Selected Papers of

FREDERICK SANGER

(with commentaries)

World Scientific Series in 20th Century Biology

Published

Vol. 1 Selected Papers of Frederick Sanger (with Commentaries)
edited by F. Sander and M. Dowding

Forthcoming

Vol. 2 A Portrait of the Immune System: Scientific Publications of Niels Kaj Jerne
edited by I. Lefkovits

Vol. 3 Selected Papers of Rita Levi-Montalcini (with Commentaries)
edited by R. Levi-Montalcini

World Scientific Series in 20th Century Biology – Vol. 1

Selected Papers of

FREDERICK SANGER

(with commentaries)

Edited by

Frederick Sanger
Margaret Dowding

Retired from the Medical Research Council Laboratory of Molecular Biology, Cambridge, England

World Scientific
Singapore • New Jersey • London • Hong Kong

Published by

World Scientific Publishing Co Pte Ltd
P O Box 128, Farrer Road, Singapore 912805
USA office: Suite 1B, 1060 Main Street, River Edge, NJ 07661
UK office: 57 Shelton Street, Covent Garden, London WC2H 9HE

Library of Congress Cataloging-in-Publication Data
Selected papers of Frederick Sanger : with commentaries / edited by Frederick Sanger, Margaret Dowding.
p. cm. -- (World Scientific series in 20th century biology ; vol. 1)
Includes bibliographical references.
ISBN 9810224303
1. Amino acid sequence. 2. Nucleotide sequence. I. Sanger, Frederick, 1918–
II. Dowding, Margaret. III. Series.
QP551.S44 1996
574.87'328--dc20 96-33754
CIP

British Library Cataloguing-in-Publication Data
A catalogue record for this book is available from the British Library.

The editors and publisher would like to thank the following publishers for their assistance and their permission to reproduce the articles found in this volume:

Academic Press (*Arch. Biochem. Biophys.*, *J. Mol. Biol.*), Annual Reviews Inc (*Ann. Rev. Biochem.*), Elsevier Science Publishers B. V. (*Biochim. Biophys. Acta*, *FEBS Lett.*), Macmillan Magazines (*Nature*), National Academy of Sciences of the United States of America, Nobel Foundation, Portland Press (*Biochem. J.*), Springer-Verlag (*Eur. J. Biochem.*)

Printed in Singapore.

Frederick Sanger

Frederick Sanger was born in 1918 in Rendcomb, a small village in the Cotswold Hills, England, where his father (also named Frederick) was the medical practitioner. He married Joan Howe in 1940 and they have two sons and a daughter. He graduated from Cambridge University in 1940, and obtained his Ph.D. in 1943 for work on the metabolism of lysine under the supervision of Dr A. Neuberger in the Department of Biochemistry, Cambridge University. He continued doing research in the Department until 1962, when he moved to the British Medical Research Council's Laboratory of Molecular Biology in Cambridge. This research is described in the following pages. He retired in 1983. His work was supported by a Beit Memorial Fellowship for Medical Research from 1944 to 1951 and thereafter by the Medical Research Council.

Dr Sanger was elected a Fellow of the Royal Society in 1954, awarded their Royal Medal in 1969, and the Copley Medal in 1977, and he gave the Croonian Lecture in 1975. He has honorary degrees from the Universities of Leicester, Oxford, Strasbourg and Cambridge. He has won many prestigious prizes and awards, including the Nobel Prize for Chemistry in 1958 and 1980.

Dr Sanger was made a C.B.E. (Commander of the Order of the British Empire) in 1963, a C.H. (Companion of Honour) in 1981, and was awarded the O.M. (Order of Merit) in 1986.

CONTENTS

PART II. RNA

PREFACE

For this collection I have selected what I consider to be my most important papers, most of which describe original research by myself and my colleagues. I have never written a book, nor have I done many review articles. The last paper in this volume is a more general and personal account of the whole of the work.

Most of my research has been concerned with "sequencing"; that is, determining the sequences of amino acids in proteins and of nucleotides in nucleic acids. While the sequences we have determined have been of interest in themselves, I consider that the most important contribution has been the development of methods as these have opened up new areas of research for ourselves and for others.

I have divided this collection of papers into three groups. The first is concerned with proteins and describes the development of methods for determining amino acid sequences and their application to insulin. The second describes methods for sequencing RNA and the determination of the nucleotide sequence of 5 S ribosomal RNA and a region in a bacteriophage that codes for a protein. The third section is on DNA sequencing and its applications. It may now seem that this was a logical and planned progression, suggesting that I had from the beginning decided to devote myself to sequencing and that all my research had been directed to this. However, this was not really the case. I have never regarded sequencing as an end in itself, but as a step towards a greater understanding of living matter. Also, I have rarely made extensive plans for my research but have been guided to each new experiment more by its intrinsic interest and technical feasibility than as the result of any long term aim. I must admit too that my choice has often depended on which experiment would be the most fun to do. The experiments reported in these papers are only a small fraction of those that my colleagues and I did; most were complete failures, inconclusive, or only preliminary. On the whole the sequencing ones were the most successful and worth publishing. And with more success and experience we became more interested in sequencing and devoted more time to it. Thus it has been a journey of exploration through unknown territory without a clear objective rather than one with a fixed destination.

Frederick Sanger
Cambridge, England

PART I

PROTEINS

Introduction

I started working on proteins in 1943, having completed my Ph.D. thesis, which was largely concerned with amino acid metabolism, under Dr. A. Neuberger. I joined the group of Professor A.C. Chibnall, who had moved to Cambridge as the new head of the Biochemistry Department, and it was mainly his influence that led me into the field of protein structure and thence to sequencing.

At that time almost nothing was known about sequencing. In fact the position of only one amino acid in any protein was known: this was a phenylalanyl residue at the N-terminus of insulin which had been detected by H. Jensen and E.A. Evans[1] in 1935. All of the 20 amino acids found as the main components of proteins had already been identified, and their relative abundance in individual proteins was being extensively studied — especially by Chibnall and his group. It was fairly generally agreed that the amino acids were linked by peptide bonds, according to the theory of Fischer and Hofmeister. Much of this basic work had been done by Fischer and his school in Germany. They were able to isolate and characterize small peptides from partial acid hydrolysates of silk fibroin. This was the main evidence for the peptide hypothesis, but it also shows that they were already interested in protein sequencing and realized its importance. This method of partial degradation was the standard way of studying the structure of large organic molecules, and it is the method that we used in our initial work. The main technical problem with this approach was the fractionation of the smaller degradation products, especially in the case of large protein molecules that gave very complex mixtures of degradation products. In fact progress in sequencing has been largely dependent on the development of improved fractionation methods.

Although at the time most workers in the field accepted the peptide hypothesis, there were some doubts and other possibilities were entertained. Other products of partial hydrolysis were the diketopiperazines (or cyclic dipeptides), and it was considered that they might be the basic components of proteins. The main adherents of this view were Abderhalden and his school. Another possible structure (the cyclol hypothesis) was put forward by Wrinch[2] and caused considerable interest at the time.

Another point of some contention was whether proteins could be regarded as pure chemical entities with all the molecules of one protein being identical, or whether they were heterogeneous mixtures of closely related molecules. It was not known how proteins were synthesized. The most popular theory was that it was some sort of reversal of the action of the proteolytic enzymes, and it was difficult to see how this could be completely specific. On the whole most people in the field assumed that proteins could be treated as pure chemicals. As there was no clear evidence to the contrary this was the safest assumption. Clearly the only way to prove or disapprove this was by doing chemical studies.

In 1943 proteins were already considered to be the most important components of living matter and were being actively studied in many laboratories from various points of view. There was much work on their physical properties, particularly their molecular weights, which could be determined by the recently introduced ultracentrifuge. There was considerable interest in the possible use of X-rays to study their structure, and several groups were attempting this. However, little progress was being made, although some rather wild claims were being put forward. The most vocal group was that of Astbury, who proposed a structure for the folding of the peptide chains based on studies of keratin. As a student I had learned this structure and was rather disillusioned when the following year he brought out an entirely different one. I think that by 1943 there was less optimism about the X-ray approach, although we fully realized that if the technical problems could be overcome it would be a most powerful method, and indeed some people doubted whether it was worth trying to determine sequences by chemical methods as in theory this could be done by X-rays. It is interesting to note that even today, when protein structures are routinely studied by X-rays, it is still necessary to use chemical means to determine their sequence.

Chemical work on proteins was concerned with amino acid analysis and peptide synthesis. Analysis was of interest from a nutritional viewpoint as well as being basic to an understanding of structure. Different methods were used for different amino acids and most were somewhat tedious and of varying accuracy. There were specific colorimetric methods for some of the amino acids, such as tryptophan and arginine; others, such as glutamic and aspartic acid, had to be separated out by chemical means and then weighed. Chibnall and his colleagues had devoted much time and effort to these methods and their analyses were probably the most reliable available, especially of insulin. There was at that time a good deal of rather wild speculation about amino acid composition, particularly a suggestion of some sort of regular arrangement along the peptide chains. As much data was being produced with not very exciting implications, I think it was tempting to see if one could find a general pattern in the results. The most popular theory was that of Bergmann and Nieman,[3] who suggested that the amino acids were arranged at regular intervals along the chains and that the number of residues of each amino acid could always be expressed by the formula $2^x \times 3^y$. It was difficult to refute this, as most small numbers fit this theory and high numbers could not usually be determined with sufficient accuracy. However, Chibnall felt that his results were sufficiently reliable and he was the first to challenge the theory.

The most influential group then working was that of Max Bergmann at the Rockefeller Institute in New York. He was generally considered to have succeeded to the crown of Emil Fischer, whose student he had been. He was best known for his work on peptide synthesis, in particular the introduction of the "carbobenzoxy" method. Together with his collaborators, Stein and Moore, he developed an ingenious new general method for amino acid analysis, known as the "solubility product method".[4] It was based on the fact that the solubility of a compound in a solution will depend on how much of that compound is already present in the solution. When

I joined Chibnall's group he asked me to try out this method. It seemed applicable to the mono amino acids, which were especially difficult when one was using other methods. It was, however, rather tedious, entailing the very accurate weighing of many small samples. Micro-balances were not as sophisticated as they are now and one had to "count the swings" — a very slow process. It soon became clear to me that it was not a practical procedure and was no match for the partition chromatography that was also being used in our lab. I think I am probably the only person outside Bergmann's laboratory to have ever used the solubility product method.

Comments on Papers 1 to 19

Paper 1

This was the first paper that I published independently, although the subject was suggested to me by Professor Chibnall, who gave me considerable guidance. He was too busy as the new head of the laboratory to take any active part in the research, but we had frequent discussions. I think he would have been justified in having his name on the paper but this was not his policy and he never even suggested it. In this respect I was lucky as it gave me a good start to my career.

Chibnall and his colleagues had carried out a careful amino acid analysis of insulin. They had chosen insulin both because of its medical importance and because it could be purchased in a pure form. Another advantage was its small size (M.W. 6,000), though this was not known at the time as it aggregated in solution and estimates were considerably higher.

Chibnall and Rees had estimated the free amino groups of insulin by the van Slyke method (a manometric technique that measured the nitrogen evolved on treatment with nitrous acid). Most of the free amino groups of proteins are due to the ϵ-amino groups of the lysine residues; however, there were many more free amino groups in insulin than lysine residues and this suggested that these were free α-amino groups at the ends of the polypeptide chains. Chibnall suggested that I try to identify and estimate them. As described in the paper, I tried several approaches and eventually came up with the dinitrophenyl (or "DNP") method.

I consider that the most important contribution to the success of this method was the use of partition chromatography, which had just been devised by Gordon, Martin and Synge[5] and used for fractionating and estimating the acetyl derivatives of the amino acids. This was an enormous advance in fractionation methods. Prior to this the mono amino acids could only be separated by tedious fractional distillation of their esters. Now a small amount of their acetyl derivatives could be put on a column and separated out in a single simple procedure. G.R. Tristram, who was working in Chibnall's group, was already using the method to analyse for the mono amino acids and was a great help to me. The DNP derivatives were particularly suitable for this chromatographic method as they were coloured and could be seen moving down the columns and estimated colorimetrically.

The method gave a clear result with insulin and was later applied to other proteins and was useful in our later work with peptides. It was the preferred method for end group analysis until it was superseded, especially by the Edman phenylisothiocyanate technique. It is interesting to note that Edman was working on this at the same time as I was developing the DNP method: however, it was not applied to proteins till considerably later. I think the main advantage of the DNP method was that the derivatives were coloured whereas the phenylthiohydantoins could only be detected in ultraviolet light, and there were no good micro-detection systems or reliable fraction collectors at that time.

There was some delay in the publication of this paper: when I submitted it to the *Biochemical Journal* it was sent to a distinguished organic chemist to be refereed, and he felt that I had not adequately characterized the end groups. I had simply identified them by their chromatographic behaviour — by comparing it with that of the DNP derivative of all the amino acids found in insulin. However, there was a firm tradition amongst chemists that a substance was not characterized unless a melting point and elementary analysis were recorded. I had therefore to do a relatively large scale isolation of the DNP derivatives from insulin in order to get the necessary data. Fortunately, in later end group studies such rigorous characterization was not considered necessary.

Paper 2

From the determination of the terminal residues we concluded that insulin was composed of four polypeptide chains, assuming a molecular weight of 12,000. Actually the molecular weight obtained by ultra-centrifugation was 48,000 whereas the X-ray evidence suggested 36,000. Using osmotic pressure methods H. Gutfreund found evidence of dissociation and suggested the value of 12,000 for the minimum subunit. In fact the value was later shown to be 6,000.[6] However, at this stage of the work we believed there were four chains. Although there was no direct evidence as to how the chains were joined together, we assumed it was by the disulphide bonds of the cystine residues, and so we tried to separate the chains by breaking these bonds. The most specific and gentle method of doing this would have been by reduction: however, we did not have any success with this, though later it was used successfully by other workers. We resorted to a more violent reaction, using performic acid, which converted the cystine to cysteic acid. Performic acid also reacts with two other amino acids: methionine and tryptophan. We were lucky that neither was present in insulin. The strongly acidic $-SO_3H$ group was very stable and its presence was helpful for fractionation of the chains, and later for small peptides. The chains were too large to be separated by the new counter-current methods and we used the classical method of fractional precipitation. Fraction A (the glycyl chain) was very acidic and was the easiest to isolate. Fraction B was more difficult. The yields were rather small, so that we could not be sure that the whole of insulin was represented in the two chains, though in fact it was.

Paper 3

One problem in applying the DNP method to insulin was that DNP-glycine was the most acid-labile of the DNP amino acids so that low yields were obtained and one had to apply a somewhat uncertain correction factor. In order to overcome this, shorter times of hydrolysis were used. Under these conditions lower yields of DNP-phenylalanine were obtained and there was a new, unidentified band on the column, which yielded DNP-phenylalanine on further hydrolysis. Clearly this was a peptide, suggesting the approach to sequencing reported in this paper. Once more fractionation was of importance, and again a new technique from A.J.P Martin's group was a decisive factor. This was paper chromatography — a form of partition chromatography but more versatile and particularly applicable to free amino acids in small amounts, which was what we needed to identify the amino acids present in our peptides.

Although this paper is less widely cited than paper 1, I feel that it is probably more important — for two reasons. It was the first time that a peptide sequence had been determined at a specific site in a protein and so represented a significant turning point in sequencing as well as in my career. Small peptides had been isolated from silk fibroin by Fischer's group, but their position was not known. It also established that proteins could be regarded as pure chemicals rather than heterogeneous mixtures. In our work we found no evidence of any fragments that did not fit into the sequences determined, which made it most probable that each position in a protein chain was occupied by one unique amino acid, and that the method of protein synthesis (about which almost nothing was known at the time) must be a very specific process. It is interesting that I do not discuss this aspect in the paper. It may be that this was because I had always assumed that proteins were pure chemicals, or perhaps I was just being cautious. It is easier to be dogmatic 40 years later.

Although the results reported in this paper were fairly significant, it was not difficult to do the work, which was completed fairly quickly. It was also enjoyable and exciting to see the results coming out from essentially simple experiments, and this certainly fostered my interest in sequencing, which lasted throughout the rest of my work.

In this paper I was still taking 12,000 as the molecular weight of insulin and concluded that the two phenylalanyl chains were identical, as were the two glycyl chains. I do mention the possibility of a subunit size of 6,000, but reject it as it does not fit in with analytical results obtained in the same laboratory.

In most of my work on insulin I was fortunate to have very little competition. It seemed strange to me that hardly anyone else felt it was sufficiently important to be working on, whereas most of my friends complained of having to compete with other laboratories. However, I was surprised to hear that D.W. Woolley had isolated DNP peptides from insulin and his results were entirely different from mine. Woolley was a blind biochemist working at the Rockefeller Institute and was well known for his work on growth factors. He worked with technical assistants but did most of the thinking and planning. Clearly he had the same idea as I had but something went wrong at the practical level. I think it is probable that he was dealing with tyrosine peptides. Fluorodinitrobenzene reacts with the hydroxyl group of tyrosine to give a colourless derivative.

Papers 4 to 8

Having obtained in a pure form from insulin the two fractions representing the two chains of approximately 20 and 30 residues, I thought it would be worth subjecting them to partial acid hydrolysis to see if I could identify peptide sequences. The general approach was not new. Emil Fischer's group had used it, but we had much better techniques available, particularly the fractionation methods of Martin's group: partition chromatography, paper chromatography, and now ion exchange chromatography and zone electrophoresis in gels. I think there is no doubt that my work owes more to Martin than to anyone else. He was extremely inventive and as near to being a genius as anyone I have met. I also found him very exciting and inspiring to talk to when I met him occasionally at meetings. He was somewhat erratic and impractical, but fortunately had more practical collaborators. Consden *et al.*[7] had already determined the sequence of the pentapeptide Gramicidin S and had demonstrated the use of paper chromatography for peptide fractionation, but nothing as large as the insulin chains had been attempted.

I started working on the glycyl chain as it was smaller, and had some encouraging results. I was then joined for a year by Hans Tuppy, a post-doctoral fellow from Austria. He was a very enthusiastic and hard worker as well as being technically skilful. He took over the phenylalanyl chain and by the end of the year had almost completed the whole sequence.

At first we were reluctant to use enzymic hydrolysis as it was thought that proteolytic enzymes might synthesize peptide bonds. Enzyme reactions were known to be reversible, and in fact synthesis had been observed under special circumstances, especially by J.S. Fruton, who was interested in the possibility that biosynthesis of proteins might be due to the reversed action of proteolytic enzymes. However, when we did not seem to be getting further with our acid hydrolysates we decided to risk using enzymes, and this made it possible to complete the phenylalanyl chain. There was no evidence of any synthesis of peptide bonds by the enzymes, nor has any been found since, and subsequently the use of enzymes became the preferred method for breaking down protein chains. In fact, as shown in paper 8, acid can cause a rearrangement of peptide sequence.

After Tuppy left I was joined by Ted Thompson from Australia and we completed the glycyl chain. Although it is the shorter of the two chains it proved more difficult. The phenylalanyl chain had a number of amino acids that occurred only once or twice and this helped considerably in interpreting the data, whereas the glycyl chain had fewer, and glutamic acid, cysteic acid and leucine accounted for about half the amino acids present.

Paper 9

Before this I had not thought in terms of trying to work out the complete primary structure of insulin, but when we had succeeded in determining the whole amino acid sequence there was clearly some pressure to continue work on the molecule to a stage where we could write a complete formula. We could not of course expect to do anything about the secondary structure.

There were essentially two problems still to be resolved: the distribution of the amide groups and the positions of the disulphide bonds that held the two chains together. Both proved particularly difficult and we would probably not have attempted them but for the challenge of achieving the first determination of a protein structure.

This paper is concerned with the amide groups. These were split off in acid digests but were left intact by enzymes, so we needed to isolate peptides from enzymic digests containing only one of the acidic amino acids — glutamic or aspartic. The enzymes we had used for sequencing had too narrow a specificity and we had to investigate other non-specific enzymes. Particularly difficult was the Glu–Glu sequence near the N-terminus of the glycyl chain. It was the last to be established and probably took as long to do as all the rest of the amide groups. The amide was eventually identified on the second Glu, thanks to the persistence of Ted Thompson. Few amino acids can have generated so much strong language.

Papers 10 and 11

For the purposes of sequence determination, insulin was a good protein to have chosen, largely because of the shortness of the chains; however, from the point of view of identifying its cross-linkages it was a bad one. Few other proteins have such a high content of cystine, and the two consecutive Cys residues in the glycyl chain constituted a particular difficulty. Our approach was to isolate peptides from intact insulin which contained cystine residues (that is, with –S–S– bridges intact) and to oxidize them with performic acid and see which of the already known cysteic acid peptides were produced. However, we could make no sense of our initial results. We eventually found that a disulphide interchange reaction was taking place, and so had to embark on a fairly detailed study of this reaction before we found suitable conditions for hydrolysis without this disulphide interchange.

By this time it had been established that the molecular weight of insulin was approximately 6,000 rather than 12,000, so there were only the two chains, as we would have expected. It should be pointed out that our results could fit a molecule of four chains — that is, two identical halves joined together by the two disulphide bonds, which in the two-chain structure form the ring.

Another problem was the difficulty of fractionating the cystine peptides, many of which were rather large and did not form well-defined spots on paper chromatography. Here again another new method proved useful. This was paper ionophoresis. Martin's group had pioneered the use of "zone" ionophoresis in silica gel, but had not used paper; various other workers introduced paper ionophoresis. One of the main problems was to prevent evaporation of water from the paper because of the heat generated by the electric current. This was elegantly overcome by Michl — whose method we used — by immersing the whole paper in an immiscible organic solvent.

Paper 11 was essentially the climax of our work on insulin. It was the first protein to be sequenced and we had developed methods and demonstrated their applicability. I think this stimulated interest in sequencing in other laboratories, and our methods were used extensively until better ones were available. Our techniques depended very much on fractionation systems using paper; their very high resolving power and ease of operation made them particularly suitable for the preliminary probing experiments, of which most of our work consisted. When one has not done something before, it is necessary to try many different ways of tackling it. However, the methods were not very quantitative and for later studies on larger proteins there was a greater emphasis on the development of more quantitative systems. Columns replaced paper and more automation was introduced, enabling the use of techniques that were not available when we were working on insulin. The most decisive advance in this direction was introduced by Moore and Stein, who used ion exchange columns and sophisticated fraction collectors to determine the sequence of ribonuclease.[8]

Papers 12 to 14

Having completed the structural work on insulin, I now had to decide what to do next. The insulin work was really initiated by Chibnall, and each subsequent experiment had been determined by what had gone before. Now I was on my own and was more or less free to work on what I chose. I was obviously interested in sequencing but did not wish simply to start on another protein, and I could see no way to attack nucleic acids. Some people suggested that I should work on carbohydrates, but this did not appeal to me.

Probably the main reason for studying the structure of proteins had been the hope that it would throw some light on their function, so it seemed that the logical thing to do next would be to study insulin function in relation to its structure. At this time there was much interest in enzymes, especially at the Cambridge University Biochemistry Laboratory, where I worked. Emil Fischer had originally put forward his "lock and key" hypothesis and it was assumed that an enzyme had an "active centre" composed of a few amino acids specifically arranged to fit the substrate. Presumably hormones must have a similar specific arrangement of amino acids to carry out their various functions, and so, having determined the sequence of insulin, perhaps we could try to find out something about its activity. We initiated a number of experiments to this end, but most did not come to anything.

One possible approach to this general problem was to compare the structure of insulins from different species. The assumption was that residues that were important for biological function would be conserved whereas variations might be expected to occur in the less important residues. A preliminary experiment in 1949 showed that there were indeed differences in the glycyl chain, and by applying the sequencing procedures we had developed it was shown where they were. More species were studied later by L.F. Smith.[9] Human insulin was found to be the same as pig insulin except that it had no Ala at the C-terminus of the phenylalanyl chain.

The differences between insulins were small and did not tell us much about function. However, this was the first comparative sequence analysis study of a protein. Since then species differences in many other proteins have been, and are still being, studied. Most proteins show greater variation than was found in insulin, and the results have often contributed to an understanding of function.

Papers 16 to 18

Following the structural work on insulin, we had become interested in the relationship of the amino acid arrangement to its biological function. Although we made little progress with insulin, it had fostered a more general concern with active centres, and these three papers describe experiments with proteolytic enzymes. The main importance of these papers is in the development of new methods rather than in the results obtained.

Sequencing of larger proteins by the partial degradation approach was a somewhat lengthy procedure and involved extensive amino acid analyses of the end products. I became interested in the possibility of sequencing by labelling a specific amino acid and deducing the sequence around it from the behaviour of its breakdown products. It was hoped that such an approach would avoid these extensive analyses and facilitate the study of specific regions in larger proteins. Radioactive isotopes were becoming available and their possible use for this approach appealed to me, largely because of their ease of detection by autoradiography. The proteolytic enzymes trypsin and chymotrypsin were ideal for testing this out since they were already known to have a serine residue in their active site that could be specifically labelled with the P-containing inhibitor. Thus, using ^{32}P we would have just one residue labelled.

Paper 16 describes a rather simple but conclusive experiment showing that elastase has the same sequence around its active serine as trypsin and chymotrypsin. Paper 17 is more concerned with the development of methods for working out sequences around the labelled residue. For this we used mainly paper ionophoresis for fractionations and direct autoradiography to detect the position of the products. This was contrary to the standard method of sequencing, which was usually done with columns and fraction collectors. I was perhaps somewhat prejudiced in my preference for paper methods rather than column ones. They had served me well for insulin and it seemed to me that you could get more information in a shorter time, which was useful — at least for preliminary exploratory experiments.

We had something of a setback with this work as we found too many products, which was very puzzling. However, as described in paper 17, we were able to work out what was happening: there was a side reaction during acid hydrolysis resulting in a rearrangement of aspartic acid — which had not been identified with less sensitive techniques.

This general approach of labelling with ^{32}P and using paper fractionations to determine sequences, without resorting to a complete analysis of every degradation product, proved particularly useful in our later work on nucleic acids.

Paper 18 describes the successful application of the method to a bacterial protease.

Paper 19

C. Milstein was working in the enzyme division of the Biochemistry Laboratory, studying the enzyme phosphoglucomutase. This contained a phosphate group that was involved in its activity. He told me about this and that it could be labelled with ^{32}P, so we decided to apply our new approach to its active centre. The presence of histidine caused additional complications, and further developments in the techniques were made before establishing the pentapeptide sequence. Although this was the only actual collaboration I had with Milstein, we have worked closely together ever since, while developing our own separate lines of research, and I owe a great deal to his help, expertise and encouragement.

References

1. H. Jensen and E.A. Evans, *J. Biol. Chem.* **108**, 1 (1935).
2. D. Wrinch, *Philos. Mag.* **26**, 313 (1938).
3. M. Bergmann and C. Niemann, *J. Biol. Chem.* **118**, 301 (1937). See also: J.S. Fruton, *A Skeptical Biochemist* (Harvard University Press, 1992), p. 48.
4. W. H. Stein, S. Moore and M. Bergmann, *J. Biol. Chem.* **143**, 121 (1941).
5. A.H. Gordon, A.J.P. Martin and R.L.M. Synge, *Biochem. J.* **37**, 79 (1943).
6. E.J. Harfenist and L.C. Craig, *J. Am. Chem. Soc.* **74**, 3087 (1952).

7. R. Consden, A.H. Gordon, A.J.P. Martin and R.L.M. Synge, *Biochem. J.* **41**, 596 (1947).
8. C.H.W. Hirs, S. Moore and W.H. Stein, *J. Biol. Chem.* **235**, 633 (1960).
9. D.S.H.W. Nicol and L.F. Smith, *Nature* **187**, 483 (1960).

[FROM THE BIOCHEMICAL JOURNAL, VOL. 39, No. 5, pp. 507–515, 1945]

The Free Amino Groups of Insulin

BY F. SANGER (Beit Memorial Fellow), *Biochemical Laboratory, Cambridge*

(*Received* 31 *August* 1945)

That the free amino groups in proteins may be the ε-amino group of lysine was first suggested by Skraup & Kaas (1906), who failed to isolate lysine from deaminized proteins. Van Slyke & Birchard (1913) later found an apparent equality between the free amino-N of proteins and one-half of their total lysine-N, which suggested that the free amino groups were due exclusively to such a source. Methods were then developed to confirm the presence of the free ε-amino group of lysine in proteins; these consisted in treating a protein with a reagent that would react with free amino groups to give derivatives that were stable to acid hydrolysis, so that on hydrolysis of the substituted protein the derivative of lysine could be isolated. The most satisfactory was that of Gurin & Clarke (1934), who treated gelatin with benzenesulphonyl chloride, hydrolyzed the resulting benzenesulphonyl gelatin and isolated ε-benzenesulphonyllysine from the hydrolysate.

In some proteins, and particularly in insulin, the number of free amino groups (Van Slyke) is far in excess of that which can now be ascribed to lysine, which suggests that the protein must contain residues of certain amino-acids which are condensed in such a way that their α-amino groups remain free. Jensen & Evans (1935) have, in fact, been able to isolate the phenylhydantoin of phenylalanine from a hydrolysate of insulin that had been treated with phenyl*iso*cyanate, thus demonstrating that some free amino groups of insulin are present on phenylalanyl residues. Chibnall (1942) has suggested that the free amino groups of proteins over and above the ε-amino groups of lysine are due to terminal residues of polypeptide chains, and that the number of these groups must therefore give a measure of the number of polypeptide chains in the protein.

We found that methanesulphonyl amino-acids were stable to acid hydrolysis and that they could be fractionated by partition chromatography (Gordon, Martin & Synge, 1943) with an indicator in the aqueous phase of the column. The rates at which the bands moved on the columns were similar to those of the acetamino-acids. The method was satisfactory when applied to synthetic peptides, but with insulin the number of terminal groups estimated by titration of the methanesulphonyl amino-acids was far less than that suggested by the method of Van Slyke. The procedure was accordingly abandoned.

Abderhalden & Stix (1923) attempted to use 2:4-dinitrochlorobenzene (DNCB) for the identification of the terminal groups of a partial hydrolysate of silk fibroin. They did not meet with much success, chiefly owing to the presence of anhydrides in the hydrolysate and the difficulties of separating the products. It seemed, nevertheless, worth while to investigate this reagent, especially as all the 2:4-dinitrophenyl-amino-acids (referred to henceforth as DNP-amino-acids) produced are bright yellow, thereby facilitating chromatographic separation. DNCB will not react with amino-acids in $NaHCO_3$ solution unless heat is applied, and this brings about a certain amount of hydrolysis of the protein. Fortunately, however, the corresponding fluoro-compound, 2:4-dinitrofluorobenzene (DNFB) was found to react readily at room temperature, and the use of this has met with considerable success, for the DNP-amino-acids produced can be estimated colorimetrically and separated almost completely from one another by partition chromatography. The solvent systems normally used for separating the acetyl-derivatives were not entirely satisfactory for the DNP-monoamino-acids, and several new systems had to be introduced; nevertheless, the method finally adopted embraced all amino-acids, though this was not possible with the methanesulphonyl derivatives.

The method as applied to insulin consisted of three stages. In the first the protein was treated with DNFB, hydrolyzed and the resulting coloured compounds separated chromatographically. The identification of these was based on band rates, and was confirmed by mixed chromatograms. Secondly, knowing which DNP-derivatives were present, one could assess the amount of each with reasonable accuracy by separating the fraction quantitatively and estimating the material present colorimetrically, using the pure DNP-amino-acid as a standard. Thirdly, the whole operation was carried out on a larger scale, so that the DNP-amino-acids could be isolated and satisfactorily characterized. The procedure should be applicable to the identification and estimation of the terminal residues in all peptides and proteins.

EXPERIMENTAL

Preparation of DNP-amino-acids

In order to work out a comprehensive chromatographic method for the identification of any DNP-amino-acid by its behaviour on a partition column, it was necessary to prepare the derivatives of all the naturally occurring amino-acids. Not all of these were obtained in pure form or completely characterized, as this was unnecessary for a study of their chromatographic behaviour; nevertheless, all the derivatives corresponding to those obtained from treated insulin were carefully purified, so that they could be used as standards of reference.

The derivatives of glycine, *dl*-alanine, *dl*-valine, *dl*-leucine and *l*-asparagine were prepared by the method of Abderhalden & Blumberg (1910). By a similar method, the derivatives of *dl*-methionine, *l*-proline, *l*-tryptophan, *dl*-aspartic acid, *l*-cystine and *l*-lysine were prepared but not completely characterized. The methods of purification and characterization of these compounds will be published later.

N-2:4-*Dinitrophenyl*-l-*phenylalanine*. 0·2 g. *l*-Phenylalanine and 0·4 g. $NaHCO_3$ were dissolved in 5 ml. water and to this was added a solution of 0·4 g. (0·28 ml.) DNFB in 10 ml. ethanol. The mixture was shaken for 2 hr. at room temperature, concentrated to remove ethanol, dissolved in water and extracted with ether to remove excess DNFB. The aqueous solution was then acidified, which caused the separation of an oil that immediately solidified in an amorphous form. The N-2:4-*dinitrophenyl*-l-*phenylalanine* was recrystallized twice from aqueous methanol. Yield, 0·27 g.; m.p. 186°. (Found: C, 54·2; H, 4·2; N, 12·6%. $C_{15}H_{13}O_6N_3$ requires C, 54·4; H, 3·9; N, 12·7%.)

The derivatives of *dl*-serine, *dl*-threonine, *l*-hydroxyproline, *l*-glutamic acid and *l*-arginine were prepared in a similar way using DNFB. Their purification and characterization will be published later. The acid equiv. wt. (277) of the serine derivative showed that it was mono-DNP-serine (271), and that DNFB does not react with aliphatic hydroxyl groups.

ε-N-2:4-*Dinitrophenyl*-l-*lysine*. Several attempts to prepare this compound by the action of DNCB or DNFB on the copper complex of lysine proved unsuccessful. It was eventually prepared from the α-acetyl-compound as follows. 0·48 g. α-acetyl-*l*-lysine (Neuberger & Sanger, 1943) and 0·75 g. $NaHCO_3$ were dissolved in 3 ml. water. To this was added a solution of 0·5 g. DNCB in 10 ml. ethanol. The mixture was then heated under reflux on a water-bath for 4 hr. The ethanol was removed by evaporation *in vacuo* and the residue dissolved in water, filtered to remove excess DNCB and acidified with HCl while hot. An oil separated, which was extracted into chloroform. The chloroform solution, after drying with anhyd. Na_2SO_4, was taken to dryness, leaving an oil. On rubbing this with dry ether the *α-acetyl-ε-DNP*-l-*lysine* crystallized and could be recrystallized from a mixture of ethanol and ether: m.p. 110°; yield 0·7 g. (75% theoretical). 0·2 g. of the compound was boiled under reflux for 3 hr. with 5 ml. 20% HCl. On cooling, the *ε-DNP*-l-*lysine hydrochloride* separated as crystals containing 1 mol. of water of crystallization. After standing overnight it was filtered off and washed with very little water and ethanol and then thoroughly with ether. It was recrystallized from dilute HCl; m.p. 186°. (Found: C, 39·0; H, 5·5; N, 15·6; Cl, 9·5%. $C_{12}H_{16}O_6N_4.HCl.H_2O$ requires C, 39·5; H, 5·2; N, 15·4; Cl, 9·8%.)

α-N-2:4-*Dinitrophenyl*-l-*lysine*. It is generally believed that the lysine residues in proteins are combined through the α-amino groups while the ε-amino groups are free. One would thus expect to isolate only ε-DNP-lysine from a hydrolysate of a DNP-protein. It is not impossible, however, that some, at least, of the lysine residues in proteins are combined through the ε-amino group, in which case *α-DNP-lysine* would be formed on hydrolysis of a DNP-protein. This second derivative was accordingly synthesized as follows. 1·25 g. ε-benzoyl-*l*-lysine were treated for 4 hr. with 1·0 g. DNCB as described above. On acidification an oil separated which solidified on standing. It weighed 2·0 g. 0·5 g. of this *α-DNP-ε-benzoyl*-l-*lysine* was boiled under reflux for 3 days with a mixture of 5 ml. acetic acid and 5 ml. conc. HCl. The mixture was then evaporated to dryness, taken up in water and filtered to remove unchanged substance. On neutralization with pyridine the α-DNP-*l*-lysine crystallized out in 60% yield; m.p. 260° (decomp.). (Found: C, 45·6; H, 5·3; N, 17·5%. $C_{12}H_{16}O_6N_4$ requires C, 46·2; H, 5·1; N, 17·9%.)

Tyrosine derivatives. When *l*-tyrosine was treated with DNCB in excess two coloured compounds were produced, which moved down a 1% butanol-chloroform column at $R=0{\cdot}5$ and 0·2 respectively. With DNFB only the faster-moving compound was produced. The acid equiv. wt. (519) indicated that this was *O*:*N*-diDNP-*l*-tyrosine (513), which had been prepared by Abderhalden & Stix (1923). The slower-moving compound was presumably *N*-DNP-*l*-tyrosine, since O-*DNP*-l-*tyrosine* is colourless. The monoDNP-tyrosine prepared by Abderhalden & Stix (1923) was presumably the *O*-DNP-derivative in an impure form.

O-2:4-*Dinitrophenyl*-l-*tyrosine*. Since DNFB reacts with the hydroxyl group of tyrosine, a hydrolysate of a DNP-protein would be expected to yield *O*-DNP-*l*-tyrosine. It was accordingly necessary to synthesize this latter compound. 0·55 g. *N*-acetyl-*l*-tyrosine (du Vigneaud & Meyer, 1932) was treated for 4 hr. with 2·0 g. DNCB as described above. The solution was only faintly coloured at the end of the reaction, in contrast to the bright colours obtained when DNCB reacts with amino groups. On acidification an oil was produced which partly crystallized. From aqueous ethanol the N-*acetyl*-O-*DNP*-l-*tyrosine* crystallized in white needles; m.p. 194°; yield 0·1 g. The product was boiled for 3 hr. under reflux with 20 % HCl; on cooling and evaporating to a small volume, a precipitate was obtained which was filtered off, dissolved in warm dilute HNO_3 and neutralized with pyridine while hot. *O*-DNP-*l*-tyrosine crystallized in white needles containing 1 mol. of water of crystallization; m.p. 202° (decomp.). (Found: C, 49·9; H, 4·2; N, 11·9%. $C_{15}H_{13}O_7N_3.H_2O$ requires C, 49·4; H, 4·1; N, 11·6%.)

Reaction of DNFB with histidine. By the action of DNCB on histidine Abderhalden & Blumberg (1910) obtained two compounds containing one and two DNP-groups respectively. I find that on treatment with excess DNFB only the diDNP-histidine is obtained, indicating that DNFB reacts with the imidazole ring. As this is free in proteins it seemed probable that a monoDNP-histidine would be formed on hydrolysis of a DNP-protein. In an attempt to prepare this compound, α-acetylhistidine (Bergmann & Zervas, 1928) was treated with DNFB, but there was very little reaction and only a small amount of a yellow oil was obtained. A portion of this was hydrolyzed with 20 % HCl and the products studied chromatographically. Two compounds were obtained. The first of these was not extracted from acid solution into ether and moved fast on a 17 % butanol-chloroform column; it was not observed when the hydrolysis products of DNP-insulin were treated. The second was ether-soluble and moved fast on a chloroform column. It seemed probable that this latter might be 2:4-dinitroaniline, derived by breakdown of the substituted histidine, and it was indeed found to have the same *R* values on ethanol-ligroin and acetone-cyclohexane columns as the aniline derivative itself. This same band was obtained on hydrolysis of both diDNP-histidine and DNP-insulin. These results indicate that DNFB reacts only slowly with α-acetylhistidine, in contrast to the almost quantitative reaction with histidine itself. It is interesting to note that benzoylation of the imidazole group of histidine is affected by the presence of other polar groups in the molecule (Gerngross, 1919) and that the same appears to be true on reaction with DNFB. To ascertain how histidine would react if present in a polypeptide chain with its carboxyl as well as its amino group in peptide linkage, the action of DNFB on α-benzoylhistidine methyl ester (Gerngross, 1919) was studied; the reaction was very slow and the same products were given on hydrolysis as with α-acetylhistidine. Finally, to ascertain how histidine would react if present as a terminal residue in a peptide chain with only the amino group free, the reaction with histidine methyl ester (Pauly, 1904) was also studied. In this case the product obtained on hydrolysis consisted entirely of the diDNP-derivative, indicating easy reaction with the imidazole group. To summarize, it may be inferred that when histidine is present in the body of a polypeptide chain of a protein or as a terminal residue with its carboxyl group free, the imidazole group will react slowly with DNFB so that on hydrolysis some 2:4-dinitroaniline will be obtained together with a possible unknown breakdown product which is insoluble in ether and moves fast on a 17 % butanol-chloroform column. If histidine is present as a terminal residue with its amino group free the diDNP-derivative will be obtained.

Reaction of DNFB with —SH *groups.* Saunders (1934) showed that DNCB reacts more readily with —SH groups than with amino groups, so that one is led to infer that when a protein containing —SH groups is treated with DNFB and hydrolyzed, *S*-DNP-cysteine would be produced. Saunders (1934) prepared a derivative by the action of DNCB on cysteine, which he stated crystallized with difficulty. I have repeated the preparation and find that the product does not give a positive ninhydrin reaction and is extracted from dilute HCl into ether. On a 3 % butanol-chloroform column it forms two bands having $R=1·2$ and 0·3 respectively. It would appear, therefore, that the condensation does not run smoothly, and the product does not exhibit the expected properties.

To obtain information as to the compound to be expected from DNP-proteins, a small amount of the *S*-DNP-derivative of reduced glutathione was prepared and hydrolyzed for 6 hr. with 20 % HCl. *S*-DNP-glutathione is a well-characterized product of known structure (Saunders, 1934). The hydrolysate appeared to contain one substance which moved fast down a 66 % methyl ethyl ketone (M.E.K.)-ether column and at $R=0·08$ in 17 % butanol-chloroform. It was not extracted from dilute HCl into ether. It is suggested that this band represents *S*-DNP-cysteine, and that the substances prepared by the action of DNCB on cysteine are formed from *S*-DNP-cysteine by the action of alkali. More experimentation is necessary to verify this, but it appears that the product obtained from glutathione would also be obtained from a protein.

Chromatographic separation of DNP-derivatives

In general the partition chromatographic method of Gordon *et al.* (1943) was employed to separate the DNP-derivatives, using a stationary aqueous phase adsorbed on the silica gel and a moving organic phase. It was, of course, unnecessary to use an indicator in the aqueous phase as the derivatives themselves are coloured. To describe such solvent systems the terminology of the above-mentioned workers is followed, i.e. 1 % butanol-chloroform refers to a solution in chloroform of butanol, 1 % (by vol.) subsequently saturated with water. In some cases glycol or an aqueous organic solvent such as ethanol or acetone was employed as the stationary phase, and a non-polar solvent in equilibrium with it as the moving phase. Such systems are especially useful for the separation of the mono-amino-acid derivatives. To prepare an ethanol-ligroin column, 1 vol. water, 1 vol. ethanol and 10 vol. ligroin (b.p. 80–100°) are shaken together until the two phases are in equilibrium. The lower aqueous phase is used as the stationary phase of the column, 1 ml. being added to each 2 g. dry silica, and the upper ethanol-ligroin phase is used as the moving phase of the column. A methanol-carbon tetrachloride column is prepared as above from 1 vol. water, 1 vol. methanol and 15 vol. carbon tetrachloride; and an acetone-cyclohexane column from 1 vol. water, 1 vol. acetone and 10 vol. cyclohexane. To prepare a glycol-benzene column, glycol and benzene are shaken together; 1 ml. of the glycol layer is added to each 1 g. dry silica, and the benzene layer used as the moving phase.

In Table 1 are shown the R values for the various ether-soluble DNP-amino-acids using various solvent systems. These values have been found to vary considerably with such factors as the particular batch of silica or solvent and with the distance travelled down the column. It is probable that there is a certain amount of adsorption of the compounds on the silica. The figures in Table 1, therefore, do no more than indicate the relative rates of the different derivatives on the columns, and R values alone cannot be used for identification unless a parallel experiment with a sample of the authentic derivative itself is run.

The data given in Table 1 show that all the amino-acid derivatives listed can be separated by one or other of the solvent systems employed, the sole exception being leucine and *iso*leucine. The method may thus be capable of estimating the component amino-acids given on hydrolysis of a protein, but the more quantitative aspects of such a procedure have not yet been fully explored. All the derivatives shown in Table 1 are extracted from acid solution into ether, and move fast on an ether column. This fact is made use of to separate them from the acid-soluble derivatives, the R values for which on various solvents are given in Table 2.

Table 1. *Band rates of ether-soluble DNP-amino-acids*

(Figures are values of R—Gordon *et al.* 1943.)

Stationary phase ...	Water		Water	Water	Water	Ethanol	Methanol	Acetone	Glycol
Mobile phase ...	Butanol-chloroform			5 %	33 %		Carbon		
	3 %	1 %	Chloroform	propanol-cyclohexane	ether-ligroin	Ligroin	tetrachloride	Cyclohexane	Benzene
2:4-Dinitroaniline	—	—	Fast	—	—	0·35	—	0·8	—
DNP-leucine	—	Fast	0·45	Fast	Fast	0·75	0·6	0·5	0·65
DNP-*iso*leucine	—	Fast	0·45	Fast	Fast	0·8	0·6	0·55	0·6
DNP-valine	—	Fast	0·4	Fast	Fast	0·55	0·5	0·4	—
DNP-phenylalanine	—	Fast	0·2	Fast	Fast	0·5	0·4	0·25	—
DNP-methionine	—	Fast	0·25	Fast	1·0	0·3	0·4	—	—
DNP-proline	Fast	0·7	0·2	1·0	0·9	0·2	0·3	0·4	0·3
DNP-alanine	Fast	0·7	0·15	1·0	1·0	0·25	0·25	0·4	0·2
DNP-tryptophan	—	—	0·06	0·9	0·8	0·1	—	—	—
DNP-glycine	—	0·5	0·08	0·6	0·6	0·1	—	—	0·07
diDNP-lysine	—	0·5	0·06	Insol.	0·25	Insol.	—	—	0·3
diDNP-tyrosine	—	0·5	0·08	Insol.	0·25	—	—	—	0·5
DNP-hydroxyproline	—	0·13	0·02	—	—	—	—	—	—
DNP-threonine	—	0·07	—	0·2	0·2	—	—	—	—
DNP-serine	—	0·05	—	Insol.	0·1	—	—	—	—
DNP-glutamic acid	0·4	0·03	—	0·35	0·3	—	—	—	—
DNP-aspartic acid	0·4	0·02	—	0·2	0·2	—	—	—	—
diDNP-cystine	0·35	—	—	Insol.	0·0	—	—	—	—

Table 2. *Band rates of acid-soluble DNP-amino-acids*

(Figures are values of R—Gordon *et al.* 1943.)

	Developing solvents		
	Methyl ethyl ketone-ether	Butanol-chloroform	
DNP-amino-acids	66 %	17 %	30 %
diDNP-histidine	Fast	Fast	—
S-DNP-cysteine	Fast	0·08	Fast
α-DNP-arginine	0·35	—	0·2
ε-DNP-lysine	0·20	—	0·15
α-DNP-lysine	0·11	—	—

Attempts to separate the DNP-derivatives by adsorption chromatography have not been very successful. When magnesium oxide or alumina were used as adsorbents decomposition took place. It was found possible, nevertheless, to separate the glycine and leucine derivatives in acid solution on a column of talc and in ether solution on a column of calcium carbonate, in neither case, however, so effectively as in the above-mentioned partition chromatography. Likewise the use of partition chromatography using filter paper (Consden, Gordon & Martin, 1944) or starch (Synge, 1944) was not very satisfactory, due to 'tailing' of the spots or bands.

Estimation of DNP-amino-acids

The DNP-amino-acids, dissolved in N-HCl (about 0·1–0·3 mg. amino-acid-N/100 ml. N-HCl), were estimated colorimetrically with a photoelectric absorptiometer. In the case of DNP-phenylalanine the solubility in this solvent was too low, and 1 % $NaHCO_3$ had to be used. Standard curves were plotted for the derivatives of the naturally occurring amino-acids. These varied considerably from one to another, and differences were noted between the curves representing the *l* and *dl* samples of the same amino-acid. In all cases the colour intensities do not obey Beer's law.

Stability of DNP-derivatives to acid hydrolysis

Since the liberation of DNP-amino-acids from a DNP-protein requires acid hydrolysis, it was necessary to investigate the stability of the former products during such treatment. To effect this, a standard solution of the derivative was made up

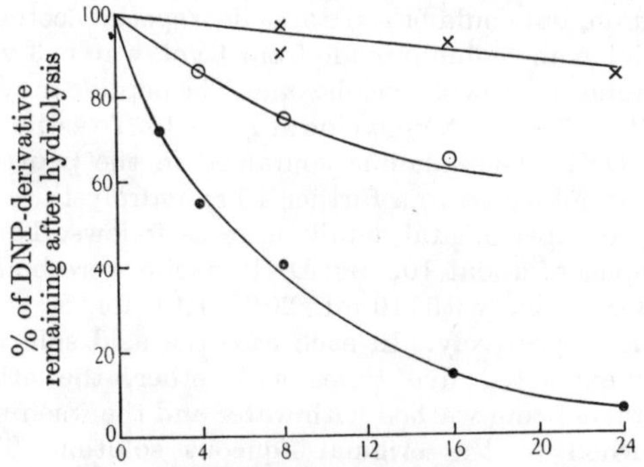

Fig. 1. Rate of hydrolysis of DNP-amino-acids with boiling 20 % (w/v) HCl. • DNP-glycine; ⊙ DNP-phenylalanine; × ε-DNP-lysine.

and samples were hydrolyzed with 20 % HCl for varying periods. These were then taken to dryness, purified on a suitable column and the amount of unchanged derivative estimated. The results of such experiments with DNP-phenylalanine, DNP-glycine and ε-DNP-lysine are shown in Fig. 1.

Approximate experiments with the DNP-derivatives of alanine, valine, arginine, serine, tryptophan, aspartic acid and diDNP-lysine indicated less than 20 % breakdown in 8 hr. It would thus appear that, with the single exception of DNP-glycine, all the derivatives exhibit satisfactory stability for the purpose intended.

Preparation of DNP-insulin

0·5 g. crystalline insulin and 0·5 g. $NaHCO_3$ were dissolved in 5 ml. water; 10 ml. ethanol and 0·5 ml. DNFB were then added and the mixture mechanically shaken for 2 hr. The DNP-insulin, which had precipitated as an insoluble yellow powder, was centrifuged down and washed with water, ethanol and ether, and air dried.

For the experiments which follow it was necessary to know the insulin content of a given sample of DNP-insulin. This could not be ascertained directly by a determination of N, as the nitro-groups would interfere, and it was therefore computed indirectly from an estimation of the amide-N. To do this a sample of the DNP-insulin was boiled for 3 hr. with 2 N-HCl and, after neutralization, the liberated NH_3 estimated in the micro-Kjeldahl apparatus using a borate buffer of pH 8·5. A control experiment with asparagine showed that treatment with DNFB did not alter the amide-N content as determined in this way. In one particular experiment, where all the values quoted below are calculated on a moisture-free basis, the sample of DNP-insulin contained 1·07 % amide-N and 0·04 % amino-N (Van Slyke). Since insulin itself contains 1·38 % amide-N (Chibnall, 1942; Rees, unpublished), it follows that 100 mg. of the DNP-insulin correspond to 77·2 mg. of insulin. If the amino-N (Van Slyke) value for insulin be taken as 0·85 % (Chibnall, 1942) it would appear that 93 % of the free amino groups of the insulin have been substituted.

Identification of the terminal residues in insulin

100 mg. DNP-insulin were boiled under reflux for 8 hr. with 10 ml. 20 % (w/w) HCl. After cooling it was extracted three times with ether. This separated the hydrolysate into two fractions; the ether extract, which contained the DNP-derivatives listed in Table 1, and the aqueous solution which contained free amino-acids and any of the DNP-derivatives listed in Table 2.

The ether extract was taken to dryness, and the material passed through an ether column to remove HCl, which would cause adsorption of the derivatives at the top of the subsequent column. The material was then fractionated on a chloroform column. There was a faint band that moved fast and so could not be an amino-acid derivative; it seems probable that it was 2:4-dinitroaniline, derived from the breakdown of a substituted histidine derivative. It could not be separated from 2:4-dinitroaniline by the use of an ethanol-ligroin or an acetone-cyclohexane column. The rest of the colour was in two bands having $R = 0{\cdot}2$ and $0{\cdot}1$ respectively. Each was 'run out' of the column separately, taken to dryness and samples used for the following experiments.

Fraction 1 (R in chloroform $= 0{\cdot}2$) formed one band having $R = 0{\cdot}5$ on an ethanol-ligroin column, and one band having $R = 0{\cdot}3$ on an acetone-cyclohexane column. These rates suggested that the material was DNP-phenylalanine, which was confirmed by the following mixed chromatograms. A sample mixed with DNP-leucine formed two bands having $R = 0{\cdot}6$ and $0{\cdot}4$ respectively on an ethanol-ligroin column. Another sample mixed with DNP-alanine formed two bands on an ethanol-ligroin column having $R = 0{\cdot}4$ and $0{\cdot}2$ respectively. A third sample mixed with DNP-valine formed two bands having $R = 0{\cdot}5$ and $0{\cdot}3$ respectively on an acetone-cyclohexane column. On mixing with authentic DNP-phenylalanine only one band was obtained on each of the above columns. Since the original hydrolysis was carried out for only 8 hr. it was possible that either band could represent a DNP-derivative of a peptide. In order to show that fraction 1 was not a peptide, a sample was hydrolyzed for a further 8 hr. with 20% HCl, extracted with ether and analyzed on the various columns given above. Only one band was obtained and the R values were unchanged. If it had been a peptide one would have expected to find evidence for a new band, representing the terminal group. It is, of course, possible that fraction 1 contained a DNP-phenylalanyl-peptide that is inseparable from DNP-phenylalanine on the columns used, but this seems unlikely and does not alter the fact that DNP-phenylalanine is present.

A sample of fraction 2 (R in chloroform $= 0{\cdot}1$) on a 5% propanol-cyclohexane column gave one band having $R = 0{\cdot}6$, and on a 33% ether-ligroin column one band having $R = 0{\cdot}55$. These rates suggested that the material was DNP-glycine and this was confirmed by the following mixed chromatograms. A sample mixed with DNP-alanine gave two bands having $R = 0{\cdot}3$ and $0{\cdot}2$ respectively on a chloroform column. Mixed with diDNP-lysine, two bands having $R = 0{\cdot}6$ and $0{\cdot}2$ respectively were formed on a 33% ether-ligroin column. Mixed with DNP-tryptophan two bands were formed which moved at $R = 0{\cdot}8$ and $0{\cdot}5$ respectively in 5% propanol-cyclohexane. When mixed with DNP-glycine only one band was obtained on each of the above columns. Further hydrolysis of the material brought about no change, showing that it was not a DNP-peptide.

After thorough ether extraction the original aqueous solution was still brightly coloured, indicative of substances listed in Table 2. The presence of ε-DNP-lysine was to be expected, and in order to ascertain what others might be there, a sample of the acid solution was taken to dryness and its behaviour on various columns studied. On a 66% M.E.K.-ether column one band was formed, having $R = 0{\cdot}2$. In 30% butanol-chloroform it moved at $R = 0{\cdot}2$. When mixed with α-DNP-arginine, two bands having $R = 0{\cdot}35$ and $0{\cdot}2$ respectively were formed on a 66% M.E.K.-ether column. These facts indicated that ε-DNP-lysine was the only coloured DNP-derivative in the aqueous solution.

The above experiments showed that the hydrolysate of DNP-insulin contained DNP-phenylalanine, DNP-glycine, ε-DNP-lysine, and no other coloured DNP-derivative.

Estimation of the amino-acids representing terminal residues in insulin

After the identity of the DNP-amino-acids produced on hydrolysis of DNP-insulin had been established, the next step was to estimate them quantitatively. It can be seen from Fig. 1 that DNP-glycine is relatively unstable to acid hydrolysis, so that if a time period of as long as 8 hr. is used a rather large correction for breakdown must be applied. Estimations were accordingly made after shorter periods of hydrolysis. In each case it was necessary to test the various fractions for the presence of peptides as described in the previous section. No evidence for these was obtained after hydrolysis for 4 hr. but after only 2 hr. a phenylalanyl-peptide was detected in fraction 2. This was inseparable from DNP-glycine on a chloroform column, but could be separated on a special acetone-cyclohexane column made from 1 vol. water, 3 vol. acetone and 10 vol. cyclohexane. The peptide moved at $R = 0{\cdot}7$ and DNP-glycine at $R = 0{\cdot}45$. To estimate the DNP-phenylalanine contained in the peptide, it was subjected to a further 4 hr. hydrolysis.

The experimental details were as follows. Three samples of about 100 mg. DNP-insulin were boiled under reflux with 10 ml. 20% HCl for 8, 4 or 2 hr. respectively. In each case the acid solution was extracted five times with ether, the ether extracts being washed with water and the washings returned to the original aqueous solution. The collected ether extracts were reduced to dryness, passed through an ether column prepared from 2 g. silica, evaporated to dryness and fractionated on a 6 g. chloroform column. The two fractions representing DNP-phenylalanine and DNP-glycine were separately collected and taken to dryness. In the 2 hr. hydrolysate the DNP-glycine fraction was purified on an acetone-cyclohexane column (containing 3 vol. acetone). The DNP-phenylalanine fraction was dissolved in 1% $NaHCO_3$, made up to 50 ml. and estimated colorimetrically. The DNP-glycine fraction was made up to 25 ml. with N-HCl. As standards the pure derivatives of the naturally occurring amino-acids were used.

The aqueous solution, after extraction with ether, was taken to dryness and subjected to a further 8 hr. hydrolysis with 20% HCl to break down any lysine peptides; this should not greatly affect the lysine estimation as the ε-DNP-derivative is relatively stable (Fig. 1). The acid solution was then made up to 100 ml., 10 ml. samples taken, reduced to dryness and passed through a 66% M.E.K.-ether column. The band was 'run out', taken to dryness and made up with N-HCl to 25 ml. for estimation.

Table 3. *Estimation of the free amino groups of insulin*

Time of hydrolysis (hr.)	Fraction	Amino-N as % insulin-N	Amino-N corrected for hydrolysis	No. of groups per insulin submolecule (mol. wt. 12,000)
8	Glycine	0·67	1·65	2·2
	Phenylalanine	1·00	1·35	1·8
	Lysine	1·16	1·3	1·7
4	Glycine	0·86	1·5	2·0
	Phenylalanine	1·05	1·2	1·6
	Lysine	1·32	1·4	1·9
2	Glycine	1·13	1·55	2·1
	Phenylalanine	1·07	—	—
	Phenylalanine from peptide	0·32	—	—
	Total phenylalanine	1·39	1·5	2·0
	Lysine	1·43	1·55	2·1
Mean	Glycine	—	—	2·1
	Phenylalanine	—	—	1·8
	Lysine	—	—	1·9

The results of the three experiments are set out in Table 3. The values given in column 3 represent the free amino-N of the amino-acid as a percentage of the original insulin-N. In column 4 these values are corrected by means of the data given in Fig. 1 for the breakdown which the derivatives are assumed to have undergone as a result of the hydrolysis; the corrections may not be strictly valid as the stability of the derivatives while still condensed in the protein or in peptide split-products may not be the same as that found for the derivatives themselves. In column 5 the corrected values are presented in terms of amino-acid residues per insulin submolecule of mol. wt. 12,000 (Chibnall, unpublished). It can be seen that within the limits of experimental error there are two residues of each amino-acid per insulin submolecule.

Isolation of DNP-amino-acids representing the terminal residues in insulin

To confirm the identity of the three DNP-derivatives obtained from insulin it was necessary to carry out an experiment on a larger scale and isolate the compounds concerned in a pure form. A 4 hr. period of hydrolysis was chosen for reasons given above.

1·03 g. air-dried DNP-insulin (corresponding to 0·755 g. moisture-free insulin) were boiled under reflux with 100 ml. 20% HCl for 4 hr. and the cooled hydrolysate extracted with ether as described above. The ether extract was taken to dryness and the material first passed through an ether column and then fractionated on a chloroform column prepared from 40 g. silica. The DNP-phenylalanine and DNP-glycine fractions were collected and worked up as follows.

N-2:4-*Dinitrophenyl*-l-*phenylalanine*. The chloroform solution was taken to dryness and the material passed through a 20 g. acetone-cyclohexane column to ensure purification. The coloured solution was taken to dryness and dissolved in a minimum of warm dilute $NaHCO_3$ solution. This solution was acidified and after standing for a few days crystals of DNP-phenylalanine had separated. The crude product weighed 14·0 mg. (38% of the material present as estimated colorimetrically). It was recrystallized from aqueous methanol and then from a mixture of acetone and cyclohexane, yielding 10·1 mg. of material which melted at 186° alone or when mixed with authentic DNP-phenylalanine. (Found: C, 55·2; H, 4·1%. Calc. for $C_{15}H_{13}O_6N_3$: C, 54·4; H, 3·9%.)

N-2:4-*Dinitrophenylglycine*. After evaporation of the chloroform solution, the whole fraction was purified on an acetone-cyclohexane column (containing 3 vol. acetone). The DNP-glycine fraction was taken to dryness, and crystallized from hot water; yield 10·5 mg. Another 5·3 mg. were obtained by working over the mother liquors, making a total yield of 15·8 mg. (90% of the amount estimated). Abderhalden & Blumberg (1910) report a m.p. of 205° for DNP-glycine. The purest samples I have synthesized usually decompose between 195 and 202° according to the rate of heating. To make the necessary comparison three melting-point tubes containing synthetic DNP-glycine, the above crystalline material and a mixture of the two respectively were heated side by side in the melting-point apparatus. They all decomposed simultaneously at 195°. (Found: C, 40·0; H, 2·9%. Calc. for $C_8H_7O_6N_3$: C, 39·8; H, 2·9%.)

ε-N-2:4-*Dinitrophenyl*-l-*lysine*. The aqueous solution containing the ε-DNP-lysine was taken to dryness and hydrolyzed for a further 20 hr. with 20% HCl. It was then evaporated to a small volume and allowed to stand overnight. An amorphous yellowish grey precipitate had formed, which was filtered off. The nature of this precipitate, which weighed 100 mg., is not yet clear. The filtrate was then taken to dryness. A column was next prepared, using 40 g. silica, on which were adsorbed 20 ml. N-HCl, 66 % M.E.K.-ether being used as the mobile phase. The ε-DNP-lysine fraction was then dissolved in 2 ml. water, which were mixed with 4 g. dry silica, suspended in the solvent, poured on to the previously prepared column and developed with 66 % M.E.K.-ether. This procedure was found more satisfactory than the normal one in the presence of large amounts of free amino-acids. As it passed out of the column the yellow material was collected in three fractions. Each of these was separately evaporated to dryness and dissolved in warm dilute HCl, filtered and allowed to evaporate slowly in a desiccator. The first fraction produced well-formed crystals of the hydrochloride, the second rather impure crystals and the third fraction would not crystallize. This suggested the presence of some colourless impurity that moves on the column at a rate somewhat slower than ε-DNP-lysine; it may have been *O*-DNP-tyrosine. On working over the various fractions, and applying chromatography to those known to be somewhat impure, there were finally obtained 10·2 mg. of pure crystals and another 14 mg. of cruder crystals, making a total yield of 24 mg. (55 % of the amount estimated). The pure material melted at 186°, alone and when mixed with authentic ε-DNP-lysine hydrochloride. (Found: C, 39·6; H, 5·4 %. Calc. for $C_{12}H_{16}O_6N_4$. $HCl.H_2O$: C, 39·5; H, 5·2 %.)

DISCUSSION

Under the conditions used in the above-mentioned experiments DNFB reacts with amino, phenolic-hydroxyl, thiol and possibly imidazole groups. The *O*-DNP-tyrosine formed by the reaction with phenolic-hydroxyl groups is colourless and thus does not interfere with the estimation of the coloured amino-derivatives. The nature of the reaction with —SH groups is not quite clear, but it appears that *S*-DNP-cysteine should be distinguishable from other DNP-derivatives chromatographically. It was not present in the insulin experiment. The reaction with the histidine imidazole group appears to be very slow, unless histidine is present as a terminal residue with its amino group free.

The results with insulin show that in every insulin submolecule of mol. wt. 12,000, two of the free amino groups are located on glycine residues, two on phenylalanine residues and two represent the ε-amino groups of lysine residues. The presence of free amino groups of phenylalanine has already been demonstrated by Jensen & Evans (1935), though not quantitatively, but terminal glycine residues have not been previously reported, nor has it been shown before that the ε-amino groups of lysine are free. The present results indicate that there are six free amino groups per submolecule of insulin. From amino-N determinations by the Van Slyke method, Chibnall (1942) suggested the presence of 21 free amino groups per molecule of mol. wt. 36,000 or seven per submolecule of 12,000. This rather higher value can probably be accounted for by the presence of glycine as a terminal residue, since glycyl-peptides have been shown to give abnormally high amino-N values by the Van Slyke method (Abderhalden & Van Slyke, 1911; Schmidt, 1929). According to Schmidt, glycylglycine gives about 135 % of the theoretical value in 10 min. so that two terminal glycine residues would react as 2·7 amino groups. Using these enhanced values for glycyl-peptides, and those of Greenstein (1933) for lysyl-peptides, it is possible to compute the course of the reaction of insulin under the Van Slyke conditions. As shown in Table 4, the agreement between the computed and experimental values is reasonable.

Table 4. *Van Slyke amino-N of insulin*

Reaction time (min.)	Apparent no. of amino groups per submolecule of mol. wt. 12,000: 2 glycyl-residues	2 phenylanalyl-residues	2 lysyl-residues	Total	Found by Van Slyke method
6	2·5	2·0	1·9	6·4	5·9
11	2·7	2·0	2·0	6·7	6·7
30	2·7	2·0	2·0	6·7	7·1

The presence of four free α-amino groups suggests that the submolecule is built up of four open polypeptide chains bound together by cross-linkages, presumably chiefly —S—S— linkages. It is, of course, possible that other chains may be present in the form of a ring structure with no free amino groups.

SUMMARY

1. A new method is described for the identification and estimation of the free amino groups of proteins and peptides by the formation of derivatives with 2:4-dinitrofluorobenzene.

2. The method has been applied to insulin. Assuming a minimum mol. wt. of 12,000, it is shown that six free amino groups are present; two of these are located on glycine residues, two on phenylalanine residues and two represent the ε-amino groups of lysine.

3. The results suggest that the insulin submolecule of mol. wt. 12,000 is made up of four open peptide chains, two of these having terminal glycyl-

residues and the other two terminal phenylalanyl-residues respectively.

I wish to express my thanks to Dr B. C. Saunders for kindly preparing the 2:4-dinitrofluorobenzene for me, to Prof. A. C. Chibnall for his advice and encouragement, and to the Medical Research Council for maintenance and expenses grants during the initial stages of this work.

REFERENCES

Abderhalden, E. & Blumberg, P. (1910). *Hoppe-Seyl. Z.* **65**, 318.

Abderhalden, E. & Stix, W. (1923). *Hoppe-Seyl. Z.* **129**, 143.

Abderhalden, E. & Van Slyke, D. D. (1911). *Hoppe-Seyl. Z.* **74**, 505.

Bergmann, M. & Zervas, L. (1928). *Biochem. Z.* **203**, 284.

Chibnall, A. C. (1942). *Proc. Roy. Soc.* B, **131**, 136.

Consden, R., Gordon, A. H. & Martin, A. J. P. (1944). *Biochem. J.* **38**, 224.

Gerngross, O. (1919). *Hoppe-Seyl. Z.* **108**, 50.

Gordon, A. H., Martin, A. J. P. & Synge, R. L. M. (1943). *Biochem. J.* **37**, 79.

Greenstein, J. P. (1933). *J. biol. Chem.* **101**, 603.

Gurin, S. & Clarke, H. T. (1934). *J. biol. Chem.* **107**, 395.

Jensen, H. & Evans, E. A. (1935). *J. biol. Chem.* **108**, 1.

Neuberger, A. & Sanger, F. (1943). *Biochem. J.* **37**, 515.

Pauly, H. (1904). *Hoppe-Seyl. Z.* **42**, 514.

Saunders, B. C. (1934). *Biochem. J.* **28**, 1977.

Schmidt, C. L. A. (1929). *J. biol. Chem.* **82**, 587.

Skraup, Z. H. & Kaas, K. (1906). *Liebigs Ann.* **351**, 379.

Synge, R. L. M. (1944). *Biochem. J.* **38**, 285.

Van Slyke, D. D. & Birchard, F. J. (1913). *J. biol. Chem.* **16**, 539.

du Vigneaud, V. & Meyer, C. E. (1932). *J. biol. Chem.* **98**, 295.

[From THE BIOCHEMICAL JOURNAL, Vol. 44, No. 1, pp. 126–128, 1949]

Fractionation of Oxidized Insulin

By F. SANGER (Beit Memorial Fellow), *Biochemical Laboratory, University of Cambridge*

(*Received* 18 *May* 1948)

From the determination of the terminal residues of insulin it was suggested that the submolecule of molecular weight 12,000 is made up of four open peptide chains bound together by disulphide linkages (Sanger, 1945). Two of these chains have glycine and two have phenylalanine as terminal residues. By the action of performic acid it is possible to split insulin into its separate polypeptide chains by conversion of the cystine to cysteic acid residues (Sanger, 1947). The object of the present work was to separate the chains in a pure form, and the first problem was to find a suitable analytical method to follow the course of fractionation. Electrophoretic analysis gave rather variable results; this was largely due to the low molecular weight of the fractions which led to rapid spreading of the boundary by diffusion and did not allow preliminary dialysis. Later the method of end-group assay (Sanger, 1945; Porter & Sanger, 1948) was used in an attempt to obtain fractions containing only glycine or only phenylalanine terminal residues. This method of characterization does not necessarily indicate complete homogeneity, but the fractions are at least representative of the two types of chains in the insulin molecule. In this paper the preparation and some preliminary observations on the properties of two fractions are reported; the one, fraction *A*, containing only glycine terminal residues and no basic amino-acids; the other, fraction *B*, containing 97% phenylalanine terminal residues.

EXPERIMENTAL

Oxidation of insulin

The method was essentially that of Toennies & Homiller (1942), who showed that within 1 hr. cystine consumes the theoretical 5 atoms of oxygen, and that after 2 hr. a more general oxidation of a number of amino-acids takes place. It was thus desirable to use a time of oxidation which was the minimum required to split the S–S bridges completely. Using paper chromatography (Consden, Gordon & Martin, 1946*b*) it was found that the reaction with free cystine was complete in 5 min., but with insulin 15 min. were required. The method finally used was as follows: 0·25 g. insulin was dissolved in 9 ml. formic acid, 1 ml. 30% (w/w) H_2O_2 was added, and the mixture allowed to stand for 15 min. at room temperature. Water (10 ml.) was added and the mixture evaporated *in vacuo* to a small volume (1–2 ml.). The oxidized protein was then precipitated by a large volume of acetone, centrifuged, washed with acetone till free of formic acid and dried in air; yield, 0·25 g.

Fractionation of the oxidized insulin

Of the various fractionation procedures tried, the most satisfactory were the conventional methods of precipitation by alteration of pH or addition of salt or ethanol. In the salting-out procedures, however, it was difficult to remove the salt since the peptides were too small to be dialyzed. This was overcome by using a volatile salt, ammonium acetate, which could be removed by evaporation.

Freshly oxidized insulin (1 g.) was dissolved in 12 ml. 0·1 M-NH_3, and 0·1 M-acetic acid was added to bring the pH of the solution to 6·5. The precipitate was centrifuged down and used in the preparation of fraction *B*. The solution was used for the preparation of fraction *A*.

Fraction A. The above solution was brought to pH 4·5 with 0·1 M-acetic acid and, after removal of the precipitate (fraction *M*), was taken almost to dryness *in vacuo*. It was then transferred to a centrifuge tube with a minimum volume (about 3 ml.) of water, and an equal volume of 50% (w/v) ammonium acetate, which had been brought to pH 5·5 with glacial acetic acid, was added. This brought about the separation of a small precipitate which was discarded. The water and ammonium acetate were then removed by leaving the solution in a high vacuum over H_2SO_4 and NaOH till it reached constant weight. The residue was a white powder; yield, 0·25–0·32 g.

Fraction B. The material precipitated at pH 6·5 from 1 g. oxidized insulin was washed with 10 ml. 0·01 M-acetic acid, dissolved in 5 ml. 0·1 M-HCl and 40 ml. absolute ethanol added. This brought about the separation of a precipitate (fraction *X*) often in the form of a gel, which was centrifuged off and washed well with 80% ethanol. The combined supernatant solution and washings were taken almost to dryness *in vacuo* and precipitated by a large volume of acetone; yield, 0·14–0·21 g. The yield of fraction *X* was 0·1–0·2 g. Various other methods were studied as a means of fractionating the oxidized insulin, but none was found more suitable than the above procedure.

Fraction *A* could be prepared satisfactorily by ionophoresis. Using ammonium acetate buffer, according to method C 1 of Consden, Gordon & Martin (1946*a*), it moved towards the anode as a discrete band, and could be identified by applying the Pauly diazo reaction to a print. After removal of the ammonium acetate *in vacuo* and elution with water the fraction was shown to contain only glycine terminal residues. Fraction *B* moved only slowly at various pH values and formed badly tailing bands indicating that adsorption on the gel had taken place. It was difficult to elute from the gel, and only a small yield was obtained of material containing only phenylalanine terminal residues.

The method of adsorption analysis (Tiselius, 1947) could be used as an analytical method (Tiselius & Sanger, 1947), but it was not found possible to use elution or displacement analysis to fractionate the material.

Using paper chromatography (Consden, Gordon & Martin, 1944), it was difficult to find a sufficiently sensitive colour test that could be applied to the paper. A weakly positive ninhydrin test was given by fraction *B*, as by insulin, but was not given by fraction *A*. The most satisfactory test was the Pauly reaction, which is given by histidine and tyrosine residues. Using this test and relatively high concentrations of the peptides it was found that they formed fast-moving tailing spots with phenol and collidine, and did not move with butanol or butanol-acetic acid.

Properties of the fractions

Fraction A. Fraction *A* is the most soluble material in the oxidized insulin. It is not precipitated at any pH value in low salt concentrations, or by any concentrations of sodium chloride or ammonium acetate, but it can be precipitated by high concentrations of ammonium sulphate.

The material, prepared as described above, contained 2·12 % NH_3 bound in salt linkage presumably to the sulphonic acid groups. Excluding this N, the N content of the dry protein was 12·80 % and the amide N 15·15 % of the total N. End-group analysis showed almost only glycine and less than 1 % phenylalanine terminal residues. The 2:4-dinitrophenyl (DNP) derivative which was prepared in the usual manner was slightly soluble in acid. Excess 1:2:4-fluorodinitrobenzene was removed as follows: The reaction mixture was taken to a small volume *in vacuo*, and the residue suspended in water and extracted well with ether. On acidification the DNP derivative was partly precipitated. Both the precipitate and the solution yielded only DNP-glycine on hydrolysis.

Paper chromatography (Consden *et al.* 1944) of a hydrolysate showed the following amino-acids to be present: leucine, isoleucine, valine, tyrosine, alanine, glycine, serine and glutamic, aspartic and cysteic acids. The following, though present in insulin, were absent from fraction *A*: lysine, arginine, histidine, threonine and phenylalanine. No proline spot could be detected, but its absence is not absolutely certain, as the ninhydrin test is not so sensitive as with other amino-acids. The absence of arginine and histidine was confirmed by the methods of Macpherson (1946) and of lysine by the absence of N^5-DNP-lysine in a hydrolysate of the DNP derivative. The absence of threonine was demonstrated by Mr M. W. Rees using the periodate technique (Rees, 1946). A very rough analysis was carried out by a method similar to the 'spot-dilution' technique of Polson, Mosley & Wyckoff (1947). All the amino-acids present could be separated using phenol-0·3 % NH_3-coal gas as solvent; standard solutions of amino-acids were run parallel with the hydrolysate of fraction *A* and the colour and size of the spots compared. The results are shown in Table 1. They are expressed as the number of residues of amino-acids/mol. of mol. wt. 2500, assuming fraction *A* to be homogeneous. While this is probably not true, the figures do give an approximate estimate of the amino-acid distribution within the limits of the method.

Table 1. *Amino-acid composition of fraction A*

Amino-acid	No. of residues/mol. of mol. wt. 2500	Amino-acid	No. of residues/mol. of mol. wt. 2500
Leucine	3	Glycine	1
Isoleucine	1	Serine	2
Valine	2	Glutamic acid	4
Tyrosine	2	Aspartic acid	2
Alanine	1	Cysteic acid	4

In developing a new method for the determination of sedimentation constants of substances of low molecular weight in the ultracentrifuge, Gutfreund & Ogston (1948) studied a rather cruder preparation of fraction *A* in which the ammonium acetate precipitation was omitted. They considered it to be relatively homogeneous and to have a mol. wt. of 2900.

Fraction B. Fraction *B* is readily precipitated near pH 6 and also by low concentrations of salt. Thus a solution in 0·03 M-HCl is precipitated by 1·5 % NaCl. Electrophoretic analysis at pH 8 showed one main boundary and a trace of a faster moving component. On ionophoresis at pH 7 it appeared to be slightly more basic than unchanged insulin, in spite of its cysteic acid content, suggesting a high content of basic amino-acids.

End-group analysis showed about 97 % phenylalanine and about 3 % glycine terminal residues. It has not yet been possible to obtain a preparation completely free of glycine terminal residues. Paper chromatography showed the presence of all the amino-acids that are found in insulin, that is to say, all the commonly occurring amino-acids except tryptophan, methionine and hydroxyproline.

In the ultracentrifuge (Gutfreund & Ogston, 1948) this preparation appeared to be less homogeneous than fraction *A* and to have a mean mol. wt. 7000.

DISCUSSION

By a variety of methods, including precipitation at pH 6 or with 3–30 % NaCl or ionophoresis at pH 7, it is possible to separate the oxidized insulin into two crude fractions: an acidic fraction *A*, with glycine as terminal residues and a basic fraction *B*, with phenylalanine as terminal residues. This does suggest that there are essentially two types of peptide chains in insulin, although the different chains of each type are probably not identical. Since the yield of the pure fraction *A* is usually greater than 25 % of the oxidized insulin, it follows that, if there are two glycyl chains in insulin, then they must both be present in this fraction, so that, besides having the same terminal residue and no basic amino-acids, they must also both be free from threonine and phenylalanine and have the same molecular weight and electrophoretic mobility.

Besides fractions *A* and *B*, the only other significant fractions are fractions *X* and *M*. It seems likely that both of the latter contain essentially the same material, since by using a lower pH for the initial precipitation, the yield of fraction *X* is increased at the expense of fraction *M* and vice versa. The sum of the yields of the two fractions is 20–30 % of the oxidized insulin, and they contain about equal amounts of the two terminal residues. By repeating the ethanol precipitation on either *X* or *M* a certain amount of fraction *B* can be prepared from them, but it seems probable that they consist chiefly of a mixture of incompletely oxidized and over oxidized insulin. Fraction *X* tends to darken on standing, suggesting the presence of oxidation products of tyrosine. The yield of these fractions is greater if an oxidation time of 5 min. is used instead of 15 min., although it is not possible to lower the yield by further oxidation. While the possibility cannot be excluded that these fractions contain another chain of the insulin molecule, it seems likely that all the chains are actually represented in the purified fractions *A* and *B*. Thus it appears that the insulin submolecule is built up of two types of peptide chains, a basic type and an acidic type. It is possible that this particular structure may account for some of the properties of insulin.

SUMMARY

Two fractions have been prepared from insulin that has been oxidized by performic acid:

A, an acidic fraction containing glycine as terminal residues and no arginine, histidine, lysine, phenylalanine or threonine.

B, a basic fraction containing phenylalanine as terminal residues and all the amino-acids present in insulin.

I wish to express my thanks to Mr M. W. Rees for the threonine analysis and to Prof. A. C. Chibnall for his advice and encouragement.

REFERENCES

Consden, R., Gordon, A. H. & Martin, A. J. P. (1944). *Biochem. J.* **38**, 224.
Consden, R., Gordon, A. H. & Martin, A. J. P. (1946*a*). *Biochem. J.* **40**, 33.
Consden, R., Gordon, A. H. & Martin, A. J. P. (1946*b*). *Biochem. J.* **40**, 580.
Gutfreund, H. & Ogston, A. G. (1948). *Biochem. J.* **44**, 163.
Macpherson, H. T. (1946). *Biochem. J.* **40**, 470.
Polson, A., Mosley, V. M. & Wyckoff, R. W. G. (1947). *Science*, **105**, 603.
Porter, R. R. & Sanger, F. (1948). *Biochem. J.* **42**, 287.
Rees, M. W. (1946). *Biochem. J.* **40**, 362.
Sanger, F. (1945). *Biochem. J.* **39**, 507.
Sanger, F. (1947). *Nature, Lond.*, **160**, 295.
Tiselius, A. (1947). *Advanc. prot. Chem.* **3**, 67.
Tiselius, A. & Sanger, F. (1947). *Nature, Lond.*, **160**, 433.
Toennies, G. & Homiller, R. P. (1942). *J. Amer. chem. Soc.* **64**, 3054.

[From THE BIOCHEMICAL JOURNAL, Vol. 45, No. 5, pp. 563–574, 1949]

The Terminal Peptides of Insulin

By F. SANGER (Beit Memorial Fellow)
Biochemical Laboratory, University of Cambridge

(*Received* 1 *June* 1949)

One of the outstanding problems of protein chemistry is the elucidation of the relative positions occupied by the amino-acid residues in the protein molecule. A method for the identification of those terminal amino-acids in which the α-amino group is free has already been applied to several proteins (Sanger, 1945; Porter & Sanger, 1948). The protein is treated with 1:2:4-fluorodinitrobenzene (FDNB) and submitted to complete hydrolysis. Thereafter the *N*-2:4-dinitrophenyl (DNP) derivatives of the amino-acids are separated from each other and quantitatively estimated. By this means insulin has been shown to contain glycine and phenylalanine in the terminal position, each present to the extent of two molecules per insulin molecule of molecular weight 12,000.

The method can be adapted in determining the sequence of amino-acids which occupy positions near to the terminal amino-acid residues. For this purpose the protein is submitted to partial hydrolysis leading to the liberation of a series of *N*-2:4-dinitrophenyl (DNP) peptides. These differ from the other products of partial hydrolysis in that they are acids which can be extracted from acid solution by organic solvents and can thus be obtained relatively free from other unsubstituted peptides. The peptide mixture so produced is relatively simple and can be fractionated by chromatography on silica gel. Determination of the structure of the individual *N*-2:4-dinitrophenyl peptides then reveals the order of the amino-acid residues in proximity to the free α-amino groups. The present paper describes the results obtained by applying the method to insulin. The nature and order of the amino-acid residues contiguous to the lysyl residues have also been determined.

Since insulin contains four free α-amino groups, it was concluded that the molecule is built up of four open polypeptide chains joined together by –S–S– bridges. These –S–S– bridges could be broken by oxidation (Sanger, 1949*a*), and on subsequent separation two main products could be obtained; an acidic fraction, *A*, which had only glycyl terminal residues, and a basic fraction, *B*, which had phenylalanyl terminal residues. For the identification of peptides described below these fractions have been used in place of the intact insulin so as to reduce the complexity of the peptide mixture dealt with.

METHODS

Materials

Insulin. The insulin used in all this work and in the earlier work (Sanger, 1945, 1949*a*) was crystalline bovine insulin having a S content of 3·2%. Fractions *A* and *B* of oxidized insulin and their DNP derivatives (referred to as DNP-*A* and DNP-*B* respectively) were prepared as previously described (Sanger, 1949*a*).

Silica gel. A number of workers who have used the silica gel solvent system procedure originally described have stated in print (Consden, Gordon, Martin & Synge, 1947; Phillips & Stephen, 1948), or in private communication, that they have experienced difficulty in that the bands have moved too rapidly down the columns. This was probably due to the properties of the silica used, although the factors enumerated below may also be contributory: (i) *Unwashed chloroform.* All $CHCl_3$ should be thoroughly washed with water to remove the ethanol added as a stabilizer. If unwashed $CHCl_3$ is used, all DNP amino-acids will travel fast on a column of silica gel. (ii) *Overloading of the column.* If too high a concentration of DNP derivative is applied to a column it will form a fast-moving band which 'tails' badly. No more than about 2 μmol. of each DNP derivative should be used on a column of diameter of 1 cm. to obtain satisfactory bands. Nevertheless, the optimum amount will depend on the particular experiment. To obtain the best separations, as are needed in the identification of a DNP amino-acid by mixed chromatograms, the smallest amount (0·1–0·2 μmol.) that can conveniently be seen on the column should be used. In quantitative estimations, a higher concentration minimizes the importance of incidental losses which are probably due to irreversible adsorption on the silica. The maximum concentration that will give a satisfactory separation should be used. In the case of the DNP peptides the limit is often set by the solubility of the derivative in the moving phase. (iii) *Water content of the silica.* The *R* values (Martin & Synge, 1941) of the derivatives vary considerably with the amount of water added to the silica, a high water content causing bands to move too rapidly. The actual capacity of the silica appears to vary from batch to batch, so that the optimal amount of water is best determined by experiments with known DNP derivatives.

All these factors apply also to the fractionation of peptides, as described below, and should be considered before discarding a gel as unsuitable for use with DNP derivatives. All gels prepared in this laboratory have been found to be satisfactory, though *R* values vary considerably from one batch to another. The gels were prepared by the method of Gordon, Martin & Synge (1943) from commercial waterglass obtained in 5 gallon drums from Joseph Crosfield Ltd., Warrington. It would seem that the source of the waterglass is more important than the actual method of preparation.

Recently Middlebrook (1949) has worked out a method for fractionating the DNP amino-acids using buffered columns (Moyle, Baldwin & Scarisbrick, 1948), which may be generally reproducible on any batch of gel. With peptides we have found that the bands tend to 'tail' rather badly on buffered columns, though they were useful for peptides that could not be dissolved easily in the solvents used on unbuffered columns, and satisfactory fractionations could often be carried out.

Fractionation of DNP peptides

The methods of fractionation of the DNP peptides are essentially the same as those used for the fractionation of DNP amino-acids (Sanger, 1945; Porter & Sanger, 1948). For a preliminary extraction of the terminal DNP peptides from other amphoteric peptides the following procedure was used. The acidic solution of the partial hydrolysate was extracted three times with equal volumes of ethyl acetate, and the combined ethyl acetate solutions extracted three times with a 1% solution of $NaHCO_3$, each bicarbonate extract being washed twice in two separating funnels with ethyl acetate. In this way one has a simple three tube counter current procedure of the type used by Craig (1944). The three bicarbonate solutions were combined, acidified and extracted again using similar counter-current procedure with ethyl acetate. This final ethyl acetate extract should contain only acidic substances. After taking to dryness the peptide mixture is fractionated on a suitable column. To describe the various solvent systems used the method and codes of Tristram (1946) are followed (Table 1). It has been found that the most generally useful solvents for the separation of the acidic terminal DNP peptides are the $CHCl_3$-*n*-butanol mixtures. The chromatographic purity of each band was usually established by running it on columns with other solvent systems.

Table 1. *Solvent systems used on silica chromatograms*

Solvent systems	Stationary phase (ml./g. silica given in brackets)	Moving phase (saturated with stationary phase)
$CHCl_3$	Water (0·5)	$CHCl_3$
CB_1	Water (0·5)	$CHCl_3$-1% (v/v) *n*-butanol
CB_5	Water (0·5)	$CHCl_3$-5% (v/v) *n*-butanol
Etc.		
Ea-HCl	N-HCl (0·5)	Ethyl acetate
Ea-pH 7	3 M-phosphate buffer, pH 7 (0·6)	Ethyl acetate
M_{66}	N-HCl (0·5)	Ether-66% (v/v) methylethylketone
M_{66}-pH 6·5	3 M-phosphate buffer, pH 6·5 (0·6)	Ether-66% (v/v) methylethylketone
Etc.		

Acetone-*cyclo*hexane: prepared from 1 vol. water, 3 vol. acetone and 10 vol. *cyclo*hexane (Sanger, 1945).

Sometimes it is found that a soluble compound, after being passed through a column and taken to dryness, has become apparently insoluble. This is believed to be due to the formation of salts, and the material can generally be brought into solution once again by adding a drop of conc. HCl and a solvent in which it is readily soluble such as ether or ethyl acetate. Such a mixture is then taken to dryness and rapidly dissolved in the required solvent. This technique has also proved useful for dissolving DNP peptides (such as *A* 5) which are themselves very insoluble.

It was found that the yellow peptides containing N^5-DNP-lysine could be fractionated from one another very successfully on suitable silica columns, on which they form sharp well-defined bands. Such bands, however, are often contaminated with other colourless peptides, which can be separated by adsorption chromatography on talc from acid solution. The DNP derivatives are adsorbed firmly on the talc, whereas the simple peptides and amino-acids usually pass straight through the column.

To prepare the column, talc (5 g.) was washed several times by decantation with N-HCl. The suspension was then poured into a chromatogram tube of 1 cm. diameter and allowed to settle under slight suction. The peptide mixture was then put on the column. The DNP derivatives were adsorbed in a band at the top. After washing with about 40 ml. N-HCl, they were eluted with a mixture of 4 parts ethanol and 1 part N-HCl, in which they moved rapidly down the column. This technique could be applied to the separated DNP-lysyl peptides or to the aqueous solution of the hydrolysate before fractionation.

When it was necessary to defer the working up of a DNP peptide fraction it was usually dissolved or suspended in dilute HCl and kept in the dark. No destruction of DNP derivatives preserved in this way has been observed.

Estimation of DNP peptides

The yellow DNP amino-acids and peptides were estimated colorimetrically using the Beckman spectrophotometer. Fig. 1 shows the spectral absorption curves for

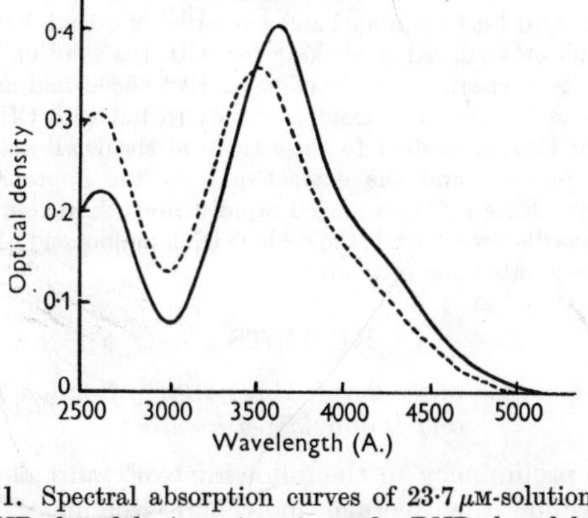

Fig. 1. Spectral absorption curves of 23·7 μM-solutions of DNP-phenylalanine (——) and DNP-phenylalanyl-valine (- - - - -) in 1% $NaHCO_3$.

DNP-phenylalanine and DNP-phenylalanyl-valine (prepared from DNP-insulin, as described below) in 1% $NaHCO_3$. It was found that the higher peptides of DNP-phenylalanine (*B* 3 and *B* 4) gave absorption curves identical with that for DNP-phenylalanyl-valine at wavelengths above 3000 A. To obtain these curves the concentration of DNP peptide was estimated from the amount of DNP-phenylalanine produced on hydrolysis. The absorption curve for DNP-glycine was almost identical with that of DNP-phenylalanine, and the curve for DNP-glycyl-glycine identical with that for DNP-phenylalanyl-valine. The DNP-glycyl peptides (*A* 4 and *A* 5) separated in the present work gave the same curve as DNP-glycyl-glycine at wavelengths >3300 A. All these DNP derivatives give identical readings at 3500 A. and the estimations have been done at this wavelength. In most experiments readings were also taken at 3900 A. and agreement between the results at the two wavelengths indicated the absence of other absorbing impurities. Table 2 shows the optical densities of 20 μM-solutions of the

Table 2. *Optical densities of 20 μM-solutions of DNP amino-acids and peptides*

DNP derivative	Solvent	Optical density (log I_0/I) 3500 A.	3900 A.
DNP-glycine	1% $NaHCO_3$	0·309	0·210
DNP-phenylalanine	1% $NaHCO_3$	0·313	0·214
DNP-glycyl-glycine	1% $NaHCO_3$	0·316	0·173
DNP-phenylalanyl-valine	1% $NaHCO_3$	0·310	0·178
N^5-DNP-lysine	N-HCl	0·296	0·204
O-DNP-tyrosine	N-HCl	0·058	0·0

different DNP derivatives. Beer's law is obeyed in solutions of concentrations <50 μM. Standard curves for a number of other DNP amino-acids in 1% $NaHCO_3$ have also been determined for a wavelength of 3500 A. and have been found to differ only slightly from those of the above derivatives.

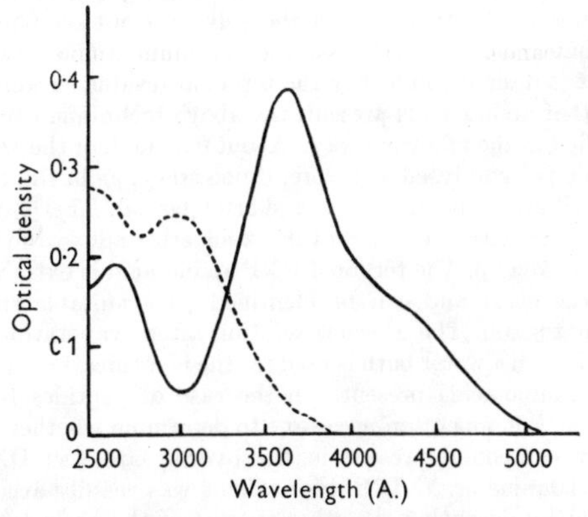

Fig. 2. Spectral absorption curves of 22 μM-solutions of N^5-DNP-lysine (——) and *O*-DNP-tyrosine (- - - - -) in N-HCl.

The determinations of the peptides containing N^5-DNP-lysine were carried out in N-HCl solutions, since N^5-DNP-lysine is unstable in alkaline solutions. The absorption curves for N^5-DNP-lysine and *O*-DNP-tyrosine are shown in Fig. 2. The curves for the lysyl peptides (*L* 2, *L* 3 and *L* 4) described below were identical with that obtained for N^5-DNP-lysine at wavelengths >3300 A. In order to avoid interference from *O*-DNP-tyrosine, estimations of N^5-DNP-lysine and its peptides were carried out using a wavelength of 3900 A.

Identification of amino-acids present in DNP peptides

The amino-acids present in the DNP peptides were identified by paper chromatography (Consden, Gordon & Martin, 1944, 1947). After purification of the band on a suitable column, a sample containing 0·1–0·5 μmol. peptide was taken to dryness in a 50 ml. round-bottomed flask; 5 drops of 5·7 N-HCl were added and the mixture refluxed for 1 min. on a microflame to dissolve the peptide and wash down the sides of the flask. The contents were sucked into a small capillary tube using a piece of valve rubber as described by Consden, Gordon & Martin (1947). The capillary was sealed off and incubated at 105° for 24 hr. After cooling, the hydrolysate was transferred to a polythene strip and the HCl removed in a desiccator. A suitable portion of the residue was then taken for identification of the amino-acids by paper chromatography. Since only a few amino-acids were present in each peptide, it was best to use one-dimensional chromatograms on which known amino-acids were run parallel as controls. In the case of fraction *A* all the amino-acids present, except leucine and isoleucine, could be separated using the system phenol-0·3% NH_3-coal gas. Leucine and isoleucine were differentiated using 1:1 *n*-butanol-benzyl alcohol-HCN. All the amino-acids from fraction *B* could be identified using the above mixtures and *n*-butanol-acetic acid (Partridge, 1948). This latter solvent has been found to be very useful for amino-acid identification and to replace collidine for most purposes, though the R_F values vary considerably with the age of the mixture, and it is always advisable to run known amino-acids parallel to the unknowns. N^5-DNP-lysine and *O*-DNP-tyrosine can also be identified using this solvent, whereas they both run with the solvent front on phenol chromatograms.

If it is desired to identify the terminal residue as well as the other amino-acids present, the above technique can be modified in the following way. About 0·5 μmol. of the DNP peptide is hydrolysed as before, if necessary, as in the case of DNP-glycyl peptides, for a shorter period. The hydrolysate is transferred to a test tube, 2 ml. ether added and the tube shaken up. The terminal DNP amino-acid is extracted into the ether and can be identified on a suitable silica chromatogram. The aqueous solution after evaporation of the ether on a water bath is used for the identification of the other amino-acids present. In the case of peptides from fraction *B* it was often necessary to determine whether the yellow compound present in the hydrolysate was DNP-phenylalanine or N^5-DNP-lysine. This was readily accomplished by decanting the ether extract of the hydrolysate into another test tube and shaking it with a few drops of $NaHCO_3$ solution. N^5-DNP-lysine remains in the acid hydrolysate in the first test tube, whereas DNP-phenylalanine passes into the $NaHCO_3$ solution in the other test tube.

Estimation of amino-acids present in DNP peptides

In order to determine the amount of each amino-acid residue present in a peptide it was necessary to use an approximate micromethod, which could differentiate between one, two or three amino-acid residues per molecule. Quantitative paper chromatography, as used in the 'spot dilution' method of Polson, Mosley & Wyckoff (1947), was found to be very suitable and sufficiently accurate for the purpose.

After estimation of the DNP peptide in 1% $NaHCO_3$ a measured fraction (containing 0·5–1·0 μmol.) was acidified and extracted three times with ethyl acetate, each extract being washed with water to remove salts. The combined ethyl acetate solutions were taken to dryness *in vacuo* and the residue boiled under reflux with 5·7 N-HCl for a sufficient time to liberate all the terminal residue (2 hr. in the case of DNP-glycyl peptides and 4 hr. for DNP-phenylalanyl peptides). After cooling, the hydrolysate was extracted with ether, each extract being washed with water. The DNP amino-acid which was present in the ether extract was purified on a $CHCl_3$ column and estimated. The combined aqueous solution and washings were taken to dryness and hydrolysed for a further 20 hr. with 5·7 N-HCl. Excess HCl was removed by repeated evaporation *in vacuo* and the residue made to a suitable volume. Samples corresponding to various amounts between 0·05 and 0·005 μmol. peptide were then applied to a one-dimensional chromatogram and similar amounts of a standard solution of amino-acids were applied alongside the unknowns. After development of the chromatogram in a suitable solvent and treatment with ninhydrin the amount of each amino-acid could be determined by comparing the colour and size of the spots. In the case of the peptides of DNP-glycine and DNP-phenylalanine, phenol-0·3% NH_3-coal gas was a suitable solvent, and for the N^5-DNP-lysine peptides *n*-butanol-acetic acid was used.

With certain peptides it was desirable to confirm the estimations by a more accurate method and it was found that, where only a few known amino-acids were present, they could be estimated with reasonable accuracy by conversion to the DNP derivatives. The following technique was used. The aqueous hydrolysate from about 1·0 μmol. of DNP peptide, after extraction of the terminal DNP amino-acid, was taken to dryness and dissolved in 1 ml. water. $NaHCO_3$ (0·1 g.) was added and a solution of 0·1 ml. FDNB in 2 ml. ethanol. After shaking for 2 hr. the solution was taken to dryness, a few drops conc. HCl added and again taken to dryness. The residue was extracted with $CHCl_3$, CB_1 or CB_3, according to the nature of the DNP amino-acids present, and the extract put on the appropriate column. Excess FDNB moved rapidly down these columns and was discarded, while the desired DNP amino-acid bands were separated and estimated.

RESULTS

Estimation of the terminal residues of fraction A and B of oxidized insulin

As a preliminary to the following work, and also to determine the average molecular weight of the peptide chains of fractions *A* and *B* of oxidized insulin it was necessary to estimate the terminal residues of these fractions as accurately as possible.

The methods employed were those described in an earlier paper on insulin (Sanger, 1945); about 20 mg. of material were used in each experiment, which was carried out in duplicate. Since the main error is probably due to the correction factor that has to be applied for the breakdown of the DNP amino-acids during hydrolysis, the times of hydrolysis were kept short, and small amounts of peptides present were estimated separately. In the earlier work the amount of protein present in a DNP protein was estimated from the content of amide N. This required more material than was available, so the molecular weight of the peptide was calculated from the molecular weight of the DNP derivative by assuming that one DNP group (mol.wt. 167) is attached to every free amino group and to every tyrosine and histidine residue (Porter, unpublished work). In the case of insulin this method gave the same results as the amide method.

Fraction A. For the estimation of the DNP-glycine in DNP-*A* a time of hydrolysis of 2 hr. was used and the DNP-glycine was purified on a $CHCl_3$ column. A small amount of a peptide band (probably *A*4) was adsorbed tightly at the top of the column. This was eluted with CB_{17} and estimated separately. The results are shown in Table 3. The

Table 3. *Terminal residues of fractions A and B*

(Results corrected for breakdown during hydrolysis.)

	Fraction	μmol. DNP amino-acid/ 100 mg. DNP derivative	Mol. wt. of DNP derivative
A	DNP-glycine	27	—
	DNP-glycyl peptide	2	—
	Total	29	3450
B	DNP-phenylalanine	17	—
	DNP-phenylalanyl peptides	1·5	—
	Total	18·5	5100
	DNP-glycine	1	—
	N^5-DNP-lysine	19·5	4800

molecular weight obtained for DNP-*A* is 3450. Fraction *A* contains one free amino group and two tyrosine residues. It has no lysine or histidine (Sanger, 1949*a*), so that the DNP-*A* should contain three DNP groups, giving an additional molecular weight of 500. The molecular weight of the average chain in fraction *A* is thus 2950, which agrees well with the value of 2900 determined in the ultracentrifuge (Gutfreund & Ogston, 1949). This value corresponds to a molecular weight of 2750 for the individual chains in the intact insulin, the difference being due to the oxygen taken up during the treatment with performic acid.

Fraction B. DNP-*B* was hydrolysed for 4 hr. in boiling 5·7 N-HCl and the hydrolysate extracted with ether. The ether extract was used for the estimation of DNP-phenylalanine and the aqueous solution for N^5-DNP-lysine. The ethereal solution was taken to dryness and the residue put on a $CHCl_3$ column. Three bands were obtained. The main DNP-phenylalanine band was estimated directly. This was followed by a fainter band containing peptide *B*2 and a small amount of DNP-glycine, which is usually present in preparations of fraction *B*. These two were separated on an acetone-*cyclo*hexane column. There was also a faint band remaining at the top of the $CHCl_3$ column, which consisted of higher DNP-phenylalanyl peptides and could be eluted with CB_{17}.

The aqueous solution, after extraction with ether, was hydrolysed for a further 20 hr. and the N^5-DNP-lysine purified on an M_{66} column and estimated.

The results for DNP-*B*, set out in Table 3, lead to an average molecular weight of 4950 on the assumption that the small amount of DNP-glycyl peptide has the same molecular weight as DNP-*A*. Fraction *B* contains two free amino groups, three tyrosine residues (estimated by difference between the tyrosine contents of insulin and fraction *A*) and two histidine residues (estimated directly on fraction *B* using histidine decarboxylase; Gale, 1945). DNP-*B* should therefore contain seven DNP groups and the molecular weight of fraction *B* would be 3800, and of the intact chains in insulin 3700. The molecular weight as determined in the ultracentrifuge was 7000 (Gutfreund & Ogston, 1949), so that it would appear that either two phenylalanyl peptide chains are joined together by some non-oxidizable cross-linkage or that in solutions of fraction *B* aggregates are formed containing two molecules. Since non-oxidizable cross-linkages are unknown, while aggregation reactions of insulin are well known (summarized by Sanger, 1949*b*), the latter explanation would seem the more probable

Isolation of DNP-phenylalanyl-valine from DNP-insulin

When DNP-insulin was hydrolysed for 2 hr. with boiling 5·7 N-HCl, a peptide band was present that moved at the same rate as DNP-glycine on a $CHCl_3$ column, but could be separated using an acetone-*cyclo*hexane column. In the experiment described by Sanger (1945), in which the DNP-amino-acids were isolated from 1·03 g. DNP-insulin, the DNP-glycine and the peptide were fractionated and the solution of the peptide in acetone-*cyclo*hexane allowed to stand overnight. It crystallized as needles as the acetone evaporated (yield 8·1 mg.).

A sample was hydrolysed and DNP-phenylalanine identified as the terminal residue. Paper chromatography showed that valine and no other amino-acid was present. By quantitative paper chromatography it was shown that there was one residue of valine per molecule of peptide and this was confirmed by estimation as the DNP derivative after purification on a $CHCl_3$ column. (Found: 3·3 % valine N as percentage of peptide; calc. for DNP-phenylalanyl-valine, 3·2; for DNP-phenylalanyl-valyl-valine, 5·3 %.) The only possible structure for this peptide is therefore DNP-phenylalanyl-valine.

Identification of DNP peptides from fraction B

In the initial experiments, DNP-*B* was hydrolysed for 8 days with conc. HCl at 37°.

DNP-phenylalanyl peptides. The material extracted from the hydrolysate by ethyl acetate was first put through a CB_{17} column. Most of the colour travelled fast and only a few faint bands could be distinguished on the column. These were not investigated in detail. The fast-moving material was then fractionated on a CB_5 column. Three main bands were formed; *B* 3, *B* 4 and a band that moved fast; there were also a number of fainter bands moving more slowly than *B* 4 and a fairly strong one that remained at the top of the column and could be eluted with CB_{17}. Bands *B* 3 and *B* 4 were shown to be homogeneous on Ea-pH 7 columns ($R = 0·25$ and 0·1 respectively). The material that moved fast in CB_5 consisted of DNP-phenylalanine, DNP-phenylalanyl-valine (*B* 2) and a small amount of DNP-glycine. These were separated using a $CHCl_3$ column, and the second band refractionated on acetone-*cyclo*hexane. In some experiments a band was present that was adsorbed at the top of the $CHCl_3$ column and usually split into a number of faint bands on development with CB_1. It contained a rather complex mixture of amino-acids. These bands were believed to be due to some secondary products (possibly esters), formed from the DNP peptides during the fractionation. They did not occur in most experiments, and the same phenomenon was sometimes observed with DNP-*A*.

Table 4. *Peptides containing DNP-phenylalanine*

Peptide	Column used for separation	*R* value	Amino-acids present
B 1	$CHCl_3$	0·4	None
B 2	$CHCl_3$	0·2	Valine
	Acetone-*cyclo*hexane	0·7	
B 3	CB_5	0·6	Valine, aspartic acid
B 4	CB_5	0·3	Valine, aspartic acid, glutamic acid

The amino-acids identified in the main fractions are shown in Table 4. In each case the terminal residue was DNP-phenylalanine, and quantitative paper chromatography showed that the amino-acids were present in equimolecular proportions in all the peptides. The regularly increasing complexity of these peptides suggested that they were all derived from the same peptide chain containing the terminal sequence DNP-phenylalanyl-valyl-aspartyl-glutamic acid, and this was confirmed by the following partial hydrolysis experiments.

Peptide *B* 3 (1·5 μmol.) was boiled under reflux with 5·7 N-HCl for 1 hr. and the degradation products studied chromatographically. The results are shown in Table 5. The only products of hydrolysis were DNP-phenylalanine and DNP-phenylalanyl-valine, which confirms the structure DNP-phenylalanyl-valyl-aspartic acid.

Table 5. *Partial hydrolysis of peptide* B 3

Peptide	Column used for separation	*R* value	Amino-acids present
B 3·1	$CHCl_3$	0·4	None
B 3·2	$CHCl_3$	0·2	Valine
	Acetone-*cyclo*hexane	0·7	
B 3·3	CB_5	0·6	Valine, aspartic acid

The partial hydrolysis of peptide *B* 4 was carried out for 14 days at 37° in conc. HCl. All the bands *B* 1, *B* 2, *B* 3 and unchanged *B* 4 were identified and shown to contain the same amino-acids as in Table 4. Since *B* 4 gives rise to peptide *B* 3, its structure can only be DNP-phenylalanyl-valyl-aspartyl-glutamic acid.

Peptides containing N^5*-DNP-lysine.* After extraction with ethyl acetate the aqueous solution from the hydrolysate of DNP-*B* was taken to dryness and fractionated on an M_{66} column. Three main bands were formed, and the faster one was refractionated on M_{66}-pH 6·5 to give two bands (*L* 1, *L* 2). There was also a fainter fast band that appeared to consist of higher peptides of N^5-DNP-lysine. Band *L* 2 was shown to be homogeneous on Ea-HCl ($R = 0·2$) and CB_{30} ($R = 0·15$) columns, and bands *L* 3 and *L* 4 were homogeneous on CB_{30} ($R = 0·07$ and 0·04 respectively). Each peptide was purified by chromatography on talc, and the amino-acids identified and shown to be present in equimolecular amounts. The results are shown in Table 6. The terminal residues were determined by deamination with nitrosyl chloride (Consden, Gordon & Martin, 1947), and confirmed by the DNP technique. It was found that the latter method gave a rather more clear-cut result, whereas the former was much simpler to carry out. In a preliminary report of this work

Table 6. *Peptides containing* N^5*-DNP-lysine*

Peptide	*R* values: On M_{66}	*R* values: On M_{66}-pH 6·5	Amino-acids present	Terminal residue
L 1	0·6	0·3	N^5-DNP-lysine	
L 2	0·6	0·1	N^5-DNP-lysine, alanine	N^5-DNP-lysine
L 3	0·2	—	N^5-DNP-lysine, threonine, proline	Threonine
L 4	0·13	—	N^5-DNP-lysine, threonine, proline, alanine	Threonine

(Sanger, 1948) the peptide *L*4 was described as threonyl-(N^5-DNP)-lysyl-alanine. The presence of proline had been overlooked in the paper chromatography, the yellow spot being confused with that due to N^5-DNP-lysine. It can, however, be clearly detected if the solvent *n*-butanol-acetic acid is used.

The results in Table 6 show that peptide *L*2 is N^5-DNP-lysyl-alanine. Peptides *L*3 and *L*4 were hydrolysed for 11 days with conc. HCl at 37°. The hydrolysate of *L*3 contained only the bands *L*1 and *L*3, and a faint band that moved at the same rate as *L*2 on M_{66}-pH 6·5 and contained N^5-DNP-lysine and proline. From peptide *L*4 all the bands *L*1, *L*2, *L*3 and *L*4 were obtained and were shown to contain the expected amino-acids. Since *L*4 has a terminal threonyl residue and gives rise to N^5-DNP-lysyl-alanine (*L*2) and a peptide (*L*3), containing threonine, proline and DNP-lysine, the only possible structure that can be assigned to it is threonyl-prolyl-(N^5-DNP-)-lysyl-alanine, and *L*3 must be threonyl-prolyl-(N^5-DNP-)-lysine.

Estimation of peptides containing DNP-phenylalanine and N^5-DNP-lysine in insulin

Having established the presence of the two peptide sequences, phenylalanyl-valyl-aspartyl-glutamic acid and threonyl-prolyl-lysyl-alanine, in fraction *B* of the oxidized insulin, the next problem was to estimate the yields of the various peptides from insulin, and to discover what proportion of the lysyl residues and the terminal phenylalanyl residues of insulin was present in these peptide sequences. When DNP-insulin was hydrolysed for 8 days with conc. HCl no peptides could be detected other than those which were obtained from DNP-*B* and very small amounts of DNP-glycyl peptides which did not interfere with the estimation of peptides containing DNP-phenylalanine and N^5-DNP-lysine. Almost all the terminal glycine was present as DNP-glycine itself, owing to the great lability of the glycyl peptide bond in strong acid.

About 100 mg. DNP-insulin were hydrolysed for 8 days at 37° with conc. HCl, the products fractionated as previously described for DNP-*B* and estimated colorimetrically. Peptides *B*5, *B*6 and *B*7 were obtained from bands that moved slower than *B*4 on CB_5. It was shown that each gave rise to band *B*4 on partial hydrolysis. *B*8 was composed of a number of faint bands that moved slowly on CB_{17}. These were eluted from the column with 80% (v/v) acetone, hydrolysed and the DNP-phenylalanine estimated after purification on a $CHCl_3$ column. The yields of the various peptides are shown in Table 7, expressed as moles of peptide as a percentage of the total DNP-phenylalanine in the original DNP-insulin, assuming there are exactly two phenylalanyl terminal residues per insulin molecule of molecular weight 12,000. It can be seen that 92% of the total DNP-phenylalanine is accounted for as compounds that fit into the peptide sequence DNP-phenylalanyl-valyl-aspartyl-glutamic acid, and it would seem evident that in fact all the terminal phenylalanyl residues are present in this sequence. An actual estimation of the amount of the peptide *B*4 in DNP-insulin can, however, be obtained by studying the hydrolysis of the peptide itself under

Table 7. *Yields of DNP-phenylalanyl peptides from an 8-day hydrolysate of DNP-insulin*

(Results expressed as mol. of DNP peptide as percentage of the total terminal phenylalanyl residues.)

	Peptide	Yield (%)	
*B*1	DNP-phenylalanine		13
*B*2	DNP-phenylalanyl-valine	16	
*B*3	DNP-phenylalanyl-valyl-aspartic acid	13	
*B*4	DNP-phenylalanyl-valyl-aspartyl-glutamic acid	30	
	Total known peptides (*B*2, *B*3, *B*4)		59
*B*5		6	
*B*6		4	
*B*7		10	
	Total higher peptides from the same sequence		20
*B*8	Unknown mixture of peptides		6
	Total		98

Table 8. *Yields of DNP peptides from an 8-day hydrolysate of peptide* B4

Peptide	Yield from *B*4 (%)	Yield from DNP-insulin (%)
*B*1	13	13
*B*2	14	16
*B*3	12	13
*B*4	55	50–56

the conditions that were used for hydrolysing the DNP-insulin. Thus, if all the DNP-phenylalanine in DNP-insulin is combined in the single tetrapeptide sequence, then the partial hydrolysis of peptide *B*4 should give the same products in the same proportions as are obtained from DNP-insulin. This assumes that the stability of the various peptide bonds is the same whether the DNP-peptide is free or combined in the protein. While this may not be completely true it is probably a safe approximation. Table 8 shows the yields of DNP-peptides from peptide *B*4 (2 μmol.) that had been hydrolysed for 8 days in conc. HCl at 37°, compared with the yields from DNP-insulin. The figure for the amount of *B*4 obtained from DNP-insulin includes higher peptides containing this sequence. It can be seen that within the limits of the methods the two sets of figures agree. The degree of accuracy of these results probably indicates that not more than about 10% of the terminal phenylalanine of insulin is present in sequences other than that detected here. Thus, not only do these results prove that both the phenylalanine terminal residues in the insulin molecule of molecular weight 12,000 are combined in the same tetrapeptide sequence, but they do justify the assumption of this value for the molecular weight.

Table 9. *Yields of peptides containing* N⁵-*DNP-lysine from an 8-day hydrolysate of DNP-insulin*

(Results expressed as mol. of peptide as percentage of the total lysine.)

	Peptide	Yield from DNP-insulin (%)	Yield from *L* 4 (%)
L 1	N^5-DNP-lysine	14	14
L 2	N^5-DNP-lysyl-alanine	19	23
L 3	Threonyl-prolyl-(N^5-DNP)-lysine	32	32
L 4	Threonyl-prolyl-(N^5-DNP)-lysyl-alanine	23	21
	Total known peptides (*L* 2, *L* 3, *L* 4)	74	
L 5	Unknown	6	
	Total	94	90

Table 9 shows the yields of the various peptides containing N^5-DNP-lysine obtained from DNP-insulin and from a similar hydrolysate of peptide *L* 4. *L* 5 is the uninvestigated band that moved fast on an M_{66} column. The N^5-DNP-lysine in it was estimated after complete hydrolysis. The yields of the peptides from DNP-insulin make it very improbable that there is a second type of peptide containing lysine and agreement between the two sets of figures confirms this.

Identification of DNP-glycyl peptides from fraction A

Hydrolysis in concentrated hydrochloric acid. Owing to the great lability to strong acid of the peptide bond involving the carboxyl group of the terminal glycyl residues, partial hydrolysates of DNP-*A* carried out in conc. HCl contained large amounts of DNP-glycine and of long peptides that could not be investigated easily, and only poor yields of small peptides that could readily be studied. Nevertheless, it was possible to determine the structure of three of these small peptides, and as most of the work was carried out on such hydrolysates, the results are reported here, although much more satisfactory yields were obtained at a later stage when the hydrolysis was carried out in dilute acid.

DNP-*A* was hydrolysed for 24 hr. with conc. HCl at 37°, and the material that was extracted with ethyl acetate first put on a CB_{17} column. Besides the main fast-moving band there were a number of fainter bands. The strongest of these, which travelled at a gradually decreasing rate down the column, gave rise to the amino-acids leucine, serine, glutamic acid and *O*-DNP-tyrosine on hydrolysis. A peptide containing these amino-acids had been reported by Woolley (1948), and it was at first assumed that the two were the same (Sanger, 1948). On quantitative estimation, however, it was found that there were about five residues of each of the above amino-acids per residue of DNP-glycine, so that the fraction was a mixture of a small amount of a DNP-glycyl peptide (probably *A* 5), which was responsible for the colour of the band on the column, and a large amount of a non-terminal peptide containing the above four amino-acids. The exact structure of this peptide has not yet been determined, but its presence does indicate that DNP-peptides, separated by the above methods, may be contaminated by non-terminal peptides.

The coloured material that moved fast on the CB_{17} column was next put on a CB_5 column. This separated a band (*A* 4) which was run out by development with CB_8. The fast-moving material from this column was fractionated on a $CHCl_3$ column. A very strong DNP-glycine band (*A* 1) was present, and this was followed by two peptide bands (*A* 2 and *A* 3) which were developed with $CB_{0{\cdot}5}$ and CB_1. Band *A* 2 was relatively faint and sometimes difficult to detect. Bands *A* 3 and *A* 4 were homogeneous on Ea-pH 7 columns ($R = 0{\cdot}4$ and $0{\cdot}1$ respectively) and band *A* 4 was homogeneous on M_{66}-pH 7 ($R = 0{\cdot}4$) and M_{66}-pH 8 ($R = 0{\cdot}07$).

Table 10. *Peptides of DNP-glycine*

Peptide	Column used for separation	*R* value	Amino-acids present
A 1*	$CHCl_3$	0·2	—
A 2	$CHCl_3$	0·06	Isoleucine
	$CB_{0{\cdot}5}$	0·2	
	CB_1	0·6	
A 3	$CHCl_3$	0·03	Isoleucine, valine
	$CB_{0{\cdot}5}$	0·1	
	CB_1	0·35	
A 4	CB_5	0·25	Isoleucine, valine, glutamic acid
	CB_8	0·8	

* DNP-glycine.

Table 10 shows the amino-acid composition of the three bands. In each case the terminal residue was DNP-glycine. Estimation showed that the amino-acids in *A* 3 and *A* 4 were present in equimolecular proportions to the DNP-glycine. The composition of the peptides suggests that they are all derived from the same peptide chain containing the terminal sequence DNP-glycyl-isoleucyl-valyl-glutamic acid and that this was the structure of *A* 4 was shown by partial hydrolysis. A sample was refluxed for 12 hr. with 0·1 N-HCl. The hydrolysate contained the peptide bands *A* 2 and *A* 3 and isoleucine and valine were identified in the *A* 3.

Hydrolysis in hydrochloric acid (0·1 N). In an attempt to find a more specific method for hydrolysing proteins, the action of boiling 0·1 N-HCl on DNP-*A* was studied. It was found that the specificity of this hydrolytic agent was very different from that of concentrated HCl, and that the smaller DNP-peptides were produced in much better yields. It seems probable that the hydrolysis of these oxidation products in dilute acid is analogous to the hydrolyses catalysed by long-chain sulphonic acids (Steinhardt & Fugitt, 1942), the peptide chains containing $-SO_3H$ groups acting as long-chain anions and catalysing a particular type of hydrolysis which is characterized by the increased lability for the amide groups and no doubt a corresponding difference in specificity towards other peptide bonds. That the $-SO_3H$ groups were catalysing the hydrolysis is indicated by the difference in the rate of hydrolysis of insulin and oxidized insulin illustrated in Fig. 3. No such difference was found when the hydrolyses were carried out in conc. HCl in which the $-SO_3H$ groups are uncharged. In these experiments the liberation of free amino groups was followed using the Van Slyke apparatus.

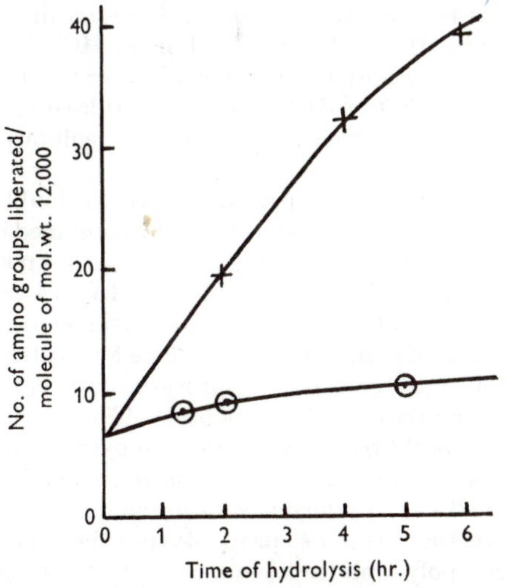

Fig. 3. Rate of hydrolysis of insulin (⊙) and oxidized insulin (+) in boiling 0·1 N-HCl.

It should be noted that the formation of diketopiperazines in dilute acid has been demonstrated (Abderhalden & Komm, 1924), so their presence was a possibility in these experiments, though the temperature employed was probably too low. The DNP derivatives could not undergo any rearrangement of this type as they contain no free amino group, and there was no evidence of any contamination of the DNP peptide bands with diketopiperazines.

In a hydrolysate of DNP-*A* that had been boiled under reflux for 12 hr. with 0·1 N-HCl there was present, besides the three previously described peptides, considerable amounts of another peptide (*A* 5) which moved with $R=0{\cdot}9$ on CB_{17} and was best purified on CB_{15} ($R=$approx. 0·4), though it was extremely insoluble, and only faintly coloured bands could be obtained. On an Ea-pH 6 column it gave an apparently homogeneous band with $R=0{\cdot}7$, though it 'tailed' rather badly on all buffered columns. This peptide contained DNP-glycine and the amino-acids isoleucine, valine and glutamic acid only. Quantitative paper chromatography showed that there were two residues of glutamic acid to every one residue of DNP-glycine, isoleucine and valine, and this was confirmed by estimation of the glutamic acid as the DNP derivative. No amide group was present, since no NH_3 was formed on hydrolysing for 4 hr. with boiling 2 N-HCl. Table 11 shows the results of an

Table 11. *Partial hydrolysis products of peptide* A 5

Peptide	Column used for separation	*R* value	Amino-acids present	Yield of peptide (%)
A 5·1	$CHCl_3$	0·1	—	29
A 5·2	CB_1	0·6	—	4
A 5·3	CB_1	0·35	Isoleucine, valine	10
A 5·4	CB_5	0·25	Isoleucine, valine, glutamic acid	19
A 5·5	CB_{15}	0·4	Isoleucine, valine, glutamic acid	27

experiment in which *A* 5 was hydrolysed for 12 hr. with boiling 0·1 N-HCl and the hydrolysis products investigated. It can be seen that it breaks down to give peptides *A* 2, *A* 3 and *A* 4, therefore the only possible structure is DNP-glycyl-isoleucyl-valyl-glutamyl-glutamic acid.

Estimation of the DNP-glycyl peptides from DNP-insulin

Since DNP-insulin is insoluble in 0·1 N-HCl it was not possible to obtain a significant yield of DNP peptides by hydrolysing it directly. However, if the insulin was first oxidized and then treated with FDNB, there was very little loss of material, and the DNP-*A* present was largely soluble and gave good yields of the peptides.

In order to obtain the maximum possible yields of peptides at the expense of the yield of DNP-glycine, the hydrolysis mixture was extracted from time to time during the course of hydrolysis, so that the larger extractable peptides (*A* 4 and *A* 5) were removed and not broken down to DNP-glycine. Also unidentified fractions were hydrolysed further to give the known peptides.

As a preliminary, the breakdown of DNP-*B* under these conditions was studied. Owing to its insolubility it was not extensively hydrolysed. The main breakdown products were peptide *B* 3 and a little *B* 2. There was also a small amount of an unidentified DNP-phenylalanyl peptide that contaminated *A* 5.

Insulin (0·2 g.) was oxidized with performic acid for 30 min. (Sanger, 1949*a*). The product was precipitated with acetone and treated with FDNB for 2 hr. in the usual manner. To remove excess FDNB about 50 ml. ether and 5 ml. water were added to the reaction mixture and, after shaking and allowing to settle, the ether layer was decanted and the aqueous layer and insoluble material were extracted twice more with ether in the same way. It was then acidified with HCl and taken to dryness *in vacuo*. Excess HCl was removed by repeated evaporation *in vacuo* with water; 50 ml. 0·1 N-HCl were added and the mixture boiled under reflux for 3 hr. It was then cooled, and extracted twice with an equal volume of ethyl acetate. After removal of dissolved ethyl acetate from the aqueous layer by evaporation *in vacuo*, it was boiled for another 3 hr. and again extracted. This process was repeated six times in all, making a total time of hydrolysis of 18 hr. The ethyl acetate solutions were taken to dryness after each extraction, and the residues stored with a little water. They were all combined and the DNP peptides purified by extraction into bicarbonate as previously described. About 3 % of the DNP-glycine was lost during this treatment. The final extract containing the terminal DNP peptides was taken to dryness, the residue dissolved in 80 % ethanol and samples taken for the analyses. Several replicate analyses were performed, but not all peptides were estimated in each case.

Peptide *A* 5 was separated on a CB_{15} column. It was usually contaminated with a small amount of a DNP-phenylalanyl peptide, which was estimated from the amount of DNP-phenylalanine produced on hydrolysis of the fraction. Peptide *A* 4 was fractionated on a CB_5 column and was free from any DNP-phenylalanyl peptides. Band *B* 3 and a faint band *B* 4 moved in front of the *A* 4 band on this column, and there were traces of other unidentified DNP-phenylalanyl peptides moving more slowly. Pure bands, *A* 2 and *A* 3, were separated from a $CHCl_3$ column which was developed with CB_1. The DNP-glycine fraction obtained from the $CHCl_3$ column contained peptide *B* 2 which was separated on an acetone-*cyclo*hexane column. The yields of the various derivatives are shown in Table 12. The figures are given as percentages of the terminal glycyl residues of the original insulin taken. The corresponding yields of the DNP-phenylalanyl peptides *B* 1, *B* 2, *B* 3 and *B* 4 were 3, 9, 20 and 3 % respectively.

On the CB_{15} column much colour was present in a number of slow-moving bands which were eluted together from the column with 80 % acetone and were designated fraction *Y*. This fraction contained about 20 % of the DNP-glycine as estimated after hydrolysis. A sample was hydrolysed for 12 hr. with boiling 0·1 N-HCl, and the products estimated as for the main hydrolysate. It gave rise to the expected peptides showing that the DNP-glycine present in it was largely in the form of higher peptides of the same sequence. The yields are shown in column 4 of Table 12.

Another 8 % of the DNP-glycine was in the insoluble residue (fraction *R*) from the original dilute acid hydrolysate. This was probably present as the DNP derivative of some incompletely oxidized insulin. In order to hydrolyse it further it was first brought into solution by incubating for 1 hr. at 37° with conc. HCl. After removal of the HCl *in vacuo*, the residue was hydrolysed for a further two periods of 12 hr. with 0·1 N-HCl, and the peptides estimated as before. The yields are shown in column 5 of Table 12.

The total yields of peptides leave no doubt that both the terminal glycyl residues are combined in the same terminal peptide sequence. It was not possible to determine the amount of peptides *A* 4 or *A* 5 directly by studying their breakdown under identical conditions of hydrolysis, as was done for peptides *B* 4 and *L* 4, since the type of hydrolysis used here could not be reproduced on a peptide. The unchanged peptide would be removed on each extraction with ethyl acetate, whereas long peptides from DNP-*A* remained in the aqueous solution and were hydrolysed further. It is also probable that the long-chain charged polypeptides have a catalytic effect on the hydrolysis, which it would be difficult to reproduce exactly on a peptide. The sum of the yields of *A* 4 and *A* 5 (55 %) does, however, give direct proof that over

Table 12. *Yields of DNP-glycyl peptides from DNP-insulin*

(Results expressed as mol. of peptide as percentage of the terminal glycyl residues in the original insulin.)

Peptide	Yield from main hydrolysate	Mean	Yield from fraction *Y* (%)	Yield from fraction *R* (%)	Total yield (%)
A 1	17·5, 16	17	3·3	1·3	22
A 2	1·4	1 }	2·6	0·5	9
A 3	4, 5	5 }			
A 4	11, 11	11	2·9	1·1	15
A 5	37, 30, 38, 35	35	3·5	1·9	40
Total known peptides					64
Total					86

half of the terminal glycyl residues are in the form of the sequence glycyl-isoleucyl-valyl-glutamic acid. Table 11 shows that neither of these peptides is completely stable to this type of hydrolysis, so that the actual amount present must be considerably greater.

DISCUSSION

It has been shown that both the terminal phenylalanyl residues are combined in the insulin molecule in the form of the single tetrapeptide sequence phenylalanyl-valyl-aspartyl-glutamic acid, and that both the lysyl residues which are in the same chains are in the form of the sequence threonyl-prolyl-lysyl-alanine. Also both the terminal glycyl residues of the other two chains are present in the pentapeptide sequence glycyl-isoleucyl-valyl-glutamyl-glutamic acid. The simplest explanation of these results is that the two phenylalanyl chains are identical, and that the two glycyl chains are identical. In the phenylalanyl chains, which contain about thirty amino-acid residues, eight of the positions relative to the terminal residues and the lysyl residues are occupied by the same amino-acids in both chains. It is, of course, possible that they might differ in the nature of one or a few residues in a position in the chain not investigated in this work. As nothing is known about the principles which govern the arrangement of amino-acids along a peptide chain, such a possibility cannot be excluded. However, if one assumes that each position in the chain can be occupied by any of the sixteen different amino-acids present in these chains, the chance that two different chains would contain the same terminal tetrapeptide is about 1 in 50,000. Certain limits are obviously imposed by the specificities of the mechanisms responsible for protein synthesis. Investigations of the free amino groups of a number of proteins by the DNP technique (summarized by Sanger, 1949*c*) have shown that the terminal position in the protein chains may be occupied by a variety of different amino-acids. There appears to be no principle that defines the nature of the residue occupying this position in different proteins and it would seem probable that this would apply to other positions in the molecule. It would thus seem a reasonable conclusion from the above results that the insulin molecule is built up of two pairs of identical chains. However, if this were indeed the case, then the simplest asymmetric unit of insulin should have a molecular weight of 6000 and all the analytical figures for the amino-acids should fit such a unit. The most recent data on the amino-acid composition of insulin have been collected by Tristram (1949). While for most amino-acids the number of residues per unit of molecular weight 6000 approximates to a whole number within the limits of the methods of assay, the values for proline and isoleucine would indicate about 1·5 and 1·3 residues respectively. It is very unlikely that these estimations are more than 10% in error, so that they do cast considerable doubt on the above conclusion, and all that can be definitely stated is that the insulin molecule is built up of two pairs of very similar chains.

In contrast to the results of this work are the results obtained by Woolley (1949). Using a method similar in principle to that used here, he has reported the isolation of a number of peptides from a tryptic hydrolysate of DNP-insulin, which he believed to be DNP-glycyl peptides. He was unable, however, to demonstrate the presence of significant amounts of DNP-glycine on hydrolysis of the peptides concerned, although it could be detected in a hydrolysate of DNP-insulin. This and the results reported in the present paper would seem to indicate that the peptides isolated by Woolley were not DNP-glycyl peptides. He used ethyl acetate and butanol extracts which would be expected to contain non-terminal peptides, though it is difficult to say what the 'chromophore' group in these peptides could have been.

The methods of investigation described in this paper should be generally applicable to all proteins and peptides that contain a free amino group. Besides giving information about the structure of the proteins, the method can probably be applied as a sensitive test for the identity and homogeneity of proteins. A simple experiment in which DNP-peptide bands are separated and the amino-acids present in them identified can readily be carried out, and it is extremely unlikely that two different proteins would give the same result, probably less likely than that they would appear homogeneous by any of the known physicochemical methods of determining protein purity.

SUMMARY

1. The dinitrophenyl method for the identification of the terminal residues of proteins has been extended by the use of partial hydrolysis to the identification and estimation of terminal peptides.

2. The method was applied to insulin; and it was shown that both the terminal phenylalanyl residues are present in the form of the tetrapeptide sequence phenylalanyl-valyl-aspartyl-glutamic acid, both the lysyl residues which are in the same chains are present in the sequence threonyl-prolyl-lysyl-alanine and both the terminal glycyl residues are present in the pentapeptide sequence glycyl-isoleucyl-valyl-glutamyl-glutamic acid.

3. It is concluded that the insulin molecule is built up of two pairs of very similar, if not identical, polypeptide chains.

I wish to express my thanks to Prof. A. C. Chibnall, F.R.S. for his advice and encouragement.

REFERENCES

Abderhalden, E. & Komm, E. (1924). *Hoppe-Seyl. Z.* **134**, 121.
Consden, R., Gordon, A. H. & Martin, A. J. P. (1944). *Biochem. J.* **38**, 224.
Consden, R., Gordon, A. H. & Martin, A. J. P. (1947). *Biochem. J.* **41**, 590.
Consden, R., Gordon, A. H., Martin, A. J. P. & Synge, R. L. M. (1947). *Biochem. J.* **41**, 596.
Craig, L. C. (1944). *J. biol. Chem.* **155**, 519.
Gale, E. F. (1945). *Biochem. J.* **39**, 46.
Gordon, A. H., Martin, A. J. P. & Synge, R. L. M. (1943). *Biochem. J.* **37**, 79.
Gutfreund, H. & Ogston, A. G. (1949). *Biochem. J.* **44**, 163.
Martin, A. J. P. & Synge, R. L. M. (1941). *Biochem. J*, **35**, 1358.
Middlebrook, W. R. (1949) *Nature, Lond.*, **164**, 501.
Moyle, V., Baldwin, E. & Scarisbrick, R. (1948). *Biochem. J.* **43**, 308.
Partridge, S. M. (1948). *Biochem. J.* **42**, 238.
Phillips, D. M. P. & Stephen, J. M. L. (1948). *Nature, Lond.*, **162**, 152.
Polson, A., Mosley, V. M. & Wyckoff, R. W. G. (1947). *Science*, **105**, 603.
Porter, R. R. & Sanger, F. (1948). *Biochem. J.* **42**, 287.
Sanger, F. (1945). *Biochem. J.* **39**, 507.
Sanger, F. (1948). *Nature, Lond.*, **162**, 491.
Sanger, F. (1949*a*). *Biochem. J.* **44**, 126.
Sanger, F. (1949*b*). *Ann. Rep. chem. Soc.* **45**, 283.
Sanger, F. (1949*c*). *Biochem. Soc. Symp.* **3**, 21.
Steinhardt, J. & Fugitt, C. H. (1942). *Bur. Stand. J. Res., Wash.*, **29**, 315.
Tristram, G. R. (1946). *Biochem. J.* **40**, 721.
Tristram, G. R. (1949). *Advanc. protein. Chem.* **5** (in the Press).
Woolley, D. W. (1948). *Fed. Proc.* **7**, 200.
Woolley, D. W. (1949). *J. biol. Chem.* **179**, 593.

[From THE BIOCHEMICAL JOURNAL, Vol. 49, No. 4, pp. 463–481, 1951]

The Amino-acid Sequence in the Phenylalanyl Chain of Insulin

1. THE IDENTIFICATION OF LOWER PEPTIDES FROM PARTIAL HYDROLYSATES

By F. SANGER (Beit Memorial Fellow) and H. TUPPY*
Biochemical Laboratory, University of Cambridge

(*Received* 17 *January* 1951)

When insulin is oxidized with performic acid, the —S—S— bridges of the cystine residues are broken by conversion to $—SO_3H$ groups (Sanger, 1949*a*) and the molecule is split into its separate polypeptide chains. From the oxidized insulin two fractions could be isolated: an acidic fraction *A*, which contained only glycyl *N*-terminal residues (see below) and no basic amino-acids, and a basic fraction *B*, having phenylalanyl *N*-terminal residues. These appeared to be the only significant fractions present.

From a study of the partial hydrolysis products of the dinitrophenyl (DNP) derivatives of the two fractions it was possible to determine the sequence of the amino-acids adjoining the *N*-terminal residues and adjoining the lysine residues (Sanger, 1949*b*). In the case of fraction *B* the terminal sequence was shown to be Phe.Val.Asp.Glu and the lysine residues were present in the sequence Thr.Pro.Lys.Ala (abbreviations for amino-acids are given in Table 1 below). No DNP-peptides were present which did not fit into these sequences, and from an estimation of the yields of the DNP-peptides produced on partial hydrolysis of DNP-insulin it was concluded that all the *N*-terminal phenylalanyl residues of insulin and all the lysine residues were present in the above two sequences respectively, and hence that there was only one type of phenylalanyl polypeptide chain in insulin. Similar though rather less clear-cut results were obtained for fraction *A*. Assuming a molecular weight of 12,000, it was concluded from these experiments that insulin is built up of two identical phenylalanyl polypeptide chains and two identical glycyl chains, these four chains being joined together by six —S—S— bridges. Various possible structures for the molecule were suggested (Sanger, 1949*c*).

The results with the terminal peptides gave a preliminary indication that the fractions *A* and *B* are both relatively homogeneous preparations of molecules having a single polypeptide chain and containing approximately 20 and 30 amino-acid residues respectively. It thus seemed that an investigation of the smaller peptides produced on partial hydrolysis might yield considerable information about the overall amino-acid sequence in these fractions. Consden, Gordon & Martin (1947) have described a method for the fractionation of lower peptides using paper chromatography which was successfully used to determine the pentapeptide sequence of 'gramicidin S' (Consden, Gordon, Martin & Synge, 1947). The present paper describes the application of this technique to partial acid hydrolysates of fraction *B* and the determination of a number of amino-acid sequences.

Throughout this paper the abbreviations for the amino-acid residues suggested by Brand & Edsall (1947) are used. These are listed in Table 1. In referring to peptides of known structure, the abbreviations for the residues are joined by a full stop (e.g. glycylalanine is written Gly.Ala). When two or more residues are included in square brackets the order is unknown. Thus, for instance, Gly.[Ala, Leu] refers to a peptide or peptides containing glycine, alanine and leucine, in which the free amino group is on the glycine residue, but the relative order of the alanine and leucine residues is unknown. The residues having the free α-amino groups in peptides will be referred to as *N*-terminal residues, those with free α-carboxyl groups as *C*-terminal residues.

Table 1. *Abbreviations for amino-acid residues*

Amino-acid	Abbreviation
Cysteic acid	$CySO_3H$
Aspartic acid	Asp
Glutamic acid	Glu
Serine	Ser
Glycine	Gly
Threonine	Thr
Alanine	Ala
Tyrosine	Tyr
Valine	Val
Leucine	Leu
Phenylalanine	Phe
Proline	Pro
Histidine	His
Lysine	Lys
Arginine	Arg
Ornithine	Orn

* Present address: II. Chemisches Universitätslaboratorium, Vienna, Austria.

METHODS

Fraction *B* was prepared from crystalline ox insulin as previously described (Sanger, 1949*a*).

Hydrolysis and preliminary group separations

Since the partial hydrolysates were too complex for direct fractionation by paper chromatography, it was necessary to carry out certain preliminary group separations to divide them into a number of simpler fractions containing fewer peptides. For this purpose it is advantageous to use methods depending on physical properties of the peptides different from those responsible for the separations on paper chromatograms. We have used adsorption on charcoal, separation on synthetic ion-exchange resins and ionophoretic methods using three- or four-compartment cells. Ionophoresis in silica jelly, which has been successfully used by Consden, Gordon & Martin (1949) to fractionate acidic peptides, was not used since difficulties are encountered with peptides containing basic or aromatic amino-acids, both of which are present in fraction *B*.

A number of experiments have been carried out using different methods both with the object of obtaining different peptides and also to investigate the various techniques. Three such experiments with hydrolysates made with 11 N-HCl are described below. Probably none of them represents the most satisfactory method of fractionating the mixture, but they do indicate the types of separations that can be obtained. The first experiment may be regarded as a preliminary one and the others, in which fewer fractions were taken, were largely confirmatory. Since sufficient peptides were not obtained from the 11 N-HCl hydrolysates to work out the entire sequence of amino-acids for fraction *B*, other methods of hydrolysis were studied. Experiments employing dilute acid (*B*4) and alkali (*B*5) for hydrolysis are also described.

*Experiment B*1. Fig. 1 summarizes the various stages of the fractionation used in this experiment. Fraction *B* (200 mg.) was hydrolysed for 4 days at 37° in 11 N-HCl, and excess HCl was removed by evaporation *in vacuo*. The hydrolysate was then transferred to a column of Amberlite IR-4B (British Drug Houses Ltd.) which had been equilibrated to pH 3·3, as described by Consden, Gordon & Martin (1948). The column, which was 10 cm. high and had a diameter of 1 cm., was washed with 200 ml. dil. HCl at pH 3·3, and the effluent, which contained neutral and basic peptides, was taken to dryness. The acidic peptides were eluted from the column with 100 ml. N-HCl and the solution taken to dryness and refractionated on a column of Amberlite IR-4B (10 cm. high) which had been equilibrated to pH 2·6. This column was developed with dil. HCl at pH 2·6. The first 100 ml. of effluent contained mainly peptides of aspartic and glutamic acids and was referred to as fraction *B*1β; the peptides of cysteic acid were collected in the next 200 ml. of effluent, which was fraction *B*1α. A further fraction could be eluted from the column with N-HCl. This contained chiefly free cysteic acid together with a few peptides.

The mixture of neutral and basic peptides obtained from the Amberlite pH 3·3 column was dissolved in a small volume of 5% (w/v) aqueous acetic acid and transferred to a column of charcoal (activated charcoal B.D.H. previously washed with 20% acetic acid), of diameter 1 cm. and height 10 cm., which was washed with 150 ml. 5% acetic acid. The effluent containing unadsorbed peptides was collected and taken to dryness. A further adsorbed fraction (*B*1γ) was obtained by washing the charcoal with a solution of 5% (w/v) phenol in 20% (w/v) aqueous acetic acid. The object of this treatment was to separate a fraction containing peptides of aromatic amino-acids (Tiselius, Drake & Hagdahl, 1947). In fact most of these peptides had already been lost by adsorption on the Amberlite, and this fraction contained other more strongly adsorbed peptides. Nevertheless, quite a valuable fractionation was obtained. The unadsorbed fraction was finally separated into a neutral (*B*1δ) and a basic (*B*1ε) fraction by ionophoresis in a simple three-compartment cell of the type used by Macpherson (1946).

*Experiment B*2. In this experiment we simply separated the hydrolysate into three fractions using a four-compartment ionophoresis cell that was designed by Dr R. L. M. Synge (1951) and which he kindly described to us. This is essentially the cell of Theorell & Åkeson (1942), to

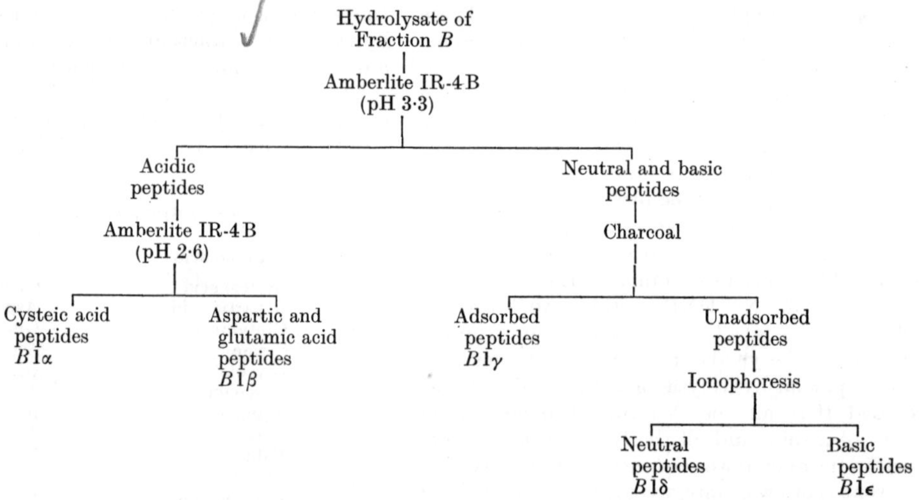

Fig. 1. Preliminary group separations in Exp. *B*1.

which is added an extra anode compartment containing H_2SO_4 as in the desalting apparatus described by Consden, Gordon & Martin (1947). In this way the acid fraction is prevented from coming into contact with the anode, where decompositions are likely to take place, especially if chlorine is being evolved.

Fig. 2 shows a diagram of the cell as used in the present experiment. It differs only in certain details from that of Synge. Membrane *a* is of cellophan, and membranes *b* and *c* of formolized gelatin (Synge has membrane *b* of formolized sheep-skin parchment and *c* of cellophan). The cell was connected to 200 V. d.c. mains with a 40 W. filament lamp in series and was cooled during the run.

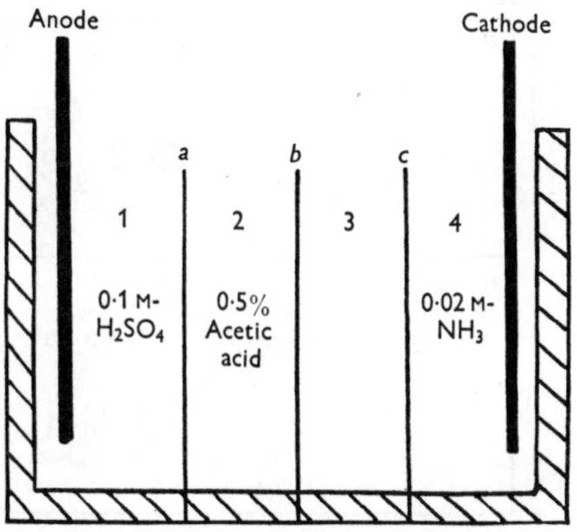

Fig. 2. Diagram of four-compartment ionophoresis cell. (For explanation see text.)

Fraction *B* (70 mg.) was hydrolysed 3 days at 37° in 11 N-HCl. After removal of excess HCl, the hydrolysate was adjusted to pH 5 and put in compartment 3 of the above cell. On applying the voltage an initial current of 110 ma. was produced and this rose to a maximum of 170 ma. and then slowly dropped. The pH of compartment 3 (which was stirred during the run) was maintained at about pH 4·5 (green to bromocresol green) by the addition of NH_3. It is necessary to add considerable amounts of NH_3 which leads to an increase in the pH of compartment 4, which may prevent the weakly basic histidine peptides from entering this compartment. It is therefore necessary to change the aqueous NH_3 solution in compartment 4 twice during the run. This also prevents some decomposition of arginine peptides, which seems to occur on prolonged ionophoresis. The separation was continued for about 1 hr. when the current had dropped to 25–30 ma. The contents of compartments 2, 3 and 4 were collected and taken to dryness. The acidic fraction from compartment 2 is referred to as *B*2α, the neutral one from compartment 3 as *B*2β and the basic one from compartment 4 as *B*2γ.

*Experiment B*3. Fraction *B* (113 mg.) was hydrolysed in 11 N-HCl at 37° for 3 days. After removal of the HCl *in vacuo*, the residue was dissolved in 5 ml. water and shaken with 100 mg. charcoal for 20 min. The charcoal was filtered off and washed with 10 ml. 5% (w/v) aqueous acetic acid. The filtrate was then fractionated into acidic (*B*3α), neutral (*B*3β) and basic (*B*3γ) fractions in the four-compartment cell as described in Exp. *B*2. A further fraction (*B*3δ) was eluted from the charcoal with 20 ml. 5% phenol in 20% acetic acid.

*Experiment B*4. Fraction *B* (50 mg.) was boiled for 24 hr. under reflux with 0·1 N-HCl (50 ml.). After removal of the HCl *in vacuo* the residue was subjected to ionophoresis as described in Exp. *B*2, and acidic (*B*4α), neutral (*B*4β) and basic (*B*4γ) fractions were obtained.

*Experiment B*5. Fraction *B* (40 mg.) was heated at 105° in a sealed tube for 5 hr. with 5 ml. 0·2 N-NaOH. After neutralization with H_2SO_4 the mixture was desalted according to Consden, Gordon & Martin (1947). It was then subjected to ionophoresis as above.

Paper chromatography of peptides

Samples (corresponding to about 10 mg. of the original fraction *B*) of the various fractions obtained after group separations were subjected to two-dimensional chromatography on Whatman no. 4 filter paper. The various solvent systems used are described below:

Phenol-0·3% NH_3. 0·3% (w/v) Aqueous NH_3 (1 l.) was put in the bottom of the chromatography tank. Phenol saturated with water was used in the trough, and the tank was filled with coal gas.

Cresol-0·03% NH_3. As above, but containing 1 l. 0·03% (w/v) aqueous NH_3 and developed with *m*-cresol.

Collidine. The collidine used in this work gave considerably lower R_F values for the amino-acids than that used by Consden, Gordon & Martin (1944, see Dent, 1948).

Butanol-acetic acid. Prepared according to Partridge (1948) by mixing 5 vol. water, 4 vol. *n*-butanol and 1 vol. acetic acid. The mixture was usually kept a few days before use, and as no special precautions were taken to ensure that all samples were equally esterified the R_F values were not always constant and consequently have not been recorded. However, in Figs. 5–16, the distances of the peptide spots from the origin are given.

Other solvents used were prepared as described by Consden *et al.* (1944).

As a rule the chromatograms were developed first with phenol or cresol in the long direction of the paper and in the second direction with another solvent (usually butanol-acetic acid). If phenol was used as the second solvent, the fluorescence of the dried spots could not be seen. For certain slow-moving peptides it was necessary to run the solvent considerably farther than the end of the filter paper. This was accomplished by attaching a second folded sheet of filter paper to the front of the chromatogram. Where this was done the number of lengths of filter paper through which the solvent ran is indicated in Figs. 5–16. After development, the chromatograms were dried, heated at 100° for 15–30 min. to bring out the fluorescence of the spots (Phillips, 1948) and examined under the ultraviolet lamp. The fluorescent areas were marked on the paper. As observed by Jones (1949), the fluorescence was found to depend to a certain extent on the particular batch of filter paper used. In order to obtain the maximum effect it was necessary to avoid any contact of the papers with phenolic vapours as, for instance, in the drier. Where good fluorescent spots were seen they could be cut out directly without treatment with ninhydrin. Usually at least one replicate sheet was treated with ninhydrin to locate the peptides, and this could be used as a reference for cutting out the other sheets on which fluorescent spots were marked. 0·025% (w/v) Ninhydrin in water-saturated *n*-butanol was used to spray the sheets. This gave a sufficient colour development with very little destruction of the peptide, so that the eluates from the ninhydrin-treated spots could be used to identify the amino-acids produced on hydrolysis. In certain cases, depending on the amino-acids concerned, there appeared to be a slight destruction of the *N*-terminal residue, but this was usually negligible. Where the peptide was to be deaminated or treated with fluorodinitrobenzene (FDNB), the use of ninhydrin was avoided and eluates from spots cut from untreated sheets were used. It was found that, if eluates from spots which had been treated with 0·025% ninhydrin were subjected to deamination, all the amino-acids present were almost completely destroyed.

Elution of peptides

For the elution of the peptides from the paper a combination of the methods of Consden, Gordon & Martin (1947) and of Dent (1947) was found convenient.

The spots were cut out and the cut trimmed to a point at one end. The other end was held between two microscope slides, which were placed in a trough of water as shown in Fig. 3. The solution passed into a capillary tube which was

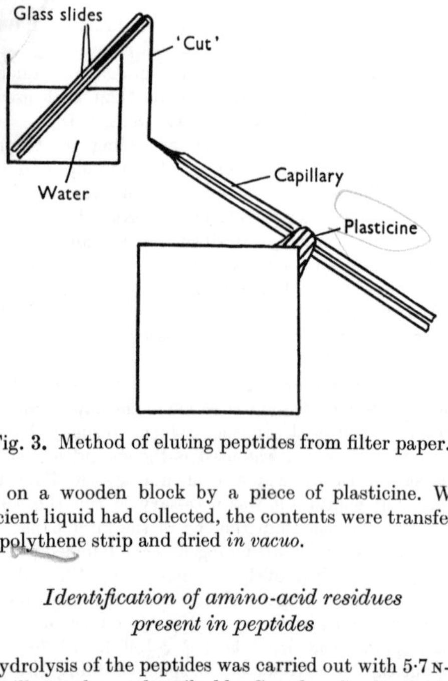

Fig. 3. Method of eluting peptides from filter paper.

held on a wooden block by a piece of plasticine. When sufficient liquid had collected, the contents were transferred to a polythene strip and dried *in vacuo*.

Identification of amino-acid residues present in peptides

Hydrolysis of the peptides was carried out with 5·7 N-HCl in capillary tubes as described by Consden, Gordon & Martin (1947), and the resulting amino-acids identified by paper chromatography. Fraction *B* contains the amino-acids listed in Table 14. It was shown by chromatography with benzyl alcohol that isoleucine is absent.

On account of the limited chromatographic apparatus available, and the labour involved, it was desirable to use one-dimensional chromatograms as far as possible to identify the amino-acids present. In the earlier experiments phenol-0·3% NH_3 was used for the preliminary identification of all peptide hydrolysates. Cysteic acid, aspartic acid, glutamic acid, serine, glycine, threonine and alanine can all be unequivocally identified in this solvent. The other faster moving amino-acids cannot be so easily distinguished especially if several of them are present. They may be more satisfactorily identified if phenol-0·03% NH_3 is used. The basic amino-acids move slower on this solvent and may frequently be recognized. The identification of all these faster moving amino-acids was always confirmed using some other solvent system or on a two-dimensional chromatogram.

For the complete analysis of a peptide two-dimensional chromatograms were run on half sheets (46 × 28 cm.) of Whatman no. 1 filter paper, which were developed in the long direction with phenol-0·3% NH_3 and in the short with butanol-acetic acid. All the amino-acids present in fraction *B* may be recognized on such a chromatogram (Fig. 4*b*).

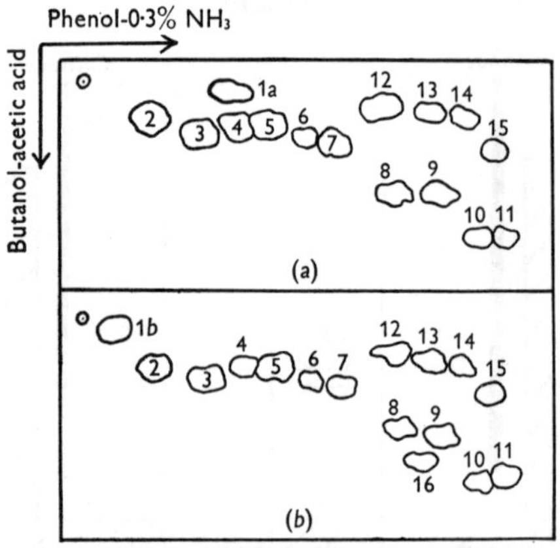

Fig. 4. Chromatograms of hydrolysates of insulin (*a*) and fraction *B* (*b*) run on half sheets of Whatman no. 1 filter paper. The identities of the spots are as follows: 1*a*, cystine; 1*b*, cysteic acid; 2, aspartic acid; 3, glutamic acid; 4, serine; 5, glycine; 6, threonine; 7, alanine; 8, tyrosine; 9, valine; 10, leucine; 11, phenylalanine; 12, histidine; 13, lysine; 14, arginine; 15, proline; 16, TyrX.

On hydrolysis of certain aromatic peptides a new spot was observed moving slightly faster than tyrosine in phenol and almost as fast as leucine in butanol-acetic acid. It gave a positive reaction with diazotized sulphanilic acid. In many cases pairs of peptides were found, one containing tyrosine and the other this new amino-acid. It is thus considered to be a breakdown product of tyrosine produced during the oxidation with performic acid, and will be referred to as TyrX. Fig. 4*b* shows a chromatogram of a hydrolysate of fraction *B* and Fig. 4*a* one of a hydrolysate of insulin. The TyrX spot (16) can be seen only in Fig. 4*b* below the tyrosine and valine spots, thus showing that it has been formed during the preparation of fraction *B*.

Since histidine gives a relatively weak spot with ninhydrin, it is possible to overlook it in certain peptides. The following modification of the Pauli test which avoids the use of aqueous solutions was found to be suitable for the identification of histidine and histidine peptides on chromatograms.

A mixture of equal volumes of 1% (w/v) *p*-anisidine in ethanolic 0·11 N-HCl and 10% (v/v) amyl nitrite in ethanol was allowed to stand a few minutes and then sprayed on to the paper. The sheets were dried in air and the colour developed with NH_3 vapour or by spraying on ethanolic KOH solution. Histidine and histidine peptides give a dark-red spot, whereas tyrosine gives a faint orange colour.

Determination of the structure of peptides

Deamination. In order to determine the *N*-terminal residues of the peptides we have deaminated them with nitrosyl chloride as described by Consden, Gordon & Martin (1947). In some experiments the reaction was carried out at 37° for 10 min., in others at room temperature for 30 min. No difference was noticed between the two procedures. For most peptides the method gave satisfactory results, though it was rarely possible to destroy the *N*-terminal residue completely. Certain amino-acids such as leucine and glutamic acid seemed to be especially resistant to this treatment, and if these were present as *N*-terminal residues, clear-cut deamination could rarely be obtained. Results with peptides containing tyrosine or arginine were also difficult to interpret.

DNP method. For most of the peptides the *N*-terminal residue has also been determined by the DNP technique, which usually gives clear-cut results, though it is rather more laborious. In some cases the DNP-derivative of the terminal residue was identified directly by chromatography using the micro method previously described (Sanger, 1949*b*). As this requires rather more material than is usually available from a paper chromatogram, it is often convenient to use an alternative procedure, and to identify the amino-acids present after hydrolysis of the DNP-peptide. Although traces of free amino-acids are formed on hydrolysis of DNP amino-acids (Consden, Gordon, Martin & Synge, 1947), it was found that they are sufficiently stable to ensure that the terminal residue is almost completely absent. The results are usually much more clear-cut than when the deamination method is used.

In this method the peptide is treated with FDNB in the usual way and the excess reagent removed by extraction of the alkaline solution with ether. It is then necessary to separate the DNP-peptide from the salt present. This may be done by extracting it from acid solution into ethyl acetate, or if it cannot be extracted by adsorbing it on a talc column from acid or neutral solution (Sanger, 1949*b*). The DNP peptide is then hydrolysed for 20 hr. with 5·7 N-HCl at 105° and the amino-acids remaining identified. In the case of peptides containing glycine as *N*-terminal residue a certain amount of breakdown of the DNP-glycine to free glycine was observed so that the results were not so clear. Better results were obtained if a shorter time of hydrolysis was used. Lysine, tyrosine and histidine, if present in a peptide with their amino groups blocked, are converted to their monoDNP derivatives which can be identified on paper chromatograms. Table 2 shows R_F values for these with butanol-acetic acid which had been prepared 24 hr. before use. The values for α-*N*-DNP-arginine, DNP-cysteic acid and for valine and leucine are given for comparison. Although *O*-DNP-tyrosine and *im*-DNP-histidine are colourless, they show up as dark spots if the papers are examined in ultraviolet light. This effect would seem to be due to quenching the fluorescence of the paper. Peptides containing *im*-DNP-histidine may be successfully fractionated on paper chromatograms with solvents such as collidine and pentanol.

RESULTS

The results obtained with the various fractions are given in Figs. 5–16 and Tables 3–13.

Figs. 5–16 are diagrams of the two-dimensional chromatograms, from which the peptide spots were eluted, and Tables 3–13 give the results obtained with the eluted peptides. In column 2 of these tables are listed the amino-acids identified on complete hydrolysis, and the approximate strength is indicated in column 3. The ×'s have the same significance as in the paper of Consden *et al.* (1949), and ? represents a very weak spot. Column 4 shows the strength of the amino-acids remaining after deamination with NOCl and hydrolysis of the peptide and column 5 that remaining after hydrolysis of the DNP peptide. In the last column are given the structures of the peptides, as far as they can be deduced from the data given in the tables. As far as possible all the results with one peptide are given together, although they were often obtained in different experiments, and sometimes from the same peptide in a different fraction. Most of the results recorded have been obtained in several experiments. Where deamination or the DNP method has been used only the most satisfactory and clear-cut results have been recorded. While such a procedure does not necessarily give a true indication of the efficiencies of the methods it is probably justified, as much more can be learnt from such a result than from an indefinite one. This is especially true for the deamination technique where several experiments were usually carried out on each peptide.

Besides the peptides shown in Fig. 5 and Table 3, fraction *B* 1α contained traces of $CySO_3H$.Ala and Ser.Val.$CySO_3H$. These peptides were probably derived from the small amount of fraction *A* contained in the preparation of fraction *B*, since they were found in very high concentration in fraction *A*. The presence of Ser.Val.$CySO_3H$ is suggested by a trace of serine that was sometimes noticed in spot 3. With this fraction there was frequently overlapping between spots 3 and 4 and between spots 5 and 6. When the eluate from the Amberlite-pH 2·6 column was collected in several successive fractions, a slower moving fraction was obtained which contained peptides 4 and 6 without the corresponding peptides containing glycine. The smaller peptides are retained more strongly on the resin than the larger.

Two chromatograms of fraction *B* 1γ are illustrated in Figs. 7 and 8, since certain peptides are separated better using cresol and others with phenol. The combined results are given in Table 5.

Table 2. *R_F values of water-soluble DNP derivatives*

(Whatman no. 4 filter paper; butanol-acetic acid.)

	R_F	Colour of spot	Ninhydrin reaction
O-DNP-Tyrosine	0·84	Colourless	Purple
α-*N*-DNP-Arginine	0·81	Yellow	No reaction
ε-*N*-DNP-Lysine	0·77	Yellow	Brown
Leucine	0·67	—	—
im-DNP-Histidine	0·57	Faint yellow	Brown
Valine	0·49	—	—
DNP-Cysteic acid	0·42	Yellow	—

The results with fraction $B1\epsilon$ are recorded with the results on fraction $B2\gamma$, which gave a more complete mixture of basic peptides. $B1\epsilon$ was identical with $B2\gamma$ except that

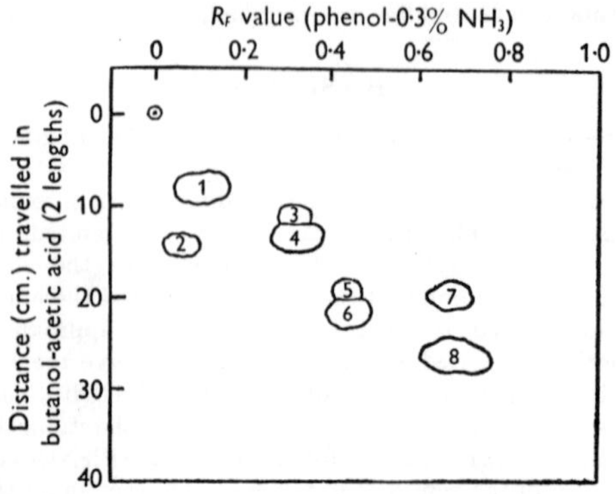

Fig. 5. Chromatogram of fraction $B1\alpha$ (see Table 3). In Figs. 5–16 the fraction was applied in the position marked by the small dotted circle, and the chromatogram run first in the direction represented horizontally, which, except in Fig. 15, was the longer dimension of the filter paper.

Ser.[His, Leu] ($B2\gamma7$) and His.Leu ($B2\gamma8$) were absent, since they had been adsorbed on charcoal and appeared in fraction $B1\gamma$.

The results with fractions $B2\alpha$ and $B2\beta$ (Figs. 10–12; Tables 7 and 8) which are largely confirmatory, are included to show the type of separation that can be obtained with mixtures of about twenty peptides, and also to show that very few peptides were overlooked in the preliminary experiments. The cresol chromatogram of fraction $B2\alpha$ (Fig. 11) is included to demonstrate the positions of the fast-moving spots 21 and 22, which did not appear on the phenol chromatogram.

A large number of solvents was studied for fractionating the basic peptides, most of which move fast on systems such as phenol-0·3% NH_3 and slow on butanol-acetic acid or collidine. Cresol-0·03% NH_3 was eventually found to give good separations (Fig. 13) and is to be recommended for the fractionation of basic peptides and also of larger peptides.

In one experiment fraction $B2\gamma$ was subjected to a further ionophoresis to obtain a fraction containing only arginine peptides. This was carried out in the four-compartment cell (Fig. 2). Compartment 2 contained 1% (w/v) acetic acid and compartment 4 0·1 M-aqueous NH_3. Compartment 3 was maintained at a pink reaction to phenolphthalein. The maximum current was 120 ma. and dropped to 30 ma. The basic fraction (from compartment 4) contained only arginine and Arg.Gly ($B2\gamma5$). The fraction from compartment 3 contained traces of these besides the other basic peptides

Considerable difficulty was experienced in fractionating the fast-moving neutral peptides, which appeared in spots $B2\beta24$ and $B2\beta25$ in Exp. $B2$. The most satisfactory solvents

Table 3. *Peptides from fraction* $B1\alpha$

Spot no. (Fig. 5)	Amino-acids present	Strength of amino-acid after: Hydrolysis	Deamination and hydrolysis	DNP treatment and hydrolysis	Structure
1	$CySO_3H$	× × × ×	?	—	$CySO_3H$.Gly
	Gly	× × × ×	× × × ×	× × ×	
2	Asp	×			[Asp, Glu]
	Glu	×			
3	$CySO_3H$	× ×		×	Val.[$CySO_3H$, Gly]
	Gly	×		×	
	Val	× ×		—	
4	$CySO_3H$	× × ×	× ×	× ×	Val.$CySO_3H$
	Val	× × ×	?	—	
5	$CySO_3H$	×	×	×	Leu.[$CySO_3H$, Gly]
	Gly	×	?	?	
	Leu	×	—	—	
6	$CySO_3H$	× × ×	× × ×	× × ×	Leu.$CySO_3H$
	Leu	× × ×	× ×	—	
7	$CySO_3H$	× ×	×	× ×	Leu.[$CySO_3H$, Gly, Val]
	Gly	× ×	×	× ×	
	Val	× ×	×	× ×	
	Leu	× ×	?	—	
8	$CySO_3H$	× × × ×	× × ×	× × ×	Leu.[$CySO_3H$, Val]
	Val	× × × ×	× × ×	× × ×	
	Leu	× × × ×	×	—	

Spot 1. Yellow colour with ninhydrin which on further heating became grey and then violet.

Spot 8. Eluates of spot 8 from two chromatograms were hydrolysed 4 days at 37° in 11 N-HCl and investigated on a phenol-0·3% NH_3 chromatogram. Besides spots corresponding to cysteic acid, valine, leucine and the unchanged peptide there was a spot having R_F 0·32 which contained valine and cysteic acid and a spot moving faster than leucine which contained valine. Since valine is present in both these dipeptides it must be in the middle of peptide 8, whose structure is therefore Leu.Val.$CySO_3H$.

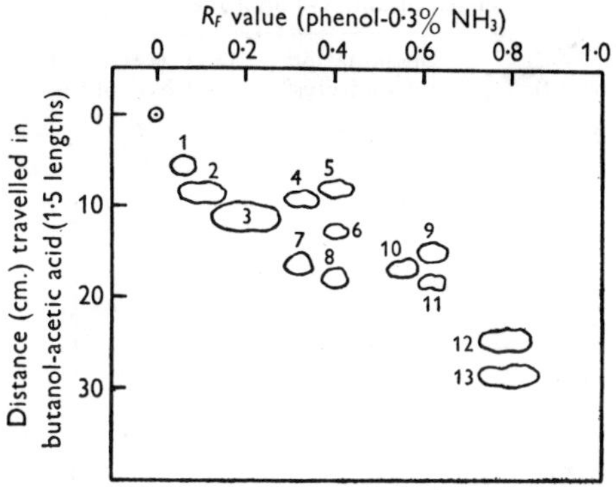

Fig. 6. Chromatogram of fraction $B1\beta$ (see Table 4).

Table 4. *Peptides from fraction* $B1\beta$

Spot no. (Fig. 6)	Amino-acids present	Strength of amino-acid after: Hydrolysis	Deamination and hydrolysis	DNP treatment and hydrolysis	Structure
1	$CySO_3H$	× ×			$CySO_3H$.Gly ($B1\alpha1$)
	Gly	× ×			
2	Asp	× × × ×			Aspartic acid
3	Glu	× × × × ×			Glutamic acid
4	$CySO_3H$	×			Val.[$CySO_3H$, Gly] ($B1\alpha3$)
	Gly	× ×			
	Val	×			
5	$CySO_3H$	×			[$CySO_3H$, Glu, His, Leu]
	Glu	×			
	His	×			
	Leu	×			
6	$CySO_3H$	×			Leu.[$CySO_3H$, Gly] ($B1\alpha5$)
	Gly	×			
	Leu	×			
7	Glu	×			Glu.Ala ($B1\delta4$)
	Ala	×			
8	Asp	×	?	×	Val.Asp
	Val	×	—	—	
9	$CySO_3H$	×			Leu.[$CySO_3H$, Gly, Val] ($B1\alpha7$)
	Gly	×			
	Val	×			
	Leu	×			
10	Asp	× ×	×	× ×	Val.[Asp, Glu]
	Glu	× ×	×	× ×	
	Val	× ×	—	—	
11	$CySO_3H$	×			Leu.[$CySO_3H$, Val] ($B1\alpha8$)
	Val	×			
	Leu	×			
12	Asp	× × × ×		× × ×	Phe.[Asp, Glu, Val]
	Glu	× × × ×		× × × ×	
	Val	× × × ×		× × × ×	
	Phe	× × × ×		×	
13	Asp	× × × ×		× ×	Phe.[Asp, Val]
	Val	× × × ×		× ×	
	Phe	× × × ×		—	

Table 5. *Peptides from fraction B* 1γ

Spot no. (Figs. 7 and 8)	Amino-acids present	Strength of amino-acid after: Hydrolysis	Deamination and hydrolysis	DNP treatment and hydrolysis	Structure
1	Glu	× ×		?	Glu.His
	His	× ×		×*	
2	Glu	× ×			[Glu, Gly, Arg]
	Gly	× ×			
	Arg	× ×			
3	Glu	× × ×	×		Glu.Arg + spot 2
	Gly	×	?		
	Arg	× × ×	× × ×†		
4	$CySO_3H$	× × ×	× ×	× ×	His.[$CySO_3H$, Leu]
	His	× × ×	?	—	
	Leu	× × ×	× × ×	× ×	
5	Thr	× × × ×			Thr.[Ala, Lys, Pro] ($B2\gamma6$)
	Ala	× × × ×			
	Lys	× × × ×			
	Pro	× × ×			
6	Asp	×		×	Phe.[Asp, Glu, His, Val]
	Glu	×		×	
	His	×		‡	
	Val	×		×	
	Phe	×		—	
7	Glu	× ×			[Glu, His, Leu]
	His	× ×			
	Leu	× ×			
8	Leu	× ×			His.Leu ($B2\gamma8$)
	His	× ×			
9	Glu	×			Val.Glu ($B1\delta7$)
	Val	×			
10	Glu	×	×		Val.[Ala, Glu]
	Ala	×	×		
	Val	×	—		
11	Tyr	×			Tyrosine
12*a*	Gly	×			Gly.Phe
	Phe	×			
12*b*	Asp	× ×		×	See below
	Glu	× ×		×	
	Gly	× ×		?	
	Val	× ×		×	
	Phe	× × ×		× ×	
13	Leu	×			Leucine + phenylalanine
	Phe	× ×			
14	Glu	× × ×			Leu.[Glu, Val] ($B2\beta14$)
	Val	× × ×			
	Leu	× × ×			
15	Glu	×			Leu.[Glu, Ala, Val] ($B2\beta15$)
	Ala	×			
	Val	×			
	Leu	×			
16	Ala	× ×			Ala.Leu ($B1\delta11$)
	Leu	× ×			
17	Unidentified mixture of fast-moving peptides, containing Phe, Leu, Val, Tyr, Ala				

* DNP derivative.

† Arginine decomposition product (R_F, 0·73 with phenol).

‡ *im*-DNP-Histidine not distinguished on this phenol chromatogram.

Spots 2 and 3 appear to be mixtures of a tripeptide [Glu, Gly, Arg] and a dipeptide [Glu, Arg], the latter moving slightly faster on butanol-acetic acid.

Spot 7, obtained from phenol chromatograms, sometimes contains serine probably due to the presence of Ser.[His, Leu] ($B2\gamma7$).

Spot 8 sometimes contains glutamic acid from spot 7.

Spot 12*a* gives a yellow colour with ninhydrin, suggesting glycine is terminal. The DNP treatment was done only on the mixture in spot 12*b*, which clearly demonstrates that glycine must be terminal.

Spot 12*b* is a mixture of 12*a* with another peptide, probably Phe.[Asp, Glu, Val] ($B1\beta12$). This latter did not appear on the cresol chromatogram (Fig. 8).

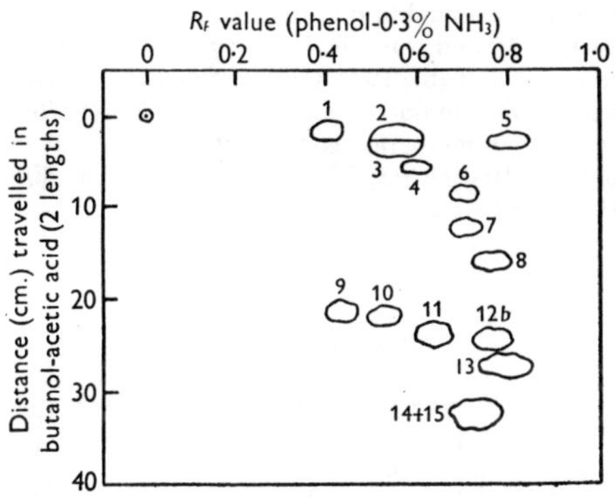

Fig. 7. Chromatogram of fraction $B1\gamma$ using phenol-0·3% NH_3 (see Table 5).

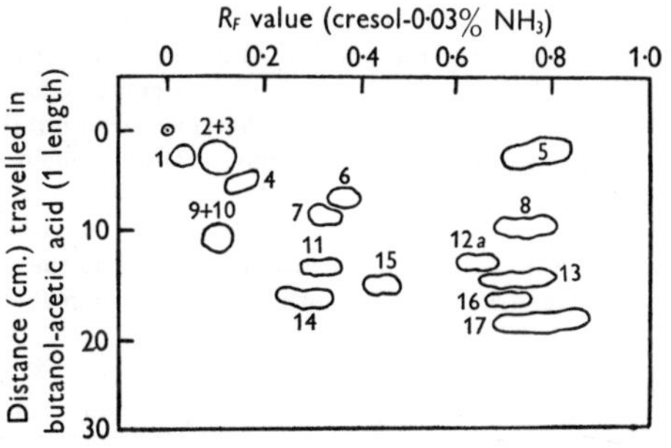

Fig. 8. Chromatogram of fraction $B1\gamma$ using cresol-0·03% NH_3 (see Table 5).

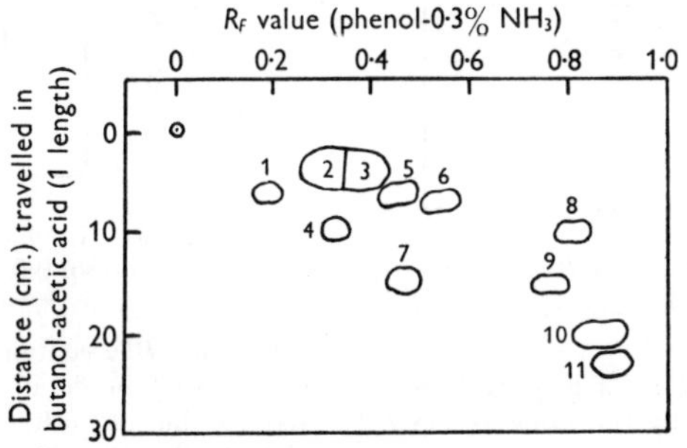

Fig. 9. Chromatogram of fraction $B1\delta$ (see Table 6).

Table 6. *Peptides from fraction* $B1\delta$

Spot no. (Fig. 9)	Amino-acids present	Strength of amino-acid after: Hydrolysis	Deamination and hydrolysis	DNP treatment and hydrolysis	Structure
1	Glu	×	×	×	Gly.Glu
	Gly	×	—	?	
2	Ser	× × ×			Serine
	Gly	×			
3	Gly	× × × ×			Glycine
4	Glu	×	?		Glu.Ala
	Ala	×	×		
5	Thr	×			Threonine
6	Ala	× × × ×			Alanine
7	Glu	× ×	× ×	× ×	Val.Glu
	Val	× ×	—	?	
8	Thr	× ×	?	—	Thr.Pro
	Pro	× ×	× ×	× × ×	
9	Val	× ×			Valine
10	Leu	× × × ×			Leucine
11	Leu	× × × ×	× × × ×	× × × ×	Ala.Leu
	Ala	× × × ×	× ×	×	

Spot 1 gives a yellow colour with ninhydrin.
Spot 7 gives a very weak reaction with ninhydrin and a relatively strong fluorescence.
Spot 8 gives a strong yellow colour with ninhydrin.
Spot 10 is often contaminated with traces of serine, which is located in the slower moving parts of the spot. If *m*-cresol is used instead of phenol the leucine spot is pure and the serine can be found together with leucine in a slower moving faint spot (R_F, 0·57). This is probably Ser.Leu derived from fraction *A*, from which it can be obtained in large amounts.
Spot 11 is often contaminated with threonine, which we have not yet been able to account for satisfactorily.

were collidine *tert.*-amyl alcohol (2-methylbutan-2-ol) and benzyl alcohol, but there seemed to be too many of these closely related peptides to give clear results. The best results were obtained with fraction $B3\beta$ (Fig. 14, Table 10), from which some of the more strongly adsorbed peptides had been removed by charcoal treatment. The very large number of these peptides containing tyrosine, valine, leucine and phenylalanine is partly due to the partial breakdown of tyrosine to TyrX, which means that two spots are obtained for each tyrosine peptide. It also suggests that these non-polar residues are grouped together in the fraction *B* chains.

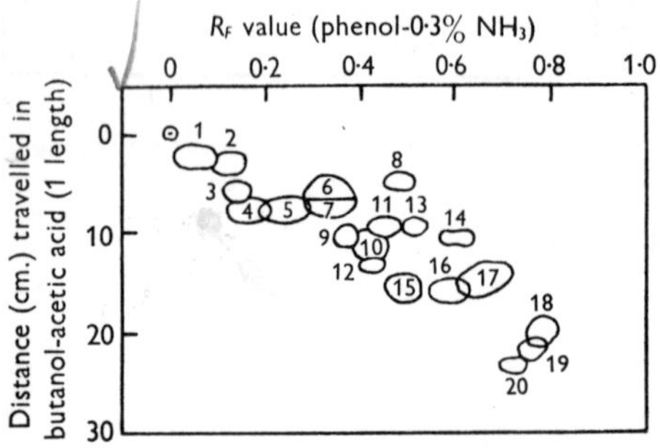

Fig. 10. Chromatogram of fraction $B2\alpha$ using phenol-0·3% NH_3 (see Table 7).

The results with the other fractions of Exp. *B*3 are not recorded, since they were essentially the same as the *B*2 fractions. Fraction $B3\delta$, which was adsorbed on charcoal, contained mainly [$CySO_3H$, Tyr, Val, Leu] ($B2\alpha21$) and Phe.[Asp, Glu, Val] ($B1\beta12$) and fast-moving peptides containing tyrosine, valine, leucine and phenylalanine.

The peptides found in fraction $B4\alpha$ are listed in Table 11. Fractions $B4\beta$ and $B4\gamma$ yielded a few new peptides, which are recorded in Table 12. Other peptides are not recorded. Spot $B4\beta1$ was obtained from a cresol-0·03% NH_3/butanol-acetic acid chromatogram and ran at the same rate in both solvents as His.[$CySO_3H$, Leu] (Peptide $B2\beta6$, Fig. 12). Peptide $B4\beta2$ was a fast-moving spot on a collidine, *tert.*-amyl alcohol chromatogram. The peptides from $B4\gamma$ were obtained from the chromatogram illustrated in Fig. 15.

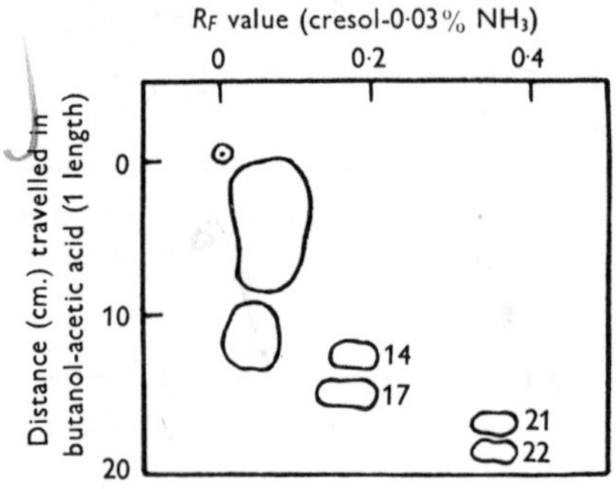

Fig. 11. Chromatogram of fraction $B2\alpha$ using cresol-0·03% NH_3 (see Table 7).

Fraction $B5\alpha$ contained very little material and was not investigated in detail. $B5\beta$ contained free glycine and alanine (but no other free amino-acids), Gly [Glu, Orn] ($B5\gamma1$), Gly Glu ($B1\delta1$) and a large amount of fast-moving peptides. There was also a yellow spot ($R_F = 0·55$ in cresol-0·03% NH_3 and a little faster than alanine on butanol-acetic acid) which gave glycine and proline on hydrolysis. It

Table 7. *Peptides from fraction B 2α*

Spot no. (Figs. 10 and 11)	Strength	Structure	Identical with
1	× × × ×	Cysteic acid	—
2	× × × ×	$CySO_3H$.Gly	$B1\alpha1$
3	× ×	Aspartic acid	—
4	× ×	Gly.Glu	$B1\delta1$
5	× × ×	Glutamic acid	—
6	× ×	Val.[$CySO_3H$, Gly]	$B1\alpha3$
7	×	Val.$CySO_3H$	$B1\alpha4$
8	×	[$CySO_3H$, Glu, His, Leu]	$B1\beta5$
9	× ×	Glu.Ala	$B1\delta4$
10	× × ×	Leu.$CySO_3H$	$B1\alpha6$
11	× × ×	Leu.[$CySO_3H$, Gly]	$B1\alpha5$
12	× ×	Val.Asp	$B1\beta8$
13	×	Alanine	—
14	× × ×	Leu.[$CySO_3H$, Gly, Val]	$B1\alpha7$
15	× × × ×	Val.Glu	$B1\delta7$
16	×	Val.[Ala, Glu]	$B1\gamma10$
17	× × × ×	Leu.[$CySO_3H$, Val]	$B1\alpha8$
18	× ×	Phe.[Asp, Glu, Val]	$B1\beta12$
19	×	Leu.[Glu, Ala, Val]	$B2\beta15$
20	× ×	Leu.[Glu, Val]	$B2\beta14$
21	×	[$CySO_3H$, TyrX, Val, Leu]	—
22	×	[$CySO_3H$, TyrXX, Val, Leu]	—

Spot 8 sometimes contains a trace of glycine.

Spot 22. Besides cysteic acid, valine and leucine, a new spot was present that had R_F, 0·53 (phenol-0·3% NH_3) and moved as fast as leucine on butanol-acetic acid. It was not detected in any other peptides and is assumed to be another breakdown product of tyrosine (TyrXX).

is assumed that this was formed by the action of the alkali on Thr.Pro ($B1\delta8$). Wieland & Wirth (1949) have shown that glycine is formed from threonine by the action of alkali.

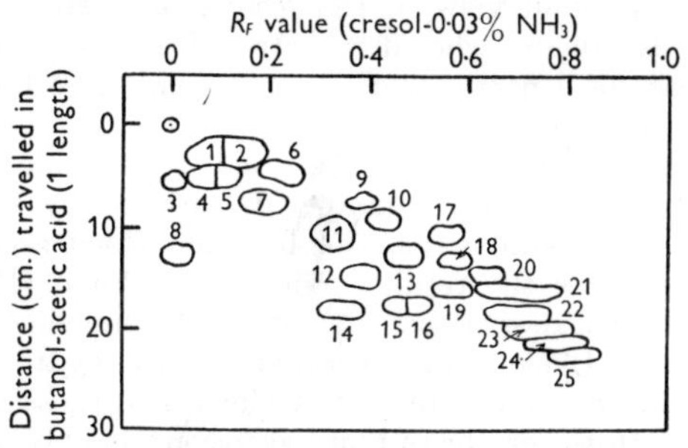

Fig. 12. Chromatogram of fraction $B2\beta$ (see Table 8).

The results with fraction $B5\gamma$ are recorded in Fig. 16 and Table 13. Besides the peptides shown there were a few unidentified spots that moved faster on butanol-acetic acid. The amino-acid referred to as Orn moved at the same R_F values as ornithine in phenol and butanol-acetic acid and is assumed to be a breakdown product of arginine.

DISCUSSION

Methods of hydrolysis

The present results support the conclusion that concentrated hydrochloric acid at low temperatures is the most generally useful method of partial hydrolysis (Synge, 1943). No apparent destruction of amino-acids occurs, and a wide variety of lower peptides is produced. The disadvantage of using only one method is that it is difficult to obtain peptides which contain the most labile bonds, and which are required to work out a complete sequence of amino-acids. The bonds that are most labile under these conditions are those involving the amino groups of serine and threonine residues (Desnuelle & Casal, 1948) and no peptides have been found containing these bonds intact. Bonds involving glycine residues are also somewhat labile, but some such peptides were obtained. Other methods of hydrolysis were therefore studied in the hope of supplementing the results from the hydrolysates made with concentrated hydrochloric acid and also in the hope of obtaining a rather less complex mixture of the fast-moving neutral peptides, suitable for fractionation. While neither of these objectives was attained, a few new peptides were detected.

In studying the terminal peptides (Sanger, 1949*b*) it was found that glycine peptides were relatively more stable to dilute acid than to concentrated acid. This is also apparent from the present work. If fraction $B4\alpha$ (Table 11) is compared with fraction $B1\alpha$ (Table 3) it will be seen that the dilute-acid hydrolysate contains a higher proportion of the peptides in which the $CySO_3H$.Gly bond is intact. Similarly, [$CySO_3H$, Gly, His, Leu] ($B4\beta1$) occurs in

Table 8. *Peptides from fraction $B2\beta$*

Spot no. (Fig. 12)	Strength	Structure	Identical with
1	× ×	Glu.His	$B1\gamma1$
2	× × × ×	Glu.Arg + [Glu, Gly, Arg]	$B1\gamma2 + B1\gamma3$
3	×	Gly.Glu	$B1\delta1$
4	× × × ×	Serine + glycine	—
5	× × ×	Glycine + threonine	—
6	× ×	His.[$CySO_3H$, Leu]	$B1\gamma4$
7	× × × ×	Alanine	—
8	×	Val.Glu + Val.[Ala, Glu]	$B1\gamma9 + B1\gamma10$
9	×	Phe.[Asp, Glu, His, Val]	$B1\gamma6$
10	× ×	[Glu, Ser, His, Val, Leu]	$B4\gamma3$
11	× ×	Tyrosine	—
12	×	TyrX	—
13	× ×	Valine	—
14	× × × ×	Leu.[Glu, Val]	—
15	× ×	Leu.[Glu, Ala, Val]	—
16	× ×	Phe.[Asp, Val]	$B1\beta13$
17	× × ×	Thr.Pro	$B1\delta8$
18	× ×	[Glu, His(?)Val, Leu]	—
19	× ×	Phe.[Asp, Glu, Val]	$B1\beta12$
20	× ×	Gly.Phe	$B1\gamma12a$
21	× × × ×	Phenylalanine + leucine	—
22	× × × ×	Ala.Leu	$B1\delta11$
23	× × ×	Ala.[Tyr, Leu]	$B3\beta9$

Spot 6 contains a trace of glycine, probably due to [His, Leu, $CySO_3H$, Gly] ($B4\beta1$).

Spot 10. Glutamic acid, histidine and leucine are slightly stronger so probably contains peptide $B1\gamma7$ ([Glu, His, Leu]).

Spot 14. After DNP treatment only valine and glutamic acid were present.

Spot 15. *N*-terminal residue determined on mixture in $B3\beta4$.

Spot 18. The histidine was not clearly distinguished on this phenol chromatogram as it comes very close to the valine spot, but it is probably significant since the peptide cannot be [Glu, Val, Leu] which is in spot 14.

Spots 24, 25. Unidentified mixtures of peptides containing tyrosine, valine, leucine and phenylalanine.

the dilute-acid hydrolysate, whereas the corresponding peptide without glycine ($B\,2\beta6$) is obtained with concentrated hydrochloric acid hydrolysis. The presence of several new histidine peptides in fraction

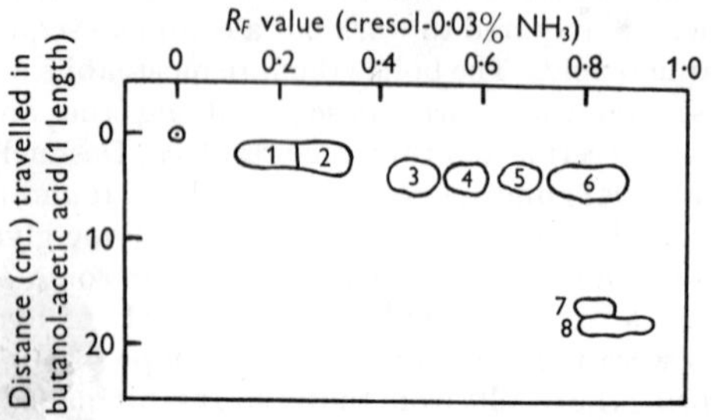

Fig. 13. Chromatogram of fraction $B\,2\gamma$ (see Table 9).

$B\,4\gamma$ (Table 12) suggests that bonds involving histidine may also be relatively more stable to dilute acid. The bonds involving the amino groups of serine and threonine, however, appear to be labile to dilute acid also, and no peptides were obtained in which they were intact. The large amount of free aspartic acid ($B\,4\alpha2$), compared with only traces of free cysteic acid (possibly present in spot $B\,4\alpha1$), is interesting in view of the results of Partridge & Davis (1950), who demonstrated a preferential release of aspartic acid when proteins were boiled with weak acids.

The hydrolysis in alkali led to considerable destruction of the amino-acids, apparently while they were still in peptide linkage, which greatly complicates the results. Wieland & Wirth (1949) found no destruction of serine or threonine on heating 12 hr. with saturated barium hydroxide, conditions which are more violent than those used here. However, the residues seem considerably more labile when in peptide linkage. No free serine or threonine was detected, and the Gly.Pro was almost certainly derived from Thr.Pro. Arginine appears to be converted almost quantitatively to ornithine since no free arginine was detected, whereas free ornithine and a peptide containing ornithine ($B\,5\gamma1$) were present.

Table 9. *Peptides from fraction B 2γ*

Spot no. (Fig. 13)	Amino-acids present	Strength of amino-acid after: Hydrolysis	Deamination and hydrolysis	DNP treatment and hydrolysis	Structure
1	Ser	× × ×			2 + lysine
	Lys	×			
	His	× × ×			
2	Ser	× × × ×	×	—	Ser.His
	His	× × × ×	× × ×	× ×*	
3	His	× ×			Arginine + histidine
	Arg	× × ×			
4	Ala	× × ×		× ×	Lys.Ala
	Lys	× × ×		—	
5	Gly	× × ×		× ×	Arg.Gly
	Arg	× × ×		—	
6	Thr	× × × × ×		×	Thr.[Ala, Lys, Pro]
	Ala	× × × × ×		× × ×	
	Lys	× × × × ×		× × ×*	
	Pro	× × × × ×		× × ×	
7	Ser	× ×		—	Ser.[His, Leu]
	His	× ×		×*	
	Leu	× ×		×	
8	His	× × ×	?	—	His.Leu
	Leu	× × ×	× ×	× ×	

* DNP derivative.

This fraction also contained traces of free neutral amino-acids, which we have always found in the basic fractions from the ionophoresis. These are not recorded.

Spot 1. Part of the Ser.His spot contains free lysine. These two are well separated if 0·3% NH_3 is used with the cresol, when the lysine travels much faster. However, the other peptides are badly separated and only Ser.His appears as a single pure spot.

Spot 4. Hydrolysis of the DNP derivative gave bis-DNP-lysine.

Spot 5. The DNP derivative of this basic peptide could be extracted (though not rapidly) into ethyl acetate from acid solution, in contrast to DNP-arginine. It was not well adsorbed on talc from acid solution, but could be adsorbed better from neutral solution.

Spot 6. Contains traces of glycine which has probably arisen by partial breakdown of the *N*-terminal threonine.

Methods of group separation

Probably none of the experiments described represent the most satisfactory procedure for fractionating the mixture concerned. Exp. *B* 1, in which most of the aromatic peptides were lost by adsorption on the resin, was designed primarily to separate cysteic acid peptides and it was only later decided to study the other fractions.

The four-compartment cell of Synge was found to give very satisfactory separations into acidic, basic and neutral peptides with only a little overlapping. Under the conditions described many of the simple peptides of glutamic acid were found in the neutral fraction ($B2\beta$) as well as in the acidic fraction ($B2\alpha$). By maintaining the specimen compartment at a slightly higher pH this overlapping could be avoided and all the glutamic acid peptides were then found in the acidic fraction, but under these conditions the neutral fraction contained histidine peptides.

The ion-exchange resin gave a good separation of peptides of cysteic acid, which are less strongly adsorbed than free cysteic acid. No clear separation of peptides containing aromatic residues was obtained by adsorption on charcoal. Several other peptides were also strongly adsorbed. Most of these

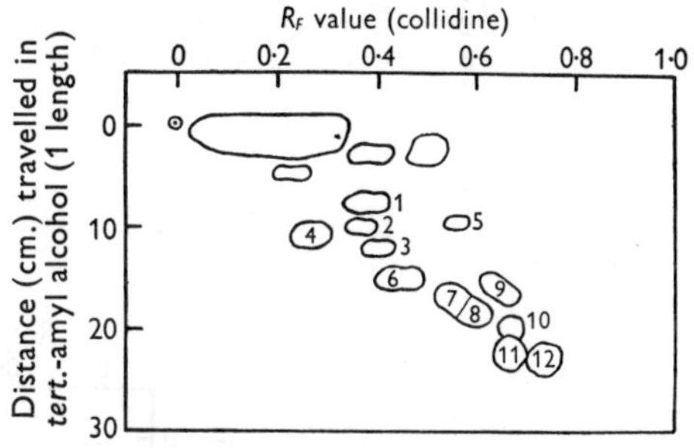

Fig. 14. Chromatogram of fraction $B3\beta$ (see Table 10).

Table 10. *Peptides from fraction B* 3β

Spot no. (Fig. 14)	Amino-acids present	Strength of amino-acid after: Hydrolysis	Strength of amino-acid after: DNP treatment and hydrolysis	Structure
1	Asp	× ×	×	Phe.[Asp, Glu, Val] ($B1\beta12$) + Gly.Phe ($B1\gamma12a$)
	Glu	× ×	×	
	Gly	× ×	?	
	Val	× ×	×	
	Phe	× × ×	×	
2	Asp	× ×		Phe.[Asp, Val] ($B1\beta13$)
	Val	× ×		
	Phe	× ×		
3	Ala	× × × × ×		Ala.Leu ($B1\delta11$)
	Leu	× × × × ×		
4	Glu	× × × ×	× × ×	Leu.[Glu, Val] ($B2\beta14$) + Leu.[Glu, Ala, Val] ($B2\beta15$)
	Ala	× ×	×	
	Val	× × × ×	× × ×	
	Leu	× × × ×	?	
5	TyrX	× ×		TyrX
6	Phe	× × ×		Phenylalanine + leucine
	Leu	× × × ×		
7	Val	× × ×	× × ×	Leu.Val
	Leu	× × ×	×	
8	Val	× ×		[Val, Phe] ($B4\beta2$) + 7
	Leu	× ×		
	Phe	×		
9	Ala	× ×	—	Ala.[Tyr, Leu]
	Tyr	× ×	?*	
	Leu	× × ×	×	
10	TyrX + Tyr	× ×	?	[TyrX, Leu] [Tyr, Leu]
	Leu	× ×	×	
11	Tyr	?		Mixture
	Val	×		
	Leu	× ×		
	Phe	× ×		
12	Tyr	×		[Tyr, Val, Leu]
	Val	× ×		
	Leu	× ×		

* DNP derivative.

Table 11. *Peptides from fraction B* 4α

Peptide	Strength	Structure	Identical with
1	× × ×	$CySO_3H$.Gly	*B* 1α1
2	× × × ×	Aspartic acid	—
3	× × ×	Glutamic acid	—
4	× ×	Val.[$CySO_3H$, Gly]	*B* 1α3
5	× ×	Leu.[$CySO_3H$, Gly]	*B* 1α5
6	×	Leu.$CySO_3H$	*B* 1α6
7	× ×	Leu.[$CySO_3H$, Gly, Val]	*B* 1α7
8	×	Leu.[Val, $CySO_3H$]	*B* 1α8

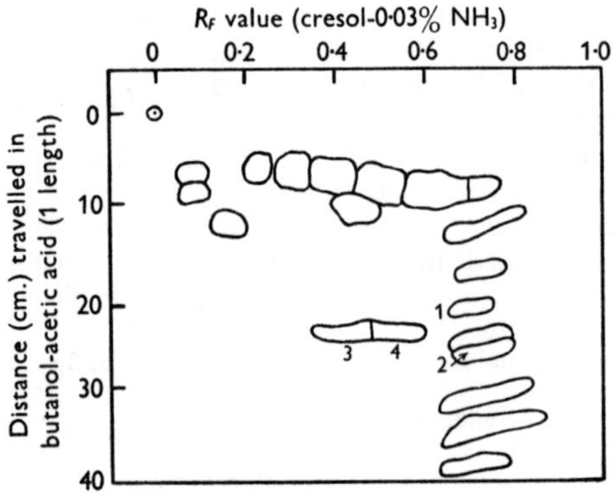

Fig. 15. Chromatogram of fraction *B* 4γ. The chromatogram was run first down the short direction of the filter paper with cresol-0·03% NH_3 and then down the long direction with butanol-acetic acid (see Table 12).

Table 12. *Peptides from fractions B* 4β *and B* 4γ

Peptide	Amino-acids present	Strength of amino-acid after: Hydrolysis	Strength of amino-acid after: DNP treatment and hydrolysis	Structure
β1	$CySO_3H$	× × ×		[$CySO_3H$, Gly, His, Leu]
	Gly	× × ×		
	His	× × ×		
	Leu	× × ×		
β2	Val	× × ×		[Val, Phe]
	Phe	× ×		
γ1*	Ser	× ×		Ser.[His, Leu] (*B* 2γ7)
	His	× ×		
	Leu	× × ×		
γ2*	Ser	×	?	Ser.[His, Val, Leu]
	His	× ×	†	
	Val	× ×	× ×	
	Leu	× ×	× ×	
γ3*	Glu	× ×		[Glu, Ser, His, Val, Leu]
	Ser	× ×		
	His	× ×		
	Val	× × ×		
	Leu	× × ×		
γ4*	Glu	× ×	×	Ser.[Glu, Ala, His, Val, Leu]
	Ser	×	?	
	Ala	× ×	×	
	His	× ×	†	
	Val	× ×	×	
	Leu	× ×	×	

* Spot nos. as in Fig. 15. These gave yellow colour with ninhydrin.
† Run on a phenol chromatogram, so *im*-DNP-His not observed.

contained both an acidic and a basic residue. Thus [Glu.Arg] ($B\,1\gamma3$) and Glu.His ($B\,1\gamma1$) were adsorbed, whereas Arg.Gly ($B\,2\gamma5$) and Ser.[His, Leu]

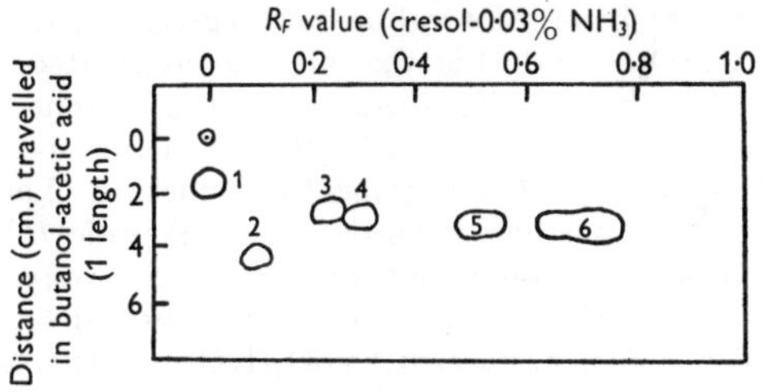

Fig. 16. Chromatogram of fraction $B\,5\gamma$ (see Table 13).

($B\,2\gamma7$) were not. Similarly, it was found that the DNP derivative of Glu.Arg could be adsorbed from acid solution on to talc, whereas the derivative of Arg.Gly was only weakly adsorbed, in contrast to the behaviour of most DNP peptides.

Probably the best method of carrying out the preliminary group separations on these hydrolysates would be to obtain first the acidic, neutral and basic fractions by ionophoresis. The acidic one could be further fractionated first with charcoal and then on the ion-exchange resin to separate peptides of cysteic acid. The neutral one could be separated into two fractions by charcoal adsorption, and the basic one could be subjected to a second ionophoresis at a higher pH to separate arginine peptides.

From the paper chromatograms it may be concluded that in general mixtures containing up to about twenty peptides (fractions $B\,2\alpha$ and $B\,2\beta$) may be satisfactorily fractionated, though for preliminary work it is preferable to have fewer present.

Structure of the peptides

The most probable composition of fraction *B* is given in Table 14. This is obtained from a consideration of the composition of insulin (Tristram, 1949) and of the very approximate analysis of fraction *A* (Sanger, 1949*a*), and may therefore not be very accurate. The value for cysteic acid was determined directly on fraction *B* by Rees by determination of the S content, and the figures for histidine, arginine, lysine and threonine are reliable since fraction *A* contains none of these residues.

N-*Terminal peptide*. The results with the DNP peptides demonstrated the presence of the *N*-terminal sequence, Phe.Val.Asp.Glu (Sanger, 1949*b*). The present results offer ample confirmation for this sequence, which may be deduced as follows:

Table 14 indicates that fraction *B* contains only one aspartic acid residue, so that all peptides con-

Table 13. *Peptides from fraction* $B\,5\gamma$

Spot no. (Fig. 16)	Amino-acids present	Strength of amino-acid after		Structure
		Hydrolysis	DNP treatment and hydrolysis	
1	Glu	× × ×	× × ×	Gly.[Glu, Orn]
	Gly	× × ×	×	
	Orn	× × × ×	× × ×*	
2	Gly	×		Glycine
3	Orn	×		Ornithine
4	Lys	×		Lysine
5	Ala	× ×		Lys.Ala ($B\,2\gamma4$)
	Lys	× ×		
6	Ala	× × ×	× × ×	Pro.[Ala, Lys]
	Lys	× × ×	× ×*	
	Pro	× × ×	?	

* DNP derivative.

Spot 1. DNP derivative hydrolysed 6 hr. This peptide was also found in the neutral fraction ($B\,5\beta$), which is where one would expect to find it.

Spot 6. DNP derivative hydrolysed 12 hr. with 11 N-HCl at 105°.

Table 14. *Probable amino-acid composition of fraction B*

Amino-acid	No. of residues/mol.
Cysteic acid	2
Aspartic acid	1
Glutamic acid	3–4
Serine	1
Glycine	3
Threonine	1
Alanine	2
Tyrosine	2
Valine	2
Leucine	3–4
Phenylalanine	3
Proline	1
Histidine	2
Lysine	1
Arginine	1

taining aspartic acid should fit into a single sequence. Six such peptides have been found: [Asp, Glu] (*B* 1α2), Val.Asp (*B* 1β8), Val.[Asp, Glu] (*B* 1β10), Phe.[Val, Asp, Glu] (*B* 1β12), Phe.[Val, Asp] (*B* 1β13) and Phe.[Asp, Glu, His, Val] (*B* 1γ6). It can be seen that the structure of the first five peptides can only be explained on the basis of the above sequence. Phe.Val (*B* 4β2) is also derived from this sequence.

Peptide *B* 1γ6 moves at different rates on the chromatograms from the other peptides of aspartic acid so must contain histidine as an integral part and not as an impurity. It can also be seen that none of the other histidine peptides would be in this position. Its structure must therefore be Phe.Val.-Asp.Glu.His. This is also supported by the identification of Glu.His (*B* 1γ1). The fact that no peptides of aspartic acid have been found that do not fit this sequence confirms that fraction *B* contains only one aspartic acid residue.

Peptides of lysine. With the DNP method it was shown that the lysine was present in the sequence Thr.Pro.Lys.Ala, and this sequence is also confirmed in the present work by the identification of the following peptides: Thr.Pro (*B* 1δ8), Lys.Ala (*B* 2γ4), Thr.[Pro, Lys, Ala] (*B* 2γ6), Pro.[Lys, Ala] (*B* 5γ6). Since fraction *B* only contains one lysine and one threonine residue this establishes the sequence. No other peptides containing proline have been found, indicating that there is also only one proline residue. It appears that the Thr.Pro bond is unusually resistant to concentrated hydrochloric acid since large amounts of Thr.Pro and the tetrapeptide were found, whereas only traces of free proline could be detected. The tripeptide Pro.Lys.-Ala was not found in the concentrated acid hydrolysate, whereas it was in the alkaline hydrolysate.

Peptides of arginine. Two dipeptides of arginine were detected: Glu.Arg (*B* 1γ3) and Arg.Gly (*B* 2γ5). Since there is only one arginine residue this establishes the sequence Glu.Arg.Gly. In peptide *B* 5γ1 (Gly.[Glu, Orn]) the ornithine is evidently derived from arginine. Since glycine is the *N*-terminal residue its structure must be Gly.Glu.Orn, and the order in fraction *B* is therefore Gly.Glu.-Arg.Gly. Gly.Glu (*B* 1δ1) was detected also and appears to originate from this sequence. This structure could give rise to two tripeptides containing glycine, glutamic acid and arginine. Peptide *B* 1γ2 is either one or both of these. Difficulties in deaminating this peptide may have been due to its being a mixture.

Peptides of cysteic acid. Three dipeptides containing cysteic acid have been detected: $CySO_3H$.Gly (*B* 1α1), Val.$CySO_3H$ (*B* 1α4) and Leu.$CySO_3H$ (*B* 1α6). Since fraction *B* contains only two cysteic acid residues, all peptides in which cysteic acid is not *N*-terminal must therefore contain the sequences Val.$CySO_3H$ or Leu.$CySO_3H$. Peptide *B* 1α3 is therefore Val.$CySO_3H$.Gly and peptide *B* 1α5 is Leu.$CySO_3H$.Gly. The only possible structure for *B* 1α8, which contains valine but no glycine is Leu.-Val.$CySO_3H$ and this was also demonstrated by partial hydrolysis. This establishes the presence of the sequence Leu.Val.$CySO_3H$.Gly in fraction *B*, and peptide *B* 1α7 clearly has this structure. Leu.-Val (*B* 3β7) has also been detected. Peptide *B* 1γ4 which contains no valine or glycine can only be His.Leu.$CySO_3H$. This establishes a second tetrapeptide sequence His.Leu.$CySO_3H$.Gly, and peptide *B* 4β1 presumably has this structure. The presence of His.Leu (*B* 2γ8) is further evidence for this sequence. Another peptide containing cysteic acid is *B* 1β5 ([$CySO_3H$, Glu, His, Leu]) and the question arises as to whether it is a pure peptide or could be a mixture. The cysteic acid cannot be present as the free amino-acid, which moves slower on phenol; therefore it must be in a peptide. No valine or glycine is present so that it must be present in the sequence Leu.$CySO_3H$. The dipeptide Leu.$CySO_3H$ goes considerably faster on butanol-acetic acid and would appear below spot 6 (Fig. 6), so that the cysteic acid in spot 5 must be present in the form His.Leu.-$CySO_3H$. This tripeptide (*B* 1γ4) goes faster on phenol (Fig. 7) so that the cysteic acid must be present in the tetrapeptide Glu.His.Leu.$CySO_3H$. In support of this structure we have also found the peptides Glu.His (*B* 1γ1), and [Glu, His, Leu] (*B* 1γ7). The sequence in fraction *B* is therefore Glu.His.Leu.$CySO_3H$.Gly. The fact that spot *B* 2α8 ([$CySO_3H$, Glu, His, Leu]) sometimes contains a trace of glycine suggests that the pentapeptide may also be present.

Peptide *B* 2α21 ([$CySO_3H$, Tyr, Val, Leu]) moves considerably faster on cresol than any of the other cysteic acid peptides including Leu.Val.$CySO_3H$ (*B* 2α17) (Fig. 11). The tyrosine must therefore be present as part of the peptide, and since it contains no glycine or histidine the cysteic acid must be in the sequence Leu.Val.$CySO_3H$ and must be the *C*-terminal residue. The structure of the peptide is thus Tyr.Leu.Val.$CySO_3H$, and the order in fraction *B* must be Tyr.Leu.Val.$CySO_3H$.Gly. This sequence would be expected to give rise to Tyr.Leu and Tyr.Leu.Val. Spots *B* 3β10 ([Tyr, Leu]) and *B* 3β12 ([Tyr, Val, Leu]) probably contain these peptides, though it is not certain that they are not mixtures.

Peptides of valine and alanine. According to Table 14 fraction *B* should only contain two valine residues. One of these is present in the *N*-terminal sequence Phe.Val.Asp.Glu.His, another in the sequence Tyr.Leu.Val.$CySO_3H$.Gly. However there are other peptides containing valine which do not fit these sequences. It is therefore necessary to assume either that fraction *B* contains three valine

Table 15. *Peptides obtained from phenylalanyl chain of insulin (fraction B)*

Dipeptides	Phe.Val (*B* 4β2); Asp.Glu (*B* 1α2); His.Leu (*B* 2γ8); CySO$_3$H.Gly (*B* 1α1); Val.Asp (*B* 1β8); Glu.His (*B* 1γ1); Leu.CySO$_3$H (*B* 1α6)	Thr.Pro (*B* 1δ8); Lys.Ala (*B* 2γ4)	Gly.Glu (*B* 1δ1); Arg.Gly (*B* 2γ5); Glu.Arg (*B* 1γ3)	Tyr.Leu (*B* 3β10); Val.CySO$_3$H (*B* 1α4); Leu.Val (*B* 3β7); CySO$_3$H.Gly (*B* 1α1)	Ser.His (*B* 2γ2); Leu.Val (*B* 3β7); Glu.Ala (*B* 1δ4); His.Leu (*B* 2γ8); Val.Glu (*B* 1δ7)	Ala.Leu (*B* 1δ11); Gly.Phe (*B* 1γ12)
Tripeptides	Phe.Val.Asp (*B* 1β13); His.Leu.CySO$_3$H (*B* 1γ4); Val.Asp.Glu (*B* 1β10); Leu.CySO$_3$H.Gly (*B* 1α5); Glu.His.Leu (*B* 1γ7)	Pro.Lys.Ala (*B* 5γ6)	Gly.Glu.Orn (*B* 5γ1); [Glu, Arg, Gly] (*B* 1γ2)	Tyr.Leu.Val (*B* 3β12); Leu.Val.CySO$_3$H (*B* 1α8); Val.CySO$_3$H.Gly (*B* 1α3)	Ser.His.Leu (*B* 2γ7); Val.Glu.Ala (*B* 1γ10); Leu.Val.Glu (*B* 2β14)	Ala.[Tyr, Leu] (*B* 3β9)
Higher peptides	Phe.Val.Asp.Glu (*B* 1β12); His.Leu.CySO$_3$H.Gly (*B* 4β1); Phe.Val.Asp.Glu.His (*B* 1γ6); Glu.His.Leu.CySO$_3$H (*B* 1β5)	Thr.Pro.Lys.Ala (*B* 2γ6)		Tyr.Leu.Val.CySO$_3$H (*B* 2α21); Leu.Val.CySO$_3$H.Gly (*B* 1α7)	Ser.His.Leu.Val (*B* 4γ2); Leu.Val.Glu.Ala (*B* 2β15); His.Leu.Val.Glu (*B* 2β18); Ser.His.Leu.Val.Glu (*B* 4γ3); Ser.His.Leu.Val.Glu.Ala (*B* 4γ4)	
Sequence in fraction *B*	Phe.Val.Asp.Glu.His.Leu.CySO$_3$H.Gly	Thr.Pro.Lys.Ala	Gly.Glu.Arg.Gly	Tyr.Leu.Val.CySO$_3$H.Gly	Ser.His.Leu.Val.Glu.Ala	Others

residues or else that it is not homogeneous. The former explanation is preferred.

Besides the sequences already considered, valine is present in the following peptides: Val.Glu (*B* 1δ7), Val.[Ala, Glu] (*B* 1γ10), Leu.[Glu, Val] (*B* 2β14) and Leu.[Val, Glu, Ala] (*B* 2β15). The only sequence fitting all these peptides is Leu.Val.Glu.Ala. This sequence may also be deduced from a consideration of the peptides of alanine. Fraction *B* contains two alanine residues; one is at the end of the sequence Thr.Pro.Lys.Ala. Other peptides are Ala.Leu (*B* 1δ11), Glu.Ala (*B* 1δ4), Val.[Ala, Glu] (*B* 1γ10), Leu.[Val, Glu, Ala] (*B* 2β15) and Ala.[Tyr, Leu] (*B* 3β9). Any peptide containing alanine, which is not *N*-terminal, must either contain Lys.Ala or Glu.Ala. *B* 1γ10 is therefore Val.Glu.Ala and *B* 2β15 is Leu.Val.Glu.Ala. The position of Ala.Leu or of Ala.[Tyr, Leu] cannot be deduced and they need not necessarily be from the same sequence.

Peptides of histidine. Two sequences containing histidine have already been considered: Phe.Val.-Asp.Glu.His and Glu.His.Leu.CySO$_3$H.Gly. Besides Glu.His and His.Leu, the only other dipeptide of histidine is Ser.His (*B* 2γ2). Since there are only two histidine residues and one serine residue in fraction *B*, the tripeptide *B* 2γ7 (Ser.[His, Leu]) can only be Ser.His.Leu. The above two pentapeptide sequences, which both contain Glu.His but not Ser.His must therefore be parts of the same sequence, so that the actual order in fraction *B* is Phe.Val.Asp.Glu.His.Leu.CySO$_3$H.Gly. Further evidence for this sequence was obtained by an extension of the DNP technique. A dilute-acid hydrolysate of the DNP derivative of fraction *B* was fractionated on a buffered Hyflo column using ethyl acetate. Two slow-moving bands were present; one contained DNP-phenylalanine, valine, aspartic acid, glutamic acid, *im*-DNP-histidine and leucine and the second contained these six residues together with cysteic acid and glycine. This in itself could not be considered as direct evidence as they could have been mixtures of terminal DNP-phenylalanyl peptides with peptides of *im*-DNP-histidine, but it adds support to the above conclusion. The establishment of this sequence depends chiefly on the two large peptides *B* 1γ6 (Phe.[Val, Asp, Glu, His]) and *B* 1β5 ([Glu, His, Leu, CySO$_3$H]) which are present in rather small amounts and therefore cannot be regarded as completely proved. Ample confirmation was, however, obtained from the results with enzymic hydrolysates (Sanger & Tuppy, 1951).

Further peptides containing histidine were present in the dilute-acid hydrolysate (fraction *B* 4γ). Peptide *B* 4γ2 (Ser.[His, Val, Leu]) must contain the sequence Ser.His.Leu and is therefore Ser.His.-Leu.Val. The four peptides of fraction *B* 4γ, shown in Fig. 15, gave a yellow-grey colour with ninhydrin which is characteristic of peptides containing serine as the *N*-terminal residue, and the DNP results showed this to be the case for *B* 4γ4. Thus the structure of *B* 4γ3 must be Ser.His.Leu.Val.Glu and of *B* 4γ4 Ser.His.Leu.Val.Glu.Ala. This sequence may also be derived by considering the valine peptides. The three valine residues are present as Phe.Val.Asp, Tyr.Leu.Val.CySO$_3$H, Leu.Val.Glu.Ala. The latter is the only one to which Ser.His.Leu.Val could be combined, so that the order in fraction *B* must be Ser.His.Leu.-Val.Glu.Ala. Spot *B* 2β18 ([Glu, His, Val, Leu]) is probably His.Leu.Val.Glu derived from this sequence.

Other peptides. In Table 15 are listed all the peptides detected in this work and the sequences which have been deduced in the foregoing discussion. We have included all peptides found in the hydrolysates with the exception of some of the fast-moving peptides such as are present in spots *B* 2β24 and *B* 2β25, which could not be identified. Apart from these sequences there are only three peptides which are not included in the five sequences listed. These are Ala.Leu, Gly.Phe, and Ala.[Tyr, Leu].

According to Table 14 all the amino-acids of fraction *B* can be accounted for in the above sequences except for one tyrosine, two phenylalanine and possibly one glutamic acid and one leucine residue.

From Table 15 it can be seen that of the various dipeptides that would be produced on breakdown of the five sequences, only one (Pro.Lys) was not detected. These results are in complete accord with the view that there is only one polypeptide chain in fraction *B*. All peptide spots present in detectable amounts have been taken into account and no peptides have been found that contradict this view. This would be almost impossible if there were two different chains in fraction *B* or if there was a significant amount of impurity.

SUMMARY

1. Partial hydrolysates of fraction *B* of oxidized insulin (Sanger, 1949*a*) have been fractionated by paper chromatography, and the structure of the resulting peptides determined.

2. It is concluded that the following amino-acid sequences are present in this fraction (abbreviations are given in Table 1, p. 463): Phe.Val.Asp.Glu.-His.Leu.CySO$_3$H.Gly, Thr.Pro.Lys.Ala, Gly.-Glu.Arg.Gly, Tyr.Leu.Val.CySO$_3$H.Gly and Ser.-His.Leu.Val.Glu.Ala.

We wish to express our thanks to the Medical Research Council for a grant to cover the cost of the insulin and to Dr R. L. M. Synge for giving us details of the four-compartment ionophoresis cell prior to publication. One of us (H.T.) wishes to thank the British Council for a Scholarship.

REFERENCES

Brand, E. & Edsall, J. T. (1947). *Ann. Rev. Biochem.* **16**, 224.
Consden, R., Gordon, A. H. & Martin, A. J. P. (1944). *Biochem. J.* **38**, 224.
Consden, R., Gordon, A. H. & Martin, A. J. P. (1947). *Biochem. J.* **41**, 590.
Consden, R., Gordon, A. H. & Martin, A. J. P. (1948). *Biochem. J.* **42**, 443.
Consden, R., Gordon, A. H. & Martin, A. J. P. (1949). *Biochem. J.* **44**, 548.
Consden, R., Gordon, A. H., Martin, A. J. P. & Synge, R. L. M. (1947). *Biochem. J.* **41**, 596.
Dent, C. E. (1947). *Biochem. J.* **41**, 240.
Dent, C. E. (1948). *Biochem. J.* **43**, 169.
Desnuelle, P. & Casal, A. (1948). *Biochim. Biophys. Acta*, **2**, 64.
Jones, T. S. G. (1949). *Discuss. Faraday Soc.* **7**, 285.
Macpherson, H. T. (1946). *Biochem. J.* **40**, 470.
Partridge, S. M. (1948). *Biochem. J.* **42**, 238.
Partridge, S. M. & Davis, H. F. (1950). *Nature, Lond.*, **165**, 62.
Phillips, D. M. P. (1948). *Nature, Lond.*, **161**, 53.
Sanger, F. (1949*a*). *Biochem. J.* **44**, 126.
Sanger, F. (1949*b*). *Biochem. J.* **45**, 563.
Sanger, F. (1949*c*). *Cold Spr. Harb. Sym. quant. Biol.* **14**, 153.
Sanger, F. & Tuppy, H. (1951). *Biochem. J.* **49**, 481.
Synge, R. L. M. (1943). *Chem. Rev.* **32**, 135.
Synge, R. L. M. (1951). *Biochem. J.* (in the Press).
Theorell, H. & Åkeson, Å. (1942). *Ark. Kemi Min. Geol.* **16** A, no. 8.
Tiselius, A., Drake, B. & Hagdahl, L. (1947). *Experientia*, **3**, 651.
Tristram, G. R. (1949). *Advanc. prot. Chem.* **5**, 83.
Wieland, T. & Wirth, L. (1949). *Ber. dtsch. chem. Ges.* **82**, 468.

[From THE BIOCHEMICAL JOURNAL, Vol. 49, No. 4, pp. 481–490, 1951]

The Amino-acid Sequence in the Phenylalanyl Chain of Insulin

2. THE INVESTIGATION OF PEPTIDES FROM ENZYMIC HYDROLYSATES

By F. SANGER (Beit Memorial Fellow) and H. TUPPY*
Biochemical Laboratory, University of Cambridge

(*Received* 17 *January* 1951)

In the previous paper (Sanger & Tuppy, 1951) a large number of simple peptides in partial hydrolysates of fraction *B* of oxidized insulin were identified and several amino-acid sequences were deduced from them. It was not possible to determine the complete structure of the polypeptide chain, partly on account of the difficulty of fractionating the less polar peptides containing the aromatic amino-acids and partly because of the relatively great lability of certain bonds such as those involving the amino groups of the hydroxyamino-acids.

In the present paper are described investigations of the peptides produced on hydrolysis of fraction *B* with the proteolytic enzymes, pepsin, chymotrypsin and trypsin. These peptides, which are mostly larger than those present in the acid and alkaline hydrolysates, could be successfully fractionated by paper chromatography, and with the knowledge presented in the previous paper it was possible to determine the structure of most of them and to work out the complete sequence of fraction *B* and hence of the phenylalanyl chain of insulin.

The abbreviations, etc., defined in the previous paper (Sanger & Tuppy, 1951) will be used throughout and the peptides with reference numbers starting *B* 1, *B* 2, *B* 3, *B* 4, *B* 5 refer to those described in that paper.

* Present address: II. Chemisches Universitätslaboratorium, Vienna, Austria.

METHODS

Enzymes

Crystalline pig pepsin was obtained from Armour Laboratories. The chymotrypsin was prepared and several times recrystallized by Mr J. Lightbown of the National Institute for Medical Research, Mill Hill, London. The trypsin was a crystalline preparation from Armour Laboratories. It was recrystallized three times according to the method of Kunitz & Northrop (1936).

Hydrolyses

Experiment Bp (pepsin). Fraction *B* (30 mg.) was dissolved in 3 ml. 0·01 M-HCl. A solution of 0·3 mg. pepsin in 0·3 ml. 0·01 M-HCl was added, and the mixture incubated at 37° for 24 hr. After boiling to inactivate the pepsin, the solution was taken to dryness *in vacuo* to remove excess HCl and subjected to ionophoresis in the four-compartment cell of Synge described in the previous paper. This separated the hydrolysate into acid, neutral and basic fractions.

Experiment Bc (chymotrypsin). Fraction *B* (50 mg.) was dissolved in 5 ml. water and brought to pH 7·5–8·0 by the addition of $NaHCO_3$. To this was added 2·3 mg. chymotrypsin, the pH again adjusted and the solution incubated at 25° for 24 hr. It was then brought to about pH 3 by addition of HCl and taken to dryness in a desiccator. The hydrolysate was again separated into acidic, neutral and basic fractions by ionophoresis.

Experiment Bt (trypsin). Fraction *B* (25 mg.) was dissolved in 2·5 ml. water and the pH adjusted to 7·5 with $NaHCO_3$. Trypsin (1·2 mg.) dissolved in 1·2 ml. water was added and the mixture incubated at 37° for 24 hr. On acidification a precipitate formed which was removed by centrifugation. It yielded on acid hydrolysis all the amino-acids that are present in fraction *B* and was not investigated further. The solution was then taken to dryness *in vacuo*.

Investigation of the peptides

The techniques used to investigate the peptides were the same as those described in the previous paper. Each fraction was studied on two-dimensional chromatograms using Whatman no. 4 or 3 filter papers. The latter is thicker and is useful when larger amounts of material are required. Butanol-acetic acid was found to be the most generally useful solvent for fractionating these large peptides, and this was usually preceded by cresol. Although in certain cases (e.g. Fig. 2) there were no apparent separations with the first solvent, it was found that much better results were obtained on two-dimensional than on one-dimensional chromatograms. Probably the first solvent removes traces of salt and other impurities and distributes each spot over a wider area.

With these larger peptides the amino-acids present were identified using two-dimensional chromatograms and *N*-terminal residues with the dinitrophenyl (DNP) technique. This was usually done both by identification of the free amino-acids and of the DNP-amino-acid present in a hydrolysate of the DNP-peptide. Due to the small amounts of peptides present it was not usually possible to identify the DNP-amino-acid completely on a number of different columns (Sanger, 1945), but it was usually put on a chromatogram that would distinguish it from other amino-acids likely to be present as terminal residues in the peptide.

Several of the peptides were subjected to further hydrolysis with a different enzyme. This was carried out in the following way: eluates of spots, usually from two chromatograms on Whatman no. 4 filter paper, were transferred to a polythene strip and taken to dryness in a desiccator. A solution (0·1 ml.) containing 0·1 mg. of the enzyme adjusted to a suitable pH was then added. The pepsin was dissolved in 0·01 N-HCl and the trypsin was dissolved in water and adjusted to pH 7·5 with $NaHCO_3$. The drops were then transferred to capillary tubes and incubated at 37° for 24 hr. After putting on to polythene strips again they were acidified, taken to dryness and fractionated on a suitable two-dimensional chromatogram.

RESULTS

Experiment Bp (pepsin)

The results obtained with the neutral and basic fractions of the hydrolysate are recorded in Figs. 1 and 2 and Table 1. The acidic fraction was investigated in three of the experiments. No significant amounts of peptides were obtained in two cases, but in the third peptides 1, 2, 4, 11, 12 and 13 (Table 1) could be detected. The different results were probably due to slight differences of pH in the middle compartment during ionophoresis.

Spot 1. Contains traces of Val.Glu which can be separated on a phenol chromatogram.

Eluates of spot 1 from two phenol/butanol-acetic acid chromatograms were hydrolysed 48 hr. in 10 N-HCl at 37° and the hydrolysate fractionated on a phenol/butanol-acetic acid chromatogram. Besides free amino-acids the peptides Glu.Ala (×), Val.Glu (× ×) and unchanged tripeptide (× × ×) were identified. This demonstrates that its structure must be Val.Glu.Ala. It is the same as spot *B*1γ10 and has the same R_F values on phenol and cresol. The Val.Glu present has the same R_F value as peptide *B*1γ9. This indicates that they are probably not present as amides but with the γ-carboxyl group of glutamic acid free.

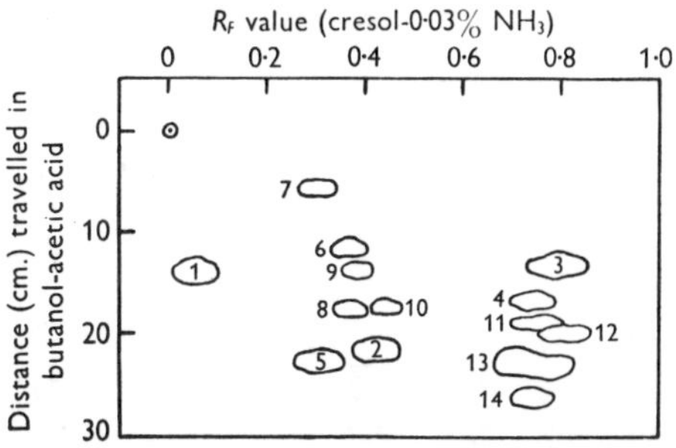

Fig. 1. Chromatogram of the neutral fraction of the peptic hydrolysate (*Bp*). In Figs. 1–6 the fraction was applied in the position marked by the small dotted circle, and the chromatogram run first in the direction represented horizontally, which except in Figs. 2 and 4 was the longer dimension of the filter paper. In each case both solvents were run for one length of the filter paper only (see Table 1).

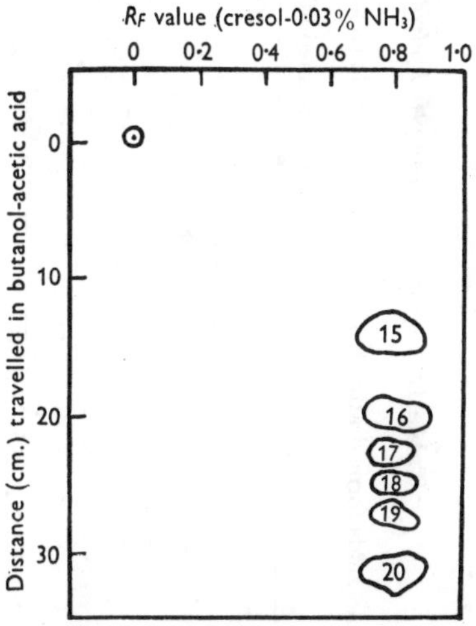

Fig. 2. Chromatogram of the basic fraction of the peptic hydrolysate (*Bp*). The chromatogram was run first down the short dimension of the filter paper with cresol-0·03 % NH_3 (see Table 1).

Spot 2. This is not the same as spot *B*2β15, which contains the same amino-acids, but has leucine as the *N*-terminal residue.

Eluates from two chromatograms were hydrolysed 3 days in 10 N-HCl at 37° and fractionated on a cresol/butanol-acetic acid chromatogram. Besides free amino-acids and small amounts of tripeptides, the dipeptides Val Glu and Ala.Leu were identified and were present in considerable amounts. From this result it is concluded that the tetrapeptide is Val.Glu.Ala.Leu.

Spot 3. The results in Table 1 indicate that phenylalanine is the *N*-terminal residue. This was confirmed by showing

Table 1. *Peptides from peptic hydrolysate* (*Bp*)

(Amino-acids formed after hydrolysis of the peptide (*H*) or of the DNP-peptide (*D*).)

Spot no. (Figs. 1 and 2)	*H* or *D*	Amino-acids formed after hydrolysis															
		$CySO_3H$	Asp	Glu	Ser	Gly	Thr	Ala	Lys	Tyr	TyrX	His	Val	Leu	Arg	Phe	Pro
1	*H*	—	—	× × ×	—	—	—	× × ×	—	—	—	—	× × ×	—	—	—	—
	D	—	—	× ×	—	—	—	× ×	—	—	—	—	—	—	—	—	—
2	*H*	—	—	× ×	—	—	—	× ×	—	—	—	—	× ×	× ×	—	—	—
	D	—	—	× ×	—	—	—	× ×	—	—	—	—	?	× ×	—	—	—
3	*H*	× ×	× ×	× ×	× ×	× ×	—	?	—	—	—	× × ×	× ×	× × ×	—	× ×	—
	D	×	×	×	×	× ×	—	?	—	—	—	×*	×	× ×	—	?	—
4	*H*	× ×	—	× ×	—	× × ×	—	—	—	—	—	—	× ×	× ×	× ×	× ×	—
5	*H*	—	—	×	—	—	—	—	—	—	—	—	×	×	—	—	—
6	*H*	?	—	?	—	×	—	—	—	—	—	—	?	?	?	—	—
7	*H*	—	?	?	—	—	—	—	—	—	—	—	?	—	—	—	—
8	*H*	—	×	×	—	—	—	—	—	—	—	—	×	—	—	?	—
9	*H*	—	—	—	—	—	—	—	—	×	—	—	—	—	—	—	—
10	*H*	—	—	—	—	—	—	—	—	—	× ×	—	—	—	—	—	—
11	*H*	× ×	—	× ×	—	× × ×	—	—	—	—	—	—	× ×	× ×	× ×	× × ×	—
12	*H*	× ×	—	× ×	—	× × ×	—	?	—	× ×	—	—	× ×	× ×	× ×	× ×	—
13	*H*	× ×	—	× ×	—	× × ×	—	(× ×)	—	(×)	—	—	× ×	× ×	× ×	× ×	—
14	*H*	—	—	—	—	—	—	×	—	×	—	—	—	× ×	—	× ×	—
15	*H*	—	—	—	—	—	× ×	× ×	× ×	× ×	—	—	—	—	—	—	× ×
	D	—	—	—	—	—	× ×	× ×	×*	—	—	—	—	—	—	—	×
16	*H*	—	—	—	—	—	× × ×	× × ×	× × ×	—	× ×	—	—	—	—	—	× ×
	D	—	—	—	—	—	× ×	× ×	×*	—	—	—	—	—	—	—	×
17	*H*	× ×	—	—	× ×	× ×	—	—	—	—	—	× × ×	—	× × ×	—	—	—
18	*H*	× ×	× ×	× ×	× ×	× × ×	—	—	—	—	—	× × ×	× ×	× × ×	—	×	—
19	*H*	×	×	×	×	×	?	×	×	—	?	×	×	×	—	×	?
20	*H*	×	?	×	×	×	?	×	×	—	?	×	?	×	—	×	?

* DNP derivative.

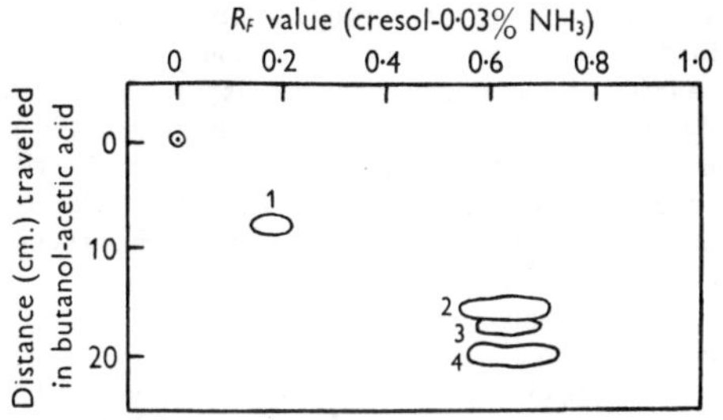

Fig. 3. Chromatogram of the neutral fraction of the chymotryptic hydrolysate (*Bc*) (see Table 2).

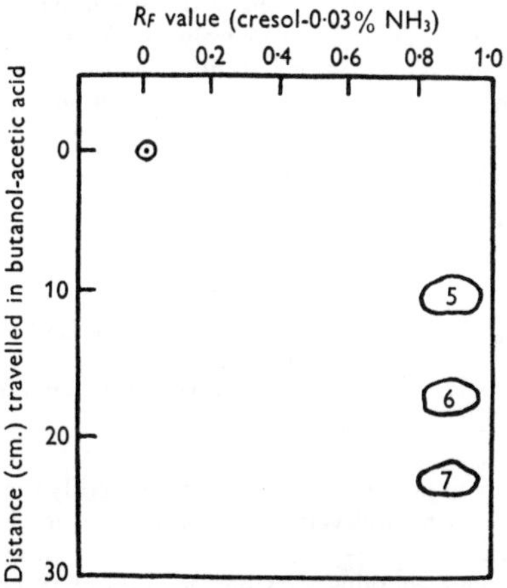

Fig. 4. Chromatogram of the basic fraction of the chymotryptic hydrolysate (*Bc*), run first down the short side of the paper with cresol-0·03 % NH_3 (see Table 2).

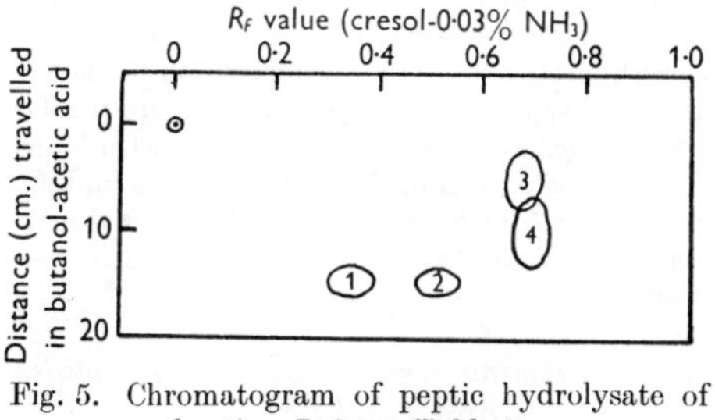

Fig. 5. Chromatogram of peptic hydrolysate of fraction *Bc*2 (see Table 2).

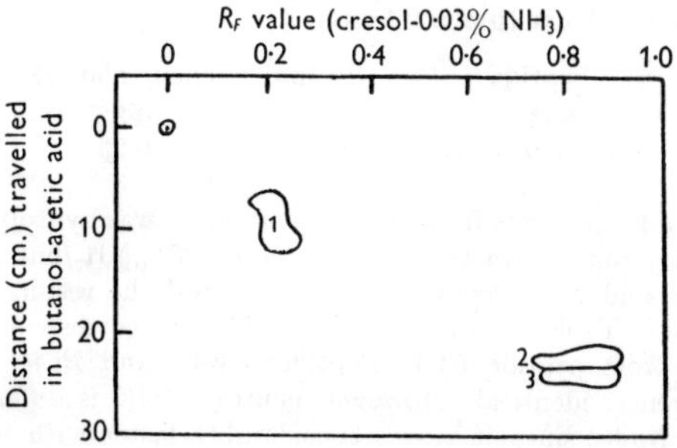

Fig. 6. Chromatogram of the tryptic hydrolysate (*Bt*) (see Table 2).

Table 2. *Peptides from chymotryptic (Bc) and tryptic (Bt) hydrolysates*

Spot no. (Figs. 3, 4, 5 and 6)	Amino-acids formed after hydrolysis															
	$CySO_3H$	Asp	Glu	Ser	Gly	Thr	Ala	Lys	Tyr	TyrX	His	Val	Leu	Arg	Phe	Pro
*Bc*1	×	—	×	—	×	—	—	—	—	—	—	×	×	×	—	—
*Bc*1 (DNP)	×	—	×	—	×	—	—	—	—	—	—	×	—	×	—	—
*Bc*2	× ×	× ×	× × ×	× ×	× × ×	—	× × ×	—	(?)	(×)	× × ×	× × ×	× × ×	—	×	—
*Bc*3	× ×	× ×	× × ×	× ×	× × ×	—	× × ×	—	(×)	(× ×)	× × ×	× × ×	× × ×	—	× ×	—
*Bc*4	× ×	—	× × ×	—	× × × ×	—	—	—	—	—	—	× × ×	× × ×	× × ×	× × ×	—
*Bc*5	—	—	—	—	×	× × ×	× × ×	× × ×	—	—	—	—	—	—	—	× × ×
*Bc*6	—	—	—	—	—	× × × ×	× × × ×	× × × ×	× × ×	—	—	—	—	—	—	× × ×
*Bc*7	—	—	—	—	—	× ×	× × ×	× × ×	—	× ×	—	—	—	—	—	× ×
*Bc*2*p*1	—	—	×	—	—	—	×	—	—	—	—	×	×	—	—	—
*Bc*2*p*2	—	—	× ×	—	—	—	× ×	—	×	—	—	× ×	× ×	—	—	—
*Bc*2*p*3	× ×	× ×	× ×	× ×	× ×	—	—	—	—	—	× ×	× ×	× ×	—	× ×	—
*Bc*2*p*4	× ×	× ×	× ×	× ×	× ×	—	× ×	—	—	—	× ×	× ×	× ×	—	× ×	—
*Bt*1	—	—	—	—	—	—	× × ×	—	—	—	—	—	—	—	—	—
*Bt*2	—	—	—	—	× ×	× ×	—	× ×	× ×	—	—	—	—	—	× × ×	× ×
*Bt*3	—	—	—	—	×	×	—	×	—	×	—	—	—	—	× ×	×
*Bt*2*p*1	—	—	—	—	—	× ×	—	× ×	× ×	—	—	—	—	—	—	×
*Bt*2*p*2	—	—	—	—	× ×	—	—	—	—	—	—	—	—	—	× ×	—

that the DNP-amino-acid produced on hydrolysis of the DNP-peptide moved at the same *R* value as DNP-phenylalanine on an ethanol-ligroin column (Sanger, 1945). This distinguishes it from the DNP derivatives of the other amino-acids present in spot 3, except possibly glycine.

Spot 4. On hydrolysis of the DNP derivative, a DNP-amino-acid was produced that was identified as DNP-leucine or DNP-valine. Chromatography of the remaining amino-acids with benzyl alcohol showed strong spots for valine and phenylalanine and only a weak leucine spot, indicating that leucine was the *N*-terminal residue.

Eluates from two chromatograms were subjected to hydrolysis with trypsin. The hydrolysate was fractionated on a half sheet of Whatman no. 1 filter paper using cresol-0·03 % NH_3/butanol-acetic acid. Only two spots were present, whose R_F values and compositions are shown in Table 3. *Bp*4*t*2 is the same as spot *B* 1γ12*a* (Gly.Phe). Further results with this peptide are given in connexion with peptide *Bc*4.

Spot 1. When a different preparation of chymotrypsin (obtained by activation of chymotrypsinogen with trypsin) was used this spot was much stronger. It was also present in hydrolysates obtained by the use of unrecrystallized trypsin, and thus appears to be produced by the combined action of chymotrypsin and trypsin on fraction *B*. The small amount present in this experiment may be due to traces of trypsin in the preparation.

Spot 2. This sometimes contains varying amounts of tyrosine and TyrX.

A hydrolysate of the DNP derivative was put on a benzyl alcohol chromatogram. It gave a weak phenylalanine spot but strong spots for leucine and valine, showing phenylalanine to be the *N*-terminal residue.

The eluates of spot 2 from two chromatograms were treated with pepsin and the hydrolysate fractionated on a half sheet of Whatman no. 1 filter paper using cresol/butanol-acetic acid. The results are shown in Fig. 5 and Table 2. Spots *Bc*2*p*1 and *Bc*2*p*3 are apparently the same as *Bp*2 and *Bp*3 respectively, which are found in the peptic hydrolysate.

Table 3. *Tryptic hydrolysis of fractions Bp*4 *and Bc*4

Fraction	R_F value (cresol-0·03 % NH_3)	Distance in butanol-acetic acid (cm.)	Strength	Composition
*Bp*4*t*1	0·19	5·6	× × ×	$CySO_3H$, Glu, Gly, Val, Leu, Arg
*Bp*4*t*2	0·66	14·4	× × ×	Gly, Phe
*Bc*4*t*1	0·21	6·9	× ×	$CySO_3H$, Glu, Gly, Val, Leu, Arg
*Bc*4*t*2	0·87	20·8	× ×	Gly, Phe

Spot 5. A small amount of Leu.Val.Glu identical with peptide *B*2β14.

Spot 7. Trace of Val.Asp.Glu.

Spot 8. Phe.Val.Asp.Glu. Peptides 7 and 8 probably contain glutamine and asparagine residues (Sanger, 1949*b*) and are therefore not identical with the corresponding peptides from the acid hydrolysates (*B*1β10, *B*1β12 respectively).

Spots 9 *and* 10. Free tyrosine and TyrX respectively.

Spots 11 *and* 12. In most experiments these two spots were not clearly separated and their composition varied considerably. The results are from a single experiment. Usually they contain free phenylalanine and leucine and at least two peptides, which contain arginine but no aspartic acid and have variable amounts of tyrosine.

Spot 13. Contains arginine but no aspartic acid. The amounts of tyrosine and alanine varied somewhat from one experiment to another. Alanine varies independently of the other amino-acids so is probably part of a low-molecular peptide (e.g. Ala.Leu). Tyrosine is probably connected with the larger peptide.

Spot 14. Probably a mixture of fast-moving small peptides.

Spot 15. The presence of tyrosine as an *N*-terminal residue was confirmed by showing that hydrolysis of the DNP derivative gave rise to a DNP-amino-acid that moved on a $CHCl_3$ column at the same rate as *ON*-bis-DNP-tyrosine. This distinguishes it from DNP-threonine.

Spot 16. The same as 15, but containing TyrX instead of tyrosine.

Spot 18. The same as spot 3 (neutral fraction).

Experiment Bc (chymotrypsin)

The results obtained with the chymotryptic hydrolysate are given in Figs. 3 and 4 and Table 2.

Spot 3. This contains varying amounts of tyrosine and TyrX, showing that it is impure.

A hydrolysate of the DNP derivative was shown on a benzyl alcohol chromatogram to contain a trace of phenylalanine and large amounts of leucine and valine, indicating that phenylalanine was the *N*-terminal residue.

A peptic hydrolysate gave a chromatogram similar to that obtained with spot *Bc*2 (Fig. 5). Spots corresponding to *Bc*2*p*1 and *Bc*2*p*2 were present but only one other large spot in a position corresponding to that of *Bc*2*p*3 and *Bc*2*p*4. This contained all the amino-acids of the original material, and was thus probably a mixture of *Bc*2*p*3 and *Bc*2*p*4 with unhydrolysed *Bc*3.

Table 4. *Amino-acid composition of peptides Bp*4*t*2 *and Bc*4*t*2

(Results expressed as μmol. amino-acid in peptide originating from 10 mg. fraction *B*.)

Peptide	Glycine	Phenylalanine
*Bp*4*t*2	0·3	0·3
*Bc*4*t*2	0·1	0·23

Spot 4. An eluate from one chromatogram was hydrolysed with trypsin and fractionated on cresol-0·03 %NH_3/butanol-acetic acid. Two spots were obtained and the results are shown in Table 3 together with those for the corresponding spots from peptide *Bp*4. Peptides *Bp*4*t*1 and *Bc*4*t*1 are apparently identical. However, peptide *Bc*4*t*2 is different from *Bp*4*t*2 since it moves considerably faster with both solvents. It gives a yellow colour with ninhydrin suggesting that glycine is the *N*-terminal residue. In Table 4 are shown the results of an experiment in which the glycine and phenylalanine in the two peptides were estimated by semi-quantitative paper chromatography. Known quantities of authentic amino-acid were run in parallel on a chromatogram and the strength of the ninhydrin spots compared visually. Whereas *Bp*4*t*2 contains the two amino-acids in equal proportions, *Bc*4*t*2 has two phenylalanine residues to one glycine residue.

Experiment Bt (trypsin)

The soluble material from the tryptic hydrolysate was fractionated directly without any preliminary group separation. Fig. 6 shows a typical chromatogram. Only three significant spots were present, and their composition is shown in Table 2.

Spot 1. Contained only alanine, and its R_F values and strong ninhydrin reaction indicated that it was the free amino-acid.

Spot 2. The slower moving parts of the spot frequently contained traces of glutamic acid, alanine, valine and leucine, which do not appear to be an integral part of the peptide. It gave a yellow colour with ninhydrin, suggesting that glycine or threonine was the *N*-terminal residue. Hydrolysis of the DNP derivative gave a DNP-amino-acid that moved on a $CHCl_3$ column at the same rate as DNP-glycine (or bis-DNP-lysine or *ON*-bis-DNP-tyrosine) and the spot for free glycine was considerably weaker than in the hydrolysate of the untreated peptide. It was thus concluded that glycine is the *N*-terminal residue.

Eluates from two chromatograms were subjected to hydrolysis with pepsin, and the hydrolysate was fractionated on a collidine/benzyl alcohol chromatogram. Only two spots were identified (see Table 2). *Bt*2*p*1 had $R_F = 0{\cdot}35$ in collidine and had travelled 7 cm. in benzyl alcohol. *Bt*2*p*2 had $R_F = 0{\cdot}46$ and had travelled 11·5 cm. It is not clear from these results whether peptide *Bt*2*p*2 is identical with *Bp*4*t*2 ([Gly, Phe]) or with *Bc*4*t*2 ([Gly, Phe, Phe]).

Spot 3. The same as spot 2, but containing TyrX instead of tyrosine.

DISCUSSION

With the rather larger peptides present in these enzymic hydrolysates, it was important to know if certain spots contained only one peptide or whether they could be mixtures. Although direct proof of homogeneity is difficult the following factors could be taken into account in assessing the purity of a spot:

(1) Constant amino-acid composition over different parts of the spot and of the same spot from different chromatograms.

(2) Comparison of the R_F value of the spot with that to be expected if all the amino-acids were present in a single peptide and with the R_F values of possible components of a mixture (see Martin, 1949).

(3) The presence of only one *N*-terminal residue is probably the best evidence of purity, since it is unlikely that two different peptides would have the same *N*-terminal residue.

(4) Consideration of the sequences previously shown to be present in fraction *B* would often indicate whether a given spot was likely to be pure or impure.

Structure of the peptides

In the previous paper the sequences Phe.Val.Asp.Glu.His.Leu.$CySO_3H$.Gly, Thr.Pro.Lys.Ala, Gly.Glu.Arg.Gly, Tyr.Leu.Val.$CySO_3H$.Gly and Ser.His.Leu.Val.Glu.Ala were shown to be present in fraction *B*. These sequences account for all the residues present except two phenylalanine, one tyrosine and possibly one leucine and one glutamic acid residue. Other peptides present were Ala.Leu, Gly.Phe and Ala.[Tyr, Leu].

Peptide *Bp* 3 contains aspartic acid and hence at least part of the *N*-terminal sequence Phe.Val.Asp.Glu.His.Leu.$CySO_3H$.Gly, in which the only aspartic acid residue in fraction *B* is present. It has a phenylalanine *N*-terminal residue and contains all the amino-acids present in this sequence and also serine. There is only one serine residue in fraction *B* and this is present in the sequence Ser.His.Leu.Val.Glu.Ala, so that this establishes a connexion between these two sequences. None of the other sequences above could come between the glycine and serine residues, since peptide *Bp* 3 contains no threonine, proline, lysine, arginine or tyrosine. The possibility that a phenylalanine, leucine or glutamic acid residue may come between the glycine and serine should be considered since all of these residues in fraction *B* have not yet been accounted for. Peptide *Bp* 17, which has cysteic acid, glycine, serine, histidine and leucine but no valine, must contain parts of both the above two sequences. It contains no phenylalanine or glutamic acid, so that neither of these two can be between the glycine and serine residue. It is very unlikely that there is a leucine residue here since Gly.Leu has never been observed, whereas the similar Gly.Phe was detected. Also it will be shown below that all the leucine residues of fraction *B* can be accounted for elsewhere. Thus it may be concluded that the *N*-terminal peptide sequence of fraction *B* is Phe.Val.Asp.Glu.His.Leu.$CySO_3H$.Gly.Ser.His.Leu.Val.Glu.Ala. The Gly.Ser bond would be extremely labile and was therefore not found intact in any of the peptides obtained from acid or alkaline hydrolysis. The exact structure of peptide *Bp* 3 cannot be deduced from the present results, but it most probably contains the first eleven residues of the above sequence, since the presence of Val.Glu.Ala.Leu (*Bp* 2) shows that the Leu.Val bond is split by pepsin. It must contain at least the first nine residues as it has serine but not more than thirteen residues as alanine is absent. Peptide *Bp* 17 is most probably His.Leu.$CySO_3H$.Gly.Ser.His.Leu. This is suggested by the presence of Phe.Val.Asp.Glu (*Bp* 8) and Val.Asp.Glu (*Bp* 7) in the peptic hydrolysate, which demonstrates a lability of the Glu.His bond to pepsin.

Peptide *Bp* 2 was shown by partial hydrolysis to be Val.Glu.Ala.Leu. This must be joined to the above

N-terminal sequence, since the other alanine residue of fraction *B* is in the form Lys.Ala. Ala.Leu (*B* 1δ11) was found in the acid hydrolysate. Further evidence for this *N*-terminal sequence is derived from peptides *Bc* 2 and *Bc* 3. These contain the amino-acids of the above sequence and a phenylalanyl *N*-terminal residue. They also have alanine and varying amounts of tyrosine, which shows they are longer than *Bp* 3. On hydrolysis with pepsin *Bc* 2 (Fig. 5, Table 2) gives a peptide (*Bc* 2*p*3) which appears to be identical with *Bp* 3 and also Val.Glu.Ala.Leu (*Bc* 2*p*1) identical with *Bp* 2. It also gives a peptide [Glu, Ala, Tyr, Val, Leu] (*Bc* 2*p*2). This can only be Val.Glu.Ala.Leu.Tyr and establishes the position of tyrosine on the *N*-terminal sequence, which must therefore be Phe.Val.Asp.Glu.His.Leu.$CySO_3H$.-Gly.Ser.His.Leu.Val.Glu.Ala.Leu.Tyr. This will be referred to as sequence 1. Further evidence for this structure comes from peptide *B* 3β9, which must be Ala.Leu.Tyr. Leu.Tyr is probably present in spot *B* 3β10. The indefinite result with the DNP method suggests both Leu.Tyr and Tyr.Leu are present.

The variation in the amounts of tyrosine in spots *Bc* 2 and *Bc* 3 suggests that they are probably mixtures of closely related peptides containing the *N*-terminal sequence and varying amounts of leucine, tyrosine and TyrX. It is probable that pepsin can split the bonds on both sides of this tyrosine residue.

Fraction *B* contains three glycine residues, one of which is present in the above sequence; others are in the two sequences Tyr.Leu.Val.$CySO_3H$.Gly and Gly.Glu.Arg.Gly, which must therefore be parts of the same sequence. Thus the sequence in fraction *B* must be Tyr.Leu.Val.$CySO_3H$.Gly.Glu.Arg.Gly. The only other peptide containing glycine is Gly.Phe (*B* 1γ12*a*) which was obtained from the acid hydrolysate. It can only be combined to the *C*-terminal glycine of the above octapeptide establishing the sequence Tyr.Leu.Val.$CySO_3H$.-Gly.Glu.Arg.Gly.Phe.

A number of peptides from the pepsin and chymotrypsin hydrolysates offer further evidence for this sequence. Peptide *Bp* 4 contains arginine as well as valine and cysteic acid and has leucine as the *N*-terminal residue, thus demonstrating the connexion between the two sequences Tyr.Leu.Val.$CySO_3H$.-Gly and Gly.Glu.Arg.Gly. On hydrolysis with trypsin two peptides are produced. One (*Bp* 4*t*2) is Gly.Phe, the other (*Bp* 4*t*1) contains all the amino-acids of *Bp* 4 except phenylalanine. It is probably identical with peptide *Bc* 1, which was very strong when what appears to be a mixture of trypsin and chymotrypsin was used for hydrolysis. It contains a leucine *N*-terminal residue and can only be Leu.Val.$CySO_3H$.Gly.Glu.Arg; *Bp* 4 must be Leu.Val.$CySO_3H$.Gly.Glu.Arg.Gly.Phe, the trypsin splitting after the arginine residue.

Peptide *Bc* 4 is also derived from this sequence and contains the same amino-acids as *Bp* 4. On hydrolysis with trypsin it gives the same hexapeptide Leu.Val.$CySO_3H$.Gly.Glu.Arg (*Bc* 4*t*1), but instead of Gly.Phe it gives a spot (*Bc* 4*t*2) moving considerably faster, but containing only these two amino-acids. A semi-quantitative analysis (Table 4) indicated that the ratio of glycine to phenylalanine in this peptide is 1:2, and the only possible conclusion is that it is Gly.Phe.Phe and that peptide *Bc* 4 is Leu.Val.$CySO_3H$.Gly.Glu.Arg.Gly.Phe.-Phe. The sequence in fraction *B* is thus Tyr.Leu.-Val.$CySO_3H$.Gly.Glu.Arg.Gly.Phe.Phe, which will be referred to as sequence 2.

Spots *Bp* 11, *Bp* 12, *Bp* 13 are probably mixtures of various peptides containing this sequence, with varying amounts of tyrosine, alanine, leucine and phenylalanine. On both sides of this sequence there are several bonds that may be split by pepsin, so that one would expect to find a large number of closely related peptides. Peptide *Bp* 6 is probably a trace of Leu.Val.$CySO_3H$.Gly.Glu.Arg.Gly. It is unlikely that pepsin would split the bond next to arginine, whereas it would be expected to split the bond before the phenylalanine residue.

The only other sequence deduced from the acid hydrolysis experiments that we have not so far considered is Thr.Pro.Lys.Ala, which is present as the tetrapeptide in spot *Bc* 5. Peptides *Bc* 6, *Bc* 7, *Bp* 15, *Bp* 16 all contain the same four residues together with tyrosine or TyrX which is their *N*-terminal residue. Their structure must therefore be Tyr.Thr.Pro.-Lys.Ala. Peptide *Bt* 2 contains threonine, proline and lysine which are unique to this sequence and also phenylalanine, which must be derived from the *C*-terminal end of sequence 2, thus establishing a connexion between these two sequences. Peptide *Bt* 2 has a glycyl *N*-terminal residue and must therefore be Gly.Phe.Phe.Tyr.Thr.Pro.Lys. On hydrolysis with pepsin it gives Tyr.Thr.Pro.Lys (*Bt* 2*p*1) and a peptide (*Bt* 2*p*2) containing glycine and phenylalanine, probably Gly.Phe.Phe. Sequence 2 must now be Tyr.Leu.Val.$CySO_3H$.Gly.Glu.Arg.Gly.-Phe.Phe.Tyr.Thr.Pro.Lys.Ala.

Structure of fraction B

Fraction *B* contains only two tyrosine residues. One is accounted for within sequence 2, the other is present both as the *C*-terminal residue of sequence 1 and as the *N*-terminal residue in sequence 2. The two sequences must therefore have this tyrosine residue in common, and the structure of fraction *B* must be Phe.Val.Asp.Glu.His.Leu.$CySO_3H$.Gly.Ser.-His.Leu.Val.Glu.Ala.Leu.Tyr.Leu.Val.$CySO_3H$.-Gly.Glu.Arg.Gly.Phe.Phe.Tyr.Thr.Pro.Lys.Ala. This structure has alanine as a *C*-terminal residue. This confirms the results of Lens (1949), who studied the action of carboxypeptidases on insulin and showed that alanine was present as a *C*-terminal residue. Fromageot, Jutisz, Meyer & Pénasse (1950) and Chibnall & Rees (1951) have also independently identified alanine as the *C*-terminal residue in insulin using their respective methods of reduction to β-amino alcohols. The liberation of free alanine (*Bt* 1) by the action of trypsin on fraction *B* also supports this conclusion. Studies with synthetic peptides have shown that trypsin splits bonds involving the carboxyl groups of arginine or lysine. It will therefore be expected to split the bond before the alanine residue, but not that after it. Free alanine could only be produced if it is already present as a *C*-terminal residue.

Table 5, which summarizes the foregoing discussion, gives the structure of fraction *B* and of the various peptides that have played a major part in the elucidation of the sequence. The residues have been numbered for convenience in the subsequent discussion. The phenylalanyl chains of insulin have the same structure as fraction *B* except that the $CySO_3H$ residues are replaced by CyS (half-cystine) residues.

The above sequence explains the difficulties experienced in fractionating the fast-moving aromatic peptides (Sanger & Tuppy, 1951). A large proportion of the least polar residues are concentrated together in the two sequences Leu.Tyr.Leu.Val and Phe.Phe.Tyr, which can be expected to give rise on partial hydrolysis to a large number of closely related peptides having similar R_F values. The difficulty is increased by the presence of TyrX.

Further evidence for the presence of the sequence Phe.Phe was obtained from a chromatogram of an aromatic fraction which was prepared from a 9-day low-temperature acid hydrolysate of insulin by adsorption on charcoal. A peptide spot was identified which had $R_F = 0{\cdot}61$ on collidine and 0·54 on benzyl alcohol and contained only phenylalanine. The free amino-acid had $R_F = 0{\cdot}40$ on collidine, 0·30 on benzyl alcohol on the same chromatogram. The only possible explanation is that the faster moving spot was Phe.Phe.

Since synthetic reactions have been shown to take place in the presence of proteolytic enzymes, the possibility that some of the peptides identified could have been formed by rearrangement of peptide bonds during the course of hydrolysis should be considered. While such a rearrangement would seem rather unlikely, we have attempted as far as possible to obtain evidence for the presence of each bond, either by identifying it in a peptide obtained by acid hydrolysis or else by identifying it in peptides produced by the action of at least two different enzymes. It is extremely unlikely that the same bond would be synthesized in both cases. The only bond to which this does not apply is the Phe.Tyr bond (positions 25, 26). This was intact only in the peptide *Bt* 2, but it

Table 5. *The structure of fraction B of oxidized insulin*

Sequences deduced from lower peptides (Sanger & Tuppy, 1951)	Phe.Val.Asp.Glu.His.Leu.$CySO_3H$.Gly; Ser.His.Leu.Val.Glu.Ala; Ala.Leu (*B*δ11); Ala.Leu.Tyr (*B*3β9); Tyr.Leu.Val.$CySO_3H$.Gly; Gly.Glu.Arg.Gly; Gly.Phe (*B*1γ12*a*); Thr.Pro.Lys.Ala
Peptides recognized in peptic hydrolysate	Phe.Val.Asp.Glu.His.Leu.$CySO_3H$.Gly.Ser.His.Leu (*Bp* 3); His.Leu.$CySO_3H$.Gly.Ser.His.Leu (*Bp* 17); Val.Glu.Ala.Leu (*Bp* 2); Leu.Val.$CySO_3H$.Gly.Glu.Arg.Gly.Phe (*Bp* 4); Tyr.Thr.Pro.Lys.Ala (*Bp* 15)
Peptides recognized in chymotryptic hydrolysate	Val.Glu.Ala.Leu.Tyr (*Bc* 2*p* 2); Leu.Val.$CySO_3H$.Gly.Glu.Arg.Gly.Phe.Phe (*Bc* 4); Tyr.Thr.Pro.Lys.Ala (*Bc* 7)
Peptides recognized in tryptic hydrolysate	Gly.Phe.Phe.Tyr.Thr.Pro.Lys (*Bt* 2); Ala (*Bt* 1)
Structure of fraction *B*	Phe.Val.Asp.Glu.His.Leu.$CySO_3H$.Gly.Ser.His.Leu.Val.Glu.Ala.Leu.Tyr.Leu.Val.$CySO_3H$.Gly.Glu.Arg.Gly.Phe.Phe.Tyr.Thr.Pro.Lys.Ala (residues numbered 1–30)
Bonds split by pepsin	⇡ 1–2; ⇡ 4–5; ↑ 11–12; ⇡ 13–14; ⇡ 14–15; ↑ 16–17; ⇡ 23–24; ↑ 24–25; ↑ 25–26
Bonds split by chymotrypsin	⇡ 15–16; ↑ 16–17; ↑ 25–26; ↑ 26–27
Bonds split by trypsin	↑ 22–23; ↑ 29–30

↑ indicates major sites of action of enzymes. ⇡ indicates other bonds split by enzymes.

is very improbable that such a bond would be formed by the action of trypsin, since the specificity for synthetic reactions should be similar to that for hydrolytic reactions. Moreover, if any extensive rearrangements had occurred it would almost certainly have been impossible to interpret the results in terms of a single amino-acid sequence for fraction *B*.

The question of location of the amide groups is not completely settled by the present work. The results with the *N*-terminal residues (Sanger, 1949*b*) showed that the aspartic and glutamic acid residues in positions 3 and 4 are in the form of amides, since several different DNP-peptides containing the four residues phenylalanine, valine, aspartic and glutamic acids were detected. According to the amide analysis of fraction *A* (Sanger, 1949*a*) and of insulin (Rees, 1946), there would appear to be only two amide groups in fraction *B*, so that the other two glutamic acid residues (positions 13, 21) would be expected to have free γ-carboxyl groups. If an amide residue is present in a protein it would be expected to be found still intact in most peptides derived from an enzymic hydrolysate, but with the amide group removed in the corresponding peptides from an acid hydrolysate. The peptides Val.Glu.Ala (*Bp* 1), Val.Glu (also in *Bp* 1) and Leu.Val.Glu (*Bp* 5) derived from the peptic hydrolysate all have the same R_F values as the corresponding peptides from the acid hydrolysates ($B1\gamma10$, $B1\gamma9$, $B2\beta14$ respectively), which suggests that they are not present as amides and that the γ-carboxyl group of the glutamic acid residue in position 13 is free in insulin. No such evidence is available for the glutamic acid residue in position 21, but the fact that peptide *Bp* 4 (Leu.Val.$CySO_3H$.Gly.Glu.Arg.-Gly.Phe) is sometimes found in the acidic fraction from the ionophoresis suggests that the γ-carboxyl group of the glutamic acid residue is also free here.

In recent years several hypotheses of protein structure have been advanced which are based on the assumption of some type of periodic arrangement of amino-acids along the polypeptide chains. It is tempting to assume that this arrangement, which is determined by the unknown mechanisms of protein synthesis, may be rather simple and may follow certain easily discernible principles. An examination of the structure of fraction *B*, however, fails to reveal any simple periodic arrangement of the residues, nor is it possible to formulate any general principles which might govern the order of amino-acids along the protein chains. It may be that when much more extensive data are available such principles may become apparent; meanwhile two interesting features of the structure, which may or may not be significant, seem worth noting.

The presence of three aromatic residues joined together in the sequence Phe.Phe.Tyr is of interest and suggests that there may be a certain tendency for the less polar residues to be grouped together. The sequence Leu.Tyr.Leu.Val is another similar grouping. Such an arrangement might be of importance in connexion with the physical or biological properties of the insulin.

In the structure given for fraction *B* there are three dipeptide sequences which occur twice in the chain. These are His.Leu (positions 5–6, 10–11), $CySO_3H$.Gly (positions 7–8, 19–20) and Leu.Val (positions 11–12, 17–18). This is more than one would expect from a mere random arrangement and might suggest a certain limitation in the mechanisms by which proteins are synthesized or it might indicate that dipeptide units are the essential building stones of proteins. It is interesting to note that one of these sequences (CyS.Gly) also occurs in glutathione.

It is difficult to say if either of these features will prove to have general significance. They certainly would not be expected if the arrangement were entirely random, but it is probably more pertinent to consider how many of the possible arrangements of the amino-acids found in fraction *B* would not show some interesting or unusual features of this type.

Action of the proteolytic enzymes on fraction B

Since the purpose of this investigation was primarily to determine the structure of fraction *B*, all the precautions necessary for a careful study of the specificity of proteolytic enzymes have not been taken, nor has the exact structure of all the peptides found been determined. Although it is believed that all the enzymes used are essentially homogeneous, no rigorous test of purity was used, nor was it possible to control accurately the pH in the very small volumes of the tryptic and chymotryptic hydrolysis mixtures. However, certain conclusions concerning the action of the enzymes can profitably be drawn, since this is the first time that such a large polypeptide has been studied. In Table 5 are indicated the various linkages that are split by the three enzymes used. It will be seen that the specificities of trypsin and chymotrypsin, which were defined by Bergmann and his associates using synthetic substrates, are almost completely applicable to this large polypeptide and thus probably to proteins also.

The only two bonds that are split to any extent by trypsin are those involving the carboxyl groups of the arginine and lysine residues. An exactly similar specificity was found by Bergmann, Fruton & Pollok (1939) and by Hofmann & Bergmann (1939) using synthetic substrates.

According to Bergmann & Fruton (1937) chymotrypsin splits those bonds involving the carboxyl groups of the aromatic residues phenylalanine and tyrosine. A typical substrate was *N*-carbobenzyl-oxytyrosylglycine amide. In fraction *B* three of the aromatic residues are grouped together in the sequence Phe.Phe.Tyr. Since chymotrypsin does not split bonds which are adjacent to a free carboxyl group, the manner in which it would split this sequence would be expected to be rather complex. The present results indicate that the main splitting takes place after the two tyrosine residues (positions 16, 26) and after the third phenylalanine residue (position 25). The small amount of tyrosine in peptide *Bc* 2 suggests that some splitting of the Leu.Tyr bond (positions 15, 16) also occurs.

Pepsin appears to have a considerably wider range of specificity than the other two enzymes. Fruton & Bergmann (1939) showed that bonds involving the amino groups of the aromatic residues were readily split in synthetic substrates. Harington & Pitt-Rivers (1944) demonstrated that Tyr.CySH is also hydrolysed. Here the bond involves the carboxyl group of the tyrosine residue. The present results also suggest that splitting takes place on both sides of the aromatic residues, though here again the situation is rather complex, since in the two sequences Ala.Leu.Tyr.Leu.Val and Phe.Phe.Tyr most of the bonds would be susceptible to hydrolysis. It does not, however, appear that the specificity is limited to bonds adjacent to aromatic residues. Thus the Leu.-Val bond (positions 11, 12) is quite strongly hydrolysed. This is evident since Val.Glu.Ala.Leu (*Bp* 2) is present in considerable amounts. Val.Glu.Ala (*Bp* 1) is also present indicating that the Ala.Leu bond is also broken. Another unexpected finding was that the Glu.His bond (positions **4**, **5**) is broken. This was evident from the presence of Phe.Val.Asp.Glu (*Bp* 8) and of Val.Asp.Glu (*Bp* 7) in the hydrolysate. As already mentioned, the glutamic acid here is probably in the form of an amide so that the actual link split is glutaminylhistidine. Apart from this, it could be said that pepsin splits those bonds adjacent to tyrosine, phenylalanine and leucine residues. It would appear to have a specificity for the more fat-soluble parts of the peptide chain. The finding of a rather wide specificity is in agreement with the results of Desnuelle, Rovery & Bonjour (1950), who studied the *N*-terminal residues liberated on treatment of ovalbumin and horse globin with pepsin.

COOH

SUMMARY

1. Fraction *B* of oxidized insulin was subjected to hydrolysis by pepsin, trypsin and chymotrypsin. The resulting peptides were fractionated by paper chromatography and their structure was investigated.

2. It is concluded that the structure of the phenylalanyl chains of insulin is Phe.Val.Asp(-NH_2). Glu(-NH_2). His. Leu. CyS. Gly. Ser. His.- Leu. Val. Glu. Ala. Leu. Tyr. Leu. Val. CyS.- Gly. Glu. Arg. Gly. Phe. Phe. Tyr. Thr. Pro. Lys.- Ala.

We wish to express our thanks to Mr J. Lightbown for a gift of chymotrypsin, and to the Medical Research Council for a grant to cover the cost of the insulin. One of us (H.T.) wishes to thank the British Council for a scholarship.

REFERENCES

Bergmann, M. & Fruton, J. S. (1937). *J. biol. Chem.* **118**, 405.
Bergmann, M., Fruton, J. S. & Pollok, H. (1939). *J. biol. Chem.* **127**, 643.
Chibnall, A. C. & Rees, M. W. (1951). *Biochem. J.* **48**, xlvii.
Desnuelle, P., Rovery, M. & Bonjour, G. (1950). *Biochim. Biophys. Acta*, **5**, 116.
Fromageot, C., Jutisz, M., Meyer, D. & Pénasse, L. (1950). *Biochim. Biophys. Acta*, **6**, 283.
Fruton, J. S. & Bergmann, M. (1939). *J. biol. Chem.* **127**, 627.
Harington, C. R. & Pitt-Rivers, R. (1944). *Nature, Lond.*, **154**, 301.
Hofmann, K. & Bergmann, M. (1939). *J. biol. Chem.* **130**, 81.
Kunitz, M. & Northrop, J. H. (1936). *J. gen. Physiol.* **19**, 991.
Lens, J. (1949). *Biochim. Biophys. Acta*, **3**, 367.
Martin, A. J. P. (1949). *Biochem. Soc. Symp.* **3**, 4.
Rees, M. W. (1946). *Biochem. J.* **40**, 632.
Sanger, F. (1945). *Biochem. J.* **39**, 507.
Sanger, F. (1949*a*). *Biochem. J.* **44**, 126.
Sanger, F. (1949*b*). *Biochem. J.* **45**, 563.
Sanger, F. & Tuppy, H. (1951). *Biochem. J.* **49**, 463.

[From THE BIOCHEMICAL JOURNAL, Vol. 53, No. 3, pp. 353–366, 1953]
 PRINTED IN GREAT BRITAIN

The Amino-acid Sequence in the Glycyl Chain of Insulin

1. THE IDENTIFICATION OF LOWER PEPTIDES FROM PARTIAL HYDROLYSATES

By F. SANGER* AND E. O. P. THOMPSON†
Biochemical Laboratory, University of Cambridge

(*Received* 16 *July* 1952)

Insulin has been shown to be composed of two types of polypeptide chains held together by —S—S— bridges. These chains may be separated after breaking the —S—S— bridges by oxidation to $—SO_3H$ groups (Sanger, 1949*a*). The more acidic glycyl chain is found in the fraction (*A*) which is soluble at pH 6·5 and the more basic phenylalanyl chain in the insoluble fraction (*B*). By making use of the methods of Consden, Gordon & Martin (1947) for fractionating peptides on paper chromatograms, Sanger & Tuppy (1951*a*, *b*) were able to determine the complete amino-acid sequence of the phenylalanyl chain. This and the following paper describe a similar investigation on fraction *A* to determine the amino-acid sequence in the glycyl chain.

End-group analysis of fraction *A* showed that it contained about twenty amino-acid residues (Sanger, 1949*a*) and this was in reasonable agreement with the value of 2900 for the molecular weight which was obtained in the ultracentrifuge (Gutfreund & Ogston, 1949). The amino-acid composition was considerably simpler than that of insulin or fraction *B* since it contained no arginine, histidine, lysine, phenylalanine, threonine or proline. By studying the dinitrophenyl (DNP) peptides produced on partial hydrolysis of the DNP derivative of fraction *A* it was shown that the *N*-terminal sequence was Gly.Ileu.Val.Glu.Glu (Sanger, 1949*b*).

The abbreviations used and the methods of writing peptide structures are the same as those given by Sanger & Tuppy (1951*a*).

METHODS

Preparation of fraction A

When the fractions of oxidized insulin were prepared as previously described (Sanger, 1949*a*) the tyrosine residues were to some extent oxidized to a product of unknown structure, which was distinguished from tyrosine by its R_F value in butanol-acetic acid and was referred to as TyrX (Sanger & Tuppy, 1951*a*). This increased the complexity of the chromatograms since, for every tyrosine-containing peptide, two spots were present, one with tyrosine and one with TyrX. In the original method a large excess of performic acid was used to oxidize the cystine residues. It has now been found that if only a slight excess is used the formation of TyrX can be avoided. The yield of the fraction *A* is somewhat lower, but for most purposes this drawback is offset by the advantage of having a simpler peptide mixture.

Insulin (1 g.) was dissolved in 8 ml. formic acid and treated with 0·48 ml. 30% (w/w) H_2O_2 (1·6 equiv. on the basis of the S content). After standing for 1 hr. 8 ml. water were added to decompose the performic acid. Subsequent treatment and preparation of the fractions was as described before (Sanger, 1949*a*). The yield of fraction *A* was 0·16–0·22 g. and of fraction *B* 0·20–0·27 g. No TyrX could be detected in either fraction.

Hydrolysis and preliminary fractionation

In the experiments described in this paper, hydrolysis was carried out in 12 N-HCl at 37°, the most generally useful conditions for obtaining a large number of small peptides. Hydrolysates prepared by the action of 0·1 N-HCl or N-NaOH at 100° were also studied, but they are not described as no new peptides were found in them, and they were not investigated in great detail.

To effect preliminary group fractionation of the hydrolysate prior to paper chromatography several methods have been studied in different experiments. An 'aromatic' fraction (A_{1a}) could be obtained by adsorption on charcoal. This method did not give a completely clear-cut separation since the fraction contained considerable quantities of the longer non-aromatic peptides besides those of tyrosine. However, it was sufficiently simple to study by paper chromatography and the method was useful for the initial removal of aromatic peptides, which do not fractionate well by ionophoresis in silica gel.

About half the residues in fraction *A* are acidic and there are no basic amino-acids. The majority of the peptides are therefore acidic and it was found that the most useful methods for group separations were those which fractionate the peptides according to the different number of charged groups they contain.

In a preliminary experiment the peptides were fractionated on the ion-exchange resin, Amberlite IR-4B (Consden, Gordon & Martin, 1948; Sanger & Tuppy, 1951*a*). This separated the hydrolysate into three clear-cut fractions containing peptides of cysteic acid, other acidic peptides and neutral peptides, respectively. In general, however, ionophoretic methods were preferred since they are easier to carry out and separate the hydrolysate into a larger number of simpler fractions. Although better separations are usually obtained by ionophoresis on filter paper, it is not convenient to work with sufficient material on this support-

* Member of the scientific staff of the Medical Research Council.

† Present address: Wool Textile Research Laboratories, 343 Royal Parade, Parkville, Victoria, Australia.

ing medium. We have therefore used silica jelly as in the original method of Consden, Gordon & Martin (1946). Difficulties due to adsorption are encountered in this method with peptides containing basic or aromatic amino-acids; the former were absent from fraction *A*, the latter had been removed by adsorption on charcoal.

Experiment A_1. This is given as a typical experiment in which all the peptides detected are recorded. It is described as a single experiment, although much of the data was in fact obtained using other fractions from experiments in which slightly different methods for preliminary separations

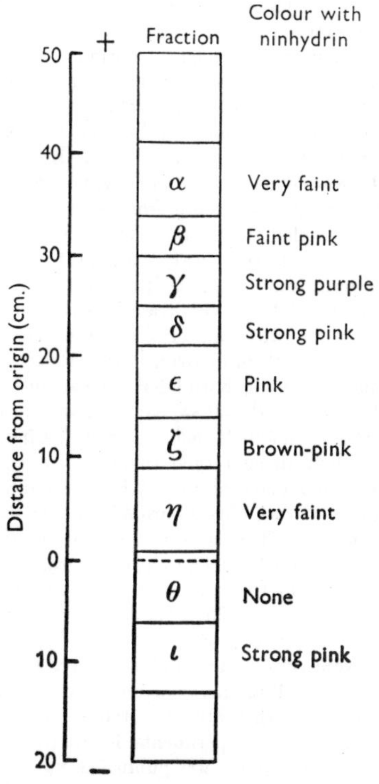

Fig. 1. Ionophoretic fractionation of unadsorbed peptides from partial hydrolysate of fraction *A* in 0·05 M-ammonium acetate (Exp. A_1).

had been employed. Fraction *A* (100 mg.) was hydrolysed for 3 days at 37° with 5 ml. 12 N-HCl. After removal of the excess acid by repeated evaporation *in vacuo*, the residue was dissolved in 4 ml. 5 % (w/v) acetic acid and stirred for 20 min. with 150 mg. charcoal (British Drug Houses Ltd., previously washed well with 20 % (w/v) acetic acid). The charcoal was filtered off and washed with 2 ml. 5 % acetic acid. The combined filtrates, containing the unadsorbed peptides, were taken to dryness and fractionated by ionophoresis as described below. The adsorbed peptides were eluted from the charcoal with 50 ml. 5 % phenol in 20 % acetic acid to give fraction A_{1a}. For the ionophoresis of the unadsorbed peptides the method (C 1) of Consden *et al.* (1946) was used. The jelly, which was strengthened with filter-paper pulp, was prepared by neutralizing 140 ml. 8 % sodium silicate solution with 5 % acetic acid (115 ml.). After setting it was washed overnight with 0·05 M-ammonium acetate to replace the Na^+ ions by NH_4^+ ions. The material to be fractionated was applied in a gutter near the cathode. The jelly was covered by a glass lid weighted to prevent uneven shrinking and puddling of the electrolyte. A potential of 220 V. was applied for 19 hr. At the end of this period a 'print' of the jelly was taken on to a sheet of filter paper, which was sprayed first with 1 % methanolic KOH to remove NH_3, and then with 5 % methanolic acetic acid followed by 0·1 % ninhydrin in butanol. As the resolution was satisfactory, the jelly was cut out according to the marker sheet and the cuts dried *in vacuo* over H_2SO_4 and NaOH. The amino-acids and peptides were eluted from the dried jelly with 5 % acetic acid. The solution was then filtered and evaporated to dryness. Fig. 1 is a diagram of the ionophoresis and shows the bands obtained. Each fraction was then investigated by paper chromatography.

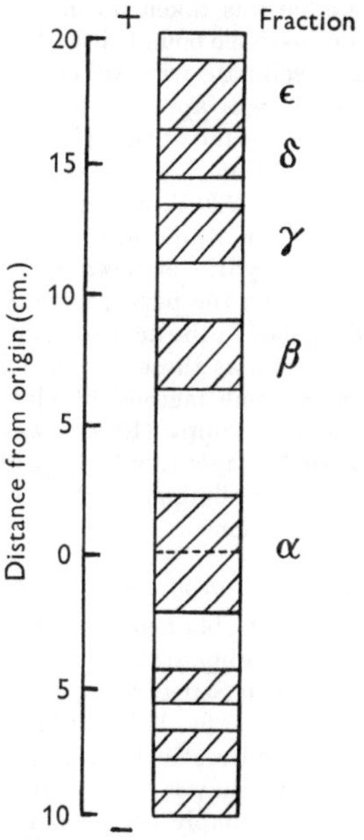

Fig. 2. Ionophoretic fractionation of cysteic acid peptides from partial hydrolysate of fraction *A* in 0·2 N-acetic acid (Exp. A_2).

Experiment A_2. Several peptides containing more than one cysteic acid residue were present in the hydrolysate of fraction *A*, and these could not easily be fractionated on paper chromatograms as they moved so slowly in all solvents studied. An attempt was therefore made to separate them by ionophoresis. Since the —SO_3H group is more acidic (pK = approx. 1·3; Andrews & Schmidt, 1927) than the —COOH group (pK = 2·5) it is possible, by choosing an electrolyte of suitable pH, to find conditions where the latter are virtually uncharged whereas the former are ionized. Consden *et al.* (1946) made use of this property to separate cysteic acid from aspartic and glutamic acids with 0·02 N-HCl as electrolyte. We have found that good separations of cysteic acid peptides can be obtained by ionophoresis in 0·2 N-acetic acid (pH 2·75). Under these conditions all peptides with no —SO_3H group move towards the cathode (due to the fully charged —NH_3 group); those with one —SO_3H group are approximately neutral and a very clear-cut fraction ($A_{2\alpha}$) with only this type of peptide can be obtained. Peptides with more than one cysteic acid residue move towards the anode, and in the case of those from fraction *A* can readily be fractionated. Free cysteic acid, which contains an unusually strong α-COOH group, also moves towards the anode.

Ionophoresis on paper was found to be very effective for this purpose, and the technique of Durrum (1950) was superior to several others that were investigated. This is probably due largely to the flow gradient of electrolyte up the paper, which tends to sharpen the bands. Since acetic acid is not itself an effective buffer it was important to remove free and bound acid from the hydrolysate and to apply the material at a relatively low concentration.

Fraction *A* was hydrolysed for 2 days with 12 N-HCl at 37°, and the excess acid removed by repeated evaporation *in vacuo*. The residue was taken up in water and excess Ag_2O added to eliminate the bound HCl. After centrifuging and washing the precipitate with water, the solution was taken to dryness. It was applied in a concentration of 100 mg./ml. as a narrow band at the apex of the filter paper (Whatman no. 3, 56 × 25 cm.). After wetting the paper with electrolyte a potential of 220 V. was applied for 16 hr. The distribution of the bands, which was revealed by spraying a marker strip with ninhydrin, is shown in Fig. 2. They were cut out and eluted from the paper, taken to dryness and investigated as described in the next section.

For large 'cuts', such as those obtained in this experiment, a more convenient method of elution than that described by Sanger & Tuppy (1951*a*) was to allow the eluates to drop on to a sheet of celluloid or directly on to the polythene strip, thereby eliminating the need for constant watching during the elution.

Examination of the peptides

The various fractions obtained in experiment A_1 were investigated by paper chromatography using essentially the same methods reported by Sanger & Tuppy (1951*a*). Usually Whatman no. 3 filter paper has been used since it has a high capacity and gives good resolutions. The solvent system phenol-0·3% NH_3, which was used extensively, will be referred to throughout simply as 'phenol', and *m*-cresol-0·3% NH_3 will be referred to as 'cresol'. With the exception of leucine and isoleucine, all the amino-acids present in fraction *A* (see Table 13) can be separated by chromatography using phenol. One-dimensional chromatograms were therefore used to identify the amino-acids present in the hydrolysates of the various peptides. Butanol-benzyl alcohol-HCN was used to distinguish between leucine and isoleucine.

To determine the *N*-terminal residues of the peptides, the DNP method has been used in most cases. Sanger & Tuppy (1951*a*) carried out the condensation between the peptide and 1:2:4-fluorodinitrobenzene (FDNB) in the presence of $NaHCO_3$. The salt had to be removed before paper chromatography could be carried out and this was achieved by extraction of the DNP peptide into ethyl acetate. In order to avoid this somewhat tedious step the reaction is now carried out in the presence of trimethylamine, which may be removed *in vacuo*. The identity of the *N*-terminal residue was deduced by Sanger & Tuppy from its absence in the hydrolysate of the DNP peptide. This clearly is less decisive than the identification of the actual DNP amino-acid present, and interpretation is especially difficult where tyrosine is present. In most cases we have therefore identified the DNP derivatives using buffered paper chromatograms (Blackburn & Lowther, 1951; Biserte & Osteaux, 1951; Monnier & Penasse, 1950). Sufficient material for such chromatograms could usually be obtained and, since the amino-acids present in the peptides were already known, it was generally possible to obtain an unequivocal result.

The peptide eluted from two chromatograms on Whatman no. 3 filter paper was dissolved in 0·1 ml. 1% trimethylamine and to this was added a solution of 0·01 ml. FDNB in 0·2 ml. ethanol. After standing for 2 hr., a few drops of water and trimethylamine solution were added and the excess FDNB extracted three times with ether. The residue, after evaporation of the aqueous solution to dryness, was taken up in three drops 5·7 N-HCl into a capillary tube. After 8 hr. heating at 105°, the yellow hydrolysate was diluted with twice its volume of water and the DNP amino-acid extracted into ether (three times). Both the ether extract and the aqueous solution were evaporated to dryness *in vacuo*. In general, the residue from the aqueous solution was chromatographed in phenol. DNP-cysteic acid and *O*-DNP-tyrosine are the only water-soluble DNP derivatives encountered with fraction *A*. DNP-cysteic acid has an R_F value of 0·52. *O*-DNP-tyrosine travels at the solvent front and because of the dirty material collected there and its weak grey colour with ninhydrin, it is not easy to detect. Using butanol-acetic acid, both DNP-cysteic acid (R_F = 0·46) and *O*-DNP-tyrosine (R_F = 0·84) can be detected, but overlapping of the amino-acids, aspartic acid, serine, glycine, and possibly glutamic acid, then occurs.

The DNP amino-acids can mostly be resolved using *tert.*-amyl alcohol as solvent (Blackburn & Lowther, 1951). The paper (Whatman no. 1) and solvent were both treated with a phthalate buffer at pH 6. The chromatograms were preferably developed in the dark at a constant temperature (20°) (Biserte & Osteaux, 1951). The R_F values obtained with this system were not constant and control spots of DNP amino-acids were always run simultaneously. Unless otherwise stated, the DNP amino-acids recorded in Tables 1–10 were identified using this solvent system.

In some of the earlier experiments, before the description of the above solvent system was published, use was made of the same solvent but a different buffer mixture (phosphate, pH 6·75) which had been developed by Partridge & Davis (unpublished). This did not give such good results as the phthalate buffer. Another solvent mixture suggested by Partridge & Davis was prepared from equal volumes of water, benzene and acetic acid, and could be used with unbuffered paper. It was useful for the DNP amino-acids with low R_F values, such as DNP-cysteic acid, DNP-aspartic acid, DNP-glutamic acid and DNP-serine.

A disadvantage with paper chromatograms of DNP derivatives are the yellow artifacts which are also formed. The main ones are probably 2:4-dinitroaniline and 2:4-dinitrophenol. The dinitroaniline travels near the front and does not complicate the interpretation of results. The dinitrophenol spot is usually very strong and is accompanied by two other spots which travel at very similar rates to some of the DNP amino-acids. These artifact spots can be distinguished and removed, however, by exposure of the paper to HCl vapour, when they lose their yellow colour.

The identification of bis-DNP-tyrosine and some of the faster-moving DNP derivatives was somewhat unsatis-

Table 1. *Peptides from fraction* A_{1a}

Spot no. (Figs. 3 and 4)	Amino-acids present	Strength of amino-acid after: Hydrolysis	Strength of amino-acid after: DNP treatment	DNP amino-acid identified	Structure*
1	$CySO_3H$	× ×			[$CySO_3H$, Glu, Ala]
	Glu	×			
	Ala	×			
2	$CySO_3H$	× × ×			[$CySO_3H$, Asp, Glu, Tyr]
	Asp	× × ×			
	Glu	× × ×			
	Tyr	× ×			
3	$CySO_3H$	×			[$CySO_3H$, Asp, Glu, Tyr]
	Asp	×			
	Glu	×			
	Tyr	×			
4	$CySO_3H$	× ×	× ×	Asp†	Asp.[$CySO_3H$, Tyr]
	Asp	× ×	?		
	Tyr	× ×	×‡		
5	$CySO_3H$	× × ×	× ×	Asp	Asp.[$CySO_3H$, Asp, Tyr]
	Asp	× × ×	×		
	Tyr	× × ×	×‡		
6	$CySO_3H$	× ×	×	Asp	Asp.[$CySO_3H$, Asp, Tyr]
	Asp	× ×	×		
	Tyr	× ×	×‡		
7	$CySO_3H$	× ×	× ×	Tyr	Tyr.$CySO_3H$
	Tyr	× ×	?‡		
8	$CySO_3H$	× ×		Gly	Gly.[$CySO_3H$, Glu, Val, Leu]
	Glu	× ×			
	Gly	×			
	Val	× ×			
	Leu	× ×			
9	Asp	×	?	Asp	Asp.Tyr
	Tyr	×	×‡		
10	Asp	× ×	?	Asp	Asp.Tyr
	Tyr	× ×	×‡		
11	Asp	× ×	× ×	Glu	Glu.[Asp, Tyr]
	Glu	× ×	×		
	Tyr	× ×	×‡		
12	Glu	× ×	× ×	Tyr	Tyr.Glu
	Tyr	× ×	?‡		
13	Tyr	× × ×			Tyrosine
14	Asp	× ×			[Asp, Tyr]
	Tyr	× ×			
15	Glu	× × ×			[Glu, Gly, Val, Leu]
	Gly	×			
	Val	× ×			
	Leu	× ×			
16	Glu	× × ×			[Glu, Val, Leu]
	Val	× ×			
	Leu	× ×			
17	Glu	× ×			[Glu, Tyr, Leu]
	Tyr	× ×			
	Leu	× ×			
18	Glu	×			Gly.[Glu, Val, Leu] ($A_{1\eta 3}$)
	Gly	×			
	Val	×			
	Leu	×			

Table 1 (*cont.*)

Spot no. (Figs. 3 and 4)	Amino-acids present	Strength of amino-acid after: Hydrolysis	Strength of amino-acid after: DNP treatment	DNP amino-acid identified	Structure*
19	Ser	× × ×	×	Ser†	Ser.[Tyr, Leu]
	Tyr	× × ×	×‡		
	Leu	× × ×	× ×		
20	Glu	× ×	×	Leu	Leu.[Glu, Tyr]
	Tyr	× ×	×‡		
	Leu	× ×	—		
21	Tyr	× ×			[Tyr, Leu]
	Leu	× ×			
22	Glu	×	×	Ser	Ser.[Glu, Tyr, Leu]
	Ser	×	?		
	Tyr	×	§		
	Leu	×	×		
23	Glu	× ×			Leu.[Glu, Val] ($A_{1\eta 4}$)
	Val	× ×			
	Leu	× ×			

* Square brackets indicate that the sequence of amino-acids inside the brackets has not been established.
† Identified using benzene-acetic acid-water.
‡ Identified as *O*-DNP-tyrosine.
§ *O*-DNP-tyrosine not identified on this phenol chromatogram.
Spots 8, 15, 18 and 19. Yellow colour with ninhydrin.

Spots 7 and 8 overlapped considerably, but as spot 8 gave an initial yellow colour with ninhydrin the separated areas of the spots could be detected and cut out. Better separated using collidine (see Fig. 13). After DNP treatment spot 7 gave a little DNP-glycine in addition to bis-DNP-tyrosine. Spots 11 and 12 often occurred as one spot and resolution could sometimes be achieved by attaching another sheet of filter paper to the bottom of the chromatogram and allowing the butanol-acetic acid to run for a longer time.

Spot 19. Yellow colour with ninhydrin. Sometimes difficult to locate as it was contaminated with red material washed down behind butanol-acetic acid front.

Spot 23 usually contained some tyrosine indicating contamination with some other peptide.

factory on the above solvent systems, since streaking often occurs near the front of the solvent and weak spots are difficult to detect. Better results were obtained with a reversed phase chromatogram using silicone-treated paper (Kritchevsky & Tiselius, 1951). A suitable solvent system was found to be $CHCl_3$ (10 vol.), *n*-propanol (10 vol.) and 1·5 N-acetic acid (6 vol.). After applying the spots to the paper, it was suspended in the tank with the organic phase (lower) and equilibrated for 3 hr. The paper was then dipped in the aqueous phase and the chromatogram allowed to develop. Bis-DNP-tyrosine and bis-DNP-lysine did not move, DNP-leucine ($R_F=0{\cdot}11$), DNP-valine ($R_F=0{\cdot}25$) and DNP-phenylalanine ($R_F=0{\cdot}13$) moved slowly while all other DNP amino-acids and the 'artifact' bands ran faster. The R_F values changed from experiment to experiment especially with the time of equilibration. The DNP amino-acids were therefore always identified by comparison with marker spots rather than by R_F values.

In earlier experiments many of the peptides were subjected to deamination by the method of Consden *et al.* (1947), but these are not recorded since clearer results have since been obtained with the DNP method. Tyrosine was never detected after deamination, but if present in a peptide in a position other than *N*-terminal, two new spots were present, which moved on a phenol chromatogram slightly more slowly than serine and slightly faster than alanine, respectively. These spots, which are possibly nitroso derivatives of tyrosine, were pale yellow before treatment with ninhydrin and gave a greyish colour after treatment. If tyrosine was present as the *N*-terminal residue, these spots were absent or relatively weak and this fact could be used as a fairly reliable guide to the structure of aromatic peptides.

RESULTS

Experiment A_1

The results obtained with the various fractions of experiment A_1 are given in Figs. 3–12 and Tables 1–10. Figs. 3–12 are diagrams of the two-dimensional chromatograms. Tables 1–10 give the results obtained with the eluted peptides. In all cases column 2 lists the amino-acids given on complete hydrolysis and the approximate strength is indicated in column 3. (The ×'s have the same significance as in the previous paper (Sanger & Tuppy, 1951*a*).) No distinction is made in these tables between leucine and isoleucine which are both referred to as Leu. Column 4 gives the strength of the amino-acid remaining in the aqueous solution after extraction of the hydrolysed DNP peptide solution with ether, and column 5 the DNP amino-acid identified in the ether extract or, in the case of DNP-cysteic acid, in the aqueous solution. The last column gives the structures of the peptides as far as they can be deduced from the data given in the tables.

Two chromatograms of fraction A_{1a} are shown. In one (Fig. 3) the initial solvent was phenol, and in the other (Fig. 4) cresol. With the latter solvent better resolution of some of the faster-moving peptides was obtained. No attempt was made to identify the slower-moving spots.

Besides the peptides given in Table 1, fraction A_{1a} contained appreciable amounts of Ser.Val.$CySO_3H$ ($A_{1\zeta 2}$), Leu.Glu.Leu ($A_{1\delta 6}$) and the free amino-acids cysteic acid, glycine and alanine. These were not present in earlier experiments where less charcoal (100 mg. charcoal/100 mg. fraction A hydrolysate) was used to adsorb the aromatic peptides.

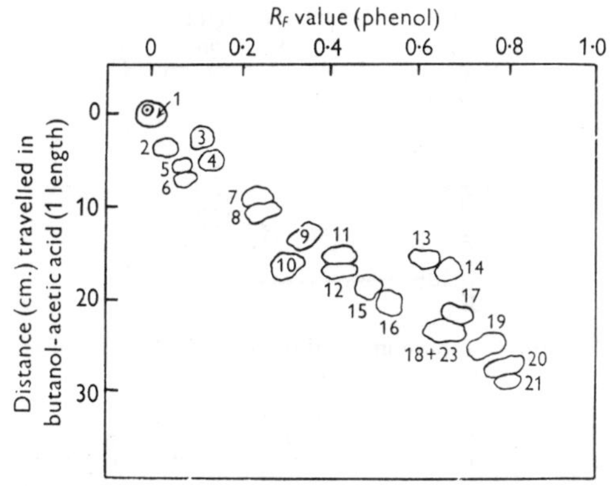

Fig. 3. Chromatogram of fraction A_{1a} using phenol (see Table 1). In Figs. 3–14 the fraction was applied in the position marked by the small dotted circle, and run first in the direction represented horizontally, which was the longer dimension of the filter paper.

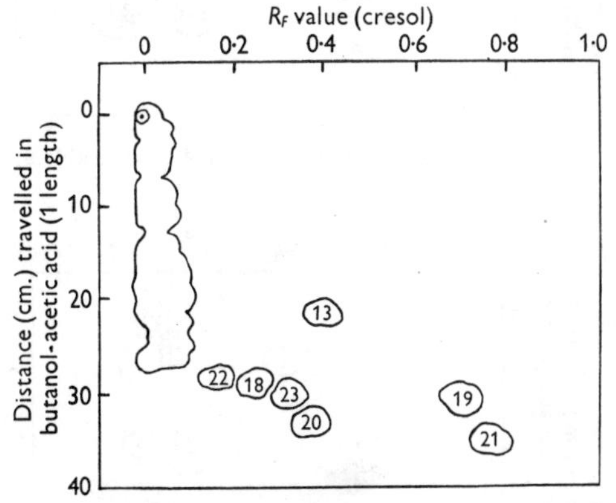

Fig. 4. Chromatogram of fraction A_{1a} using cresol (see Table 1).

The duplication of many of the spots containing aspartic acid and tyrosine may be due to the presence of small amounts of TyrX, different amide contents, or due to different contents of aspartic acid residues. At least three spots (9, 10 and 14) were obtained which gave on hydrolysis aspartic acid and tyrosine and analysed as Asp.Tyr. This suggested that the aspartic acid residue concerned was derived initially from an asparagine residue and spot 14 may in fact be Asp(-NH_2).Tyr, as it gives a yellow colour with ninhydrin similar to asparagine. Again, spots 4–6 seem identical in amino-acid composition. The DNP treatment suggests that spots 5 and 6 differ from spot 4, in having a greater proportion of aspartic acid. Spots 1–6 were sometimes eluted together from a phenol chromatogram, which had been developed only one length in each direction, and the material re-run on a second chromatogram using phenol (2 lengths) and butanol-acetic acid ($1\frac{1}{2}$ or 2 lengths). Although much better resolutions were obtained in this way, there was never sufficient material available for a satisfactory end-group determination.

The chromatograms obtained with the ionophoretic fractions ($A_{1\alpha}$–$A_{1\iota}$) are much simpler than that obtained with the aromatic fraction (A_{1a}), only a few peptides being present in each. The results with fraction $A_{1\theta}$ are not recorded since it only contained traces of the monoamino-acids and Gly.[Glu, Val, Leu] ($A_{1\eta 3}$).

Partial hydrolysis of peptide A_{1a19}. Peptide A_{1a19} was further characterized by hydrolysing a sample with 12N-HCl for 3 days at 37°. The hydrolysate was analysed on a strip of Whatman no. 4 filter paper which was developed in only one direction with butanol-acetic acid (1 length). In addition to unchanged peptide and free amino-acids, a yellow spot ($R_F = 0{\cdot}65$) was obtained which yielded only serine and leucine on hydrolysis. It gave a yellow colour with ninhydrin, indicating that the serine was N-terminal and that it was therefore Ser.Leu (peptide $A_{1\iota 7}$). This established the structure of the original tripeptide (A_{1a19}) as Ser.Leu.Tyr. A spot very close to the unchanged tripeptide gave mainly leucine and tyrosine on hydrolysis but also contained some serine. It was probably Leu.Tyr mixed with a little Ser.Leu.Tyr.

Partial hydrolysis of peptide $A_{1\delta 6}$. This spot had a much lower R_F value in phenol than Glu.Leu ($A_{1\epsilon 8}$) or Leu.Glu ($A_{1\epsilon 7}$). The strength of the amino-acids after DNP treatment suggested two glutamic residues were present. Samples of this peptide were collected and hydrolysed 3 days with 12N-HCl at 37°. The hydrolysate was chromatographed using phenol (1 length) and butanol-acetic acid (1 length) as solvents. Samples of Leu.Glu and Glu.Leu were run as markers during the development with butanol-acetic acid. The results are shown in Table 6. Spot 2 was unchanged tripeptide while spots 3 and 4 corresponded to Leu.Glu and Glu.Leu respectively. This establishes the structure of $A_{1\delta 6}$ as Glu.Leu.Glu since it has glutamic acid as the N-terminal residue and gives Glu.Leu and Leu.Glu on partial hydrolysis.

Experiment A_2

Fraction $A_{2\alpha}$ (Fig. 2), which contained only peptides with one cysteic acid residue, was fractionated by chromatography and the results are shown in Fig. 13 and Table 11.

Samples of fraction $A_{2\beta}$ were chromatographed using phenol (2 lengths) followed by butanol-acetic acid (2 lengths) or collidine (1 length). In both cases

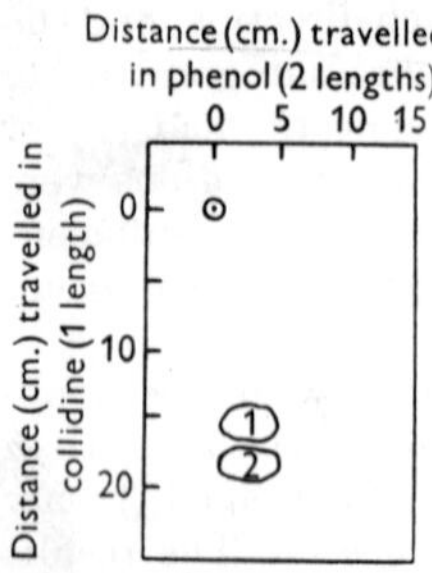

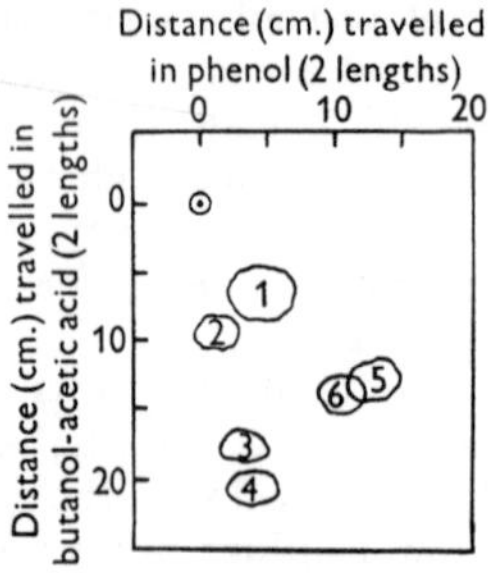

Fig. 5. Fig. 6.

Fig. 5. Chromatogram of fraction $A_{1\alpha}$ (see Table 2). When the R_F was low a second folded sheet was clamped to the lower edge of the original sheet and the distance of the spots from the original is then given instead of R_F.

Fig. 6. Chromatogram of fraction $A_{1\beta}$ (see Table 3).

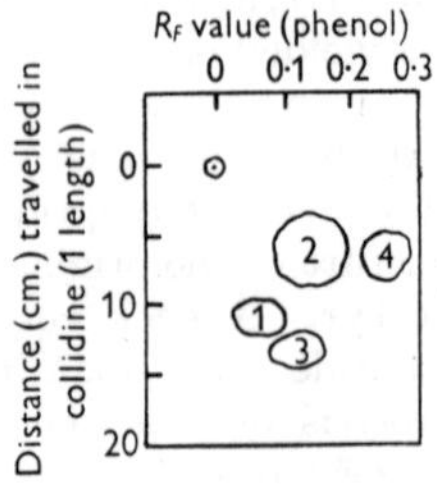

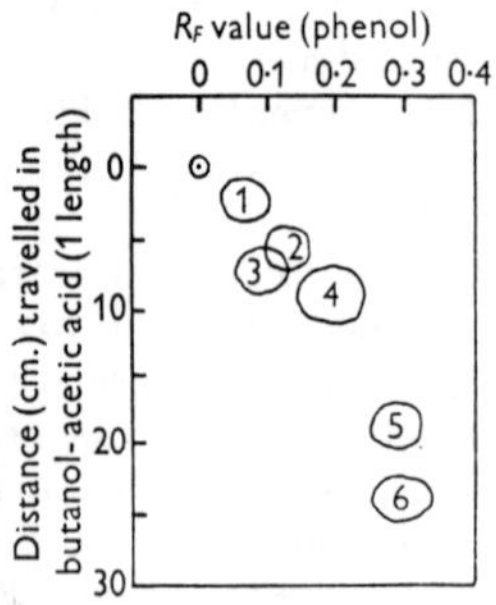

Fig. 7. Fig. 8.

Fig. 7. Chromatogram of fraction $A_{1\gamma}$ (see Table 4).

Fig. 8. Chromatogram of fraction $A_{1\delta}$ (see Table 5).

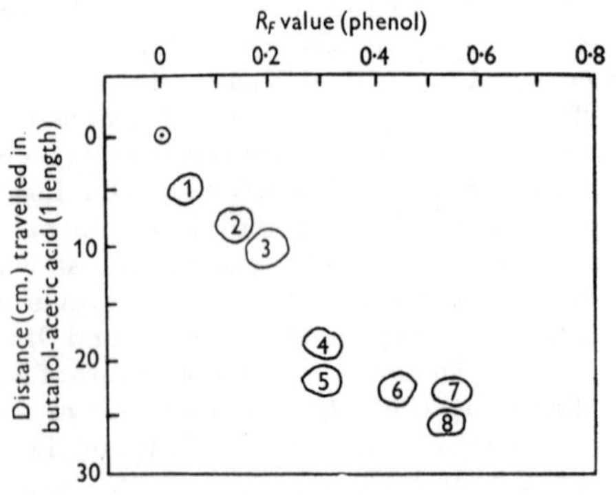

Fig. 9. Chromatogram of fraction $A_{1\epsilon}$ (see Table 7).

one main spot ($A_{2\beta 1}$) was obtained with smaller amounts of three other spots. The main spot gave a yellow colour with ninhydrin and had moved 9 cm. on phenol and 19 cm. on collidine or 12 cm. on phenol and 7·5 cm. on butanol-acetic acid. It contained $CySO_3H$ (× ×), Glu (× ×), Gly (×), Ala (× ×), Val (×), Leu (×).

Fraction $A_{2\gamma}$ was again submitted to ionophoresis using the same conditions to remove some contamination from fraction $A_{2\delta}$. This gave a pure

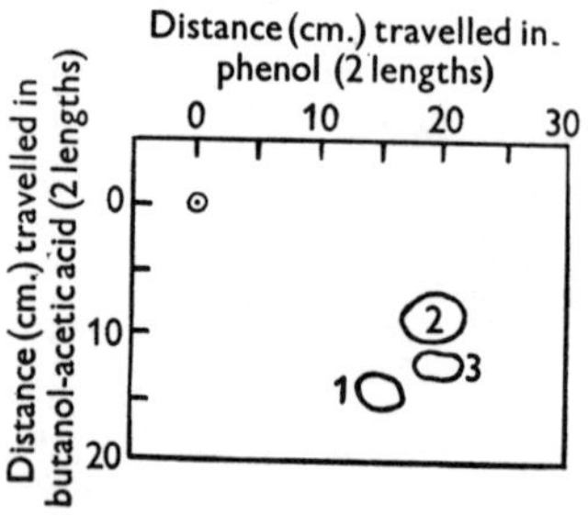

Fig. 10. Chromatogram of fraction $A_{1\zeta}$ (see Table 8).

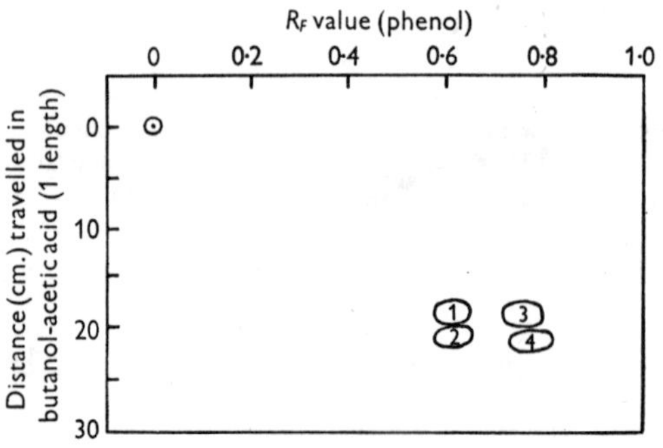

Fig. 11. Chromatogram of fraction $A_{1\eta}$ (see Table 9).

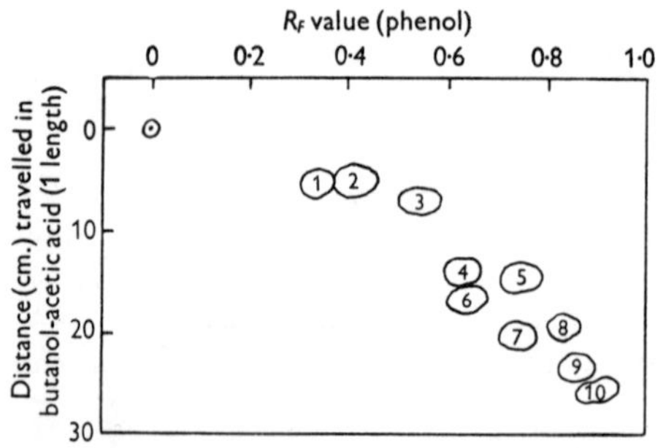

Fig. 12. Chromatogram of fraction $A_{1\iota}$ (see Table 10).

peptide ($A_{2\gamma 1}$) which contained $CySO_3H$ (× × ×), Glu (× ×), Ala (× ×). Glutamic acid was shown to be the *N*-terminal residue, indicating the structure Glu.[$CySO_3H$, $CySO_3H$, Ala].

A sample of this peptide was hydrolysed 3 days with 12 N-HCl at 37°, and the hydrolysate chromatographed on a sheet of Whatman no. 4 paper. The results are shown in Fig. 14 and Table 12. Glu.$CySO_3H$ was obtained so that the structure of $A_{2\gamma 1}$ is Glu.$CySO_3H$.[$CySO_3H$, Ala].

Table 2. *Peptides from fraction* $A_{1\alpha}$

Spot no. (Fig. 5)	Amino-acids present	Strength of amino-acid after: Hydrolysis	DNP treatment	DNP amino-acid identified	Structure
1	$CySO_3H$ Glu	× × × × × ×	× × × ×	Glu*	Glu. $CySO_3H$
2	$CySO_3H$ Asp	× × × × × ×	× × × ×	$CySO_3H$†	$CySO_3H$. Asp

* Identified using benzene-water-acetic acid as solvent.
† Identified in aqueous layer using phenol.

Table 3. *Peptides from fraction* $A_{1\beta}$

Spot no. (Fig. 6)	Amino-acids present	Strength of amino-acid after: Hydrolysis	DNP treatment	DNP amino-acid identified	Structure
1	$CySO_3H$	× × × ×			Cysteic acid
2	$CySO_3H$ Glu	× ×			Glu. $CySO_3H$ ($A_{1\alpha 1}$)
3	Glu Asp	× × × × × ×	× × × ×	Glu*	Glu. Asp
4	Glu	× × ×			See footnote
5	Asp	× ×			Aspartic acid
6	$CySO_3H$ Ala	× ×			$CySO_3H$. Ala ($A_{1\gamma 3}$)

* Identified using benzene-water-acetic acid as solvent.

Spot 1. Spread over extensive area, due to salt or overloading.

Spot 4. Contained only glutamic acid but moves at a very different rate from free glutamic acid or glutamine. It is presumably Glu. Glu known to be present in the *N*-terminal sequence. Its rate, relative to cysteic acid, was similar to that found for the synthetic material by Consden, Gordon & Martin (1949), and the ionophoretic rate was faster than that of glutamic acid, in agreement with the results of these workers. On paper ionophoresis using 0·05 M-ammonium acetate, glutamic acid travelled 6·1 cm. while spot 4 travelled 7·7 cm.

Table 4. *Peptides from fraction* $A_{1\gamma}$

Spot no. (Fig. 7)	Amino-acids present	Strength of amino-acid after: Hydrolysis	DNP treatment	DNP amino-acid identified	Structure
1	$CySO_3H$	× ×			Cysteic acid
2	Asp	× × ×			Aspartic acid
3	$CySO_3H$ Ala	× × × × × ×	× × × ×	$CySO_3H$*	$CySO_3H$. Ala
4	Glu	× ×			Glutamic acid

* Identified in aqueous phase using butanol-acetic acid.

Spots 2 and 3 were not well separated when butanol-acetic acid was used as the second solvent.

Spot 3. Yellow colour with ninhydrin turning to grey on longer heating.

Table 5. *Peptides from fraction* $A_{1\delta}$

Spot no. (Fig. 8)	Amino-acids present	Strength of amino-acid after: Hydrolysis	DNP treatment	DNP amino-acid identified	Structure
1	$CySO_3H$	× ×			Cysteic acid
2	Asp	× ×			Aspartic acid
3	$CySO_3H$ Ala	× × × ×			$CySO_3H$. Ala ($A_{1\gamma 3}$)
4	Glu	× × × ×			Glutamic acid
5	Asp Glu Leu	× × × × × ×	× × × —	Leu	Leu. [Asp, Glu]
6	Glu Leu	× × × × × × ×	× × × ×	Glu	Glu. [Glu, Leu]

Table 6. *Partial hydrolysate of* $A_{1\delta 6}$

Spot no.	R_F in phenol	Distance moved in butanol-acetic acid (cm.)	Amino-acids present	Strength
1	0·18	7·3	Glu	×
2	0·26	18·2	Glu	× ×
			Leu	×
3	0·53	19	Glu	× ×
			Leu	× ×
4	0·53	21	Glu	×
			Leu	×
5	0·78	20	Leu	×

All leucine spots from the chromatogram of these peptide hydrolysates were cut out, washed with acetone to remove excess ninhydrin, eluted and the material rechromatographed using butanol-benzyl alcohol as solvent. Only leucine, and no isoleucine, was present.

Table 7. *Peptides from fraction* $A_{1\epsilon}$

Spot no. (Fig. 9)	Amino-acids present	Strength of amino-acid after: Hydrolysis	DNP treatment	DNP amino-acid identified	Structure
1	$CySO_3H$	×			Cysteic acid
2	$CySO_3H$	×			$CySO_3H$.Ala ($A_{1\gamma 3}$) + aspartic acid
	Asp	×			
	Ala	×			
3	$CySO_3H$	×			[$CySO_3H$, Val] ($A_{1\zeta 3}$) + glutamic acid
	Glu	× ×			
	Val	×			
4	Asp	× ×			Leu.[Asp, Glu] ($A_{1\delta 5}$)
	Glu	× × ×			
	Leu	× ×			
5	Glu	× × ×			Glu.[Glu, Leu] ($A_{1\delta 6}$)
	Leu	× ×			
6	Glu	× × ×			Val.Glu + A_{1a16} (see footnote)
	Val	× ×			
	Leu	×			
7	Glu	× × ×	× × ×	Leu	Leu.Glu
	Leu	× × ×	×		
8	Glu	× × ×	×	Glu	Glu.Leu
	Leu	× × ×	× × ×		

Spot 6 is probably a mixture of peptide A_{1a16} ([Glu, Val, Leu]) with Val.Glu. In another experiment, in which more charcoal (approx. 1 g./100 mg. hydrolysate) had been used to remove the 'aromatic' fraction, Val.Glu was identified as a single spot in this position on a chromatogram from a similar ionophoretic fraction. Deamination with nitrosyl chloride indicated valine was *N*-terminal.

Table 8. *Peptides from fraction* $A_{1\zeta}$

Spot no. (Fig. 10)	Amino-acids present	Strength of amino-acid after: Hydrolysis	DNP treatment	DNP amino-acid identified	Structure
1	Glu	×			Glutamic acid
2	$CySO_3H$	× × ×	× × ×	Ser	Ser.[$CySO_3H$, Val]
	Ser	× × ×	×		
	Val	× × ×	× × ×		
3	$CySO_3H$	×			[$CySO_3H$, Val]
	Val	×			

Besides the peptides given above, fraction $A_{1\zeta}$ also contained traces of Leu.Glu ($A_{1\epsilon 7}$) and Glu.Leu ($A_{1\epsilon 8}$) which were detected on a chromatogram in which each solvent was run only one length.

Spot 2. Yellow colour with ninhydrin.

Spot 3 gave a very faint colour with ninhydrin, and could more easily be detected by its fluorescence in ultraviolet light. Deamination with nitrosyl chloride indicated that valine was probably the *N*-terminal residue.

Table 9. *Peptides from fraction* $A_{1\eta}$

Spot no. (Fig. 11)	Amino-acids present	Strength of amino-acid after: Hydrolysis	DNP treatment	Structure
1	Glu	×		Leu.Glu ($A_{1\epsilon7}$)
	Leu	×		
2	Glu	×		Glu.Leu ($A_{1\epsilon8}$)
	Leu	×		
3	Glu	×	×	Gly.[Glu, Val, Leu]
	Gly	×	—	
	Val	×	×	
	Leu	×	×	
4	Glu	× ×	× ×	Leu.[Glu, Val]
	Val	× ×	× ×	
	Leu	× ×	?	

Spot 3. Yellow colour with ninhydrin.

Table 10. *Peptides from fraction* $A_{1\iota}$

Spot no. (Fig. 12)	Amino-acids present	Strength of amino-acid after: Hydrolysis	DNP treatment	DNP amino-acid identified	Structure
1	Ser	× ×			Serine
2	Gly	× ×			Glycine
3	Ala	× × ×			Alanine
4	Tyr	× ×			Tyrosine
5	Val	× ×			Valine
6	Ser	× ×	?	Ser	Ser.Val
	Val	× ×	× ×		
7	Ser	× ×	×	Ser	Ser.Leu
	Leu	× ×	× × ×		
8	Leu	× ×			Leucine
9	Ser	×			Ser.Leu.Tyr ($A_{1\alpha19}$)
	Tyr	×			
	Leu	×			
10	Leu	× ×	?	Leu	Ileu.Val
	Val	× ×	× ×		

Spots 6, 7 and 9. Yellow colour with ninhydrin.
Spot 7 was shown to contain leucine and not isoleucine using butanol-benzyl alcohol.

Table 11. *Peptides from fraction* $A_{2\alpha}$

Spot. no. (Fig. 13)	Strength	Structure	Identical with
1	× ×	Glu.$CySO_3H$	$A_{1\alpha1}$
2	× ×	$CySO_3H$.Asp	$A_{1\alpha2}$
3	×	[$CySO_3H$, Asp, Glu, Tyr]	$A_{1\alpha2}$, $A_{1\alpha3}$
4	× ×	[$CySO_3H$, Asp, Tyr]	$A_{1\alpha4}$, $A_{1\alpha5}$
5	× × ×	$CySO_3H$.Ala	$A_{1\gamma3}$
6	× ×	Tyr.$CySO_3H$	$A_{1\alpha7}$
7	×	Gly.[$CySO_3H$, Glu, Val, Leu]	$A_{1\alpha8}$
8	×	[$CySO_3H$, Glu, Val, Leu]	—
9	× × × ×	Ser.[$CySO_3H$, Val]	$A_{1\zeta2}$

Spots 7 and 8. The glutamic acid spot was stronger than the other amino-acid spots.
Spots 7 and 9. Yellow colours with ninhydrin.
Spot 8. Blue colour with ninhydrin.

Fraction $A_{2\delta}$ was chromatographed using phenol (2 lengths) and collidine (1 length). Two spots were obtained. One ($A_{2\delta 2}$) was free cysteic acid. It had travelled 6·5 cm. in phenol and 16·5 cm. in collidine. The other ($A_{2\delta 1}$) had travelled 1·4 cm. in phenol and 20·7 cm. in collidine. It contained cysteic acid and alanine, and hydrolysis of the DNP derivative gave DNP-$CySO_3H$ (× ×), $CySO_3H$ (× ×), Ala (× ×). Its structure was therefore $CySO_3H$.[$CySO_3H$, Ala].

Fraction $A_{2\epsilon}$ gave only free cysteic acid on hydrolysis.

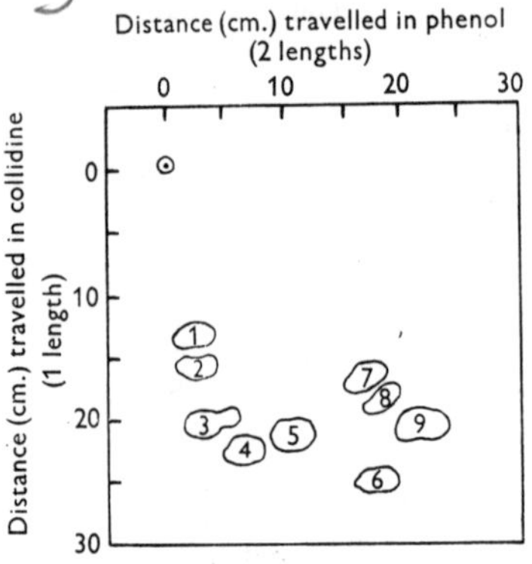

Fig. 13. Chromatogram of fraction $A_{2\alpha}$ (see Table 11).

DISCUSSION

Since the amino-acid content of fraction *B* is accurately known from the structure determined by Sanger & Tuppy (1951*b*) it is possible to calculate the composition of fraction *A* as the difference between the analytical values for insulin and those for fraction *B*. Using the values collected by Tristram (1949) for the composition of insulin, the figures shown in column 2 of Table 13 are obtained for fraction *A*. In the third column are shown the values obtained in an approximate analysis done by paper chromatography (Sanger, 1949*a*).

Peptides of tyrosine

The various tyrosine peptides are all found in fraction A_{1a}. There are four dipeptides containing tyrosine: Tyr.Glu (A_{1a12}), Asp.Tyr (A_{1a9}, A_{1a10}, A_{1a14}), Tyr.$CySO_3H$ (A_{1a7}) and [Tyr, Leu] (A_{1a21}). Since there are only two tyrosine residues in fraction *A* the structure of A_{1a21} must be Leu.Tyr. This is confirmed by the structure of A_{1a19} which was shown by partial hydrolysis to be Ser.Leu.Tyr. Peptide A_{1a20} (Leu.[Glu, Tyr]) cannot be Leu.Glu.-Tyr since Glu.Tyr has not been identified, whereas Tyr.Glu (A_{1a12}) has. It must therefore be Leu.Tyr.-Glu and the sequence in fraction *A* becomes Ser.Leu.Tyr.Glu. Peptide A_{1a22} presumably has this structure.

A DNP peptide containing those four amino-acids was detected as a contaminant of a DNP-glycyl peptide in partial hydrolysates of the DNP derivative of fraction *A* (Woolley, 1948; Sanger, 1949*b*). It presumably had the structure Ser.Leu.-(*O*-DNP) Tyr.Glu.

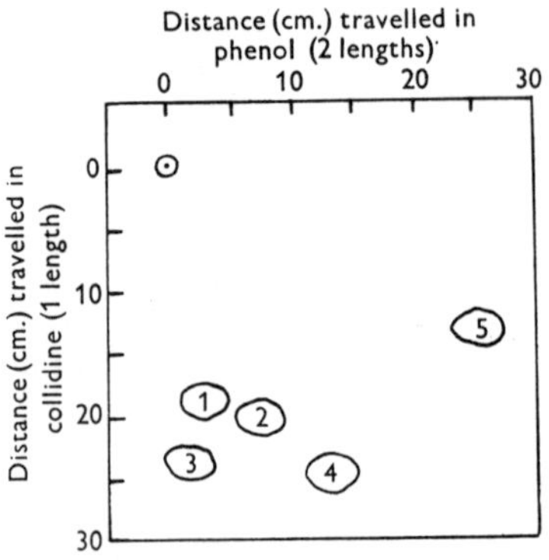

Fig. 14. Chromatogram of partial hydrolysate of peptide $A_{2\gamma 1}$ (see Table 12).

Table 12. *Partial hydrolysis of peptide $A_{2\gamma 1}$*

Spot. no. (Fig. 14)	Strength	Structure	Identical with
1	× ×	Glu.$CySO_3H$	A_{1a1}
2	×	Cysteic acid	—
3	× ×	$CySO_3H$.[$CySO_3H$, Ala]	$A_{2\delta 1}$
4	× ×	$CySO_3H$.Ala	$A_{1\gamma 3}$
5	×	Glutamic acid	—

Table 13. *Probable amino-acid composition of fraction A*

	No. of residues/mol.	
Amino-acid	Calculated from insulin – fraction *B*	Found by approx. analysis
Cysteic acid	4	4
Aspartic acid	2	2
Glutamic acid	4*	4
Serine	2	2
Glycine	0–1	1
Alanine	1	1
Tyrosine	2	2
Valine	1	2
Leucine	2	3
Isoleucine	1–2	1

* The figures given by Tristram would indicate a value of 4·5 for the glutamic acid content. Recent analyses (Chibnall & Rees, 1951) make it almost certain that 4 is the correct figure.

Two other dipeptides (Asp.Tyr and Tyr.$CySO_3H$) are not included in the above sequence and must therefore contain the other tyrosine residue. The sequence is therefore Asp.Tyr.$CySO_3H$ and peptide A_{1a4} (Asp.[$CySO_3H$, Tyr]) clearly has this structure. Peptide A_{1a11} has an *N*-terminal glutamic acid residue and also contains aspartic acid and tyrosine.

Its structure is therefore Glu.Asp.Tyr and the sequence in fraction *A* is Glu.Asp.Tyr.$CySO_3H$. This is confirmed by the presence of Glu.Asp ($A_{1\beta 3}$). The presence of free aspartic acid in the hydrolysate of the DNP derivatives of peptides A_{1a5} and A_{1a6} was taken as evidence that there was a second aspartic acid in these peptides. This would make the above sequence Glu.Asp.Tyr.$CySO_3H$.-Asp. However, the evidence is too slight to be certain of this.

N-*terminal sequence*

Using the DNP technique it was shown that the *N*-terminal sequence was Gly.Ileu.Val.Glu.Glu (Sanger, 1949*b*). Confirmation of this structure is obtained from a number of peptides which are listed in Table 14. Owing to the stability of bonds involving isoleucine and valine many large peptides from this sequence are present, and are mostly found in the fraction adsorbed on charcoal (A_{1a}). The two peptides containing one glutamic acid residue (Ileu.Val.Glu and Gly.Ileu.Val.Glu) move rather rapidly on the phenol chromatogram in spot A_{1a18}. The two slower-moving spots (A_{1a16}, A_{1a15}) containing these same amino-acids presumably have two glutamic acid residues.

Since there is only one glycine residue in fraction *A*, any peptide containing glycine must be derived from this *N*-terminal sequence. Such a peptide is A_{1a8}. It contains the amino-acids of this sequence and also cysteic acid. Since it moves relatively slowly on a chromatogram, and is found in fraction $A_{2\alpha}$ (spot $A_{2\alpha 7}$) on ionophoresis in acetic acid, the cysteic acid must be an integral part of the peptide, whose structure is therefore Gly.Ileu.Val.Glu.-Glu.$CySO_3H$. The presence of Glu.$CySO_3H$ ($A_{1\alpha 1}$) confirms this sequence and peptide $A_{2\alpha 8}$ is presumably Ileu.Val.Glu.Glu.$CySO_3H$. Although the results hitherto discussed do not preclude the presence of another residue of leucine, valine or glutamic acid between the second glutamic acid and the cysteic acid residue in this sequence, it will be seen below that this is in fact not possible.

Peptides of alanine

There is only one alanine residue in fraction *A* and only one dipeptide was found. This was $CySO_3H$.Ala ($A_{1\gamma 3}$). Alanine was also found in peptides that moved towards the anode in Exp. A_2, that is to say in peptides containing two cysteic acid residues. Peptide $A_{2\delta 1}$ was $CySO_3H$.[$CySO_3H$, Ala] and peptide $A_{2\gamma 1}$ Glu.$CySO_3H$.[$CySO_3H$, Ala]. Clearly two sequences are possible Glu.$CySO_3H$.-$CySO_3H$.Ala or Glu.$CySO_3H$.Ala.$CySO_3H$. If the second structure were correct Ala.$CySO_3H$ should be detected in the hydrolysate. Consden & Gordon (1950) have shown it to be stable, for it was isolated from a 10-day partial hydrolysate of wool. They also showed that it could be separated from $CySO_3H$.Ala by ionophoresis or chromatography with collidine. In spite of many attempts, no evidence for the presence of this peptide in the fraction *A* hydrolysate could be obtained and it is therefore concluded that the former sequence (Glu.$CySO_3H$.$CySO_3H$.Ala) is correct. Similarly, Glu.$CySO_3H$.Ala, which would be expected from the second sequence, was never detected. Unfortunately it has not been possible to obtain direct evidence for the presence of $CySO_3H$.$CySO_3H$. It would be expected to move slowly on chromatograms and is probably present in spot A_{1a1} with $CySO_3H$.$CySO_3H$.Ala and Glu.$CySO_3H$.$CySO_3H$.-Ala. On ionophoresis in acetic acid it is probably in fraction A_{26} with cysteic acid.

The fact that Glu.$CySO_3H$ is present in the *N*-terminal position of the above sequence (Glu.-$CySO_3H$.$CySO_3H$.Ala) and in the *C*-terminal position in the sequence Gly.Ileu.Val.Glu.Glu.-$CySO_3H$ suggests that the two are joined together. This is confirmed from the composition of peptide $A_{2\beta 1}$, which contained the amino-acids of the *N*-terminal sequence and also alanine. Since it moved towards the anode on ionophoresis in acetic acid, it must contain the two adjacent cysteic acid residues and its structure is therefore Gly.Ileu.Val.Glu.-Glu.$CySO_3H$.$CySO_3H$.Ala.

Other peptides

Three sequences have so far been deduced:

(1) Gly.Ileu.Val.Glu.Glu.$CySO_3H$.$CySO_3H$.Ala.
(2) Ser.Leu.Tyr.Glu.
(3) Glu.Asp.Tyr.$CySO_3H$.

Apart from the valine peptides derived from sequence 1 there are three peptides containing valine: Ser.Val ($A_{1\epsilon 6}$), Val.$CySO_3H$ ($A_{1\zeta 3}$) and Ser.[$CySO_3H$, Val] ($A_{1\zeta 2}$). Since there are certainly not more than two valine residues in fraction *A* these must all come from the same sequence: Ser.Val.$CySO_3H$. This structure may also be derived from considering the peptides of serine.

Aspartic acid is present in sequence 3 and also in the peptides $A_{1\delta 5}$ (Leu.[Asp, Glu]) and $A_{1\alpha 2}$ ($CySO_3H$.Asp). Since there are only two aspartic acid residues in fraction *A* and since the above two peptides cannot come from the same sequence, peptide $A_{1\delta 5}$ must be joined to the *N*-terminal end of sequence 3, which thus becomes Leu.Glu.Asp.-Tyr.$CySO_3H$. Leu.Glu is found in peptide $A_{1\epsilon 7}$.

Peptide $A_{1\delta 6}$ was shown by partial hydrolysis to be Glu.Leu.Glu. Since there are only four glutamic acid residues this must be joined on to sequence 3 to give Glu.Leu.Glu.Asp.Tyr.$CySO_3H$. Any other way of fitting the sequences together would require at least five glutamic acid residues.

Arthur C. Clarke
British Interplanetary Society

Table 14. *Peptides obtained in partial hydrolysate of fraction A of insulin*

Dipeptides	Ileu.Val $(A_{1\iota 10})$; Glu.$CySO_3H$ $(A_{1\alpha 1})$	Ser.Leu $(A_{1\iota 7})$; Glu.Leu $(A_{1\epsilon 8})$; Asp.Tyr (A_{1a9})	Ser.Val $(A_{1\iota 6})$	$CySO_3H$.Asp $(A_{1\alpha 2})$
	Val.Glu $(A_{1\epsilon 6})$	Leu.Tyr (A_{1a21}); Leu.Glu $(A_{1\epsilon 7})$; Tyr.$CySO_3H$ (A_{1a7})	Val.$CySO_3H$ $(A_{1\zeta 3})$	
	Glu.Glu $(A_{1\beta 4})$; $CySO_3H$.Ala $(A_{1\gamma 3})$	Tyr.Glu (A_{1a12}); Glu.Asp $(A_{1\beta 3})$		
Higher peptides	Ileu.Val.Glu $(A_{1\eta 4})$; $CySO_3H$.$CySO_3H$.Ala $(A_{2\delta 1})$	Ser.Leu.Tyr (A_{1a19}); Leu.Glu.Asp $(A_{1\delta 5})$	Ser.Val.$CySO_3H$ $(A_{1\zeta 2})$	
	Gly.Ileu.Val.Glu $(A_{1\eta 3})$	Leu.Tyr.Glu (A_{1a20}); Glu.Asp.Tyr (A_{1a11})		
	Glu.$CySO_3H$.$CySO_3H$.Ala $(A_{2\gamma 1})$			
	Ileu.Val.Glu.Glu (A_{1a16})	Asp.Tyr.$CySO_3H$ (A_{1a4})		
	Gly.Ileu.Val.Glu.Glu (A_{1a15})	Glu.Leu.Glu $(A_{1\delta 6})$		
	Ileu.Val.Glu.Glu.$CySO_3H$ $(A_{2\alpha 8})$	Ser.Leu.Tyr.Glu (A_{1a22}); Glu.Asp.Tyr.$CySO_3H$ (A_{1a2})		
	Gly.Ileu.Val.Glu.Glu.$CySO_3H$ $(A_{1a8}, A_{2\alpha 7})$			
	Gly.Ileu.Val.Glu.Glu.$CySO_3H$.$CySO_3H$.Ala $(A_{2\beta 1})$			
Sequences in fraction *A*	Gly.Ileu.Val.Glu.Glu.$CySO_3H$.$CySO_3H$.Ala	Ser.Leu.Tyr.Glu.Leu.Glu.Asp.Tyr.$CySO_3H$	Ser.Val.$CySO_3H$	$CySO_3H$.Asp

In the three sequences deduced there are still five glutamic acid residues so that one must be common to two sequences. This can only be the *C*-terminal residue of sequence 2 and the *N*-terminal residue of sequence 3, which are therefore joined together, and the structure in fraction *A* becomes Ser.Leu.Tyr.-Glu.Leu.Glu.Asp.Tyr.$CySO_3H$. Both the 'Leu' residues in this sequence are leucine and not isoleucine since peptides $A_{1\iota 7}$ (Ser.Leu) and $A_{1\delta 6}$ (Glu.Leu.Glu) contain no isoleucine. The presence of peptide A_{1a17} ([Glu, Tyr, Leu]), which was different from A_{1a20} (Leu.Tyr.Glu), can readily be understood from this sequence. A number of structures are possible, such as Tyr.Glu.Leu or Leu.Tyr.Glu.Leu.Glu.

All the peptides encountered in this work are now accounted for in terms of the four sequences:

(1) Gly.Ileu.Val.Glu.Glu.$CySO_3H$.$CySO_3H$.Ala.
(2) Ser.Leu.Tyr.Glu.Leu.Glu.Asp.Tyr.$CySO_3H$.
(3) Ser.Val.$CySO_3H$.
(4) $CySO_3H$.Asp.

All the amino-acids known to be present in fraction *A* are present in these sequences and there is one too many cysteic acid residues. The $CySO_3H$.-Asp must therefore be joined to the *C*-terminal end of sequence 2 or 3, but it is not possible to decide conclusively which.

In conclusion, the various peptides identified are listed in Table **14** with the sequences deduced from them.

SUMMARY

1. Fraction *A* of oxidized insulin has been subjected to partial hydrolysis with acid and the resulting peptides fractionated by charcoal adsorption, ionophoresis and paper chromatography.

2. It is concluded that the following sequences are present in this fraction:

Gly.Ileu.Val.Glu.Glu.$CySO_3H$.$CySO_3H$.Ala,
Ser.Leu.Tyr.Glu.Leu.Glu.Asp.Tyr.$CySO_3H$,
Ser.Val.$CySO_3H$ and $CySO_3H$.Asp.

One of us (E. O. P. T.) wishes to thank the Commonwealth Government of Australia for a Commonwealth Scientific and Industrial Research Organization studentship. Part of the work described in this paper was carried out during the tenure by F.S. of a Beit Memorial Fellowship, which is gratefully acknowledged.

REFERENCES

Andrews, S. & Schmidt, C. L. A. (1927). *J. biol. Chem.* **73**, 651.
Biserte, G. & Osteaux, R. (1951). *Bull. Soc. Chim. biol., Paris*, **33**, 50.
Blackburn, S. & Lowther, A. G. (1951). *Biochem. J.* **48**, 126.
Chibnall, A. C. & Rees, M. W. (1951). *Biochem. J.* **48**, xlvii.
Consden, R. & Gordon, A. H. (1950). *Biochem. J.* **46**, 8.
Consden, R., Gordon, A. H. & Martin, A. J. P. (1946). *Biochem. J.* **40**, 33.
Consden, R., Gordon, A. H. & Martin, A. J. P. (1947). *Biochem. J.* **41**, 590.
Consden, R., Gordon, A. H. & Martin, A. J. P. (1948). *Biochem. J.* **42**, 443.
Consden, R., Gordon, A. H. & Martin, A. J. P. (1949). *Biochem. J.* **44**, 548.
Durrum, E. L. (1950). *J. Amer. chem. Soc.* **72**, 2943.
Gutfreund, H. & Ogston, A. G. (1949). *Biochem. J.* **44**, 163.
Kritchevsky, T. H. & Tiselius, A. (1951). *Science*, **114**, 299.
Monnier, R. & Penasse, L. (1950). *C.R. Acad. Sci., Paris*, **230**, 1176.
Sanger, F. (1949*a*). *Biochem. J.* **44**, 126.
Sanger, F. (1949*b*). *Biochem. J.* **45**, 563.
Sanger, F. & Tuppy, H. (1951*a*). *Biochem. J.* **49**, 463.
Sanger, F. & Tuppy, H. (1951*b*). *Biochem. J.* **49**, 481.
Tristram, G. R. (1949). *Advanc. Protein Chem.* **5**, 83.
Woolley, D. W. (1948). *Fed. Proc.* **7**, 200.

[FROM THE BIOCHEMICAL JOURNAL, VOL. 53, No. 3, pp. 366–374, 1953]

The Amino-acid Sequence in the Glycyl Chain of Insulin

2. THE INVESTIGATION OF PEPTIDES FROM ENZYMIC HYDROLYSATES

BY F. SANGER* AND E. O. P. THOMPSON†
Biochemical Laboratory, University of Cambridge

(*Received* 16 *July* 1952)

In the previous paper (Sanger & Thompson, 1953*a*) four sequences were shown to be present in fraction *A* of oxidized insulin. It was found, as in the case of fraction *B* (Sanger & Tuppy, 1951*a*), that the complete sequence could not be derived from the small peptides produced on acid hydrolysis, due largely to the great lability of the bonds involving the amino groups of the serine residues. It was therefore necessary to investigate the action of proteolytic enzymes, and the present paper describes the results obtained with peptic and chymotryptic hydrolysates. Trypsin was found to be without action on fraction *A*.

* Member of the scientific staff of the Medical Research Council.

† Present address: Wool Textile Research Laboratories, 343 Royal Parade, Parkville, Victoria, Australia.

Although, in fact, it was only necessary to consider the structure of one of the peptides (A_{p5}) to work out the sequence of fraction A, the hydrolysates were investigated in considerable detail, since confirmation was desired and since the results throw some light on the specificity of the enzymes concerned. Also a number of the peptides from these hydrolysates were required to determine the position of the amide groups (Sanger & Thompson, 1953*b*). The action of carboxypeptidase on fractions A and B, insulin, and acetyl insulin has also been studied since this was used to confirm the presence of asparagine as the C-terminal residue of the glycyl chain.

The peptides with reference numbers starting A_1 or A_2 refer to those described in the previous paper.

METHODS

Fraction A was prepared free of TyrX as previously described (Sanger & Thompson, 1953*a*). The pepsin used was crystalline pig pepsin (Armour Laboratories). The chymotrypsin was prepared by Mr J. Lightbown of the National Institute for Medical Research. The carboxypeptidase was a commercial preparation obtained from Armour Laboratories. Acetylinsulin was prepared by the method of Hughes (1947).

Experiment A_p (pepsin)

In a typical experiment 50 mg. fraction A were dissolved in 5 ml. 0·1 N-acetic acid and 2 mg. pepsin added. The mixture was incubated at 37° for 48 hr. After boiling to inactivate the enzyme, the solution was centrifuged to remove denatured pepsin and evaporated to dryness *in vacuo*. In some experiments the digestion was carried out in 0·01 N-HCl, but no difference in the peptides obtained could be detected.

No preliminary fractionation was necessary and samples of the hydrolysate were investigated directly on paper chromatograms. Where Whatman no. 4 filter paper was used an amount equivalent to 5 mg. of the original fraction A was applied, and for Whatman no. 3 paper 10–12·5 mg.

The peptides were investigated essentially by the methods described in the previous paper.

Experiment A_c (chymotrypsin)

Fraction A (50 mg.) was dissolved in 5 ml. water, 2·5 mg. chymotrypsin were added and the pH adjusted to 7·5 with dilute NH_3. The mixture was incubated at 37° for 24 hr. It was boiled to inactivate the enzyme and evaporated to dryness *in vacuo*. Before application to a chromatogram, the residue was taken up in a small volume of water, and centrifuged to remove insoluble material.

Experiments with carboxypeptidase

In preliminary experiments it was evident that the carboxypeptidase preparation contained traces of chymotrypsin, which were splitting fraction A in the chain. In order to avoid this the enzyme was first incubated with di*iso*propyl fluorophosphonate, which is a specific inhibitor for chymotrypsin (Jansen, Nutting, Jang & Balls, 1949). The action of carboxypeptidase was not affected by this reagent. In order to remove the long-chain material, which frequently interfered with the paper chromatograms, the molecular sieve procedure of Thompson & Partridge (1952) was used.

The di*iso*propyl fluorophosphonate was kept as a 0·1 M solution in dry *iso*propanol. When required for use it was brought to 0·001 M by the addition of water. Of a suspension of carboxypeptidase in water (10–15 mg./ml.) 0·05 ml. was run into 1·45 ml. of this inhibitor solution, and, after adjusting the pH to 7·5, the mixture was incubated at 37° for 30 min. The solution was then added to the substrate (fraction A, 10 mg.; fraction B, 15 mg.; insulin and acetyl insulin, 25 mg.). Samples (0·3 ml.) were withdrawn at various times with a vacuum pipette. These were heated in a boiling-water bath for 5 min. to inactivate the enzyme. To each were added approximately 50 mg. sulphonated polystyrene resin, which was in bead form (15–30 mesh/in.) and had been regenerated before use by successive washings with 2 N-HCl, water, 2 N-NaOH, water, 2 N-HCl and water (Thompson & Partridge, 1952). The mixture was shaken for 1 hr., the supernatant solution decanted and the beads washed twice with distilled water. The amino-acids were then displaced from the resin with 5 N-NH_3 (0·2 ml.) and investigated on paper chromatograms.

RESULTS

Experiment A_p (pepsin)

Two chromatograms of the peptic hydrolysate are shown in Figs. 1 and 2. When each solvent was run only one length of the paper (Fig. 1) spots 3–9 were not sufficiently resolved to give clear results. It was therefore necessary to clamp a second folded sheet of paper on to the bottom edge during each run so that the chromatograms were developed two lengths in each direction (Fig. 2). The combined results are shown in Table 1.

Partial hydrolysis of peptides A_{p5}, A_{p14}, A_{p15}. Eluates of spot A_{p5} from two chromatograms on Whatman no. 3 paper, were hydrolysed with 12 N-HCl at 37° for 3 days and the hydrolysate was chromatographed on Whatman no. 4 paper. The results are shown in Fig. 3 and Table 2.

Eluates of spots A_{p14} and A_{p15} were treated with 1:2:4-fluorodinitrobenzene and the dinitrophenyl (DNP) derivatives hydrolysed 3 days with 12 N-HCl at 37°. The hydrolysates were extracted with ether, and in both cases bis-DNP-tyrosine was identified in the extract by using reversed phase paper chromatography. The aqueous solutions were chromatographed on Whatman no. 4 paper. Peptide A_{p14} gave on partial hydrolysis Glu.Leu ($A_{1\epsilon8}$) Leu.Glu ($A_{1\epsilon7}$) and Glu.Leu.Glu ($A_{1\delta6}$) in addition to the free amino-acids and consequently had the structure Tyr.Glu.Leu.Glu. Peptide A_{p15} gave only free amino-acids and Glu.Leu ($A_{1\epsilon8}$) on partial hydrolysis and must therefore have been Tyr.Glu.Leu.

Table 1. *Peptides from peptic hydrolysate* (A_p) *of fraction A of oxidized insulin*

Spot no. (Figs. 1 and 2)	Amino-acids identified	Strength	DNP amino-acids identified	Structure*
1	$CySO_3H$	× × ×		
	Asp	× × × ×	Glu	Glu.[$CySO_3H$, Asp, Tyr]
	Glu	× × ×		
	Tyr	× × ×		
2	$CySO_3H$	×		
	Asp	× ×	Glu	Glu.[$CySO_3H$, Asp, Tyr]
	Glu	×		
	Tyr	×		
3	$CySO_3H$	× × ×		
	Asp	× × × ×	Asp	Asp.[$CySO_3H$, Asp, Tyr]
	Tyr	× × ×		
4	$CySO_3H$	×		
	Asp	× ×	Asp	Asp.[$CySO_3H$, Asp, Tyr]
	Tyr	×		
5	$CySO_3H$	× × × × ×		
	Glu	× × × × ×		
	Ser	× × × × ×		
	Gly	× × × ×	Gly	Gly.[$CySO_3H$, Glu, Ser, Ala, Val, Leu]
	Ala	× × × ×		
	Val	× × × × ×		
	Leu	× × × × ×		
6	$CySO_3H$	× ×		
	Asp	× ×		
	Glu	× ×	Leu†	Leu.[$CySO_3H$, Asp, Glu, Tyr]
	Tyr	× ×		
	Leu	× ×		
7	$CySO_3H$	×		
	Asp	× ×		
	Glu	×		[$CySO_3H$, Asp, Glu, Tyr, Leu]
	Tyr	×		
	Leu	×		
8	Glu	×		Glutamic acid
9	$CySO_3H$	× ×		
	Glu	× ×		
	Ser	× ×	Gly	Gly.[$CySO_3H$, Glu, Ser, Ala, Val, Leu]
	Gly	× ×		
	Ala	× ×		
	Val	× ×		
	Leu	× ×		
10	Glu	× ×		[Glu, Tyr]
	Tyr	× ×		
11	Glu	× ×		[Glu, Tyr]
	Tyr	× ×		
12	Tyr	×		Tyrosine
13	Glu	× ×		Leu.Glu
	Leu	×		
14	Glu	× × × ×		
	Tyr	× × ×	Tyr†	Tyr.[Glu, Leu]
	Leu	× × ×		
15	Glu	× × ×		
	Tyr	× × ×	Tyr†	Tyr.[Glu, Leu]
	Leu	× × ×		
16	Glu	×		
	Gly	×		Gly.Ileu.Val.Glu ($A_{1\eta 3}$)
	Val	×		
	Leu	×		

* Where amino-acids are enclosed in square brackets the sequence has not been directly established.
† Identified by reversed-phase paper chromatography.
Spot 5. Yellow colour with ninhydrin.
Spot 9. Red-yellow colour with ninhydrin.
Spot 13. The aqueous solution of a hydrolysate of the DNP derivative contained Glu (× ×) Leu (×) suggesting it is Leu.Glu.

Experiment A_c (chymotrypsin)

Fig. 4 shows a chromatogram of the chymotryptic hydrolysate. Since there was considerable overlapping of some of the spots, each one was purified by ionophoresis on paper in 0·05 M-ammonium acetate using the technique of Durrum (1950). With the exception of spot (6 + 7), only one main component was found in each case and the amino-acids given on hydrolysis are recorded in Table 3. In the case of spot (6 + 7) two bands were visible after ionophoresis when the paper was examined in ultraviolet light. One (A_{p6}) was neutral and the other (A_{p7}) was acidic. A_{p6} gave a yellow colour with ninhydrin, whereas no colour was obtained with A_{p7}. The composition of the two is shown in Table 3.

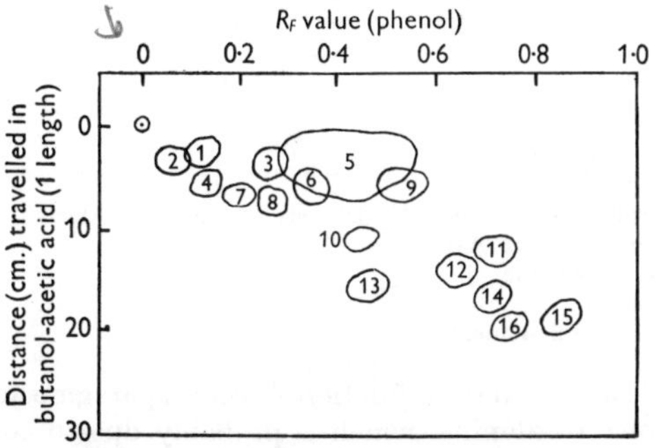

Fig. 1. Chromatogram of peptic hydrolysate of fraction A of oxidized insulin (Exp. A_p) in which both solvents were run one length of paper (see Table 1).

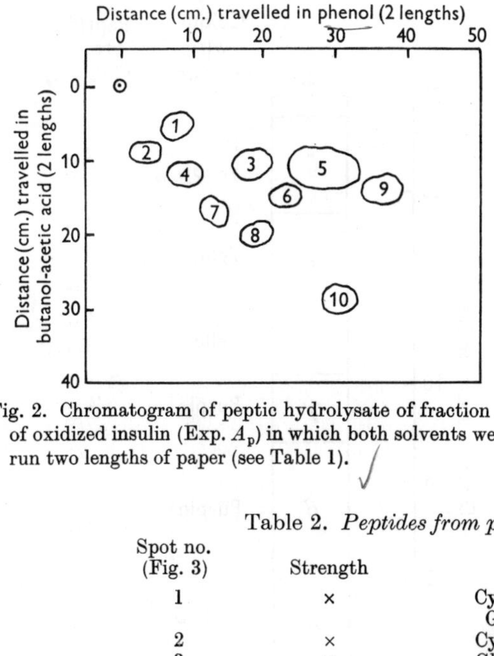

Fig. 2. Chromatogram of peptic hydrolysate of fraction A of oxidized insulin (Exp. A_p) in which both solvents were run two lengths of paper (see Table 1).

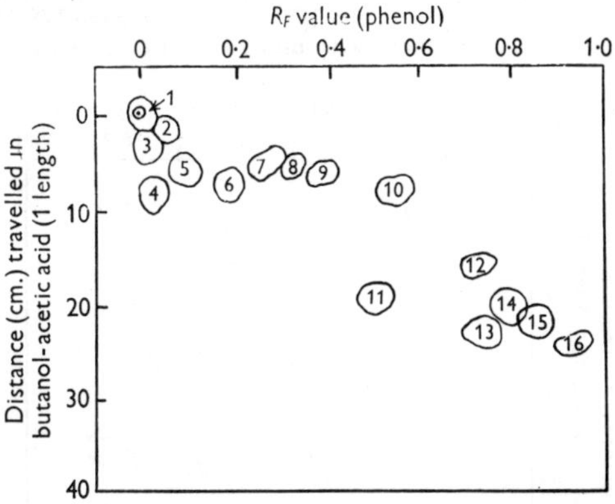

Fig. 3. Chromatogram of partial acid hydrolysate of peptide A_{p5} (see Table 2).

In later experiments, when further samples of these peptides were required, it was more convenient to run a large-scale ionophoresis first, followed by two-dimensional chromatography of the bands obtained. Fig. 5 is a diagram of such an ionophoresis, showing the relative positions of the different peptides.

Table 2. *Peptides from partial hydrolysate of peptide A_{p5}*

Spot no. (Fig. 3)	Strength	Structure	Identical with
1	×	$CySO_3H.CySO_3H.Ala$ + $Glu.CySO_3H.CySO_3H.Ala$	$A_{2\delta 1}$ + $A_{2\gamma 1}$
2	×	Cysteic acid	—
3	×	$Glu.CySO_3H$	$A_{1\alpha 1}$
4	×	Glu.Glu	$A_{1\beta 4}$
5	× ×	$CySO_3H.Ala$	$A_{1\gamma 3}$
6	×	Glutamic acid	—
7	× × ×	$Ser.Val.CySO_3H$	$A_{1\zeta 2}$
8	× ×	Serine	—
9	× ×	Glycine	—
10	× ×	Alanine	—
11	× ×	Gly.Ileu.Val.Glu.Glu. + Ileu.Val.Glu.Glu	$A_{1\alpha 15}$ $A_{1\alpha 16}$
12	×	Ser.Val	$A_{1\iota 6}$
13	×	Ileu.Val.Glu	$A_{1\eta 4}$
14	× ×	Ser.Leu	$A_{1\iota 7}$
15	×	Leucine + isoleucine	—
16	?	Ileu.Val	$A_{1\iota 10}$

Table 3. *Peptides from chymotryptic hydrolysate (A_c) of fraction A of oxidized insulin*

Spot no. (Fig. 4)	Amino-acids formed after hydrolysis								
	$CySO_3H$	Asp	Glu	Ser	Gly	Ala	Tyr	Val	Leu
1	×××	×××	—	—	—	—	—	—	—
2	×××	—	×××	×××	×××	×××	—	×××	××
3	××	×××	××××	—	—	—	××	—	××
4	××××	—	××××	××××	××××	××××	×××	××××	××××
5	—	××	×××	—	—	—	××	—	××
6	—	—	—	××	—	—	×××	—	×××
7	—	××	×××	—	—	—	××	—	××

Spot 1. Yellow colour with ninhydrin, turning to purple on prolonged heating. Hydrolysis of the DNP derivative gives DNP-cysteic acid and aspartic acids. Its structure is therefore $CySO_3H$.Asp.
Spot 2. Yellow colour with ninhydrin. Contained isoleucine but no leucine.
Spot 3. Hydrolysis of the DNP derivative gave DNP-glutamic acid.
Spot 4. Yellow colour with ninhydrin. Contained both leucine and isoleucine.

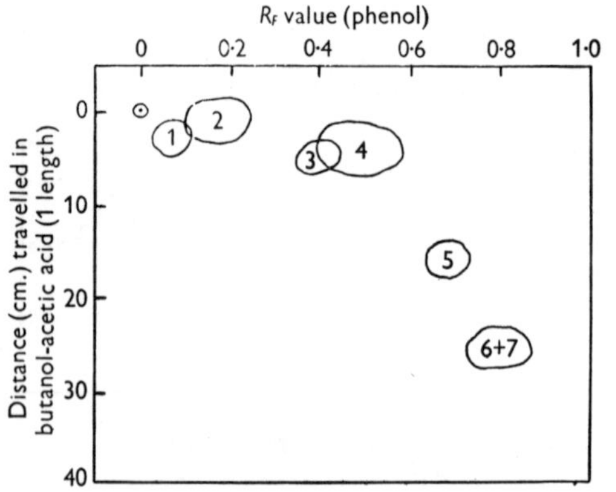

Fig. 4. Chromatogram of chymotryptic hydrolysate of fraction *A* of oxidized insulin (Exp. A_c) (see Table 3).

Table 4. *Action of carboxypeptidase on insulin derivatives*

Substrate	Time of incubation at 37°	Free amino-acids liberated	Strength
Fraction *B*	30 min.	Ala	×××××
Fraction *A*	1 hr.	Asp (-NH_2)	×
		Ala	?
Insulin	1·5 hr.	Ala	×××××
		Asp (-NH_2)	××
		Asp	?
	8 hr.	Ala	×××××
		Asp (-NH_2)	×××
		Asp	×
Acetylinsulin	4 hr.	Ala	×××××
		Asp (-NH_2)	××××
		Asp	×

Carboxypeptidase experiments

The results obtained with carboxypeptidase are summarized in Table 4. Free alanine was liberated very rapidly from fraction *B* and insulin, as has been shown by Lens (1949). No evidence could be found of free lysine which is the residue adjacent to alanine (Sanger & Tuppy, 1951*b*). The first amino-acids liberated from fraction *A* were asparagine and a trace of alanine, which is probably due to contamination with fraction *B*. On prolonged incubation traces of glutamic acid, serine, tyrosine and leucine were liberated. Better yields of asparagine were obtained from insulin. This effect may be due to inhibition of the enzyme by the —SO_3H group of the adjacent cysteic acid residue in fraction *A*. An approximate estimation by the spot-dilution technique of Polson, Mosley & Wyckoff (1947) indicated

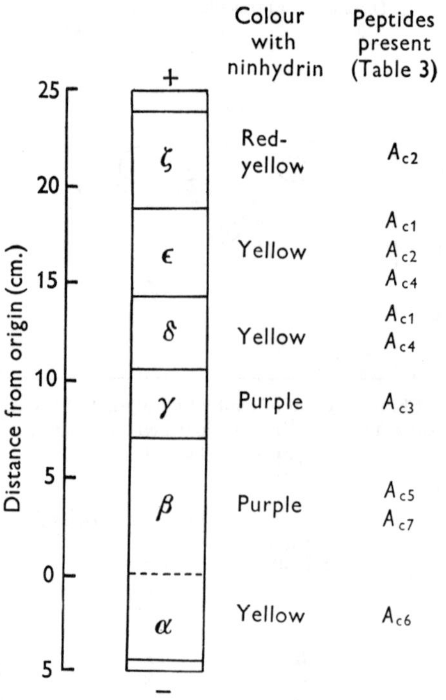

Fig. 5. Ionophoresis of chymotryptic hydrolysate of fraction *A* of oxidized insulin in 0·05 M-ammonium acetate (see Table 3).

that after 8 hr. incubation the yield of asparagine from insulin was 25–30 % of the theoretical value, assuming two *C*-terminal asparagine residues per molecule of mol.wt. 12000.

Acetylinsulin was investigated in the hope that the lysine residue from the phenylalanyl chain could be split off as ε-acetyllysine. This expectation was not realized, but a much better yield of asparagine was obtained. After a reaction period of 4 hr. the yields of both alanine and asparagine were approximately 90 % of the theoretical value. Aspartic acid was also obtained in a yield of less than 10 %. It is probably a secondary product derived from asparagine, although control experiments in which asparagine was incubated with carboxypeptidase showed that the breakdown was very small and would hardly account for the amounts of aspartic acid formed in the experiments described above.

DISCUSSION

Amino-acid sequence of fraction A

In the previous paper four sequences were deduced as being present in fraction *A*. These were:

(1) Gly.Ileu.Val.Glu.Glu.$CySO_3H$.$CySO_3H$.Ala (*N*-terminal sequence).

(2) Ser.Leu.Tyr.Glu.Leu.Glu.Asp.Tyr.$CySO_3H$.

(3) Ser.Val.$CySO_3H$.

(4) $CySO_3H$.Asp.

The peptide present in the highest concentration in the peptic hydrolysate was peptide A_{p5}. It contained all the amino-acids of fraction *A* except aspartic acid and tyrosine, and had a glycine *N*-terminal residue. On partial hydrolysis with acid it gave the peptides listed in Table 2. Of these, all but peptides 7 (Ser.Val.$CySO_3H$), 12 (Ser.Val) and 14 (Ser.Leu) are derived from sequence 1. Since peptide A_{p5} gives Ser.Leu on hydrolysis but contains no tyrosine, this leucine residue must be present as the *C*-terminal residue. It contains therefore sequence 1 as its *N*-terminal sequence, Ser.Leu as its *C*-terminal sequence and Ser.Val.$CySO_3H$ which must be within the chain. Neither $CySO_3H$.Asp nor the remainder of sequence 2 can be present in A_{p5}, so that its structure must be Gly.Ileu.Val.Glu.Glu.$CySO_3H$.$CySO_3H$.Ala.Ser.Val.$CySO_3H$.Ser.Leu. Having established this sequence containing three cysteic acid residues there is only one such residue to which $CySO_3H$.Asp can be attached and the structure of fraction *A* must be Gly.Ileu.Val.Glu.Glu.$CySO_3H$.$CySO_3H$.Ala.Ser.Val.$CySO_3H$.Ser.Leu.Tyr.Glu.Leu.Glu.Asp.Tyr.$CySO_3H$.Asp.

For simplicity in deducing this sequence, considerable use has been made of the figures given in the previous paper (Sanger & Thompson, 1953*a*, Table 13) for the amino-acid composition of fraction *A*. It could, however, have been deduced as the simplest structure which would fit all the experimental results obtained, and, in fact, was originally worked out in this way. For instance, it could be shown that all the tyrosine-containing peptides could not be explained on the basis of one sequence, but could be explained on the basis of two sequences. This is fairly conclusive evidence that there are two tyrosine residues. If there were three it is unlikely that all the peptides could be fitted into two sequences. While the structure deduced is the simplest one that will fit the experimental results, a few more complicated ones are possible but are ruled out by the analytical figures. For instance, if there were an extra serine residue between positions 8 and 9 (Table 5) this might not have been detected, but the analytical results show that there are only two such residues in the chain. It is seen that an exact amino-acid analysis of the chain may in this way be worked out from the nature of the peptides without carrying out any quantitative estimation. It is interesting that the very approximate analysis of fraction *A* (Table 13, Sanger & Thompson, 1953*a*) which was done by paper chromatography is in fact correct except for the leucine value which should be 2 and not 3.

The fact that it is possible to deduce this single sequence proves that fraction *A* is an essentially homogeneous peptide and therefore that there is only one type of glycyl chain in insulin.

As was the case with the phenylalanyl chain, there is no obvious periodicity of the amino-acids in this chain. In fraction *B* it was noted that the aromatic and less polar residues were grouped together. There is no such arrangement in fraction *A*, so that this cannot be regarded as a general property of proteins. In fraction *B* there were three dipeptide sequences which occurred twice in the chain, and this was considerably more than would be expected on a random arrangement. No dipeptide sequences occur twice in the glycyl chain, although there are now six which occur twice in the whole insulin. These are Val.Glu (*A* 3–4, *B* 12–13), Val.$CySO_3H$ (*A* 10–11, *B* 18–19), Leu.Tyr (*A* 13–14, *B* 15–16), His.Leu (*B* 5–6, *B* 10–11), Gly.$CySO_3H$ (*B* 7–8, *B* 19–20) and Leu.Val (*B* 11–12, *B* 17–18). The sequence CyS.Gly, which occurs in glutathione and twice in the phenylalanyl chain, is not found in the glycyl chain. It would thus seem that no general conclusions can be drawn from these results concerning the general principles which govern the arrangement of amino-acid residues in protein chains. In fact, it would seem more probable that there are no such principles, but that each protein has its own unique arrangement; an arrangement which endows it with its particular properties and specificities and fits it for the function that it performs in nature.

Table 5. *Peptides obtained by enzymic hydrolysis of fraction A of oxidized insulin*

Sequences deduced from lower peptides (Sanger & Thompson, 1953*a*)	Gly.Ileu.Val.Glu.Glu.CySO$_3$H.CySO$_3$H.Ala
	Ser.Val.CySO$_3$H
	Ser.Leu.Tyr.Glu.Leu.Glu.Asp.Tyr.CySO$_3$H
	CySO$_3$H.Asp
Peptides from peptic hydrolysate	Gly.Ileu.Val.Glu.Glu.CySO$_3$H.CySO$_3$H.Ala.Ser.Val.CySO$_3$H.Ser.Leu (A_{p5})
	Leu.Glu.Asp.Tyr.CySO$_3$H.Asp (A_{p6}, A_{p7})
	Gly.Ileu.Val.Glu.Glu.CySO$_3$H.CySO$_3$H.Ala.Ser.Val (A_{p9})
	Tyr.Glu (A_{p10}, A_{p11})
	Glu.Asp.Tyr.CySO$_3$H.Asp (A_{p1}, A_{p2})
	Gly.Ileu.Val.Glu (A_{p16})
	Tyr.Glu.Leu (A_{p15})
	Asp.Tyr.CySO$_3$H.Asp (A_{p3}, A_{p4})
	Tyr.Glu.Leu.Glu (A_{p14})
	Leu.Glu (A_{p13})
Peptides from chymotryptic hydrolysate	Gly.Ileu.Val.Glu.Glu.CySO$_3$H.CySO$_3$H.Ala.Ser.Val.CySO$_3$H.Ser.Leu.Tyr (A_{c4})
	CySO$_3$H.Asp (A_{c1})
	Gly.Ileu.Val.Glu.Glu.CySO$_3$H.CySO$_3$H.Ala.Ser.Val.CySO$_3$H (A_{c2})
	Glu.Leu.Glu.Asp.Tyr.CySO$_3$H.Asp (A_{c3})
	Ser.Leu.Tyr (A_{c6})
	Glu.Leu.Glu.Asp.Tyr (A_{c5}, A_{c7})
Structure of fraction *A*	Gly.Ileu.Val.Glu.Glu.CySO$_3$H.CySO$_3$H.Ala.Ser.Val.CySO$_3$H.Ser.Leu.Tyr.Glu.Leu.Glu.Asp.Tyr.CySO$_3$H.Asp
	1 2 3 4 5 6 7 8 9 10 11 12 13 14 15 16 17 18 19 20 21
Bonds split by pepsin	4 ⇡ 5; 10 ⇡ 11; 13 ↑ 14; 14 ⇡ 15; 15 ⇡ 16; 16 ↑ 17; 17 ↑ 18
Bonds split by chymotrypsin	11 ⇡ 12; 14 ↑ 15; 19 ⇡ 20

↑, major sites of action of enzymes; ⇡, other bonds split by enzymes.

Structure of the peptides

Having established the sequence of residues in fraction A it is now possible to consider the peptides identified in the various hydrolysates. Their structures are given in Table 5. The fact that these can all be determined in this way provides confirmatory evidence for the main sequence.

Since pepsin splits bonds involving the amino groups of tyrosine residues in synthetic peptides (Fruton & Bergmann, 1939), it was expected that splitting of fraction A would occur at position 18–19 to give Tyr.CySO$_3$H.Asp. Although peptides containing these three amino-acids were present (A_{p3}, A_{p4}), they had aspartic acid and not tyrosine as the N-terminal residues. Splitting by pepsin therefore occurred at bond 17–18 (Glu.Asp) and no evidence could be obtained for splitting of bond 18–19. Further considerations on the specificity of pepsin will be presented in a later paper (Sanger & Thompson, 1953*b*), when the positions of the amide groups have been established.

The presence of duplicate spots of the peptides Asp.Tyr.CySO$_3$H.Asp (A_{p3}, A_{p4}), Glu.Asp.Tyr.-CySO$_3$H.Asp (A_{p1}, A_{p2}) and Leu.Asp.Tyr.-CySO$_3$H.Asp (A_{p6}, A_{p7}) is probably due to deamidation. Amide groups are present on the aspartic acid residues (Sanger & Thompson, 1953*b*) and some hydrolysis probably occurred during the digestion with pepsin in acid solution (cf. Melville, 1935). Similarly, peptides A_{p10} and A_{p11} (Tyr.Glu) differ by an amide group.

Partial acid hydrolysis of peptide A_{p9} on a rather small amount of material suggested the structure shown in Table 5, since it gave the peptides from the N-terminal sequence and Ser.Val but no Ser.Val.-CySO$_3$H or Ser.Leu.

In the chymotryptic hydrolysate two peptides (A_{c5}, A_{c7}) containing the same amino-acids (Asp, Glu, Tyr, Leu) are present. A_{c7} moves faster on the chromatograms than A_{c5} and gives no colour with ninhydrin, suggesting that it has no free α-amino group. Both are derived from the sequence Glu.Leu.Glu.Asp.Tyr and evidence to be presented later (Sanger & Thompson, 1953*b*) indicates that this N-terminal glutamic acid residue is present in the form of an amide. Glutamine residues containing a free α-amino group are particularly labile and readily cyclize to give the corresponding pyrollidonecarboxylic acid derivative (Melville, 1935). It seems probable that peptide A_{c7} is such a derivative formed from A_{c5} during the digestion or subsequent operations.

Peptide A_{c6} can only be Ser.Leu.Tyr identical with A_{1a19}, indicating that splitting with chymotrypsin has occurred at bond 11–12 (CySO$_3$H.Ser). This is confirmed by the presence of peptide A_{c2} which contains isoleucine and the other amino-acids from the N-terminal sequence but no leucine. Since the splitting of this bond was rather unexpected, control experiments were carried out to show that the effect was due to chymotrypsin and not to the experimental conditions used. An end-group determination on the fraction A used showed that there were no free amino groups of serine present, glycine being the only N-terminal residue. Samples of fraction A were then treated in the usual manner with chymotrypsin which had been previously incubated at 37° for 30 min. with a 0·001 M solution of the specific inhibitor di*iso*propylfluorophosphonate. Chromatography of the evaporated solutions showed no spot corresponding to peptide A_{c2}. Only one large area (R_F in phenol = 0·25) was obtained on the chromatograms which gave a yellow colour with ninhydrin and on hydrolysis gave all the amino-acids of fraction A. It was obviously unchanged fraction A. It is therefore concluded that the splitting of this bond is due to the action of chymotrypsin or possibly of another enzyme that is very closely associated with it and which is inhibited by di*iso*propyl fluorophosphonate.

Further evidence for this splitting was provided in an experiment in which peptide A_{p5} was treated with chymotrypsin. Ser.Leu was identified as the only neutral component after ionophoresis in 0·05 M-ammonium acetate. The bands moving towards the anode were separated by chromatography and shown to be similar in R_F values and composition to A_{c2} and A_{p5} (unchanged peptide).

The C-terminal residue of fraction A

Peptide A_{c1} contains the same amino-acid sequence as peptide $A_{1\alpha2}$ (CySO$_3$H.Asp). However, it was found to behave differently on ionophoresis in 0·05 M-ammonium acetate. Thus after ionophoresis for 6 hr. with a potential of 220 V., $A_{1\alpha2}$ had moved 8·4 cm., whereas A_{c1} had only moved 5·1 cm. This clearly shows that A_{c1} is an amide. Its structure is therefore CySO$_3$H.Asp(-NH$_2$) and the C-terminal residue of fraction A is asparagine. Although there is now considerable confirmatory evidence for this (Chibnall & Rees, 1952; Harris, 1952), it had been suggested at the time this sequence was first being worked out, that glycine was present as a C-terminal residue in insulin, presumably on the glycyl chains (Fromageot, Jutisz, Meyer & Penasse, 1950; Chibnall & Rees, 1951). As a result of more recent work Chibnall & Rees (private communication) have now withdrawn this claim.

The analytical results on fraction A indicated that there was only one glycine residue and this appeared to be the N-terminal residue. This was supported by the results of Butler, Phillips, Stephen & Creeth (1950), which we have been able to confirm. They showed that no glycine was present in a hydrolysate of the DNP derivative of fraction *A*, showing that the only glycine residues are *N*-terminal.

The results with carboxypeptidase support the conclusion that asparagine is the *C*-terminal residue of the glycyl chain. In the case of acetylinsulin the liberation was approximately quantitative, assuming that it is present in both the glycyl chains, if one assumes that the molecular weight is 12000.

SUMMARY

1. Fraction *A* of oxidized insulin was hydrolysed with pepsin and chymotrypsin. The resulting peptides were fractionated and their structure investigated.

2. It is concluded that the amino-acid sequence in the glycyl chain of insulin is Gly.Ileu.Val.Glu.-Glu.CyS.CyS.Ala.Ser.Val.CyS.Ser.Leu.-Tyr.Glu.Leu.Glu.Asp.Tyr.CyS.Asp.

We wish to express our thanks to Mr J. Lightbown for the chymotrypsin and to Dr B. C. Saunders for the di*iso*propyl fluorophosphonate.

One of us (E. O. P. T.) wishes to thank the Commonwealth Government of Australia for a Commonwealth Scientific and Industrial Research Organization studentship.

Part of the work described in this paper was carried out during the tenure by F.S. of a Beit Memorial Fellowship.

REFERENCES

Butler, J. A. V., Phillips, D. M. P., Stephen, J. M. L. & Creeth, J. M. (1950). *Biochem. J.* **46**, 74.

Chibnall, A. C. & Rees, M. W. (1951). *Biochem. J.* **48**, xlvii.

Chibnall, A. C. & Rees, M. W. (1952). *Biochem. J.* **52**, iii.

Durrum, E. L. (1950). *J. Amer. chem. Soc.* **72**, 2943.

Fromageot, C., Jutisz, M., Meyer, D. & Penasse, L. (1950). *Biochim. biophys. Acta*, **6**, 283.

Fruton, J. S. & Bergmann, M. (1939). *J. biol. Chem.* **127**, 627.

Harris, J. I. (1952). *J. Amer. chem. Soc.* **74**, 2944.

Hughes, W. R. L. (1947). Cited by Herriot, R. M. in *Advanc. Protein Chem.* **3**, 169.

Jansen, E. F., Nutting, M. D., Jang, R. & Balls, A. K. (1949). *J. biol. Chem.* **179**, 189.

Lens, J. (1949). *Biochim. biophys. Acta*, **3**, 367.

Melville, J. (1935). *Biochem. J.* **29**, 179.

Polson, A., Mosley, Y. M. & Wyckoff, R. W. G. (1947). *Science*, **105**, 603.

Sanger, F. & Thompson, E. O. P. (1953*a*). *Biochem. J.* **53**, 353.

Sanger, F. & Thompson, E. O. P. (1953*b*). (Unpublished.)

Sanger, F. & Tuppy, H. (1951*a*). *Biochem. J.* **49**, 463.

Sanger, F. & Tuppy, H. (1951*b*). *Biochem. J.* **49**, 481.

Thompson, A. R. & Partridge, S. M. (1952). *Nature, Lond.*, **169**, 495.

VOL. **9** (1952) BIOCHIMICA ET BIOPHYSICA ACTA 225

THE INVERSION OF A DIPEPTIDE SEQUENCE DURING HYDROLYSIS IN DILUTE ACID

by

F. SANGER AND E. O. P. THOMPSON

Biochemical Laboratory, Cambridge (England)

In deducing the amino-acid sequence of a protein or polypeptide chain from the structure of the peptide fragments produced on partial hydrolysis, it is essential that the sequence present in the breakdown product is the same as was present in the original molecule. It has been pointed out[1] that an inversion of the order of amino-acids in a dipeptide could theoretically occur via the diketopiperazine. Thus if a dipeptide AB, which is less stable than the corresponding peptide BA, is exposed to conditions which catalyse ring formation, a conversion of AB to BA might be expected according to the equation;

$$AB \rightleftharpoons \begin{bmatrix} A \\ B \end{bmatrix} \rightleftharpoons BA$$

where $\begin{bmatrix} A \\ B \end{bmatrix}$ represents the diketopiperazine. Such inversion could occur without the accumulation of the intermediate diketopiperazine, since in general diketopiperazines are more labile than the corresponding dipeptides. In order to see if this reaction does in fact take place we have studied the effect of various conditions of "hydrolysis" on the dipeptide glycylvaline (Gly.Val), which is considerably less stable than the corresponding inverted peptide valylglycine (Val.Gly)[2,3]. ABDERHALDEN AND KOMM[4] showed that ring formation occurs at high temperatures in the presence of dilute acid, and the following experiment demonstrates that inversion can take place when hydrolysis is carried out in 0.1 *N* HCl at 100°.

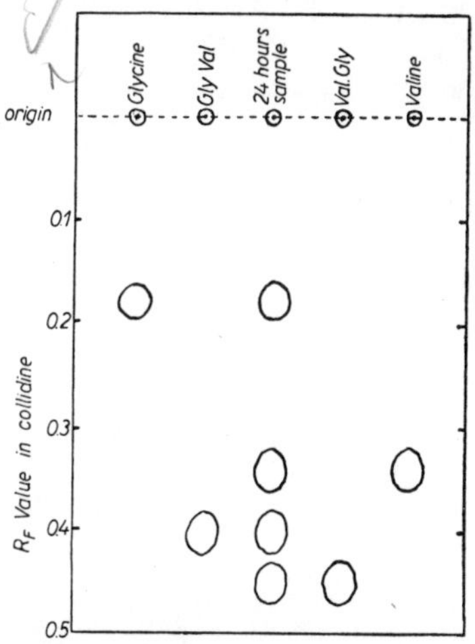

Fig. 1. Paper chromatogram of "partial hydrolysate" of Gly.Val.

Gly.Val (1 mg) was boiled under reflux with 0.1 *N* HCl (1 ml). Samples (0.1 ml) were withdrawn after 6, 11, 23 and 30 hours, evaporated to dryness on polythene and applied to a paper chromatogram together with control samples of Gly.Val and Val.Gly. The chromatogram was developed with collidine and better resolution could be obtained by attaching a folded half-sheet of filter paper to the bottom of the chromatogram so that the effective distance run by the solvent front was 1½ lengths. Fig. 1 is a diagram of the resulting chromatogram of the 24 hour sample and controls after spraying with ninhydrin. Besides spots corresponding in R_F value to glycine (R_F = 0.18), valine (R_F = 0.34) and unchanged Gly.Val (R_F = 0.40), a spot was present which had the same R_F value (0.45) as Val.Gly. Samples of the two faster-moving spots were collected from replicate chromatograms, using marker strips (sprayed with ninhydrin) from the same paper to locate the spots. The eluates of these spots were treated with 1:2:4-fluorodinitrobenzene as described by SANGER AND TUPPY[5]. The dinitrophenyl (DNP) peptides were extracted from the acidified reaction mixture into ether, and after removal of the solvent subjected to hydrolysis in boiling 5.7 *N* HCl for 3 hours. The hydrolysates were extracted with ether, and the DNP-amino-acids present in the extract identified using a buffered paper chromatogram developed with tertiary amyl alcohol[6]. The amino-acids remaining in the aqueous solution were identified on a phenol chromatogram. The results are given in Table I. They clearly show that the peptide with R_F = 0.45 was Val.Gly and that an inversion of the peptide Gly.Val had occurred during "hydrolysis" in boiling 0.1 *N* HCl.

These experiments indicate the need for caution in interpreting the results of partial hydrolysis experiments using dilute acid. During work on insulin[5,7] this type of hydrolysis has been used since it shows a rather different specificity from that shown by concentrated acids. However, no fundamental conclusions have been drawn from the structure of dipeptides identified in such a hydrolysate, and in fact no evidence of any inversion was obtained. Partial acid hydrolysis is normally carried out with 12 *N* HCl at 37° and under these conditions no inversion of Gly.Val could be demonstrated.

TABLE I

PROPERTIES OF HYDROLYSATES OF DNP-PEPTIDES

R_F value of spot	*Products of hydrolysis of DPN peptide*		
	Ether extract	*Aqueous residue*	
0.40	DNP-glycine	Glycine (?)	valine (xx)
0.45	DNP-valine	Glycine (x)	valine (?)

REFERENCES

[1] F. SANGER, *Advances in Prot. Chem.*, 7 (1952) 1.
[2] P. A. LEVENE, P. E. STEIGER, AND A. ROTHEN, *J. Biol. Chem.*, 97 (1932) 717.
[3] R. L. M. SYNGE, *Biochem. J.*, 39 (1945) 351.
[4] E. ABDERHALDEN AND E. KOMM, *Z. Physiol. Chem.*, 139 (1924) 147.
[5] F. SANGER AND H. TUPPY, *Biochem. J.*, 49 (1951) 463.
[6] S. BLACKBURN AND A. G. LOWTHER, *Biochem. J.*, 48 (1951) 126.
[7] F. SANGER, *Biochem. J.*, 45 (1949) 563.

Received June 19th, 1952

[FROM THE BIOCHEMICAL JOURNAL, VOL. 59, No. 3, pp. 509-518, 1955]

The Amide Groups of Insulin

BY F. SANGER,* E. O. P. THOMPSON† AND RUTH KITAI
Department of Biochemistry, University of Cambridge

(*Received* 6 *September* 1954)

Previous work has shown that insulin consists of two types of polypeptide chains the amino acid sequences of which have been determined by partial hydrolysis experiments (Sanger & Tuppy, 1951*a*, *b*; Sanger & Thompson, 1953*a*, *b*). For an assumed molecular weight of approximately 6000, which is the simplest chemical unit consisting of one phenylalanyl and one glycyl chain, there are six amide groups present (Rees, 1946). From a quantitative study of the products of reduction of insulin ester with lithium borohydride Chibnall & Rees (1952) showed that the three aspartic acids, one of which is

* Member of the scientific staff of the Medical Research Council.

† Present address: Wool Textile Research Laboratories, 343 Royal Parade Parkville, Victoria, Australia.

the *C*-terminal residue of the glycyl chain, are all present as amides, whereas three glutamic acid residues are amidized and the other four have free carboxyl groups. The following paper describes experiments to locate these amide groups on the individual dicarboxylic acid residues. To achieve this, use has been made of the fact that amide groups are split rapidly by the action of acid, but are not in general attacked by proteolytic enzymes. Thus peptides obtained from enzymic digests contain the original amide groups intact, whereas those obtained from acid hydrolysates do not. Two methods are thus available for identifying amide groups. In the first, the ionophoretic rates of the peptides from enzymic digests are compared with the rates of other similar peptides obtained from acid and enzymic hydrolysates. From these results it is possible to calculate the number of charged groups on the peptides and hence the number of carboxyl groups that are masked as amides. In the second method the number of amide groups present in the peptides from enzymic digests is determined directly by an estimation of the ammonia produced by hydrolysis with acid.

A number of peptides were identified from peptic and chymotryptic hydrolysates of fraction *A* (Sanger & Thompson, 1953*b*) and these were used to locate most of the amide groups. However, other peptides were required, especially those containing only one of the two glutamic acid residues in positions 4 and 5 (see Table 7), and so hydrolysates obtained by the action of two other enzymes (papain and mould protease) were investigated and the results are reported. The mould protease used was first isolated in a crystalline form by Crewther & Lennox (1950), but was not considered to be a pure substance. It is extremely active though rather unspecific in its attack.

The abbreviations and reference numbers to the peptides are those reported in previous papers (Sanger & Tuppy, 1951*a*, *b*; Sanger & Thompson, 1953*a*, *b*). Pyr is used to denote a residue of pyrrolidone carboxylic acid (5-oxopyrrolidine-2-carboxylic acid).

EXPERIMENTAL

Action of mould protease on fraction A (*Expt.* *A*m). The mould protease used was a crystalline sample obtained through Dr R. R. Porter of the National Institute for Medical Research, Mill Hill, London.

A solution of fraction *A* (30 mg.) in 5 ml. water was brought to pH 8 and 1·5 mg. mould protease were added. After incubating at 37° for 24 hr. the mixture was boiled to inactivate the enzyme and taken to dryness *in vacuo*. Samples (5 or 10 mg.), after centrifuging to remove insoluble matter, were fractionated on Whatman no. 4 or no. 3 paper chromatograms, using phenol–0·3% NH_3 followed by butanol–acetic acid as solvents and the peptide spots eluted and subjected to hydrolysis and end-group determination as described in previous papers (Sanger & Tuppy, 1951*a*, *b*; Sanger & Thompson, 1953*a*, *b*).

Action of papain on fraction A (*Expt.* *A*pa). The papain used was a commercial preparation supplied by British Drug Houses Ltd. A sample (16 mg.) was activated before use by incubation for 1 hr. with 1·6 ml. 1% NaCN solution at 37°. The activated enzyme (0·5 ml.) was then added to a solution of 50 mg. fraction *A* in 2 ml. water. The pH was adjusted to 7 with dilute NH_3 and the mixture incubated at 37° for 24 hr. Subsequent working up was as described for mould protease.

Preparation of peptide Ap5 by ionophoresis. Peptide *A*p5, which consists of the first thirteen residues of the glycyl chain (Sanger & Thompson, 1953*b*) was prepared from insulin by making use of its high mobility as an acid in 0·2N acetic acid, owing to its content of three cysteic acid residues.

Insulin (50 mg.) in 10 ml. 0·01N-HCl was treated with 1·0 mg. pepsin and incubated at 37° for 24 hr. After being taken to dryness, the residue was oxidized with performic acid (Sanger, 1949*a*). The formic acid was removed *in vacuo*, the residue dissolved in 1·5 ml. water and applied along the centre line of a sheet of Whatman no. 3 filter paper. It was then subjected to ionophoresis in an apparatus similar to that of Durrum (1950). A potential of 220 v was applied for 21 hr. A strip was cut from the paper and tested with ninhydrin. The *A*p5 was present in a band which stained rather weakly with ninhydrin and had moved 12·5 cm. towards the anode. From a ninhydrin determination on a hydrolysate of the eluted material it appeared that the yield of *A*p5 was about 30% of the theoretical yield from insulin. In several experiments a faint slower moving band could be distinguished, which appears to be peptide *A*p9 (the first ten residues of the glycyl chain). For further purification the *A*p5 was subjected to ionophoresis at pH 3·7 in the apparatus of Michl (1951). The material obtained from 50 mg. insulin was put on a 18 cm.-wide strip of no. 3 filter paper; the ionophoresis was continued for 3·5 hr. at 1500 v. The main *A*p5 band had moved 19 cm. towards the anode. Several fainter slow moving bands were also present.

Ionophoresis of peptides. The ionophoretic rates of peptides have been compared using the apparatus of Durrum (1950). The electrolyte used was 0·05M ammonium acetate (pH 6·8). The peptides obtained from paper chromatograms were applied to a sheet of Whatman no. 1 filter paper either directly by the 'strip transfer' method, or after elution from the paper and drying in a desiccator. The potential used was 220 v; since the rates were not reproducible from experiment to experiment, only those done on the same paper were compared. If the original peptide spot had been treated with ninhydrin, the purple colouring matter usually migrated slightly towards the anode and was distinguished from the colour given by neutral spots on subsequent spraying, since they moved towards the cathode with the electroendosmotic flow.

Micro amide determinations on peptides. Estimations of amide N on the peptides eluted from paper chromatograms were carried out by the standard procedure (see Bailey, 1937), the ammonia being estimated by the micro-distillation method of Conway (1950). In order to determine the amount of peptide present in the samples, portions were subjected to hydrolysis, and the total amino acids estimated by the ninhydrin method of Moore & Stein (1948).

Samples of the peptides required for analysis were collected from several chromatograms on Whatman no. 3 filter paper, or from ionophoresis experiments. If the spots had been located with dilute ninhydrin, the 'cuts' were thoroughly washed in acetone before elution. The peptide sample was dissolved in a small amount of water and divided into portions which were transferred to small flat glass dishes or Polythene strips. In general, one-sixth was taken for estimation of the amount of peptide present and two one-third samples for amide estimation.

The smaller portion for estimation of total peptide was taken to dryness in a desiccator, dissolved in 5·7N-HCl, transferred to a capillary tube, and incubated at 105° for 24 hr. The hydrolysate was then transferred to Polythene and the excess HCl removed by repeated evaporation in a desiccator. In order to remove any NH_3, which would react with the ninhydrin reagent, 0·1 ml. 0·1M-K_2CO_3 was added to the residue on the Polythene strip, which was placed *in vacuo* for 2 hr. The NH_3-free hydrolysate was then dissolved in 5 ml. water and portions taken for estimation by the ninhydrin method using 1 ml. of ninhydrin reagent and a 30 min. reaction period (Moore & Stein, 1951).

The larger samples of the eluted peptides for amide estimations were taken to dryness, 0·1 ml. 0·1M-K_2CO_3 added and free NH_3 was removed by evaporation in a desiccator for 2 hr. The dry samples were then dissolved in 0·2 ml. 2N-HCl, transferred to capillary tubes and incubated at 105° for 3 hr. Reagent blanks were carried out simultaneously. The hydrolysed samples were divided into portions, each containing *ca.* 5 μg. NH_3-N, which was determined by the standard Conway method. 0·0005N-HCl was used in the centre compartment and back titration was carried out with 0·0025N-$Ba(OH)_2$.

RESULTS

Ionophoresis of peptic hydrolysate of fraction B (*B*p). In order to obtain samples of peptides from fraction *B* for amide estimations, 50 mg. of a peptic

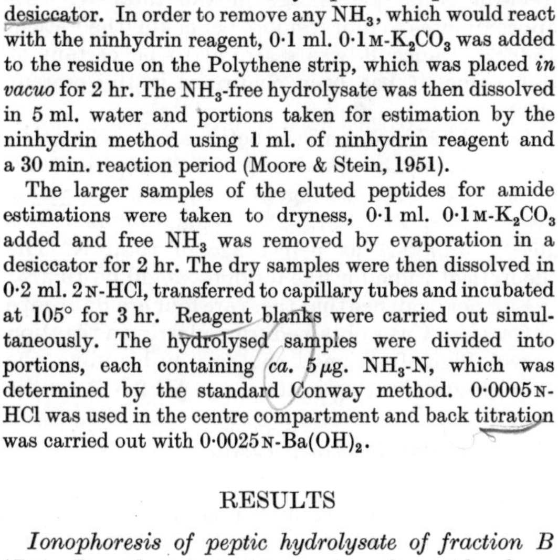

Fig. 1. Ionophoresis of peptic hydrolysate of fraction *B* in 0·05M ammonium acetate. 220 v, 20 hr.

hydrolysate were fractionated by ionophoresis in 0·05M ammonium acetate using a 25 cm.-wide sheet of Whatman no. 3 paper. A number of clear bands were obtained and were cut out as shown in Fig. 1. Paper chromatography of hydrolysates of samples of the fractions indicated the composition shown and with the exception of band γ seemed to be reasonably pure.

Action of mould protease on fraction A. The results obtained with the mould protease hydrolysates of fraction *A* are summarized in Fig. 2 and Table 1. All peptides were hydrolysed to identify the amino acids. In Tables 1–4 the approximate relative amount (strength) of the peptide spots is indicated in the second columns. This was determined by the strength of the ninhydrin colour reaction of the amino acids obtained on hydrolysis of the peptide. The ×'s have the same significance as in previous papers (Sanger & Tuppy, 1951*a*, *b*; Sanger & Thompson, 1953*a*, *b*). The *N*-terminal residues of some of the peptides were determined by hydrolysis of their DNP-derivatives. The experimental results are summarized in column 3 (composition) and in column 4 are shown the most probable structures. These are deduced from a consideration of the data in column 3, the structure of fraction *A* and the known points of splitting by the enzyme, as determined from the structure of the other peptides present. Also the structures of some of the peptides have been inferred by comparison of their R_F values and ionophoretic rates with those of peptides of known structure.

Action of papain on fraction A (*A*pa). The results obtained with the papain hydrolysate are summarized in Fig. 3 and Table 2. Better resolution of the slower-moving peptides (spot nos. 1–9) was obtained by running both solvents two lengths of the paper. The results in Table 2 have the same significance as those in Table 1 and the probable structures were deduced in the same manner.

Peptide *A*pa1, which contained no free α-amino group, was detected by examination of the chromatogram in ultraviolet light. It was subjected to partial hydrolysis with 12N-HCl at 37° for 3 days and the peptides listed in Table 3 were identified. There was also present a spot (R_F in phenol, 0·7; R_F in butanol–acetic acid, 0·68) which gave no colour with ninhydrin, but gave glutamic acid on hydrolysis. This was most probably pyrrolidone carboxylic acid, which was shown to have these R_F values. This and the fact that peptide *A*pa1 contains no free amino group indicate that it is a pyrrolidone carboxylic acid derivative, and the partial hydrolysis experiments show that its structure is that given in Table 2.

No satisfactory explanation can be offered for the presence of spot *A*pa9, which does not fit into the structure of fraction *A*. On ionophoresis it moved toward the anode, suggesting the glutamic acid residue is not present as an amide. The yellow colour with ninhydrin suggests that serine is *N*-terminal. It is tentatively suggested that synthesis may have occurred under the influence of the papain, but it is hoped to investigate this effect further.

*Action of papain on A*p5. Peptide *A*p5 purified by a single ionophoresis, was hydrolysed with cyanide-activated papain as described above and the

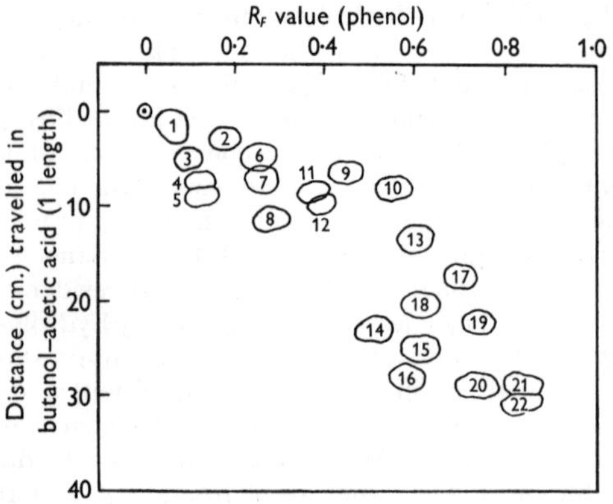

Fig. 2. Chromatogram of mould protease hydrolysate of fraction *A* (Expt. *A*m) (see Table 1).

hydrolysate subjected to ionophoresis in 0·2 M acetic acid overnight at 220 v. Test strips were coloured with ninhydrin, and by the chlorination procedure of Rydon & Smith (1952). The distribution of the bands is shown in Fig. 4, and their composition in Table 4. Since band 5 gave no colour with ninhydrin but a strong colour by the chlorination method and contained glutamic acid, it is presumed to be a pyrrolidonoyl peptide. The presence of free tyrosine (band 2) is probably due to some impurity in the *A*p5.

Total amides of fraction A. Sanger (1949*a*) obtained a value of 15·15% for the amide-N of fraction *A* as percentage of the total N. This approximates to four residues per chain (theoretical 16·0%). Since it was essential to be certain of this value, it has been confirmed in the present work.

Bound NH_3 was removed before amide estimation by dissolving the fraction *A* (15–20 mg.) in 1 ml. 0·1 M-K_2CO_3 and leaving in a vacuum desiccator over H_2SO_4 for 4 hr. The value obtained was 16·7% amide-N as percentage of total N. The theoretical value for five residues would be 19·2%.

Further evidence that there are four amide groups in fraction *A* was obtained from plotting a titration curve. Since fraction *A* contains no histidine, the only groups titrating in the range pH 2·5–7 are the

Table 1. *Peptides from mould protease hydrolysates of fraction A* (*A*m)

Spot no. (Fig. 2)	Approx. strength of peptide	Composition*	Probable structure
1	× × ×	Glu.[$CySO_3H$, Ser, Ala, Val]	Glu.$CySO_3H$.$CySO_3H$.Ala.Ser.Val.$CySO_3H$
2	× × ×	[$CySO_3H$, Ser, Ala, Val]	$CySO_3H$.Ala.Ser.Val.$CySO_3H$
3	×	[$CySO_3H$, Glu, Ser, Ala]	Mixture
4	× ×	[$CySO_3H$, Asp, Tyr]	Tyr.$CySO_3H$.Asp
5	×	[$CySO_3H$, Asp, Tyr]	—
6	× × × ×	[$CySO_3H$, Asp, Tyr]	Asp.Tyr.$CySO_3H$.Asp
7	×	[$CySO_3H$, Asp, Tyr]	—
8	×	—	Glutamic acid
9	× ×	[Asp]	Asparagine
10	× ×	[Glu]	Glutamine
11	{ × × ×	[Ser]	Serine
	×	[$CySO_3H$, Val]	Val.$CySO_3H$ and/or Ser.Val.$CySO_3H$
12	× ×	[$CySO_3H$, Glu, Gly, Val, Leu]	Gly.Ileu.Val.Glu.Glu.$CySO_3H$
13	×	—	Alanine
14	× × ×	Val.Glu	Val.Glu
15	× ×	Glu.[Glu, Leu]	Glu.Leu.Glu
16	× × ×	Leu.Glu	Leu.Glu
17	× ×	[Glu, Tyr]	Tyr.Glu
18	× ×	—	Tyrosine
19	×	[Glu, Val]	?
20	× ×	[Glu, Gly, Val, Leu]	Gly.Ileu.Val.Glu
21	× × ×	—	Leucine
22	× × ×	[Ser, Tyr, Leu]	Ser.Leu.Tyr

* Leucine and isoleucine were not distinguished. Both are recorded as Leu.

Spot 1. Faint colour with ninhydrin. *Spots* 1 *and* 2. May have an extra serine in the *C*-terminal position. *Spot* 3. Contained both glutamic acid and cysteic acid as *N*-terminal residues, so it was clearly a mixture. *Spot* 4. Grey colour with ninhydrin similar to tyrosine, suggesting tyrosine is *N*-terminal. *Spot* 6. Similar R_F value and colour with ninhydrin (yellow) to *A*p3 and *A*pa3 (Asp.Tyr.$CySO_3H$.Asp). *Spots* 5 *and* 7. May be deamidated forms of *A*m4 and *A*m6 or one could be Asp.Tyr.$CySO_3H$. *Spot* 9. Yellow-brown with ninhydrin. *Spots* 12 *and* 22. Yellow with ninhydrin. *Spot* 20. On the chromatogram was a mixture with a tyrosine-containing peptide. It was purified by ionophoresis, yellow colour with ninhydrin. Same R_F value as peptide *A*1η3 (Gly.Ileu.Val.Glu).

Table 2. *Peptides from papain hydrolysate of fraction A* (*A*pa)

Spot no. (Fig. 3)	Approx. strength of peptide	Composition*	Probable structure
1	× × ×	[$CySO_3H$, Glu, Ser, Ala, Val]	Pyr.$CySO_3H$.$CySO_3H$.Ala.Ser.Val.$CySO_3H$
2	× ×	[$CySO_3H$, Asp, Glu, Tyr]	Glu.Asp.Tyr.$CySO_3H$.Asp
3	× ×	[$CySO_3H$, Asp, Tyr]	Asp.Tyr.$CySO_3H$.Asp
4	×	[$CySO_3H$, Asp, Tyr]	—
5	× × ×	[$CySO_3H$, Asp, Glu, Tyr, Leu]	Leu.Glu.Asp.Tyr.$CySO_3H$.Asp
6	×	[$CySO_3H$, Asp, Glu, Tyr, Leu]	—
7	×	—	Glutamic acid
8	× ×	—	Glycine
9	×	[Glu, Ser, Tyr]	?
10	× ×	—	Tyrosine
11	× × ×	[Glu, Tyr]	Tyr.Glu
12	×	[Glu, Tyr, Leu]	Tyr.Glu.Leu.Glu
13	×	[Glu, Ser, Tyr, Leu]	Ser.Leu.Tyr.Glu
14	×	[Glu, Tyr, Leu]	Tyr.Glu.Leu
15	× × × ×	[Glu, Gly, Val, Leu]	Gly.Ileu.Val.Glu
16	× × ×	Leu, Glu	Leu.Glu
17	× ×	[Glu, Val, Leu]	Ileu.Val.Glu
18	× ×	[Ser, Leu]	Ser.Leu
19	× ×	[Glu, Leu]	Pyr.Leu

* Leucine and isoleucine were not distinguished. Both are recorded as Leu.

Spot 1. No colour with ninhydrin. Located by slight fluorescence in ultraviolet light. It gave no DNP derivative, indicating no free —NH_2 group. *Spot* 2. R_F values and ionophoresis rates identical with those of peptide *A*p1. *Spot* 3. Yellow with ninhydrin. R_F values and ionophoretic rates identical with those of peptide *A*p3. *Spots* 4 *and* 6. Only detected when solvents run 2 lengths. May be deamidated forms of 3 and 5 or may have other structures. *Spot* 5. R_F values and ionophoretic rates identical with those of peptide *A*p6. *Spot* 8. Gives traces of serine, valine and cysteic acid after hydrolysis. *Spots* 9, 13 *and* 18. Yellow with ninhydrin. *Spot* 10. Corresponds in composition to spots *A*p12 and *A*m18 while spot 11 corresponds to spots *A*p11 and *A*m17, but the rates of movement in phenol and butanol–acetic acid were reversed. *Spot* 12. Obtained from the chromatogram was contaminated with spot 15, from which it could be separated by ionophoresis. Is probably identical with *A*p14. *Spot* 14. Probably identical with *A*p15. *Spot* 15. Strong grey colour with ninhydrin. *Spot* 19. No visible colour with ninhydrin. Located by fluorescence. Its R_F value in phenol is greater than the known peptides of leucine and glutamic acid. It is probably a pyrrolidone carboxylic acid derivative formed from an *N*-terminal glutamine residue.

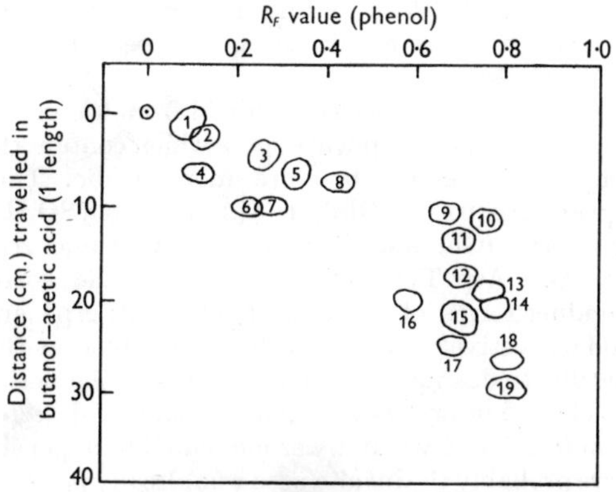

Fig. 3. Chromatogram of papain hydrolysate of fraction *A* (Expt. *A*pa) (see Table 2).

Table 3. *Partial hydrolysis products of peptide A*pa1

Peptide	Strength
$CySO_3H$.$CySO_3H$.Ala (*A*2δ1)	× ×
Cysteic acid	× ×
Glu.$CySO_3H$ (*A*1α1)	×
$CySO_3H$.Ala (*A*1γ3)	× ×
Glutamic acid	× ×
Ser.Val.$CySO_3H$ (*A*1ζ2)	× × × ×
Serine	×
Alanine	× × ×

carboxyl groups. The —SO_3H groups are considerably stronger and are probably completely ionized in this range. Fraction *A* (20·1 mg.) from which the bound acetate had been removed by washing well with 0·1 N-HCl in acetone, was titrated to pH 2·27 with 0·025 N-HCl. It was then titrated back to pH 7 with 0·025 N-NaOH, taking the usual precautions. The results (Fig. 5) are corrected for the activity of HCl in water. Although there was considerable scatter at the lower pH values, it is evident that there are three free —COOH groups on fraction *A*. One of these is the free α-COOH group, so that two of the six dicarboxylic amino acids have free ω-COOH groups and the other four are present as amides.

Ionophoretic rates of peptides. Table 5 shows the distance moved by the various peptides on ionophoresis in 0·05 M ammonium acetate. In this table the results listed in each column were obtained in the same experiment and so are directly comparable. Results from different experiments are not strictly comparable, although in general a peptide moving slightly towards the cathode is neutral and peptides moving towards the anode are acidic.

Table 6 shows the results of the amide determinations on the peptides produced by enzymic hydrolysis of fractions *A* and *B*.

Table 4. *Action of papain on peptide Ap*5 (*Ap*5pa)

Band no. (Fig. 4)	Strength	Composition	Probable structure
1	× × ×	[Glu, Gly, Val, Leu or Ileu]	Gly.Ileu.Val.Glu
2	×	[Tyr]	Free tyrosine
3	× × ×	[$CySO_3H$, Ser, Val, Leu or Ileu]	Val.$CySO_3H$.Ser.Leu
4	×	[$CySO_3H$, Glu, Ser, Gly, Ala, Val, Leu or Ileu]	Unchanged *Ap*5
5	× × × ×	[$CySO_3H$, Glu, Ser, Ala]	Pyr.$CySO_3H$.$CySO_3H$.Ala.Ser

Band 3. Also contained traces of glutamic acid.

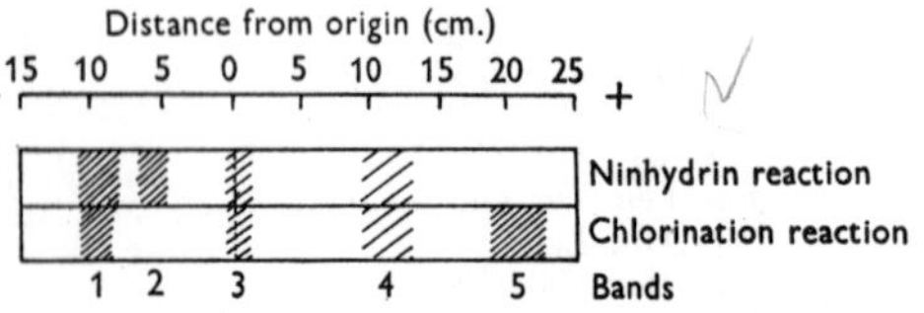

Fig. 4. Ionophoresis of papain hydrolysate of peptide *Ap*5 in 0·2M acetic acid, 220V, 16 hr. (Expt. *Ap*5pa) (see Table 4).

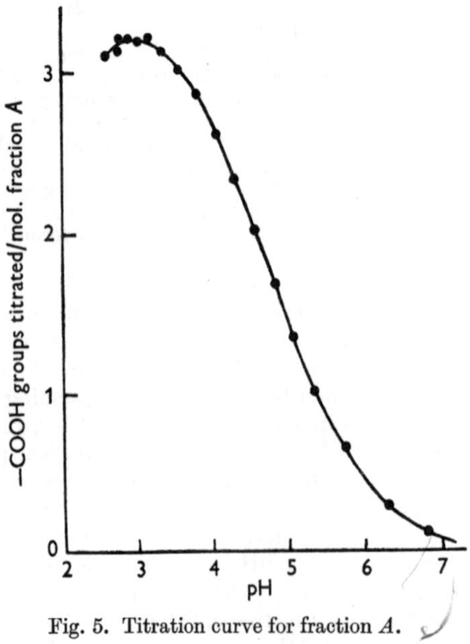

Fig. 5. Titration curve for fraction *A*.

DISCUSSION

Amide groups in the phenylalanyl chain

The amide contents of the peptides from fraction *B* (Table 6) were obtained using the fractions obtained by ionophoresis (Fig. 1), which may not have been completely pure. There may therefore be some error in the results but they are sufficiently accurate for the present purposes.

Sanger (1949*b*) showed that the aspartic and glutamic acid residues in positions 3 and 4 respectively were both in the form of amides, since several different *N*-terminal DNP-peptides containing the same amino acids could be obtained. This is supported by the amide content of peptide *Bp*3, which gives a value approximating to the expected two residues.

Sanger & Tuppy (1951*b*) concluded that the glutamic acid residue in position 13 (see Table 7) contained a free γ-COOH group, since peptides *Bp*1 (Val.Glu.Ala + Val.Glu) and *Bp*5 (Leu.Val.Glu) from the peptic hydrolysates had the same R_F values as the corresponding peptides from the acid hydrolysates. This is confirmed by the fact that the peptides *Bp*1 and *Bp*2 (Val.Glu.Ala.Leu) are acidic (Fig. 1) and contain no amide groups (Table 6).

The remaining glutamic acid residue (position 21) is present in peptide *Bp*4 (Leu.Val.$CySO_3H$.Gly.Glu.Arg.Gly.Phe). This is slightly acidic (Fig. 1) and contains no amide groups, so that the glutamic acid residue must have a free γ-COOH group.

Amide groups in the glycyl chains

It has already been shown that the *C*-terminal residue is asparagine (Harris, 1952; Sanger & Thompson, 1953*b*) and this is confirmed by the presence of one amide group in peptide *Ac*1 ($CySO_3H$.Asp).

Several of the peptides obtained from the peptic hydrolysate have duplicate spots which contain the same amino acids, but are more acidic. Thus peptides *Ap*10 (Tyr.Glu), *Ap*4 (Asp.Tyr.$CySO_3H$.Asp), *Ap*2 (Glu.Asp.Tyr.$CySO_3H$.Asp) and *Ap*7 (Leu.Glu.Asp.Tyr.$CySO_3H$.Asp) are the corresponding forms of peptides *Ap*11, *Ap*3, *Ap*1 and *Ap*6 respectively. It was at first thought that this was due to deamidation occurring under the action of pepsin, but no loss of amide-N could be detected from fraction *A* when it was incubated with pepsin. More probably the insulin used was already partially deamidated, since Harfenist (1953) has shown that certain preparations contain a component having one less amide group than the normal insulin.

Peptides *Ap*11 and *Am*17 both contain the sequence Tyr.Glu and are neutral. They are therefore Tyr.$GluNH_2$ and the glutamic acid residue in position 15 is present as an amide. There is considerable confirmation for this. Peptides *Ap*15, *Apa*14 (Tyr.Glu.Leu) are also neutral and *Ap*15 contains one amide group. Peptide *Am*15 (Glu.Leu.Glu) moves more slowly on ionophoresis than Leu.Glu (*Am*16) and must therefore be $GluNH_2$.Leu.Glu.

Table 5. *Ionophoretic rates of peptides in* 0·05 M *ammonium acetate*

The figures listed in each column represent the results obtained in a single experiment.

Peptide	Spot no.	Distance moved towards anode (cm.)								
		(15)*	(10)*	(15)*	(15)*	(14)*	(8·5)*	(15)*	(18)*	(18)*
Tyr.Glu	*Ap*11	—	−1	−2	—	—	—	—	—	—
	*Ap*10	—	—	5·5	—	—	—	—	—	—
	*Am*17	—	—	—	—	—	−1	—	—	—
Tyr.Glu.Leu	*Ap*15	—	−1	−2	—	—	—	—	—	—
	*Apa*14	—	—	—	—	—	—	−1	—	—
Tyr.Glu.Leu.Glu	*Ap*14	—	4·4	—	—	—	—	—	—	—
	*Apa*12	—	—	—	—	—	—	3·6	—	8
Leu.Glu	*Apa*16	—	—	—	—	—	—	6	—	—
	*Am*16	—	—	—	—	8·5	—	—	—	—
Glu.Leu.Glu	*Am*15	—	—	—	—	7	—	—	—	—
Asp.Tyr.CySO$_3$H.Asp	*Ap*3	8·7	7	—	8·2	—	—	—	—	—
	*Ap*4	15	—	14	—	—	—	—	—	—
	*Apa*3	—	—	—	8	—	—	—	—	—
Glu.Asp.Tyr.CySO$_3$H.Asp	*Ap*1	12·3	9	—	11·6	—	—	—	—	—
	*Ap*2	18·7	—	—	—	—	—	—	—	—
	*Apa*2	—	—	—	11·7	—	—	—	—	—
Leu.Glu.Asp.Tyr.-CySO$_3$H.Asp	*Ap*6	—	—	9	9·5	—	—	—	—	—
	*Ap*7	15	—	—	—	—	—	—	—	—
	*Apa*5	—	—	—	9·7	—	—	—	8	—
Gly.Ileu.Val.Glu	*Ap*16	—	—	4·3	—	—	—	—	—	—
	*Apa*15	—	—	—	—	—	—	4·3	—	—
	*Am*20	—	—	—	—	—	5	—	—	—
Ileu.Val.Glu	*Apa*17	—	—	—	6	—	—	—	—	—
Val.Glu	*Am*14	—	—	—	9	9·5	—	—	—	—
Ser.Leu.Tyr.Glu	*Apa*13	—	—	—	—	—	—	−1	—	—
Ser.Leu.Tyr	*Am*22	—	—	—	—	—	−1	—	—	—
Glu.Leu.Glu.Asp.Tyr.-CySO$_3$H.Asp	*Ac*3	—	—	—	—	—	—	—	6·7	—
Glu.Leu.Glu.Asp.Tyr	*Ac*5	—	—	—	—	—	—	—	—	6·7
Glutamine	*Am*10	—	—	—	—	−1	—	—	—	—
Asparagine	*Am*9	—	—	—	—	−1	—	—	—	—
Glutamic acid	—	14·5	—	16	—	—	—	—	—	—

* Time of ionophoresis (hr.).

Table 6. *Amide determinations of peptides*

Peptide	Sequence	No. of estimations averaged	Amide groups (residues/mol.)
*Bp*1	Val.Glu.Ala	2	0·15
*Bp*2	Val.Glu.Ala.Leu	2	0·15
*Bp*4	Leu.Val.CySO$_3$H.Gly.Glu.Arg.Gly.Phe	3	0·13
*Bp*3	Phe.Val.Asp.Glu.His.Leu.CySO$_3$H.Gly.Ser.His.Leu	2	1·7
*Ac*1	CySO$_3$H.Asp	1	0·7
*Ap*15	Tyr.Glu.Leu	2	0·98
*Ap*14	Tyr.Glu.Leu.Glu	2	1·03
*Ap*3	Asp.Tyr.CySO$_3$H.Asp	4	2·10
*Ap*1	Glu.Asp.Tyr.CySO$_3$H.Asp	7	1·94
*Ap*5*pa*1	Gly.Ileu.Val.Glu	2	0·15
*Ap*5	Gly.Ileu.Val.Glu.Glu.CySO$_3$H.CySO$_3$H.Ala.Ser.Val.-CySO$_3$H.Ser.Leu	17	1·16

Similarly, Glu . Leu . Glu . Asp . Tyr . $CySO_3H$. Asp (*A*c3) moves slower than Leu . Glu . Asp . Tyr . - $CySO_3H$. Asp (*A*pa5).

In the chymotryptic digests two peptides (*A*c5, *A*c7) were isolated having the same amino acid composition but one (*A*c7) gave no colour with ninhydrin. It was pointed out that this was presumably due to the lability of an *N*-terminal glutaminyl residue which tends to cyclize to a pyrrolidone carboxylic acid residue. Thus *A*c5 is probably $GluNH_2$. Leu . Glu . Asp . Tyr and *A*c7 Pyr . Leu . Glu . Asp . Tyr. Also Pyr . Leu (*A*pa19) was found in the papain hydrolysate.

Whereas peptide *A*p15 (Tyr . $GluNH_2$. Leu) was found to be neutral, peptide *A*p14 (Tyr . Glu . Leu . - Glu) is acidic, which suggests the second glutamic acid residue has a free carboxyl group. This is supported by the fact that Leu . Glu (*A*pa16, *A*m16) and Glu . Leu . Glu (*A*m15) are acidic. The glutamic acid residue in position 17 therefore contains a free γ-COOH group. Amide determination also showed that peptide *A*p14 (Tyr . $GluNH_2$. Leu . Glu) had only one amide group, as did *A*p15 (Tyr . $GluNH_2$. Leu). Peptides *A*p1 and *A*pa2 (Glu . Asp . Tyr . - $CySO_3H$. Asp) differ from peptides *A*p3 and *A*pa3 (Asp . Tyr . $CySO_3H$. Asp) only by one glutamic acid residue but are more acidic on ionophoresis, which confirms that this glutamic acid residue (position 17) is not amidized. Similarly peptide *A*p6 (Leu . Glu . - Asp . Tyr . $CySO_3H$. Asp) is less acidic than *A*p1 and more acidic than *A*p3.

Peptide *A*p3 (Asp . Tyr . $CySO_3H$. Asp) contains two amide groups (Table 6). These can only be located on the aspartic acid residues so that both (positions 18 and 21) must be present as asparagine. Similarly *A*p1 (Glu . Asp . Tyr . $CySO_3H$. Asp) contains two amide residues, and the glutamic acid residue has a free γ-COOH group. Further evidence for an amide in position 18 is that Glu . Leu . Glu . - Asp . Tyr (*A*c5) moves slower on ionophoresis than Tyr . Glu . Leu . Glu (*A*p14). These two peptides differ only by the presence of the aspartic acid residue in question, although they contain a different tyrosine residue.

Three enzymes have split the bond between the two glutamic acid residues in positions 4 and 5. Thus, papain gave a high yield of peptide *A*pa15 (Gly . Ileu . Val . Glu). On ionophoresis it moved towards the anode, showing that the glutamic acid was not present as an amide. Similarly, *A*p16 and *A*m20, which have the same sequence as *A*pa15 are acidic and NH_3 determination showed there was no amide in *A*p5pa1. Peptides *A*pa17 (Ileu . Val . Glu) and *A*m14 (Val . Glu) are also acidic. Peptide *A*p5pa5 was identified as a pyrrolidonoyl peptide (Pyr . $CySO_3H$. $CySO_3H$. Ala . Ser) indicating that the residue in position 5 is glutamine.

From these results it was expected that peptide *A*p5 (Gly . Ileu . Val . Glu . Glu . $CySO_3H$. $CySO_3H$. - Ala . Ser . Val . $CySO_3H$. Ser . Leu) would contain one amide group (on position 5). However, innumerable determinations gave values which varied somewhat but were considerably closer to two than to one. Also a few determinations on peptides *A*p7, *A*c2 and *A*c4 gave similar high results. The *A*p5 was prepared either from chromatograms or more often by a single ionophoresis in 0·2 M acetic acid. Although the apparent absence of aspartic acid and tyrosine from hydrolysates suggested that the *A*p5 was reasonably pure, it now seems that the high amide value was due to the presence of an amide-rich impurity, since when *A*p5 was further purified by high-voltage ionophoresis at pH 3·7, values approximating to one were obtained. The figures averaged in Table 6 are all for such purified material. This impurity appears to travel slightly slower than *A*p5 in the 0·2 M acetic acid ionophoresis since a cut taken behind the *A*p5 band gave a higher amide value than the main band. It seems probable that it is largely unchanged fraction *A* and possibly

Table 7. *Amide distribution in insulin*

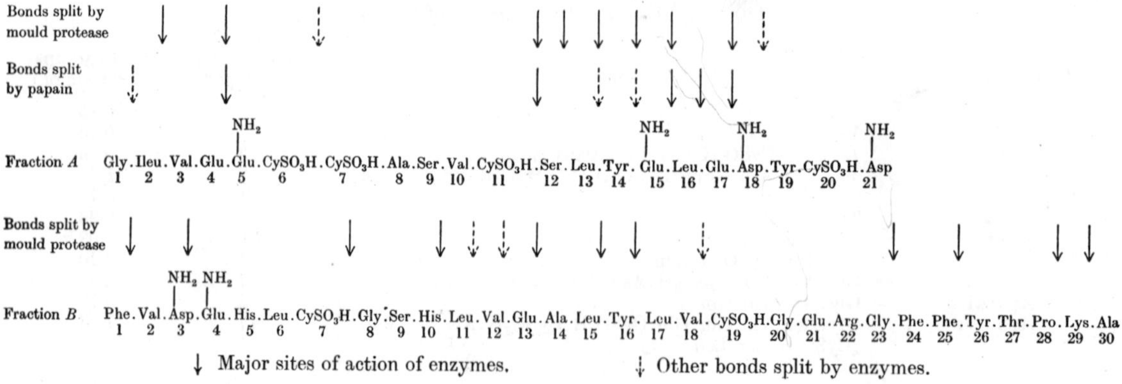

other similar large peptides. Fraction *A* has a comparatively high amide content and a small amount present in the *A*p5 would cause considerable error. In our experience it forms very long spread-out streaks both on ionophoresis and chromatography and, if present, would be expected to contaminate *A*p5 and other peptides (*A*p9, *A*c2, *A*c4) which move at similar rates.

On the basis of these combined results it is concluded that the distribution of amide groups in insulin is as shown in Table 7. This agrees well with the results of Chibnall & Rees (1952), who showed that all three aspartic acid and three of the glutamic acid residues are amidized.

The possibility of iso*glutamine residues*

Kandel, Kandel, Kovacs & Bruckner (1954) have recently suggested that insulin contains an *iso*-glutamine residue. This is based on the finding that when insulin is subjected to the Hofmann degradation and hydrolysed, a substance is produced which gives a 2:4-dinitrophenylhydrazone that moves at the same rate on a butanol–3 % NH_3 paper chromatogram as the dinitrophenylhydrazone of 4-oxobutyric acid. This product would be obtained if an *iso*glutaminyl residue were present in the protein. We have repeated the above experiment and have confirmed the presence of a spot that moves on a butanol–3 % NH_3 chromatogram at the same rate as the synthetic dinitrophenylhydrazone of 4-oxobutyric acid. There were, however, rather more spots on the chromatogram than the two recorded by Kandel *et al.* Samples of the dinitrophenylhydrazone spot were eluted from the paper and reinvestigated by chromatography on *tert.*-amyl alcohol–phthalate (pH 6) buffer (Blackburn & Lowther, 1951) and in the 'toluene' solvent of Biserte & Osteux (1951). With the former solvent two strong spots were produced which had moved 6 and 11 cm. down the paper respectively and a fainter spot that had moved 19 cm. A control spot of the synthetic hydrazone had moved 19 cm. Similar results were obtained with the 'toluene' solvent. It thus seems that the spot that was identified by Kandel *et al.* as the above hydrazone, was in fact a mixture of several substances, one of which may have been the hydrazone. It is present in only small quantities and would probably account for about 1–2 % of one residue in insulin (mol.wt. 5700). In view of these results and the known complexity of the reactions of hypochlorite with amino acids, peptides and proteins (see e.g. Goldschmidt & Strauss, 1930 and previous papers) it seems that there is no clear evidence for the existence of *iso*glutamine residues in insulin.

The fact that the residues in positions *A*5 and *A*15 are converted to a pyrrolidonoyl group when they are the *N*-terminal residues of a peptide (e.g. in peptides *A*pa1 and *A*c7, respectively) indicates that they are glutaminyl residues. An *iso*glutaminyl residue would not decompose to a pyrrolidonoyl peptide but would give the amide of pyrrolidone carboxylic acid. No such evidence exists for the residue in position *B*4, but a study of the Hofmann degradation of fractions *A* and *B* of oxidized insulin showed that the spot moving at the same rate as the dinitrophenylhydrazone of 4-oxobutyric acid was produced from both fractions.

Action of the proteolytic enzymes

The main sites at which splitting occurred under the action of papain and mould protease are indicated in Table 7. The sites of action of mould protease on fraction *B* are based on results of a single experiment and should not be regarded as conclusive and are not comprehensive. Both enzymes show a wide specificity which is somewhat similar to that of pepsin, though the relative intensities of splitting of the various bonds was markedly different for the three enzymes. Mould protease carried the digestion of fraction *A* further than the other enzymes. For structural studies in the future, this enzyme may prove useful, for example, in degrading longer chains obtained by the use of other enzymes.

It is doubtful if the preparations of the two proteinases represent pure enzymes. In the case of papain we have also studied the action of an activated sample of crystalline mercuripapain prepared by Kimmel & Smith (1954). The specificity appeared to be essentially the same as that found with the crude commercial sample.

It is interesting that papain breaks peptide *A*p5 in a different way from fraction *A*. In fraction *A* strong splitting occurs at the $CySO_3H$.Ser bond (position 11–12), whereas this bond was stable in *A*p5 and splitting occurred largely at the Ser.Val bond (position 9–10). This result could be explained if papain was inhibited by a free carboxyl group on the second residue from the bond. Other explanations are possible, but it does illustrate that the specificity of proteolytic enzymes is not conditioned solely by the nature of the residues involved in the bond in question.

SUMMARY

1. Fraction *A* of oxidized insulin was hydrolysed with mould protease and cyanide-activated papain and the resulting peptides studied. Both enzymes showed a rather wide specificity.

2. The positions of the amide groups of insulin were determined by estimating the relative ionophoretic mobilities and the amide contents of peptides from enzymic hydrolysates of fractions *A* and *B* of oxidized insulin. The results are summarized in Table 7.

We wish to thank Dr R. R. Porter for the mould protease and Dr Emil L. Smith for the crystalline mercuripapain.

One of us (E. O. P. T.) wishes to thank the Commonwealth Government of Australia for a Commonwealth Scientific and Industrial Research Organization studentship.

REFERENCES

Bailey, K. (1937). *Biochem. J.* **31**, 1406.
Biserte, G. & Osteux, R. (1951). *Bull. Soc. Chim. biol., Paris*, **33**, 50.
Blackburn, S. & Lowther, A. G. (1951). *Biochem. J.* **48**, 126.
Chibnall, A. C. & Rees, M. W. (1952). *Biochem. J.* **52**, iii.
Conway, E. J. (1950). *Microdiffusion Analysis and Volumetric Error*. London: Lockwood.
Crewther, W. G. & Lennox, F. G. (1950). *Nature, Lond.*, **165**, 680.
Durrum, E. L. (1950). *J. Amer. chem. Soc.* **72**, 2943.
Goldschmidt, S. & Strauss, K. (1930). *Ber. dtsch. chem. Ges.* **63**, 1218.
Harfenist, E. J. (1953). *J. Amer. chem. Soc.* **75**, 5528.
Harris, J. I. (1952). *J. Amer. chem. Soc.* **74**, 2944.
Kandel, I., Kandel, M., Kovacs, J. & Bruckner, V. (1954). *Naturwissenschaften*, **41**, 281.
Kimmel, J. R. & Smith, E. L. (1954). *J. biol. Chem.* **207**, 515.
Michl, H. (1951). *Monatshefte*, **82**, 489.
Moore, S. & Stein, W. H. (1948). *J. biol. Chem.* **176**, 367.
Moore, S. & Stein, W. H. (1951). *J. biol. Chem.* **192**, 663.
Rees, M. W. (1946). *Biochem. J.* **40**, 632.
Rydon, H. N. & Smith, P. W. G. (1952). *Nature, Lond.*, **169**, 922.
Sanger, F. (1949*a*). *Biochem. J.* **44**, 126.
Sanger, F. (1949*b*). *Biochem. J.* **45**, 563.
Sanger, F. & Thompson, E. O. P. (1953*a*). *Biochem. J.* **53**, 353.
Sanger, F. & Thompson, E. O. P. (1953*b*). *Biochem. J.* **53**, 366.
Sanger, F. & Tuppy, H. (1951*a*). *Biochem. J.* **49**, 463.
Sanger, F. & Tuppy, H. (1951*b*). *Biochem. J.* **49**, 481.

[From THE BIOCHEMICAL JOURNAL, Vol. 60, No. 4, pp. 535–540, 1955]

Disulphide Interchange Reactions

By A. P. RYLE and F. SANGER*
Department of Biochemistry, University of Cambridge

(*Received* 26 *November* 1954)

In studies designed to determine the distribution of the disulphide bridges, insulin was hydrolysed with cold concentrated hydrochloric acid and many more cystine peptides were obtained than could be accounted for by a unique structure for insulin. This result led to the suspicion that an interchange reaction took place at the disulphide bonds, and this suspicion was confirmed by experiments with model disulphides (Sanger, 1953). This paper reports the results of some further studies on the disulphide interchange reaction in acid, neutral and slightly alkaline solutions, which were made with the aim of finding how, and under what conditions, the interchange occurred, and particularly how it might be prevented during studies on the arrangement of the disulphide bonds of proteins.

In acid solution the reaction has been studied using a model system consisting of L-cystine and *NN'*-bis-2:4-dinitrophenyl-L-cystine (bisDNP-cystine). The coloured product of the reaction, unlike bisDNP-cystine, is not extracted from acid solution by ether, so that the progress of the reaction can readily be followed spectrophotometrically. It was assumed to be *N*-2:4-dinitrophenylcystine (mono-DNP-cystine), and this has now been confirmed by isolation of the crystalline compound. The above model system could not be used for quantitative studies in neutral solutions because of the insolubility of cystine, which was therefore replaced by the more soluble cystylbisglycine or by oxidized glutathione (GSSG). A preliminary report of this work has been published (Ryle & Sanger, 1954).

EXPERIMENTAL

Materials

BisDNP-L-cystine was prepared by the method of Porter & Sanger (1948) and L-cystylbisglycine by the method of Bailey (1950). The insulin used was batch 9011 G obtained from Boots Pure Drug Co., Nottingham.

Product of reaction between cystine and bisDNP-cystine. L-Cystine (100 mg.) and bisDNP-cystine (500 mg.) were dissolved in a mixture of 10 ml. 12N-HCl and 20 ml. acetic acid and allowed to stand at room temp. for 3–4 days. From part of the resulting mixture a small quantity of the product was obtained in crystalline form in the following way. The solution was diluted with 2 vol. of water, extracted with ether to remove unchanged bisDNP-cystine and concentrated to a small volume by evaporation *in vacuo* at 40°. A column of talc (Hopkins and Williams Purified B.P.C., previously washed with N-HCl) 3·5 cm. high and 4·5 cm. in diameter was prepared and washed with 0·1N acetic acid under applied pressure 5–10 cm. Hg to hasten the flow. The yellow solution was poured on to the top of the column and washed through with 0·1N acetic acid. Although the yellow material did not stick fast to the talc, the emerging liquid was found to be free of chloride before it reached the bottom of the column. At this stage the eluting liquid was changed to ethanol–N acetic acid (80:20, by vol.) and the yellow solution was collected as it came off the column. This solution was extracted with ether to remove ethanol and any bisDNP-cystine not already removed, and after concentration by evaporation *in vacuo* at 40° yellow crystals appeared on cooling. A second crop of crystals was obtained by further concentration of the mother liquor. The crystals were quickly washed with 0·2N acetic acid and dried in a vacuum desiccator. Under the microscope they appeared as bunches of fine yellow needles, m.p. 179° (uncorr.).

The preparation and characterization of monoDNP-L-cystine (m.p. 187°) has now been described by Bettelheim (1955).

The crystalline material was chromatographed on paper together with a sample of monoDNP-cystine kindly given us by Dr V. du Vigneaud. The unknown material moved as a single yellow spot which turned brown on treatment with ninhydrin. In descending runs (Whatman no. 1 paper) R_F values were: 0·64 in phenol–0·3% aq. NH_3 (Sanger & Tuppy, 1951); 0·84 in butanol–acetic acid–water (Partridge, 1948); 0·07 in *tert.*-amyl alcohol–phthalate buffer (pH 6) (Blackburn & Lowther, 1951). Both the authentic and the unknown material moved at the same speed in each case.

A small amount of the material was oxidized with 0·1 ml. performic acid solution (30% (w/v) H_2O_2–formic acid, 1:9 by vol.) at room temp. for 15 min. The formic acid was removed by repeated drying *in vacuo* and the residue chromatographed in the same three systems as before. In each case one yellow spot which did not react with ninhydrin and one ninhydrin-reactive spot appeared. These spots ran at the same speeds as DNP-cysteic acid (Sanger & Thompson, 1953) and cysteic acid respectively (see Table 1). The product is thus confirmed to be monoDNP-cystine.

The light absorption of the material in N-HCl and in 1% (w/v) $NaHCO_3$ was measured in the Beckman spectrophotometer. Beer's law was obeyed at the concentrations studied (less than 50 μM). The absorption is close to that of ε-DNP-lysine (Sanger, 1949). Absorption curves for acid and alkaline solutions are shown in Fig. 1.

The course of the interchange reaction in acid solution

The model system of cystine and bisDNP-cystine was used. In the experiments in HCl solution the reactants were dissolved separately in 12N-HCl and on mixing were

* Member of the Scientific Staff of the Medical Research Council.

Table 1. *R_F values of oxidation products of monoDNP-cystine*

Solvent system	Authentic DNP-cysteic acid	Yellow spot	Authentic cysteic acid	Ninhydrin-reactive spot
Phenol–0·3% aq. NH_3 (Sanger & Tuppy, 1951)	0·50	0·50	0·08	0·08
Butanol–acetic acid–water (Partridge, 1948)	0·73	0·74	0·16	0·15
tert.-Amyl alcohol–phthalate buffer (pH 6) (Blackburn & Lowther, 1951)	0·02	0·02	0·00	0·00

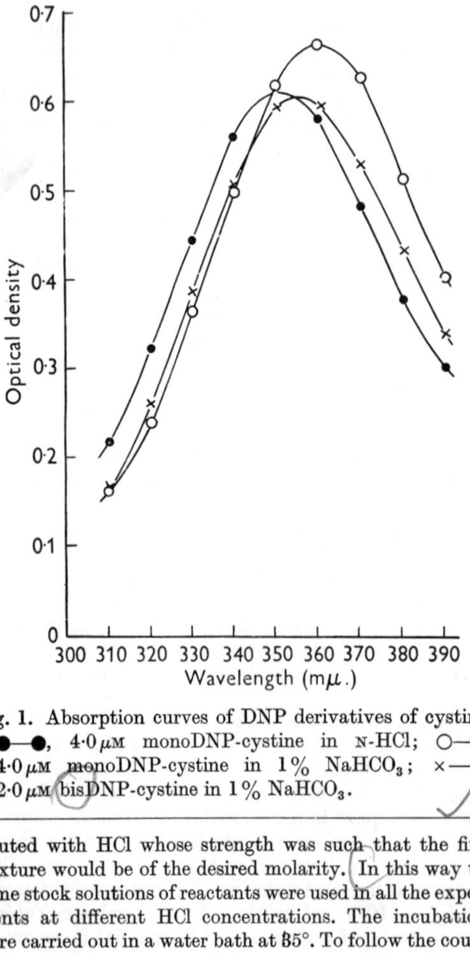

Fig. 1. Absorption curves of DNP derivatives of cystine: ●—●, 4·0 μM monoDNP-cystine in N-HCl; ○—○, 4·0 μM monoDNP-cystine in 1% $NaHCO_3$; ×—×, 2·0 μM bisDNP-cystine in 1% $NaHCO_3$.

diluted with HCl whose strength was such that the final mixture would be of the desired molarity. In this way the same stock solutions of reactants were used in all the experiments at different HCl concentrations. The incubations were carried out in a water bath at 35°. To follow the course of the reaction samples (2 ml.) were taken from time to time, pipetted into 4 ml. of water and shaken with 2 ml. ether. The ether was sucked off and the extraction repeated with a further 2 ml. ether. Trial experiments showed that this procedure was sufficient to remove the bisDNP-cystine from the aqueous phase. A sample from this phase was then diluted to a suitable volume and the absorption at 350 mμ. of the resulting solution was read in the Beckman spectrophotometer against water. The amount of monoDNP-cystine formed in each ml. of reaction solution was calculated from the absorption.

The course of the reaction in neutral and slightly alkaline solutions

The low solubility of cystine in neutral solution prevented its use in these experiments, so cystylbisglycine was used instead. The reactions were carried out in buffer solutions in a water bath at 35°, and were followed by the same means as in the experiments in acid solution except that the samples were pipetted into 2 vol. of N-HCl instead of into water.

The experiments in the absence of oxygen were carried out in Thunberg tubes in which the solutions of the two reactants were evacuated before being mixed. The tubes were re-evacuated after each sample had been taken.

RESULTS

Disulphide interchange in acid solution

Effect of acid. Fig. 2 shows the course of the reaction in various concentrations of HCl at 35°. It is very rapid in conc. HCl but falls off markedly as

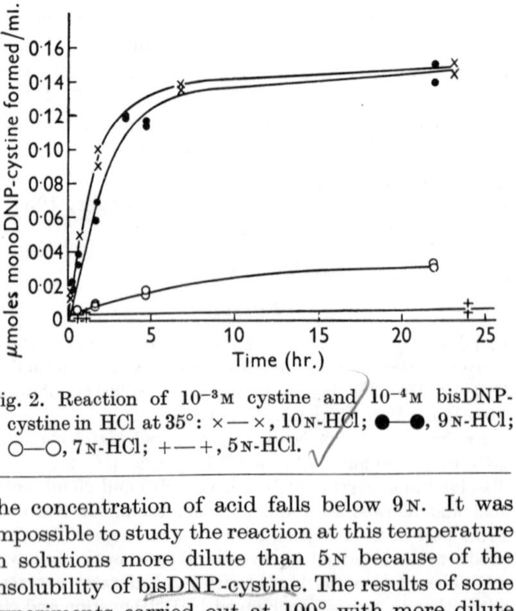

Fig. 2. Reaction of 10^{-3}M cystine and 10^{-4}M bisDNP-cystine in HCl at 35°: ×—×, 10N-HCl; ●—●, 9N-HCl; ○—○, 7N-HCl; +—+, 5N-HCl.

the concentration of acid falls below 9N. It was impossible to study the reaction at this temperature in solutions more dilute than 5N because of the insolubility of bisDNP-cystine. The results of some experiments carried out at 100° with more dilute acid are given in Table 2. Here excess cystine was used and the results are expressed as percentage of the maximum interchange reaction theoretically possible if all the bisDNP-cystine were converted into monoDNP-cystine. In conc. acid the reaction is much more rapid in HCl than in corresponding strengths of H_2SO_4, though in dilute acids the reverse is the case. In 0·01 N acids or in water the reaction was more rapid than in 0·1 N acid but there was some production of H_2S, suggesting that other reactions were taking place. The reaction is at a minimum in 2 N-H_2SO_4 or in approx. 0·5 N-HCl.

Table 2. *Interchange reaction in various concentrations of* HCl *and* H_2SO_4 *at* 100°

3×10^{-4} M bisDNP-cystine refluxed 2 hr. in acid with 10^{-2} M cystine. Figures are % theoretically possible interchange reaction.

Normality of acid	% interchange in HCl	% interchange in H_2SO_4
0·1	4·6	18
0·5	—	3·3
1·0	4·7	2·8
2·0	—	1·5
3·0	—	2·3
5·0	26	4·7

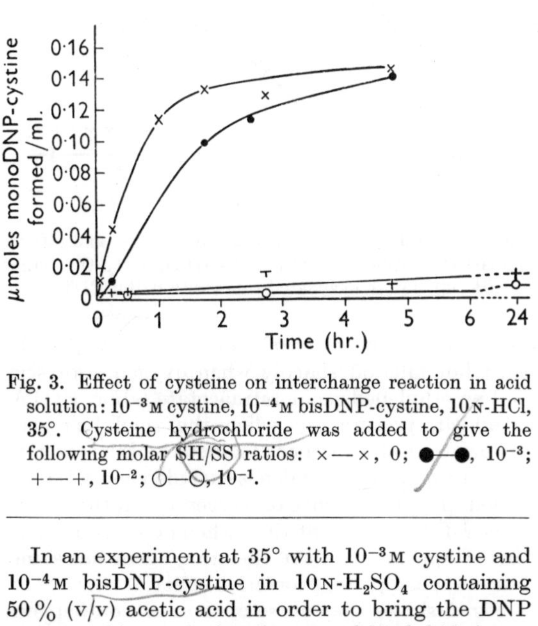

Fig. 3. Effect of cysteine on interchange reaction in acid solution: 10^{-3} M cystine, 10^{-4} M bisDNP-cystine, 10 N-HCl, 35°. Cysteine hydrochloride was added to give the following molar SH/SS ratios: ×—×, 0; ●—●, 10^{-3}; +—+, 10^{-2}; ○—○, 10^{-1}.

In an experiment at 35° with 10^{-3} M cystine and 10^{-4} M bisDNP-cystine in 10 N-H_2SO_4 containing 50 % (v/v) acetic acid in order to bring the DNP compound into solution the rate of disulphide interchange was found to be low, 10 % of the bisDNP-cystine having reacted after 24 hr. and 45 % after 5 days.

Effect of added thiol. Huggins, Tapley & Jensen (1951) described the gelation of various proteins in conc. urea solutions and produced evidence indicating that it was brought about by the rearrangement of some of the disulphide bonds of the protein, the rearrangement occurring by reaction between disulphide groups and thiols:

$$R^1SSR^2 + R^3SH \rightleftharpoons R^1SSR^3 + R^2SH.$$

It seemed possible that the disulphide interchange reaction studied here might occur by the same means. If this were so, the addition of thiols should accelerate the interchange reaction. Fig. 3 shows the effect on the reaction in 10 N-HCl of adding varying quantities of cysteine hydrochloride. It is apparent that, rather than accelerating the reaction, cysteine markedly inhibits it.

Conditions for protein hydrolysis. The above results suggest that the interchange reaction is slower in high concentrations of H_2SO_4 than in corresponding concentrations of HCl and that it is markedly inhibited by thiols. In order to determine how much interchange occurred during various conditions for the hydrolysis of insulin, samples of insulin (10 mg./ml.) were allowed to hydrolyse in the presence of bisDNP-cystine (0·02 M), and the amount of coloured material remaining in the aqueous solutions after ether extraction was taken as a measure of the extent of the interchange reaction. This colour is due to a mixture of peptides of the type

$$\begin{array}{c} \text{DNP-Cy} \\ | \\ \text{S} \\ | \\ \text{S} \\ | \\ \text{R}_1.\text{Cy}.\text{R}_2 \end{array}$$

The results are given in Table 3 as the number of disulphide bonds of insulin that have reacted, assuming that when one such bond has reacted two DNP groups appear in the aqueous phase. It can be seen from Table 3 that no appreciable interchange occurs in 10 days when the hydrolysis is carried out at 37° with 10 N-H_2SO_4 in the presence of thiol, and there is only a very little after hydrolysis in the same solution at 100° for 2 hr. These two sets of conditions were used for the hydrolyses of insulin to determine the arrangement of the disulphide bonds (Ryle, Sanger, Smith & Kitai, 1955).

Similar experiments with cystine and bisDNP-cystine yielded similar results. The interchange reaction at 35° between 10^{-3} M cystine and 10^{-4} M bisDNP-cystine in 10 N-H_2SO_4 in 50 % acetic acid was completely inhibited by $1·1 \times 10^{-5}$ M thioglycollic acid, no reaction being detectable after 5 days. At 100° $1·1 \times 10^{-4}$ M thioglycollic acid provided complete inhibition of interchange over 1 hr., while in the same time in the presence of $1·1 \times 10^{-5}$ M thioglycollic acid 15 % interchange occurred.

Table 3. *Interchange reaction occurring during the hydrolysis of insulin*

0·02 M bisDNP-cystine; 10 mg. insulin/ml.

Acid	Temp. (°)	Additions	Time	Approx. no. —S.S— bonds reacting/ molecule insulin (mol.wt. 6000)
12 N-HCl*	37	—	1 hr.	0·3
			2 hr.	0·65
			5 hr.	1·2
5·7 N-HCl in 50% (v/v) acetic acid*	37	10^{-3} M cysteine	1 day	0·35
			4 days	1·0
			10 days	1·35
10 N-H_2SO_4 in 50% (v/v) acetic acid	37	2×10^{-3} M cysteine	5 days	0
			10 days	0·01
10 N-H_2SO_4 in 30% (v/v) acetic acid	100	—	30 min.	0·35
			90 min.	0·7
10 N-H_2SO_4 in 30% (v/v) acetic acid	100	5×10^{-3} M thioglycollic acid	35 min.	0·05
			120 min.	0·15

* 2×10^{-3} M-bisDNP-cystine; 1 mg. insulin/ml.

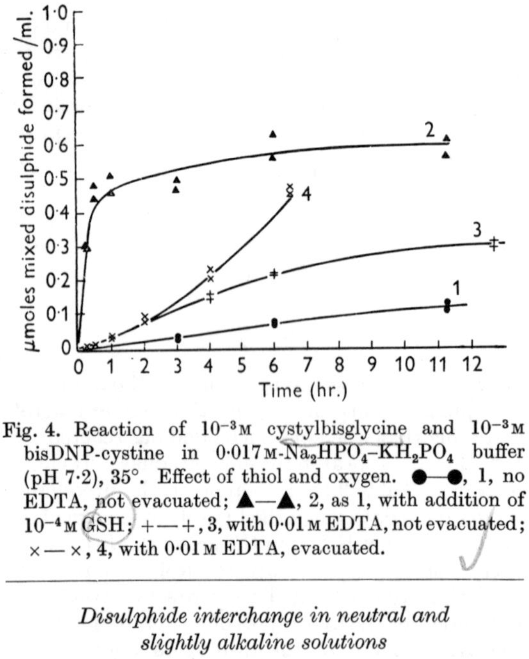

Fig. 4. Reaction of 10^{-3} M cystylbisglycine and 10^{-3} M bisDNP-cystine in 0·017 M-Na_2HPO_4–KH_2PO_4 buffer (pH 7·2), 35°. Effect of thiol and oxygen. ●—●, 1, no EDTA, not evacuated; ▲—▲, 2, as 1, with addition of 10^{-4} M GSH; +—+, 3, with 0·01 M EDTA, not evacuated; ×—×, 4, with 0·01 M EDTA, evacuated.

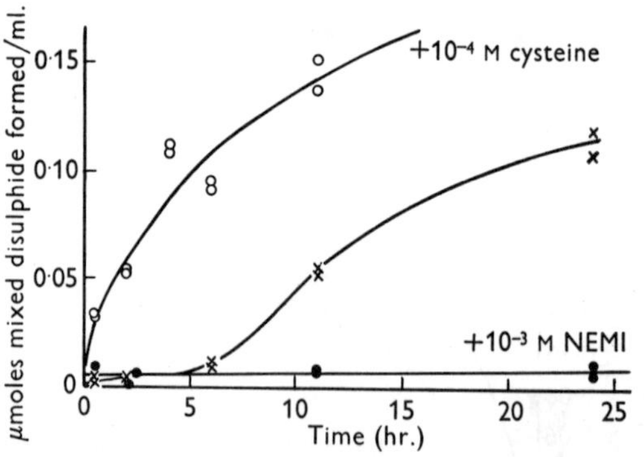

Fig. 5. Reaction of 10^{-3} M cystylbisglycine and 10^{-4} M bisDNP-cystine in 0·017 M-Na_2HPO_4–KH_2PO_4 buffer (pH 7·2), 35°, in evacuated tubes with 0·01 M EDTA.

Disulphide interchange in neutral and slightly alkaline solutions

Some preliminary experiments indicated that in neutral solution the disulphide interchange occurring between bisDNP-cystine and cystylbisglycine was catalysed by the addition of thiols, and since thiols are oxidized by oxygen in neutral solution in the presence of traces of heavy metal it seemed possible that the reaction might proceed faster if oxygen were removed, or oxidation prevented, by the addition of some chelating agent such as ethylenediaminetetraacetate (Versene).

In Fig. 4 are shown the results of some experiments which indicate that this is the case. The reaction in the presence of added thiol is rapid at the start but falls off sharply when oxygen is present and no ethylenediaminetetraacetate (abbr. EDTA) is added, presumably when the thiol has been oxidized. In the absence of added thiol the reaction is most rapid when oxygen is excluded and EDTA is added. In the presence of oxygen it is rather faster when EDTA is present than when none is added, so that it appears that the chelating agent has some, but not a complete, protective effect.

Fig. 5 shows the course of the reaction between 10^{-3} M cystylbisglycine and 10^{-4} M bisDNP-cystine (one-tenth of the concentration of the DNP compound used before). Catalysis by added thiol is again observed, and almost complete inhibition by *N*-ethylmaleimide (NEMI).

Fig. 6 shows the course of the reaction between GSSG and bisDNP-cystine. The rate is greater at the higher pH, and the reaction is almost completely inhibited by the addition of *p*-chloromercuribenzoate (*p*CMB).

A curious feature of the interchange reaction in neutral solution in the absence of oxygen was the lag phase of somewhat variable duration occurring at the start of the reaction. In Fig. 7 are shown the results of four experiments which are identical, except that in two tubes freshly prepared solutions of the reactants were used, while in the other two solutions which had stood under vacuum at 37° for 16 hr. were used. In the experiments with these 'aged' solutions there is no lag phase.

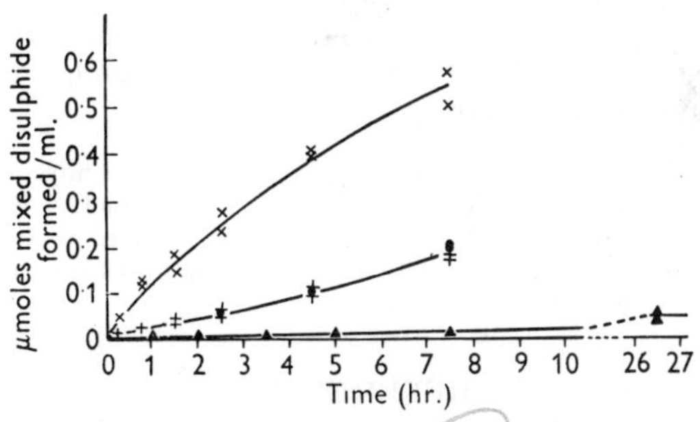

Fig. 6. Reaction of $0{\cdot}8 \times 10^{-3}$M GSSG with $0{\cdot}8 \times 10^{-3}$M bisDNP-cystine in 0·017 M-Na_2HPO_4–KH_2PO_4 buffer at 35° in evacuated tubes. ×—×, pH 8·0 with 0·01 M EDTA; ●—●, pH 7·2 with 0·01 M EDTA; +—+, pH 7·2, without EDTA; ▲—▲, pH 7·2, with $0{\cdot}8 \times 10^{-4}$M *p*-chloromercuribenzoate.

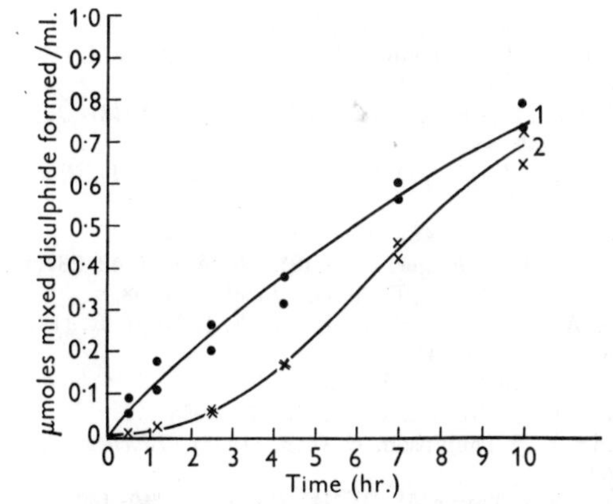

Fig. 7. Reaction of 10^{-3}M cystylbisglycine and 10^{-3}M bisDNP-cystine in 0·017 M-Na_2HPO_4–KH_2PO_4 buffer (pH 7·2), 0·01 M EDTA, 35°; ●—●, 'aged' solutions of disulphides; ×—×, fresh solutions of disulphides.

DISCUSSION

The most striking feature of the results reported above is the difference in the response of the disulphide interchange in acid and neutral solutions to the addition of thiols. The finding that thiols inhibit the reaction in acid solution and catalyse it in neutral solution must indicate that the mechanism under the two conditions is different. This conclusion accords well with the finding that the rate passes through a minimum in dilute acid and is greater in strongly acid or neutral solutions (Table 2).

The inhibition of the interchange reaction in neutral solution by NEMI which combines rapidly and quantitatively with thiols (Friedmann, Marriann & Simon-Reuss, 1949), by *p*CMB, which similarly removes thiols from solution (Hellerman, Chinard & Deitz, 1943), and by oxygen, and its catalysis by added thiol indicates that it is here carried by thiol groups in the same manner as the gelation reaction of Huggins *et al.* (1951). A catalytic amount of thiol must be produced from the disulphides, presumably by hydrolytic fission (Cecil, 1950), so that the interchange reaction would proceed as follows:

$$R^1SSR^1 + OH^- \rightleftharpoons R^1S^- + R^1SOH$$
$$R^2SSR^2 + R^1S^- \rightleftharpoons R^1SSR^2 + R^2S^-$$
$$R^2S^- + R^1SSR^1 \rightleftharpoons R^1SSR^2 + R^1S^- \text{ etc.}$$

At higher pH values the concentration of RS^- produced by the initial reaction with OH^- would be greater, so that the overall interchange reaction would be found to proceed faster, as observed in Fig. 6.

If it is assumed that the hydrolytic attack on the disulphides is slow, this mechanism may explain the lag phase observed in experiments in which fresh solutions of the disulphides were employed, as opposed to those in which 'aged' solutions were used (Fig. 7): in the experiments with fresh solution thiols are slowly formed from the disulphides during the lag phase, while in the experiments with 'aged' solutions the thiols will have been formed before the start of the experiment.

The inhibition of the interchange by thiol-binding reagents is most striking and, as described in the succeeding paper, such reagents have been employed in preventing interchange in neutral solutions during studies on protein disulphides.

It may not be out of place to speculate on the role which disulphide interchange reactions play in living systems. In the experiments described above a quite rapid interchange took place under neutral conditions in the presence of 10^{-4}M glutathione (3 mg./100 ml.) or cysteine. In the blood the concentration of glutathione found by Benedict & Gottschall (1933) is about 40 mg./100 ml. corresponding to about 80 mg./100 ml. for the red blood corpuscles in which all the glutathione is found. Bartlett & Stevenson (1954) give a lower figure, about 23 mg./100 ml., for whole blood and about 170 and 15 mg./100 g. wet weight for the liver and muscle, respectively, of growing rats. If it is assumed that the glutathione is distributed uniformly through these two last tissues its concentration ranges from 0·5 to 5·0 mM—rather higher than that used in the experiments described above.

The conditions in the body are therefore such as could promote a rapid disulphide interchange, so that any small disulphides would be expected to react in this way. Whether such interchange can occur within and between protein molecules *in vivo* will depend on the rigidity of the protein structure and on the extent to which the disulphide bonds are accessible to glutathione. While little is known of the latter factor, the observations of Huggins *et al.* (1951) on the urea-induced gelation of albumins (a reaction which is essentially a disulphide interchange reaction) indicate that with these proteins interchange can only occur when the hydrogen bonding of the protein has been shattered by the high concentration of urea. If, as seems possible, other proteins behave in the same way, their disulphide bonds should be stable under physiological conditions, so that the interchange reaction may not be of importance.

Most of the small sulphur-containing molecules, such as glutathione and coenzyme A, are found in the reduced form, so are unlikely to play any except a catalytic part in disulphide interchange. The fact that these thiols are found side by side with proteins containing disulphide bonds is probably another reflexion of the stability conferred on the protein disulphides by the configuration of the molecule, maintained by other bonds within and between the polypeptide chains. Whilst it seems that in general proteins are unlikely to take part in disulphide interchange reactions under physiological conditions, it still remains possible that there are special cases in which such reactions are of importance.

No conclusions can be drawn here about the mechanism of the interchange reaction in acid solutions, but the most striking features here are the inhibition by thiols and the great speed of the reaction in 10N-HCl compared with that in 10N-H_2SO_4 or 5N-HCl. It should be pointed out that the activity of hydrochloric acid rises very rapidly with increasing concentration, so that that of the 10N acid is much greater than that of sulphuric acid of the same strength.

SUMMARY

1. A disulphide interchange reaction of the type

$$R^1SSR^1 + R^2SSR^2 \rightleftharpoons 2R^1SSR^2$$

has been studied in acid, neutral and slightly alkaline solutions.

2. The reaction is inhibited in acid solution by thiols.

3. The reaction in neutral and alkaline solution is catalysed by thiols and inhibited by thiol-binding reagents.

4. A mechanism for the reaction in neutral and alkaline solutions is suggested.

5. Suitable conditions have been determined for the hydrolysis of proteins without disulphide interchange occurring.

We wish to thank Dr V. du Vigneaud for a sample of monoDNP-cystine and Dr J. L. Bailey for help and advice in preparing cystylbisglycine. A.P.R. is indebted to the Medical Research Council for a scholarship for training in research methods.

REFERENCES

Bailey, J. L. (1950). *J. chem. Soc.* p. 3461.

Bartlett, P. D. & Stevenson, A. (1954). *Fed. Proc.* **13**, 594.

Benedict, S. R. & Gottschall, G. (1933). *J. biol. Chem.* **99**, 729.

Bettelheim, F. R. (1955). *J. biol. Chem.* **212**, 235.

Blackburn, S. & Lowther, A. G. (1951). *Biochem. J.* **48**, 126.

Cecil, R. (1950). *Biochem. J.* **47**, 572.

Friedmann, E., Marriann, D. H. & Simon-Reuss, I. (1949). *Brit. J. Pharmacol.* **4**, 105.

Hellerman, L., Chinard, F. P. & Deitz, V. R. (1943). *J. biol. Chem.* **147**, 443.

Huggins, C., Tapley, D. F. & Jensen, E. V. (1951). *Nature, Lond.*, **167**, 592.

Partridge, S. M. (1948). *Biochem. J.* **42**, 238.

Porter, R. R. & Sanger, F. (1948). *Biochem. J.* **42**, 287.

Ryle, A. P. & Sanger, F. (1954). *Biochem. J.* **58**, v.

Ryle, A. P., Sanger, F., Smith, L. F. & Kitai, R. (1955). *Biochem. J.* **60**, 541.

Sanger, F. (1949). *Biochem. J.* **45**, 563.

Sanger, F. (1953). *Nature, Lond.*, **171**, 1025.

Sanger, F. & Thompson, E. O. P. (1953). *Biochem. J.* **53**, 353.

Sanger, F. & Tuppy, H. (1951). *Biochem. J.* **49**, 463.

[From THE BIOCHEMICAL JOURNAL, Vol. 60, No. 4, pp. 541–556, 1955]

The Disulphide Bonds of Insulin

By A. P. RYLE, F. SANGER,* L. F. SMITH* and RUTH KITAI
Department of Biochemistry, University of Cambridge

(*Received* 26 *November* 1954)

In order to deduce the unique structure of ox insulin it is necessary to know its molecular weight. In previous papers a value of 12 000 was assumed, since physical measurements suggested that this was the weight of the smallest unit that existed in solution. Recently, however, Harfenist & Craig (1952) have used a new chemical method and have found a value of approximately 6000. It is difficult to see how this result could be wrong and some recent physical measurements have shown that dissociation into units of molecular weight lower than 12 000 definitely occurs (Fredericq, 1953; Kupke & Linderstrøm-Lang, 1954). It may thus safely be concluded that the molecular weight of insulin is 5734, on the basis of condensation of the amino acids present with normal elimination of water. This molecule is composed of two polypeptide chains joined together by the disulphide bridges of three cystine residues. Treatment with performic acid splits the insulin to two fractions *A* and *B*, which are the oxidized forms of the glycyl and the phenylalanyl chain respectively (Sanger, 1949*a*). The sequence of amino acids in these two polypeptide chains has been determined by partial hydrolysis methods (Sanger, 1949*b*; Sanger & Tuppy, 1951*a*, *b*; Sanger & Thompson, 1953*a*, *b*; Sanger, Thompson & Kitai, 1955) and is shown in Table 1. Fraction *A* contains four and fraction *B* two cysteic acid residues, originating from the three cystine residues of insulin. The purpose of the present study was to find which half-cystine residues are joined together in intact insulin.

For this it was necessary to identify peptides containing cystine residues and to determine their structure. The procedure used may be summarized as follows:

(1) Partial hydrolysis of insulin under conditions where the disulphide bonds remained intact.

(2) Fractionation of cystine peptides from one another. It was not, however, necessary at this stage to separate them from other peptides not containing cystine, since these were separated from the cysteic acid peptides during stage 4.

(3) Oxidation of cystine peptides to cysteic acid peptides.

(4) Fractionation of cysteic acid peptides.

(5) Identification of the cysteic acid peptides from the amino acids produced on hydrolysis. These peptides had already been obtained from partial hydrolysates of the fractions *A* and *B*, so that their structures were known and they could be completely identified from their hydrolysis products.

From the cysteic acid peptides produced the structure of the original cystine peptides could be deduced and hence the distribution of the disulphide bonds in insulin.

During stages (1) and (2) it was essential to avoid any re-arrangement of the disulphide bonds, and this was the chief difficulty in the work. Initially, hydrolysis was carried out in concentrated hydrochloric acid and the fact that it was impossible to interpret the results in terms of a unique structure for insulin suggested that a random re-arrangement of the disulphide bonds had taken place. This was confirmed by studies in model systems and the reaction has been further studied (Ryle & Sanger, 1955) and shown to occur in acid and neutral solution by different mechanisms. The reaction in neutral solution was catalysed by thiol compounds and could be prevented to some extent by thiol inhibitors such as *N*-ethylmaleimide (NEMI). It was thus possible to use proteolytic enzymes for partial hydrolysis and the position of one disulphide bridge was found in this way. Experiments using chymotrypsin and a crude pancreatic extract are described. However, no enzyme could be found that would break between the two half-cystine residues in positions *A* 6 and *A* 7 (Table 1) and acid hydrolysis had to be used to locate the remaining two disulphide bridges.

Various conditions of acid hydrolysis were tried but invariably led to re-arrangement until it was found that, unlike the neutral reaction, the acid one was inhibited by thiol compounds. It also appeared to be slower in H_2SO_4 than in HCl of the same concentration (Ryle & Sanger, 1955). A rather long time of hydrolysis was necessary to avoid the production of large amounts of peptides containing the *A* 6.*A* 7 sequence (Table 1), as they do not provide the information required to establish the position of the bridges.

It was also necessary to avoid re-arrangement during the fractionation of the cystine peptides. Initially paper chromatography was used, but it was

* Members of the Scientific Staff of the Medical Research Council.

Table 1. *The structure of fractions A and B of oxidized insulin*

Fraction *A* (Sanger & Thompson, 1953 *a*, *b*; Sanger *et al.* 1955)

	1	2	3	4	5	6	7	8	9	10	11	12	13	14	15	16	17	18	19	20	21
					NH_2										NH_2			NH_2			NH_2
	Gly	Ileu	Val	Glu	Glu	$CySO_3H$	$CySO_3H$	Ala	Ser	Val	$CySO_3H$	Ser	Leu	Tyr	Glu	Leu	Glu	Asp	Tyr	$CySO_3H$	Asp

Fraction *B* (Sanger & Tuppy, 1951 *a*, *b*)

	1	2	3	4	5	6	7	8	9	10	11	12	13	14	15	16	17	18	19	20	21	22	23	24	25	26	27	28	29	30
			NH_2	NH_2																										
	Phe	Val	Asp	Glu	His	Leu	$CySO_3H$	Gly	Ser	His	Leu	Val	Glu	Ala	Leu	Tyr	Leu	Val	$CySO_3H$	Gly	Glu	Arg	Gly	Phe	Phe	Tyr	Thr	Pro	Lys	Ala

difficult to obtain sufficient material, and frequently re-arrangement appeared to have occurred. The best results were obtained with paper ionophoresis and in the most recent experiments this method was used almost entirely.

The peptides with reference numbers starting *A* or *B* are those obtained in previous papers as follows: *A* 1, *A* 2, Sanger & Thompson (1953*a*); *A*p, *A*c, Sanger & Thompson (1953*b*); *B* 1, *B* 2, *B* 3, *B* 4, *B* 5, Sanger & Tuppy (1951*a*); *B*p, *B*c, *B*t, Sanger & Tuppy (1951*b*).

A preliminary communication of the present work was made by Sanger, Smith & Kitai (1954).

MATERIALS

The insulin used throughout this work was 6-times recrystallized cattle insulin (batch 9011G), obtained from Boots Pure Drug Co., Nottingham. Crystalline chymotrypsin was obtained from Worthington Biochemical Sales Co., Freehold, New Jersey, U.S.A.

METHODS

Ionophoresis

The method used was essentially that of Michl (1951) in which the ionophoresis is carried out at high potential gradients in pyridine–acetic acid buffers on filter papers

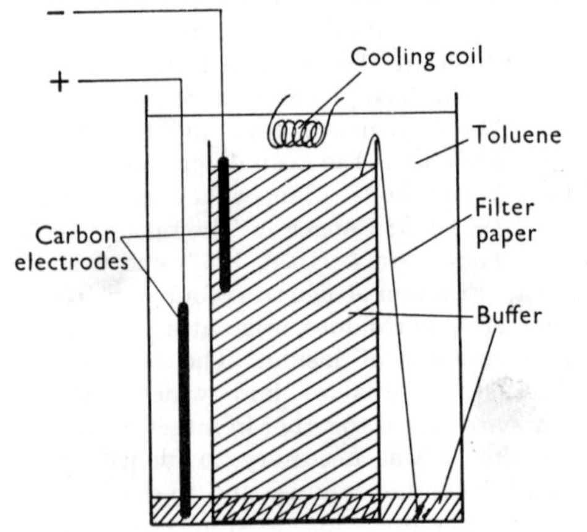

Fig. 1. High-voltage ionophoresis apparatus (Michl,1951).

completely immersed in toluene. This last prevents evaporation and cools the paper. It was found that sharper bands were obtained by this method than by other methods studied and the short times required for separations minimized the danger of breakdown or re-arrangement of cystine peptides during ionophoresis.

The apparatus, which is shown in Fig. 1, was made from two battery jars, one fitting inside the other. By having the level of the buffer solution higher in the cathode than in the anode vessel, electroendosmotic flow was to some extent counterbalanced by hydrostatic flow. Carbon rods were used for the electrodes and were connected to platinum wires underneath the toluene. Suitably shaped glass rods were used to prevent the paper from touching the sides of the vessels, and the whole apparatus was covered with a glass lid.

Since the buffer (pyridine–acetic acid) is somewhat soluble in the toluene and large volumes were involved, it was not convenient to keep changing the buffer for different experiments. Two units were therefore employed; the one contained buffer at pH 6·5 (10 vol. pyridine: 0·4 vol. acetic acid: 90 vol. water), the other at pH 3·6 (1 vol. pyridine: 10 vol. acetic acid: 89 vol. water). In some of the earlier experiments 0·2 M acetic acid (pH 2·75) was used as the electrolyte. A paper up to 20 cm. wide could be used in the apparatus. The material to be fractionated was applied as a thin line across the dry paper to within about 1 cm. of each edge. It was allowed to dry and then laid on a glass plate with a glass rod underneath the line of application. The paper was wetted with the buffer to within about 1 cm. of the applied material and the buffer allowed to flow in towards the line of application from both sides, thus concentrating the material in a thinner line. It was essential to ensure that the buffer flowed in evenly and at the same rate from both sides and to be certain that the paper was completely wetted before putting in the apparatus. Where rather large concentrations of peptides are applied they sometimes wet very slowly. If a spot is not wetted by the buffer it is wetted by the toluene and causes an unevenness of flow. After blotting the paper to remove excess buffer it was put in the apparatus.

For most purposes a potential of 1500 v was applied. Using a 20 cm. wide Whatman no. 3 filter paper, a current of 60–70 mA was produced with the buffer at pH 3·6 and of 40–45 mA at pH 6·5. Owing to differences in design of the apparatuses the potential gradients produced were 29 and 33 v/cm. respectively. With no. 4 papers about half these currents are obtained. At the higher currents there is considerable heating, and it was necessary to have a cooling coil in the toluene. To find the position occupied by the peptides after ionophoresis, marker strips were cut from the paper and tested with a suitable dipping reagent, as below.

Ninhydrin. A 0·25% solution in acetone was used (Toennies & Kolb, 1951).

CN–nitroprusside. To test for the presence of cystine peptides the CN–nitroprusside test of Toennies & Kolb (1951) was used. This was found to be more sensitive than other tests investigated, though the colour was rather transitory, and it was necessary to record the position of the bands fairly rapidly.

Tyrosine test. To test for tyrosine peptides the method of Acher & Crocker (1952) was used.

Chymotryptic hydrolysis (*Expt. I*c)

Insulin (100 mg.) was suspended in 10 ml. water and dissolved by the addition of NH_3 to give a final pH of 8. Chymotrypsin (4 mg.) and 1 mg. of NEMI were added and the mixture was incubated at 37° for 24 hr. The solution was then brought to pH 5·7 by the addition of dilute acetic acid. This produced a precipitate, which will be referred to as the 'chymotryptic core'. It is probably essentially the same as the 'core' studied by Butler, Dodds, Phillips & Stephen (1948) which was obtained from a chymotryptic hydrolysate by precipitation with trichloroacetic acid. The yield of 'core' was about 65 mg. It was further purified by dissolving in dilute NH_3 and precipitating at pH 5·7. In certain experiments the 'core' did not precipitate immediately on bringing the hydrolysate to pH 5·7 and it was necessary to concentrate to a smaller volume. However, once the precipitate had been obtained, it could not be redissolved near pH 5·7.

The soluble material from the hydrolysate was freeze-dried before being subjected to fractionation by ionophoresis.

Hydrolysis with a crude pancreatic extract (*Expt. I*x)

The preparation used was a crude 'trypsin' (British Drug Houses, no. 557). It apparently contained several other enzymes, since several bonds which were outside the specificity range of trypsin or chymotrypsin were extensively hydrolysed.

Insulin (300 mg.) was dissolved in 18 ml. 0·01 N-HCl, 1 mg. of NEMI was added and the pH was adjusted to 7·8 by the addition of 0·1 N-NH_3. 1·2 mg. of the 'trypsin' preparation was then added in solution in a little water, and the pH was re-adjusted to 7·8. After 4 hr. incubation at 37° the pH had fallen considerably and was brought back to 7·8. After a further 17 hr. incubation the pH was 6·8 and the solution was very cloudy. The digest was then brought to pH 6 by the addition of 0·2 N acetic acid and centrifuged to remove the precipitated material. On concentrating the supernatant solution *in vacuo* a further small precipitate appeared, but this was ignored, and the whole of the concentrate was applied across a 25 cm. wide strip of Whatman no. 3 paper and was subjected to ionophoresis at pH 6·5 (in buffer containing approx. 10^{-3} M NEMI) at 1500 v for 12 hr.

Acid hydrolyses

A number of different experiments were carried out using acid with various conditions of hydrolysis and fractionation. Two such experiments will be described which incorporate the various techniques used and results obtained. In the first (*I*1) 10 N-H_2SO_4 at 100° was used and the peptides were separated into two groups (*I*1α and *I*1β) by ion-exchange chromatography and were then fractionated by a number of different ionophoreses. In the second (*I*2) the hydrolysis was carried out at 37° and the neutral cystine peptides were separated by two-dimensional ionophoresis.

*Experiment I*1. The 'chymotryptic core' (100 mg.) was dissolved in a mixture of 5 ml. 20 N-H_2SO_4, 3 ml. glacial acetic acid and 2 ml. water. Thioglycollic acid (3 μl.) was added and the mixture heated on a boiling-water bath for 45 min.

Removal of the H_2SO_4 with Ba caused extensive losses of the cystine peptides and was much more conveniently carried out using the basic ion-exchange resin Amberlite IR-4B, 44–100 mesh/in. (manufactured by The Rohm and Haas Co., Philadelphia, U.S.A.) in the acetate form. A column was prepared 8 cm. high in a tube of radius 2 cm. having a tap at the bottom. The resin was equilibrated against 20% (v/v) aqueous acetic acid and the hydrolysate poured on the column. To prevent uneven flow the top half of the column was stirred with a long glass rod so that most of the H_2SO_4 was initially adsorbed at the top of the column. The column was then allowed to flow and was developed with 20% acetic acid. The effluent was collected till a drop on a filter paper no longer gave the ninhydrin reaction.

75–100 ml. was usually sufficient. In some earlier experiments 5% acetic acid was used for elution, but the cystine peptides were considerably retarded and a greater volume of effluent was required. The solution was concentrated to 10–20 ml. *in vacuo* in the rotary evaporator of Craig, Gregory & Hausmann (1950), using a solid CO_2–methyl cellosolve mixture to cool the condensing bulb. The thioglycollic acid and remaining acetic acid were extracted with ether and a small amount of NEMI (2–3 mg.) was added.

The solution was brought to pH 5 with aqueous NH_3 and put on a column of Amberlite IR-4B in the acetate form, which had been washed with distilled water until the effluent attained pH 3–4. The column was developed with water. Part of the hydrolysate was not retained by the column and was collected and freeze-dried ($I1\beta$). The peptides retained on the column were removed by eluting with 20% (v/v) aqueous acetic acid, the solution concentrated, extracted with ether and freeze-dried ($I1\alpha$).

Experiment I2. Insulin (100 mg.) was dissolved in a mixture of 7 ml. 20N-H_2SO_4 and 7 ml. acetic acid containing 6 μl. thioglycollic acid and kept at 37° for 17 days. H_2SO_4 was removed on Amberlite IR-4B as described in Expt. *I*1 and the material freeze-dried in the presence of NEMI.

Two-dimensional ionophoresis. Hydrolysate from Expt. *I*2 (50 mg.) was dissolved in 0·15 ml. water and applied on a line near to the anode end of a 18 cm. wide strip of no. 3 filter paper for ionophoresis at pH 3·6. A potential of 1500 v was applied for 5·5 hr. After drying the paper, a strip was cut from one edge and the position of the cystine peptides located by the CN–nitroprusside test. The appearance of this test strip is shown in Fig. 2. The most slowly

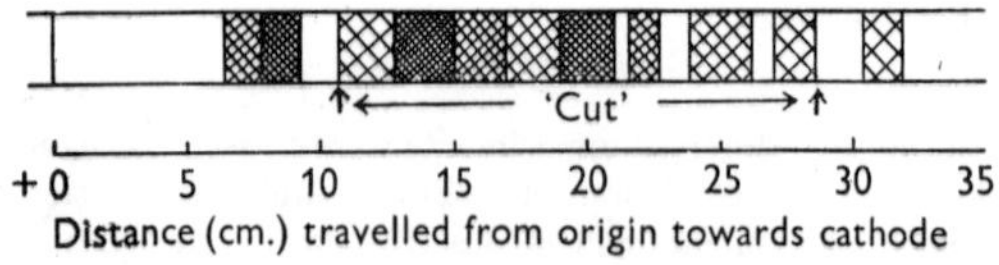

Fig. 2. Ionophoresis of partial hydrolysate of insulin (Expt. *I*2): 50 mg./18 cm. wide no. 3 paper; pH 3·6; 29 v/cm.; 5·5 hr. In this and subsequent figures representing ionophoretic separations the different types of shading indicate the colour reaction used to locate the bands as follows: ▨, ninhydrin reaction; ▩, CN–nitroprusside reaction; ▤, tyrosine reaction. The depth of shading represents the approximate strength of the colour reaction.

migrating material was largely free cystine, so was not fractionated further. The part of the paper between 10·6 and 28·6 cm. from the origin was then cut out and material eluted from it on to a second no. 3 filter paper as shown in Fig. 3. In this method one edge of the 'cut' was put in the trough of 10% acetic acid, which washed the peptide bands down near the 'front'. In order to keep this front as even as possible it was sometimes necessary to pipette drops of acetic acid on to the paper behind it. The other edge of the 'cut' was clamped down on the second paper with a glass plate so that only the extreme edge was in contact. When the acetic acid front had flowed into the second paper to form a band 1–2 cm. wide, the elution was stopped, the band dried off and then subjected to ionophoresis at pH 6·5 for 10 hr. using 1500 v. A small amount of NEMI was added to the buffer just before wetting the paper, to inhibit interchange reactions.

At the end of the ionophoresis the paper was hung up in air for about 10 min. to allow the surface toluene to drop off and evaporate. It was then laid on a flat sheet of glass. A dry sheet of no. 1 paper was laid on top of it and rapidly pressed down with another glass plate to make even contact and thus take a 'print'. Both papers were then dried. The no. 1 paper was treated with the CN–nitroprusside reagent, and from the positions of the spots, the cystine peptides could be located on the no. 3 paper, cut out and eluted.

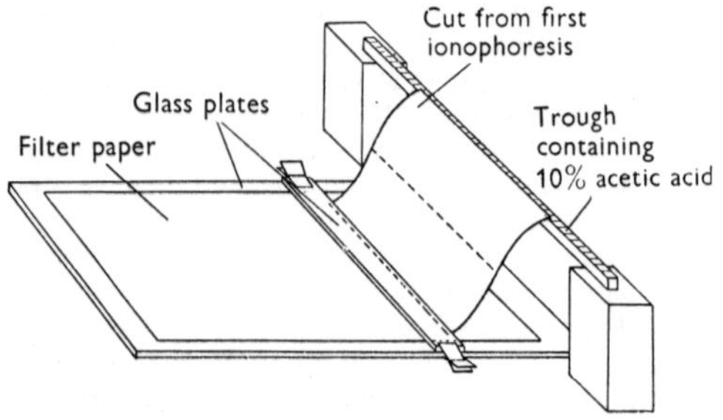

Fig. 3. Elution of peptide material during two-dimensional ionophoresis.

Structure of cystine peptides

To determine the structure of the cystine peptides they were oxidized with performic acid to the corresponding peptides of cysteic acid, which were then fractionated and subjected to complete hydrolysis. Previously the fractionation had been carried out by paper ionophoresis in acetic acid by which means all the cysteic acid peptides, except those containing a basic amino acid, could be completely separated from other peptides (Sanger & Thompson, 1953*a*). More satisfactory results have now been obtained by carrying out the ionophoresis in pyridine–acetate buffers at pH 3·6. Here some fractionation of the individual cysteic acid peptides is obtained, which simplifies their identification.

The areas occupied by cystine peptides were cut out of the papers and eluted with 10% acetic acid. The eluates were put in small test tubes and taken to dryness in a desiccator. A few drops of a solution of performic acid prepared by mixing 1 vol. 33% (w/w) H_2O_2 with 9 vol. formic acid, were added. Oxidation was allowed to proceed for 15–30 min., a few drops of water were added and the solutions taken to dryness in a desiccator. Each fraction was then transferred to a polythene strip with a small volume of water and taken to dryness again to ensure removal of formic acid.

The residues were dissolved in about 10 μl. water and small portions (each about 2 μl.) were transferred with a capillary tube to a sheet of no. 4 paper near to the cathode end and subjected to ionophoresis at pH 3·6 for 2 hr. at 1500 v. The papers were dried and developed with ninhydrin.

From this preliminary ionophoresis some information could be obtained about the cysteic acid peptides present and it could be decided which fractions should be investigated further and at what concentration to carry out the ionophoresis. Thus, for instance, peptides containing two cysteic acid residues could readily be identified (see Figs. 14, 21) and since such peptides could not provide the required information, they were usually not investigated further. On the other hand, fractions giving two peptides each containing one residue of cysteic acid were of particular interest.

For further investigation of the cysteic acid peptides, the whole fractions were put on paper as a line and subjected to ionophoresis as above. A strip was cut from the edge and treated with ninhydrin to locate the position of the peptides, which were then cut out, eluted and hydrolysed. The resulting amino acids were identified by paper chromatography using phenol–0·3 % aqueous NH_3.

RESULTS

Chymotryptic hydrolysate (*Expt. Ic*)

Fig. 4 shows the distribution of peptide bands after ionophoresis of the soluble fraction (30 mg.) from the chymotryptic hydrolysate.

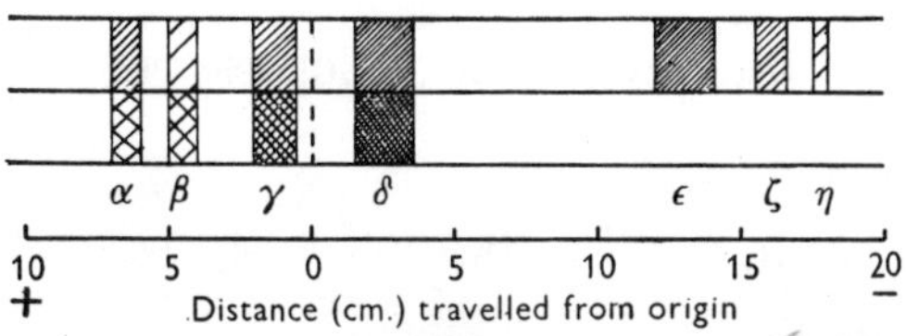

Fig. 4. Ionophoresis of peptides from soluble fraction of chymotryptic hydrolysate (Expt. *Ic*): 30 mg. on 18 cm. wide no. 3 paper; pH 6·5; 33 v/cm.; 2 hr.

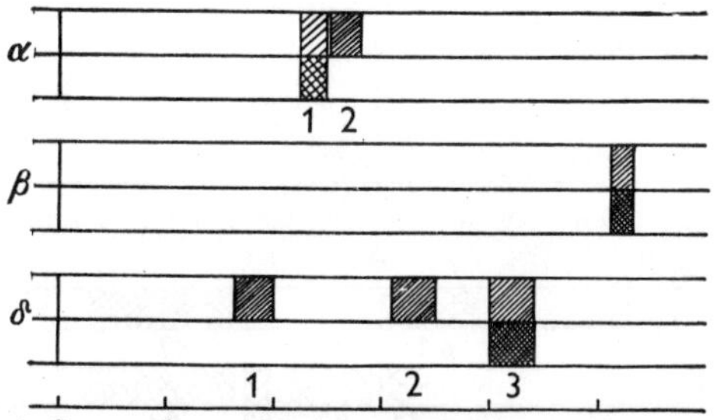

Fig. 5. Ionophoresis of fractions *Icα*, *Icβ* and *Icδ* (Fig. 4): no. 3 paper; pH 3·6; 29 v/cm.; 7 hr.

Band ϵ, which contained no cystine, was refractionated on a two-dimensional paper chromatogram using phenol–0·3 % NH_3/butanol–acetic acid. One main spot was present which proved to be Tyr.Thr.Pro.Lys.Ala (Table 2). Band ζ was similarly shown to be Thr.Pro.Lys.Ala and band η Pro.Lys.Ala. This last peptide had not been obtained from the chymotryptic treatment of fraction *B* (Sanger & Tuppy, 1951*b*). Its presence is surprising since it would not be expected from the known specificity of chymotrypsin.

Bands α, β and δ were further purified by ionophoresis at pH 3·6 (Fig. 5). Both δ1 and δ2 gave phenylalanine and tyrosine on hydrolysis. Band δ2 probably contains the free amino acids, since they move at the same rate on ionophoresis at pH 3·6. Band δ1 is probably a dipeptide Phe.Tyr, which would be expected to travel slower than the free amino acids. This dipeptide was not detected in the chymotryptic hydrolysate of fraction *B*, but might be expected to occur from the known specificity of chymotrypsin.

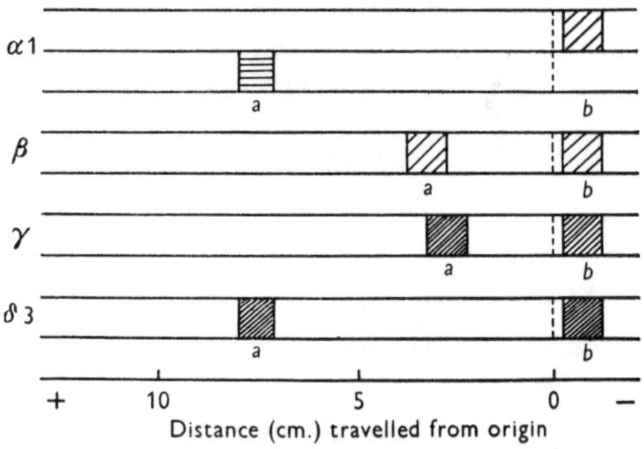

Fig. 6. Ionophoresis of oxidation products of bands α1, β, γ and δ3 (Figs. 4, 5): no. 4 paper; pH 3·6; 29 v/cm.; 2 hr. (see Table 2).

The four cystine peptides (α1, β, γ, and δ3) were oxidized and the resulting cysteic acid peptides fractionated by ionophoresis at pH 3·6 (Fig. 6). The results are summarized in Table 2. All four gave the same peptide (*b*, Fig. 6) which was neutral at pH 3·6 and was characterized by the presence of arginine and a large amount of glycine. Its structure can only be Leu.Val.$CySO_3H$.Gly.Glu.Arg.Gly.Phe.-Phe, identical with peptide *Bc*4.

The three acidic peptides α1a, βa and γa gave the same amino acids on hydrolysis but moved at different rates at pH 3·6. Band γa was by far the strongest and is probably $GluNH_2$.Leu.Glu.Asp-NH_2.Tyr.$CySO_3H$.$AspNH_2$ (*Ac*3) which would be expected to be a major component. Band α1a gave no ninhydrin reaction, suggesting that it was a pyrrolidonoyl derivative identical with γa except that the *N*-terminal glutamine residue had cyclized. This would account for α1 moving faster towards the anode than the other acidic cystine peptides as it has one less amino group. Band βa probably differs from γa only in having one amide group less and therefore being slightly more acidic.

On oxidation the only neutral cystine peptide (δ3) gave a neutral peptide (δ3b) identical with α1b, βb and γb, and an acidic peptide δ3a having the structure $CySO_3H$.$AspNH_2$ (*Ac*1).

Table 2. *Peptides from chymotryptic hydrolysate of insulin* (*Expt. I*c)

Band (Figs. 4, 5, 6)	Amino acids of oxidized peptides	Probable structure
α1	a: $CySO_3H$, Asp, Glu, Tyr, Leu b: $CySO_3H$, Glu, Gly, Val, Leu, Phe, Arg	NH_2 (on Asp), NH_2 (on Asp) Pyr*. Leu. Glu. Asp. Tyr. Cy. Asp S S Leu. Val. Cy. Gly. Glu. Arg. Gly. Phe. Phe
α2	Asp, Glu, Leu, Tyr (weak Val)	NH_2 (on Glu), NH_2 (on Asp) Glu. Leu. Glu. Asp. Tyr (*A*c5)
β	a: $CySO_3H$, Asp, Glu, Tyr, Leu b: $CySO_3H$, Glu, Gly, Val, Leu, Phe, Arg	See text
γ	a: $CySO_3H$, Asp, Glu, Tyr, Leu b: $CySO_3H$, Glu, Gly, Val, Leu, Phe, Arg	NH_2 (on Glu), NH_2 (on Asp), NH_2 (on Asp) Glu. Leu. Glu. Asp. Tyr. Cy. Asp S S Leu. Val. Cy. Gly. Glu. Arg. Gly. Phe. Phe
δ1	Tyr, Phe	Phe. Tyr
δ2	Tyr, Phe	Phe + Tyr
δ3	a: $CySO_3H$, Asp b: $CySO_3H$, Glu, Gly, Val, Leu, Phe, Arg	NH_2 (on Asp) Cy. Asp S S Leu. Val. Cy. Gly. Glu. Arg. Gly. Phe. Phe
ε	Thr, Ala, Tyr, Pro, Lys	Tyr. Thr. Pro. Lys. Ala (*B*c7)
ζ	Thr, Ala, Pro, Lys	Thr. Pro. Lys. Ala (*B*c5)
η	Ala, Pro, Lys	Pro. Lys. Ala

* Pyrrolidonoyl.

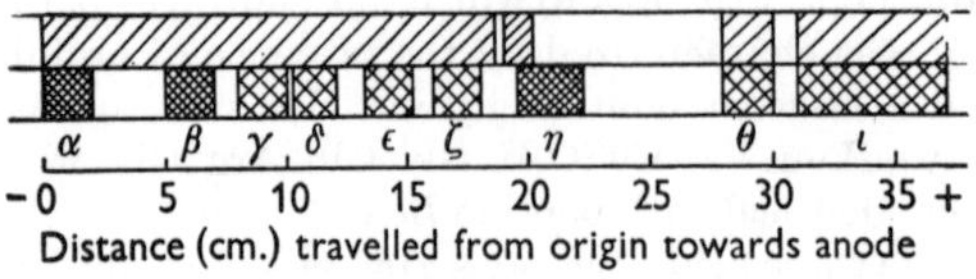

Fig. 7. Ionophoresis of crude 'tryptic' hydrolysate of insulin (Expt. *I*x). Soluble material from hydrolysate of 300 mg. insulin on 25 cm. wide no. 3 paper; pH 6·5; 33 v/cm.; 12 hr.

Hydrolysis with crude pancreatic extract (*Expt. I*x)

Fig. 7 shows the positions of the bands obtained by ionophoresis at pH 6·5 of the crude 'trypsin' hydrolysate. Band η, which had moved ahead of most of the ninhydrin-reactive material, was thought to be fairly pure and was oxidized directly. When part of the material from bands α–ζ and θ and ι was oxidized several cysteic acid peptides were obtained from most of the bands, indicating that they contained more than one cystine peptide.

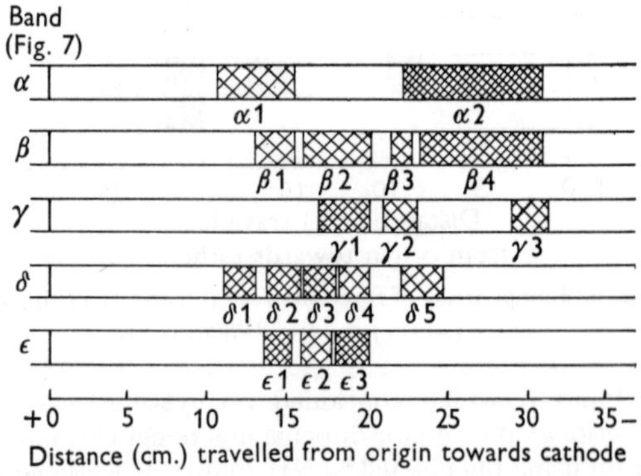

Fig. 8. Purification of bands α–ε from Expt. *I*x (Fig. 7) by ionophoresis: pH 3·6; 29 v/cm.; 6 hr.

The material from band ζ behaved in the same way as that from band η, and bands θ and ι were considered too weak to give any results after further manipulations, so these three bands were not further dealt with.

Table 3. *Cysteic acid peptides obtained from crude 'tryptic' hydrolysate of insulin* (*Expt. I*x)

Band (Figs. 7–9)	Amino acids present in cysteic acid peptide	Assumed structure of cysteic acid peptide
β4a	$CySO_3H$, Glu, Gly, Val, Leu	Leu.Val.$CySO_3H$.Gly.Glu
β4b	$CySO_3H$, Asp	$CySO_3H$.$AspNH_2$
γ2a	$CySO_3H$, Gly, Val, Leu	Leu.Val.$CySO_3H$.Gly
γ2b	$CySO_3H$, Asp	$CySO_3H$.$AspNH_2$
δ1a	$CySO_3H$, Ser, Ala, Val	Same as ηb
δ1b	$CySO_3H$	Cysteic acid
δ2a	$CySO_3H$, Ser, Gly	?
δ2b	$CySO_3H$, Ser, Gly, Leu	Leu.$CySO_3H$.Gly.Ser
δ2c	$CySO_3H$, Ser, Ala, Val	Same as ηb
δ2d	$CySO_3H$, Ser, Gly	?
ηa	$CySO_3H$, Ser, Gly	$CySO_3H$.Gly.Ser
ηb	$CySO_3H$, Ser, Ala, Val	$CySO_3H$.$CySO_3H$.Ala.Ser.Val.$CySO_3H$
ηc	$CySO_3H$, Ser, Gly, Ala	?

The remainder of the material from bands α–ε was purified by ionophoresis at pH 3·6. The separated cystine peptides, detected by the CN–nitroprusside reagent (Fig. 8) were eluted, oxidized and run again at pH 3·6 for the separation of the cysteic acid peptides whose positions are shown in Fig. 9.

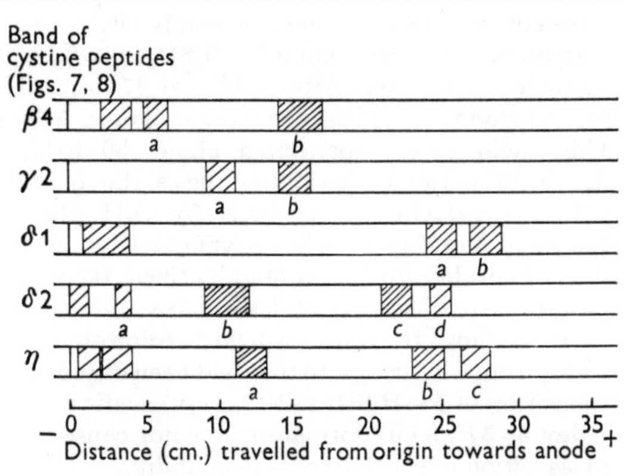

Fig. 9. Ionophoresis of cysteic acid peptides from Expt. *I*x (Figs. 7, 8): pH 3·6; 29 v/cm., approx. 2 hr. Fractions β4, γ2 and δ1 were run for slightly longer than the other fractions (see Table 3).

The cysteic acid peptides strong enough to give results were eluted and hydrolysed, except in those cases where one of a pair of bands could be identified as cysteic acid by its speed on ionophoresis. The amino acids found in the hydrolysates and the probable identification of the cysteic acid peptides are shown in Table 3. Certain peptides (e.g. β2 and γ1) gave results indicating that they were probably complex mixtures and that no definite conclusions could be drawn from them. The results with these peptides are not recorded.

Peptides β4b and γ2b were identified as $CySO_3H$.$AspNH_2$ (identical with *A*c1). γ2a can only be Leu.Val.$CySO_3H$.Gly (positions *B*17–20), since these four amino acids do not occur together elsewhere in the molecule, so that peptide γ2 must be

Leu.Val.Cy.Gly
S
S
Cy.$AspNH_2$.

By analogy it is concluded that β4 is

Leu.Val.Cy.Gly.Glu
S
S
Cy.$AspNH_2$,

though from its amino acid composition β4a could be Gly.Ileu.Val.Glu.Glu.$CySO_3H$ (positions *A* 1–6). Bands δ1a, δ2c and ηb all contain the same characteristic peptide which is present in large amounts in this hydrolysate. Judging from its ionophoretic mobility it is very acidic and has a high content of cysteic acid. In order to determine its structure it was subjected to partial hydrolysis.

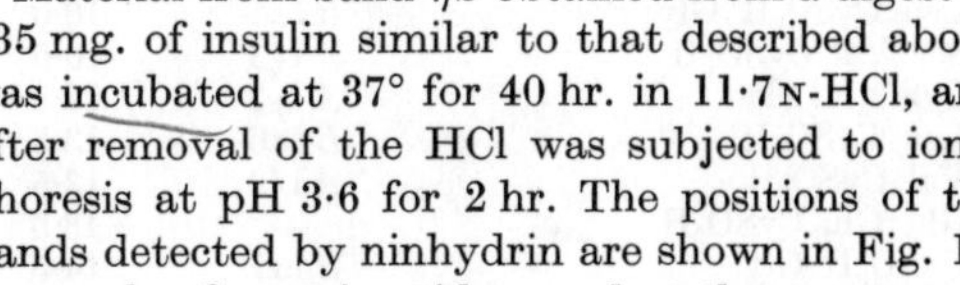

Fig. 10. Ionophoresis of partial hydrolysate of peptide *I*xηb: pH 3·6; 29 v/cm.; 2 hr.

Material from band ηb obtained from a digest of 135 mg. of insulin similar to that described above was incubated at 37° for 40 hr. in 11·7 N-HCl, and after removal of the HCl was subjected to ionophoresis at pH 3·6 for 2 hr. The positions of the bands detected by ninhydrin are shown in Fig. 10. A sample of cysteic acid moved at the same speed as band 6. The identities of the components of the various bands are given in Table 4. These show that ηb has the structure $CySO_3H$.$CySO_3H$.Ala.Ser.Val.$CySO_3H$ (from positions *A* 6–11). It is probable that it does not contain the serine residue from position *A* 12, since such a peptide would be expected to give large amounts of free serine on acid hydrolysis, whereas only small amounts were obtained from peptide ηb.

Table 4. *Peptides obtained from partial hydrolysate of peptide Ixηb*

Band (Fig. 10)	Structure
1	Alanine
2	Serine
3	Ser.Val.$CySO_3H$ (*A* 1ζ3)
4	$CySO_3H$.Ala (*A* 1γ3)
5	$CySO_3H$.$CySO_3H$.Ala (*A* 2δ1)
6	Cysteic acid
7	$CySO_3H$.$CySO_3H$

The other main cysteic acid peptide from the cystine peptide η is ηa([$CySO_3H$, Ser, Gly]). These three residues only occur together in positions *B* 7–9 so that it must be $CySO_3H$.Gly.Ser. How ηb and ηa are linked together is not clear from these results since ηb contains three cysteic acid residues.

The two main oxidation products of peptide δ2 are $CySO_3H$.$CySO_3H$.Ala.Ser.Val.$CySO_3H$ (δ2c) and δ2b which must be Leu.$CySO_3H$.Gly.Ser. Peptides δ2a, δ2d, and ηc were present in only small amounts and their composition is not certain. Their ionophoretic mobilities make it unlikely that δ2a and δ2b are $CySO_3H$.Gly.Ser (ηa).

Acid hydrolysis

In choosing conditions for acid hydrolysis a number of factors had to be considered. In Table 5 are listed the various peptides of cysteic acid encountered after oxidation of a partial hydrolysate of the 'chymotryptic core'. From these it is possible to calculate that ninety-five different cystine peptides would be expected before oxidation. Of these, twelve contain a half-cystine residue (CyS) on one side of the —S.S— bond, and provide no information about the distribution of —S.S— bonds. Seventy-two contain the sequence Cy.Cy intact and are also of no use, whereas eleven contain a peptide sequence on both sides of the —S.S— bond and are of the type required in the present study. In order to avoid the very complicated mixture that would be produced if there was much of the Cy.Cy sequence left intact it was necessary to hydrolyse for as long as possible, but if the hydrolysis were carried out for too long the peptides containing a CyS residue would predominate and also there would be more danger of the interchange reaction occurring.

Table 5. *Cysteic acid peptides from oxidized partial hydrolysate of 'chymotryptic core' of insulin*

Peptide	Reference no. in previous papers	Distance moved towards anode on ionophoresis at pH 3·6 (Fig. 14) (cm.)
His.Leu.$CySO_3H$	*B* 1γ4	−1·0
His.Leu.$CySO_3H$.Gly	*B* 4β1	−1·0
Gly.Ileu.Val.Glu.Glu.$CySO_3H$	*A* 2α7	9·0
Leu.$CySO_3H$.Gly	*B* 1α5	10·0
Ser.Val.$CySO_3H$	*A* 1ζ3	10·5
$CySO_3H$.Ala	*A* 1γ3	12·5
Leu.$CySO_3H$	*B* 1α6	13·5
Val.$CySO_3H$	*A* 1ζ2	13·5
Glu.Glu.$CySO_3H$	—	13·5
$CySO_3H$.Gly	*B* 1α1	14·5
Glu.$CySO_3H$	*A* 1α1	15·5
Glu.$CySO_3H$.$CySO_3H$.Ala	*A* 2γ1	19·5
$CySO_3H$.$CySO_3H$.Ala	*A* 2γ1	22·5
Glu.$CySO_3H$.$CySO_3H$	—	24·0
Cysteic acid	—	28·0
$CySO_3H$.$CySO_3H$	—	30·0

Using model systems it was shown that less interchange occurred in H_2SO_4 than in HCl of corresponding concentration (Ryle & Sanger, 1955). Although at high concentrations H_2SO_4 is less effective as a hydrolytic agent, it appeared that it would nevertheless be an advantage to use it. Insulin is insoluble in aqueous H_2SO_4 so that it was necessary to add acetic acid to dissolve it.

In an initial experiment 3·7 N-H_2SO_4 in 25 % (v/v) acetic acid was used. After 2 days at 37° a precipitate started to form and was separated after 5 days. This 'acid core' represented about 30–40 % of the insulin and on oxidation gave the peptides Gly.Ileu.Val.Glu.Glu.$CySO_3H$.$CySO_3H$.Ala and Phe.Val.Asp.Glu.His.Leu.$CySO_3H$.Gly, besides larger peptides which were not identified. It seems to be a mixture of large peptides which are less soluble than insulin. It could also be precipitated with water from a hydrolysate that had been obtained by the action of 5 N-H_2SO_4 in 50 % (v/v) acetic acid for 1 day at 37°, so its formation was not caused by a mass action effect due to its insolubility.

In later experiments 10 N-H_2SO_4 in 50 % (v/v) acetic acid at 37° was used for hydrolysis. The rate of hydrolysis of insulin in this medium is given in Table 6.

Table 6. *Rate of hydrolysis of insulin in 10 N sulphuric acid*

Amino N (Van Slyke nitrous acid method, 11 min. reaction) as % of value after hydrolysis for 48 hr. at 105° (see Peters & Van Slyke, 1932).

Reagent	Temp. (°)	Time	% hydrolysis
10 N-H_2SO_4 in 50 % (v/v) acetic acid	37	4 days	26
		10 days	37
		14 days	49
10 N-H_2SO_4 in 30 % (v/v) acetic acid	100	23 min.	40
		35 min.	66
		60 min.	80
		11 hr.	105

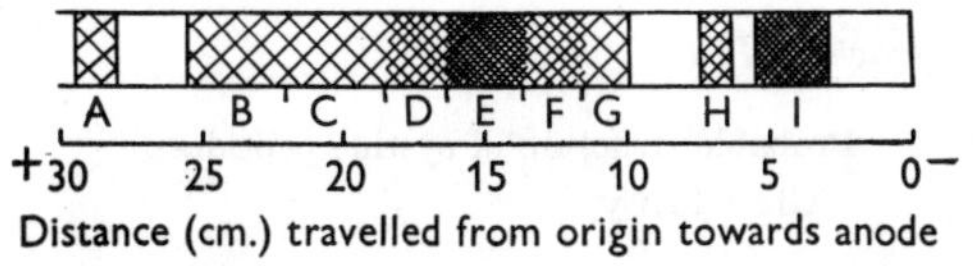

Fig. 11. Ionophoresis of fraction *I*1α: 90 mg. on 20 cm. wide no. 3 paper; pH 6·5; 33 v/cm.; 1·75 hr.

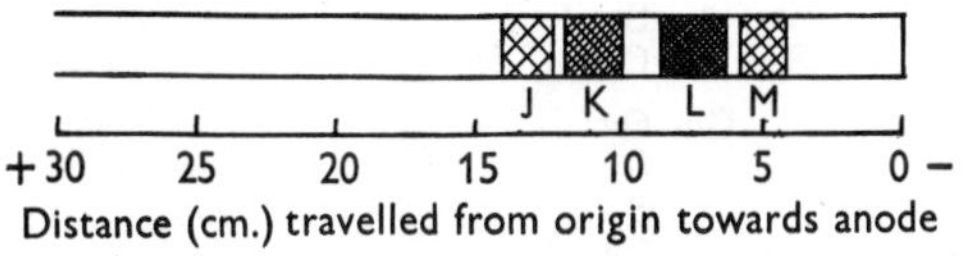

Fig. 12. Fractionation of band *I*1αI (Fig. 11) by ionophoresis: no. 3 paper; pH 6·5; 33 v/cm.; 12 hr.

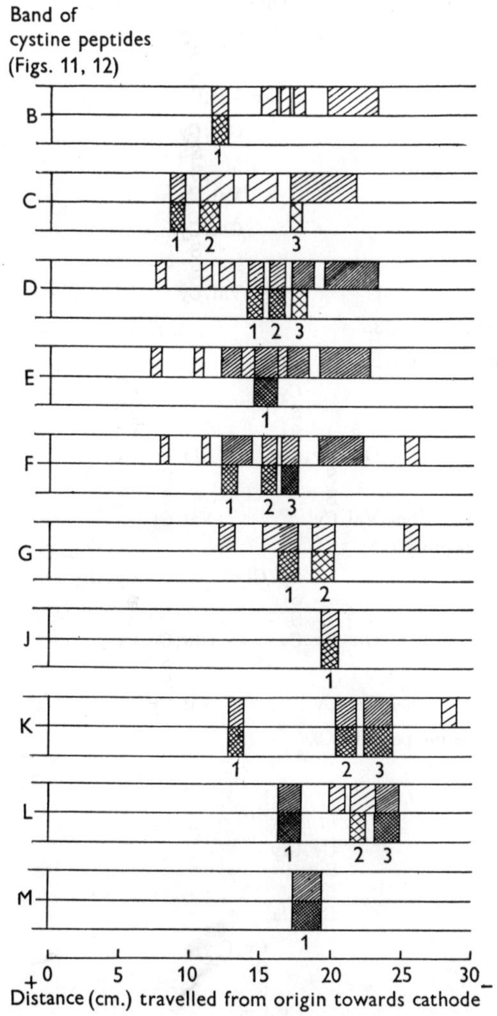

Fig. 13. Further ionophoretic fractionation of peptides from *I*1α (Figs. 11, 12): pH 3·6; 29 v/cm.; 6 hr.

To obtain sufficiently simple peptides at 37° it was necessary to continue the hydrolysis for at least 14 days, so in some experiments to save time a temperature of 100° was used. The corresponding rates of breakdown are also shown in Table 6. There was slightly more disulphide interchange at this higher temperature, but it did not significantly affect the results (see Ryle & Sanger, 1955).

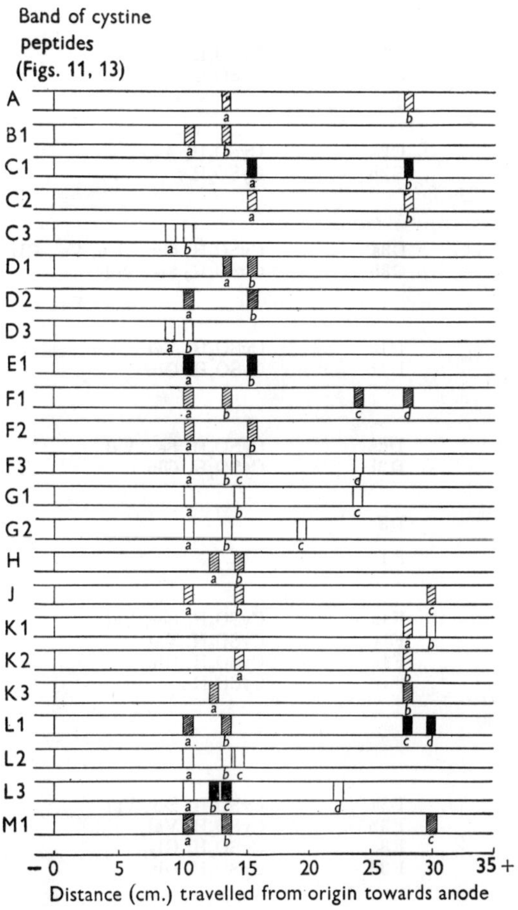

Fig. 14. Ionophoresis of cysteic acid peptides from *I*1α (Figs. 11–13): no. 4 paper; pH 3·6; 29 v/cm.; 2 hr. (see Table 7). All bands coloured with ninhydrin.

*Experiment I*1. Using Amberlite IR-4B at pH 3–4 the hydrolysate was separated into two fractions *I*1α and *I*1β. The former, which was adsorbed on the column, contained acidic and some neutral cystine peptides, whereas fraction *I*1β contained the remaining neutral and basic peptides. Most of the neutral non-cystine peptides were in fraction *I*1β as would be expected (Consden, Gordon & Martin, 1948).

Table 7. *Peptides from fraction I1α*

Band (Fig. 14)	Amino acids of oxidized peptides	Probable structure of cystine peptides
Aa Ab	$CySO_3H$, Glu $CySO_3H$	Glu.Glu.Cy S S Cy
B1a B1b	$CySO_3H$, Ser, Val $CySO_3H$, Glu	Glu.Glu.Cy S S Ser.Val.Cy
C1a C1b	$CySO_3H$, Glu $CySO_3H$	Glu.Cy S S Cy
C2a C2b	$CySO_3H$, Asp $CySO_3H$	Cy S S Cy.Asp
C3a C3b	$CySO_3H$, Glu, Ileu, Val, Gly $CySO_3H$, Ser, Val	Gly.Ileu.Val.Glu.Glu.Cy S S Ser.Val.Cy
D1a D1b	$CySO_3H$, Val $CySO_3H$, Glu	Glu.Cy S S Val.Cy
D2a D2b	$CySO_3H$, Ser, Val $CySO_3H$, Glu	Glu.Cy S S Ser.Val.Cy
D3		Same as peptide C3.
E1		Same as peptide D2.
F1a F1b F1c F1d	$CySO_3H$, Ser, Val $CySO_3H$, Val $CySO_3H$, Glu $CySO_3H$	Cy S S Glu.Cy.Cy S S Ser.Val.Cy + Cy S S Glu.Cy.Cy S S Val.Cy
F2		Same as peptide D2.
F3a F3b F3c F3d	$CySO_3H$, Ser, Val $CySO_3H$, Val $CySO_3H$, Gly $CySO_3H$, Glu	Cy.Gly S S Glu.Cy.Cy S S Ser.Val.Cy + Cy.Gly S S Glu.Cy.Cy S S Val.Cy
G1		Same as serine peptide F3.
G2a G2b G2c	$CySO_3H$, Ser, Val $CySO_3H$, Glu, Val $CySO_3H$, Glu, Ala	?
Ha Hb	$CySO_3H$, Ala $CySO_3H$, Gly	Cy.Gly S S Cy.Ala
Ja Jb Jc	$CySO_3H$, Ser, Val $CySO_3H$, Gly $CySO_3H$	Cy.Gly S S Cy.Cy S S Ser.Val.Cy

Table 7 (*cont.*)

Band (Fig. 14)	Amino acids of oxidized peptides	Probable structure of cystine peptides
K1a K1b	$CySO_3H$ $CySO_3H$	?
K2a K2b	$CySO_3H$, Gly $CySO_3H$	Cy.Gly S S Cy
K3a K3b	$CySO_3H$, Ala $CySO_3H$	Cy S S Cy.Ala
L1a L1b L1c L1d	$CySO_3H$, Ser, Val $CySO_3H$, Val $CySO_3H$ $CySO_3H$	Cy S S Cy.Cy S S Ser.Val.Cy + Cy S S Cy.Cy S S Val.Cy
L2a L2b L2c	$CySO_3H$, Ser, Val $CySO_3H$, Val, Leu $CySO_3H$, Gly	?
L3a L3b L3c L3d	$CySO_3H$, Ser, Val $CySO_3H$, Ala $CySO_3H$, Leu $CySO_3H$, Ala	Mainly Leu.Cy S S Cy.Ala (*I*2H)
M1a M1b M1c	$CySO_3H$, Ser, Val $CySO_3H$, Leu (weak Val) $CySO_3H$	Leu.Cy S S Cy.Cy S S Ser.Val.Cy + Leu.Cy S S Cy.Cy S S Val.Cy

Fig. 11 shows the ionophoretic separation of fraction *I*1α at pH 6·5. The main neutral band *I*1αI was eluted and refractionated at the same pH for a longer time (12 hr.) when it separated into a number of different fractions (Fig. 12). The acidic cystine peptides (*B–G*) were not adequately separated and were refractionated by ionophoresis at pH 3·6, as also were most of the neutral peptides (Fig. 13). Fig. 14 shows the ionophoretic separations at pH 3·6 of the oxidation products of the various cystine peptides. Not all the ionophoreses were in fact run for exactly 2 hr. but the diagram has been drawn as if they were, so that all samples of a particular peptide are found on the same vertical line. After ionophoresis the various peptides were identified by their amino acid composition and the results are given in Table 7.

The various peptides encountered in this work are listed in Table 5 together with the distance they move on ionophoresis at pH 3·6 under the conditions of Fig. 14. A number of these peptides had not been encountered in previous work with the oxidized chains, and were only revealed by the more efficient ionophoretic method used here. Thus Glu.Glu.$CySO_3H$ (*I*1αAa) was identified as a peptide containing a high proportion of glutamic acid with cysteic acid and moving considerably slower on ionophoresis than Glu.$CySO_3H$. Glu.$CySO_3H$.$CySO_3H$ (*I*1αF1c, *I*1αF3d) was a peptide moving at the characteristic rate of peptides having two residues of cysteic acid, and $CySO_3H$.$CySO_3H$ (*I*1αL1d, *I*1αM1c, etc.) was found as a band moving faster than free cysteic acid but giving only cysteic acid on hydrolysis.

From their ionophoretic rates and composition it was possible to identify the cysteic acid peptides, and the structures of the cystine peptides given in Table 7 were deduced from them. Where a peptide with two cysteic acid residues was produced on oxidation it was not possible to deduce the structure of the original cystine peptide from the experimental results and the formulae given were derived when the structure of insulin was known.

During the purification of the cystine peptides at pH 3·6 (Fig. 13), weak CN–nitroprusside-reacting bands were frequently observed, moving towards the anode or slowly towards the cathode. Their acidity suggested that they were cysteic acid peptides. As they also contained an intact —S.S— bond, they were probably derived from the peptides containing the Cy.Cy sequence by partial oxidation during the purification procedure. *I*1αK1 is probably such a peptide. The others were too faint to investigate.

Fig. 15 shows the ionophoresis of fraction *I*1β.

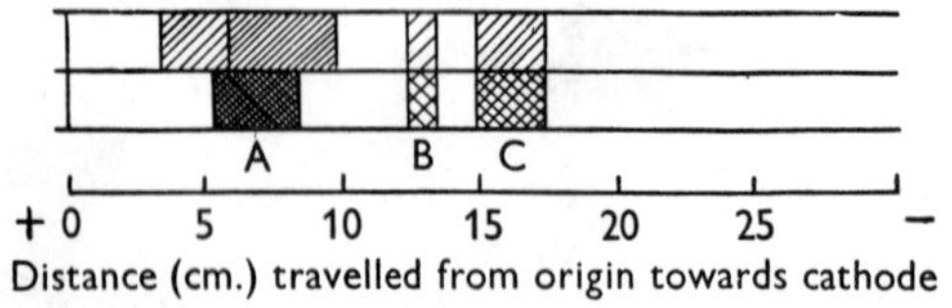

Fig. 15. Ionophoresis of fraction *I*1β: no. 3 paper; pH 6·5; 33 v/cm.; 6 hr.

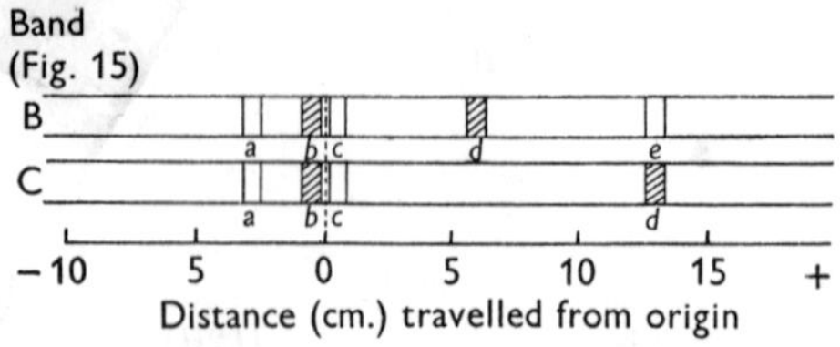

Fig. 16. Ionophoresis of oxidation products of fractions *I*1βB and *I*1βC (Fig. 15): no. 4 paper; pH 3·6; 29 v/cm.; 1 hr. (see Table 8).

The two basic peptides B and C were not present in sufficient concentration to refractionate, so were oxidized directly and subjected to ionophoresis at pH 3·6 (Fig. 16). The results are given in Table 8. It is probable that neither of these were pure but consisted of the peptides shown in Table 8.

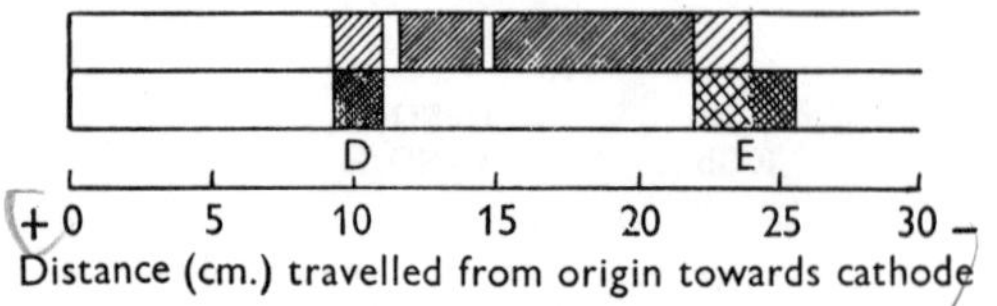

Fig. 17. Ionophoresis of fraction *I*1βA (Fig. 15): no. 3 paper; pH 3·6; 29 v/cm.; 7 hr.

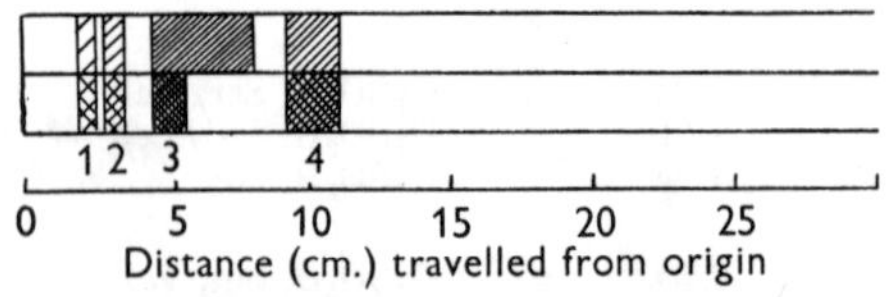

Fig. 18. One-dimensional chromatogram of fraction *I*1βE (Fig. 17): no. 3 paper; *n*-butanol–acetic acid–water (4:1:5, by vol.). Run for 36 hr. at 25° and solvent allowed to drip off the bottom of the paper.

Table 8. *Peptides from fraction I*1β

Band (Figs. 16, 19)	Amino acids of oxidized peptides	Structure of cystine peptide
Ba	No $CySO_3H$	His.Leu.Cy—S—S—Cy.Ala + His.Leu.Cy.Gly—S—S—Cy.Ala
Bb	$CySO_3H$, Gly, His, Leu	
Bc	No $CySO_3H$	
Bd	$CySO_3H$, Ala	
Be	$CySO_3H$	
Ca	No $CySO_3H$	His.Leu.Cy—S—S—Cy + His.Leu.Cy.Gly—S—S—Cy
Cb	$CySO_3H$, His, Leu, (weak Gly)	
Cc	No $CySO_3H$	
Cd	$CySO_3H$	
D	$CySO_3H$	Cy—S—S—Cy
E1a	$CySO_3H$, Glu, Ser, Gly (all very weak)	?
E1b	$CySO_3H$	
E2a*	$CySO_3H$, Ala	Cy—S—S—Ala.Cy
E2b	$CySO_3H$	
E3a	$CySO_3H$, Ser, Val	Cy—S—S—Ser.Val.Cy
E3b	$CySO_3H$	
E4a	$CySO_3H$, Leu	Leu.Cy—S—S—Cy
E4b	$CySO_3H$	

* Peptide E2a moved considerably faster on ionophoresis at pH 3·7 (Fig. 19) than $CySO_3H$.Ala (e.g. in peptide *I*1αK3a). It seems probable that it is Ala.$CySO_3H$ formed by inversion during hydrolysis. This reaction has in fact been shown to occur under similar conditions (Tuppy & Bodo, 1954).

The main neutral band $I1\beta$A was refractionated by ionophoresis at pH 3·6 (Fig. 17) to give two bands D and E. D was free cystine, whereas E was further fractionated by one-dimensional paper chromatography in butanol–acetic acid (Fig. 18). Fig. 19 shows the ionophoresis of the oxidized peptides and the results are listed in Table 8.

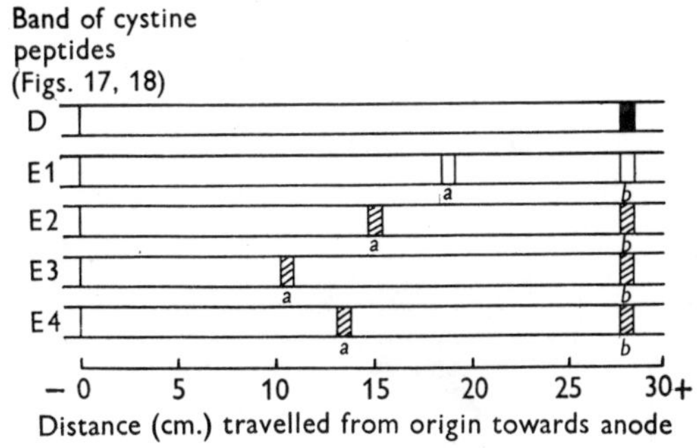

Fig. 19. Ionophoresis of oxidized peptides from $I1\beta$A (Figs. 15, 17, 18): no. 4 paper; pH 3·6; 29 v/cm.; 2 hr. (see Table 8). All bands coloured with ninhydrin.

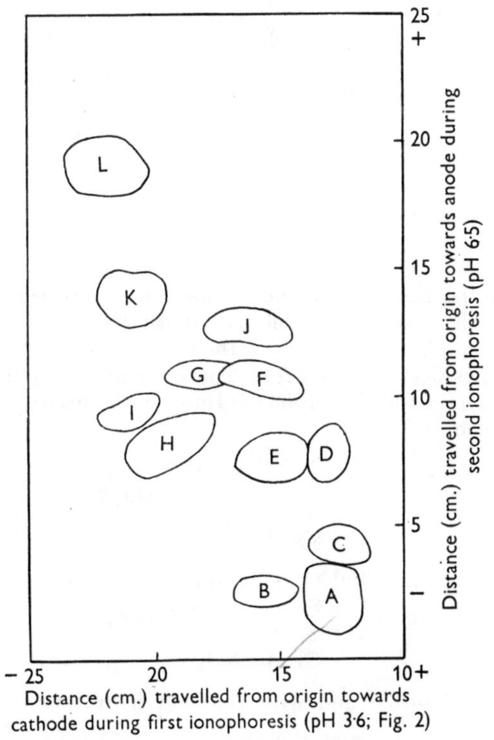

Fig. 20. Diagram of two-dimensional ionophoresis of partial acid hydrolysate of insulin (Expt. $I2$). First run as shown in Fig. 2. Second run: pH 6·5; 33 v/cm.; 10 hr.; spots located by CN–nitroprusside test.

Experiment I2. Fig. 20 shows the distribution of the cystine peptides in the two-dimensional ionophoresis experiment, and Fig. 21 is a diagram of the ionophoreses of their oxidation products at pH 3·6 showing the distribution of the cysteic acid peptides produced. Some of these were eluted, hydrolysed and their amino acid composition determined. The results and the deductions made are summarized in Table 9. The probable structure of most of the other peptides could be deduced from the position of the cysteic acid peptides on the pH 3·6 ionophoresis.

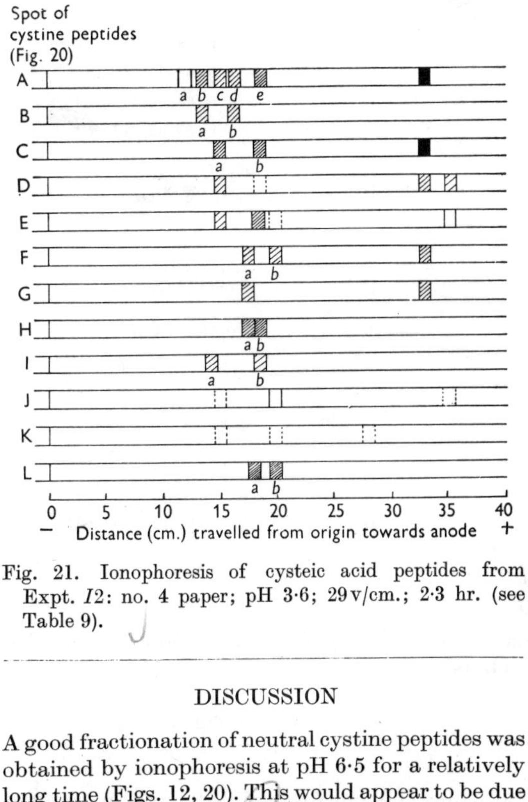

Fig. 21. Ionophoresis of cysteic acid peptides from Expt. $I2$: no. 4 paper; pH 3·6; 29 v/cm.; 2·3 hr. (see Table 9).

DISCUSSION

A good fractionation of neutral cystine peptides was obtained by ionophoresis at pH 6·5 for a relatively long time (Figs. 12, 20). This would appear to be due to the unusually low pK of the amino groups of cystine residues in peptides. Greenstein, Klemperer & Wyman (1939) found values of 6·36 and 6·95 for the pK' values of the amino groups of cystylbisglycine, whereas the corresponding values for bisglycylcystine were more normal (7·94 and 7·94). Thus peptides such as

Cy.Ala
S
S (Fig. 20, spot L)
Cy.Gly

are somewhat acidic at pH 6·5 owing to the amino groups being partially discharged and

Table 9. *Peptides from Expt. I2*

Band (Figs. 20, 21)	Structure of cysteic acid peptide	Structure of cystine peptide
Aa	Tyr.Leu.Val.CySO$_3$H (*B*2α21)	Mixture of peptides of type Cy S S X.Cy
Ab	Leu.Val.CySO$_3$H (*B*1α8)	
Ac	Ser.Val.CySO$_3$H	
Ad	Tyr.CySO$_3$H (*A*1a7)	
Ae	Leu.CySO$_3$H + Val.CySO$_3$H	
Ba	Leu.Val.CySO$_3$H (*B*1α8)	Leu.Val.Cy S S Tyr.Cy
Bb	Tyr.CySO$_3$H (*A*1a7)	
Ca	Ser.Val.CySO$_3$H	Peptide *I*1βE3 + Val.Cy S S Cy
Cb	Val.CySO$_3$H	
D	Same as band *I*1αL1	
E	Same as band *I*1αM1	
F	Same as peptide *I*1αK2 + *I*1αK3	
G	Same as peptide *I*1αK3	
Ha	CySO$_3$H.Ala	Cy.Ala S S Leu.Cy
Hb	Leu.CySO$_3$H	
Ia	Leu.CySO$_3$H.Gly	Cy.Ala S S Leu.Cy.Gly
Ib	CySO$_3$H.Ala	
J	Peptide *I*1αJ	
K	—*	Ser.Val.Cy S S Probably Cy.Cy S S Cy.Gly
L	Same as peptide *I*1αH	

* The cysteic acid peptides were not hydrolysed, but the structure was deduced from their ionophoretic rates.

move relatively rapidly on ionophoresis, peptides such as

Cy.Ala
S
S (Fig. 20, spot H)
Leu.Cy

containing only one free amino group of cystine are less acidic, whereas those like

Leu.Val.Cy
S
S (Fig. 20, spot B)
Tyr.Cy

are almost neutral. In this way the neutral cystine peptides could largely be separated from each other and also from other neutral peptides not containing cystine.

Two main cystine-containing peptides (*I*cδ3 and *I*cγ, Table 2) were present in the water-soluble fraction of the chymotryptic hydrolysate. Their oxidation products had been encountered in studies on the fractions *A* and *B*, and their structures, which are shown in Table 2, establish the presence of a disulphide bridge joining the half-cystine residues in positions *A* 20 and *B* 19. This is confirmed by the results with the crude 'tryptic' hydrolysate, in which

Leu.Val.Cy.Gly
S
S (*I*xγ2)
Cy.AspNH$_2$

and

Leu.Val.Cy.Gly.Glu
S
S (*I*xβ4)
Cy.AspNH$_2$

were detected and with the acid hydrolysate in which

Leu.Val.Cy
S
S (*I*2B)
Tyr.Cy

was found. In the chymotryptic hydrolysate, the remaining cystine residues are present in the insoluble 'core' and in the crude 'tryptic' hydrolysate they are present in peptides $Ix\eta$ and $Ix\delta 2$ in which CyS.Gly.Ser and Leu.CyS.Gly.Ser respectively are coupled with CyS.CyS.Ala.Ser.Val.-CyS. Neither enzyme preparation split between the three CyS residues of the glycyl chain.

A number of neutral peptides from the acid hydrolysates gave $CySO_3H$.Ala on oxidation. These were:

Cy.Gly
S
S (*I*1αH)
Cy.Ala

Leu.Cy
S
S (*I*2H)
Cy.Ala

and

Leu.Cy.Gly
S
S (*I*2I).
Cy.Ala

$CySO_3H$.Ala was found in the basic fraction *I*1βB linked with His.Leu.$CySO_3H$ and His.Leu.-$CySO_3H$.Gly. All these peptides establish that there is a disulphide bond between positions *A* 7 and *B* 7.

This only leaves the half-cystine residues in positions *A* 6 and *A* 11 to be considered and they must therefore be joined together. This is proved by the presence of relatively high concentrations of the acidic peptides

Ser.Val.Cy
S
S (*I*1αD2)
Glu.Cy

and

Val.Cy
S
S (*I*1αD1)
Glu.Cy

and confirmed by the peptides

Glu.Glu.Cy
S
S (*I*1αB1)
Ser.Val.Cy

and

Gly.Ileu.Val.Glu.Glu.Cy
S
S (*I*1αC3)
Ser.Val.Cy

A considerable number of peptides containing the Cy.Cy sequence were detected. Since their structures could all be explained on the basis of the above results, they added confirmation and the fact that no peptides were detected which could not be explained showed that no appreciable disulphide interchange had occurred during the experiments.

The results are summarized in Table 10, which gives the complete structure of insulin. This structure contains a ring, in which the disulphide bond in positions *A* 6–*A* 11 is included. Clearly to form such a bond the polypeptide chain must be folded in such a way that the two half-cystine residues are close together and this fact must be

Table 10. *The structure of insulin*

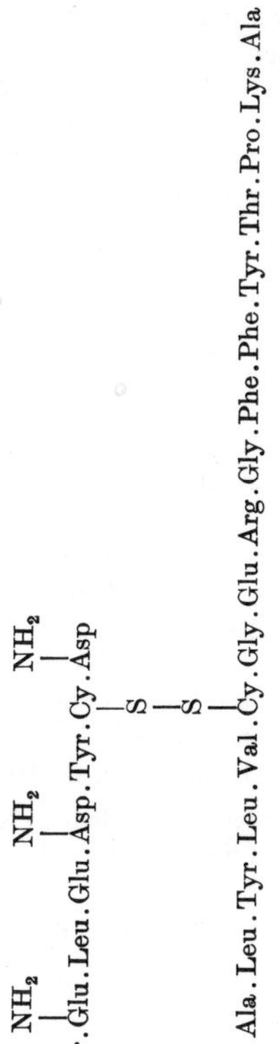

taken into account when considering possible configurations for the polypeptide chain in this area. It is interesting that the ring is the same size as the similar disulphide ring found in oxytocin (Tuppy & Michl, 1953; du Vigneaud, Ressler & Trippett, 1953) and vasopressin (Acher & Chauvet, 1953; du Vigneaud, Lawler & Popenoe, 1953), which suggests that it may have a possible structural or biological significance.

SUMMARY

1. Insulin was subjected to partial hydrolysis with chymotrypsin, with a crude pancreatic extract and with acid under conditions in which the disulphide bonds were stable.

2. Cystine-containing peptides in the hydrolysates were separated and their structure determined after oxidation to cysteic acid peptides. Paper ionophoresis at high potential gradients in pyridine-acetic acid buffers was found useful for the separations.

3. From the structure of the cystine peptides the distribution of the disulphide bonds of insulin was deduced and is shown in Table 10.

A.P.R. is indebted to the Medical Research Council for a Scholarship for training in research methods.

REFERENCES

Acher, R. & Chauvet, J. (1953). *Biochim. biophys. Acta*, **12**, 487.

Acher, R. & Crocker, C. (1952). *Biochim. biophys. Acta*, **9**, 704.

Butler, J. A. V., Dodds, E. C., Phillips, D. M. P. & Stephen, J. M. L. (1948). *Biochem. J.* **42**, 116.

Consden, R., Gordon, A. H. & Martin, A. J. P. (1948). *Biochem. J.* **42**, 443.

Craig, L. C., Gregory, J. D. & Hausmann, W. (1950). *Analyt. Chem.* **22**, 1462.

Du Vigneaud, V., Lawler, H. C. & Popenoe, E. A. (1953). *J. Amer. chem. Soc.* **75**, 4880.

Du Vigneaud, V., Ressler, C. & Trippett, S. (1953). *J. biol. Chem.* **205**, 949.

Fredericq, E. (1953). *Nature, Lond.*, **171**, 570.

Greenstein, J. P., Klemperer, F. W. & Wyman, J. (1939). *J. biol. Chem.* **129**, 681.

Harfenist, E. J. & Craig, L. C. (1952). *J. Amer. chem. Soc.* **74**, 3087.

Kupke, D. W. & Linderstrøm-Lang, K. (1954). *Biochim. biophys. Acta*, **13**, 153.

Michl, H. (1951). *Mh. Chem.* **82**, 489.

Peters, J. P. & Van Slyke, D. D. (1932). *Quantitative Clinical Chemistry*. Vol. 2, p. 385. London: Baillière, Tindall and Cox.

Ryle, A. P. & Sanger, F. (1955). *Biochem. J.* **60**, 535.

Sanger, F. (1949*a*). *Biochem. J.* **44**, 126.

Sanger, F. (1949*b*). *Biochem. J.* **45**, 563.

Sanger, F., Smith, L. F. & Kitai, R. (1954). *Biochem. J.* **58**, vi.

Sanger, F. & Thompson, E. O. P. (1953*a*). *Biochem. J.* **53**, 353.

Sanger, F. & Thompson, E. O. P. (1953*b*). *Biochem. J.* **53**, 366.

Sanger, F., Thompson, E. O. P. & Kitai, R. (1955). *Biochem. J.* **59**, 509.

Sanger, F. & Tuppy, H. (1951*a*). *Biochem. J.* **49**, 463.

Sanger, F. & Tuppy, H. (1951*b*). *Biochem. J.* **49**, 481.

Toennies, G. & Kolb, J. J. (1951). *Analyt. Chem.* **23**, 823.

Tuppy, H. & Bodo, G. (1954). *Mh. Chem.* **85**, 807.

Tuppy, H. & Michl, H. (1953). *Mh. Chem.* **84**, 1011.

(*Reprinted from Nature, Vol. 164, p. 529, September 24, 1949*)

Species Differences in Insulins

FROM a study of a number of crystalline insulins from different animal species, Scott and Fisher[1] found that they all had the same microscopic appearance, sulphur content and physiological activity; and Wasserman and Mirsky[2] could find no immunological differences using sensitization and complement fixation experiments. It has thus often been concluded that different insulins are chemically identical, although almost all other proteins have been found to show species differences. The following experiment shows, however, that there are certain definite differences in the detailed chemical structure of insulins from different animal species.

Source of insulin	Serine	Glycine	Threonine	Alanine
Ox	++	+	—	++
Pig	++	+	+	+
Sheep	+	++	—	++

Samples of insulins derived from the ox, pig and sheep were oxidized with performic acid and the *A* fractions prepared[3]. There were no differences in the course of fractionation and separation into the two main fractions, *A* and *B*, indicating the same general overall structure. The *A* fractions were hydrolysed and the amino-acids identified and estimated semi-quantitatively by paper chromatography. There were considerable differences in the contents of the amino-acids, serine, glycine, threonine and alanine, but not of the other amino-acids. The most significant of these differences, which are summarized in the accompanying table, is the presence of threonine in the fraction *A* of pig insulin and its absence from the others.

F. SANGER
(Beit Memorial Research Fellow)
Biochemical Laboratory,
Cambridge.
Aug. 11.

[1] Scott, D. A., and Fisher, A. M., *Trans. Roy. Soc. Canada*, **34**, (V) 137 (1940).

[2] Wasserman, P., and Mirsky, I. A., *Endocrin.*, **31**, 115 (1942).

[3] Sanger, F., *Biochem. J.*, **44**, 126 (1949).

Printed in Great Britain by Fisher, Knight & Co., Ltd., St. Albans.

[From THE BIOCHEMICAL JOURNAL, Vol. 60, No. 4, pp. 556–565, 1955]

The Structure of Pig and Sheep Insulins

By H. BROWN,* F. SANGER† and RUTH KITAI
Biochemical Laboratory, University of Cambridge

(*Received* 26 *November* 1954)

Insulins from a variety of different animal species show the same biological activity (Scott & Fisher, 1940) and immunological behaviour (Wasserman & Mirsky, 1942). They have the same crystalline form and mixtures of different insulins behave as a single substance in the phase-rule solubility test (Lens & Evertzen, 1952). However, in a preliminary chemical study of pig and sheep insulin it was shown that, whereas the general structure was similar, there were certain differences in individual amino acid residues (Sanger, 1949*b*). Harfenist & Craig (1952) have recently analysed the different insulins, and found differences in the contents of the amino acid residues given in Table 1. No differences were found for the other amino acids.

The complete sequence of amino acids in cattle insulin has recently been determined in this laboratory (Sanger & Tuppy, 1951*a*, *b*; Sanger & Thompson, 1953*a*, *b*), and the present paper describes similar studies on pig and sheep insulins. Since the larger part of the molecule was the same for all three species it was justifiable not to determine the amino acid sequence unequivocally in each case, but only to identify lower peptides embodying each residue in the molecule. Where

* Damon Runyon Cancer Research Fellow. Present address: Department of Surgery, University of Wisconsin, U.S.A.

† Member of the Scientific Staff of the Medical Research Council.

these were identical with the corresponding peptides from cattle insulin it could be assumed that they were derived from identical sequences. Any differences in the original insulins were readily evident from the peptides produced.

Table 1. *Analysis of different insulins* (*Harfenist & Craig*, 1952)

Results expressed as amino acid residues/molecule of insulin.

Amino acid	Cattle insulin (mol.wt. 5734)	Sheep insulin (mol.wt. 5704)	Pig insulin (mol.wt. 5778)
Serine	2·89	2·07	2·79
Threonine	0·97	0·96	1·77
Glycine	3·94	4·70	3·94
Alanine	2·91	2·99	2·17
Valine	4·68	4·80	3·68
Isoleucine	0·66	0·69	1·54

After splitting the insulins into the glycyl (fraction *A*) and phenylalanyl (fraction *B*) chains by oxidation (Sanger, 1949*a*) each chain was hydrolysed with enzymes. The resulting large peptides were fractionated by paper ionophoresis and subjected to complete hydrolysis. Some were also subjected to partial acid hydrolysis to give small peptides, which were separated on paper chromatograms. In this way it was possible to obtain peptides representing all of the residues in the insulin chains.

Peptides with reference numbers starting *A* and *B* refer to those obtained from cattle insulin and described by Sanger & Thompson (1953*a*, *b*) and Sanger & Tuppy (1951*a*, *b*).

MATERIALS

Various preparations of pig insulin have been used throughout this work:

(1) Crystalline pig insulin obtained from Dr J. Lens of N. V. Organon, Holland, as used by Lens & Evertzen (1952).

(2) Crystalline pig insulin, N. V. Philips-Roxane, Weesp, Holland.

(3) Novo pig insulin (Novo Terapeutisk A/S, Copenhagen), several-times recrystallized.

(4) Crystalline insulin (pig pancreas), batch no. E 178, Allen and Hanburys Ltd. and The British Drug Houses Ltd.

(5) A freeze-dried preparation of pig insulin obtained from Eli Lilly Company, Indianapolis, U.S.A.

In most of the work reported in this paper sample 2 was used, though some of the earlier preliminary experiments were carried out with sample 5. The other samples were used only in the experiment recorded in Fig. 5.

The sheep insulin used was crystalline insulin (sheep pancreas) batch E 73, Allen and Hanburys Ltd. and The British Drug Houses Ltd.

The cattle insulin was batch 9011 G (six-times recrystallized), Boots Pure Drug Company (Nottingham).

METHODS

Oxidation

Oxidation and fractionation of the oxidized insulin was carried out as described for cattle insulin (Sanger, 1949*a*). The fractions were found to behave similarly except that the fraction *A* of pig insulin was largely insoluble in 25% (w/v) ammonium acetate and was precipitated during the final stage of the purification. This precipitate, which contained no detectable phenylalanyl *N*-terminal residue, was used as fraction *A* in most of the following work. The fractions *A* and *B* from pig and sheep insulins are referred to below as P*A*, P*B*, and S*A*, S*B*, respectively.

Enzymic hydrolyses

The treatment of the fractions with proteolytic enzymes was carried out as described in previous papers (Sanger & Tuppy, 1951*b*; Sanger & Thompson, 1953*b*). The hydrolysates were fractionated by ionophoresis under toluene in the apparatus of Michl (1951) (see Ryle, Sanger, Smith & Kitai, 1955). Approx. 10 mg. of hydrolysate were fractionated on a 18 cm. wide sheet of Whatman no. 3 filter paper. After ionophoresis and drying, marker strips were cut out and tested with ninhydrin (Toennies & Kolb, 1951), the Sakaguchi reaction as modified by Jepson & Smith (1953) and by the following modification of the Pauly test.

Pauly test on filter paper. Equal volumes of 5% (w/v) $NaNO_2$ and of 1% (w/v) sulphanilic acid in 10% (w/v) HCl were mixed and applied to a clean strip of no. 3 filter paper on a glass plate, so that the paper was wet all over but no puddles were formed. The strip to be tested was then pressed on to the wet strip with another glass plate so that the reagent was evenly blotted on to it. It was removed and allowed to dry in air for 10 min. It was then treated in the same way with a solution of 20% (w/v) Na_2CO_3. Histidine peptides give a bright orange red, tyrosine peptides a brownish red. This test is more sensitive than the one used previously (Sanger & Tuppy, 1951*a*), although rather more tedious. The above 'blotting' technique has been found to be a convenient method of testing papers with water-soluble reagents.

Peptides studied. Peptides P*A*p5 and S*A*p5 were prepared by ionophoresis in 0·2 N acetic acid of pepsin hydrolysates of the corresponding oxidized insulin as described by Sanger, Thompson & Kitai (1955).

Partial acid hydrolysis of peptides from enzymic hydrolysates

Material eluted from an ionophoretic fractionation of 10–20 mg. of one of the oxidized fractions was subjected to partial hydrolysis in 12 N-HCl for 3 days at 37°. After removal of HCl the small peptides were fractionated by two-dimensional chromatography on a sheet of Whatman no. 4 filter paper using phenol–3% aq. NH_3 and *n*-butanol–acetic acid–water (4:1:5, by vol.) as solvents. The spots were located by spraying with 0·025% ninhydrin in *n*-butanol and their amino acid composition determined after complete hydrolysis in the usual manner (Sanger & Tuppy, 1951*a*).

RESULTS

N-*Terminal residues*

By the method of forming 2:4-dinitrophenyl (DNP) derivatives (Sanger, 1945) it was shown that both insulins contained glycine and phenylalanine as *N*-terminal residues.

The phenylalanyl chains (fractions P*B*, S*B*)

Peptic hydrolysates (P*B*p and S*B*p). Fig. 1 shows an ionophoretic separation of the peptic hydrolysates of the *B* fractions of cattle (*B*p), pig (P*B*p) and sheep (S*B*p) insulin. It can be seen that the hydrolysates appear indistinguishable, indicating at least that the distribution of charged groups and of groups attacked by pepsin are the same for all three fractions. In fact, their structures were found to be identical. For this reason only the results with fraction P*B* will be described, since fraction S*B* was investigated by essentially the same methods and gave the same results. Table 2 shows the amino acids given on hydrolysis of the various peptides from P*B*p and the structure of the corresponding peptides from cattle insulin, or, if the peptide had not previously been encountered, the presumed structure.

Peptides P*B*p7 and P*B*p10 were subjected to further partial hydrolysis with acid. The results are given in Tables 3 and 4. The R_F values on the chromatograms were the same as those for the corresponding peptides from cattle insulin, which are also listed in the Tables.

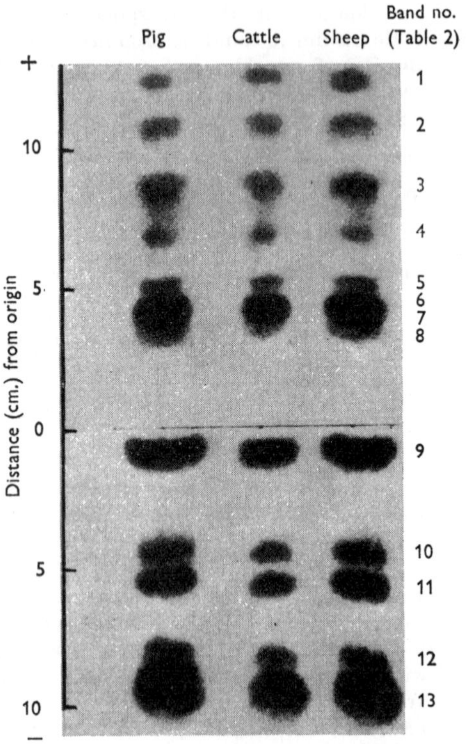

Fig. 1. Ionophoretic separations of peptic hydrolysates of *B* fractions of cattle (*B*p), pig (P*B*p) and sheep (S*B*p) insulins. Pyridine–acetate buffer (pH 6·5); 28 v/cm.; 3 hr. (cf. Ryle *et al.* 1955). Coloured with ninhydrin.

Table 2. *Peptides from peptic hydrolysate of fraction* P*B*

Peptide (Fig. 1)	Amino acids found on hydrolysis	Corresponding peptide from cattle insulin (Sanger & Tuppy, 1951*b*)
1	Glu, Val	Val.Glu (in *B*p1)
2	Glu, Ala, Val	Val.Glu.Ala (*B*p1)
3	Glu, Ala, Val, Leu	Val.Glu.Ala.Leu (*B*p2) (see footnote)
4	Glu, Ala, Val, Leu	See footnote
5	As for 6	See footnote
6	$CySO_3H$, Glu, Gly, Val, Leu, Phe, Arg	Leu.Val.$CySO_3H$.Gly.Glu.Arg.Gly.Phe (*B*p4)
7	$CySO_3H$, Glu, Gly, Tyr, Val, Leu, Phe, Arg	Tyr.Leu.Val.$CySO_3H$.Gly.Glu.Arg.Gly.Phe
8	As for 6	See footnote
9	All residues found in fraction B	Mixture of neutral peptides and amino acids
10	$CySO_3H$, Ser, Gly, Leu, His	His.Leu.$CySO_3H$.Gly.Ser.His.Leu (*B*p17)
11	$CySO_3H$, Asp, Glu, Ser, Gly, Val, Leu, Phe, His	Phe.Val.Asp.Glu.His.Leu.$CySO_3H$.Gly.Ser.His.Leu (*B*p3)
12	Thr, Ala, Tyr, Pro, Lys (Phe)	See footnote
13	Thr, Ala, Tyr, Pro, Lys	Tyr.Thr.Pro.Lys.Ala (*B*p15)

3. Partial acid hydrolysis on a small amount revealed the presence of [Ala, Leu] (P*B*p3a) corresponding to Ala.Leu (*B*1δ11).

4. This was present in only trace amounts, so analysis was not certain. It may contain tyrosine, in which case it could be Val.Glu.Ala.Leu.Tyr.

5 and 8. These were present in small amounts and probably differ from 6 and 7 by containing an extra residue on the *N*- or *C*-terminal position.

12. Present only in traces. In one experiment with P*B*p phenylalanine was detected, suggesting the structure Phe.Tyr.Thr.Pro.Lys.Ala, but there was not sufficient material to confirm this.

Table 3. *Peptides from partial acid hydrolysate of peptide* P*B*p7

Peptide	Amino acids found on hydrolysis	Corresponding peptide from cattle insulin
a	$CySO_3H$, Gly	$CySO_3H$.Gly (*B*1α1)
b	Glu, Gly	Gly.Glu (*B*1δ1)
c	$CySO_3H$, Glu, Gly, Val	Val.$CySO_3H$.Gly (*B*1α3) +free glutamic acid
d	$CySO_3H$, Val	Val.$CySO_3H$ (*B*1α4)
e	Glu, Arg	Glu.Arg (*B*1γ3)
f	Glu, Gly, Arg	[Glu, Arg, Gly] (*B*1γ2)
g	Gly, Arg	Arg.Gly (*B*2γ5)
h	$CySO_3H$, Val, Leu	Leu.Val.$CySO_3H$ (*B*1α8)
i	$CySO_3H$, Gly, Val, Leu	Leu.Val.$CySO_3H$.Gly (*B*1α7)
j	$CySO_3H$, Tyr, Val, Leu	Tyr.Leu.Val.$CySO_3H$ (*B*2α21)
k	Gly, Phe	Gly.Phe (*B*1γ12)

Table 4. *Peptides from partial acid hydrolysate of peptide* P*B*p10

Peptide	Amino acids found on hydrolysis	Corresponding peptide from cattle insulin
a	$CySO_3H$, Gly	$CySO_3H$.Gly (*B*1α1)
b	$CySO_3H$, Leu	Leu.$CySO_3H$ (*B*1α6)
c	$CySO_3H$, Leu, His	His.Leu.$CySO_3H$ (*B*1γ4)
d	Ser, His	Ser.His (*B*2γ2)
e	Ser, Leu, His	Ser.His.Leu (*B*2γ7)
f	Leu, His	His.Leu (*B*2γ8)

Tryptic hydrolysates (P*B*t and S*B*t). Samples (30 mg.) of fractions P*B* and S*B* were treated for 24 hr. with trypsin at pH 8, and the water-soluble material subjected to ionophoresis at pH 6 on a 20 cm. wide strip of no. 3 filter paper for 3 hr. at 1500 v. Two ninhydrin-reacting bands were present, one (P*B*t1) had moved 6 cm. towards the cathode and was free alanine, the other (P*B*t2) had moved 17·5 cm. towards the cathode and gave on hydrolysis [Gly, Thr, Tyr, Phe, Pro, Lys]. It corresponded to the cattle peptide (*B*t2) Gly.Phe.Phe.Tyr.Thr.Pro.Lys, and was identical from both pig and sheep material. In order to determine the number of phenylalanine residues present samples of the hydrolysates were treated with fluorodinitrobenzene and assayed by the method of Levy (1954). It was found that there was approximately twice as much phenylalanine present as glycine or threonine.

The glycyl chain of pig insulin (fraction P*A)*

The results with fraction P*A* will be described as single experiments in which peptides derived from enzymic hydrolysates are subjected to further hydrolysis with acid, although in fact many of the peptides were obtained initially from direct acid hydrolysis of the fraction *A*.

Partial hydrolysis of peptide P*A*p5. Peptide P*A*p5 derived from 25 mg. pig insulin was subjected to partial acid hydrolysis and the hydrolysate fractionated by paper chromatography in the usual manner. Fig. 2 is a diagram of the chromatogram, and in Table 5 are given the analyses of the various spots.

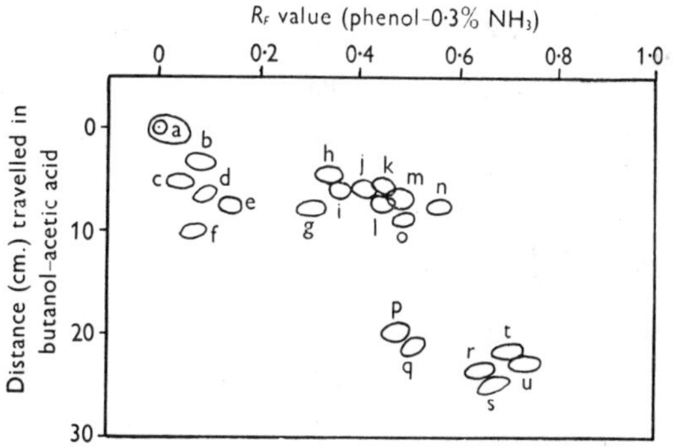

Fig. 2. Chromatogram of partial acid hydrolysate of peptide P*A*p5. The material was applied in the position marked ⊙ and run first in phenol–0·3 % NH_3 and then in butanol–acetic acid (solvent front moved approx. 40 cm.) (see Table 5).

Chymotryptic hydrolysate. (P*A*c) Fraction P*A* (10 mg.) was treated with chymotrypsin (0·5 mg.) at pH 7·5 and the hydrolysate subjected to ionophoresis at pH 6·5 for 3 hr. using 1000 v. The bands were distributed as shown in Fig. 3 and their composition is given in Table 6. Peptides 1, 2 and 3 were subjected to partial acid hydrolysis. Peptide 1 gave rise to the same peptides that were obtained from P*A*p5 (Table 5) together with [Ser, Tyr, Leu] corresponding to *A*1a19 (Ser.Leu.Tyr) and [$CySO_3H$, Asp] presumably due to the presence of $CySO_3H$.Asp (*A*c1) in band P*A*c1. The peptides detected in the hydrolysate of P*A*c2 are listed in Table 7. With the exception of a, b and d the same peptides were obtained from P*A*c3.

The glycyl chain of sheep insulin (fraction S*A)*

Partial hydrolysis of peptide S*A*p5. In Fig. 4 and Table 8 are shown the results obtained with a partial acid hydrolysate of peptide S*A*p5 prepared

Table 5. *Peptides from partial acid hydrolysate of peptide* P*A*p5

Peptide (Fig. 2)	Amino acids found on hydrolysis	Strength	Corresponding peptide from cattle insulin
a	$CySO_3H$ Glu Ala	× × × × × ?	Glu . $CySO_3H$. $CySO_3H$
b	$CySO_3H$	× × ×	Cysteic acid
c	$CySO_3H$ Glu	× × × ×	Glu . $CySO_3H$ ($A\,1\alpha1$)
d	$CySO_3H$ Thr	× ×	—
e	$CySO_3H$ Ala	? ?	$CySO_3H$. Ala ($A\,1\gamma3$)
f	Glu	×	Glu . Glu ($A\,1\beta4$)
g	Glu	× × ×	Free glutamic acid
h	$CySO_3H$ Ser Val	× × ×	Ser . Val . $CySO_3H$ ($A\,1\zeta2$)
i	Ser	× ×	Free serine
j*	Gly	× × ×	Free glycine
k*	$CySO_3H$ Ser Thr Ileu	× × × × × × × × × × × ×	—
l*	$CySO_3H$ Ser Ileu	× × × × × × × × ×	—
m*	Thr	× ×	Free threonine
n	Ala	×	Free alanine
o	$CySO_3H$ Ileu or Leu	× ×	—
p	Glu Gly Val Ileu	× × × × ×	Gly . Ileu . Val . Glu . Glu ($A\,1a15$)
q	Glu Val Ileu	× × × ×	Ileu . Val . Glu . Glu ($A\,1a16$)
r	Glu Gly Val Ileu	× × × ×	Gly . Ileu . Val . Glu ($A\,1\eta3$)
s	Glu Val Ileu	× × ×	Ileu . Val . Glu ($A\,1\eta4$)
t	Ser Leu	× × × × × × × ×	Ser . Leu ($A\,1\iota7$)
u	Leu + Ileu	× × ×	Free leucine and isoleucine

* Spots j, k, l and m overlapped considerably on this chromatogram, but could be obtained pure by developing longer with both solvents. Peptides k and l were in fact first obtained pure from a fraction of cysteic acid peptides prepared by ionophoresis in 0·2 N acetic acid of a partial acid hydrolysate of fraction P*A* (as in Expt. *A* 2 of Sanger & Thompson, 1953*a*).

in the same way as the partial hydrolysate of P*A*p5 (above).

Chymotryptic hydrolysate (S*A*c). Hydrolysis of fraction S*A* with chymotrypsin yielded peptides similar to those obtained from fraction P*A* (Fig. 3), and partial hydrolysis of bands S*A*c2 and S*A*c3 produced the same peptides as P*A*c2 and P*A*c3 respectively (Table 7). Partial hydrolysis of S*A*c1 gave rise to the same peptides as S*A*p5 (Table 8) together with Ser.Leu.Tyr and $CySO_3H$.Asp.

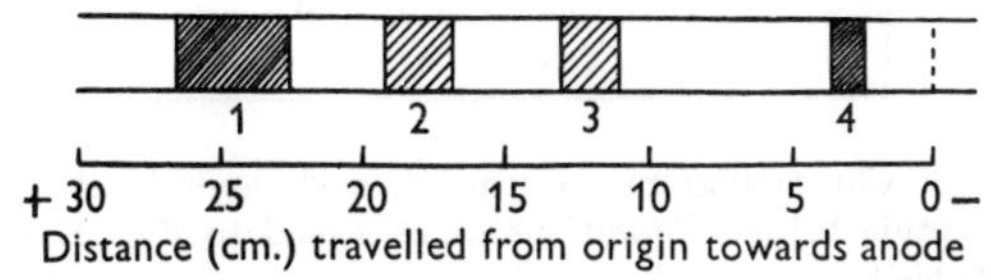

Fig. 3. Ionophoresis of chymotryptic hydrolysate of fraction *A* of pig insulin (P*A*c): pyridine–acetate buffer, pH 6·5; 28 v/cm.; 3 hr.; bands located by ninhydrin reaction (see Table 6).

summarized in Table 9, where the peptides identified in the present work are indicated by underlining the corresponding sequence in the formula for cattle fraction *B*. All these peptides were in fact derived from peptic hydrolysates of fraction *B* (P*B*p and S*B*p). The reference numbers refer to those used in Tables 2–4.

In order to show that a particular position in the chain is occupied by the same amino acid residue in the pig or sheep as in the cattle insulin, we have considered it necessary to detect the amino acid in a characteristic small peptide. Such a peptide must have a sequence which only occurs once in the chain and must contain only one residue of the amino acid concerned. This does not give an absolute proof of the structure, but it would be an extreme coincidence if any difference were not detected. In the following discussion only the pig insulin will be considered. The results with the sheep were identical.

Table 6. *Peptides from chymotryptic hydrolysate of fraction* P*A*

Peptide (Fig. 3)	Amino acids found on hydrolysis	Corresponding peptide in cattle insulin
1	$CySO_3H$, Asp, Glu, Ser, Gly, Thr, Ala*, Tyr, Val, Leu, Ileu	Mixture of *A*c4, *A*c2, *A*c1
2	$CySO_3H$, Asp, Glu, Tyr, Leu	Glu.Leu.Glu.Asp.Tyr.$CySO_3H$.Asp (*A*c3)
3	Asp, Glu, Tyr, Leu	Glu.Leu.Glu.Asp.Tyr (*A*c5)
4	Ser, Tyr, Leu	Ser.Leu.Tyr (*A*c6)

* Alanine present only in trace amounts.

Table 7. *Peptides from partial acid hydrolysate of peptide* P*A*c2

Peptide	Amino acids found on hydrolysis	Corresponding peptide from cattle insulin (Sanger & Thompson, 1953*a*)
a	$CySO_3H$, Asp	$CySO_3H$.Asp (*A* 1α2)
b	$CySO_3H$, Asp, Tyr	(Asp).Tyr.$CySO_3H$.(Asp) (*A* 1a4, *A* 1a5)
c	Asp, Glu	Glu.Asp (*A* 1β2)
d	$CySO_3H$, Tyr	Tyr.$CySO_3H$ (*A* 1a7)
e	Asp, Glu, Leu	Leu.Glu.Asp (*A* 1δ5)
f	Asp, Glu, Tyr	Glu.Asp.Tyr (*A* 1a11)
g	Asp, Tyr	Asp.Tyr (*A* 1a9)
h	Glu, Leu	Glu.Leu.Glu (*A* 1δ6)
i	Glu, Leu	Leu.Glu (*A* 1ε7)
j	Glu, Leu	Glu.Leu (*A* 1ε8)

DISCUSSION

The phenylalanyl chains

Fig. 1 shows that the ionophoretic separations of the peptic hydrolysates of the *B* fractions from cattle, pig and sheep insulins are identical, suggesting that their structures are the same. This is confirmed by the fact that all the small peptides found on partial hydrolysis of the pig and sheep material had also been obtained from cattle insulin. These results are

Position 1 is occupied by phenylalanine, since this is the *N*-terminal residue determined by the DNP method. Peptide P*B*p11 (Table 2) differs from peptide P*B*p10 in containing a single residue of phenylalanine, valine, aspartic and glutamic acids. These are the first four residues of the chain of cattle insulin and must therefore also be the first four of fraction P*B*. If the relative order of the residues was different, it would not have been detected, but this would seem to be rather an unlikely coincidence.

Positions 5–11 are well covered by the small peptides from partial hydrolysis of P*B*p10 (Table 4). It is interesting to note that the structure of peptide P*B*p10 could not have been determined from the peptides isolated, since the simpler structure Ser.His.Leu.$CySO_3H$.Gly would in fact fit the results.

Residues 12–15 are found in small peptides from the peptic hydrolysate, and residues 16–24 in peptides from the partial hydrolysate of P*B*p7. A characteristic peptide present in large amounts in the peptic hydrolysate is P*B*p13, which represents residues 26–30. Each residue occurs only once in the peptide, which is the same for cattle and pig insulin, provided there is no reversal of order. This leaves only residue 25 unaccounted for. However, that this is phenylalanine may be concluded from the fact that peptide P*B*t2 contains only the expected amino acids and has twice as much phenylalanine as glycine or threonine. It may thus be concluded that the phenylalanyl chains of pig and sheep insulins are identical with the phenylalanyl chain of cattle insulin.

The glycyl chain of pig insulin

The results with the *A* fractions are summarized in Table 10. Small peptides were obtained to cover the whole of the chain and establish that positions 1–7 and 11–21 are occupied by the same amino acids in pig and cattle insulin.

Ignoring for the present trace amounts, certain peptides which had been found in cattle material

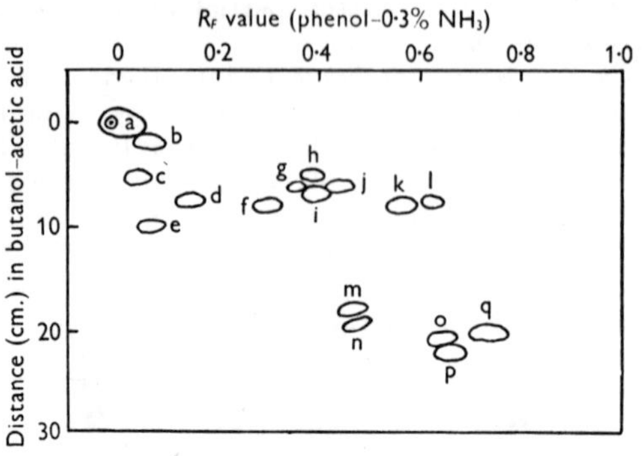

Fig. 4. Chromatogram of partial acid hydrolysate of peptide S*A*p5. Conditions as in legend to Fig. 2 (see Table 8).

Table 8. *Peptides from partial acid hydrolysate of peptide* S*A*p5

Peptide (Fig. 4)	Amino acids found on hydrolysis	Strength	Corresponding peptide from cattle insulin
a	$CySO_3H$ Glu Ala	× × × × × × × ×	Mixture (*A* 2γ1, *A* 2δ1, etc.)
b	$CySO_3H$ Ala	× × × ×	Free cysteic acid + $CySO_3H.CySO_3H$.Ala (*A* 2δ1)
c	$CySO_3H$ Glu	× × × ×	Glu.$CySO_3H$ (*A* 1α1)
d	$CySO_3H$ Ala	× × × ×	$CySO_3H$.Ala (*A* 1γ3)
e	Glu	×	Glu.Glu (*A* 1β4)
f	Glu	× × ×	Free glutamic acid
g	Ser	×	Free serine
h	$CySO_3H$ Ser Val	× × ×	Ser.Val.$CySO_3H$ (*A* 1ζ2)
i	$CySO_3H$ Gly Val	× × × × × × × × × × × ×	—
j	Gly	× ×	Free glycine
k	Ala	× × ×	Free alanine
l	Gly Ala	× × × ×	—
m	Glu Gly Val Leu or Ileu	× × × × ×	Gly.Ileu.Val.Glu.Glu (*A* 1a15)
n	Glu Val Leu or Ileu	× × × ×	Ileu.Val.Glu.Glu (*A* 1a16)
o	Glu Gly Val Leu or Ileu	× × × ×	Gly.Ileu.Val.Glu (*A* 1η3)
p	Glu Val Leu or Ileu	× × ×	Ileu.Val.Glu (*A* 1η4)
q	Ser Leu or Ileu	× × × × × × ×	Ser.Leu (*A* 1ι7) + leucine and isoleucine

Table 9. *The structure of the phenylalanyl chains of pig and sheep insulins*

In the reference numbers, *PB* or *SB* is omitted, only the suffixes being given.

Structure of fraction *B* of cattle insulin:

Phe.Val.Asp(NH_2).Glu(NH_2).His.Leu.$CySO_3H$.Gly.Ser.His.Leu.Val.Glu.Ala.Leu.Tyr.Leu.Val.$CySO_3H$.Gly.Glu.Arg.Gly.Phe.Phe.Tyr.Thr.Pro.Lys.Ala

1 2 3 4 5 6 7 8 9 10 11 12 13 14 15 16 17 18 19 20 21 22 23 24 25 26 27 28 29 30

Corresponding peptides from fraction *B* of pig and sheep insulins:

p10a; p1; p7d; p7e

p10b; p10d; p3a; p7a; p7g

p10c; p10e; p7b; p7k

p2; p7c; p13

p3; p7h

p7i; t2

p7j

Table 10. *The structure of the glycyl chains of pig and sheep insulins*

Structure of fraction *A* of cattle insulin:

Gly.Ileu.Val.Glu.Glu(NH_2).$CySO_3H$.$CySO_3H$.Ala.Ser.Val.$CySO_3H$.Ser.Leu.Tyr.Glu(NH_2).Leu.Glu.Asp(NH_2).Tyr.$CySO_3H$.Asp(NH_2)

1 2 3 4 5 6 7 8 9 10 11 12 13 14 15 16 17 18 19 20 21

Corresponding peptides from fraction *A* of pig insulin (Fraction P*A*):

P*A*p5f; P*A*p5t; P*A*c2j; P*A*c2g

P*A*p5s; P*A*p5c; P*A*c2c; P*A*c2a

P*A*p5q; P*A*c2i; P*A*c2d

P*A*p5p; P*A*c4; P*A*c2e

P*A*p5r; P*A*p5a; P*A*c2f

P*A*c2h

Other peptides found in fraction P*A*:

$CySO_3H$.Thr (P*A*p5d)

Ileu.$CySO_3H$ (P*A*p5o)

Ser.Ileu.$CySO_3H$ (P*A*p5l)

Thr.Ser.Ileu.$CySO_3H$ (P*A*p5k)

Structure of fraction *A* of pig insulin:

Gly.Ileu.Val.Glu.Glu(NH_2).$CySO_3H$.$CySO_3H$.Thr.Ser.Ileu.$CySO_3H$.Ser.Leu.Tyr.Glu(NH_2).Leu.Glu.Asp(NH_2).Tyr.$CySO_3H$.Asp(NH_2)

Other peptides found in fraction *A* of sheep insulin (fraction S*A*):

$CySO_3H$.Ala (S*A*p5d)

Gly.Val.$CySO_3H$ (S*A*p5i)

Ala.Gly (S*A*p5l)

Structure of fraction *A* of sheep insulin:

Gly.Ileu.Val.Glu.Glu(NH_2).$CySO_3H$.$CySO_3H$.Ala.Gly.Val.$CySO_3H$.Ser.Leu.Tyr.Glu(NH_2).Leu.Glu.Asp(NH_2).Tyr.$CySO_3H$.Asp(NH_2)

were absent from pig insulin. These were $CySO_3H$.-Ala, Val.$CySO_3H$ and Ser.Val.$CySO_3H$. In their places are found [$CySO_3H$, Thr] (*PA*p5d), [$CySO_3H$, Ileu] (*PA*p5o), [$CySO_3H$, Ser, Ileu] (*PA*p5l) and [$CySO_3H$, Ser, Thr, Ileu] (*PA*p5k). These sequences can only be explained if the amino acid sequence in positions 7–11 is $CySO_3H$.Thr.Ser.Ileu.$CySO_3H$ and the structure of the peptides is as shown in Table 10. Theoretically the reverse order is possible but such a structure would not give large amounts of [$CySO_3H$, Ser, Ileu] and [$CySO_3H$, Ser, Thr, Ileu] owing to the lability of bonds involving the amino groups of hydroxyamino acids. In this connexion it is interesting that the Thr.Ser bond found here is considerably more stable than the Ala.Ser bond which was never found intact after acid hydrolysis of cattle material.

The glycyl chain of sheep insulin

The sequence of amino acids in positions 1–7 and 11–21 is the same in the sheep insulin as for the cattle and pig insulin. $CySO_3H$.Ala (*SA*p5d) is also found, as it was found in cattle material, so that position 8 is occupied by alanine. The main difference between the sheep and cattle fraction *A* is that Ser.Val.$CySO_3H$ is virtually absent and in its place is found [$CySO_3H$, Gly, Val] (*SA*p5i), whose structure is thus most probably Gly.Val.$CySO_3H$. This is confirmed by the presence of the new peptide Ala.Gly (*SA*p5l). The sequence in positions 7–11 is thus $CySO_3H$.Ala.Gly.Val.$CySO_3H$.

General

Whereas the main peptides produced on partial hydrolysis of pig or sheep insulin were those listed in Table 10 which were characteristic of the particular species, it was always found that small amounts of the peptides characteristic of cattle insulin were also present. Thus, for instance, Ser.Ileu.$CySO_3H$ was obtained in high concentrations from fraction P*A*, but a small amount of Ser.Val.$CySO_3H$ was present in spot P*A*p5h. Similarly, preparations of fraction P*A* or P*A*p5 always contained some alanine.

To test if this was characteristic of all preparations of pig insulin peptide P*A*p5 was prepared from a number of different samples of pig insulin; the peptides were hydrolysed and the amino acids present separated on a phenol chromatogram. The results are shown in Fig. 5. All samples do in fact contain alanine, though in varying amounts. Sample 1, from which alanine is almost absent, was prepared by Lens & Evertzen (1952) and is probably the most rigorously purified. The other samples were commercially produced. Most of the work reported here was done with sample 2, which contains the most alanine of any. It seems probable that this phenomenon is due to contamination of the preparations with small amounts of cattle insulin. On the other hand, it is possible that the pig and sheep do actually produce more than one type of insulin. This effect has never been observed with cattle insulin. Within the limits of the methods used there is no threonine in the cattle fraction *A* or peptide *A*p5 and neither Ser.Ileu.$CySO_3H$ nor Gly.Val.$CySO_3H$ have been found, so that it would be strange if cattle produce a single unique insulin and pigs and sheep produce more than one.

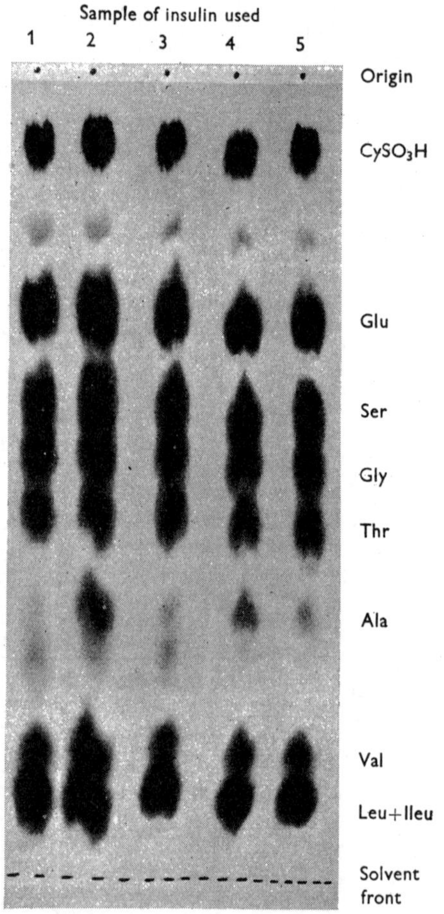

Fig. 5. Paper chromatograms on phenol–0·3% (w/v) aqu. NH_3 of hydrolysates of peptide P*A*p5 prepared from various samples of pig insulin. The chromatograms have been overloaded to show the presence of alanine.

It may, however, be noted that all the work on cattle insulin from this laboratory was done with insulin from the same source (six-times recrystallized insulin, Batches 9633B (Sanger & Tuppy, 1951*a*, *b*, and earlier papers) and 9011G (Sanger & Thompson, 1953*a*, *b* and later papers) from Boots Pure Drug Co., Nottingham). It is thus conceivable that if traces of insulins having a different structure were present, they might have been lost during the recrystallizations. It is hoped to investigate this phenomenon further.

The above results are in qualitative agreement with the analyses of Harfenist & Craig (Table 1). The pig insulin differs from the cattle insulin in that an alanine and a valine residue are replaced by a threonine and an isoleucine residue respectively and the sheep insulin is the same as the cattle except that glycine replaces serine.

It is interesting that the only structural differences are in those residues which lie within the disulphide ring in insulin (Ryle *et al.* 1955). The fact that insulin is readily inactivated by any treatment that affects the disulphide bonds suggests that they may be concerned in its biological activity and it might seem that the part of the glycyl chain between positions *A*6–11, which contains three half-cystine residues, would be a physiologically important part of the molecule, perhaps the 'active centre'. However, it is clear from the present results that the activity of insulin is not dependent on the unique structure of the residues in these positions, since they can vary without any change in activity.

SUMMARY

1. The structures of pig and sheep insulin have been determined by subjecting them to partial hydrolysis and comparing the peptides produced with those obtained from cattle insulin.

2. The amino acid sequence in the phenylalanine chains is identical for all three insulins.

3. The glycyl chains are the same except for three residues in positions 8–10 inclusive. In the cattle insulin this sequence is Ala.Ser.Val, in the pig insulin Thr.Ser.Ileu and in the sheep insulin Ala.Gly.Val.

Some of the experiments reported here were carried out by advanced students during a postgraduate course in biochemistry. We wish to express our gratitude to them and to Dr J. Lens of N. V. Organon, Dr J. G. Faber of N. V. Philips-Roxane, Dr O. K. Behrens of the Lilly Research Laboratories and Mr F. A. Robinson of Allen and Hanburys Ltd. for samples of pig and sheep insulin.

REFERENCES

Harfenist, E. J. & Craig, L. C. (1952). *J. Amer. chem. Soc.* **74**, 4216.
Jepson, J. B. & Smith, I. (1953). *Nature, Lond.*, **172**, 1100.
Lens, J. & Evertzen, A. (1952). *Biochim. biophys. Acta*, **8**, 332.
Levy, A. L. (1954). *Nature, Lond.*, **174**, 126.
Michl, H. (1951). *Mh. Chem.* **82**, 489.
Ryle, A. P., Sanger, F., Smith, L. F. & Kitai, R. (1955). *Biochem. J.* **60**, 541.
Sanger, F. (1945). *Biochem. J.* **39**, 507.
Sanger, F. (1949*a*). *Biochem. J.* **44**, 126.
Sanger, F. (1949*b*). *Nature, Lond.*, **164**, 529.
Sanger, F. & Thompson, E. O. P. (1953*a*). *Biochem. J.* **53**, 353.
Sanger, F. & Thompson, E. O. P. (1953*b*). *Biochem. J.* **53**, 366.
Sanger, F., Thompson, E. O. P. & Kitai, R. (1955). *Biochem. J.* **59**, 509.
Sanger, F. & Tuppy, H. (1951*a*). *Biochem. J.* **49**, 463.
Sanger, F. & Tuppy, H. (1951*b*). *Biochem. J.* **49**, 481.
Scott, D. A. & Fisher, A. M. (1940). *Trans. Roy. Soc. Can.* **34** (V), 137.
Toennies, G. & Kolb, J. J. (1951). *Analyt. Chem.* **23**, 823.
Wasserman, P. & Mirsky, I. A. (1942). *Endocrinology*, **31**, 115.

Harold Kroto talked about him

Am Cancer Soc
Am Lung Assoc
Am Heart Assoc
anxiolytic

Reprinted from ARCHIVES OF BIOCHEMISTRY AND BIOPHYSICS, Volume 65, No. 1, November, 1956
Academic Press Inc. *Printed in U.S.A.*

ARCHIVES OF BIOCHEMISTRY AND BIOPHYSICS **65**, 427–438 (1956)

Species Differences in Insulin

J. I. Harris,[1] F. Sanger[1] and M. A. Naughton

From the Department of Biochemistry, University of Cambridge, England

Received July 3, 1956

INTRODUCTION

In a previous paper (1), the amino acid sequences in pig and sheep insulins were compared with that of cattle insulin (Fig. 1). It was found that all three proteins had a very similar structure. The phenylalanyl chains were identical, and the only differences found were in the three residues occupying positions 8, 9, and 10 in the glycyl chain. The sequence in cattle insulin was Ala.Ser.Val, in pig insulin Thr.Ser.Ileu, and in sheep insulin Ala.Gly.Val. The three residues are present in the part of the molecule that forms a ring containing the intrachain disulfide bridge joining positions 6 and 11 in the glycyl chain, a part of the molecule that might be expected to play an important role in the physiological activity of the hormone. Since the three species studied are rather closely related, it was of interest to study insulins from more distantly related species to see if the same type of differences are found. The following paper describes the results obtained with insulin from horse and whale.

MATERIALS

Horse insulin was prepared in the Nordisk Insulin Laboratory, Copenhagen, and obtained through Dr. H. Neurath; the whale insulin was prepared by Dr. K. Nagasawa of the National Hygienic Laboratory, Tokyo, and obtained through Dr. A. M. Fisher.

METHODS AND RESULTS

The methods used were essentially the same as those used for the study of pig and sheep insulin (1). The insulins were first separated into the separate chains (fractions A and B) by oxidation in performic acid. Each fraction was then subjected to hydrolysis by proteolytic enzymes, the resulting peptides fractionated by paper ionophoresis, and some of them subjected to a further partial hydrolysis

[1] Members of the Scientific Staff of the Medical Research Council.

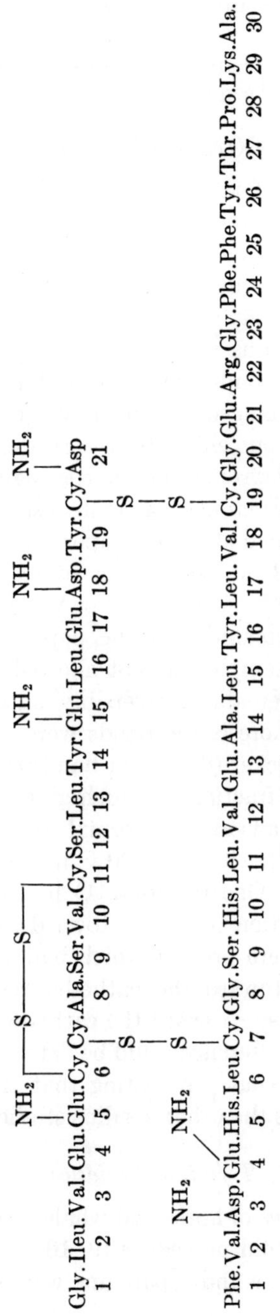

FIG. 1. The Structure of cattle insulin.

428

with acid. The results were compared with those obtained in corresponding experiments with cattle, pig, and sheep material.

Fractionation of Oxidized Insulin by Ionophoresis

Originally, separation of the fractions A and B was carried out by precipitation methods (2), and this is still the most convenient method if more than a few milligrams are required. Recently the fractions have been separated by countercurrent distribution (3, 4) and by partition chromatography (5). For analytical work and for preparing small quantities of the A and B chains we have used paper ionophoresis in 20% formic acid. Under these conditions fraction A moves toward the anode and fraction B and unchanged insulin move toward the cathode. The separations were carried out either by the method of Durrum (6) or of Michl (7), the latter giving somewhat sharper bands.

Insulin (0.4 mg.) was treated in a small test tube with 0.02 ml. performic acid, which had stood at room temperature for 1 hr. after preparation. After 15 min., 0.02 ml. water was added, and the material was applied, as a streak 3 cm. long to a sheet of Whatman No. 4 filter paper, for ionophoresis. The remainder of the paper was wetted with 20% formic acid so that the sample was not allowed to become dry. If this occurs, some material remains irreversibly adsorbed at the point of application. After ionophoresis the bands were located by the tyrosine reagent of Acher and Crocker (8). In our hands this has proved to be the most convenient test for fraction A, which gives a very weak color with ninhydrin and does not adsorb the normal dyes used for locating proteins. After ionophoresis for 6 hr. at 220 v. in the apparatus of Durrum, two bands were present. One (fraction A) had moved 3.6 cm. toward the anode, the other (fraction B) 2.9 cm. toward the cathode. Frequently two other faint bands were present which had moved 1.5 cm. toward the anode and 1.4 cm. toward the cathode, respectively. Unchanged insulin moved slightly faster toward the cathode than fraction B.

Using this method no difference could be detected between the insulins from cattle, horse, and whale, indicating that the latter two have the same general structure as the other insulins studied.

Terminal Residues

The N-terminal residues of horse and whale insulins were investigated by the phenylthiohydantoin procedure (9, 10), and found to be glycine and phenylalanine. Alanine and asparagine were shown to be the C-terminal amino acids by the use of carboxypeptidase (11). The same terminal residues occur in cattle, pig, and sheep insulins (1), although according to a recent report (12) the C-terminal alanine of the B chain does not occur in fish insulin.

The Phenylalanyl Chains (*Fraction B*)

The B fractions from horse and whale insulins were studied as previously described for cattle, pig, and sheep insulins (1). Hydrolysis was carried out with pepsin, the resulting peptides were fractionated by paper ionophoresis at pH 6.5, and the amino acid content of complete acid hydrolyzates was determined by two-dimensional paper chromatography. Peptides Bp7 (Tyr.Leu.Val.$CySO_3H$.Gly.Glu.Arg.Gly.Phe.) and Bp10 (His.Leu.$CySO_3H$.Gly.Ser.His.) were subjected to partial acid hydrolysis, and all of the peptide fragments which were obtained could be fitted into the sequences previously established.

From these results it was concluded that the amino acid sequence of the phenylalanyl chain in both horse and whale insulins is identical with that previously reported for cattle, pig, and sheep insulins (1).

The Glycyl Chains: (*Fraction A*)

Samples of the A fractions from cattle, horse, and whale insulins were subjected to complete acid hydrolysis (5.7 *N* HCl, 24 hr. at 105°) and investigated by chromatography on paper. The results obtained are summarized in Table I.

Action of Chymotrypsin. Samples (10 mg., 4 μmoles) of each of the A fractions, prepared from cattle, horse, and whale insulins, were allowed to react with chymotrypsin (0.1 mg.) at pH 8.0 and 30°. The amount of alkali necessary to keep the pH of the reaction mixture constant was recorded by means of a Jacobsen-Léonis autotitrator. The base consumed during the reaction was recorded as a function of time. As shown in Fig. 2, there is a rapid initial reaction during which 3.5 μmoles of base was consumed in the first 5 min.; subsequently the reaction becomes very slow, an additional 5.2 μmoles of base being consumed during the course of the next 16 hr.

A sample of the digestion mixture was taken after 5 min., adjusted to pH 3 to prevent further enzyme action, and subjected to ionophoresis for 2 hr. at pH 6.5 and a potential gradient of 30 v./cm. In this way the three A fractions were each resolved into two components 1 and 2 (Fig. 3a). 1 and 2 derived from cattle fraction A were eluted, hydrolyzed

TABLE I

The Composition of Fraction A from Cattle, Horse and Whale Insulin

The hydrolyzates were analyzed by two-dimensional chromatography using *n*-butanol–benzyl alcohol (1:1 *v/v*) on buffered paper (13) and phenol–0.3% NH_3.

Amino acid	Cattle	Horse	Whale
Cysteic acid	xxx	xxx	xxx
Aspartic acid	x	xx	xx
Glutamic acid	xxx	xxx	xxx
Serine	xx	x	xx
Glycine	x	xx	x
Threonine	—	x	x
Alanine	x	?	?
Tyrosine	xx	xx	xx
Valine	xx	x	x
Leucine	xx	xx	xxx (Leucine + Isoleucine)
Isoleucine	x	xx	

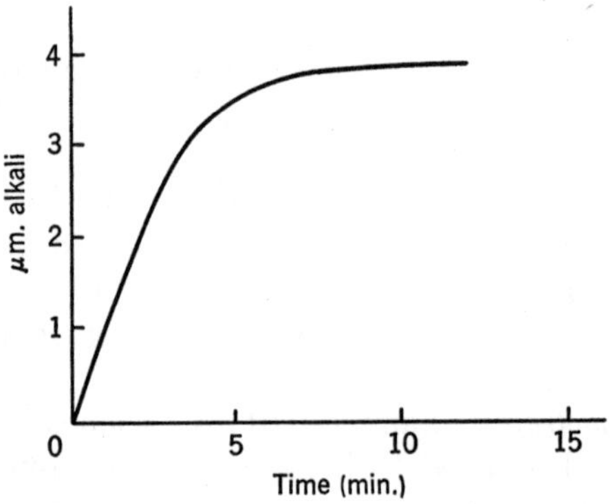

FIG. 2. The uptake of alkali during the reaction of fraction A of insulin (10 mg.) with chymotrypsin (0.1 mg.) at pH 8.0 and 30°.

with acid, and examined for constituent amino acids by two-dimensional paper chromatography. The results revealed 2 to be the N-terminal peptide, Al–14 and 1 to be the C-terminal peptide Al4–21, which are thus shown to be identical with peptides Ac4 and Ac3, respectively, reported in a previous paper (14). It follows that the Tyr.Glu(NH_2) bond in the A chain (Al4–15) is very rapidly and specifically split by chymotrypsin.

After 16 hr. a sample of the chymotryptic digest gave two additional components when subjected to ionophoresis as described above (Fig. 3b); the ionophoretic pattern was identical with that described previously

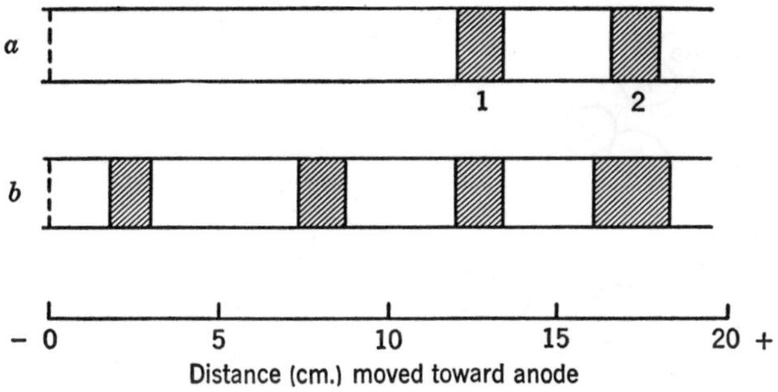

FIG. 3. Diagram of the ionophoresis of chymotryptic hydrolyzates of fraction A. Pyridine–acetate buffer pH 6.5 (10 vol. pyridine–0.4 vol. acetic acid–90 vol. water). 30 v./cm.; 2 hr. (*a*) 5-min. hydrolyzate; (*b*) 16-hr. hydrolyzate.

(1), showing that splitting had occurred at the other two susceptible positions (A11–12 and A19–20) in the A chain (Fig. 1).

Partial Acid Hydrolysis. By partial hydrolysis with acid it was shown that peptide Ac3 from whale and horse material had the same structure as that from cattle insulin. Differences were however found in peptide Ac4 from the various species.

Samples of peptide Ac4 from horse and whale insulins were hydrolyzed for 3 days at 37° with 12 *N* HCl, and the hydrolyzates were fractionated by two-dimensional paper chromatography using phenol–0.3 % NH_3 and butanol–acetic acid as solvents.

Figure 4 is a diagram of the chromatogram of the horse material, and the compositions of the various peptides are recorded in Table II. A number of peptides derived from positions A15–21 (e.g., Glu.Leu.Glu, Asp.Tyr, etc.) were also present on this chromatogram. This was presumably due to contamination of peptide Ac4 with some peptide Ac3 or with unchanged fraction A. They are not recorded in Fig. 4. With the exception of *i* and *j* all the peptides were similar to those obtained from cattle insulin, and were identified from their compositions and positions on the chromatogram. The peptides $CySO_3H$.Ala, Ser.Val, and Ser.Val.-$CySO_3H$ which are derived from cattle insulin were not found on the chromatogram, indicating that the sequence Ala.Ser.Val is not present in positions A8–10 of horse insulin.

Spot *i* gave an initial yellow color with ninhydrin. After hydrolysis of the DNP-derivative, DNP-threonine and free glycine were identified, showing that its structure was Thr.Gly.

Spot *j* ([$CySO_3H$, Ileu]) was also obtained from a partial hydrolyzate

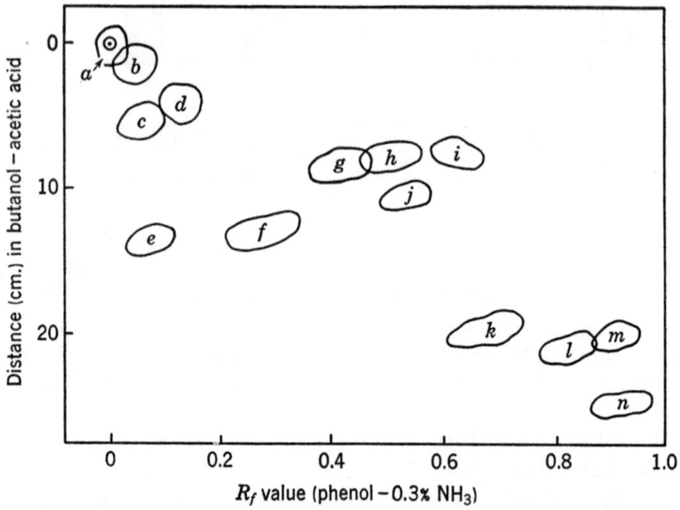

FIG. 4. Chromatogram of partial acid hydrolyzate of peptide Ac4 from horse insulin. Run first in phenol–0.3% NH_3 and then in *n*-butanol–acetic acid–water (4:1:5, by vol.)

TABLE II

Peptides Identified in a Partial Hydrolyzate of Peptide Ac4 from Horse Insulin

(See Fig. 4)

a. Glu.$CySO_3H$.$CySO_3H$
b. Cysteic acid
c. Glu.$CySO_3H$
d. Probably Glu.Ileu.Val.Glu.Glu.$CySO_3H$.$CySO_3H$
e. Glu.Glu
f. Glutamic acid
g. Gly.Ileu.Val.Glu.Glu.$CySO_3H$
h. Serine and glycine + trace of unidentified cysteic acid peptide.
i. [Gly, Thr]
j. [$CySO_3H$, Ileu]
k. Ileu.Val.Glu.Glu
l. Ileu.Val.Glu + Ser.Leu.Tyr
m. Leucine
n. Leu.Tyr

of fraction A by ionophoresis at pH 3.5 (2 hr., 2000 v. applied at the rate of 0.5 mg. hydrolyzate/cm. on Whatman No. 3 paper). The presence of isoleucine was established using *n*-butanol–benyzl alcohol (1:1 *v*/*v*) on buffered paper (13).

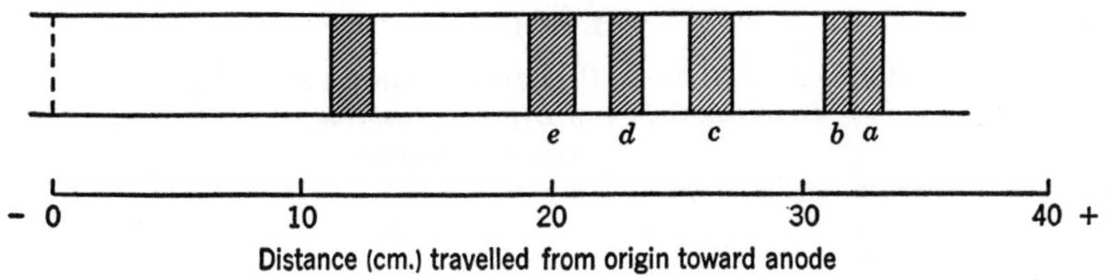

FIG. 5. Diagram of ionophoresis of partial acid hydrolyzate of peptide Ac4 from whale insulin. Pyridine–acetate buffer pH 3.5 (1 vol. pyridine–10 vol. acetic acid–89 vol. water); 2000 v.; 3 hr.

TABLE. III

Peptides Identified in Partial Hydrolyzate of Peptide Ac4 from Whale Insulin

(See Fig. 5)

Band (Fig. 5)	Amino acids found on hydrolysis	N-terminal residue (DNP method)	Structure of peptides
a.	$CySO_3H$, Asp	—	$CySO_3H$.Asp
b.	$CySO_3H$, Asp, Glu	—	$CySO_3H$.Asp + Glu.$CySO_3H$
c.	$CySO_3H$, Ileu	Ileu	Ileu.$CySO_3H$
d.	Asp	—	Aspartic acid
e.	$CySO_3H$, Ser, Ileu	Ser	Ser[$CySO_3H$,Ileu]

From these results it may be concluded that the sequence Thr.Gly.-Ileu.$CySO_3H$ is present in positions A8–11 of the glycyl chain of horse insulin.

The results with the partial acid hydrolyzate of peptide Ac4 from whale insulin were similar to those for horse insulin (Fig. 4), except that spots *i* and *j* were absent, and in approximately the position of spot *i* was a spot having the composition [$CySO_3H$, Ser, Thr, Ileu]. This was similar to spot PAp5k obtained from pig insulin (1) and presumably has the structure Thr.Ser.Ileu.$CySO_3H$. More definite evidence for this sequence however came from an investigation of the peptides obtained by ionophoresis at pH 3.5 of the partial hydrolyzate.

Figure 5 is a diagram of the ionophoresis that had been carried out for 3 hours at 2000 v., and the composition of the peptides is given in Table III. From these results it may be concluded that the sequence in positions A8–11 in whale insulin is Thr.Ser.Ileu.$CySO_3H$.

DISCUSSION

The results obtained with horse and whale insulins are very similar to those obtained previously with cattle, pig, and sheep insulins (1). In

TABLE IV

The Amino Acid Sequence Contained within the Intrachain Disulfide Bridge of Different Insulins

Species	Sequence in positions 7	8	9	10
Cattle	$CySO_3H$.	Ala.	Ser.	Val.
Pig	$CySO_3H$.	Thr.	Ser.	Ileu.
Sheep	$CySO_3H$.	Ala.	Gly.	Val.
Horse	$CySO_3H$.	Thr.	Gly.	Ileu.
Whale	$CySO_3H$.	Thr.	Ser.	Ileu.

all five species the only differences are confined to the three amino acid residues which occur within the intrachain disulfide bridge, that is, in positions 8–10 of the A (glycyl) chain; the remainder of the A chain and the entire B (phenylalanyl) chain appear to have identical sequences. The sequences which have been found for positions A7–10 are listed in Table IV.

In all the five species studied, it may be seen that position 8 is occupied either by alanine or threonine, position 9 by serine or glycine, and position 10 by valine or isoleucine. It emerges that whale insulin has the same structure as pig insulin. If we generalize on the basis of five different examples, it would appear that there are only three positions in the insulin molecule in which species differences can occur, and that each of these positions is occupied by one of two amino acids. This limits the number of possible structures to eight. Hence it is not surprising that after investigating five insulins from different species, two are found to have a common structure, and it seems likely that the identity of pig and whale insulins should be ascribed to coincidence rather than to any unexpected phylogenic relationship.

Although other examples of species differences in proteins have now been shown to occur, the various replacements which have been reported do not in general seem to conform to any predictable and consistent pattern. Thus although another example of an Ala-Thr replacement has been demonstrated in the case of serum albumins from human and bovine sources, respectively (15), an alanine residue which occurs in a peptide isolated from pig, horse, cattle, and salmon cytochromes is found to be replaced by a serine residue in the corresponding peptide isolated from chicken cytochrome (16). Similar observations have been made in the case of different strains of tobacco mosaic virus. The protein moieties prepared from four different strains of the virus were found to contain the same C-terminal amino acid sequence, Thr.Ser.Gly.Pro.Ala.Thr; in a fifth strain, however, the C-terminal sequence was found to be Thr.-(Ala, Thr).Pro.Ala.Thr, indicating a double replacement in adjacent positions, which in view of the unestablished (Ala, Thr) sequence involves either the replacement of Ala by Ser and of Thr by Gly, or the replacement of Ala by Gly and Thr by Ser (17). The N-terminal amino sequences of hemoglobins from different species show that Val-Met, Gly-Ser and Glu-Asp replacements have occurred (18, 19). Species differences have also been found in polypeptide hormones from the pituitary. Thus in the case of the corticotropins (ACTH), the sequence Leu.-Ala in pig corticotropin (20) is replaced by Ala.Ser in the sheep hormone (21). Whereas the replacement of alanine by serine is in no way unusual, the replacement of its adjacent leucine by alanine is unexpected, and even results in the elimination of the Glu(NH_2).Leu bond which in pig corticotropin is a major site of cleavage by pepsin (20). This change in structure, however, has no significant effect on the biological activities of the two corticotropins since the part of the molecule in which it occurs does not appear to be essential for activity (20, 21).

In the case of vasopressin, the penultimate arginine residue in cattle vasopressin (22, 23) is replaced by lysine in the pig hormone (22):

```
Cy.Tyr.Phe.Glu(NH2).Asp(NH2).Cy.Pro.Arg.Gly(NH2)
|                              |        Lys
|____________S—S_______________|
```

In contrast to insulin, the amino acids within the intrachain disulfide structure are not changed in the peptides from the two species. It may also be significant that a change in the same penultimate position, in this case the replacement of the basic amino acid by leucine (coupled with the replacement of phenylalanine by isoleucine), changes the vasopressin structure into that of oxytocin (24, 25) which has very different biological properties.

To return to insulin, it is of considerable interest that in all the five species studied, structural differences are confined to three of the four amino acids which occur within the intrachain disulfide ring. This would imply that during evolutionary development successful mutations have occurred only in those genes which control the structure of these three sites. One possible explanation would be that the whole of the rest of the molecule is essential for biological activity and that any change in another residue would be fatal. This however seems unlikely since there is considerable evidence to show that certain other parts of the molecule can be modified without loss of activity. For example, complete activity is maintained when the C-terminal alanine residue of the B chain is removed by carboxypeptidase (26, 27), and also when the free amino groups in the molecule are blocked by acetylation (28).

The location of the three sites responsible for the species differences may be determined by the relative "instability" of the genes controlling the synthesis of this part of the molecule, so that they would have a greater tendency to undergo successful mutations than those controlling the synthesis of the remainder of the molecule. In this connection Linderstrøm-Lang (29) has proposed a model for the insulin molecule based on the Pauling α-helix. In this model it was suggested that the insulin molecule could be represented by two α-helices; a right-handed helix to represent the B chain and a left-handed helix to represent the A chain. The part of the A chain between the intrachain disulfide bridge (A6–11, Fig. 1) could not be accommodated into an undistorted α-helix, and it was therefore suggested that it forms a "loop" outside the main A chain helix. The fact that all the species differences in insulin appear to be confined to the loop may be explained on the basis of Linderstrøm-Lang's suggestion (29), based on the results of deuterium-exchange studies (30), that the "loop" amino acids would have greater mobility and accessibility than the other amino acids which are stabilized between the interchain S—S bridges. It is conceivable that this additional "degree of freedom" may in some way be reflected in the genetic factors which control the synthesis of this part of the molecule. Clearly such ideas are highly speculative, but it does seem that more extensive studies of species differences in the amino acid sequences of polypeptide chains may lead to interesting conclusions concerning evolutionary trends in protein biosynthesis.

ACKNOWLEDGMENTS

We wish to thank Dr. H. Neurath, Dr. C. H. Hagedorn, Dr. K. Nagasawa, and Dr. A. M. Fischer for their kindness in providing the materials for this investigation.

SUMMARY

The structure of horse and whale insulins have been determined and compared with other insulins previously investigated.

The only differences found were in the three residues occupying positions 8–10 in the glycyl (A) chain. The sequence in horse insulin is Thr. Gly.Ileu and in whale insulin is Thr.Ser.Ileu.

References

1. Brown, H., Sanger, F., and Kitai, R., *Biochem. J.* **60**, 556 (1955).
2. Sanger, F., *Biochem. J.* **44**, 126 (1949).
3. Pierce, J. G., *J. Am. Chem. Soc.* **77**, 184, (1955).
4. Craig, L. C., *Ciba Foundation Colloquia on Endocrinol.* **9**, 104 (1956).
5. Andersen, W., *Compt. rend. trav. lab. Carlsberg. Sér. chim.* **29**, 49 (1954).
6. Durrum, E. L., *J. Am. Chem. Soc.* **72**, 2943 (1950).
7. Michl, H., *Monatsh.* **82**, 489 (1951).
8. Acher, R., and Crocker, C., *Biochim. et Biophys. Acta* **9**, 704 (1952).
9. Edman, P., *Acta Chem. Scand.* **4**, 283 (1950).
10. Fraenkel-Conrat, H., *J. Am. Chem. Soc.* **76**, 3606 (1954).
11. Harris, J. I., "Symposium on Peptide Chemistry," *Chem. Soc. Spec. Publ.* **2**, 71 (1955).
12. Akabori, S., *J. Japan. Biochem. Soc.* **25**, 243 (1953).
13. McFarren, E. F., *Anal. Chem.* **23**, 168 (1951).
14. Sanger, F., and Thompson, E. O. P., *Biochem. J.* **53**, 366 (1953).
15. Thompson, E. O. P., *J. Biol. Chem.* **208**, 565 (1954).
16. Tuppy, H., and Paleus, S., *Acta Chem. Scand.* **9**, 353 (1955).
17. Niu, C. I., and Fraenkel-Conrat, H., *Arch. Biochem. and Biophys.* **59**, 538 (1955).
18. Porter, R. R., and Sanger, F., *Biochem. J.* **42**, 287 (1948).
19. Ozawa, H., and Satake, K., *J. Biochem.* (Japan) **42**, 641 (1955).
20. Bell, P. H., *J. Am. Chem. Soc.* **76**, 5565 (1954).
21. Li, C. H., Geschwind, I. I., Cole, R. D., Raacke, I. D., Harris, J. I., and Dixon, J. S., *Nature* **176**, 687 (1955).
22. Vigneaud, V. du, Lawler, H. C., and Popenoe, E. A., *J. Am. Chem. Soc.* **75**, 4879 (1953).
23. Acher, R., and Chauvet, J., *Biochim. et Biophys. Acta* **12**, 487 (1953).
24. Tuppy, H., and Michl, H., *Monatsh.* **84**, 1011 (1953).
25. Vigneaud, V. du, Ressler, C., and Trippett, J., *J. Biol. Chem.* **205**, 949 (1953).
26. Harris, J. I., and Li, C. H., *J. Am. Chem. Soc.* **74**, 2945 (1952).
27. Nicol, D., and Smith, L. F., *Biochem. J.* **64**, 17p. (1956).
28. Stern, K. G., and White, A., *J. Biol. Chem.* **122**, 371 (1938).
29. Linderstrøm-Lang, K., "Symposium on Peptide Chemistry," *Chem. Soc. Spec. Publ.* **2**, 1 (1955).
30. Hvidt, Aa., and Linderstrøm-Lang, K., *Compt. rend. trav. lab. Carlsberg. Sér. chim.* **29**, 385 (1955).

LES PRIX NOBEL

EN 1958

THE CHEMISTRY OF INSULIN.

BY

F. SANGER

STOCKHOLM 1959

THE CHEMISTRY OF INSULIN.

by

F. Sanger.

Nobel lecture, December 1, 1958.

It is great pleasure and privilege for me to give an account of my work on protein structure and I am deeply sensitive of the great honour that has been done to me in recognising my work in this way. Since the work on insulin has extended over about 12 years it will be necessary to give a somewhat simplified account and to omit most of the work that did not contribute directly to the main problem, the determination of the structure of a protein.

In 1943 the basic principles of protein chemistry were firmly established. It was known that all proteins were built up from amino acid residues bound together by peptide bonds to form long polypeptide chains. Twenty different amino acids are found in most mammalian proteins and by analytical procedures it was possible to say with reasonable accuracy how many residues of each one was present in a given protein. Practically nothing, however, was known about the relative order in which these residues were arranged in the molecules. This order seemed to be of particular importance, since although all proteins contained approximately the same amino acids they differed markedly in both physical and biological properties. It was thus concluded that these differences were dependent on the different arrangement of the amino acid residues in the molecules. Although very little was known about amino acid sequence, there was much speculation in this field. The most widely discussed theory was that of BERGMANN and NIEMANN who suggested that the amino acids were arranged in a periodic fashion, the residues of one type of amino acid occurring at regular intervals along the chain. On the other extreme there were those who suggested that a pure protein was not a chemical individual in the classical sense but consisted of a random mixture of similar individuals.

Due largely to the work of CHIBNALL and his colleagues (1) insulin had been studied in considerable detail. It had a somewhat simpler composition than most proteins, in that two of the commonly occurring amino acids, tryptophan and methionine were absent and an accurate analysis was available. Moreover, using the van Slyke procedure, Chibnall had shown that insulin was peculiar in having a high content of free α-amino groups. This indicated that it was composed of relatively short polypeptide chains since free α-amino groups would be found only on those residues (the N-terminal residues) which were present at one end of a chain. Thus the number of chains could be determined from the number of these N-terminal residues. The nature of one of these N-terminal residues was in fact known. JENSEN & EVANS (2) had shown that the phenylhydantoin of phenylalanine could be isolated from an acid hydrolysate of insulin that had been treated with phenylisocyanate, thus indicating that phenylalanine was at the end of one of the chains. At that time this was the only case where the position of an amino acid in a protein was known.

There was considerable doubt about the actual molecular weight of insulin and hence the number of amino acid residues present. Values varying from 36 000 to 48 000 were reported by physical methods but it was shown by GUTFREUND (3) that these high values were due to aggregation and it was suggested that the real molecular weight or subunit was 12 000. This indicated that there were about 100 residues in the molecule. More recently HARFENIST & CRAIG (4) have shown that the actual value is about 6 000; however during most of our work it was believed to be 12 000.

In order to study in more detail the free amino groups of insulin and other proteins, a general method for labelling them was worked out (5). This was the dinitrophenyl (or DNP) method. The reagent used was 1 : 2 : 4 fluorodinitrobenzene (FDNB) which reacts with the free amino groups of a protein or peptide to form a DNP derivative:

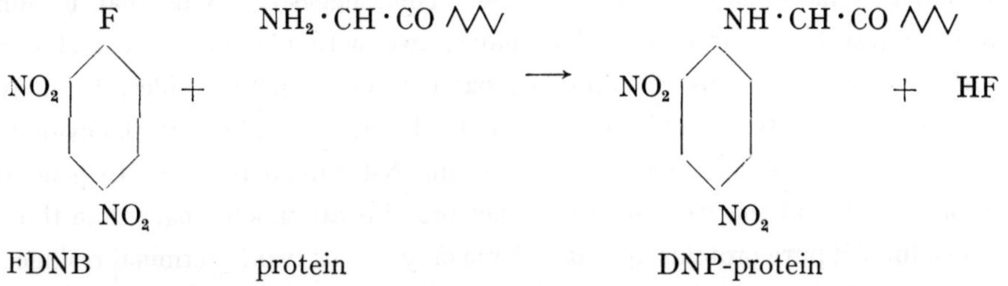

FDNB protein DNP-protein

The reaction takes place under mildly alkaline conditions which normally do not cause any breakage of the peptide bonds.

The DNP-protein is then subjected to hydrolysis with acid which splits the peptide bonds in the chain, leaving the N-terminal residue in the form of its DNP-derivative.

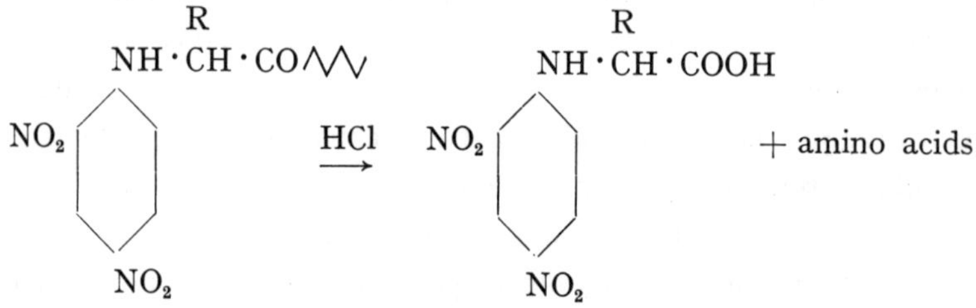

DNP-protein DNP-amino acid

The DNP-amino acids are bright yellow substances and can be separated from the unsubstituted amino acids by extraction with ether. They could be fractionated by partition chromatography, a method which had just been introduced by GORDON, MARTIN & SYNGE (6) at that time. The DNP-amino acids could then be identified by comparison of their chromatographic rates with those of synthetic DNP-derivatives. In the original work on insulin silica gel chromatography was used, though more recently other systems particularly paper chromatography have been found more satisfactory. Having separated and identified the DNP derivatives they could be estimated colourimetrically.

When the method was applied to insulin three yellow DNP-derivatives were found in the hydrolysate of the DNP-insulin. One of these was not extracted into ether and was ε-DNP-lysine which was formed by reaction of the FDNB with the free ε-amino group of lysine residues which are bound normally within the polypeptide chain. The others were identified as DNP-phenylalanine and DNP-glycine, and estimation showed that there were two residues of each assuming a molecular weight of 12 000. This suggested to us that insulin was composed of four polypeptide chains, two with phenylalanine and two with glycine end groups. This method has now been applied widely to many proteins and peptides, and together with the Edman phenylisothiocyanate method is the standard method for studying N-terminal residues. In general it has been found that the chains of other proteins are much longer than those of insulin. All pure proteins appear to have only one or two N-terminal residues.

It seemed probable that the chains of insulin were joined together by the disulphide bridges of cystine residues. Insulin is relatively rich in cystine and this was the only type of cross-linkage that was definitely known to occur in proteins. It was thus next attempted to separate the peptide chains by splitting the disulphide bridges. Earlier attempts to do this by reduction to — SH derivatives had not proved successful and had given rise to insoluble products which were probably due to some type of polymerisation. More satisfactory results were obtained by oxidation with performic acid (7). The cystine residues were converted to cysteic acid residues thus breaking the cross-links:

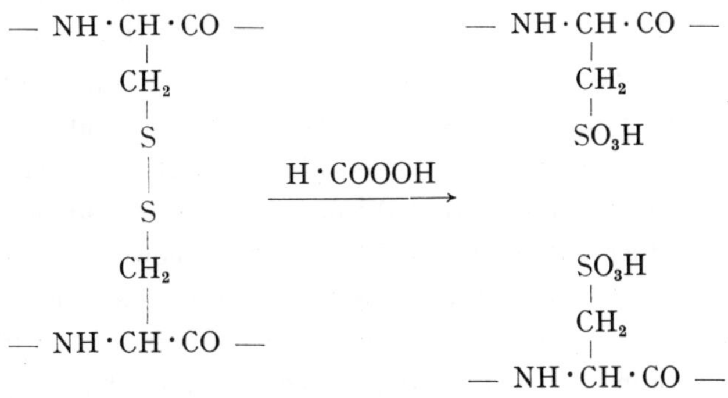

cystine residue (CyS S Cy) — 2 cysteic acid residues ($CySO_3H$)

Performic acid also reacts with methionine and tryptophan residues, the two amino acids which fortunately were absent from insulin.

From the oxidized insulin two fractions could be separated by precipitation methods. One (fraction A) contained glycine and the other (fraction B) phenylalanine N-terminal residues. Fraction A was acidic and had a simpler composition than insulin, in that the six amino acids lysine, arginine, histidine, phenylalanine, threonine and proline were absent from it. It thus had no basic amino acids, which were found only in fraction B. From a quantitative determination of the end groups it was concluded that fraction A contained about 20 residues per chain, four of these being cysteic acid and fraction B had 30 residues, two of which were cysteic acid. Since the yield of each fraction was greater than 50 % in terms of the N-terminal residues present and since they appeared to be homogeneous it seemed likely that there was only one type of glycyl chain and one type of phenylalanyl chain. This was confirmed by a study of the N-terminal sequences (8).

When the DNP derivative of fraction B was subjected to complete acid hydrolysis DNP-phenylalanine was produced. If however it was subjected to a milder acid treatment so that only a fraction of the peptide bonds were split, DNP-phenylalanyl peptides were produced which contained the amino acid residues near to the N-terminal end and by an analysis of these peptides it was possible to determine the N-terminal sequence to four or five residues along the chain. The results with fraction B are shown in Table 1. It was concluded from these results that all the N-terminal phenylalanine residues of insulin were present in the sequence Phe·Val·Asp·Glu. This suggested that if there were in fact two phenylalanyl chains, then these two were identical. Similar results were obtained with fraction A, and it was shown that the N-terminal glycine residues were present in the sequence Gly·Ileu·Val·Glu·Glu.

These results, besides giving information about the position of certain residues in the polypeptide chains, showed for the first time that the molecule was composed of only two types of chains and that if the molecular weight was 12 000 as was then believed, then the molecule was built up of two identical halves. The other alternative, which was later shown to be the case, was that the actual molecular weight was 6 000. In any case the structural problem was somewhat simplified since we were now concerned with determining the sequence in two chains containing 20 and 30 residues respectively.

The main technical problem was the fractionation of the extremely complex mixtures that resulted from partial hydrolysis of a protein. However CONSDEN, GORDON, MARTIN & SYNGE (9) had shown that small peptides could be well fractionated by paper chromatography and had determined the sequence in the pentapeptide 'gramicidin S' from the composition of peptides produced on acid hydrolysis.

Table 1.

Peptide	Products of complete hydrolysis of peptide	Products of partial hydrolysis	Structure	Yield from DNP insulin[1]
B1	DNP-phenylalanine	—	DNP-Phe	13
B2	DNP-phenylalanine Valine	B1	DNP-Phe·Val	16
B3	DNP-phenylalanine Valine Aspartic acid	B1, B2	DNP-Phe·Val·Asp	13
B4	DNP-phenylalanine Valine Aspartic acid Glutamic acid	B1, B2, B3	DNP-Phe·Val. Asp·Glu	30
Other Bands giving B4 on partial hydrolysis				20
			Total	92

[1] Moles peptide as per cent of total N-terminal phenylalanine residues of insulin.

At this point (1949) I was joined by Dr HANS TUPPY who came to work in Cambridge for a year. Although we did not seriously envisage the possibility of being able to determine the whole sequence of one of the chains within a year, it was considered worth-while to investigate the small peptides from an acid hydrolysate using essentially the methods that had been applied to 'gramicidin S'. Studies were initiated on both the chains at the same time but it soon became clear that there would be more difficulties with fraction A although it was the shorter chain and the work on fraction B progressed so favourably and TUPPY worked so hard that by the end of the year we were virtually able to deduce the whole of the sequence of its 30 residues (10).

Fraction B was subjected to partial hydrolysis with acid. Since the mixture was too complex for direct analysis by paper chromatography it was necessary to carry out certain preliminary group separations in order to obtain fractions containing 5—20 peptides that could then be separated on paper. This was accomplished by ionophoresis, ion exchange chromatography and adsorption on charcoal. These simplified mixtures were then fractionated by two-dimensional paper chromatography. The peptide spots were cut out and the material eluted from the paper, subjected to complete hydrolysis and analysed for its constituent amino acids. Another sample of the peptide was then investigated by the DNP technique to determine the N-terminal residue. Table 2 illustrates the results obtained with a very acidic fraction obtained by ion-exchange chromatography. This contained only peptides of cysteic acid. Since there are only two such residues in fraction B all these peptides must fit into two sequences. The way in which the two sequences Leu · $CySO_3H$ · Gly and Leu · Val · $CySO_3H$ · Gly were deduced from the results obtained with the peptides is illustrated in the table.

Table 2. Cysteic acid peptides identified in a partial acid hydrolysate of fraction B

(The inclusion of residues in brackets indicates that their relative order is not known.)

	$CySO_3H$ · Gly	$CySO_3H$ · Gly
	Val · $CySO_3H$	Leu · $CySO_3H$
	Val · ($CySO_3H$, Gly)	Leu · ($CySO_3H$, Gly)
	Leu · (Val, $CySO_3H$)	
	Leu · (Val, $CySO_3H$, Gly)	
Sequences deduced	Leu · Val $CySO_3H$ · Gly	Leu · $CySO_3H$ · Gly

In this way about 45 peptides were identified in various fractions of the partial acid hydrolysate and the following five sequences were deduced as being present in the phenylalanine chain.

1. Phe · Val · Asp · Glu · His · Leu · $CySO_3H$ · Gly (N-terminal sequence).
2. Gly · Glu · Arg · Gly.
3. Thr · Pro · Lys · Ala.
4. Tyr · Leu · Val · $CySO_3H$ · Gly.
5. Ser · His · Leu · Val · Glu · Ala.

These five sequences contain all but four of the amino acid residues of fraction B. It was not possible to determine from the small peptides derived from acid hydrolysates the position of the remaining four residues or how the above five sequences were joined together. There were two reasons for this. Firstly there was considerable technical difficulty in fractionating the peptides containing two or more of the non-polar residues such as tyrosine or leucine. It happened that these residues were grouped together in the chain (see below) and gave rise to a mixture of peptides that moved fast on paper chromatograms and were not well resolved. The second difficulty was due to the great lability to acid of the bonds involving the amino groups of the serine and threonine residue. It was never possible to find a peptide containing this bond and hence to know what residue preceded the serine and threonine.

It was thus necessary to use another method of hydrolysis that would show a different specificity from concentrated acid. Hydrolysates prepared by the action of dilute acid at high temperatures or of alkali were studied but yielded little further information. Much more successful however was the use of proteolytic enzymes (11). Initially we had refrained from using them since it was considered that they might bring about re-arrangement of the peptide bonds by transpeptidation or actual reversal of hydrolysis. Subsequent work has however shown that this is not a very serious danger and in fact proteolytic enzymes are the most useful hydrolytic agent for studies of amino acid sequences.

Proteolytic enzymes are much more specific than is acid since only a few of the peptide bonds are susceptible. They give rise to larger peptides which in general are more difficult to fractionate by paper chromatography. However there are relatively few of them so that the mixtures are less complex. In this initial work we used essentially the same methods for studying the enzymic peptides as we had used for the acid ones, depending largely on paper chromatography for fractionation, although more recently it has been shown that better separations can be obtained by ion-exchange chromatography and by ionophoresis.

As an example we may consider a peptide Bp3 obtained by the action of pepsin. It had the following composition Phe($CySO_3H$, Asp, Glu, Ser, Gly, Val, Leu, His) of which the most important components are aspartic acid and serine since they occur only once in the chain. Aspartic acid is present only in the N-terminal sequence (1) and serine is in sequence 5. This shows that all of sequence 1 and at least the N-terminal part of sequence 5 is present in peptide Bp3. That none of the other sequences are present follows from the fact that Bp3 contains no arginine (sequence 2), threonine, proline, or lysine (sequence 3) or tyrosine (sequence 4). One may thus conclude that the two sequences are joined together. By studying other peptides obtained by the action of pepsin, trypsin and chymotrypsin it was possible to find out how the various sequences were arranged and to deduce the complete sequence of the phenylalanyl chain which is shown below:

Phe · Val · Asp · Glu · His · Leu · $CySO_3H$ · Gly · Ser · His · Leu · Val · Glu · Ala · Leu · Tyr · Leu · Val · $CySO_3H$ · Gly · Glu · Arg · Gly · Phe · Phe · Tyr · Thr · Pro · · Lys · Ala.

In this work many more peptides were studied from both acid and enzymic hydrolysates than were actually necessary to deduce the sequence. This was considered essential since the methods used were new and were qualitative rather than quantitative. The fact that all the peptides fitted into the unique sequence given above added further proof to its validity.

Essentially similar methods were used to determine the sequence of fraction A (12). Although the shorter of the two chains, the determination of its structure was more difficult. Fraction B contains several residues that occur only once in the molecule and this helps considerably in interpreting the results, whereas fraction A has only a few such residues and these are all near one end. Also fraction A is much less susceptible to enzymic hydrolysis. It is not attacked by trypsin and there is a sequence of thirteen residues which is not split by chymotrypsin or pepsin either. Considerable difficulty was at first experienced with the cysteic acid peptides. Fraction A contains the sequence $CySO_3H$ · $CySO_3H$ and this gave rise to a number of very water-soluble peptides which would not fractionate easily by paper chromatography. However it was found that by paper ionophoresis at pH 2.75 they could be well separated since they were the only acidic peptides present. At this pH — COOH groups

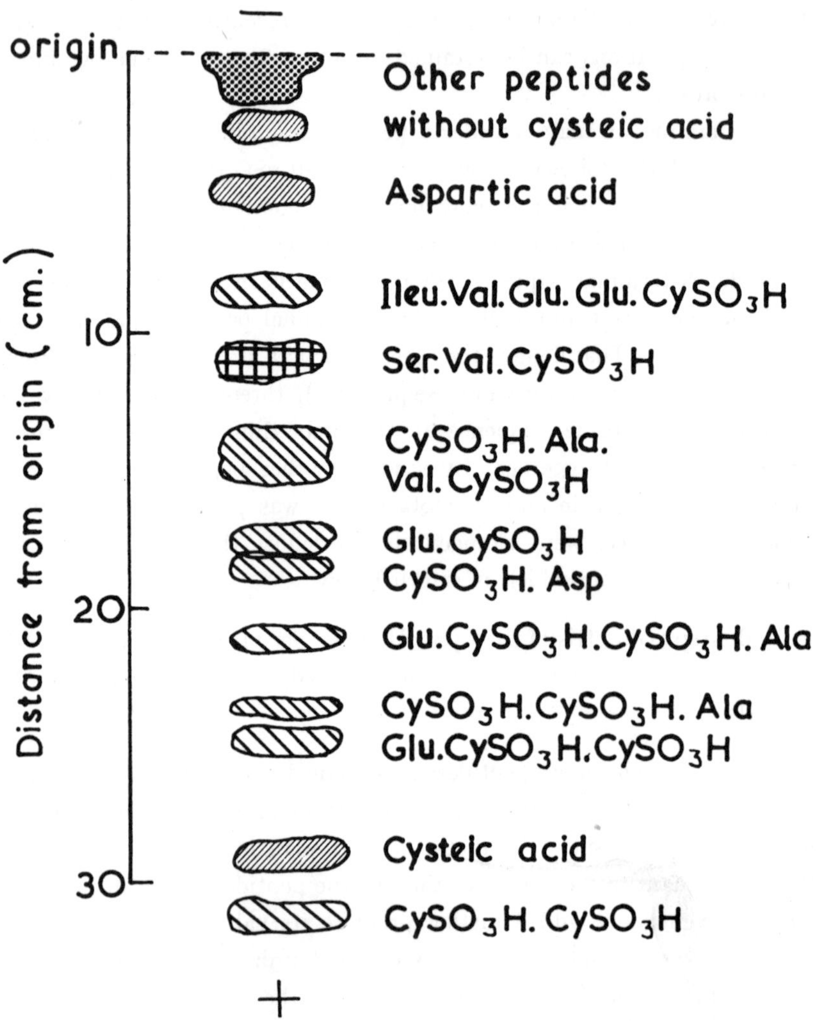

Fig. 1. Ionophoresis of partial acid hydrolysate of fraction A at pH 3,5 showing separation of cysteic acid peptides.

are uncharged, — SO_3H groups carry a negative and — NH_2 groups a positive charge. Peptides without cysteic acid were all positively charged, those with one cysteic acid were neutral and could be separated as a group and those with two cysteic acids were negatively charged. If a slightly higher pH (3,5) is used for the ionophoresis, the — COOH groups become slightly charged and all the peptides containing one cysteic acid residue move slowly towards the anode and can be fractionated in this way. This method was found very useful for the separation and identification of cysteic acid peptides. Fig. 1 is a tracing of an ionogram of an acid hydrolysate of fraction A carried out in this way.

The sequence of fraction A was found to be:

Gly · Ileu · Val · Glu · Glu · $CySO_3H$ · $CySO_3H$ · Ala · Ser · Val · $CySO_3H$ · Ser · Leu · Tyr · Glu · Leu · Glu · Asp · Tyr · $CySO_3H$ · Asp.

When a protein is hydrolysed with strong acid, it gives rise not only to amino acids but also to a certain amount of ammonia. This is present in the form of amide groups on some of the aspartic and glutamic acid residues. It was thus necessary to determine the position of these groups (13). This was done by studying the ionophoretic rates and amide contents of peptides derived from enzymic hydrolysates, since the amide groups are not split off by enzymes, whereas they are by acid. The position of the amide groups are indicated in fig. 2 by the symbols $\underset{|}{NH_2}$.

Having determined the structure of the two chains of insulin the only remaining problem was to find how the disulphide bridges were arranged. About this time it was shown by HARFENIST & CRAIG that the molecular weight of insulin was of the order of 6 000, so that it consisted of two chains containing three disulphide bridges, and not of four chains as we had originally thought. The fact that fraction A contained four cysteic acid residues whereas fraction B had only two indicated that two bridges must connect the two chains together and one must form an intrachain bridge connecting one part of the A chain with another part of the same chain.

In order to determine the distribution of the disulphide bridges, it was necessary to isolate from unoxidized insulin peptides containing intact cystine residues. These could then be oxidized to give cysteic acid peptides which could be recognised since they had been found in the hydrolysates of the oxidized chains. However an unexpected difficulty arose, in that during hydrolysis a reaction occurred which caused a random rearrangement of the disulphide bonds, so that cystine peptides were isolated which were not actual fragments of the original insulin and it would have appeared from the results that every half-cystine was combined to every other half-cystine residue.

This disulphide interchange reaction could be demonstrated and studied using as a model system a mixture of cystine and bis-DNP cystine, which reacted together to give mono-DNP-cystine

$$\begin{array}{r}\text{DNP — Cy — S}\\ |\\ \text{DNP — Cy — S}\end{array} + \begin{array}{r}\text{Cy—S}\\ |\\ \text{Cy—S}\end{array} \rightleftarrows 2\ \begin{array}{r}\text{DNP — Cy—S}\\ |\\ \text{Cy—S}\end{array}$$

An ether-soluble coloured substance was thus converted to a water-soluble coloured substance and the course of the reaction could be studied by measuring the distribution of colour between ether and water.

It was found that there were two types of disulphide interchange reactions (14). One took place in neutral and alkaline solution and was catalysed by — SH compounds. It is probably due to initial hydrolysis of the disulphide which then catalyses a chain reaction:

$$R_1\,SSR_2 \xrightarrow{OH^-} R_1S^- + R_2SOH$$

$$R_1S^- + R_3SSR_4 \longrightarrow R_1SSR_3 + R_4S^- \qquad \text{etc.}$$

In neutral conditions the reaction could be inhibited by — SH inhibitors so that it was possible to use enzymic hydrolysis to obtain cystine peptides (15). Thus for instance with chymotrypsin a peptide was obtained which on oxidation gave the two cysteic acid peptides $CySO_3H$ · $AspNH_2$ and Leu · Val · $CySO_3H$ · Gly · Glu · Arg · Gly · Phe · Phe. The structure of the cystine peptide was thus:

```
           Cy · AspNH2
           |
           S
           |
           S
           |
Leu · Val · Cy · Gly · Glu · Arg · Gly · Phe · Phe
```

establishing the presence of a disulphide bridge between the two half-cystine residues nearest the C-terminal ends of the two chains.

It was not, however, possible to determine the positions of the other two disulphide bonds using enzymic hydrolysis, since no enzyme would split between the two consecutive half-cystine residues of the A chain. It was therefore necessary to re-investigate the possibility of using acid hydrolysis.

The disulphide interchange reaction that occurred in acid solution was found to be different from that occurring in neutral and alkaline solution and instead of being catalysed by — SH compounds was actually inhibited by them. Not only did this show that a different reaction was involved but it also made it possible to prevent it occurring during acid hydrolysis. Thus when insulin was treated with concentrated acid to which a small amount of thioglycolic acid was added cystine peptides could be isolated which were in fact true breakdown products and from which the distribution of the remaining two disulphide bonds could be deduced. These are shown in Fig. 2, which shows the complete structure of insulin.

145

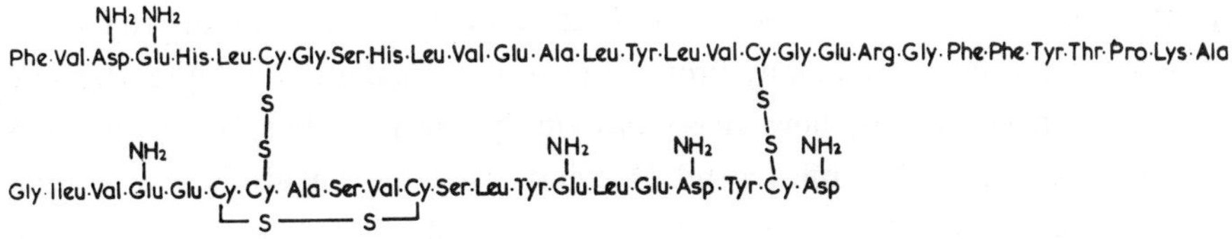

Fig. 2. The structure of insulin.

Of the various theories concerned with protein chemistry our results supported only the classical peptide hypothesis of HOFMEISTER and FISCHER. The fact that all our results could be explained on this theory added further proof, if any were necessary, to its validity. They also showed that proteins are definite chemical substances possessing a unique structure in which each position in the chain is occupied by one and only one amino acid residue.

Examination of the sequences of the two chains reveals no evidence of periodicity of any kind nor does there seem to be any basic principle which determines the arrangement of the residues. They seem to be put together in a random order, but nevertheless a unique and most significant order, since on it must depend the important physiological action of the hormone.

As yet little is known about the relationship of the physiological action of insulin to its chemical structure. One approach to this problem was to study the insulins from different animal species (16, 17). Since all insulins show the same activity it could be concluded that differences would be found only in parts of the molecule that were not important for activity.

All the above results were obtained on cattle insulin. When insulins from four other species were studied by essentially the same methods it was found that the whole of the B chain was identical in all species and the only differences were found in the three amino acids contained within the disulphide ring of the A chain, which in the cattle are Ala·Ser·Val and in the other species are as follows:

Pig	Thr·Ser·Ileu
Sheep	Ala·Gly·Val
Horse	Thr·Gly·Ileu
Whale	Thr·Ser·Ileu

These results suggest that the exact structure of the residues in this position is not important for biological activity, but it does not follow that the whole of the rest of the molecule is important.

The determination of the structure of insulin clearly opens up the way to similar studies on other proteins and already such studies are going on in a number of laboratories. These studies are aimed at determining the exact chemical structure of the many proteins that go to make up living matter and hence at understanding how these proteins perform their specific functions on which the processes of life depend. One may also hope that studies on proteins may reveal changes that take place in disease and that our efforts may be of more practical use to humanity.

REFERENCES.

1. A. C. CHIBNALL, Proc. Roy. Soc. B, *131*, 136 (1942).
2. H. JENSEN and E. A. EVANS, J. Biol. Chem. *108*, 1 (1935).
3. H. GUTFREUND, Biochem. J. *42*, 544 (1948).
4. E. J. HARFENIST and L. C. CRAIG, J. Amer. Chem. Soc. *74*, 3087 (1952).
5. F. SANGER, Biochem. J. *39*, 507 (1945).
6. A. H. GORDON, A. J. P. MARTIN and R. L. M. SYNGE, Biochem. J. *37*, 79 (1943).
7. F. SANGER, Biochem. J. *44*, 126 (1949).
8. F. SANGER, Biochem. J. *45*, 563 (1949).
9. R. CONSDEN, A. H. GORDON, A. J. P. MARTIN and R. L. M. SYNGE, Biochem. J. *41*, 596 (1947).
10. F. SANGER and H. TUPPY, Biochem. J. *49*, 463 (1951).
11. — — Biochem. J. *49*, 481 (1951).
12. — and E. O. P. THOMPSON, Biochem. J. *53*, 353, 366 (1953).
13. — — and R. KITAI, Biochem. J. *59*, 509 (1955).
14. A. P. RYLE and F. SANGER, Biochem. J. *60*, 535 (1955).
15. — — L. F. SMITH and R. KITAI, Biochem. J. *60*, 541 (1955).
16. H. BROWN, F. SANGER and R. KITAI, Biochem. J. *60*, 556 (1955).
17. J. I. HARRIS, F. SANGER and M. A. NAUGHTON, Arch. Biochem. Biophys. *65*, 427 (1956).

Stockholm 1959. Kungl. Boktr. P. A. Norstedt & Söner

Reprinted from
Biochim. Biophys. Acta

VOL. **34** (1959) 243

The amino acid sequence around the reactive serine of elastase

The group of hydrolytic enzymes which are characterized by being inhibited by DFP contain an unusually reactive serine residue. In the case of trypsin[1] and chymotrypsin[2,3] this serine is present in the sequence Gly·Asp·Ser·Gly· and in other enzymes of the group[4-6] it appears to be present in this or a very similar sequence. Elastase has recently been shown[7] to have an entirely different specificity from trypsin or chymotrypsin and to be inhibited by DFP. The following experiment shows that it too has a similar sequence around its reactive serine residue.

Elastase was prepared by the method of LEWIS, WILLIAMS AND BRINK[8], and purified on a carboxymethylcellulose column[7]. 40 mg of the purified material were dissolved in 25 ml 0.067 *M* phosphate buffer (pH 8.0) and treated with 1.25 ml 0.02 *M* DFP in isopropanol. (The DFP contained 3.5 mC ^{32}P/mg). After incubation for 3 h at 37° the mixture was dialyzed overnight against 5 l 0.01 *M* acetic acid at 2° and freeze-dried. A sample of the di-isopropyl phosphoryl-elastase was then run on a carboxymethylcellulose column. It emerged as a radioactive peak which moved in the same position as unchanged elastase but had no activity against elastin. From the specific radioactivity of the purified material it was calculated that it was 90 % pure assuming a molecular weight[8] of 25,000 and that one mole DFP reacted with one mole elastase.

The di-isopropyl phosphoryl derivatives of trypsin and chymotrypsin were prepared similarly but without chromatographic purification. Samples (0.1 mg) of the derivatives were incubated for 3 days at 37° with 12 *N* HCl. After removal of the HCl *in vacuo*, the hydrolysates were subjected to ionophoresis (40 V/cm, 2 h) on Whatman No. 52 paper in pyridine–acetate buffer, pH 3.5, in the apparatus of MICHL[9]. Each sample was applied over 2 cm of the paper and there was slight overlapping (about 0.2 cm) between the different samples. A similar hydrolysate of ^{32}P-labelled ovalbumin was applied for comparison.

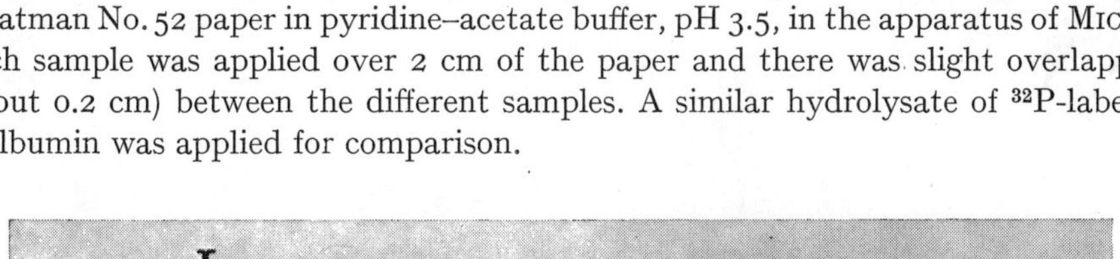

Fig. 1. Radioautograph of ionogram of partial acid hydrolysates of di-isopropyl [^{32}P]phosphoryl derivations of chymotrypsin, elastase, trypsin and of ^{32}P-labelled ovalbumin.

Abbreviation: DFP, di-isopropyl phosphorofluoridate.

VOL. **34** (1959) SHORT COMMUNICATIONS 244

Fig. 1 is a radioautograph of the ionogram. The free phosphate band has been allowed to run off the anode end of the paper. The exact correspondence of the radioactive bands from the three di-isopropyl phosphoryl-enzymes indicates that the hydrolysates contain identical peptides of phosphoserine which are entirely different from those derived from ovalbumin. The structure of the different peptide bands is not certain at present but is under investigation. From the number present and their identity in the three enzymes it is probably safe to conclude that elastase, like trypsin and chymotrypsin contains the sequence Gly·Asp·Ser·Gly· around its reactive serine residue.

Department of Biochemistry, University of Cambridge (Great Britain)

B. S. HARTLEY
M. A. NAUGHTON
F. SANGER*

1 G. H. DIXON, D. L. KAUFFMAN AND H. NEURATH, *J. Am. Chem. Soc.*, 80 (1958) 1260.
2 F. TURBA AND G. GUNDLACH, *Biochem. Z.*, 327 (1955) 186.
3 N. K. SCHAFFER, L. SIMEL, S. HARSHMAN, R. R. ENGLE AND R. W. DRISKO, *J. Biol. Chem.*, 225 (1957) 197.
4 P. E. KOSHLAND AND M. J. ERWIN, *J. Am. Chem. Soc.*, 79 (1957) 2657.
5 R. A. OOSTERBAHN, H. S. JANSZ AND J. A. COHEN, *Biochim. Biophys. Acta*, 20 (1956) 402.
6 J. A. GLADNER AND K. LAKI, *J. Am. Chem. Soc.*, 80 (1958) 1264.
7 M. A. NAUGHTON AND F. SANGER, *Biochem. J.*, 70 (1958) 4 P.
8 U. J. LEWIS, D. E. WILLIAMS AND N. G. BRINK, *J. Biol. Chem.*, 222 (1956) 705.
9 H. MICHL, *Monatsh. Chem.*, 82 (1951) 489.

Received November 27th, 1958

* Member of the external staff of the Medical Research Council.

Printed in The Netherlands

Biochem. J. (1960) **77**, 149
Printed in Great Britain

The Amino Acid Sequence around the Reactive Serine Residue of some Proteolytic Enzymes

By M. A. NAUGHTON,* F. SANGER,† B. S. HARTLEY and D. C. SHAW
Department of Biochemistry, University of Cambridge

(*Received* 22 *February* 1960)

The three endopeptidases of the pancreas, chymotrypsin, trypsin and elastase, are stoicheiometrically inhibited by di*iso*propyl phosphorofluoridate, which combines with an unusually reactive serine residue in the enzymes. In chymotrypsin and trypsin this residue has been shown to be present in the sequence Gly.Asp.Ser.Gly (Turba & Gundlach, 1955; Schaffer, Simet, Harshman, Engle & Drisko, 1957; Oosterbaan, Kunst, Van Rotterdam & Cohen, 1958; Dixon, Kauffman & Neurath, 1958). [For definitions of the abbreviations used for amino acids in this paper see *Biochem. J.* (1957), **66**, 6.] That elastase contains the same sequence was concluded from experiments in which a partial acid hydrolysate of the di*iso*propoxy[^{32}P]phosphinyl derivative of elastase was subjected to paper ionophoresis and shown to contain the same radioactive peptides as did similar hydrolysates of the di*iso*propoxy[^{32}P]-phosphinyl derivatives of chymotrypsin and trypsin (Hartley, Naughton & Sanger, 1959). The present paper describes some further experiments with two-dimensional ionophoresis which confirm the above-mentioned result.

Considerable difficulties were encountered when we attempted to identify the various radioactive peptide spots present in these hydrolysates. The

* Present address: Department of Biology, Massachusetts Institute of Technology, Cambridge 39, Mass., U.S.A.

† Member of the external staff of the Medical Research Council.

mixture was much more complex than would have been expected and the results could not be explained on the simple assumption that the only reaction taking place during the partial hydrolysis of a protein by acid was the splitting of peptide and amide bonds. Thus, for instance, when hydrolysis was carried out for 3 days in 12N-hydrochloric acid, at least three peptides were present which appeared to have the structure aspartyl–serine phosphate (Asp.SerP) and were interconvertible on further hydrolysis. Another three appeared to have the structure Asp.SerP.Gly. Since the serine phosphate residue in the radioactive proteins was present as the di*iso*propyl derivative heterogeneity was to be expected if the removal of the *iso*propyl groups by acid was not complete (Schaffer *et al.* 1957), and in fact small amounts of mono*iso*propyl derivatives were detected. The main difficulties, however, appeared to be due to some interconversions of the aspartyl residues and the results could best be explained by assuming that the reactions shown in Scheme 1 were taking place.

on partial hydrolysis and the effect of degradations such as acid hydrolysis and the Edman procedure on these peptides.

Thus, for instance, it should be possible to determine the number of residues in a labelled peptide from the products it gives on partial acid hydrolysis. A dipeptide should give only two labelled products, unchanged dipeptide and the free amino acid. A tripeptide in which the labelled residue is terminal would give itself, the amino acid and one new dipeptide. A tripeptide with the labelled residue in the middle should give two dipeptides and so forth. Similarly, some of the interrelationships of the different peptides may be deduced by studying the products produced after removal of the *N*-terminal residues by the Edman procedure. It was considered that the present system, in which a single residue in a known sequence is labelled, would be a good one on which to work out the methods. We have thus studied the effect of partial acid hydrolysis on the various bands in some detail and these results have contributed considerably to their identification.

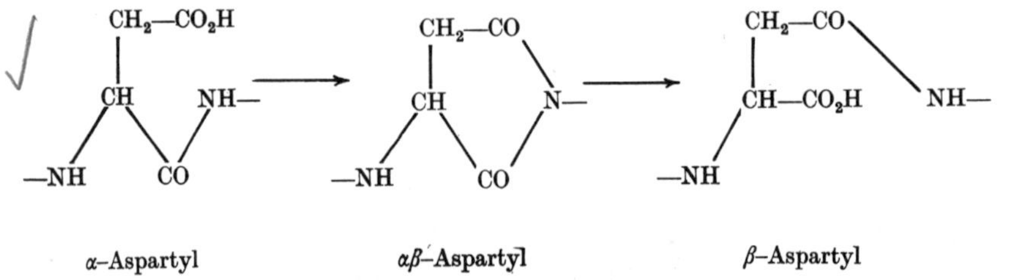

Scheme 1

That the reaction in Scheme 1 can occur during acid hydrolysis was shown by Swallow & Abraham (1958), who isolated ε-(αβ-aspartyl)lysine [referred to by them as ε-(aminosuccinyl)lysine] from a partial hydrolysate of bacitracin A. In model experiments they also showed that ε-(α-aspartyl)-lysine and ε-(β-aspartyl)lysine were interconvertible but were both largely converted into the αβ-aspartyl derivative. They found that an αβ-aspartyl derivative was formed from α-(α-aspartyl)-lysine but was not as stable as the ε-(αβ-aspartyl)-lysine. The present results agree with those of Swallow & Abraham, and suggest that the above-mentioned reactions will be of general occurrence in experiments where partial acid hydrolysis is used on proteins or peptides containing aspartic acid.

One of the objects of this investigation was to try to develop methods for identifying the amino acid sequence in the vicinity of a known isotopically labelled residue when only small amounts of material are available. Such methods would involve studying the electrophoretic and chromatographic rates of the radioactive peptides produced

Where enough material was available we have identified the amino acids present in the peptides after complete hydrolysis and in some cases have determined the *N*-terminal residues by the 2:4-dinitrophenyl technique.

EXPERIMENTAL

The following abbreviations for amino acid residues will be used in this paper: SerP, *O*-phosphoserine; SerIP, mono*iso*propyl phosphoserine; SerDIP, di*iso*propyl-phosphoserine.

Materials

Trypsin and chymotrypsin were crystalline salt-free preparations from Worthington Biochemical Corp. Elastase was prepared by a method to be described (Naughton & Sanger, 1960).

Di*iso*propyl [^{32}P]phosphorofluoridate (DF^{32}P) was a gift from Dr R. Davies and usually contained between 1 and 30 μc of ^{32}P/mg.

*Preparation of di*iso*propoxy[^{32}P]phosphinyl enzymes*

Various preparations of the di*iso*propoxy[^{32}P]phosphinyl (formerly 'di*iso*propylphosphoryl') derivatives were made. A satisfactory procedure was as follows.

The enzyme (25 mg.) was dissolved in 5 ml. of 0·1 M-phosphate buffer, pH 8·0, and treated with 0·63 ml. of 0·02 M-DF^{32}P in propan-2-ol. After incubation at room temperature for 3hr. the mixture was dialysed against two changes of 5 l. of 0·01 M-acetic acid at 2° and freeze-dried.

*Hydrolysis of di*isopropoxy[^{32}P]*phosphinyl enzymes and peptides*

Three methods of hydrolysis were used: (*a*) hydrolysis with 12 N-HCl at 37°; (*b*) hydrolysis with 5·7 N-HCl at 100° in a boiling-water bath; (*c*) hydrolysis with 5·7 N-HCl at 37°.

In general, where only small amounts of protein or peptides were to be hydrolysed the reaction was carried out in small test tubes with about 0·2 ml. of HCl. The HCl was then removed *in vacuo* over NaOH, water was added and the residue dried off about three times. In some experiments redistilled 5·7 N-HCl was used, but this made no apparent difference to the course of hydrolysis. As a general approximation it would appear that treatment for 10 min. by method (*b*) is equivalent to 24 hr. by method (*a*).

Ionophoresis of peptides

Paper ionophoresis was normally carried out by the method of Michl (1951) (Ryle, Sanger, Smith & Kitai, 1955). Except where otherwise stated, the separations were done on Whatman no. 52 filter paper with a voltage gradient of 40 v/cm. The most useful buffer for the fractionation of the phosphoserine peptides was pyridine–acetate, pH 3·5 (pyridine–acetic acid–water; 1:10:189, by vol.). The relative rates of the different peptides are very sensitive to slight changes in pH and it will be noted that the fractionations obtained in this work are rather different from those reported previously (Hartley *et al.* 1959). Although the buffer used was the same it seems that the actual pH on the paper will depend on the amounts of pyridine and acetic acid dissolved in the toluene which will be in equilibrium with the buffer on the paper. In the present experiments the toluene contained 0·3% (v/v) of acetic acid. The buffer at pH 6·5 contained pyridine–acetic acid–water (10:0·4:90, by vol.) and the toluene used in the tank with it contained 8% (v/v) of pyridine.

Each peptide fraction to be analysed, which should not contain more than about 0·1 mg. of material, was applied as a thin line over 2 cm. to the dry paper. Samples to be compared, as for instance in comparing the hydrolysates of the di*iso*propoxyphosphinyl derivatives (e.g. Fig. 11), or where peptides were run with a control mixture for identification, were applied with no gap between them but with slight overlapping (about 0·2 cm.). In this way a rigorous test for the identity of any two bands is obtained, since identical substances should give a continuous band between the two samples. After application of the samples the rest of the paper was wetted with buffer in such a way that it flowed evenly in towards the origin from each side, thus sharpening the bands of the applied samples.

In order to obtain satisfactory radioautographs each sample applied should contain at least 1 μmc of ^{32}P. To prepare the pH: mobility curves (Fig. 10) ionophoresis was carried out between cooled plates (Gross, 1955), since there was less flow of buffer into the paper from the electrode vessels than in the toluene-cooled system.

Two-dimensional ionophoresis. In order to prepare comparative two-dimensional ionophoresis patterns as shown in Fig. 12, the following technique was found useful. The samples to be compared were applied side by side and run in one dimension; the paper was then dried and a radioautograph was prepared (Fig. 11). A strip 0·5 cm. wide was then cut from the part of the paper that was to be run in a second dimension and the peptides were transferred to another sheet of paper by blotting. To do this a pad of clean filter paper slightly larger than the strip was wetted with water and blotted so that there was no excess of moisture. The strip was then wetted by pressing on to the pad with a thin glass plate. It was then transferred and pressed firmly on to the origin line of a clean sheet of paper. The process was repeated once or twice. This method transferred about 50% of the material to the new paper, which was then subjected to ionophoresis in a different buffer. An alternative method of doing two-dimensional ionophoresis which gives rather compact spots is to use cellulose acetate membrane (Kohn, 1957) as support during the first run. The cellulose acetate strip was wetted with buffer (pH 3·5) before application of the sample. Otherwise it was run under toluene in the same manner as the paper strips and a radioautograph prepared. The bands were sharp but usually rather uneven (see Fig. 5*a*). There was considerably more electroendosmosis than on paper and the relative rates of some of the peptides were slightly different although the buffer was the same. A strip about 0·5 cm. wide was cut from the cellulose acetate at a place where the bands were running evenly. A sheet of Whatman

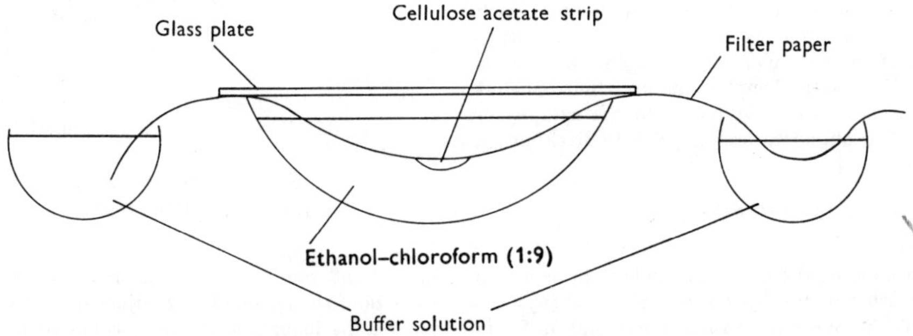

Fig. 1. Method of transferring material from a cellulose acetate strip to filter paper for two-dimensional ionophoresis.

no. 52 paper was wetted with buffer at pH 6·5 and the cellulose acetate strip pressed on to the origin line so that it was wetted with the buffer. While in this position it was wetted with a small amount of ethanol–chloroform (1:9) from a pipette. This was allowed to evaporate, thus sticking the cellulose acetate to the paper. The whole sheet was then inverted and the part around the origin line was soaked in ethanol–chloroform (1:9) as shown in Fig. 1. The paper was kept wet by immersion of its ends in troughs containing the buffer solution. The cellulose acetate dissolved in the ethanol–chloroform, leaving the radioactive spots on the paper. After soaking for about 2 hr. the paper was transferred to the ionophoresis tank for running at pH 6·5. Fig. 5*b* is a radioautograph of a paper on which a two-dimensional ionophoresis had been carried out in this way.

Isolation of peptides

In most experiments in which the peptides were studied by radioautography only small amounts of the hydrolysates were used and no preliminary separation of the phosphoserine peptides was necessary. However, in order to obtain sufficient amounts of peptides to determine their amino acid content and end groups, two experiments were done on a somewhat larger scale. Di*iso*propoxy[^{32}P]phosphinyl-chymotrypsin (100–200 mg.) was hydrolysed with acid (method *a* for 2 days or method *b* for 20 min.). The HCl was removed *in vacuo* with a rotary evaporator and the residue was dissolved in 2 ml. of water and applied to a column (2 cm. diam.) of Dowex-50 (10 g., in the acid form). The column was developed with water. The material coming through with the solvent front was collected and contained over 90% of the added radioactivity. This treatment, which removes the majority of the peptides not containing serine phosphate (Flavin, 1954), was necessary to avoid overloading during ionophoresis. After repeated evaporation *in vacuo* the peptide mixture was applied as a 35 cm. band 10 cm. from the cathode end of a sheet of Whatman no. 52 paper. Ionophoresis was carried out at pH 3·5 at 40 v/cm. for 2·5 hr. A radioautograph was prepared, a part of which is shown in Fig. 4. The bands were cut out and eluted with water. Each peptide was purified by ionophoresis at pH 6·5. Samples were taken for hydrolysis and for end-group determination by the 2:4-dinitrophenyl technique. Where possible 0·1 μmole of peptide was used for hydrolysis. The amounts of the various peptides were estimated from the radioactivity.

The amino acids in the hydrolysates were identified by ionophoresis in 2·5% (v/v) formic acid for 25 min. at 80 v/cm. This system gives very sharp bands and clearly separates all the amino acids expected in these peptides (Ser, Asp, Gly, Glu, Ala). *N*-Terminal residues were determined by the 2:4-dinitrophenyl technique, dinitrophenyl-amino acids being identified by paper chromatography on *tert.*-amyl alcohol–phthalate buffer, pH 5·0 (Blackburn & Lowther, 1951).

Edman degradation

In the Edman degradation it is usual to identify the *N*-terminal residue as the phenylthiohydantoin or as the amino acid missing from a hydrolysate of the phenylthiocarbamyl peptide. We have used a somewhat different approach and have identified the radioactive product produced. This gives considerable information about the relationships between the different peptides. Various modifications of the original Edman method have been employed. Most of the results reported here were obtained by the procedure described below, which is a combination of various published methods (Edman, 1950; Fraenkel-Conrat & Harris, 1954).

The radioactive peptide (containing at least 1 μmc of ^{32}P) was taken to dryness in a small test tube and dissolved in 0·1 ml. of water. Phenyl*iso*thiocyanate–pyridine (0·1 ml.; 1:9, v/v) was added and the mixture shaken at room temperature for 3–5 hr. Benzene (2 ml.) and a few drops of water were added, and the mixture was shaken and the benzene layer removed. The extraction was repeated and the aqueous residue taken to dryness over H_2SO_4 in a desiccator. After addition of water and evaporation two or three times, 0·2 ml. of 3 N-HCl was added and the mixture allowed to stand for 5 hr. at room temperature. The HCl was removed *in vacuo* over NaOH and the residue subjected to ionophoresis at pH 3·5. A partial hydrolysate of di*iso*propoxy[^{32}P]phosphinyl-chymotrypsin was run in parallel as a control in order to be able to identify the product.

We have also used the method of Sjöquist (1957) on some of the peptides. Little difference was noticed between the methods, but it was observed that some peptides gave much clearer results than others. Those containing *iso*propyl groups appeared to react better than those with unesterified phosphate. When peptides containing *N*-terminal serine phosphate (e.g. peptide 10 and some peptides from ovalbumin; F. Sanger, unpublished work) were subjected to this procedure, the only product present in significant amounts migrated at the same rate as inorganic phosphate. It is unlikely that the phenylthiohydantoin of serine phosphate, which is the expected product, would move so fast and it appears that the compound is unstable and breaks down to give phosphate. Similarly, peptide 11 (SerIP.Gly) breaks down to give *iso*propyl phosphate (band 2).

Radioautographs

The filter paper to be radioautographed was stuck with cellulose self-adhesive tape into a cardboard filing folder, one side of which was covered with lead sheeting (0·25 mm. thick). One or two sheets of Ilford Industrial G X-ray film were then stuck in the folder, which was kept in a light-proof box until ready for developing. Where individual bands from a sample put on over 2 cm. paper contained 1 μmc of ^{32}P the film could be developed after about 24 hr. If the bands contained 0·1 μmc about 10–15 days were required.

In order to mark the radioautographs and to be able to align them with the filter papers, the papers were marked with a radioactive pencil: commercial propelling pencil leads were immersed overnight in radioactive residues (^{14}C or ^{35}S) of high specific activity and then dried in an oven at 105°. The leads were then used in a propelling pencil.

*Time course of hydrolysis of di*iso*propoxy[^{32}P]phosphinyl enzymes*

The di*iso*propoxy[^{32}P]phosphinyl enzyme was dissolved in HCl and samples were measured into small test tubes, the size of the sample being adjusted so that the amounts put on the ionophoresis paper would be constant (about 1 μmc of ^{32}P/cm.). In most experiments each sample contained about 2 μmc and was put on the ionophoresis paper over a 2 cm. width; however, in the experiment illustrated in Fig. 2 the size of the samples varied. Each sample was incubated at 37° or 100° for the appropriate period and after removal of the HCl *in vacuo* they were subjected to ionophoresis at pH 3·5, being put on side by side in the appropriate order (see Figs. 2 and 3).

RESULTS

Most of the results reported in this section were obtained with chymotrypsin or trypsin. However, they apply equally well to elastase since no major differences could be detected in the radioactive peptides obtained from the di*iso*propoxy[^{32}P]-phosphinyl derivatives of the three enzymes (see below).

*Course of hydrolysis of di*iso*propoxy*[^{32}P]*phosphinyl enzymes*

Fig. 2 shows the course of hydrolysis of di*iso*-propoxy[^{32}P]phosphinyl-trypsin by method (*b*). It can be seen that the main stable derivatives are in bands 1, 4 and 10. Bands, 3, 6, 7 and 9B are also relatively stable but are present in smaller amounts. It seemed that a period of hydrolysis of 20–30 min. gave a suitable mixture of peptides, and most of the work was carried out under these conditions. In some experiments method (*a*) was used and it was found that a suitable time of hydrolysis was 2–3 days. There were no apparent qualitative differences between the two methods of hydrolysis, so that to save time method (*b*) was normally used.

In an attempt to identify larger peptides, hydrolysates prepared by method (*c*) were also studied. The course of this reaction is shown in Fig. 3 and appears to be somewhat different from that with methods (*a*) and (*b*). A 3-day period of hydrolysis gave relatively high concentrations of a number of bands, such as 7B, 9A, 11, 14, 15 and 18, which were absent or present only in small amounts in the other more complete hydrolysates, and this time of hydrolysis was then used to study these slower-moving peptides.

Identification of radioactive bands

Fig. 4 shows a radioautograph of a typical ionophoresis experiment of a partial acid hydrolysate of di*iso*propoxy[^{32}P]phosphinyl-chymotrypsin. On running again at pH 6·5, most of the bands were essentially pure though there were usually smaller amounts of other contaminants. Fig. 5 shows a two-dimensional ionophoresis run first on cellulose acetate at pH 3·5 and then on paper at pH 6·5, as described in the Experimental section, and the numbering of the various peptide spots is

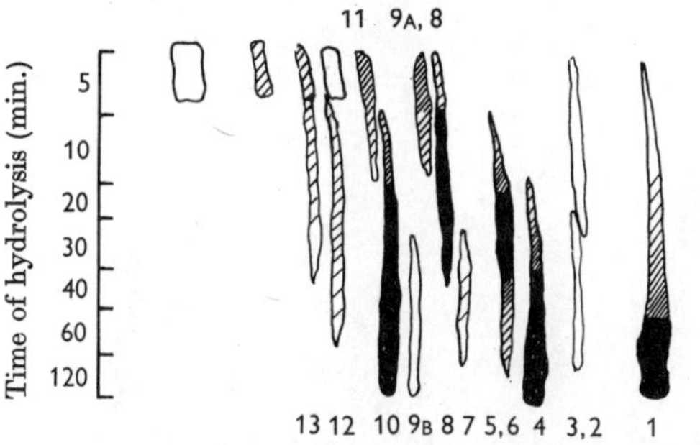

Fig. 2. Time course of hydrolysis of di*iso*propoxy[^{32}P]-phosphinyl-trypsin by method (*b*). Samples of di*iso*propoxy[^{32}P]phosphinyl-trypsin were hydrolysed for different periods and subjected to ionophoresis at pH 3·5, 40 v/cm., 2 hr.

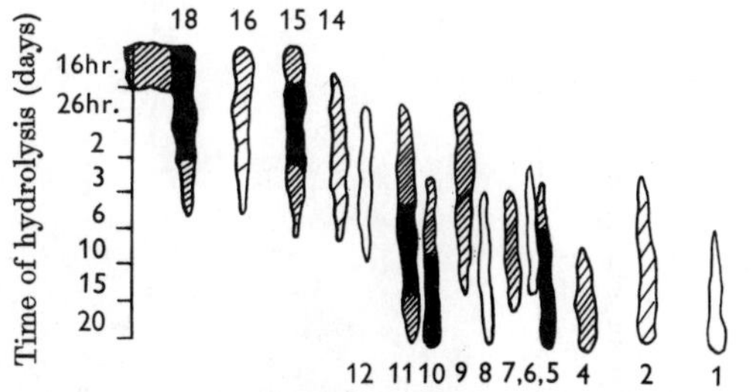

Fig. 3. Time course of hydrolysis of di*iso*propoxy[^{32}P]-phosphinyl-trypsin by method (*c*). Samples were hydrolysed for different periods and subjected to ionophoresis at pH 3·5, 40 v/cm., 2 hr.

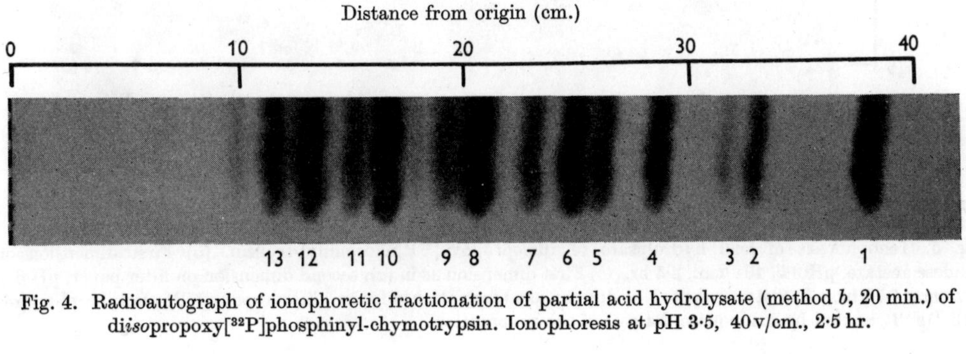

Fig. 4. Radioautograph of ionophoretic fractionation of partial acid hydrolysate (method *b*, 20 min.) of di*iso*propoxy[^{32}P]phosphinyl-chymotrypsin. Ionophoresis at pH 3·5, 40 v/cm., 2·5 hr.

shown in Fig. 6. This is a diagram of a two-dimensional ionophoresis pattern carried out entirely on filter paper and is thus slightly different from Fig. 5. It incorporates all the main peptides studied in this work, including those in the hydrolysate obtained by method (*c*) (3 days).

The yields of peptides from a hydrolysate (method *a*, 2 days) of di*iso*propoxy[^{32}P]phosphinyl-chymotrypsin are given in Table 1. The total recovery of ^{32}P is low (62 %) so that there have been considerable losses, probably mainly during elution of the peptides from the filter paper and running again at pH 6·5.

The results of the large-scale experiments, in which some of the peptides were hydrolysed to identify the amino acids present, and in which the end groups of some peptides were determined, are shown in Table 2. This method of analysis clearly requires that the peptides should be free of any contaminating non-radioactive peptides. Most of these should have been removed on the Dowex-50 column and should not move as acids at pH 3·5.

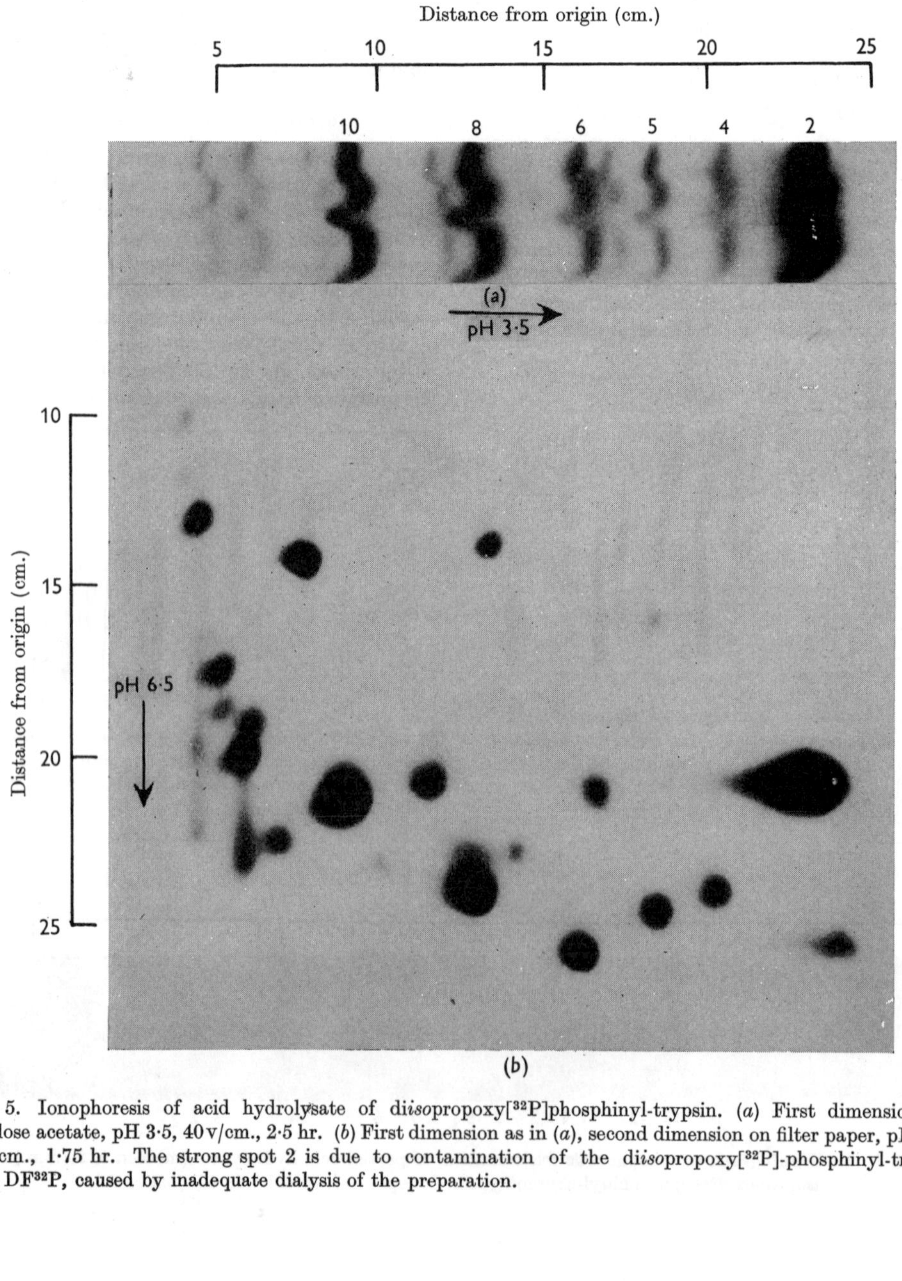

Fig. 5. Ionophoresis of acid hydrolysate of di*iso*propoxy[^{32}P]phosphinyl-trypsin. (*a*) First dimension on cellulose acetate, pH 3·5, 40 v/cm., 2·5 hr. (*b*) First dimension as in (*a*), second dimension on filter paper, pH 6·5, 40 v/cm., 1·75 hr. The strong spot 2 is due to contamination of the di*iso*propoxy[^{32}P]-phosphinyl-trypsin with DF^{32}P, caused by inadequate dialysis of the preparation.

However, the presence of glutamic acid in the hydrolysate of band 5A before purification at pH 2·1 indicates that some contamination may take place. It was at first considered that peptide 5A was [Asp, SerP, Gly, Glu]. However, this could not be reconciled with the partial-hydrolysis results, so that further purification was attempted.

Each band was partially hydrolysed (usually by method *b*, 30 min.) and subjected to ionophoresis at pH 3·5. A sample of a hydrolysate of di*iso*propoxy[^{32}P]phosphinyl-chymotrypsin was run adjacent to each peptide hydrolysate as a control in order to be able to identify the various degradation products. The results are given in Table 3. The yields given were estimated by eye from the darkening of the radioautographs and indicate the relative amount of each product obtained from a given quantity of each peptide band. The letters 'tr' indicate the presence of a small but significant amount of the product. A small amount of free phosphate (band 1) was obtained from each band but this is not included in the Table 3.

Since the course of hydrolysis by method (*c*) is somewhat different, the main products obtained by this method were also subjected to a further hydrolysis for 3 days to show their inter-relationships. These results are given in Table 4.

Samples of some of the peptides were subjected to the Edman degradation procedure and the product was identified by ionophoresis at pH 3·5. The results were not always entirely clear-cut. Unchanged material was frequently present and varying amounts of free phosphate. The results of experiments where the main product of the degradation was clear are given in Table 3.

Two of the bands could be identified directly by comparison of their rates with marker compounds of known structure. These were band 1, which was identified as free phosphate, and band 4, which was serine phosphate. Band 2 is present together with band 1 in a mild acid hydrolysate (e.g. method *c*, 1–3 days) of DF^{32}P. On hydrolysis it gives rise to band 1 only and appears to be mono*iso*propyl phosphate.

The only peptide band that was readily identifiable was band 10, which behaved essentially as a dipeptide on partial acid hydrolysis, giving only serine phosphate in significant amounts. It was

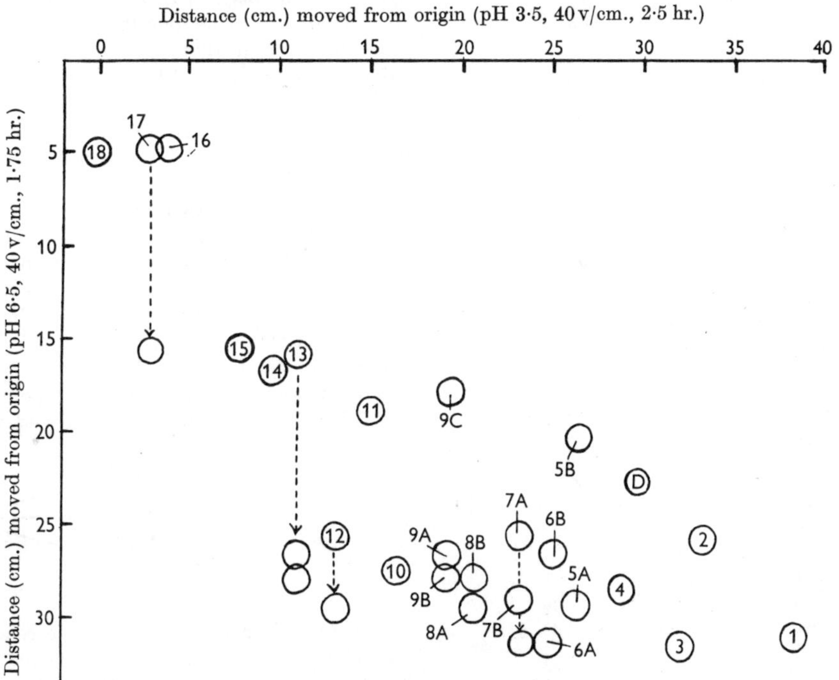

Fig. 6. Diagram showing the migration rates at pH 3·5 and pH 6·5 of radioactive compounds obtained from partial acid hydrolysates of di*iso*propoxy[^{32}P]phosphinyl-chymotrypsin. Ionophoresis on Whatman no. 52 filter paper. The materials were applied near to the cathode end of the paper. There was considerable flow of buffer into the paper from both ends during the ionophoresis. Spot D shows the position occupied by di*iso*propyl phosphate. The broken arrows indicate the changes that take place during running at pH 6·5.

the most stable peptide (Fig. 2) and gave serine and glycine on complete hydrolysis. It was clearly SerP.Gly.

Although the main product of partial hydrolysis of SerP.Gly was serine phosphate, two other bands were given in small amounts. They moved at the rates of bands 9B and 10*x* (Fig. 7). Peptide 9B appears during the later stages of hydrolysis by method (*b*) and is very stable. On partial hydrolysis it gives band 4 and a trace of band 10. It is formed in larger amounts if peptide 10 is hydrolysed in more dilute acid (e.g. 3N-HCl, 100°; Fig. 7, no. 4). On Edman degradation it gave serine phosphate. It seems probable that it is Gly.SerP formed by inversion of SerP.Gly via the intermediate anhydride. This reaction has been shown to occur during hydrolysis in 2N-HCl (Schaffer, Harshman & Engle, 1955) and also appears to occur in 5·7N-HCl at 100°, though it was probably not significant when method (*a*) was used.

The nature of band 10*x* is unknown. Its yield varies considerably and is greatly increased if the partial hydrolysis is carried out in the presence of a small amount of glucose (Fig. 7, no. 5) or of filter paper. It would thus appear to be an artifact formed from SerP.Gly due to the presence of impurities (probably carbohydrate) extracted from the filter paper. Its slow rate of migration at pH 3·5 suggests decarboxylation or else that one of the acidic groups is substituted.

The results with the other main peptide bands could not readily be explained in any simple way. A number of bands were interconvertible. For instance, peptide 5A on partial hydrolysis gave 12 and 12 gave 5A. When certain bands were run again as controls at pH 3·5 they were found to be converted into other bands and these conversions could be brought about by incubation overnight at pH 6·5 (Fig. 8). Thus 12 was completely converted into 5A, 13 which appears to be a mixture was converted into two peptides moving at the rates of 6B and 8B respectively. Band 7 was partly converted into band 3 and was partly unchanged. It seemed to contain two compounds: 7A, which is converted into 3 at pH 6·5, and 7B, which is stable at pH 6·5. The decomposition of bands 12 and 13 can be seen in Fig. 5 by the streaks originating from spots 12 and 13. Evidently in this experiment the conversion was taking place during the ionophoresis at pH 6·5.

The various interconversions of peptides 3, 6A, 7A, 5A, 8A, 12 and 10 are summarized in Scheme 2.

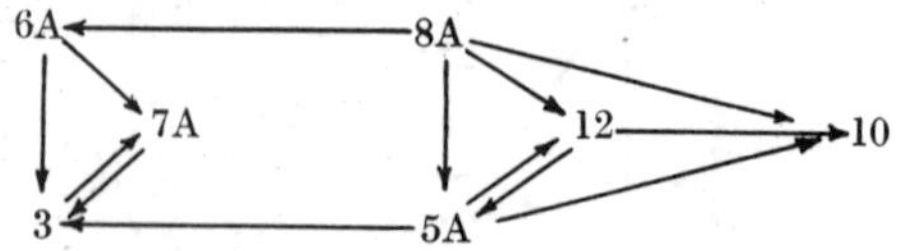

Scheme 2

The fact that 3, 6A and 7A are related and on partial hydrolysis give only serine phosphate and one another suggested that they were all different forms of the same dipeptide, presumably Asp.SerP. Peptides 5A, 8A and 12 were also similarly related and broke down to give 10 (SerP.Gly) and different forms of Asp.SerP. They thus appeared to be the corresponding three forms of the tripeptide Asp.SerP.Gly.

Table 1. *Yield of peptides from hydrolysate (method* a, *2 days) of di*iso*propoxy[^{32}P]phosphinyl-chymotrypsin*

Yield is expressed as ^{32}P recovered in peptide band as percentage of ^{32}P of original di*iso*propoxy[^{32}P]phosphinyl-chymotrypsin.

Band	Yield
1	5·0
2	1·5
3	2·4
4	3·0
5A	11·0
5B	0·2
6A	1·5
6B	1·0
7 (A + B)	5·0
8A	9·0
8B	1·4
9 (A + B + C)	3·0
10	10·7
11	1·6
12	2·3
13 (A + B)	3·2
Total yield	61·8

Table 2. *Amino acid composition and* N*-terminal residues of peptides from di*iso*propoxy[^{32}P]phosphinyl-chymotrypsin*

Band no. (Fig. 6)	Amino acids present	Approx. strength	*N*-Terminal residue
1	None	—	—
2	None	—	—
4	Ser	× × ×	—
	SerP	×	—
	Glu	tr	—
5A*	Ser	× ×	Asp
	Gly	× ×	—
	Asp	× ×	—
6A	Ser	× ×	Asp
	Asp	× ×	—
8A	Ser	× ×	Asp
	Gly	× ×	—
	Asp	× ×	—
	Glu	tr	—
10A	Ser	× ×	—
	Gly	× ×	—

* Band 5A, after purification by ionophoresis at pH 3·5 and 6·5, contained glutamic acid (× × ×). This was not present after further ionophoresis in 2·5% (v/v) formic acid (pH 2·1).

Table 3. *Partial hydrolysis and Edman degradations of serine phosphate peptides from di*isopropoxy[^{32}P]*phosphinyl-chymotrypsin*

Band no. (Fig. 6)	Peptide bands given after hydrolysis (method *b*, 30 min.) 3	4	5	6	7	8	10	12	Other bands	Main product of Edman degradation	Probable structure
3	× ×	× ×	—	—	×	—	—	—	—	—	β-Asp.SerP
4	—	×	—	—	—	—	—	—	—	1	SerP
5A	× ×	× ×	× × ×	—	×	—	× × ×	×	—	—	β-Asp.SerP.Gly
5B	—	×	—	—	—	—	—	—	—	—	SerIP
6A	tr	× ×	—	× ×	tr	—	—	—	—	4	α-Asp.SerP
6B	×	× ×	×	—	tr	—	× ×	tr	—	—	β-Asp.SerIP.Gly
7A*	× ×	× × ×	—	—	×	—	—	—	—	—	αβ-Asp.SerP
7B*	tr	× × ×	—	× ×	tr	—	—	—	—	5B	α-Asp.SerIP
8A	tr	× ×	tr	× ×	tr	× × ×	× × ×	tr	—	10	α-Asp.SerP.Gly
8B	×	× ×	× ×	—	tr	×	× × ×	×	13 (×)	5†	Gly.β-Asp.SerP.Gly?
9A	tr	× × ×	tr	× ×	tr	× ×	× × ×	tr	—	11	α-Asp.SerIP.Gly
9B	—	× ×	—	—	—	—	tr	—	9 (× ×)	4	Gly.SerP
9C	×	× × ×	—	—	× ×	—	—	—	9 (× ×)	—	?
10	—	×	—	—	—	—	× × ×	—	9B (tr), 10*x* (tr)	1	SerP.Gly
11	—	×	—	—	—	—	× × ×	—	9B (tr), 10*x* (tr)	2	SerIP.Gly
12	× ×	× ×	× × ×	—	×	tr	× × ×	×	—	—	αβ-Asp.SerP.Gly
13A‡	×	× ×	×	—	×	—	× ×	tr	13 (tr)	—	αβ-Asp.SerIP.Gly
13B‡	×	× ×	×	—	×	tr	× ×	×	13 (×)	—	Gly.αβ-Asp.SerP.Gly?
14	tr	× × ×	—	× ×	tr	—	—	—	—	16	α-Asp.SerDIP
15	tr	× ×	tr	×	tr	×	× ×	tr	9A (tr)	18	α-Asp.SerDIP.Gly
16	—	×	—	—	—	—	—	—	—	—	SerDIP
17	tr	× ×	tr	×	×	tr	× ×	tr	—	—	αβ-Asp.SerDIP.Gly?
18	—	×	—	—	—	—	× × ×	—	2 (× ×)	—	SerDIP.Gly

* Band 7 was incubated at pH 6·5 overnight and refractionated at pH 3·5. 7A was the fast-moving band (3), 7B the slower-moving one. 7B was obtained in higher yield directly from the 3-day hydrolysate by method (*c*).

† Result of a single experiment. There was a relatively large amount of unchanged peptide present.

‡ Band 13 was incubated at pH 6·5 overnight and refractionated at pH 3·5. 13A moved at the rate of band 6, 13B at the rate of band 8.

Table 4. *Partial hydrolysis (method* c) *of peptides from di*isopropoxy[^{32}P]*phosphinyl-chymotrypsin*

Band no. (Fig. 6)	Peptide bands given after partial hydrolysis (method *c*, 3 days): 5*	7B	9A	10	11	14	15	17	18	Other bands
7B	tr	××	—	—	—	—	—	—	—	—
9A	tr	×	××	—	×	—	—	—	—	8 (tr), 13 (tr)
11	tr	—	—	tr	××	—	—	—	—	2 (tr)
14	×	×	—	—	—	××	—	—	—	—
15	tr	×	×	tr	×	×	××	tr	×	2 (tr)
18	—	—	—	tr	××	—	—	—	×	2 (××)

* Bands 5 and 6 were not well resolved. It is probable that the band in this position is 5B though it could in some cases have been 5A, 6A or 6B.

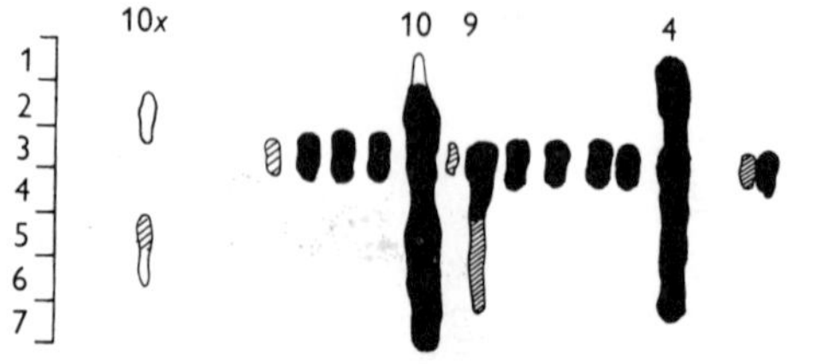

Fig. 7. Breakdown of peptide 10 under various conditions of hydrolysis. Samples of peptide 10 (containing ^{32}P, 5 μmc) hydrolysed as below and subjected to ionophoresis at pH 3·5, 40 v/cm., 2·5 hr. 1, Method (*b*), 3 hr.; 2, method (*a*), 3 days; 3, control hydrolysate (method *b*, 30 min.) of di*iso*propoxy[^{32}P]phosphinyl-chymotrypsin; 4, 3N-HCl, 100°, 6 hr.; 5, as for 6 with a trace of glucose added; 6, method (*b*), 35 min.; 7, untreated.

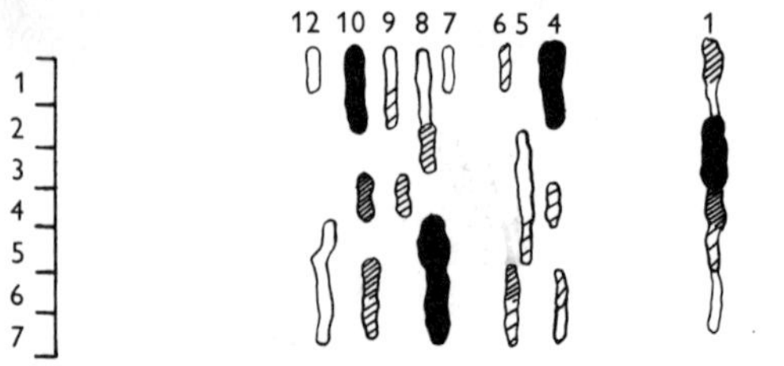

Fig. 9. Breakdown of peptide 8 under various conditions. Samples of peptide 8 (containing ^{32}P, 1 μmc) were treated as below and subjected to ionophoresis at pH 3·5, 40 v/cm., 2 hr. 1, Method (*b*), 55 min.; 2, 2N-HCl, 100°, 2 hr.; 3, water, 100°, 6 hr.; 4, 20% (v/v) formic acid, 100, 4 hr.; 5, thionyl chloride, 18°, 18 hr.; 6, 6N-HCl, 120°, 3 min.; 7, method (*b*), 25 min.

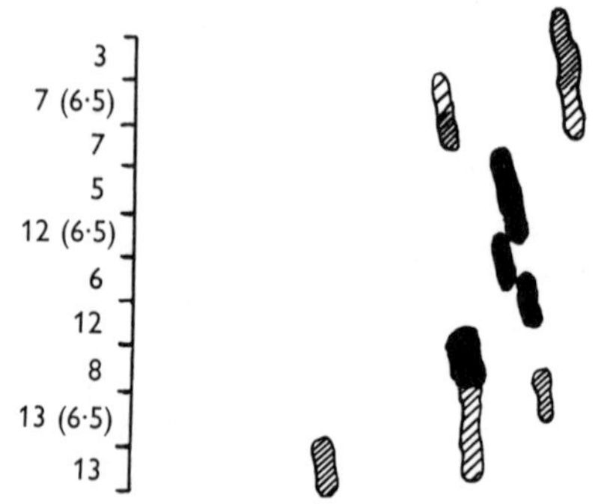

Fig. 8. Decomposition of peptides 7, 12, 13 after incubation at pH 6·5, for 16 hr. Other markers were eluted from papers on which they had been fractionated by ionophoresis, dried down and run again (1 μmc per sample). Ionophoresis: pH 3·5, 40 v/cm., 2 hr.

These results are best explained by assuming that 6A and 8A are normal α-aspartyl peptides, 7A and 12 the relatively unstable ring forms (αβ-aspartyl derivatives) and 3 and 5A the β-aspartyl peptides. The assignment of the β form to 3 and 5A was at first based largely on their greater ionophoretic rate at pH 3·5. Since the aspartic acid is *N*-terminal in both peptides, the aspartic acid carboxyl group in the β form will be adjacent to an α-amino group and will have a lower p*K* than the normal β-carboxyl group and hence move faster at pH 3·5. Peptide 3 is formed during the later stages of hydrolysis and seems to be relatively stable. John & Young (1954) showed that α-Asp.Val can be converted into β-Asp.Val by heating in water at 100° for 6 hr. Under these conditions considerable conversion of 8A into 5A and of 6A into 3 occurred (see Fig. 9). The various results given in Tables 2–4 are in agreement with the above interpretation and can thus be regarded as confirmation of it.

Probably the best method of characterizing the various aspartyl derivatives would be by their titration curves (see Swallow & Abraham, 1958). Since we had insufficient material for this we attempted to determine the p*K* values of the various groups by measuring the ionophoretic mobility of the peptides at different pH values. If the mobility is plotted against pH, the resulting curve should be analogous to a titration curve. In order to compensate for electroendosmosis, for the flow of buffer into the paper from the electrode vessels and for variations in conditions of running, the results were expressed relative to a marker of known titration curve. For this purpose serine phosphate was used and the radioactive material (from band 4) was run together with the peptides at each pH value. A glucose marker was also used to determine the displacement of the origin. A titration curve of serine phosphate was then prepared in the normal way, and the net charge was calculated for the different pH values. From this the theoretical mobility of serine phosphate was determined, assuming a value at pH 5·1 of 13 cm. This was the experimental value found after ionophoresis for 1·5 hr. at 30 v/cm. The relative mobilities for the peptides were calculated as

$$\frac{\text{Distance of peptide from glucose marker}}{\text{Distance of serine phosphate from glucose marker}} \times \text{theoretical mobility for serine phosphate.}$$

The results are, then, the distance travelled by the peptide in 1·5 hr. at 30 v/cm. under the experimental conditions (heating etc.) that prevailed during the experiment at pH 5·1. Since the absolute mobilities could not be determined it was considered preferable to express the results in these arbitrary units.

Some of the curves so obtained are shown in Fig. 10. Although they cannot be regarded as very accurate owing to the limitations of the method, they do give an indication of the p*K* values of the various groups involved and clearly show that one of the carboxyl groups and the amino groups in 3 and 5A are considerably stronger than the corresponding groups in 6A and 8A. Unfortunately 7A and 12 were too unstable at most pH values to determine their mobilities. One of the p*K* values of the phosphate group is about 5·5–6·0 (Fölsch & Österberg, 1959), and this is absent from peptide 11, thus confirming that it is a mono*iso*propyl derivative (see below). The p*K* values of the amino groups decrease with increased peptide size (4 > 6A > 8A), as would be expected.

Fig. 9 shows the results of various treatments on peptide 8 (α-Asp.SerP.Gly). The main products of hydrolysis are normally 4, 6, 10 and 12 (e.g. in sample 7), the relative amounts of each depending on the time of hydrolysis. Treatment with 20% (v/v) formic acid (sample 4) leads almost exclusively to peptide 10 (SerP.Gly) and its inversion product 9B. This is in agreement with the results of Partridge & Davis (1950), who showed a preferential splitting of bonds involving aspartyl residues in weak acid.

The action of thionyl chloride was studied to see if it converted an α-aspartyl peptide (e.g. peptide 8) into the αβ form (e.g. peptide 12). This does take place to a small extent, though most of 8 was unchanged under the conditions used. A more appreciable conversion of peptide 6 into 7 was brought about by treatment with thionyl chloride at 100° for 15 min. This can be regarded as further support for the conclusion that 7A and 12 are αβ forms.

A number of peptides, especially in the hydrolysate prepared by method (*c*) for 3 days, contain the phosphate as the mono*iso*propyl or di*iso*propyl derivatives. These can be recognized by their relative rates of migration at pH 3·5 and 6·5, since the second ionization constant (p*K* 5·5–6·0) of the phosphate is blocked in them. Also, on treatment with alkali (4% triethylamine, 30 min., 105°) the mono*iso*propyl derivatives gave mono*iso*propyl phosphate (band 2), which could be identified. The following bands were shown in this way to be mono*iso*propyl derivatives: 5B, 7B, 9A, 11. Their probable structure was deduced from the products of their hydrolysis and is shown in Table 3. Peptides 14, 15, 16, 17 and 18 appeared from their behaviour on degradation to be di*iso*propyl derivatives. When they were treated with triethylamine and the product was subjected to iono-

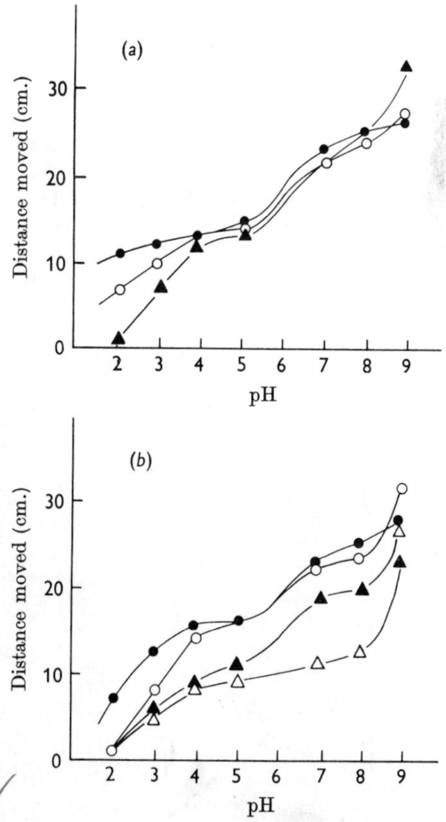

Fig. 10. Mobility–pH curves for ^{32}P-containing bands from partial hydrolysate of di*iso*propoxy[^{32}P]phosphinyl-chymotrypsin. The curves are plotted relative to the theoretical distance of migration of serine phosphate at 30 v/cm. for 1·5 hr. (see text). (*a*) ●, Serine phosphate (theoretical curve); ○, band 5A; ▲, band 8A. (*b*) ●, Band 3; ○, band 6A; ▲, band 10A; △, band 11.

phoresis at pH 3·5 under toluene the radioactivity was almost completely lost. This also occurred when DF^{32}P was treated in the same way. If, however, the ionophoresis was done between cooled plates a band was obtained from the alkaline degradation of DF^{32}P and of peptide 15 which moved slightly faster than serine phosphate. It is presumably di*iso*propyl phosphate. It appears that it is extracted into toluene, possibly in the form of its triethylamine salt.

Schaffer *et al.* (1957) identified the peptide Gly.Asp.SerP.Gly in hydrolysates of di*iso*propoxy[^{32}P]phosphinyl-chymotrypsin (method *a*, 3 days). It seems most likely that band 8B has this structure, although it was present in rather small amounts and was not completely characterized. It is clearly in the β form and 13B is the corresponding $\alpha\beta$ form. Schaffer & Lang (1959) have reported the identification of a labelled peptide having the amino acids [Ser.Gly.Glu.Ala] (see also Turba & Gundlach, 1955). We have been unable to detect any peptide that could have this structure. There are a number of weak bands that have not been thoroughly characterized. The most likely structure for 9C is $\alpha\beta$-Asp.SerIP. On running again at pH 3·5 and on incubation with alkali it gives rise to a peptide moving between bands 3 and 4, which could be β-Asp.SerIP.

When band 10 (SerP.Gly) was repurified by ionophoresis at pH 6·5 a small amount of a rapidly moving band was always present. This proved to be free phosphate. Similarly, when bands 11 (SerIP.Gly) and 18 (SerDIP.Gly) were purified at pH 6·5, faster-moving bands were present which could have been *iso*propyl phosphate and di*iso*propyl phosphate respectively. It would thus appear that these peptides are somewhat unstable to the conditions employed. There were several other bands present in small amounts whose structures could not be determined. The results with these are not recorded here.

Comparison of ^{32}P*-containing peptides from the di*iso*propoxy*[^{32}P]*phosphinyl derivatives of chymotrypsin, trypsin and elastase*

Samples of the di*iso*propoxy[^{32}P]phosphinyl derivatives of trypsin, chymotrypsin and elastase were hydrolysed for 20 min. by method (*b*) and subjected to ionophoresis in parallel together at pH 3·5. Fig. 11 is a radioautograph of the paper which shows no difference between the three enzymes. By the use of the blotting method two-dimensional ionophoresis patterns were prepared from the above-mentioned parallel one-dimensional ones and these are shown in Fig. 12. A shorter-time hydrolysis (10 min.) of di*iso*propoxy[^{32}P]phosphinyl-elastase is also shown. These results reveal no differences in the ^{32}P-containing peptides from the three enzymes.

DISCUSSION

The above-described results show that the complexity of the hydrolysates of the di*iso*propoxy-[^{32}P]phosphinyl-enzymes is largely due to two factors: the incomplete hydrolysis of the *iso*propyl

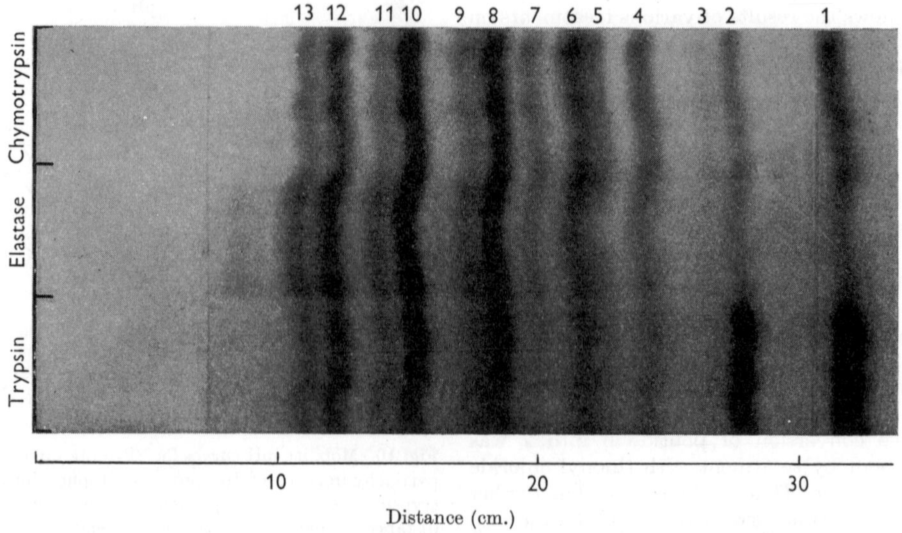

Fig. 11. Ionophoresis of partial hydrolysates (method *b*, 20 min.) of the di*iso*propoxy[^{32}P]phosphinyl derivatives of chymotrypsin, elastase and trypsin, pH 3·5, 40 v/cm., 2 hr. The strong bands 1 and 2 from the trypsin derivative are derived from DF^{32}P, which contaminated the preparation.

ester groups and the aspartyl-interconversion reaction. It seems likely that this interconversion is a general reaction and is to be expected in experiments where proteins or peptides containing aspartic acid are subjected to hydrolysis with acid. It probably occurred during the hydrolysis of fraction A of oxidized insulin with acid (Sanger & Thompson, 1953), since three peptides (Ala9, Ala10, Ala14) with the apparent structure Asp.Tyr were detected. They were explained as being due to the presence of an amide group and to TyrX (chlorotyrosine; Thompson, 1954). It is unlikely that amide groups would in fact be stable under the conditions of hydrolysis and a more likely explanation is that peptide Ala14, whieh gave a yellow colour with ninhydrin, was the β-Asp.Tyr. John & Young (1954) report that β-aspartyl peptides give a yellow ninhydrin colour.

Although Schaffer *et al.* (1957) did not detect the interconversion in their experiments, which were done under the same conditions as we have used, it is probable that it occurred and that their peptide 2 was the same as our peptide 5A (β-Asp.SerP.Gly), whereas peptide 3 was our 8A. They found that although both reacted with phenyl *iso*thiocyanate in the Edman procedure, the

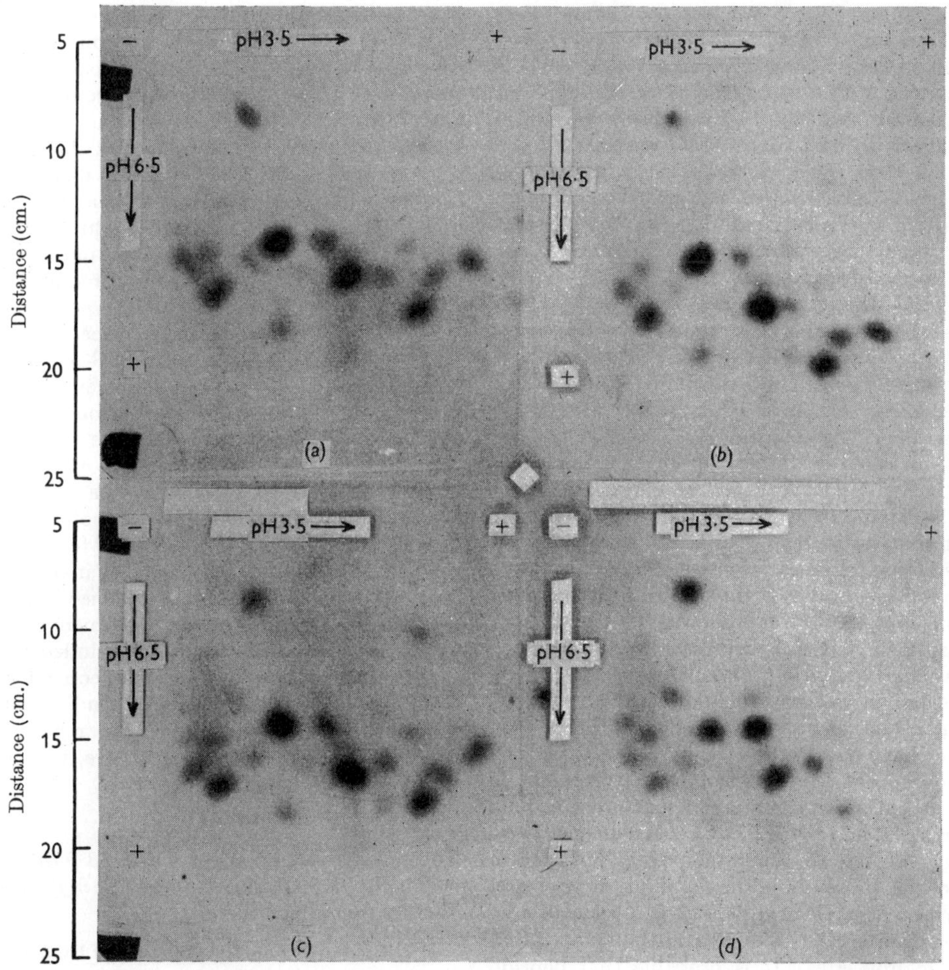

Fig. 12. Comparative two-dimensional ionophoresis patterns of partial hydrolysates (method *b*) of the di*iso*propoxy[^{32}P]phosphinyl derivatives of trypsin, chymotrypsin and elastase. First dimension, pH 3·5, as in Fig. 11; second dimension, pH 6·5, 40 v/cm., 1 hr. Transfer to second dimension was by the blotting technique (see text). (*a*) Di*iso*propoxyphosphinyl-trypsin, 20 min. hydrolysate. (*b*) Di*iso*propoxyphosphinyl-chymotrypsin, 20 min. hydrolysate. (*c*) Di*iso*propoxyphosphinyl-elastase, 20 min. hydrolysate. (*d*) Di*iso*propoxyphosphinyl-elastase, 10 min. hydrolysate.

phenylthiohydantoin of aspartic acid was liberated only from peptide 3 and not from peptide 2, as would be expected if the latter was a β-aspartyl peptide.

Various methods of hydrolysis were studied (e.g. Fig. 9) in an attempt to find conditions in which the interconversions of aspartic acid peptides could be avoided. No such conditions could be found but the reaction seems to be less marked in 5·7 N-hydrochloric acid than in 12 N-hydrochloric acid. The main difference between hydrolyses in 12 N-hydrochloric acid (method *a*) and in 5·7 N-hydrochloric acid (method *c*) at 37° is in the stability of the *iso*propyl groups, which are much more resistant under the latter conditions. There was no evidence of any significant amounts of peptides larger than tripeptides in the 3-days hydrolysate (method *c*). Only relatively small amounts of the β and $\alpha\beta$ forms of the aspartyl peptides were present. It also appears that the interconversion reaction is less marked when hydrolysis is carried out by method *b* than when carried out by method *a*. There was no evidence of any inversion of a dipeptide sequence with method (*c*) so that it seems that in general this would be a better method to use if side reactions are to be avoided.

From the identity of the radioautographs of the ionograms of partial hydrolysates of the di*iso*propoxy[^{32}P]phosphinyl derivatives of elastase, chymotrypsin and trypsin, it was concluded that elastase, like the other enzymes, contained the sequence Gly.Asp.SerP.Gly (Hartley *et al.* 1959). This conclusion was based on the assumption that only three peptides (Asp.SerP, SerP.Gly and Asp.SerP.Gly) would be derived from the sequence Asp.SerP.Gly. In fact six to seven identical bands were obtained from the three enzymes. Hence it was assumed that they must have more than three residues in common; Schaffer *et al.* (1957) had shown the presence of Gly.Asp.SerP.Gly in hydrolysates of the type we were using. Clearly in view of the rearrangement reactions of the aspartyl peptides, the above conclusions are no longer valid. In fact in the present work we have shown that the hydrolysates of the three di*iso*propoxy[^{32}P]phosphinyl enzymes have about 20 peptides in common but can only conclude for certain that they all have the sequence Asp.Ser.Gly. Owing to the interconversion reaction and to the presence of mono- and di-*iso*propyl esters, each aspartic acid peptide can occur in nine forms, which greatly complicates the composition of the hydrolysates. It is probable that peptide 8B is Gly.Asp.SerP.Gly, and it seems to be present in the hydrolysates of di*iso*propoxy[^{32}P]-phosphinyl-chymotrypsin, -trypsin and -elastase (Fig. 12), suggesting they have the tetrapeptide sequence in common. However, the amount of this peptide is very small and its structure uncertain so that the suggestion cannot be regarded as conclusive.

The possibility has to be considered whether the aspartyl residue in the intact proteins could be the β or $\alpha\beta$ form rather than the normal α form. This seems unlikely since the proportion of the β forms increases with time of hydrolysis and the conversion of a β form into an α form (e.g. 5A $\rightarrow$ 8A) has not been observed on partial acid hydrolysis of the isolated peptides, whereas the reverse reaction does take place (e.g. 8A $\rightarrow$ 5A). On the other hand the yields of the β and $\alpha\beta$ forms do seem to be unexpectedly high. Table 5 shows the relative yields of bands 8A (α-Asp.SerP.Gly), 5A (β-Asp.SerP.Gly) and 12 ($\alpha\beta$-Asp.SerP.Gly) from a partial hydrolysate of di*iso*propoxy[^{32}P]phosphinyl-chymotrypsin and from a partial hydrolysate of peptide 8A. It can be seen that the relative yields of the β and $\alpha\beta$ forms are greater from the protein than from the isolated peptide. This can probably best be explained by assuming that the conversion is more rapid if the α-amino group of the aspartyl residue is blocked as in a larger peptide, so that most of the reaction takes place during the early stages of hydrolysis of the protein and the β forms (e.g. 3 and 5A) are probably produced largely from the β and $\alpha\beta$ forms of larger peptides rather than from the corresponding α forms (6A and 8A). In this connexion if 8B is Gly.β-Asp.SerP.Gly and 13B its $\alpha\beta$ form, there is no evidence of the α form which suggests that the longer period of survival of the larger peptide in the acid has led to its complete conversion. Whereas it seems most likely that the natural form of the aspartyl residue is the α form, the possibility cannot be excluded that it may be partly in the β or $\alpha\beta$ form. It seems unlikely that the reaction could be involved in the enzymic activity of the proteins, since it appears to be brought about by acid treatment; however, again the possibility cannot be excluded.

It must, of course, be more than coincidence that the three proteolytic enzymes studied have the sequence Asp.Ser.Gly around the reactive serine residue. It seems that there are two possible explanations for this.

Table 5. *Yields of the α, β and $\alpha\beta$ forms of Asp.SerP.Gly from diisopropoxyphosphinyl-chymotrypsin and from α-Asp.SerP.Gly*

Band no. (Fig. 6)	Relative yield of band in hydrolysate (method *b*, 30 min.): Di*iso*propoxy-[^{32}P]phosphinyl-chymotrypsin	Band 8A
5A	36	12
8A	100	100
12	38	12

One is that the sequence is essential for enzymic activity, in which case some function must be ascribed to the aspartyl and glycyl residues for the catalytic activity, presumably in activating the serine residue. Rydon (1958) has suggested a mechanism involving the aspartyl residue, though it is more generally assumed that the activity of the serine residue is due to its spatial proximity to a specific histidine residue. No function has yet been suggested for the glycyl residue. It could perhaps be steric; for instance, it might be that a larger side chain would sterically hinder the catalytic activity. It seems possible, however, that the similarities in the structures of trypsin and chymotrypsin may extend over a larger part of the molecule (see Šorm *et al.* 1957) and this could hardly be explained on the basis of the catalytic activity. An alternative explanation is in the biological origin of the three proteins and it seems likely that they are in fact 'descended' from a common protein and have developed their specific sequences by independent mutations. Thus it may be that some more primitive organism contained a single protease and that, during the evolutionary process, three enzymes with different specificities (trypsin, chymotrypsin and elastase) have developed from it. Thus one can envisage evolution of proteins as taking place within a species, as well as between species. The nature and course of these changes could probably be revealed by a detailed study of the sequence of these and related proteins. A similar situation is found with the pituitary hormones: vasopressin and oxytocin on the one hand (du Vigneaud, Lawler & Popenoe, 1953) and adrenocorticotrophic hormone and melanophore-stimulating hormone on the other (Harris & Roos, 1956). Here again proteins or peptides from the same organ which have different biological activities have similarities in amino acid sequence and this may be expected to be a general phenomenon as more proteins are studied.

SUMMARY

1. The di*iso*propoxy[^{32}P]phosphinyl derivatives of chymotrypsin, trypsin, and elastase were subjected to partial acid hydrolysis and the products investigated after ionophoretic fractionation.

2. During acid hydrolysis an interconversion of the aspartyl peptides occurs in which the normal α form is converted into the $\alpha\beta$ and β forms.

3. By comparison of the radioactive peptides in the hydrolysates of the di*iso*propoxy[^{32}P]phosphinyl derivatives of the three enzymes it is concluded that elastase, like trypsin and chymotrypsin, contains the sequence Asp.Ser.Gly around its reactive serine residue.

We wish to thank Dr D. R. Davies of the Ministry of Supply Chemical Defence Experimental Establishment for generous gifts of DF^{32}P and Dr W. C. Frith of Courtaulds Ltd. for samples of cellulose acetate membrane. D.C.S. is on leave from C.S.I.R.O. Wool Research Laboratories, Australia, and gratefully acknowledges the award of a Hackett Studentship from the University of Western Australia. We thank Dr S. Bernard for valuable discussion in connexion with this work.

REFERENCES

Blackburn, S. & Lowther, A. G. (1951). *Biochem. J.* **48**, 126.

Dixon, G. H., Kauffman, D. L. & Neurath, H. (1958). *J. Amer. chem. Soc.* **80**, 1260.

du Vigneaud, V., Lawler, H. C. & Popenoe, E. A. (1953). *J. Amer. chem. Soc.* **75**, 4880.

Edman, P. (1950). *Acta chem. scand.* **4**, 277.

Flavin, M. (1954). *J. biol. Chem.* **210**, 771.

Fölsch, G. & Österberg, R. (1959). *J. biol. Chem.* **234**, 2298.

Fraenkel-Conrat, H. & Harris, J. I. (1954). *J. Amer. chem. Soc.* **76**, 6058.

Gross, D. (1955). *Nature, Lond.*, **176**, 72.

Harris, J. I. & Roos, P. (1956). *Nature, Lond.*, **178**, 90.

Hartley, B. S., Naughton, M. A. & Sanger, F. (1959). *Biochim. biophys. Acta*, **34**, 243.

John, W. D. & Young, G. T. (1954). *J. chem. Soc.* p. 2870.

Kohn, J. (1957). *Clin. chim. Acta*, **2**, 297.

Michl, H. (1951). *Mh. Chem.* **82**, 489.

Naughton, M. A. & Sanger, F. (1960). *Biochem. J.* (in the Press).

Oosterbaan, R. A., Kunst, P., Van Rotterdam, J. & Cohen, J. A. (1958). *Biochim. biophys Acta*, **27**, 549.

Partridge, S. M. & Davis, H. F. (1950). *Nature, Lond.*, **165**, 62.

Rydon, H. N. (1958). *Nature, Lond.*, **182**, 928.

Ryle, A. P., Sanger, F., Smith, L. F. & Kitai, R. (1955). *Biochem. J.* **60**, 541.

Sanger, F. & Thompson, E. O. P. (1953). *Biochem. J.* **53**, 353.

Schaffer, N. K., Harshman, S. & Engle, R. R. (1955). *J. biol. Chem.* **214**, 799.

Schaffer, N. K. & Lang, R. P. (1959). *Fed. Proc.* **18**, 317.

Schaffer, N. K., Simet, L., Harshman, S., Engle, R. R. & Drisko, R. W. (1957). *J. biol. Chem.* **225**, 197.

Sjöquist, J. (1957). *Ark. Kemi*, **11**, 129.

Šorm, F., Keil, B., Holeyšorský, V., Knesslová, V., Kostka, V., Mäsiar, P., Meloun, B., Mikeš, O., Tomasek, V. & Vaněček, J. (1957). *Coll. Trav. chim. Tchécosl.* **22**, 1310.

Swallow, D. L. & Abraham, E. P. (1958). *Biochem. J.* **70**, 364.

Thompson, E. O. P. (1954). *Biochim. biophys. Acta*, **15**, 440.

Turba, F. & Gundlach, G. (1955). *Biochem. Z.* **327**, 186.

fakleo
NPS
v100 c s t i

(Reprinted from Nature, Vol. 187, *No.* 4740, *pp.* 872–873, *September* 3, 1960)

Amino-acid Sequence about the Reactive Serine of a Proteolytic Enzyme from *Bacillus subtilis*

An increasing number of hydrolytic enzymes have been reported to be inhibited by di*iso*propylphosphorofluoridate. Where the resulting di*iso*propylphospho-enzyme has been investigated, the phosphate was isolated in the form of phospho-serine peptides and the amino-acid sequence about the reactive serine has been determined in several cases. Trypsin[1] chymotrypsin[2–4], elastase[5,6] and thrombin[7] have the sequence Gly.Asp.Ser.Gly., while horse liver ali esterase[8] and pseudocholinesterase[9] have Gly.Glu.Ser.-Ala.Gly. The proteolytic enzyme subtilisin was isolated from a strain of *B. subtilis* by Güntelberg and Ottesen[10] and they recorded its inhibition by di*iso*propylphosphorofluoridate, while Matsubara[11] obtained a phospho-peptide from strain *N*[1] but did not report any sequence. The present communication deals with an investigation of the enzyme from Novo Terapeutisk Laboratorium known as 'bacterial trypsin', which is similar to, but distinguishable from, the original subtilisin.

A sample was treated with di*iso*propylphosphorofluoridate labelled with phosphorus-32 at *p*H 7·4 and a partial acid hydrolysate was subjected to ionophoresis on Whatman No. 52 paper at *p*H 3·5 in parallel with similarly treated chymotrypsin and liver ali esterase (sample kindly provided by Dr. E. C. Webb). The radioautograph of the ionogram showed that serine phosphate occurred in all three; but the pattern of phospho-peptides from bacterial protease was entirely different from that obtained from either chymotrypsin[6] or liver ali esterase and therefore there must be a different sequence about the labelled site. Repetition of the hydrolysis of samples of di*iso*-propylphospho-enzyme, under the same conditions, gave markedly different proportions of the bands within the pattern. The proportions were also affected by subsequent air oxidation and thioglycollic acid reduction, and some individual bands were found to be interconvertible by oxidation or reduction. This suggested that there was, near the labelled serine, an amino-acid which could undergo reversible air oxidation. It seemed probable that this was methionine, which can be readily oxidized

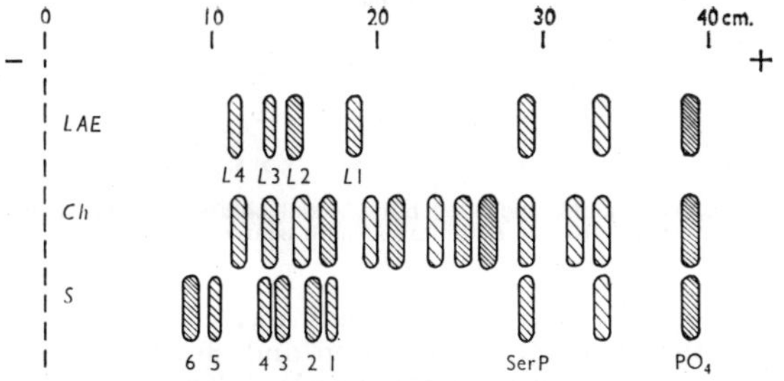

Fig. 1. Diagram of radioautograph of ionogram (*p*H 3·5, 40 V./cm., 2 hr.) of partial acid hydrolysis of di*iso*propylphospho-enzyme derivatives of liver ali esterase (*LAE*), chymotrypsin (*Ch*) and oxidized bacterial protease (*S*). The *LAE* bands are considered to be: *L*1, Glu.SerP.; *L*2, SerP.Ala.; *L*3, Glu.SerP.-Ala; *L*4, Gly.Glu.SerP.Ala. The interpretation of the bands from chymotrypsin is being reported elsewhere (ref. 6)

to the sulphoxide[12]. The di*iso*propylphospho-enzyme was therefore oxidized with performic acid to convert any methionine into the stable sulphone, and a simplified, reproducible pattern was obtained on hydrolysis. This pattern, which is also obtained if the previous hydrolysates were oxidized with performic acid prior to ionophoresis, is shown in Fig. 1 together with those from di*iso*propylphospho-chymotrypsin and di*iso*propylphospho-liver ali esterase.

The interrelationships of these oxidized bands (numbered in diagram) were determined by study of their behaviour to partial acid hydrolysis and Edman degradation[6] and their chromatographic and electrophoretic mobilities. Additional information about the N-terminal residue was obtained by treating the peptide with sodium periodate solution and observing whether the electrophoretic movement at *p*H 3·5 was affected. N-terminal serine and threonine residues should be oxidized to give a glyoxylic acid residue by loss of ammonia and formaldehyde or acetaldehyde,

Table 1

Peptide	Products of partial acid hydrolysis (excluding PO_4 and SerP)	Main radioactive product of Edman degradation	Effect of sodium periodate*
1	1	PO_4	−
2	2	SerP	+
3	1 and 2	1	+
5	1	PO_4	−
6	1, 2, 3 and 5	5	+

* + indicates that the peptide moved faster after treatment; − shows the peptide was not affected.

Peptide 4 is probably a larger peptide the structure of which has not yet been determined.

Table 2

Peptide	Thr	Ser	Met SO_2	Ala	Asp	Glu	Gly
1	—	×××	××	—	××	*tr*	*tr*
2	×××	×××	—	—	*tr*	*tr*	*tr*
3	×××	×××	×××	—	×	×	×
5	*tr*	×××	××	×××	*tr*	*tr*	*tr*
6	×××	×××	×××	×××	*tr*	—	—

Relative amounts indicated by ×'s and obtained by visual comparison of intensities. *tr* indicates small contamination.

respectively[13]; so that the radioactive peptide would lose the terminal amino-group and should therefore move faster towards the anode. The results are summarized in Table 1.

A larger sample (50 mgm.) was then treated with di*iso*propylphosphorofluoridate labelled with phosphorus-32, oxidized with performic acid, hydrolysed and the radioactive peptides isolated by paper ionophoresis after removal of the neutral and basic peptides on a column of 'Dowex 50'. The amino-acid composition of each of these, after total hydrolysis, was determined on paper and is presented in Table 2.

Contamination by non-radioactive, acidic peptides was not great and so from Table 2 and the determined interrelationships of the peptides, represented in Fig. 2, these experiments indicated the sequence Thr.Ser.Met.Ala.

It is interesting that the sequence about the reactive serine in this enzyme is totally different from those determined for other di*iso*propylphosphoenzymes. They had in common an acidic amino-acid, the serine and a small neutral aliphatic residue. Comparing subtilisin with chymotrypsin, trypsin, elastase and thrombin, which are all endo-peptidases with different specificities, it would appear that the sequence immediately about the serine is not responsible for the unusual reactivity towards di*iso*propylphosphorofluoridate which results in the inhibition of the proteolytic activity. Some other grouping, possibly including histidine[14], which is spatially close

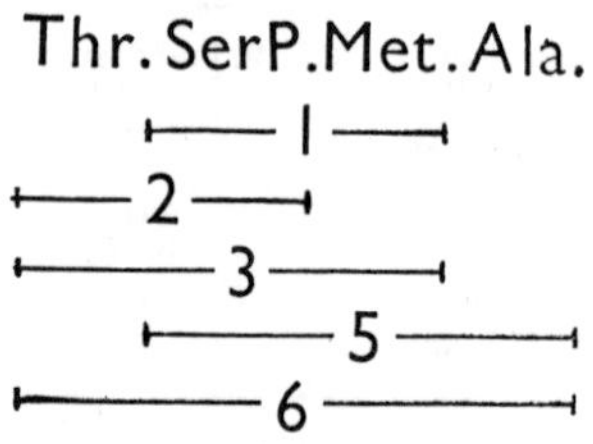

Fig. 2. Interrelationships of peptides. SerP indicates O-phosphoserine

due to folding of the chain, probably activates the serine.

F. SANGER*
D. C. SHAW†

Department of Biochemistry,
University of Cambridge.

* Member of the external staff of the Medical Research Council.

† Hackett Student from the University of Western Australia and on leave from C.S.I.R.O. Wool Research Laboratories, Australia.

[1] Dixon, G. H., Kauffman, D. L., and Neurath, H., *J. Amer. Chem. Soc.*, **80**, 1260 (1958).

[2] Turba, F., and Gundlach, G., *Biochim. Z.*, **327**, 186 (1955).

[3] Schaffer, N. K., Simet, L., Harshman, S., Engle, R. R., and Drisko, R. W., *J. Biol. Chem.*, **225**, 197 (1957).

[4] Oosterbaan, R. A., Kunst, P., Van Rotterdam, J., and Cohen, J. A., *Biochim. Biophys. Acta*, **27**, 549 (1958).

[5] Hartley, B. S., Naughton, M. A., and Sanger, F., *Biochim. Biophys. Acta*, **34**, 243 (1959).

[6] Naughton, M. A., Sanger, F., Hartley, B. S., and Shaw, D. C., *Biochem. J.*, **77**, 149 (1960).

[7] Gladner, J. A., and Laki, K., *J. Amer. Chem. Soc.*, **80**, 1264 (1958).

[8] Jansz, H. S., Posthumus, C. H., and Cohen, J. A., *Biochim. Biophys. Acta*, **33**, 396 (1959).

[9] Jansz, H. S., Brons, D., and Warringa, M. G. P. J., *Biochim. Biophys. Acta*, **34**, 573 (1959).

[10] Güntelberg, A. V., and Ottesen, M., *C.R. Lab. Carlsberg, Ser. Chim.*, **29**, 36 (1953).

[11] Matsubara, H., *J. Biochem. (Japan)*, **46**, 107 (1959).

[12] Harris, J. I., and Roos, P., *Biochem. J.*, **71**, 434 (1959).

[13] Nicolet, B. H., and Shinn, L. A., *J. Amer. Chem. Soc.*, **61**, 161 (1939).

[14] Discussed in Symposium on Enzyme Reaction Mechanisms, *J. Cell. and Comp. Physiol.*, **54**, Supp. 1 (1959).

Printed in Great Britain by Fisher, Knight & Co., Ltd., St. Albans

Biochem. J. (1961) **79**, 456
Printed in Great Britain

An Amino Acid Sequence in the Active Centre of Phosphoglucomutase

By C. MILSTEIN* and F. SANGER†
Department of Biochemistry, University of Cambridge

(*Received* 17 *August* 1960)

The presence of phosphate in phosphoglucomutase was first demonstrated by Jagannathan & Luck (1949), who also showed that it may readily be exchanged with phosphate in the substrates and that it is thus probably involved in the active centre of the enzyme. Anderson & Jollès (1957) observed that it was not released from the protein even under strong acid conditions and were able to isolate serine phosphate from the partially hydrolysed enzyme. This offered the possibility of studying the amino acid sequence around the point of attachment of the phosphate by isotopic techniques. In a preliminary study of the peptides produced from a partial hydrolysate of phosphoglucomutase labelled with ^{32}P, Koshland & Erwin (1957) and Koshland, Ray & Erwin (1958) suggested that the sequence around the serine phosphate (SerP) residue was Asp.SerP.Gly.Glu.Ala.-Val. [For definitions of the abbreviations of amino acids used in this paper see *Biochem. J.* (1957), **66**, 6.] This was very similar to that found in trypsin and chymotrypsin around the serine residue that can react with di*iso*propyl phosphorofluoridate. The sequence Asp.Ser.Gly thus appeared to be a common feature of enzymes with rather dissimilar functions, and these conclusions have been discussed by several authors in connexion with the problem of the relationship between structure and function (Lumry, 1959; Koshland, 1959; Linderstrøm-Lang & Schellman, 1959; Dixon, Neurath & Pechère, 1958). In the work of Koshland & Erwin evidence for the purity of the peptides studied was lacking, and, by comparing partial hydrolysates of [^{32}P]phosphoglucomutase with similar hydrolysates of labelled chymotrypsin, we were able to show (Milstein & Sanger, 1960) that the sequence around the serine phosphate residue in phosphoglucomutase was not Asp.SerP.Gly. The present paper describes the determination of a pentapeptide sequence around the serine phosphate residue.

In most experiments with peptides of serine phosphate it is usually possible to separate them from other peptides by virtue of their strong acid group by ion-exchange methods (Flavin, 1954) or by ionophoresis at an acid pH. Since most of the labelled peptides in a partial hydrolysate of [^{32}P]-phosphoglucomutase contain a basic residue, they are neutral and purification is much more difficult. We have thus determined as much as possible of the sequence by isotopic methods, since it is relatively easy to separate the labelled peptides from one another but very difficult to separate them from the very complex mixture of non-labelled peptides that is derived from such a large protein. Some such methods have been described previously (Naughton, Sanger, Hartley & Shaw, 1960; Sanger & Shaw, 1960; Milstein & Sanger, 1960). The interrelationships of the various peptides (Tables 3 and 4) were first determined by studying the effect of partial acid hydrolysis and Edman degradation on them. The presence of a histidine residue after the serine phosphate and of an acidic residue next to the histidine was concluded from the ionophoretic rates of the peptides at different pH values.

Another method for the identification of histidine peptides was also developed by making use of the lability of the imidazole ring to photo-oxidation. Weil, Gordon & Buchert (1951) showed that some amino acids can be oxidized when they are exposed under a visible light in the presence of methylene

* Present address: Instituto Nacional de Microbiología, Av. Velez Sarsfield 563, Buenos Aires, Argentina.
† Member of the External Staff of the Medical Research Council.

blue and that peptide bonds do not participate in the photochemical action of the dye. Of the amino acids studied, the most reactive appeared to be histidine, tyrosine, tryptophan, methionine and cysteine. It was suggested that, during the reaction, the imidazole ring breaks, with formation of an aldehyde group. It would therefore be expected that the electrophoretic mobility of a peptide containing histidine will change after exposure to photo-oxidation. Since no other basic amino acid should react under those conditions, differentiation of histidine peptides from other basic amino acids containing peptides is possible.

Another specific test that proved useful was the use of periodate (Sanger & Shaw, 1960). If the ionophoretic rate of a peptide was altered after treatment with periodate it could be concluded that it contained serine or threonine in the N-terminal position.

Further information about the individual residues present in the peptides could be obtained by studying their R_F values on paper chromatography. Pardee (1951) showed that the chromatographic rate of a peptide in a given solvent system can be related to the chromatographic rate of the amino acid components by the following formula:

$$\boldsymbol{R}T \ln (1/R_{F(p)} - 1) = (n-1)A + B + \Sigma \boldsymbol{R}T \ln (1/R_{F(a)} - 1) \quad (1)$$

in which $\boldsymbol{R}$ is the molar gas constant, T the absolute temperature, n the number of amino acid residues in the peptide, and A and B are constants. $R_{F(p)}$ and $R_{F(a)}$ are the R_F values of the peptide and the amino acids respectively. This relationship has been shown to fit well with the experimental data obtained on different solvent systems with 46 peptides (Moore & Baker, 1958). These authors reported no significant differences for groups of peptides in which the sequence of amino acids was altered.

From this it follows that if, for a given solvent system, the values of the constants and the R_F values of a peptide with one unknown amino acid are known, the R_F of the unknown amino acid could be easily deduced. The R_F values of an unknown amino acid will not in most cases be enough to characterize it. However, from the combination of R_F data on more than one system and information obtained by mobility studies and specific reactions, it should be possible to obtain valuable information leading to the resolution of the sequence around one known amino acid. With these techniques it could be concluded that the most probable sequence was Thr.Ala.SerP.His.-(Glu or Asp). In order to confirm this and to identify the acidic amino acid it was necessary to purify certain of the peptides and identify the amino acids present by established techniques. A rather extensive purification was found to be necessary, but since the size and approximate composition of the peptides was known it was easy to ascertain when they were pure. From these results it was shown that the sequence was Thr.Ala.SerP.His.Asp.

METHODS

Phosphoglucomutase (PGM) was prepared by the method of Najjar (1948) and dialysed against distilled water before use. Trypsin and chymotrypsin were crystalline salt-free preparations from Worthington Biochemical Corp. Elastase was a gift from M. A. Naughton (Naughton & Sanger, 1960).

Preparation of [^{32}P]phosphoglucomutase

Two methods were used for the preparation of [^{32}P]PGM. In method 1 PGM was labelled by allowing it to react with glucose 6-[^{32}P]phosphate, which was prepared through the hexokinase reaction. A mitochondrial suspension was used to generate adenosine [^{32}P]triphosphate from [^{32}P]phosphate. In method 2 glucose 1-[^{32}P]phosphate was prepared by the action of potato phosphorylase on starch in the presence of [^{32}P]phosphate, and allowed to react with the PGM.

Method 1. Each 8·2 ml. of the incubation mixture contained (amounts in μmoles unless otherwise stated): $MgSO_4$, 60; glucose, 225; 2-amino-2-hydroxymethylpropane-1:3-diol (tris)–HCl buffer, 125; glutamate or proline, 75; KCl, 250; adenosine triphosphate (ATP), 2; $H_3{}^{32}PO_4$ (carrier-free), 0·2–2 mc; fresh mitochondria suspension, 0·75 ml.; and excess of hexokinase. The mitochondria were freshly prepared from sucrose homogenates of rat liver and tested before use. Yeast hexokinase was prepared according to Berger, Slein, Colowick & Cori (1946), by using the preparation of stage 5. The total volume of the incubation mixture was 8·2 ml. and the pH 7·3. After the addition of the hexokinase, the mixture was kept for 40 min. at room temperature. Histidine (75μmoles, pH 7·5) was then added together with a dialysis bag containing a solution of PGM in a small volume (0·5–5 ml.); the whole was allowed to equilibrate in the cold overnight. The dialysis bag was then transferred to a 5 l. beaker and dialysed against distilled water with stirring. After 12 hr. the water was changed, and the dialysis repeated five times, unless otherwise indicated. Even after additional changes of the water, considerable amounts of radioactive phosphate and glucose phosphate were released when the protein was subjected to tryptic digestion or partial acid hydrolysis and to ionophoresis at pH 6·5 or 3·5. This preparation was used for the structural studies, unless otherwise indicated. The specific activity of the protein ranged from about 0·05 to 0·5μc/mg.

As a control in one experiment this procedure was repeated exactly as above, except that PGM was denatured before use. The denatured PGM was prepared by precipitating 2 mg. of protein with 10% trichloroacetic acid and then dialysing it against 1·5 l. of distilled water. Native enzyme (5 mg.) was included in the same experiment. After the usual procedure (except that a smaller amount of ^{32}P was used) the native enzyme contained 7·5μmc/mg. of PGM, whereas the denatured one contained 0·3μmc/mg. of protein. As will be seen below, in the denatured preparation, only phosphate and glucose phosphate were detected

dr Ram

after partial acid hydrolysis or tryptic digestion, whereas the native preparation contained phosphopeptides in addition.

Method 2. A crude phosphorylase preparation was prepared by blending 20 g. of potatoes with 20 ml. of water. Kaolin (0·6 g.) was added and the mixture centrifuged. To 5 ml. of 2·5% starch solution were added 0·5 ml. of 0·2M-phosphate buffer, pH 7·0, 1·5 ml. of the above supernatant solution, a few drops of toluene and 1 mc of $H_3{}^{32}PO_4$ (carrier free). The mixture was kept at room temperature for 8 hr. A solution of 5 mg. of PGM in 1·0 ml. of water containing histidine (0·04M, pH 7·5) and $MgSO_4$ (2·5 mM) was dialysed overnight against the above mixture. The dialysis bag was then placed in a 5 l. beaker and dialysed against several changes of water.

Tryptic hydrolysis of [^{32}P]phosphoglucomutase

[^{32}P]PGM was twice dialysed against 40% urea and then against water until the urea was removed. Under these conditions the denatured PGM was precipitated. The precipitate was suspended in $(NH_4)_2CO_3$ (0·5%) to produce a 0·25% suspension, and crystalline trypsin (7·5μg./ml.) was added. Incubation was carried out at 37° with occasional stirring. Samples were removed at various times and subjected to ionophoresis after removal of the $(NH_4)_2CO_3$ *in vacuo*.

Partial acid hydrolysis. Two methods of partial acid hydrolysis were used:

(*a*) 5·7N-HCl at 100°. In general the reaction was carried out in small test tubes with approx. 0·2 ml. of acid. The time of hydrolysis was 30 min. unless otherwise indicated.

(*b*) 12 N-HCl at 37°. This method was used principally when larger amounts of protein were hydrolysed. Between 20 and 70 mg. of protein was incubated with 3–10 ml. of 12 N-HCl, for 30–40 hr. at 37° in tubes fitted with ground-glass stoppers.

Ionophoretic fractionation of peptides

The peptides were fractionated by high-voltage paper ionophoresis as previously described (Michl, 1951; Naughton *et al.* 1960). The buffer at pH 4·0 was prepared by adding pyridine to the buffer at pH 3·5. Unless otherwise stated, Whatman no. 52 paper was used with voltage gradients of 40 v/cm.

The pH–mobility curves (Fig. 9) were determined in the apparatus of Gross (1955) with Whatman no. 1 filter paper and 50 v/cm. (Naughton *et al.* 1960).

In order to standardize the positions of peptide spots and the conditions of each ionophoretic run, it was found convenient to use a mixture of coloured markers, which were usually applied at one side of the paper and run in parallel with the radioactive peptides. A suitable mixture for running at pH 3·5 or 6·5 towards the anode was made by using the following dyes: Pappenheim's panoptic stain; isamine blue; xylene cyanol FF; phenol red; methyl orange; safranine; Bordeaux; methyl green; light green SF.

For two-dimensional ionophoresis the following method, which was preferable to those previously described (Naughton *et al.* 1960), was used. Ionophoresis was carried normally in the first direction and the paper was radioautographed. A strip containing the peptides was cut out and sewn with a sewing machine to a second strip of paper. The paper was then wetted with the buffer to be used in the second dimension in such a way that it flowed evenly towards the strip from each side, thus sharpening the bands. The paper was then subjected to ionophoresis in the second dimension.

Paper chromatography

Two systems were used for studying the R_F values of the peptides: butan-1-ol-acetic acid-water-pyridine (30:6:24:20, by vol.) (Waley & Watson, 1953); phenol–water: phenol saturated with water at 25°. Whatman no. 52 filter paper was used with both systems.

Photo-oxidation technique for the detection of histidine peptides

The method was tested first with pure amino acids. The following amino acids were applied on a sheet of Whatman no. 52 paper, in two series one next to the other: tryptophan, tyrosine, arginine, lysine, histidine and methionine. The paper was moistened with pyridine–acetate buffer (10% of pyridine, 0·4% of acetic acid, pH 6·5) and then with methylene blue (0·2% in absolute ethanol). The paper was then covered with a polythene sheet, and one of the series of amino acids covered with a black paper to use as control. The paper was pressed between glass plates and the part containing both series of amino acids put under a 150 w lamp at 30 cm. for 2 hr. It was then dried in the dark, the amino acids were subjected to ionophoresis at pH 2·1 for 20 min. and the paper was sprayed with ninhydrin. The methylene blue, which covers part of the sheet, does not seem to interfere with the ninhydrin reaction and the spots can easily be located. It was observed (Fig. 1) that tryptophan, present in the control, completely disappeared after photo-oxidation. Histidine, which ran together with the basic amino acids in the control, is slightly slower than tyrosine after the treatment. No difference was detected in the other amino acids tested.

The test for histidine peptides was made as follows: a partial acid hydrolysate of [^{32}P]PGM containing about 2·5μmc of ^{32}P was applied on Whatman no. 52 paper in a 1·5 cm. wide band and subjected to ionophoresis at pH 4·0

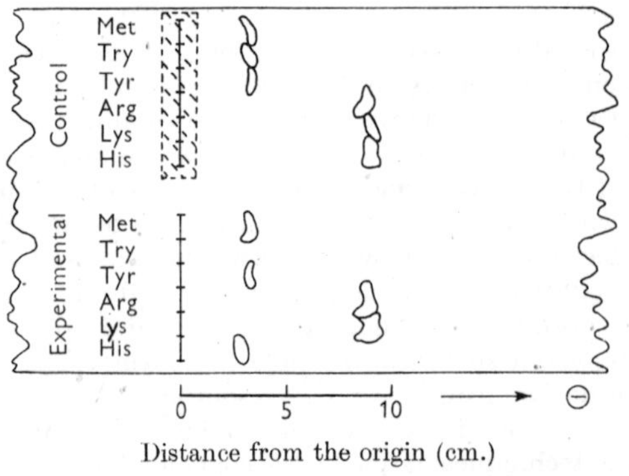

Fig. 1. Effect of photo-oxidation on the electrophoretic mobilities of amino acids at pH 2·1. Before electrophoresis both series of amino acids were photo-oxidized as indicated in the Methods section. During photo-oxidation the shaded area was covered with a black paper.

for 2 hr. It was then radioautographed, and a strip was cut to include peptides 7, 6, 5 and 4, and part of the glucose [^{32}P]phosphate to be used as a marker. The strip was moistened with buffer solution and methylene blue, and placed under the 150 w lamp in a dark room. Another strip of the same paper containing histidine, lysine and arginine, covered and uncovered with black paper, was subjected to the same treatment and used as further control. Both strips were placed together under the same light, and exposed for 80 min. After being dried in the dark, they were sewn with a sewing machine on a sheet of Whatman no. 52 paper at the same distance from the edge of the paper as the hydrolysate was applied in the first run. Ionophoresis was repeated under the same conditions (pH 4·0, 2 hr.).

Direct identification of amino acids in peptides

In order to obtain sufficient quantities of the peptides to identify the amino acids present in them directly, larger-scale experiments were carried out: [^{32}P]PGM (70 mg.) was hydrolysed for 2 days at 37° in 12N-HCl and passed through a column (20 cm. × 0·8 cm. diam.) of Dowex 50X4 (200–400 mesh). About 50% of the radioactivity passed straight through the column when washed with water and appeared in the first 20 ml. of eluate. This was collected, freeze-dried and labelled F_1. The column was then developed with ammonium acetate solution (2%, pH 3·7), which eluted about 30% of the radioactivity. This fraction, which contained most of the material in the partial hydrolysate, was freeze-dried and labelled F_3. No more radioactivity could be eluted from the column by increasing the pH to 7·0.

Samples of fractions F_1 and F_3 were subjected to ionophoresis at pH 4·0 and 6·5. The bands obtained were designated by numbers as shown in Fig. 4. F_1 contained the acidic components, namely bands 2 (serine phosphate), 4 and 5 as well as 1 (phosphate) and 3 (glucose phosphate). F_3 contained the 'neutral' components (bands 6 and 7). This fractionation on Dowex 50 was very useful for the purification of peptide 5 but unnecessary for the 'neutral' peptides.

All the peptides were purified further by paper ionophoresis and chromatography (Table 6), generally with the two-dimensional method. Whatman no. 52 paper was used where possible, but, when the amount of material was so large that it would have caused overloading on no. 52, it was applied to no. 3 MM paper.

The first run of fraction F_3, or when the Dowex 50 step was omitted, was made on Whatman no. 3 MM paper (peptides 6 and 7, Table 6) and the bands were eluted and dried. In other cases the samples were applied as 2 cm. bands and run in two dimensions. After location of the spots by radioautography they were cut out and were either eluted and hydrolysed for amino acid determination or sewn on another sheet of paper and again run in two directions.

The amino acids were identified by two-dimensional paper electrophoresis (pH 2·1; 30 min.; 80 v/cm.) and chromatography in butanol–acetic acid–water (3:1:1, by vol.) for 3 hr. (L. F. Smith, unpublished work), except for 7A and 7C (Table 6), in which case only electrophoresis was used.

RESULTS

Labelling of phosphoglucomutase

According to Kennedy & Koshland (1957) PGM is inhibited by di*iso*propyl phosphorofluoridate (DFP). An attempt was therefore made to label the enzyme by treatment with [^{32}P]DFP. The conditions used were similar to those used by the above authors for obtaining 100% inhibition of PGM in 15 min.: PGM was incubated with [^{32}P]-DFP (1·0 mM), cysteine (25 mM), $MgSO_4$ (1·5 mM) and glucose 1-phosphate (5 mM). The pH was adjusted to 8·2. After 3 hr. the mixture was dialysed extensively against water. A partial acid hydrolysate of the freeze-dried protein was examined by ionophoresis at pH 3·5 followed by radioautography. No 'phosphopeptides' or 'phospho-amino acids' were detected but only inorganic phosphate and a trace of mono*iso*propyl phosphate. A second experiment in which the concentration of the DFP was increased ten times, the incubation period increased to 12 hr. and the enzyme precipitated with HCl before dialysis to avoid the possibility of removal of the phosphate during dialysis, was also negative.

Labelling the PGM by exchange of phosphate with labelled substrate proved much more successful. According to Kennedy & Koshland (1957) their preparation contained about 1·5 μmc/mg. of protein, by using as starting material 1 mc of ^{32}P. This would have been insufficient for our purposes. With the techniques described above, it was possible to prepare a labelled protein containing as much as 0·5 μc/mg. This offered the possibility of using minute amounts of protein for sequence studies.

Stability of the bound ^{32}P *in labelled phosphoglucomutase.* The [^{32}P]PGM prepared by these methods still contained some diffusible ^{32}P. When dialysed five times against 100 ml. of water for 24 hr., some radioactivity still diffused out. The material after such treatment was dialysed under various conditions and the amount of radioactivity that diffused out was measured (Table 1). Three separate experi-

Table 1. *Removal of* ^{32}P *from* [^{32}P]*phosphoglucomutase by dialysis*

Expt.	Dialysis against	Amount diffused out (%)
A	Water	5·6
	Water	6·1
	Water	6·0
B	0·1 mM-Phosphate (pH 6·0)	1·5
	4·0 mM-Phosphate (pH 6·2)	3·7
	Water	1·0
C	40% Urea	7·1
	Water	1·3
	Water	1·2

ments were performed, in which samples containing about 50 μmc of [^{32}P]PGM were dialysed against 25 ml. of the indicated solution for 24 hr. The radioactivity of the dialysate was determined and the dialysis was repeated with a further 25 ml. of solution. In all cases some radioactivity diffused out, although in a small proportion compared with the total amount. When the urea-treated enzyme (Table 1, Expt. C) was digested with trypsin and the hydrolysate subjected to ionophoresis at pH 6·5, a spot corresponding to phosphate and amounting to about 30% of the total remaining radioactivity was obtained. This suggests that there is still some labile phosphate remaining inside the sac even after very extensive dialyses, namely three times against 5000 vol. of water (12 hr. each with stirring), five times against 100 vol. of water (24 hr. each), once against 16 vol. of 40% urea (24 hr.) and finally twice against 16 vol. of water for 24 hr.

Tryptic hydrolysate of [^{32}P]phosphoglucomutase

Fig. 2 indicates the time course for the hydrolysis of [^{32}P]PGM with trypsin. The results clearly suggest that band B (not necessarily a single radioactive peptide) is the primary breakdown product formed by the splitting of very susceptible bonds, the reaction being complete when all the denatured material has come into solution (usually less than 4 hr.). A second slower reaction takes place and a radioactive peptide is released from band B which moves quicker towards the anode at pH 6·5 (band A). This second reaction hydrolyses band B after incubation for 17 hr. A third product, band C, begins to be apparent at that time. Trypsin usually contains traces of other pancreatic enzymes, especially chymotrypsin, and it is possible that band C is due to the action of such a contaminant on band A or B. Partial acid

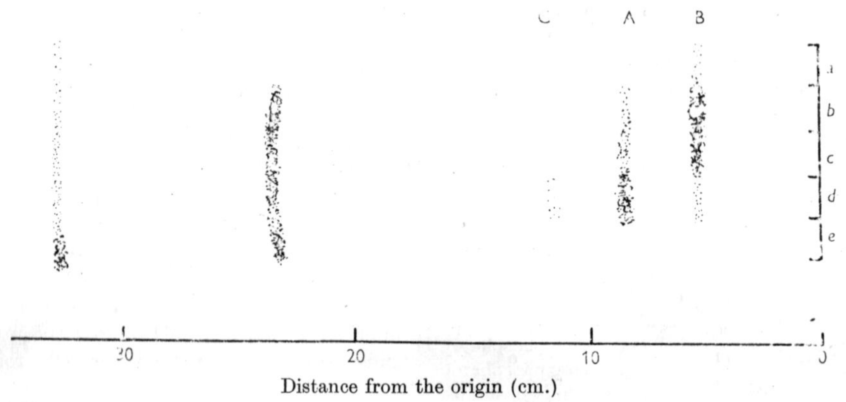

Fig. 2. Tracing of radioautograph obtained after ionophoresis (pH 6·5; 40 v/cm.; 2 hr.) of tryptic digest of [^{32}P]PGM. Conditions of the digestion were as described in text. Times of digestion: *a*, 1½ hr.; *b*, 4 hr.; *c*, 7 hr.; *d*, 17 hr.; *e*, markers of bands 1 and 3 (Fig. 4).

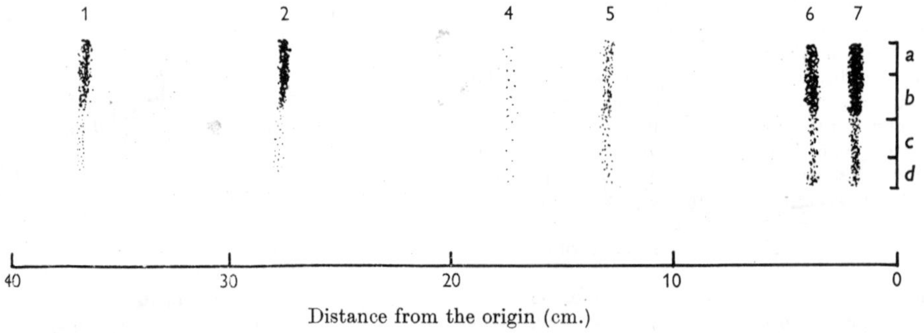

Fig. 3. Tracing of radioautograph obtained after ionophoresis (pH 3·5; 40 v/cm.; 2 hr.) of partial acid hydrolysates of tryptic peptides (Fig. 2) derived from [^{32}P]PGM prepared by methods 1 and 2. *a*, Band B (method 1); *b*, band A (method 1); *c*, band A (method 2); *d*, band B (method 2).

hydrolysates of the three bands were made and since no difference of pattern was observed, either the pure bands or a mixture of them was used in further partial hydrolysis studies.

The [32P]PGM prepared by method 2 was also subjected to hydrolysis by trypsin. Fig. 3 shows an ionogram of the partial acid hydrolysate of the tryptic bands, which was run together with a similar hydrolysate of [32P]PGM prepared by method 1. The exact matching of the bands indicates that both methods of labelling are equivalent.

As a control a sample of PGM that had been denatured before the labelling procedure was also subjected to tryptic digestion. The small amount of radioactivity which remained associated with the protein after dialysis gave after digestion a spot moving with phosphate and a trace of a second spot moving with glucose phosphate on ionophoresis at pH 6·5.

Partial acid hydrolysis of [32P]*phosphoglucomutase*

Fig. 4 shows the pattern obtained when [32P]-PGM was subjected to partial acid hydrolysis (5·7N-HCl; 100°; 30 min.) and ionophoresis at pH 3·5. Band A from the tryptic digest (Fig. 2) was similarly hydrolysed and run side by side. As a control, denatured PGM was also subjected to the same treatment. For comparison the pattern obtained from di*iso*propoxy[32P]phosphinylchymotrypsin, which had been subjected to the same treatment, is shown. It is quite clear that the main features in the patterns of PGM and chymotrypsin are different, so that it may be concluded that PGM does not contain the sequence Gly.Asp.SerP.Gly which is in chymotrypsin.

Band 1 runs in the position of inorganic orthophosphate and band 2 was identified as serine phosphate by comparison with a known marker on ionophoresis at pH 3·5 and 2·1.

Band 3 appeared to be an impurity, which can be released during the isolation of the tryptic peptide. A sample eluted from band 3 was run in parallel with glucose 6-phosphate in ionophoresis at pH 3·5. The paper was treated with phosphate reagent (Burrows, Grylls & Harrison, 1952) and then radioautographed. There was a satisfactory match between the blue colour and the radioactive spot. It thus seems likely that it is a glucose phosphate, and the known stability of the phosphate residues would suggest that it is glucose 6-phosphate rather than glucose 1-phosphate.

Bands 6 and 7 are mixtures of a number of components which can be separated at pH 2·1 or 6·5. The ionophoresis of the peptide mixture at pH 6·5 is shown in Fig. 5. Fig. 6 shows a two-dimensional ionogram run at pH 3·5 and 6·5, and shows the relationships between the patterns obtained at the two pH values. At pH 6·5 bands 4 and 5 are contaminated by a 'tailing' of band 6A and can best be purified at pH 3·5, whereas the other peptides are present as mixtures (bands 6 and 7) at pH 3·5 and are best purified at pH 6·5. The proportions of the peptides in hydrolysates obtained by two different methods were determined by measuring the relative degree of darkness of the radioautographs of the ionograms run at pH 6·5. The measurements were carried out in a Chromatograph automatic recording reflective densitometer (Joyce, Loebel and Co. Ltd., A8, Princeway Team Valley, Gateshead, Co. Durham) and the results are shown in Table 2.

The time course of the hydrolysis was followed by ionophoresis at pH 3·5 and 6·5. A sample of the tryptic bands was hydrolysed with 5·7N-HCl in a boiling-water bath and samples were taken out at different times and subjected to ionophoresis at pH 3·5. The relative amounts of each peptide were determined by densitometry of the radioautographs. Fig. 7 shows clearly that bands 7, 2 and 1 increased continuously. Bands 5 and 6, on

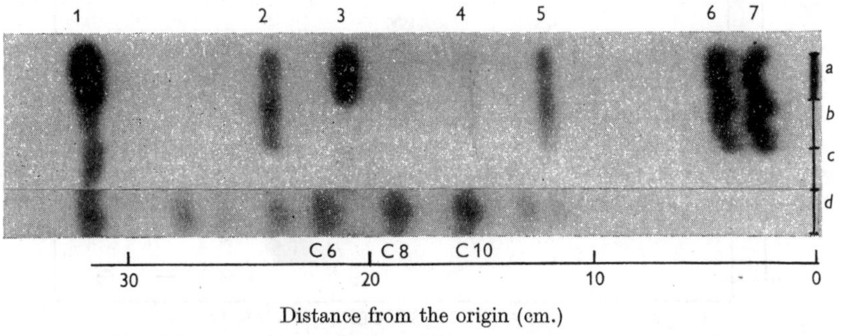

Fig. 4. Radioautograph obtained after ionophoresis (pH 3·5; 40 v/cm.; 2 hr.) of partial acid hydrolysates of 32P derivatives. *a*, [32P]PGM; *b*, band A (Fig. 2); *c*, denatured PGM subjected to the same labelling and subsequent treatment as *a*; *d*, di*iso*propoxy[32P]phosphinylchymotrypsin; the Asp.SerP (C6), Asp.SerP.Gly (C8), and SerP.Gly (C10) bands are indicated (Naughton, Sanger, Hartley & Shaw, 1960).

the other hand, pass through a maximum. Band 4 does not seem to change much after reaching a low level. The time course of the hydrolysis was also followed at pH 6·5. In this case two different hydrolysis conditions were used: 5·7 N-HCl and 12 N-HCl. The time course of the hydrolysis is more or less similar for both conditions, when 10 min. hydrolysis in the 5·7 N-HCl is compared with 24 hr. in 12 N-HCl as suggested by Naughton *et al.* (1960). Bands 6 A and 7 A are the main ones present after 30 min. in 5·7 N-HCl (or 72 hr. in 12 N). Band 7 A is, however, still increasing whereas band 6 A is beginning to diminish. Both seem to be quite stable peptides, especially the former.

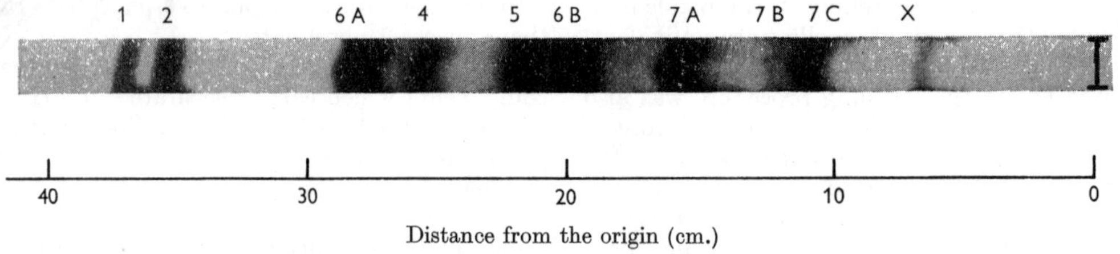

Fig. 5. Radioautograph obtained after ionophoresis (pH 6·5; 40 v/cm.; 2 hr.) of the partial acid hydrolysate of band B (Fig. 2).

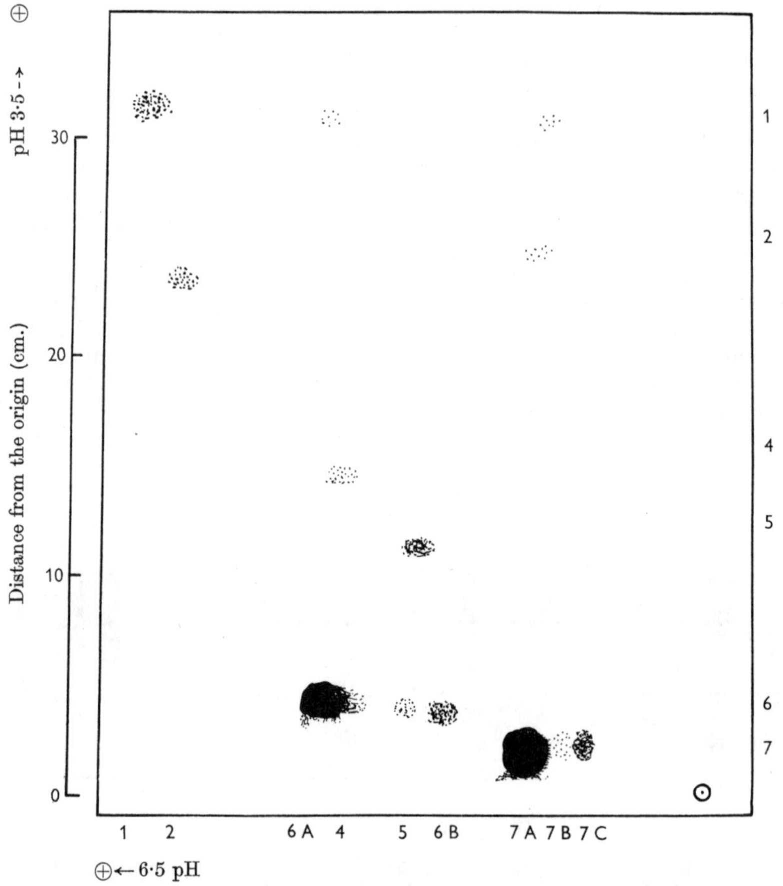

Fig. 6. Tracing of radioautograph of a two-dimensional ionogram (pH 3·5 and 6·5; 40 v/cm.; 2 hr. and 110 min. respectively) of the partial acid hydrolysate of a mixture of isolated bands from the tryptic digest of [^{32}P]PGM.

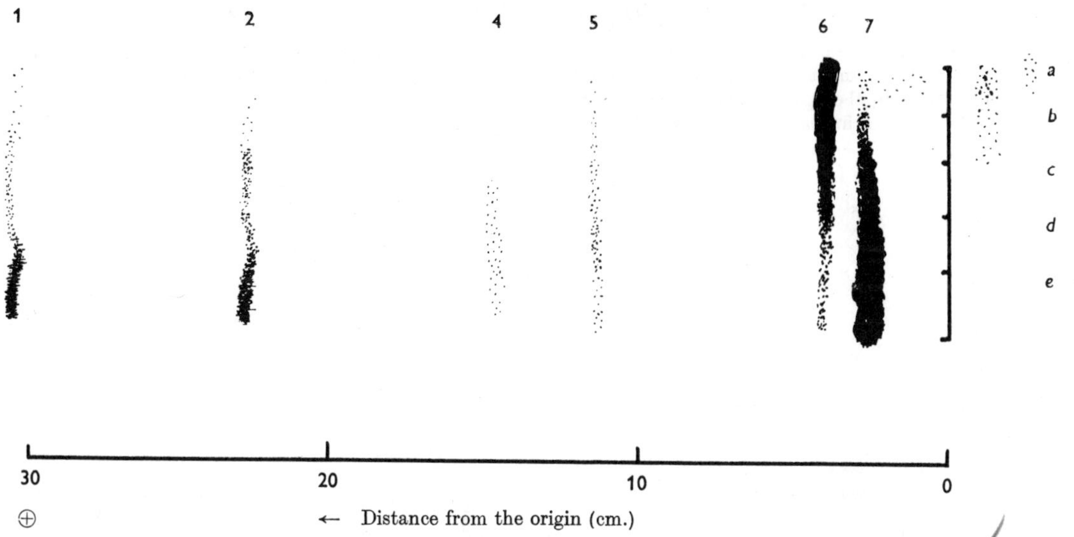

Fig. 7 (*a*). Tracing of radioautograph obtained after ionophoresis (pH 3·5; 40 v/cm.; 2 hr.) of the partial hydrolysis of band A (Fig. 2) for different times: *a*, 5 min.; *b*, 10 min.; *c*, 20 min.; *d*, 30 min.; *e*, 45 min.

Table 2. *Yields of peptides (%) from partial acid hydrolysis of* [^{32}P]*phosphoglucomutase*

Band no.	5·7 N-HCl for 30 min. at 100°	12 N-HCl for 3 days at 37°
1	14·0	5·5
2	15·6	11·6
4*	2·5	2·6
5	8·2	10·8
6A	18·4	26·8
6B	3·9	4·6
7A	24·4	28·6
7B	1·7	1·2
7C	6·2	7·7
Others	4·7	1·2

* Contaminated with material running as band 6 at pH 3·5.

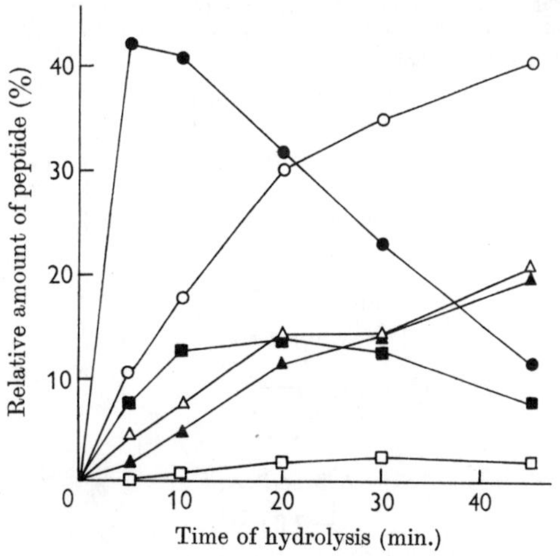

Fig. 7 (*b*). Relative amounts of each band during the hydrolysis, calculated by densitometry of the radioautograph [Fig. 7 (*a*)]. △, Band 1; ▲, band 2; □, band 4; ■, band 5; ●, band 6; ○, band 7.

Structure of the peptides

In order to determine the interrelations of the various bands they were subjected to a further partial acid hydrolysis (5·7 N-HCl; 100°; 45 min.) and the products identified by ionophoresis together with markers of the original partial acid hydrolysate. The results are shown in Table 3. Table 4 shows the products produced when the various bands were subjected to the Edman degradation (Naughton *et al.* 1960).

Peptide 7A gives only serine phosphate on partial hydrolysis, suggesting that it is a dipeptide, and, since phosphate and not serine phosphate was obtained as the main product after the Edman degradation, the serine phosphate is *N*-terminal. Peptide 6A also has *N*-terminal serine phosphate and gives 7A on partial hydrolysis, indicating that it is the tripeptide formed by addition of a residue to the *C*-terminal end of 7A. Peptide 5 gives 4 and serine phosphate on partial hydrolysis. It does not give 7A and therefore the serine phosphate must be the *C*-terminal residue. It is converted into 4 by the Edman degradation. It is thus most probably a tripeptide and peptide 4 the other dipeptide. The yield of band 4 is very low and it was not possible to carry out the re-hydrolysis and Edman degradations with it. The various relationships of the

Table 3. *Products of the rehydrolysis of radioactive bands*

Bands 1, 2, 4 and 5 were isolated by ionophoresis at pH 3·5, the other bands at pH 6·5, and hydrolysed in 5·7 N-HCl for 45 min. at 100°. They were again subjected to ionophoresis at the same pH used for their isolation and identified by their position with respect to markers of the original partial hydrolysate. The relative amount of each product is indicated with crosses; tr., trace.

Band rehydrolysed (Figs. 4 and 5) ...	1	2	4	5	6A	7A	7B	7C	X
1	×	.	.	.	.	.	.	.	.
2	×	× ×	.	.	.	.	.	.	.
4	×	× ×	× × ×	.	.	.	.	.	.
5	× ×	× × ×	×	×	.	.	.	.	.
6A	×	× ×	.	.	× ×	× × × ×	.	.	.
6B	×	× ×	×	× ×	× ×	× × × ×	tr.	tr.	.
7A	×	×	.	.	.	× × × ×	.	.	.
7B	×	× ×	× × ×	.	.	× × × ×	× × ×	.	.
7C	×	× × ×	×	×	.	× × × ×	×	×	.
X	×	×	.	.	.	× × × ×	.	.	× ×

Table 4. *Products of the Edman degradation of radioactive bands*

Band 5 was isolated at pH 3·5 and bands 6A and 7A at pH 6·5 as indicated in Figs. 4 and 5 respectively. The bands were eluted and subjected to the Edman degradation followed by ionophoresis at the same pH used for their isolation. Identification of the products was made by their position with respect to markers of the original partial hydrolysate. The relative amount of the products is indicated with crosses; tr., trace.

Band	Products
5	1 (×)+4 (× × ×)+5 (× ×)
7A	1 (× × × ×)+2 (tr.)+a* (× ×)+7A (× ×)
6A	1 (× × ×)+2 (tr.)+6A (× × × ×)

* Unknown spot, running between 5 and 7A.

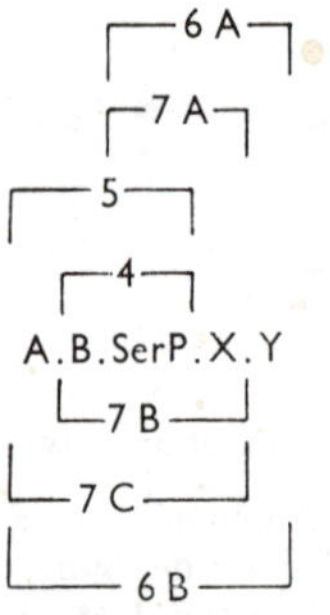

Fig. 8. Diagram showing the interrelationships of peptides of the partial acid hydrolysate.

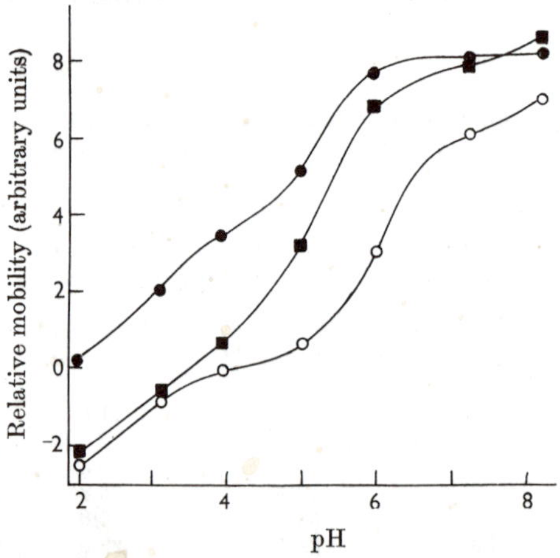

Fig. 9. Mobilities at different pH values of some of the peptides isolated as shown in Fig. 5. For comparison, the mobility of SerP.Gly (●) is also included. Mobilities are expressed relative to a standard serine phosphate marker. The points are experimental values and the curves are theoretically calculated from the following p*K* values: Peptide 7A (○); 1·1 (PO_4), 2·5 (CO_2H), 5·7 and 6·2 (PO_4 and basic), 8·8 (α-NH_2); Peptide 6A (■); 1·1 (PO_4), 2·5 (CO_2H), 4·5 (CO_2H), 5·7 and 5·7 (PO_4 and basic), 8·8 (α-NH_2).

peptides are shown in Fig. 8, where the unknown amino acid residues are indicated by A, B, X and Y. It can be seen that all possible partial hydrolysis products of the pentapeptide sequence have been detected with the exception of the tetrapeptide B.SerP.X.Y.

Fig. 9 shows the pH–mobility curves for peptides 6A and 7A and a SerP.Gly marker. The results were calculated relative to a serine phosphate marker in the following way: a titration curve of serine phosphate was drawn, and, assuming mobility to be proportional to net charge, an arbitrary mobility curve was obtained by substituting the alkali volume by an arbitrary centimetre scale. Glucose markers were used to correct for the displacement of the origin due to endosmosis. The relative mobility for a peptide at a given pH is equal to: (Distance of peptide from glucose marker)/(distance of SerP from glucose marker) × arbitrary mobility of SerP at same pH (Naughton *et al.* 1960). Values for band 3 (glucose phosphate) were also determined and showed good agreement with the theoretical curve calculated from the known pK values. At pH 4·0, SerP.Gly has one negative charge, whereas the dipeptide 7A is neutral. This suggests that the unknown amino acid called X in Fig. 8 is a basic one. Comparing the increase in rates of SerP.Gly and peptide 7A between pH 4·0 and pH 7·2, it can be seen that although the molecular weight of peptide 7A is greater, its mobility has increased more than the increase in mobility of SerP.Gly, suggesting another group ionizing near the range of the second phosphate. Fig. 9 also shows the theoretical curve, based on ionizable groups in peptide 7A, including a basic group with a pK of 5·7 or 6·2. It can be seen that this curve fits the experimental data. The tripeptide 6A is more acidic than 7A at pH values higher than 3·5. At pH 4·5, when the second phosphate group is completely discharged, the tripeptide is clearly acidic. The inclusion of an acidic group with pK 4·5, and slight changes in the pK of the phosphate or basic, or both, groups of peptide 7A, fit in with the experimental data. The mobility data, strongly suggest then, the presence of a basic amino acid (pK between 5 and 7) and an acidic amino acid (pK about 4·5) on the C-terminal side of serine phosphate. Peptides 4 and 5 run at both pH values (3·5 and 6·5), as would be expected for a di- and tri-peptide having neutral amino acids and a C-terminal serine phosphate.

The above curves indicate that residue X (Fig. 8) is a basic amino acid having a pK about 6·5 and therefore histidine. This was further confirmed by photo-oxidation (see below) and by the following experiment: band 7 (3 μmc) isolated by ionophoresis at pH 3·5 was added to 3 mg. of a hydrolysate (5·7N-HCl; 100°; 1 hr.) of unlabelled PGM and subjected to ionophoresis at pH 3·5 (2 hr.; Whatman no. 1 paper). The radioactive band was cut out, eluted and re-run at pH 6·5 for 90 min. and then at pH 2·1 for 50 min. The paper was then radioautographed and sprayed with Pauly reagent (Fraenkel-Conrat & Singer, 1956). A faint but distinct red spot fell in exactly the same position as the radioactivity. Since the possibility of a histidine peptide contamination having the same mobility as peptide 7A in the three different systems used seems unlikely, the assumption that SerP.His is one of the dipeptides of the partial hydrolysis of PGM seems reasonable.

Fig. 10 shows the results of the photo-oxidation experiment. The pattern obtained in the first run is shown at the origin. After photo-oxidation and re-running in the second dimension but with the same buffer, glucose phosphate and peptides 4 and 5 are mostly unchanged, whereas most of peptides 6 and 7 are in a new position, indicating an increase in negative charge. Some of the unchanged peptides are still present. Band 6 shows two unchanged spots, and two corresponding faster-moving spots (presumably 6A, 6B and their oxidized derivatives) and some free phosphate. Band 7 gave one spot in the position of the unchanged material and several other spots. The strongest one, running in a position between 4 and 5, is probably the oxidized 7A, and the slower-moving spots the other peptides present in 7 which can be separated at pH 6·5. There are also a phosphate spot and three very faint spots whose nature is unknown. It is quite clear from this experiment that the main components in 6 and 7 are histidine peptides. The presence of spots in the positions of the unchanged peptides from 6 and 7 suggests that in these conditions the reaction is not quantitative. In this respect it has to be considered that 10% of unchanged histidine would hardly be detected in the ninhydrin control experiment, but would give a very clear spot in the radioactive experiment.

In order to identify the presence of N-terminal serine or threonine residues, peptides 4 and 5 were

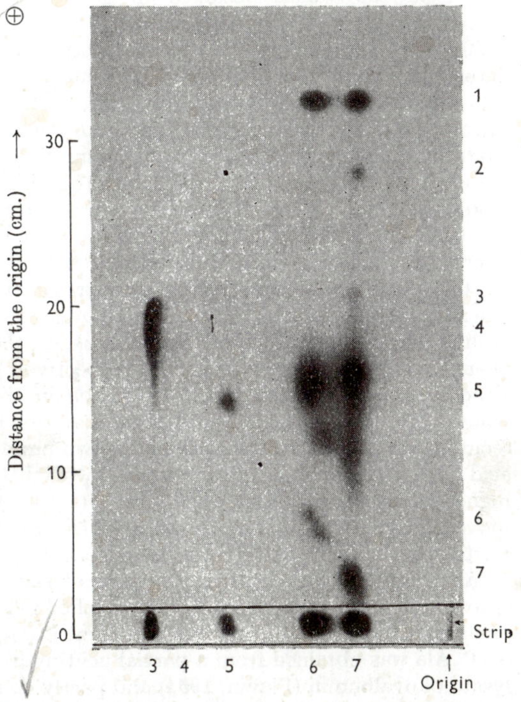

Fig. 10. Radioautograph obtained after ionophoresis of the photo-oxidized peptides. A paper strip obtained by ionophoresis (pH 3·5; 40 v/cm.; 2 hr.) was subjected to photo-oxidation and then sewn to another sheet of Whatman no. 52 filter paper and subjected to ionophoresis in the second dimension with the same conditions. The photograph also shows the radioautograph of the photo-oxidized strip, and the position in which it was sewn before ionophoresis in the second dimension.

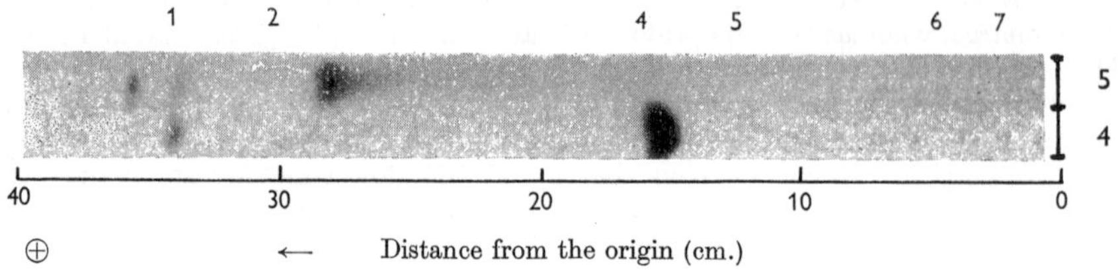

Fig. 11. Radioautograph obtained after ionophoresis (pH 3·5; 40 v/cm.; 2 hr.) of the periodate-treated peptides 4 and 5 from Fig. 3.

Table 5 shows the R_F values in one solvent system for a number of peptides of known constitution and for the main radioactive bands. The values of the constant $(n-1)A+B$ [equation (1)] for the known peptides have also been calculated. For the butanol–acetic acid–water–pyridine system the values for the constant are within the expected limit of errors, whereas for the phenol–water system there was much more spreading of the values. Much more reliance therefore can be placed on the values obtained in the former system.

The R_F of peptide 4 in the butanol–acetic acid–water–pyridine system was 0·06. Assuming a value of -1240 for the constant $(n-1)A+B$ it can be calculated from equation (1) that the R_F of the unknown amino acid B is 0·20. If we assume that the error in R_F is not more than 0·05 (Moore & Baker, 1958), the R_F of B must lie between 0·15 and 0·25. The only neutral amino acids to be considered are alanine, proline and possibly threonine (R_F 0·23). The results with the periodate treatment indicate that it is not threonine. Alanine and proline have very different R_F values on the phenol–water system (0·49 and 0·81 respectively), although there was considerable variation in the values of $(n-1)A+B(-300 \pm 260)$ in this system it seems reasonably certain that the unknown amino acid is not proline. Proline has an R_F slightly higher than leucine, whereas peptide 4 (R_F 0·06) moves much more slowly than Leu.SerP (R_F 0·37). Further evidence that peptide 4 was Ala.SerP was obtained by comparison with a sample of radioactive Ala.SerP prepared by another method from ovalbumin (D. C. Shaw, personal communication). SerP.Ala was obtained from a partial acid hydrolysate of ovalbumin (Flavin, 1954) and partly converted into Ala.SerP by inversion of the dipeptide sequence in hot dilute acid. This material ran at the same rate as peptide 4 on ionophoresis at pH 6·5 for 1 hr. 45 min. at 40 v/cm.

Assuming that peptide 4 was Ala.SerP, it was calculated that the R_F of residue A (Fig. 8) was 0·27. The periodate reaction indicated that it is either serine or threonine: serine has R_F 0·12 and threonine 0·23, suggesting that the residue is most

Table 5. *Chromatographic constants of amino acids and peptides in butanol–acetic acid–water–pyridine*

R_F values were obtained after development for 48 hr. Values of $A+B$ and of $2A+B$ were calculated from equation (1).

Sample	R_F	$A+B$
Arginine	0·13	.
Histidine	0·13	.
Glutamic acid	0·14	.
Aspartic acid	0·09	.
Proline	0·25	.
Glycine	0·12	.
Alanine	0·19	.
Valine	0·36	.
Leucine	0·55	.
Threonine	0·23	.
Serine	0·12	.
Serine phosphate	0·03	.
Leu.SerP	0·18	−1010
Gly.SerP	0·03	−1140
SerP.Gly	0·04	−1290
SerP.Ala	0·07	−1360
Ala.SerP	0·06	−1250
Glu.SerP	0·05	−1360
Thr.SerP	0·05	−1325
Mean value ($A+B$)		−1240 ± 125
Peptide 7A (SerP.His)	0·04	−1270
		$2A+B$
Leu.Gly.Gly	0·40	−1940
Gly.Gly.Gly	0·09	−2050
Mean value ($2A+B$)		−1990
Peptide 4	0·06	.
Peptide 5	0·08	.
Peptide 6A	0·03	.

treated with periodate (Sanger & Shaw, 1960). The result is shown in Fig. 11. Peptide 5 has completely disappeared and has been replaced by a main spot moving faster than peptide 4 and two minor spots. Peptide 4, on the other hand, has remained largely unchanged. This indicates that peptide 5 has a *N*-terminal serine or threonine residue, whereas peptide 4 has not. The sequence around the serine phosphate residue may thus now be written:

$$\begin{pmatrix}\text{Thr}\\ \text{Ser}\end{pmatrix}.\text{B}.\text{SerP}.\text{His}.\begin{pmatrix}\text{Asp}\\ \text{Glu}\end{pmatrix}$$

where B is some neutral amino acid other than threonine or serine.

Table 6. *Amino acid composition of purified peptides*

Figures in parentheses represent the approximate amount present as estimated visually from the strength of the ninhydrin colour; tr., trace.

	Purification procedure					
	Chromatography on Dowex 50	Ionophoresis pH			Chromatography in butanol–acetic acid–water (3:1:1, by vol.)	
Peptide		2·1	3·5	6·5		Amino acid composition
5	+	.	+	.	.	Glu (3), Ser (2), Ala (2), Thr (1), Val (1), Ileu (1), Leu (1)
5	+	+	+	.	+	Ala (1), Ser (1), Thr (1)
7A	+	+	+	+	.	His (2), Ser (2), Glu (1), Asp (1), Thr (tr.), Ala (tr.)
7B	+	+	+	+	.	His (1), Ala (1), Ser (1), Thr (tr.), Glu (tr.), Asp (tr.)
6A	.	+	+	+	+	His (1), Ser (1), Asp (1), Glu (tr.), Ala (tr.), Gly (tr.)
6A	+	+	+	+	.	His (1), Ser (1), Asp (1), Gly (1), Glu (0·5)

probably threonine. From the R_F of peptide 6A in butanol–acetic acid–water–pyridine a value of 0·18 for the acidic residue (Y in Fig. 8) was calculated. Glutamic acid has R_F 0·14 and aspartic acid 0·09. This result would suggest that glutamic acid is the most probable. However, further work (see below) showed that residue Y was in fact aspartic acid. It seems probable that equation (1) will be less reliable where amino acids with charged side chains are concerned, since variations in pK values between amino acids and peptides would be expected to affect the partition coefficients, whereas this is not taken account of in the equation.

It may be concluded from the above results that the sequence around the serine phosphate residue is most probably:

$$\text{Thr.Ala.SerP.His.}\begin{pmatrix}\text{Asp}\\\text{Glu}\end{pmatrix}.$$

Table 6 shows the results of the large-scale experiments in which the amino acids present in the peptides were identified by the standard ninhydrin technique. Several preliminary attempts failed to give pure peptides and extensive purification was necessary. Glutamic acid, aspartic acid and glycine were often present as impurities. The spot identified as serine phosphate (band 2) contained more glutamic acid than serine after purification by ionophoresis at pH 3·5 and 6·5.

Since the pentapeptide sequence could be deduced from the composition of the two tripeptides 5 and 6A, special care was taken to obtain these in a pure form, and the results with these two (Table 6) were clear and demonstrate that the pentapeptide sequence is Thr.Ala.SerP.His.Asp. The results with the other peptides are consistent with this sequence, although in most cases they are clearly not completely pure. Thus peptide 7A contained glutamic and aspartic acids in about half the amount of the true components, histidine and serine.

DISCUSSION

In this work we have attempted as far as possible to use methods that depend only on radioactive techniques for the identification of peptides and their degradation products rather than those that depend on the more conventional ninhydrin reaction. Such methods which can be employed on a much smaller scale have the advantage that they can be applied to isotopic peptides that are contaminated with other non-isotopic substances provided they are free from other isotopic substances. The extreme difficulty of purifying peptides from an acid hydrolysate of a large protein is evident from this work, and, if the ninhydrin method had been used by itself, it would have been very difficult to derive the correct sequence. On the other hand, it is relatively easy to obtain isotopic peptides free from one another, and it is hoped that the techniques described here may prove of value in other cases where it is desired to know the sequence around a given labelled amino acid residue.

The method of deduction of the interrelationships of the various peptides from the partial acid hydrolysate (Table 5) depends on the assumption that all the peptide bonds are to some extent hydrolysed by acid. Thus, for instance, peptide 7A was said to be a dipeptide since it gave only serine phosphate on partial hydrolysis. It could possibly have been a tripeptide SerP.X.Y where the bond X–Y was extremely stable so that SerP.X was never produced in appreciable amounts. In fact

Table 7. *Amino acid sequences around reactive serine in enzymes*

Enzyme	Sequence	Reference
Trypsin	Gly.Asp.Ser.Gly	Dixon, Kauffman & Neurath (1958)
Chymotrypsin		Turba & Gundlach (1955); Schaffer *et al.* (1957)
Elastase		Hartley, Naughton & Sanger (1959)
Thrombin		Gladner & Laki (1958)
Liver ali-esterase	Gly.Glu.Ser	Jansz, Posthumus & Cohen (1959)
Pseudocholinesterase		Jansz, Brons & Warringa (1959)
Subtilisin	Thr.Ser.Met.Ala	Sanger & Shaw (1960)
Phosphorylase A	Lys.Glu-NH_2.Ileu.SerP.Val.Arg	Fisher, Graves, Grittenden & Krebs (1959)
Phosphoglucomutase	Thr.Ala.SerP.His.Asp	This paper

the further confirmatory experiments did show that all the deductions made from the partial hydrolysis experiments were correct, and whether or not such methods will be entirely reliable can only be decided by further experience. From the partial acid hydrolysis the relationships of the peptides could have been as shown or in the reverse order. The Edman degradation studies showed which was the correct order and confirmed some of the relationships.

The pH–mobility curves have proved useful for identifying charged residues in peptides. It was hoped that they could be used further, for instance to obtain information about the exact pK values of amino and carboxyl groups and about the size of residues involved in peptides. However, preliminary experiments with known serine phosphate peptides and other neutral peptides suggested that accurate results could not easily be obtained. It appeared that other factors, such as adsorption on the paper and hydration, may determine the rate of migration at different pH values, since many of the pH–mobility curves obtained did not agree with any predicted formula (e.g. Alberty, 1953). Thus for instance at all pH values the difference in mobility between SerP.Gly and SerP.Ala was considerably greater than would be expected. At high pH values the migration of serine phosphate peptides was frequently slower than expected, suggesting increased hydration with the increased charge. Perhaps by a more intensive study of the relationships of the structures of peptides to their ionophoretic mobilities, it may be possible to devise methods for identifying residues in peptides in this way. However, at present it seems that more reliable information can be obtained by the use of specific reactions, such as the periodate and the photo-oxidation reaction, and by studying the R_F values of the peptides as described above. With the two chromatographic systems we have used, it was not possible to identify with certainty all the amino acids present in the peptides, though a large number of possibilities could be eliminated. In order to make such a method completely reliable, it would be necessary to use more chromatographic systems and to have more known peptides to use as markers. The formula of Pardee (1951) is not completely accurate and certain peptides may be exceptional (Moore & Baker, 1958). It is to be hoped that with further experience it may be possible to improve the reliability and to develop it as a general method of identifying the amino acid sequence around a given labelled amino acid residue.

The results reported here demonstrate that the structure around the serine phosphate residue is Thr.Ala.SerP.His.Asp (or Asp-NH_2). There seems little doubt that the serine phosphate is involved in the 'active centre' of the enzyme since the phosphate can be exchanged with phosphate in the substrates and, also, since the inactive enzyme does not become labelled. Whether or not the other residues in this sequence are concerned in the enzymic activity cannot be said; but the presence of a histidine residue next to the serine phosphate is of particular interest, since there is some evidence that such a residue is involved in the activity (Koshland *et al.* 1958). There are only three histidine residues in PGM, but whether in fact this particular one is involved in the activity cannot be said with certainty at present.

It is clear that the amino acid sequence around the serine phosphate residue of PGM is in no way similar to that found around the reactive serine residue in other enzymes which react with di*iso*propyl phosphorofluoridate. Some sequences in the immediate vicinity of reactive serine or serine phosphate residues in various enzymes are listed in Table 7. Although there are similarities between different enzymes as between trypsin and chymotrypsin or between liver ali-esterase and pseudocholinesterase, these similarities are not as universal as has sometimes been suggested and cannot at present be made the basis of any general hypothesis.

SUMMARY

1. Phosphoglucomutase labelled with ^{32}P was prepared by two methods involving reaction of the enzyme with labelled substrates.

2. The [^{32}P]phosphoglucomutase was subjected to partial hydrolysis with trypsin and with acid and the [^{32}P]peptides were purified by ionophoresis.

3. The interrelationships of the peptides in the acid hydrolysate were determined by subjecting them to further partial acid hydrolysis and to the Edman degradation procedure.

4. The individual residues in these peptides were identified from the ionophoretic and chromatographic mobilities of the peptides, by specific reactions and from the amino acid composition of highly purified peptides.

5. It was concluded that the amino acid sequence around the serine phosphate (SerP) residue was Thr.Ala.SerP.His.Asp (or Asp-NH_2).

We are grateful to Dr G. Fölsch for a gift of Leu.SerP and Gly.SerP and to Miss R. M. Carpenter for the mitochondria and hexokinase. One of us (C.M.) wishes to acknowledge a scholarship from the British Council.

REFERENCES

Alberty, R. A. (1953). In *The Proteins*, vol. 1A, p. 461. Ed. by Neurath, H. & Bailey, K. New York: Academic Press Inc.

Anderson, L. & Jollès, G. R. (1957). *Arch. Biochem. Biophys.* **70**, 121.

Berger, L., Slein, M. W., Colowick, S. P. & Cori, C. F. (1946). *J. gen. Physiol.* **29**, 379.

Burrows, S., Grylls, F. S. M. & Harrison, J. S. (1952). *Nature, Lond.*, **170**, 800.

Dixon, G. H., Kauffman, D. L. & Neurath, H. (1958). *J. biol. Chem.* **233**, 1373.

Dixon, G. H. S., Neurath, H. & Pechère, J. P. (1958). *Annu. Rev. Biochem.* **27**, 489.

Fisher, E. H., Graves, D. J., Grittenden, E. R. S. & Krebs, E. G. (1959). *J. biol. Chem.* **234**, 1698.

Flavin, M. (1954). *J. biol. Chem.* **210**, 771.

Fraenkel-Conrat, J. & Singer, B. (1956). *Arch. Biochem. Biophys.* **65**, 296.

Gladner, J. A. & Laki, K. (1958). *J. Amer. chem. Soc.* **80**, 1263.

Gross, D. (1955). *Nature, Lond.*, **176**, 72.

Hartley, B. S., Naughton, M. A. & Sanger, F. (1959). *Biochim. biophys. Acta*, **34**, 244.

Jagannathan, V. & Luck, J. M. (1949). *J. biol. Chem.* **179**, 569.

Jansz, H. S., Brons, D. & Warringa, M. G. P. J. (1959). *Biochim. biophys. Acta*, **34**, 573.

Jansz, H. S., Posthumus, C. H. & Cohen, J. A. (1959). *Biochim. biophys. Acta*, **33**, 387.

Kennedy, E. P. & Koshland, D. E. (1957). *J. biol. Chem.* **228**, 419.

Koshland, D. E. (1959). In *The Enzymes*, vol. 1, p. 318. Ed. by Boyer, P. D., Lardy, H. & Myrbäck, K. New York: Academic Press Inc.

Koshland, D. E. & Erwin, M. J. (1957). *J. Amer. chem. Soc.* **79**, 2657.

Koshland, D. E., Ray, W. J. & Erwin, M. J. (1958). *Fed. Proc.* **17**, 1147.

Linderstrøm-Lang, K. V. & Schellman, J. A. (1959). In *The Enzymes*, vol. 1, p. 498. Ed. by Boyer, P. D., Lardy, H. & Myrbäck, K. New York: Academic Press Inc.

Lumry, R. (1959). In *The Enzymes*, vol. 1, p. 196. Ed. by Boyer, P. D., Lardy, H. & Myrbäck, K. New York: Academic Press Inc.

Michl, H. (1951). *Mh. Chem.* **82**, 489.

Milstein, C. & Sanger, F. (1960). *Biochim. biophys. Acta*, **42**, 173.

Moore, T. B. & Baker, C. G. (1958). *J. Chromat.* **1**, 513.

Najjar, V. (1948). *J. biol. Chem.* **175**, 281.

Naughton, M. A. & Sanger, F. (1960). *Biochem. J.* **78**, 156.

Naughton, M. A., Sanger, F., Hartley, B. S. & Shaw, D. C. (1960). *Biochem. J.* **77**, 149.

Pardee, A. B. (1951). *J. biol. Chem.* **190**, 757.

Sanger, F. & Shaw, D. C. (1960). *Nature, Lond.*, **187**, 872.

Schäffer, N. K., Simet, L., Harshman, S., Engle, R. R. & Drisko, R. W. (1957). *J. biol. Chem.* **225**, 197.

Turba, F. & Gundlach, G. (1955). *Biochem. Z.* **327**, 186.

Waley, S. G. & Watson, J. (1953). *Biochem. J.* **55**, 328.

Weil, L., Gordon, W. G. & Buchert, A. R. (1951). *Arch. Biochem. Biophys.* **33**, 90.

PART II

RNA

Introduction

By 1960 much more was known about the role of nucleic acids and their importance in biochemistry than had been in the early 1940s, and with my experience and interest in sequencing I was anxious to see if it was possible to sequence them. However, this had hitherto seemed an impossible task and nothing had been done. The main problem was the absence of any small nucleic acid on which to develop methods. With proteins we had insulin, which enabled us to start on them, but the only known nucleic acids were very large and probably mostly mixtures. The main stimulus to start work on RNA was the discovery of S-RNA (now called tRNA) in 1958 by Zamecnik and his colleagues.[1] These were small (about 75 nucleotides) but were present as mixtures, so the first thing to do was to obtain a pure one.

The first person in my group to work on RNA was L.F. Smith, who had spent a sabbatical in Zamecnik's laboratory, and he started trying to purify tRNA when he returned. However, he left the department before he got very far, and others continued. Another important event that directed our activities towards nucleic acids was the arrival of G.G. Brownlee as a Ph.D. student. I asked him whether he would like to work on proteins or RNA, and being of an adventurous nature he chose RNA. Our attempts to purify tRNA went rather slowly at first. However, Holley and his colleagues[2] were more successful and obtained pure alanine-tRNA by counter-current fractionation. By 1965 they were able to sequence it using procedures very similar to those applied to protein sequencing. Besides working on tRNA we were doing preliminary experiments on nucleotide fractionation as this would be of paramount importance to any sequence studies. Also, it seemed to me that different methods from those used for proteins would be needed as the chemistry was very different; and the fact that there were only four mononucleotides to be considered compared to the 20 amino acids meant that the interpretation of results of partial degradation would be more difficult, but the analysis of end products would be easier. Paper 21, which was our first contribution to RNA sequencing, describes a new method for fractionating small nucleotides. The success of this work convinced us that we could make useful contributions and stimulated a complete change from proteins to nucleic acids.

All my previous work had been done in the Biochemistry Department of Cambridge University, but in 1962 I moved to the new Laboratory of Molecular Biology of the Medical Research Council. Here I joined the group headed by Max Perutz, which had been working in the University's Physics Department. It included John Kendrew, Francis Crick, Sydney Brenner, Hugh Huxley and John Smith. Perutz was head of the new lab, which was composed of three divisions. I was head of the Division of Protein Chemistry, later to be renamed Protein and Nucleic Acid Chemistry. This move had two main advantages from the point of view of my work on RNA. Some of my new colleagues had worked on, or were interested in, nucleic acids; in particular John Smith was the nucleic acid expert and was very helpful to me. Also, I had more facilities and space and could take on more collaborators.

I have never worked with many collaborators as I prefer to do experiments myself with just a few close colleagues. Most of the other members of the division worked independently on protein subjects with their own small groups.

Comments on Papers 20 to 27

Paper 20

This paper was somewhat of a side issue to our main objective of trying to sequence RNA. In our studies on the active centres of enzymes (papers 16 to 19) we had been able to determine sequences around an amino acid that had been labelled by reaction with a radioactive compound. Could such an approach be used on RNA? Our idea was to label tRNA with a radioactive amino acid (its normal substrate) and then use similar techniques to determine the sequence of nucleotides around the binding site. We chose methionine since it contained sulphur and could be labelled with ^{35}S — a convenient radioactive isotope. In an initial experiment using paper ionophoresis we observed an unexpected spot. Usually such spots turn out to be artefacts, but Kjeld Marcker was not satisfied with that explanation and was determined to find out what it was. This took us some time. Formyl derivatives were not known in natural products and this possibility had not occurred to us until I was looking through a volume of Emil Fischer's collected papers (which I had been given by his son Hans Fischer) and found that he had synthesized formyl amino acids.

Marcker later extended this work[3] and showed that the formyl-methionine is incorporated into the N-terminal position of proteins in *E. coli* and is important for initiation of protein synthesis. All proteins are initiated in this way. In some the formyl group is later removed; in others the methionine is also removed. In contrast with *E. coli*, protein initiation in higher organisms does not require the formyl group on the initiation methionine, with the exception of mitochondria, which do use formyl-methionine. Thus our "artefact spot" led to the discovery of the method of protein initiation.

Paper 21

By the early 1960s protein sequencing was done in the main using column techniques on ion exchange resins, both for fractionating the peptides and for estimating the amino acids derived from them. These methods had almost completely replaced the two-dimensional paper fractionation methods that we had used earlier. I still had a preference for paper techniques, especially for preliminary experiments, as they were quicker and, in general, gave more information — though of a qualitative rather than quantitative nature. Paper methods had been widely used for fractionating mononucleotides, which were detected under ultraviolet light. This method was not very sensitive and I found it impossible to see any distinct spots from partial digests of RNA, so it seemed that a more sensitive means of detection was needed. We had been labelling proteins with ^{32}P and were impressed with its sensitivity and ease of detection using autoradiography. All mononucleotides contained phosphorous, which could be replaced by ^{32}P through biosynthesis, so this seemed ideal for the purpose. This greatly facilitated all our work on RNA and was used in all our subsequent studies on nucleic acids. Only recently has ^{32}P-labelling been replaced to a great extent by fluorescent labelling for sequencing. This of course requires more sophisticated equipment than was available to us.

As we had no small pure RNA to study we used ribosomal RNA for our preliminary work. We could not expect to get any useful sequence information from such large molecules, but it was readily prepared in a radioactive form and was ideal for working out fractionation techniques. We tested many of the standard systems but with most of them the oligonucleotides streaked very badly. Eventually we came up with the two-dimensional system described here and this formed the basis for most of our subsequent work on RNA.

We had originally thought that nucleic acids would be more difficult to sequence than proteins because they contained only four mononucleotides. However, we soon found that we could take advantage of this fact to devise simpler methods for identifying and sequencing the products of partial hydrolysis. In our work with labelled proteins we had tried to derive information about the amino acids in peptides from the behaviour of the peptide under various conditions on the separation systems, thus avoiding tedious analyses of every peptide. This had been quite difficult as it was necessary to differentiate between 20 amino acids. Using a similar approach with RNA was more rewarding as we only had to differentiate between the four mononucleotides. In this way we were able to identify small oligonucleotides directly from their position on the two-dimensional fractionation system, and devised a rapid method for determining their sequences using an exonuclease.

In retrospect I consider that this was my most important paper on RNA since it opened up a somewhat new approach that was more suitable than the methods used by Holley. It also convinced us that nucleic acid sequencing was feasible, and by this time I was fully committed to the change from proteins to RNA.

Papers 22 to 25

Having developed methods for working on nucleotides we were keen to try them out on a suitable RNA. Holley had already shown that tRNAs could be sequenced. A special feature of them is that they contain minor bases, many of which occur only once in the molecule. This makes it easier to find overlaps between the degradation products, thus facilitating working out the sequence. Most other RNAs do not contain minor bases, so that any general method would be required to work without them. In 1963 Rosset and Monier had isolated a small RNA (5 S RNA) containing 120 nucleotides from ribosomes, and this seemed suitable for trying out our new methods. It was larger than tRNA and contained no minor bases. Papers 22 and 23 describe the determination of the complete sequence, most of which was done by Brownlee as part of his Ph.D. work. Barrell and I were more concerned with the further development of methods. Those described in paper 21 turned out not to be adequate for determining the complete sequence and further ways had to be found, including a new fractionation system; this we called homochromatography. This was used to fractionate the larger oligonucleotides obtained from partial digests.

At this time 5 S RNA was the largest nucleic acid to have been sequenced, and in fact the only one without minor bases, so this was something of a milestone.

We had done quite a lot of work trying to get a pure tRNA and had eventually succeeded in obtaining pure phenylalanine tRNA by binding it specifically to polyuridylic acid. Paper 24 describes its sequence. By this time several other tRNAs had also been sequenced.

Paper 25 describes an improved method of homochromatography which enabled us to obtain somewhat larger fragments.

Papers 26 and 27

When we started working on RNA one of the most exciting prospects was to "break the code". The hope was to find the sequence of a messenger RNA that was coding for a known sequence of amino acids in a protein. With this in mind I put quite a lot of effort into trying to isolate a messenger RNA from hen oviducts, but met with little success. A more promising source of coding-RNA was that of bacteriophages, which could be readily prepared in a radioactive form and, in the case of R17, the amino acid sequence of the coat protein was known. The main problem was, however, that it was over 3,000 nucleotides long — so how could we begin to sequence it? Thus far our methods had been applicable to 5 S RNA of 120 nucleotides, but it was difficult to see how they could be extended to 3,000. Our main method for breaking down RNA was ribonuclease T_1, which split specifically at G residues. 5 S RNA contained about 30 of these, but R17 would have about 700 so that any digestion would be extremely complex. It was, however, possible to separate out a few of the larger T_1 products using the homochromatography system (paper 25), and these were sequenced by Peter Jeppesen.[4] To get any farther we would have to carry out partial hydrolysis, and this would be expected to give an extremely complex mixture of larger fragments. We were very surprised therefore when Jerry Adams, a postdoctoral fellow from Harvard, ran a partial digest on acrylamide gel ionophoresis and obtained an autoradiograph showing a number of sharp, well-separated bands representing fragments about 50 nucleotides long. The first one we sequenced proved to be coding for a region of the coat protein. By this time the genetic code had been determined by an entirely different and less direct method; however, our results did at least give independent confirmation of the code, and had opened up the possibility of doing larger RNA sequences. One surprising feature of the fragments was that they formed base-pairing loops. This explains why we were able to isolate single pure products. The double-stranded RNA in the loops is resistant to ribonuclease T_1 so that splitting occurs only in single-stranded regions between the loops, giving a small number of large fragments.

By now there was considerable interest in sequencing RNA, and the work of two other laboratories was of special interest to us. The group of Walter Fiers in Belgium[5] was studying the related RNA bacteriophage MS2. They were mostly using our methods and eventually completed the whole of the 3,000 residues. We did not continue further with R17 as we were becoming more interested in DNA. The other group was that of Charles Weissmann.[6] They worked on another bacteriophage, Q_β. This contains a replicase that makes a complementary copy of the RNA starting from its $3'$ end. In this way they could make short stretches of radioactive RNA for sequencing. I was particularly interested in this work and made use of the copying approach in our DNA studies.

References

1. M.B. Hoagland, M.L. Stephenson, J.F. Scott, L.I. Hecht and P.C. Zamecnik, *J. Biol. Chem.* **231**, 241 (1958).
2. R.W. Holley, J. Agpar, G.A. Everett, J.T. Madison, M. Marquisee, *et al.*, *Science* **147**, 1462 (1965).
3. B.F. Clark and K.A. Marcker, *J. Mol. Biol.* **17**, 394 (1966).
4. P.G.N. Jeppesen, *Biochem. J.* **124**, 357 (1971).
5. R. De Wachter, A. Vandenberghe, R. Contreres and W. Fiers, *Proc. Natl. Acad. Sci. U.S.A.* **68**, 585 (1971).
6. M.A. Billeter, J.E. Dahlberg, H.M. Goodman, J. Hindley and C. Weissmann, *Nature* **224**, 1083 (1969).

J. Mol. Biol. (1964) 8, 835–840

N-Formyl-methionyl-S-RNA

K. Marcker and F. Sanger
Medical Research Council Laboratory of Molecular Biology
Hills Road, Cambridge, England

(*Received 17 February 1964*)

The reaction between methionine and S-RNA of *Escherichia coli* has been investigated. It has been demonstrated that, after the initial attachment to its specific S-RNA, the free α-amino group of the attached methionine may become formylated.

1. Introduction

The general role of S-RNA in protein biosynthesis is now well established. The amino acid is first activated by interaction with an enzyme and ATP to form a bound amino acyl-AMP. It is then transferred from this compound to form an ester linkage on the 2′- or 3′-hydroxyl group of the ribosyl moiety of the terminal adenosyl unit of S-RNA. This reaction is catalysed by the same enzyme which activates the amino acid. The amino acyl-S-RNA then reacts with the ribosomes, where the amino acid is incorporated into a nascent protein, starting from the amino end.

Zachau, Acs & Lipmann (1958) showed that the adenosyl ester of leucine was liberated when leucyl-S-RNA was digested with ribonuclease, thus demonstrating that ribonuclease acted upon the amino acyl-S-RNA in the following way:

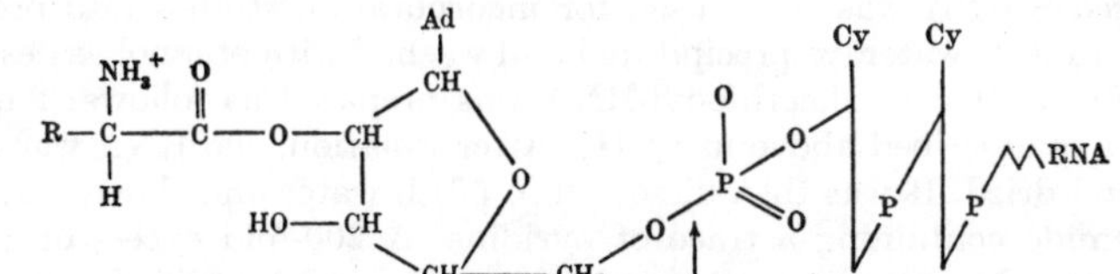

The authors have carried out similar experiments with [^{35}S]methionine in order to characterize the complex formed between methionine and its specific S-RNA. The adenosyl esters liberated by ribonuclease digestion were separated from the other nucleotides by high-voltage ionophoresis at pH 3·5. Besides the normal methionine ester, a second radioactive substance was present which was identified as the adenosyl ester of N-formyl-methionine. This indicates that the reaction between methionine and its specific S-RNA leads to the formation of the normal methionyl-S-RNA ester and, surprisingly, to N-formyl-methionyl-S-RNA ester. It is also shown that the formylation of the free α-amino group of methionine only takes place after methionine has become esterified to the S-RNA.

2. Materials and Methods

S-RNA was prepared from frozen cells of *E. coli* strain B according to the method of Zubay (1962). The supernatant solution containing the amino acid activating enzymes was prepared by grinding frozen cells with twice their weight of alumina and extracting with 2·5 vol. of 0·001 M-tris buffer pH 7·8 containing 0·01 M-magnesium acetate, 0·006 M-mercaptoethanol and 1 to 2 μg/ml. DNase. All manipulations were carried out at

4°C. The crude extract was clarified by two centrifugations (10,000 rev./min for 20 min). The ribosome-free supernatant solution was obtained from the clarified extract by 2 hr centrifugation at 105,000 ***g***. The top two-thirds of the supernatant solution were withdrawn and stored in small tubes at −20°C without prior dialysis.

Methionine labelled with ^{35}S was made according to the method of Sanger, Bretscher & Hocquard (1963). The ATP, GTP and PEP† were all used as their sodium salts. Pyruvate kinase (*Krystallsuspension*) was purchased from Boehringer and Soehne. The RNase used was a lyophilized Worthington preparation. The S-RNA labelled with ^{14}C in the terminal adenosine was a gift from Mr. M. Bretscher and had a specific activity of 9·3 mc/ m-mole (Bretscher, 1963). Formyl-DL-methionine (m.p. 100°C) was prepared by the method of Windus & Marvel (1931). Two sets of conditions were used for charging S-RNA with methionine:

(1) 0·2 ml. M-tris (pH 7·4) 0·15 M-$MgCl_2$ buffer, 0·1 ml. 0·1 M-ATP, 40 μl. 2 M-$NH_4OOC.CH_3$, approximately 5 μc [^{35}S]methionine, 4 to 5 mg S-RNA, 40 to 50 μl. *E. coli* supernatant solution, water to 2 ml. The mixture was then incubated for 5 min at 37°C. When using these conditions, it was found that mainly [^{35}S]methionyl-S-RNA was formed.

(2) 0·2 ml. M-tris (pH 7·4) 0·15 M-$MgCl_2$ buffer, 0·1 ml. 0·1 M-ATP, 40 μl. 2 M-$NH_4OOC.CH_3$, approximately 5 μl. [^{35}S]methionine, 4 to 5 mg S-RNA, 200 to 400 μl. *E. coli* supernatant solution, water to 2 ml.; incubation for 10 min at 37°C. Under these conditions it was found that a high proportion of the methionyl-S-RNA was transformed into formyl[^{35}S]methionyl-S-RNA, although there were considerable differences between different batches of *E. coli* supernatant solution. After the incubation, the mixture was chilled in ice and 0·1 vol. of 20% CH_3COOK (pH 5) was added. It was then shaken in the cold with 2 ml. water-saturated phenol and the RNA in the aqueous layer was precipitated by addition of 3 vol. ethanol. The precipitate was washed once with ethanol. If the charged S-RNA was to be used for characterization purposes, it was then dissolved in a small amount of water and incubated for 5 min at room temperature with 5 to 10 μg RNase. The mixture was put on Whatman no. 3 MM paper and subjected to ionophoresis for 1 hr at pH 3·5 and 3000 v. Finally, a radioautograph was prepared, usually by leaving the paper in contact with X-ray film for one night.

If the charged S-RNA was to be used for incorporation studies into proteins, it was re-dissolved in 1 ml. of water, re-precipitated and washed with ethanol as described above.

Artificially formylated [^{35}S]methionyl-RNA was prepared as follows: 2 mg of S-RNA were incubated as described above using (1). After isolation, the RNA was washed twice with ethanol and dried. It was then dissolved in 50 μl. water and slowly put into 1 ml. of dimethylformamide containing a trace of pyridine. A 200-fold excess of *p*-nitrophenylformate (prepared by coupling *p*-nitrophenol with formic acid in the presence of dicyclohexylcarbodiimide) in 1 ml. of dimethylformamide was then added. The mixture was shaken for 1 hr at room temperature, and the S-RNA was isolated by addition of 0·1 vol. 20% potassium acetate and 3 vol. ethanol.

To study the transfer of methionine from its S-RNA derivatives into protein, the following incubation mixture was used: 50 μl. M-tris buffer (pH 7·8), 40 μl. 0·2 M-$Mg(CH_3COO)_2$, 50 μl. 2 M-$NH_4OOC.CH_3$, 10 μl. M-β-mercaptoethanol, 10 μl. 0·1 M-ATP, 20 μl. 0·01 M-GTP, 80 μl. 0·05 M-PEP, 20 μl. pyruvate kinase suspension (2 mg/ml.). To each tube was added 100 μl. of an undialysed S-30 fraction prepared from freshly grown *E. coli* cells according to the method of Nirenberg & Matthaei (1961). Finally, the respective RNA's were added and the final volume made up to 0·5 ml. by the addition of water. After incubation for 15 min at 37°C, an equal volume of 10% TCA was added. The precipitate was then heated for 15 min at 90 to 95°C and filtered through a Millipore filter. It was washed several times with 5% TCA and twice with 1 : 1 ethanol–ether and finally plated and counted.

3. Results

Plate I shows a typical radioautograph of a ribonuclease digest of S-RNA labelled with [^{35}S]methionine. There are two distinct bands (A and B) travelling towards the cathode, in addition to free methionine. That the presence of these two bands

† Abbreviations used: PEP, phosphoenolpyruvate; TCA, trichloroacetic acid.

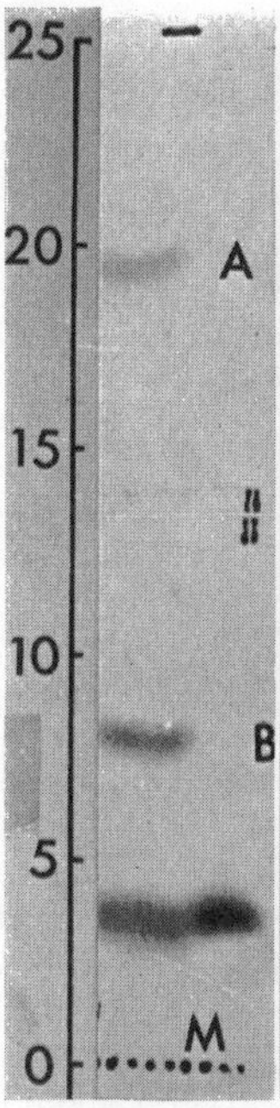

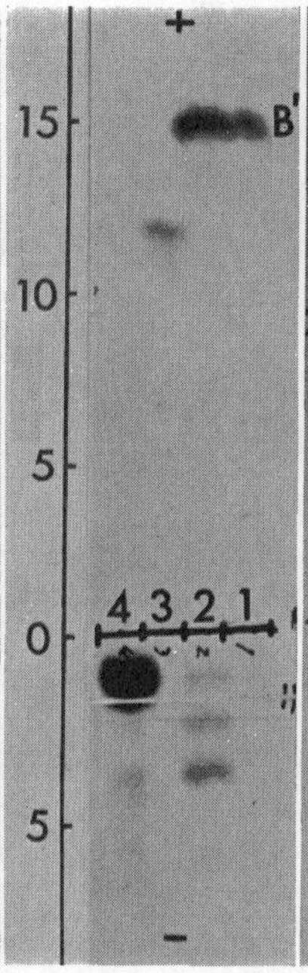

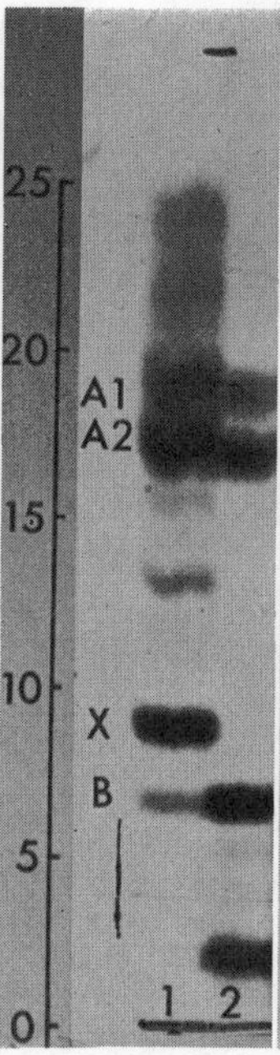

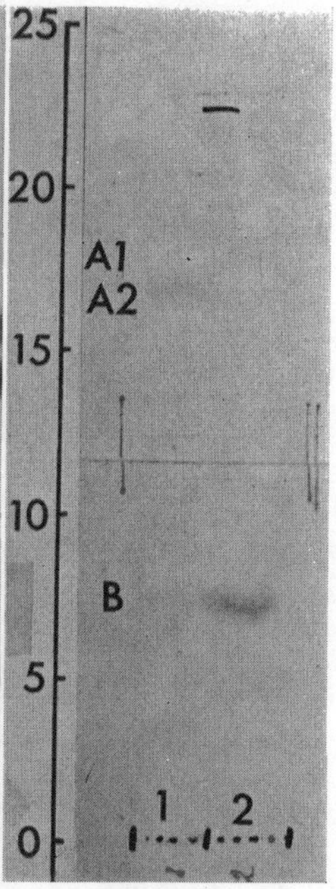

PLATE I PLATE II PLATE III PLATE IV

PLATE I. Ribonuclease digest of methionyl-S-RNA. 2 mg *E. coli* S-RNA were incubated for 10 min at 37°C in 1 ml. of the reaction mixture described in the Materials and Methods section. The amount of supernatant solution was 100 μl. The S-RNA was then isolated and digested with RNase as described. The digest was subjected to ionophoresis for 1 hr pH 3·5, 3000 v. After drying, the paper was radioautographed overnight. M refers to free methionine used as marker.

PLATE II. Radioautograph of ionogram (pH 3·5, 3000 v, 1 hr) of: (1) band B (Plate I) eluted and incubated 1 hr in dilute NH_3 (pH 10·5) at 37°C; (2) formyl[^{35}S]methionine made according to Sheehan & Young (1958); (3) *N*-acetyl[^{35}S]methionine made by acetylation of [^{35}S]methionine with acetyl chloride; (4) [^{35}S]methionine.

PLATE III. Radioautograph of ionogram (pH 3·5, 3000 v, 1 hr) of: (1) ribonuclease digest of methionyl-S-RNA carrying a ^{14}C label in the terminal adenosine position: [^{14}C]S-RNA (0·25 μc) was incubated for 10 min at 37°C in a mixture containing tris, $MgCl_2$, ATP and $NH_4OOC.CH_3$ in the concentrations described under Materials and Methods, 5 μl. 0·1 M-methionine and 75 μl. *E. coli* supernatant solution. The RNA was isolated and digested with ribonuclease as described; (2) ribonuclease digest of [^{35}S]methionyl-S-RNA, prepared as above but using [^{35}S]methionine (0·5 μc) and unlabelled S-RNA. Band X has been eluted and shown to be free adenosine.

PLATE IV. Radioautograph of ionogram (pH 3·5, 3000 v, 1 hr) of: (1) ribonuclease digest of [^{35}S]methionyl-S-RNA, prepared as described under (1) of Materials and Methods; (2) ribonuclease digest of [^{35}S]methionyl-S-RNA from the same preparation as above after formylation with *p*-nitrophenyl formate.

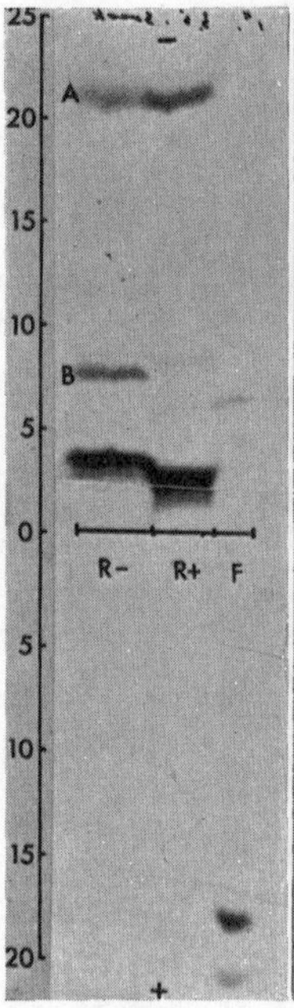

PLATE V

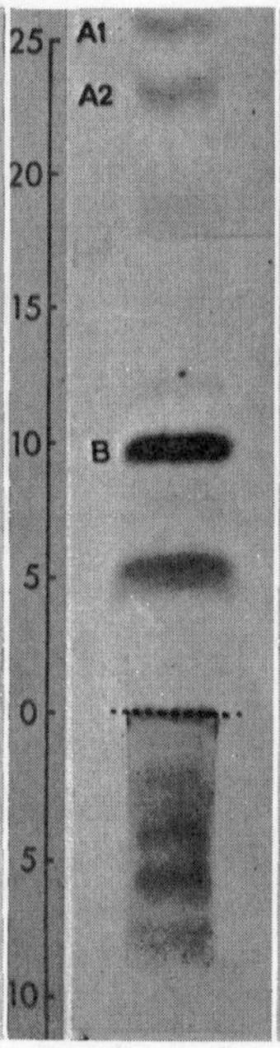

PLATE VI

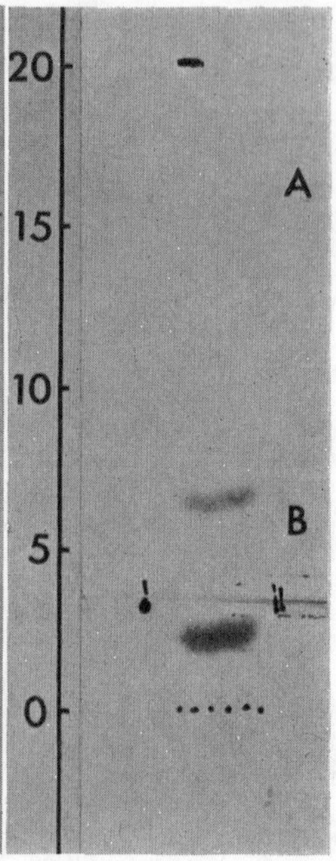

PLATE VII

PLATE V. Effect of ribonuclease on the formation of formyl-methionyl-S-RNA.

(R+), medium preincubated with ribonuclease. A mixture containing 10 μl. M-tris buffer (pH 7·4), 0·15 M-$MgCl_2$, 5 μl. 0·1 M-ATP, 2 μl. 2 M-$NH_4OOC.CH_3$, [^{35}S]methionine (0·5 μc), 0·5 mg S-RNA, water to 100 μl. was incubated for 5 min at 37°C with 10 μg RNase. 20 μl. *E. coli* supernatant solution containing 100 μg RNase/ml. was then added. The mixture was incubated for an additional 10 min at 37°C and applied directly to the paper.

(R–), incubation as above except that ribonuclease was omitted in every step, but after the final incubation 10 μg RNase was added and the mixture applied to the paper.

(F), a synthetic mixture obtained by formylation of [^{32}S]methionine, containing formyl-methionine and formyl-methionine sulphoxide. Ionophoresis at pH 3·5, 3000 v, 1 hr.

Band A in both R− and R+ is believed to be largely *S*-adenosyl methionine. Its mobility does not alter on treatment at pH 10·5. Band B in R− was eluted and shown to give *N*-formyl-methionine after treatment at pH 10·5.

PLATE VI. Ribonuclease digest of [^{35}S]methionyl-S-RNA isolated from growing *E. coli*. *E. coli* strain B cells were grown overnight in the synthetic medium described by McQuillen (1955). Next morning, growth was restarted by addition of 10% glucose to a concentration of 0·8%. When the cells were in the log phase, 100 μc $^{35}SO_4^{=}$ were added. After 3 min, chloromycetin was added to a final concentration of 50 μg/ml. (This was found to lead to an accumulation of counts in the S-RNA fraction.) After an additional 3 min, the cells were chilled in ice, centrifuged and washed once with 0·001 M-tris (pH 7·4), 0·01 M-$Mg(CH_3COO)_2$ buffer. The cells were then resuspended in approximately 2 ml. of the buffer, and 1 mg of solid lysozyme was added. The suspension was left for 5 to 10 min at 4°C, then frozen and thawed 5 to 6 times, 4 ml. of water-saturated phenol were added and the mixture shaken for 15 min at 4°C. The S-RNA was isolated from the aqueous phase, digested with RNase and subjected to ionophoresis at pH 3·5 as described in the Materials and Methods section. Band B has been eluted and found to contain *N*-formyl-methionine.

PLATE VII. 5 mg S-RNA were incubated as described in method (2) of Materials and Methods. The amount of supernatant solution was 400 μl. After 30 min incubation at 37°C, a sample of 0·2 ml. was withdrawn and precipitated without phenol-shaking by 0·1 vol. 20% CH_3COOK (pH 5) and 3 vol. ethanol. The isolated S-RNA was then digested with RNase and run at pH 3·5, 3000 v for 1 hr.

was not due to incomplete splitting of the methionyl-S-RNA by ribonuclease has been shown by using ^{32}P-labelled S-RNA charged with unlabelled methionine. Ionophoresis of a ribonuclease digest of this S-RNA under the same conditions did not give any bands corresponding to those obtained from [^{35}S]methionyl-S-RNA. Thus neither A nor B contained phosphorus, whereas any incomplete digestion product would contain phosphorus.

The two bands were eluted from the paper and incubated with dilute ammonia at pH 10·5 for one hour at 37°C. Band A was converted to methionine. This was consistent with the known alkali-lability of the amino acyl-S-RNA ester linkage and suggested that it was the methionyl adenosine ester. As shown below (Plate III) this band contains the terminal adenosine residue of the RNA.

In many experiments (e.g. Plates III and VI) there are two bands close together near the position of band A, the relative amounts varying considerably from one experiment to another. Both behave similarly on treatment at pH 10·5. The reason why two compounds are obtained is not entirely clear. In the case of leucyl-adenosine ester, only one such compound was demonstrated. We have found that ribonuclease digests of other amino acid S-RNA's such as the proline and phenylalanine RNA's give two similar bands. It seems probable that the effect is due to a migration of the amino acid between the 2′- and 3′-hydroxyl groups of the ribosyl moiety of the terminal adenosine. Rammler & Khorana (1963) have recently shown that such isomers do in fact become rearranged in the above way, and they also found that it was possible to separate them on silicic acid columns.

The substance from band B, on treatment at pH 10·5, did not give methionine but a compound B′ which by ionophoresis was found to be rather acidic at pH 3·5 (Plate II). This showed that it had gained an additional negative charge by the alkali treatment, since before it was slightly basic. It was found, however, that by mild acidic hydrolysis (15 min 100°C, N-HCl) B′ was converted quantitatively into methionine. This suggested that it was a derivative of methionine in which the α-amino group was blocked. It was therefore subjected to ionophoresis at various pH values in parallel with synthetic *N*-acetyl methionine. At pH 2·1 it was neutral, but at pH 3·5, 6·5 or 8·9 it travelled ahead of *N*-acetyl methionine, and the distance between the two bands was approximately constant at these pH values. Attempts to acetylate it did not alter its ionophoretic mobility, thus showing that it did not have a free α-amino group. Since it moved faster than *N*-acetyl methionine on ionophoresis at pH 3·5, it seemed most likely that it was *N*-formyl methionine. This was especially supported by the finding that the derivative was converted into methionine by mild acidic hydrolysis. The acidic lability of the *N*-formyl group has long been known (Fischer & Warburg, 1905). In fact, this group has been used in peptide synthesis as an amino-blocking group because of its easy removal by mild acidic hydrolysis. Compound B′ was therefore submitted to ionophoresis at pH 3·5 together with synthetic *N*-formyl[^{35}S]methionine. The result is shown in Plate II. As can be seen from the radioautograph, the synthetic *N*-formyl-methionine and the unknown compound B′ do travel together, confirming that the latter is indeed *N*-formyl-methionine. As further confirmation, samples of band B′ were mixed with synthetic non-radioactive formyl-DL-methionine and subjected to ionophoresis at pH 3·5 and 6·5 and to chromatography in water-saturated benzyl alcohol. After preparing radioautographs, the papers were developed with the platinic iodide reagent (Toennies & Kolb, 1951) to locate the synthetic compound. In each case it was found to be co-extensive with the radioactive material. In some experiments a band moving slightly faster than band B′ at pH 3·5 was present. This appeared to be formyl-methionine sulphoxide.

That *N*-formyl-methionine is attached to S-RNA is shown by the following facts. *N*-formyl-methionine is released by a very mild alkali treatment from compound B (Plate II), thus showing that it is bound through its carboxyl group by a very alkali-labile bond. When undigested *N*-formyl-methionyl-RNA was subjected to ionophoresis at pH 3·5, most of the radioactivity stayed at the origin, with streaking extending towards the cathode. No radioactivity could be found corresponding to the bands shown in Plate I. Finally, when S-RNA labelled with ^{14}C in the terminal adenosine unit was charged with unlabelled methionine, digested with ribonuclease and subjected to ionophoresis together with a ribonuclease digest of [^{35}S]methionyl-S-RNA, band B, as well as the two forms of band A, was found to contain ^{14}C-labelled adenosine, as shown in Plate III. When [^{35}S]methionyl-S-RNA was formylated in dimethyl-formamide by *p*-nitrophenyl-formate ester and subsequently digested with ribonuclease, it was found that the two A band substances were converted into the band B substance, as shown in Plate IV.

The above results indicate that *N*-formyl-methionyl-S-RNA may be formed in cell-free systems from *E. coli*. There are two possible ways in which this compound could be made: (1) by formylation of methionine followed by combination of the *N*-formyl-methionine with S-RNA; (2) by formylation of methionyl-S-RNA. In order to distinguish between these two possibilities, the effect of pretreating the components of the incubation mixture with ribonuclease was studied. After the incubation, the whole mixture was submitted to ionophoresis at pH 3·5 as usual, together with a control in which ribonuclease was added after the 10 minutes incubation. As shown in Plate V, no compound corresponding to *N*-formyl-methionyl adenosine ester (band B) was present. Furthermore, no free *N*-formyl-methionine could be found. The same result was obtained when the S-RNA was omitted from the reaction mixture. Attempts to label the S-RNA with synthetic formyl[^{35}S]methionine did not lead to any significant incorporation. It therefore appears that reaction (1) can be ruled out and that formylation takes place on the methionyl-S-RNA.

It was of interest to see whether or not the same compound could be isolated from growing cells. Accordingly *E. coli* cells brought into the log phase were grown for a short time with [^{35}S]methionine or $^{35}SO_4^{=}$ in the growth medium. The S-RNA was then isolated, digested with ribonuclease and subjected to ionophoresis at pH 3·5. The result of such an experiment is shown in Plate VI. As can be seen, the same three bands appear as were found in the *in vitro* system. The radioactivity in band B has definitely been shown to be due to *N*-formyl-methionine; it can therefore be concluded that the same compound is formed in growing cells.

Attempts have been made to see whether or not other amino acids are formylated on their respective S-RNA's. So far, all these attempts have failed both *in vivo* and *in vitro*, with the possible exception of cysteine. A ribonuclease digest of S-RNA isolated from a $^{35}SO_4^{=}$ grown methionine-requiring *E. coli* mutant, which should not contain radioactive methionine, did give a band with the same mobility as band B; but the band has not yet been positively shown to contain *N*-formyl-cysteine. The fact that the formyl derivative of other amino acyl-S-RNA's could not be detected cannot of course be regarded as evidence that the reaction does not occur in these cases, since the derivatives may be too rapidly destroyed.

Experiments have also been undertaken to see if organisms other than *E. coli* make *N*-formyl-methionyl-S-RNA. This has definitely been shown to be the case for yeast. S-RNA isolated from yeast grown on $^{35}SO_4^{=}$ was treated with ribonuclease and shown to give rise to bands A and B. S-RNA from rat liver or hen oviduct that had been incubated with [^{35}S]methionine, gave only band A (methionyl adenosine ester) and no detectable amount of the formyl derivative.

Since it is possible that the formation of *N*-formyl-methionyl-S-RNA is connected in some way with protein biosynthesis, experiments were undertaken in order to see whether or not methionine in the form of *N*-formyl-methionyl-S-RNA was incorporated into protein. Before doing so, it was thought necessary to develop a method for obtaining *N*-formyl-methionyl-S-RNA essentially free from methionyl-S-RNA. This could best be done by using more *E. coli* supernatant solution and incubating for a longer time, or by carrying out the incubation as described under (2) and then isolating the S-RNA and re-incubating once more without the addition of free methionine. Plate VII shows a ribonuclease digest of formyl-methionyl-S-RNA prepared by one of the methods described above. It is seen that it is possible to convert methionyl-S-RNA almost completely into *N*-formyl-methionyl-S-RNA.

Incorporation studies were carried out using an *E. coli* cell-free system (Nirenberg & Matthaei, 1961). Such systems would incorporate about 2 to 8% of methionine when present as the S-RNA derivative. Samples of S-RNA containing predominantly methionine or formyl-methionine were prepared as above. Incorporation into hot TCA-insoluble material was obtained in both cases, and in general methionine was somewhat more efficiently incorporated when it was in the form of formyl-methionyl-S-RNA than when in the form of methionyl-S-RNA. For example, in a typical experiment in which 10,000 counts per minute of the S-RNA's were used, 379 counts were incorporated from the methionine derivative and 621 from the formyl methionine derivative. This incorporation is dependent on the energy system, since no significant incorporation occurred when this was omitted, and is inhibited by addition of puromycin at a concentration (5×10^{-4} M) which is known to inhibit protein biosynthesis. It was also shown that the transfer from formyl-methionyl-S-RNA was not due to stimulation of the incorporation of the small amount of contaminating methionyl-S-RNA by some unknown factor present in the preparation, since no stimulation from [^{35}S]methionyl-S-RNA was obtained by adding a non-radioactive formyl-methionyl-S-RNA preparation.

4. Discussion

The above results indicate that *N*-formyl-methionyl-S-RNA is formed by an enzymic reaction from methionyl-S-RNA. It is also found in growing cells of *E. coli* and yeast, the two derivatives being present in about the same concentration, as measured by the yields of the adenosyl derivatives. This may, however, be deceptive, as the *N*-formyl derivative is more stable than the methionyl-S-RNA and therefore its relative yield may appear higher due to greater losses of the methionine derivative during isolation.

It is highly probable that the unknown compound attached to S-RNA isolated from $^{35}SO_4$-grown *E. coli* cells by Aronson *et al.* (1959) is identical with *N*-formyl-methionine. By incubation of the isolated S-RNA with alkali, they found that in addition to [^{35}S]methionine another radioactive compound was liberated, which on chromatography in butanol–formic acid travelled with the front. Although no further attempt at identification was undertaken, its chromatographic behaviour is consistent with the idea that the unknown compound is in fact *N*-formyl-methionine, which under those conditions would be expected to be non-polar.

Whether or not all the methionyl-S-RNA can be formylated is not certain. Preparations could be made which contained almost exclusively the formyl derivative (as shown in Plate VII); however, relatively long incubation periods were used and it may be that the more labile methionyl-S-RNA was broken down, leaving only the formyl-methionyl-S-RNA. The nature of the formyl donor has not been investigated in detail. It has been found that addition of [^{14}C]sodium formate to the incubation mixture does not lead to the incorporation of any radioactive formyl groups into formyl-methionyl-S-RNA. It is, therefore, unlikely that the formyl donor is readily generated from free formate under the conditions used. However, further work is clearly necessary on this point.

Since the reaction involves S-RNA, which is generally considered to be an essential intermediate in the synthesis of protein, the question arises as to whether the formylation has any role in protein biosynthesis. Studies with the cell-free system suggested that the formyl-methionyl-S-RNA was incorporated as well as, if not better than, the methionyl-S-RNA into the hot TCA precipitate. Experiments are in progress to decide whether this incorporation does in fact constitute the true formation of peptide linkages. If this is so, it is difficult to avoid the conclusion that the formyl-methionyl-S-RNA is in fact involved in the normal pathway of incorporation of methionine into protein. It is felt, however, that this would be a premature conclusion and that further work is necessary to evaluate the biological significance of the formylation reaction.

We are very grateful to various members of this laboratory for advice and help during the course of this work, and especially to Mr. M. S. Bretscher and Dr. R. Monroe. One of the authors (K. M.) wishes to thank the Carlsberg–Wellcome Foundation for a scholarship.

REFERENCES

Aronson, A. I., Bolton, E. T., Britten, R. S., Cowie, D. B., Duerksen, J. D., McCarthy, B. J., McQuillen, K. & Roberts, R. B. (1959). *Yearb. Carneg. Instn*, **59**, 270.
Bretscher, M. S. (1963). *J. Mol. Biol.* **7**, 446.
Fischer, E. & Warburg O. (1905). *Ber.* **38**, 3997.
McQuillen, K. (1955). *Biochim. biophys. Acta*, **17**, 382.
Nirenberg, M. W. & Matthaei, H. J. (1961). *Proc. Nat. Acad. Sci., Wash.* **47**, 1588.
Rammler D. H. & Khorana H. G. (1963). *J. Amer. Chem. Soc.* **85**, 1997.
Sanger, F., Bretscher, M. S. & Hocquard, E. J. (1963). *J. Mol. Biol.* **8**, 38.
Sheehan, J. C. & Young, D. D. H. (1958). *J. Amer. Chem. Soc.* **80**, 1154.
Toennies, G. & Kolb, J. J. (1951). *Analyt. Chem.* **23**, 823.
Windus, W. & Marvel, C. S. (1931). *J. Amer. Chem. Soc.* **53**, 3490.
Zachau, H. G., Acs, G. & Lipmann, F. (1958). *Proc. Nat. Acad. Sci., Wash.* **44**, 885.
Zubay, G. (1962). *J. Mol. Biol.* **4**, 347.

J. Mol. Biol. (1965) **13**, 373–398

A Two-dimensional Fractionation Procedure for Radioactive Nucleotides

F. Sanger, G. G. Brownlee and B. G. Barrell

Medical Research Council Laboratory of Molecular Biology
Cambridge, England

(*Received 21 April 1965*)

A method is described for the two-dimensional fractionation of ribonuclease digests of ^{32}P-labelled RNA. High-voltage ionophoresis is used in both dimensions. The first is on cellulose acetate at pH 3·5, the second on DEAE-paper at an acid pH. The method is capable of resolving the di- and tri- and most of the tetranucleotides in digests prepared by the action of ribonuclease T_1 or pancreatic ribonuclease. It has been applied to the 16 s and 23 s components of ribosomal RNA which show significant quantitative differences, and to sRNA from *Escherichia coli* and from yeast. A method is described for the determination of the sequence of a nucleotide by partial digestion with spleen phosphodiesterase.

1. Introduction

Progress in the determination of amino acid sequences in proteins has been largely dependent on the development of methods for the fractionation of the peptides produced by partial hydrolysis. Similarly, the study of nucleotide sequences in nucleic acids is likely to be limited by the techniques available for the fractionation of oligonucleotides. The methods most used at present are the two-dimensional procedure of Rushizky & Knight (1960), which uses ionophoresis and chromatography on paper (Rushizky & Sober, 1962; Armstrong, Hagopian, Ingram, Sjöquist & Sjöquist, 1964), and ion-exchange column chromatography, particularly on DEAE-cellulose or DEAE-Sephadex (Khorana & Vizsolyi, 1961; Staehelin, 1961,1964*a*; Rushizky, Bartos & Sober, 1964). Although the latter method is probably the most efficient separation procedure and is the most widely employed, paper techniques are often preferable from the point of view of speed and ease of operation. The advantages of both techniques may be combined by the use of ion-exchange papers. Tyndall, Jacobson & Teeter (1964) have described a two-dimensional technique using chromatography on DEAE-paper at two different pH values. We have found that good separations may be obtained by high-voltage ionophoresis on DEAE-paper at an acid pH, and this system is used as the second dimension for the two-dimensional technique. It appears that fractionation is due both to ion-exchange and to electrophoretic effects. The positive charges on the paper cause a rapid electro-endosmotic flow of buffer from the cathode to the anode which carries the nucleotides through the paper, thus subjecting them to ion-exchange chromatography. Superimposed on this is the electrophoretic fractionation which is probably an important factor in the fractionation, since considerably better separations are obtained by this technique than by simple chromatography on the DEAE-paper. The ionophoresis essentially opposes the ion-exchange effect, the more acidic components moving faster by ionophoresis but slower by ion-exchange.

In conjunction with this method, we have used ionophoresis on cellulose acetate as the basis for a two-dimensional technique. Ionophoresis of nucleotides on paper is somewhat limited by a tendency of the larger ones to streak. Previous experiments with peptides had indicated that considerably less streaking occurred if cellulose acetate was used instead of paper, and the same was found to be true for nucleotides. A disadvantage of the use of cellulose acetate for many purposes is that it is only suitable for small amounts of materials. Nucleotides are normally detected by their absorption of ultraviolet light, and this limits the scale on which one can work. In order to lower this scale, we have studied nucleotides which have been labelled with ^{32}P and have detected and estimated them by radioautography and counting techniques. Many nucleic acids can be prepared in a radioactive form by biological labelling and some, such as messenger RNA, can be studied only in this form.

The following paper describes the two-dimensional fractionation procedure and its application to ribonuclease digests of the two main components of ribosomal RNA from *Escherichia coli*. This material, which may readily be prepared, has been used to study the potentialities of the method and to determine the position of the different nucleotides on the two-dimensional fingerprint. We have studied digests prepared by the action of ribonuclease T_1 (Sato & Egami, 1957) because of its high specificity, splitting only at G† residues, but have also done preliminary work with pancreatic ribonuclease. Digests of mixed sRNA have also been studied. These are complicated by the presence of minor components and have not been investigated so completely. During this work a new minor component was detected and identified as the alkali degradation product of dihydrouridylic acid. In the meantime, the presence of dihydrouridylic acid as a component of sRNA has been reported by Madison & Holley (1965).

The structure of the nucleotides has been determined by conventional methods and by a new technique involving partial digestion with spleen phosphodiesterase. A somewhat similar method using venom phosphodiesterase was recently described by Holley, Madison & Zamir (1964). Spleen phosphodiesterase splits off 3′-mononucleotides sequentially from the 5′ terminal end of an oligonucleotide (Razzell & Khorana, 1961). If we consider an oligonucleotide obtained from a ribonuclease T_1 digest and suppose its structure is ABCG, where A, B and C are unknown residues, on partial digestion with spleen phosphodiesterase the following degradation products will be found: BCG, CG and mononucleotides. It has been found that on the DEAE-paper–pH 1·9 system used for the two-dimensional procedure, any nucleotide will move faster than a corresponding nucleotide having the same structure but with one extra residue added to its 5′ terminal end; i.e., BCG will move faster than ABCG, etc. Consequently, the various degradation products of an oligonucleotide will be arranged in order of their size on fractionation in this system, the larger ones moving slower than the smaller ones. The distance between any two products which differ by only one residue will depend on the nature of that residue. Thus from these distances one may deduce the nature of the various residues, and hence it is frequently possible to determine a complete sequence from a single degradation and ionophoretic fractionation.

† Abbreviations used: The letters A, C, G and U will be used for 3′ nucleotide residues. Thus G refers to guanosine-3′-phosphate and ACG refers to the trinucleotide ApCpGp (Shuster, Khorana & Heppel, 1959). The residue bearing the free 5′-hydroxyl group will be referred to as the 5′ terminal residue and that with the 3′-monoesterified phosphate as the 3′ terminal residue.

2. Materials and Methods

(a) *Materials*

Most of the samples of cellulose acetate electrophoresis strips were obtained from British Celanese Ltd., Plastics Division, through the kindness of Mr Prescott and Mr Hughes. Different batches varied considerably in their resolving power, and this appeared to be related to the method of production. Batch 1625 was a particularly suitable one, and material similar to this may now be obtained in sheets 95 cm × 25 cm from Oxo Ltd., Oxoid Division, 20 Southwark Bridge Road, London, S.E.1.

The DEAE-paper was Whatman Chromedia DE81 (originally known as DE20).

Ribonuclease T_1 was prepared by Mr R. E. Offord of this laboratory by the method of Rushizky & Sober (1962). Pancreatic ribonuclease was obtained from the Nutritional Biochemicals Corporation. Bacterial alkaline phosphomonoesterase, snake venom phosphodiesterase, micrococcal nuclease and spleen phosphodiesterase were obtained from the Worthington Biochemical Corporation.

In order to follow the course of an ionophoresis, it was convenient to apply a mixture of coloured dyes to be run side by side with the unknowns to act as markers. The dye mixture used in this work contained equal volumes of 1% Xylene Cyanol F.F (blue), 2% Orange G (yellow) and 1% Acid Fuchsin (pink) (all from George T. Gurr, Ltd., London, S.W.6).

(b) *Ionophoresis*

High-voltage ionophoresis was normally carried out by the method of Michl (1951). This, however, could not be used with the DEAE-paper, which is very fragile and difficult to handle when wet. Satisfactory results were obtained in an apparatus similar to that used by Naughton & Hagopian (1962) in which the paper is supported on a rack that can be lifted in and out of the tank. The centre partition was 15 cm high and the rods on the rack were arranged to accommodate papers 57 or 85 cm long.

The following buffer systems were used: pH 3·5, 0·5% pyridine–5% acetic acid (v/v); pH 1·9, 2·5% formic acid–8·7% acetic acid (v/v); pH 6·5, 10% pyridine–0·3% acetic acid (v/v). For the pH 6·5 system the coolant was toluene; for the other systems it was white spirit 100 (Esso Petroleum Company).

For ionophoresis at intermediate values of pH, a water-cooled flat plate apparatus, as described by Gross (1961), was used. The buffer systems were 5% acetic acid adjusted with ammonia to pH 3·5; 1% acetic acid adjusted with ammonia to pH 4·1; 0·5% acetic acid adjusted with ammonia to pH 4·4; 0·25% acetic acid adjusted with ammonia to pH 4·9; and 1·7% *N*-ethyl morpholine adjusted with acetic acid to pH 7·9. The separations took about 1 hr at 80 v/cm.

Due to the small amounts of nucleotides being studied, it was necessary to take all possible precautions against contamination with nucleases (Apgar, Holley & Merrill, 1962). Rubber gloves were worn when there was danger of contamination from the fingers. Small test tubes and centrifuge tubes were treated with silicone and washed with glass-distilled water. Non-radioactive "carrier" nucleotides were usually added to isolated samples to avoid too low concentrations.

(c) *Preparation of ^{32}P-ribosomal RNA*

The 16 s and 23 s components of ribosomal RNA were prepared from *E. coli* (strain MRE600) that had been grown on 100 ml. of the medium of Garen & Levinthal (1960), which contained 10^{-4} M-β-glycerophosphate and about 10 mc carrier-free [^{32}P]phosphate. This organism, which was kindly supplied by Dr H. E. Wade, of M.R.E. Experimental Station, Porton, Wilts., is reported to be free of ribonuclease.

A large inoculum was used, and after about 5 hr growth the bacteria were collected in a thick-walled centrifuge tube and frozen. They were then thawed and ground up for 5 min at 0°C with alumina, using a glass rod with a diameter about half that of the inside of the centrifuge tube. The broken cells were then extracted with 2 ml. of buffer (0·01 M-magnesium acetate, 0·01 M-tris–acetate (pH 7·4), 0·006 M-mercaptoethanol). After sedimenting the debris at 10,000 *g* for 10 min, the supernatant solution was centrifuged for 90 min at 39,000 rev./min in the Spinco model L ultracentrifuge. The ribosomal pellet, which was just visible, was dissolved in about 0·1 ml. 0·1 M-LiCl, 0·01 M-tris buffer (pH 7·4) containing 0·5% sodium dodecyl sulphate. It was then subjected to density-gradient centrifugation at 10°C in 5 ml. of the above medium using a gradient from 5 to 20% sucrose. It was centrifuged for 4 hr at 39,000 rev./min in the SW39 rotor. Fractions were collected and samples counted. The peak fractions for the 23 s and 16 s components were combined and the RNA precipitated with 2·5 vol. ethanol. After standing in the deep-freeze overnight, the precipitates were collected. In some experiments the peaks were re-run on the same density-gradient system to ensure purification. The ethanol precipitates were then treated with 1 ml. 2% sodium acetate (adjusted to pH 5·4) and 1 ml. water-saturated phenol. After shaking for 1 hr at room temperature, the aqueous layer was collected and the RNA precipitated with 2·5 vol. ethanol. It was washed with 75% ethanol and with absolute ethanol and dried in a desiccator. The residue was then dissolved in 0·1 ml. water and kept frozen until used. The yields varied considerably, but on the average the final solutions contained about 1 μc ^{32}P and 1 μg RNA per μl.

(d) *Preparation of ^{32}P-sRNA*

(i) *From* E. coli

E. coli was grown on 5 to 20 mc carrier-free ^{32}P on the medium described above. After centrifugation, the cells were transferred in a volume of 1 ml. to a small centrifuge tube and 1 ml. water-saturated phenol and 0·5 mg carrier sRNA were added. The tube was tightly stoppered and shaken vigorously at 5°C for 1 hr. After centrifugation, the aqueous layer was separated and treated with 0·1 vol. 2 M-sodium acetate (pH 5) and 2 vol. 95% ethanol. The precipitate was collected after standing in the cold, washed with 95% ethanol and ether and dried. It was then dissolved in 0·1 ml. 0·1 M-tris–HCl (pH 8·0), 0·1 M-LiCl, 0·5% sodium dodecyl sulphate and subjected to density-gradient centrifugation in 2·5 ml. of the above medium at 10°C using a gradient from 5 to 20% sucrose for 10 hr at 39,000 rev./min. Ten-drop fractions were collected and the position of the 4 s peak found by counting 0·1-μl. samples. The main peak sedimented with a maximum near the middle of the gradient. The peak tubes were pooled, and the RNA was precipitated with 2 vol. ethanol and collected after standing overnight at 4°C. The yield of [^{32}P]-sRNA was between 1 and 2% (counts) of the radioactive material added to the medium. The method is modified from that described by Gilbert (1963).

(ii) *From yeast*

Yeast was grown in a medium which was modified from that of Williams & Dawson (1952). $(NH_4)_2HPO_4$ and KH_2PO_4 in their medium were replaced by triammonium citrate (3·25 g/l.). The medium was adjusted to pH 5·2 with citric acid. 10 mc of [^{32}P]phosphate and 1 to 5 mg bakers' yeast were added to 50 ml. medium which was incubated for about 6 hr at 30°C with efficient aeration. The uptake of radioactivity from the solution into the yeast was 95% complete. The cells were collected by centrifugation, 1 mg yeast sRNA was added as carrier and the [^{32}P]sRNA was isolated by the method of Holley *et al.* (1961). The yield was about 1% (counts) of the radioactive material added to the medium.

(e) *Preparation of [^{14}C]methyl-labelled sRNA*

E. coli strain K12, *RC^{rel}*, *met$^-$*, *leu$^-$*, *thr$^-$* was used (CB3 of Dr S. Brenner). This is a multiple auxotrophic derivative of K12 W6, an RNA relaxed strain. Borek, Ryan & Rockenbach (1955) described the properties of K12 W6; Stent & Brenner (1961) prepared the multiple auxotrophic derivative of it. CB3 was grown overnight on 100 ml. of M9 (which contained/l. 5·8 g Na_2HPO_4, 3·0 g KH_2PO_4, 0·5 g NaCl, 1·0 g NH_4Cl, 0·02% glucose and 10^{-3} M-$MgSO_4$), supplemented with 2 μg/ml. L-methionine, 2 μg/ml. L-threonine and 2 μg/ml. L-leucine. Under these conditions glucose is limiting. In the morning, 2 ml. of 20% glucose was added, and growth started exponentially almost immediately. After 2 hr the cells reached an optical density at 550 mμ of 0·5 and were harvested by centrifugation, washed and resuspended in M9 with no amino acids added. The cells were bubbled for 15 min, after which 10 μc of [^{14}C]methyl-labelled methionine (30 μc/μmole) was added. Growth was allowed to go on for another 3 hr, when the turbidity had reached

0·95 and the uptake (counts) of radioactive material was 40%. [^{14}C]methyl-labelled sRNA was prepared by the method of Holley *et al.* (1961) and the yield was about 1·5% (counts) of the radioactive material added to the medium.

(f) *Enzymic digestion*

In preliminary experiments with ribonuclease T_1, difficulty was experienced in obtaining reproducible fingerprints, and it was necessary to define the hydrolysis conditions rather closely. If too mild conditions were used, cyclic nucleotides were produced which led to a duplication of each spot. Under too violent conditions, there was some splitting of bonds not involving G residues. This may have been partly due to contaminants in the ribonuclease T_1 preparation, but was probably more often due to contaminants in the RNA preparations. In order to minimize this latter effect, it was considered advisable to digest with relatively high concentrations of enzyme for a short time. The following conditions were found to give satisfactory results with ribosomal RNA and sRNA: enzyme/substrate ratio 1:20, 0·02 M-tris buffer (pH 7·4) containing 0·002 M-EDTA (neutralized). For amounts of RNA from 5 to 20 μg the digestion was carried out in a volume of 5 to 10 μl. in a small capillary tube. The digest could then be applied directly to the cellulose acetate for fractionation. Incubation was for 30 min at 37°C. The same conditions were also used for digestion with pancreatic ribonuclease.

(g) *Fractionation procedure*

The digest, which should not contain more than 0·1 mg nucleotides in a volume of less than 10 μl., was first fractionated by high-voltage ionophoresis at pH 3·5 according to the method of Michl (1951) on strips of cellulose acetate. The strip should be about 3 cm wide and the length will depend on the fractionation required. To obtain a fingerprint with all the smaller nucleotides, as in Plates I and II, the strip should be about 60 cm long. For better fractionation of larger nucleotides, as in Plates III and IV, strips of 95 cm can be used in conjunction with the "moving up" technique described below. The cellulose acetate was first wetted with the buffer. To avoid the inclusion of air bubbles, this should be done from one side by floating it on the buffer contained in a Petri dish. The point of application, which should be about 10 cm from the cathode end of the strip, was blotted and the digest applied as a spot and allowed to soak in. Samples of the coloured marker were applied each side of the digest. Care had to be taken to avoid the strip drying out while the sample was being applied. This was done by having the ends dipping in the buffer or covered with wet Kleenex tissue-paper. Excess buffer was then removed from the strip by blotting and the strip was rapidly put into the ionophoresis tank with the point of application near the negative electrode vessel, which was at the top of the tank. In the tanks normally used, the distance between the two electrode vessels was about 50 cm. With the longer strips the lower end was allowed to lie in the buffer at the bottom of the tank (anode compartment). Ionophoresis was carried out at 3000 v. For the shorter strips it was continued until the yellow marker had almost reached the anode buffer (about 2 hr). It was found that the majority of the nucleotides migrate slower than the pink and faster than the blue marker. In order to achieve better resolution in the area between these two markers, the following "moving up" technique was used with the longer strips. Ionophoresis was continued until the pink marker was near the anode buffer. The strip was then taken out and excess buffer blotted off the end that was lying in the anode buffer. It was then replaced in the tank so that the blue marker was just outside the buffer in the cathode compartment. The part of the strip behind the blue marker, which contained the origin, could be cut off. Ionophoresis was then continued until the pink marker had again approached the buffer. The blue marker had by then moved away from the cathode end and the moving up procedure could be repeated. The process was continued until the pink marker was near the end of the strip and there was a distance of about 40 cm between the two markers.

Fractionation in the second dimension was carried out on sheets of DEAE-paper. Normally sheets 57 cm × 46 cm were used, but it may be obtained in rolls, and longer sheets (e.g. 85 cm) have also been used with satisfactory results. This material is very fragile and difficult to handle when wet. The cellulose acetate, on the other hand, is easy to handle when wet but brittle when dry. Attempts to sew the cellulose acetate strip on to the DEAE-paper were thus unsuccessful. A good transfer of the material from the cellulose acetate could be obtained, however, by the following blotting procedure. A sheet of the DEAE-paper was laid on a glass plate. The strip of cellulose acetate on which the nucleotides had been fractionated was removed from the tank and hung up so that the excess white spirit dripped off. Before the buffer dried, the wet strip was laid on the DEAE-cellulose sheet about 10 cm from one of the short sides. A pad of four strips (46 cm × 2 cm) of Whatman 3 MM paper that had been soaked in water was then put on top of the cellulose acetate strip and a glass plate placed on top to press the strips together evenly. Water from the paper pad passes through the cellulose acetate and into the DEAE-paper, carrying the nucleotides with it. Being acidic, they are held on to the DEAE-paper by ion-exchange and remain in the position in which they are first washed on. One thus obtains a good transfer of the fractionated nucleotides without any smearing of the spots as usually occurs with blotting techniques. By this means a strip of the DEAE-paper about 6 cm wide was wetted. To ensure a more complete transfer, more water could be added from a pipette to the paper pad while still in position. In a control experiment, it was shown that 90 to 95% of a ribonuclease T_1 digest of RNA could be transferred from cellulose acetate to DEAE-paper by this technique. After removal of the pad and cellulose acetate strip, the position of the coloured spots was marked on the DEAE-paper, which was hung up to dry in a current of warm air.

After drying, the DEAE-paper was wetted with the pH 1·9 buffer as follows. Wetting was started on either side of the line on which the nucleotides were applied, and the fronts of the solution were allowed to meet along this line. After wetting the rest of the cathode end of the paper (i.e. one-third of the total length of paper), it was put on the rack for running in the special ionophoresis apparatus. The rest of the paper was then wetted and the rack lowered slowly into the tank with the applied nucleotides near the cathode compartment. The system has a high conductivity and there is considerable heating if very high voltages are used. To obtain a fingerprint as in Plate I, about 1500 v was used until the blue marker was almost at the top of the rack (about 4 hr). For better separations, as in Plate III, the ionophoresis was carried out overnight at 1500 v, by which time the blue marker had usually run off the end of the paper. The DEAE-paper was finally dried in air while still on the rack. To prepare radioautographs, the paper was marked with radioactive ink (^{35}S) and put in a folder with one or two sheets of Ilford industrial G X-ray film. The folders were covered on one side with a thin sheet (0·5 mm) of lead so that they could be stacked together in a light-proof cabinet. Where more than 1 μc of [^{32}P]RNA had been used, the film could be developed after 1 day and there would be sufficient material to study the structure of the nucleotides. If it was only desired to obtain a fingerprint, less material could be used (about 0·01 μc for a short fractionation as in Plate I) and development was after 1 to 2 weeks.

The fractionation of a T_1 ribonuclease digest of [^{14}C]methyl-labelled sRNA was done exactly as for [^{32}P]sRNA, which was always run in parallel with it.

The fractionation of an alkaline hydrolysate of [^{14}C]methyl-labelled sRNA was done on a two-dimensional system which differed from that described above. The digest was loaded as a spot on to Whatman 3 MM paper and run at pH 3·5 for 2 hr at 60 v/cm, after which the strip of paper (bounded on each side by the blue marker) was sewn on to a sheet of Whatman no. 1 paper for chromatography in the second dimension using propan-2-ol–HCl–water as the developer. After running for 24 hr, the paper was dried and a radioautograph made. A hydrolysate of [^{32}P]sRNA was run in parallel.

(h) *Preparation of ^{32}P-β-ureidopropionic acid* N-*ribotide*

The method used was adapted from that published by Cohn & Doherty (1956) in that the intermediate, dihydrouridylic acid, was not isolated. A solution containing 1 mg of rhodium catalyst and 1 μc of [^{32}P]uridylic acid in 0·5 ml. of 0·01 M-sodium acetate (pH 5·0) was bubbled with hydrogen for 2 hr. The catalyst was removed by centrifugation and the supernatant fraction transferred on to a piece of polythene and dried down in a vacuum desiccator. It was taken up in about 0·1 ml. of 0·2 N-NaOH and incubated at 37°C for 1 hr. The solution was then loaded on to the origin of the Whatman 3 MM paper for ionophoresis at pH 4·4.

(i) *Structure of the nucleotides*

The positions of the nucleotides on the DEAE-paper was determined from the radioautograph, using the marks from the radioactive ink to line up the film with the paper. The spots were cut out and eluted with alkaline triethylamine carbonate prepared as follows: CO_2 was passed into a mixture of 70 ml. water and 30 ml. redistilled triethylamine until saturated. More triethylamine was then added to about pH 10. In order to minimize losses due to the very small amounts of nucleotides present, non-radioactive nucleotides were added as carrier to the eluting medium. These were prepared as follows: 1·0 g yeast RNA was treated with 25 ml. 1·0 N-KOH at 37°C for 20 min. The KOH was then neutralized with perchloric acid and the precipitate removed by centrifugation. The supernatant solution was made up to 50 ml. and thus contained 20 mg nucleotides/ml. 2 ml. of this was added to each 100 ml. of the eluting mixture.

Elution was carried out as described by Sanger & Tuppy (1951), except that thin-walled capillary tubes were used. These were prepared by drawing out melting point tubes (Gray, 1964). Elution was very rapid by this method and was complete within 20 min. The eluted material was usually put on to polythene sheets and taken to dryness in a desiccator in a partial vacuum. If a good vacuum is put on the desiccator, the triethylamine bubbles badly. In order to ensure complete removal of the triethylamine carbonate, water was added to the spots and they were taken to dryness several times.

(i) *Composition*

In order to identify the mononucleotides present in an oligonucleotide, it was digested with about 10 μl. 0·2 N-NaOH at 37°C for 16 hr. This was done in drawn-out melting point tubes. The solution was drawn into the tube using a piece of polythene tubing and the ends were sealed off. After incubation, the digest was applied to Whatman no. 52 paper for ionophoresis at pH 3·5 and 60 v/cm for 1 to 1½ hr. Each digest was applied over 2 cm of the paper with 1-cm gaps between them. The four mononucleotides were well separated on this system and the composition of each digest could be determined from the radioautograph. With nucleotides from ribonuclease T_1 digests, G was present as a single spot, but with those from pancreatic ribonuclease digests it usually formed two spots, which were presumably the 2′- and 3′-phosphates. With simple nucleotides it was usually possible to deduce how many residues of each mononucleotide were present by estimating the intensity of the bands on the radioautograph; however, this was usually checked by direct estimation on the scintillation counter. Areas of equal size for the various bands were cut out from the paper, using the radioautograph as a guide. The area was usually 1·8 cm × 2·6 cm and was chosen to take in the whole of each spot. If the spots contained less than about 100 cts/min the results were not very reliable, but above this about 10% accuracy was achieved.

sRNA contains a number of minor components. Many of these separated from the four major components on the pH 3·5 ionophoresis. They were investigated further by paper chromatography using the following systems on Whatman no. 1 paper: propan-2-ol (680 ml.), concn HCl (176 ml.) and water to 1 l. (Wyatt, 1951); and propan-2-ol (70 ml.), water (30 ml.), concn NH_3 (1 ml.) (Markham & Smith, 1952).

(ii) *Digestion of oligonucleotides with enzymes*

To determine the sequence of residues in oligonucleotides, they were further degraded with nucleases and the products identified. The most useful enzyme for the investigation of nucleotides from ribonuclease T_1 digests was pancreatic ribonuclease and, conversely, ribonuclease T_1 was used to study nucleotides from pancreatic ribonuclease digests. In both cases the products that can be obtained are the same and consist of the mononucleotides G, C and U and oligonucleotides with various numbers of A residues terminated at the 3′ end by G, C or U. Most of these could be satisfactorily separated by ionophoresis at pH 3·5 (Fig. 1).

Pancreatic ribonuclease. Samples of the nucleotides that had been dried down on polythene were treated with 5 to 10 μl. of 0·001 M-EDTA–0·01 M-tris buffer (pH 7·4) containing 0·2 mg pancreatic ribonuclease/ml. and transferred to capillary tubes which were sealed

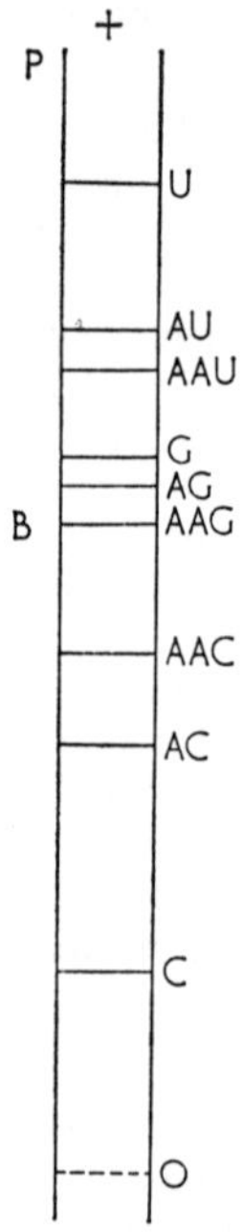

FIG. 1. Diagram showing the position of products from pancreatic digests of nucleotides from ribonuclease T_1 digests on ionophoresis at pH 3·5. B and P show the positions of the blue and pink markers, respectively.

at the drawn out end and incubated at 37°C for 30 min. The samples were then treated with 0·1 N-HCl to break down any cyclic phosphates as follows. 2 μl. 0·5 N-HCl was put on a polythene strip. The tip of the capillary was broken and the digest squirted out, mixed with the HCl and sucked back into the capillary, which was resealed and incubated at 37°C for 1 hr. The digest was then transferred to Whatman no. 52 paper and subjected to ionophoresis at pH 3·5 at 3 kv until the pink marker was near the end of the paper. The products could usually be identified by their position (Fig. 1), but if there was any doubt the nucleotide was eluted with water and subjected to alkaline hydrolysis for analysis. This was most frequently necessary with AG, AAG and AAAG, which are only just resolved on the ionophoresis.

Ribonuclease T_1. Products from pancreatic ribonuclease digests were digested with ribonuclease T_1 for 30 min at 37°C, using the following mixture: 0·001 M-EDTA–0·01 M-tris buffer (pH 7·4), 0·1 mg ribonuclease T_1/ml.

Micrococcal nuclease. Digestion was carried out in 0·05 M-borate buffer (pH 8·8) containing 0·01 M-$CaCl_2$, with an enzyme concentration of 0·1 mg/ml. Incubation was for 2 hr at 37°C and the products were fractionated at pH 3·5. Since the specificity of this enzyme is not clearly defined (Cunningham, Catlin & Privat de Garilhe, 1956; Reddi, 1959), it was not possible to identify the products from their positions, and it was necessary to analyse them and in certain cases to determine the terminal residues as described below with phosphomonoesterase and phosphodiesterase.

(iii) *Determination of terminal residues*

The nucleotides were first treated with bacterial alkaline phosphomonoesterase to remove the terminal 3′-phosphate group and the product purified by ionophoresis at pH 6·5. The 3′-terminal residue was usually known, due to the specificity of the ribonucleases; but if it was desired to determine it, the dephosphorylated nucleotide was digested with alkali. The 3′-terminal residue would be present as a nucleoside and would not be seen on the radioautograph. It was thus identified as the mononucleotide that was present in an alkaline digest of the oligonucleotide, but not in that of the dephosphorylated product.

To identify the 5′ terminal residue, the dephosphorylated product was treated with snake venom phosphodiesterase. This breaks the 3′ phosphoester bonds (Razzell & Khorana, 1959) so that the 5′ terminal residue is now present as a nucleoside and is identified as the residue that is missing from the digest.

The digestion with phosphomonoesterase was carried out for 30 min at 37°C in capillary tubes with 5 to 10 μl. of 0·005 M-$MgCl_2$–0·1 M-NH_4HCO_3 containing 0·1 mg enzyme/ml. It was then taken to dryness to remove the ammonium bicarbonate and subjected to ionophoresis at pH 6·5 (50 v/cm, 1·5 hr) on Whatman no. 52 paper. The bands were located by radioautography, cut out and eluted with water containing carrier nucleotides. If more than one main band was present, they were also cut out and studied.

Treatment of the dephosphorylated product or a sample of it with venom phosphodiesterase was carried out with 5 to 10 μl. of the following solution: 0·01 M-magnesium acetate–0·025 M-tris buffer (pH 8·5)–0·1 mg enzyme/ml. After incubation for 2 hr at 37°C, the 5′-mononucleotides were separated by ionophoresis at pH 3·5.

(iv) *Partial digestion with spleen phosphodiesterase*

In order to know the position of the unchanged material on the final DEAE-paper ionophoresis, each nucleotide was divided into two samples: a large sample which was digested, and a smaller one containing about half as much material which was used as a control.

Conditions of digestion were varied somewhat for different nucleotides and sometimes samples were studied under different conditions. The following conditions were found to be the most generally useful and were used in most preliminary experiments with products from ribonuclease T_1 digests. The nucleotide was treated with 5 to 10 μl. of the following mixture in a capillary tube for 30 min at 37°C: 0·1 M-ammonium acetate (pH 5·7), 0·002 M-EDTA, 0·05% Tween 80, 0·2 mg enzyme/ml. For nucleotides containing more than one C residue, digestion was usually carried out for 60 min. Nucleotides from pancreatic ribonuclease digests were digested more rapidly, and for satisfactory results an enzyme concentration of 0·1 mg/ml. was used.

After digestion, the material was applied directly to DEAE-paper. The untreated sample was applied next to it, with no space between. A sample of an unfractionated digest of ribosomal RNA was also applied to each paper as well as the coloured markers. After wetting with the pH 1·9 mixture, ionophoresis was carried out as described above until the blue marker had reached the top of the rack (about 4 hr at 1·5 kv).

3. Results

(a) *Ribonuclease T_1 digests of ribosomal RNA*

Plates I to IV show two-dimensional fractionations of ribonuclease T_1 digests of the 16 s and 23 s ribosomal components. The fractionations shown in Plates I and II have been run for shorter times to include all the small nucleotides, except G, whereas those in Plates III and IV have been run for longer times to obtain better fractionation of the larger nucleotides. The system of numbering the spots is shown in Fig. 2 and the results of the analyses of the different nucleotides are summarized in Table 1. These combined results were obtained in several different experiments. It was found that the distribution of the spots was reproducible if identical conditions of digestion were used. All the nucleotides recorded in Table 1 have G as the 3′ terminal residue, indicating good specificity of ribonuclease T_1. Occasionally nucleotides not containing a G residue were obtained. This was probably due to contamination with other nucleases. Thus, in the fingerprint obtained with the 23 s material there were a number of faint spots which were not present in the 16 s material. Most of these gave only A

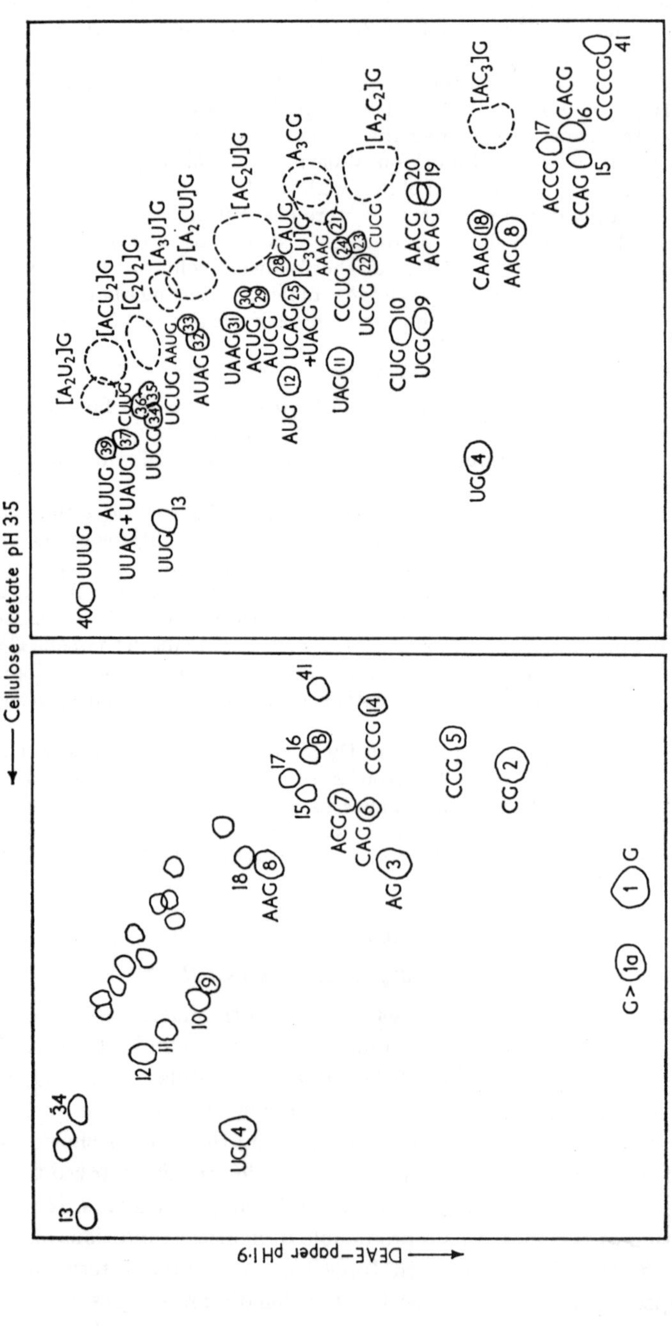

FIG. 2. Diagram showing the position of nucleotides from a ribonuclease T_1 digest of RNA on the two-dimensional system. (a) Illustrates a fractionation that has been run for a short time on the DEAE-paper; (b) one that has been run for a longer time. B marks the position of the blue marker.

TABLE 1

Nucleotides from ribonuclease T_1 digests

Spot no. (Fig. 2)	Composition† C	A	U	5′ terminal‡ residue	Products obtained with pancreatic ribonuclease	Products obtained with micrococcal nuclease	Structure deduced
1	—	—	—				G
1a	—	—	—				G>
2	*x*	—	—	C			CG
3	—	0·9	—	A			AG
4	—	—	*x*	U			UG
5	2·1	—	—				CCG
6	0·9	0·7	—		C, AG		CAG
7	1·0	1·0	—		G, AC	G, AC	ACG
8	—	1·9	—	A	unchanged		AAG
9	0·95	—	0·9	U			UCG
10	1·1	—	0·9	C			CUG
11	—	0·9	0·85	U	U, AG		UAG
12	—	1·2	1·0	A	G, AU		AUG
13	—	—	1·9				UUG
14	2·9	—	—				CCCG
15	*xx*	*x*	—	C	C, AG		CCAG
16	*xx*	*x*	—	C	C, G, AC		CACG
17	1·9	1·1	—	A	C, G, AC		ACCG
18	1·0	2·1	—		C, AAG	AG, C, A	CAAG
19	1·0	2·0	—		G, AC, AG, AAC		ACAG+AACG
21	—	3·0	—		unchanged		AAAG
22	*xx*	—	*x*	U		C, G, UC, CC	UCCG
23 }	1·9	—	1·0 {	C		C, G, CG, CU	CUCG
24 }				C		CC, UG	CCUG
25	1·2	1·1	1·0	U	C, U, AG, AC		UCAG+UACG
28	1·15	1·0	1·0	C	C, G, AU		CAUG
29	1·0	1·0	1·1	A	C, U, G, AC, AU		ACUG+AUCG
31	—	2·2	1·2		U, AAG		UAAG
32 }	—	2·1	1·0 {		AU, AG		AUAG
33 }					G, AAU		AAUG
34 }			{	U		C, G, UC, UU, CG	UUCG
35 }	1·1	—	2·0 {	U		C, U, UC, UG	UCUG
36 }			{	C		C, U, G, CU, UG	CUUG
37	—	1·0	1·9		U, G, AU, AG		UUAG+UAUG
39	—	*x*	*xx*	A	U, G, AU		AUUG
40	—	—	3·1				UUUG
41	4·0	—	—				CCCCG

† Compositions are expressed as the yield of a given mononucleotide relative to one mole of G and were estimated in the scintillation counter as described in the text. Where the results are given in *x*'s instead of figures, the composition was determined only by inspection of the radio-autograph. All nucleotides contained G, which is not recorded.

‡ 5′ terminal residues (column 5) were determined by degradation with phosphomonoesterase and phosphodiesterase as described in the text.

1a. On alkaline digestion gave a double spot corresponding to the 2′- and 3′-guanylic acids.

25 may contain a small amount of C terminal residue.

34, 35, 36. There was considerable overlapping between these components. The sequences of the (CU) dinucleotides from the micrococcal nuclease digest were determined using phosphomono-esterase and phosphodiesterase.

and U on hydrolysis. This suggested that the 23 s RNA was slightly contaminated with a nuclease which was not present in the 16 s RNA.

Using an enzyme to substrate ratio of 1:20, the only 2′,3′-cyclic phosphate present was that of guanylic acid (spot 1a, Fig. 2), which appears to be more stable than the cyclic phosphates of the oligonucleotides. When less enzyme was used, cyclic forms of other nucleotides were present, which led to a considerably more complex pattern. The cyclic forms moved faster than the non-cyclic phosphates on the cellulose acetate ionophoresis dimension, but at approximately the same rate in the DEAE-paper dimension.

The results obtained by pancreatic ribonuclease digestion were very clear and gave considerable information about structure. Some difficulty was experienced in determining the 5′ terminal residues using phosphomonoesterase and phosphodiesterase, since the various preparations of phosphomonoesterase which were used appeared to contain a nuclease, so that on purification of the dephosphorylated product several spots were frequently present and it was necessary to study them all to decide which was the required one.

The probable structures of the components of the different spots are shown in the last column of Table 1, and these results are summarized in Fig. 2. The di- and trinucleotides are all readily separated by this technique. Of the 27 possible tetranucleotides, 10 are present as single pure spots. Very slight overlapping usually occurs between CCUG(23) and CUCG(24) and between AAUG(32) and AUAG(33). The following pairs occur in single spots, where there is slight evidence of separation; AUCG(29) and ACUG(30), ACAG(19) and AACG(20). No separation has been observed between the pairs UCAG(25) and UACG(26), and UUAG(37) and UAUG(38). The three isomers having the composition $(CU_2)G$ (34–36) are in a large spot which shows partial resolution of all three. We have not yet detected the nucleotide CUAG in ribosomal RNA, which would be expected to occur in or near spot 28. In both the 16 s and the 23 s material, spot 28 appears to contain only CAUG, giving AU and no AG when treated with pancreatic ribonuclease. It is possible that CAUG may run with spot 25.

The composition of the pentanucleotide spots has been determined in many cases, but the sequences have not been studied in detail and it seems probable that most spots are mixtures of isomers. There are 81 possible pentanucleotides, and for some compositions, e.g. $(C_2AU)G$, there are 12 possible isomers. There is a considerable spread of isomers, as is indicated by the difference in the patterns in the pentanucleotide area obtained from the 16 s and 23 s components. The areas occupied by the different isomers are indicated in Fig. 2 with dashed lines.

The position of a nucleotide on the fingerprint is determined largely by its composition. Figure 3 shows the relationship between the composition of a nucleotide and its position. It was prepared as follows. The "centre of gravity" of each set of isomers was first marked with a point. Lines were then drawn to link up the points representing nucleotides that differed only in the number of C residues they contained. Thus, for instance, a line was drawn joining UG, (CU)G, $(C_2U)G$ and $(C_3U)G$, etc. Similarly, lines were drawn connecting nucleotides that differed only in A residues. The mononucleotide that has the greatest effect on mobility in the DEAE-paper dimension is U, so that the fingerprint may be regarded as being composed of three sections representing nucleotides with two, one or no U residues, respectively. The lines joining the spots form three graticules corresponding to the different sections, and one axis on each

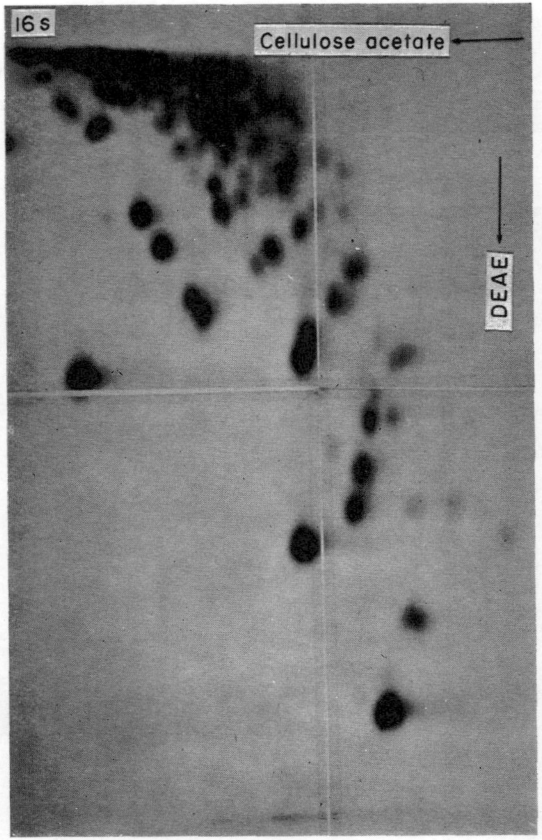

Plate I

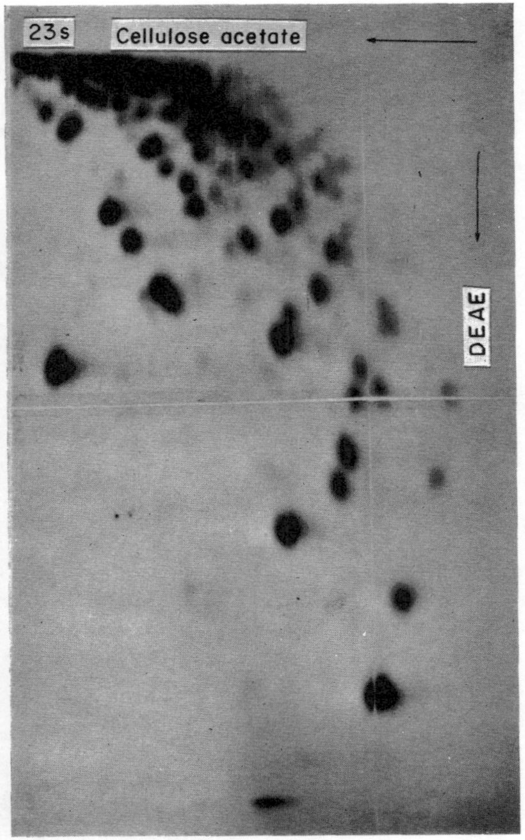

Plate II

Plate I. Radioautograph of a two-dimensional fractionation (cellulose acetate pH 3·5, 3 kv, 2 hr; DEAE-paper pH 1·9, 1·5 kv, 4 hr) of a ribonuclease T_1 digest of the 16 s component of ribosomal RNA.

Plate II. Radioautograph of a two-dimensional fractionation (as in Plate I) of a ribonuclease T_1 digest of the 23 s component of ribosomal RNA.

[facing p. 384

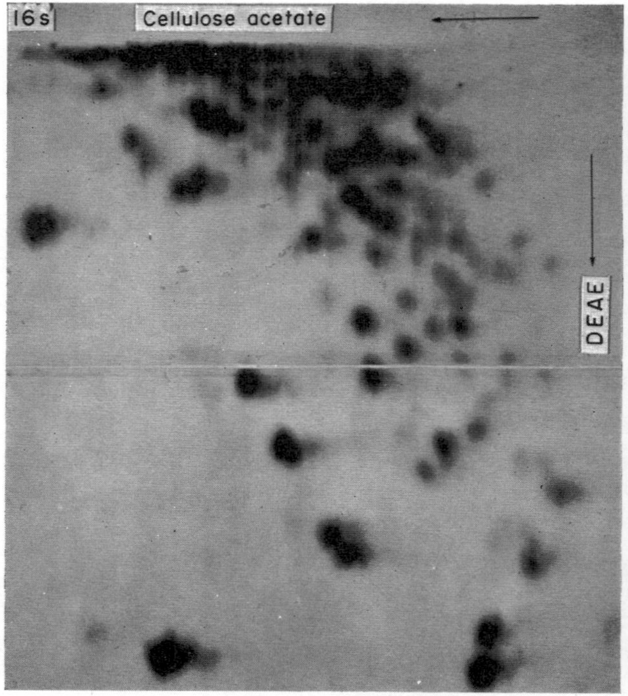

Plate III

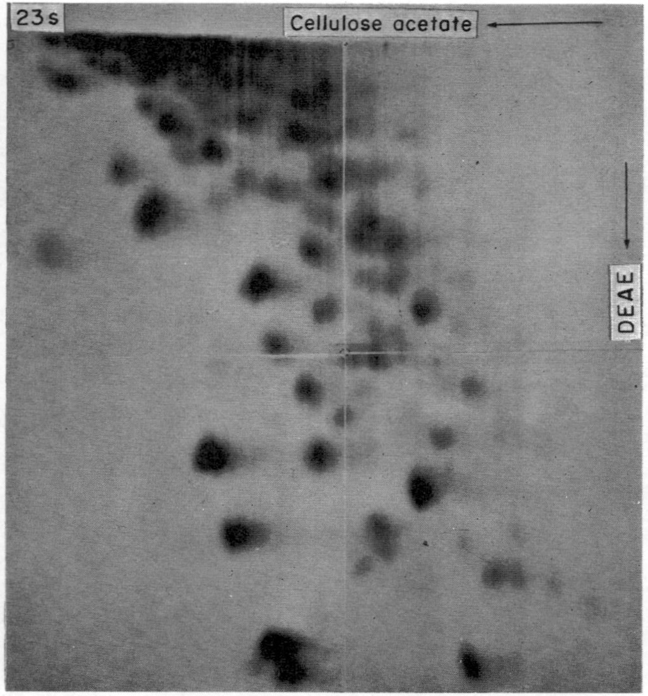

Plate IV

Plate III. Radioautograph of a two-dimensional fractionation (cellulose acetate pH 3·5, 3 kv with "moving up" as described in text; DEAE-paper pH 1·9, 1·5 kv, 16 hr) of a ribonuclease T_1 digest of the 16 s component of ribosomal RNA.

Plate IV. Radioautograph of a two-dimensional fractionation (as in Plate III) of a ribonuclease T_1 digest of the 23 s component of ribosomal RNA.

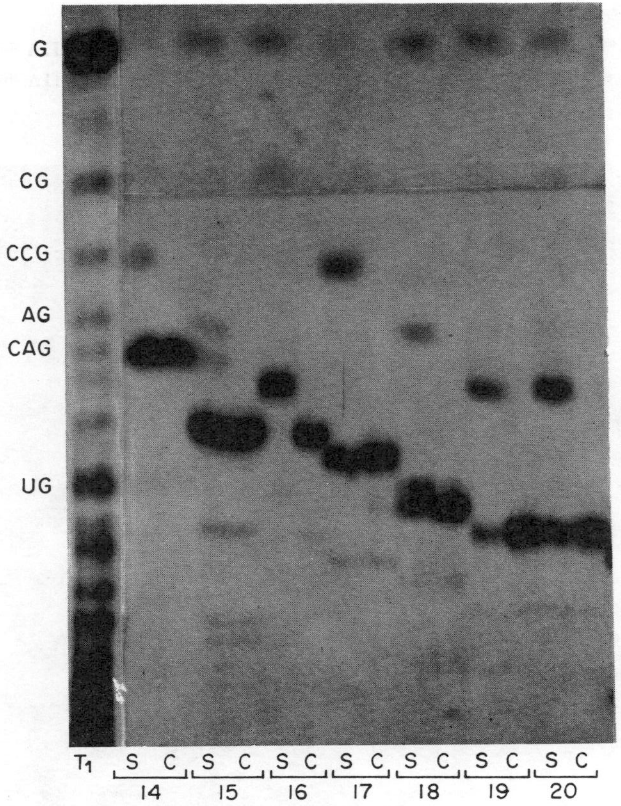

PLATE V

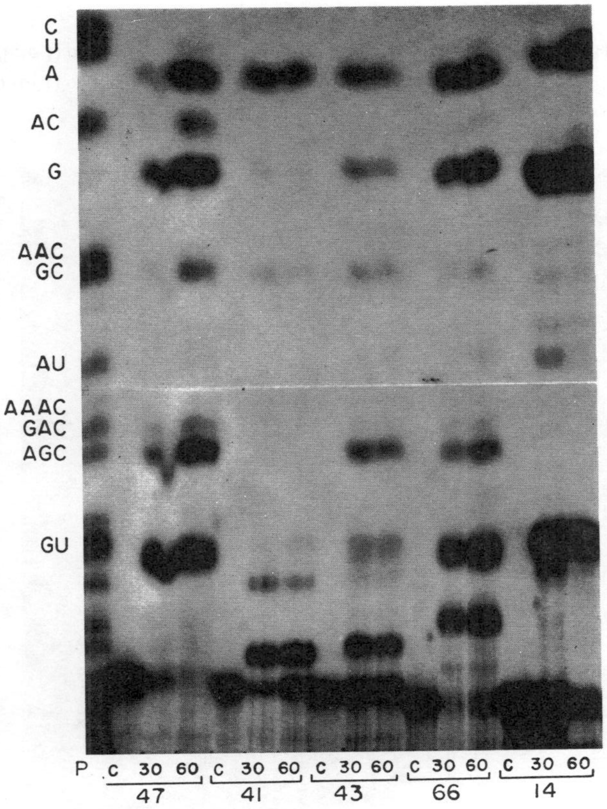

PLATE VI

PLATE V. Radioautograph of an ionogram on DEAE-paper (pH 1·9, 1·5 kv, 4 hr) of the products from partial digestion with spleen phosphodiesterase of nucleotides from a ribonuclease T_1 digest. T_1 is a ribonuclease digest; C is the untreated nucleotide; S is a sample that has been treated with spleen phosphodiesterase as described in the text.

PLATE VI. Radioautograph of an ionogram (as in Plate V) of the products of partial digestion with spleen phosphodiesterase of some nucleotides from a pancreatic ribonuclease digest of ribosomal RNA.

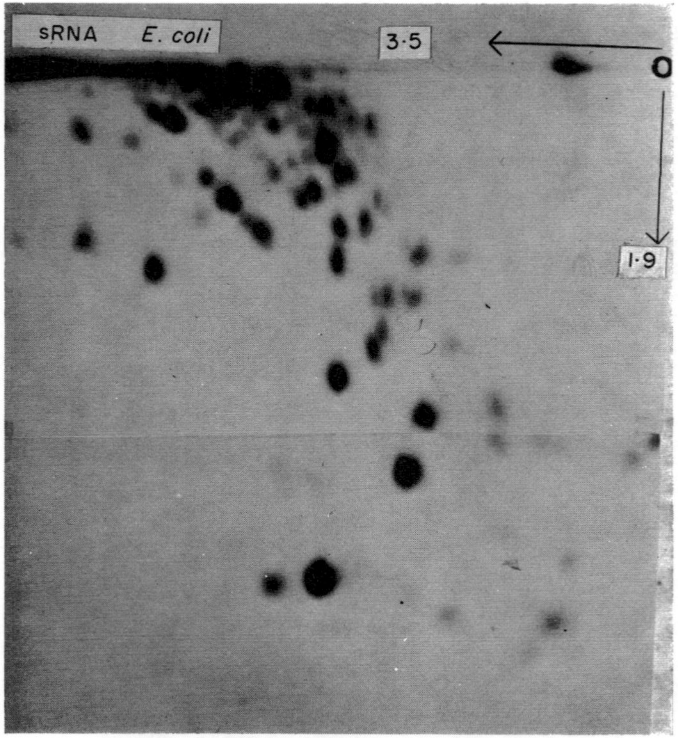

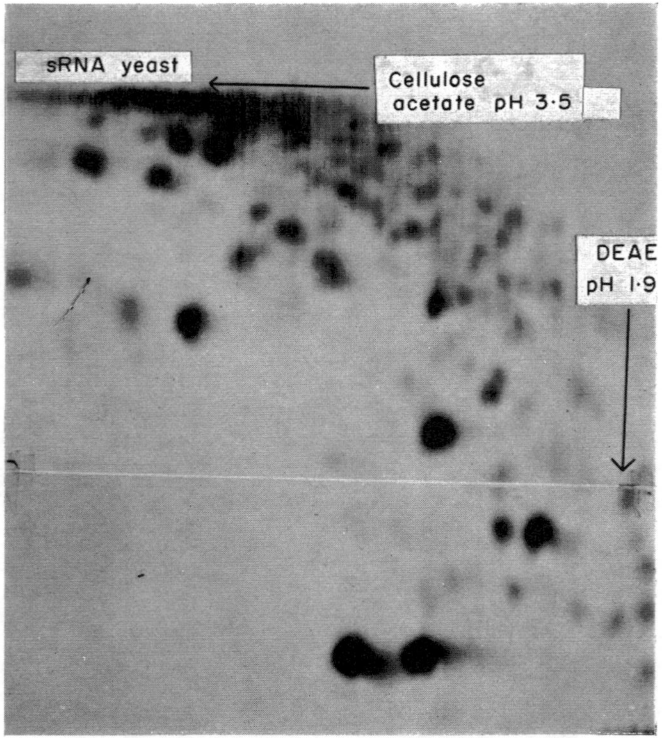

PLATE VII

PLATE VIII

PLATE VII. Radioautograph of a two-dimensional fractionation (as in Plate I) of a ribonuclease T_1 digest of sRNA from *E. coli*.

PLATE VIII. Radioautograph of a two-dimensional fractionation (as in Plate I) of a ribonuclease T_1 digest of sRNA from bakers' yeast.

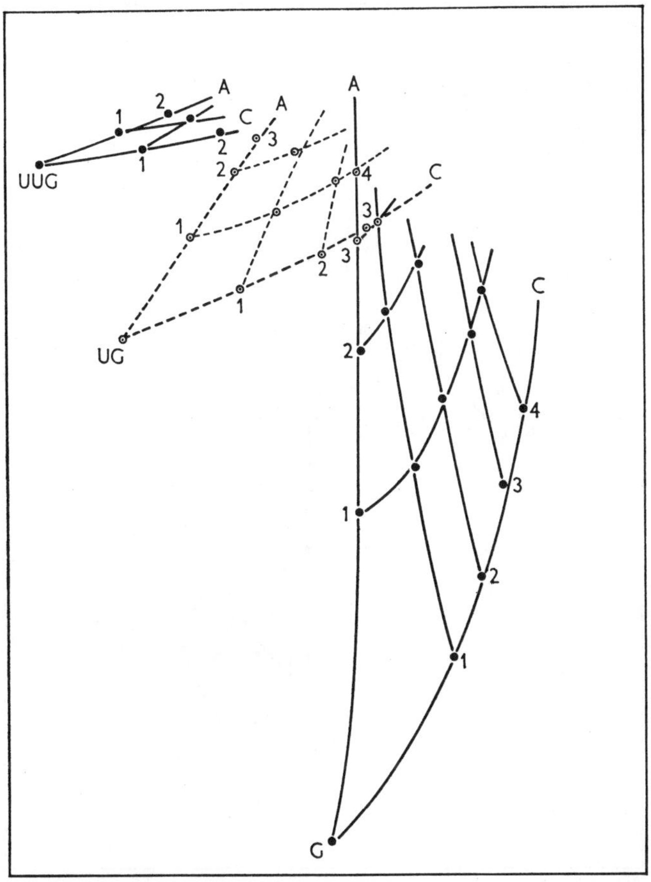

FIG. 3. Diagram illustrating the relationship between the composition of a nucleotide and its position on the two-dimensional system using ionophoresis at pH 1·9 for the DEAE-paper dimension. For explanation see text.

graticule represents the number of A residues, whereas the other gives the number of C residues. In this way the probable composition of a nucleotide may be determined from its position on the map. In the pentanucleotide area there is considerable overlapping between the different graticules. Thus, for instance $(A_3C)G$ is in the same area as $(C_3U)G$. This overlapping may be almost completely avoided by carrying out the fractionation on DEAE-paper using between 7 and 10% formic acid instead of the pH 1·9 mixture (2·5% formic acid, 8·7% acetic acid). This increases the separation of the sections containing different numbers of U residues, and a distribution as shown in Fig. 4 is obtained. The resolution of isomers of the faster moving spots is probably less efficient on this system, although it is probably better for the slow-moving ones. The system has not yet been investigated in sufficient detail to say precisely to what extent the composition of a nucleotide can be determined from its position on this map, and it seems probable that some overlapping may occur between the different isomers. Thus the tetranucleotide CAAG comes very close to the trinucleotide AAG, and it seems that larger nucleotides that bear this type of relationship to one another may also overlap.

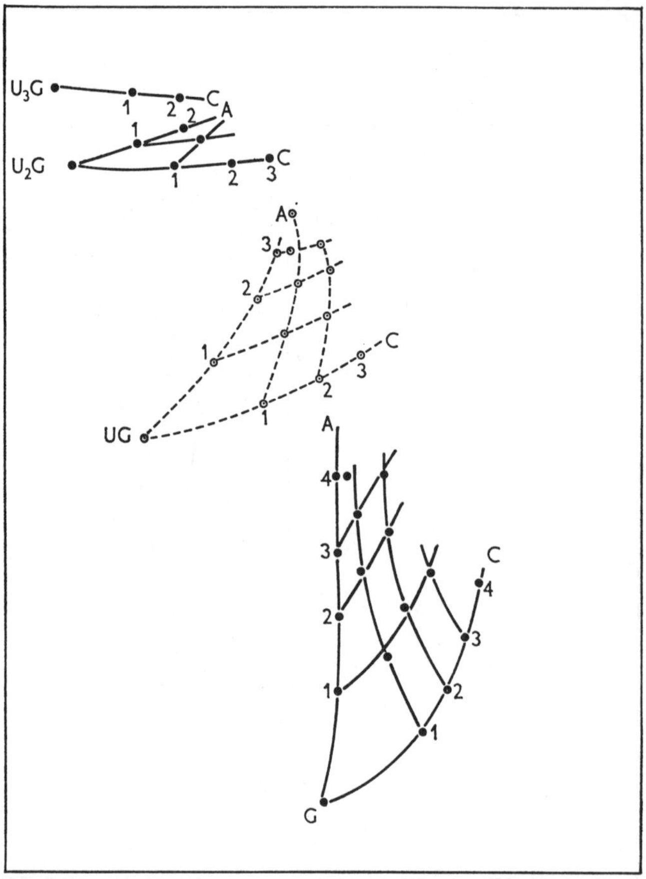

FIG. 4. Diagram as in Fig. 3 but using 10% formic acid for the DEAE-paper dimension.

The fact that there is a separation of isomers indicates that the sequence affects the behaviour on the two-dimensional system. One generalization that seems possible is that isomers which have A as the 5′ terminal residue move slower in the DEAE dimension than those having other 5′ terminal residues.

(b) *Partial digestion with spleen phosphodiesterase*

The results listed in Table 1 were obtained before the introduction of the method using spleen phosphodiesterase. This method has been applied to most of the spots the sequence of which had been determined by the normal methods, and the results have been used as confirmation and to demonstrate the general applicability of the method. Some of the results are illustrated in Plate V and Fig. 5.

If we consider the nucleotide ABCG, where A, B and C are unknown residues, its breakdown products will be BCG and CG. Suppose the distance moved from the origin by ABCG on the DEAE-paper is y and the distance between it and its first degradation product (BCG) is x, we can then define a value M for ABCG as x/y. This value depends on the nature of the 5′ terminal residue A. For all the nucleotides studied from the ribonuclease T_1 digests, it has been found that the values of M lie within the limits

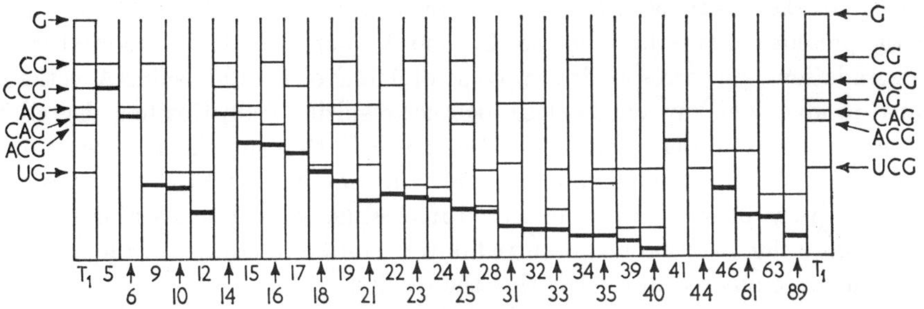

FIG. 5. Diagram illustrating the fractionation on DEAE-cellulose (pH 1·9) of the products of partial digestion with spleen phosphodiesterase of oligonucleotides from ribonuclease T_1 digests of ribosomal RNA. The thick line represents the position of the unchanged nucleotide which was run as a control. Mononucleotides are not included.

shown in Table 2 and there is no overlapping between the values for the different residues; thus the various residues may be identified directly from the values of *M*. The dinucleotides and many of the trinucleotides can be identified from their position on the DEAE-paper by comparison with the ribonuclease T_1 digest that is run as a marker.

TABLE 2

Values of M *on DEAE-paper ionophoresis (pH 1·9) for nucleotides from ribonuclease T_1 digests*

5′ terminal residue	Range of *M*-values
C	0·05–0·3
A	0·5 –1·0
U	1·5 –2·5

As examples, we may consider the results with the three isomers (C_2A)G, numbers 15, 16 and 17 (see Plate V and Fig. 5). The value of *M* for the first degradation step for 16 is 0·18. This falls within the limits for C (Table 2), which is therefore the 5′ terminal residue. The value of *M* for the second step is 0·58, establishing it as an A residue. The product produced at this stage is rapidly moving and can only be the dinucleotide CG, thus confirming the structure CACG. For no. 15, the values of *M* for the first two steps are 0·22 and 0·07, respectively, and the dinucleotide moves at the rate of AG, thus establishing the sequence CCAG. Nucleotide no. 17 gives only one visible degradation product under the conditions used. This is related to the unchanged material by a value of 0·64 and occupies a unique position which identifies it as the trinucleotide CCG, thus confirming the structure ACCG.

For most of the nucleotides studied it was possible to deduce the complete sequence by this method. Difficulties might be experienced if the nucleotides studied were mixtures. Thus the results with no. 19 could be interpreted in various ways; from the values for *M* of the different bands, it could for instance fit a sequence ACCACG. This, however, is clearly impossible from its position on the fingerprint and is precluded by a knowledge of the positions of the different nucleotides on the DEAE-paper system. Thus the presence of two dinucleotides CG and AG indicates that it is a mixture. The first degradation step corresponds to a value of 0·64 for M, establishing A as the only terminal residue. The two bands close together resulting from this first degradation can be identified from their position as CAG and ACG, thus showing it to be a mixture of ACAG and AACG. Similarly, spot 25 is a mixture of UACG and UCAG. Occasionally extra bands were present in the untreated nucleotide control, suggesting that some degradation or side reaction had occurred during elution or storage of the nucleotide. Such bands were ignored in interpreting the results.

One limitation of the method lies in the different rates at which the different residues are released by the enzyme, U residues being split off more rapidly than A, and A more rapidly than C. Thus it may be that not all the possible degradation products of a nucleotide are present in a particular digest; this may lead to a misinterpretation in the case of larger nucleotides. For this reason it is advisable not to rely completely on this method, but to have some other data such as a composition on all nucleotides. Thus in Fig. 5, CG was not observed in nucleotides 17 and 22, UG was not observed in 24 and in 32 the intermediate UAG was missing. Of the pentanucleotides studied, no. 41 loses a 5′ terminal residue giving a band which corresponds to band 14 (CCCG) and no further breakdown products under the conditions used; it is therefore CCCCG. Similarly 44 is ACCCG, 46 is AACCG, the final product being CCG, which is uniquely identified from its position, 63 is AUCCG and 89 UUCCG.

(c) *Pancreatic ribonuclease digests of ribosomal RNA*

Digests prepared by the action of pancreatic ribonuclease have not been studied in such detail as those obtained with ribonuclease T_1. However, the simplicity in the composition of the nucleotides and the smaller number of larger ones made a study of these digests somewhat simpler. Figure 6 shows the pattern of spots obtained, and the sequence determinations are summarized in Table 3. The results were obtained from various digests of ribosomal RNA components. Of the 16 tetranucleotides, 10 have been identified as single pure spots. The following pairs of isomers show partial separation but there is usually some overlapping: AAGU(18) and AGAU(20), GAGC(25) and GGAC(27), GAGU(26) and GGAU(28).

The determination of the 5′ terminal residue by digestion with phosphomonoesterase and venom phosphodiesterase was more generally satisfactory than with the nucleotides from ribonuclease T_1 digests, since only one product was usually obtained after phosphomonoesterase treatment. This suggested that the contaminant in the preparations of phosphomonoesterase causing degradation of the nucleotides in ribonuclease T_1 digests had a specificity similar to pancreatic ribonuclease. In fact it was always possible to identify the 5′ terminal residues from the positions of the nucleotides on the fingerprint, since isomers with A end groups move slower on the DEAE-paper than those with G end groups.

The degradation method with spleen phosphodiesterase also gave rather clear results. In this case one is only concerned with A and G residues, and these are both split off at a similar speed, so that a complete spectrum of degradation products is usually produced. For the pancreatic ribonuclease digestion products, the values of M associated with an A 5′ terminal residue varied from 0·4 to 1·1 and for a G terminal residue from 1·2 to 3·1. There is a wide spread and the limits are rather close; thus for values between 0·9 to 1·5 it is advisable to have confirmation from other data. The

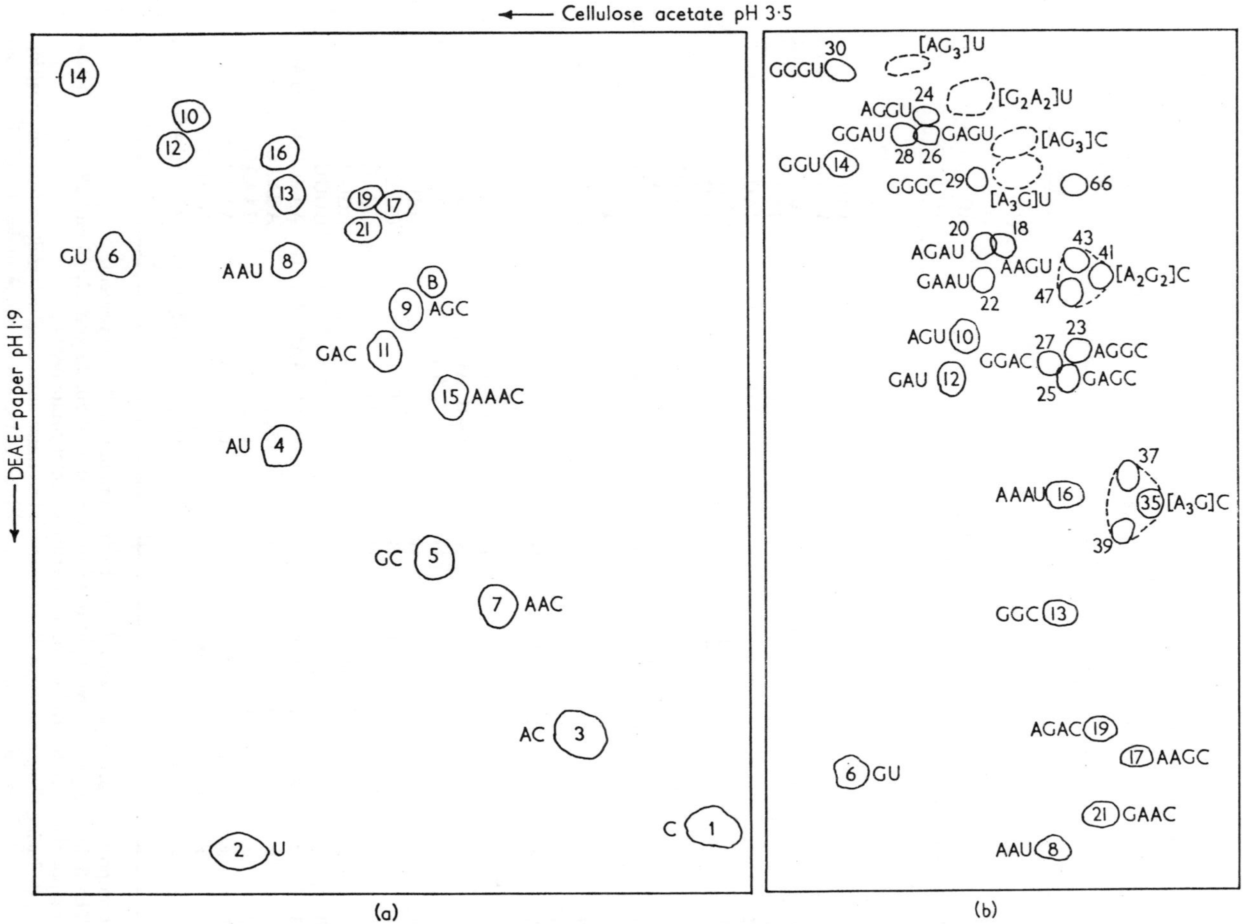

FIG. 6. Diagram showing the position of nucleotides from a pancreatic ribonuclease digest on the two-dimensional system. (a) Illustrates a fractionation that has been run for a short time on the DEAE-paper dimension; (b) one that has been run for a long time. B marks the position of the blue marker.

TABLE 3

Nucleotides from pancreatic ribonuclease digests

Spot no. (Fig. 6)	Composition† C	U	A	G	5′ terminal residue	Products obtained with ribonuclease T_1	Structure deduced
1	*x*	—	—	—			C
2	—	*x*	—	—			U
3	*x*	—	*x*	—			AC
4	—	*x*	*x*	—			AU
5	*x*	—	—	*x*			GC
6	—	*x*	—	*x*			GU
7	*x*	—	1·9	—			AAC
8	—	*x*	1·7	—			AAU
9	*x*	—	*x*	*x*		C, AG	AGC
10	—	*x*	1·0	1·0	A	U, AG	AGU
11	*x*	—	*x*	*x*		G, AC	GAC
12	—	*x*	*x*	*x*	G	G, AU	GAU
13	*x*	—	—	2·0			GGC
14	—	*x*	—	1·7			GGU
15	*x*	—	2·6	—			AAAC
16	—	*x*	2·8	—			AAAU
17	*x*	—	1·7	1·0	A	C, AAG	AAGC
18‡	—	*x*	2·1	1·0		U, AAG	AAGU
19	*x*	—	*xx*	*x*	A	AG, AC	AGAC
20‡	—	*x*	2·1	1·0		AG, AU	AGAU
21	*x*	—	1·9	0·9	G	G, AAC	GAAC
22	—	*x*	2·0	1·2		G, AAU	GAAU
23	*x*	—	1·0	1·9	A	C, G, AG	AGGC
24	—	*x*	*x*	*xx*	A	U, G, AG	AGGU
25‡	*x*	—	1·1	2·0	G	C, G, AG	GAGC
26‡	—	*x*	1·0	1·9	G		GAGU
27‡	*x*	—	1·1	2·0	G	G, AC	GGAC
28‡	—	*x*	1·0	1·9	G	G, AU	GGAU
29	*x*	—	—	*xxx*			GGGC
30	—	*x*	—	2·7			GGGU
33	*x*	—	2·9	0·9		C, AC, AAG, AAAG	AAAGC+AAGAC
37	*x*	—	*xxx*	*x*		AG, AAC	AGAAC
39	*x*	—	2·6	1·0		G, AAAC	GAAAC
41	*x*	—	2·0	1·9		G, C, AAG	AAGGC§
43	*x*	—	*xx*	*xx*		C, AG	AGAGC
47	*x*	—	2·1	1·9		G, AC, AG, C, AAG	GAGAC+GAAGC§

† Compositions are expressed relative to the amount of C or U present.

‡ The following pairs show slight overlapping: 18 and 20, 25 and 27, 26 and 28. In each case the composition was determined on a mixture of both components.

§ In deducing these structures, the 5′ terminal residues have been determined from the position on the DEAE-paper, assuming that isomers with an A terminal residue move slower than those with a G.

extreme values are found among the smaller nucleotides, which in any event are best identified from their position on the DEAE-paper.

Some results with some larger nucleotides are shown in Plate VI. The values of *M* for the various steps in no. 41 are 0·46, 0·68 and 1·8. This latter step gives rise to the dinucleotide GC, demonstrating the sequence AAGGC. Two dinucleotides, AC and GC, are present in the digest of no. 47, indicating it is a mixture. The first two degradation steps are clear, giving GA (*M* being 1·7, 0·52). Two bands are then produced close together which correspond to the positions of GAC and AGC on the marker. It thus appears to be a mixture of GAGAC and GAAGC. Number 43 could be interpreted as AAAAGC; however, this would not fit with its position and it is probable that the third band from the origin is due to contamination with spot 21 and that the actual sequence is AGAGC. Number 66 appears to analyse clearly as the hexanucleotide GAAAGC.

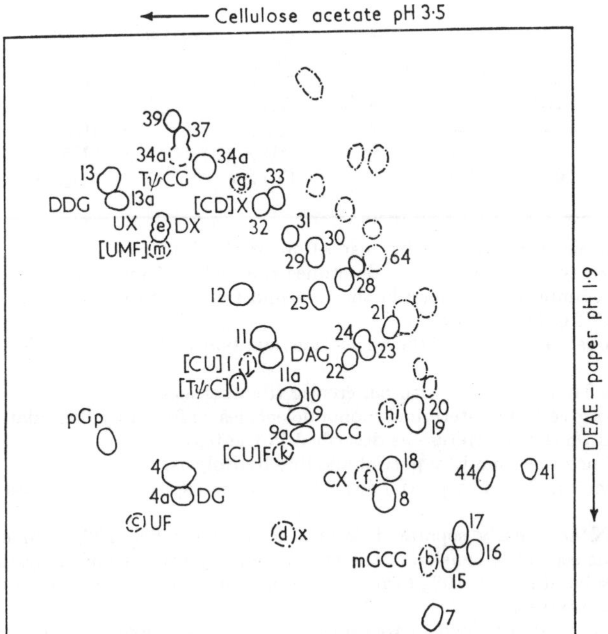

FIG. 7. Diagram showing the position of nucleotides from a ribonuclease T_1 digest of sRNA on the two-dimensional system. The sequence or composition is shown for those nucleotides, symbolized by letters a to m, containing minor bases. The numbered spots refer to nucleotides observed in digests of ribosomal RNA, the sequence of which is shown in Fig. 2. Those not mentioned there are referred to in the text.

(———) Indicates that the nucleotide was detected in both yeast and *E. coli*, (– – –) that it was detected only in yeast, and (– · – · –) that it was seen only in *E. coli*. See Table 4 and text for further explanation.

TABLE 4

Nucleotides containing minor bases from ribonuclease T_1 digests of sRNA

Spot no. (Fig. 7)	Composition† C	A	G	U	Minor nucleotides‡	Structure deduced	Source (yeast (y) or *E. coli* (c))
4a	—	—	x	—	(Y) x	DG	y+c
9a	x	—	x	—	(Y) x	DCG	y+c
11a	—	x	x	—	(Y) x	DAG	y+c
13a	—	—	x	—	(Y) xx	DDG	y+c
34a	x	—	x	—	(T) x (ψ) x	TψCG	y+c
34a>	x	—	x	—	(T) x (ψ) x	TψCG>	y+c
b	x	—	x	—	(mG) x	mGCG	c
c	—	—	—	x	(F) x	UF	y
d	—	—	—	—	(X)	X	c
e	—	—	—	x	(X) xx	UX	y+c
e	—	—	—	—	(Y) x, (X) xx	DX	y+c
f	1·0	—	—	—	(X) 1·8	CX	c
g	1·0	—	—	—	(X) 1·7, (Y) 1·0	(CD)X	c
h	x	—	x	x	(mG) x	—	c
i	x	—	—	—	(T) x, (ψ) x	TψC	y+c
j	x	—	—	x	(I) x	(CU)I	y
k	x	—	—	x	(F) x	(CU)F	y
m	—	—	—	x	(M) x, (F) x	(UMF)	y
z	—	0·9	1·0	—	(Y) 2·1	(D_2A)G	c

† Compositions were determined by inspection of the intensity of blackening of the radio-autograph and are symbolized by x's. Sometimes the yields of mononucleotides were estimated in a scintillation counter as described in the text and the result expressed per one mole of G, or per one mole of C if G was absent.

‡ The most probable structure of the nucleotides symbolized by the letters is recorded in Table 5 and in the text.

4a gives D (see Table 5) and G with pancreatic ribonuclease.

9a, 11a and 13a give D as 5′-terminal mononucleotides as found by degradation with phosphomonoesterase and phosphodiesterase as described in the text.

34a. This sequence was found by partial alkaline hydrolysis.

b. This sequence was found by partial degradation with a spleen phosphodiesterase as described in the text.

e. UX and DX are clearly separated from one another when 10% formic acid is used for ionophoresis in the second dimension of the two-dimensional fractionation system.

g and z were isolated from a 10% formic acid run as in e above. z is not shown in Fig. 7.

h is probably a mixture.

i. Sequence was not determined; it was presumed to be a breakdown product of 34a.

j and k. I and F are presumed to be 3′-terminal from published observations on the specificity of T_1 ribonuclease for minor nucleotides (Holley *et al.*, 1963; Staehelin, 1964*b*).

(d) *sRNA*

Plates VII and VIII show two-dimensional fractionations of ribonuclease T_1 digests of mixed sRNA from *E. coli* and from yeast. They show considerable differences from ribosomal RNA. There are those spots, identified by letters of the alphabet in Fig. 7 and Table 4, which are not present in ribosomal RNA and are oligonucleotides containing minor bases, and also there are certain of the larger oligonucleotides that are particularly strong, suggesting that there are some sequences of bases common to many of the sRNA molecules.

(i) *Minor components*

Those minor mononucleotides which separated from G and U on ionophoresis at pH 3·5 were further analysed by paper chromatography. The properties of these various components are listed in Table 5. Identification is based on their mobility on ionophoresis at pH 3·5, on their chromatographic behaviour and on the published composition of sRNA. T and ψ are ribothymidylic acid and pseudo-uridylic acid. The most probable structure for X is 2′-*O*-methylguanylyl guanylic acid.

TABLE 5

Ionophoretic and chromatographic properties of minor nucleotides and guanylic acid

Symbol	R_U† on ionophoresis at pH 3·5	R_U† on chromatograms Propan-2-ol–HCl	R_U† on chromatograms Propan-2-ol–NH_3	Probable structure (see text)
Y	1·07	0·93, 1·1	0·63	β-Ureidopropionyl *N*-ribotide
D	1·0	0·90	—	4,5-Dihydrouridylic acid
T	0·98	1·1	1·7	Thymidylic acid
ψ	0·98	0·79	0·63	Pseudouridylic acid
I	0·90	0·54	—	Inosinic acid
X	0·89	0·22	0·20	2′-*O*-Methylguanylyl guanylic acid
mG	0·82	0·64	0·57	—
F	0·76	0·64	—	N^2-Dimethylguanylic acid
G	0·73	0·55	0·50	Guanylic acid
M	0·70	0·72	—	N^1-, or N^2-Methylguanylic acid
A_2	0·44	0·83	2·0	6-Methyladenylic acid
A_1	0·25	0·68	1·5	2-Methyladenylic acid

† R_U is the distance moved relative to uridylic acid.

Dihydrouridylic acid. One of the minor components (Y) appeared to be present in relatively large amounts and did not have the properties of any of the then known minor components. This was identified as a breakdown product of dihydrouridylic acid, the presence of which in sRNA has recently been reported by Madison & Holley (1965). Our identification is briefly reported here since the methods used are somewhat different from those of Madison & Holley and confirm their work.

Figure 8 shows the pH–mobility curve for Y plotted relative to U. The best fit to the experimental values, indicated by ×'s, was the theoretical titration curve of an acidic group of p*K* 4·1. These results suggested that Y had a carboxyl group of p*K* about 4·1.

Compound Y was resolved into two spots on the propan-2-ol–HCl system. The faster and stronger spot was identified as a mixture of ribose phosphate and inorganic

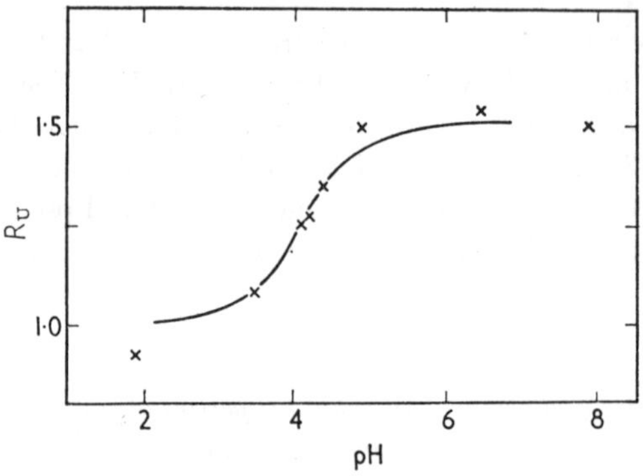

FIG. 8. pH–mobility curve for Y. The mobility is expressed as an R_U value which is the distance of Y from the origin divided by the distance of U. The curve is a theoretical titration curve of an acidic group of p*K* 4·1.

phosphate by ionophoresis at pH 3·5. The slower and weaker spot was shown to re-run in the same position on chromatography and to run in a position between uridylic acid and compound Y on ionophoresis at pH 3·5. The results suggested Y was acid-labile, and gave ribose phosphate as the major product and an unidentified substance as the minor product as soon as the acid developer touched Y on the paper. The purine nucleotides, adenylic acid and guanylic acid, do not cleave under these conditions. Compound Y also degraded slightly to ribose phosphate on ionophoresis at pH 1·9. Again this was interpreted as a partial acid hydrolysis occurring because of the acidity of the buffer used for the ionophoresis.

It was of interest to establish whether Y had a methyl group. An alkaline digest of mixed [^{14}C]methyl-labelled sRNA from *E. coli* was fractionated on the two-dimensional system described in section 2(g). No radioactive spot was seen in the known position of Y. The results suggested that Y was not methylated.

The presence of a carboxyl group in compound Y and its acid lability suggested that Y might be a nucleotide with the purine or pyrimidine ring opened by alkali.

Further evidence that Y was a degradation product was obtained from a study of the digestion products obtained from nucleotide 11a. With alkali, 11a gave A, G and Y, but on digestion with pancreatic ribonuclease it gave AG and a band moving at the same rate as U on ionophoresis at pH 3·5 and slower than U on paper chromatography in propan-2-ol–HCl. This band was referred to as D. It was clear that D was very similar to U, since trinucleotides containing it were always found on the two-dimensional system very close to corresponding trinucleotides in which the D was replaced by U. Also D was not methylated since the D-containing spots from a ribonuclease T_1 digest of [^{14}C]methyl-labelled sRNA were not radioactive.

These results suggested that D was dihydrouridylic acid, which decomposes with alkali to give β-ureidopropionic acid *N*-ribosyl phosphate (Y) (Cohn & Doherty, 1956). To confirm this, [^{32}P]β-ureidopropionic acid *N*-ribosyl phosphate was prepared from [^{32}P]U by the method of Cohn & Doherty (1956) and shown to have the same ionophoretic mobility at pH 4·4 and 3·5, and the same chromatographic behaviour on propan-2-ol–HCl–water or propan-2-ol–NH_3–water as compound Y. Cohn & Doherty also reported that it is labile to acid, which fits with the properties of Y.

Compound X. Compound X is a dinucleotide because equimolar amounts (that is equal counts) of inorganic phosphate and a new band in the expected position on ionophoresis of 2′-*O*-methylguanylyl guanosine are formed after digestion with bacterial alkaline phosphomonoesterase. This new band gave guanosine 5′-phosphate after hydrolysis with snake venom phosphodiesterase. X is methylated, because a compound with its properties was isolated in good yield from an alkaline digest of [^{14}C]methyl-labelled sRNA from *E. coli* (strain CB3). Also free X occurred on the two-dimensional separation system in a position compatible with this structure. When X occurred in oligonucleotides isolated from a ribonuclease T_1 digest of *E. coli* sRNA (see Table 4), it had twice as much radioactive material in it as the other mononucleotides. This suggested there were two phosphate groups in X.

Smith & Dunn (1959) first reported the occurrence of 2′(3)-*O*-methylribose in alkaline-resistant dinucleotides isolated from the RNA of rat liver and plants. Hall (1963) reported that all four of the 2′-*O*-methylribosides were present in soluble RNA from *E. coli*, but that 2′-*O*-methylguanosine was the commonest of the four and isolated it in a yield of 0·1%. We would add to these reports that we have not detected any other alkali-resistant dinucleotides in the products of a T_1 ribonuclease digest of sRNA from *E. coli* or yeast. Presumably they must be present in very much smaller amounts than 2′-*O*-methylguanylyl guanylic acid.

Other minor components. We are less sure of the identity of mG. It is acid-labile and methylated, and may possibly be 4-amino-5(*N*-methyl)-formamido-isocytosine ribotide 1, which is the alkaline degradation product of N^1-methyl guanylic acid.

Other minor mononucleotides have been detected in oligonucleotides isolated from a T_1 digest of [^{14}C]methyl-labelled sRNA and in an alkaline hydrolysis of [^{14}C]methyl-labelled sRNA. These are listed in Table 5 as A_1, which is probably 2-methyladenylic acid, and A_2, which is probably 6-methyladenylic acid (Dunn, Smith & Spahr, 1960). These components are indistinguishable from C and A in their R_U values (the distance moved relative to uridylic acid) on ionophoresis at pH 3·5 and are therefore missed in digests of [^{32}P]sRNA.

F, I and M have been detected in sRNA of yeast and are assigned from their chromatographic properties (Holley *et al.*, 1963), as N^2-dimethylguanylic acid, inosinic acid, and N^1- or N^2-monomethylguanylic acid.

Table 4 and Fig. 7 show only those oligonucleotides which have been separated in a pure state. There are many others that are present as mixtures. For instance, a T_1 ribonuclease digest of [^{14}C]methyl-labelled sRNA from *E. coli* shows that there are methylated nucleotides covering all those areas of the fingerprint where phosphate-labelled nucleotides are found.

sRNA from yeast and *E. coli* show many similarities to one another. Both contain the same dihydrouridylic acid-containing nucleotides and the sequence TψCG. Dihydrouridylic acid occurs in relatively high proportions in sRNA from both sources. It is probably about as common as pseudouridylic acid in *E. coli*. In addition to those T_1 sequences recorded in Table 4, it has been detected in the sequences GGD and AGD from pancreatic ribonuclease digests. TψCG is remarkable in that it gives the strongest radioactive spot excepting guanylic acid in a T_1 digest of sRNA. The sequence GTψCG must be common to many or all of the different amino acid-accepting species of sRNA in both yeast and *E. coli* (Holley, Everett, Madison, Marquisee & Zamir, 1964, *Abstr. 6th Int. Congr. Biochem.*, section 1, p. 9). Similarly, the commonest sequence containing dihydrouridylic acid is DAG. It is stronger than UAG and about one-third as strong as TψCG and would therefore appear to be present in many different sRNA's.

(ii) *Other nucleotides*

A comparison of a T_1 digest of sRNA (Plates VII and VIII) with ribosomal RNA (Plates I and II) shows that some of the larger oligonucleotides are particularly strong, and others, present in ribosomal RNA, are absent from sRNA. The following are some examples seen in sRNA from *E. coli*: spot 64 (Fig. 7) is mainly CUCAG, and the other 11 isomers of the composition $(C_2UA)G$ are absent or in much smaller amounts. Spot 44 is CACCG and is the only nucleotide of the composition $(C_3A)G$ present. Of the (CAU)G isomers, UCAG(25) is the commonest and UACG(26) and CUAG have not been detected. The three isomers $(AC_2)G$ (15, 16, 17) form a triangle in ribosomal RNA, but ACCG(17) is weak in *E. coli* sRNA. Even among the trinucleotides, the amounts of the two isomers are unequal in certain cases and there are nearly twice as much CUG as UCG and twice as much CAG as ACG. We suspect from these results that there are some sequences which are common even in the mixed sRNA. GCUCAG, for example, must be present in more than half the different sRNA's in *E. coli*. It is not common, however, in sRNA from yeast, and this is one of the most obvious distinctions between them.

4. Discussion

The fractionation procedure described in this paper is capable of distinguishing the di-, tri- and most of the tetranucleotides in digests prepared by the action of ribonuclease T_1 or pancreatic ribonuclease. The resolution of larger nucleotides depends on the complexity of the mixture. It should thus prove a useful tool in the study of sequences in smaller RNA's. Most of the information required about a nucleotide mixture may be obtained by running it on the two-dimensional system using 10% formic acid for the DEAE-paper dimension (Fig. 4). The approximate composition of a spot is determined from its position, and the sequence may be determined by partial degradation with spleen phosphodiesterase.

The method may also be used as a fingerprinting technique to characterize RNA's and perhaps to detect small differences as between species and mutants. A comparison of the patterns obtained for the ribonuclease T_1 digests of the two ribosomal RNA components (Plates I to IV) shows definite differences among the tetranucleotides. Thus UAAG(31) is almost completely absent in the 16 s component but present in the 23 s, whereas AUUG(39) is present in the 16 s but absent from the 23 s. The three isomers $(AC_2)G$ (15, 16, 17) form a triangle with each one in about equal strength in the 23 s component; but in the 16 s component CCAG is much stronger than the other two. The general pattern of spots is very different in the pentanucleotide areas, indicating that there must be many different sequences. That this is the type of behaviour that would be expected is illustrated by the following very approximate calculations.

The yield of G from a ribonuclease T_1 digest will depend on the frequency of the sequence GG. If, as a rough approximation, one assumes an RNA to be a random arrangement of the 4 mononucleotides present in equal amounts, the sequence G will occur once every 4 residues and GG once every 16 residues. A sequence such as GAG, which gives rise to the dinucleotide AG, will occur once every 64 residues. Similarly, a trinucleotide will be obtained once from a run of 250 residues, a tetranucleotide once from 1000 residues, and a pentanucleotide once from every 4000 residues. The 16 s and 23 s components contain about 1600 and 3200 residues, respectively; thus the average yield of a tetranucleotide would be 2 to 3 moles and of a pentanucleotide would be less than one. It is thus not surprising that certain tetranucleotides and many of the pentanucleotides are absent from the digests. This illustrates that the method can be an effective fingerprinting method for RNA's of this size, since characteristic patterns are obtained.

The conditions for the DEAE-paper ionophoresis are rather acidic and some depurination of the nucleotides might be expected. That this cannot be very extensive is apparent from the fact that all expected nucleotides are found in the digests. It would seem that the effective pH on an ion-exchange material is not necessarily that of the buffer with which it is washed. Occasionally after a long run at a high voltage gradient the tank may become warm and streaking may be observed behind some of the spots; this may be due to depurination.

We wish to thank Mr J. R. Prescott and Mr R. P. Hughes of British Celanese Ltd. for generous supplies of cellulose acetate, Dr F. Egami for a gift of ribonuclease T_1, Mr R. E. Offord for the preparation of ribonuclease T_1 and Drs D. B. Dunn and J. D. Smith for discussions and helpful advice.

REFERENCES

Apgar, J., Holley, R. W. & Merrill, S. H. (1962). *J. Biol. Chem.* **237**, 796.
Armstrong, A., Hagopian, H., Ingram, V. M., Sjöquist, I. & Sjöquist, J. (1964). *Biochemistry*, **3**, 1194.
Borek, E., Ryan, A. & Rockenbach, J. (1955). *J. Bacteriol*, **69**, 460.
Cohn, W. E. & Doherty, D. G. (1956). *J. Amer. Chem. Soc.* **78**, 2863.
Cunningham, L., Catlin, B. W. & Privat de Garilhe, M. (1956). *J. Amer. Chem. Soc.* **78**, 4642.
Dunn, D. B., Smith, J. D. & Spahr, P. F. (1960). *J. Mol. Biol.* **2**, 113.
Garen, A. & Levinthal, C. (1960). *Biochim. biophys. Acta*, **38**, 470.
Gilbert, W. (1963). *J. Mol. Biol.* **6**, 389.
Gray, W. R. (1964). Ph.D. Dissertation, Cambridge.
Gross, D. (1961). *J. Chromatography*, **5**, 194.
Hall, R. H. (1963). *Biochem. Biophys. Res. Comm.* **12**, 429.
Holley, R. W., Apgar, J., Doctor, B. P., Farrow, J., Marini, M. A. & Merrill, S. H. (1961). *J. Biol. Chem.* **236**, 200.
Holley, R. W., Apgar, J., Everett, G. A., Madison, J. T., Merrill, S. H. & Zamir, A. (1963). *Cold Spr. Harb. Symp. Quant. Biol.* **28**, 117.
Holley, R. W., Madison, J. T. & Zamir, A. (1964). *Biochem. Biophys. Res. Comm.* **17**, 389.
Khorana, H. G. & Vizsolyi, J. P. (1961). *J. Amer. Chem. Soc.* **83**, 675.
Madison, J. T. & Holley, R. W. (1965). *Biochem. Biophys. Res. Comm.* **18**, 153.
Markham, R. & Smith, J. D. (1952). *Biochem. J.* **52**, 552, 558.
Michl, H. (1951). *Monatsh. Chem.* **82**, 489.
Naughton, M. A. & Hagopian, H. (1962). *Analyt. Biochem.* **3**, 276.
Razzell, W. E. & Khorana, H. G. (1959). *J. Biol. Chem.* **234**, 2114.
Razzell, W. E. & Khorana, H. G. (1961). *J. Biol. Chem.* **236**, 1144.
Reddi, K. K. (1959). *Biochim. biophys. Acta*, **36**, 132.
Rushizky, G. W., Bartos, E. M. & Sober, H. A. (1964). *Biochemistry*, **3**, 626.
Rushizky, G. W. & Knight, C. A. (1960). *Virology*, **11**, 236.
Rushizky, G. W. & Sober, H. A. (1962). *J. Biol. Chem.* **237**, 834.
Sanger, F. & Tuppy, H. (1951). *Biochem. J.* **49**, 463.
Sato, K. & Egami, F. (1957). *J. Biochem., Tokyo*, **44**, 753.
Shuster, L., Khorana, H. G. & Heppel, L. A. (1959). *Biochim. biophys. Acta*, **33**, 452.
Smith, J. D. & Dunn, D. B. (1959). *Biochim. biophys. Acta*, **31**, 573.

Staehelin, M. (1961). *Biochim. biophys. Acta*, **49**, 11.
Staehelin, M. (1964*a*). *J. Mol. Biol.* **8**, 470.
Staehelin, M. (1964*b*). *Biochim. biophys. Acta*, **87**, 493.
Stent, G. S. & Brenner, S. (1961). *Proc. Nat. Acad. Sci., Wash.* **47**, 2005.
Tyndall, R. L., Jacobson, K. B. & Teeter, E. (1964). *Biochim. biophys. Acta*, **87**, 335.
Williams, R. B. & Dawson, R. M. C. (1952). *Biochem. J.* **52**, 314.
Wyatt, G. R. (1951). *Biochem. J.* **48**, 584.

J. Mol. Biol. (1967) **23**, 337–353

Nucleotide Sequences from the Low Molecular Weight Ribosomal RNA of *Escherichia coli*

G. G. Brownlee and F. Sanger

Medical Research Council Laboratory of Molecular Biology
Hills Road, Cambridge, England

(*Received 24 August 1966*)

Nucleotides from a ribonuclease digest of ^{32}P-labelled 5 s† ribosomal RNA of *Escherichia coli* have been studied by two-dimensional ionophoresis on modified celluloses. All the nucleotides from ribonuclease T_1 and pancreatic ribonuclease digests were separated and their structure was determined. The yields of the smaller nucleotides were reproducible and greater than 70%. 5 s RNA consists of about 115 nucleotides of which pUG is 5′-terminal and $GCAU_{OH}$ is 3′-terminal.

1. Introduction

A two-dimensional procedure for fractionating radioactive nucleotides has been described (Sanger, Brownlee & Barrell, 1965; Sanger & Brownlee, 1966). Previously the method was applied to the study of complex mixtures of nucleotides from high molecular weight ribosomal RNA and mixed transfer RNA. We were interested to see to what extent the technique could be applied to the determination of nucleotide sequences in RNA, and this paper reports its application to the fractionation and sequential analysis of nucleotides from 5 s ribosomal RNA.

The preparation and some properties of 5 s RNA of *Escherichia coli* were reported by Rosset & Monier (1963) and Rosset, Monier & Julien (1964). Since then, these authors and others have shown that a similar RNA is present in a variety of organisms (Marcot-Queiroz, Julien, Rosset & Monier, 1965; Comb, Sarkar, DeVallet & Pinzino, 1965; Galibert, Larsen, Lelong & Boiron, 1965; Sarkar & Comb, 1965; Hindley, 1966, personal communication; Schleich & Goldstein, 1966; Jackson, 1966). The evidence suggests that it is a low molecular weight RNA which exists with the two species of high molecular weight RNA (15 to 18 s, 23 to 28 s) as an integral structural component of all ribosomes.

In the two-dimensional procedure, the first dimension is done by ionophoresis at pH 3·5 on cellulose acetate and the second dimension by ionophoresis on DEAE-cellulose either at pH 1·9 or in 7% formic acid. The use of 7% formic acid in the second dimension gives better results than the pH 1·9 mixture in the fractionation of nucleotides containing more than one U‡ residue, and is more generally useful.

† The low molecular weight ribosomal RNA was conveniently termed "5 s" RNA by Rosset *et al.* (1964) to distinguish it from "4 s" or transfer RNA. This terminology will be used throughout this paper, although it is not meant to imply that the value of *S* is exactly 5·0.

‡ The letters A, C, G and U refer to adenosine, cytosine, guanine, and uridine 3′-phosphate. Thus a combination such as CCG is the nucleotide CpCpGp (Shuster, Khorana & Heppel, 1959). However, if it is stated that CCG is dephosphorylated, or if it is derived from a digest in which bacterial alkaline phosphatase has been used, it refers to the nucleotide CpCpG.

To improve the resolution of nucleotides rich in U further, we have prepared combined digests with ribonuclease T_1 and alkaline phosphatase. Removal of the 3′-terminal phosphate group of nucleotides lowers their net negative charge, so that all those containing U residues move faster than when they have a 3′-terminal phosphate (Fig. 1). However, those nucleotides that do not contain a U residue move more slowly on the 7% formic acid system after dephosphorylation than before. This is caused by the complex nature of the fractionation, which is a combination of ionophoresis and ion-exchange.

Another advantage of using dephosphorylated nucleotides is that they are susceptible to digestion with snake venom phosphodiesterase, and it was possible to use partial digestion with this enzyme (Holley, Madison & Zamir, 1964) for sequence determination from the 3′-end. This usually gave clearer results for larger nucleotides than the use of spleen phosphodiesterase, since it is less specific for individual mononucleotides.

2. Materials and Methods

Most of the materials and methods, including the two-dimensional fractional procedure, used in this work were described in our previous papers (Sanger *et al.*, 1965; Sanger & Brownlee, 1967). Only new or significant modifications to methods previously described are included here.

(a) *New materials*

Bacterial alkaline phosphatase was prepared by Dr C. Milstein of this laboratory and was a highly purified preparation made by the method of Pigretti & Milstein (1965). Ribonuclease T_1 was bought from Sankyo Ltd., Tokyo, Japan.

(b) *Preparation of* ^{32}*P-labelled 5 s RNA*

^{32}P-labelled *E. coli* (strain MRE600) was prepared as follows: 5 ml. of an overnight culture, grown in the low-phosphate medium of Garen & Levinthal (1960) under conditions where glucose was limiting (0·02%), was added to 50 ml. of the above medium containing 0·2% glucose and 10 mc carrier-free [^{32}P]phosphate (Radiochemical Centre, England, PBS1). A parallel control flask was also inoculated, and both were incubated at 37°C with efficient aeration. Under these conditions, growth started without a lag and the ^{32}P-labelled cells were harvested while still in the exponential growth phase, when the control culture had reached an optical density of 0·5 at 550 mμ. Sometimes a control culture was not set up and growth was stopped when about 90% of the [^{32}P]phosphate had been incorporated into the cells. Growth took 2 to 3 hr, with a doubling time of about 35 min. Two methods have been used to purify 5 s RNA from the cells harvested by centrifugation.

Method 1 was a modification of that of Rosset *et al.* (1964). The cells were ground with about twice their weight of alumina in a centrifuge tube at 0°C and then extracted twice with 1 ml. of buffer (0·01 M-Tris acetate (pH 7·4), 0·01 M-magnesium acetate). After removing the debris by centrifugation at 8000 rev./min for 20 min, the combined supernatant fractions were centrifuged for 90 min at 39,000 rev./min in the Spinco model L ultracentrifuge. The ribosomes were then washed in the following way. The pellet, which was just visible, was rinsed twice and then homogenized with a small glass rod in the same 2-ml. centrifuge tube in 1 ml. of buffer containing a low magnesium concentration (0·01 M-Tris acetate (pH 7·4), 10^{-4} M-magnesium acetate). This took about 15 min at 0°C. The undissolved particles and debris were removed by centrifugation at 5000 rev./min for 5 min in the Spinco model L, and the supernatant fractions carefully layered on to 1 to 2 ml. of 5% sucrose in 0·01 M-Tris acetate (pH 7·4), 10^{-4} M-magnesium acetate and centrifuged for 6 hr at 39,000 rev./min in the SW39 rotor at 0°C. The supernatant fraction was discarded and the ribosomal pellet, which was not visible, was rinsed with water. This was a convenient stage to freeze the washed ribosomes to −20°C. The material left in the tube was dissolved in 0·1 ml. of 0·5% sodium dodecyl sulphate and layered on to 4 ml. of a 5 to 20% sucrose gradient containing 0·5% sodium dodecyl sulphate, 0·1 M-lithium chloride and 0·1 M-Tris acetate (pH 8·0) and centrifuged in the SW39 rotor for 6 hr at 39,000 rev./min at 10°C. Seven-drop fractions were collected by piercing the side of the tube half-way down with a fine needle and the 5 s peak was located by counting 0·1-μl. portions. A second smaller peak of lower *S* value which was not hydrolysed by either RNase or DNase was also present. The peak tubes were pooled and shaken vigorously with an equal volume of phenol for 1 hr at room temperature. After separating the layers and re-extracting the phenol layer with half its volume of 0·1 M-sodium chloride, the combined aqueous layers were precipitated with 2 vol. of ethanol. After standing overnight at −20°C, the precipitate was collected by centrifugation, dried in a vacuum desiccator and dissolved in 0·1 ml. water. It was stored at −20°C. The yield was about 1% of the radioactive material added, with a specific activity of about 1 μc (10^6 counts/min)/μg RNA.

Method 2. The RNA extracted by vigorously shaking the cells with phenol at 4°C for 1 hr was fractionated by disc electrophoresis in polyacrylamide gels (Richards, Coll & Gratzer, 1965). This method of preparation of 5 s RNA was developed and described by Dr J. Hindley (manuscript in preparation,) and is therefore described here only in outline. The 5 s RNA moves as a sharp band on disc electrophoresis, more slowly than the more diffuse transfer RNA region but well separated from it. Apart from transfer RNA, it is the next major component in RNA prepared by cold phenol extraction of cells. After electrophoresis, staining with acridine orange and destaining in the cold room, that part of the gel with the 5 s RNA band was cut out and homogenized with enough solid urea to make a thick paste and the RNA extracted by shaking for 1 hr at room temperature with equal volumes of an alkaline buffer (0·1 M-Tris base, 0·01 M-EDTA, 1·0 M-sodium chloride) and phenol. Although most of the orange colour was extracted from the gel into the phenol layer, the proportion of RNA extracted into the aqueous layer was variable but always more than half. That remaining in the gel was estimated, in one experiment, as 49%, which was the soluble radioactive material after incubating the gel in 20% hydrogen peroxide at 37°C for 24 hr. The final yield of 5 s RNA recovered from the aqueous layer by precipitation with 2 vol. of ethanol was about 1·5% of the initial radioactive material.

Either method can give an equally good preparation but the second is quicker and easier to carry out.

(c) *Enzymic digestion*

The enzymic digestions are conveniently done as follows. A sample is dried in a small silicone-treated tube and then dissolved in about 2 μl. of a solution of an appropriate amount (Sanger *et al.*, 1965) of carrier transfer RNA (General Biochemicals, U.S.A.), enzyme and buffer. It is then incubated in the tip of a drawn-out capillary tube and loaded on cellulose acetate for fractionation in the first dimension of the two-dimensional system.

For the combined ribonuclease T_1 and phosphatase digestion, incubation was for 60 min at 37°C using an enzyme to substrate (5 s-RNA) ratio of 1 to 20 for ribonuclease T_1 and 1 to 5 for bacterial alkaline phosphatase in 0·01 M-Tris (pH 8·0).

(d) *Structure of the nucleotides*

The sequence of an oligonucleotide was found by enzymic degradation with nucleases followed by fractionation by paper ionophoresis and location of the products by radioautography. The procedures described below were all small-scale methods and incubation was done in capillary tubes in a volume of 5 to 10 μl. The products of a digest were identified either from their position on ionophoresis, or were further analysed usually by alkaline hydrolysis.

(i) *Determination of 5′-terminal residue*

Dephosphorylated nucleotides from combined ribonuclease T_1 and alkaline phosphatase digests were digested with 0·1 mg of snake venom phosphodiesterase/ml. in 0·02 M-Tris chloride (pH 8·9) and 0·01 M-magnesium chloride for 1 hr at 37°C, and the 5′-mononucleotides fractionated by paper ionophoresis at pH 3·5. The 5′-terminal residue is liberated as a nucleoside, and is thus identified as that residue which is missing from a venom phosphodiesterase digest but present in an alkaline digest.

(ii) *Secondary splits*

Nucleotides from T_1 ribonuclease digests were incubated with 0·1 mg of ribonuclease T_1/ml. in 0·01 M-Tris chloride (pH 7·5), 0·001 M-EDTA for 30 min at 37°C and fractionated by ionophoresis on DEAE-paper at pH 1·9 until the blue marker, xylene cyanol F.F. (see Sanger *et al.*, 1965) had reached the top of the rack.

(iii) *Partial digestion with snake venom phosphodiesterase*

In preliminary experiments, the dephosphorylated nucleotide was incubated for 30 min at 37°C with 0·01 mg of enzyme/ml. in 0·02 M-Tris chloride (pH 8·5), 0·01 M-magnesium chloride and fractionated by ionophoresis on DEAE-paper at pH 3·5 (0·5% pyridine–5% acetic acid, v/v). However, the conditions of digestion varied according to the size of the nucleotide, so that 60 min incubation was needed for a decanucleotide whereas only 30 min was needed for a hexanucleotide. Sometimes it was found convenient to digest the sample for 30 min and load half on the paper for ionophoresis. The remainder was incubated for a further 30 min and loaded next to the first sample. A control, to which no enzyme had been added, was run beside the sample.

3. Results

(a) *Ribonuclease T_1 digests*

Plate I shows a two-dimensional fractionation of a ribonuclease T_1 digest of 5 S RNA and a map which shows the system of numbering of the spots. This fractionation includes all the T_1 nucleotides except spot 19, which is the 3′-terminal nucleotide CAU_{OH}. Plate II shows a two-dimensional fractionation of a combined ribonuclease T_1 and alkaline phosphatase digest and a map giving the position of dephosphorylated nucleotides and their sequences. CG, CCG and AG are not included in this fingerprint.

Preliminary information about the composition of the nucleotides can be deduced from their position on the "fingerprint" by reference to Fig. 1. Figure 1(a) shows the relationship between the composition of a nucleotide from a ribonuclease T_1 digest

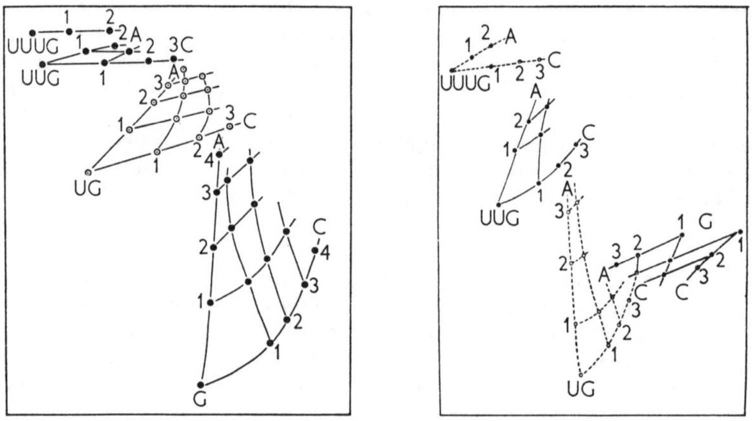

(a) (b)

Fig. 1. Diagram illustrating the relationship between the composition of a nucleotide and its position on the two-dimensional fractionation system using 7% formic acid in the second dimension: (a) for nucleotides from ribonuclease T_1 digests; (b) for dephosphorylated nucleotides from a combined ribonuclease T_1 and alkaline phosphatase digest.

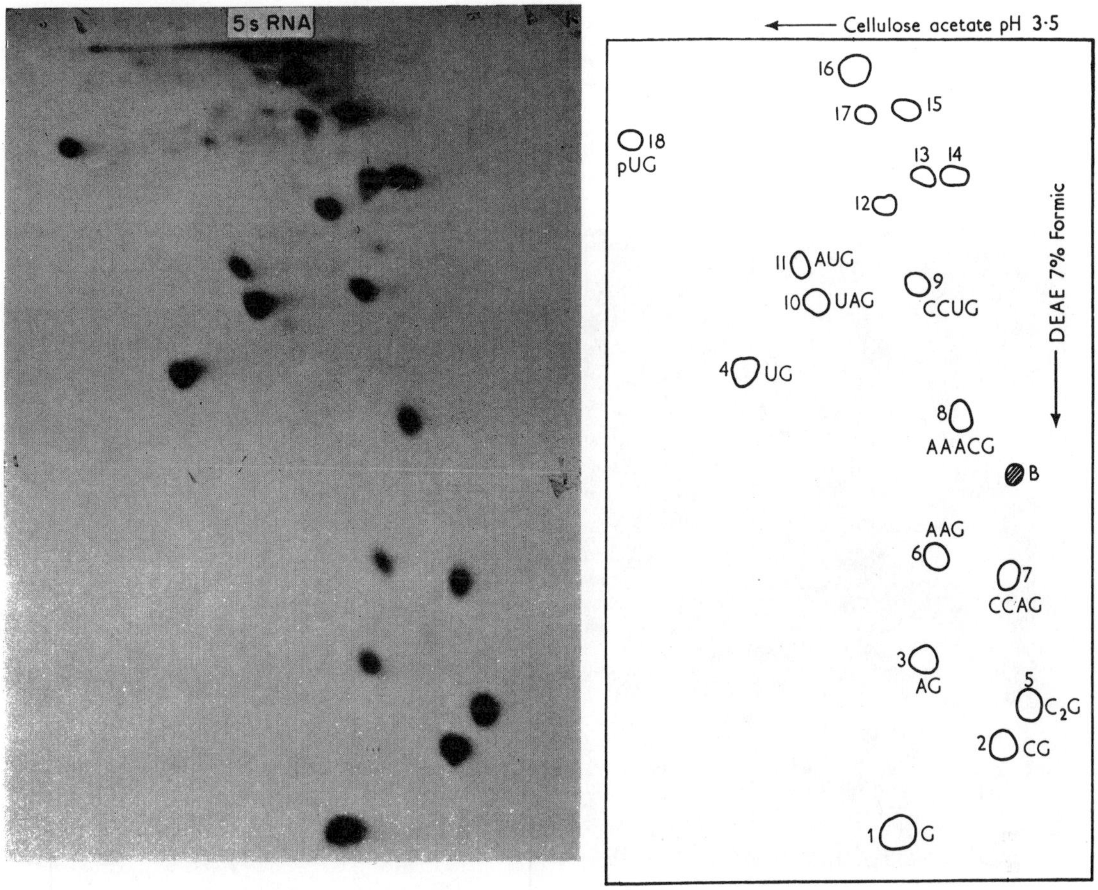

(a) (b)

PLATE I. A two-dimensional fractionation of a ribonuclease T_1 digest of 5 s RNA.

(a) Radioautograph and (b) a map of the major spots with their identifying number and the sequence of the smaller nucleotides. B is the blue marker. Spot 19, CAU_{OH}, was not included in this fingerprint—it is the slowest nucleotide in the first dimension but the fastest nucleotide in the second dimension.

[facing p. 340

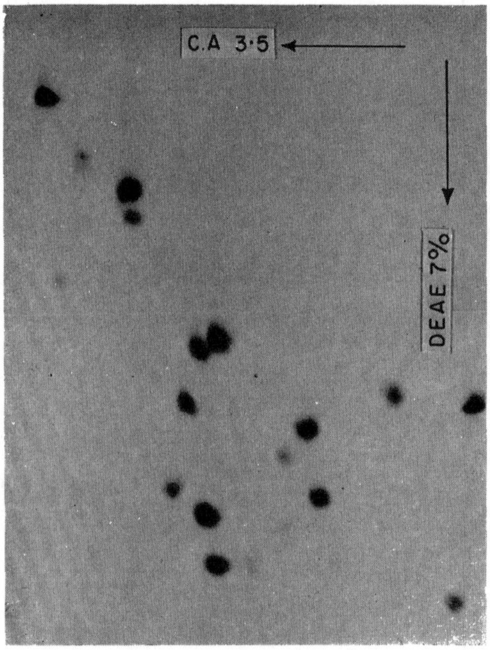

(a)

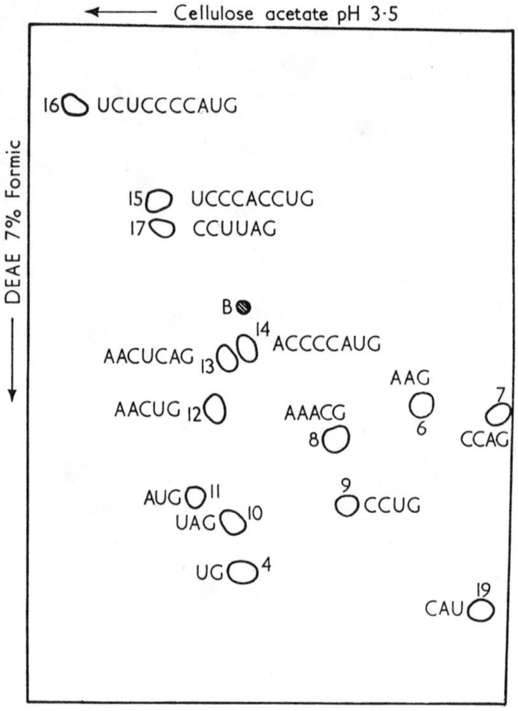

(b)

Plate II. A two-dimensional fractionation of a combined ribonuclease T_1 and bacterial alkaline phosphatase digest of 5 s RNA.

(a) Radioautograph and (b) a map of the major spots with their identifying numbers and sequence. CG, CCG and AG were not included in this fingerprint. Their position can be seen in Fig. 1(b). B is the blue marker.

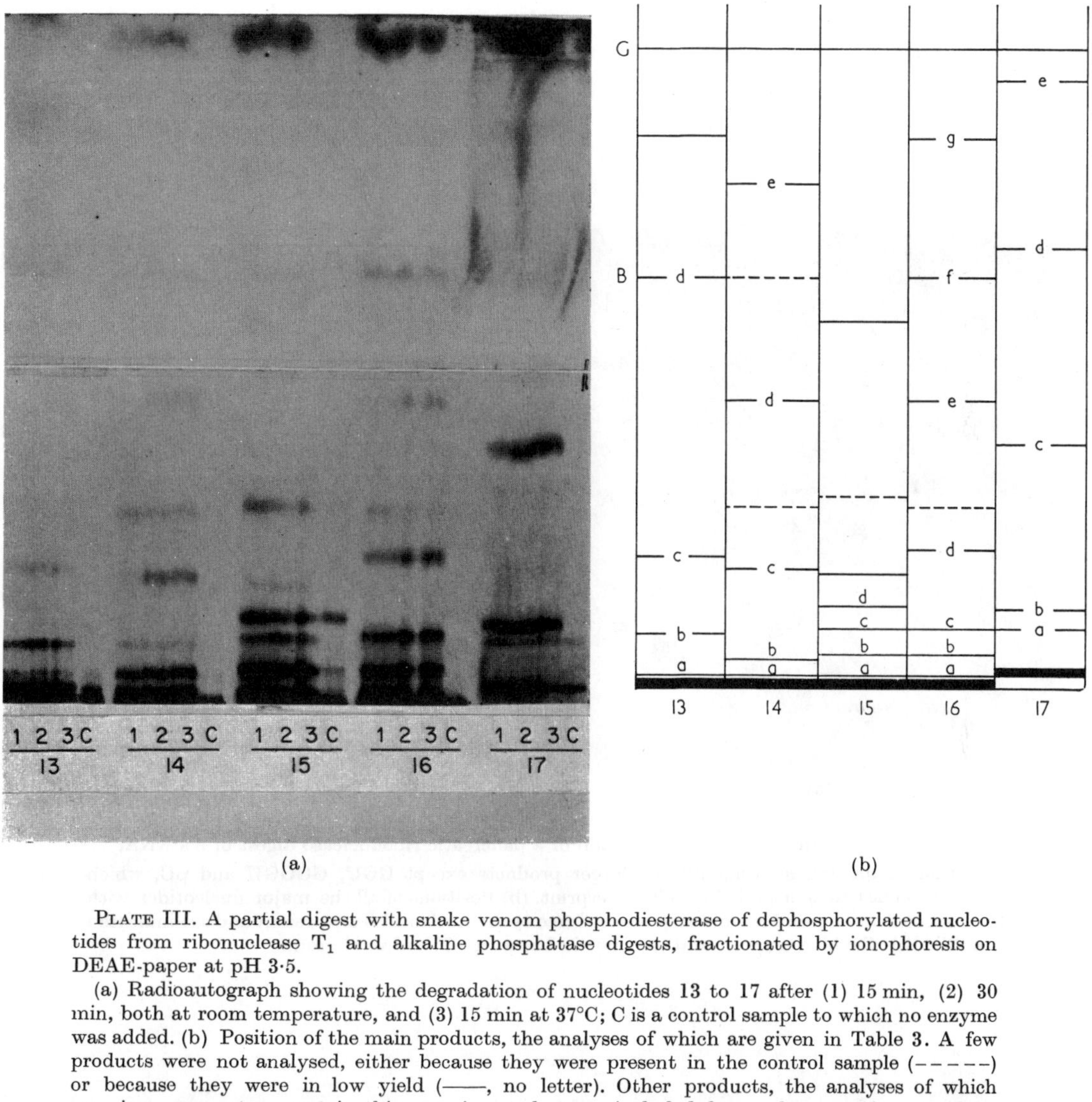

(a) (b)

PLATE III. A partial digest with snake venom phosphodiesterase of dephosphorylated nucleotides from ribonuclease T_1 and alkaline phosphatase digests, fractionated by ionophoresis on DEAE-paper at pH 3·5.

(a) Radioautograph showing the degradation of nucleotides 13 to 17 after (1) 15 min, (2) 30 min, both at room temperature, and (3) 15 min at 37°C; C is a control sample to which no enzyme was added. (b) Position of the main products, the analyses of which are given in Table 3. A few products were not analysed, either because they were present in the control sample (- - - - -) or because they were in low yield (——, no letter). Other products, the analyses of which are given, are not present in this experiment, but are included from other experiments where longer digestion times were used. B is the blue marker and the thick line is unchanged material. G is free guanosine 5′-phosphate.

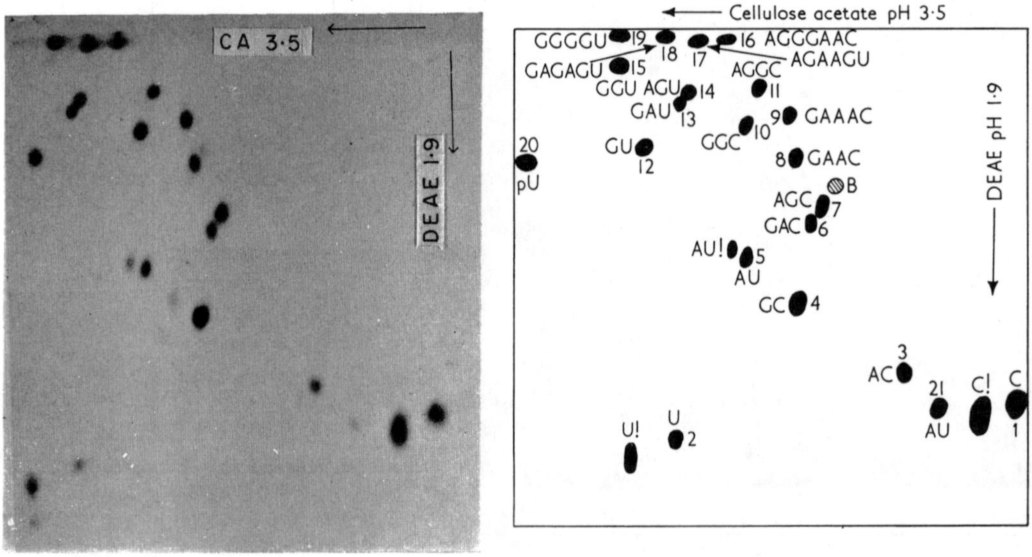

(a) (b)

Plate IV. A two-dimensional fractionation of a pancreatic ribonuclease digest of 5 s RNA.

(a) Radioautograph showing all the larger products except GGU, GGGGU and pU, which have run too fast to be included in this fingerprint. (b) Positions of all the major nucleotides with their indentification numbers and sequences. Some cyclic nucleotides are present and are indicated by a "!"; B is the blue marker, and spot no. 21 has a free 3′-hydroxyl group.

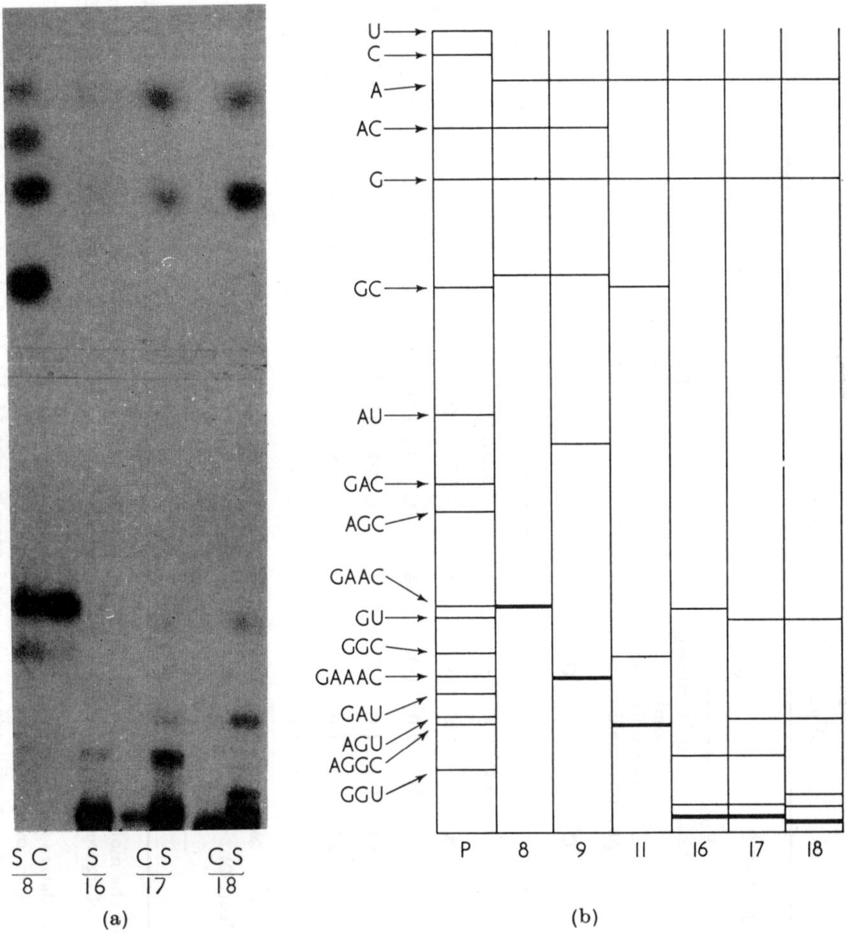

PLATE V. A partial digest with spleen phosphodiesterase of certain pancreatic nucleotide fractionated by ionophoresis on DEAE-paper at pH 1·9.

(a) Radioautograph: S is the sample and C the control. (b) Position of the partial products, with the unchanged material shown by a thick line. P is a pancreatic ribonuclease digest of 5 s RNA run as a marker.

TABLE 1

Nucleotides from ribonuclease T_1 digests

Spot no. (Plates I & II)	Composition† Alkaline digest of nucleotides from ribonuclease T_1 digests				Venom phosphodiesterase digest of nucleotides from combined ribonuclease T_1 and phosphatase digests				5′-Terminal residue deduced	Pancreatic ribonuclease products‡	Second ribonuclease T_1 products (Table 2)	Structure deduced
	C	A	G	U	C	A	G	U				
8	1·1	2·9	1·0	—	x	xx	x	—	A	AAAC, G		AAACG
12	1·1	2·1	1·0	1·0	x	x	x	x	A	AAC, U, G		AACUG
13	2·1	3·0	1·0	1·1	2·2	1·8	1·0	1·0	A	C (1·3), AAC (1·0), AG (1·0), U (1·2)	(C_2A_2U), AG	AAC(UC)AG
14	3·9	2·2	1·0	1·1	3·4	1·2	1·0	1·0	A	C (3·1), AC (1·0), G (1·1), AU (0·9)	$(A\underline{\underline{C}})$, AUG	ACCCCAUG
15	4·6	1·1	1·0	2·0	4·7	1·0	1·0	1·0	U	C (3·8), AC (1·0), G (1·3), U (0·9)	(C_3U), ACCUG	UCCCACCUG
16	4·8	1·0	1·0	3·0	5·1	0·9	1·0	1·9	U	C (4·9), G (1·0), AU (1·0), U (2·0)	$(C_{4-5}U_2)$, AUG	$U(C_5U)AUG$
17	2·0	1·0	1·0	1·9	x	x	x	xx	C	C (2·5), AG (1·0), U (2·4)	(C_2U_2), AG	$C(CU_2)AG$
18	—	—	1·0	—pUp§(0·9)		—				—	—	pUG
19	x	x	—	—‖	—	x	—	x	C	—	—	CAU_{OH}

† The compositions are expressed as the molar yield of a given nucleotide relative to one mole of G and were estimated in a scintillation counter. Where the results are given as "x's" instead of figures, the composition was estimated by visual inspection of the radioautograph. The figures are an average of at least three separate experiments.

‡ The yields are expressed as moles relative to 1 mole of the dinucleotide released in each case.

§ Its mobility relative to U = 1 on paper ionophoresis at pH 3·5 is 1·64.

‖ Its composition, found by a complete digest with venom phosphodiesterase, was (A, U), thus showing it had a 3′-terminal U residue with a free 3′-hydroxyl group.

and its position, and was prepared by analysing the products of digestion of 16 s and 23 s ribosomal RNA and joining up the points with increasing numbers of C and A residues (Sanger *et al.*, 1965). Figure 1(b) shows a similar diagram for the dephosphorylated nucleotides derived from a combined ribonuclease T_1 and alkaline phosphatase digest.

The sequence of the smaller nucleotides was worked out as previously described. The sequence of the larger nucleotides was determined in two stages. Table 1 summarizes the results of the first stage, which includes the identification of the 5′-terminal residue and the products of pancreatic ribonuclease digests and of the secondary splits. The secondary splits were obtained by re-incubation of the nucleotides with a high concentration of ribonuclease T_1, but it is uncertain whether the effect was always due to the ribonuclease T_1 or to another contaminating nuclease. The results were sufficient to deduce the sequence of all the nucleotides except no. 13, 16 and 17. The determination of the sequence by partial digestion with spleen phosphodiesterase, which had proved useful in the case of smaller nucleotides isolated from 16 s and 23 s ribosomal RNA (Sanger *et al.*, 1965), gave ambiguous results when applied to the larger nucleotides 13, 14, 15, 16 and 17 from 5 s RNA, probably because secondary splits were occurring. However, this method confirmed the sequence of no. 8 and 12 and established A as 5′-terminal in no. 13 and 14.

(i) *Secondary splits*

Figure 2 shows the secondary splits of the larger nucleotides isolated from ribonuclease T_1 digests on further incubation with ribonuclease T_1. The analyses of the

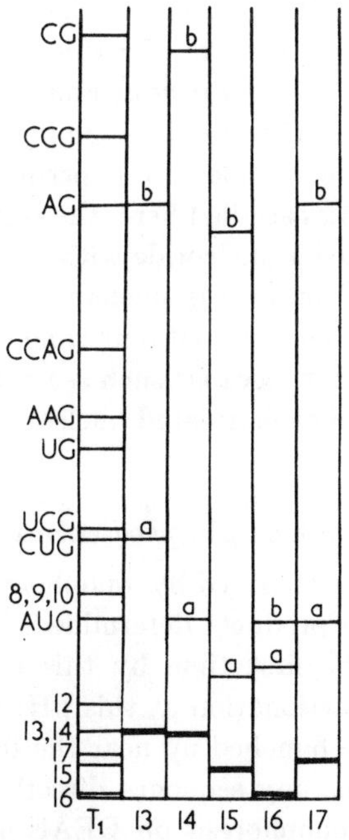

FIG. 2. Diagram showing the secondary splits of some ribonuclease T_1 nucleotides (see Plate I) fractionated by ionophoresis on DEAE-paper at pH 1·9. T_1 is a marker ribonuclease T_1 digest of 5 s RNA with the products identified by number (see Plate I) or sequence. The thick line shows the position of unchanged nucleotides.

TABLE 2

Products of secondary splits of nucleotides from ribonuclease T_1 digests

Band no. (Fig. 2)	Composition†				Probable structure
	C	A	G	U	
13a§	1·9	1·8	—	1·0	(C_2A_2U)
b	—	x	x	—	AG
14a	—	x	x	x	AUG
b	3·8	1·0	—	—	(C_4A)
15a	2·0	1·0	0·7	0·8	ACCUG‡
b	3·4	—	—	1·0	(C_3U)
16a‖	4·3	—	—	2·0	($C_{4-5}U_2$)
b	—	1·0	1·0	1·0	AUG
17a	2·0	—	—	2·0	(C_2U_2)
b	—	x	x	—	AG

† The yields were expressed relative to the residue shown in 1·0 or 2·0 amounts.
‡ Found by partial digestion with spleen phosphodiesterase.
§ Average of three separate experiments.
‖ Average of two separate experiments.

products of the split, labelled a and b, are collected in Table 2. Information from both the composition and from the position of the nucleotide on ionophoresis was used in determining the probable structure. For this purpose it is useful to remember that a U residue behaves much like a G residue and thus (C_3U) is in much the same position as (C_3G) would be, and (C_2U_2) is in a similar region to (C_2U)G. The extent of secondary splits varied from experiment to experiment and also depended on the nucleotide in question. In all cases studied here, the split occurred on the 5′-terminal side of an A residue and liberated a nucleotide with a 3′-terminal phosphate. Usually less than a quarter of the nucleotide was unchanged. It seemed possible that the breakdown was caused by a nuclease contaminating the sample rather than ribonuclease T_1 or a contaminant of it, because such secondary splits were also detected, but to a lesser extent, when dephosphorylated nucleotides were incubated with snake venom phosphodiesterase.

(ii) *Partial digestion with snake venom phosphodiesterase*

Initially the partial digest was analysed by ionophoresis on DEAE-paper at pH 1·9 and the composition of all the products determined. The sequence of spots 13, 16 and 17 was determined for the first time by this method. However, there were certain disadvantages in the fractionation at this pH: the products had a relatively high mobility and tended to be bunched up near the free mononucleotides, and also it was not possible to determine any sequence directly from the mobility values of the products. An analysis by ionophoresis on DEAE-paper at pH 3·5 proved to be more useful in this respect.

Plate III(a) shows a partial venom digest of spots 13 to 17 fractionated by ionophoresis on DEAE-paper at pH 3·5, and Plate III(b) a line diagram of the pro-

Table 3

Nucleotides from partial digests with snake venom phosphodiesterase fractionated by ionophoresis on DEAE-paper at pH 3·5

Band no. (Plate III)	Composition† Alkaline hydrolysis	Composition† Venom phosphodiesterase	Value of M (see text)	Probable structure
13a	$\underline{C}$ $\underline{A}$ U		2·9	A(AC)UCA
b	C $\underline{A}$ U	$\underline{C}$ A U	0·9	A(AC)UC
c	C $\underline{A}$		1·7	A(AC)U
d	—		—	A(AC)
14a	$\underline{C}$ A		1·9	(ACC)CCAU
b	$\underline{\underline{C}}$ A		2·7	(ACC)CCA
c	$\underline{\underline{C}}$ A		0·9	(ACC)CC
d	$\underline{C}$ A		0·6	(ACC)C
e	—		—	(ACC)
15a	$\underline{\underline{C}}$ A U		1·7	(UCCCA)CCU
b	$\underline{\underline{C}}$ A U		0·9	(UCCCA)CC
c	$\underline{\underline{C}}$ A U		—	(UCCCA)C
d§	$\underline{\underline{C}}$ U		—	(UCCC)p‡
16a	$\underline{\underline{C}}$ U		2·5	(UC)UCCCCA
b	$\underline{C}$ U		0·9	(UC)UCCCC
c	$\underline{C}$ U		0·9	(UC)UCCC
d	C U		0·9	(UC)UCC
e	C $\underline{U}$		0·8	(UC)UC
f§	A U		—	AUG
g	C U		—	(UC)U
17a§	C U		—	CCUUp‡
b	C U		2·1	CCUUA
c	$\underline{C}$ U		1·4	CCUU
d§	A		—	AG
e	C & $\underline{G}$	C U & G	—	CCU

† Compositions were estimated by visual inspection of the radioautograph and are expressed relative to the nucleotide present in lowest amounts. If a nucleotide is once underlined there are two residues, if twice underlined either three or more. It was not possible to estimate with any degree of accuracy ratios greater than 3 to 1.

‡ These nucleotides have a 3′-terminal phosphate; others are dephosphorylated.

§ These nucleotides were present in the control sample, to which no venom phosphodiesterase had been added, and were the same as the products of secondary splits recorded in Table 2.

ducts obtained. The analysis of the degradation products is given in Table 3. In the interpretation of the composition of the various products of digestion by alkaline hydrolysis, it is important to remember that the 3′-terminal residue is released as a nucleoside and is therefore not detected. As an illustration, consider the degradation products of spot 16 in Table 3 which are labelled a to g. Assuming that the nucleotides fractionate in order of size, where the smallest ones move fastest, then 16g (C,U) is the trinucleotide (UC)X, where X is unknown. 16f (A,U) is obviously unrelated to the degradation product already considered and is AUG, which is liberated by secondary splitting (see Fig. 2 and Table 2). 16e (C,$\underline{U}$) differs from 16g in having an extra U residue and is therefore (UC)UX. Therefore 16g must be (UC)U. Similarly 16d (C,U) is probably (C_2U_2), which differs from 16e by a C residue and is therefore (UC)UCX. 16a, b and c looked as though they differed from 16d by having more C's. The sequence of 16a is therefore (UC)UCCCCX. With the additional information in Table 1, the structure of 16 can be written as UCUCCCCAUG.

Our previous experience in the analysis of partial digests of nucleotides with spleen phosphodiesterase on DEAE pH 1·9 ionophoresis showed that it was possible to deduce the structure of a nucleotide directly from the mobility of its degradation products. A value of *M* was defined and its value for any one split was characteristic of the nucleotide released. In a similar approach to this, the values of *M* for the splits shown in Plate III on DEAE pH 3·5 ionophoresis have been collected in Table 3. The value of *M* is defined (as before) as equal to x/y, where y is the distance of any nucleotide from the origin of the DEAE-paper and x is the distance between it and its first degradation product.

For the limited number of nucleotides studied, it has been found that the values of *M* lie within the limits shown in Table 4, with the restriction that values of *M* for the splits where the product of the split is faster than the blue marker are not included.

TABLE 4

Values of M *on DEAE-paper ionophoresis (pH 3·5) for nucleotides from combined ribonuclease* T_1 *and alkaline phosphatase digests*†

3′-Terminal residue	Range of *M*-values	No. of observations
C	0·6 to 1·2	13
A	2·1 to 2·9	4
U	1·7 to 1·9	3
G	2·6 to 4·4	6

† The values of *M* for splits in which the product of the degradation is faster than the blue marker are not included.

There is no overlapping between the values for C, A and U, although the values for A and U are rather close to one another and may possibly overlap. Thus in no. 13 four bands, a to d, were recognized, of which an analysis was obtained on a, b and c. 13c (C,$\underline{A}$) was related to 13b (C,$\underline{A}$,U) by the loss of a U; 13b was related to 13a (C,$\underline{A}$,U) by the loss of a C, and 13a was related to unchanged material (C_2A_3U) by the loss of an A. Thus the 5′ nucleotide must be released in the following order –A,C,U– and the sequence of 13 must be (AAC)UCAG. This same information can be read off from the values of M for the first three splits, all of which occur proximal to the blue marker. The first is an A (M=2·9), the second a C (M=0·9), and the third a U (M=1·7).

For spot 14, the values of M suggested that nucleotides were released in the order –U, A, C–, and with the additional information from the composition of the products, the structure (ACC)CCAUG was deduced.

For spot 15, the values of M suggested that nucleotides were released in the order –U, C–, and knowing the composition of the products the structure (UCCCA)CCUG was deduced.

Spot 17 was particularly labile and more than half was degraded in the untreated control sample to give the same products as were described under secondary splits. Bands 17b, c and e were not present in the control. 17b (C,U), which by its position is likely to be (C_2U_2), was related to unchanged material (C_2U_2A) by the loss of an A and to 17c (C̲,U) by the addition of a U. 17e (C), which runs with free G, is related to 17c by the loss of another U. The 5′-mononucleotides are therefore released in the order –A,U,U– and the sequence is therefore CCUUAG. A complete venom phosphodiesterase digest of 17e gave (CU) and confirmed its sequence CCU.

One advantage of this method of sequence analysis compared with analysis by partial digests with spleen phosphodiesterase is that it is possible to get all possible degradation products. This must mean that the different 5′-mononucleotides are released at fairly similar rates by venom phosphodiesterase, whereas with spleen phosphodiesterase C residues are released much more slowly than the others (Sanger *et al.*, 1965). Table 5 is an attempt to summarize the evidence for the base sequence of the larger nucleotides from ribonuclease T_1 digests. This type of notation is derived from that used in sequence analysis of peptides and is a useful way of emphasizing the variety of methods which can be used. Nevertheless, it is impossible to summarize completely all the evidence for a particular structure, which can be checked by the reader only by careful inspection of that part of Tables 1 to 4 which relates to the nucleotide in question.

(b) *Pancreatic ribonuclease digest*

Plate IV shows the fractionation of a pancreatic ribonuclease digest of 5 s RNA on the two-dimensional system, using pH 1·9 in the second dimension. Like the ribonuclease T_1 digests, there is no difficulty in separating all the products in a pure form even in a fairly short separation. The results of the sequence analysis are given in Plate IV(b).

All the nucleotides except 16, 17, 18, 19, 20 and 21 have previously been found and analysed in digests of high molecular weight ribosomal RNA, and details of the method of sequence analysis are not repeated here. Table 6 summarizes the initial analysis by alkali and digestion with ribonuclease T_1. As G, AG, AAG did not separate very well on the system (paper ionophoresis at pH 3·5) used to analyse the products of ribonuclease T_1 digestion, this area was eluted and run on DEAE-paper ionophoresis at pH 1·9, when they clearly separated according to size. A single analysis suggested that spot 19 might be G_3U, but its position indicated that it was probably G_4U. This illustrates that the position of a nucleotide on the fingerprint is a more reliable indication of its composition than quantitative analysis.

TABLE 5

Summary of the evidence for the base sequence of the larger nucleotides from ribonuclease T_1 digests

Spot no. (Plates I & II)	Sequence†
13	A A C U C A G
14	A C C C C A U G
15	U C C C A C C U G
16	U C U C C C C A U G
17	C C U U A G

† The symbols and the method of digestion from which the sequence information was derived are as follows:

(a) ⟶, Partial digestion with spleen phosphodiesterase; (b) ⟵, partial digestion with snake venom phosphodiesterase; (c) S↓, a secondary split at the point indicated; (d) —— (S), a fragment of known composition from a secondary split; (e) - - ⟶ (S), a fragment from a secondary split the sequence of which is known from its mobility on ionophoresis; (f) - - ⟶ (P), a pancreatic ribonuclease fragment of known sequence; (g) —, the 5′-terminal nucleotide as found from the difference in the composition by alkaline hydrolysis and hydrolysis with snake venom phosphodiesterase; (h) - - ⟶ (Sp), a fragment from a partial digestion with spleen phosphodiesterase the sequence of which is known from its position on ionophoresis.

TABLE 6

Nucleotides from pancreatic ribonuclease digest

Spot no. (Plate IV)	Composition†	Products from ribonuclease T_1‡	Structure deduced
16	$\underline{\underline{A}}\ \underline{\underline{G}}\ C$	AAC, AG, $\underline{G}$	(AG,G,G)AAC
17	$\underline{\underline{A}}\ \underline{G}\ U$	AG, AAG, U	(AG,AAG)U
18	$\underline{A}\ \underline{\underline{G}}\ U$	$\underline{AG}$, G, U	(AG,AG,G)U
19	$\underline{\underline{G}}\ U$		$G_{3-4}U$
20	pU		pU
21	A		AU_{OH}

† See footnotes to Table 3.
‡ The yield of di- or trinucleotides is expressed as a molar yield. Thus in 16, although the intensity of blackening of the radioautograph by G and AG is similar, the molar yield of G is twice AG.

Degradation with spleen phosphodiesterase (Plate V) gave the sequences of 16, 17, 18 and 19. The values of M for a degradation step where A is removed are in the range 0·4 to 1·1, and for G, 1·2 to 3·1. Thus for no. 16, the values of M for the various steps are 0·8(A), 2·3(G) and 2·0(G), with the last step giving rise to a band in the position of GAAC. Number 16 is therefore AGGGAAC. Number 17 analyses as AGAAGU and no. 18 as GAGAGU. The degradation of no. 19 is not shown, but the values of M for the first two steps are 2·7 and 3·0, the latter step giving a band in the GGU position. Number 19 is therefore GGGGU. These results were in agreement with the results of Table 6.

The ease of analysis by degradation with spleen phosphodiesterase made it possible to carry out the complete sequence analysis of the pancreatic ribonuclease products much more quickly than the analysis of nucleotides from ribonuclease T_1 digests.

(c) *Relative molar yields of nucleotides*

Table 7 shows some estimates of the relative molar yields of the nucleotides from both ribonuclease T_1 and pancreatic ribonuclease digests of 5 s RNA. These were calculated as follows. After fractionation, the radioactivity of each nucleotide was estimated in a liquid-scintillation system by direct measurement of the counts/minute in the area of paper which contained the nucleotide. The counts/minute divided by the number of phosphate residues in the nucleotide in question was expressed relative to AG = 2·0 in ribonuclease T_1 digests and to AU = 2·0 in pancreatic ribonuclease digests. These dinucleotides were taken as a standard for reference, because there was evidence from the sequence information present in the other digest that this should represent the actual molar yield. Thus AG in a ribonuclease T_1 digest derives from a GAG sequence in the nucleic acid. The only two GAG sequences appear in the same pancreatic nucleotide, GAGAGU. Similarly, the 2 AU residues in a pancreatic ribonuclease digest are present in nucleotides 14 and 16 of the ribonuclease T_1 digest.

To test the two-dimensional fractionation system, portions of a ribonuclease T_1 digest were fractionated in three different ways. In experiments 1 and 2, the digest was fractionated, on one dimension only, by ionophoresis on DEAE-paper at pH 1·9; in experiment 3, fractionation was on the usual two-dimensional system using pH 1·9 for the second dimension; and in experiment 4 the digest was fractionated on a two-dimensional system which used ionophoresis on DEAE-paper in 7·5% triethylamine carbonate (pH 9 to 10) in the second dimension. The individual yields using these various methods (Table 7) are all within 15% of the mean value, and most are much closer than this to the mean. Thus the losses, at least for the smaller nucleotides no. 1 to 7, on transfer of material from the cellulose acetate to DEAE-paper must be negligible. Also nucleotides cannot be appreciably degraded by fractionation on DEAE-paper at pH 1·9, as the yields are similar to those obtained by fractionation on DEAE-paper at pH 9 to 10. However, the yields of the larger nucleotides (penta- to decanucleotides) were sometimes lower than 1·0. This probably reflected several factors, among them a loss on transfer from cellulose acetate to DEAE-paper as well as a low yield because the digestion with ribonuclease T_1 had not gone to completion, leaving some nucleotides with cyclic phosphate which could not always be included in the figures given in Table 7. Experiment 5 (the same fractionation as that shown in Plate I) and experiment 6 were separate digests of another preparation of 5 s RNA fractionated using 7% formic acid in the second dimension of the two-dimensional system. There were no cyclic nucleotides present, whereas in other experiments (e.g. experiment 3) up to half the free G content was present as G cyclic phosphate. The yields relative to AG were about 20% higher than in experiments 1 to 4 for most nucleotides; e.g. CG, CCG, UG, CCAG, CCUG: although they were the same as in experiments 1 to 4 for AAG, AAACG and AUG. Probably the yields are high relative to AG, because AG itself as well as AAG, AAACG and AUG are relatively unstable to overdigestion with ribonuclease T_1 (or to a contaminant of it) and degrade further. This interpretation accounts for the larger nucleotides being recovered in an apparent yield which is greater than 1. If the yields are calculated relative to UG = 4·0 in experiment 5 or 6, then the yields of the larger nucleotides do more nearly approximate to the expected yield. Thus the yields of 15, 16 and 17 corrected in this way are 0·9, 1·0 and 0·4, respectively.

In summary, the differences between experiments 5 and 6 and 1 to 4 are caused by the different conditions of digestion, rather than by variations in yields caused by experimental error in the method of fractionation. In general, it would appear that the main limitation in the use of the method for quantitative work will be in the control of the hydrolytic procedure.

Experiments 7 to 10 (Table 7) give some quantitative analyses, derived from different digestions of the same preparation, of the products of pancreatic ribonuclease digestion of 5 s RNA. On the basis of the present evidence, the yields of some nucleotides, e.g. GU and GGU, were ambiguous and were equal to 2·5. However, the yields of other nucleotides such as GC and C were surprisingly consistent. It is possible that the variation in the GU type was a result of some wrong splits occurring, because G cyclic phosphate could usually be detected on "fingerprints". Another difficulty was that the yields of one T_1 nucleotide (no. 17) and one pancreatic nucleotide (no. 16) were consistently in 0·5 molar yields relative to other larger nucleotides. The significance of this result is not clear.

TABLE 7

Molar yields of nucleotides

Spot no. (Plates I & II)	Sequence	Ribonuclease T_1 digests Molar yield† 1‡	2	3	4	5	6	Average molar yield of exp. 1 to 4	Corrected§ molar yield
1	G	12·8	12·8	10·4	11·2	13·7	12·6	11·8	10
2	CG	4·3	4·2	4·3	4·4	6·1	7·8	4·3	4
3	AG	2·00	2·00	2·00	2·00	2·00	2·00	2·00	2
4	UG	3·5	4·0	4·0	3·4	4·7	5·0	3·7	4
5	CCG	2·7	2·7	2·9	2·9	3·8	3·8	2·8	3
6	AAG	1·3	1·1	1·1	1·1	0·9	0·9	1·2	1
7	CCAG	0·7	0·7	0·7	0·9	1·3	1·4	0·8	1
8	AAACG			0·7	0·7	0·8	0·8	0·7	1
9	CCUG			0·7	0·8	1·1	1·3	0·8	1
10	UAG			2·9	2·8	3·4	3·7	2·9	3
11	AUG			0·9	1·0	1·0	1·0	1·0	1
12	AACUG			0·8	0·7	1·1	1·0	0·8	1
13	AACUCAG			0·6	0·8	0·7	0·9	0·7	1
14	ACCCCAUG			0·5	0·7	0·9	0·9	0·6	1
15	UCCCACCUG			0·6	0·7	1·1	1·0	0·7	1
16	UCUCCCCAUG			0·4	0·6	1·2	1·3	0·5	1
17	CCUUAG			0·3		0·5	0·5	0·3	0·5
18	pUG			0·9	0·5	1·2		0·7	1
19	CAU_{OH}			0·7				0·7	1
Others	U(CU)G			0·1				0·1	0
,,	CUG			0·3	0·4	0·0	0·1	0·4	0
,,	UCG			0·2	0·3	0·0	0·0	0·3	0
								Total =	115

Table 7 continued]

Spot no. (Plate V)	Sequence	Pancreatic ribonuclease digests: Molar yield† 7	8	9	10	Average molar yield	Corrected§ molar yield
1	C	17·6	20·8	1·9		19·1	19
2	U	4·0	4·8	4·8		4·5	5 to 6
3	AC	1·1	1·1	1·2		1·1	1
4	GC	7·0	7·0	7·1		7·0	7
5	AU	2·00	2·00	2·00	2·00	2·00	2
6	GAC	1·1	1·0			1·1	1
7	AGC	1·9	2·0			2·0	2
8	GAAC	1·0	0·9			1·0	1
9	GAAAC	1·0	0·8			0·9	1
10	GGC	2·0	1·7			1·9	2
11	AGGC	0·8	0·6			0·7	1
12	GU	2·1	2·8	2·4	2·3	2·4	2 to 3
13	GAU	1·0	1·0	1·0		1·0	1
14	AGU	1·2	1·2	1·0		1·1	1
15	GGU			2·4	2·6	2·5	2 to 3
16	AGGGAAC	0·4				0·4	0·5
17	AGAAGU	1·0	0·8			0·9	1
18	GAGAGU	1·0	0·7			0·9	1
19	GGGGU						1
20	$_pU$						1
21	AU_{OH}	0·8	0·9				1
Others	A_2C	0·05					0
						Total =	113 to 117

† Expressed relative to AG or AU taken as 2·00.

‡ Experiments 1 to 4 were separate fractionations of the same digest. 1 and 2 were fractionated on DEAE-cellulose at pH 1·9, 3 was fractionated on the usual two-dimensional system using the pH 1·9 mixture for the second dimension, and 4 was fractionated on a two-dimensional system which used ionophoresis on DEAE-paper using 7·5% triethylamine bicarbonate pH 9 to 10 in the second dimension. Experiments 5 and 6 were separate digests and fractionations of another preparation. Experiments 7, 8, 9 and 10 were separate digests and fractionations of a third preparation. All were prepared by polyacrylamide gel ionophoresis. Gaps in the Table indicate that no quantitative estimate was made in that particular experiment.

§ The observed molar yield for G can be corrected, knowing the sequence and yields of the pancreatic nucleotides. Similarly, the yield of C can be checked, knowing the sequence and yields of the T_1-nucleotides. For further discussion see text.

The corrected molar yields for each digest are derived from the average molar yields by taking into account any relevant sequence information from the other digest. For example, the average yield of G in ribonuclease products (taking GGU equal to 2·5 and AGGGAAC equal to 0·5) is 9·5. Therefore the corrected molar yield for G is 10. The average yield of U in pancreatic ribonuclease digests is 4·5; but because the number of pyrimidine-U sequences in the ribonuclease T_1 digests equals 5·5, the figure for yield of U is corrected to 5 or 6. Other figures needed no correction. Thus the average yield of C was 19 and the number of pyrimidine-C sequences in T_1-nucleotides was also 19. Having made corrections of this type, one can test whether they are likely to be correct. Thus, any pancreatic nucleotide ending in GU should be present as a T_1-nucleotide starting in U and there are nine in each digest. However, there are twelve sequences ending in GC in pancreatic ribonuclease digests but only eleven T_1-nucleotides starting in C. There are many other sequences that also agree, such as the number of AG, AU, AAC and AAAC sequences. Another test is to add up all the nucleotides to estimate the number of mononucleotide residues; this comes to 115 in the ribonuclease T_1 digest when no. 17 is counted in half-molar amounts, and to 113 to 117 in the pancreatic ribonuclease digest counting no. 16 as half a mole. Thus there is good, although not exact, agreement in the yields of pancreatic and ribonuclease T_1 sequences.

Listed in Table 7 are some other nucleotides which are present in much lower amounts than the numbered nucleotides. Thus, in ribonuclease T_1 digests U(CU)G, CUG and UCG are present, presumably derived from an RNA contaminating the preparation. Also, as can be seen from Plate I, there is a background blackening of the radioautograph on or near the origin of the second dimension, which is probably derived from contaminating material. However, in good preparations (experiments 5 and 6) the yield of these contaminants is less than 0·1 mole, so that the preparation of 5 s RNA is probably about 90% pure. The contaminating RNA does not give nucleotide-containing rare bases and is thus not a transfer RNA type. The fact that the contaminants are characteristic suggests it is another distinct RNA of a similar size to 5 s RNA.

4. Discussion

When the two-dimensional ionophoretic method of fractionation was applied to the examination of radioactive nucleotides from ^{32}P-labelled 5 s RNA of *E. coli*, all the 19 nucleotides in a ribonuclease T_1 digest and all the 21 nucleotides in a pancreatic ribonuclease digest were separated from one another. Thus, if the mixture of nucleotides to be separated is not too complex, the resolution of the method is good. Also, the quantitative yields of nucleotides were reproducible and approximated to the expected integral figures.

Various methods were used in the sequential analysis of the oligonucleotides, but the most generally useful way was partially to degrade them with exonucleases and isolate and analyse all the products. Thus, sequential analyses of oligonucleotides from pancreatic ribonuclease digests were conveniently and rapidly carried out using spleen phosphodiesterase, and nucleotides from combined ribonuclease T_1 and phosphatase digests were analysed using snake venom phosphodiesterase. The specific degradation, called secondary splitting, was a useful new method of analysing nucleotides from ribonuclease T_1 digests.

The results from the two different nuclease digests confirmed one another, but the sequence information was insufficient to allow us to connect any of the larger fragments from the two digests. However, the results show that this RNA is about 115 residues long, with pUG as 5′-terminal and $GCAU_{OH}$ as 3′-terminal. We are at present investigating new methods for fractionating larger nucleotides derived from partial ribonuclease T_1 digests, which is the way by which Holley and his collaborators (Holley *et al.*, 1965) determined the structure of alanine transfer RNA of yeast. By using ^{32}P-labelled RNA, it is possible to isolate by a fairly simple method sufficient material to study structure. The procedure is less time consuming and has, up to a certain limit of size, higher resolution than the column fractionation methods which are currently used to obtain sufficient material to study non-radioactive RNA.

This study thus provides a basis of experience for finding the sequence of nucleotides from ^{32}P-labelled purified transfer RNA, as well as a basis for defining structural differences between related mutants or different species of RNA.

REFERENCES

Comb, D. G., Sarkar, N., DeVallet, J. & Pinzino, C. J. (1965). *J. Mol. Biol.* **12**, 509.

Galibert, F., Larsen, C. J., Lelong, J. C. & Boiron, M. (1965). *Nature*, **207**, 1039.

Garen, A. & Levinthal, C. (1960). *Biochim. biophys. Acta*, **38**, 470.

Holley, R. W., Apgar, J., Everett, G. A., Madison, J. T., Marquisee, M., Merrill, S. H., Penswick, J. R. & Zamir, A. (1965). *Science*, **147**, 1462.

Holley, R. W., Madison, J. T. & Zamir, A. (1964). *Biochem. Biophys. Res. Comm.* **17**, 389.

Jackson, R. (1966). Ph.D. dissertation, Cambridge.

Marcot-Queiroz, J., Julien, J., Rosset, R. & Monier, R. (1965). *Bull. Soc. Chim. Biol.* **47**, 183.

Pigretti, M. M. & Milstein, C. (1965). *Biochem. J.* **94**, 106.

Richards, E. G., Coll, J. A. & Gratzer, W. B. (1965). *Analyt. Biochem.* **12**, 452.

Rosset, R. & Monier, R. (1963). *Biochim. biophys. Acta*, **68**, 653.

Rosset, R., Monier, R. & Julien, J. (1964). *Bull. Soc. Chim. Biol.* **46**, 87.

Sanger, F. & Brownlee, G. G. (1967). In *Methods in Enzymology—Nucleic Acids*, ed. by L. Grossman & K. Moldave. New York: Academic Press, in the press.

Sanger, F., Brownlee, G. G. & Barrell, B. G. (1965). *J. Mol. Biol.* **13**, 373.

Sarkar, N. & Comb, D. G. (1965). *Fed. Proc.* **24**, 292.

Schleich, T. & Goldstein, J. (1966). *J. Mol. Biol.* **15**, 136.

Shuster, L., Khorana, H. G. & Heppel, L. A. (1959). *Biochim. biophys. Acta*, **33**, 452.

J. Mol. Biol. (1968) **34**, 379–412

The Sequence of 5 s Ribosomal Ribonucleic Acid

G. G. Brownlee, F. Sanger and B. G. Barrell

Medical Research Council Laboratory of Molecular Biology
Hills Road, Cambridge, England

(*Received 26 January 1968*)

The nucleotide sequence of the low molecular weight ribosomal RNA (5 s RNA) of *Escherichia coli* has been studied using ^{32}P-labelled RNA and paper fractionation techniques. This is shown to be: pUGCCUGGCGGCCGUAGCGCGGUGGUC-CCACCUGACCCCAUGCCGAACUCAGAAGUGAAACGCCGUAGCGCCGAUGG-UAGUGUGGGGUCUCCCCAUGCGAGAGUAGGGAACUGCCAGGCAU$_{OH}$.† There are two major species of 5 s RNA which differ by a base in only one position. A new two-dimensional procedure suitable for fractionating oligonucleotides up to 25 residues long is described.

1. Introduction

Previous work on nucleotide sequences of RNA has been largely confined to the transfer RNA's and the structure of a number of them has been completely elucidated (Holley *et al.*, 1965; Zachau, Dütting & Feldmann, 1966; Madison, Everett & Kung, 1966; RajBhandary *et al.*, 1967; Bayev *et al.*, 1967). The oligonucleotides were fractionated on columns of DEAE-cellulose or DEAE-Sephadex and were detected by their ultraviolet absorption. Recently we have described methods for the study of sequences in ^{32}P-labelled nucleotides using methods of fractionation on paper (Sanger, Brownlee & Barrell, 1965). The greatly increased sensitivity of the isotopic techniques makes it possible to obtain sufficient material from a two-dimensional paper fractionation to determine sequences in the separated oligonucleotides. The method has been applied to the low molecular weight ribosomal RNA of *Escherichia coli* (5 s RNA) (Brownlee & Sanger, 1967) and the structures of all the end-products of ribonuclease T_1 and ribonuclease A (pancreatic ribonuclease) digestion have been determined (Table 1). These will be referred to as the T_1 and A end-products, respectively.

The 5 s RNA was originally described by Rosset & Monier (1963) and appears to be an integral structural component of all ribosomes. In contrast to transfer-RNA it contains no "minor" bases. Our previous results suggested that it had about 115 nucleotides of which pU-G was 5′-terminal and G-C-A-U$_{OH}$ 3′-terminal. From the structure of the end-products of digestion with the two ribonucleases it was not possible to obtain significant overlapping sequences; i.e. the sequence of the A end-products could not be used to join together any two T_1 end-products. Thus it was

† In this paper one-letter symbols for the nucleosides have been used as recommended in *J. Biol. Chem.* (1966), **241**, 527. Similarly an abbreviation such as 7MeG has been used for 7-methylguanosine. We have also followed these recommendations by using hyphens to replace p in short known sequences but for reasons of space have omitted the hyphens altogether in known sequences of more than six nucleosides. Unless otherwise stated the oligonucleotide fragments mentioned in this paper have a 3′-terminal phosphate group and a free 5′-terminal hydroxyl group.

TABLE 1

End-products of ribonuclease digestion of 5 s RNA (MRE600)

Ribonuclease T_1				Ribonuclease A			
		Yield				Yield	
Identification no.	Sequence	Found†	Calc. from final structure	Identification no.	Sequence	Found‡	Calc. from final structure
t1	G	12·3	11	a1	C	19·1	19
t2	C-G	5·2	5	a2	U	4·5	6
t3	A-G	2·00	2	a3	A-C	1·1	1
t4	U-G	4·1	4	a4	G-C	7·1	7
t5	C-C-G	3·1	3·5	a5	A-U	2·0	2
t6	A-A-G	1·1	1	a6	G-A-C	1·1	1
t7	C-C-A-G	1·0	1	a7	A-G-C	2·0	2
t8	A-A-A-C-G	0·8	1	a8	G-A-A-C	1·0	1
t9	C-C-U-G	1·0	1	a9	G-A-A-A-C	0·9	1
t10	U-A-G	3·3	3·5	a10	G-G-C	2·0	2
t11	A-U-G	1·0	1	a11	A-G-G-C	0·8	1
t12	A-A-C-U-G	0·9	1	a12	G-U	2·8	2·5
t13	AACUCAG	0·8	1	a13	G-A-U	1·0	1
t14	ACCCCAUG	0·8	1	a14	A-G-U	1·2	1
t15	UCCCACCUG	0·9	1	a15	G-G-U	2·9	3
t16	UCUCCCCAUG	0·9	1	a16	AGGGAAC	0·6	1
t17	C-C-U-U-A-G	0·4	0·5	a17	A-G-A-A-G-U	1·1	1
t18	pU-G	0·7	1	a18	G-A-G-A-G-U	1·0	1
t19	C-A-U_{OH}	0·7	1	a19	G-G-G-G-U	0·7	1
				a20	pU	0·9	1
				a21	A-U_{OH}	0·9	1

† Yields were obtained by measuring the radioactivity in the area of the paper containing the nucleotide, using a liquid-scintillation counter (Brownlee & Sanger, 1967) and the results are expressed relative to A-G = 2·00. They are unbiased averages of four experiments, with the exception of t16 and t17 where there were only three measurements.

‡ These results are the average of at least seven experiments and were expressed relative to $\frac{(\text{G-G-C}) + (\text{A-G-C}) + (\text{A-U})}{3} = 2{\cdot}00$.

necessary to isolate other larger degradation products in order to proceed further with the sequence determination.

In the original work on alanine transfer-RNA Holley *et al.* (1965) used partial digestion with ribonuclease T_1 to obtain longer oligonucleotides and complete the sequence. Other workers have used this method and partial digestion with ribonuclease A. We have also applied these methods of digestion extensively to the ^{32}P-labelled 5 s RNA. Under conditions of partial hydrolysis both enzymes split preferentially at single-stranded regions and most of the longer products are from the more stable double-stranded regions and are about 15 to 25 residues long. Initially we attempted to fractionate them on the two-dimensional system using 7% formic acid for the DEAE-paper dimension (Sanger *et al.*, 1965). However, most of the large nucleotides remained at the origin on the DEAE-paper although they separated well on the cellulose acetate ionophoresis dimension, especially when urea was added to the buffer. It was thus necessary to find an alternative fractionation method for the second dimension which would be suitable for nucleotides of this size. The best results were obtained using chromatography on DEAE-paper, rather than ionophoresis. In order to prevent the large nucleotides being adsorbed too strongly on the DEAE-paper, a high concentration of an anion was necessary. This was not possible with ionophoretic systems as it increased the conductivity too much with consequent overheating. In the normal column chromatographic systems with DEAE-cellulose, the necessary anion concentration is achieved by using a sodium chloride gradient to displace the nucleotides from the DEAE groups. A technique for applying this method to DEAE-paper and thus using it as a second dimension, following cellulose acetate ionophoresis, has been developed by B. G. Barrell & R. E. Offord (manuscript in preparation) and was applied to partial digests of the 5 s RNA. Although the spots were somewhat streaky, fractionation was usually adequate and much of the initial work on partial digests was done in this way.

More recently we have developed an alternative procedure (referred to as homochromatography) in which a high concentration of a mixture of non-radioactive nucleotides is used to develop the chromatogram and displace the radioactive nucleotides successively from the DEAE-paper. In the conventional procedure elution is effected by a gradient of sodium chloride, the different nucleotides being displaced by different chloride ion concentrations. An alternative procedure is to displace the radioactive nucleotides by a series of anions of different valency or affinity for the DEAE groups. Such a system is produced when a concentrated mixture of nucleotides (or of any other suitable anions of different valency) is applied to the end of a DEAE chromatogram. The nucleotides saturate the DEAE groups and displace one another producing a series of fronts. The smaller ones with lower valency or affinity are displaced by the larger ones and therefore move faster. The radioactive nucleotides move with the different fronts and are fractionated according to their affinity for the DEAE. The principle of the method is illustrated in Figure 1 and the actual experimental technique is described in the Materials and Methods section of this paper.

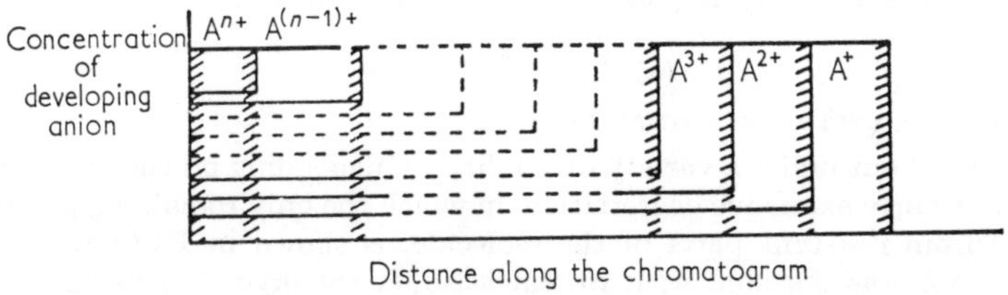

FIG. 1. Diagram illustrating homochromatography.

A mixture of radioactive nucleotides is applied at one end of a DEAE-paper chromatogram (left-hand end of Figure), which is developed with a concentrated mixture of anions (non-radioactive nucleotides, A^{+}, A^{2+}, A^{3+} . . . $A^{(n-1)+}$, A^{n+}). The anions saturate the DEAE-paper and displace one another to form a series of fronts on which the non-radioactive nucleotides (indicated by the shaded areas) are carried.

Using this method as a two-dimensional system combined with cellulose acetate ionophoresis, it was possible to obtain reasonable separations of nucleotides of up to about 25 residues. Longer ones remained at the origin of the DEAE-paper.

From partial digests with ribonuclease T_1 and A, most of the sequence could be deduced, particularly in the double-stranded more stable regions; however, other methods were necessary to complete the whole sequence. Partial digests with other ribonucleases were studied and the most useful of these was the acid ribonuclease from spleen (Bernardi & Bernardi, 1966). Although it has no absolute specificity it shows a preference for splitting at adenosine residues and this gives somewhat different products from those given by the other two ribonucleases, and a number of new sequences were determined.

Two methods were used in which the specificity of the enzyme was limited by chemical modification of the RNA. Such methods were particularly suitable for studying single-stranded regions which were difficult to determine using normal partial digestion. Gilham (1962) described a water-soluble carbodi-imide derivative that will react with uridine and guanosine but not with cytidine residues. The substituted uridine is not susceptible to ribonuclease A so that specific splitting at cytidine residues can be achieved (Lee, Ho & Gilham, 1965). This method was applied to the 5 s RNA and a number of useful sequences could be derived.

When RNA is methylated with dimethyl sulphate, methylation occurs preferentially at G residues to give 7-methylguanosine, which is resistant to ribonuclease T_1 (Lawley & Brookes, 1963; Brimacombe, Griffin, Haines, Haslam & Reese, 1965). To obtain partial digestion products containing 7MeG residues, the RNA was partially methylated and then completely digested with ribonuclease T_1. Although the complexity of the digests was increased by a number of side reactions, the nucleotides were particularly suitable for fractionation by the two-dimensional procedure due to the extra positive charge of 7MeG at acid pH values, and a number of overlaps from the single-stranded regions could be obtained.

By a combination of these various methods it was possible to obtain enough partial digestion products to determine the complete sequence, which has been reported in a previous communication (Brownlee, Sanger & Barrell, 1967). The present paper describes the methods and the derivation of the sequence in detail.

2. Materials and Methods

Most of the materials and methods, including the preparation of ^{32}P-labelled 5 s RNA and the standard two-dimensional ionophoretic fractionation procedure, have been described in our previous papers (Sanger *et al.*, 1965; Brownlee & Sanger, 1967).

(a) *Enzymic digestion*

(i) *Partial digestion with ribonucleases T_1 and A*

This was carried out under a variety of conditions, depending on the extent of degradation required. A fairly extensive degradation, in which the only remaining partial products were derived from resistant parts of the molecule, is shown in Plate I. About 1 μc of ^{32}P-labelled RNA was digested with ribonuclease T_1 (Sankyo Co.) using an enzyme to substrate ratio of 1 : 100 in 0·01 M-Tris chloride, pH 7·5, 0·02 M-magnesium chloride for 30 min at 0°C. Similar conditions were suitable for studying fairly complete ribonuclease A digests. For more partial digests smaller weights of enzyme were used so that partial products from the more labile parts of the molecule were present in addition to the more resistant fragments seen at higher enzyme concentrations. Plate II shows a ribonuclease A digest using an enzyme to substrate ratio of 1 : 500. Many of the ribonuclease T_1

digests were prepared in 0·01 M-EDTA instead of magnesium chloride. The mildest digestion conditions used were 1 : 1000 ribonuclease T_1 in 0·01 M-EDTA and otherwise as above.

(ii) *Partial digestion with spleen acid ribonuclease*

The acid ribonuclease was a gift from Dr G. Bernardi and was obtained as a 0·2% solution in 0·1 M-sodium acetate, pH 5·0. An amount of 5 s RNA containing 1 to 10 μc of ^{32}P and 5 to 10 μg RNA was incubated with 2 μl. of enzyme for 30 min at 37°C.

(iii) *Partial digestion with* B. subtilis *ribonuclease*

Bacillus subtilis ribonuclease was kindly provided by Dr S. Nishimura. 2 μc of 5 s RNA and 10 μg carrier tRNA was digested with 2 μl. of a solution of 0·1 mg/ml. enzyme in 0·05 M-Tris chloride, pH 8·5 for 30 min at 0°C.

In all cases digestion was carried out in the smallest possible volume, usually 1 or 2 μl., in the tip of a drawn-out capillary tube and applied directly onto the cellulose acetate for fractionation in the first dimension of the two-dimensional system.

(b) *Fractionation of partial digests*

(i) *First dimension*

In the previously published method (Sanger *et al.*, 1965) an enzyme digest is fractionated by ionophoresis on strips of cellulose acetate (Oxoid) at pH 3·5 using 0·5% pyridine–5% acetic acid–water as the buffer. However, rather bad streaking of the larger oligonucleotides was apparent with some batches of cellulose acetate, which could be prevented by using a buffer containing 7 M-urea. Thus for partial digests 5% acetic acid–7 M-urea, adjusted to pH 3·5 with pyridine, was used both to wet the strip and in the two electrode compartments. 85 cm long runs were carried out for 2 to 3 hr at 6 to 8 kv in an apparatus similar to that used by Naughton & Hagopian (1962), in which the cellulose acetate was supported on a rack that could be lifted in and out of the tank. Cooling was not necessary because of the low current used. Most partial products moved between the slowest pink (acid fuchsin) and the blue (xylene cyanol F.F.) dyes run as markers, so that this region was normally used for the second dimension.

(ii) *Second dimension—homochromatography*

The nucleotides were transferred, after fractionation in the first dimension, onto a sheet of DEAE-paper (Whatman DE81, 47 cm × 57 cm) 5 cm from one of the short ends of the paper. After drying, the paper was folded in the middle over a supporting rack and lowered into a chromatography tank, previously saturated with water, and allowed to equilibrate for 2 to 3 hr. 50 ml. of the developing mixture of nucleotides (see below) was then placed in a trough inside the tank and the chromatogram allowed to develop by ascending chromatography for 16 to 24 hr at room temperature. The blue dye, used as a marker, had an R_F of 0·4 to 0·5 and the yellow an R_F of about 0·25. The paper was normally dried on the rack in air and a radioautograph prepared. No nucleotide ran faster than the blue, whilst most partial products were slower than the yellow. Because of the low R_F values of the larger oligonucleotides, the origin may be placed at 1 to 2 cm from one end of the chromatogram and developed as above. However, although the larger nucleotides moved further off the origin, they were more streaky and the results were probably less useful.

A commercial preparation of yeast ribonucleic acid, dissolved in 9 M-urea, was a convenient source of a mixture of oligonucleotides. Thus *homomixture a* was prepared by dissolving 10 g of yeast RNA (British Drug Houses Ltd.) in 200 ml. of 9 M-urea and adjusting the pH to 7·5 with 10 N-KOH. The final volume was about 205 ml. This mixture gave the separations shown in Plates I and II. Another mixture, in which the RNA was partially hydrolysed with alkali, was also used, especially in the acid ribonuclease experiments. Thus *homomixture b* was prepared by suspending 10 g of yeast RNA in 100 ml. of water at room temperature and adding 10 ml. of 10 N-KOH with stirring. After 30 min at room temperature the solution was cooled in ice and adjusted to pH 6·5 with 36% perchloric acid. After standing at 0°C for 1 hr, the potassium perchlorate was centrifuged off and the solution saturated with urea at 0°C.

(c) *Analysis of partial products*

Spots were marked out, numbered, and cut out of the paper, as described before (Sanger *et al.*, 1965). In order to remove the urea, the pieces of paper were washed, with stirring, in a large volume of ethanol for about 2 hr at room temperature. After allowing the ethanol to evaporate, the nucleotides could then be eluted with 30% triethylamine carbonate, pH 9 to 10, as before (Sanger *et al.*, 1965). Normally the material was divided into two, one-half for complete digestion with approximately 10 μl. of ribonuclease T_1 (0·2 mg/ml.) in 0·01 M-Tris chloride (pH 7·5), 0·01 M-EDTA for 4 hr at 37°C in a capillary tube, and the other half for digestion with ribonuclease A (0·2 mg/ml.) in the same buffer as above for 2 hr at 37°C. These digests were analysed by ionophoresis on long sheets (85 cm) of DEAE-paper using the pH 1·9 system. Most of the end-products (see Table 1) could be tentatively identified from their position relative to the products of marker ribonuclease digests of 5 s RNA which were always run in parallel (Fig. 2). The bands were then eluted and their compositions determined by alkaline hydrolysis. This distinguished between products having similar mobilities on the DEAE–pH 1·9 system. Some products occupied positions which did not correspond to bands in the marker digest—for example (CU_2) and (C_2U) which were generated by ribonuclease T_1 digests of some partial ribonuclease A products.

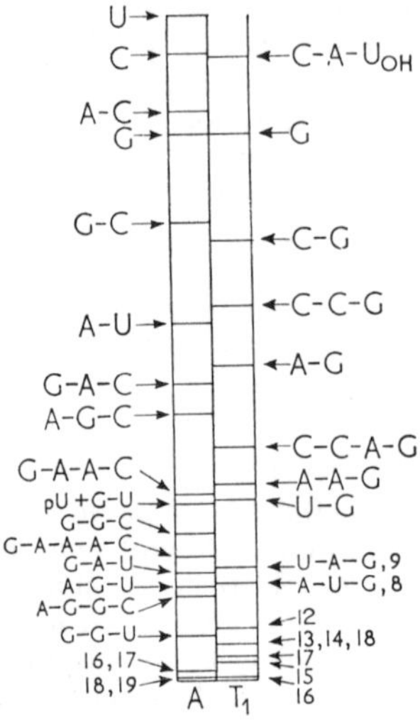

FIG. 2. Diagram showing the position of the products of complete ribonuclease T_1 and A digestions of 5 s RNA, fractionated by ionophoresis on DEAE-paper at pH 1·9. The numbered sequences are identified in Table 1.

(d) *Partial methylation with dimethyl sulphate*

In each experiment 5 s RNA containing 1 to 10 μc ^{32}P was used. Non-radioactive RNA was added to give a total of about 20 μg RNA. It was dissolved in 80 μl. 5% sodium acetate adjusted to pH 6·8. 2 μl. dimethyl sulphate were added and the mixture incubated at room temperature with occasional shaking for 20 min. 100 μl. water and 20 μl. 20% sodium acetate (pH 5·4) were added and the RNA precipitated with 450 μl. ethanol. After standing at −20°C for several hours the precipitate was centrifuged, washed with ethanol and dried *in vacuo*. A small sample was usually hydrolysed with alkali and the digest subjected to ionophoresis at pH 3·5 to determine the extent of methylation of the guanosine. N^7-Methylguanylic acid is decomposed by alkali to give predominantly 4-amino-5(N-methyl)-formamido-isocytosine ribotide (Lawley & Wallick, 1957), which has $R_U = 0{\cdot}82$ (R_U = mobility relative to uridine 3′-phosphate) on ionophoresis at pH 3·5 (Sanger *et al.*, 1965). There is also another degradation product which has an

$R_U = 1{\cdot}2$, and is present in variable amounts. Under the above reaction conditions about 30 to 50% of the G residues were usually methylated and this material was suitable for digestion with ribonuclease T_1. To the remainder of the methylated RNA was added 1 μl. of a 0·1% solution of ribonuclease T_1 and 2 μl. 0·1 M-Tris buffer (pH 7·4) containing 0·01 M-EDTA (neutralized) and the mixture was incubated at 37°C for 30 min.

Fractionation was carried out by the standard two-dimensional technique. A long strip (85 cm) of cellulose acetate was used for the first dimension. After running it was cut at the position of the blue marker and the two pieces blotted onto two separate sheets of DEAE-paper (85 cm long). The one containing material that had moved more slowly than the blue marker on cellulose acetate was run in the pH 1·9 system, and the faster moving sample run on either the pH 1·9 or the 7% formic acid system. The sequences of the 7MeG-containing nucleotides were largely determined by digestion with ribonuclease A and ionophoresis of the products at pH 3·5 on Whatman no. 52 paper. The positions of the main degradation products in this system are summarized in Fig. 3. C and 7MeG-C are not resolved so the C band was usually cut out, sewed onto another sheet of paper and subjected to ionophoresis at pH 6·5 (10% pyridine–0·3% acetic acid–water). 7MeG-C moves at 0·4 times the rate of C at this pH.

Several attempts to determine the structure of these 7MeG-containing nucleotides by partial digestion with spleen phosphodiesterase were unsuccessful.

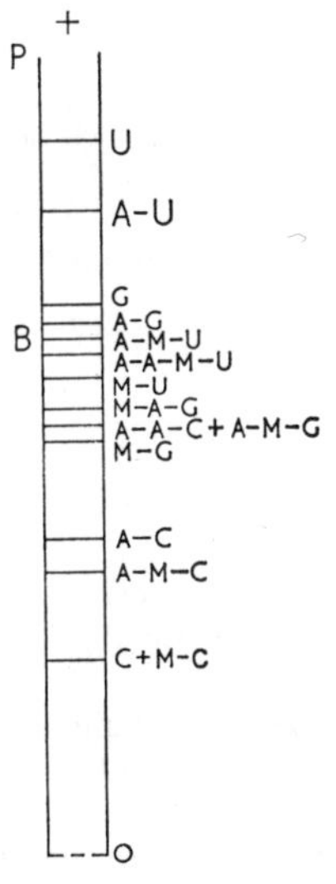

FIG. 3. Diagram showing the position of the products of ribonuclease A digests of nucleotides from ribonuclease T_1 digests of partially methylated 5 s RNA, fractionated by paper ionophoresis at pH 3·5. B and P show the positions of the blue and pink markers and M refers to N^7-methylguanosine. O is origin.

(e) *Reaction with a water-soluble carbodi-imide derivative*

The reaction of *N*-cyclohexyl-*N′*-(β-morpholinyl-(4)-ethyl) carbodi-imide-methyl-*p*-toluene sulphonate with uridine is shown in Fig. 4 (Gilham, 1962). Some reaction with guanosine also occurs. Ribonuclease A cleaves only adjacent to unmodified pyrimidines so that partial products are produced. Two methods for applying this method to 5 s RNA are described. In the first a partial reaction with CMCT† was carried out followed by complete digestion with ribonuclease A. The products were then fractionated directly on the standard two-dimensional ionophoretic system. They were easily separated from unmodified nucleotides because of their extra positive charge. In the second method a

† Abbreviation used: CMCT, *N*-cyclohexyl-*N′*-(β-morpholinyl-(4)-ethyl) carbodi-imide-methyl-*p*-toluene sulphonate.

more complete reaction with CMCT was used and the products of ribonuclease A digestion were fractionated after removal of the blocking group with ammonia. This necessitated the prior inactivation of the ribonuclease A which was done by treatment with mercapto-ethanol and pronase.

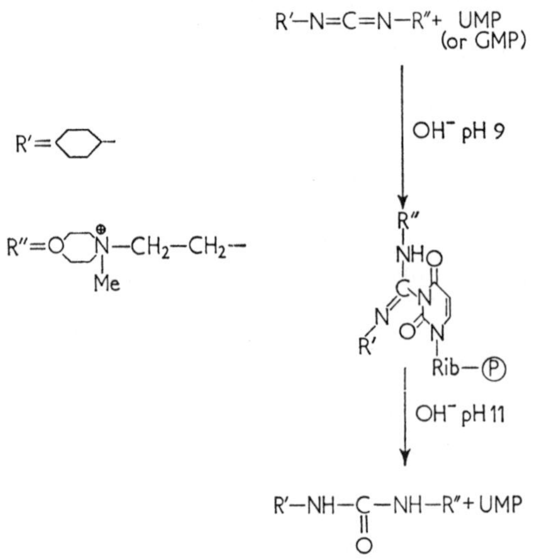

FIG. 4. The reaction of a water-soluble carbodi-imide derivative with uridylic (and guanylic) acids.

(i) *Partial reaction*

The reaction mixture (10 to 50 μl.) contained 1 to 2 μg and approximately 1 μc of 5 s RNA, between 4·0 and 7·5 mg/ml. CMCT (freshly dissolved) and 0·02 M-Tris chloride, 0·002 M-EDTA, at pH 8·9. After 16 hr at 37°C the RNA was precipitated by the addition of one-tenth volume of 1·0 M-NaCl and 2 vol. ethanol. After recovering the precipitate by centrifugation, the partially modified RNA was digested with ribonuclease A (Worthington or Sigma) using an enzyme to substrate ratio of 1 : 20 in 0·01 M-Tris chloride, 0·01 M-EDTA, pH 7·5, for 30 min at 37°C and fractionated using the standard two-dimensional system. Adequate separation was obtained by fractionation on short lengths (57 cm) of cellulose acetate. Nucleotides from the cathode end as far as the blue marker were transferred onto DEAE-paper (57 cm long) for fractionation in the second dimension, using the pH 1·9 system.

New spots on the "fingerprint", containing a modified G or U, were eluted and incubated with 10 to 20 μl. 0·2 N-ammonia for 6 hr at 37°C to remove the blocking group. After incubation the solution was dried down on polythene, wetted with water and each spot divided into three parts for analysis (i) by alkaline hydrolysis to establish a composition, (ii) by further digestion with ribonuclease A, and (iii) by digestion with ribonuclease T_1. The products of enzymic digestion were fractionated by ionophoresis on DEAE-paper using the pH 1·9 system. Marker ribonucleases T_1 and A digests of 5 s RNA were run in parallel with the samples, and the products of digestion were identified by their position relative to these markers (see Fig. 2) and from a knowledge of the composition of the original oligonucleotide.

(ii) *Complete reaction*

The procedure was that of Lee *et al.* (1965). 5 s RNA was reacted with CMCT under the same conditions as for the partial reaction except that the concentration of CMCT was 100 mg/ml. After precipitation of the RNA and hydrolysis with ribonuclease A the solution was dried down on polythene and mixed with 5 μl. mercaptoethanol and 5 μl.

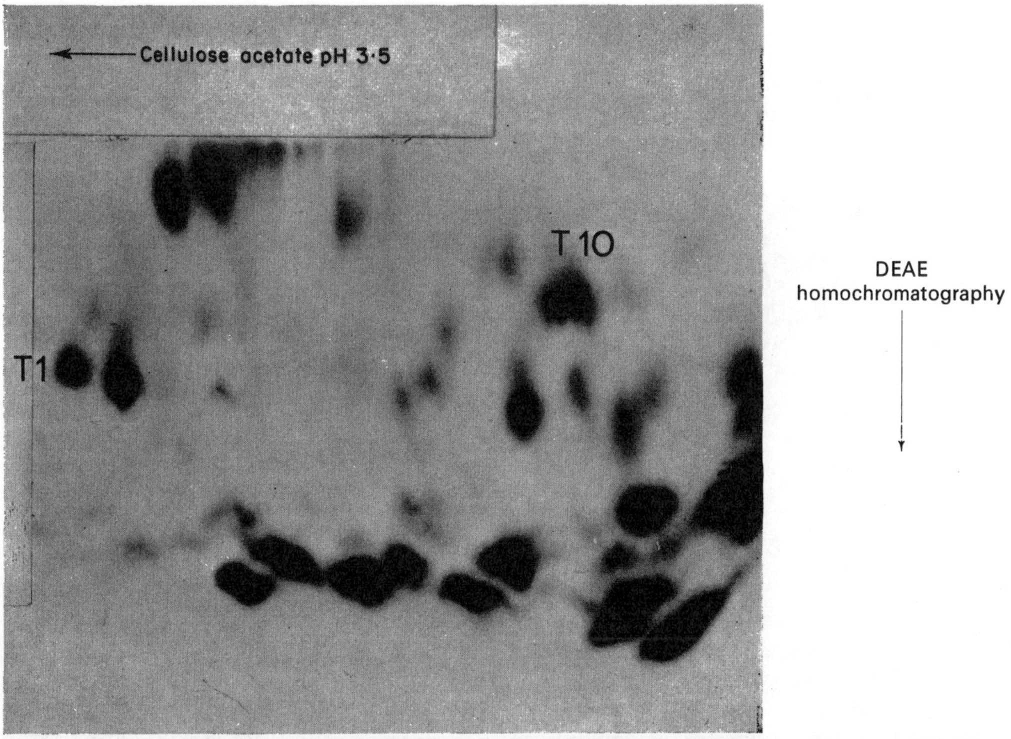

Plate I

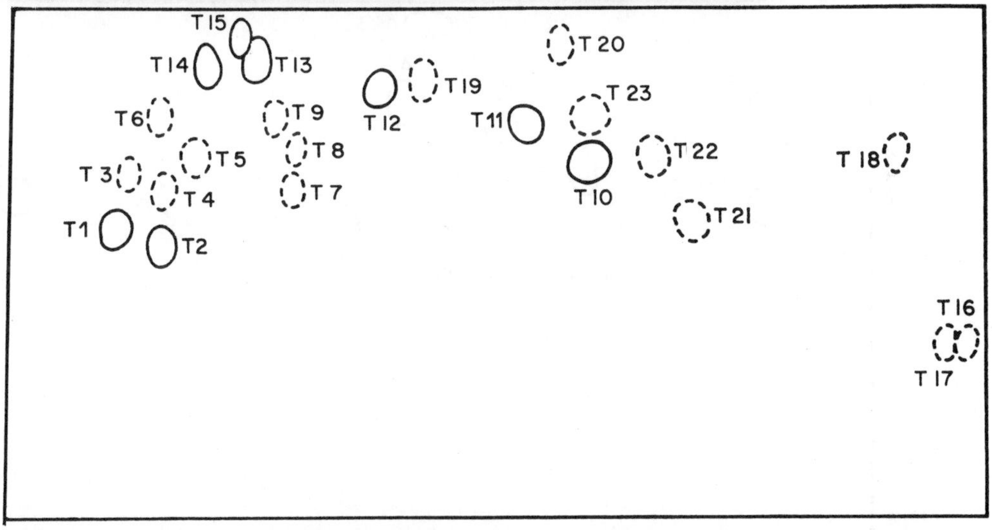

Fig. 5

Plate I and Fig. 5. Radioautograph and diagram of a two-dimensional fractionation of a partial ribonuclease T_1 (1 : 100 for 30 min. at 0°C) digest of 5 s RNA using ionophoresis on cellulose acetate at pH 3·5 in 7 M-urea in the first dimension; and using *mixture a* for homochromatography in the second dimension (see Materials and Methods). Those spots drawn in full circles are visible in the Plate while those in broken circles were isolated in other experiments. Those oligonucleotides moving fast in the second dimension are end-products of T_1-ribonuclease digestion and are not included in the diagram. Analyses are given in Table 2.

[*facing p. 386*

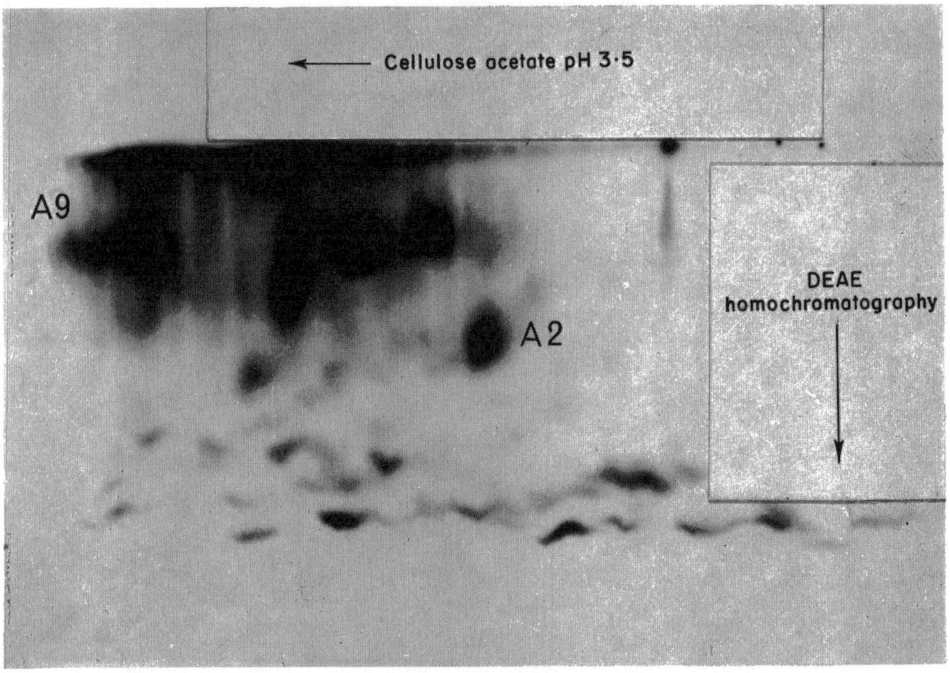

PLATE II

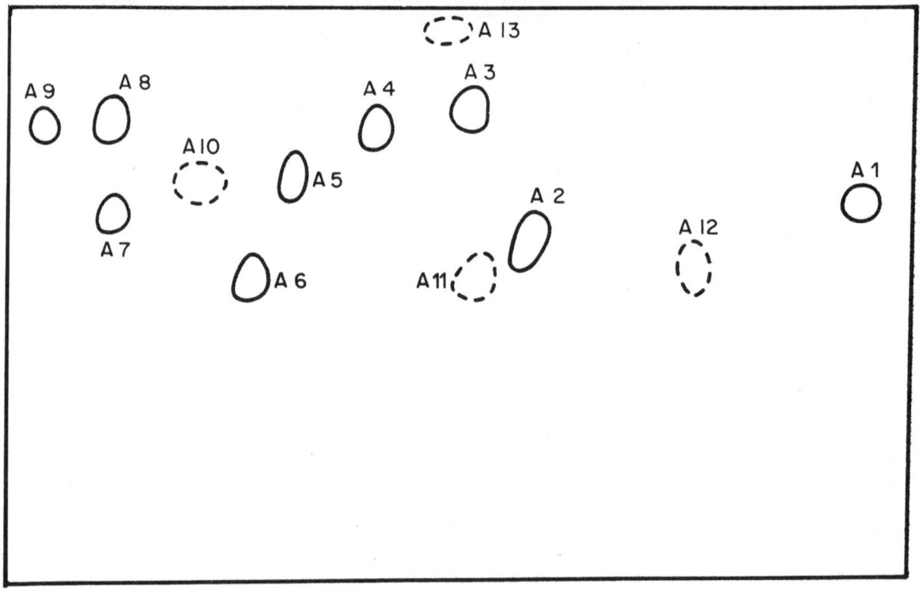

FIG. 6

PLATE II and FIG. 6. Radioautograph and diagram of a partial ribonuclease A (1: 500 for 30 min at 0°C) digest 5 s RNA fractionated as described in Plate I. The full circles in the diagram are visible in the Plate; the broken circles are the positions of nucleotides isolated in other experiments. Analyses are given in Table 3.

dimethylformamide and incubated in a sealed capillary tube for 8 hr at 37°C in order to reduce the ribonuclease. After evaporating down again on polythene the residue was dissolved in 10 μl. of 0·5 mg/ml. pronase (Calbiochem) and incubated at 37°C for 16 hr in order to hydrolyse the ribonuclease. It was then mixed with an equal volume of concentrated ammonia for 2 hr at room temperature to remove the blocking groups. After drying-off the ammonia, the residue was ready for fractionation on the standard two-dimensional system.

3. Results

Table 1 is a summary of previous results of the sequences present in 5 s RNA, *E. coli* strain MRE600. The experimental yields, which are an average of many experiments, compare well with the yield calculated from the final structure (Figs 14 and 15).

(a) *Partial ribonuclease T_1 and A digestion*

Plate I shows a two-dimensional fractionation of a partial ribonuclease T_1 digest (1 to 100) and Plate II a less complete ribonuclease A digest (1 to 500) fractionated by ionophoresis on cellulose acetate at pH 3·5 in the first dimension and homochromatography on DEAE-paper using *homomixture a* in the second dimension. The spots near the front are the end-products of digestion and the slower-moving and somewhat less sharp spots are the partial products. The numbering scheme for these spots and for others isolated in other experiments is shown in Figures 5 and 6, and their analysis is given in Tables 2 and 3. In a few cases, indicated by an underlined sequence in the Tables, it was possible to deduce the structure of the partial products from their analyses. In most cases, however, this was not possible and results from other experiments had to be used. For this reason the derivation of the sequences of the nucleotides in these Tables is considered, after all the results have been presented, in the Discussion.

TABLE 2

Analysis of the products of partial ribonuclease T_1 digestion

Spot no. (Fig. 5)	Products of further ribonuclease digestion†		Probable structure of nucleotides‡
	T_1	A	
T1	UCUCCCCAUG U-G G	G-G-G-G-U G-U A-U G	UGUGGGGUCUCCCCAUG
T2	UCUCCCCAUG U-G G	G-G-G-G-U A-U G	UGGGGUCUCCCCAUG
T3	UCUCCCCAUG U-A-G U-G G	G-G-G-G-U A-G-U G-U A-U G	UAGUGUGGGGUCUCCCCAUG

Table 2 continued]

Spot no. (Fig. 5)	Products of further ribonuclease digestion†		Probable structure of nucleotides‡
	T_1	A	
	UCUCCCCAUG	G-G-G-G-U	
	U-A-G	A-G-U	
T4	U-G	G-U	UAGUGUGGGGUCUCCCCAUGCG
	C-G	A-U	
	G	G-C	
		G	
	UCUCCCCAUG	G-G-G-G-U	
	U-A-G	G_2,A	
T5	U-G	A-G-U	UAGUGUGGGGUCUCCCCAUGCGAG
	A-G	G-U	
	C-G	A-U	
	G	G-C	
	UCUCCCCAUG	G-G-G-G-U	
	U-A-G	G-G-U	
	A-U-G	A-G-U	
T6	U-G	G-U	AUGGUAGUGUGGGGUCUCCCCAUG
	G	A-U	
		G	
	pU-G	G-G-C	
	C-C-U-G	pU	
T7§	C-A-G	A-G	pUGCCUGGCGGCAG
	C-G	G-C	
	G		
	pU-G	G-G-C	
	C-C-U-G	pU	
T8	C-C-G	G-C	pUGCCUGGCGGCCG
	C-G	G	
	G		
	C-C-U-U-A-G	G-G-C	
	pU-G	pU	
T9‖	C-C-U-G	A-G	pUGCCUGGCGGCCUUAG
	C-G	G-C	
	G		

Table 2 continued]

Spot no. (Fig. 5)	Products of further ribonuclease digestion†		Probable structure of nucleotides‡
	T_1	A	
T10	UCCCACCUG ACCCCAUG	G-A-C A-U A-C G	UCCCACCUGACCCCAUG
T11	UCCCACCUG ACCCCAUG G	G-U G-A-C A-U A-C G	GUCCCACCUGACCCCAUG
T12	UCCCACCUG ACCCCAUG U-G G	G-G-U G-A-C A-U A-C G	UGGUCCCACCUGACCCCAUG
T13	UCCCACCUG ACCCCAUG U-G G	G-G-U G-U G-A-C A-U A-C G	GUGGUCCCACCUGACCCCAUG
T14	UCCCACCUG ACCCCAUG U-G C-G G	G-G-U G-A-C A-U A-C G	CGGUGGUCCCACCUGACCCCAUG
T15	UCCCACCUG ACCCCAUG U-G C-G G	G-G-U G-A-C A-U G-C A-C G	CGCGGUGGUCCCACCUGACCCCAUG
T16	AACUCAG C-C-G	G-A-A-C A-G	CCGAACUCAG
T17	AACUCAG A-A-G	A_3,G_2 A-A-C	AACUCAGAAG

Table 2 continued]

Spot no. (Fig. 5)	Products of further ribonuclease digestion† T_1	A	Probable structure of nucleotides‡
T18	AACUCAG A-A-G C-C-G	A_3,G_2 G-A-A-C	CCGAACUCAGAAG
T19	A-A-A-C-G U-A-G C-C-G C-G	G-U A-G-C A-A-A-C G-C	AAACGCCGUAGCGCCG
T20	AACUCAG A-A-A-C-G U-A-G U-G A-A-G C-C-G A-G	A-G-A-A-G-U G-A-A-A-C G-A-A-C G-U	CCGAACUCAGAAGUGAAACGCCGUAG
T21	A-A-C-U-G C-C-A-G C-A-U_{OH} G	A-G-G-C G-C A-A-C A-U_{OH}	AACUGCCAGGCAU_{OH}
T22	A-A-C-U-G C-C-A-G C-A-U_{OH} G	A-G-G-C G-A-A-C G-C A-U_{OH}	GAACUGCCAGGCAU_{OH}
T23	A-A-C-U-G C-C-A-G	G-G-A-A-C A-G-G-C G-C	GGAACUGCCAGGCAU_{OH}

† These sequences were identified partly from a knowledge of the position of the end-products of ribonuclease digestion on DEAE-paper ionophoresis at pH 1·9 (Fig. 2) and partly from the results of further alkaline hydrolysis. Approximate yields of nucleotides were estimated by visual inspection of the radioautographs. If a nucleotide is underlined once there are 2 moles present; if twice there are more than 2 moles. Mononucleotides were frequently run off the end of the paper and are therefore not usually included in the Table. In some cases, e.g. T23, C-A-U_{OH} and A-U_{OH} were also run off.

‡ The structures given are those that can be deduced from the experimental results given in the Table and from the T_1 and A end-products, and also from a knowledge of other partial products (see text). Where a sequence is underlined it was deduced uniquely from its further degradation products and without reference to other partial products.

§ Present in 5 s RNA from strain CA265.

|| Present in 5 s RNA from strain MRE600 only.

TABLE 3

Analysis of products of partial ribonuclease A digestion

Spot no. (Fig. 6)	Products of further ribonuclease digestion†		Probable structure of nucleotides‡
	T_1	A	
A1	(C_2, A,U)G A,C_4	G-A-C A-C	ACCUGACCCC
A2	A-U-G C_2, A_2,U C-C-G	G-A-A-C A-U G-C G	AUGCCGAACUC
A3	A-A-C-U-G C-C-A-G A-G G C-A-U_{OH}	AGGGAAC A-G-G-C G-C	AGGGAACUGCCAGGCAU_{OH}
A4	A-A-A-C-G U-G A-A-G A-G C-C-G	A-G-A-A-G-U G-A-A-A-C G-U G-C	AGAAGUGAAACGCCGU
A5	A-U-G A-G C-C-G C-G G	G-G-U G-A-U A-G-C G-C	AGCGCCGAUGGU
A6	A-U-G A-G C-G G	G-A-G-A-G-U A-U G-C	AUGCGAGAGU
A7	pU-G C-C-U-G C-A-G C-G G	A-G-U G-G-C G-U pU G-C	pUGCCUGGCGGCAGU
A8	pU-G C-C-U-G C-C-G C-G G	G-G-C G-U pU G-C	pUGCCUGGCGGCCGU

Table 3 continued]

Spot no. (Fig. 6)	Products of further ribonuclease digestion†		Probable structure of nucleotides‡
	T_1	A	
A9	UCUCCCCAUG U-G A-G C-G G	G-G-G-G-U G-A-G-A-G-U A-G-U G-U A-U G-C	AGUGUGGGGUCUCCCCAUGCGAGAGU
A10	pU-G C-C-U-G C-G G	pU G-G-C G-C	pUGCCUGGCGGC
A11	A-U-G C-C-G G	G-G-U G-A-U G-C	GCCGAUGGU
A12	A-A-A-C-G C-C-G G	G-A-A-A-C G-U G-C	GAAACGCCGU
A13	AACUCAG A-A-A-C-G A-U-G A-A-G U-G C-C-G	A-G-A-A-G-U G-A-A-A-C G-A-A-C G-C	AUGCCGAACUCAGAAGUGAAACGC
A14§	pU-G C-C-U-G C_2,U	G-G-C pU G-C	pUGCCUGGCGGCCU
A15§	C,U_2 U-G A-G G	G-G-G-G-U A-G-U G-U	AGUGUGGGGUCU
A16§	U-G A-G U	A-G-U G-U	A-G-U-G-U
A17§	C,U_2 G	G-G-G-G-U	GGGGUCU

† and ‡ See footnotes to Table 2.
§ Isolated by salt gradient chromatography on DEAE-paper.

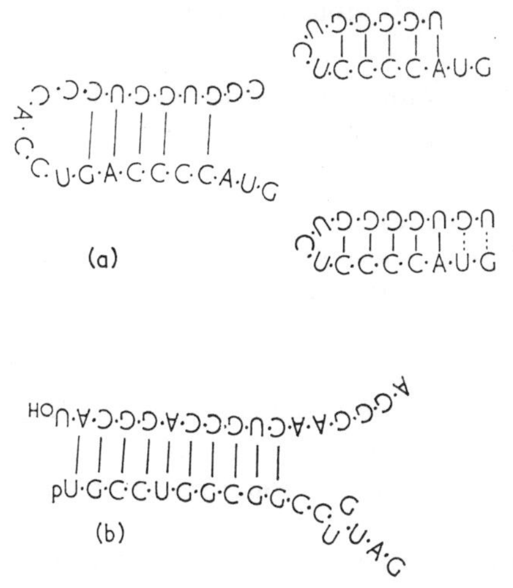

FIG. 7. (a) Shows a possible base-pairing scheme for some partial T_1 oligonucleotides and (b) a possible base pairing between the two ends of 5 s RNA. A single dash indicates G—C or A—U base pairs; broken dashes are G--U pairs.

Fractionation on DEAE-paper by homochromatography is largely according to size. Nevertheless, some nucleotides moved faster than their size would suggest. Thus in Figure 5, T1 and T2 (17 and 15 residues long, respectively) had roughly the same mobility as each other, and were faster than T10 (17 long). This may be explained if we assume that T1 and T2 adopted a compact hairpin structure by base-pairing and so interacted less strongly with the DEAE groups, whereas T10 did not (see Figure 7). In the experiments using the salt gradient chromatography, T12 and T14 were faster than T10 although they were larger in size. This observation was interpreted, again, by the formation of a hairpin loop (Fig. 7). However, the pairing must be less stable than in T1 because these pairs are broken in the homochromatography where T10, 12 and 14 are fractionated according to size (Fig. 5).

Most partial products observed under fairly extensive degradation conditions were related to T1 or T10 or were derived from the two ends of the molecule. This suggested that both ends were somewhat resistant to enzymic cleavage, and the fact that their sequences are complementary suggests that they pair together to form a stable hydrogen-bonded structure as shown in Figure 7(b). The products of fairly extensive ribonuclease A digestion also included resistant fragments from the two ends of the molecule (e.g. A3 and A8) in good yield. This supported the idea that the specificity of hydrolysis was again dependent on some secondary structure and is consistent with the scheme in Figure 7(b). Such pairing between the two ends of an RNA also seems to be a general feature of tRNA (see e.g. Holley *et al.*, 1965).

With ribonuclease A, and in contrast to ribonuclease T_1, breaks occurred at the pyrimidine residues in loops as well as in the single-stranded regions between the base-paired regions of the molecule. Thus fragments were present which derived from half of a hairpin loop (e.g. A1 and A15, and others such as UGACCCCAUGC and CUGACCCC isolated by the salt gradient technique, not mentioned in Table 3). With less extensive degradation fragments other than those already mentioned were isolated, e.g. A4, A5, T16, T17, and probably derive from non-helical regions of the molecule.

(b) *Partial acid ribonuclease digestion*

Figure 8 shows a diagram of a "fingerprint" of 5 s RNA partially digested with acid ribonuclease. The sequences of the nucleotides were determined by digestion with ribonucleases T_1 and A. The results for some of the smaller ones are included in the Figure and for the larger ones are given in Table 4. In some experiments urea was added to the pH 3·5 buffer for the cellulose acetate ionophoresis. This increased the sharpness of the spots and slightly altered the pattern. A large amount of material was still stuck at the origin on the DEAE 7% formic acid run, and in some experiments this was cut out, sewed onto another sheet of DEAE paper and run on the homochromatography system. Figure 9 and Table 5 summarize the combined results from several experiments in which the conditions for the homochromatography were varied somewhat.

Under the conditions used there was a considerable amount of cyclic phosphates. The enzyme appears to have no absolute specificity for any one nucleotide but most splits seem to have occurred after A residues, some after G residues and a few after C.

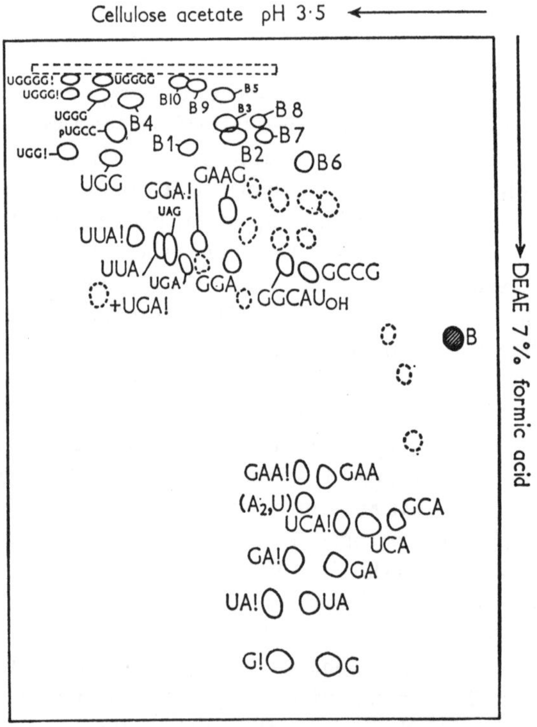

FIG. 8. Diagram of a two-dimensional fractionation of a partial acid ribonuclease digest of 5 s RNA fractionated by ionophoresis on cellulose acetate at pH 3·5 and on DEAE-paper using 7% formic acid.

The area on the origin of the DEAE-paper, included in the dashed lines, was cut out and subjected to homochromatography (Fig. 9). The analyses of spots B1 to B9 are given in Table 4. For lack of space, the customary abbreviation > p has not been used and in its place ! indicates an oligonucleotide terminated by a 2′,3′-cyclic phosphate. For the same reason hyphens have been omitted, UGGG etc. indicating known sequences. B is the blue marker and dashed circles are nucleotides which were not sequenced.

TABLE 4

Nucleotides from acid ribonuclease digest isolated on the 7% formic acid system

Spot no. (Fig. 8)	Products of ribonuclease digestion† T$_1$	Products of ribonuclease digestion† A	Structure of nucleotide Deduced‡	Structure of nucleotide Probable§
B1	U-G C-G A	G-A G-C U	U-G-C-G-A	
B2	U-G C-C-G A-A	A_2,G G-C U C	UGCCGAA	
B3	U-G C-C-G A-A-C	(A_2,G)C G-C U C	UGCCGAAC	
B4	U-G C-G G	G-G-C G U	U-G-G-C-G	
B5	U-A-G C-C-G C-G A	A-G-C G-A G-C U C	UAGC(C,G)CGA	UAGCGCCGA
B6‖		A-A-A-C G-C G C	AAACGCCG	
B7	C-G C-C-G G A	G-A G-C C	GC(C,G)CGA	GCGCCGA
B8	A-G C-G C-C-G A	A-G-C G-A G-C C	AGC(C,G)CGA	AGCGCCGA
B9	A-A-A-C-G U-G A-A-A-C¶ C-C-G G¶	G-A-A-A-C G-C G C U	UG(AAACG,CCG)	UGAAACGCCG

† See footnote † to Table 2.

‡ The structures given are those that can be deduced from the experimental results given in the Table and from a knowledge of the T$_1$ and A end-products. Inclusion of nucleotides in brackets indicates that their relative order could not be deduced.

§ These sequences are deduced by making use of results from other partial digestion products and the derivation is described in the text. Where there are many possible structures from the experimental results only the final deduced sequence is given.

‖ Only detected in one experiment.

¶ Present in low yield and probably derived by further degradation of A-A-A-C-G.

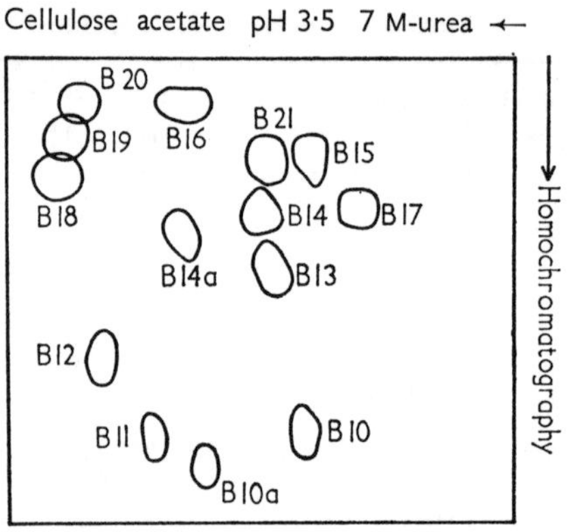

FIG. 9. Diagram illustrating a two-dimensional fractionation of a partial acid ribonuclease digest of 5 s RNA using ionophoresis on cellulose acetate at pH 3·5 in 7 M-urea for the first dimension and homochromatography (*mixture b*, see Materials and Methods) for the second dimension.

The upper border of the diagram is the origin of the second dimension. In some experiments the material left on the origin after ionophoresis on DEAE-paper using 7% formic acid (see Fig 8) was sewed onto another sheet of DEAE-paper and subjected to homochromatography; in others the chromatogram was run directly after the ionophoresis on cellulose acetate. Spots B19, B20, B21 were obtained using a *mixture a*, modified to contain twice as much nucleic acid, for homochromatography. The radioactive nucleotides had moved considerably further although the spots were rather "streaky". Spots B10a and B14a gave the same sequence as B10 and B14, respectively, and are probably cyclic forms (see Table 5).

(c) *Partial methylation*

Plate III, Figures 10 and 11, and Table 6 summarize the combined results from six separate experiments. There was some variation between experiments, probably due to differences in extent of methylation of G and to side reactions. In Table 6 the sequences deduced are expressed in three ways. Column 4 gives the sequences that can be deduced uniquely from the experimental results given in the Table. Column 5 makes use of a knowledge of the end-products of ribonuclease T_1 digestion. Thus, for instance, the results with spot M18 give the structure [U,(A,G)C]G indicating four possible sequences:

1. U-A-G-C-G
2. U-G-A-C-G
3. A-G-C-U-G
4. G-A-C-U-G

Sequence 2 cannot occur in the 5 s RNA since A-C-G is not a T_1 end-product; similarly 3 and 4 are not possible since neither C-U-G nor A-C-U-G are T_1 end-products. This leaves only sequence 1, U-A-G-C-G, which is therefore given in column 5. In certain cases a unique structure could not be deduced in this way but only by making use of results from other partial digestion products. These sequences are given in the final column of Table 6.

There were a number of spots whose structure could not be interpreted and these are not included in the Table. Their presence was partly due to incomplete resolution of the nucleotides, and probably partly due to unidentified side reactions of the methylation. In certain cases two spots appeared to have the same structure. In most cases this was probably due to methylation of adenosine residues, which would

TABLE 5

Nucleotides from acid ribonuclease digests isolated by homochromatography

Spot no. (Fig. 9)	Products of ribonuclease digestion: T_1	Products of ribonuclease digestion: A	Probable structure of nucleotide[a]
B10	U-G A-G C-G	G_3,A_2 G-C	UGCGAGAG
B11	UCUCCCCA UCUCCCC[b] U-G G	G-G-G-G-U A C U	UGGGGUCUCCCCA + UGGGGUCUCCCC
B12	UCUCCCCA UCUCCCC[b] U-G G	G-G-G-G-U G-U A C U	UGUGGGGUCUCCCCA + UGUGGGGUCUCCCC
B13	pU-G C-C-U-G C-G G C-C	pU G-G-C G-C C U	pUGCCUGGCGGCC
B14	pU-G C-C-U-G C-C-G C-G G	pU G-G-C G-C G C U	pUGCCUGGCGGCCG
B15	A-A-C-U-G U-A-G C-C-A-G A-G[c] G	AGGGAAC A-G-G-C G-C C U	UAGGGAACUGCCAGGCA(U_{OH})
B16	UCCCACCUG U-A-G U-G A-C-C-C-C-A C-G G	G-G-U A-G-C G-A-C G-C A-C	UAGCGCGGUGGUCCCACCUGACCCCA

Table 5 continued]

Spot no. (Fig. 9)	Products of ribonuclease digestion T_1	A	Probable structure of nucleotide[a]
B17	AACUCAG U-G A-A-G C-C-G	A-G-A-A-G[d] G-A-A-C G-C	UGCCGAACUCAGAAG
B18[e]	UCUCCCCA U-A-G U-G G	G-G-G-G-U A-G-U G-U	UAGUGUGGGGUCUCCCCA
B19[f]	UCUCCCCA U-A-G U-G G	G-G-G-G-U G-G-U A-G-U G-U	UGGUAGUGUGGGGUCUCCCCA
B20[f]	UCUCCCCAUG U-A-G U-G G	G-G-G-G-U G-G-U A-G-U G-U A-U	UGGUAGUGUGGGGUCUCCCCAUG
B21[g]	pU-G C-C-U-G U-A-G U-G C-C-G C-G G	G-G-U G-G-C G-U pU A-G-C G-C C U	pUGCCUGGCGGCCGUAGCGCGGUG

[a] The structures given are deduced from the experimental results shown in the Table and from a knowledge of the T_1 and A end-products and of other partial products the derivation of which is described in the text. The sequences underlined could be deduced uniquely from the results in the Table and a knowledge of the T_1 and A end-products.

[b] Present in less than molar amounts.

[c] Probably derived from partial breakdown of C-C-A-G.

[d] Analysed as A_3,G_2. An examination of A end products shows that A-G-A-A-G is the only possible sequence.

[e] Contained also traces of UCUCCCCAUG, A-G and G-G-U.

[f] B19 and B20 were contaminated with each other.

[g] Contained also traces of A-A-G and several unidentified products.

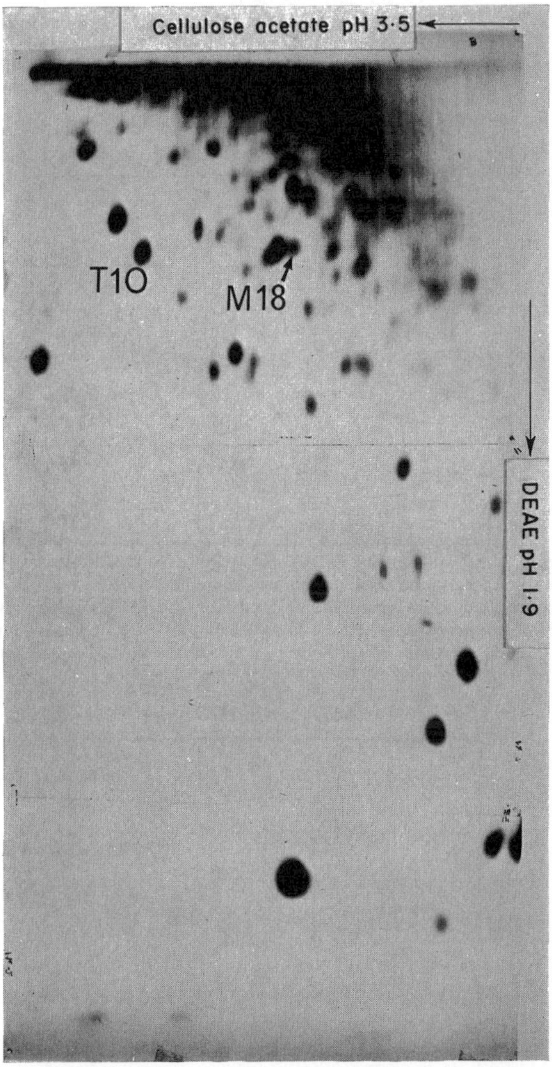

PLATE III. Radioautograph of a two-dimensional separation of a ribonuclease T_1 digest of partially methylated 5 s RNA. Fractionation was by ionophoresis on cellulose acetate, pH 3·5, and on DEAE-paper at pH 1·9. The spots are numbered and identified on the left side of Fig. 10 and the analyses are given in Table 6.

[*facing p. 399*

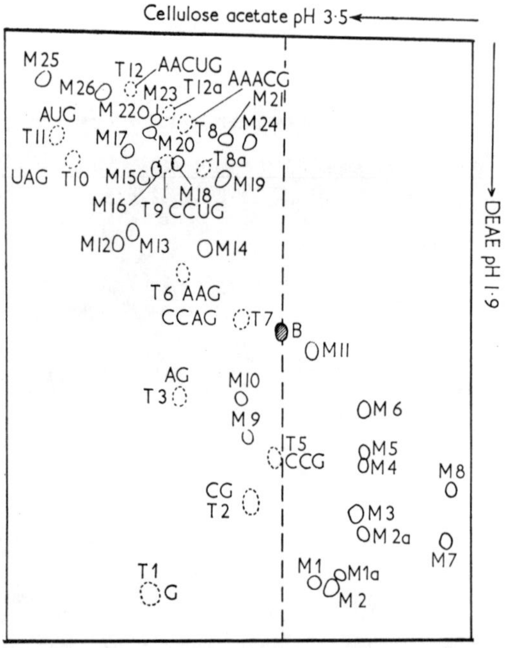

Fig. 10. Diagram of a two-dimensional fractionation of a ribonuclease T_1 digest of partially methylated 5 s RNA fractionated by ionophoresis on cellulose acetate at pH 3·5.

The cellulose acetate was then cut at the blue marker and the two halves were fractionated by ionophoresis on two separate sheets of DEAE-paper at pH 1·9. The analyses of the M-containing spots is given in Table 6 (see also Plate III). B is the blue marker. Hyphens have been omitted from the known nucleotide sequences because of lack of space.

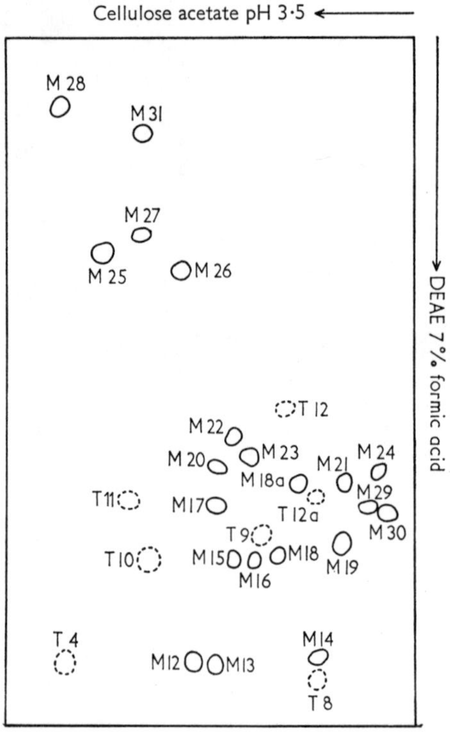

Fig. 11. Diagram of a two-dimensional fractionation of a ribonuclease T_1 digest of partially methylated 5 s RNA fractionated by ionophoresis on cellulose acetate, pH 3·5, and on DEAE-paper using 7% formic acid. The M-containing spots are identified in Table 6 and the T spots in Table 1.

TABLE 6

Nucleotides from ribonuclease T_1 digests of partially methylated 5 s RNA

Spot no. (Figs 10 and 11)	Composition[a] C	A	U	M[b]	G	Products obtained with ribonuclease A	Sequence deduced from results	from end-products	from other partial products
M1				×	×	M-G	G-G		
M1a				×	×	M-G	G-G		
M2						C M-G	C-G-G		
M2a						C M-G	C-G-G		
M3	×			×	×	M-C G	G-C-G		
M4	2·1			0·8	×	C M-C G	(C,G)C-G	} C-G-C-G and/or G-C-C-G[c]	
M5	××			×	×	C M-C G	(C,G)C-G		
M6	×××			×	×	C M-C G	(C_2,G)C-G	C(C,G)C-G	C-G-C-C-G
M7	1·0			1·5	1	M-C M-G	G-C-G-G		
M8	2·1			1·7	1	C M-C M-G	(C,G)C-G-G	C-G-C-G-G or G-C-C-G-G	C-G-C-G-G
M9		×		×	×	(A,M)G	(A,G)G[d]		
M10	×	×		×	×	C (A,M)G	C(A,G)G	C-G-A-G	
M11	××	×		×	×	C (A,M)G	C-C(A,G)G	C-C-A-G-G or C-C-G-A-G	C-C-A-G-G
M12			×	×	×	U M-G	U-G-G		
M13			×	×	×	M-U G	G-U-G		
M14						M-U M-G	G-U-G-G		
M15		×	×	×	×	U (A,M)G	U(A,G)G	U-A-G-G or U-G-A-G	U-A-G-G
M16		×	×	×	×	M-U A-G	G-U-A-G		
M17		1·0	1·2	0·9	1	A-U M-G	A-U-G-G		

M18	×	×	×	×	×	U (A,M)C G	[U,(A,G)C]G	U-A-G-C-G	
M18a						U (A,M)C G	[U,(A,G)C]G	U-A-G-C-G	
M19	1·9		1·0	1·0	1	C U M-G	(C_2,U)G-G	C-C-U-G-G	
M20	××	×	×	×	×	C (A,M)U G	$[C_{1-2},(A,G)U]$G	(C)C-G-A-U-G[e]	C-C-G-A-U-G
M21	××	×	×	×	×	C M-U A-G[f]	$(C_2,$G-U)A-G	C-C-G-U-A-G	
M22		××	×	×	×	A(A,M)U[g] G	A(A,G)U-G		A-A-G-U-G
M23		××	×	×	×	(A,M)U A-G	(A,G)U-A-G	A-G-U-A-G	
M24	×(×)	×	×	××	×	C (A,M)C M-U G			CCGUAGCG[h]
M25			××	×	×	U M-U G	(G,U)U-G	U-G-U-G	
M26			××	××	×	U M-U M-G	(G,U)U-G-G	U-G-U-G-G	
M27						U (A,M)U G	[(A,G),U]U-G	U-A-G-U-G or U-G-A-U-G	U-A-G-U-G
M28						pU C U M-C G	pU(C,U,G-C)G	pU-G-C-C-U-G	
M29	×	××	×	×	×	U $(A_2,$M)C G	$[U,(A_2,G)C]$G	G-A-A-C-U-G	
M30[i]						C U M-C (A,M)C G		UAGCGCCG or UAGCCGCG or AGCGCCUG	UAGCGCCG
M31[i]						U M-U (A,M)U G			UAGUGUG

[a] Compositions are expressed as the yield of mononucleotide relative to one mole of G. Where the results are given in ×'s, the composition was estimated by inspection of the radioautograph. Where numbers are given it was determined in the scintillation counter.

[b] In this Table M is 7-methylguanylic acid. It was identified by the presence of its degradation product, 4-amino-5(*N*-methyl)-formamido-isocytosine ribotide in alkaline digests. Its yield was not quantitative and allowance was made for this in estimating the yield of M.

[c] The presence of two spots, giving the same products with ribonuclease A, suggested that both C-G-C-G and G-C-C-G are present. However, this is not conclusive as certain nucleotides (e.g. M-G and C-M-G) appear to be present in two spots.

[d] If splitting has occurred only at G residues this would have come from the sequence G-A-G-G or G-G-A-G, neither of which is found in ribonuclease A end products.

[e] It was uncertain whether this was C-G-A-U-G or C-C-G-A-U-G. The position on the two-dimensional system suggested one C whereas the analysis suggested two.

[f] Analysis of the ribonuclease A products on the scintillation counter gave 2·4 moles C to one of M-U and A-G.

[g] Partial digestion with spleen phosphodiesterase indicated A as the 5′-terminal residue.

[h] Various other sequences (e.g. CGUAGCG, CGUAGCCG) were possible depending on the analysis for C, which was uncertain.

[i] These nucleotides were detected in only one experiment.

be expected to occur to some extent to give 1-methyladenylic acid (Lawley & Brookes, 1963). On alkaline hydrolysis 1-methyladenylic acid is converted to 6-methyladenylic acid, which is not separated from adenosine on ionophoresis at pH 3·5. It could, however, be separated by chromatography on propan-2-ol–HCl (Wyatt, 1951). Methylation of adenosine probably accounts for the presence of two spots with the apparent sequence A-A-A-C-G (T8, T8a), A-A-C-U-G (T12, T12a) and U-A-7MeG-C-G (M18, M18a). It would not, however, account for two spots with the apparent sequence 7MeG-G (M1, M1a) and C-7MeG-G (M2, M2a). The nature of the side reaction involved here is not clear.

(d) *Reaction with water soluble carbodi-imide*

(i) *Partial reaction*

Figure 12 is a diagram summarizing the results of six different ribonuclease A digests of CMCT-modified 5 s RNA run on the standard two-dimensional system. The "blocked" oligonucleotides (e.g. G-U-A-G-C) lie to the right of the unmodified nucleotides because of their extra positive charge at pH 3·5. Their mobility in the second dimension is also considerably greater than a corresponding unmodified

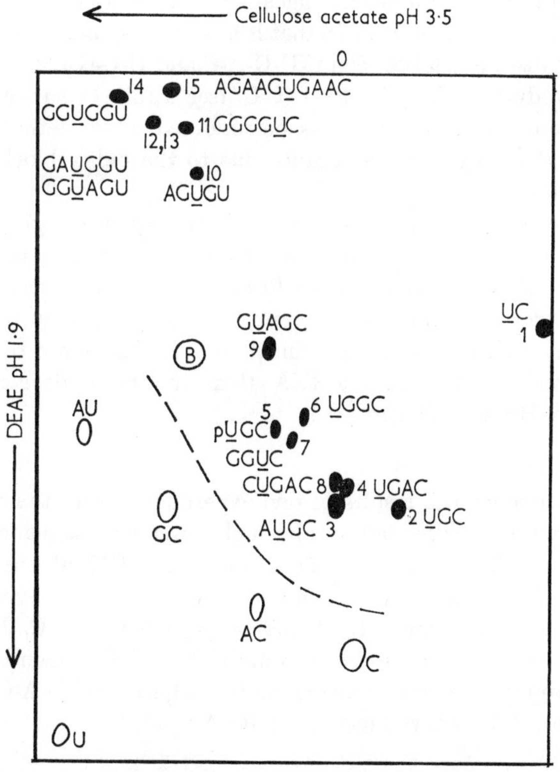

FIG. 12. Diagram showing the position of all CMCT-blocked nucleotides containing more than one pyrimidine residue isolated by partial reaction with CMCT followed by ribonuclease A digestion.

Fractionation was by ionophoresis on cellulose acetate, pH 3·5, and DEAE-paper at pH 1·9. The underlining indicates the modified residue. Some unmodified nucleotides (open circles) are included for reference. B is the blue marker and O marks the origin of the first dimension. The analyses are given in Table 7. Hyphens have been omitted from the known nucleotide sequences because of lack of space.

oligonucleotide and is consistent with the addition of two positive charges at pH 1·9. The two charges are on the quaternary ammonium group of the morpholine group and, presumably, on the substituted guanidino group of the CMCT-base adduct (see Fig. 4).

The pattern of spots in Figure 12 was not entirely reproducible although most spots were consistently present. However, C15 was usually weak and C7, C11 and C13 were only detected once. Some modified nucleotides with only one pyrimidine residue were present, usually in lower amounts than those shown in Figure 12; for example, AGGGAAC, G-A-G-A-G-U and A-G-A-A-G-U. They were detected in positions on the "fingerprint" consistent with the modification of one of their G residues.

Table 7 gives the evidence for the sequences of the nucleotides from Figure 12. The sequence of many spots was found merely from a knowledge of the ribonuclease A end-products. Thus C15 gives A-G-A-A-G-U and G-A-A-A-C. Because of the specificity of the blocking reaction for U residues, the sequence must be AGAAGUGAAAC. This was confirmed by the presence of U-G in the ribonuclease T_1 digest. Because the blocking of U residues was only partial, some products were obtained having no C's; for example, A-G-U-G-U (C10). Among these types of partial products G-G-U-G-G-U (C14) was present in very high yield, suggesting that its internal U was more reactive than its 3′-terminal U. C8 is unusual in that it has two C's, and, although the sequence analysis does not distinguish between C-U-G-A-C and U-C-G-A-C, the fact that there is no T_1 end-product ending in Py-U-C-G (see Table 1) supports the sequence C-U-G-A-C. Ribonuclease A must thus be hindered in its action adjacent to this 5′-terminal C, perhaps by steric hindrance due to the bulky blocking group on the adjacent U residue.

From these results, it is clear that fewer new sequences are generated by the partial carbodi-imide method than by the partial methylation method. Although both U and G residues may be modified, the former must react considerably faster (see Gilham, 1962) under our conditions. Moreover, not all residues react equally well. This is consistent with a chemical modification occurring more readily in the more open, single-stranded parts of 5 s RNA than in the hydrogen-bonded regions (Augusti-Tocco & Brown, 1965).

(ii) *Complete reaction*

Figure 13 summarizes the results of one experiment using the complete reaction technique. This method gave similar, although fewer new sequences, to the partial reaction method. The only new sequence, numbered C16 at the end of Table 7, analysed as C-U-G-G-C. Nevertheless, a useful feature of this type of analysis was that all the sequences of Figure 13 should be preceded by a C. This proved to be correct except for A-G-C, which derived from two U-A-G-C sequences in the molecule. Also, A-U-G-C appeared twice as strong as the adjacent U-G-A-C, suggesting that there could be two A-U-G-C residues in 5 s RNA.

4. Discussion

In the following discussion it is assumed that during partial digestion with ribonuclease T_1 only the bonds following G residues are split, so that any nucleotide isolated from such a digest can be considered to be preceded by a G on its 5′ end. Similarly it will be assumed that nucleotides isolated from ribonuclease A digests are preceded by a pyrimidine nucleotide.

Since the products isolated by the partial methylation method are obtained from a ribonuclease T_1 digest, it is to be expected that they are preceded by a G residue and this appears to be justified except in the case of nucleotide M9 (A,7MeG)G, which was present in small amounts. There is thus some slight doubt in coming to the above conclusion and no sequence is based solely on it.

TABLE 7

Nucleotides isolated by ribonuclease A digestion of 5 s RNA partially reacted with CMCT

Spot no. (Fig. 12)	Composition	Products of further enzymic digest after removal of CMCT†		Structure deduced
		RNase A	RNase T_1	
C1	C U			U-C
C2	C G U	G-C U	U-G C	U-G-C
C3	C A G U	A-U G-C	A-U-G C	A-U-G-C
C4	C A G U	G-A-C U	U-G A-C	U-G-A-C
C5	C G pU	pU G-C		pU-G-C
C6	C $\underline{G}$ U	G-G-C U	U-G G C	U-G-G-C
C7	C $\underline{G}$ U	G-G-U C	(U,C) $\underline{G}$	G-G-U-C
C8		G-A-C C U	(C,U)G A-C	C-U-G-A-C
C9	C A $\underline{G}$ U	G-U A-G-C	U-A-G G C	G-U-A-G-C
C10	A $\underline{G}$ $\underline{U}$	A-G-U G-U	U-G A-G U	A-G-U-G-U
C11	C $\underline{\underline{G}}$ U	G-G-G-G-U C G-G-G-U‡ G-G-U‡	$\underline{\underline{G}}$ (U,C)	G-G-G-G-U-C
C12	A $\underline{\underline{G}}$ $\underline{U}$	G-G-U G-A-U	A-U-G $\underline{G}$ U	G-A-U-G-G-U
C13§	A $\underline{\underline{G}}$ $\underline{U}$	G-G-U A-G-U G-A-U G-U‡	A-U-G U-A-G $\underline{G}$ U	G-G-U-A-G-U (and C12)
C14	$\underline{G}$ U	G-G-U G-U‡ G‡ U‡	U-G $\underline{\underline{G}}$ U	G-G-U-G-G-U
C15	C $\underline{\underline{A}}$ $\underline{\underline{G}}$ U	A-G-A-A-G-U G-A-A-A-C A-A-A-C‡ A-A-C‡ A-C‡	U-G A-A-G A-A-A-C A-G	AGAAGUGAAAC
C16‖		G-G-C C U		C-U-G-G-C

† The nucleotides were identified by their position relative to marker (RNase A and T_1) digests of 5 s RNA (Fig. 2) on DEAE-ionophoresis at pH 1·9. (U-C) lies between G and C-G. A-A-A-C runs with A-U; and A-A-C with G-C.

‡ Present in smaller amounts and probably the products of over-digestion with enzyme.

§ Contaminated by C12.

‖ Isolated by the "complete reaction" method. See text and Fig. 13.

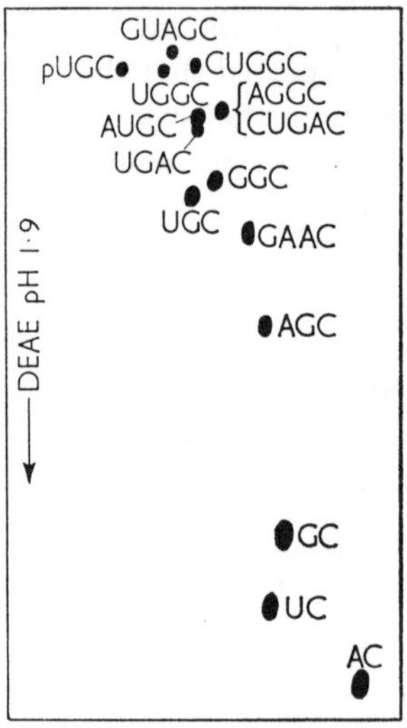

FIG. 13. Diagram of a two-dimensional fractionation as in Fig. 12 ribonuclease A digest of 5 s RNA after "complete" reaction with CMCT and then removal of the "blocking" group.

5′-terminal sequence (1)

pU-G is the 5′-terminal sequence. Nucleotide M28 (pU[C,U,G-C]G) contained the pU and must therefore start pU-G. There are then two sequences possible from the experimental results: pU-G-C-U-C-G or pU-G-C-C-U-G. However, the former is impossible as C-U-C-G is not a T_1 end-product, so that the 5′-terminal sequence becomes pU-G-C-C-U-G. pU-G-C (C5) was found by the partial carbodi-imide method and pU-G and C-C-U-G were found together in many partial digestion products (T7–T9, A7, A8, A10), all of which also give C-G with ribonuclease T_1 and two moles of G-G-C with ribonuclease A. Because there are only two G-G-C sequences in *E. coli* strain MRE600, both must be in this 5′-terminal sequence. The sequence U-G-G-C (C6) and C-U-G-G-C (C16), both of which contain G-G-C, must thus be included in the above nucleotides. This makes the only possible common sequence pUGCCUGG-CGGC. This is confirmed by M7 (G-C-G-G), which is derived from the sequence G-G-C-G-G, since it comes from a T_1 digest. A10 has this common sequence, pUGCCUGGCGGC, B13 has an additional C, and T8 gives an additional C-C-G with ribonuclease T_1 and must be pUGCCUGGCGGCCG. T9 gives the same degradation products as T8, but instead of C-C-G has C-C-U-U-A-G. It must come from the 5′ end of the molecule but its presence cannot be explained on the basis of a unique sequence and it is concluded that there are two different molecules of 5 s RNA, one with the sequence pUGCCUGGCGGCCG (1) and another with the sequence pUGCCUGGCGGCCUUAG (1a). This heterogeneity, which in this case applies only to the strain MRE600, will be discussed further below. Nucleotide A14 also confirms the latter sequence and A8 extends (1) by a U. By analogy with sequence 1a, sequence 1 can be expected to be pUGCCUGGCGGCCGUAG . . . and this is supported by M21 (C-C-G-U-A-G). Sequence 1a is not found in other strains studied, including CA265 on which much of the subsequent work was carried out. In this strain, instead, the sequence pUGCCUGGCGGCAG (T7) was present which differed from T8 by having a C-A-G instead of C-C-G. A7 extended the sequence T7 by a U residue, giving pUGCCUGGCGGCAGU (1b). Thus only sequence 1 is common to MRE600 and CA265 and for the purposes of the following discussion 1a and 1b will be ignored.

3′-terminal sequence (5)

T21 contains the 3′-terminal sequence and analyses uniquely as follows: A-A-C-U-G is given in a ribonuclease T_1 digest and must come from the sequence G-A-A-C-U-G as it is a T_1 end-product (t12). Since A-A-C and not G-A-A-C is in the ribonuclease A digest of T21, A-A-C-U-G must be the 5′ end and C-A-U_{OH} is the 3′ end. C-C-A-G, from the ribonuclease T_1 digest, and A-G-G-C, from the ribonuclease A digest, overlap to give C-C-A-G-G-C. These three sequences account for all the digestion products except G-C and must be joined together in the sequence AACUGCCAGG-CAU_{OH}. A3 gives A-A-C-U-G with ribonuclease T_1 and AGGGAAC with ribonuclease A and shows that the two must overlap, which extends the 3′-terminal sequence to AGGGAACUGCCAGGCAU_{OH}. B15 gives the degradation products expected from this sequence but has U-A-G rather than A-G in the ribonuclease T_1 digest, thus extending the sequence to UAGGGAACUGCCAGGCAU_{OH} . . . (5).

Sequence 4

T2 gives UCUCCCCAUG (t16), U-G and G with ribonuclease T_1 and G-G-G-G-U and A-U as the main ribonuclease A products. C11 has the sequence G-G-G-G-U-C so that G-G-G-G-U must overlap with the 5′ end of t16, and since T2 gives no G-U with ribonuclease A the U-G must join to the 5′ end of G-G-G-G-U to give the sequence UGGGGUCUCCCCAUG . . . (a). A15 contains the G-G-G-G-U sequence. It gives (C,U_2) with ribonuclease T_1, which must represent part of the t16 sequence, so that the 3′ end of A15 is UGGGGUCU. A-G-U must be the 5′ end since A-G is obtained in the ribonuclease T_1 digest. (If it were preceded by a G or A, A-G-U would not be given by ribonuclease A; if it were preceded by C or U, A-G would not be given by ribonuclease T_1.) A15 also gives G-U with ribonuclease A and two moles of U-G with ribonuclease T_1, and this can only be explained by the sequence AGUGUGGGGUCU. A16, A17 and C10 confirm this sequence. This extends sequence (a) above to AGUGUGGGGUCUCCCCAUG (b). T1, B11 and B12 agree with and confirm this sequence. T3 and B18 give U-A-G with ribonuclease T_1 establishing a G-U on the 5′ end of (b). T4 differs from T3 by the addition of a C-G and G-C, thus adding an extra C-G on the 3′ end of (b), and T5 gives A-G with ribonuclease T_1 and G_2,A with ribonuclease A, adding an A-G to the 3′ end to give GUAGUGUGGGGUCUCCCC-AUGCGAG (c). U-G-C-G-A (B1) confirms the 3′ end. UGCGAGAG (B10) extends it by a further A-G. A-G-A-G can only come from the ribonuclease A end-product G-A-G-A-G-U (a18) giving GUAGUGUGGGGUCUCCCCAUGCGAGAGU (d). This is nicely confirmed by A6, AUGCGAGAGU. a18 is the only sequence ending in G-A-G-U so that M23 (A-G-U-A-G), which should be derived from G-A-G-U-A-G, must overlap with it, extending sequence (d) to GUAGUGUGGGGUCUCCCCAUGCGAGAGUAG . . . (4).

Sequence 2

Spot T10 gives only two products (t14, t15) with ribonuclease T_1 and the presence of G-A-C in the ribonuclease A digest establishes their order as UCCCACCUGA-CCCCAUG (e). t14 and t15 are also present in a series of progressively larger fragments, T11 to T15. T11 differs from T10 by the addition of a G in the ribonuclease T_1 digest and a G-U in the ribonuclease A digest. The G must therefore be added to the 5′ end of (e). T12 differs from T11 by an extra U-G in the ribonuclease T_1 digest and G-G-U instead of G-U in the ribonuclease A digest. That G-G-U is on the 5′ side of sequence (e) may be concluded from B16 which contains G-G-U and t15 but only the 5′ part of t14 (A-C-C-C-C-A). T12 is therefore UGGUCCCACCUGACCCCAUG (f). T13 has an extra G-U compared with T12, which therefore adds a G to the 5′ end of (f), and T14 differs from T13 by a C-G in the ribonuclease T_1 digest and by having two G-G-U sequences and no G-U in the ribonuclease A digest. Because there is no G-C in the ribonuclease A digest C-G must be 5′-terminal and the sequence is CGGUGGUCCC-ACCUGACCCCAUG (g). Other sequences containing G-G-U are M14 (G-G-U-G-G) and C14 (pyGGUGGU), which confirm that the two G-G-U's are consecutive.

Besides giving two G-G-U's, t15 and A-C-C-C-C-A, B16 gives U-A-G and two C-G's with ribonuclease T_1 and A-G-C and G-C with ribonuclease A. It gives no G-U with ribonuclease A, so U-A-G must be the 5′ end and overlaps with A-G-C to give U-A-G-C. To accommodate two C-G's in B16 the sequence of (g) must be extended to give UAGCGCGGUGGUCCCACCUGACCCCAUG (h). This is confirmed by M8 which is derived from G-C-G-C-G-G. T15 is also consistent with (h).

Spot B2 analyses uniquely for UGCCGAA and B3 for UGCCGAAC. There are two possible T_1 end-products starting A-A-C that could be involved in the 3′ end of this sequence—A-A-C-U-G (t12) and AACUCAG (t13). t12 is already involved in the 3′-terminal sequence (5) so that t13 should overlap with the UGCCGAAC to give UGCCGAACUCAG. This is confirmed by the isolation of the unique sequences AUGCCGAACUC (A2) and CCGAACUCAG (T16) which extend the above sequence to pyAUGCCGAACUCAG (i). There are only two T_1 end-products ending in py-A-U-G; these are t16 which is already accounted for in sequence 4 and therefore could not overlap with (i), and t14, which is the 3′ end of sequence (h). Thus (h) and (i) must overlap to give UAGCGCGGUGGUCCCACCUGACCCCAUGCCGAACUCAG (j). T17 and T18 analysed uniquely as AACUCAGAAG and CCGAACUCAGAAG, respectively, and established an extra A-A-G on the 3′ end of (j). This was confirmed also by B17 which analysed uniquely as the sequence UGCCGAACUCAGAAG. A-G-A-A-G is derived from a17 (A-G-A-A-G-U) and this is joined to a9 (G-A-A-A-C) in nucleotide C15, which is AGAAGUGAAAC. A-A-A-C is present in the T_1 end-product A-A-A-C-G, and B6 and B9 show that A-A-A-C-G is associated with C-C-G giving AGAAGUG-AAACGCCG (k). A12 and A4 allow (k) to be extended by a U, and T20 extends (k) by a U-A-G. Thus (j) may be extended to UAGCGCGGUGGUCCCACCUGACCCC-AUGCCGAACUCAGAAGUGAAACGCCGUAG . . . (2).

Sequence 3

C12 is pyGAUGGU. M16 (G-U-A-G) is derived from G-G-U-A-G and overlaps with C12 to give pyGAUGGUAG (l) since of the three G-G-U's two are already accounted for in sequence 2.

A11 gives A-U-G, C-C-G and G with ribonuclease T_1 and G-G-U, G-A-U and G-C with ribonuclease A and is therefore GCCGAUGGU. This extends (l) to GCCGAUGGUAG (m).

B5, 7 and 8 terminate at their 3′ end in C-G-A and the only sequence with which they can overlap is (m). Also A5 must overlap with (m), which can thus be extended to give UAGCGCCGAUGGUAG . . . (3).

All the nucleotides present in a complete ribonuclease T_1 or A digest of 5 s RNA are now included in the five sequences listed in Table 8. These five sequences represented a stage in the chronological development of the work at which progress became particularly difficult. It will be seen that all the sequences except sequence 1 have G-U-A-G at the 5′ end, and all except sequence 5 have it at the 3′ end. There are thus four G-U-A-G sequences present in the 5 s RNA and the remaining problem was to determine which ones overlapped. All occurred in the more labile single-stranded regions of the molecule, so that it was difficult to obtain the necessary overlapping nucleotides, which would have to contain a characteristic sequence on either side of a G-U-A-G. This was eventually accomplished by a more detailed study of the partial digests to include as far as possible larger nucleotides and ones that were present in relatively small amounts. During this work considerable confirmation of the above five sequences was obtained and the necessary overlaps found.

TABLE 8

1.	pUGCCUGGCGGCCGUAG
2.	GUAGCGCGGUGGUCCCACCUGACCCCAUGCCGAACUCAGAAGUGAAACGCCGUAG
3.	GUAGCGCCGAUGGUAG
4.	GUAGUGUGGGGUCUCCCCAUGCGAGAGUAG
5.	GUAGGGAACUGCCAGGCAU$_{OH}$

T6 gave t16, U-G and U-A-G with ribonuclease T_1 and G-G-G-G-U with ribonuclease A, indicating that the 5′ end of sequence 4 is involved. It also gave A-U-G with ribonuclease T_1 and G-G-U with ribonuclease A so that the 3′ end of sequence 3 is involved, and therefore sequences 3 and 4 must overlap.

T19 was obtained in both a partial T_1 digest and a *Bacillus subtilis* ribonuclease digest. It contained A-A-A-C-G, U-A-G, C-G and C-C-G (two moles), indicating the 3′ end of sequence 2 was present. It also gave A-G-C with ribonuclease A. There are only two A-G-C sequences and one is on the 5′ end of this same sequence (2) and cannot be involved here. The other is at the 5′ end of sequence 3 and must therefore be the one concerned in this nucleotide. This establishes an overlap between sequence 2 and sequence 3.

Thus, having joined together 2, 3 and 4 to give a single sequence, there are only three sequences left—1 which is 5′ terminal, 5 which is 3′ terminal, and 2-3-4 which must be in the middle. These three can now be overlapped at their G-U-A-G's to give the complete sequence shown in Figure 14. B21, which contains the various products from sequence 1, gives A-G-C and G-G-U with ribonuclease A and thus confirms the overlap between sequence 1 and sequence 2. No clear confirmation has been obtained for the overlap between sequences 4 and 5.

In the absence of an exact knowledge of the function of the 5 s RNA, a discussion of the significance of the sequence can be little more than speculative: however, a number of interesting points may be considered.

In both strains of *E. coli* that we have studied there appear to be two forms of 5 s RNA present in about equal amounts. The sequence given in Figure 14 is common to both species. Strain MRE600 contains a second form in which the G in position 13 is changed to a U and CA265 one in which the C in position 12 is changed to an A. Since a batch of *E. coli* grown from a single colony contained both forms they are

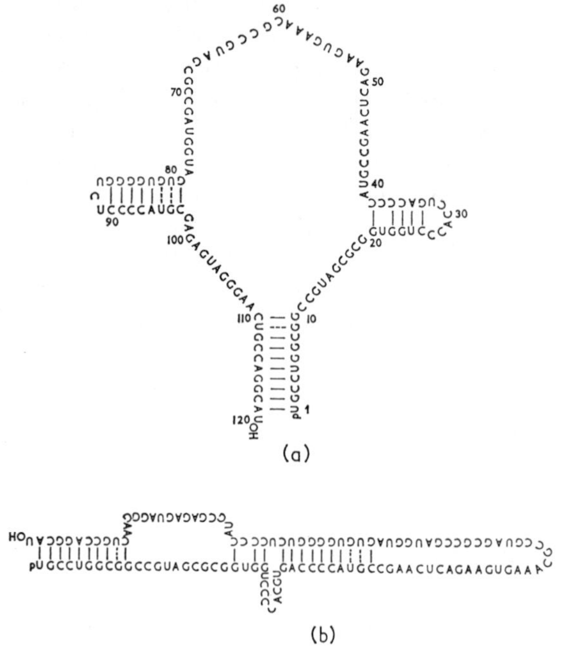

FIG. 14. The sequence of 5 s RNA of *E. coli*.

(a) and (b) are two possible base-paired structures. A solid line indicates a standard (G—C or A—U) pair and a dashed line a G- -U pair; for example between residues 81 and 95. Other more extensively base-paired schemes have been suggested (Cantor, 1967; Boedtker & Kelling, 1967).

presumably under the control of two separate genes. From hybridization studies Morell, Smith, Dubnau & Marmur (1967) suggest that there are a number of genes controlling the production of 5 s RNA, so that it is not surprising to find two forms. Indeed one would have expected to find many more differences and it seems more surprising that it was possible to determine a unique sequence. Preliminary studies on some of the minor components of digests of 5 s RNA suggest that there may be other forms of 5 s RNA with one or two base changes present in smaller amounts (10 to 20%).

Certain regions of the sequence were particularly resistant to digestion with ribonucleases and this was presumed to be due to their being involved in hydrogen-bonded double-stranded structures. Results by other workers on transfer RNA molecules have shown that this is the most likely explanation for resistance (Holley *et al.*, 1965; Zachau *et al.*, 1966), various loops being formed by suitable base-pairing sequences. This conclusion is strengthened by the finding that the sequences in the resistant regions are such as to allow plausible stable base-paired structures, and the most probable form of this base-pairing has been suggested (Brownlee *et al.*, 1967) and is shown in Figure 14(a). It has less base pairing than the transfer RNA's but, in common with them, shows pairing between the two ends of the molecule and other sequences in loops. It should, however, be pointed out that other structures, such as that shown in Figure 14(b), are possible. The various areas concerned are homologous (see below) as well as complementary and can thus be paired together in various ways. From the anomalously high chromatographic rate of nucleotides containing the sequences UGGGGUCUCCCCAUG and UGGUCCCACCUGACCCCAUG, it was suggested that they were isolated as "hairpin" loops (as in Fig. 7(a)), which were present as such in the intact 5 s RNA. However, it is also possible that these loops are formed after liberation of the nucleotide in the partial digest and a structure such as that in Figure 14(b) is not excluded. Both models probably indicate a minimum secondary structure of 5 s RNA and other less stable helical regions could be present. Cantor (1967) has recently studied the optical rotatory dispersion of the 5 s RNA and has suggested that there is more base pairing than is shown in the above models. A model proposed by him on the basis of maximum base pairing is difficult to reconcile with the partial digestion results, but illustrates the possibility of obtaining more base pairing in a three-dimensional model

Certain homologies in the sequence have been pointed out (Brownlee *et al.*, 1967) and are illustrated in Figure 15. It was suggested that this may have been due to gene

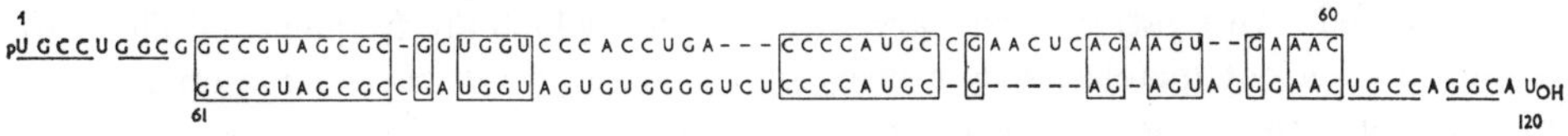

Fig. 15. Homologies between the two halves of the sequences of 5 s RNA.

Residues are numbered as in Fig. 14 and homologies shown by the boxed areas. Dashes are where gaps have to be left in the sequence in order to maximize these homologies. The underlining shows similarities between the two ends of the molecule.

duplication during evolution of the 5 s RNA. The homology of the two ends of the molecule is more difficult to account for. It may represent a second duplication event, or it could have arisen by a single duplication if the original DNA were assumed to be cyclic. One must also seek some explanation for the fact that homology is shown in certain regions but that between them there is no homology. Presumably the preservation of the homologous sequences against mutations could best be explained if they are essential for the function of the molecule. From this one can speculate on a symmetrical function for the molecule; i.e. it could have two similar active centres or binding sites. On this basis the possibility cannot be excluded that the homologies may have arisen by convergent evolution from a single gene rather than by divergent evolution from a duplicated structure.

It is interesting to note that the sequence G-A-A-C occurs twice in the molecule. This is complementary to the sequence G-T-Ψ-C which has been found to be a common feature of all transfer RNA's so far studied, and has been considered as a possible ribosome binding site. If this were significant it could implicate 5 s RNA in binding transfer RNA to the ribosome. It should, however, be pointed out that there are likely to be many other G-A-A-C sequences in the larger ribosomal RNA and that the G of the G-T-Ψ-C is normally considered to be involved in base pairing in the transfer RNA. Also the two G-A-A-C sequences in the 5 s RNA do not come into equivalent positions in the homologous sequences shown in Figure 15.

Forget & Weissman (1967) have recently determined the sequence of the 5 s RNA from human carcinoma (KB) cells. Although it shows a certain amount of homology with the *E. coli* material it is surprisingly different and does not show many of the features discussed above. The main common features are that both 5 s RNA's are 120 residues long and both probably show base pairing between the two ends. Other possible base-pairing residues seem to differ in the two molecules. The KB cell material contains repeating sequences but different ones and in different positions from those found in *E. coli* material. Again the KB cell material contains the sequence G-A-A-C but in a different position from that in the *E. coli*.

This work has been largely concerned with investigating methods for the determination of sequences in RNA. As found by other workers, the most successful technique for degradation is partial digestion with ribonuclease and we would include the spleen acid ribonuclease as being a useful enzyme besides ribonucleases T_1 and A. All three enzymes split preferentially in certain regions which are believed to be single-stranded: however, their different specificities do make it possible to obtain the necessary overlaps; thus the main products obtained in all partial digests are nucleotides of about 15 to 30 residues long containing the stable (double-stranded) regions intact but differing in the distance they extend into the single-stranded region at either end. In fact it would probably have been possible to determine the complete sequence using only the above three enzymes. However, this would have required a more detailed investigation of the larger nucleotides, which are frequently difficult to obtain in sufficient quantities. For studying the structure of these large nucleotides we have only used digestion with ribonucleases T_1 and A. The end-products have already been well characterized and could usually be identified in a one-dimensional ionophoretic separation on DEAE-paper. In most cases the results of the two digests did not give a unique sequence for a single oligonucleotide and this was usually deduced from other partial degradation products. In this respect the smaller nucleotides obtained from the two chemical modification procedures were useful. Both methods were somewhat limited by the presence of side reactions, but the products were relatively easy to fractionate and the specificity was different from that shown by direct partial digestion in that there was no preferential splitting in single-stranded regions.

The main technical problem has been the fractionation of the oligonucleotides and we have been particularly interested in studying paper techniques because of their rapidity and ease of handling. The original two-dimensional method (Sanger *et al.*, 1965) was satisfactory for nucleotides of up to about 10 to 15 residues long, depending on the number of U's and G's they contained. The separations obtained are probably as good as those obtained by column procedures and are more rapid. Larger nucleotides, however, were not well separated and an alternative second dimension was necessary. Chromatography on DEAE-paper and particularly the homochromatography method fulfilled this need as far as the present work on 5 s RNA was concerned. It was possible to purify nucleotides of up to about 25 to 30 residues long, provided they were not present in too complex mixtures. Larger nucleotides still remained adsorbed to the origin of the DEAE-paper. It would clearly have simplified the problem if these large nucleotides could be well fractionated, and for studying longer RNA's such methods will probably be necessary.

REFERENCES

Augusti-Tocco, G. & Brown, G. L. (1965). *Nature*, **206**, 683.

Bayev, A., Venkstern, T., Mirzabekov, A., Krutilina, A., Axelrod, V., Li, L. & Engelhardt, V. (1967). *Proc. 4th Fed. Eur. Biochem. Soc.*, Oslo, in the press.

Bernardi, A. & Bernardi, G. (1966). *Biochim. biophys. Acta*, **129**, 23.

Boedtker, H. & Kelling, D. G. (1967). *Biochem. Biophys. Res. Comm.* **29**, 758.

Brimacombe, R. L. C., Griffin, B. E., Haines, J. A., Haslam, W. J. & Reese, C. M. (1965). *Biochemistry*, **4**, 2452.

Brownlee, G. G. & Sanger, F. (1967). *J. Mol. Biol.* **23**, 337.
Brownlee, G. G., Sanger, F. & Barrell, B. G. (1967). *Nature*, **215**, 735.
Cantor, C. R. (1967). *Nature*, **216**, 513.
Forget, B. G. & Weissman, S. H. (1967). *Science*, **158**, 1695.
Gilham, P. T. (1962). *J. Amer. Chem. Soc.* **84**, 687.
Holley, R. W., Apgar, J., Everett, G. A., Madison, J. T., Marquisee, M., Merrill, S. H., Penswick, J. R. & Zamir, A. (1965). *Science*, **147**, 1462.
Lawley, P. D. & Brookes, P. (1963). *Biochem. J.* **89**, 127.
Lawley, P. D. & Wallick, C. A. (1957). *Chem. & Ind.*, p. 633.
Lee, J. C., Ho, N. W. Y. & Gilham, P. T. (1965). *Biochim. biophys. Acta*, **95**, 503.
Madison, J. T., Everett, G. A. & Kung, H. (1966). *Science*, **153**, 531.
Morell, P., Smith, I., Dubnau, D. & Marmur, J. (1967). *Biochemistry*, **6**, 258.
Naughton, M. A. & Hagopian, H. (1962). *Analyt. Biochem.* **3**, 276.
RajBhandary, U. L., Chang, S. H., Stuart, A., Faulkner, R. D., Hoskinson, R. M. & Khorana, H. G. (1967). *Proc. Nat. Acad. Sci., Wash.* **57**, 751.
Rosset, R. & Monier, R. (1963). *Biochim. biophys. Acta*, **68**, 653.
Sanger, F., Brownlee, G. G. & Barrell, B. G. (1965). *J. Mol. Biol.* **13**, 373.
Wyatt, G. R. (1951). *Biochem. J.* **48**, 584.
Zachau, H. G., Dütting, D. & Feldmann, H. (1966). *Hoppe Seyler's Z. Physiol. Chem.* **347**, 212.

Volume 3, number 4 FEBS LETTERS June 1969

THE SEQUENCE OF PHENYLALANINE tRNA FROM *E.COLI*

B.G.BARRELL and F.SANGER
MRC Laboratory of Molecular Biology, Cambridge, England

Received 21 April 1969

Gassen and Uziel [1] have recently proposed a structure for phenylalanine tRNA from *E.coli*. We have also been studying the sequence of this tRNA but have obtained certain results that are in conflict with the above sequence.

^{32}P-labelled $tRNA^{Phe}$ was prepared from *E.coli* grown on ^{32}P-phosphate [2]. Initially it was purified by making use of its specific binding to ribosomes in the presence of polyuridylic acid [3] and much of the preliminary work was done on this material. The yield of pure $tRNA^{Phe}$ was rather low with this method and in later work purification was effected using reversed phase partition chromatography [4].

Figs. 1 and 3 show the two-dimensional fractionation [2] of a ribonuclease Tl and a pancreatic ribonuclease digest of the $tRNA^{Phe}$. The sequences of the nucleotides are given in the accompanying diagrams. X appears to be the same unknown unstable nucleotide that was found in $tRNA^{Met}_{M}$ [5]. 2mt6iA appears to be the same minor component that is found in $tRNA^{Tyr}$ and is therefore tentatively identified as 2-thiomethyl-6-isopentenyl adenylic acid [6].

Digests prepared by partial digestion with ribonuclease Tl, pancreatic ribonuclease and *B.subtilis* ribonuclease were also studied and fractionated on the homochromatography system [7]. The results are summarized in fig. 2 which shows the oligonucleotides that were obtained and the sequence derived from them. The sequence is deduced as the only possible one that fits our experimental results; however in certain cases, particularly in the D loop (positions 10–20), the conclusions depend on the results with only one partial digestion product (indicated by —x— in fig. 2) and there is insufficient confirmatory evidence to establish the sequence without making use of analogies with other tRNAs.

Yaniv et al. [8] have shown that in $tRNA^{Val}_{1}$ the 4-tU in position 8 can react under the influence of UV light with a C residue in position 13 to form a covalent bond. A similar reaction also occurs in $tRNA^{Phe}$ [8] and our results indicated that a stable cross linkage was formed during some of the fractionation procedures between the nucleotide A4tUAG and CUCAG. This caused some difficulty in interpreting the results.

The above sequence can be written in the "cloverleaf" form (fig. 4) and is found to conform to the common features of other tRNAs. The sequence GAA is in the anticodon position. This is the expected anticodon for the two known codons for phenylalanine, UUC and UUU. It is followed by a minor nucleotide that is probably the same as that found in $tRNA^{Tyr}$.

The sequence at the 3′ end (positions 48–76) agrees with that reported by Gassen and Uziel. They do not record the base X and there are several differences in the anticodon loop* and the D loop.

Abbreviations: $tRNA^{Phe}$, phenylalanine accepting transfer RNA; G, U, A, C, T, Ψ, the 3′-phosphates of guanosine, uridine, adenosine, cytidine, ribosylthymine and pseudouridine (5-ribosyluracil) respectively; D, dihydrouridine 3′-phosphate; 7mG, 7-methylguanosine 3′-phosphate; 4-tU, 4-thio-uridine 3′-phosphate; 2mt6iA, probably 2-thiomethyl-6-isopentenyl adenosine 3′-phosphate.

* Uziel and Gassen (Biochemistry, in press) have recently identified an anticodon sequence identical to the one we report.

Volume 3, number 4 FEBS LETTERS June 1969

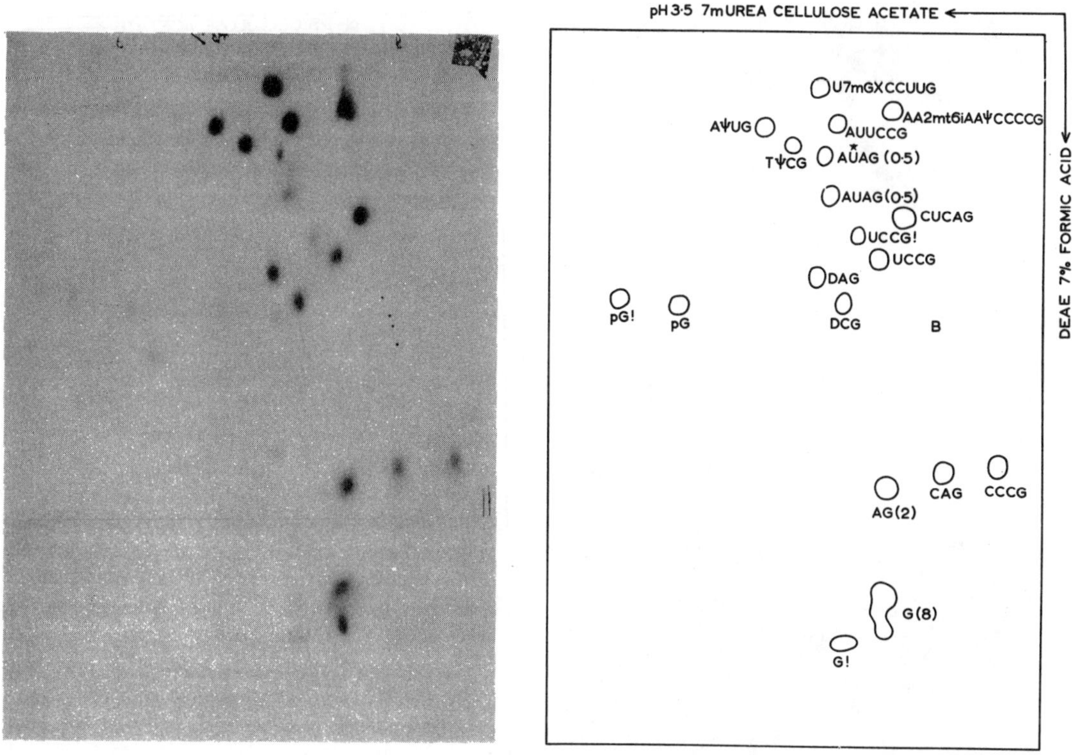

Fig. 1. A two-dimensional fractionation of a ribonuclease Tl digest of $tRNA^{Phe}$. Radioautograph and diagram showing the sequences of the nucleotides. B is the blue marker. $CACCA_{OH}$ is not included in this fingerprint. Some cyclic nucleotides are present and are indicated by "!". Molar yields of nucleotides are given in brackets. Where the nucleotides occur in one molar yield no figures are given. AU*AG is probably A4-tUAG which has been converted to AUAG while the 7% formic acid on the paper was drying. AUAG and the streak immediately above it was probably formed from A4-tUAG when the paper was wetted and electrophoresed in 7% formic acid. The streaking of the G is due to incomplete removal of the urea after electrophoresis in the first dimension.

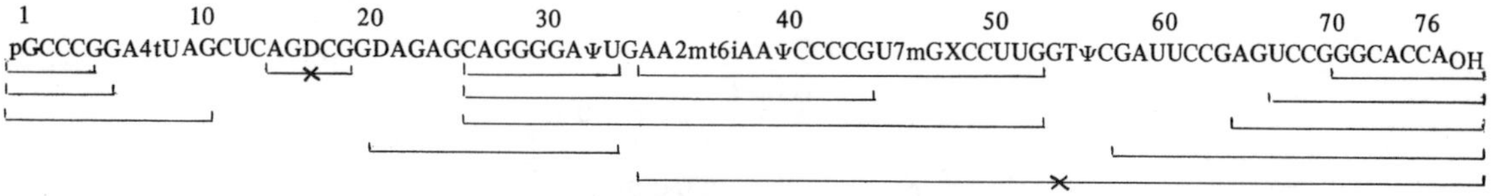

Fig. 2. Nucleotide sequence of $tRNA^{Phe}$ showing the products obtained by limited digestion with ribonuclease Tl, pancreatic ribonuclease and *B.subtilis* ribonuclease. (—x—) indicates that the product was only obtained once. All the others have been isolated several times.

Volume 3, number 4 FEBS LETTERS June 1969

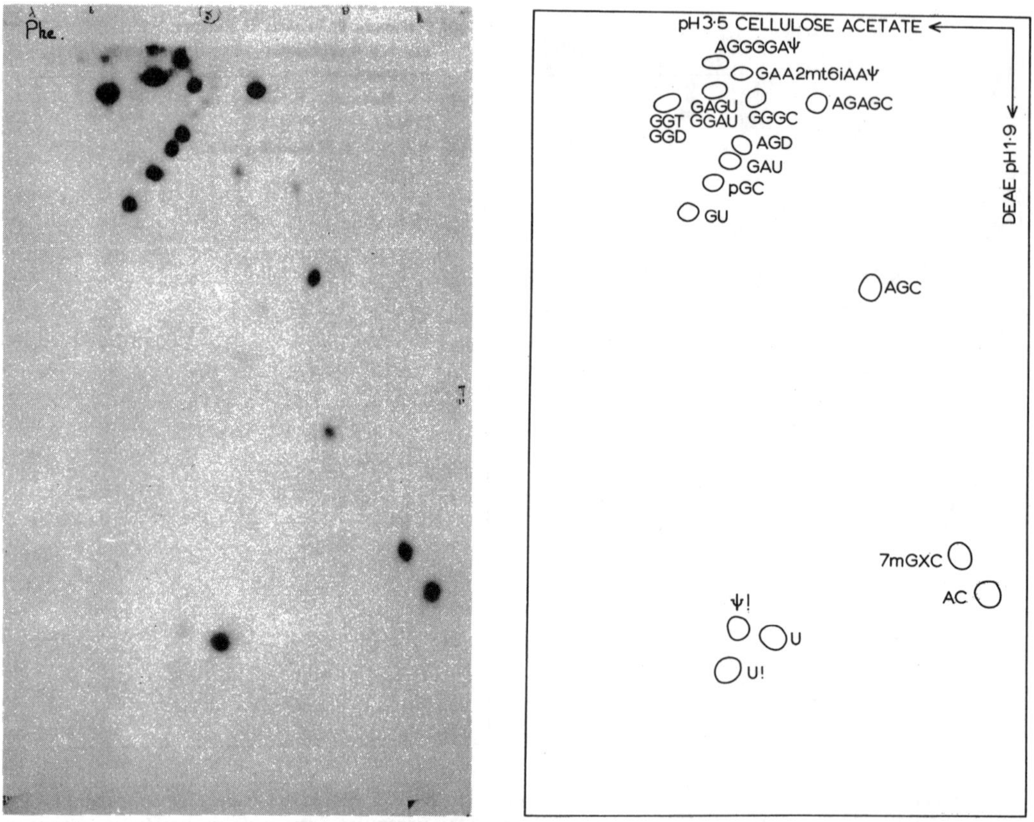

Fig. 3. A two-dimensional fractionation of a pancreatic ribonuclease digest of tRNA^{Phe}. Radioautograph and diagram showing the sequences of the nucleotides. C and C> are not included in this fingerprint. AGGGGAΨ is always obtained in low yield and some of the minor spots on the fingerprint are due to breakdown of this nucleotide. GGAU is also obtained in low yield and is formed from GGA4-tU after electrophoresis at acid pH.

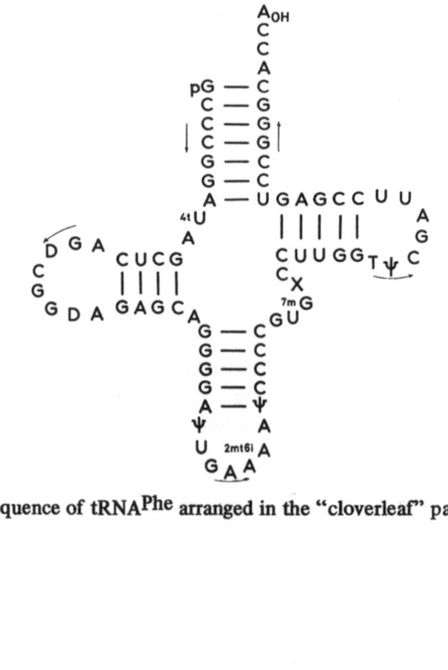

Fig. 4. The nucleotide sequence of tRNA^{Phe} arranged in the "cloverleaf" pattern typical of other tRNAs.

Volume 3, number 4 FEBS LETTERS June 1969

References

[1] H.G.Gassen and M.Uziel, Abstracts FEBS Madrid (1969) 128.
[2] F.Sanger, G.G.Brownlee and B.G.Barrell, J. Mol. Biol. 13 (1965) 373.
[3] M.Nirenberg and P.Leder, Science 145 (1964) 1399.
[4] A.D.Kelmers, G.D.Novelli and M.P.Stulberg, J. Biol. Chem. 240 (1965) 3979.
[5] S.Cory, K.A.Marcker, S.K.Dube and B.F.C.Clark, Nature 220 (1968) 1039.
[6] F.Harada, H.J.Gross, F.Kimura, S.H.Chang, S.Nishimura and U.L.Rajbhandary, Biochem. Biophys. Res. Commun. 33 (1968) 299.
[7] G.G.Brownlee, F.Sanger and B.G.Barrell, J. Mol. Biol. 34 (1968) 379.
[8] M.Yaniv, B.G.Barrell and A.Favre, to be published.

European J. Biochem. 11 (1969) 395—399

Chromatography of ^{32}P-Labelled Oligonucleotides on Thin Layers of DEAE-Cellulose

G. G. Brownlee and F. Sanger

Medical Research Council Laboratory of Molecular Biology, Cambridge

(Received July 19/September 15, 1969)

A two-dimensional fractionation procedure has been developed for separating radioactively-labelled oligonucleotides of up to 50 residues long, using uniformly ^{32}P-labelled 5S RNA of *Escherichia coli* as a model compound. The method uses ionophoresis on cellulose acetate at pH 3.5 in the first dimension; and ascending chromatography with a concentrated mixture of oligonucleotides on thin layers of mixed DEAE-cellulose and cellulose in the second dimension.

We have previously described two separate two-dimensional fractionation systems, using modified papers, suitable for studying enzymatic digests of ^{32}P-labelled RNA. The first [1] is a two-dimensional ionophoretic procedure for studying the end-products of digestion with ribonuclease T_1 or pancreatic ribonuclease. The second procedure [2] uses both ionophoresis and chromatography and is suitable for studying the products of partial enzymatic digestion of purified low-molecular weight RNA, such as 5S RNA or transfer RNA. The second dimension is a type of displacement chromatography by which unlabelled oligonucleotides are used to displace labelled oligonucleotides fixed on DEAE-paper, and we have referred to this as "homochromatography". Homochromatography on DEAE-paper is limited to the fractionation of oligonucleotides of less than 25 residues in length, those larger than this remaining at, or near, the origin. It was of some importance to develop methods for extending this technique to larger fragments, especially for the determination of the sequence of molecules larger than tRNA or 5S RNA. A method suitable for fragments of up to 50 residues long was therefore developed using 5S RNA of known sequence [2] as a model compound, and this is described in this paper. It uses ionophoresis on cellulose acetate at pH 3.5 for the first dimension and homochromatography on thin layers of mixed DEAE-cellulose and cellulose in the second dimension. This method may also be used with advantage for separating smaller oligonucleotides (10—30 residues in length) as these are better resolved and show less streaking than on the DEAE-paper system.

Unusual Abbreviations. tRNA = transfer RNA; G>p = guanosine-2′,3′-cyclic phosphate; Cp, Ap, Gp and Up the 3′-phosphates of cytidine, adenosine, guanosine and uridine, respectively.

Enzymes. T_1-ribonuclease (EC 2.7.7.26); pancreatic ribonuclease (EC 2.7.7.16).

MATERIALS AND METHODS

Preparation of Thin Layers

The Desaga basic thin layer equipment and the adsorbants were purchased from Camlab (Glass) Ltd. (Milton Road, Cambridge). A slurry of DEAE-cellulose (MN 300 DEAE) and cellulose (MN 300 cellulose) was prepared sufficient for four long glass plates (20×40 cm) or eight short (20×20 cm) plates, as follows. For the "1:10" plates, 1.5 g DEAE-cellulose and 15 g cellulose were mixed and then added, with stirring, to 102 ml distilled water until the mixture was fairly evenly dispersed and wetted. The slurry was then homogenized thoroughly in a fast electric blender for about 3 min to break up aggregated particles. It was then de-aerated thoroughly and 50 ml poured into the spreader (giving a fixed 250 μ layer) which was then passed fairly rapidly across the plates positioned on the template. The other half of the slurry could be used when the spreader was reclosed and positioned to spread the remaining glass plates on a second template. The plates were separated slightly and allowed to dry at room temperature. On inspection they should have no streaks in the direction of spreading, which are usually caused by air bubbles trapped in the slurry and prevent even flow; nor should lumps be present, which are aggregated particles reflecting incomplete homogenization. Slight ridges perpendicular to the direction of spreading are less significant and indicate that the layer was not spread evenly (usually too slowly) or that the slurry was too dilute. Although perfect plates probably give the best separations, we have not observed markedly inferior fractionations with plates that were somewhat imperfect in any of the above respects. Plates containing a higher proportion of DEAE-cellulose to cellulose, in a ratio of 1 part to 7.5 parts instead of 1:10, were also prepared. Finally a razor blade was used to define a sharp edge to the thin layers.

Fractionation Procedure

^{32}P-labelled 5S RNA was extracted from *E. coli* (CA265) grown in the presence of [^{32}P]phosphate and was purified by acrylamide gel electrophoresis on 10 % acrylamide slabs [3]. A different preparation purified by successive column chromatography on DEAE-Sephadex [4] and on a reversed phase column [5] was kindly provided by Dr. J. Abelson and was used for the fractionation shown in Fig. 3.

Partial T_1-ribonuclease (Sankyo, Japan) digests were prepared in 2—5 µl in 0.02 M Tris-chloride, 0.02 M magnesium chloride, pH 7.5, incubated for 30 min at 0° and fractionated on cellulose acetate strips (Oxoid or Schleicher & Schüll) at pH 3.5 in the first dimension, as before [1]. The solution used to wet the strips (5 % acetic acid—7 M urea, adjusted to pH 3.5 with pyridine) has a very weak buffering capacity and the pH was found to increase at room temperature. It was therefore stored at 4° and rejected when the pH rose higher than 3.7. Nucleotides were blotted from the cellulose acetate to the plates as follows. The strip was removed from the tank used for the first dimensional run and monitored with a portable Geiger counter to find the position of the oligonucleotides, which in partial digests was rather fast, usually in the region just slower than the major red, acid fuchsin, marker. This part of the strip was then placed on the thin layer plate, 3 cm from one of the short ends, perpendicular to the direction of spreading of the plate. Three moist, but not too wet, strips of Whatman 3 MM paper were carefully placed on top followed by a glass plate to maintain efficient contact. Water thus flowed into the layer, transferring the oligonucleotides onto the DEAE-cellulose layer. Care is needed in this procedure to obtain efficient transfer and we prefer to use strips that still have some "white spirit" on their surface, rather than let the buffer in the strip dry out. Excess water or pressure applied to the glass plate does not help the transfer but rather tends to dislodge the layer. Transfer is not entirely quantitative and is probably poorer for the larger fragments. However over 80 % of the total material is usually transferred. Because only 20 cm was available for the second dimension we preferred to use short runs in the first dimension on approx. 50 cm cellulose acetate strips for about 1 h, applying 6 kV. This spread out the partial products less well than the longer runs on 85 cm strips, but allowed them to be included in a single second-dimensional run.

Fractionation in the second dimension was carried out at 50—60° in an oven using one or other of the homomixtures [2] described below. Which particular mixture was used depended on whether the largest fragments, 15 to 40—50 residues long, were to be isolated (homomixture *a*); an intermediate size, in the range of 10 to approx. 25 residues (homomixture *b*); or even smaller oligonucleotides, 1 to approx. 15 (homomixture *c*). The thin-layer tank, used for the short plate, or the tall tank (formed by inverting one thin-layer tank on top of another) used for long plates, are equilibrated with approx. 100 ml of the homomixture at 60° in the oven. The plate is also equilibrated at 60°, but outside the tank. Before starting the ascending homochromatography, the plate is briefly chromatographed with distilled water until the front is about 5 cm up the plate. This is designed to wash excess urea, transferred from the first dimension, away from the origin which otherwise interferes in the fractionation, especially of faster moving oligonucleotides (as can be seen in Fig. 3, which was not washed). An alternative procedure, which is quicker and therefore results in less cooling of the plate, is to spray the region of the origin with distilled water. After this the plate is transferred into the equilibrated tank and ascending homochromatography continued until the front reaches the top of the plate. This usually takes about 2 h for short plates and 5 h for long ones, although we have noticed considerable variation in running time, which we cannot account for.

The plate is then dried, marked with red ink containing [^{35}S]sulphate, and a radioautograph prepared by exposing to an X-ray film in a lead-lined folder. It was convenient to prepare two separate exposures (one for an elution template and the other for the records) by exposing two films one on top of the other.

Homomixtures were prepared as follows:

Homomixture a was essentially as before [2] and was prepared by dissolving 10 g of yeast RNA (British Drug Houses Ltd.) in 200 ml of 7 M urea and adjusting to pH 7.5 with 10 N KOH. The mixture was kept frozen and was used in the separation shown in Fig. 3. It has also been used to isolate a T_1-oligonucleotide (21 residues long) from R 17 bacteriophage RNA [6].

Homomixture b was a preparation of homomixture *a* dialysed against 7 M urea for 2—3 h at 4°. In this time there was approximately 10 % increase in volume. It was used for the fractionation shown in Fig. 2.

Homomixture c was a dialysed and hydrolysed homomixture prepared by dissolving 10 g of yeast RNA in 100 ml of 1 N KOH and hydrolysing it for 15 min at room temperature. The solution was then neutralized to pH 7.5 with concentrated HCl and dialysed against distilled water for 2—4 h. 84 g of urea was then added and the volume made up to 200 ml with distilled water to give a 5 % homomixture in 7 M urea. Homomixture *c* may be diluted with 7 M urea to give a 3 % mixture, which gives somewhat sharper spots than the 5 % mixture, although the R_F values are lower.

Homomixture *c* was not used in the experiments quoted in this paper but is described here for completeness. It is useful for the fractionation of the oligonucleotides encountered in complete T_1-ribonuclease digests of RNA, when it is usually used as a 3% mixture on 1:7.5 plates. Fragments from 1 to about 15 residues long are separated from one another approximately according to chain length. The resolution of the smaller oligonucleotides is generally poorer than on the standard two-dimensional ionophoretic fractionation system [1], although fractionation of those in the range 10—15 residues is usually rather better. Longer ones are better separated using homomixture *b* or homomixture *a*.

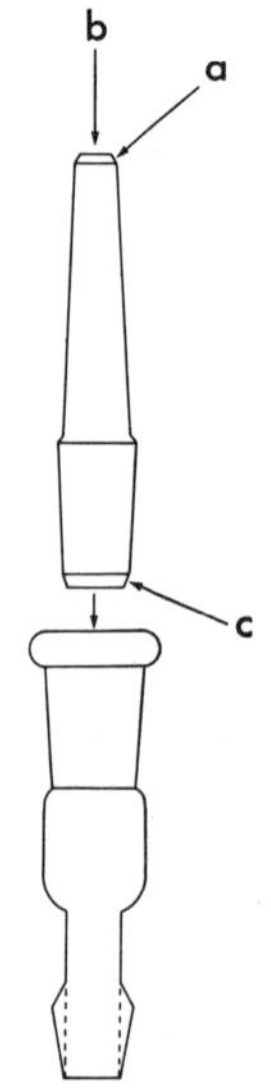

Fig. 1. *Diagram of a thin layer elution device.* This has a B10/19 ground glass joint which can be connected to a vacuum line. The upper part has an orifice "b" of approx. 2 mm internal diameter and is ground on its circumference to give a sharp edge, "a", suitable for scraping and dislodging the thin layer. A sintered disc of porosity 1 or 2 is welded in at "c". Device available from T. W. Wingent Ltd. (Milton, Cambridge, England)

Elution and Analysis

Those spots to be eluted were marked and numbered on the radioautograph that showed the [^{35}S]sulphate ink marks, and holes were cut out with a razor blade corresponding to these spots. This radioautograph then functioned as a "template" and was stuck with sellotape onto the thin layer, aligning it by means of the ink marks. The thin layer spots thus exposed through the holes of this template could then be scraped off, using the elution device shown in Fig. 1 which is connected to a vacuum line. The cellulose is thus sucked up through the narrow orifice (b) and is trapped in the sinter (c). The porosity of the sinter should be number 1 or 2. If it is too course the sinter lets some of the smaller particles through in the succeeding washing and elution procedure, whilst if it is too fine particles do not easily suck up and subsequent elution is very tedious. Ten to twenty samples were usually eluted, using separate top sections of the elution device and the same bottom section. These were then inverted and held in position in a stand with clips. Urea was first removed from the samples by washing with 95% ethanol before eluting oligonucleotides with triethylamine carbonate, pH 10.0 [1]. Two or three drops (0.2 ml) were collected into small silicone-treated [1] test tubes, to which a further two or three drops of water was added, and the triethylamine carbonate removed *in vacuo* using three successive washes with water.

The analysis of partial digestion products by further treatment with T_1 and pancreatic ribonuclease was as before [2] except that lower enzyme concentrations and shorter digestion times could be used as less carrier nucleotide was present on the DEAE-thin layers than on the DEAE-paper. Suitable conditions for both T_1 and pancreatic ribonuclease were 10 μl of 0.1 mg/ml of enzyme and digestion was for 30 min at 37°. Conditions for other sequence procedures of spots from thin layeis are presented by Adams *et al.* [6].

RESULTS

Fig. 2 shows in its upper half a radioautograph of a partial T_1-ribonuclease digest (1:4000 for 30 min at 0°) of 5S RNA using a long 1:10 plate and homomixture *b* for the second dimension. Although there is a large number of products they are resolved from one other as discrete spots, except for the faster moving and smaller products which move as bands, and the slower products where the mixture is too complex to resolve. The sequence of some of the fragments was determined by further T_1-ribonuclease digestion and by reference to the known sequence of 5S RNA [2] reproduced in Fig. 4. In all cases identical partial digestion fragments had been previously isolated by homochromatography on DEAE-paper, so that there was no difficulty in identification. These sequences are recorded by the residue number (see Fig. 4) of the two ends of the fragment in the lower half of Fig. 2. Oligonucleotides from 13 to 28 residues long were isolated and this suggests that the method is suitable for fractionation of oligonucleotides of this size range, assuming the mixture to be fractionated is no more complex than in Fig. 2. If it is simpler the size range of fractionation may well be extended to fragments over 30 residues long.

Fig. 3 shows a less extensive but simpler T_1-ribonuclease digestion of 5S RNA which had been purified by column chromatography. Although the conditions of digestion are more violent than those of the previous experiment, the digestion is less complete, which presumably reflects differences in the con-

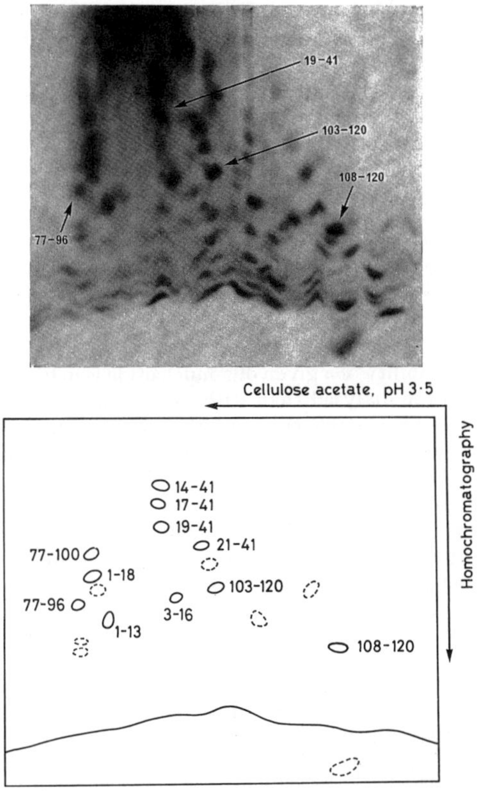

Fig. 2. *Radioautograph and diagram of a two-dimensional fractionation of a partial T_1-ribonuclease digest (1:4000, 30 min at 0°) of 5S RNA using a 50 cm cellulose acetate strip (Schleicher & Schüll) in the first dimension at pH 3.5 in 7 M urea; and using homomixture* b *for homochromatography on a long, 1:10, thin layer plate in the second dimension.* The numbers on the diagram are an abbreviation for the sequence which can be read off by referring to Fig. 4. The diagram and radioautograph are aligned vertically to allow the reader to recognize which spots on the radioautograph the numbers on the diagram refer. To aid identification a few are marked directly on the radioautograph. Dotted circles are included in the diagram to show some fairly strong spots that were not eluted. The major nucleotide front has an R_F of approx. 0.5 relative to the solvent front. The single nucleotide ahead of this is G>p

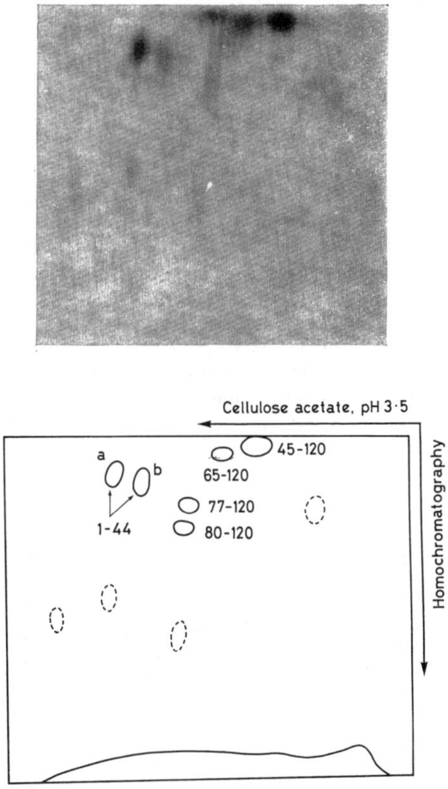

Fig. 3. *Radioautograph and diagram of a two-dimensional fractionation of a partial T_1-ribonuclease digest (1:500, 30 min at 0°)* of *5S RNA (purified by column chromatography).* The sample was fractionated as in Fig. 2 except for the following: "Oxoid" cellulose acetate (57 cm) was used in the first dimension, and homomixture *a* and a short plate in the second dimension. The meaning of symbols is as in the legend to Fig. 2 and the text. The wavy line at the bottom of the diagram marks the nucleotide front which in this case is close to the solvent front

figuration of the RNA prepared by two different methods. Fractionation in this case was on a short 1:10 plate using homomixture *a* in the second dimension. The faster moving products are very streaky as in this experiment no attempt was made to chromatograph urea away from the origin before homochromatography (see Methods section). Nevertheless some of the slower moving spots were eluted and their sequences determined by further digestion with T_1 and pancreatic ribonuclease.

As is shown in the lower half of Fig. 3, oligonucleotides of from 41 to 76 residues long were isolated pure. These fragments had not been isolated in our previous work and this provided confirmation that the sequence shown in Fig. 4 was correct. Two alternative forms of the sequence of residues 1 to 44 were isolated, marked "a" and "b" on Fig. 3, which on analysis were found to differ only in the residue at position 12. This was adenine in spot "a" and cytosine in spot "b". These separate fragments derive from the two major species of 5S RNA and are presumed to separate in the first dimension on the basis of differences in shape, as the single base change would probably not be sufficient to alter mobility on the basis of a charge difference. The unequal yields of the two fragments presumably reflect a different susceptibility of the two forms of 5S RNA to cleavage with T_1 ribonuclease under the conditions used, perhaps because of slight differences in shape. Thus, using this system, fragments of up to 45 residues long are well separated, and even larger

```
1                  10  C                 20                      30
pU-G-C-C-U-G-G-C-G-G-C- -G-U-A-G-C-G-C-G-G-U-G-G-U-C-C-C-A-C-
                       A
                    40                      50                        60
C-U-G-A-C-C-C-C-A-U-G-C-C-G-A-A-C-U-C-A-G-A-A-G-U-G-A-A-A-C-
                    70                      80                        90
G-C-C-G-U-A-G-C-G-C-C-G-A-U-G-G-U-A-G-U-G-U-G-G-G-G-U-C-U-C-
                    100                     110                      120
C-C-C-A-U-G-C-G-A-G-A-G-U-A-G-G-G-A-A-C-U-G-C-C-A-G-G-C-A-U OH
```

Fig. 4. *The nucleotide sequence of 5 S RNA of* E. coli *(strain CA265)*. Both cytosine and adenine occur in position 12, in approximately equal yields, as deduced from the yields of the T_1-ribonuclease end-product, CpApGp, which is 0.5 relative to other end-products present in 1.0 relative molar yields

fragments may be isolated if the mixture is not too complex. Fragments smaller than 15 residues long move at, or near, the nucleotide front and are not well fractionated.

DISCUSSION

Thin layer techniques are well known to offer advantages over paper fractionation systems [7]. The operator has strict control over the composition, the thickness and uniformity of the system, and usually considerably better separations are observed in shorter times than with paper. In the work reported here the technique allowed us to reduce the "effective" capacity of the DEAE-cellulose by mixing it with an excess of cellulose. We were thus able to isolate larger fragments by homochromatography on DEAE-thin layer than was possible on DEAE-paper. Moreover, the resolution obtained was better as the spots were less diffuse. In comparing the homomixtures described, mixtures *b* and *c*, which are both dialysed mixtures, give better resolution than mixture *a*, which usually gives slightly more streaky, though acceptable, results. The first two mixtures are thus preferable for separating mixtures of fragments of less than 30 residues long, while for larger fragments mixture *a* is best. The reason for the improved resolution with the dialysed mixtures may be due to removal of salt, as the conditions of dialysis do not result in the loss of a significant amount of nucleotide material from the mixtures. This would explain why the mobility of a given oligonucleotide is in fact slower with the dialysed than with the undialysed mixture. Thus it is possible that with the homomixtures *b* and *c* a true displacement chromatography is occurring.

The fractionation system described in this paper should be useful for separating fragments of ribonucleic acids of unknown sequence, especially for the larger RNA molecules where rather complex mixtures of products may be obtained.

The expert assistance of Mrs. E. Cartwright is gratefully acknowledged.

REFERENCES

1. Sanger, F., Brownlee, G. G., and Barrell, B. G., *J. Mol. Biol.* 13 (1965) 373.
2. Brownlee, G. G., Sanger, F., and Barrell, B. G., *J. Mol. Biol.* 34 (1968) 379.
3. Peacock, A. C., and Dingman, C. W., *Biochemistry*, 6 (1967) 1818.
4. Nishimura, S., Harada, F., Narushima, U., and Seno, T., *Biochim. Biophys. Acta*, 142 (1967) 133.
5. Kelmers, A. D., Novelli, G. D., and Stulberg, M. P., *J. Chem.* 240 (1965) 3979.
6. Adams, J. M., Jeppesen, P. G. N., Sanger, F., and Barrell, B. G., *Nature*, 223 (1969) 1009.
7. Randerath, K., and Randerath, E., In *Methods in Enzymology, Nucleic Acids* (edited by L. Grossmann and K. Moldave), Academic Press, London-New York 1967, Vol. XII, part. A, p. 323.

G. G. Brownlee and F. Sanger
M.R.C. Laboratory of Molecular Biology
Hills Road, Cambridge, CB2 2QH, Great Britain

(*Reprinted from Nature, Vol.* 223, *No.* 5210, *pp.* 1009–1014, *September* 6, 1969)

Nucleotide Sequence from the Coat Protein Cistron of R17 Bacteriophage RNA

by
J. M. ADAMS*
P. G. N. JEPPESEN
F. SANGER
B. G. BARRELL
MRC Laboratory of Molecular Biology, Cambridge

The sequence of fifty-seven nucleotides in the coat protein cistron of phage R17 RNA directly confirms the genetic code, shows that the code used by the phage is degenerate and suggests that highly ordered base-paired structures exist in this RNA. Such base-paired loops may be involved in regulation of cistron expression and packing of the RNA in the phage particle.

ALTHOUGH the nature of the genetic code is well established, it has not been possible until now to determine by chemical means a sequence from a messenger RNA and to show that it is related by the code to the sequence of amino-acids in the protein that it specifies. The best characterized messenger RNAs that can be obtained in a pure form are the single-stranded RNAs containing about 3,300 nucleotide residues[1] isolated from RNA bacteriophages, such as R17, f2 and MS2. The nucleotide sequences at the ends of these molecules[2–6] have been determined and, for MS2 RNA, the sequences of the products of pancreatic ribonuclease digestion[7]. R17 RNA codes for three proteins[8], one of which is the phage coat protein of known amino-acid sequence[9]. Here we report a nucleotide sequence from the coat protein cistron of R17 RNA.

In this laboratory we have developed fractionation methods for ^{32}P-labelled oligonucleotides[10–12] which have been applied in the determination of the nucleotide sequences of *t*RNAs[13] and the 5*S* ribosomal RNA[12], which is 120 nucleotides long. The method used for separating nucleotides up to about ten residues in length is ionophoresis on a two-dimensional system using cellulose acetate in one dimension and DEAE-paper in the other[10]. Longer oligonucleotides are not well fractionated by the ionophoresis on DEAE-paper; however, those up to about thirty residues in length can be separated by a form of displacement chromatography on DEAE-paper (termed "homochromatography")[12], in which a mixture of non-radioactive nucleotides is used to develop the chromatogram and fractionate the radioactive nucleotides.

Recently we have found that even larger oligonucleotides can be resolved with homochromatography on a thin layer of DEAE-cellulose instead of on DEAE-paper; nucleotides up to fifty residues long can be fractionated very well by a new two-dimensional system—ionophoresis on cellulose acetate followed by thin layer homochromatography[14]. When R17 RNA was digested completely with T_1 ribonuclease a few large oligonucleotides were produced that could be purified by this two-dimensional system. One of these proved to be of particular interest because its nucleotide sequence corresponds to an amino-acid sequence in the coat protein of the phage.

Large Oligonucleotides from a Complete T_1 Ribonuclease Digest of R17 RNA

R17 RNA, labelled uniformly with ^{32}P, was digested with T_1 ribonuclease and the digest fractionated on a two-dimensional system using ionophoresis at *p*H 3·5 in 7 M urea on cellulose acetate as the first dimension and thin layer chromatography on DEAE-cellulose as the second dimension[14]. Fig. 1 is an autoradiograph of the fractionation. The smaller oligonucleotides are not well resolved but certain of the larger products are clearly separated from the others. These large oligonucleotides were isolated from the chromatogram and subjected to preliminary sequence analysis as follows.

First, base compositions were determined by hydrolysing a portion of each oligonucleotide with alkali; the resulting 3′-mononucleotides were fractionated by electrophoresis on paper at *p*H 3·5, located by autoradiography, and then measured by scintillation counting. A second portion of each oligonucleotide was digested with pancreatic ribonuclease and the digestion products separated by electrophoresis on DEAE-paper at *p*H 3·5 (Fig. 2). Most of the products could be identified uniquely from their position on the paper. Where confirmation was necessary, the composition of each product was analysed by alkaline hydrolysis.

Using the genetic code, one can write a set of nucleotide sequences corresponding to the known amino-acid sequence of the R17 coat protein. By comparing the information about each of the large oligonucleotides obtained as given here with the possible nucleotide sequences for the coat

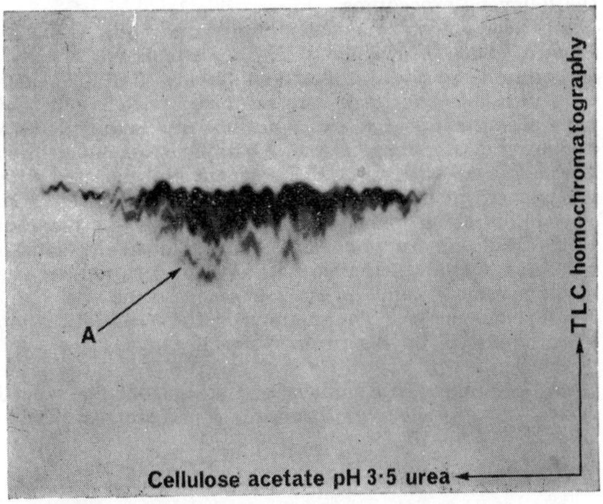

Fig. 1. A two-dimensional fractionation of a ribonuclease T_1 digest of R17 RNA. Uniformly ^{32}P-labelled bacteriophage R17 RNA was prepared as described by Dahlberg[2]. 20 μg (about 10^7 c.p.m.) was digested with 1 μg of ribonuclease T_1 dissolved in 3 μl. of 0·01 M *tris*-HCl, *p*H 7·4, containing 0·001 M EDTA, at 37° C for 30 min, fractionated on a two-dimensional system[14] using homomixture *a* for homochromatography with DEAE-cellulose and cellulose in the ratio of 1 : 7·5 on a short plate.

* Present address: Institut de Biologie Moléculaire, 1211 Genève 4, Switzerland.

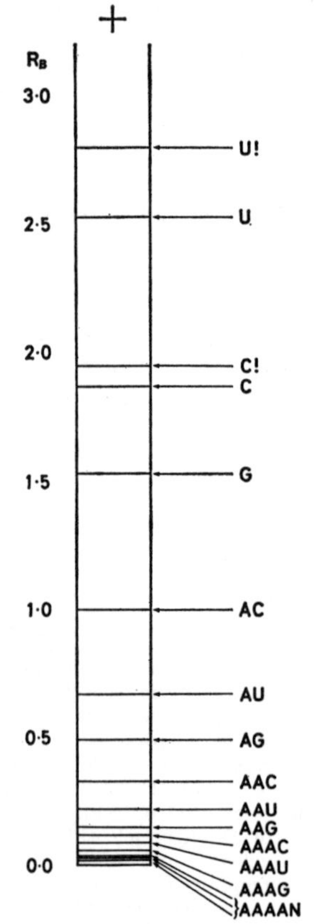

Fig. 2. Fractionation of pancreatic ribonuclease digestion products on DEAE-paper at *p*H 3·5.

protein, we found that oligonucleotide A could have come from the coat protein cistron.

Sequence of Oligonucleotide A

To establish the sequence of oligonucleotide A we used the following procedures.

(*a*) Reaction with a carbodiimide reagent: Gilham[15] showed that N-cyclohexyl-N′-(β-morpholinyl-(4)-ethyl) carbodiimide-methyl-p-toluene sulphonate (CMCT) reacts with uridylate and guanylate residues in RNA and that pancreatic ribonuclease can hydrolyse the modified RNA only after cytidylate residues. A sample of oligonucleotide A was treated with the CMCT reagent and digested with pancreatic ribonuclease; the digest was subjected to electrophoresis on paper at *p*H 3·5. Fig. 3 is a diagram of the resulting fractionation. The modified nucleotides were then eluted, incubated with ammonia to remove the blocking groups, and digested a second time with pancreatic ribonuclease. The products of the second digestion were identified by electrophoresis on paper at *p*H 3·5 (Table 1).

(*b*) Digestion with ribonuclease U_2: this enzyme, which was kindly provided by Professor F. Egami, is specific for purines when it is used at a low concentration[16]. Thus the enzyme hydrolyses oligonucleotides from a T_1 ribonuclease digest only after adenylate residues. A portion of oligonucleotide A was digested with ribonuclease U_2 and the products were fractionated by electrophoresis on DEAE-paper at *p*H 1·9. The base compositions of the products were determined by alkaline hydrolysis (Table 1). Certain products appear in pairs, differing only in the number of adenylate residues. This reflects the increased resistance of adenylate tracts to attack by ribonuclease U_2 and is useful for obtaining overlapping sequences with the pancreatic ribonuclease products.

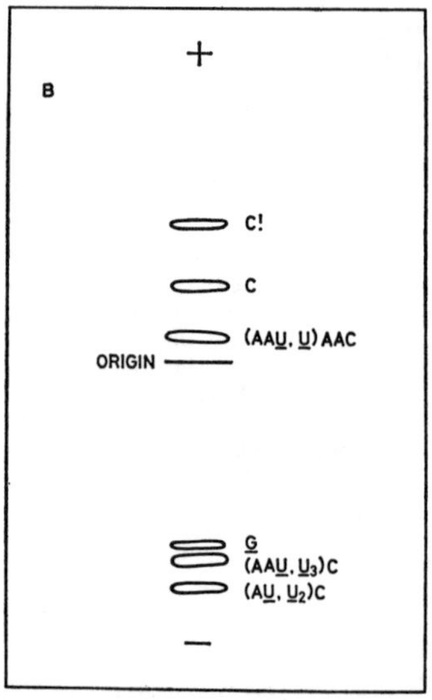

Fig. 3. Fractionation of the pancreatic ribonuclease products from carbodiimide-blocked oligonucleotide A. A sample of oligonucleotide A was incubated overnight at 37° C with 10 μl. of a 100 mg/ml. solution of CMCT dissolved in 0·01 M *tris*-HCl, *p*H 8·9, containing 0·001 M EDTA. 10 μl. of a 0·1 mg/ml. solution of pancreatic ribonuclease in 0·05 M *tris*-HCl, *p*H 7·4, containing 0·001 M EDTA, was then added and the mixture incubated for a further 30 min at 37° C. The digest was then applied as a 2 cm streak at the centre of a 57 cm strip of Whatman 3*MM* paper. Electrophoresis was then carried out in *p*H 3·5 pyridine acetate buffer for 1 h at 3 kV.

Combining the results obtained with ribonuclease U_2 and pancreatic ribonuclease (both before and after blocking with CMCT), a unique sequence can be constructed for oligonucleotide A (Fig. 4). If this nucleotide sequence acts as a messenger for the synthesis of protein, it can code for three possible amino-acid sequences, depending on the phase in which the message is read. These three sequences are also shown in Fig. 4. Amino-acid sequences 1 and 2 are not found in the coat protein sequence but sequence 3 is present at positions 89 (Glu) to 95 (Phe)[9]. Because a sequence of seven amino-acids is statistically a very rare structure, it is considered highly likely that oligonucleotide A is a fragment from the coat protein cistron of R17 RNA. To extend the knowledge of the nucleotide sequence in this region of the RNA we tried to find a larger fragment of the phage RNA that contains oligonucleotide A.

Sequence of nucleotide A (G) A A U U A A C U A U U C C A A U U U U C G

Possible amino-acid sequences coded for by nucleotide A:

Asn . — . Leu . Phe . Gln . Phe . Ser .
Ile . Asn . Tyr . Ser . Asn . Phe . Arg .
Glu . Leu . Thr . Ile . Pro . Ile . Phe .

Fig. 4. The nucleotide sequence of oligonucleotide A and the three possible amino-acid sequences for which it can code.

Table 1. SEQUENCE ANALYSIS OF OLIGONUCLEOTIDE A

(*a*) Base composition (G = 1·0)	(*b*) Pancreatic ribonuclease digestion products	(*c*) CMCT blocked pancreatic ribonuclease digestion products	(*d*) U_2 ribonuclease digestion products
U 8·6	2AAU*	C	A
		(AA$\dot{U}$,$\dot{U}$)AAC	AA
G 1·0	AAC		(C,U)A
		$\dot{G}$	
A 6·5	AU		UUA
		(AA$\dot{U}$,U_3)C.	
C 3·9	G		UUAA
	3C*	(A$\dot{U}$,$\dot{U}_2$)C	
	6U*		
			(U_2,C_2)A
			(U_2,C_2)AA
			(U_4,C)G

* These estimations were based on a visual examination of the autoradiograph together with the base composition determined in (*a*).

Oligonucleotide A was located on the thin layer of DEAE-cellulose in Fig. 1 by autoradiography, and isolated in the following manner. The DEAE-cellulose in the region of the spot was scraped off, washed with 95 per cent ethanol to remove the urea, and the nucleotide was eluted with 30 per cent triethylamine carbonate, *p*H 10·5, in a total volume of about 200 μl. The triethylamine carbonate was removed by evaporation under vacuum and the oligonucleotide subjected to preliminary sequence analysis. (*a*) The base composition was determined by hydrolysis with 5–10 μl. of 0·2 M NaOH; the products were separated by electrophoresis on paper at *p*H 3·5 (ref. 10), located by autoradiography and quantitated by scintillation counting. (*b*) The oligonucleotide was digested for 30 min at 37° C in 5–10 μl. pancreatic ribonuclease (0·1 mg/ml. enzyme in 0·01 M *tris*-HCl, *p*H 7·5, and 0·001 M EDTA). The products were separated by electrophoresis on DEAE-paper at *p*H 3·5 (Fig. 2) and identified by alkaline hydrolysis and/or mobility. (*c*) The modified nucleotide was eluted from the various bands in Fig. 3, dried, and incubated overnight with 0·2 M ammonia to remove the blocking groups. After the ammonia had been dried off, the nucleotides were digested with 10 μl. of 0·1 mg/ml. pancreatic ribonuclease for 30 min at 37° C in buffer containing 0·01 M *tris*-HCl and 0·001 M EDTA, and the products fractionated by electrophoresis on paper at *p*H 3·5. $\dot{U}$ indicates the modified uridylic acid residue. (*d*) A portion of oligonucleotide A was incubated at 37° C for 2 h with 10 μl. of ribonuclease U_2 (0·1 units/ml. in 0·05 M sodium acetate, *p*H 4·5, containing 0·002 M EDTA and 0·1 mg/ml. bovine serum albumin).

Partial T_1 Ribonuclease Digest of R17 RNA

When a partial enzymic digest of ribosomal RNA is electrophoresed on a polyacrylamide gel a number of discrete bands are found[17]. We tried this approach for making specific fragments of R17 RNA. Samples of ^{32}P-labelled R17 RNA were digested with various amounts of ribonuclease T_1 at 0° C in a buffer of high ionic strength, and the partial digests were electrophoresed on a long flat slab of 12·5 per cent polyacrylamide gel by a modification of the method of Peacock and Dingman[18] which was developed in this laboratory with G. G. Brownlee. (A flat slab is particularly suitable for autoradiography and also for comparing different samples on the same gel.) Fig. 5 shows an autoradiograph of the fractionation obtained. In the undigested control sample virtually all the RNA remains at the origin because it is too large to penetrate the gel. With increasing amounts of added enzyme, however, more and more bands appear and there is a progressive increase in the amounts of the smaller, faster-moving fragments. As many as forty discrete bands can be seen in the more extensively digested samples. The RNA fragments in these bands range in size from guanosine monophosphate, in the fastest moving band, to fragments more than 300 nucleotides long near the top of the gel.

This experiment shows that there is an extremely wide range in the rate at which T_1 ribonuclease splits different guanylate residues in the molecule, presumably because of the structure of the RNA. Moreover, it shows that gel electrophoresis is capable of resolving many of the fragments that result from this very specific hydrolysis. Recently Gould, Pinder and Matthews[20] have reported that a partial digest of the RNA from the related phage μ2 also contains a number of specific fragments and we find that this is the case with f2 phage RNA as well. Gesteland and Boedtker[1] have shown by physical studies that R17 RNA has an unusually compact structure at a high ionic strength. The highly specific fragmentation found in the present experiments indicates that phage RNAs have highly organized secondary structures.

The fragmentation of the RNA was usually sufficiently reproducible in different experiments, using different pre-

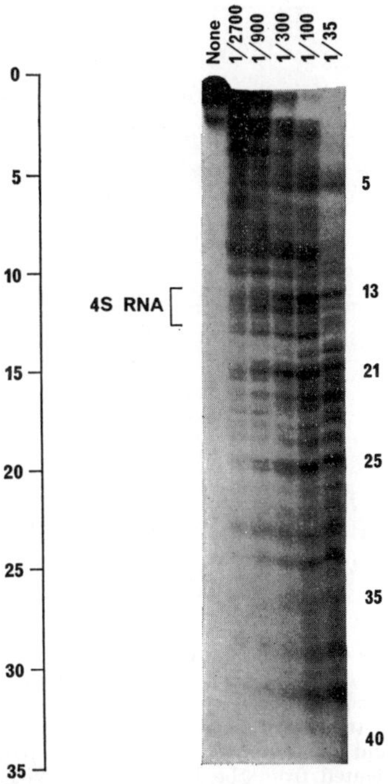

Fig. 5. Gel electrophoresis of a partial digest of R17 RNA with ribonuclease T_1. Samples (20 μg) of ^{32}P-labelled R17 RNA (specific activity 1·5 μCi/μg RNA) were digested with different amounts of ribonuclease T_1 for 1 h at 0° C in 25 μl. of a buffer consisting of 0·2 M NaCl, 0·02 M magnesium acetate, and 0·05 M *tris*-HCl (*p*H 7·5); the enzyme–substrate ratios are shown at the top of the figure. The reaction mixtures were extracted with phenol at 0° C in capillary tubes and then extracted three times with ether. The samples were made 10 per cent with respect to sucrose with a 50 per cent sucrose solution containing bromophenol blue, loaded into the sample wells of a vertical flat slab of 12·5 per cent polyacrylamide gel, and subjected to electrophoresis at 400 V and 40 mA for 16 h at 2°–4° C. At the end of the run the front glass plate was pulled away from the gel and the gel was marked with radioactive ink, covered with cellophane, and subjected to autoradiography for 30 min. The gel was prepared in the following manner. The mould consisted of two thoroughly cleaned glass plates (40 cm × 20 cm × 0·4 cm) separated at their sides by two 'Vaseline'-coated 'Perspex' spacers (40 cm × 2 cm × 0·3 cm) and sealed at the bottom with plasticine. The assembly was held together at the sides by strong spring clips. The gel was made from 12·1 per cent acrylamide and 0·4 per cent bisacrylamide (both recrystallized as described by Loening[19]) and 0·04 M *tris*-acetate, *p*H 8·3, which was also the buffer used in the reservoirs. The gel solution (250 ml.) was de-aerated, 0·6 ml. of freshly prepared 10 per cent ammonium persulphate and 0·25 ml. of N,N,N,N-tetramethylethylenediamine were added, and the solution was poured quickly into the mould. The sample wells were formed by pushing a slotted strip of 'Perspex' (15 cm × 3 cm × 0·3 cm) into the top of the gel. After gelation was complete (in about 5 min) the well-former and plasticine were removed and the bottom of the assembly was placed in a 2 l. reservoir; the top of the gel was connected to another 2 l. reservoir by a short length of Whatman 3*MM* paper. The buffer in the reservoirs was changed once during a run. On the right of the figure are shown the numbers of some of the bands that were studied. The position to which a 4*S*-RNA marker ran is also indicated.

parations of RNA or enzyme, for each band to be identified simply from the overall band pattern. To isolate enough of the fragments to characterize them, preparative digests were made with up to 5 mCi of ^{32}P-labelled R17 RNA and the digests were loaded across the width of a flat slab gel (20 cm × 40 cm × 0·3 cm). In these experiments we chose digestion conditions that would give primarily fragments of a size (up to about 200 nucleotides in length) suitable for sequence analysis. Appropriate conditions were digestion for 1 h at 0° C with an enzyme to substrate ratio of 1 : 100 (w/w) or 10 h at 0° C with a ratio of 1 : 1,000.

Because we wished to study the fragments from different gel bands, it was necessary to find a convenient method of extracting the RNA. The following method was found satisfactory. The bands were cut from the gel using the autoradiograph as a guide, broken into small pieces, and put into cylindrical tubes (10 cm × 1 cm)

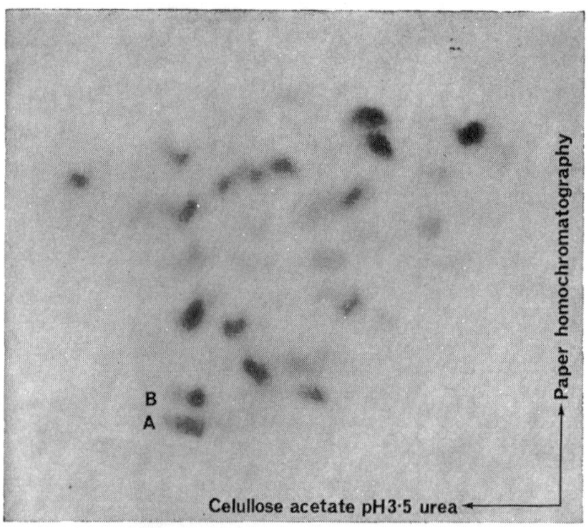

Fig. 6. A two-dimensional homochromatography fingerprint of a ribonuclease T_1 digest of gel band 19. (Indicating oligonucleotides A and B.)

having one end covered with DEAE-paper supported on cellophane. The tubes were then placed between two reservoirs containing 0·04 M *tris*-acetate, *p*H 8·3. When subjected to electrophoresis the RNA migrated out of the gel and was trapped by the DEAE-paper; it could then be eluted from the DEAE-paper with 30 per cent triethylamine carbonate (*p*H 9·7) as described earlier[10]. To get an estimate of the purity of fragments isolated from the gel, the RNA from the different bands was digested with T_1 ribonuclease and fingerprinted using ionophoresis on cellulose acetate at *p*H 3·5 as the first dimension and homochromatography on DEAE-paper as the second (homomixture *c* of ref. 14 but with a 30 min alkaline hydrolysis). The complexity of the fingerprints suggested that many of the bands in the upper part of the gel (greater than eighty nucleotides in length) contain

Table 2. ANALYSIS OF THE NUCLEOTIDE SEQUENCE OF OLIGONUCLEOTIDE B

(*a*) Pancreatic ribonuclease digestion products	(*b*) CMCT blocked pancreatic ribonuclease digestion products	(*c*) U_2 ribonuclease digestion products	Deduced sequence
AAAU	UAC	A	UACUUAAAUAUG
AU	(U_2,AAAU,AU)G	AA	
AC		2UA	
3U		(C,U_2)A	
G		(C,U_2)AA	
		UG	

Oligonucleotide B was eluted from the homochromatography fingerprint (Fig. 6). After drying off under vacuum the following digestion procedures were used. (*a*) A sample was incubated for 2 h at 37° C with 10 μl. of a 0·2 mg/ml. solution of pancreatic ribonuclease dissolved in 0·01 M *tris*-HCl, *p*H 7·4, containing 0·001 M EDTA. (*b*) A further sample was incubated overnight at 37° C with 10 μl. of 100 mg/ml. solution of CMCT dissolved in 0·01 M *tris*-HCl, *p*H 8·9, containing 0·001 M EDTA. 10 μl. of a 0·2 mg/ml. solution of pancreatic ribonuclease dissolved in 0·05 M *tris*-HCl, *p*H 7·4, 0·001 M EDTA was then added, and the mixture incubated for a further 2 h at 37° C. (*c*) A third sample was incubated overnight at 37° C with 10 μl. of a solution of ribonuclease U_2 containing 1 unit/ml. dissolved in 0·05 M sodium acetate buffer, *p*H 4·5, containing 0·002 M EDTA and 0·1 mg/ml. bovine serum albumin. The products of these procedures were fractionated and characterized as described for oligonucleotide A in Table 1.

relatively homogeneous fragments while those nearer the bottom of the gel contain several components.

RNA Fragment containing Oligonucleotide A

To determine which of the fragments from the gel contained oligonucleotide A, the large T1 ribonuclease products from each fingerprint were digested with pancreatic ribonuclease; oligonucleotide A was found in the fingerprints of the material from bands 19 and 21. A fingerprint of band 19 is shown in Fig. 6. Another large nucleotide (B) was also present in these bands and by the procedures described its sequence was shown to be UACUUAAAUAUG (Table 2). This sequence can also code for three amino-acid sequences and one of these is Tyr-Leu-Asn-Met, which is found in the coat protein just before the sequence coded for by nucleotide A. Band 21, which also contained nucleotides A and B, was subjected to further analysis. Its position on the acrylamide gel suggested that it contained a fragment in the size range of 50–60 residues. The fingerprint of the material from the band was, however, considerably more complex than one would expect for a single component of this size

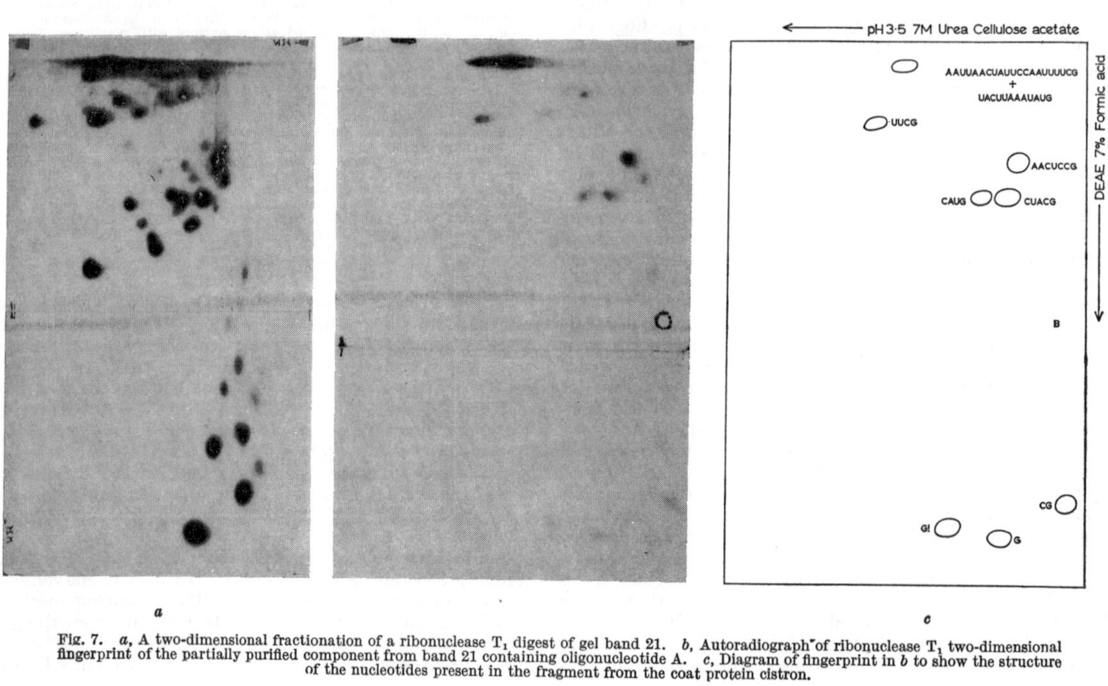

a *c*

Fig. 7. *a*, A two-dimensional fractionation of a ribonuclease T_1 digest of gel band 21. *b*, Autoradiograph of ribonuclease T_1 two-dimensional fingerprint of the partially purified component from band 21 containing oligonucleotide A. *c*, Diagram of fingerprint in *b* to show the structure of the nucleotides present in the fragment from the coat protein cistron.

Table 3. ANALYSIS OF NUCLEOTIDE SEQUENCES OF REMAINING T_1 RIBONUCLEASE OLIGONUCLEOTIDES IN THE FRAGMENT OF R17 RNA (FIG. 7)

(*a*) Pancreatic ribonuclease digestion products	(*b*) CMCT blocked pancreatic ribonuclease digestion products	(*c*) U_2 ribonuclease digestion products	Deduced sequence
AAC U 2C G	AAC C U̇C Ġ	A AA (C_3,U)G	AAC(UC,C)G*
AC U C G	C U̇AC Ġ . .	(C,U)A CG	CUACG
2U C G	UUC Ġ	—	UUCG
AU C G	C . . AUG	CA UG	CAUG
	—	—	CG
	—	—	G

The oligonucleotides from the two-dimensional fingerprint shown in Fig. 7*a* were eluted[10,11]. Samples were subjected to the following procedures. (*a*) Incubation for 30 min at 37° C with 10 μl. of a 0·1 mg/ml. solution of pancreatic ribonuclease in 0·01 M *tris*-HCl, *p*H 7·4, containing 0·001 M EDTA. (*b*) Incubation overnight at 37° C with 10 μl. of a 20 mg/ml. solution of CMCT dissolved in 0·01 M *tris*-HCl, *p*H 8·9, containing 0·001 M EDTA, followed by the addition of 10 μl. of a 0·1 mg/ml. solution of pancreatic ribonuclease in 0·05 M *tris*-HCl, *p*H 7·4, 0·001 M EDTA, and incubation for a further 30 min at 37° C. (*c*) Incubation for 2 h at 37° C with 10 μl. of a 0·1 unit/ml. solution of U_2 ribonuclease in 0·05 M sodium acetate buffer, *p*H 4·5, containing 0·002 M EDTA and 0·1 mg/ml. bovine serum albumin. The products of these digestion procedures were fractionated and characterized by the usual methods.

* The sequence of this oligonucleotide could not be unambiguously assigned with the information presented in the table. In addition, a partial snake venom diesterase digestion was performed: a sample of the oligonucleotide was incubated for 30 min at 37° C with 10 μl. of a 0·1 mg/ml. solution of bacterial alkaline phosphatase dissolved in 0·05 M *tris*-HCl, *p*H 8·9, containing 0·01 M $MgCl_2$ and 1 mg/ml. carrier s-RNA. Following this, 10 μl. of a 0·05 mg/ml. solution of snake venom diesterase in the same buffer was added and digestion allowed to continue for 0 min (a control), 15 min and 30 min at room temperature. The digests were applied to DEAE-paper and the products fractionated by electrophoresis at *p*H 1·9. The following partial digestion products were identified: $(A_2C_2U)C_{OH}$

$(A_2CU)C_{OH}$

$(A_2C)U_{OH}$

The sequence was therefore shown to be AACUCCG.

and therefore we concluded that band 21 contained a mixture of at least two—and probably more—fragments of approximately the same size. In an attempt to separate the component containing oligonucleotide A from the others the material from band 21 was subjected to electrophoresis on cellulose acetate in 7 M urea (*p*H 3·5) at 6 kV for 3 h. After the run the RNA fragments were blotted on to DEAE-paper and an autoradiograph was made. Three main components were evident and each was eluted and digested with T_1 ribonuclease. The digests were fractionated by ionophoresis on the standard two-dimensional system: cellulose acetate in 7 M urea (*p*H 3·5) followed by DEAE-paper in 7 per cent formic acid[10]. Fig. 7*b* shows the fingerprint of the component containing nucleotides A and B (these nucleotides do not separate in this system) compared with the fingerprint of the unfractionated band 21 (Fig. 7*a*). The purification was still not complete but was sufficient to show that the spots indicated in Fig. 7*c* all come from the same fragment. The sequences of the small oligonucleotides from these spots were determined by standard procedures (Table 3).

To determine the order in which the oligonucleotides occur in the fragment, the RNA was partially digested with T_1 ribonuclease. The partial digestion products were fractionated using electrophoresis on cellulose acetate in 7 M urea (*p*H 3·5) as a first dimension, followed by either electrophoresis on DEAE-paper in 7 per cent formic acid or thin layer homochromatography as a second dimension. The former system was used to isolate the smaller products and the latter to isolate the larger ones. Each partial digestion product was then characterized by further digestion with T_1 ribonuclease and pancreatic ribonuclease.

The analysis of the partial digestion products (Table 4) permits the derivation of the complete sequence of the fragment, which is shown in Fig 8. It can be seen that this sequence of fifty-seven nucleotides is related by the genetic code to the sequence of amino-acids in positions 81–99 of the coat protein[9].

Table 4. ANALYSIS OF PARTIAL T_1 RIBONUCLEASE DIGESTION PRODUCTS FROM THE FRAGMENT OF R17 RNA

	Ribonuclease T_1 digestion products	Pancreatic ribonuclease digestion products	Deduced sequence
$T_{III}17$	UUCG CG	GU G 2C U	CGUUCG
T_I3	CAUG UUCG CG G	GGC GU AU G 2C U	CAUGGCGUUCG
$T_{II}11$	CUACG AACUCCG	GAAC G AC 2U 3C	CUACGAACUCCG
$T_{II}3$	AAUUAACUAUUCCAAUUUUCG AACUCCG CUACG	GAAC 2AAU AC GC AAC G AC U ≡ C =	AAUUAACUAUUCCAAUUUUCGCUACGAACUCCG
T_I6	AAUUAACUAUUCCAAUUUUCG UACUUAAAUAUG UUCG CUACG CAUG CG 2G	GGAAU AAAU GGC 2GU AAU 3AU GC 2AC U ≡ C =	CAUGGCGUUCGUACUUAAAUAUGGAAUUAACUAUUCCAAUUUUCGCUACG

A sample of the material from gel band 21 was digested with 1/500 of its weight of T_1 ribonuclease for 15 min at 0° C in about 3 μl. of 0·01 M *tris*-HCl, *p*H 7·4. The partial digestion products were fractionated (see text) and then characterized by further digestion with both T_1 ribonuclease and pancreatic ribonuclease. The structure of each nucleotide was deduced from its degradation products and from the structures listed above it in the table. Other partial degradation products were present which did not give rise to the T_1 ribonuclease digestion products listed in Table 3 and were therefore derived from the other components of gel band 21.

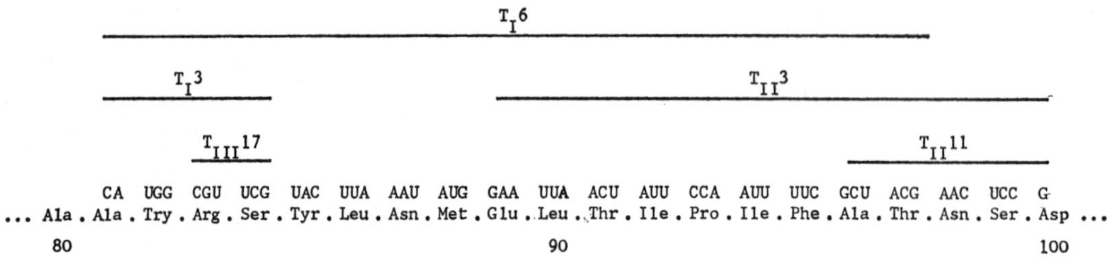

Fig. 8. Nucleotide sequence of the fragment from the coat protein cistron of R17 RNA, indicating partial T_1 digestion products and the corresponding amino-acid sequence of the coat protein.

This is the first time that a sequence from a messenger RNA has been determined by chemical means and shown to correspond to the sequence of amino-acids in the protein for which it codes; the results can be regarded as one of the most direct confirmations of the correctness of the genetic code. It is also of interest to see which codons are actually used by this bacteriophage. Table 5 shows the genetic code, in which the codons found in the above sequence are indicated by underlining the amino-acids concerned. Six amino-acids are found twice in the sequence. Two of these (Leu and Ile) are specified both times by the same codon; however, the other four (Thr, Ser, Asn, Ala) are coded for by two different codons. The data are not sufficient to make any generalizations but at least it may be concluded that the code used by the bacteriophage is degenerate.

Table 5. THE GENETIC CODE

2nd→ 1st ↓	U	C	A	G	3rd ↓
	Phe	Ser	Tyr	Cys	U
U	Phe	Ser	Tyr	Cys	C
	Leu	Ser	—	—	A
	Leu	Ser	—	Try	G
	Leu	Pro	His	Arg	U
C	Leu	Pro	His	Arg	C
	Leu	Pro	Gln	Arg	A
	Leu	Pro	Gln	Arg	G
	Ile	Thr	Asn	Ser	U
A	Ile	Thr	Asn	Ser	C
	Ile	Thr	Lys	Arg	A
	Met	Thr	Lys	Arg	G
	Val	Ala	Asp	Gly	U
G	Val	Ala	Asp	Gly	C
	Val	Ala	Glu	Gly	A
	Val	Ala	Glu	Gly	G

Codons found in the nucleotide sequence of the fragment from the coat protein cistron of R17 RNA have been indicated by underlining the amino-acids concerned.

Secondary Structure of the Fragment

An interesting feature of the sequence is that it can be written in the form of a simple loop showing considerable base-pairing (Fig. 9). Of the twenty-four pairs in this structure nineteen are complementary. This is very unlikely to occur by chance and therefore we believe that the sequence most probably occurs in a double helical configuration in the virus. In this structure all the guanylate residues in the sequence are involved in base pairs and would thus be expected to be resistant to T_1 ribonuclease. This would explain the presence of this fragment in the partial digest of the whole molecule. The unexpected specificity of the partial hydrolysis of R17 RNA suggests that other such highly ordered base-paired structures exist in the RNA; these may be important in the packing of the RNA into the virus particle and may also be involved in the regulation of cistron expression.

It thus appears that the sequence of a messenger RNA, at least in phage RNA, is determined not only by the need to specify an amino-acid sequence but also by its need to assume a particular secondary structure. In Fig. 9 the phasing of the codons is indicated by dots. It can be seen that the third positions do not come opposite one another. Codons that differ only in the third position often code for the same amino-acid. Thus mutations occurring in two-thirds of the base pairs could change the RNA secondary structure without altering the amino-acid sequence of the protein that is synthesized. It may be that this is one of the functions of the degeneracy of the code.

Because protein biosynthesis depends on the recognition of codons by the anticodon on *t*RNAs, it seems that the messenger RNA must be single-stranded during translation. Thus the finding of a double-stranded structure in a messenger RNA suggests that the protein-synthesizing mechanism must be capable of unfolding such a structure. Similarly the phage RNA synthetase must be able to unfold the RNA during transcription.

Fig. 9. Secondary structure of the fragment from the coat protein cistron of R17 RNA.

We thank Mr A. Coulson for his skilful technical assistance, Dr G. G. Brownlee for much helpful advice during the course of this work and Professor F. Egami for a gift of ribonuclease U_2. J. M. A. thanks the Helen Hay Whitney Foundation for a fellowship and P. G. N. J. the Medical Research Council for a scholarship.

Received July 28, 1969.

1. Gesteland, R. F., and Boedtker, H., *J. Mol. Biol.*, **8**, 496 (1964).
2. Dahlberg, J. E., *Nature*, **220**, 548 (1968).
3. De Wachter, R., and Fiers, W., *J. Mol. Biol.*, **30**, 507 (1967).
4. Weith, H. L., and Gilham, P. T., *J. Amer. Chem. Soc.*, **89**, 5473 (1967).
5. Roblin, R., *J. Mol. Biol.*, **31**, 51 (1968).
6. De Wachter, R., Verhassel, J-P., and Fiers, W., *FEBS Lett.*, **1**, 93 (1968).
7. Min Jou, W., and Fiers, W., *J. Mol. Biol.*, **40**, 187 (1969).
8. Gussin, G. N., *J. Mol. Biol.*, **21**, 435 (1966).
9. Weber, K., *Biochemistry*, **6**, 3144 (1967).
10. Sanger, F., Brownlee, G. G., and Barrell, B. G., *J. Mol. Biol.*, **13**, 373 (1965).
11. Brownlee, G. G., and Sanger, F., *J. Mol. Biol.*, **23**, 337 (1967).
12. Brownlee, G. G., Sanger, F., and Barrell, B. G., *J. Mol. Biol.*, **34**, 379 (1968).
13. Goodman, H. M., Abelson, J., Landy, A., Brenner, S., and Smith, J. D., *Nature*, **217**, 1019 (1968). Dube, S. K., Marcker, K. A., Clark, B. F. C., and Cory, S., *Nature*, **218**, 232 (1968). Cory, S., Marcker, K. A., Dube, S. K., and Clark, B. F. C., *Nature*, **220**, 1039 (1968). Barrell, B. G., and Sanger, F., *FEBS Lett.*, **3**, 275 (1969). Yaniv, M., and Barrell, B. G., *Nature*, **222**, 278 (1969).
14. Brownlee, G. G., and Sanger, F., *Europ. J. Biochem.* (in the press).
15. Gilham, P. T., *J. Amer. Chem. Soc.*, **84**, 687 (1962).
16. Arima, T., Uchida, T., and Egami, F., *Biochem. J.*, **106**, 609 (1968).
17. Gould, H. J., *J. Mol. Biol.*, **29**, 307 (1967).
18. Peacock, A. C., and Dingman, C. W., *Biochemistry*, **6**, 1818 (1967).
19. Loening, U. E., *Biochem. J.*, **102**, 251 (1967).
20. Gould, H. J., Pinder, J. C., and Matthews, H. R., *Analyt. Biochem.*, **29**, 1 (1969).

Printed in Great Britain by Fisher, Knight & Co., Ltd., St. Albans.

Biochem. J. (1972) **128**, 993–1006
Printed in Great Britain

993

Nucleotide Sequences of Two Fragments from the Coat-Protein Cistron of Bacteriophage R17 Ribonucleic Acid

By P. G. N. JEPPESEN,* B. G. BARRELL, F. SANGER and A. R. COULSON
Medical Research Council Laboratory of Molecular Biology,
Hills Road, Cambridge CB2 2OH, U.K.

(*Received* 21 *January* 1972)

Bacteriophage R17 RNA was labelled with ^{32}P and was subjected to partial digestion with ribonuclease T_1. The products were fractionated by ionophoresis on polyacrylamide gel. Two fragments were purified and their nucleotide sequences determined by methods involving complete and further partial digestion with ribonucleases A and T_1. Fragment 20 had a sequence that coded for the amino acids in positions 32–53 of the coat protein of the bacteriophage. Fragment 20X, on further purification in 7M-urea, gave rise to two smaller nucleotides whose sequences coded for the amino acids in positions 56–66 and 67–76 of the coat protein. The sequence of the two fragments was such that they could be written in the form of loops stabilized by base-pairing.

The genome of the *Escherichia coli* bacteriophage R17 is a single-stranded linear RNA molecule comprising approx. 3500 nucleotides, and it acts as a messenger coding for the synthesis of three proteins: the coat protein, which is the main protein component of the virus; the A protein or 'maturation' protein, which is also present in the mature virus; and the RNA replicase, which is responsible for copying the viral genome but is not present in the virus particle. The virion components are present in the proportions RNA:coat protein:A protein 1:180:1 (Gussin, 1966; Jeppesen *et al.*, 1970*a*). The amino acid sequence of the coat protein has been determined by Weber (1967), but little is known about the sequences of the other two proteins.

When ^{32}P-labelled bacteriophage R17 RNA was digested with ribonuclease T_1, a number of large oligonucleotides (15–24 residues long) could be separated in a pure form by a two-dimensional system involving ionophoresis on cellulose acetate and thin-layer 'homochromatography' on DEAE-cellulose (Brownlee & Sanger, 1969; Jeppesen, 1971). The sequences of these oligonucleotides were determined and some of them contained sequences that were related by the genetic code to amino acid sequences in the coat protein. Larger degradation products of the RNA could be obtained by partial digestion with ribonuclease T_1 and fractionation by ionophoresis on polyacrylamide gels (Adams *et al.*, 1969) (Plate 1). One of the bands from the fractionation (band 21) contained an RNA fragment of 57 residues whose sequence was determined, and this was shown to code for the amino acids in positions 81–100 of the coat protein.

Most of the other bands have now been investigated in more detail, and two more (bands 20 and 20X) have been found to belong to the coat-protein cistron and their sequences have been determined. A preliminary account of this work has been published (Jeppesen *et al.*, 1970*b*).

Methods

Preparation of the ^{32}P-labelled bacteriophage R17 RNA, partial ribonuclease T_1 digestion, and fractionation of the products by ionophoresis on polyacrylamide gel were as described previously (Adams *et al.*, 1969).

Previously RNA had been eluted from the gel by a method involving ionophoresis on to DEAE-cellulose paper and subsequent elution from the DEAE-cellulose paper with triethylamine carbonate. Although this method was relatively efficient and easy to carry out on a large number of samples simultaneously, the products obtained were frequently found to be degraded. The reason for this is not clear; it may have been due to nuclease contamination in the large volume of buffer used in the reservoir, or in the DEAE-cellulose paper, or to slight alkaline hydrolysis during elution from the DEAE-cellulose paper. The following method, although more tedious, usually yielded intact RNA and has been generally adopted.

The gel (vol. approx. 1 ml) was roughly broken up by extruding it through the orifice of a 2 ml plastic disposable hypodermic syringe into 2 ml of redistilled, water-saturated phenol contained in a 10 ml silicone-treated glass homogenizer tube. Then 2 ml of 0.1 M-NaCl, buffered with 0.04 M-tris–HCl, pH 8.3,

* Present address: Cold Spring Harbor Laboratory, Cold Spring Harbor, N.Y. 11724, U.S.A.

containing 0.002M-EDTA, was added and the gel was finely homogenized for 1–5 min with a Teflon homogenizer rotated by an electric motor. (All of these procedures were carried out at 0–4°C.) The homogenate was kept at 0°C for about 30min with occasional mixing, after which the phases were separated by brief low-speed centrifugation and the aqueous phase was removed. The phenol/polyacrylamide slurry was then washed twice in a similar manner with 1 ml portions of buffer, and the combined aqueous extracts, contained in a silicone-treated 15ml glass centrifuge tube, were precipitated by the addition of 2.5vol. of ethanol and being left at −20°C overnight. The precipitate was then collected by centrifugation at 8000rev./min for 15min at 4°C (MSE High-Speed 18 centrifuge, 8×50 rotor). The phenol, which was added primarily to remove any contaminating nuclease activity that might have been present, was also found to increase the efficiency of extraction from the gel, compared with experiments performed without the addition of phenol, and recovery of radioactivity could be as high as 80–85%. There was often a certain amount of soluble (i.e. low-molecular-weight) polyacrylamide present in gels, which was also extracted by the above procedure and which co-precipitated with the RNA on addition of ethanol. This could be an advantage in that low concentrations of RNA could be precipitated with the polyacrylamide acting as a carrier: however, subsequent operations involving solution of the RNA in small volumes (e.g. for re-running extracts on cellulose acetate strips) were made more difficult because of the gelatinous nature of the dissolved polymer. The following additional acid-precipitation step, which does not co-precipitate polyacrylamide, has been used to purify RNA extracted from gels that contained a large amount of soluble polyacrylamide (as was apparent by a heavy white ethanol precipitate). The first ethanol precipitate was redissolved in 1 ml of water at 0°C, and carrier RNA was added if necessary to bring the total RNA concentration to 1.0 E_{260}unit/ml. Then 0.1ml of 10% (v/v) $HClO_4$ was added and the whole was well mixed, left for 10min at 0°C, and centrifuged at 8000rev./min for 15min at 4°C. The supernatant was poured off and the pellet was resuspended in 1.0ml of 2% (w/v) sodium acetate, pH5.0, after which the RNA was reprecipitated with ethanol as described above. Precipitation of ^{32}P-labelled RNA fragments with $HClO_4$ in this manner was estimated at better than 90% efficient and has not been shown to damage the RNA in any detectable way.

In all the two-dimensional fractionation systems used, the first dimension was by ionophoresis on cellulose acetate at pH3.5 in 7M-urea as described by Brownlee & Sanger (1969). In the t.l.c. homochromatography systems the second dimension was either in homomixture *b* (Brownlee & Sanger, 1969) or in homomixture *c*, which was prepared as in the above paper to give a final concentration of 3% (w/v) RNA.

Products of complete ribonuclease T_1 digestion were usually fractionated with homomixture *c*. Products of partial digestion were fractionated with homomixture *b*. After elution the products of partial digestion were divided into two samples. One sample was subjected to complete digestion with ribonuclease T_1 (enzyme:RNA ratio 1:20, 30min at 37°C in 5–10μl of 0.01M-tris–HCl buffer, pH7.4, containing 0.001M-EDTA) and the products were fractionated one-dimensionally by ionophoresis on DEAE-cellulose paper in 7% (v/v) formic acid. They could be frequently identified from their mobility on this system, but this was normally confirmed by further degradation with ribonuclease A and fractionation of the products by ionophoresis on DEAE-cellulose paper at pH3.5. A second sample of the partial ribonuclease T_1 digestion products was treated with ribonuclease A (conditions were as for ribonuclease T_1)

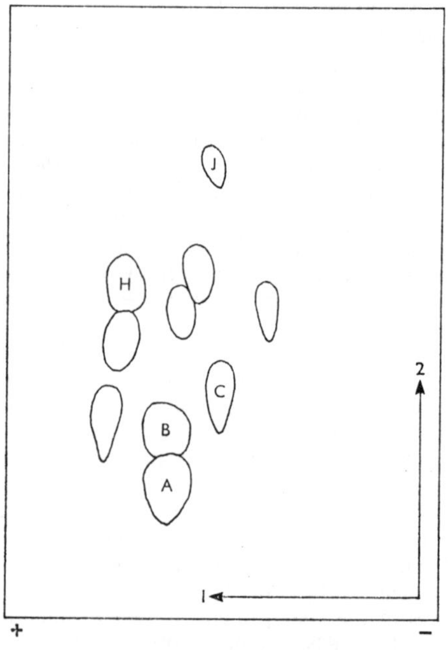

Fig. 1. *Re-fractionation of the material extracted from gel band 20X by t.l.c. homochromatography by using homomixture b*

The ethanol precipitate of the material extracted from the gel was dissolved in approx. 5μl of 7M-urea, pH3.5, and was applied to the cellulose acetate first dimension. The origin for the first dimension is to the right of the figure, as it is not transferred to the t.l.c. plate (Sanger *et al.*, 1965).

Table 1. *Sequence analysis of ribonuclease T_1 oligonucleotides from coat-protein-cistron RNA fragments contained in band* 20*X*

Spots C and H (Fig. 1) were eluted and digested with ribonuclease T_1 and the products were fractionated by one-dimensional ionophoresis on DEAE-cellulose paper in 7% formic acid. Sequence analyses were performed as described by Jeppesen (1971). The values in parentheses indicate molar yields of oligonucleotides to the nearest whole number, where these are not unity. Where yields of products were estimated visually from the radioautograph, rather than determined by scintillation counting, this is indicated by underlining; for example, (—)and (═) would mean approximately twice and three times the molar yield respectively. -U̇- and -Ġ- represent carbodiimide-modified uridylate and guanylate residues respectively.

Oligo-nucleotide	Ribonuclease A digestion products	CD products (Jeppesen, 1971)	Ribonuclease U_2 digestion products	Deduced sequence
T_1-oligonucleotides from spot C				
h	←—— (see Table 1 of Jeppesen, 1971)		——→	C-A-A-A-U-A-C-A-C-C-A-U-U-A-A-A-Gp
1	A-A-Gp Cp(2) Up	U̇-A-A-Ġp $\underline{\text{Cp}}$	—	C-C-U-A-A-Gp
2	Gp Cp Up	U̇-Cp Ġp	—	U-C-Gp
3	Gp Up	—	—	U-Gp
4	A-Gp	—	—	A-Gp
5	Gp	—	—	Gp
T_1-oligonucleotides from spot H				
6*	A-A-Cp A-Gp Cp(2) Up	U̇-Cp A-Ġp Cp A-A-Cp	(U,C_2)A-Gp (U,C_2)Ap C-A-Ap C-Ap Gp Ap	C-A-A-C-U-C-A-Gp
7*	A-Cp Gp Up	U̇-Ġp A-Cp	—	A-C-U-Gp
8	Gp Up(2)	—	—	U-U-Gp
9	A-Gp Up	—	—	U-A-Gp
10*	Gp Up	—	—	$\underline{\underline{\text{U-Gp}}}$
11*	Gp	—	—	$\underline{\underline{\text{Gp}}}$

* These oligonucleotides were also found in a ribonuclease T_1 digest of spot J (Fig. 1) from the re-fractionation of band 20X (U-Gp and Gp were in approximate molar yield in this case).

and the products were fractionated by ionophoresis on DEAE-cellulose paper at pH 1.9 (Sanger *et al.*, 1965) and identified by further digestion with ribonuclease T_1.

Other methods of sequence analysis were as described previously (Adams *et al.*, 1969; Jeppesen, 1971).

Results

Plate 1 shows a radioautograph of a preparative ionophoretic separation in polyacrylamide of a partial ribonuclease T_1 digest of bacteriophage R17 RNA. The RNA corresponding to gel bands was extracted and ribonuclease T_1 'fingerprints' were made by using the thin-layer homochromatography system

Table 2. *Analysis of partial ribonuclease T_1 digestion products of spot C from band* 20X (*Fig.* 2)

The spots numbered in Fig. 2 were eluted and analysed as described in Table 1 and by Jeppesen (1971).

Oligo-nucleotide (Fig. 2)	Ribonuclease T_1 digestion products	Ribonuclease A digestion products	Deduced sequence
*t*13	U-Gp Gp	G-Up Gp	G-U-Gp
*t*12	U-C-Gp A-Gp	G-A-Gp Cp Up	U-C-G-A-Gp
*t*10	C-C-U-A-A-Gp U-Gp	A-A-Gp G-Cp Cp Up	U-G-C-C-U-A-A-Gp
*t*9	C-C-U-A-A-Gp U-Gp Gp	A-A-Gp G-Up G-Cp Cp Up	G-U-G-C-C-U-A-A-Gp
*t*8	C-C-U-A-A-Gp U-Gp A-Gp Gp	A-G-G-Up A-A-Gp G-Cp Cp Up	A-G-G-U-G-C-C-U-A-A-Gp
*t*7	C-C-U-A-A-Gp U-C-Gp U-Gp A-Gp Gp	G-A-G-G-Up A-A-Gp G-Cp Cp Up	U-C-G-A-G-G-U-G-C-C-U-A-A-Gp

Table 2—*continued*

*t*5	C-A-A-A-U-A-C-A-C-C-A-U-U-A-A-A-Gp	A-A-A-G-Up	C-A-A-A-U-A-C-A-C-C-A-U-U-A-A-A-G-U-C-Gp
	U-C-Gp	A-A-A-Up	
		A-Cp	
		A-Up	
		Gp	
		Cp	
		Up	
*t*4	C-A-A-A-U-A-C-A-C-C-A-U-U-A-A-A-Gp	A-A-A-G-Up	C-A-A-A-U-A-C-A-C-C-A-U-U-A-A-A-G-U-C-G-A-Gp
	U-C-Gp	A-A-A-Up	
	A-Gp	G-A-Gp	
		A-Cp	
		A-Up	
		Cp	
		Up	

Summary of partial digestion products

Deduced sequence ... C-A-A-A-U-A-C-A-C-C-A-U-U-A-A-A-G-U-C-G-A-G-G-U-G-C-C-U-A-A-Gp

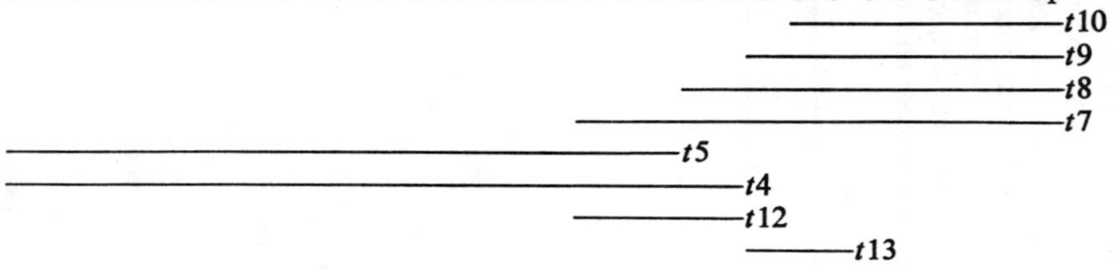

(Jeppesen, 1971). The band pattern was found to be reasonably reproducible from one preparative gel to the next, although there were often slight differences in relative mobilities of bands (as characterized by their 'fingerprints'), which were probably due to slight variations in polymerization between gels. However, after initial screening experiments in which each of the bands was 'fingerprinted', the similarities of successive gels made it possible in most cases to recognize particular bands from the gel pattern alone, although this was always confirmed by subsequent 'fingerprinting'.

The 'fingerprints' of the gel bands were considerably simpler than those for the whole bacteriophage R17 RNA, so that more smaller oligonucleotides were obtained in a pure form. Thus, whereas in the whole bacteriophage R17 RNA 'fingerprint' most oligonucleotides smaller than 12 residues were impure, in the isolated gel bands oligonucleotides larger than about eight residues were usually pure. Initially the RNA from spots was subjected to digestion with pancreatic ribonuclease and the products were separated by ionophoresis on DEAE-cellulose paper at pH3.5 (Adams *et al.*, 1969). From these results it was usually possible to determine if any of the products could be derived from the coat-protein cistron (as illustrated for band 20 in Fig. 6) or from other known parts of the RNA, such as the ribosomal binding sites (Steitz, 1969) or the ends of the RNA (Adams & Cory, 1970).

Band 21 had already been shown to be derived from the coat-protein cistron, and its sequence has been described (Adams *et al.*, 1969). On digestion with ribonuclease T_1, band 20X gave rise to oligonucleotide *h*, which had been found to have a nucleotide sequence corresponding to the amino acid sequence in position 57–61 of the coat protein (Jeppesen, 1971). Band 20, however, did not contain any of the larger oligonucleotides that had been identified in the ribonuclease T_1 digest of the whole bacteriophage R17 RNA, but preliminary sequence results on a number of the smaller oligonucleotides suggested that they could be derived from a section of the coat-protein cistron (Fig. 6). The sequences of the RNA fragments contained in these bands were therefore investigated in further detail.

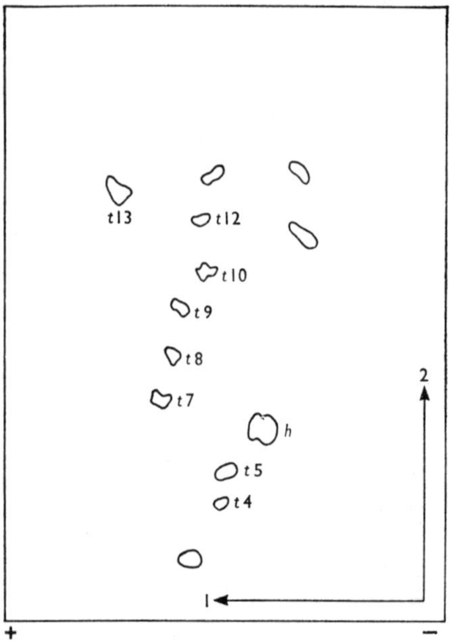

Fig. 2. *Partial ribonuclease T_1 digest of component C from gel band 20X (Fig. 1) fractionated by t.l.c. homochromatography by using homomixture b*

Digestion was with an enzyme:RNA ratio of 1:1000 in 5μl of 0.01M-tris–HCl buffer, pH7.4, containing 0.001M-EDTA, for 15min at 0°C. Analyses of the products are given in Table 2.

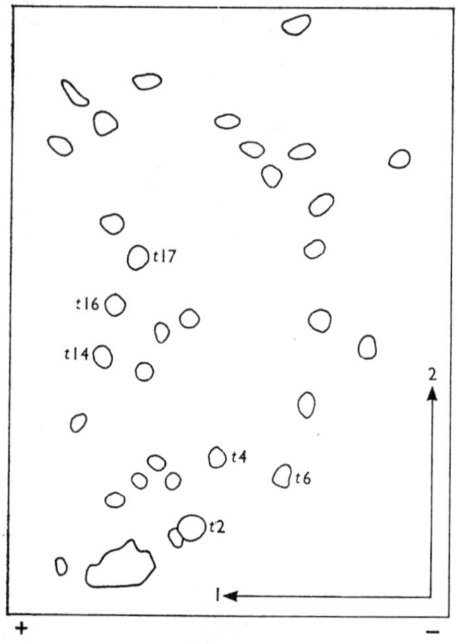

Fig. 3. *Partial ribonuclease T_1 digest of component H from gel band 20X (Fig. 1) fractionated by t.l.c. homochromatography by using homomixture c*

Digestion was done as in Fig. 2 with an enzyme:RNA ratio of 1:5000.

The Biochemical Journal, Vol. 128, *No.* 5 Plate 1

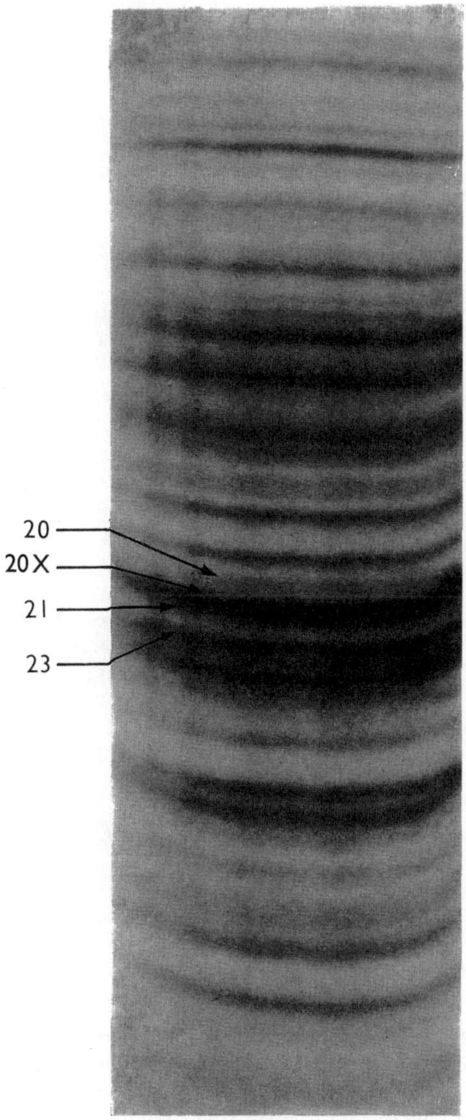

EXPLANATION OF PLATE I

*Radioautograph of a polyacrylamide-gel ionophoresis of a partial ribonuclease T_1 digest of bacteriophage R*17 *RNA*

Digestion was done with an enzyme:RNA ratio of 1:500 in 0.04M-tris–acetate–0.02M-magnesium acetate buffer, pH7.4, for 16h at 0°C. Ionophoresis was in a 12.5% polyacrylamide gel as described by Adams *et al.* (1969). The positions of bands 20, 20X, 21 and 23 are indicated. Band 20X is not visible as a separate band, but represents the trailing part of the strong band indicated as 21.

P. G. N. JEPPESEN, B. G. BARRELL, F. SANGER AND A. R. COULSON (*Facing p.* 998)

The Biochemical Journal, Vol. 128, *No.* 5 Plate 2

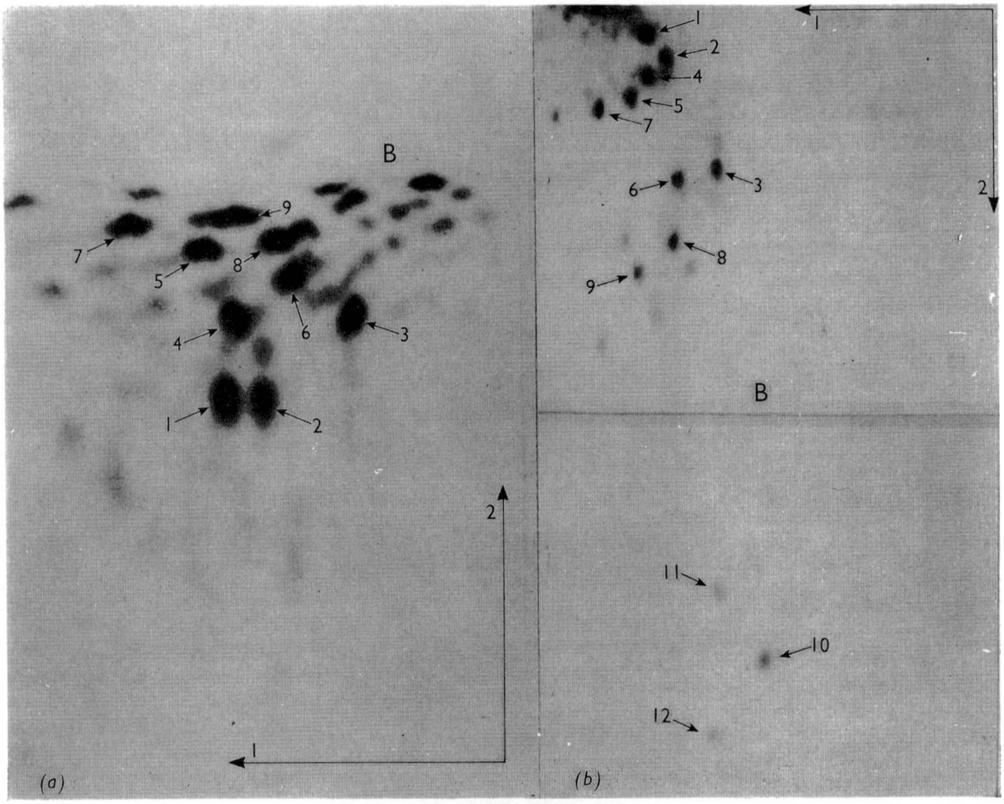

EXPLANATION OF PLATE 2

Radioautographs of two-dimensional fractionations of complete ribonuclease T_1 digests of the material extracted from gel band 20 (*Plate* 1)

Digestion was done with an enzyme:RNA ratio 1:20 in 0.01 M-tris–HCl buffer, pH 7.4, containing 0.001 M-EDTA for 30min at 37°C. (*a*) T.l.c. homochromatography system by using homomixture *c*; (*b*) ionophoresis on DEAE-cellulose paper in 7% formic acid as the second dimension. Analyses of the spots are given in Table 4. B marks the position of the blue marker (Sanger *et al.*, 1965).

P. G. N. JEPPESEN, B. G. BARRELL, F. SANGER AND A. R. COULSON

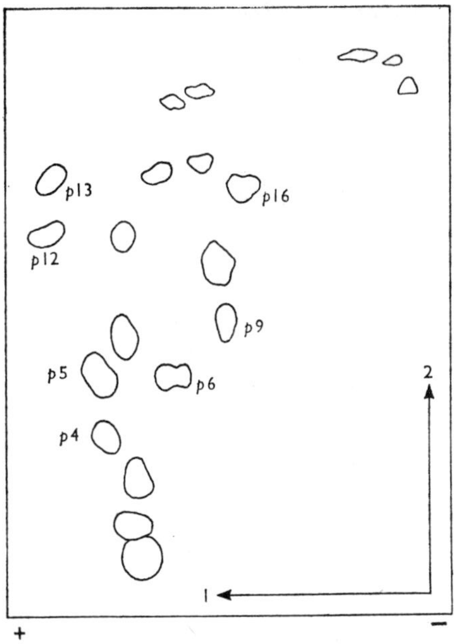

Fig. 4. *Partial ribonuclease A digest of component H from gel band 20X (Fig. 1) fractionated by t.l.c. homochromatography by using homomixture b*

Digestion was done as in Fig. 2 with an enzyme:RNA ratio of 1:1000.

Band 20X

The 'fingerprint' of a ribonuclease T_1 digest of band 20X showed that more than one component was present. In particular it was heavily contaminated with the material from band 21, which gave rise to oligonucleotide *a* (Jeppesen, 1971). To purify the component that gave rise to oligonucleotide *h*, the two-dimensional homochromatography re-fractionation procedure (Brownlee & Sanger, 1969) was used, and the result is shown in Fig. 1. Each of the major spots was eluted. The 'fingerprints' of the material from spots A and B, which corresponded to fragments of around 60 base residues, contained no trace of oligonucleotide *h*. However, when ribonuclease T_1 digests were made of samples of the faster-moving spots and the products were separated by one-dimensional ionophoresis on DEAE-cellulose paper in 7% formic acid, oligonucleotide *h* was found among the products of spot C, characterized by analysis of its ribonuclease A digestion products. Table 1 lists the ribonuclease T_1 products of spot C together with their sequence-analysis results. As for the component from band 21 (Adams *et al.*, 1969), these ribonuclease T_1 products could be ordered by referring to the amino acid sequence of the coat protein (see Fig. 5), but again the order was rigorously confirmed by partial ribonuclease T_1 digestion and characterization of the products (Fig. 2 and Table 2).

The paradox posed by this result was that C, a fragment of RNA only 31 nucleotides long, should run on the polyacrylamide gel more slowly than the 57-nucleotide component in band 21. However, analysis of the other major components of band 20X suggested the explanation, discussed below, that the ribonuclease T_1 digestion products of spot H (Table 1) could be arranged in such a way that if a single extra guanylate residue were inserted at the beginning they could code for the polypeptide sequence directly following that coded by component C (see Fig. 5). The order was established by partial digestions with ribonucleases T_1 and A, and analysis of the partial-digestion products as described above (Figs. 3 and 4 and Table 3). Another spot in Fig. 1, J, contained a part of sequence H (Fig. 5).

Therefore, by combining the sequences of spots C and H, and inserting the missing guanylate residue, the complete nucleotide sequence coding for amino acids 57–75 of the coat protein could be constructed (Fig. 5), although it must be emphasized that sequences C and H were never isolated covalently linked, and therefore the overlapping sequence has not been directly confirmed. Sequences C and H together constitute 58 base residues, a number consistent with the mobility of band 20X on the polyacrylamide gel. It was concluded therefore that the two migrate as a single species on the gel, held together by some interaction (e.g. hydrogen-bonded secondary structure) that is stable under the conditions of gel ionophoresis, but that is broken by the denaturing conditions of the two-dimensional re-fractionation procedure (i.e. the presence of 7M-urea.) This is readily understood if the structure is written in the form of a 'hairpin loop' (see Fig. 10 and the Discussion section).

Band 20

The ribonuclease T_1 'fingerprints' of band 20, shown in Plate 2, revealed it to be a fairly pure single major component. A preliminary analysis of the major ribonuclease T_1 products by ribonuclease A digestion indicated the possibility that they were all derived from a region of the coat-protein cistron coding for amino acids, approximately in positions 30–50 of the protein chain. This is shown by reference to Fig. 6, which illustrates the general method used for preliminary screening of ribonuclease T_1 digestion products to see if they could be derived from the coat-protein cistron. The ribonuclease A digestion products of ribonuclease T_1 oligonucleotides from band 20 are listed in column 2 of Table 4. Oligonucleotide 2 contains the sequences A-Cp, A-A-A-Gp, Cp and

Table 3. *Analysis of partial digestion products of spot H from band 20X*

The spots numbered in Figs. 3 and 4 were eluted and analysed as described in Table 1 and by Jeppesen (1971).

Oligo-nucleotide	Ribonuclease T_1 digestion products	Ribonuclease A digestion products	Deduced sequence
Partial ribonuclease T_1 digestion products (Fig. 3):			
*t*4	C-A-A-C-U-C-A-Gp U-Gp Gp	G-G-Cp A-A-Cp A-Gp Cp Up	U-G-G-C-A-A-C-U-C-A-Gp
*t*6	C-A-A-C-U-C-A-Gp A-C-U-Gp	A-G-A-Cp A-A-Cp Gp Cp Up	C-A-A-C-U-C-A-G-A-C-U-Gp
*t*2	C-A-A-C-U-C-A-Gp A-C-U-Gp U-U-Gp	A-G-A-Cp A-A-Cp G-Up Gp Cp Up	C-A-A-C-U-C-A-G-A-C-U-G-U-U-Gp
*t*14	A-C-U-Gp U-U-Gp Gp	G-Up G-Gp A-Cp Up	A-C-U-G-U-U-G-Gp
*t*17	U-A-Gp U-Gp	G-Up A-Gp Up	U-G-U-A-Gp
*t*16	U-A-Gp U-Gp Gp	G-Up A-Gp	G-U-G-U-A-Gp
Partial ribonuclease A digestion products (Fig. 4):			
*p*13	U-Gp Gp Up	G-G-Up G-Up	G-G-U-G-G-U-G-Up
*p*12	U-U-Gp U-Gp Gp Up	G-G-Up G-Up Up	G-U-U-G-G-U-G-G-U-G-Up
*p*16	C-A-A-C-U-Cp U-Gp Gp	G-G-Cp A-A-Cp Cp Up	U-G-G-C-A-A-C-U-Cp
*p*9	C-A-A-C-U-C-A-Gp U-Gp A-C-Up Gp	G-G-Cp A-G-A-Cp A-A-Cp Cp Up	U-G-G-C-A-A-C-U-C-A-G-A-C-Up

Table 3—*continued*

Oligonucleotide	Ribonuclease T_1 digestion products	Ribonuclease A digestion products	Deduced sequence
Partial ribonuclease A digestion products (Fig. 4) (cont.):			
*p*6	C-A-A-C-U-C-A-Gp	G-G-Cp	U-G-G-C-A-A-C-U-C-A-G-A-C-U-G-Up
	A-C-U-Gp	A-G-A-Cp	
	U-Gp	A-A-Cp	
	Up	G-Up	
	Gp	Cp	
		Up	
*p*5	U-U-Gp	G-G-Up	A-G-A-C-U-G-U-U-G-G-U-G-G-U-G-Up
	A-C-U-Gp	G-Up	
	U-Gp	A-G-A-Cp	
	A-Gp	Up	
	Gp		
	Up		
*p*4	U-U-Gp	G-G-Up	A-G-A-C-U-G-U-U-G-G-U-G-G-U-G-U-A-Gp
	A-C-U-Gp	G-Up	
	U-A-Gp	A-G-A-Cp	
	U-Gp	Up	
	Gp	A-Gp	
	A-Gp		

Summary of partial digestion products

```
                                         ————————————————————t14
                     ——————————————————————————————————————t2
                     ——————————————————————————t6                 ———————————————t16
              ———————————————————————t4                             ——————————————t17
Deduced sequence ... U-G-G-C-A-A-C-U-C-A-G-A-C-U-G-U-U-G-G-U-G-G-U-G-U-A-Gp
              ——————————————————p16                  ——————————————————p13
                                                 ——————————————————————p12
              ——————————————————————————————p9
              ——————————————————————————————————p6
                                   ——————————————————————————————————————p5
                                   ——————————————————————————————————————————p4
```

```
                                                          J
                                                 ——————————————————
                       C                                  H
      ——————————————————————————————————————     ————————————————————————————————
(G)C AAA UAC ACC AUU AAA GUC GAG GUG CCU AAG (G)UG GCA ACU CAG ACU GUU GGU GGU GUA G

... Arg-Lys-Tyr-Thr-Ile-Lys-Val-Glu-Val-Pro-Lys-  Val-Ala-Thr-Gln-Thr-Val-Gly-Gly-Val-Glu ...

    56          60                                           70                   76
```

Fig. 5. *Amino acid sequence of the coat protein from positions* 56–76 *and corresponding nucleotide sequence found in the RNA fragments C, H and J from band* 20*X*

For details see Tables 1, 2 and 3, and Figs. 2, 3 and 4.

Up (2mol). To see if it could be derived from the coat-protein cistron, a list of possible ribonuclease-digestion products was prepared, as in Fig. 6, but for the whole cistron. If we just consider the section shown in Fig. 6, there are two positions where A-A-A-Gp could occur (corresponding to positions 43–44 and 50–51 of the amino acid sequence). Both could occur in a ribonuclease T_1 digestion product having A-Cp, but the

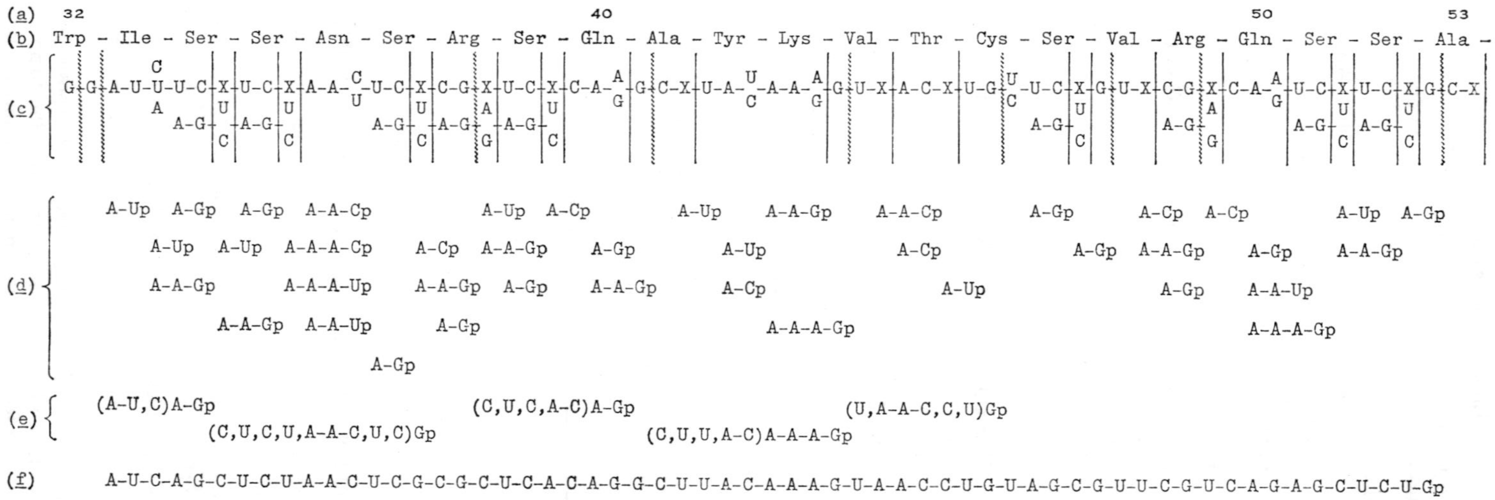

Fig. 6. *Method for screening ribonuclease T_1 digestion products to test whether they could be derived from the coat-protein cistron*

The method is applied to the oligonucleotides from band 20. (*a*) Position of amino acid residue in coat protein; (*b*) part of the amino acid sequence of coat protein; (*c*) corresponding nucleotide sequence predicted from the genetic code: ⦚ indicates an obligatory ribonuclease T_1 split, | indicates a possible ribonuclease T_1 split; (*d*) possible ribonuclease A digestion products of ribonuclease T_1 oligonucleotides from the RNA; (*e*) ribonuclease T_1 oligonucleotides from band 20 (Plate 2) characterized by their ribonuclease A digestion products (Table 4, column 2), indicating possible sequences that are compatible with (*c*) and (*d*); (*f*) nucleotide sequence of band 20 fragment as finally deduced (see Tables 4, 5, 6 and 7).

Table 4. *Sequence analysis of ribonuclease T_1 digestion products of the coat-protein-cistron RNA fragments contained in band* 20

The major spots numbered in Plate 2 were eluted and analysed as described in Table 1.

Oligo-nucleotide	Ribonuclease A digestion products	CD products (Jeppesen, 1971)	Ribonuclease U_2 digestion products	Deduced sequence
1	A-A-Cp Gp Cp(3) Up(3)	U̇-A-A-Cp U̇-Cp Ġp Cp	C(U_2,C)A-Ap* C(U_2,C)Ap* (U,C_2)Gp Ap	C-U-C-U-A-A-C-U-C-Gp
2	A-A-A-Gp A-Cp Cp Up(2)	U̇-U̇-A-Cp A-A-A-Ġp Cp	C-A-A-A-Gp C-A-A-Ap C-A-Ap C-Ap (U_2,C)Ap	C-U-U-A-C-A-A-A-Gp
3	A-Gp A-Cp Cp(2) Up	U̇-Cp A-Ġp Cp A-Cp	C(U,C)Ap* C-A-Gp C-Ap Gp	C-U-C-A-C-A-Gp
4	A-A-Cp Gp Cp Up(2)	U̇-Ġp U̇-A-A-Cp Cp	(U,C_2)Gp U-A-Ap U-Ap	U-A-A-C-C-U-Gp
5	Gp Cp(2) Up(2)	U̇-Ġp U̇-Cp Cp	—	C-U-C-U-Gp*
6	A-Up A-Gp Cp	A-U̇-Cp A-Ġp	U-C-A-Gp U-C-Ap Ap	A-U-C-A-Gp
7	Gp Cp Up(2)	U̇-U̇-Cp Ġp	—	U-U-C-Gp
8	A-Gp Cp Up	U̇-Cp A-Ġp	—	U-C-A-Gp
9	A-Gp Up	—	—	U-A-Gp
10	Cp Gp	—	—	C-Gp
11	A-Gp	—	—	A-Gp
12	Gp	—	—	Gp

* The 5′-end groups in these oligonucleotides were determined by analysis of their complete snake venom phosphodiesterase-hydrolysis products after dephosphorylation (Sanger *et al.*, 1965) and comparing these with an alkaline hydrolysate.

product corresponding to positions 50–51 would not give Cp and Up with ribonuclease A, whereas the product corresponding to positions 43–44 could give Cp and Up(2), if it extended from positions 41–44. Thus oligonucleotide 2 (Table 4) could be derived from this position. This coincidence could of course be fortuitous; however, when the other oligonucleotides listed in Table 4 were examined in this way it was found that they also could be derived from the neighbouring part of the coat-protein cistron, as shown in Fig. 6(*e*).

When the material from band 20 was re-run by the

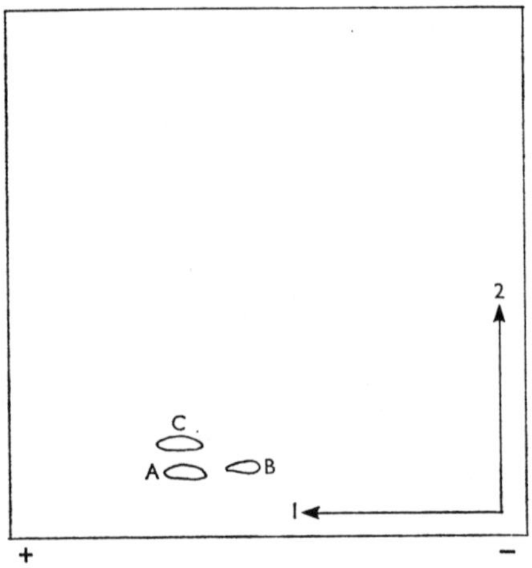

Fig. 7. *Re-fractionation of the material extracted from gel band 20 by t.l.c. homochromatography by using homomixture b*

For details see Fig. 1.

Table 5. *Ribonuclease A digestion products of the coat-protein-cistron RNA fragments contained in band 20*

Purified band 20 material (spot A, Fig. 7) was digested with ribonuclease A (enzyme:RNA ratio 1:20, 30min at 37°C, in 0.01M-tris–HCl buffer, pH7.4, containing 0.001 M-EDTA) and the products were fractionated in one dimension by DEAE-cellulose paper ionophoresis at pH.19. The sequences of the products were determined by analysis of their ribonuclease T_1 digestion products.

A-G-A-G-Cp	A-Cp (2mol)
A-A-A-G-Up	G-Cp (2mol)
A-G-G-Cp*	G-Up (3mol)
A-G-Cp (2mol)	Gp
A-A-Cp (2mol)	Cp (>3 mol)
A-Up	Up (>3 mol

* The sequence of this oligonucleotide must be as written, since there is only one A-Gp ribonuclease T_1 digestion product present in band 20 (Table 4) and this occurs in ribonuclease. A digestion product A-G-A-G-Cp.

Table 6. *Products obtained by digestion with ribonuclease T_1 and A of fragments from spots B and C (Fig. 7) from band 20*

Spots B and C were eluted from the t.l.c. plate shown in Fig. 7 and analysed as described by Jeppesen (1971).

	Ribonuclease T_1 digestion products	Ribonuclease A digestion products
Spot B	C-U-C-U-A-A-C-U-C-Gp	A-G-G-Cp
	C-U-U-A-C-A-A-A-Gp	A-A-A-Gp
	C-U-C-A-C-A-Gp	A-G-Cp
	A-U-C-A-Gp	A-A-Cp
	C-Gp	A-Up
	Gp	A-Cp (2mol)
		G-Cp (2mol)
		Cp (>3 mol)
		Up (>3 mol)
Spot C	U-A-A-C-C-U-Gp	A-G-A-G-Cp
	C-U-C-U-Gp	A-G-Cp
	U-U-C-Gp	A-A-Cp
	U-C-A-Gp	G-Up (3mol)
	U-A-Gp	Gp
	A-Gp	Cp (>3 mol)
	C-Gp	Up (>3 mol)

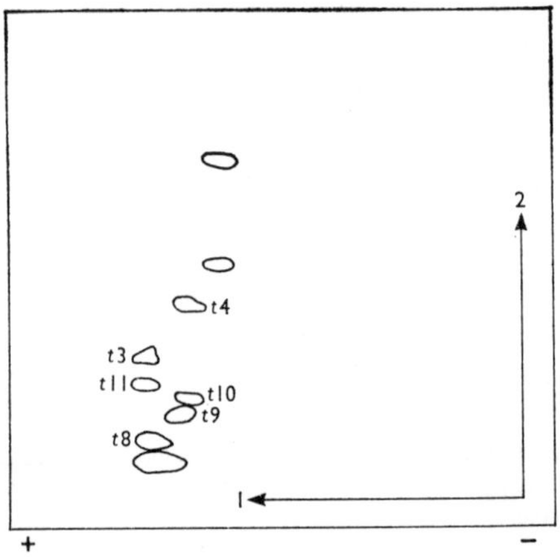

Fig. 8. *Partial ribonuclease T_1 digest of component B from gel band 20 (Fig. 7) fractionated by t.l.c. homochromatography by using homomixture b*

Digestion conditions were as in Fig. 2. The analyses of the partial digestion products are given in Table 7.

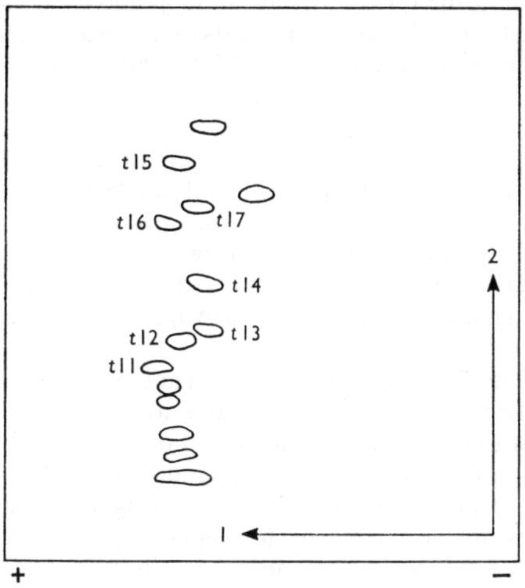

Fig. 9. *Partial ribonuclease T_1 digest of component C from gel band* 20 (*Fig.* 7) *fractionated by t.l.c. homochromatography by using homomixture b*

Digestion conditions were as in Fig. 2. The analyses of the partial digestion products are given in Table 7.

two-dimensional homochromatography system only three major components were found (Fig. 7). A ribonuclease T_1 'fingerprint' showed that spot A contained all the major-yield spots of Plate 2. The sequence of these ribonuclease T_1 products (Table 4) further supported the probability that the fragment was from the coat-protein cistron, and, together with the nucleotide sequences of the ribonuclease A-digestion products (Table 5), the complete sequence of component A could be postulated (Fig. 6).

The two other major spots obtained by re-running band 20 (spots B and C, Fig. 7) were each found by analysis of their ribonuclease T_1 and ribonuclease A digestion products to be derived from spot A by cleavage at a single site (Table 6). Whereas all the ribonuclease T_1 digestion products of spot A are found in either spot B or C, the ribonuclease A digestion product A-A-A-G-Up is not present in spot B or C, but A-A-A-Gp is found in spot B. This establishes the site of the 'hidden break' and the relative order of the two fragments. To establish the complete sequence of the fragments from spots B and C, they were subjected to partial digestion with ribonuclease T_1 (Figs. 8 and 9) and the products were analysed by complete digestion with the same enzyme. These results are summarized in Table 7.

Table 7. *Products obtained by partial digestion of spots B and C from band* 20 (*Fig.* 7) *with ribonuclease T_1*

The numbered spots from Figs. 8 and 9 were eluted and analysed by complete digestion with ribonuclease T_1 (see the text)

Deduced sequence of spot B (Fig. 8) A-U-C-A-G-C-U-C-U-A-A-C-U-C-G-C-G-C-U-C-A-C-A-G-G-C-U-U-A-C-A-A-A-Gp

Deduced sequence of spot C (Fig. 9) U-A-A-C-C-U-G-U-A-G-C-G-U-U-C-G-U-C-A-G-A-G-C-U-C-U-Gp

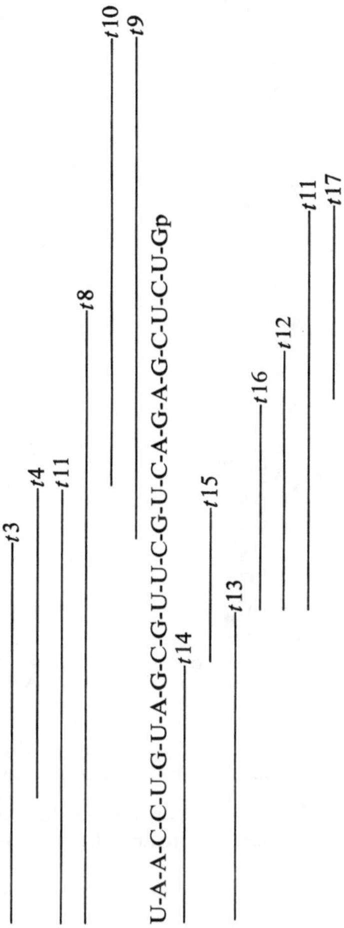

Discussion

The nucleotide sequences of the two fragments from bands 20X and 20 are summarized in Figs. 5

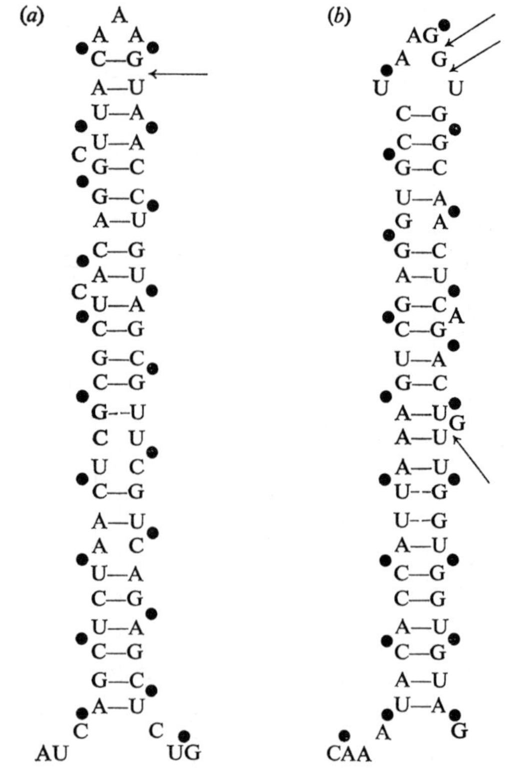

Fig. 10. *Nucleotide sequences of the fragments from bands* 20 (*a*) *and* 20*X* (*b*) *written in the form of 'hairpin loops' and showing the suggested base pairing*

Arrows indicate guanylate residues susceptible to enzyme attack. The 'reading frame' is shown by dots.

---

and 6, which also show the amino acid sequences in the coat protein for which they code. The significance of these and other sequence studies on the bacteriophage R17 RNA has been discussed previously (Jeppesen *et al.*, 1970*b*; Sanger, 1971) and will not be dealt with in detail here.

For the fragment from band 21 (Adams *et al.*, 1969), it was found that the nucleotide sequence could be written in the form of a hairpin loop and it was suggested that such a structure exists in the virus and may play an important part in the biological properties of the RNA. Both the fragments described in the present paper can also be written in the form of such loops (Fig. 10). In these cases it was found necessary to allow occasional 'looping out' of single residues to give a maximum number of base pairs (see Fresco *et al.*, 1960), but the structures thereby obtained correlate well with the observed susceptibility of certain guanylate residues to cleavage by ribonuclease T_1.

The methods used in the present and previous work have been shown to be applicable to the determination of relatively long sequences in the bacteriophage RNA and can clearly be extended considerably. Fiers *et al.* (1971) have been carrying out similar studies on the related bacteriophage MS2 and have reported the sequence of almost the whole of the coat-protein cistron.

We thank Dr. F. Egami for gifts of ribonuclease U_2 and Dr. G. Bernardi for spleen phosphodiesterase. P. G. N. J. thanks the Medical Research Council for a Scholarship.

References

Adams, J. M. & Cory, S. (1970) *Nature* (*London*) **227**, 570

Adams, J. M., Jeppesen, P. G. N., Sanger, F. & Barrell, B. G. (1969) *Nature* (*London*) **223**, 1009

Brownlee, G. G. & Sanger, F. (1969) *Eur. J. Biochem.* **11**, 395

Fiers, W., Contreras, R., De Wachter, R., Haegemen, G., Merragaert, J., Min Jou, W. & Vandenberghe, A. (1971) *Biochimie* **53**, 495

Fresco, J. R., Alberts, B. M. & Doty, P. (1960) *Nature* (*London*) **188**, 98

Gussin, G. N. (1966) *J. Mol. Biol.* **21**, 435

Jeppesen, P. G. N. (1971) *Biochem. J.* **124**, 357

Jeppesen, P. G. N. Steitz, J. A., Gesteland, R. F. & Spahr, P. F. (1970*a*) *Nature* (*London*) **226**, 230

Jeppesen, P. G. N., Nichols, J. L., Sanger, F. & Barrell, B. G. (1970*b*) *Cold Spring Harbor Symp. Quant. Biol.* **35**, 13

Sanger, F. (1971) *Biochem. J.* **124**, 833

Sanger, F., Brownlee, G. G. & Barrell, B. G. (1965) *J. Mol. Biol.* **13**, 373

Steitz, J. A. (1969) *Nature* (*London*) **224**, 957

Weber, K. (1967) *Biochemistry* **6**, 3144

PART III

DNA

Introduction

I suppose the ultimate goal for a sequencer was to sequence DNA and thus be able to read the instructions for living matter. Before about 1968 this goal seemed almost unattainable, but after our success with RNA we felt it was worth seeing what could be done with DNA. One problem was its size. There was no small DNA comparable to tRNA or 5 S RNA on which we could develop methods; the smallest were the single-stranded bacteriophages of about 5,000 residues, and most of the more interesting DNAs were in the millions. Another real problem was the lack of suitable enzymes for degrading DNA in a specific way. Our work on RNA had depended much on the enzyme ribonuclease T_1, which split specifically at G residues. There was no deoxyribonuclease of comparable specificity. The restriction enzymes, which are of course very specific deoxyribonucleases, had not been discovered at the time but these were used extensively in our later work.

At this time (the early 1970s) there were about ten people in my group working on DNA. Most were American post-doctoral fellows and needed little attention from me. Several different lines of research were taken up with some success and reported by the members of the group; the papers are not included here as my name was not on them, but they should be mentioned as they contributed to the general attack on DNA.

In 1960 Burton and Petersen[1] developed a method for depurinating DNA; this destroyed the purine nucleotides, leaving runs containing only pyrimidines (C and T). Hitherto, radioactive labelling had been done biologically, but with some nucleic acids this was difficult. To overcome this Maria Szekely[2] devised an *in vitro* labelling technique using polunucleotide kinase. This could be applied to both RNA and DNA, and the products were particularly suitable for analysis by the 3′ exonuclease digestion method (as described in paper 21). She applied the method to the depurination products of bacteriophage fd DNA and showed that their sequence could be determined after separation of their partial exonuclease digest on a two-dimensional system. This approach was developed further by Ling,[3] who determined the sequence of all the fd depurination products by a method which became known as the "wandering spot" technique and was used extensively in our later work. It was particularly useful for the depurination products where only C and T were present. It was not completely reliable if all four mononucleotides were present, but nevertheless proved very useful.

There were several phages available, but the one we chose to study predominantly was the single-stranded DNA bacteriophage ϕX174. It had been studied extensively by Sinsheimer and his colleagues and I think our main reason for choosing it was that John Sedat had worked with Sinsheimer and knew how to grow and handle it. The main initial problem was to isolate a smaller fragment from ϕX. There were a number of deoxyribonucleases known. One that proved useful was endonuclease IV, which had recently been isolated by Sadowski and Hurwitz.[4] It splits before C residues and has a high preference for single-stranded DNA.

Galibert, Sedat and Ziff[5] were able to isolate a pure fragment of 48 nucleotides from a limited digest of ϕX with this enzyme, and this was fractionated by ionophoresis on acrylamide gel. They sequenced it by depurination and by further digestion with endonuclease IV.

Robertson, Barrell, Weith and Donelson[6] used a different method to prepare a 50-nucleotide fragment from ϕX. Under suitable conditions ribosomes will bind firmly to the initiation site on a messenger RNA. That is the site where synthesis of a protein will start. As the ϕX DNA has the same sequence as the corresponding messenger RNA, ribosomes could also bind to it. The complex was then hydrolysed with pancreatic DNAase, which is a non-specific enzyme. The initiation site was protected by the bound ribosome and was isolated and sequenced using depurination, endonuclease IV and another enzyme, streptodornase, which appears to have a rather narrow specificity.

Another possible way to determine the sequence of a DNA was to copy it into RNA using RNA polymerase. This method was studied by Elizabeth Blackburn[7] using the endonuclease IV fragment prepared by Ziff *et al.* One problem was that the polymerase was non-specific as to where it started or stopped, so that a complex mixture was produced; however, by subjecting the mixture to partial digests she showed that it was possible to determine the sequence of all but the first five residues.

The work I was principally involved in employed copying with DNA polymerase and is described in papers 28 and 29.

Thus in 1973 we had determined three sequences of about 50 residues — at that time the longest DNA sequences to have been done. (There was a rumour in the lab that there was to be a reward of a three-week holiday for the first person to sequence 50 residues of DNA. As there were ten of us we all took a weekend off.)

Comments on Papers 28 to 48

Papers 28 and 29

The first DNA sequence to be determined was published by Wu and Kaiser[8] in 1968. Bacteriophage λ contains "sticky ends" — that is to say, it is a long double-stranded DNA with a short single strand (12 residues) at each end. These sticky ends were ideal substrates for DNA polymerase, and by measuring the course of incorporation of each nucleotide into these sticky ends Wu and Kaiser were able to deduce their sequence. The method was somewhat tedious and not generally applicable. Encouraged by their work and the success of Weissmann's copying techniques with RNA polymerase on Q_β, I became interested in the idea of developing a more general approach to sequencing DNA using DNA polymerase. There were two main problems. One was how to make a unique starting place for the polymerase. This required a small piece of DNA (the primer) that would hybridize onto a single-stranded bacteriophage DNA. The polymerase would add nucleotides to the 3′ end of the primer. The only way to prepare a suitable primer seemed to be by synthesis. At this time oligonucleotide synthesis was very much in its infancy. Methods were being developed, principally in the laboratory of Khorana, but they were still very specialized and we had no experience. However, I was lucky to encounter Hans Kössel at a meeting; he had worked with Khorana and knew how to synthesize oligonucleotides. He had also had the same idea as we had of making a specific primer for bacteriophage f1 that was based on what was then thought to be the known amino acid sequence of the coat protein. In fact this turned out to be wrong, but luckily the octanucleotide did prime at a specific position on the DNA. It took him and Fischer about a year to make this.

The other main problem was how to sequence the DNA produced. We were not very satisfied with the enzymes available but, being a copying technique, we were able to make use of a very elegant method proposed by Berg, Fancher and Chamberlin in 1963. This was known as "ribosubstitution" and is described in these papers.

These papers involved some new techniques that were reasonably successful. They could clearly be improved and we visualized doing longer sequences by synthesizing a new primer corresponding to the 3′ end of the deduced sequence. This process could be repeated — possibly completely around the bacteriophage ϕX genome. It was a rather wild dream that would probably be limited by the rate of synthesis of primers. However, we did not go any farther as the method was soon superseded by the much faster and simpler "plus and minus" method.

A further decanucleotide was in fact synthesized corresponding to a sequence at the ribosome binding site of ϕX (see Ref. 7). Using this Donelson *et al.*[9] determined a sequence of 40 residues, confirming the earlier results and somewhat extending the known sequence.

It can be seen that these papers are to some extent forerunners of the plus and minus method. They describe primed synthesis with DNA polymerase, and the extension of the "wandering spot" technique, which probably helped us to develop the new method (see comment on paper 30).

Paper 30

I consider this to be my most important paper, because it describes an entirely new approach to sequencing. Hitherto all our work had been based on the principle of partial degradation. The large molecules were broken down, usually with enzymes, to give smaller fragments which were fractionated and further degraded to give their sequence. As the work proceeded we became more and more concerned with trying to develop methods whereby we could deduce sequences of small fragments from their behaviour on the fractionation systems, rather than having to resort to further digestions and analyses. This approach was better suited to nucleic acids than to proteins. At this stage the method we were mostly using was the wandering spot technique. Although not always absolutely reliable it was relatively quick and simple. (Good examples of it are shown in the paper of Donelson *et al.* — Ref. 9 — and other references given in the Introduction to Part III.) The same approach could be used to follow synthesis of a sequence by DNA polymerase, as shown in Fig. 1 of paper 28. This worked on a short sequence and I became interested in trying to extend it to longer sequences. Some idea of what I was thinking about is shown in Fig. 3(c) of paper 28. One of the main problems was that there were not sufficient intermediate spots because of the variable rate of the polymerase reaction at different nucleotide positions. We were therefore doing experiments aimed at obtaining a more even distribution of intermediates. This was also what was required for the first stage of the plus and minus method, but it was a big leap from this two-dimensional idea to the final one-dimensional sequencing method.

In a more recent review (paper 48) I described how I believed the one-dimensional sequencing originated but expressed some uncertainty as to whether I was correct. I have lately been looking more carefully through my old notebooks and trying to find out what actually did happen. Whereas the experimental details and protocols are fairly clearly described, the ideas and purposes behind them are less well recorded, so that it is difficult, for instance, to know at what stage we first thought of a one-dimensional method. Another problem is that the original autoradiographs, which were the main record of results, are now lost. This may seem strange and was certainly careless, but it should be borne in mind that at the time this was just another idea and there was no strong reason to believe that it would be particularly successful. When doing research one has plenty of ideas, most of which do not come to anything, and this one was no less wild than most of them. Especially wild was the idea that we would be able to find a fractionation system that could separate according to size.

In paper 48 I suggested that the first experiments had been using what we later called the "zero" system, employing a low concentration of one triphosphate — for instance dATP — with high concentrations of the other three, so that the system would run out of dATP and all chains would eventually terminate at the residues before the As. From my notebooks I find that this was not really the case as the zero system was not used until almost a year after the first plus and minus systems.

The first definite one-dimensional experiment was in about April 1973. We had done several experiments with DNA polymerase aiming to make as complex a mixture of products as possible, and then separate this mixture from the polymerase and excess triphosphates. We had also just obtained some T4 DNA polymerase and the first experiment was done using the plus system (a single triphosphate). It thus seems that this was inspired by the results of Englund,[10] which had given us the idea of making a mixture with all the chains terminating in the same residue. The incubation mixtures were fractionated in parallel on the homochromatography system. The results were not very clear but they did "show that some things have happened". In the same experiment we had run one sample with *coli* DNA polymerase and three triphosphates ("minus C" system), and this looked somewhat better than the others. So in the next experiment we ran a complete minus system, and in later experiments the two systems together. Most were run using homochromatography, but soon some samples were run on acrylamide electrophoresis by John Donelson, who had been using this fractionation system in other experiments. These were relatively small gels and were only marginally better than homochromatography. However, when we started running longer gels they were clearly better and were then used exclusively. These early experiments were all done using rC-substituted material primed by the octamer so that we could cut out bands from the gels to see if they were what we thought they would be. Also, we already knew much of this sequence. It took us some time before we got results as clear as those shown in the paper. One of the main problems at first was that there was sometimes an apparent inversion of sequence. This was probably due to secondary structure in the template DNA and was largely overcome by using severe denaturing conditions: 8M urea and a high voltage which allowed the gel to become hot. As mentioned in the Introduction, there were still some problems, but, as the method was so much simpler and quicker than any that had gone before, we decided to adopt it.

Papers 31 to 33

These papers represent a transitional stage before we were committed to the plus and minus method. By this time restriction enzymes had been discovered and at least one (Hin_d II) was available to us, and others with different specificities became available as the work on ϕX proceeded. Restriction enzymes were of course an important contribution to DNA sequencing. When we started work on DNA we were hampered by the lack of suitable specific enzymes for breaking it down. Now we had these extremely specific ones that recognized sequences of four to six residues. Although the plus and minus method did away with much of the necessity for partial degradation, we still needed methods for the initial breaking down of the large ϕX molecule. Hin_d II was ideal for this, giving 13 fragments that we could use as primers on the single-stranded DNA.

Paper 31 extends the sequence determined by Robertson *et al.* at the ribosomal binding site at the start of gene G. As well as the plus and minus method, Elizabeth Blackburn was using the copying technique with RNA polymerase that she had pioneered (see Ref. 7) and Gillian Air was sequencing the corresponding protein.

Paper 32, which deals with the neighbouring gene F, describes use of endonuclease IV digestion and the methods pioneered by Ziff *et al.*,[11] while Blackburn used RNA polymerase on the endonuclease products.

Paper 33 uses only the plus and minus method for DNA with supporting evidence from the amino acid sequence.

Papers 34 to 36

Although paper 34 does not have my name on it, I have included it as it reports the most interesting finding in the ϕX sequence. Ever since Beadle and Tatum enunciated the hypothesis of "one gene, one enzyme" it had been assumed that the genes were arranged in linear order along the genome with one DNA sequence coding for one protein sequence. This was the basis for determining the order of genes by "mapping", which was one of the main preoccupations of geneticists.

We were fortunate to be joined for a year by Clyde Hutchison, who was an expert on ϕX and its genetics. He brought with him a mutant strain of ϕX with an altered E gene. At this stage we did not know the position of this gene on the DNA. By comparison of sequence data on the normal and mutant DNA it was found that the change in the mutant was in the region coding for gene D, demonstrating that one DNA sequence was coding for two protein sequences. This was of course always a theoretical possibility as any nucleotide sequence can code for up to three different amino acid sequences according to which phase it is read in; that is, at which point initiation takes place. However, this was thought to be unlikely as it was difficult to imagine how such an overlapping system could evolve.

Even when the overlapping of genes D and E was allowed for, there appeared to be more protein sequence than could be accounted for by the size of the DNA, so further overlapping was suspected. Paper 35 describes the identification of this between genes A and B, making use of further mutants provided by Hutchison.

By 1977 the sequence of the ϕX genome was virtually completed by the plus and minus method and we decided to publish it (paper 36) since it effectively told us all we needed to know about ϕX and demonstrated the power of the new method, without which we would certainly have taken much longer. Most of the later work was done by other members of the group as Alan Coulson and I were principally involved in developing a new method.

Papers 37 and 38

Although the plus and minus technique was much better than previous methods, there were still problems and it looked as if improvements could be made if we could find the right conditions. The main difficulty was that the strength of the different bands varied considerably, and some were completely missing. Also, on any single incubation the bands were distributed over a relatively small area of the gel. To overcome this in part, we had to carry out a preliminary incubation with samples taken out at varying times to get a wide distribution of sizes. It would have been good if we could have avoided this preliminary step, and we tried various other systems, but with little success. However, the plus and minus method was slowly improving as it was being used.

One possibility that occurred to us fairly early was to use chain-terminating inhibitors. This was largely inspired by the work of Kornberg[12] showing that dideoxythymidine triphosphate (ddTTP) was a substrate for DNA polymerase, but that once it was incorporated no further nucleotides could be added since it had no 3′ hydroxyl group. So by using ddTTP in the incubation mixture with DNA polymerase, one could clearly make a mixture of nucleotides all ending in T, which was what was needed. There was no apparent reason why it should be any better than the other methods but we were anxious to try it. The main problem was that we had no such terminating inhibitors. I wrote to Kornberg asking if he could let me have some ddTTP but he had already used up all he had. I was lucky, however, to chance on Klaus Geider at a meeting and discover that he had made some ddTTP. He kindly let me have some and we did a single incubation with it; the result was better than anything we had ever seen before. Bands of more or less equal strength were spread out over a long distance on the gel. Clearly we had to get the system going. Besides giving a much clearer result it would be quicker to do as no preliminary incubation would be necessary. The difficulty was that the other three dideoxynucleotide triphosphates had never been made, so Alan Coulson and I set about preparing them, although we had no experience of nucleotide chemistry.

My main experience in synthetic organic chemistry had occurred about 30 years earlier when I was making amino acid derivatives, and particularly the DNP amino acids. This required a good deal of crystallization, which I very much enjoyed doing. The first sign of a pure compound was to obtain pure, well-shaped crystals, and it was always exciting to come into the lab in the morning and find a flask full of beautiful crystals, though more often all one found was an oily mess at the bottom. I think the fun of preparing nice crystals was one of the factors that attracted me to chemistry when I was still studying. I was therefore rather disappointed when I found that nucleotide products did not crystallize, and that to test for purity you had to run them on some chromatography system. This certainly took some of the fun out of the work but we settled down to it and it was well worth the effort. (This observation may seem rather trivial, but I think one should not underestimate the importance of "fun" in research.)

Other possible terminating inhibitors were the arabinonucleotides, and while Alan Coulson and I were preparing the dideoxynucleotides Steve Nicklen was studying the arabinoderivatives. The C and A derivatives were available. They worked reasonably well and would certainly have been an improvement on the plus and minus method, but the dideoxys were marginally better and so we finally settled on them. This was the basis of all our subsequent work on DNA and in fact is still the predominant method in use today, though with many technical improvements.

At about this time another, similar rapid sequencing method was developed by Maxam and Gilbert.[13] This again depended on making mixtures with all the chains ending in the same nucleotide: four incubations were made corresponding to the four nucleotides. They did not use DNA polymerase but labelled the 3′ end of a fragment with $^{32}PO_4$ and then subjected the DNA to partial degradation with chemical reagents that broke the chain either specifically or almost specifically at a given residue. Different reagents were used for the four nucleotides so that four samples were produced and run on the acrylamide gels. The method worked well and the fact that most of the reagents were not absolutely specific did not matter; in fact it sometimes helped interpretation.

The method was first described in a lecture at a Gordon Conference by Gilbert, and protocols of the techniques were circulated, but it was not published till 1977. It is interesting that although they used many of our techniques they have carefully avoided giving any references to our work. The method was easier to use than the plus and minus one and was adopted by several other laboratories, especially in America. Some American labs had had difficulties with our methods, which I think was probably because the DNA polymerase they were using was less pure.

Paper 38 describes a small technical improvement that made a big difference to the work. Even thinner gels are used today.

Papers 39 and 40

Paper 36 summarized our results using the plus and minus method and gave a provisional sequence for ϕX showing the main features of the genome. We were in some doubt as to whether it was necessary to complete the sequence as this would take a lot of work. However, when we had developed the more accurate and rapid dideoxy method, and the method of Maxam and Gilbert had been described, we decided to complete the work. The two methods complemented each other, that of Maxam and Gilbert being particularly suitable for sequences adjacent to restriction enzyme sites, which were sometimes difficult with our system. Every sequence was thoroughly checked and so far no mistakes have been reported. It was by far the longest sequence to be determined at the time.

By now we were using computers for storing and analysing the data. When we started working on DNA some people suggested that we should be using computers, but I was rather reluctant as I felt we were not getting enough data to warrant that. Also, I thought it would take some of the pleasure out of it as I enjoyed looking through the sequences and seeing what could be made of them. However, as the methods became quicker and more and more data was collected it became clear that we should use a computer — at least for storing the sequence. We had found that errors were often coming in just in copying out sequences. Clearly this could be avoided by storing the sequence in a computer but none of us had any experience. However, Michael Smith had a brother working in another laboratory who was an expert and they set up our first storage of the ϕX sequence. This was useful for a while, but a bit inconvenient, especially when Mike left the lab. However, there was a great deal of computer work being done in the laboratory in connection with X-ray analysis and Bart Barrell managed to get one of their experts interested in sequences. This was Rodger Staden and he organized the storage and designed programmes for analysing data which were extremely useful to us. As these became increasingly complex he left the X-ray group and worked full-time on sequence analyses. Today software is a very important part of any major sequencing project and research workers seem to spend most of their time sitting in front of their screens while the robots run the dideoxy gels.

Paper 41

The main drawback to the dideoxy method was that it was applicable only to single-stranded DNA. This was fine with ϕX, which is naturally single-stranded, but most DNAs exist in the double-stranded form and it is frequently difficult to separate the strands. We tried various ways of overcoming this, but by far the best was a cloning procedure worked out by Gronenborn and Messing.[14] The new recombinant DNA technology, or "genetic engineering", had just been introduced, though there were rather rigorous restrictions on its use, which probably explains why we did not use it sooner. The idea had been suggested to me by Walter Schaffner in 1976 when he spent some time in our lab, but it was Messing who got the method going by preparing a suitable vector. The concept was to clone fragments of double-stranded DNA in the replicative form of a single-stranded bacteriophage. This was then grown and the single strands isolated and used as the template for DNA polymerase. As a primer for the dideoxy method a piece of DNA that corresponded to a flanking sequence in the vector was used. In this way the same primer was used for each clone. The other great advantage of this cloning procedure was that it solved the fractionation problem, which had often been the limiting factor in sequencing methods. During the ϕX work we had used restriction enzyme fragments for priming on different sites of the single-stranded template. These had to be pure, and could usually be fractionated by a single run on acrylamide gel ionophoresis. However, with larger DNAs such as bacteriophage λ this would have been a formidable task. In the new method, we used the same primer on different templates which were fractionated by cloning. Cloning is probably the ideal way of fractionating and it does not matter how complex the mixture is: you simply spread it on a plate, essentially using single molecules, which must of course be pure. This paper describes our use of the method, which was a major breakthrough in sequencing and enabled us to attack the larger DNAs of mitochondria and bacteriophage λ. The cloning procedure was not generally applicable to the method of Maxam and Gilbert, and so most laboratories switched to the cloning and dideoxy system, and it is still the method of choice today (1995).

Papers 42 to 45

Armed with the new cloning technique, we felt prepared to tackle DNAs that were double-stranded and larger than ϕX. Initially we chose mitochondrial DNA for various reasons. Its size was suitable (about 16,000 nucleotides) and there was considerable biological interest. Most of the proteins used by mitochondria are encoded by the normal nuclear DNA, but mitochondria contain a relatively small DNA of their own which codes for a few proteins as well as the tRNAs and ribosomal RNAs.

A good deal was already known about this DNA, especially from the work of Attardi and his colleagues, but no sequence had been determined. Certain properties had suggested that mitochondria were more closely related to procaryotes than to eucaryotes, and it was speculated that they had originated from some prehistoric procaryote that had had a complete genome of its own and, as it established a symbiotic relationship in eucaryotic cells, gradually lost most of its genes.

One of the most interesting findings was that mitochondria had a slightly different genetic code from the normal one. Previously it had been assumed that the code was universal, and indeed it was difficult to see how it could change during the course of evolution. Paper 42 describes this discovery, and I have again included this although my name is not on it as it represents an important achievement of the group. When we were working on these larger DNAs it was necessary for someone to collect from the various contributors the data to be put into the computer and analysed. This was usually done by Bart Barrell, who had considerable experience and was expert at studying sequences and picking out interesting features, such as tRNA genes in mitochondria. He was the first to spot the overlapping genes in ϕX (paper 34) and the altered code in mitochondria. This was a somewhat sedentary occupation and I myself preferred to be working at the bench, particularly on the development of methods.

Paper 43 describes the decoding system in mitochondria, which again is different from the normal. In fact mitochondria appear to be unique in many ways and it is difficult to relate them to normal procaryotic or eucaryotic systems.

Paper 44 describes the complete sequence of 16,569 nucleotides. This was the first human DNA sequence to be determined and can be regarded as the start of the human genome project.

Paper 45 describes parallel studies on the bovine mitochondrial DNA. This work was undertaken partly because amino acid sequence data was available on some of the proteins. This proved particularly important in connection with the coding differences.

At the same time as we were working on the mammalian DNA, Tzagoloff and his colleagues at Columbia University, New York, were working on yeast mitochondrial DNA[15] using the method of Maxam and Gilbert. We corresponded frequently with Tzagoloff, comparing our results, and this helped particularly in identifying coding regions. The yeast material was much larger than the human and was surprisingly different, although there was much homology at the sequence level.[16] Yeast also used a different genetic code from nuclear DNA, but this was also somewhat different from the human mitochondria code.[17] We also had useful correspondence with Attardi, who had been studying the messenger RNA of mitochondria.[18]

Paper 47

Bacteriophage λ was one of the organisms most extensively studied by biochemists and much was known about its genes and control mechanisms. There was therefore considerable interest in its sequence, and in fact parts had already been determined. It contained about 48,000 nucleotides, and so was a step up from mitochondria. In fact Alan Coulson and I had started work on it before we started on the mitochondria. Because it was human mitochondria that we wanted to work on, there were rather severe regulations as to the containment facilities required. We had to work in a special isolated room, and while it was being equipped we decided to try out the new cloning procedure on λ, and collected a good deal of sequence information on this before we were able to start on the mitochondria. When the work on the mitochondria was finished we returned to λ to complete it. This took somewhat longer than we had anticipated. We used the "shotgun" approach, collecting clones at random and sequencing them. In this way the work goes very rapidly at first, but as it proceeds more and more of the results are redundant and new data is obtained more slowly. Eventually it becomes necessary to use specific methods for obtaining the missing pieces, and they are much slower than the random approach.

References

1. K. Burton and G.B. Petersen, *Biochem, J.* **75**, 17 (1960).
2. M. Szèkely and F. Sanger, *J. Mol. Biol.* **43**, 607 (1969).
3. V. Ling, *J. Mol. Biol.* **64**, 87 (1972).
4. P. Sadowski and J. Hurwitz, *J. Biol. Chem.* **244**, 6192 (1969).
5. F. Galibert, J.W. Sedat and E.B. Ziff, *J. Mol. Biol.* **87**, 377 (1974).
6. H.D. Robertson, B.G. Barrell, H.L. Weith and J.E. Donelson, *Nature* **241**, 38 (1973).
7. E.H. Blackburn, *J. Mol. Biol.* **93**, 367 (1975).
8. R. Wu and A.D. Kaiser, *J. Mol. Biol.* **35**, 523 (1968).
9. J.E. Donelson, B.G. Barrell, H.L. Weith, H. Kössel and H. Schott, *Eur. J. Biochem.* **58**, 383 (1975).
10. P.T. Englund, *J. Biol. Chem.* **246**, 3269 (1971).
11. E.B. Ziff, J.W. Sedat and F. Galibert, *Nature New Biol.* **241**, 34 (1973).
12. M.B. Atkinson, M.P. Deutscher, A. Kornberg, A.F. Russell and J.G. Moffat, *Biochemistry* **8**, 4897 (1969).
13. A.M. Maxam and W. Gilbert, *Proc. Natl. Acad. Sci. U.S.A.* **74**, 560 (1977).
14. B. Gronenborn and J. Messing, *Nature* **272**, 375 (1978).
15. G. Macino, G. Coruzzi, F.G. Nobrega, M. Li and A. Tzagoloff, *Proc. Natl. Acad. Sci. U.S.A.* **76**, 3784 (1979).
16. M. Zamaroczy and G. Bernardi, *Gene* **47**, 155 (1986).
17. G. Macino and A. Tzagoloff, *Proc. Natl. Acad. Sci. U.S.A.* **76**, 131 (1979).
18. J. Montoya, D. Ojala and G. Attardi, *Nature* **290**, 465 (1981).

Reprinted from
Proc. Nat. Acad. Sci. USA
Vol. 70, No. 4, pp. 1209–1213, April 1973

Use of DNA Polymerase I Primed by a Synthetic Oligonucleotide to Determine a Nucleotide Sequence in Phage f1 DNA

(octadeoxyribonucleotide/pulse-labeling/homochromatography/intercistronic region/evolution)

F. SANGER*, J. E. DONELSON*, A. R. COULSON*, H. KÖSSEL†, AND D. FISCHER†

* Medical Research Council Laboratory of Molecular Biology, Hills Road, Cambridge, England; and † Institut für Biologie III (Genetik und Molekularbiologie), Universität Freiburg, 78 Freiburg, W. Germany

Contributed by F. Sanger, January 16, 1973

ABSTRACT **A sequence of 50 residues in f1 DNA has been determined by the extension of a chemically synthesized octadeoxyribonucleotide by *Escherichia coli* DNA polymerase I, with radioactive nucleoside triphosphates and f1 DNA template. The polymerized product was synthesized either in the presence of manganese and a mixture of ribo- and deoxyribotriphosphates or in a magnesium-containing reaction with one or more of the four triphosphates absent. The sequence determination depended largely on fractionation of the polymerized products by two-dimensional "homochromatography." This approach and the techniques for the subsequent sequence analysis should be of general use for determining other sequences of DNA. Several features of this sequence suggest that it is located in an intercistronic region of f1 DNA.**

Techniques for determining the nucleotide sequence of various radioactive RNA species are well developed (1). Since naturally occurring DNA molecules are very large, the application of many of these techniques to DNA sequence analysis involves one of two initial approaches. Either a uniformly labeled radioactive DNA molecule must be degraded to small fragments (20–100 nucleotides long) that can be isolated for sequence studies (2–4), or the radioactivity must be introduced selectively *in vitro* into a small specific region of an unlabeled DNA molecule (5–8). This report describes a method that uses the second approach with *Escherichia coli* DNA polymerase I (9).

The aminoacid sequence Trp-Met-Val in positions 26–28 of the major coat protein (49 amino acids) of bacteriophage fd has been reported (10). Since the plus strand is the "sense" strand (11), and since tryptophan and methionine are each coded for by a unique codon, the corresponding octadeoxyribonucleotide sequence A-C-C-A-T-C-C-A in the minus strand can be predicted. We have chemically synthesized this sequence on the assumption that it would serve as a unique primer, i.e., a unique starting point, for the incorporation of nucleoside triphosphates by *E. coli* DNA polymerase I, using fd or f1 plus strand as template. The sequence of nucleotides polymerized onto this unique primer–template could then be studied.

The octanucleotide functioned efficiently as a primer on f1 and fd DNA, but we soon realized that the sequence made was not that expected from the published sequence of the coat protein. However, as the priming appeared to be specific, the reaction could be used for the development of methods for DNA sequencing, and has led to the elucidation of a sequence of 50 residues in f1 DNA that may be of considerable biological interest. This paper is a preliminary account of the work; details of the sequence analyses will be published elsewhere.

Previous successful use of the pulse-labeling technique (12, 13) has been with RNA, and has relied mainly on the analysis of ribonuclease T1 digests of the synthetic product obtained after various time intervals. We have used some different approaches that depend largely on the fractionation of the synthetic products of different lengths by a two-dimensional "homochromatography" (14). This system was used directly to determine the sequence of the first 12 residues, using only nearest-neighbor analysis and depurination (15). To determine further sequences it was necessary to have a specific method of degrading the DNA. Berg *et al.* (16) have shown that it is possible to incorporate a ribonucleotide into a deoxynucleotide sequence with DNA polymerase I if manganese is present in the incubation medium instead of magnesium. If a single ribotriphosphate is used with the other three deoxyribotriphosphates, it is incorporated to give bonds that are susceptible to alkali or an appropriate ribonuclease. In theory, this reaction offers an attractive approach to sequencing DNA as different ribonucleotides can be introduced, giving rise to different products, which should overlap one another. We have, therefore, applied this method to the octanucleotide-primed synthesis and, using largely rCTP incorporation, have isolated specific fragments by digestion with pancreatic ribonuclease A and ordered them by studying their distribution in the synthetic products of differing lengths.

MATERIALS

The octadeoxyribonucleotide A-C-C-A-T-C-C-A was synthesized chemically by procedures to be presented elsewhere. DNA was extracted with phenol from phage f1 (a gift of H. D. Robertson). [α-^{32}P]Deoxynucleoside triphosphates (10–100 Ci/mmol) were synthesized by the method of Symons (17). *E. coli* DNA polymerase I was the gift of T. Jovin.

RESULTS

Octanucleotide-Primed Incorporation in the Absence of One or More Nucleotide Triphosphates. Preliminary experiments showed that the octanucleotide could serve as a primer with either fd or f1 DNA as template. In order to determine the sequence of the first few nucleotides incorporated onto the octanucleotide, various incubations were conducted in the absence of one or more nucleoside triphosphates. After a short

Abbreviations: The symbols A, C, G, and T refer to the deoxyribonucleosides, and rC and rG refer to the ribonucleosides.

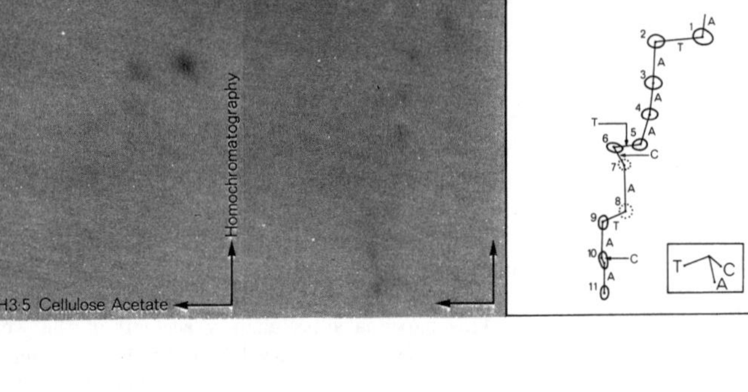

FIG. 1. Fractionation of products synthesized in the absence of one or more nucleoside triphosphates. (*a*) Radioautograph of a fractionation. The reaction mixture (10 μl) contained 0.6 μg of f1 DNA, 0.2 μg of octanucleotide in 50 mM Tris·HCl (pH 7.4), 30 mM NaCl, 8 mM $MgCl_2$, 2 mM dithiothreitol, 0.1 mM dTTP, 1–5 μM [^{32}P]dATP (50 Ci/mmol), and 1.0 unit of DNA polymerase I. After 1 min at 8°, EDTA was added to 20 mM and the reaction mixture was spotted directly on cellulose acetate for ionophoresis at pH 3.5. The second dimension was in a 3% homomixture hydrolyzed for 30 min (2). (*b*) Same as *a*, except that 0.1 mM dCTP was also present during the incubation for 2 min at 8°. (*c*) Composite diagram of *a* and *b* showing the course of synthesis of the oligonucleotide chain. The *inset* shows the predicted effect of the addition of one residue to an oligonucleotide (2, 3).

incubation, the reactions were run directly on the two-dimensional system, which resolved the [^{32}P]deoxyribotriphosphates that had not reacted from labeled oligonucleotides. In a reaction containing TTP and [^{32}P]dATP, six oligonucleotides of increasing size were observed (Fig. 1*a*) and, from their relative mobilities (2, 3) and nearest-neighbor analyses, the sequence of the first six nucleotides polymerized onto the octanucleotide could be deduced. As polymerization stopped at this point, the next position must be either a C or G residue. Fig. 1*b* shows the results of an experiment in which dCTP was also present in the reaction mixture, and Fig. 1*c* shows a stylized version of the autoradiograph. Again, from

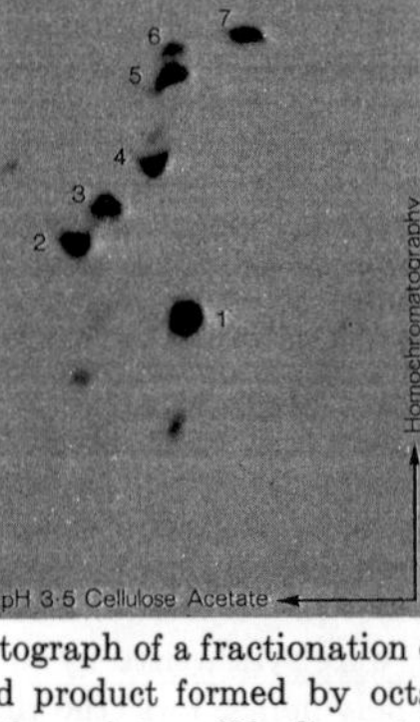

FIG. 2. Radioautograph of a fractionation of a ribonuclease A digest of the mixed product formed by octanucleotide-primed synthesis. The reaction mixture (20 μl) contained 67 mM Tris·HCl (pH 7.4), 0.67 mM $MnCl_2$, 1.5 mM 2-mercaptoethanol, 1.0 mM rCTP, 0.05 mM each of dGTP and dTTP, 1.2 μg of f1 DNA, 0.4 μg of octanucleotide, 2–4 units of DNA polymerase I, and about 0.5 μCi of [^{32}P]dATP (50 Ci/mmol). Incubation was at 25° for 10 min, and 0.13 μCi of [^{32}P] was incorporated. The products were purified by an unpublished procedure of V. Ling involving precipitation with ethanol and with HCl, and digested with ribonuclease A (enzyme to substrate ratio 1:3, 37°, 2 hr). The first dimension was separated by ionophoresis at pH 3.5 on cellulose acetate, the second in a 5% homomixture hydrolysed for 10 min (14).

nearest-neighbor analysis and depurination of the larger oligonucleotides, a sequence of 12 nucleotides could be deduced. In many experiments spots **7** and **8** (Fig. 1*c*) were much weaker than the other spots, or were not even observed; this decrease was due to variations in the asynchronous incorporation of nucleotides by the polymerase, because the relative mobilities and nearest-neighbor analyses of the larger oligonucleotides remained the same. Spot **10** is a mixture of two unresolved oligonucleotides differing by the addition of a 3′-terminal pC residue. This combination was detected by a nearest-neighbor analysis of the larger oligonucleotide in spot **11**. From position differences of these synthesis products on the fingerprint, much of their sequence could be "read off" directly, in the same manner as that of degradation products of partial exonuclease digests (2, 3).

Incorporation Experiments in the Presence of rCTP. When f1 DNA and the octanucleotide were incubated with DNA polymerase I in the presence of manganese, rCTP, and the other three deoxyribotriphosphates, rCMP was incorporated in place of dCMP, and subsequent treatment with ribonuclease A cleaved the polynucleotide chain at the rC residues. These products could then be fractionated by the two-dimensional system. Fig. 2 shows a fingerprint in which [^{32}P]dATP was used. There are seven strong spots and a number of fainter ones, which are more apparent if longer times of incubation are used. The strong spots are numbered **1–7**, and are considered here since they represent the first 50 residues incorporated onto the octanucleotide. Most of the other spots presumably represent sequences from further along the chain, but there are a few that are present as faint spots at early stages of incorporation and represent some type of artifact, perhaps a second initiation site with the octanucleotide or miscopying by the polymerase. The sequences of the seven components were determined, by nearest-neighbor analysis with different [^{32}P]deoxyribotriphosphates (5), depurination (3), and partial digestion with exonucleases (2, 3, 19), and are shown in Table 1. The partial digestion of spot **1** by spleen exonuclease showed that it contained the octanucleotide sequence, and the remaining part of the sequence confirmed that structure already described above.

Proc. Nat. Acad. Sci. USA 70 (1973)

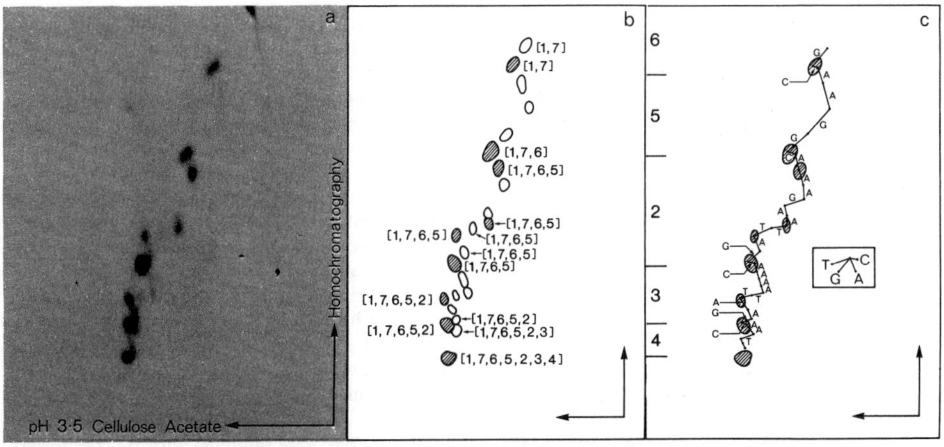

FIG. 3. Fractionation of the products synthesized by DNA polymerase I in the presence of the octanucleotide, Mn, and rCTP. (*a*) Radioautograph of a fractionation. Incubation was for 4 min at 10°, as described in Fig. 2. The product was purified by precipitation with ethanol, and applied to a cellogel strip for fractionation in the first dimension. The second dimension was in homomixture *b* (14). (*b*) Composite diagram showing spots observed in several experiments similar to that shown in *a*, but with various times of incubation and different labels (dATP and dGTP). The oligonucleotides appearing as strong spots in *a* are indicated by *shading*. The products from the spots were digested with ribonuclease A (50 μg per spot, overnight, 37°), and the oligonucleotides were identified by one-dimensional chromatography in a 3% homomixture hydrolysed for 10 min, or by ionophoresis on DEAE-paper in 7% formic acid (19), together with marker oligonucleotides 1–7 from a digest of unfractionated material. These are shown in *brackets* on the diagram. (*c*) Diagram showing the relationship between the position of the spots and the course of synthesis of the oligonucleotide chain. Each spot (Fig. 3*b*) or predicted spot is indicated by a *dot*. The dots are joined by *lines*, whose directions indicate the residue added. The *numbers* to the left of the diagram show at what level on the fingerprint the different sequences, represented by oligonucleotides 2–6 (Table 1), appear in the synthetic product.

Order of the Oligonucleotides. Various methods were investigated for determining the order of the oligonucleotides listed in Table 1. The most successful was to fractionate on the two-dimensional system the product synthesized by DNA polymerase I in the presence of Mn and rCTP. By use of a relatively strong unhydrolyzed "homomixture" (14), fragments of up to 60 residues could be fractionated and then digested with pancreatic ribonuclease A to identify the components. Fig. 3*a* shows such a fractionation.

The presence of certain relatively strong spots well separated from each other indicates that the rate of synthesis by the polymerase is not constant along the chain, but is held up at certain residues. Much of this effect is due to the rC residues, which the polymerase has difficulty in incorporating, though in some experiments there was also an apparent delay at the radioactively labeled residues—presumably because the corresponding triphosphate was present in very low concentrations. The spots from the fingerprint were eluted

TABLE 1. *Oligonucleotides isolated from a ribonuclease A digest of the mixed product containing rC*

Oligonucleotide (Fig. 2)	Sequence
1	A-C-C-A-T-C-C-A-A-T-A-A-A-T-rC(A)
2	A-A-A-G-A-A-T-T-A-G-rC(A)
3	A-A-A-A-T-T-A-A-G-rC(A)
4	A-A-T-A-A-A-G-rC(rC)
5	A-A-G-G-rC(A)
6	A-G-G-rC(A)
7	A-T-A-rC(A)

and digested with ribonuclease A; the digestion products were identified by one-dimensional chromatography. The combined results from several experiments are summarized in Fig. 3*b*. There were several products that did not correspond to any of the markers and appeared in only one or two fragments, but not in the larger fragments. These were intermediates from the 3′ end of the synthesized chain. After digestion with ribonuclease A, this 3′-hydroxyl terminus is liberated as an oligonucleotide that is different from the markers and, if it is labeled, will appear on the analysis.

From the results in Fig. 3*b* the oligonucleotides can be arranged in the order **1–7–6–5–2–3–4**. It was important to be certain that no other residues could be interspersed between the recognized oligonucleotides numbered **1–7**. Table 1 shows that all these oligonucleotides have Ap at their 5′ ends. Thus, if there were any product lying between them in the sequence, it would be labeled in an experiment where [^{32}P]dATP was used. As no such products were present, we concluded that the above oligonucleotides are linked directly in the sequence shown in Fig. 4*a*. Two further lines of evidence support this conclusion. Tentative sequences were ascribed to the products derived from an alkaline digest of the material from an incorporation with rGTP; these sequences precluded the presence of any intermediates.

Fig. 3*c* is a diagram showing the relationship between the course of synthesis of the DNA chain and the relative positions of the spots on the two-dimensional fractionation system. This diagram is based on the above results and on the expected effect of the addition of a residue to a nucleotide chain, as indicated in the inset and as already discussed above (Fig. 1). It is not possible to use the fingerprint to determine the sequence directly, since many intermediates are not seen and

a) A-C-C-A-T-C-C-A-A-T-A-A-A-T-C-A-T-A-C-A-G-G-C-
A-A-G-G-C-A-A-A-G-A-A-T-T-A-G-C-A-A-A-A-T-T-A-
A-G-C-A-A-T-A-A-A-G-C-C

b) 1 10
----G-G-C-T-T-T-A-T-T-G-C-T-
20 30
T-A-A-T-T-T-T-G-C-T-A-A-T-T-C-T-T-T-G-C-C-T-T-
40 50
G-C-C-T-G-T-A-T-G-A-T-T-T-A-T-T-G-G-A-T-G-G-T----

c) ----C-T-T-A-G-G-A-T-T-T-A-A-T-C-A-T-G-T-T-T----

FIG. 4. DNA sequence in phage f1 DNA. (*a*) Sequence of the minus strand, as determined in this paper. The first eight residues are from the synthetic octanucleotide used for priming. (*b*) The compliment of the sequence in (*a*); this is the sequence in the plus strand and in the messenger RNA. *Underlining* indicates termination codons; *broken underlining* indicates initiation codons. (*c*) Sequence around the initiating site of a structural protein of ϕX DNA (4).

the differences in position are very small for the larger products. The fact that it is possible to draw such a diagram gives useful confirmation of the sequence and precludes the possibility of any extensive sequence being missed. Certain features, such as the two –T–T– sequences, are readily recognized on the fingerprint (Fig. 3*a*).

DISCUSSION

The methods described in this paper should be of fairly general application to the determination of sequences in DNA, though no doubt each situation will present its own problems. In theory, this procedure, requires only a synthetic oligonucleotide complimentary to a known sequence and conditions where polymerase will copy faithfully. In practice, the choice of the octanucleotide primer in this work was entirely fortuitous as it does not prime in the expected position in the coat protein cistron. The recent report by Snell and Offord (20) on the aminoacid sequence of the coat protein of the related phage ZJ-2 raises some doubt as to whether the sequence in this position might in fact be Trp-Ala-Met instead of Trp-Met. Although efficient and sufficiently specific priming was obtained with the octanucleotide, it would probably be advantageous to use a longer primer. In these experiments it was necessary to use relatively high concentrations of primer (about 200-fold excess over the f1 DNA), and it is probable that other sequences may have been formed to a lesser extent as faint spots were often detected that could not be explained by the single unique sequence.

The advantages of using the method of Berg *et al.* (16) to introduce ribonucleotides, and therefore specifically susceptible bonds into the DNA, are obvious and this has been the subject of several recent communications (18, 21, 22). In agreement with the conclusions of Van de Sande *et al.* (21), we have obtained the best results with rC, and most of the sequences have been based on its use, whereas, although good incorporations were obtained with rG, the fingerprints were rather complex and somewhat variable. This may have been partly due to the nature of the sequence involved. Others (21, 22) have concluded that DNA polymerase I has difficulty in inserting two consecutive ribonucleotides. The first –C–C– is at the 3′ end of the deduced sequence, whereas the first G residues are in a –G–G–C–A–A–G–G– sequence. It may be that in a different situation the use of rG may be more advantageous. As might be expected from the large number of A residues, almost no incorporation of rA could be obtained.

An important question to be considered is whether DNA polymerase I is copying faithfully under these special conditions and, therefore, whether the sequence determined is the exact compliment of the plus strand of f1 DNA. In particular, the unusually large number of A residues present in apparently homologous sequences raised the possibility of some artifact of incorporation. Van de Sande *et al.* (21) report that misincorporation with rCTP may occur at 37°, but not at 10°. Most of our experiments were done at 25°, but all essential results have now been confirmed at 10°. Further evidence that the copying is faithful is the agreement with the results obtained under normal conditions with three deoxyribotriphosphates. However, the main evidence is that it is possible to deduce a unique sequence with no evidence of randomiza-

Mutation event	Sequence
Duplication	T G C C T
C → T	T G C C T T G C C T
Duplication	T G C T T T G C C T
Duplication	T G C T T T T T T G C C T
T → G, T → A, T → A,	T G C T T T T T T G C T T T T T G C C T
Insert A	G G C T T T A T T G C T/A T T T T G C C T
Duplication	G G C T T T A T T G C T A A T T T T G C C T
Duplication	G G C T T T A T T G C T A A T T T T G C T A A T T T T G C C T
Duplication	G G C T T T A T T G C T A A T T T T G C T A A T T T T G C C T T G C C T
Insert T	G G C T T T A T T G C/T A A T T T T G C T A A T T T T T T G C C T T G C C T
T → C	G G C T T T A T T G C T T A A T T T T G C T A A T T T T T T G C C T T G C C T
	G G C T T T A T T G C T T A A T T T T G C T A A T T C T T T G C C T T G C C T

FIG. 5. Hypothetical course of evolution of the sequence in positions 1–39 (Fig. 4*b*). *Underlining* indicates the position in which the next mutation event takes place.

tion. If misincorporation were occurring, it would not be specific and the products would be expected to be much more heterogeneous.

The sequence determined (Fig. 4*a*) corresponds to that of the minus strand of f1 DNA. The complimentary sequence, which corresponds to the plus strand and the messenger RNA, is shown in Fig. 4*b*. The last eight residues are based on the assumption that the synthetic octanucleotide used as a primer initiates by hybridizing with its exact complimentary sequence on the template DNA. This assumption seems likely, since such a small oligonucleotide would not be expected to prime unless it paired perfectly, but there is no direct proof.

The sequence is remarkable in that it is very rich in T residues and shows considerable repetition of homologous sequences. Thus, T–T–G–C is repeated four times, T-G-C-C-T twice, and T–A–A–T–T twice. The most likely explanation is the way in which the DNA has evolved. It is probable that the increase in size of a DNA molecule is brought about by duplication of existing sequences, which will then evolve by mutation to give functional DNA. Thus, at early stages of evolution one would expect to find repeated sequences. Such repetitions have been found in RNA sequences (23) and particularly in satellite DNA (24). Fig. 5 shows one way in which the sequence from position 1–39 could have evolved from the pentanucleotide T–G–C–C–T by a process involving six duplications, two insertions of single residues (which could be regarded as single residue duplications), and only five base changes. Clearly there are many other ways in which the evolution could have taken place, but this scheme does illustrate that a sequence of the type found here might be found shortly after an increase in DNA size has occurred by duplication of relatively short sequences. This mechanism would imply either that the DNA section has evolved relatively recently, or else that it acquired an essential function at an early stage in evolution and thus remained stable. This latter possibility is suggested by the further speculation given below.

Robertson *et al.* (4) have recently determined the sequence of a ribosomal binding site in ϕX 174 DNA that was found to contain the initiation point for one of the structural proteins of the virus (25). This sequence is compared in Fig. 4*c* with the sequence of f1 DNA described here, and there are certain interesting homologies. Both are rich in T residues and contain the sequence G–A–T–T–T–A shortly before an A–T–G triplet (positions 54–56). The sequence Pu–Pu–T–T–T–Pu is characteristically found in this position relative to initiation sites for several protein cistrons (4).

This finding suggests the possibility that the A–T–G of this f1 DNA fragment might be an initiation site for a protein. Moreover, the sequence contains three termination codons, T–A–A in positions 13–15 and 22–24, and T–G–A in position 43–45. All three are in phase with each other, so that only the first one could actually function as a protein chain terminator, but the other two might act as a defense mechanism against incomplete termination. A similar situation is found at the end of the coat protein cistron of phage R17 (26), which has three termination codons in phase with each other, the first two being adjacent. It is thus tempting to speculate that the sequence could represent the intercistronic space between two protein cistrons, one of which ends in the aminoacid sequence Gly–Phe–Ile–Ala and the next one of which starts with formylMet–Val. It is hoped that further work will reveal whether this is the case. If it were not, there are two possible aminoacid sequences for which the DNA fragment could code, since two of the three possible phases do not contain termination codons.

Some preliminary experiments in connection with this work were carried out with Dr. V. Ling. We thank Dr. T. Jovin for a gift of DNA polymerase I and Dr. H. D. Robertson for phage f1. J.E.D. is supported by the Helen Hay Whitney Foundation. H.K. wishes to acknowledge support from the Deutsche Forschungsgemeinschaft (SFB 46) and from EMBO.

1. Barrell, B. G. (1971) *Procedures in Nucleic Acids*, eds. Cantoni, G. L. & Davies, D. R. (Harper & Row, New York), Vol. 2, pp. 751–779.
2. Ling, V. (1972) *J. Mol. Biol.* **64**, 87–102.
3. Ziff, E. B., Sedat, J. W. & Galibert, F. (1971) *Nature New Biol.* **241**, 34–37.
4. Robertson, H. D., Barrell, B. G., Weith, H. L. & Donelson, J. E. (1973) *Nature New Biol.* **241**, 38–40.
5. Wu, R. & Taylor, E. (1971) *J. Mol. Biol.* **57**, 491–511.
6. Padmanabhan, R. & Wu, R. (1972) *J. Mol. Biol.* **65**, 447–467.
7. Padmanabhan, R., Padmanabhan, R. & Wu, R. (1972) *Biochem. Biophys. Res. Commun.* **48**, 1295–1302.
8. Oertel, W. & Schaller, H. (1972) *FEBS Lett.* **27**, 316–320.
9. Kornberg, A. (1969) *Science* **163**, 1410–1418.
10. Asbeck, V. F., Beyreuther, K., Köhler, H., Wettstein, G. & Braunitzer, G. (1969) *Hoppe-Seyler's Z. Physiol. Chem.* **350**, 1047–1066.
11. Hayashi, M., Hayashi, M. N. & Spiegelman, S. (1963) *Proc. Nat. Acad. Sci. USA* **50**, 664–671.
12. Billeter, M. A., Dahlberg, J. E., Goodman, H. M., Hindley, J. & Weissmann, C. (1969) *Nature* **224**, 1083–1086.
13. Blattner, F. R. & Dahlberg, J. E. (1972) *Nature New Biol.* **237**, 227–232.
14. Brownlee, G. G. & Sanger, F. (1969) *Eur. J. Biochem.* **11**, 395–399.
15. Burton, K. (1967) in *Methods in Enzymology*, eds. Grossman, L. & Moldave, K. (Academic Press, New York), Vol. 12, part A, pp. 222–224.
16. Berg, P., Fancher, H. & Chamberlin, M. (1963) in *Symposium on Informational Macromolecules*, eds. Vogel, H., Bryson, V. & Lampen, J. O. (Academic Press, New York & London), pp. 467–483.
17. Symons, R. H. (1969) *Biochim. Biophys. Acta* **190**, 548–550.
18. Salser, W., Fry, K., Brunk, C. & Poon, R. (1972) *Proc. Nat. Acad. Sci. USA* **69**, 238–242.
19. Sanger, F., Brownlee, G. G. & Barrell, B. G. (1965) *J. Mol. Biol.* **13**, 373–398.
20. Snell, D. T. & Offord, R. E. (1972) *Biochem. J.* **127**, 167–178.
21. Van de Sande, J. H., Loewen, P. C. & Khorana, H. G. (1972) *J. Biol. Chem.* **247**, 6140–6148.
22. Wu, R., Donelson, J. E., Padmanabhan, R. & Hamilton, R. (1972) *Bull. Inst. Pasteur, Paris* **70**, 203–233.
23. Brownlee, G. G., Sanger, F. & Barrell, B. G. (1968) *J. Mol. Biol.* **34**, 379–412.
24. Southern, E. M. (1970) *Nature* **227**, 794–798.
25. Air, G. & Bridgen, J. (1973) *Nature New Biol.* **241**, 40–41.
26. Nichols, J. L. (1970) *Nature* **225**, 147–151.

J. Mol. Biol. (1974) **90**, 315–333

Determination of a Nucleotide Sequence in Bacteriophage f1 DNA by Primed Synthesis with DNA Polymerase

F. Sanger, J. E. Donelson, A. R. Coulson

Medical Research Council Laboratory of Molecular Biology
Hills Road, Cambridge, England

H. Kössel and D. Fischer

Institut für Biologie III (Genetik und Molekularbiologie)
Universität Freiburg, 78 Freiburg, W. Germany

(Received 16 April 1974)

A method is described for the determination of nucleotide sequences in DNA by using specific oligonucleotides as primers for copying specific regions by DNA polymerase. The method was applied to bacteriophage f1 DNA using the synthetic octanucleotide A-C-C-A-T-C-C-A as primer and a sequence (sequence A) of 81 nucleotides was determined. Synthesis was carried out in the presence of manganese and with one of the deoxyribotriphosphates (dCTP or dGTP) replaced by the corresponding ribotriphosphate so that mixed oligonucleotides were found which could be specifically split at the ribonucleotide residues by the appropriate ribonuclease or by alkali. The relative order of the digestion products was determined by fractionating the undigested oligonucleotides according to size on a two-dimensional system and digesting the isolated products. In the presence of rGTP the octanucleotide appeared to prime at a second site giving rise to a second sequence (B) besides sequence A. The complementary sequence to sequence A, which corresponds to the plus strand of f1 DNA and to the messenger RNA, contains five nonsense codons, four of which are in the same phase, and two possible initiation codons. It also contains a repetitive sequence which suggests its evolutionary origin by duplication.

1. Introduction

One approach to the determination of nucleotide sequences in nucleic acids is by specific copying or repair with polymerases. This makes it possible to introduce ^{32}P-labelled nucleotides into a small region and to apply the "pulse-labelling" method. Billeter *et al.* (1969) have determined extensive sequences in bacteriophage Qβ RNA using the phage's own replicase. ^{32}P-labelled nucleotides were introduced into the 5′ end of either strand and elegant techniques have been developed for deducing the sequences thus labelled. In order to use the method for sequencing DNA it is necessary to employ a system in which the repair starts at a unique position. Blattner & Dahlberg (1972) and Lebowitz *et al.* (1971) have used RNA polymerase starting at its normal promoter sites to determine sequences in phage λ DNA. Wu and his colleagues (Wu & Kaiser, 1968; Wu & Taylor, 1971) have employed DNA polymerase to study the "sticky ends" of phage λ. These "sticky ends" are single-stranded extensions of 12 residues to the otherwise double-stranded molecule.

They form a suitable substrate for DNA polymerase, which was used to copy them in the presence of various radioactive triphosphates, and techniques were developed for determining these sequences.

An alternative approach that has been used by ourselves and by Khorana (see Loewen *et al.*, 1974) is to use a synthetic oligonucleotide to prime repair by DNA polymerase at a specific site. In our work, which was described in a preliminary communication (Sanger *et al.*, 1973) the template was the single-stranded circular DNA (6000 nucleotides) of bacteriophage f1 and the synthetic oligonucleotide was A-C-C-A-T-C-C-A. This primer was chosen on the basis of the reported amino acid sequence in the protein, in the expectation that it would prime in the cistron for this protein. Although this did not prove to be the case there was nevertheless a primer-dependent incorporation in the presence of *Escherichia coli* DNA polymerase I and nucleoside triphosphates (one of which was highly labelled with ^{32}P). The octanucleotide hybridized to a specific site on the template, which was presumably the complementary sequence or very similar to it, and nucleotides were added sequentially to its 3′ end, thus making possible a "pulse-labelling" approach to sequencing.

By using conditions in which only a few residues were added, and fractionating the products on two-dimensional "homochromatography", a series of spots was obtained from which it was possible to deduce a sequence of twelve residues. This was based largely on the relative positions of the oligonucleotide spots on the fingerprint according to the method originally used by Ling (1972) for analysing partial exonuclease digests. To extend the sequence determination farther, the "ribo-substitution" approach originally proposed by Berg *et al.* (1963) was used. The incubation with DNA polymerase was carried out in the presence of manganese, rCTP† and the three other deoxyribonucleoside triphosphates so that the sequence of the octanucleotide primer was extended by about 50 residues to give a product in which all the C residues were present as ribonucleotides. These rC bonds could then be cleaved by ribonuclease A and the specific degradation made it possible to determine a sequence of 50 residues. Other workers (Berg *et al.*, 1963; Salser *et al.*, 1972; Van de Sande *et al.*, 1972) have shown that the ribo-substitution technique may be used with ribotriphosphates other than rCTP. However, preliminary experiments with our system using rGTP substitution were very difficult to interpret. Although good primer-dependent incorporation was obtained, digestion of the product with ribonuclease T_1 or alkali gave rise to oligonucleotides that did not fit the sequence (sequence A) that had been determined using rCTP substitution. It now appears that under the conditions used for rGTP substitution at least one other octanucleotide-primed sequence (sequence B) is formed. This is evident from the experiments shown in Plates III and IV. In these experiments the products from the DNA polymerase incubations were fractionated directly without ribonuclease digestion on the two-dimensional system using homochromatography. When rCTP was used (Plate III) essentially a single series of spots was present which represented the gradual building up of sequence A on the octanucleotide. From analysis of these spots the relative order of the ribonuclease A digestion products could be deduced. However, when rGTP substitution was used (Plate VI) two separate series of spots were present, representing the building up of two different sequences. On alkali hydrolysis of the oligonucleotides from series A, products were obtained that confirmed sequence A

† Abbreviations used: as this paper is concerned mainly with DNA, the symbols A, C, G and T refer to the deoxyribon ucleosides and rC and rG to the ribonucleosides.

and made it possible to extend it by a further 31 residues. The oligonucleotides in series B represent an entirely different sequence, which was not detected in the experiments with rCTP substitution.

2. Materials and Methods

(a) *Materials*

The octadeoxyribonucleotide, A-C-C-A-T-C-C-A, was synthesized chemically by procedures described elsewhere (Schott *et al.*, 1973). Random DNA primer was the gift of E. Ziff and was prepared by digesting 2 mg calf thymus DNA/ml with 220 μg pancreatic DNAase/ml in 50 mM-Tris·HCl (pH 7·4), 5 mM-$MgCl_2$ for 2 h at 37°C.

DNA was phenol-extracted from bacteriophage f1 (a gift of H. D. Robertson).

[α-^{32}P]deoxyribonucleoside triphosphates (spec. act. = 10 to 100 mC/μmol) were synthesized using only slight modification of the method of Symons (1969). Unlabelled nucleoside triphosphates were bought from Sigma Chemical Co.

E. coli DNA polymerase I was either a gift of T. Jovin or bought from Boehringer Corporation. Nucleases were bought from commercial sources: micrococcal nuclease and pancreatic ribonuclease A from Worthington Biochemical Corporation; spleen and venom exonuclease from Boehringer Corporation; ribonuclease T_1 from Sankyo Co.

(b) *Incorporation by DNA polymerase I*

(i) *Using four deoxyribonucleoside triphosphates in the presence of magnesium*

The basic incubation reaction (10 to 100 μl) contained 0·008 μg f1 DNA/μl, 0·005 μg octanucleotide/μl in 50 mM-Tris·HCl (pH 7·4), 60 mM-KCl, 8 mM-$MgCl_2$, 4 mM-2-mercaptoethanol, 200 μM-unlabelled and 0·5 to 100 μM-[α-^{32}P]deoxyribonucleoside triphosphates and 0·1 to 0·2 unit of DNA polymerase I/μl. The reaction was terminated by adding EDTA to 20 mM or blowing into 1 ml 5% trichloroacetic acid.

(ii) *Using rCTP in the presence of manganese*

The conditions used were similar to those of Salser *et al.* (1972). The reaction mixture (10 to 100 μl) contained 67 mM-Tris (pH 7·4), 0·67 mM-$MnCl_2$, 1·5 mM-mercaptoethanol, 1·0 mM-rCTP, 0·05 mM of the two unlabelled deoxyribotriphosphates and a much lower concentration of the radioactive triphosphate (approx. 0·02 to 0·1 μCi/μl), 0·06 μg f1 DNA/μl, 0·02 μg of the octanucleotide A-C-C-A-T-C-C-A/μl, 0·1 to 0·2 unit DNA polymerase I/μl.

After incubation a small portion was used for assay of incorporation by trichloroacetic acid-precipitation and the remainder was added to 0·9 ml water, 0·3 ml water-saturated phenol and 60 μg carrier tRNA. After 10 min at 0°C the layers were separated by centrifugation and the nucleic acid in the aqueous layer precipitated by the addition of 0·1 ml 20% sodium acetate (pH 5) and 2·5 ml ethanol overnight at −20°C. The precipitate was collected by centrifugation. In some experiments this precipitate was used directly, in which case it was washed once with 1 ml ethanol and dried in a desiccator. It contained unreacted triphosphate and the product was still hybridized to the f1 DNA. In most experiments the product was further purified by an acid-precipitation step that was developed by V. Ling (personal communication). The precipitate was dissolved in 0·3 ml water and 0·2 ml N-HCl was added. After 10 min at 0°C the precipitate was centrifuged down and washed with 0·5 ml 0·2 N-HCl and with 1·0 ml ethanol and dried in a desiccator. The acid-precipitation and washing was carried out rapidly to prevent depurination. This precipitate was insoluble in water: it was dissolved in a few drops of 20% triethylamine carbonate (pH 10) and taken to dryness in a desiccator. To remove excess triethylamine carbonate a drop of water was added and again taken to dryness. The procedure was repeated 3 times. This residue was then soluble in water and could be used for rerunning or for digestion with ribonuclease A.

(iii) *Incorporation experiments with rGTP*

Incorporation of rGTP was carried out as described for rCTP except that 1·0 mM-rGTP was present and dCTP, if unlabelled, was in 0·05 mM concentration. For digestion with ribonuclease T_1 the product isolated by the above method was incubated at 37°C overnight in 5 μl of a solution containing 0·05 M-Tris·HCl (pH 7·4), 0·005 M-EDTA and 25 μg ribonuclease T_1.

(c) *Digestion of rCMP-containing oligonucleotides with ribonuclease A*

Considerably more violent conditions were required to digest these mixed oligonucleotides with ribonuclease A than are used for RNA. The product obtained by the above isolation procedure was dissolved in 5 μl of a solution containing 0·02 M-Tris·HCl (pH 7·4), 0·002 M-EDTA and 20 μg ribonuclease A and incubated at 37°C for 2 h in a sealed capillary tube. This digest was applied directly without further purification to the cellulose acetate or DEAE-thin layer plates.

(d) *Fractionations*

For most separations the two-dimensional "homochromatography" system was as described by Brownlee & Sanger (1969), using the following "homo-mixtures": homomixture B, as described by Brownlee & Sanger (1969); homomixture C, 5% RNA hydrolysed 10 min and dialysed; homomixture D, 4% RNA hydrolysed 30 min and dialysed (Jeppesen, 1971); homomixture E, as for homomixture B but using 7% unhydrolysed RNA. For the fractionation of the ribonuclease digests commercially prepared DEAE-cellulose layers (Polygram Cel300 DEAE, Machery–Nagel & Co., Düren) were used. Migration and flow rate were considerably slower on these than on the normal cellulose–DEAE-cellulose plates. Development was usually for 12 h, though this varied somewhat from batch to batch. A pad of filter paper was normally attached to the top of the plate with clips and the developing solution was allowed to soak into this until the blue marker dye (xylene cyanol FF) had not quite reached the top. The undigested products were fractionated on glass plates containing a layer of cellulose and DEAE-cellulose (7·5:1) as previously described (Brownlee & Sanger, 1969).

(e) *Sequence determination of the oligonucleotides*

Nearest-neighbour analyses were performed by digesting the eluted oligonucleotides in 10 μl 80 mM-NH_4HCO_3 (pH 8·1), 0·5 mM-$CaCl_2$ containing 1 μg micrococcal nuclease for 1 h at 37°C in a drawn-out capillary. The solution was dried on polythene sheets in a desiccator and taken up in 10 μl of 100 mM-ammonium acetate (pH 5·7), 10 mM-EDTA containing 10 μg spleen exonuclease and incubated for 4 h at 37°C in a sealed capillary. The mononucleotides were separated by ionophoresis on Whatman 3MM paper at pH 3·5.

Depurination was carried out as described by Ling (1972) and the products identified by ionophoresis on DEAE-paper at pH 3·5 (Ziff *et al.*, 1973) using a mixture of the depurination products of whole ϕX DNA as a reference. rGp residues were not degraded under the conditions used, so when ribonuclease T_1 digestion products of the material synthesized in the presence of rGTP were depurinated, oligonucleotides were obtained containing rGp at their 3′ end. The behaviour of some of these on the ionophoresis system is summarized in Fig. 1. They were normally identified by digestion to prGp with venom exonuclease under the "vigorous" conditions described by Galibert *et al.* (1974).

Partial digestion with exonucleases

(i) *Spleen exonuclease.* The conditions necessary for optimum digestion varied considerably for different nucleotides, and somewhat for different batches of enzyme. Thus it was usually preferable to carry out a preliminary analysis to help establish these conditions.

Method A is essentially the method of Sanger *et al.* (1965). Half of the material eluted from one fingerprint was dissolved in 20 μl of a solution containing 0·1 M-ammonium acetate (pH 5·7), 0·01 M-EDTA and 0·2 mg spleen exonuclease/ml and incubated at 37°C. Samples were removed at 0, 20, 40 and 90 min and applied directly to DEAE-paper to stop the reaction. The products were then fractionated by ionophoresis at pH 1·9 and the sequence deduced from *M* values and nearest-neighbour analyses of the products. These conditions were suitable for oligonucleotide C4 (Plate I). For smaller nucleotides 0·1 mg enzyme was more suitable, whereas with larger ones (e.g. nucleotide C1) there was still much undigested material present.

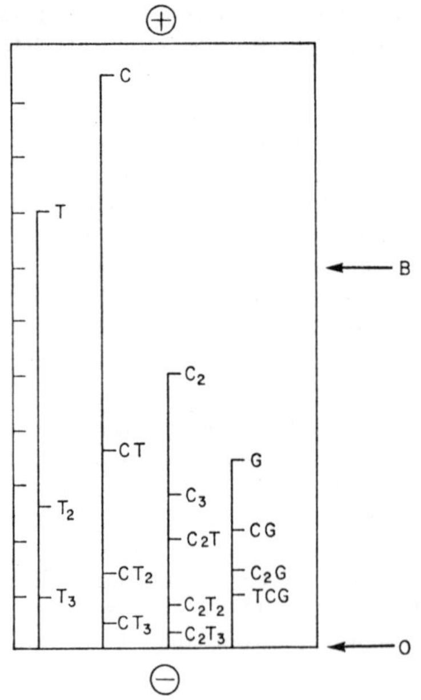

FIG. 1. Diagram showing the relative rates of migration of depurination products from mixed DNA containing rG residues on ionophoresis at pH 3·5 on DEAE-paper (see Galibert *et al.*, 1974). O marks the origin, B the position of the blue marker (xylene cyanol FF).

Method B is the method described by Ziff *et al.* (1973) in which the products are fractionated on the two-dimensional system and the sequence deduced from the relative positions of the spots.

(ii) *Venom exonuclease.* The products eluted from the thin-layer plates were digested under various conditions depending on their size, according to the method of Galibert *et al.* (1974). 30 μg of alkaline phosphatase (Whatman Biochemicals Ltd) was added to each incubation to remove the 3′ terminal phosphate groups.

(f) *Determination of the order of the nucleotides*

In order to identify the ribonuclease A digestion products of the synthesized chains containing rCp that had been fractionated on the two-dimensional system, the material was eluted from the thin-layer plate and digested with 5 μl 0·02 M-Tris·HCl, 0·002 M-EDTA and 50 μg ribonuclease A in a sealed capillary overnight at 37°C. Usually the digests were fractionated one-dimensionally using homomixture D on polygram thin-layers (as in Plate IV). Most products could be identified by their migration rates, and frequently markers of the ribonuclease A digestion products that had been isolated from digests such as those shown in Plate I were run in parallel. The identifications were normally confirmed by eluting the spots and subjecting them to "nearest-neighbour" analysis, as described above.

In certain cases ionophoresis on DEAE-paper, using 7% formic acid, was used for identification. This was particularly useful for small oligonucleotides which were not separated well on homochromatography. For some of the larger synthesized products identification was carried out on the two-dimensional system (see Plate V). This gave clearer results though it is more laborious than the one-dimensional methods.

The synthesized products containing rGp were difficult to digest with ribonuclease T_1 due partly to the large amounts of RNA present in the spots eluted from the homochromatography plates, so they were digested with alkali. The products were dissolved in 15 μl 0·3 M-NaOH and incubated in a sealed capillary tube for 15 h at 37°C. In order to remove the NaOH the digest was transferred to a 2 cm × 4 cm piece of CM-cellulose paper. The paper was eluted with water by the normal technique (Sanger *et al.*, 1965) and the eluate collected and dried down. The digestion products were identified as described for rCp-containing products using markers from a ribonuclease T_1 digest of the unfractionated material containing rGp.

3. Results

(a) *Test of priming ability of the octanucleotide*

In order to show that the octanucleotide could serve as a primer for the incorporation of nucleotide triphosphates by DNA polymerase I on f1 DNA template, acid-precipitable counts were determined at increasing incubation times. As shown in Figure 2 incorporation is primer-dependent and in the presence of random DNA primer reaches a plateau at approximately 1·2 rounds of synthesis on the circular f1 DNA template. Figure 2 shows that the octanucleotide does indeed serve as a primer for DNA polymerase I. The faster rate of appearance of acid-precipitable material with random DNA primer than with the octanucleotide suggests, as expected, that random DNA primer provides many more starting points than the octanucleotide.

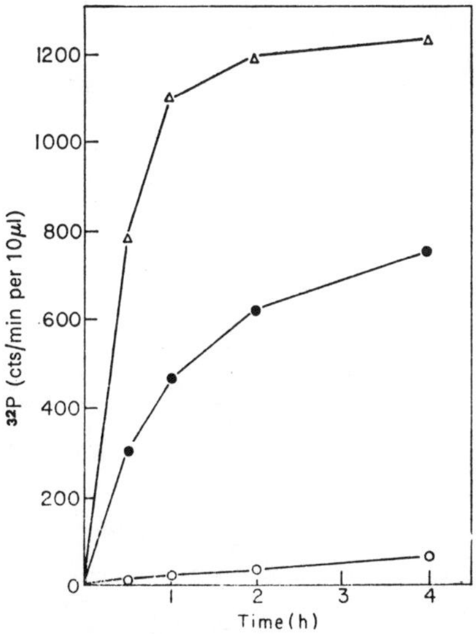

FIG. 2. Incorporation of [^{32}P]TTP into DNA by DNA polymerase at 15°C in the presence of f1 DNA, as described in Materials and Methods, section (b) (i). Incorporation was measured as trichloroacetic acid-precipitable radioactivity. —○—○—, No added primer; —△—△—, 0·5 μg random DNA primer added per 50 μl incubation; —●—●—, 0·25 μg of the octanucleotide A-C-C-A-T-C-C-A added as primer. One round of synthesis corresponds to an incorporation of 1040 cts/min per 10 μl.

(b) *Incorporation in the presence of rCTP*

Plate I shows the results of experiments in which f1 DNA and the octanucleotide were incubated with DNA polymerase in the presence of manganese, rCTP and the three other deoxyribonucleoside triphosphates, one of which (ATP or GTP) was labelled with ^{32}P. The products obtained were digested with ribonuclease A and fractionated on the two-dimensional homochromatography system. Although the

TABLE 1

Analysis of the oligonucleotides from a ribonuclease A digest of the product containing rC

Oligonucleotide (Plate I)	Labelled residue†	Nearest-neighbour analysis‡				Depurination products§
		rCp	Ap	Gp	Tp	
C1	pA	1·0	2·9	—	1·1	pTp, pTprCp
	pT	—	+	—	—	
C2	pA	1·0	2·8	0·9	1·0	pTpTp, prCp
	pG	—	+	—	—	
	pT	—	1·1	—	1·0	
C3	pA	1·0	4·2	—	1·0	pTpTp, prCp
	pG	—	+	—	—	
	pT	—	1·3	—	1·0	
C4‖	pA	—	+++	—	+	pTp
	pG	—	+	—	—	
	pT	—	+	—	—	
C5‖	pA	+	+	—	—	prCp
	pG	—	+	+	—	
C6	pA	+	—	—	—	prCp
	pG	—	+	+	—	
C7	pA	1·0	—	—	0·95	pTp, prCp
	pT	—	+	—	—	
C10	pA	+	++	+	+	pTp, pTpTpCp
	pG	—	—	—	+	Tp
	pT	—	++	—	+	
C12	pA	—	++	—	+	pTp
	pG	—	+	—	—	
	pT	1·0	0·95	—	—	
C13	pA	—	++	—	+	Tp
	pG	+	—	—	—	pTprCp
	pT	—	+	—	—	
C18	pA	+	—	—	—	TprCp¶
C19	pA	+	—	—	+	
C23	pA	—	—	—	+	pTp
	pG	—	—	+	+	pTpTp
	pT	—	—	++	+	
C24	pA	—	—	+	+	pTpTp
	pG	—	—	—	+	pTp
	pT	—	++	—	+	
C25	pA	—	+	—	++	pTp
	pG	—	—	—	+	Tp
	pT	+	+	+	—	
C26	pA	—	+	—	—	

† Where no results are given for pA, pG or pT, no spot was obtained, indicating that the oligonucleotide was not labelled by that triphosphate.

‡ The numbers give the relative ratios of the mononucleotides estimated by scintillation counting. Where only +s are given the concentration was estimated by inspection of the radio-autograph, the number of +s indicating the relative intensities.

§ Identified by ionophoresis at pH 3·5 on DEAE-paper (Ziff *et al.*, 1973).

‖ Spots C4 and C5 obtained in the experiment shown in Plate I were both mixtures. Spot C5 contained a small amount of G(A) and A(T) and spot C4 a small amount of rC(A). The results given here were from an experiment in which there was less extensive incorporation.

¶ From its position on the fingerprint and the nearest-neighbour analysis C18 could be either GC(A) or TC(A). To distinguish between these two possibilities a sample was subjected to depurination and "ionophoresed" in parallel with the untreated material on DEAE-paper at pH 3·5. It was unchanged, showing it to be TC(A).

TABLE 2

Partial digestion with exonucleases of oligonucleotides from a ribonuclease A digest of the product containing rCp

Oligonucleotide (Plate I)	Exonuclease	Labelled residue	Method of analysis (see Materials and Methods)	Sequence (the residues in brackets are the nearest-neighbours present in the residual nucleotide)
C1	S	pA	A	-A-A-T-(A,C)
	S	pA	B	A-C-C-A-T-C-C-A-A-T-A-A
	V	pT	B	-T-C
C2	S	pA	B†	A-A-A-G-A-A-(T,C)
	S	pG	B†	A-A-A-G-A-A-(A)
	V	pT	B	-A-G-C
C3	S	pA	A	-T-T-A-
	S	pA	B†	A-A-A-A-T-T-(A,C)
	V	pT		-A-A-G-C
C4	S	pA	A	A-A-T-A-
	S	pG	A	A-A-T-A-
	V	pA	A	(T)A-A-G‡-C
C5	S	pA	A	A-A-
C6	S	pA	A	A-G‡-G‡-C
C7	S	pA	A	A-
C10	S	pA	B	-G-A-T-A-A-A-T-
C12	S	pA	A	A-T-§
C13	S	pA	A	T-
C23	S	pA	B	G-G-T-T-
C24	S	pA	B	A-T-T-
C25	S	pA	B	T-G-T-A-T-
C26	S	pA	A	A-A-

S, spleen; V, venom.

† These results are from 2 experiments; one as described in method B, and another in which the products were separated by one-dimensional homochromatography and subjected to nearest-neighbour analysis.

‡ As this sequence was deduced from *M* values alone these positions could be occupied by G or T.

§ The spleen exonuclease digests of C12 and C4 were run in parallel. Unchanged C12 ran at the same rate as the first degradation product of C4 and the first product of C12 at the same rate as the second product of C4, indicating that C12 has the same sequence as C4 except that the 5′ A residue is absent.

conditions used (10 min at 25°C) were essentially the same as in the experiment described in the preliminary communication (Sanger *et al.*, 1973; Fig. 2), synthesis has gone considerably farther, giving rise to several more digestion products. In general we have found considerable variation in the extent of synthesis in different experiments, which is not fully understood. One factor seems to be the ^{32}P-labelled triphosphate used. These vary considerably from batch to batch and the longer they are stored the shorter the chains formed by DNA polymerase appear to be. Most batches of ^{32}P-labelled triphosphates inhibit polymerase when used at relatively high concentrations. The reason for this is not understood and so far it has been impossible to define conditions to obtain exactly reproducible results. Therefore it

TABLE 3

Sequence of the oligonucleotides from a ribonuclease A digest of the product containing rCp (deduced from the results given in Tables 1 and 2)

Oligonucleotide (Plate I)	Sequence (residues in brackets represent the nearest-neighbour to the 3′ terminal rC)
C1	A-C-C-A-T-C-C-A-A-T-A-A-A-T-rC(A)
C2	A-A-A-G-A-A-T-T-A-G-rC(A)
C3	A-A-A-A-T-T-A-A-G-rC(A)
C4	A-A-T-A-A-A-G-rC(rC)†
C5	A-A-G-G-rC(A)
C5′	(A_2, G_2)rC(A)
C6	A-G-G-rC(A)
C7	A-T-A-rC(A)
C10	T-G-A-T-A-A-A-T-T-rC(A)
C12	A-T-A-A-A-G-rC(T)
C13	T-A-A-A-T-rC(G)
C18	T-rC(A)‡
C19	T-A-rC(A)‡
C23	G-G-T-T-G-T-A-rC(rC)†
C24	A-T-T-A-T-G-A-rC(rC)†
C25	T-G-T-A-T-A-A-rC(T)
C26	A-A-A-A-A-rC(rC)§

† No nearest-neighbour analysis was done with rCTP, but residues were concluded to be next to rC if they were not labelled by any of the other three triphosphates.

‡ Sequence deduced partly from position on the fingerprint (Plate I(a)).

§ The number of Ap residues in C26 is not absolutely certain and is based on the position of the oligonucleotides and its first 2 spleen exonuclease digestion products on DEAE-paper ionophoresis at pH 1·9.

is usually necessary to establish conditions in a preliminary experiment with a particular batch of triphosphate. Although the incorporation in the experiment shown in Plate I was done at 25°C, in more recent experiments we have usually used lower temperatures (0°C or 10°C). Van de Sande *et al.* (1972) have shown that errors are more likely to occur at high incubation temperatures. We have found no significant differences in our system at different temperatures.

Table 1 shows the results of nearest-neighbour analysis with different [α-^{32}P]triphosphates and the depurination products of the main oligonucleotides. Table 2 gives the results obtained by partial exonuclease digestion and the deduced sequences are listed in Table 3. These results are taken from a number of experiments using different labels and incubation conditions. Results with minor spots that were not complete are not recorded. Spot 17 was a mixture, one component of which was probably T-A-G-rC(A). While all the main fragments are derived from a single sequence (sequence A), there are a few oligonucleotides that appear as faint spots at early stages of incorporation and represent some type of artefact—perhaps another site of initiation by the octanucleotide. It has not however been possible to correlate them with the sequence B obtained in rGTP incorporation experiments.

TABLE 4

Analysis of the oligonucleotides from a ribonuclease T_1 digest of the product containing rG

Oligonucleotide (Plate II)	Labelled residue	Nearest-neighbour analysis rGp	Ap	Cp	Tp	Depurination products
G1	pA	—	3·0	2·1	2·0	pTpCp, pTp, pCp
	pC	—	+	—	+	pTpCp, pCp
	pT	—	+	—	—	pTpCp, pTp
G3	pA	—	++++	++	+	pTpCp, pCpCp, pTp
	pC	—	+	+	++	pTpCp, pCpCp, pTpCprGp
G5	pA	+	+	—	++	pCprGp, pTpTpTp, (pT_2, pC_2)p
	pC	—	+	—	—	pCprGp
	pT	—	+	+	++	
G9	pA	—	4·0	1·0	0·85	pTpTp, Cp
	pC	+	—	—	—	prGp
	pT	—	1·0	—	1·1	pTpTp
G10	pA	—	3·1	1·0	1·0	pTp, Cp
	pC	+	—	—	—	
	pT	—	+	—	—	pTp
G11	pA	—	2·0	—	1·0	
	pC	—	—	—	+	pTpCprGp
	pT	—	+	+	—	
G12	pA	—	2·2	1·0	1·0	pTp
	pC	+	—	—	—	
	pT	—	+	—	—	
G13	pA	—	1·0	—	1·0	pTpTp
	pC	+	—	—	—	
	pT	—	+	—	+	pTpTp
G14†	pA	1·0	2·2	1·2	—	prGp, Cp
G14′†	pC	—	+	—	—	pCprGp
G15	pA	—	+	+	—	Cp
G16	pA	+	—	—	+	Tp, prGp
G18‡	pC	+	+	+	—	pCpCprGp
G19	pA	+	—	+	—	
	pC	—	—	+	+	T$C_3$$p_4$
	pT	—	—	+	—	
G20	pA	—	—	+	—	Cp
	pT	+	+	—	—	
G21	pT	+	—	+	—	
G22	pT	+	—	—	+	
G23	pC	+	—	—	—	
G24	pC	+	—	—	—	
G27	pA	—	—	—	+	
	pC	+	—	—	+	
	pT	—	—	+	+	

† G14 represents the product obtained after short incubation times and is not labelled by dCTP. G14′ appears after long incubation times with dCTP labelling.

‡ Where no results are given for pA, pC or pT, no spots were detected. This was usually because the oligonucleotide was not labelled by the triphosphate. In the case of G18, however, it was due to the different extents of synthesis on incubation in different experiments; i.e. in the experiment with pA, synthesis did not proceed far enough to include G18.

(c) *Incorporation with rGTP*

A large number of experiments have been carried out using rGTP incorporation. At first there was considerable variation in the fingerprints obtained and the results were very difficult to interpret. One source of this variation was due to incomplete digestion with ribonuclease T_1. The mixed products are relatively insensitive to this enzyme and high concentrations are necessary to achieve quantitative cleaving of the rGp bonds. Another source of variation in the results was differences in the batches of DNA polymerase, which probably vary in their 5′ exonuclease activity. Most of the early work using rCTP incorporation was carried out with a preparation of DNA polymerase kindly provided by Dr T. Jovin. It apparently contained very little 5′ exonuclease activity, whereas other preparations used later contained much more. To avoid this extra complication we have recently used polymerase from which the 5′ exonuclease activity has been removed by the method of Klenow & Henningsen (1970). Even so there were still considerable variations from one experiment to another—particularly in respect of the relative amounts of oligonucleotides from sequence A and sequence B. Plate II(a) shows the fingerprint of a ribonuclease T_1 digest from an experiment using [^{32}P]ATP, and Plate II(b) and (c) are from experiments using [^{32}P]CTP. In Plate II(c) the synthesis has gone considerably farther than in Plate II(b) and several extra spots are present. The difference in extent of synthesis is probably largely due to the different batches of ^{32}P used since the incubation conditions were very similar. Plate II(d) is an experiment using [^{32}P]TTP and Plate II(e) is a composite diagram showing the position of the different oligonucleotides. Table 4 gives the results of the nearest-neighbour analyses and depurination products. Table 5 shows the results of partial digestion with exonucleases and the deduced sequences are listed in Table 6. With the exception of oligonucleotide G1, the sequences given can be obtained from the experimental results. For

TABLE 5

Partial digestion with exonucleases of oligonucleotides from a ribonuclease T_1 digest of the product containing rGp

Oligonucleotide (Plate II)	Exonuclease	Labelled residue	Method of analysis (see Materials and Methods)	Sequence
G1	S	pA	B	A-C-C-A-T-
G3	S	pC	B	A-A-A-C-
	V	pC	B	-C-A-A-T-C-G
G5	S	pA	B	A-C-C-A-T-
G9	S	pC	A	-T-T-A-A-G
G10	S	pC	A	-A-A-T-A-A-A-G
G11	S	pC	B	C-T-A-A-A-T-
G12	S	pC	A	-A-T-A-A-A-G
G13	S	pC	A & B	A-A-T-T-A-G
G14′	S	pC	A	A-
G18	S	pC	A	T-A-C-C-G
G19	S	pC	A	C-C-T-
G27	S	pC	A	T-T-C-T-A-G

S, spleen; V, venom.

instance the sequence of oligonucleotide G9 was deduced as follows. The partial digestion with spleen exonuclease showed the 3′ terminal sequence T-T-A-A-G. The 5′ sequence must be C-A since depurination of the A-labelled material gave Cp. All the nearest-neighbour results are accounted for in these two terminal sequences except for three A-A sequences. These must therefore be in the centre of the molecule, establishing the sequence C-A-A-A-A-T-T-A-A-G(C). The results for oligonucleotide G1 show that it has a 5′ terminal sequence A-C-C-A-T-, 3′ terminal -G(G) and contains the sequences -A-T-C-A-, -A-C-A and $(\text{-A-T-A-})_2$ linked through A residues. This is entirely consistent with the sequence of the first 22 residues of sequence A as already determined using rCTP substitution (Sanger *et al.*, 1973). It is therefore assumed to have this sequence. It can be seen that the sequences of oligonucleotides G1, G9, G10, G13, G14, G15 and G24 are consistent with those of the major oligonucleotides (C1-C7) from the rCTP incorporations; however, the sequences of several of the strong spots (G3, G5, G6, G16, G20) are not consistent, suggesting either that there was a second octanucleotide-primed sequence or that synthesis had gone much farther with rGTP than with rCTP. The experiments described below show that the former was the case.

TABLE 6

Sequence of the oligonucleotides from a ribonuclease T_1 digest of the product containing rGp (deduced from the results given in Tables 4 and 5)

Oligonucleotide (Plate II)	Sequence deduced
G1	A-C-C-A-T-C-C-A-A-T-A-A-A-T-C-A-T-A-C-A-G(G)
G5	A-C-C-A-T-C-C-T-A-A-T-T-T-A-C-G(A)†
G9	C-A-A-A-A-T-T-A-A-rG(C)
G10	C-A-A-T-A-A-A-rG(C)
G11	C-T-A-A-A-T-C-rG(G)
G12	C-A-T-A-A-A-rG(C)
G13	A-A-T-T-A-rG(C)
G14	C-A-A-A-rG(A)
G14′	A-A-A-C-rG
G15	C-A-A-rG(rG)
G16	T-A-rG(A)‡
G18	T-A-C-C-rG(C)
G19	C-C-T-C-A-rG(A)
G20	C-A-T-rG(T)
G21	C-T-rG(T)
G22	T-T-rG(T)‡
G23	A-rG(C)‡
G24	rG(C)‡
G27	T-T-C-T-A-rG(C)

† The derivation of this sequence is discussed in the text.

‡ The identification of G23 and G24 was established by ionophoresing them on DEAE-paper at pH 1·9 with ApGp and Gp markers. The identification of G16 and G22 has been confirmed by running on this system.

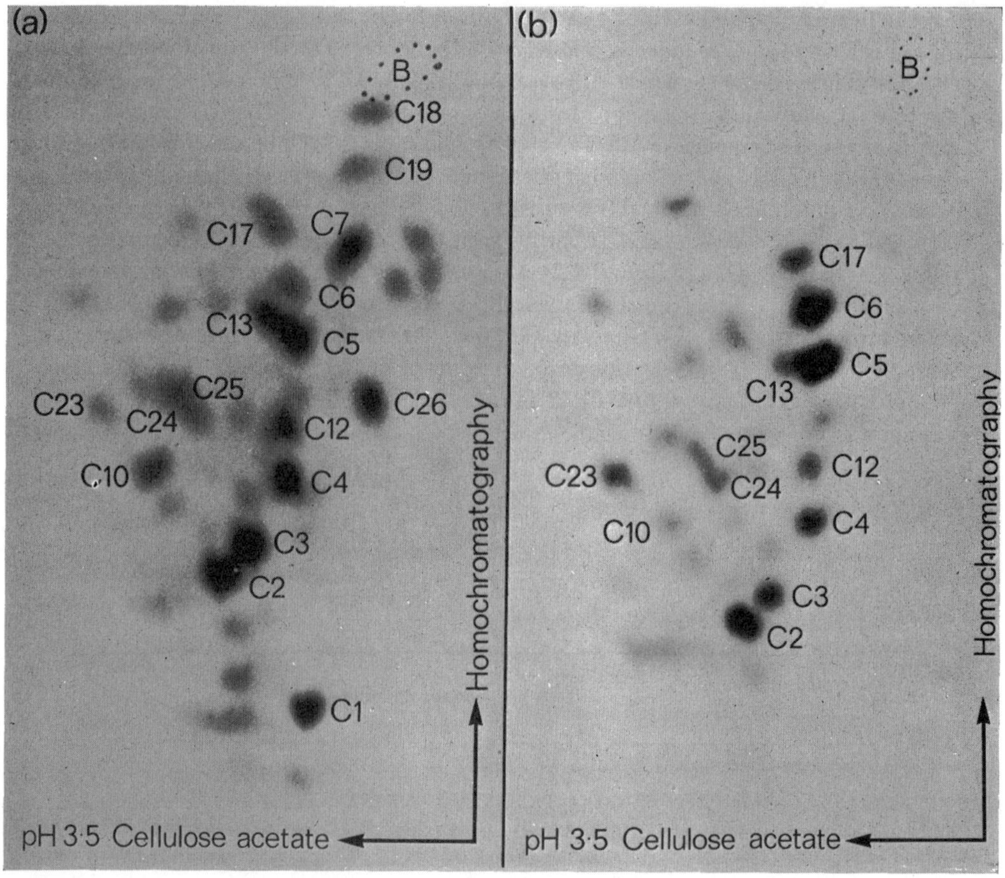

PLATE I. Radioautographs of two-dimensional fractionations of ribonuclease A digests of the mixed product formed by octanucleotide-primed synthesis with DNA polymerase I in the presence of Mn and rCTP. (a) Contained approx. 0·5 μCi [^{32}P]dATP having a specific activity of 50 mCi/μmol; (b) 0·2 μCi [^{32}P]dGTP with a similar activity. Incubation was at 25°C for 10 min and 0·13 μCi ^{32}P was incorporated in both cases. The products were purified by precipitation with ethanol and acid as described in Materials and Methods, digested with ribonuclease A and fractionated by ionophoresis on cellulose acetate in 8 M-urea at pH 3·5 (first dimension) and homochromatography using homomixture C.

[*facing p. 326*

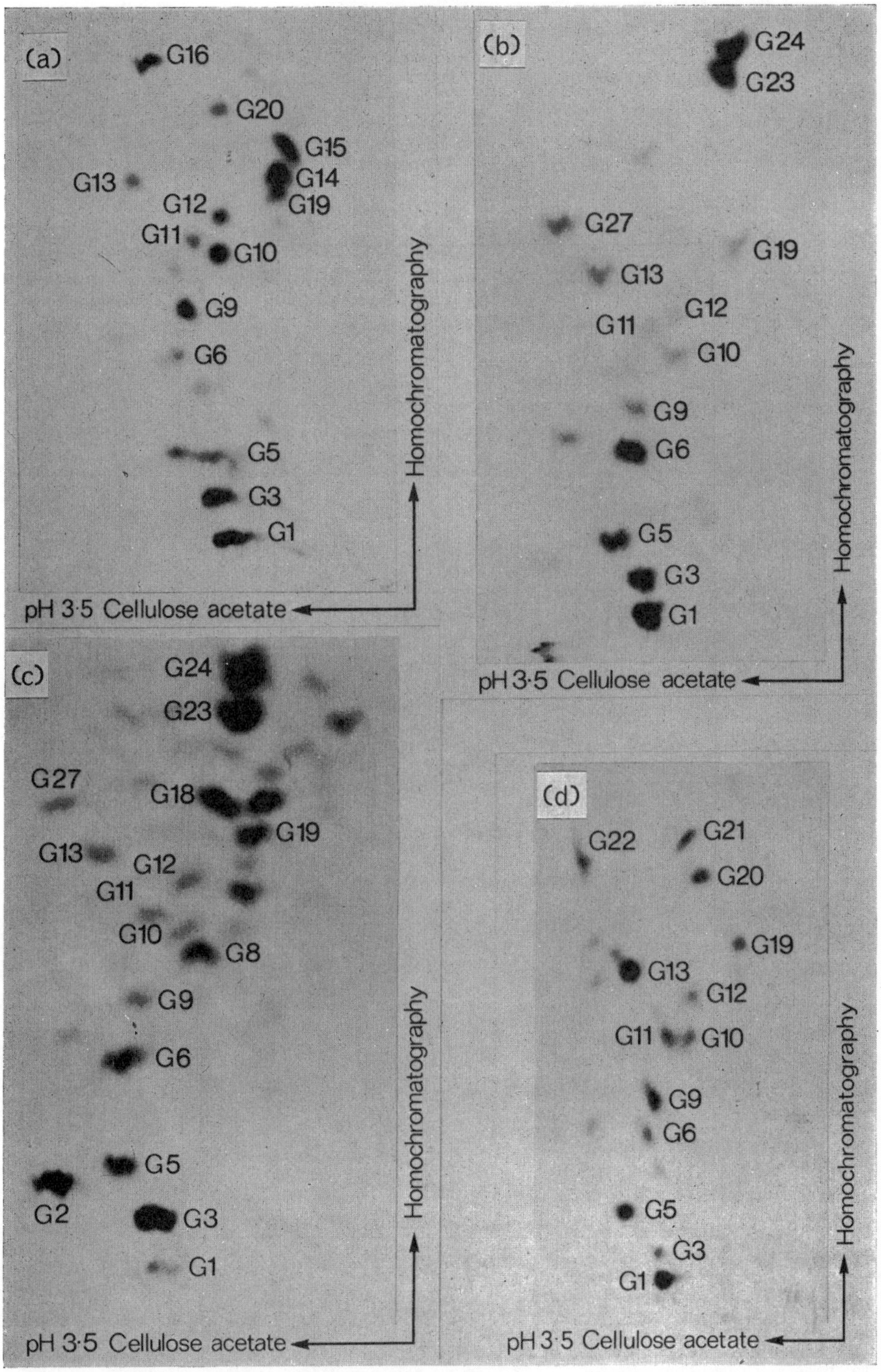

Plate II. Radioautographs of two-dimensional fractionations of ribonuclease T_1 ((a), (b) and (d)) or alkali (c) digests of the products synthesized by DNA polymerase in the presence of rGTP, as described in the text, using different ^{32}P-labelled triphosphates and incubation conditions as follows: (a) [^{32}P]dATP, 2 min, 25°C; (b) [^{32}P]dCTP, 3 min, 25°C; (c) [^{32}P]dCTP, 4 min, 25°C; (d) [^{32}P]TTP, 2 min, 25°C; (e) composite diagram showing the distribution of oligonucleotides from different experiments.

Fractionation in the second dimension was with homomixture D.

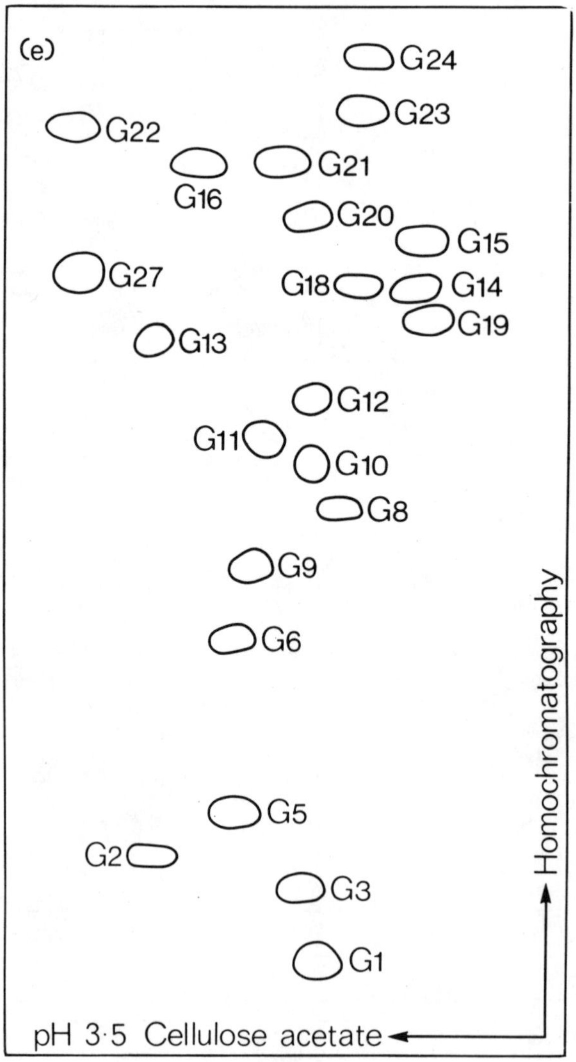
(e)
G24
G23
G22
G16
G21
G20
G15
G27
G18
G14
G19
G13
G12
G11
G10
G8
G9
G6
G5
G2
G3
G1
pH 3·5 Cellulose acetate
Homochromatography

PLATE II.

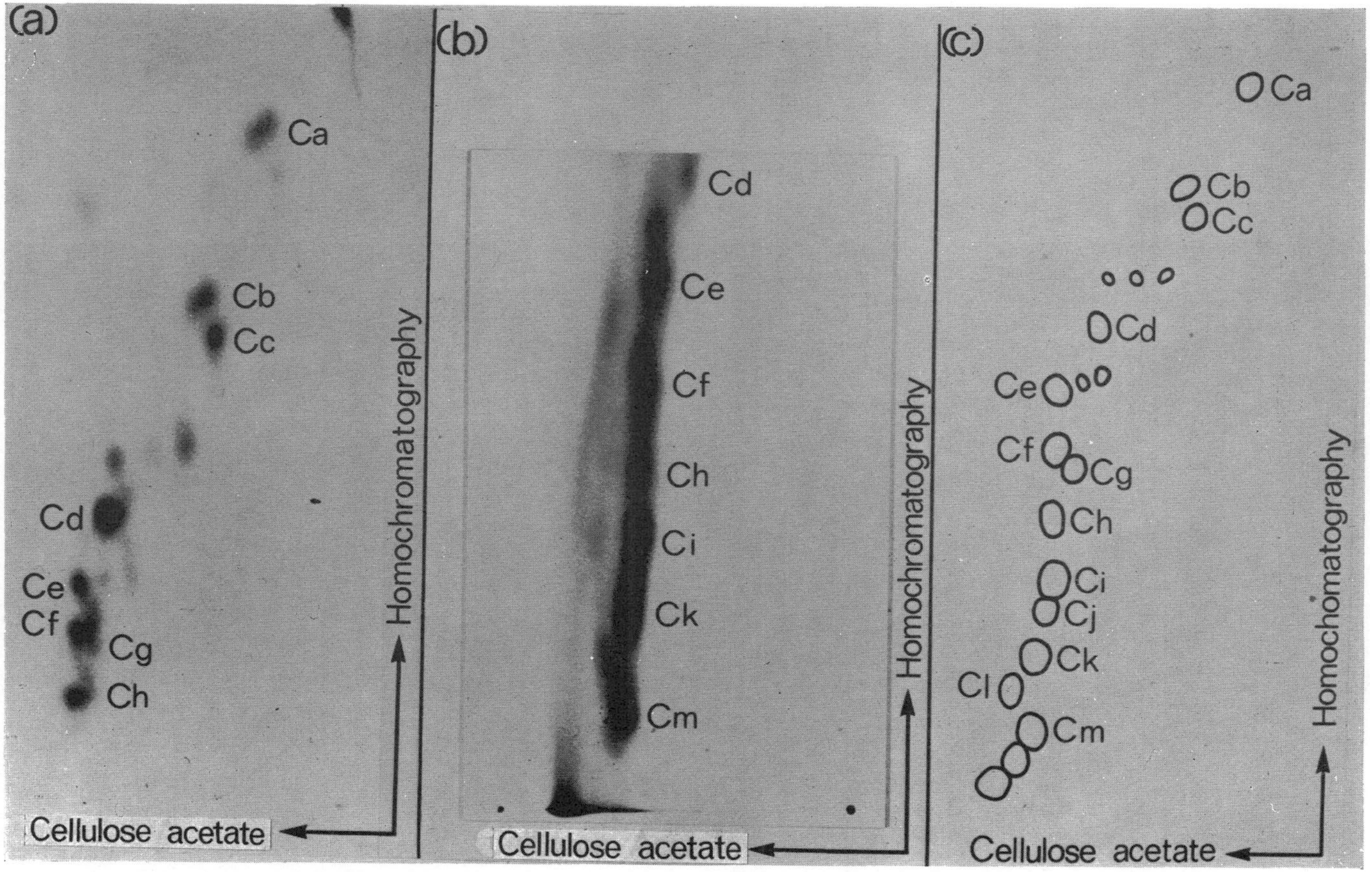

PLATE III. Two-dimensional fractionations of the unhydrolysed oligonucleotides synthesized by DNA polymerase in the presence of octanucleotide, Mn and rCTP. (a) Incubation was for 4 min at 10°C with 0·4 μCi of [^{32}P]dATP (spec. act. approx. 10 mCi/μmol): homochromatography using homomixture B. (b) Incubation for 60 min at 0°C with [^{32}P]dATP and the product fractionated using homomixture E. (c) Composite diagram showing the relative distribution of the oligonucleotides from various experiments.

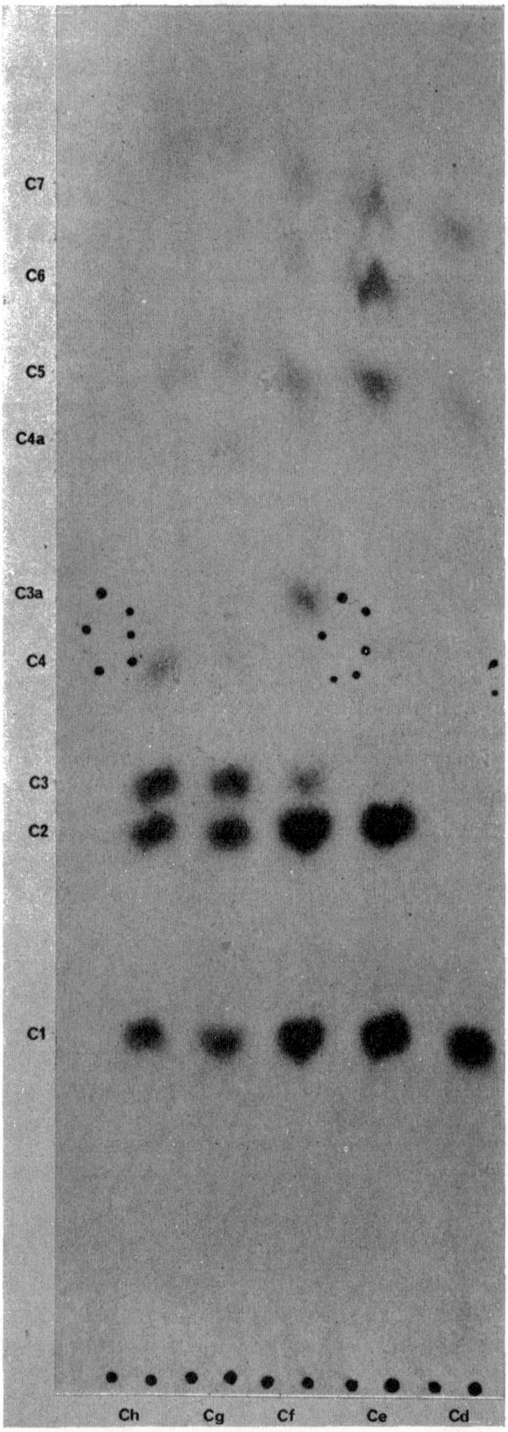

PLATE IV. An example showing the identification of the ribonuclease A digestion products of the oligonucleotides (Cd-Ch, Plate III(c)) synthesized by DNA polymerase in the presence of rCTP and [^{32}P]dATP by one-dimensional homochromatography using homomixture D (see text).

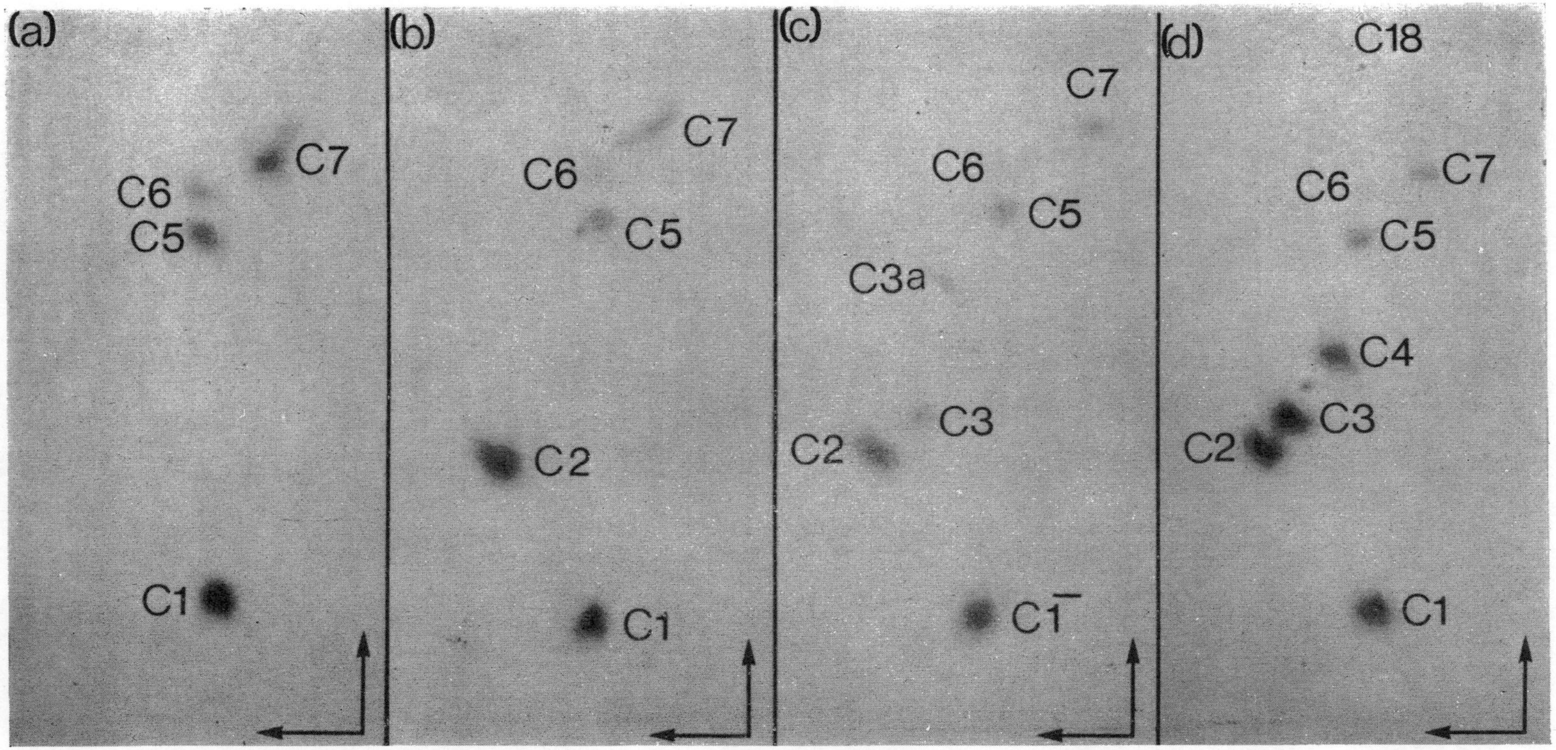

PLATE V. Two-dimensional identification of the digestion products of the oligonucleotides formed by DNA polymerase and fractionated according to size (Plates III and VI). (a) to (d) were from experiments in which the synthesis was carried out in the presence of rCTP using [^{32}P]dATP; (e) to (g) with rCTP and [^{32}P]TTP: the oligonucleotides were digested with ribonuclease A; (h) was from an experiment in which the synthesis was carried out in the presence of rGTP using [^{32}P]dCTP: digestion was with NaOH. The oligonucleotides (Plates III and VI) were (a) Cc; (b) Ce; (c) Cf; (d) Ci; (e) Ci; (f) Cl; (g) Cm; (h) Gj. Fractionation of the digests was by ionophoresis at pH 3·5 on cellulose acetate (right to left) followed by thin-layer homochromatography using homomixture D. The numbering of the spots is as in Plate I and II, C3a and C23a are products resulting from incomplete synthesis of C3 and C23 as explained in the text.

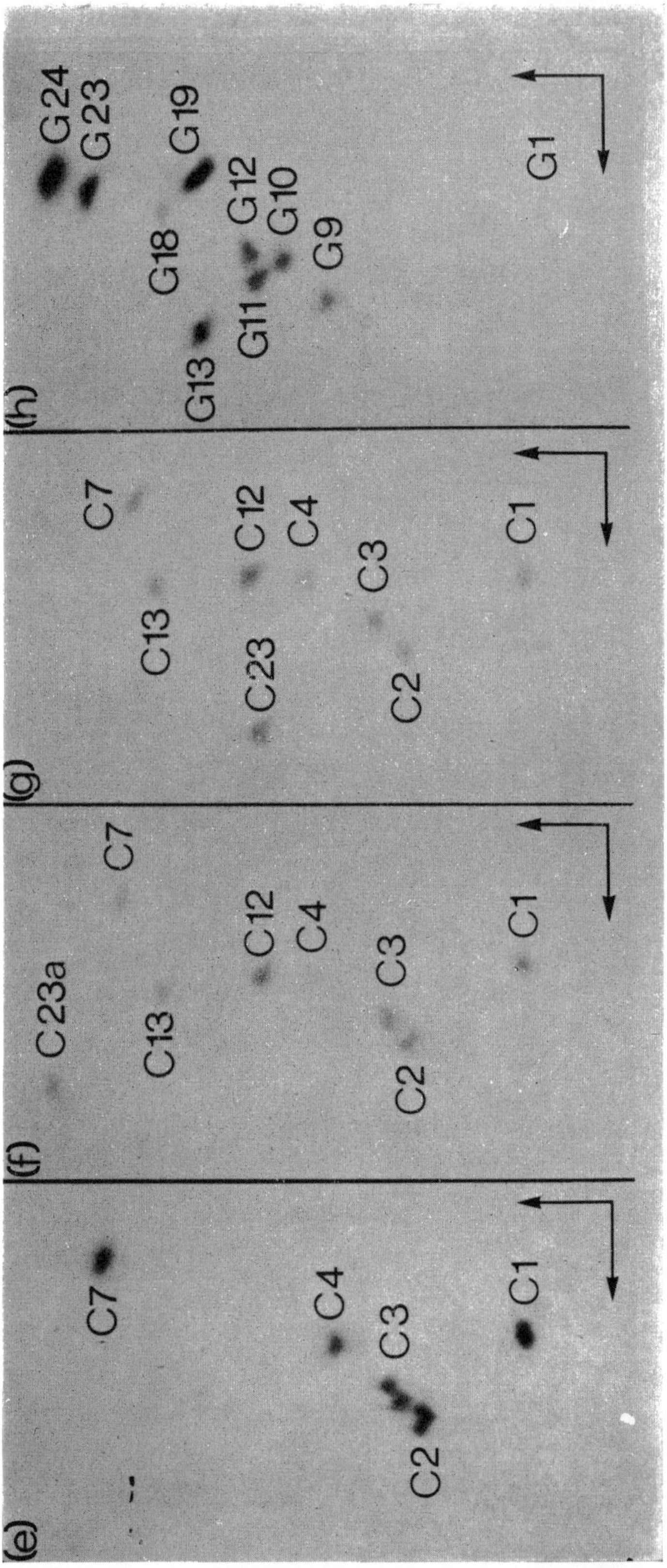

PLATE V. The experiments shown in (e) to (g) were carried out by Dr Jacob.

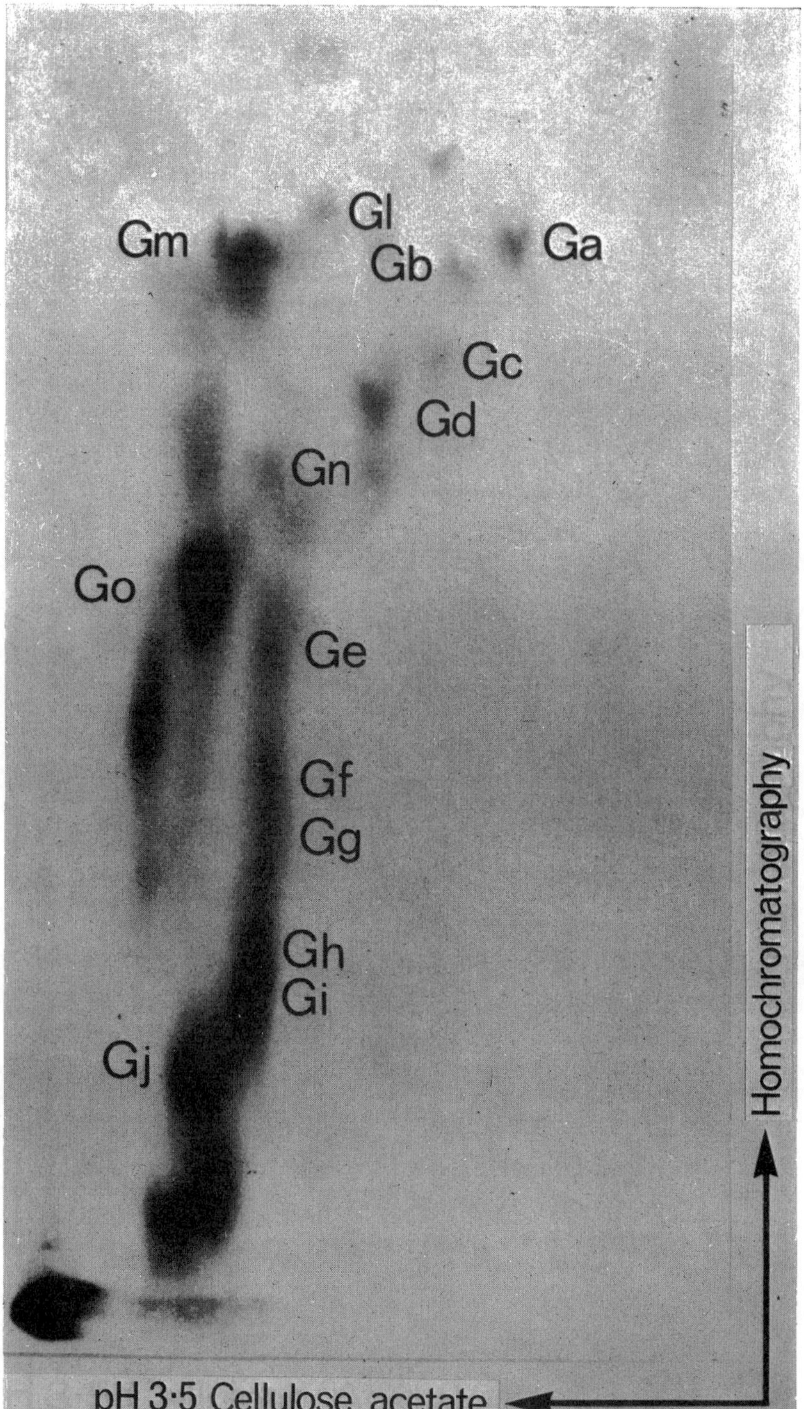

PLATE VI. Radioautograph of the two-dimensional fractionation (homomixture E) of the oligonucleotides synthesized by DNA polymerase in the presence of rGTP. Incubation was for 2 min at 25°C using [^{32}P]TTP as label.

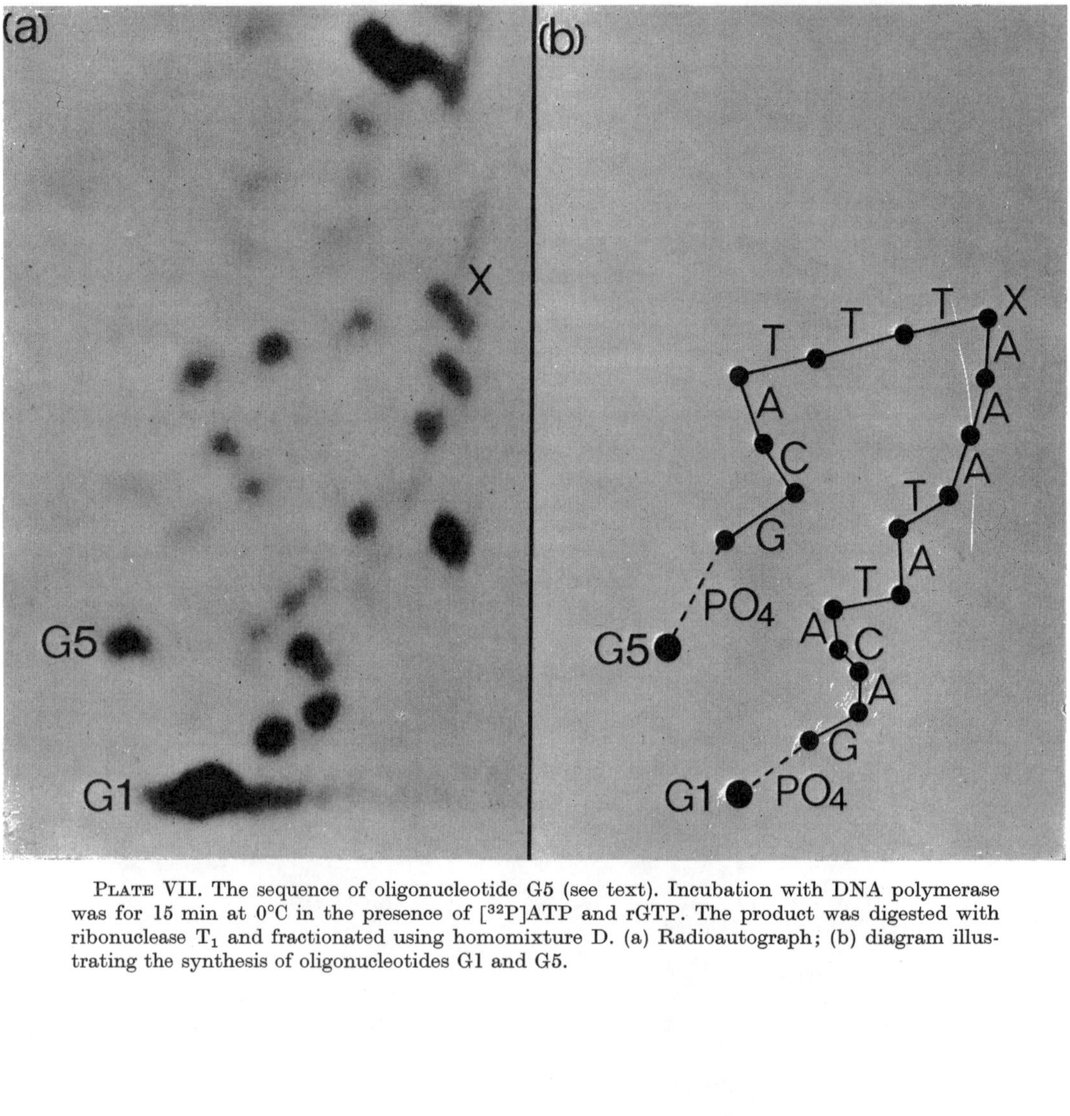

PLATE VII. The sequence of oligonucleotide G5 (see text). Incubation with DNA polymerase was for 15 min at 0°C in the presence of [^{32}P]ATP and rGTP. The product was digested with ribonuclease T_1 and fractionated using homomixture D. (a) Radioautograph; (b) diagram illustrating the synthesis of oligonucleotides G1 and G5.

(d) *Order of the oligonucleotides*

To determine the relative order of the ribonuclease A digestion products from rCTP incorporation experiments, the DNA polymerase incubation mixtures were fractionated directly on a two-dimensional homochromatography system on which fragments of about 60 to 80 residues would move away from the origin. In this way a series of spots was obtained which represented oligonucleotides of increasing size being built up from the octanucleotide primer. Since they contained rC residues they could be digested with ribonuclease A, to give products that had already been studied in the digests of the complete incubation mixture (as in Plates IV and V). The identifications were usually confirmed by nearest-neighbour analysis.

In the preliminary communication (Sanger *et al.*, 1973) we described a group of experiments in which products of up to about 60 nucleotides in length were fractionated using a 5% unhydrolysed homomixture (B, as in Plate III(a)). From this the relative order of oligonucleotides C1-C7 was determined giving the sequence of the first 50 residues synthesized onto the octanucleotide. In order to extend this sequence, longer fragments of up to about 90 residues were studied by fractionation on a 7% homomixture (E, as in Plate III(b)) and subjected to digestion with ribonuclease A. Plate III(c) is a composite diagram summarizing the results of a number of similar experiments using different ^{32}P-labelled triphosphates and different incubation conditions.

Plate IV shows an example of the identification of the ribonuclease A digestion products of spots Cd-Ch (Plate III(c)). It can be seen that, besides the recognizable products (C1-C7) that had already been detected (Plate I, Table 3), there were other

TABLE 7

Products of ribonuclease A digestion of the oligonucleotides of different length synthesized by DNA polymerase in the presence of rCTP

Oligonucleotide (Plate III)	Products of digestion with ribonuclease A† (Plate I; Tables 1 to 3)
Ca	C1, C7
Cb	C1, C7, C6
Cc	C1, C7, C6, C5
Cd	C1, C7, C6, C5
Ce	C1, C7, C6, C5, C2
Cf	C1, C7, C6, C5, C2
Cg	C1, C7, C6, C5, C2, C3
Ch	C1, C7, C6, C5, C2, C3, C4
Ci	C1, C7, C6, C5, C2, C3, C4, C18
Cj	C1, C7, C6, C5, C2, C3, C4, C18, C5′a‡
Ck	C1, C7, C6, C5, C2, C3, C4, C18, C5′, C12
Cl	C1, C7, C6, C5, C2, C3, C4, C18, C5′, C12, C13
Cm	C1, C7, C6, C5, C2, C3, C4, C18, C5′, C12, C13, C23

† Products that were present in digests of only one oligonucleotide and were identified as precursors (see text) are not included in the Table, with the exception of C5′a.

‡ C5′a is an oligonucleotide that was found slightly to the right of spot C19 (Plate I(a)) in an experiment in which [^{32}P]ATP was used. The "nearest neighbour" to pA was Gp. It was not present in digests of larger fragments.

spots present that did not correspond to known products and that did not appear again in larger oligonucleotides. Thus spot Cf gave rise to products C1, C2, C5, C6, C7, a low yield of C3 and a product, C3a, that is derived from the 3′ terminus of oligonucleotide Cf and has the sequence A-A-A-A-T-T-A-A-G_{OH}. As would be expected, C3a is not detected in spots Cg and Ch where the synthesis of the whole sequence of C3 is complete. The presence of both C3 and its smaller precursor, C3a, in spot Cf demonstrates that spot Cf really contains two lengths of synthesized chain, one ending at the pG residue and one beyond the subsequent pCp residue of C3. Similarly spot C4a from oligonucleotide Cg has the sequence A-A-T-A-A-A-G_{OH} and is an intermediate in the formation of the sequence of C4. It should be noted that these intermediate products do not have a 3′ phosphate group.

Plate V shows examples of the identification of the digestion products using the two-dimensional fractionation. This method was especially useful for the larger oligonucleotides. The combined results of these experiments are summarized in Table 7. In general the products of intermediate synthesis, such as C3a and C4a, are not included in this Table. The one exception is C5′a, found in the digest of oligonucleotide Cj, which is included since it establishes the relative order of C5′ and C12. Since these two products appear together in oligonucleotide Ck it is not clear which follows C18. C5′a is a precursor of one of them as it does not occur in larger oligonucleotides. Nearest-neighbour analysis showed that it contained the sequence GpA. This is not present in C12 but is in C5′, and therefore C5′a must be a precursor of C5′ and the relative sequence must be C18-C5′-C12.

From the result in Table 7 it can be seen that the ribonuclease A digestion products are arranged in the order: C1-C7-C6-C5-C2-C3-C4-C18-C5′-C12-C13-C23. In the case of products C1-C7 it could be concluded that no other sequence could come between those already deduced (Table 3) since oligonucleotides C2-C7 have Ap as the 5′ terminal residue, and oligonucleotides C1-C3 and C5-C7 have -rC(A) at the 3′ terminus. Therefore any product that came between them would be labelled if [^{32}P]ATP were used for incorporation and there were in fact no other such products detected. However, this is no longer true after product C4, which has -rC(rC) at its 3′ terminus. Experiments were carried out in which ATP, GTP or TTP were used as the labelled triphosphate so that most possible intermediate fragments would have been detected. However, it was not possible to use [^{32}P]rCTP for labelling as the high concentration of rCTP necessary to obtain rC incorporation inhibited the enzyme. Thus, since fragment C18 followed C4, the sequence C-C-T-C was expected, however the possibility of the sequence C-C-C-T-C could not be excluded. Early experiments involving depurination of material synthesized with all four deoxyribonucleotides suggested that such a sequence was unlikely. However, a much clearer proof of the correct sequence came from the results with rGTP substitution given below.

Another problem with the results using rCTP was that the sequence of C5′ was not known. It was only recognized in digests of oligonucleotides Cj-Cm by the higher yield of spot C5 and the presence of a G(A) nearest-neighbour in it. Since product C has rC(rC) at its 3′ end, ribonuclease A digests of the larger oligonucleotides would be expected to contain rCp. This would not have been detected with the fractionation methods normally used. However, in one experiment using [^{32}P]TTP labelling, digests of oligonucleotides Ch and Cl were subjected to ionophoresis at pH 1·9 on DEAE-paper in parallel with some marker rCp. Both digests contained rCp, showing the presence of the sequence rC-T.

TABLE 8

Products of alkali digestion of the oligonucleotides of different lengths synthesized by DNA polymerase in the presence of rGTP

Oligonucleotide (Plate VI)	Products of alkali digestion (Plate II, Tables 4 to 6)
Ga	G1a
Gb	G1b
Gc	G1
Gd	G1, G15
Ge	G1, G15, G14, G13
Gf	G1, G15, G14, G13, G9
Gg	G1, G15, G14, G13, G9, G10
Gh	G1, G15, G14, G13, G9, G10, G19
Gi	G1, G15, G14, G13, G9, G10, G19, G12
Gj	G1, G15, G14, G13, G9, G10, G19, G12, G11, G22
Gl	G5, G20
Gm	G5, G20, G16
Gn	G5, G20, G16
Go	G5, G20, G16, G3

Plate VI shows the results of some experiments in which the synthesized product containing rGp was fractionated on the two-dimensional system using a 7% homo-mixture (E). The products obtained by subsequent alkali digestion of the spots are listed in Table 8. Spots Ga-Gj gave the products (G1, G15, etc.) that would have been predicted from sequence A already deduced in the rCTP experiments, as well as some extra products in spots Gg-Gj which extended this sequence. There is however another prominent series of spots, lying to the left of the first series, which represents the building up of another oligonucleotide chain. On alkali digestion these spots give products (G5, G20, etc.) that do not fit in with sequence A.

Most of the results in Table 8 were obtained using [^{32}P]ATP and [^{32}P]TTP as labels and in the experiments with [^{32}P]CTP only a few of the spots were studied. From Table 6 it can be seen that the only products that would have been detected with [^{32}P]CTP labelling but not [^{32}P]ATP and [^{32}P]TTP labelling are G23 (A-rG) and G24 (rG). G23 was detected in digests of Gj, Gn, Go. It was not present in Ge. G24 was present in Gd and some larger oligonucleotides. From the results given in Table 8 it can be concluded that the sequence of products is: G1-(G24, G15)-(G14, G13)-G9-G10-G19-G12-(G11, G22) with G23 coming after G13 and G14.

The sequence of G5 was of particular interest since it appeared to represent the start of sequence B and this was supported by the partial digestion with spleen exonuclease. With [^{32}P]ATP labelling it gave (pT_2, pC_2)p as a depurination product, but this was not seen with [^{32}P]CTP labelling. The Cp residues must therefore have been derived from the octanucleotide primer and suggested the sequence A-C-C-A-T-C-C-T-A. Presumably the 3′ terminal pA residue of the octanucleotide was removed by the 3′ exonuclease activity of DNA polymerase and the resulting A-C-C-A-T-C-C acted as a primer for sequence B. From the nearest-neighbour and depurination results given in Table 4, G5 contains the sequences A-C-C-A-T-C-C-T-A (5′ end), A-C-G(A) (3′ end) and T-T-T-A, which must be in the centre. It also contains an

A-A sequence. Thus two sequences are possible: A-C-C-A-T-C-C-T-A-A-T-T-T-A-C-G(A) or A-C-C-A-T-C-C-T-A-T-T-T-A-A-C-G(A). The experiment shown in Plate VII shows that the former sequence is probably correct. In this experiment a mixture of relatively short oligonucleotides was synthesized in the presence of rGTP and the products were digested and fractionated on the two-dimensional system. Spots G1 and G5 are present at the bottom of the fingerprint. A series of intermediate products of synthesis leading to spot 1 can be seen to the right of the fingerprint. Their relative positions are consistent with the sequence of spot 1, which can be read off as described in the previous communication (Sanger *et al.*, 1973) and illustrated in Plate VII(b). A similar series of intermediate products can be seen leading to spot 5. These show the sequence -T-T-T-A-C-G. Spot X from which this sequence starts is common to both sequences. In the case of the sequence leading to spot 1 it is known to be A-C-C-A-T-C-C-A-A-T. It is therefore likely that the sequence leading to spot 5 has the same base composition. Of the two possible sequences for G5 discussed above, these results are consistent only with the sequence A-C-C-A-T-C-C-T-A-A-T-T-T-A-C-G. The sequence of G3 cannot be uniquely deduced from the results given.

From the results shown in Table 8 for oligonucleotide Gl-Go it can be deduced that the products occur in the sequence G5-G20-G16-G3 with G23 coming somewhere between G5 and G3, since it was found in oligonucleotides Gn and Go (see above). From the nearest-neighbour results it can only fit between G5 and G20, establishing the order G5-G23-G20-G16-G3 for sequence B. It will be seen that there are a number of products found in the ribonuclease T_1 digests (Plate II) which are not accounted for in sequence A or B. The most prominent of these is G6 which is sometimes a strong spot (e.g. in Plate II(b)) and yet was not found in alkaline digests of the main synthesized oligonucleotides. It was however sometimes found in some minor oligonucleotides which usually contained neither G1 or G5. The most likely explanation is that there is yet a third site at which the octanucleotide can act as a primer.

4. Discussion

From the results of Table 7 it was concluded that the products of ribonuclease A digestion of the oligonucleotides synthesized in the presence of rCTP are arranged in the sequence C1-C7-C6-C5-C2-C3-C4-C18-C5′-C12-C13-C23, and from Table 8 the products of alkaline digestion of the oligonucleotides synthesized with rGTP are in the sequence G1-(G24, G15)-(G14, G13)-G9-G10-G19-G12-(G11, G22). The unique sequence that can be deduced from these results is given in Figure 3. The order up to residue 58 was established from the rCTP experiments (Sanger *et al.*, 1973) and is confirmed by the rGTP experiments. Fragment G19, which follows G10, extends this sequence by establishing an overlap with the 3′ terminal -C-C sequence to include the T-C of product C18. G19 ends in C-A-G(A), so that the next product (C5′) from the rCTP experiments must start A-G-A. This establishes the sequence of C5′, which could not be deduced from the results in Tables 1 and 2, and makes it possible to continue the sequence. Product C23 ends G-T-A-C(C). This gives a good overlap with G18 (T-A-C-C-G(C)) which can thus be used to extend the sequence a further two residues, although it was not actually detected uniquely in this position by analysis of the oligonucleotides of different length. From a comparison of the fingerprints of Plate II(b) and (c) it is clear that G18 only appears when a relatively long chain is synthesized.

```
     1                 10                20
(a)  (A-C-C-A-T-C-C-A)A-T-A-A-A-T-C-A-T-A-C-A-G-G-C-A-A-G-
(b)  |-------------- C1 ------------||-C7-||-C6-||-C5-
(c)  |---------------- G1 -----------------|  |-G15-|

              30                40
     G-C-A-A-A-G-A-A-T-T-A-G-C-A-A-A-A-T-T-A-A-G-C-
     -||--------- C2 ---------||-------- C3 -------|
     |-G14-||-G13--||----------- G9 -----------||-

     50                60                70
     A-A-T-A-A-A-G-C-C-T-C-A-G-A-G-C-A-T-A-A-A-G-C-
     |----- C4 -----| |-C18-||-C5'-||----- C12 ----|
     |----- G10 -----||-- G19 --||-G23-||-- G12 ---||-

                 80
     T-A-A-A-T-C-G-G-T-T-G-T-A-C-C-G-C-
     |--- C13 ---||---- C23 ----|
     --- G11 ----| |-G22-||-G18-|
```

FIG. 3. Sequence deduced from the octanucleotide-primed synthesis on bacteriophage f1 DNA. (a) Sequence deduced; (b) products from ribonuclease A digestion of the oligonucleotide synthesized in the presence of rCTP (Tables 3 and 7). (c) Products from the alkali digestion of the oligonucleotide synthesized in the presence of rGTP (Tables 6 and 8).

The sequence shown in Figure 3 corresponds to that of the minus strand of phage f1 DNA. The complementary sequence, which is that of the plus strand and corresponds to that of the messenger RNA, is shown in Figure 4. The sequence of residues 82-89 has not been determined directly but is given as the complementary sequence to the octanucleotide primer. It is, however, possible that it may prime by imperfect base pairing to a very similar but not identical sequence.

The sequence contains three possible initiation and five possible termination codons. In the previous paper (Sanger *et al.*, 1973) it was suggested that the A-T-G in position 85-87 might act as an actual protein chain initiator since it is preceded a few residues away by the sequence Pu-Pu-T-T-T-Pu, which appears to be a characteristic of certain initiation sites (Robertson *et al.*, 1973). An unexpected finding from the extended sequence is that the A-T-G in position 23-25 is also preceded by a similar characteristic sequence G-A-T-T-T-A-G. If, however, this A-T-G were an initiator it could only synthesize the dipeptide fMet-Leu, which would be terminated by the T-G-A codon in position 29-31. Although this seems unlikely, it is not impossible.

```
1                  10                    20
G-C-G-G-T-A-C-A-A-C-C-G-A-T-T-T-A-G-C-T-T-T-A-T-G-C-

       30                40                   50
T-C-T-G-A-G-G-C-T-T-T-A-T-T-G-C-T-T-T-A-A-T-T-T-T-G-C-

                60                 70
T-A-A-T-T-C-T-T-T-G-C-C-T-T-G-C-C-T-G-T-A-T-G-A-T-T-

    80
T-A-T(T-G-G-A-T-G-G-T)
```

FIG. 4. Sequence present in the plus strand of f1 DNA. (Possible termination codons are shown by underlining; possible initiation codons by broken underlining.)

It is interesting to note that Bretscher (1969) showed the formation of formylmethionyl dipeptides when phage fd DNA was incubated with a protein-synthesizing system. However, fMet-Leu was not one of them.

In the preliminary communication it was noted that the three termination codons in positions 44-46, 53-55 and 74-76 are in the same phase, and it was suggested that the first one (44-46) could act as a protein chain terminator, with the other two codons ensuring complete termination. Such a situation of three in-phase terminators was found for the coat protein of bacteriophage R17 (Nichols, 1970). It is now seen that there is a fourth termination codon (position 29-31) in phase with the other three. Therefore the T-A-A in position 44-46 cannot act directly as a protein chain terminator but the T-G-A in position 29-31 could, with three in-phase terminators to ensure complete termination. Clearly the function of this DNA sequence cannot be deduced at present, but there are a number of interesting possibilities and it is to be expected that further work on the DNA and protein sequences of the bacteriophage will shed light on which terminators or initiators are used. If in fact the whole fragment is coding for a protein sequence this sequence can be uniquely determined since two of the phases contain terminators, the T-A-G in position 16-18 being out of phase with the other four terminators.

An interesting feature of this sequence is that it shows considerable repetition of homologous sequences, suggesting its evolutionary origin by a process of simple duplications (Sanger *et al.*, 1973). This feature appears to extend back to about residue 12. In particular the sequence G-C-T-T-T-A-T is found in positions 18-24 and 33-39. The repetitious pattern does not appear to extend to the sequence before about residue 12 and this conclusion is supported by sequences of the oligonucleotides C10, C19, C24, C25, C26 (Table 3), which presumably are found in positions before the deduced sequence. They do not show obvious signs of the homologies found in the other products.

From the results given above it may be deduced that the sequence B primed by the octanucleotide is (A-C-C-A-T-C-C)-T-A-A-T-T-T-A-C-G-A-G-C-A-T-G-T-A-G-A-A-A-C. This is however only put forward as a tentative sequence since it is felt that there is not sufficient confirmatory evidence to regard it as fully established. The complementary sequence contains a termination codon in two phases so that if it codes for a protein the amino acid sequence is uniquely defined. It is not clear why sequence B was formed only in the experiments using rGTP and was not formed in significant amounts when rCTP was used.

The method for sequencing DNA described in this paper should be generally applicable where suitable primers are available. Such primers could be synthetic or naturally occurring oligonucleotides from the opposite strand of the DNA to be sequenced (see Oertel & Schaller, 1972). The octanucleotide was probably not an ideal primer and there were special difficulties due to the repetitive nature of the main sequence (A) and the presence of a second priming site giving rise to sequence B. It would probably be better to use a somewhat longer primer and, if unique results could be obtained with both rGTP and rCTP substitution, the deduction of the sequence should be relatively rapid and simple.

We wish to thank Dr T. Jovin for a gift of DNA polymerase I. Some preliminary experiments in connection with this work were carried out with Dr V. Ling, who also developed the method for purifying the product synthesized by DNA polymerase. The experiments reported in Plate V(e) to (g) were carried out by Dr J. M. Jacob during a stay in this laboratory. One of us (J. E. D.) was supported by the Helen Hay Whitney Foundation. Another author (H. K.) wishes to acknowledge support from the Deutsche Forschungsgemeinschaft (SFB46) and from EMBO.

REFERENCES

Berg, P., Fancher, H. & Chamberlin, M. (1963). In *Symposium on Informational Macromolecules* (Vogel, H., Bryson, V. & Lampen, J. O., eds), pp. 467–483, Academic Press, New York and London.
Billeter, M. A., Dahlberg, J. E., Goodman, H. M., Hindley, J. & Weissmann, C. (1969). *Nature (London)*, **224**, 1083–1086.
Blattner, F. R. & Dahlberg, J. E. (1972). *Nature New Biol.* **237**, 227–232.
Bretscher, M. S. (1969). *Cold Spring Harbor Symp. Quant. Biol.* **34**, 651–653.
Brownlee, G. G. & Sanger, F. (1969). *Eur. J. Biochem.* **11**, 395–399.
Galibert, F., Sedat, J. W. & Ziff, E. B. (1974). *J. Mol. Biol.* **87**, 377–407.
Jeppesen, P. G. N. (1971). *Biochem. J.* **124**, 357–366.
Lebowitz, P., Weissman, S. M. & Radding, C. M. (1971). *J. Biol. Chem.* **246**, 5120–5139.
Loewen, P. C., Sekiya, T. & Khorana, H. G. (1974). *J. Biol. Chem.* **249**, 217–226.
Klenow, H. & Henningsen, I. (1970). *Proc. Nat. Acad. Sci., U.S.A.* **65**, 168–175.
Ling, V. (1972). *J. Mol. Biol.* **64**, 87–102.
Nichols, J. L. (1970). *Nature (London)*, **225**, 147–151.
Oertel, W. & Schaller, H. (1972). *FEBS Letters*, **27**, 316–320.
Robertson, H. D., Barrell, B. G., Weith, H. L. & Donelson, J. E. (1973). *Nature New Biol.* **241**, 38–40.
Salser, W., Fry, K., Brunk, C. & Poon, R. (1972). *Proc. Nat. Acad. Sci., U.S.A.* **69**, 238–242.
Sanger, F., Brownlee, G. G. & Barrell, B. G. (1965). *J. Mol. Biol.* **13**, 373–398.
Sanger, F., Donelson, J. E., Coulson, A. R., Kossel, H. & Fischer, D. (1973). *Proc. Nat. Acad. Sci., U.S.A.* **70**, 1209–1213.
Schott, H., Fischer, D. & Kossel, H. (1973). *Biochemistry*, **12**, 3447–3453.
Symons, R. H. (1969). *Biochim. Biophys. Acta*, **190**, 548–550.
Van de Sande, J. H., Loewen, P. C. & Khorana, H. G. (1972). *J. Biol. Chem.* **247**, 6140–6148.
Wu, R. & Kaiser, A. D. (1968). *J. Mol. Biol.* **35**, 523–527.
Wu, R. & Taylor, E. (1971). *J. Mol. Biol.* **57**, 491–511.
Ziff, E. B., Sedat, J. W. & Galibert, F. (1973). *Nature New Biol.* **241**, 34–37.

J. Mol. Biol. (1975) **94**, 441–448

A Rapid Method for Determining Sequences in DNA by Primed Synthesis with DNA Polymerase

F. Sanger and A. R. Coulson

Medical Research Council
Laboratory of Molecular Biology
Hills Road, Cambridge CB2 2QH, England

(Received 20 December 1974)

A simple and rapid method for determining nucleotide sequences in single-stranded DNA by primed synthesis with DNA polymerase is described. It depends on the use of *Escherichia coli* DNA polymerase I and DNA polymerase from bacteriophage T4 under conditions of different limiting nucleoside triphosphates and concurrent fractionation of the products according to size by ionophoresis on acrylamide gels. The method was used to determine two sequences in bacteriophage ϕX174 DNA using the synthetic decanucleotide A-G-A-A-A-T-A-A-A-A and a restriction enzyme digestion product as primers.

1. Introduction

In previous papers (Sanger *et al.*, 1973; Donelson, J. E., Barrell, B. G., Weith, H. L., Kössel, H. & Schott, H., unpublished data) we have described the determination of two nucleotide sequences in bacteriophage DNA using DNA polymerase primed by synthetic oligonucleotides. In this method the oligonucleotide was bound to a specific complementary region on the single-stranded DNA, and nucleoside triphosphates were added by the DNA polymerase to the 3′ end of the primer. By using ^{32}P-labelled triphosphates, a radioactive complementary copy of a defined part of the template DNA was synthesized and subjected to sequence determination, which was greatly facilitated by use of the ribosubstitution technique.

This paper describes a relatively rapid and simple alternative procedure for deducing sequences by primed synthesis with DNA polymerase.

2. Principle of the Method

Figure 1 illustrates the principle of the method by considering its application to a small hypothetical sequence in a DNA chain. DNA polymerase I is first used to extend the primer oligonucleotide and copy the template in the presence of the four deoxyribotriphosphates, one of which is labelled with ^{32}P. Ideally this synthesis should be non-synchronous and as random as possible, so that the maximum number of oligonucleotides of different length, all starting from the primer, is formed. This mixture is then purified on an agarose column to remove the excess triphosphates and samples are re-treated in various ways as follows:—

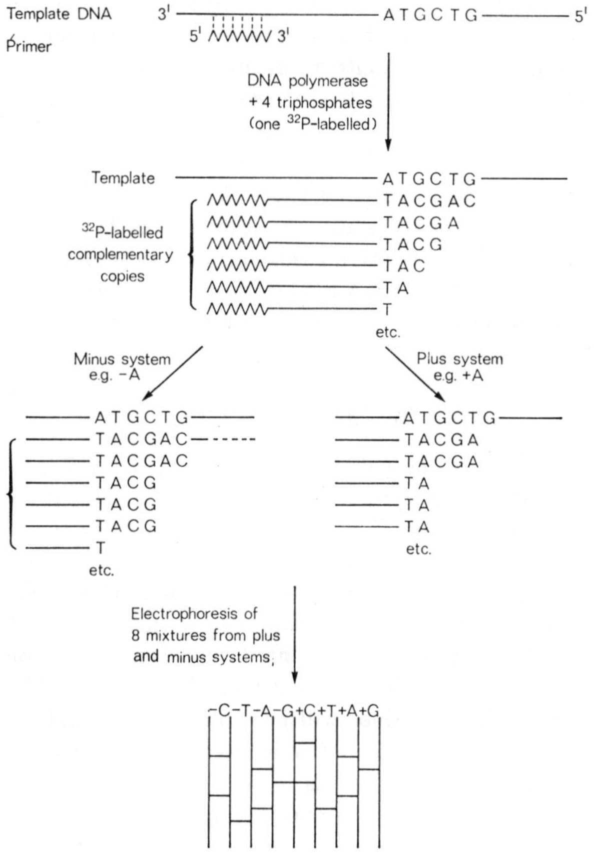

FIG. 1. The principle of the method.

(a) *The "minus" system*

In their original work on the "sticky ends" of phage λ DNA, Wu & Kaiser (1968) showed that if DNA polymerase acted in the absence of one triphosphate, synthesis would proceed accurately up to a position where the missing triphosphate should have been incorporated, and they used this principle to deduce sequences by assaying the relative amount of each nucleotide incorporated in the presence of different triphosphate mixtures (Wu & Taylor, 1971). The minus system described here uses a similar principle.

The random mixture of oligonucleotides, which is still hybridized to the template DNA, is reincubated with DNA polymerase I in the presence of three deoxyribotriphosphates. Synthesis then proceeds as far as it can on each chain: thus, if dATP is the missing triphosphate (the —A† system), each chain will terminate at its 3′ end at a position before an A residue. Separate samples are incubated, with each one of the four triphosphates missing.

† Since this paper is concerned only with DNA the symbols A, C, G and T refer to the deoxyribonucleosides.

The four incubation mixtures are then denatured to separate the newly-synthesized strands from the template, subjected to electrophoresis on acrylamide gel in the presence of 8 M-urea, and a radioautograph prepared. In this fractionation system mobility is essentially proportional to size, so that the various synthesized oligonucleotides (which have a common 5′ end) will be arranged according to size. Ideally each oligonucleotide should be separated from its neighbour, which contains one more residue. The radioautograph from the −A system will contain bands corresponding to positions before the A residues in the synthesized chain. Thus the positions of As are located. Similarly the relative positions of the other residues may be located and, ideally, the sequence of the DNA read off from the radioautograph. This system alone is usually not sufficient to establish a sequence, so a second similar system is normally used in conjunction with it.

(b) *The "plus" system*

This system makes use of the method of Englund (1971,1972) who showed that, in the presence of a single deoxyribotriphosphate, DNA polymerase from bacteriophage T4-infected *Escherichia coli* (T4 polymerase) will degrade double-stranded DNA from its 3′ end, but that this exonuclease action will stop at residues corresponding to the one triphosphate that is present. This method is applied to the random oligonucleotide mixture obtained above. Samples are incubated with T4 polymerase and a single triphosphate and then fractionated by electrophoresis on acrylamide gel. Thus in the +A system only dATP is present and all the chains will terminate with A residues. The positions of A residues will be indicated by bands on the radioautograph. Usually these will be in products one residue larger than the corresponding bands in the −A system, but if there is more than one consecutive A residue the distance between the bands in the −A and +A systems will indicate the number of such consecutive residues. In the example illustrated in Figure 1 the smallest oligonucleotide gives a band in the −T position, indicating that the next residue after its 3′ terminus will be a T. This is confirmed by the presence of a band in the +T position in the next largest oligonucleotide. The bands in the +T and −A positions in this product show that its 3′ terminus is T and the residue following is an A, thus defining the dinucleotide sequence T-A. Similarly the next largest oligonucleotide defines the dinucleotide A-C, and so establishes the sequence T-A-C.

3. Materials and Methods

(a) *Chemicals and enzymes*

^{32}P-labelled deoxyribonucleotide triphosphates were synthesized by the method of Symons (1974) or obtained from New England Nuclear at a specific activity of about 100 mCi/μmol. Phage ϕX DNA was a gift from H. L. Weith. Phage ϕX RF (replicative form) was a gift from J. W. Sedat. *E. coli* DNA polymerase (nach Klenow) was obtained from Boehringer Chemical Corporation. *Haemophilus influenzae* restriction enzyme (*Hind* II and III) was prepared by the method of Smith & Wilcox (1970), and some was a gift from T. Maniatis. T4 polymerase was prepared by the method of Goulian *et al.* (1968), and some was a gift from K. Murray and A. G. Isaksson.

Two experiments will be described here to illustrate the use of the method.

(b) *Experiment 1 (Plate I)*

The initial reaction mixture (100 μl) contained 0·02 M-Tris·HCl (pH 7·4), 0·01 M-$MgCl_2$, 0·01 M-mercaptoethanol, 0·05 mM-dATP, -dTTP and -dCTP, 10 μCi of [^{32}P]dGTP (100

Ci/mmol), 6 μg φX174 DNA, 1·0 μg of the decanucleotide primer (Schott, 1974) and 16 units DNA polymerase. Incubation was at 0°C. After 2 min 50 μl were removed and added to 5 μl 0·2 M-EDTA to terminate the reaction. After 8 min 30 μl were removed and inhibited, and the remaining 20 μl incubated for a total of 30 min. The combined incubation mixtures were then applied to a column of Agarose (Bio-Gel A-0·5 m, 200 to 400 mesh, Bio-Rad Laboratories) using a 1-ml disposable plastic pipette. The column was made up and run in 2·0 mM-Tris·HCl (pH 7·4), 0·04 mM-EDTA. The fractionation was followed using a hand radiation monitor and the front band containing the synthesized DNA collected manually. It was concentrated to about 100 μl, and 5-μl samples taken for reincubation with 5 μl of the plus and minus mixtures and 1 μl of the appropriate enzyme.

The minus mixtures contained 0·04 M-Tris·HCl (pH 7·4), 0·02 M-$MgCl_2$, 0·02 M-mercaptoethanol, and the appropriate three dNTPs in 0·02 mM concentration. Incubation with DNA polymerase (0·8 unit in 1 μl) was for 30 min at 0°C.

The plus mixtures contained 0·13 M-Tris·HCl (pH 8·0), 0·013 M-$MgCl_2$, 0·02 M-mercaptoethanol, and the one triphosphate in 0·4 mM concentration. Incubation with T4 polymerase (approx. 0·02 unit in 1 μl) was for 30 min at 37°C.

The reactions were stopped by the addition of 1 μl 0·2 M-EDTA. 25 μl of freshly deionized formamide containing 0·3% xylene cyanol FF and 0·3% bromphenol blue were added and the solutions were heated at 95 to 100°C for 3 min before layering onto the acrylamide gel.

Electrophoresis was carried out on a 15% acrylamide gel (20 cm × 40 cm × 0·1 cm) at room temperature according to the method of Peacock & Dingman (1967). The buffer in the gel and in the anode compartment was Tris–glycine (3·028 g Tris-base, 14·4 g glycine/l)–8 M-urea. The cathode contained the same buffer without urea. It was run at 400 V until the bromphenol blue marker had travelled 30 cm from the origin (approx. 16 h). The gel was covered with "Saran wrap" and radioautographed.

(c) *Experiment 2 (Plate II)*

*Hin*d fragment 1 was prepared by digestion of 100 μg of phage φX RF with *Hin*d II + III and purified by ionophoresis on a 5% acrylamide gel (Edgell *et al.*, 1972). The amount obtained from 20 μg RF was mixed with 6 μg single-stranded φX DNA in 40 μl water and heated to 95°C for 3 min. 4 μl H × 10 buffer (66 mM-Tris·HCl (pH 7·4), 66 mM-$MgCl_2$, 100 mM-mercaptoethanol, 0·5 M-NaCl) was added and it was incubated at 67°C for 4 h. 20 μl of this annealed material was then incubated with DNA polymerase under the conditions described in section (b), above, in a volume of 50 μl. The labelled triphosphate was [^{32}P]dATP. Half of the solution was incubated for 2 min at 0°C and half for 8 min. These were combined and purified on an Agarose column and the eluate concentrated to 40 μl. 2-μl samples were then taken for reincubation with 2 μl of the plus and minus mixtures and 1 μl of the enzyme solution.

The minus mixtures were prepared by mixing equal volumes of the H × 10 buffer and 0·1 mM solutions of each of the three appropriate triphosphates. The plus mixtures contained 1 vol. H × 10 buffer, 1 vol. of a 0·2 mM solution of the triphosphate and 2 vol. of water. After incubation with enzyme as described above, 1 μl *Hind* II + III enzyme was added (this was assayed as sufficient enzyme to digest 2·5 μg bacteriophage λ DNA in 1 h at 37°C) and the solutions were reincubated at 37°C for 15 min. The reactions were terminated with EDTA and fractionated on a 12% acrylamide gel. A voltage of 1000 V was applied until the bromphenol blue marker had travelled 35 cm from the origin (4 to 5 h). The gel was allowed to become relatively hot during the run and no cooling was applied.

4. Results

(a) *The sequence primed by A-G-A-A-A-T-A-A-A-A (experiment 1)*

The method was initially tested out using the primer A-G-A-A-A-T-A-A-A-A on φX DNA. The sequence of the first 41 residues primed by this decamer was already known (Robertson *et al.*, 1973; Donelson *et al.*, unpublished data) and is shown in Figure 2. Plate I shows the results of one experiment. In interpreting the results it is essential to know the relative positions of the bands from the different samples on the radioautograph. This is easy where samples are run side by side but involves careful measuring for ones further apart. Thus, to show that the bands in

```
        5         10        15        20        25
A-G-A-A-A-T-A-A-A-A-G-T-C-T-G-A-A-A-C-A-T-G-A-T-T
-------------------
      primer

        30        35        40        45        50
A-A-A-C-T-C-C-T-A-A-G-C-A-G-A-A-A-A-C-C-T-A-C-C-G-C
```

FIG. 2. The sequence of the first 41 residues primed by the decanucleotide on ϕX DNA (Donelson *et al.*, unpublished data).

the +A and —T system are both in position 20, the strong —C band in position 28 was used as a reference. The distance between this —C and the —T in position 20 was the same as the sum of the distance from the —C to the +G in position 22 and that from this +G to the +A in position 20. Plate I also shows the exact lining up of the different bands, from the radioautograph.

If we consider position 18, there are two bands present, one in the +A and one in the —C system. This indicates that the 3′ residue of the oligonucleotide in this position is an A and that the next residue in the chain will be a C. A dinucleotide, A-C, is thus defined. As would be expected, the next position contains a band in the +C column, thus confirming the identification of the C. It also has a —A, which establishes A as the next residue. This is clearly followed by a sequence T-G; however at this stage the sequence is less easy to deduce, due to the presence of artifact bands, which appear to occur in this region and which will be discussed below. From positions 23 to 51 the sequence can be read off readily, and agrees well with the known sequence. Beyond residue 51 a partial sequence can be predicted but there are some uncertain parts, especially where "runs" of the same residue are concerned. For instance, it is clear that there are runs of T and A residues in the position labelled b, suggesting a sequence A_{4-6} T_2 A_{3-4} T_{4-6}, but the exact number of residues cannot be decided. This partial sequence will be discussed in a separate paper in connection with results obtained by other techniques.

(b) *Priming with fragments from restriction enzyme digests*

Besides using synthetic oligonucleotides as primers for DNA polymerase, it is possible to use the specific fragments obtained by the action of restriction enzymes (Maniatis *et al.*, 1974). The method is essentially the same as that used for the oligonucleotide primer except that the products from the incubations with the plus and minus mixtures are digested with the restriction enzyme before applying to the acrylamide gel for electrophoresis. Thus the radioactive oligonucleotides all have the same 5′ terminus, which corresponds to the original cleavage point of the enzyme. Plate II shows the result of an experiment in which fragment 1 from a digest prepared by the action of the restriction enzyme from *H. influenzae* (*Hin*d II + III) on double-stranded ϕX DNA (Edgell *et al.*, 1972) was used as primer on the single-stranded ϕX

DNA. The results here were particularly clear-cut and it was possible to predict a sequence of 70 residues that proved to be essentially correct. This DNA sequence codes for a part of the largest coat protein (from gene *F*) of the bacteriophage and has been largely determined by Sedat J. W., Ziff, E. B. & Galibert, F. (unpublished data) by direct partial digestion methods (see Galibert *et al.*, 1974), and by Blackburn (unpublished data) using transcription methods. Data were also obtained on peptides from the protein by Air (unpublished data). Although neither method gave the complete sequence, the results from the four techniques, which depend on entirely different approaches, left no doubt as to the correct sequence. These results will be discussed in a separate communication.

5. Discussion

In order that the method should give reliable results, various criteria must be satisfied. First, the products synthesized must all have the same 5′ terminus. *E. coli* DNA polymerase I normally has 5′ exonuclease activity; however, this can be removed by digestion with subtilisin and fractionation on Sephadex (Klenow & Henningsen, 1970). Thus DNA polymerase treated in this way was used in this work.

It is essential that the oligonucleotides are fractionated according to size. In preliminary experiments in which 20% acrylamide gels were used this was found not to be the case, some oligonucleotides migrating faster than corresponding smaller ones. This happened particularly in the region marked a in Plate I. The effect has not been encountered on 12% gels, though migration rate is not always exactly proportional to size, and the effect of the addition of a single residue is not exactly constant (see below). The reason for the anomalous migratory behaviour at higher gel concentrations is not clear, though it has been noted before (see Ikemura & Dahlberg, 1973). It may be connected with the secondary structure of the oligonucleotides and is certainly much worse in non-denaturing conditions.

Ideally oligonucleotides of all possible lengths should be present in the initial product of synthesis, so that all residues are represented in the plus and minus systems. In fact it is difficult to achieve this and under all conditions studied we have found that certain products are formed in relatively high yield, whereas others are absent. This suggests that the polymerase acts at different rates at different sites. It may be that this effect is partly related to the secondary structure of the template. It can also depend on the relative concentration of the triphosphates used. Thus if a low concentration of the ^{32}P-labelled triphosphate is used, "piling-up" frequently occurs before these residues. We find that the best results are obtained if synthesis is carried out for short times with a relatively high concentration of polymerase. Nevertheless it is frequently found that some expected products are missing (e.g. the +G in position 36, Plate I), and this constitutes a limitation of the method and is one reason why it is necessary to use both the plus and minus systems.

The main difficulty with the method occurs when consecutive runs of a given mononucleotide are present. Ideally if all oligonucleotides are present in the initial product of synthesis, each component of a run should appear as a band in the minus system, though the smallest component will be the strongest since it will be formed by extension of other smaller products—whereas the larger components are merely the unchanged oligonucleotides from the initial incubation. In Plate I it is possible to see the run of three As in positions 25–27 and of four As in positions 39–42. However it

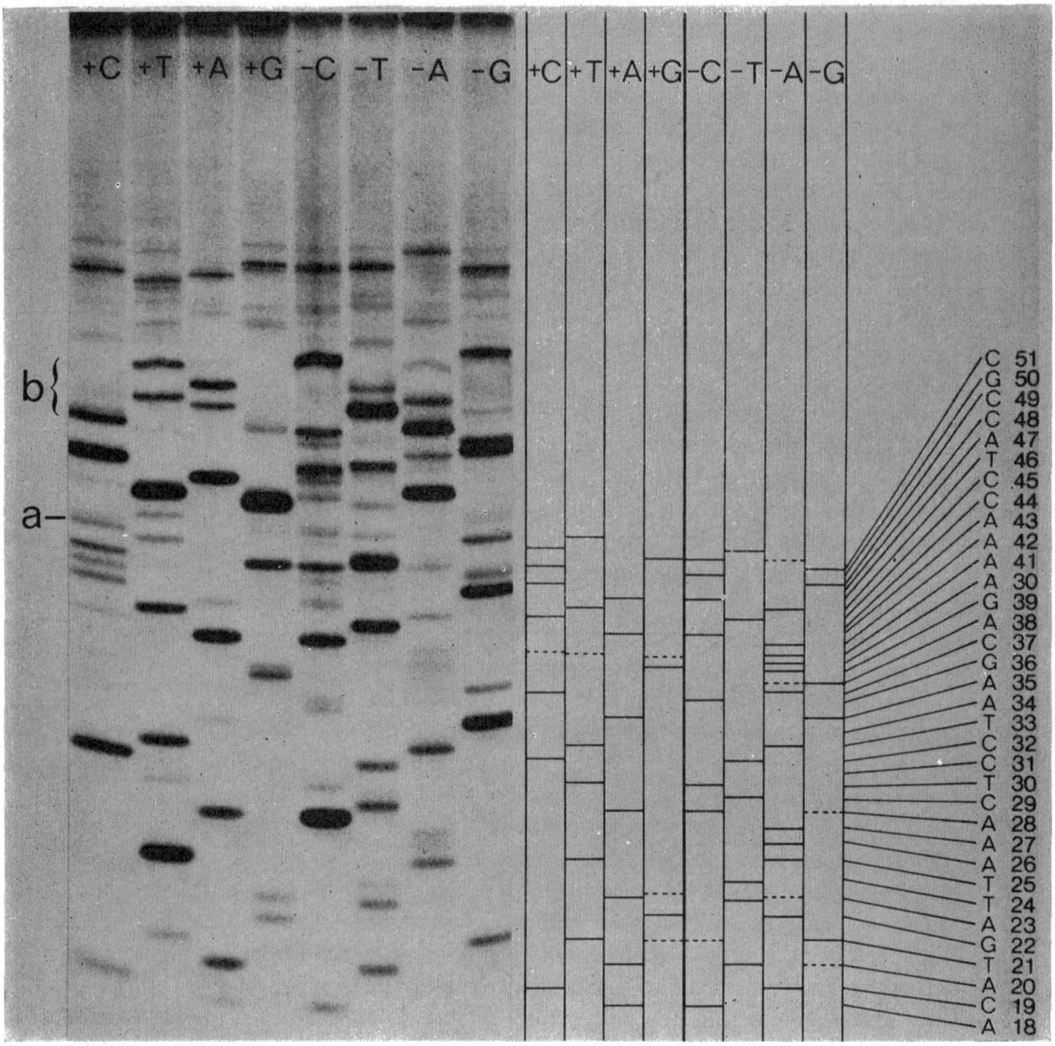

PLATE I. Radioautograph of experiment 1 (see Materials and Methods) in which the decanucleotide was used as a primer. The diagram shows the interpretation and the sequence deduced. The dashed lines represent artifact bands (see text). The bromphenol blue marker was 1 cm below the section of the gel shown in the Plate and the xylene cyanol FF marker opposite the product 50 nucleotides long.

[*facing p. 446*

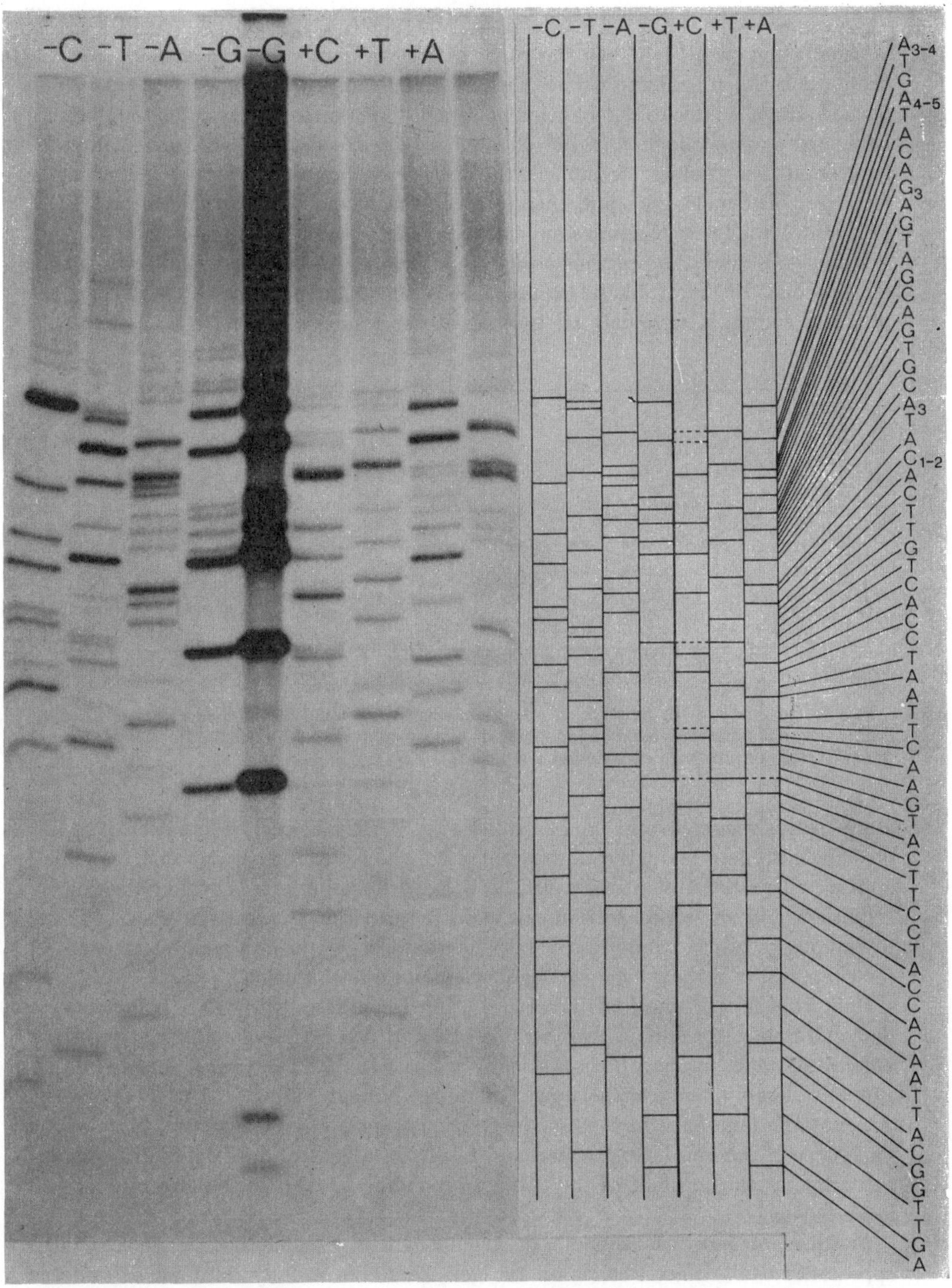

PLATE II. Radioautograph of experiment 2 (see Materials and Methods) in which *Hind* fragment 1 was used as a primer, and diagram illustrating the interpretation and sequence. The very dark centre sample labelled −G contained five times as much material as the other samples. The +G sample gave a number of artifact bands and it was not used in the interpretation. This was probably due to contamination of the +G mixture used in this experiment. A later experiment with a fresh mixture gave satisfactory results though resolution of the bands was less good.

is usually not possible to see the runs in this way and we have mainly used the distances between bands to deduce the size of a given run. In Figure 3 the change in distance travelled due to the addition of a single mononucleotide residue (the "jump") is plotted against the size of the oligonucleotides fractionated in Plates I and II. In general these values are sufficiently consistent; however they are less accurate for larger oligonucleotides and there are certain anomalies. In particular it will be seen from Plate I that position a appears to give four bands corresponding to —T,—C, +C and +G. Since the previous position contains a —G and there is a +T in a subsequent one, the most likely explanation is that the sequence is G-C-T and the products ending in G and C are moving at the same rate, although the expected

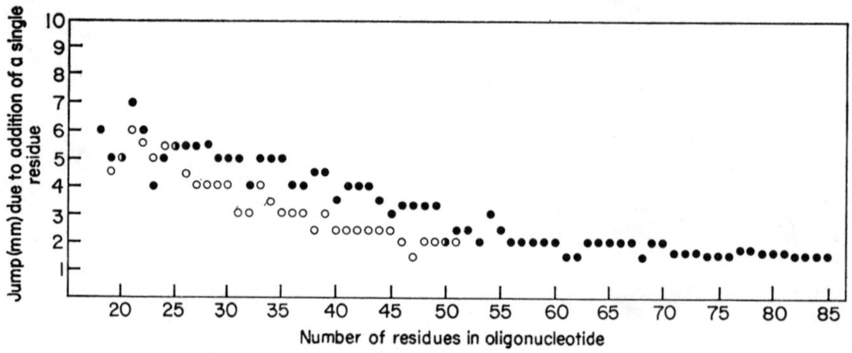

Fig. 3. The effect of the addition of a single nucleotide residue to oligonucleotides of different length on the distance travelled on ionophoresis in acrylamide gel. (○) Results from experiment 1 (Plate I); (●) results from experiment 2 (Plate II).

jump in this region should be 1·5 to 2·0 mm. Clearly in such a situation it is not possible to say how many C residues are present; however, fortunately such anomalies appear to be rare. Most of them can be avoided by using more severely denaturing conditions and low concentration gels as in the experiment shown in Plate II: thus by running the electrophoresis at 1000 V, so that the gel was at a higher temperature, some separation of the above two oligonucleotides was obtained.

One problem with methods involving primed synthesis with DNA polymerase is that one frequently finds a considerable blank incorporation of triphosphate in the absence of added primer. This is usually worse with ϕX than with phage f1 DNA and considerably worse with larger DNAs (see Maniatis *et al.*, 1974). In general it does not affect the present method very seriously since the blank seems to represent some type of random incorporation and results in a general darkening of the radioautograph at the top half of the gel, corresponding to oligonucleotides of over 100 residues long.

Another problem with the method, which is not fully understood, is the occasional appearance of "artifact" bands on the gel. These are usually faint bands and are marked as broken lines in Plates I and II. They can usually be recognized by the fact that they are not consistent in the plus and minus systems. The most serious one in Plate I is the +G band in position 23. It is clearly anomalous since there is no corresponding —G band, and two +G bands should never occur in adjacent positions. These artifact bands are usually not reproducible from one experiment to another, but their presence does emphasize the need for caution in interpreting the results. They are usually found clustered in a certain region (e.g. positions 21 to 23, Plate I) and may represent positions where the polymerases react particularly slowly—perhaps due to some secondary structure of the DNA.

The above method represents a rapid and simple technique for determining sequences in specific regions of a DNA chain, if suitable primers are available. Thus, if successfully carried out, it is possible to deduce a sequence of 50 nucleotides in a few days. At present, due to the limitations discussed above, it cannot be regarded as a completely reliable method and it is necessary to have confirmatory data such as amino acid sequences or transcription results. It has nevertheless been found extremely useful and it is hoped that its reliability will increase as more experience is gained.

Some preliminary experiments in connection with this approach were done with J. E. Donelson. We wish to thank H. Kössel and H. Schott for providing the decanucleotide primer, K. Murray and A. G. Isaksson for a gift of T4 DNA polymerase, T. Maniatis for *Hind* enzyme, H. L. Weith for ϕX DNA, and J. W. Sedat for ϕX RF DNA.

REFERENCES

Edgell, M. H., Hutchison, C. A. & Sclair, M. (1972). *J. Virol.* **9**, 574–582.
Englund, P. T. (1971). *J. Biol. Chem.* **246**, 3269–3276.
Englund, P. T. (1972). *J. Mol. Biol.* **66**, 209–224.
Galibert, F., Sedat, J. & Ziff, E. (1974). *J. Mol. Biol.* **87**, 377–407.
Goulian, M., Lucas, Z. J. & Kornberg, A. (1968). *J. Biol. Chem.* **243**, 627–638.
Ikemura, T. & Dahlberg, J. E. (1973). *J. Biol. Chem.* **248**, 5024–5032.
Klenow, H. & Henningsen, I. (1970). *Proc. Nat. Acad. Sci., U.S.A.* **65**, 168–175.
Maniatis, T., Ptashne, M., Barrell, B. G. & Donelson, J. E. (1974). *Nature (London)*, **250**, 394–397.
Peacock, A. C. & Dingman, C. W. (1967). *Biochemistry*, **6**, 1818–1827.
Robertson, H. D., Barrell, B. G., Weith, H. L. & Donelson, J. E. (1973). *Nature New Biol.* **241**, 38–40.
Sanger, F., Donelson, J. E., Coulson, A. R., Kössel, H. & Fischer, D. (1973). *Proc. Nat. Acad. Sci., U.S.A.* **70**, 1209–1213.
Schott, H. (1974). *Die Makromolekulare Chemie*, **175**, 1683–1693.
Smith, H. O. & Wilcox, K. W. (1970). *J. Mol. Biol.* **51**, 379–391.
Symons, R. H. (1974). In *Methods in Enzymology* (Grossman, L. & Moldave, K., eds), vol. 29, part E, pp. 102–115, Academic Press, New York and London.
Wu, R. & Kaiser, A. D. (1968). *J. Mol. Biol.* **35**, 523–527.
Wu, R. & Taylor, E. (1971). *J. Mol. Biol.* **57**, 491–511.

J. Mol. Biol. (1975) **96**, 703–719

The Nucleotide and Amino Acid Sequences of the N (5′) Terminal Region of Gene G of Bacteriophage ϕX174

G. M. Air, E. H. Blackburn, F. Sanger and A. R. Coulson

Medical Research Council
Laboratory of Molecular Biology
Hills Road, Cambridge CB2 2QH, England

(Received 12 March 1975)

Gene G of bacteriophage ϕX174 DNA has been characterized by three techniques. (1) Transcription by RNA polymerase of appropriate DNA fragments from restriction enzyme digestion. (2) Use of such DNA fragments as primers for DNA polymerase. (3) Amino acid sequencing of the gene product. The combined results give a DNA sequence of 195 nucleotides which codes for the N-terminal 65 amino acids of the protein sequence.

1. Introduction

Under certain conditions a complex can be formed between ribosomes and the single-stranded DNA of bacteriophage ϕX174. Robertson *et al.* (1973) showed that in this complex a sequence of about 50 nucleotides in the DNA was protected from the action of nucleases. By using DNA labelled *in vivo* with ^{32}P, the protected fragment could be isolated and its sequence determined. It was found that this sequence contained the initiation site for one of the structural proteins of the bacteriophage. This was the spike protein of molecular weight about 20,000, coded by gene G (Burgess & Denhardt, 1969), whose N-terminal sequence of nine amino acids was determined by Air & Bridgen (1973). The DNA fragment contained the sequence T-T-T-T-A-T-T-T-C-T. The complementary decanucleotide A-G-A-A-A-T-A-A-A-A has been synthesised (Schott, 1974) and used as a primer for DNA polymerase to confirm and extend the ribosome binding site sequence (Donelson, Barrell, Weith, Kössel & Schott, unpublished results).

The replicative form of ϕX174 DNA can be specifically degraded into 13 fragments (Edgell *et al.*, 1972) with the restriction endonuclease from *Haemophilus influenzae* (Hin_d II and III) and the relative order of these fragments has been determined by Jeppesen & Sanders (unpublished results) and by Lee & Sinsheimer (1974). Fiddes (personal communication) found that the above decanucleotide could act as a primer for DNA polymerase using the denatured form of Hin_d fragment 9 as a template, and the expected sequence for the binding site was formed. This established that this sequence and the initiation site for the gene G protein were in Hin_d fragment 9. The relative order of the Hin_d fragments in this region was 1-9-10-2, reading in the 5′ to 3′ direction on the viral or coding strand. Thus the nucleotide sequence coding for the gene G protein starts in fragment 9, continues through fragment 10, and terminates within fragment 2. This paper describes the further determination of the DNA sequence in this cistron up to the end of fragment 10, and the corresponding amino acid sequence of the gene G protein.

To determine the sequence of Hin_d fragment 10 we have used two methods which make use of RNA polymerase and DNA polymerase, respectively. The two different approaches complement one another very well, and together gave the complete sequence. A number of workers have determined DNA sequences by asymmetric transcription of native DNA templates *in vitro* and sequence analysis of the ^{32}P-labelled RNA (Zain *et al.*, 1974; Dhar *et al.*, 1974; Maizels, 1973). Transcription of single-stranded DNA with *Escherichia coli* RNA polymerase has been used for sequence determination, making use of RNA synthesis both with and without complementary oligonucleotide primers (Terao *et al.*, 1972; Gilbert & Maxam, 1973). Blackburn (1975) has used this method to determine sequences in a single-stranded fragment of DNA obtained by limited digestion of ϕX174 viral DNA by T4 endonuclease IV. When denatured fragment 10 was copied by similar techniques, sequences corresponding to both strands were formed. This gave a considerable amount of sequence information on this relatively small fragment (79 base-pairs), especially since overlapping sequences could be derived from complementary oligonucleotides.

Sanger & Coulson (1975) have described a relatively rapid method of determining DNA sequences with DNA polymerases using specific oligonucleotides as primers. When the single-stranded ϕX174 viral DNA is used as template and a restriction enzyme fragment as primer, the minus strand of the fragment can anneal to the ϕX174 viral strand DNA and synthesis with DNA polymerase proceeds from the 3′ end of this primer. Thus when fragment 2 was used as primer sequences corresponding to those in fragment 10 were formed, and in the single experiment described here a sequence of 60 residues was obtained. The method was also used with Hin_d fragment 10 as primer. The sequences formed were from fragment 9, and it was possible to deduce a sequence of 90 residues from the junction of fragments 9 and 10 to join the 3′ end of the binding site sequence of Robertson *et al.* (1973). There were a few doubtful residues in this DNA sequence but these were readily resolved using the amino acid sequence data.

In parallel with the work on DNA, the amino acid sequence of the gene G protein has been studied. This has been completed up to residue 67. The DNA sequence described above codes for the first 65 of these residues. 170 residues from this DNA were determined in the work described here; the first 27 have been determined by Robertson *et al.* (1973) and Donelson, Barrell, Weith, Kössel & Schott (unpublished results).

2. Materials and Methods

(a) *Transcription of* Hin$_d$ *fragment 10*

For each incubation 0·15 μg of Hin_d fragment 10 in 20 μl water was denatured by heating to 100°C for 30 s followed by rapid cooling to 0°C. To this denatured template DNA was added an equal volume of 2 times concentrated transcription reaction reagent, including RNA polymerase. Each reaction mixture (40 μl) contained 25 mM-Tris·HCl (pH 7·8), 8 mM-KCl, 10 mM-$MgCl_2$, 10 mM-β-mercaptoethanol, 0·5 mM-EDTA, 1 mM-potassium phosphate, three unlabelled ribonucleoside triphosphates (0·5 mM each) and one [α-^{32}P]ribonucleoside triphosphate (30 to 50 μM) at a specific activity of 100 Ci/mmol (New England Nuclear, U.S.A.), 2 to 3 μg *E. coli* DNA-dependent RNA polymerase (the generous gift of A. A. Travers, prepared as described by Burgess & Travers, 1971), and 0·15 μg of denatured Hin_d fragment 10. Incubations were carried out for 18 to 20 h at 20°C. Reaction was stopped by the addition of EDTA to 10 mM. Nucleic acids were precipitated by the addition of 0·5 ml water, 50 μl of 3 M-sodium acetate, 20 μg of *E. coli* tRNA (the gift of B. F. C. Clark) as carrier, and 1·5 ml of ethanol. After leaving at −70°C for 30 min, the material was centrifuged at 4°C at 2600 revs/min for 15 min, and the supernatant decanted. The precipitate was dissolved in water (50 to 100 μl), heated to 100°C for 1 min chilled rapidly and transferred to a fresh siliconized tube for lyophilization. The dry material was taken up in 10 μl of 10 mM-Tris·HCl (pH 7·4), 1 mM-EDTA containing 1 mg T_1 ribonuclease/ml (Sankyo, Tokyo, Japan) or pancreatic ribonuclease (Worthington Biochemical Corporation, N.J., U.S.A.). Digestion was carried out for 45 min at 37°C and the products of digestion were fractionated in two dimensions by electrophoresis on cellulose acetate at pH 3·5 in 7 M-urea, followed by homochromatography as the second dimension using a 3% "homomix" hydrolysed for 10 min (Brownlee & Sanger, 1969). The spots were located by autoradiography, eluted, and sequence analysis carried out as described previously (Sanger *et al.*, 1965; Brownlee & Sanger, 1967; Barrell, 1971). Nearest neighbour information was obtained by alkaline hydrolysis of the products of digestion of each oligonucleotide (Bishop *et al.*, 1968).

(b) *Primed DNA synthesis*

Hin_d fragments 2 and 10 were prepared by digestion of ϕX174 RF DNA with endonuclease R·Hin_d II and III under the conditions described by Smith & Wilcox (1970), and fractionated by electrophoresis on a 5% polyacrylamide gel. They were eluted electrophoretically from the gel as described by Galibert *et al.* (1974). The primed synthesis of DNA was carried out as described for experiment 2 (Sanger & Coulson, 1975), in which Hin_d fragment 1 was used as the primer.

(c) *Amino acid sequence*

Several grams of ϕX174 *am*3 were grown at the Microbiological Research Establishment, Porton Down, Wilts., U.K., and the frozen infected cells sent to Cambridge. Purification of the phage and preparation of the G protein was as previously described (Air & Bridgen, 1973). The protein in the pooled column fractions was *S*-carboxymethylated before dialysis and freeze-drying.

Chymotryptic digestion of a 10% suspension of the insoluble protein (20 mg) was in 1% ammonium bicarbonate at an enzyme–substrate ratio of 1:100 (w/w). After 3 h at 37°C the remaining precipitate was centrifuged off and the supernatant immediately fractionated on a column of Sephadex G50 in 50 mM-ammonium bicarbonate, monitored at 225 nm. Pooled fractions were freeze-dried and the individual peptides in each separated on Whatman 3MM paper by electrophoresis at pH 6·5 (Naughton *et al.*, 1960) followed by chromatography in butanol/acetic acid/water/pyridine (15:3:12:10) (Waley & Watson, 1953). After dipping the papers in fluorescamine (10 μg/ml in acetone containing 0·5% pyridine; Udenfriend *et al.*, 1972), the fluorescent spots were washed with acetone and the peptides eluted with the pH 6·5 electrophoresis buffer. Where necessary, further purification was by paper electrophoresis at pH 2·1 (acetic acid/formic acid/water, 4:1:45).

Tryptic digestion was in the same manner, but since few tryptic peptides are soluble the Sephadex G50 fractionation was omitted.

Cyanogen bromide cleavage was in 70% formic acid for 24 h at 4°C, using equal weights of cyanogen bromide and protein. The digest was freeze-dried, taken up in anhydrous formic acid, diluted to 50% formic acid, and fractionated on a column of Sephadex G75 in 50% formic acid. The soluble fragments were purified by paper electrophoresis and chromatography. Insoluble mixtures were fractionated on DEAE-cellulose and CM-cellulose (Whatman DE52 and CM52) in 8 M-urea buffers, but not all the fragments could be purified.

Amino acid sequences of peptides (20 to 50 nmol) were obtained by the dansyl–Edman method (Gray, 1967). Since some peptides were insoluble, or became so after some cycles of degradation, at each step the peptide was taken up in 150 μl pyridine and 150 μl water, the portion taken for dansylation, and 7·5 μl phenylisothiocyanate (PITC) (Pierce, Sequenal grade) added. For large insoluble peptides washing with benzene (3 × 1 ml) after PITC coupling then drying under a stream of nitrogen at 45°C gave a much improved degradation. Cleavage with trifluoroacetic acid was for 10 min at 45°C. The dansyl amino acids were identified by thin-layer chromatography on polyamide layers (Woods & Wang, 1967) using their solvents (1) 1·5% formic acid and (2) benzene/acetic acid (9:1). Other solvents were (3) ethyl acetate/methanol/acetic acid (10:1:1) (Crowshaw *et al.*, 1967), and, for arginine, (5) 0·05 M-trisodium phosphate/ethanol (3:1) (S. Magnusson, unpublished results). Amide residues could often be assigned from the electrophoretic mobility of the peptide.

The N-terminal sequence of the whole protein was obtained from an extended run in the Beckman 890 Sequencer as previously described (Air & Bridgen, 1973).

3. Results

(a) *Transcription of denatured* Hin_d *fragment 10*

Plate I shows autoradiographs of fingerprints of transcripts of Hin_d fragment 10, labelled separately with three of the [α^{32}P]nucleoside triphosphates and digested with T_1 ribonuclease. The sequence information obtained for the fingerprinted oligonucleotides is given in Table 1. From these data complete sequences could be deduced for all oligonucleotides except T1, T2, T3 and T19.

Plate II shows the fingerprints obtained when the transcripts, labelled separately with each of the ^{32}P nucleotides, were digested with ribonuclease A and fractionated in two dimensions. The sequences determined for all these oligonucleotides are given in Table 2. The final sequence of Hin_d fragment 10 was determined by combining the results of these experiments with the experiments in which DNA polymerase was used to extend fragment 2.

(b) *Priming with DNA polymerase*

Plate III shows the result of an experiment in which Hin_d fragment 10 was used as a primer on ϕX DNA and the "plus" and "minus" system described by Sanger & Coulson (1975) was used to determine the sequence extending from the 3′ end of the complementary strand of fragment 10. Column 1 shows the sequence predicted from this experiment and column 2 the sequence finally deduced by making use of the results from the amino acid sequence. There was some uncertainty in predicting the number of residues in 'runs' of a given nucleotide; however there was no doubt about the relative order of the nucleotides. The sequence starts at residue 21 as measured from the restriction enzyme cleavage site between fragments 9 and 10. As is usual in these experiments, no bands were obtained for the smaller oligonucleotides. This is probably due to the exonuclease activity of the DNA polymerase present during digestion with the Hin_d enzyme. The small fragments produced by the Hin_d enzyme are melted at 37°C and degraded by the 3′ exonuclease activity of the polymerase. Attempts to remove the polymerase have not so far led to successful results. To obtain residues 1 to 20 an experiment was done in which the Hin_d splitting was omitted and the products containing extended fragment 10 were fractionated directly by polyacrylamide gel ionophoresis. They could be reasonably well separated by running the ionophoresis twice the normal time, although fragment 10 is 79 residues long. In two experiments the earlier bands up to residue 21 were very weak; however they could just be seen and Figure 1 shows the combined results.

To obtain further information about the "runs" of pyrimidine residues an experiment was done on ten times the normal scale with the +G system, using [^{32}P]dATP

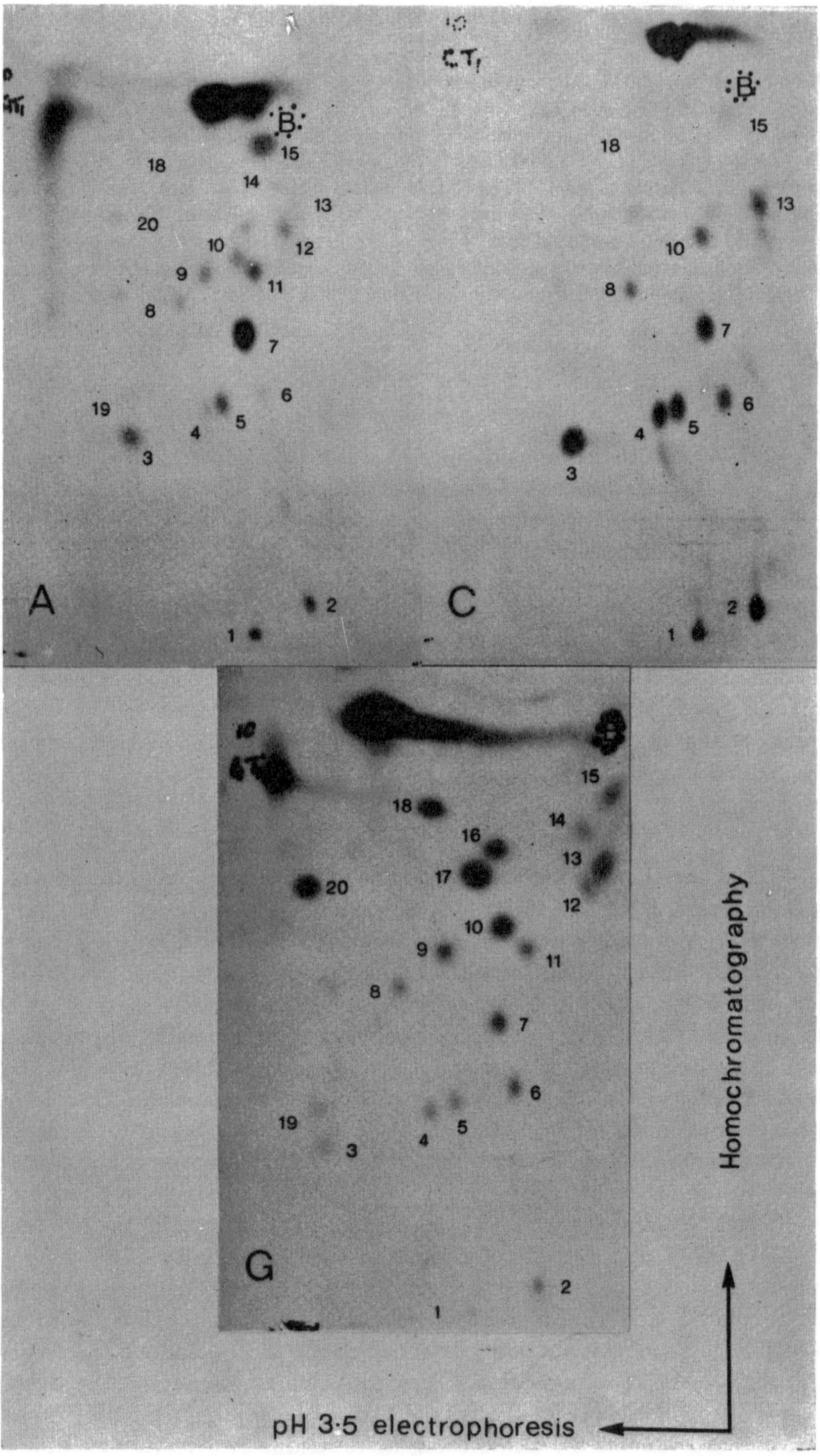

PLATE I. Autoradiographs of the products of T_1 ribonuclease digestion of *Hin*$_d$ fragment 10 transcripts fractionated as described by Brownlee & Sanger (1969). For the fingerprints shown, either [α-^{32}P]ATP (A), [α-^{32}P]CTP (C) or [α-^{32}P]GTP (G) were present in the transcription incubations to label the synthesised RNA. The dotted circles indicate the position of the xylene cyanol FF dye. Sequences of the oligonucleotides are given in Table 1.

[*facing p. 706*

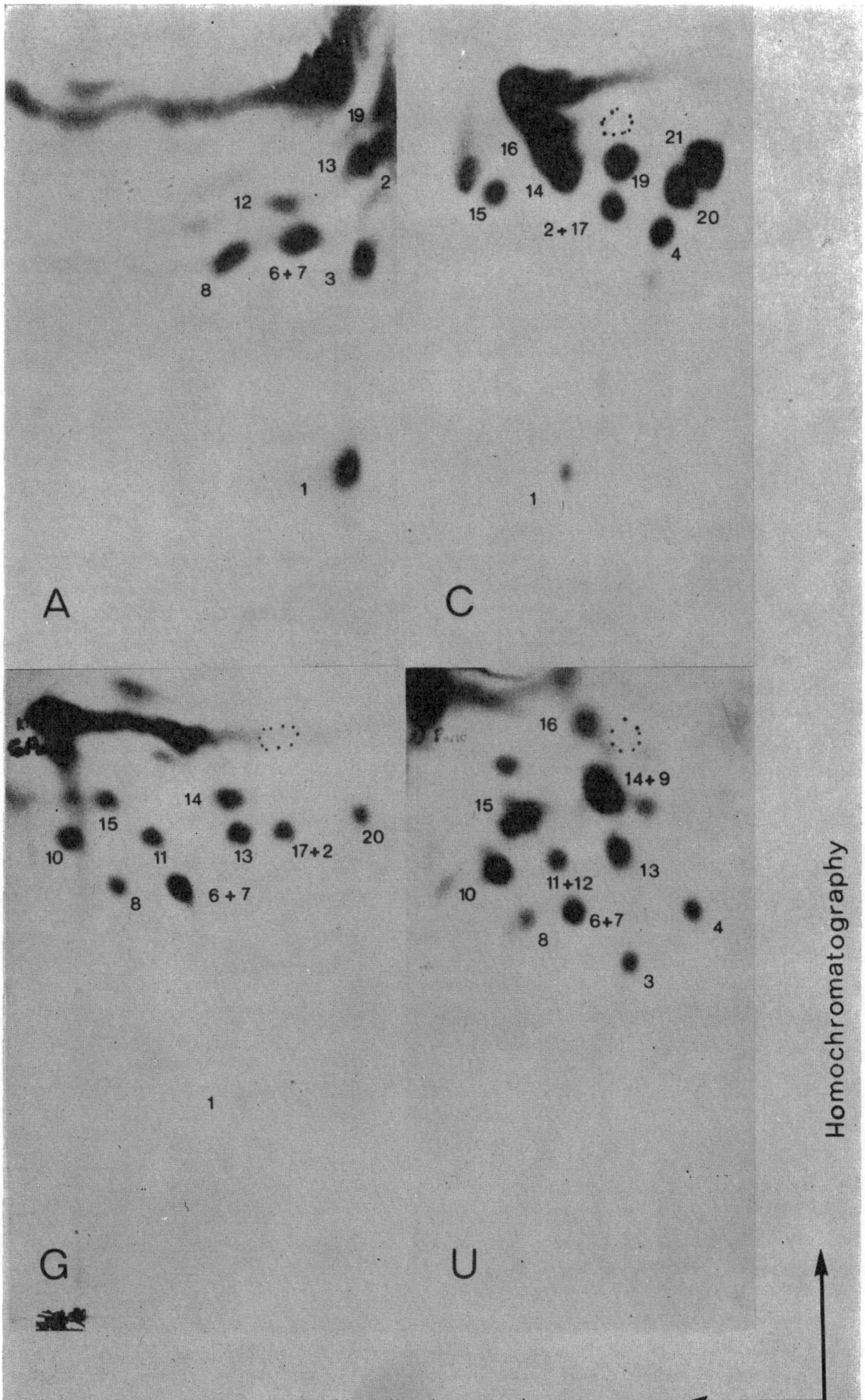

PLATE II. Autoradiographs of the ribonuclease A digestion products of Hin_d fragment 10. Each of the four ribonucleoside triphosphates, labelled at the α position with ^{32}P, was used in turn to label the transcripts, as indicated by A, C, G or U. Sequences of the oligonucleotides are given in Table 2.

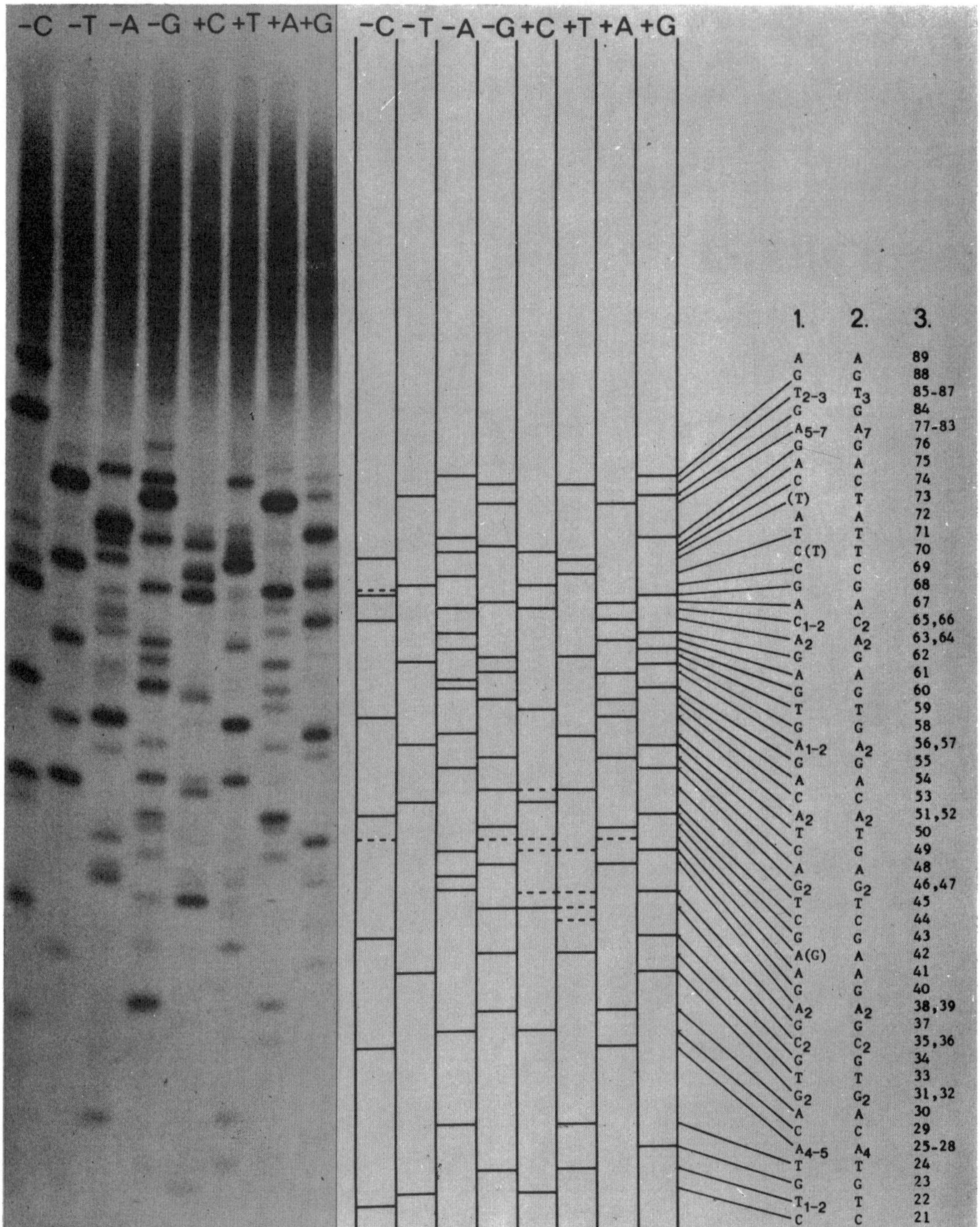

PLATE III. Autoradiograph of a polyacrylamide gel electrophoresis of samples from an experiment in which Hin_d fragment 10 was used as a primer for DNA polymerase on ϕX DNA, as described by Sanger & Coulson (1975) (experiment 2). [^{32}P]ATP was used for labelling. The diagram summarises the interpretation of the autoradiograph. The broken lines represent artefact bands (see Sanger & Coulson, 1975). Column 1 gives the sequence predicted from this experiment, column 2 the sequence finally established, and column 3 the number of residues from the 3′ terminus of the minus strand of Hin_d fragment 10.

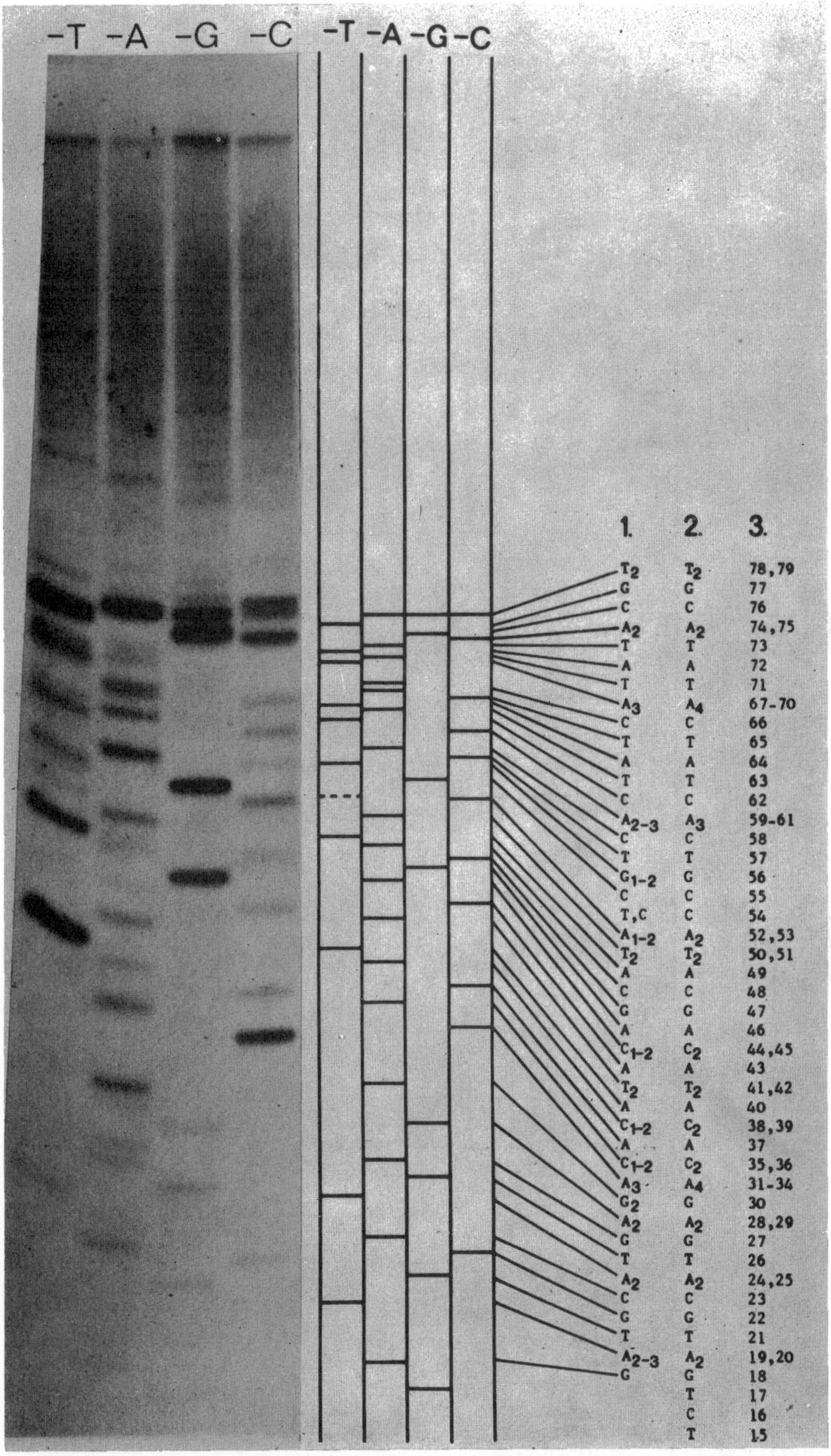

PLATE IV. Autoradiograph of a polyacrylamide gel electrophoresis of samples from an experiment in which *Hin*$_d$ fragment 2 was used as a primer for DNA polymerase, as in Plate III. Only the "minus" system was used in this experiment.

TABLE 1

Transcript of Hin$_d$ *fragment 10: products of ribonuclease* T_1 *digestion*

Oligonucleotide	Label	Pancreatic ribonuclease digestion products†	Ribonuclease U_2 digestion products identified	Deduced sequence
T1	C	A-A-A-C, A-A-C, A-U(C), U(C) + + + +		U-C-A-A-A-C-U-A-U-C-A-A-A-A-U-A-U-A-A-C-G(U)
	A	A-A-A-A-U(A), A-A-A-C (not A), + + + + + A-A-C, A-U(A), C(A), U(A) + + ++ +	U-A-A	
	U	A-A-A-C(U), A-A-A-A-U (not U), + + A-U (not U), G(U) + +	C-U-A(U), U-A(U)	
	G	A-A-C(G)		
T2	C	A-A-A-A-C(C), A-G(C), A-C(C) + + + + + + +	C-C-A-G(C), G(C)	A-A-A-A-C-C-A-C-C-A-U-U-A-C-C-A-G(C)
	A	A-A-A-A-C (not A), C(A), U(A) + + + + +++ +		
	U	A-U(U)		
	G	A-G (not G)		
T3	C	U-U(C), G(C), U(C) ++ + ++	—	U-U-U-U-C-U-U-C-A-U-U-G(C)
	A	C(A)		
	U	A-U(U), U(U), C(U)		
	G	U(G)		
T4	C	A-U(C), C(C) + +	—	U-A-U-C-C-A-U-C-U-G(A)
	A	G(A), C(A), U(A) + + +		
	U	A-U (not U), C(U) + +		
	G	U(G)		

TABLE 1—*continued*

Oligonucleotide	Label	Pancreatic ribonuclease digestion products†	Ribonuclease U_2 digestion products identified	Deduced sequence
T5	C	A-$\underset{+}{\text{U}}$(C), A-$\underset{+}{\text{C}}$ (not C)	—	A-U-A-C-A-U-C-U-G(U)
	A	A-$\underset{+}{\text{U}}$(A), A-$\underset{+}{\text{C}}$(A)		
	U	$\underset{+}{\text{G}}$(U), $\underset{+}{\text{C}}$(U)		
	G	U(G)		
T6	C	A-A-C(C)	—	C-A-U-U-A-A-C-C-G(U)
	A	A-A-$\underset{+}{\text{C}}$ (not A), $\underset{+}{\text{C}}$(A), $\underset{+}{\text{U}}$(A)		
	U	A-$\underset{++}{\text{U}}$(U), $\underset{+}{\text{G}}$(U)		
	G	C(G)		
T7	C	U(C)	—	C-A-U-U-C-A-G(A)
	A	A-$\underset{+}{\text{G}}$(A), $\underset{+}{\text{C}}$(A)		
	U	A-U(U)		
	G	A-G (not G)		
T8	C	G(C)	—	U-U-A-A-U-G(C)
	A	A-A-$\underset{+}{\text{U}}$ (not A), $\underset{+}{\text{U}}$(A)		
	U	A-A-U, U(U)		
	G	A-A-U(G)		
T9	C	—	—	U-A-A-U-G(G)
	A	A-A-$\underset{+}{\text{U}}$ (not A), $\underset{+}{\text{U}}$(A)		
	U	A-A-U (not U)		
	G	A-A-U(G), G(G)		
T10	C	G(C)	—	A-A-U-G(C) + A-U-A-G(U) (a) (b)

	A	A-A-U (not A), A-U(A)		
	U	A-A-U (not U). A-G(U), A-U (not U)		
	G	A-G (not G), A-A-U(G)		
T11	C	—	—	C-A-A-U-G(A)
	A	A-A-U (not A), C(A), G(A) + + +		
	U	A-A-U (not U)		
	G	A-A-U(G)		
T12	C	—	—	A-A-G(A)
	A	A-A-G(A)		
	U	—		
	G	A-A-G (not G)		
T13	C	A-C (not C)	—	A-C-G(G), C-A-G(A
	A	A-G(A), C(A)		
	U	—		
	G	A-C(G), G(G)		
T14	C	—	—	A-G(A)
	A	A-G(A)		
	U	—		
	G	A-G (not G)		
T15	C	G(C)		C-G(A)
	A	G(A)		(C)
	U	G(U)		(U)
	G	C(G)		
T16	C	G(C), U(C)		U-C-G(C), C-U-G(G)
	A	—		(a) (b)
	U			
	G	U(G), G(G)		
T17	C	—		A-U-G(G)
	A	—		(U)
	U	A-U (not U), G(U)		
	G	A-U(G), G(G)		

Table 1—*continued*

Oligonucleotide	Label	Pancreatic ribonuclease digestion products†	Ribonuclease U_2 digestion products identified	Deduced sequence
T18	C	G(C)	—	U-G(C)
	A	G(A)		(A)
	U	—		(G)
	G	U(G), G(G)		
T19	C	—	—	U-rich
	A	A-U(A), G(A), U(A)		
	G	—		
T20	C	—	—	U-U-U-G(A)
	A	G(A)		
	U	U(U)		
	G	U(G)		

† The symbols +, ++, etc. indicate the relative intensities of the products of pancreatic ribonuclease digestion before correction for the number of ^{32}P atoms in each product.

TABLE 2

Transcript of Hin$_d$ *fragment 10: ribonuclease A oligonucleotides*

Pancreatic ribonuclease digestion product	Sequence
P1	G-A-A-G-A-A-A-A-C(C)
P2	G-A-C(G)
P3	A-A-A-A-U(A)
P4	A-A-A-C(U)
P6	G-A-A-U(G)
P7	A-G-A-U(G)
P8	G-G-A-U(A)
P9	A-U(U)
P10	(G)a G-G-U(U)b (A)c
P11	A-G-U(U)
P12	G-A-U
P13	A-A-U(G)
P14	A-U$^{(C)}_{(U)}$
P15	G-U$^{(U)}_{(C)}$
P16	U$^{(U)}_{(C)}$
P17	A-G-C
P19	G-C$^{(A)}_{(U)}$
P20	A-A-C$^{(C)}_{(G)}$
P21	A-C(C)

labelled material, and some of the stronger bands eluted and subjected to depurination analysis as described by Galibert *et al.* (1974). These results are listed in Table **3**.

It can be seen from Plate III that there is a long run of A residues at positions 77 to 83. The following evidence indicated that it was A_7. *Hin*$_d$ fragment 9 was transcribed with RNA polymerase using [^{32}P]ATP and the product digested with pancreatic ribonuclease. An oligonucleotide was detected which contained at least ten residues, as judged from its position on the homochromatography system. With ribonuclease T_1 it gave A-G(A) and a product identified as $A_{>5}$-G. This indicated the sequence A-G-$A_{>6}$-G-Py. This long run of at least nine purine residues in the complementary strand should correspond to a long run of pyrimidine residues in the viral strand. Such a sequence was identified in the depurination products by Ling (1972) and its sequence shown to be C-T_7-C-T. This sequence was also confirmed by the amino acid sequence (position 12 to 15, Fig. 5).

Plate IV shows an experiment in which *Hin*$_d$ fragment 2 was the primer and just the "minus" system was used. The sequence was somewhat more tentative than that primed by fragment 10 (in which both "minus" and "plus" systems were used), but the results were sufficiently clear to establish the sequence in this region, which was already almost complete from the transcription data. Apart from uncertainties about the number of residues in runs, the only errors in the predicted sequence were

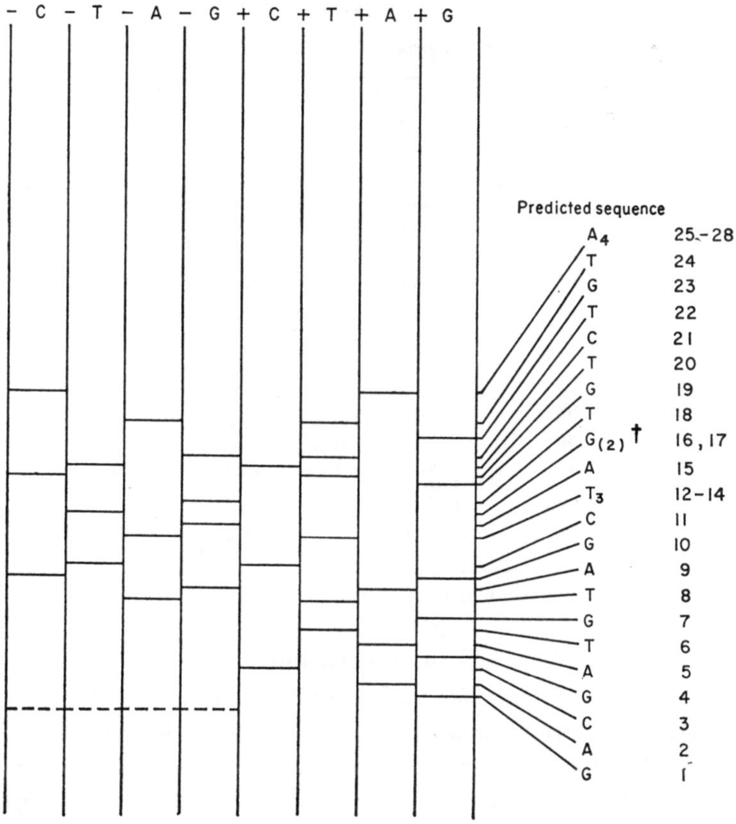

FIG. 1. Diagram showing the combined results of two experiments in which Hin_d fragment 10 was used as a primer and the final products were electrophoresed without prior digestion with the Hin_d enzyme. In one experiment [^{32}P]TTP was used and equal samples were incubated with the DNA polymerase for 2 min and 10 min, respectively. The ionophoresis was run until the xylene cyanol FF marker (the slower of the two) had run 37 cm. In the second experiment incubations were for 1 min and 3 min with [^{32}P]dATP and the marker was run 27 cm. The resolution was better in the first experiment but the slower bands were stronger in the second. The broken line represents a band seen only when the TTP was used as label. The first two bands (in the +A and +G systems) were also seen only with this label.

† The final sequence deduced differed from the predicted one only in that there were two G residues in positions 16 and 17 instead of one.

at residues 30 to 34 when the sequence G-G-A-A-A was predicted instead of G-A-A-A-A, and the possible insertion of a T at residue 54. In a second similar experiment the former sequence was clearly corrected but the T in position 54 persisted.

(c) *Amino acid sequence*

The sequences obtained from the N-terminus and the various peptides relevant to this region are shown in Figure 2: solid and broken bars indicate, respectively, complete and incomplete cleavage. No peptide overlaps joining residues 43 to 44 and 57 to 58 have been found, but the nucleic acid sequence provides these overlaps.

TABLE 3

Depurination products ([^{32}P]dATP-labelled)

Primer	Position of band (+G)	Depurination products
Plate III		
H10	23	pTp, p(C, T_3, p_3)p
	32	pTp, pCp, p(C, T_3, p_3)p
	34, 40, 42, 49	pTp, pCp, p(C, T_3, p_3)p
	62	pTp, pCp, p(C, T_3, p_3)p
	68	pTp, pCp, pCpCp, p(C, T_3, p_3)p
	76	pTp, pCp, pCpCp, p(T, C, p)p, p(C, T_2, p_2)p, p(C, T_3, p_3)p
Plate IV		
H2	18	pCp, pTp, p(C_2, T, p_2)p
	22	pCp, pTp, p(C_2, T, p_2)p
	30	pCp, pTp, p(C_2, T, p_2)p
	47	pCp, pTp, p(C_2, T, p_2)p, pTpTp, pCpCp
	56	pCp, pTp, p(C_2, T, p_2)p, pTpTp, pCpCp
	77	pCp, pTp, p(C_2, T, p_2)p, pTpTp, pCpCp, p(T, C, p)p

Underlining indicates more than one equivalent by inspection.

4. Discussion

(a) *Nucleotide sequence of* Hin$_d$ *fragment 10*

The sequences of all the products of digestion of transcripts with ribonucleases T_1 and A (Tables 1 and 2) were compared with all predicted complementary sequences. This procedure enabled a number of these oligonucleotides to be overlapped, as both DNA strands were copied, and the sequence shown in Figure 3 was completely deduced from these results. Sequences of the oligonucleotides T1, T2, T3 and T19 were deduced using the data obtained by DNA synthesis primed with *Hin*$_d$ fragment 2 (Plate IV). The predicted sequence T-C-A_{2-3}-C-T-A-T-C-A_3-T-A-T-A_2-C-G shown in Plate IV (positions 57 to 77) clearly corresponded to the oligonucleotide T1 which, on digestion with ribonuclease A, was shown to contain the sequences U(C), A-A-A-C(U), U(A), A-U(C), C(A), A-A-A-A-U(A), A-U(A), A-A-C(G) and G(U). Similarly oligonucleotide T2, which contained the sequences (deduced by digestion with ribonucleases A and U_2) A-A-A-A-C(C), C(A), A-C(C), A-U(U), A-G(G) and C-C-A-G(C), corresponded to the sequence from positions 31 to 48, as shown in Plate IV.

The oligonucleotide P1 was complementary to part of the product T3, which contained the sequences U-U-C, U-G(C), C-A and A-U-U. These sequence data, in addition to the position of T3 on the fingerprint (Plate I) (suggesting a U-rich composition) and the sequences of products T11 and T12 (Table 1), were combined with the data shown in Plate IV and the sequence U-U-U-U-C-U-U-C-A-U-U-G(C) was deduced for T3 (Fig. 4). The sequence of T19 was deduced from the data of Plate IV and was consistent with sequences obtained for complementary oligonucleotides. Finally, combining these results with the sequence data shown in Plate IV and Figure 3, the remainder of the sequence in *Hin*$_d$ fragment 10 was deduced. Figures 4 and 5 show the complete nucleotide sequence. Confirmatory results were obtained by partial digestion of the [^{32}P]GMP-labelled transcript with T_1 ribonuclease and analysis of the fractionated products as described previously (Blackburn, 1975).

```
                                                  5           10             15                20            25
Sequencer                 Met-Phe-Gln-Thr-Phe-Ile-Ser-Arg-His-Asn-Ser-Asn-Phe-Phe-Ser-Asp-Lys-Leu-Val-Leu-   -Ser-Val-   -   -
CNBr peptides
Tryptic peptides                                         |His-Asx-Ser-Asx-Phe-Phe-Ser-Asx-Lys|Leu-Val-Leu-Thr-Ser-Val-Thr
Chymotryptic peptides     Met-Phe-Gln-Thr-Phe|Ile-Ser-Arg-His-Asn-Ser-Asn-Phe-Phe|Ser-Asp-Lys-Leu-Val-Leu|Thr-Ser-Val-Thr-Pro-

                                   30            35             40              45             50
Sequencer                 Ala-   -Ser-Ala-Pro-Val-Leu-Glx
CNBr peptides
Tryptic peptides
Chymotryptic peptides     Ala-Ser-Ser-Ala-Pro-Val-Leu-Gln-Thr-Pro-Lys-Ala-Thr-Ser-Ser-Thr-Leu-Tyr|Phe-Asx-Ser-Leu-Thr-Val-Asx-

                                   55            60             65
Sequencer
CNBr peptides                                                |Asx-Thr-Ser-Val-Asx-
Tryptic peptides
Chymotryptic peptides     Ala-Gly-Asn-Gly-Gly-Phe-Leu|His-Cys-Ile-Glx-Met-Asx-Thr-Ser-Val-Asx
```

FIG. 2. Amino acid sequences of peptides and protein from the N-terminal region of ϕX174 G protein. The bars indicate sites of cleavage by enzymes or cyanogen bromide (CNBr).

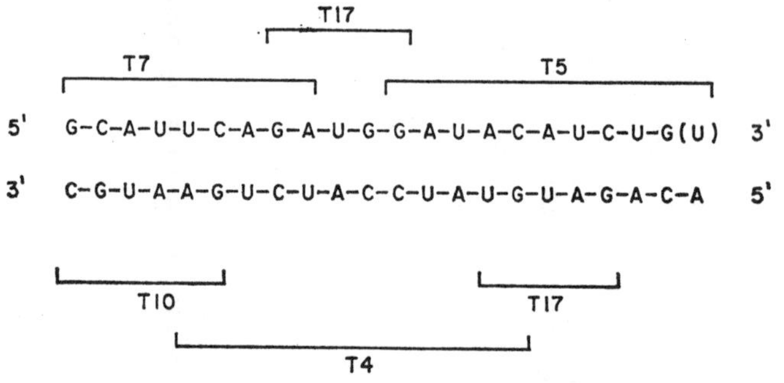

FIG. 3. A nucleotide sequence within *Hin*$_d$ fragment 10 deduced by inspection of the products of pancreatic and T_1 ribonuclease digestion of transcripts (Tables 1 and 2).

Apart from one residue at the 3′ end and one at the 5′ end, the complete sequence of fragment 10 was worked out from the two approaches to DNA sequencing. The DNA sequence was determined first and, as there are nonsense codons in two of the phases, it was possible to predict the amino acid sequence, which was then confirmed experimentally.

According to Plate IV, there appears to be an abrupt limit to the sequences at position 79, which presumably corresponds to the junction between fragments 9 and 10, so that all larger products synthesized were split by the *Hin*$_d$ enzyme. Bands were present in this position corresponding to -C, -A and -G. The next band below these was in the -T column, so that the 3′-terminal residue of the non-coding strand of fragment 10 is a T—though it is not completely clear from this result whether there are one or two T residues. That there are two T residues can be concluded from the amino acid sequence and from the known specificity of the *Hin*$_d$ enzyme. The amino acid sequence in positions 40 and 41 is Ser-Thr (Fig. 5). The only way this could be consistent with a *Hin*$_d$ site would be if the DNA sequence of the plus strand were G-T-C-X-A-C, where X is a purine. The G-T-C was established from the 5′ end of the sequence primed by fragment 10 (see below), and the A-C by positions 77 and 78 from Plate IV; i.e. the sequence G-T in the minus strand. Thus the position corresponding to X must be present in the sequence primed by fragment 2 and can only be a second T residue.

(b) *Sequence in* Hin$_d$ *fragment 9*

The sequence predicted from the experiments in which *Hin*$_d$ fragment 10 was used as a primer for DNA polymerase is given in Plate III and Figure 1. As both "plus" and "minus" systems were used there was less uncertainty than in the experiment primed by fragment 2. However confirmation was still necessary—especially for the sequence at the 5′ end (Fig. 1) since in this case the bands were very faint and the "jumps" small. Some confirmation was obtained from the depurination results (Table 3): in particular the sequence C-T-T-T-A in positions 11–15 was established by the identification of the depurination product p(C,T_3,p_3)p labelled by [^{32}P]ATP. Any ambiguities in the DNA sequence were eliminated by the amino acid sequence, with the possible exception of position 90. The last strong band seen on the radioautograph in Plate III is in the -A column (residue 89) and this is followed by a faint band in the -T column, suggesting the sequence A_{1-2}-T. As an Asn residue is present

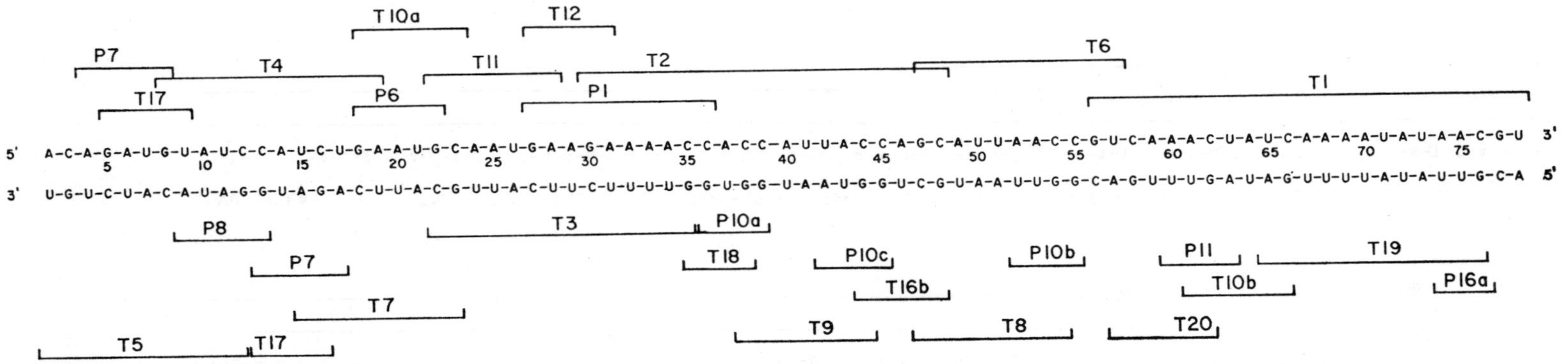

FIG. 4. Nucleotide sequence of Hin_d fragment 10 showing the ribonuclease digestion products identified from transcripts of both strands (Tables 1 and 2), The lower strand is the coding sequence. Residue numbers in the upper, complementary, strand correspond to those in Plate IV.

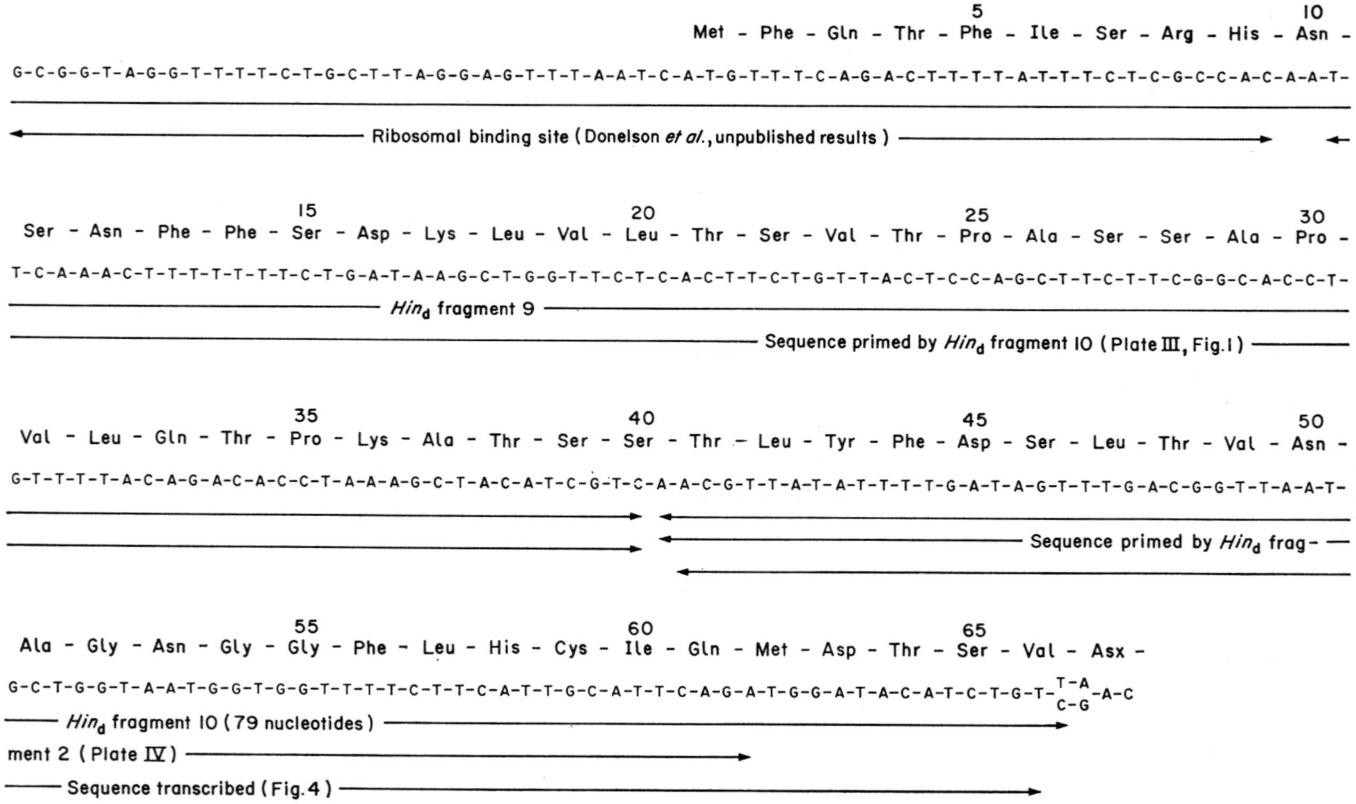

FIG. 5. The amino acid sequence of the N-terminal region of the spike protein (M_r 20,000) of ϕX174 and of the gene (G) that codes for it.

in position 10 (Fig. 5) (predicting the sequence A-T-T or G-T-T in the non-coding strand), the above result is best interpreted as A-A-T-T (residues 89 to 92)—an A being assigned to position 90.

In the experiments summarised in Figure 1 the earlier sequences were identified only from the "plus" system, and the first two residues (G-A) only when [^{32}P]TTP was used as label. Also in this experiment there was a strong band in all the "minus" systems corresponding to position 0. It seems likely that this represents intact fragment 10 that had become labelled with [^{32}P]TTP by the 3′ exonuclease and repair activity of the DNA polymerase. This supports the identification of T at the 3′ terminus of the non-coding strand of band 10, as discussed above.

(c) *The N-terminal region of gene G of bacteriophage φX174*

The combined results of this work are summarized in Figure 5, which shows the amino acid sequence of the first 67 residues of the gene G protein and the DNA sequence that codes for the first 65 of these. In this work three independent techniques have been used. As they depend on entirely different principles they complement each other very effectively. Including the sequence of the binding site determined by Robertson *et al.* (1973) and Donelson, Barrell, Weith, Kössel & Schott (unpublished results), a continuous sequence of 227 nucleotides has been established.

Of the 65 codons in this DNA sequence, it is of interest that the frequency of bases in the third position is as follows: T 58%, G 18·5%, A 15%, and C 8%. This is in marked contrast to the codons used in MS2 and R17 where the third position has U in 29%, G in 20%, A in 24% and C in 27% of the known codons (Adams *et al.*, 1969; Steitz, 1969; Min Jou *et al.*, 1972; Contreras *et al.*, 1972,1973; Jeppesen *et al.*, 1972; Haegeman & Fiers, 1973). Other coding data in φX174 will be presented in a separate paper (Air, Blackburn, Coulson, Galibert, Sanger, Sedat & Ziff, unpublished results).

The possible secondary structures that could be assumed by the single-stranded viral DNA or messenger RNA have not yet been exhaustively analysed. A preliminary inspection of the sequences shows some possible "looped" structures in Hin_d fragment 10 (for example, from residues 3 to 18 (Fig. 4)).

REFERENCES

Adams, J. M., Jeppesen, P. G. N., Sanger, F. & Barrell, B. G. (1969). *Nature (London)*, **223**, 1009–1014.

Air, G. M. & Bridgen, J. (1973). *Nature New Biol.* **241**, 40–41.

Barrell, B. G. (1971). In *Procedures in Nucleic Acid Research* (Cantoni, G. L. & Davies, D. R., eds), vol. 2, pp. 751–779, Harper & Row, New York, Evanston and London.

Blackburn, E. H. (1975). *J. Mol. Biol.* **93**, 367–374.

Bishop, D. H. L., Mills, D. R., Spiegelman, S. (1968). *Biochemistry*, **7**, 3744–3753.

Brownlee, G. G. & Sanger, F. (1967). *J. Mol. Biol.* **23**, 337–353.

Brownlee, G. G. & Sanger, F. (1969). *Eur. J. Biochem.* **11**, 395–399.

Burgess, A. B. & Denhardt, D. T. (1969). *J. Mol. Biol.* **44**, 377–386.

Burgess, R. R. & Travers, A. A. (1971). In *Procedures in Nucleic Acid Research* (Cantoni, G. L. & Davies, D. R., eds), vol. 2, pp. 851–863, Harper & Row, New York, Evanston and London.

Contreras, R., Vandenburghe, A., Volkaert, G., Min Jou, W. & Fiers, W. (1972). *FEBS Letters*, **24**, 339–342.

Contreras, R., Ysebeart, M., Min Jou, W. & Fiers, W. (1973). *Nature New Biol.* **241**, 99–101.

Crowshaw, K., Jessup, S. J. & Ramwell, P. W. (1967). *Biochem. J.* **103**, 79–84.

Dhar, R., Weissman, S. M., Zain, B. S., Pan, J. & Lewis, A. M. (1974). *Nucleic Acids Research*, **1**, 595–613.
Edgell, M. H., Hutchison, C. A. & Sclair, M. (1972). *J. Virol.* **9**, 574–582.
Galibert, F., Sedat, J. & Ziff, E. (1974). *J. Mol. Biol.* **87**, 377–407.
Gilbert, W. & Maxam, A. (1973). *Proc. Nat. Acad. Sci., U.S.A.* **70**, 3581–3584.
Gray, W. R. (1967). In *Methods in Enzymology* (Hirs, C. H. W., ed.), vol. 11, pp. 469–475, Academic Press, New York.
Haegeman, G. & Fiers, W. (1973). *Eur. J. Biochem.* **36**, 135–143.
Jeppesen, P. G. N., Barrell, B. G., Sanger, F. & Coulson, A. R. (1972). *Biochem. J.* **128**, 993–1006.
Lee, A. S. & Sinsheimer, R. L. (1974). *Proc. Nat. Acad. Sci., U.S.A.* **71**, 2882–2886.
Ling, V. (1972). *Proc. Nat. Acad. Sci., U.S.A.* **69**, 742–746.
Maizels, M. N. (1973). *Proc. Nat. Acad. Sci., U.S.A.* **70**, 3585–3589.
Min Jou, W., Haegeman, G., Ysebeart, M. & Fiers, W. (1972). *Nature (London)*, **237**, 82–88.
Naughton, M. A., Sanger, F., Hartley, B. S. & Shaw, D. C. (1960). *Biochem. J.* **77**, 149–163.
Robertson, H. D., Barrell, B. G., Weith, H. L. & Donelson, J. E. (1973). *Nature New Biol.* **241**, 38–40.
Sanger, F. & Coulson, A. R. (1975). *J. Mol. Biol.* **94**, 441–448.
Sanger, F., Brownlee, G. G. & Barrell, B. G. (1965). *J. Mol. Biol.* **13**, 373–398.
Schott, H. (1974). *Die Makromolekulare Chemie*, **175**, 1683–1693.
Smith, H. O. & Wilcox, K. W. (1970). *J. Mol. Biol.* **51**, 379–391.
Steitz, J. A. (1969). *Nature (London)*, **224**, 957–964.
Terao, T., Dahlberg, J. E. & Khorana, H. G. (1972). *J. Biol. Chem.* **247**, 6157–6166.
Udenfriend, S., Stein, S., Böhlen, P., Dairman, W., Leimgruber, W. & Weigele, M. (1972). *Science*, **178**, 871–872.
Waley, S. G. & Watson, J. (1953). *Biochem. J.* **55**, 328–337.
Woods, K. R. & Wang, K. T. (1967). *Biochim. Biophys. Acta*, **133**, 369–370.
Zain, B. S., Weissman, S. M., Dhar, R. & Pan, J. (1974). *Nucleic Acids Research*, **1**, 577–594.

J. Mol. Biol. (1976) **107**, 445–458

Gene *F* of Bacteriophage ϕX174. Correlation of Nucleotide Sequences from the DNA and Amino Acid Sequences from the Gene Product

G. M. Air, E. H. Blackburn†, A. R. Coulson, F. Galibert‡
F. Sanger, J. W. Sedat§ and E. B. Ziff¶

Medical Research Council, Laboratory of Molecular Biology
Hills Road, Cambridge CB2 2QH, England

(*Received 6 May 1976*)

Application of three different methods to obtain nucleotide sequences from ϕX174 DNA has yielded a continuous sequence of 281 nucleotides and partial sequences totalling another 263 nucleotides from adjoining regions. Simultaneous work on amino acid sequences of ϕX174 coat proteins showed that these nucleotide sequences are part of the *F* gene, coding for the largest coat protein of the bacteriophage. A total of 136 codons has been identified, of which 56% have thymine as the base in the third position of the coding triplet.

1. Introduction

(a) *The nucleotide sequences*

In developing methods to determine nucleotide sequences in ϕX174 DNA several approaches have been used in this laboratory. Essentially they are (1) partial degradation of ^{32}P-labelled single-stranded DNA with endonuclease IV or the depurination reaction (Burton & Petersen, 1960) to obtain small oligonucleotides which can then be sequenced by partial digestion with spleen and snake venom phosphodiesterases (Ling, 1972; Robertson *et al.*, 1973; Ziff *et al.*, 1973; Galibert *et al.*, 1974); (2) transcription of the DNA with RNA polymerase and sequencing the ^{32}P-labelled RNA products (Blackburn, 1975,1976); (3) copying regions of the DNA using DNA polymerase with a specific primer to give short stretches of radioactive DNA. Two approaches have been used to determine such sequences: (a) incorporation of a ribonucleotide into the DNA, thus providing sites for ribonuclease or alkali cleavage to give oligonucleotides of suitable length for sequencing (Sanger *et al.*, 1974); (b) the "plus and minus" method (Sanger & Coulson, 1975).

† Present address: Department of Biology, Yale University, New Haven, Conn. 06520, U.S.A.
‡ Present address: Laboratoire d'Hématologie Expérimentale, Centre Hayem, Hôpital Saint-Louis, Paris X, France.
§ Present address: Department of Radiobiology, Yale University Medical School, New Haven, Conn. 06520, U.S.A.
¶ Present address: Rockefeller University, 1230 York Avenue, New York, N.Y. 10021, U.S.A.

In the work summarized here these methods have been used to obtain nucleotide sequences from a region comprising about 10% of the ϕX174 genome.

Ziff *et al.* (1973) showed that when the single-stranded DNA from ϕX174 that had been labelled with ^{32}P *in vivo* was digested under mild conditions with endonuclease IV and the products fractionated by ionophoresis on acrylamide gel, a number of relatively pure fragments could be isolated. Methods were developed for studying the sequences of such fragments of ^{32}P-labelled DNA and the first sequence to be completed was that of band 6, which was 48 nucleotide residues long (Galibert *et al.*, 1974). These studies have been extended to some of the larger bands and this work is reported in an accompanying paper (Sedat *et al.*, 1976). It was found that all the fragments formed by limited endonuclease IV digestion (bands 2, 3, 5A, 5B, 6, 7 and 8) came from the same region of the DNA, and their relative order was found to be as shown in Figure 5 of Sedat *et al.* (1976). The complete sequence of bands 5A and 5B, most of band 3, and fragmentary sequences of band 2 have been determined.

Another approach to DNA sequencing is by transcription of unlabelled DNA with RNA polymerase in the presence of ^{32}P-labelled triphosphates and studying the synthesised RNA by techniques that are already available. Blackburn (1975,1976) has used this method on endonuclease IV bands 6 and 3, and the results confirm and supplement the data obtained by direct sequencing of these bands.

DNA may also be sequenced by copying with DNA polymerase and a method has been developed for determining specific sequences by using suitable primers (Sanger & Coulson, 1975). Either synthetic oligonucleotides or fragments from restriction enzyme digests can be used as primers and the technique described is rapid and simple. A sequence of 70 residues could be predicted by priming with *Hind* fragment 1 (Sanger & Coulson, 1975). This sequence corresponds to the 3′ end of endonuclease IV band 3. From the combined results of the three methods it was possible to deduce the complete sequence of band 3, although neither method alone was sufficient.

(b) *Localization of the endonuclease IV bands in fragments from restriction enzyme digestion*

In order to relate the oligonucleotides obtained by endonuclease IV partial digestion to restriction enzyme fragments, one of us (E. H. Blackburn) has studied transcription products of the fragments produced by digestion with restriction endonucleases from *Haemophilus influenzae* (*Hind*II and *Hind*III) (Smith & Wilcox, 1970; Edgell *et al.*, 1972) and *Haemophilus aegyptius* (*Hae*III) (Middleton *et al.*, 1972). Fragments *Hind* 1, *Hind* 6b, *Hae* 4 and *Hae* 8 were transcribed and the products compared with products obtained from the endonuclease IV bands. This established the relationships summarised in Figure 2.

(c) *Localization of nucleotide sequences in the genetic map*

The preliminary map of the restriction enzyme fragments relative to the genetic map of ϕX174 (Chen *et al.*, 1973; Sedat *et al.*, 1976) suggested that the sequence of the endonuclease IV fragments would probably code for the gene *E* protein or near it. However, Air (1976) has been studying the amino acid sequences of the structural proteins of ϕX174 and as soon as sufficient amino acid and DNA sequences were available it was clear that sequences in the largest coat protein (the product of gene *F*) corresponded according to the genetic code to DNA sequences in the endonuclease IV fragments. This was amply confirmed by further work. The 5′ end of endonuclease IV band 5A corresponded to the third amino acid residue of the *F* protein, thus establishing an exact relationship between the genetic map and the DNA fragments from both restriction enzyme and endonuclease IV digests.

2. Results

The detailed methods and results of the DNA and protein sequence work referred to in this paper are given in preceding papers (Galibert *et al.*, 1974; Sanger & Coulson, 1975; Blackburn, 1975) and in the accompanying papers (Air, 1976; Blackburn, 1976; Sedat *et al.*, 1976). We give here only the additional experiments used to correlate the sequence data.

(a) *Location of endonuclease IV bands in restriction enzyme fragments*

Endonuclease IV band 3 contains the sequence G-G-C-C (Sedat *et al.*, 1976) which, in double-stranded DNA, is recognised and cleaved by endonuclease *Hae*III. ϕX174 duplex DNA was therefore predicted to be cleaved at this point by the *Hae* enzyme. Endonuclease IV band 2 contains the sequence G-T-T-A-A-C near the 5′ terminus, hence a junction of two *Hind* fragments was expected at this position. To ascertain which *Hae* and *Hind* fragments these were, use was made of a number of relatively long and characteristic polypyrimidine tracts present in the endonuclease IV band sequences. In order to identify these tracts in the corresponding restriction enzyme fragments and hence to line up the three sets of fragments, purified restriction enzyme fragments were transcribed with RNA polymerase and the products digested with ribonuclease A (Fig. 1 and Table 1). The large products (polypurine tracts) were then studied by ribonuclease T_1 digestion. For any polypyrimidine tract present in the double-stranded restriction enzyme fragments there will be a complementary ribonuclease A (polypurine) product. Thus the correspondence between the transcripts of restriction fragments and polypyrimidine tracts in the endonuclease IV bands was used to establish overlaps as shown in Figure 2.

Table 1 shows that *Hind* fragment 1 contains the polypyrimidine tracts A′, B′ and C′, while the sequences D′, E′ and F′ are found in *Hind* 6a or 6b. Fragment *Hae* 4 contains the sequences A′, B′, C′ and D′.

In addition, analysis of ribonuclease T_1 oligonucleotides in transcripts of the restriction fragments *Hae* 8, *Hae* 4 and *Hind* 6a + 6b helped in assigning the endonuclease IV bands to restriction fragments. The sequence data on these oligonucleotides are given in Table 2 and the results included in Figure 2. In these experiments the *Hind* fragments 6a and 6b were not separated, but the *Hind* cleavage map of ϕX174 shows that the fragments containing the endonuclease IV bands can only be 6b (Jeppesen *et al.*, 1976).

(b) *Location of the endonuclease IV bands in the genetic map*

A genetic map of ϕX174 has been deduced by Benbow *et al.* (1972,1974) from genetic data and estimation of the sizes of the proteins coded by ϕX174. A cleavage map of the ϕX174 genome is also known in which the fragments produced by the restriction endonucleases *Hae*III and *Hind*II are accurately related to each other (Lee & Sinsheimer, 1974; Jeppesen *et al.*, 1976). The amino acid and nucleotide sequences summarized here show that the 5′ nucleotide of endonuclease IV band 5A is the third nucleotide of the triplet coding for amino acid residue number 3 of the

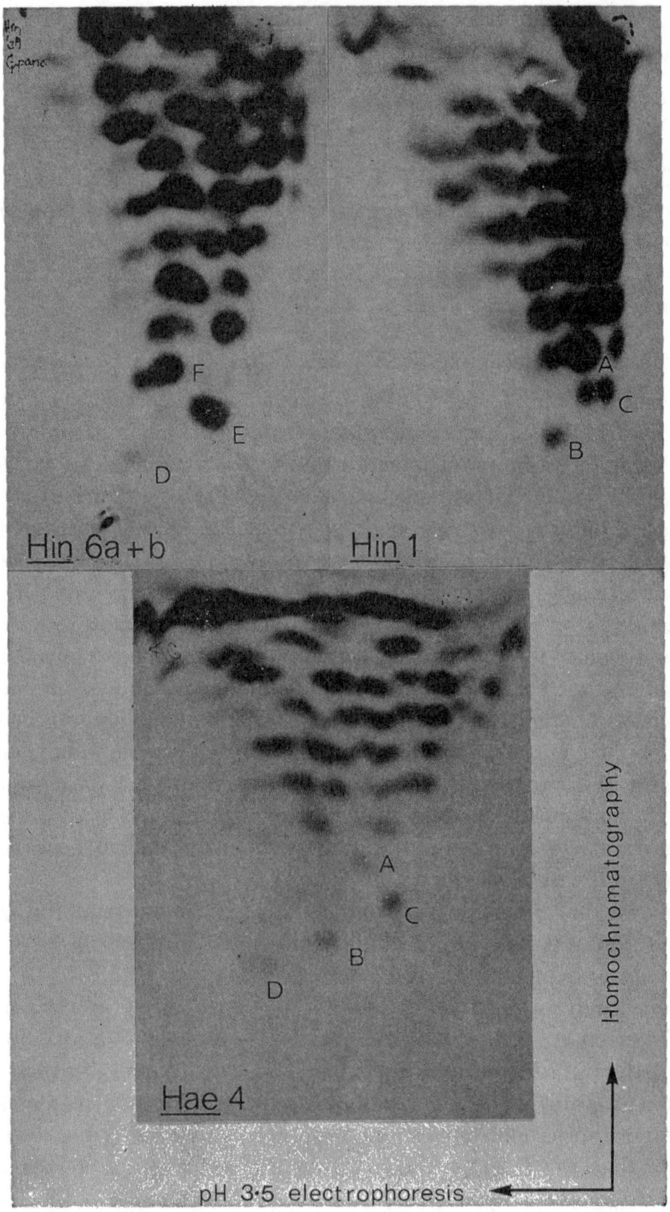

FIG. 1. Pancreatic RNAase fingerprints of transcripts of denatured fragments of ϕX174 replicative form DNA. The template fragments were produced by cutting the DNA with restriction endonuclease *Hin*dII and *Hin*dIII (Smith & Wilcox, 1970; Edgell *et al.*, 1972) or *Hae*III (Middleton *et al.*, 1972). Each DNA fragment was denatured and transcribed with *Escherichia coli* RNA polymerase, as described previously (Air *et al.*, 1975). Transcripts were labelled by incorporation of [α-^{32}P]ATP (shown for *Hind* fragment 1) and [α-^{32}P]GTP (shown for fragments *Hae* 4 and *Hind* 6a + 6b). The synthesised RNA was digested with pancreatic RNAase and the products fractionated in 2 dimensions. The lettered spots correspond to depurination products sequenced in endonuclease IV bands.

TABLE 1

Transcripts of restricted ϕX174 DNA: pancreatic ribonuclease digestion products

Oligonucleotide†	*Hind*‡ fragment	*Hae*‡ fragment	Estimated size §	T_1 ribonuclease digestion products: A-label	T_1 ribonuclease digestion products: G-label	Corresponding depurination product†	Endonuclease IV band in which depurination product found
A	1	4	7	G(A), A-G(A), A-A-A-G	(present as mixture)	C-T-T-T-C-T-C (A′)	2
B	1	4	9	A-A-G, G(A)	A-A-G, A-G, C(G)	(C-G)C-T-T-C-C-T-C-C-T (B′)	2
C	1	4	8	A-A-A-A-G, A-A-G	A-A-A-A-G, A-A-G, C(G)	(C-G)C-T-T-T-T-C-T-T (C′)	2
D	6b	4	11	A-G(A), G(A)	A-G, G(G), U(G)	(C-A)C-T-C-C-C-T-C-T-C-C-C (D′)	Junction of 2 and 3
E	6b	n.i.	9	A-A-A-G, A-G(A), G(A)	A-A-A-G, A-G, G(G), A-C(G)	(C-G)T-C-T-T-T-C-T-C-C (E′)	3
F	6b	n.i.	8	A-A-A-G, G(A)	A-A-A-G, G(G)	C-C-T-T-T-C-C-C (F′)	5A

† The polypurine tracts identified in the transcripts are designated by letters (A, B, etc.) and the corresponding polypyrimidine tracts by A′, B′, etc.
‡ Fragments designated according to size, as described by Edgell *et al.* (1972), Middleton *et al.* (1972), Jeppesen, *et al.* (1976) Lee & Sinsheimer (1974).
§ Sum of A plus G residues.
n.i., not identified.

TABLE 2

Transcripts of restricted $\phi X174$ *DNA:* T_1 *ribonuclease digestion products*

Oligonucleotide	*Hind* fragment	*Hae* fragment	Pancreatic ribonuclease digestion products of oligonucleotides (labelled with [^{32}P]CMP)†	Location in endonuclease IV bands	DNA strand used as template
1	6b	8	<u>A-C</u>, A-U(C), U(C), C(C)	Band 5a	Non-coding
2	6b	8	A-A-C, <u>A-C</u>, <u>U(C)</u>, C(C)	Overlaps bands 6 and 3	Non-coding
4	n.i.	8	A-C, U(C), C(C)	Band 3	Non-coding
6	6b	4	A-A-C, A-U(C), <u>A-C</u>, U(C), C(C)	Band 3	Coding

† An underlined product is present in more than one molar yield, as judged by inspection of the radioautograph.

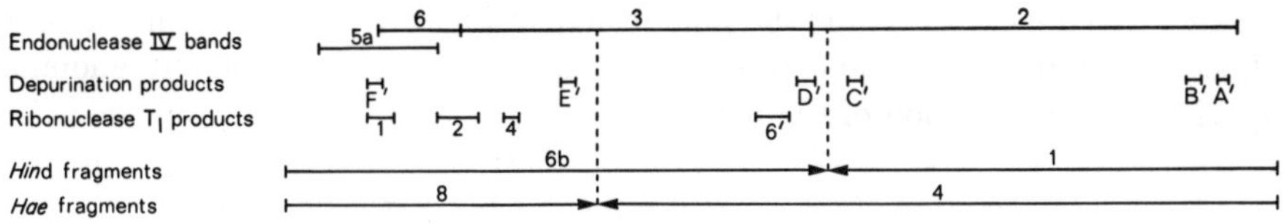

FIG. 2. Schematic representation of the positions of the large depurination products (A′ to F′) whose complements (A to F) were identified in transcripts of restriction fragments (Table 1) and of the large ribonuclease T_1 products (Table 2). The positions of the endonuclease IV bands relative to restriction fragments are shown.

F gene product. Having established the relationships between the endonuclease IV bands and the *Hind* and *Hae* restriction fragments as above, the genetic map and the *Hind*/*Hae* cleavage map can be superimposed as shown in Figure 3. The alignment in the first parts of gene *F*, as above, and gene *G* (Air *et al.*, 1975) is exact to the nucleotide level. The amino acid sequence data for the *F* protein (Air, 1976) and *G* protein (Air *et al.*, 1975, 1976) indicate that the molecular weights estimated by gel electrophoresis in the presence of sodium dodecyl sulphate are close to correct. The total length of the *J* protein is reported to be 37 amino acid residues (Edgell & Hutchison, unpublished data). The rest of the genetic map is not precise, since the protein molecular weights are not reliably known. The molecular weights used by Benbow *et al.* (1974) would require over 20% more DNA than is available, even with no allowance for intergenic sequences. It is known, however, that the origin of replication, in the *A* gene, is within *Hind* fragment 3 (Johnson & Sinsheimer, 1973) and that a major promotor site before gene *D* is in both *Hind* fragment 6 and *Hae*

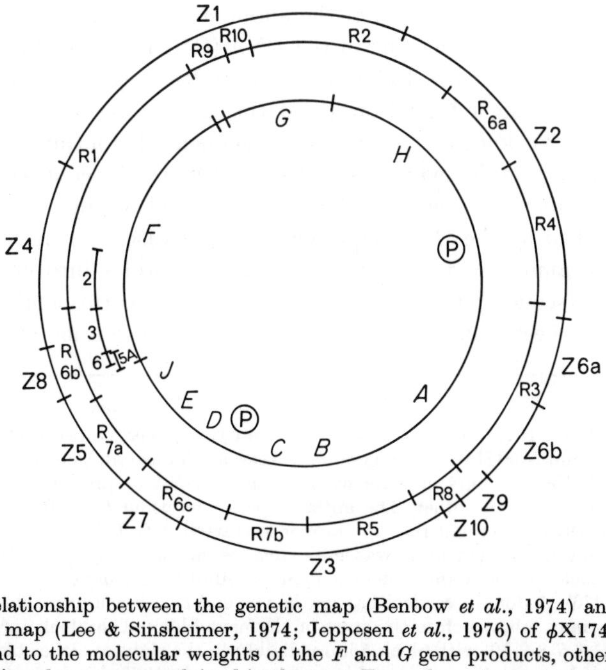

FIG. 3. The relationship between the genetic map (Benbow *et al.*, 1974) and the restriction enzyme cleavage map (Lee & Sinsheimer, 1974; Jeppesen *et al.*, 1976) of ϕX174 DNA. The gene lengths correspond to the molecular weights of the *F* and *G* gene products, otherwise the quantitative alignment is unknown, as explained in the text. From the outermost circle the maps shown are *Hae*III (Z), *Hind* (R), endonuclease IV (2, 3, 6 and 5A), and the genetic map. Promoter sites are indicated P (Chen *et al.*, 1973).

fragment 3 (Chen *et al.*, 1973) and the genetic map in Figure 3 takes these into account. We hope that further DNA and protein sequences will clarify the severe anomalies in the *H-A-B-C-D-E*-region of the map.

(c) *Location of amino acid sequences in the genetic map*

The amino acid sequence data on the gene *F* (capsid) protein is fragmentary (Air, 1976) but, when considered in conjunction with the nucleotide sequences, many overlaps are provided. The overall fitting together of the DNA and protein sequences is shown in Figure 4. As explained above, the 5′ end of endonuclease IV band and

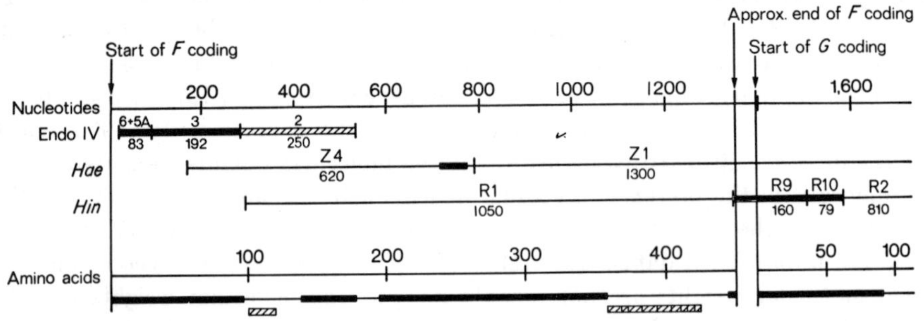

FIG 4. Alignment of the nucleotide and amino acid sequence data for the *F* and *G* genes of ϕX174 and their products. The solid bars indicate sequences completely known; partial sequences are cross-hatched. The junction between *Hind* fragments 9 and 1 is known to be in the *F–G* intergenic region (Sanger & Coulson, 1975; Fiddes, 1976), so there must be some error in the length either of fragment R1 or of the protein.

5A is part of the codon for residue 3 of the amino acid sequence. By combining the results obtained by direct sequencing of the DNA (Galibert *et al.*, 1974; Sedat *et al.*, 1976), transcription (Blackburn, 1976), and using DNA polymerase with *Hind* fragment 1 as the primer (Sanger & Coulson, 1975), the complete sequence of endonuclease IV bands 5A, 6 and 3 is known. The amino acid sequence is continuous to residue 50, but non-overlapped peptide sequences coded by the DNA sequence are available up to residue 91. Thus the overlaps between amino acid residues 50–51, 56–57 and 74–75 are established by the DNA sequence. Endonuclease IV band 3 would end in residue 94 of the amino acid sequence, and 12 nucleotides into endonuclease IV band 2 have been sequenced from this junction, but no corresponding amino acid sequences have yet been found. Sedat *et al.* (1976) have also some very preliminary,

---

FIG. 5—*continued*

pyrimidine tracts (Ling, 1972; Harbers *et al.*, 1976, and this paper). Sites of cleavage by endonuclease IV, *Hind*II and *Hae*III restriction endonucleases are indicated. The amino acid sequences are from Air (1976). Vertical bars indicate where either a DNA or protein sequence was overlapped only by reference to the other. The initial experiment using *Hind* fragment 1 as primer (Sanger & Coulson, 1975) suggested that the sequence started A-G-T-T-G. A subsequent experiment in which the initial polymerisation was less extensive indicated that this sequence should be A-G-T-G. (b) The nucleotide and amino acid sequences found for endonuclease IV band 2 (Sedat *et al.*, 1976; Air, 1976). A double vertical bar marks the end of a block of residues; an unknown number of residues may follow before the start of the next block. The blocks probably occur in the order 1-2-3-4-11-6-7-8-9-10, with block 5 immediately to the right of either block 1 or block 4. In block 8 the G residue at position 7 in Table 4 of Sedat *et al.* (1976) has been deleted, as otherwise the amino acid sequence becomes out of phase. (c) The nucleotide sequence primed by fragment *Hae* 1 and the corresponding amino acid sequence.

Endonuclease IV 5A | Endonuclease IV 6

A-C-T-G-G-C-G-C-C-G-A-G-C-G-T-A-T-G-C-C-G-C-A-T-G-A-C-C-T-T-T-|C-C-C-A-T-C-T-T-G-G-C-T-T-C-C-T-T-

10 20

Ser - Asn - Ile - Gln - Thr - Gly - Ala - Glu - Arg - Met - Pro - His - Asp - Leu - Ser - His - Leu - Gly - Phe - Leu -

Endonuclease IV 5A | Endonuclease IV 6 | Endonuclease IV 3

G-C-T-G-G-T-C-A-G-A-T-T-G-G-T-C-G-T-|C-T-T-A-T-T-A-C-C-A-T-T-T-|C-A-A-C-T-A-C-T-C-C-G-G-T-T-A-T-C-G-C-T-G-G-C-G-A-C-T-C-C-

30 40

Ala - Gly - Gln - Ile - Gly - Arg - Leu - Ile - Thr - Ile - Ser - Thr - Thr - Pro - Val - Ile - Ala - Gly - Asp - Ser -

Hae 8 | *Hae* 4

T-T-C-G-A-G-A-T-G-G-A-C-G-C-C-G-T-T-G-G-C-G-C-T-C-T-C-C-G-T-C-T-T-T-C-T-C-C-A-T-T-G-C-G-T-C-G-T-G-G-|C-C-T-T-G-C-T-A-T-T-

50 60

Phe - Glu - Met - Asp - Ala - Val - Gly - Ala - Leu - Arg | Leu - Ser - Pro - Leu - Arg - Arg | Gly - Leu - Ala - Ile -

G-A-C-T-C-T-A-C-T-G-T-A-G-A-C-A-T-T-T-T-T-A-C-T-T-T-T-T-A-T-G-T-C-C-C-T-C-A-T-C-G-T-C-A-C-G-T-T-T-A-T-G-G-T-G-A-A-C-A-G-

70 80

Asp - Ser - Thr - Val - Asp - Ile - Phe - Thr - Phe - Tyr - Val - Pro - His - Arg | His - Val - Tyr - Gly - Glu - Gln -

Endonuclease IV 3 | Endonuclease IV 2

Hind 6b | *Hind* I

T-G-G-A-T-T-A-A-G-T-T-C-A-T-G-A-A-G-G-A-T-G-G-T-G-T-T-A-A-T-G-C-C-A-C-T-C-C-T-C-T-|C-C-C-G-A-C-T-G-T-T-|A-A

90

Trp - Ile - Lys - Phe - Met - Lys - Asp - Gly - Val - Asn - Ala

(a)

*Hind*6b | *Hind* I

1: C-C-C-G-A-C-T-G-T-T-|A-A ‖ 2: C-C-A-T-G-C-C-G-C-T-T-T-T-C-T-T ‖ 3: -T-T-A-G-C-C-C-T-G-A-T-A-C-C-A-A-T-A-A-A-A-T ‖

4: C-G-T-T-T-G-T-T-T-C-A-G-G- ‖ 11: -T-A-G-C-A-G-C-T-A-T-T-T-T-A-A-G-C-G-C-C-G-T-G- ‖

B1

Met - Pro - Asp - Arg - Thr - Glu -

6: -G-C-T-A-A-C-C-C-T-A-A-T-G-A-G- | 7: -A-T-G-A-T-G-C-T-C-G-T-T-A-T-G-G | 8: -C-C-G-T-T-G-C-T-G-C-C-A-T-C-T |

B10 B20

Ala - Asn - Pro - Asn - Glu - Leu - Asp - Glx - Asx - Asp - Ala - Arg | Tyr - Gly - Phe - Arg | Cys - Cys - His - Leu -

9: C-A-A-A-A- -C-A-T-T-T-G-G-A-C-T-G-C-T | C-C-G-C-T-T-C-C-T-C-C-T- 10: -A-G-A-C-T-G-A-G-C-T-T-T-C-T-C-G-C-C- ‖

B30 B40

Lys - Asx - Ile - Trp - Thr - Ala - Pro - Leu - Pro - Pro - Glx - Thr - Glu - Leu - Ser - Arg

5: C-A-A-A-C-T-A-C-T-G-G-T-T-A-T-A-T-T-G-A ‖

(b)

A-C-C-G-T-C-C-T-T-T-A-C-T-T-G-T-C-A-T-G-C-G-C-T-C-T-A-A-T-C-T-C-T-G-G-G-C-A-T-C-T-G-G-C-T-A-T-G-A-T-G-T-T-G-A-T-G-G-A-A-C-T-G-A-C

C40 C50 C60

Arg - Pro - Leu - Leu - Val - Met - Arg - Ser - Asn - Leu - Trp - Ala - Ser - Gly - Tyr - Asp - Val - Asp - Gly - Thr - Asp

(c)

FIG. 5. Nucleotide and amino acid sequences from gene F and its product. (a) The DNA sequence obtained by direct methods of endonuclease IV bands 5A, 6 and 3 is shown as (------) (Sedat *et al.*, 1976), and those obtained by other methods are indicated as follows: (— — — —) sequences transcribed by RNA polymerase (Blackburn, 1975,1976); (.....) sequences obtained by priming with restriction fragment *Hind* 1 (Sanger & Coulson, 1975); (————) sequences of the poly-

ordered but not overlapped, sequences from endonuclease IV band 2 which total about 250 nucleotides. None of the peptide sequences available can be matched to the first half of this band, but in the second half the nucleotide and amino acid sequences provide mutual overlaps and an almost completely unambiguous sequence can be written for each of them. The details are included in Figure 5. The amino acid sequence data also yielded another continuous sequence of 142 amino acids (C1 to C70, Air, 1976; C71 to C142, unpublished data). To determine its position in the protein an experiment was done in which *Hae* fragment 1 was used as the primer in the plus and minus system (Sanger & Coulson, 1975). The results are shown in Figure 6. A sequence of 65 nucleotides was deduced which corresponded to residues

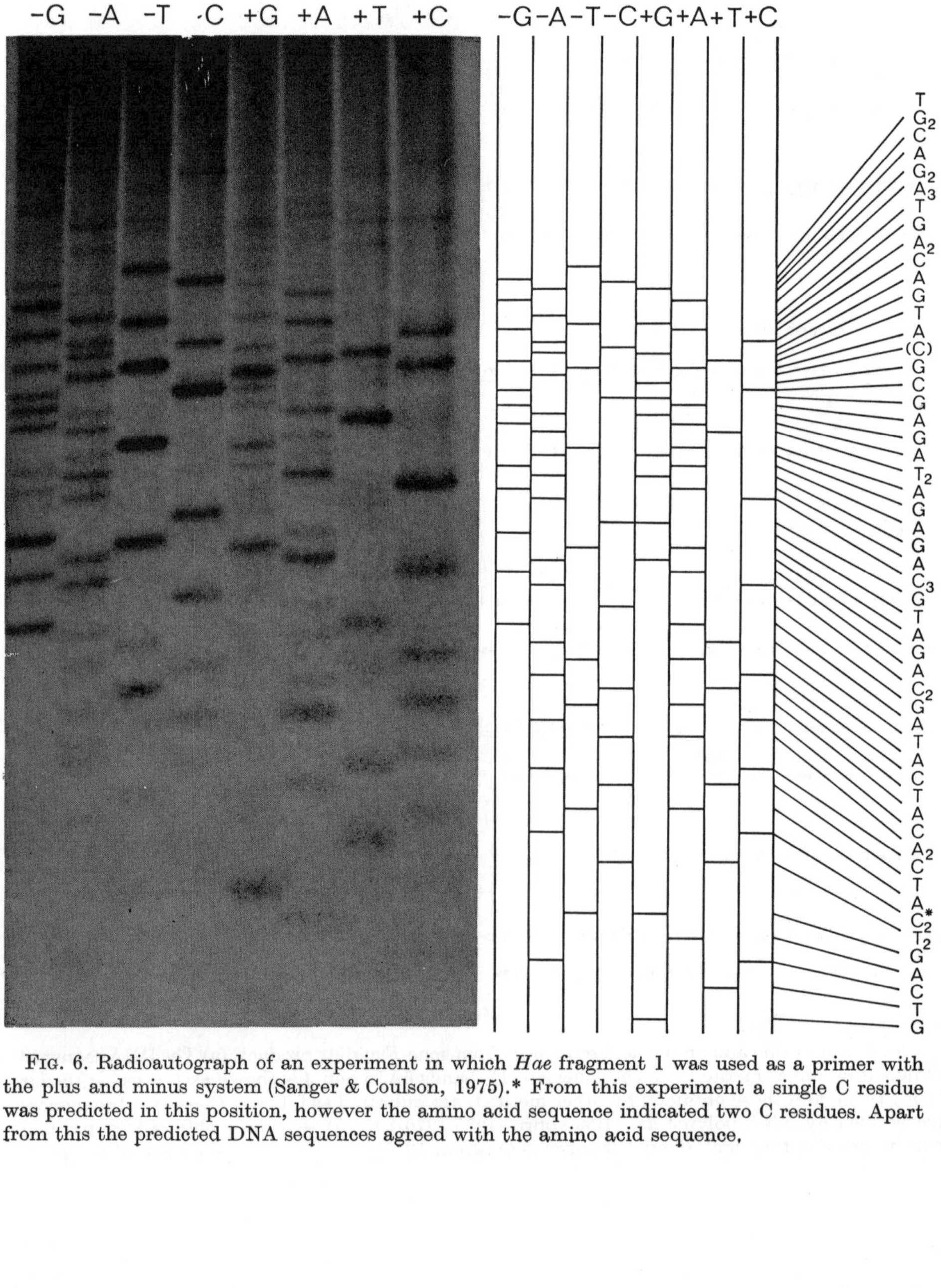

FIG. 6. Radioautograph of an experiment in which *Hae* fragment 1 was used as a primer with the plus and minus system (Sanger & Coulson, 1975).* From this experiment a single C residue was predicted in this position, however the amino acid sequence indicated two C residues. Apart from this the predicted DNA sequences agreed with the amino acid sequence.

C39 to C60 of the amino acid sequence. Thus, allowing for the 15 to 20 nucleotides closest to the *Hae* site that are not seen by this method, the amino acid sequence C1 to C142 is aligned as shown in Figure 4.

Figure 5 shows the correspondence of all nucleotide and amino acid sequence data obtained so far for the *F* gene of ϕX174 and its product. The correlated sequences total 136 codons, including assignment of several Glu, Gln, Asp and Asn residues from the nucleotide sequence that in the amino acid sequence results could only be designated as Glx or Asx (Air, 1976). Amino acid residues 63, 70 to 73, 81, B22 and C50 were given by the DNA sequence as these were not identified, or were uncertain, in the protein sequence work.

3. Discussion

In the determination of these sequences we have used several independent methods which each yielded partial information. The combined results however establish a continuous sequence of nucleotides and we have found in this work, as in the work on gene *G* and its product (Air *et al.*, 1975, 1976), that the combination of several partial results is more satisfactory than trying to obtain a complete result by a single method. This was desirable in the DNA work where copying methods were used, which might have introduced errors, although such errors have not been detected (Blackburn, 1975, 1976; Sanger & Coulson, 1975; Air *et al.*, 1975). In amino acid sequence work, where the methods are long-established and supposedly routine, the work involved in determining the final details can be extremely frustrating. In this project the provision of overlaps by the DNA sequence and being able to locate sequence fragments in the protein, as shown in Figure 4, were of great assistance in obtaining sequences in a relatively short time.

(a) *Secondary structure in* F *gene DNA*

Although we have not carried out a rigorous search for possible hydrogen-bonding into "looped" structures, there is not much secondary structure possible in these DNA sequences. In the endonuclease IV bands there are many runs of T bases but no similar runs of A bases to which they could hydrogen-bond. This absence of secondary structure is compatible with the action of endonuclease IV in excising this region specifically under conditions of maximum hydrogen-bond formation (Ziff *et al.*, 1973; Galibert *et al.*, 1974).

(b) *The coding of the* F *and* G *proteins of ϕX174*

The distribution of the 136 *F* gene codons shown in Figure 5 and the 65 codons previously identified in gene *G* (Air *et al.*, 1975) are shown in Table 3. Although the data are limited, there appears to be a significant preference for certain codons, particularly for those with T in the third position. Table 4 shows the frequency of each base in the third position and confirms that this preference is real: 56% of codons in the *F* gene end in T and 59% of those in gene *G*. The DNA of ϕX174 is in any case rich in T (32·7%), but if it is assumed that the first two positions in each codon are random (i.e. 25% of each base) and that the high T content of the DNA is only due to changes in the third position, the frequency of T in this position would be only 48%. It seems likely therefore that this is a local effect and it suggests that there are regions elsewhere in the DNA with a lower content of T.

TABLE 3

The codons identified in φX174 genes F *and* G

		F	*G*	Total			*F*	*G*	Total
Asp	GAU	1111	111	7	Val	GUU	11111	1111	9
	GAC	111111		6		GUC	11		2
Asn	AAU	111	111	6		GUA	1		1
	AAC	1	1	2		GUG			
Thr	ACU	11111111	111	11	Met	AUG	1111	11	6
	ACC	1		1	Ile	AUU	1111111	11	9
	ACA		111	3		AUC	1		1
	ACG		11	2		AUA			
Ser	UCU	11111	11111	10	Leu	UUA	1	11	3
	UCC	11		2		UUG	1	1	2
	UCA	1	11	3		CUU	111111111	1	10
	UCG		11	2		CUC	111	1	4
	AGU		1	1		CUA			
	AGC					CUG		1	1
Glu	GAA	1		1	Tyr	UAU	1111	1	5
	GAG	1111		4		UAC			
Gln	CAA				Phe	UUU	11	111111	8
	CAG	11	111	5		UUC	111		3
Pro	CCU	11111	11	7	Trp	UGG	111		3
	CCC				His	CAU	1111	1	5
	CCA	1	1	2		CAC	1	1	2
	CCG	111		3	Lys	AAA	1	1	2
Gly	GGU	1111	111	7		AAG	11	1	3
	GGC	111111		6	Arg	CGU	111111111		9
	GGA	1		1		CGC	11	1	3
	GGG					CGA			
Ala	GCU	1111111	111	10		CGG			
	GCC	111		3		AGA			
	GCA	1	1	2		AGG			
	GCG								
Cys	UGU								
	UGC	11	1	3		Total	136	65	201

TABLE 4

The frequency of each base in the third position of codons in the F *and* G *genes of φX174 and for comparison in the RNA phages MS2, R17, f2 and Qβ*

	φX174 Number		φX174 %		MS2, R17, f2, Qβ
	F	*G*	*F*	*G*	%
T	76	38	55·9	58·5	29·1
C	33	5	24·3	7·7	26·9
A	8	10	5·9	15·4	23·6
G	19	12	14·0	18·4	20·4
Total	136	65			

The data used for the MS2, R17, f2 and Qβ codons were taken from Barrell & Clark (1974) and totalled 461 codons.

The reason for the high T content is not known. It is similarly high in other single-stranded DNA phages including the filamentous viruses fd and f1. Previous studies on RNA phages (Adams *et al.*, 1969; Steitz, 1969; Min Jou *et al.*, 1972; Contreras *et al.*, 1972,1973; Jeppesen *et al.*, 1972; Haegeman & Fiers, 1973; Sanger, 1971) show no significant preference for any particular codons, in contrast to the results with ϕX174. In studies on frameshift mutants in the lysozyme gene of bacteriophage T4, in which the T (and A) content is 32·5%, a somewhat similar phenomenon is observed. Of 19 codons, seven had U and seven had A as the third base (Ocada *et al.*, 1970).

It is possible that the cause is chemical rather than biological, i.e. that there are chemical reasons for changes towards T residues to occur in these phages and, if there is no selective preference for the third position of codons, T residues would accumulate. It seems more probable however that the effect has been brought about by natural selection and is of some advantage to the organism. Either it is an advantage to have an overall high T content and changes have occurred in the variable third position to accommodate this, or it is an advantage to have T in the third position and the overall analysis is merely a result of this. In the former case the effect may be concerned with the physical constraints of the DNA and may involve the packaging into the virion, though both the filamentous and the spherical phages, in which the packaging must be very different, have a high T content. There are however other stages in the life-cycle of the organism that could equally well be affected by the physical properties of the DNA.

Denhardt & Marvin (1969) have suggested that the high T content of the single-stranded DNA phages might be explained on the basis of an alteration in the mechanism of translation on infection such that codons ending in T residues are more readily recognised than other codons. A change in the "wobble" pairing might, for instance, account for this. However, this would predict a low frequency of C in the third position, which is observed in gene G codons but not in gene F. In the F gene codons A is noticeably low in the third position, and this is true also when F and G gene codons are averaged.

In these coding sequences the out-of-phase terminator codons are not randomly distributed. They do not appear to be associated with internal ATG triplets. In spite of similarities in some of these internal ATG sequences to initiation ATG sequences (e.g. Robertson *et al.*, 1973) as noted in phage MS2 RNA sequences (Haegeman & Fiers, 1973), there is no evidence that false initiations would be corrected by rapid termination.

REFERENCES

Adams, J. M., Jeppesen, P. G. N., Sanger, F. & Barrell, B. G. (1969). *Nature (London)*, **223**, 1009–1014.

Air, G. M. (1976). *J. Mol. Biol.*

Air, G. M., Blackburn, E. H., Sanger, F. & Coulson, A. R. (1975). *J. Mol. Biol.* **96**, 703–719.

Air, G. M., Sanger, F. & Coulson, A. R. (1976). *J. Mol. Biol.* (in the press).

Barrell, B. G. & Clark, B. F. C. (1974). *Handbook of Nucleic Acid Sequences*, Joynson-Bruvvers Ltd., Oxford.

Benbow, R. M., Mayol, R. F., Picchi, J. C. & Sinsheimer, R. L. (1972). *J. Virol.* **10**, 99–114.

Benbow, R. M., Zuccarelli, A. J., Davis, G. C. & Sinsheimer, R. L. (1974). *J. Virol.* **13**, 898–907.

Blackburn, E. H. (1975). *J. Mol. Biol.* **93**, 367–374.

Blackburn, E. H. (1976). *J. Mol. Biol.* **107**, 417–431.

Burton, K. & Petersen, G. B. (1960). *Biochem. J.* **75**, 17–27.

Chen, C.-Y., Hutchison, C. A. & Edgell, M. H. (1973). *Nature New Biol.* **243**, 233–236.

Contreras, R., Vandenburghe, A., Volkaert, G., Min Jou, W. & Fiers, W. (1972). *FEBS Letters*, **24**, 339–342.

Contreras, R., Ysebaert, M., Min Jou, W. & Fiers, W. (1973). *Nature New Biol.* **241**, 99–101.

Denhardt, D. T. & Marvin, D. A. (1969). *Nature (London)*, **221**, 769–770.

Edgell, M. H., Hutchison, C. A. & Sclair, M. (1972). *J. Virol.* **9**, 574–582.

Fiddes, J. C. (1976). *J. Mol. Biol.* **107,** 1–24.

Galibert, F., Sedat, J. & Ziff, E. B. (1974). *J. Mol. Biol.* **87**, 377–407.

Haegeman, G. & Fiers, W. (1973). *Eur. J. Biochem.* **36**, 135–143.

Harbers, B., Delaney, A. D., Harbers, K. & Spencer, J. H. (1976). *Biochemistry*, **15**, 407–414.

Jeppesen, P. G. N., Barrell, B. G., Sanger, F. & Coulson, A. R. (1972). *Biochem. J.* **128**, 993–1006.

Jeppesen, P. G. N., Sanders, L. & Slocombe, P. M. (1976). *Nucleic Acids Res.* **3,** 1329–1339.

Johnson, P. H. & Sinsheimer, R. L. (1973). *Fed. Proc. Fed. Amer. Soc. Exp. Biol.* **32**, 491.

Lee, A. S. & Sinsheimer, R. L. (1974). *Proc. Nat. Acad. Sci., U.S.A.* **71**, 2882–2886.

Ling, V. (1972). *Proc. Nat. Acad. Sci., U.S.A.* **69**, 742–746.

Middleton, J. H., Edgell, M. H. & Hutchison, C. A. (1972). *J. Virol.* **10**, 42–50.

Min Jou, W., Haegeman, G., Ysebaert, M. & Fiers, W. (1972). *Nature (London)*, **237**, 82–88.

Ocada, Y., Amagase, S. & Tsugita, A. (1970). *J. Mol. Biol.* **54**, 219–246.

Robertson, H. D., Barrell, B. G., Weith, H. L. & Donelson, J. E. (1973). *Nature New Biol.* **241**, 38–40.

Sanger, F. (1971). *Biochem. J.* **124**, 833–843.

Sanger, F. & Coulson, A. R. (1975). *J. Mol. Biol.* **94**, 441–448.

Sanger, F., Donelson, J. E., Coulson, A. R., Kössel, H. & Fischer, D. (1974). *J. Mol. Biol.* **90**, 315–333.

Sedat, J., Ziff, E. B. & Galibert, F. (1976). *J. Mol. Biol.* **107,** 391–416.

Smith, H. O. & Wilcox, K. W. (1970). *J. Mol. Biol.* **51**, 379–391.

Steitz, J. A. (1969). *Nature (London)*, **224**, 957–964.

Ziff, E. B., Sedat, J. W. & Galibert, F. (1973). *Nature New Biol.* **241**, 34–37.

J. Mol. Biol. (1976) **108**, 519–533

Nucleotide and Amino Acid Sequences of Gene *G* of ϕX174

G. M. Air, F. Sanger and A. R. Coulson

Medical Research Council Laboratory of Molecular Biology
Hills Road, Cambridge CB2 2QH, England

(Received 16 July 1976)

The nucleotide sequence of the coding region of gene *G* of ϕX174 and the amino acid sequence of the *G*-coded "spike" protein of the virion have now been completed. From the 5′ A of the initiating ATG to the 3′ end of the terminator triplet, the gene consists of 528 nucleotides and codes for a protein of 175 amino acids, molecular weight 19,053.

1. Introduction

In studying ribosome binding to ϕX174 single-stranded DNA, Robertson *et al.* (1973) isolated and sequenced a ribosome-protected DNA fragment which proved to contain the initiation region for translation of a spike protein of the virion (M_r 19,000) which is coded by gene *G* (Air & Bridgen, 1973). The DNA and protein sequences were extended (Donelson *et al.*, 1975; Air *et al.*, 1975), and we have now obtained the complete nucleotide and amino acid sequences of gene *G* (528 nucleotides) and its product (175 amino acids). As in earlier work (Air *et al.*, 1975) neither the DNA nor the protein sequence was determined to absolute certainty, but any doubtful residues of one sequence have been confirmed by reference to the other. Since the emphasis has been to develop reliable methods of sequencing DNA, the nucleotide sequence has been the primary aim, with enough amino acid sequence work to make sure that the sequence was correct.

In the previous work (Robertson *et al.*, 1973; Donelson *et al.*, 1975; Air *et al.*, 1975) several methods of DNA sequencing have been employed, including incorporation of ^{32}P-labelled deoxyribonucleotide triphosphates using a synthetic primer and DNA polymerase with the single-stranded viral DNA as template. The rapidly expanding knowledge of restriction enzymes and their sites has meant that almost any region of ϕX174 DNA contains enough known restriction sites to isolate fragments suitable for sequence determination by priming methods, with no necessity for synthesising oligonucleotide primers. In the DNA sequencing described here, all the 331 residues which complete the gene *G* sequence were determined using the "plus and minus" method (Sanger & Coulson, 1975) with viral strand ϕX174 DNA as template and restriction fragments produced by the enzymes from *Haemophilus influenzae* R_f (*Hinf* I, Middleton, Stankus, Edgell, Hutchison & Barrell, unpublished data), *H. aphrophilus* (*Hap* II, Takanami, 1973), *H. haemolyticus* (*Hha* I, Roberts, unpublished data) as primers. The primers were selected by reference to the physical map of ϕX174 (Jeppesen *et al.*, 1976), part of which is shown in Figure 1. This restriction fragment map has been aligned with the genetic map in the F-G regions as previously reported (Air *et al.*, 1975,1976).

The amino acid sequence of the protein coded by gene *G* has been determined by the standard methods of trypsin, chymotrypsin and CNBr digestion and isolating and sequencing the peptides produced, except that the DNA sequence has been used to overlap peptide sequences and to confirm uncertain residues.

2. Materials and Methods

Single-stranded viral DNA was prepared from φX174 *am3cs*70 (lysis defective). Replicative form φX DNA was prepared by B. G. Barrell and J. C. Fiddes by the method of Godson & Vapnek (1973) and some was a gift from C. A. Hutchison. Restriction enzyme fragments were prepared and mapped by various members of this laboratory (N. L. Brown, B. G. Barrell, J. C. Fiddes, P. M. Slocombe). They were usually prepared by fractionating the digests on gradient polyacrylamide gels (Jeppesen, 1974) and eluting electrophoretically (Galibert *et al.*, 1974).

^{32}P-labelled deoxyribonucleotide triphosphates were obtained from New England Nuclear at a specific activity of about 100 mCi/μmol, and were used as soon after receipt as possible. It appeared that less satisfactory results were obtained with older triphosphate preparations and this was not only due to loss of specific activity.

T4 DNA polymerase was prepared by the method of Goulian *et al.* (1968). For the minus reactions we have used a mutant form of DNA polymerase I which lacks the 5′ exonuclease activity (Heijneker *et al.*, 1973).

All test tubes and reactivials used in the DNA work were siliconized before use. Those for the Edman and dansyl reactions were acid-washed.

For isolation of the phage coat proteins, cells from two 300-litre cultures of *Escherichia coli* C infected with φX174 *am3cs*70 were obtained from the Microbiological Research Establishment, Porton Down, Wiltshire, England.

Trypsin (TPCK-treated), chymotrypsin and carboxypeptidases A and B were from Worthington.

Phenylisothiocyanate, trifluoroacetic acid and butyl acetate were sequenal grade (Pierce).

(a) *DNA sequencing*

The plus and minus method used for the DNA sequencing was essentially as described by Sanger & Coulson (1975) with slight modifications. The following detailed description applies to the experiments using fragment *Hap* 5 as primer (Figs 6 and 7). The techniques used in the other experiments were very similar, the most significant difference being that restriction enzyme digestion was originally done after the plus and minus polymerase reactions. In recent experiments we have found it preferable and simpler to do it before.

After isolation of the restriction enzyme fragments they were dissolved in water such that 1 μl contained the product derived from 1 μg φX RF. 10 μl of this solution was mixed with 2 μl single-stranded φX DNA (1·2 μg), sealed in a capillary tube and heated at 100°C for 3 min. After cooling and opening the capillary, 2 μl H × 10 buffer (66 mM-Tris·HCl (pH 7·4), 66 mM-$MgCl_2$, 10 mM-dithiothreitol, 0·5 M-NaCl) and 6 μl water were added. The capillary was resealed and incubated for 4 h at 67°C. The annealed material was then added to 20 μl of a solution containing 10 μCi [^{32}P]dATP, 0·1 mM-dGTP, dCTP and TTP in H buffer. After cooling to 0°C, 2 μl DNA polymerase 1 (Boehringer, grade 1) were added. After 2 min, approximately one-third of the reaction mixture was removed and added to 5 μl 0·2 M-EDTA. One-third samples were again removed after 8 and 30 min, combined with the first sample, 25 μl phenol added, and thoroughly mixed in a 0·5-ml reactivial. This was centrifuged, the aqueous layer sucked off with a drawn-out capillary tube and washed twice with 2 ml ether to remove excess phenol. Excess ether was then evaporated by a current of air and the solution applied to a column of Sephadex G100 (Pharmacia) in a 1-ml disposable plastic serological pipette (Falcon), which was made up and run in 5 mM-Tris·HCl (pH 7·4) 0·1 mM-EDTA. These columns run much faster than the agarose columns originally used (Sanger & Coulson, 1975) and the results appear to be more satisfactory. However, care must be taken to avoid them drying out. The fractionation was followed using a hand radiation monitor. The synthesized DNA was eluted in the void volume and clearly separated from the excess triphosphate. It was collected in about 200 μl and taken to dryness in a desiccator, dissolved in 20 μl H buffer and digested with the appropriate restriction enzyme. The amount of enzyme used was sufficient to digest 3 μg DNA completely under the conditions used (usually 1 h at 37°C). In the case of *Hap* fragment 5, the enzyme *Hpa* II was used, which has the same specificity (C-C-G-G) as the *Hap* enzyme.

After digestion the solution was diluted to a suitable volume (usually about 50 μl) and 2-μl samples used in the plus and minus reactions. The yield of radioactive DNA varies considerably between experiments. In general this solution should be diluted to contain about 3 nCi/μl.

Plus mixes are 0·2 mM solutions of a single deoxyribotriphosphate in 1·5 × H buffer. Minus mixes are 0·01 mM solutions of the appropriate three triphosphates in 1·5 × H buffer.

The reaction mixtures contained 2 μl of the DNA reaction product, 2 μl of the appropriate "mix" and 1 μl DNA polymerase in a drawn-out capillary. Minus reactions were carried out at 0°C for 30 min with approximately 0·8 units of the mutant DNA polymerase (Heijneker *et al.*, 1973) and the plus reactions for 30 min at 37°C with T4 DNA polymerase (approx. 0·02 unit). At the end of the reaction 1 μl 0·2 M-EDTA and 25 μl freshly deionized formamide containing 0·03% xylene cyanol FF and 0·03% bromophenol blue were added. The solutions were heated at 100°C for 3 min and then loaded onto the polyacrylamide gel. After the electrophoresis the gel was soaked for 10 to 30 min in 10% acetic acid to fix the oligonucleotides and prevent diffusion of the bands. It was then soaked in water for 1 to 2 min before covering with Saran Wrap for autoradiography.

(b) *Preparation of polyacrylamide gels*

The plus and minus reaction products were fractionated on 12% gels in 8 M-urea at pH 8·3. To facilitate production of gels, a Perspex box with removable front (see Fig. 2) was made to hold six 20 cm × 40 cm × 1·5 mm slab gel templates, each sealed at the edges with silicone grease and waterproof tape. The gels were poured from the bottom *via* a raised funnel, with the slot formers in place. (The slots, 9 or 13 per gel, were the full thickness of the gel, i.e. 1·5 mm.)

The buffer system was Tris·borate/EDTA (pH 8·3) (TBE) (Peacock & Dingman, 1967). 10 × TBE was made by dissolving 108 g Tris base, 55 g boric acid and 9·3 g EDTA in water to 1 l.

The gel solution was made as follows. 720 g urea (Schwarz-Mann, ultra pure), 180 g acrylamide and 7·5 g bis-acrylamide (both BDH, "purified for electrophoresis") were made up to approximately 1·25 l with water and dissolved with slight warming. This solution was stirred for 30 min with approximately 50 g Amberlite MB-1 resin (BDH), which was removed by filtration through Whatman no. 1 filter paper in a glass sinter funnel. 150 ml of 10 × TBE was added, the solution made up to 1·5 l and degassed. 5 ml of 10% ammonium persulphate and 0·5 ml *N,N,N′,N′*-tetraethylmethylenediamine were added prior to pouring. The gels were removed from the box after 1 to 2 h, the slot formers removed and excess acrylamide washed off. The gels can be stored for at least 2 weeks immersed in 8 M-urea/TBE.

Prior to use about 0·5 cm of gel is removed to allow insertion of the Whatman 3MM paper wick, and the 8 M-urea/TBE storage buffer is washed out and replaced by TBE. It has been found unnecessary to have any urea in the buffer compartments of the electrophoresis apparatus.

With a gradient chamber in place of the funnel the above apparatus has been used for making gradient gels.

(c) *Protein sequencing*

The purification of the gene *G* protein and the methods used to separate and sequence the peptides produced by CNBr, trypsin and chymotrypsin were as described in previous papers (Air *et al.*, 1975; Air, 1976). The amounts of protein used to determine the *G* protein sequence were 6·5 mg for tryptic digestion, 20 mg for chymotryptic digestion and ~30 mg for CNBr cleavage. The peptides eluted from paper after electrophoresis and chromatography were freeze-dried and taken up in 200 μl water. 20 μl were taken for acid hydrolysis and amino acid analysis, and 150 μl for sequencing by the manual dansyl-Edman method as detailed previously (Air *et al.*, 1975; Air, 1976). Carboxypeptidase A and B digestions were at an enzyme : substrate ratio of 1 : 50 in 1% ammonium bicarbonate at room temperature. Portions were taken at 0, 10 and 30 min, immediately frozen in a solid CO_2/acetone bath, and lyophilised. The dry samples were taken up in sample loading buffer, centrifuged, and the supernatants analysed on the Durrum amino acid analyser.

3. Results

(a) *DNA sequence*

The only method used to study the DNA sequence was the plus and minus method using restriction enzyme fragments as primers (Sanger & Coulson, 1975). Figure 1 shows the sites of splitting of the gene *G* region with restriction enzymes based on the results of Edgell *et al.* (1972), Lee & Sinsheimer (1974) and Jeppesen *et al.* (1976).

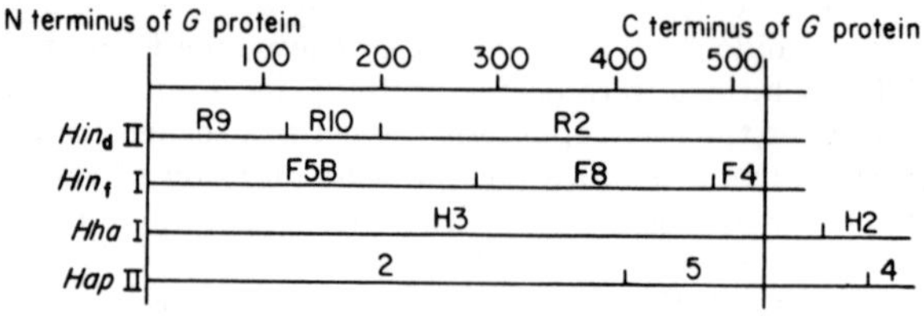

FIG. 1. The sites of restriction enzyme cleavage in the gene *G* region of φX174 as mapped by Edgell *et al.* (1972), Lee & Sinsheimer (1974), Jeppesen *et al.* (1976). The enzymes shown are *Hind* II, *Hinf* I, *Hha* I, *Hap* II, and their exact sites of cleavage are included in Fig. 10.

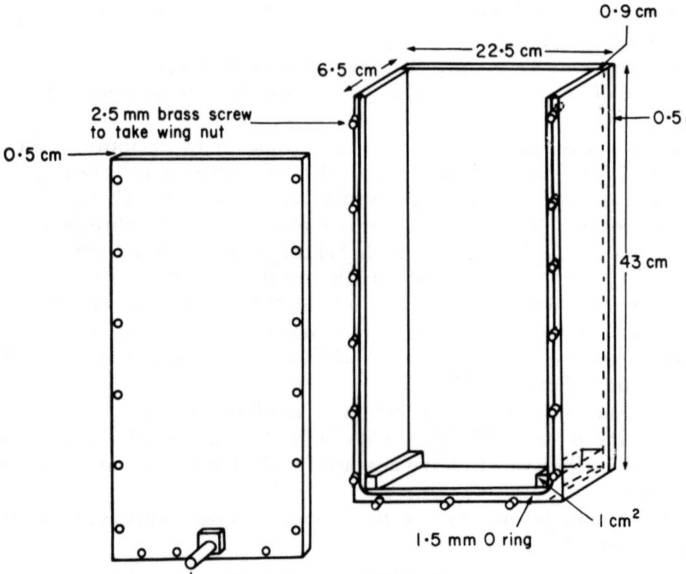

FIG. 2. Apparatus for production of 6 acrylamide gels.

Fragments *Hha* 2, *Hinf* 4, *Hap* 5 and *Hinf* 8 have been used as primers and some results are shown in Figures 4 to 8 which are radioautographs of the acrylamide gels used in the plus and minus method. Using the four primers it was possible to obtain data covering the 331 nucleotide sequences of the *G* gene that had not been sequenced in the previous paper (Air *et al.*, 1975). It will be seen that the quality of the experiments varies considerably. Thus the radioautographs for *Hha* 2 (Fig. 4) or *Hap* 5 (Fig. 6) are clear and relatively easy to read, whereas that for *Hinf* 4 is very much harder with a number of artefact bands and bands missing. In general several experiments were done with each primer until the data were sufficiently good to establish the DNA sequence together with the amino acid sequence data. The sequences finally deduced in this way are shown in the Figures.

The 3′ end of the minus strand of fragment *Hha* 2 is about 55 residues from the DNA coding for the C terminus of the *G* protein. An initial priming with this fragment suggested a sequence whose complement is shown in Figure 3. At this time the C-terminal sequence of the *G* protein had not been identified, but a number of tryptic peptides had been sequenced whose position in the protein was unknown. One of these had the sequence shown in Figure 3. The last six amino acids corresponded by the genetic code to the DNA sequence and since this was followed by the termination codon T-G-A it suggested that the peptide was the C-terminus of the protein. This was confirmed by further protein work as described below.

Nucleotide sequence deduced in preliminary experiment	A-A-A-G-G-G-A-T-T-A-T- -T-G-T-C-T-C-C-A-G-C-C-A-C-T-T-A-A-G-T-G-A
Amino acid sequence of a tryptic peptide	Glx - Ile - Ile - Cys - Leu - Glx - Pro - Leu - Lys
Corrected DNA sequence	——A——T——

FIG. 3. Identification of the termination codon of the *G* gene.

In the experiment shown in Figure 4 the acrylamide gel electrophoresis was run for a relatively long time so that the oligonucleotides in the bands at the bottom of the gel are about 60 nucleotides long. The oligonucleotide in position 85 is approximately 145 residues long. Nevertheless, bands in this area are separated and the sequences can be read with reasonable accuracy. This radioautograph is relatively dark which facilitates the reading. There are, however, a few artefacts. The most noticeable is in position 4 which contains a band in the —T system. The main reason for identifying this as an artefact is that the next oligonucleotide gives a band in the +C system. Where there are runs of a particular nucleotide in a sequence one occasionally sees bands in the minus system corresponding to each component of the run but more often only one band is seen which corresponds to the residue before the run and this is usually stronger than any subsequent bands. Thus the presence of two bands in consecutive oligonucleotides in the —T system in positions 4 and 5 is further evidence that one of them is an artefact. This same artefact was found in most other experiments in which fragment *Hha* 2 was used as a primer so that the possibility of a T residue instead of a C residue in this position could not be excluded. Nor could it be excluded by the amino acid sequence. However, when fragment *Hap* 4, which primes about 40 residues beyond the *Hha* 2 priming site, was used this artefact was completely absent. Other artefacts which could be more easily eliminated were found in the +C system in positions 14 and 41. The exact reason for these occasional artefacts

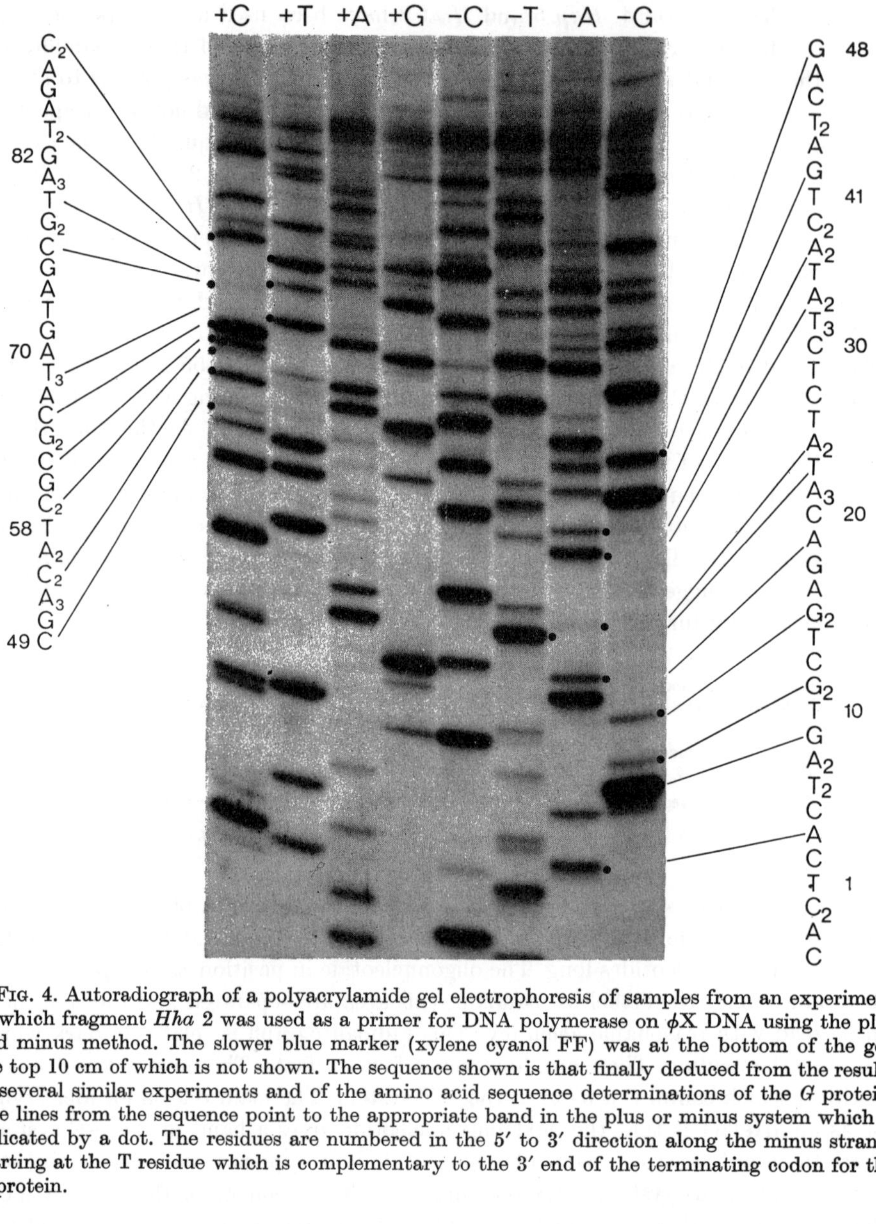

FIG. 4. Autoradiograph of a polyacrylamide gel electrophoresis of samples from an experiment in which fragment *Hha* 2 was used as a primer for DNA polymerase on ϕX DNA using the plus and minus method. The slower blue marker (xylene cyanol FF) was at the bottom of the gel, the top 10 cm of which is not shown. The sequence shown is that finally deduced from the results of several similar experiments and of the amino acid sequence determinations of the *G* protein. The lines from the sequence point to the appropriate band in the plus or minus system which is indicated by a dot. The residues are numbered in the 5′ to 3′ direction along the minus strand, starting at the T residue which is complementary to the 3′ end of the terminating codon for the *G* protein.

is unknown. They vary considerably from one experiment to another but some, such as that in position 4, are extremely persistent. There were considerably more when *Hinf* 4 was used as primer (Fig. 5), e.g. in positions 94 (+C) and 96 (+T).

Positions 42 to 46 give the sequence G-A-T-T-C which is a specificity site for *Hinf* (Hutchison & Barrell, unpublished data) and represents the junction between *Hinf* fragments 4 and 8. Fragment *Hinf* 4 was therefore used as a primer to extend the sequence (Fig. 5). By comparing the pattern of bands in Figure 5 with those at the top of Figure 4 it is clear that the same sequences are involved, thus establishing an overlap between the two sets of data. Similar overlaps can be seen between other

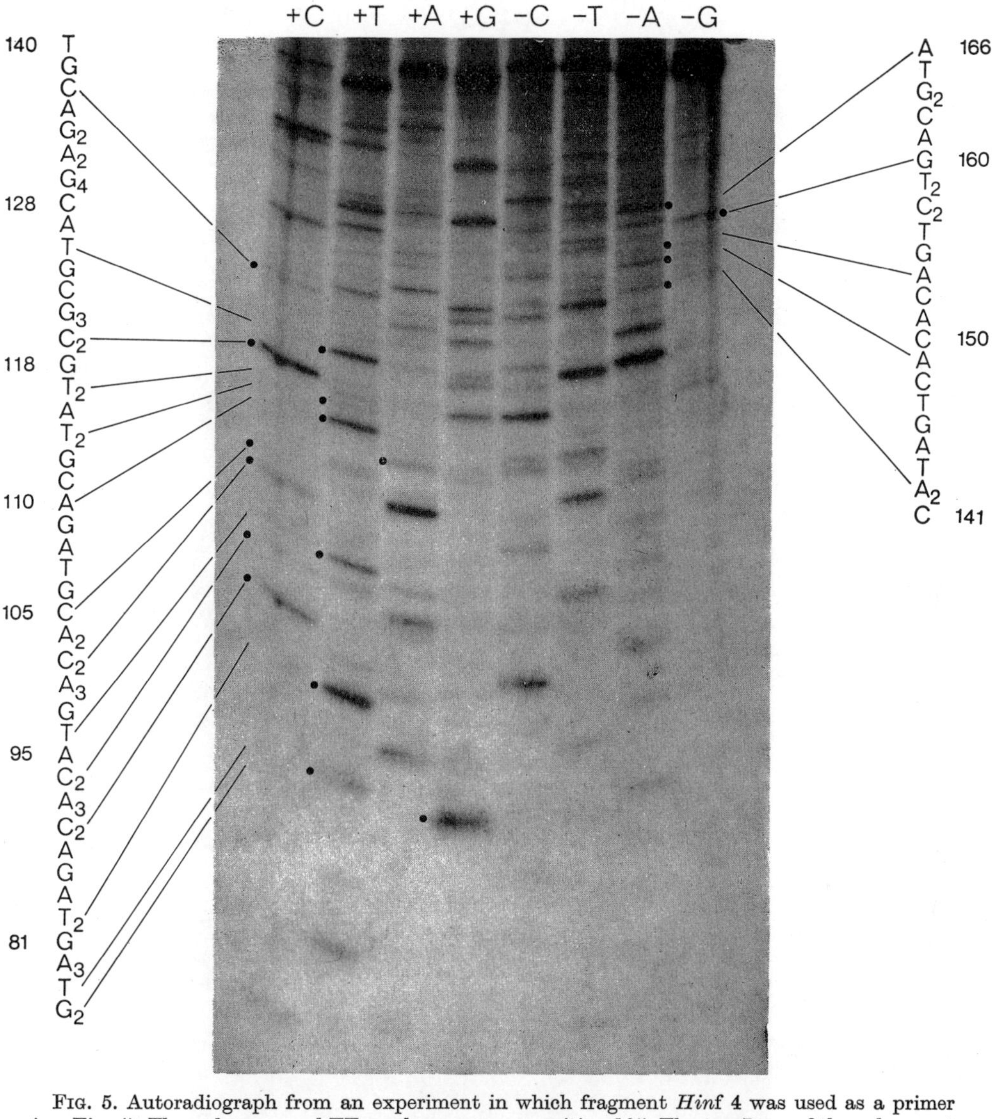

FIG. 5. Autoradiograph from an experiment in which fragment *Hinf* 4 was used as a primer (see Fig. 4). The xylene cyanol FF marker was near position 105. The top 7 cm of the gel are not shown. The results in the −G system in this experiment were unsatisfactory and were not used. The electrophoresis has run rather unevenly so that the bands lie on curves, especially at the edges. This curvature is taken into account in marking the positions of the sequences.

primed sequences. For instance, the pattern at the bottom of Figure 8 matches that at the top of Figure 6. It may be seen that the pattern of bands in any one area of a gel represents a "fingerprint" or characterization of a specific sequence and it is thus possible to use these patterns to obtain overlaps and as an alternative though somewhat more laborious method of determining a restriction enzyme fragmentation map.

The next restriction site is the *Hap* site (C-C-G-G) in position 119 to 122. Fragment *Hap* 5 was used to prime from this site (Fig. 6). Although the results in the +A system were unsatisfactory in the experiment shown, the sequence is relatively easy to read off. Positions 236 to 243 are an alternating sequence of A and G residues, in

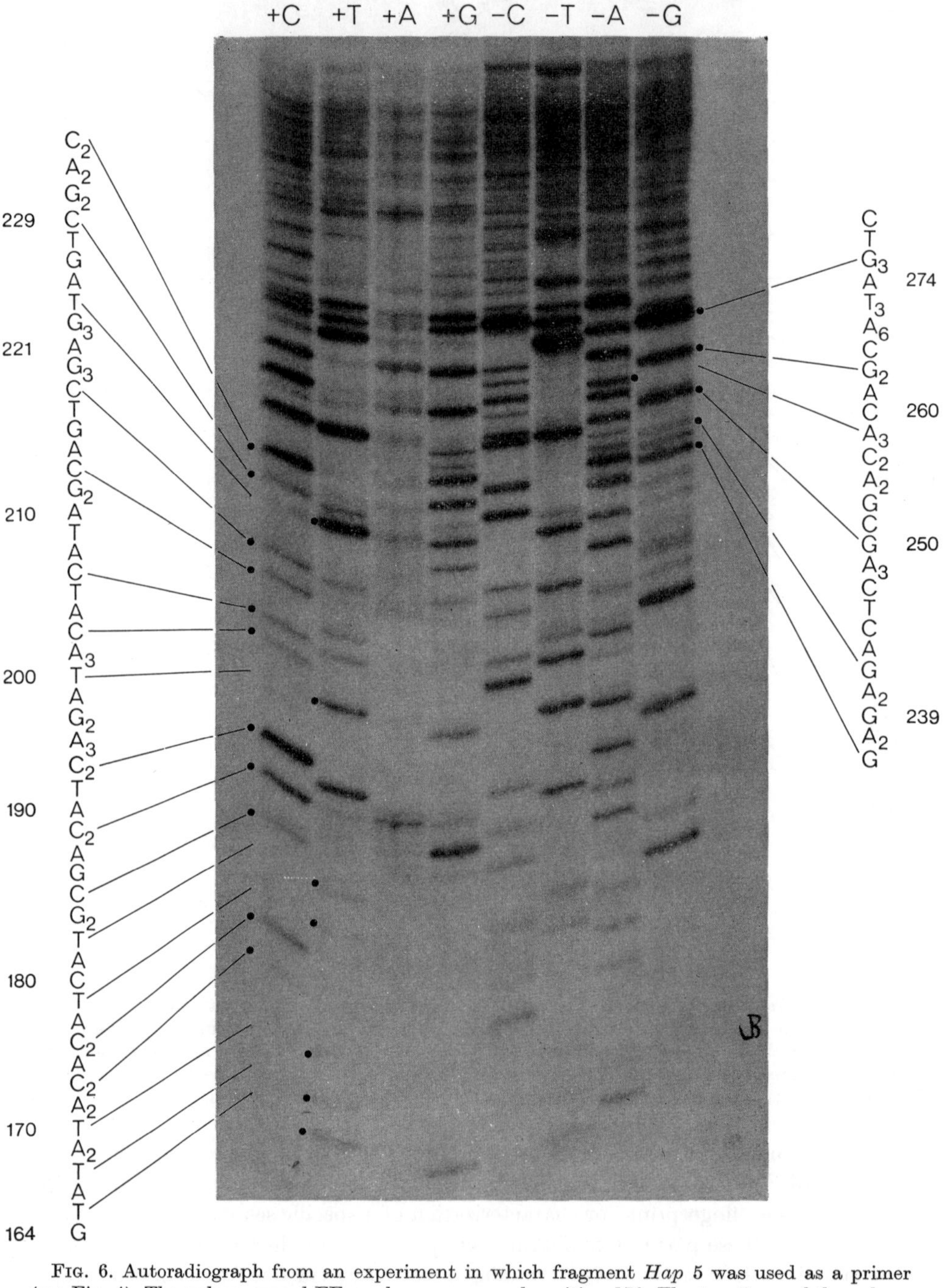

FIG. 6. Autoradiograph from an experiment in which fragment *Hap* 5 was used as a primer (see Fig. 4). The xylene cyanol FF marker was around position 170. The top 5 cm of the gel are not shown. The results with the +A system were unsatisfactory in this experiment.

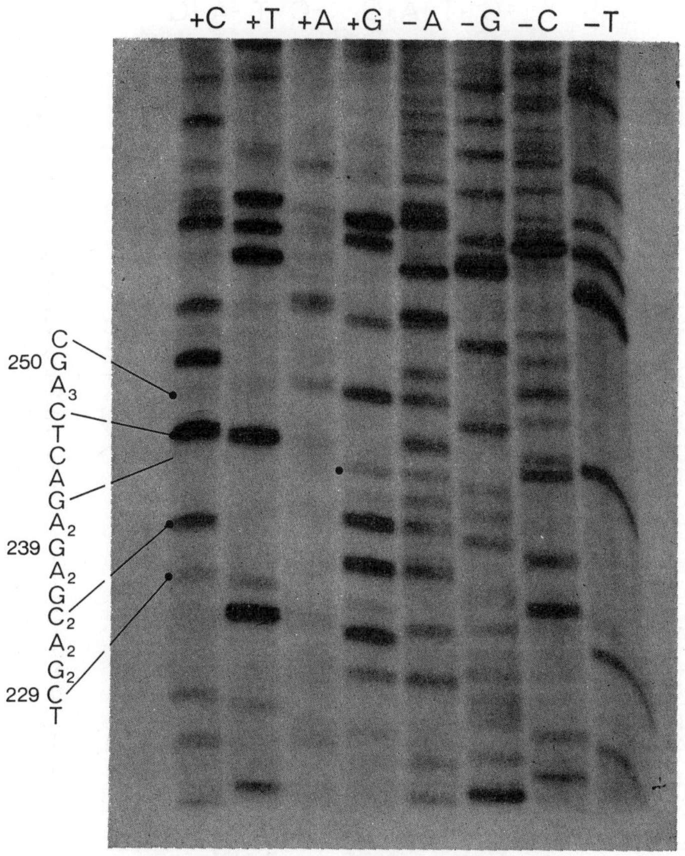

FIG. 7. The result of an experiment with fragment *Hap* 5 as in Fig. 6, except that the polyacrylamide gel electrophoresis was carried out for twice as long. The top 19 cm of the gel are not shown.

which there was some doubt of the relative frequency of each residue, and this could not be fully resolved by the amino acid sequence. Figure 7 shows an experiment similar to that shown in Figure 6 except that the electrophoresis was run for twice as long. The bands in the —A and —G systems are now further spread out and it is relatively simple to deduce the sequence. The corresponding pyrimidine tract has been identified in the plus strand by Harbers *et al.* (1976).

Figure 8 shows an experiment primed by fragment *Hin*f 8 from the site G-A-C-T-C at position 242 to 246. The sequence extends into the sequence primed by fragment *Hind* 2 from the G-T-T-G-A-C at position 328 to 333, which was reported in a previous paper (Air *et al.*, 1975).

(b) *Amino acid sequence*

After tryptic digestion of the *G* protein only seven soluble peptides were obtained. These were purified directly by paper electrophoresis and chromatography, detected with fluorescamine, eluted and sequenced as far as possible.

As would be expected, chymotryptic digestion gave a much more complex mixture of peptides, and a preliminary fractionation on Sephadex G50 in 1% ammonium

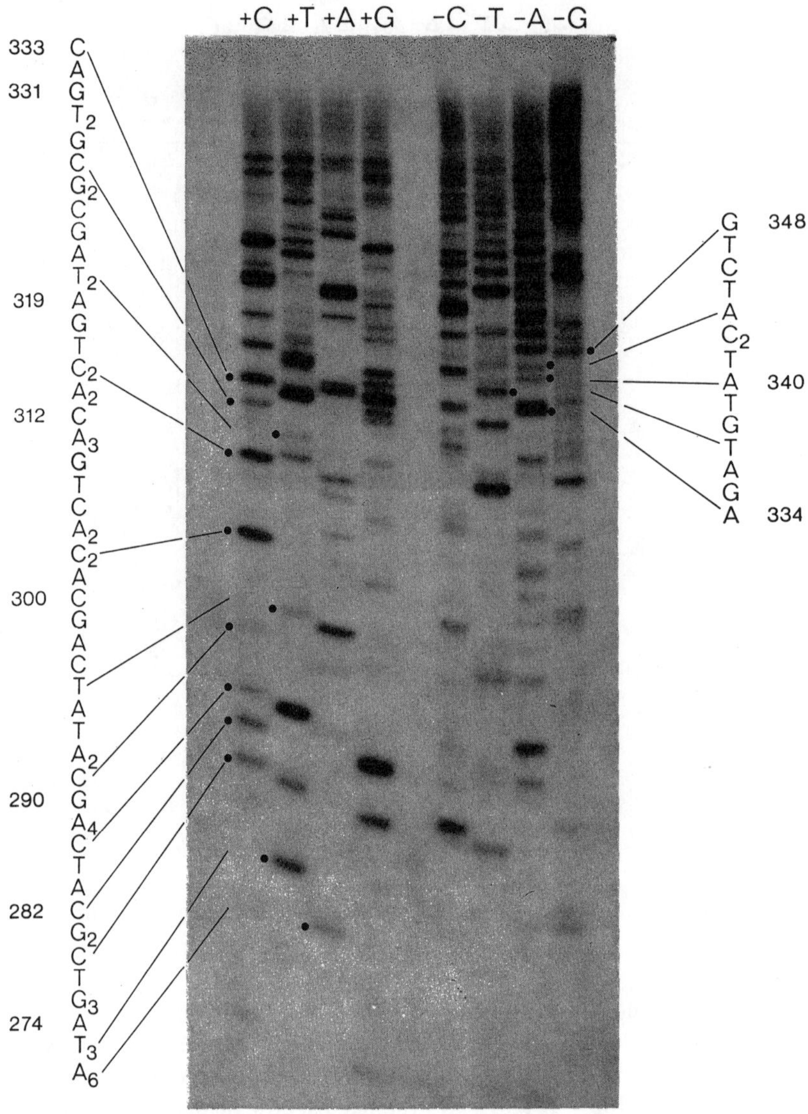

FIG. 8. Autoradiograph from an experiment in which fragment *Hinf* 8 was used as a primer (see Fig. 4). The xylene cyanol FF marker was at position 295. The top 2 cm and bottom 8 cm of the gel are not shown.

bicarbonate was carried out. Four fractions were pooled and the peptides in each resolved by paper electrophoresis and chromatography. It was not found necessary to attempt to purify any insoluble tryptic or chymotryptic peptides, particularly because the nucleotide sequence could be used to order the peptides and check that no amino acid residues were missing.

Amino acid analysis of the *G* protein showed three methionine residues, and since one is N-terminal treatment with CNBr should give three fragments. The CNBr digest was fractionated firstly on Sephadex G75 in 50% formic acid. This separated a small (30 residues) fragment later shown to be C-terminal, but the N-terminal fragment

(61 and 62 residues due to partial cleavage of the N-terminal methionine residue) and the middle fragment (83 residues) were not resolved. The latter was purified by chromatography on a column of Whatman DE52 equilibrated and loaded in 8 M-urea, 10 mM-Tris, 50 mM-mercaptoethanol (pH 8·0) with a linear gradient to 0·6 M-KCl. Tryptic digestion of the CNBr fragments yielded some peptides not seen after trypsin treatment of the whole protein (residues 116 to 124, 135 to 145, 157 to 166).

The C-terminus of this protein was not easy to characterise. All the tryptic or chymotryptic peptides isolated contained an amino acid of appropriate specificity at the C-terminus, so none could be assigned to the carboxyl end of the protein. Resolution of homoserine or homoserine lactone on the older amino acid analysers was erratic, and its absence from hydrolysates of a CNBr fragment was not a reliable indication that it contained the C-terminus of the protein. Digestion of the protein, which as purified is very insoluble, with carboxypeptidases A and B did not give a convincing result, and after solubilization by succinylation of α- and ε-amino groups (Klotz, 1967) no amino acids were released by carboxypeptidases.

As described above, in the nucleotide sequence of the gene *G* region the tryptic peptide Glx-Ile-Ile-Cys-Leu-Glx-Pro-Leu-Lys could be seen, followed by a termination codon, just where the *G* protein sequence would end if the molecular weight estimation from gel electrophoresis in sodium dodecyl sulphate (19,000) is correct. To confirm this C-terminal sequence the small CNBr fragment (residues 146 to 175) was repurified. Amino acid analysis on the Durrum analyser showed that it did not contain homoserine, therefore it is the C-terminal fragment. Treatment of the CNBr fragment with carboxypeptidases A and B released only lysine, as would be expected from a -Pro-Leu-Lys sequence. The sequence of this CNBr fragment had been

Residues 1 to 67: Air *et al.* (1975)

	65 70 75
CNBr peptides	-Met | Asx-Thr-Ser-Val-Asx-Ala-Ala-Asx-Glx-Val-Val-Ser-Val-
Tryptic peptides	
Chymotryptic peptides	-Asn | Ala-Ala-Asx-Glx (Val, Val, Ser, Val,
	80 85 90 95 100
CNBr peptides	Gly-Ala-Asx-Ile- -Phe-Asx
Tryptic peptides	| Phe-Phe-Ala-Cys-Leu-Val-Arg | Phe-Glx-Ser-Ser-Ser-Val-Pro-
Chymotryptic peptides	Gly, Ala, Asx, Ile, Ala) Phe ¦ Asp-A- -Asp-Pro (Lys, Phe, Phe) | | Val-Arg-Phe ¦ Glu-Ser-Ser-Ser-Val-Pro-
	105 110 115 120 125
CNBr peptides	
Tryptic peptides	Thr-Thr-Leu (Pro, Thr, Ala, Tyr, Asx, Val, Tyr, Pro, Leu, Asx, Gly) Arg | His-Asp-Gly-Gly-Tyr (Tyr, Thr, Val) Lys | Asp-
Chymotryptic peptides	Thr-Thr-Leu-Pro-Thr-Ala-Tyr | Asp-Val-Tyr-Pro-Leu ¦ Asp-Gly-Arg-His-Asp-Gly-Gly-Tyr-Tyr | Thr-Val-Lys-Asx-
	130 135 140 145 150
CNBr peptides	| Val- -Ser-Asx-Phe-
Tryptic peptides	Cys-Val-Thr-Ile-Asp-Val-Leu-Pro-Arg | Thr-Pro-Gly-Asx-Asx-Val-Tyr-Val-Gly-Phe-Met |
Chymotryptic peptides	Cys-Val-Thr-Ile-Asx-Val-Leu (Pro, Arg) Thr (Pro, Gly) Asx (Asx, Val, Tyr) | Val-Gly-Phe |
	155 160 165 170 175
CNBr peptides	Thr-Ala-Thr-Lys-Cys-Arg-Gly-Leu-Val-Ser-Leu-Asx-Glx-Val-Ile-Lys (Glx, Ile, Ile, Cys, Leu, Glx, Pro, Leu, Lys)
Tryptic peptides	| Cys-Arg | Gly (Leu, Val, Ser, Leu, Asn, Gln, Val, Ile) Lys | Glx-Ile-Ile-Cys-Leu-Glx-Pro-Leu-Lys |
Chymotryptic peptides	| Thr-Ala-Thr-Lys-Cys-Arg-Gly-Leu ¦ Val-Ser-Leu |

FIG. 9. Amino acid sequences of the peptides used to deduce the sequence of residues 67 to 175 of the protein coded by gene *G*. The sites of cleavage of enzymes or CNBr are indicated by vertical bars; broken bars indicate partial cleavages. Residues in parentheses were present in the amino acid composition but the sequence was not determined. The C-terminus was confirmed as described in the text.

determined by the dansyl-Edman method as far as the lysine at residue 166. The amino acid composition of the fragment showed that only the tryptic peptide Glx-Ile-Ile-Cys-Leu-Glx-Pro-Leu-Lys could be added to the sequence; this was confirmed by tryptic digestion of the fragment when this was the only peptide not contained in the known sequence of the CNBr fragment.

The amino acid sequence data used to determine the sequence of the first 67 amino acids of this protein were given previously (Air *et al.*, 1975), and in Figure 9 these data are continued for the remaining 108 residues. Although the peptide data has given overlaps all through this region many of them are dangerously short and could not be relied upon. The nucleotide sequence has provided an unambiguous confirmation of the peptide order.

The nucleotide sequence gave a Trp residue at position 147 which had not been detected in the CNBr fragment, presumably because it was destroyed during the CNBr reaction. Several amide residues were assigned by the DNA sequence, where protein sequencing gave only "Glx" or "Asx" (residues 70, 71, 78, 138, 139, 149, 167, 172) and some doubtful residues were confirmed by the DNA sequence.

The "spike" protein coded for by gene *G* of ϕX174 DNA is 175 amino acid residues in length, with a calculated molecular weight of 19,053, in good agreement with the estimate from sodium dodecyl sulphate gel electrophoresis.

4. Discussion

The results obtained by nucleotide and amino acid sequencing are combined in Figure 10, which shows the nucleotide sequence of gene *G* of ϕX174 from the 5′ A of the initiating A-T-G codon to the 3′ end of the termination codon (528 nucleotides) and the amino acid sequence of its product (175 amino acids). Also included in Figure 10 are the sites of cleavage by restriction enzymes *Hind* II, *Alu* I, *Hinf* I and *Hap* II. Four sequences previously identified as depurination products (Ling, 1972; Harbers *et al.*, 1976) are present, three at positions 129 to 137, 236 to 243, 357 to 364 of the DNA sequence and one in the position corresponding to residues 13 to 16 of the amino acid sequence.

As noted previously in ϕX174 DNA protein coding sequences (Air *et al.*, 1975, 1976), there is a preponderance of T residues in the third position of the codons. Of the 175 codons of gene *G*, T occurs as the third base in 95 of them (54·3%). The frequency of T at the first and second positions of the codons, which is dependent on (or dictates) the amino acid sequence is 25·7% and 31·4%, respectively, and in the overall composition of gene *G* 37·1%, so we are still expecting to find regions in ϕX174 DNA of low T content to bring the total base composition to the known value of 32·7% T.

A search was made for possible base-pairing between sequences within the gene but there appeared to be no significant sequences longer than would be expected for a random sequence (A. McLachlan, unpublished data). This is in contrast to the results with the RNA bacteriophages where extensive hairpin loops are found.

The protein coded by gene *G* is located in the 12 spikes of the virion, the molar amount present giving five molecules of *G* protein per spike (Burgess, 1969). Electron micrographs (J. T. Finch, J. Robertus, A. Crowther, unpublished data) show a pentamer of globular subunits which may be the individual *G* protein molecules. The rules of Chou & Fasman (1974) have been applied to the amino acid sequence

10 20
Met - Phe - Gln - Thr - Phe - Ile - Ser - Arg - His - Asn - Ser - Asn - Phe - Phe - Ser - Asp - Lys - Leu - Val - Leu -
A-T-G-T-T-T-C-A-G-A-C-T-T-T-T-A-T-T-T-C-T-C-G-C-C-A-C-A-A-T-T-C-A-A-A-C-T-T-T-T-T-T-T-C-T-G-A-T-A-A-G-C-T-G-G-T-T-C-T-C-
Alu 2 ↑ *Alu* 16

30 40
Thr - Ser - Val - Thr - Pro - Ala - Ser - Ser - Ala - Pro - Val - Leu - Gln - Thr - Pro - Lys - Ala - Thr - Ser - Ser -
A-C-T-T-C-T-G-T-T-A-C-T-C-C-A-G-C-T-T-C-T-T-C-G-G-C-A-C-C-T-G-T-T-T-T-A-C-A-G-A-C-A-C-C-T-A-A-A-G-C-T-A-C-A-T-C-G-T-C-A-
Alu 16 ↑ *Alu* 15a *Alu* 15a ↑ *Alu* 3 *Hind* 9 ↑ *Hind* 10

50 60
Thr - Leu - Tyr - Phe - Asp - Ser - Leu - Thr - Val - Asn - Ala - Gly - Asn - Gly - Gly - Phe - Leu - His - Cys - Ile -
A-C-G-T-T-A-T-A-T-T-T-T-G-A-T-A-G-T-T-T-G-A-C-G-G-T-T-A-A-T-G-C-T-G-G-T-A-A-T-G-G-T-G-G-T-T-T-T-C-T-T-C-A-T-T-G-C-A-T-T-
360 350

70 80
Gln - Met - Asp - Thr - Ser - Val - Asn - Ala - Ala - Asn - Gln - Val - Val - Ser - Val - Gly - Ala - Asp - Ile - Ala -
C-A-G-A-T-G-G-A-T-A-C-A-T-C-T-G-T-C-A-A-C-G-C-C-G-C-T-A-A-T-C-A-G-G-T-T-G-T-T-T-C-A-G-T-T-G-G-T-G-C-T-G-A-T-A-T-T-G-C-T-
340 330 320 310 300 290
Hind 10 ↑ *Hind* 2

90 100
Phe - Asp - Ala - Asp - Pro - Lys - Phe - Phe - Ala - Cys - Leu - Val - Arg - Phe - Glu - Ser - Ser - Ser - Val - Pro -
T-T-T-G-A-T-G-C-C-G-A-C-C-C-T-A-A-A-T-T-T-T-T-T-G-C-C-T-G-T-T-T-G-G-T-T-C-G-C-T-T-T-G-A-G-T-C-T-T-C-T-T-C-G-G-T-T-C-C-G-
280 270 260 250 240 230
Hinf 5b ↑ *Hinf* 8

110 120
Thr - Thr - Leu - Pro - Thr - Ala - Tyr - Asp - Val - Tyr - Pro - Leu - Asp - Gly - Arg - His - Asp - Gly - Gly - Tyr -
A-C-T-A-C-C-C-T-C-C-C-G-A-C-T-G-C-C-T-A-T-G-A-T-G-T-T-T-A-T-C-C-T-T-T-G-G-A-T-G-G-T-C-G-C-C-A-T-G-A-T-G-G-T-G-G-T-T-A-T-
220 210 200 190 180 170

130 140
Tyr - Thr - Val - Lys - Asp - Cys - Val - Thr - Ile - Asp - Val - Leu - Pro - Arg - Thr - Pro - Gly - Asn - Asn - Val -
T-A-T-A-C-C-G-T-C-A-A-G-G-A-C-T-G-T-G-T-G-A-C-T-A-T-T-G-A-C-G-T-C-C-T-T-C-C-C-C-G-T-A-C-G-C-C-G-G-C-A-A-T-A-A-C-G-T-C-
160 150 140 130 120 110
Hap 2 ↑ *Hap* 5

150 160
Tyr - Val - Gly - Phe - Met - Val - Trp - Ser - Asn - Phe - Thr - Ala - Thr - Lys - Cys - Arg - Gly - Leu - Val - Ser -
T-A-C-G-T-T-G-G-T-T-T-C-A-T-G-G-T-T-T-G-G-T-C-T-A-A-C-T-T-T-A-C-C-G-C-T-A-C-T-A-A-A-T-G-C-C-G-C-G-G-A-T-T-G-G-T-T-T-C-G-
100 90 80 70 60 50

170
Leu - Asn - Gln - Val - Ile - Lys - Glu - Ile - Ile - Cys - Leu - Gln - Pro - Leu - Lys
C-T-G-A-A-T-C-A-G-G-T-T-A-T-T-A-A-A-G-A-G-A-T-T-A-T-T-T-G-T-C-T-C-C-A-G-C-C-A-C-T-T-A-A-G-T-G-A
40 30 20 10
Hinf 8 ↑ *Hinf* 4

FIG. 10. The nucleotide sequence of gene *G* of ϕX174 and the amino acid sequence of its product. The nucleotide sequence is given as the DNA sequence of the viral (plus) strand, which is the complement of the minus strand sequence obtained in Figs 4 to 8. Restriction sites of *Hind* II, *Hinf*, *Hap* I and *Alu* I are indicated. The numbering of the amino acid residues from the N-terminus is shown above the amino acid sequence and of the nucleotide residues are shown below the corresponding sequence. They are numbered in the 3′ to 5′ direction starting from the last residue of the termination codon and correspond to the numbering used in Figs 4 to 8.

of the *G* protein and predict many, often alternating, regions of α-helix and β-sheet, consistent with a globular conformation.

Of the results described in this paper the 108 amino acids of the protein sequence were identified from 163 successful thin-layer chromatograms, whereas the 331 residues of DNA sequence could essentially have been deduced from four 12% acrylamide–urea gels, though in fact duplicate experiments were carried out. As in studies on the *F* protein of ϕX174 (Air, 1976), manual methods of amino acid sequencing have been used almost exclusively. The only method found to separate the phage coat proteins has been gel filtration in 4 M-guanidine hydrochloride or 8 M-urea, which works very well but is expensive and messy to scale up. Hence it was judged preferable to make small amounts of protein, isolate peptides rapidly by paper electrophoresis and chromatography, and sequence them manually rather than purifying the much larger amounts of peptides necessary for automated sequencing.

The basic reason why the simple plus and minus method of sequencing DNA cannot be adapted for proteins is the same reason that nucleotide sequencing was once considered intrinsically more difficult than protein sequencing, namely that the four nucleotides have a great deal of similarity and can be made to behave essentially identically as regards solubility, electrophoretic mobility, susceptibility to both polymerising and degradative enzymes, and so on. Amino acids, on the other hand, have vastly different properties. Peptides do not behave consistently as residues are added or removed, and rates of exo- or endoprotease reactions are unpredictable until the whole sequence is known. Hence the impossibility of determining the C-terminal sequence of this protein (-Pro-Leu-Lys) with carboxypeptidases and the existence of chymotryptic peptides containing three phenylalanine residues (residues 67 to 88) or three tyrosines (residues 108 to 121). Insolubility remains a serious problem in protein sequence work: if the DNA sequence had not been available for confirming overlaps attempts would have been made to separate the insoluble tryptic peptides, which contain 40% of the sequence, and there is no general method known to purify small, insoluble peptides. This solubility problem is absent in nucleotide sequencing work; all fragments from a restriction enzyme digest can be seen on an appropriate acrylamide gel in equimolar proportions so that bands containing unresolved fragments are immediately recognised as double the intensity of their neighbours. Peptides may be present in varying yields due to partial cleavage or insolubility, and since any detection methods are dependent to some extent on the amino acid composition the major products cannot be readily distinguished from the many minor ones.

The plus and minus method used to sequence the DNA of gene *G* is rapid and straightforward, but some anomalies may arise, as detailed in Results, and as yet we have not had complete confidence that very long sequences will not have errors. Sequences can be confirmed using the other well-established methods of DNA sequencing (Robertson *et al.*, 1973; Galibert *et al.*, 1974; Sanger *et al.*, 1973) or protein coding regions can be compared to the amino acid sequence (Air *et al.*, 1975,1976). As previously discussed, most uncertainties in a DNA sequence obtained by the plus–minus method are in determining the lengths of runs of a single nucleotide. Since the amino acid sequence gives the phase of the coding sequence it has been very useful in checking that the DNA sequence is of the correct length. However, as more experience is gained with the plus and minus method we are finding that this checking is becoming less necessary since most run lengths can be resolved by running the gel longer, as in Figure 7, or priming with another restriction fragment.

We wish to thank other members of this laboratory (B. G. Barrell, N. L. Brown, J. C. Fiddes, C. A. Hutchison, P. M. Slocombe and M. Smith) for help, advice and the preparation of materials, H. L. Heijneker for a gift of the mutant DNA polymerase I, R. Kamen for a gift of T4 DNA polymerase, and J. H. Spencer for making his results available to us before their publication.

REFERENCES

Air, G. M. (1976). *J. Mol. Bicl.* **107**, 433–434.

Air, G. M. & Bridgen, J. (1973). *Nature New Biol.* **241**, 40–41.

Air, G. M., Blackburn, E. H., Sanger, F. & Coulson, A. R. (1975). *J. Mol. Biol.* **96**, 703–719.

Air, G. M., Blackburn, E. H., Coulson, A. R., Galibest, F., Sanger, F., Sedat, J. W. & Ziff, E. B. (1976). *J. Mol. Biol.* **107**, 445–458.
Burgess, A. B. (1969). *Proc. Nat. Acad. Sci., U.S.A.* **64**, 613–617.
Chou, P. Y. & Fasman, G. D. (1974). *Biochemistry*, **13**, 222–245.
Donelson, J. E., Barrell, B. G., Weith, H. L., Kössel, H. & Schott, H. (1975). *Eur. J. Biochem.* **58**, 383–395.
Edgell, M. H., Hutchison, C. A. & Sclair, M. (1972). *J. Virol.* **9**, 574–582.
Galibert, F., Sedat, J. & Ziff, E. B. (1974). *J. Mol. Biol.* **87**, 377–407.
Godson, G. N. & Vapnek, D. (1973). *Biochem. Biophys. Acta*, **299**, 516–520.
Goulian, M., Lucas, Z. J. & Kornberg, A. (1968). *J. Biol. Chem.* **243**, 627–638.
Harbers, B., Delaney, A. D., Harbers, K. & Spencer, J. H. (1976). *Biochemistry*, **15**, 407–414.
Heijneker, H. L., Ellens, D. J., Tjeerde, R. H., Glickman, B. W., van Dorp, B. & Pouwels, P. H. (1973). *Mol. Gen. Genet.* **124**, 83–96.
Jeppesen, P. G. N. (1974). *Anal. Biochem.* **58**, 195–207.
Jeppesen, P. G. N., Sanders, L. & Slocombe, P. M. (1976). *Nucl. Acids Res.* **3**, 1323–1339.
Klotz, I. M. (1967). In *Methods in Enzymology* (Hirs, C. H. W., ed.), vol. 11, pp. 576–540, Academic Press, New York and London.
Lee, A. S. & Sinsheimer, R. L. (1974). *Proc. Nat. Acad. Sci., U.S.A.* **71**, 2882–2886.
Ling, V. (1972). *Proc. Nat. Acad. Sci., U.S.A.* **69**, 742–746.
Peacock, A. C. & Dingman, C. W. (1967). *Biochemistry*, **7**, 668–674.
Robertson, H. D., Barrell, B. G., Weith, H. L. & Donelson, J. E. (1973). *Nature New Biol.* **241**, 38–40.
Sanger, F. & Coulson, A. R. (1975). *J. Mol. Biol.* **94**, 441–448.
Sanger, F., Donelson, J. E., Coulson, A. R., Kössel, H. & Fischer, D. (1973). *Proc. Nat. Acad. Sci., U.S.A.* **70**, 1209–1213.
Takanami, M. (1973). *FEBS Letters*, **34**, 318–322.

(Reprinted from Nature, Vol. 264, No. 5581, pp. 34–41, November 4, 1976)

Overlapping genes in bacteriophage φX174

B. G. Barrell, G. M. Air & C. A. Hutchison III*

MRC Laboratory of Molecular Biology, Hills Road, Cambridge CB2 2QH, UK

Bacteriophage φX174 genes D and E are translated from the same DNA sequence but in different reading frames.

BACTERIOPHAGE φX174 contains a single strand of DNA approximately 5,400 nucleotides in length, which has been shown to be of the same sense as the mRNA and to code for nine cistron products. As shown in Table 1, there is an apparent discrepancy between the length of the DNA in the genome and the amount required to code separately for the proteins; significantly more DNA would be needed to code for proteins of the accepted molecular weights than is available. An almost complete sequence of φX DNA has been obtained in this laboratory using the "plus and minus" technique of Sanger and Coulson[1], confirming that the length of the DNA is not more than 5,400 nucleotides (G. M. A., B. G. B., N. L. Brown, A. R. Coulson, J. C. Fiddes, C. A. H., F. Sanger, P. M. Slocombe and M. Smith, in preparation). At the same time, a study of the amino acid sequences of the major φX coded proteins has enabled the boundaries of genes *D*, *J*, *F*, *G* and *H* to be located on the restriction map of φX DNA (Fig. 1). Table 1 shows that the molecular weights of the F, G, J and H proteins have been found to be close to correct, while the D protein molecular weight is somewhat larger than the sodium dodecyl sulphate (SDS) gel estimate. It has not yet been possible to determine directly the boundaries of genes *A*, *B* and *C* in the remaining region between gene *H* and gene *D*. The order of genes *D* to *H* is in agreement with the genetic map of φX reported by Benbow *et al.*[4] except that they located gene *E* between *D* and *J*. Here we report the DNA sequence from the promoter preceding gene *D* through to gene *J*. We find that gene *E* amber mutations lie within the DNA sequence which codes for the gene *D* protein, and interpret this to mean that genes *D* and *E* overlap on the DNA sequence. Sequences of gene *E* amber mutations show that *D* and *E* are translated in two reading frames from a common DNA sequence.

*Present address: Department of Bacteriology and Immunology and Curriculum in Genetics, University of North Carolina, Chapel Hill, North Carolina 27514

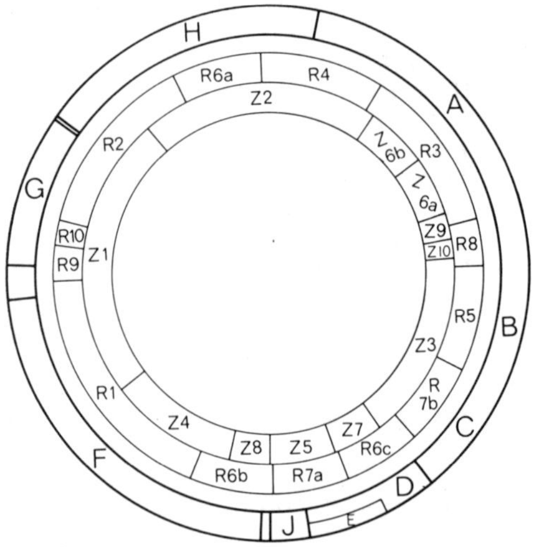

Fig. 1 A map of φX174 DNA. The boundaries of genes *D*, *E*, *J*, *F*, *G* and *H* are shown aligned with the restriction maps of *Hae*III (fragments designated Z) and *Hin*dII (fragments designated R)[2,3]. The precise location of genes *A*, *A′*, *B* and *C* have not yet been determined.

It should be emphasised that genes *D* and *E* seem perfectly independent by normal genetic criteria. The *D* protein is produced in large amounts in the infected cell. It is necessary

Table 1 φX174 coding capacity

Gene	Protein molecular weight from SDS gels*	Number of nucleotides: From protein molecular weight estimates	Number of nucleotides: From nucleotide sequence	Protein molecular weight from sequence information†
A	62,000	1,701		
(A′)	35,000	(960)	⩾1,851	67,400
B	19,000–25,000	⩾522		
C	7,000	192		
D	14,500	399	456	16,811‡
E	10,000–17,500	⩾273	(273)	9,940
J	5,000	138	114	4,097‡
F	48,000	1,317	1,287	46,700
G	19,000	522	525	19,053‡
H	37,000	1,014	981	35,500
Total		⩾6,078	5,374§	
Length of φX174 DNA		≃5,370		

*There is considerable variation in reported molecular weight estimates of φX174 coded proteins. The values given are those most frequently recorded, or recently reinvestigated[4,22,28–33].

†Protein molecular weights are calculated from the number of nucleotides using the formula:

$$\text{Protein molecular weight} = \text{No. of nucleotides}/(3 \times 0.00915)$$

‡These values are calculated from the amino acid sequences (Freymeyer *et al.*, in preparation; ref. 35 and this paper).

§This figure includes the 160 nucleotides of the intercistronic regions so far determined.

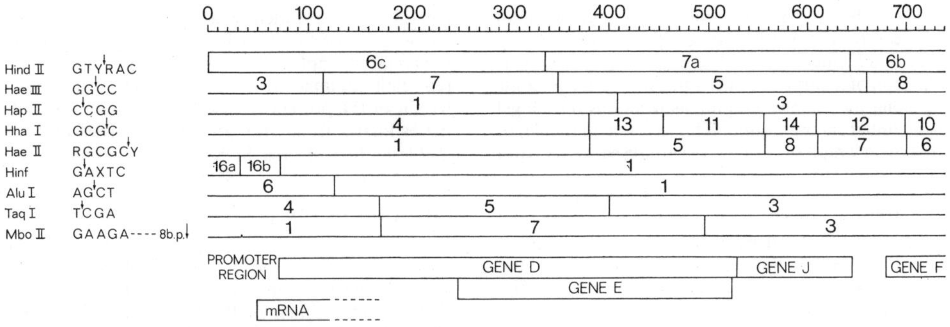

Fig. 2 Restriction map of the gene *D* region. The cleavage sites are shown of *Hin*dII, *Hae*III (refs 2 and 3), *Hap*II (ref. 7), *Hha*I, *Hae*II, *Hinf*, *Alu*I (ref. 3), *Taq*I (S. Sato, C. A. H., and J. I. Harris, in preparation), *Mbo*II (N. L. Brown, C. A. H., and M. Smith, in preparation).

for the production of viral single-stranded DNA, but its precise role in this process is unclear and there is no evidence that it is directly involved in DNA synthesis[5,6]. Gene *E* is responsible for lysis of the host cell. All nonsense mutants in gene *E* characterised so far produce completely functional full size gene *D* protein. In fact, the protein chemistry of the gene *D* protein which is reported in this paper has been carried out using material produced by an amber mutant in gene *E* (*am*3). Similarly, normal gene *E* function has been observed in the gene *D* nonsense mutants.

Nucleotide sequence analysis of the gene *D* region

The cleavage sites of ten restriction enzymes of different specificities have been mapped on φX174 RF I (double stranded) DNA. The cleavage sites of these enzymes in the gene *D* region are shown in Fig. 2.

Using the restriction fragments shown in Fig. 2 as primers with either the (+) or (−) strands of φX DNA as a template, sequences of approximately 100 residues on either side of all the restriction enzyme sites in this region were determined using the plus and minus technique of Sanger and Coulson[1]. Figure 3 shows a good example of this method using fragment *Taq* 4 as a primer on φX (−) strand DNA cleaved with *Mbo*II. A long and a short fractionation of the same reactions is shown, from which an almost unambiguous sequence of at least 126 residues can be deduced, starting 14 residues from the *Mbo*II 1/7 site.

In this way a large number of overlapping sequences from each restriction site and from both strands were obtained which was more than enough to deduce the entire sequence of the gene *D* region. Although the method is very reliable these overlapping sequences overcame some problems encountered with the plus and minus technique. One disadvantage is that the first 15 residues from the restriction site are not normally seen on the acrylamide gels. This is probably due to the instability of short products on the template which are subsequently degraded by the exonuclease activity of the DNA polymerase present during the restriction enzyme digest of the plus and minus reactions. Another problem encountered is unequal "jump sizes" on the gel between products differing by one residue in length. These occurred in specific regions of the sequence, probably due to local secondary structure at the end of products which were not completely denatured during gel electrophoresis. Electrophoresis at high current (high temperature) gives much improved separation. This

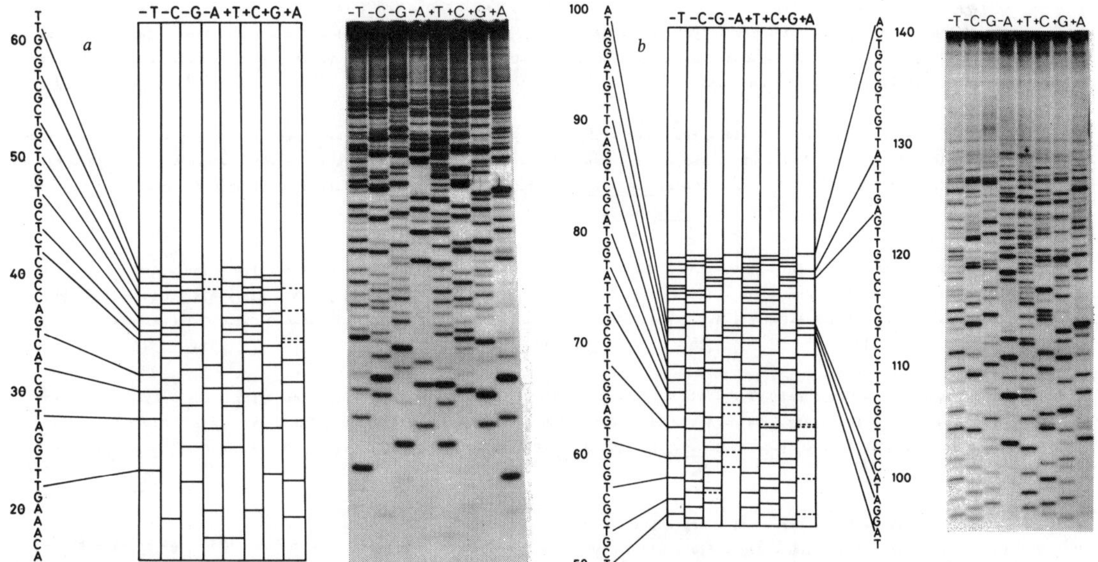

Fig. 3 *a*, An autoradiograph of a plus and minus[1] sequencing gel using fragment *Taq*4 as a primer on φX (−) strand template. The products were cleaved with *Mbo*II. A diagram showing the interpretation of the sequence is shown alongside. Electrophoresis was continued until the bromphenol blue marker reached the bottom of the gel. *b*, The same reaction as in *a* but fractionated for a longer time, such that the xylene cyanol marker migrated near to the bottom of the gel.

problem could also be overcome by priming in the same region from an adjacent restriction site but on the complementary strand. An early difficulty in deducing accurately the number of residues in a run of a particular nucleotide has been overcome by including alongside the plus and minus reactions on the gel a digest of the initial incorporation. This gives a graticule of every possible chain length of the newly synthesised DNA from the restriction enzyme site.

In some cases other techniques such as limited exonuclease digestion of rCTP or rGTP substitution products[9,10] and depurination of nick-translated restriction enzyme fragments[11] have been used to confirm the sequences deduced from the plus and minus system where they were in any doubt. The DNA sequence is shown in Fig. 4.

Amino acid sequence analysis of the D protein

A further confirmation of the DNA sequence was provided by amino acid sequence analysis of the gene *D* protein. Figure 5 shows the sequences of the protein and of the peptides derived by trypsin, chymotrypsin, thermolysin and cyanogen bromide digestion. The DNA sequence has been used for overlapping several peptides, and amino acids 58, 59, 93 and 123 were not identified in the protein work. The sequence of the first eight amino acids of this protein has been reported by Farber[6] but it should be noted that our sequence differs by a serine replacing arginine at residue 7. The calculated molecular weight of the protein is 16,811 which is somewhat higher than the estimates from SDS gel electrophoresis of 14,500 daltons (Table 1). In 8 M urea–SDS gels the molecular weight estimate is more accurate at ~16,000 (G. M. A., unpublished).

Location of the promoter and genes *D* and *J*

The amino acid sequence analysis of the gene *D* product unequivocally defines the boundaries of gene *D* on the DNA sequence since termination codons are found in phase before the initiation codon and after the C-terminal amino acid codon. This rules out the possibility of a longer precursor to the gene *D* protein. The gene *D* initiation codon lies within the *Hin*dII 6b–*Hae*III 3 region reported by Chen *et al.*[12] to contain an RNA polymerase binding site. Smith and Sinsheimer[13] have identified the *in vitro* mRNA start from this region as ppp(C) G–A–U–G–C. We find that this sequence occurs 30 residues to the left of the initiation codon of gene *D* (Fig. 4).

The termination codon of gene *D* overlaps by one nucleotide

```
                G-A-G-T-C-C-G-A-T-G-C-T-G-T-T-C-A-A-C-C-A-C-T-A-A-T-A-G-G-T-A-A-G-A-A-A-T-C-[A-T-G]-
         Hinf 16a↑16b    ↑mRNA start                                                      ↑D start

D     Ser - Gln - Val - Thr - Glu - Gln - Ser - Val - Arg - Phe - Gln - Thr - Ala - Leu - Ala - Ser - Ile - Lys - Leu - Ile -
      A-G-T-C-A-A-G-T-T-A-C-T-G-A-A-C-A-A-T-C-C-G-T-A-C-G-T-T-C-C-A-G-A-C-C-G-C-T-T-T-G-G-C-C-T-C-T-A-T-T-A-A-G-C-T-C-A-T-T-
Hinf 16b↑1           10             20             30        Hae III 3↑7          50    Alu 6↑1           60

D     Gln - Ala - Ser - Ala - Val - Leu - Asp - Leu - Thr - Glu - Asp - Asp - Phe - Asp - Phe - Leu - Thr - Ser - Asn - Lys -
      C-A-G-G-C-T-T-C-T-G-C-C-G-T-T-T-T-G-G-A-T-T-T-A-A-C-C-G-A-A-G-A-T-G-A-T-T-T-C-G-A-T-T-T-T-C-T-G-A-C-G-A-G-T-A-A-C-A-A-A-
                     70             80        Mbo II      Taq 4↑5  1↑7                       110            120

                                                                                                          E (Met)-Val-
D     Val - Trp - Ile - Ala - Thr - Asp - Arg - Ser - Arg - Ala - Arg - Arg - Cys - Val - Glu - Ala - Cys - Val - Tyr - Gly -
      G-T-T-T-G-G-A-T-T-G-C-T-A-C-T-G-A-C-C-G-C-T-C-T-C-G-T-G-C-T-C-G-T-C-G-C-T-G-C-G-T-T-G-A-G-G-C-T-T-G-C-G-T-T-T-[A-T-G]-G-T-
                     130          . 140             150             160              170       ↑E start

           Ter           Ter                                                                              Leu
            ↑             ↑                                                                                ↑
E     Arg - Trp - Thr - Leu - Trp - Asp - Thr - Leu - Ala - Phe - Leu - Leu - Leu - Leu - Ser - Leu - Leu - Leu - Pro - Ser -
D     Thr - Leu - Asp - Phe - Val - Gly - Tyr - Pro - Arg - Phe - Pro - Ala - Pro - Val - Glu - Phe - Ile - Ala - Ala - Val -
      A-C-G-C-T-G-G-A-C-T-T-T-G-T-G-G-G-A-T-A-C-C-C-T-C-G-C-T-T-T-C-C-T-G-C-T-C-C-T-G-T-T-G-A-G-T-T-T-A-T-T-G-C-T-G-C-C-G-T-C-
                ↓             ↓                              210             220             230              ↓
                A am27, amN11 A am3 , am34                                                                    T dE3 , dE5

E     Leu - Leu - Ile - Met - Phe - Ile - Pro - Ser - Thr - Phe - Lys - Arg - Pro - Val - Ser - Ser - Trp - Lys - Ala - Leu -
D     Ile - Ala - Tyr - Tyr - Val - His - Pro - Val - Asn - Ile - Gln - Thr - Ala - Cys - Leu - Ile - Met - Glu - Gly - Ala -
      A-T-T-G-C-T-T-A-T-T-A-T-G-T-T-C-A-T-C-C-C-G-T-C-A-A-C-A-T-T-C-A-A-A-C-G-G-C-C-T-G-T-C-T-C-A-T-C-A-T-G-G-A-A-G-G-C-G-C-T-
                     250         Hind 6c↑7a         270     Hae III 7↑5                                 Hha 4↑13

E     Asn - Leu - Arg - Lys - Thr - Leu - Leu - Met - Ala - Ser - Ser - Val - Arg - Leu - Lys - Pro - Leu - Asn - Cys - Ser -
D     Glu - Phe - Thr - Glu - Asn - Ile - Ile - Asn - Gly - Val - Glu - Arg - Pro - Val - Lys - Ala - Ala - Glu - Leu - Phe -
      G-A-A-T-T-T-A-C-G-G-A-A-A-A-C-A-T-T-A-T-T-A-A-T-G-G-C-G-T-C-G-A-G-C-G-T-C-C-G-G-T-T-A-A-A-G-C-C-G-C-T-G-A-A-T-T-G-T-T-C-
                     310             320       Taq 5↑3     Hap 1↑3  340                  350                360

E     Arg - Leu - Pro - Cys - Val - Tyr - Ala - Gln - Glu - Thr - Leu - Thr - Phe - Leu - Leu - Thr - Gln - Lys - Lys - Thr -
D     Ala - Phe - Thr - Leu - Arg - Val - Arg - Ala - Gly - Asn - Thr - Asp - Val - Leu - Thr - Asp - Ala - Glu - Glu - Asn -
      G-C-G-T-T-T-A-C-C-T-T-G-C-G-T-G-T-A-C-G-C-G-C-A-G-G-A-A-A-C-A-C-T-G-A-C-G-T-T-C-T-T-A-C-T-G-A-C-G-C-A-G-A-A-G-A-A-A-A-C-
                     370        Hha 13↑11         390             400                410    Mbo II      420

E     Cys - Val - Lys - Asn - Tyr - Val - Arg - Lys - Glu
                                                                 J    Ser - Lys - Gly - Lys - Lys - Arg - Ser -
D     Val - Arg - Gln - Lys - Leu - Arg - Ala - Glu - Gly - Val - Met
      G-T-G-C-G-T-C-A-A-A-A-A-T-T-A-C-G-T-G-C-G-G-A-A-G-G-A-G-[T-G-A]-T-G-[T-A-[A]-T-G]-T-C-T-A-A-A-G-G-T-A-A-A-A-A-A-C-G-T-T-C-T-G
          7↑3        430             440            ↑           ↑    ↑                     470              480
                                                  E stop    D stop  J start
```

Fig 4. The nucleotide sequence of the gene *D* region. The amino acid sequences of the D protein (see Fig. 5) and the J protein are aligned with the DNA sequence. The amino acid sequence of the gene *E* product predicted from the DNA sequence is also shown.

```
                                        5                   10                  15                  20                  25
                        Ser-Gln-Val-Thr-Glu-Gln-Ser-Val-Arg-Phe-Gln-Thr-Ala-Leu-Ala-Ser-Ile-Lys-Leu-Ile-Gln-Ala-Ser-Ala-Val-
Sequencer               Ser-Glx-Val-Thr-Glx-Glx-Ser-Val-Arg-Phe     Thr-Ala-Leu-Ala-Ser-Ile
Tryptic peptides
Chymotryptic peptides   Ser-Glx-Val-Thr-Glx-Glx-Ser-Val-Arg-Phe|Gln-Thr-Ala-Leu|Ala-Ser-Ile-Lys-Leu|Ile-Gln-Ala-Ser-Ala-Val-
Thermolytic peptides    Ser-Glx-Val-Thr-Glx-Glx-Ser|Val-Arg|                           |Ile-Lys-Leu|               |Ala-Val-

                                        30                  35                  40                  45                  50
                        Leu-Asp-Leu-Thr-Glu-Asp-Asp-Phe-Asp-Phe-Leu-Thr-Ser-Asn-Lys-Val-Trp-Ile-Ala-Thr-Asp-Arg-Ser-Arg-Ala-
Sequencer
Tryptic peptides                                                                                        |Ser-Arg|Ala-
Chymotryptic peptides   Leu|Asx-Leu-Thr-Glx-Asx-Asx-Phe(Asx,Phe,Leu)|Thr-Ser-Asn-Lys-Val-Trp|Ile-Ala-Thr-Asx
Thermolytic peptides    Leu-Asx-Leu-Thr-Glx-Asx-Asx-Phe-Asx-Phe|                          |Ile-Ala(Thr)Asx(Arg)Ser-Arg|Ala

                                        55                  60                  65                  70                  75
                        Arg-Arg-Cys-Val-Glu-Ala-Cys-Val-Tyr-Gly-Thr-Leu-Asp-Phe-Val-Gly-Tyr-Pro-Arg-Phe-Pro-Ala-Pro-Val-Glu-
Sequencer
Tryptic peptides        Arg|Arg|
Chymotryptic peptides                                    |Gly-Thr-Leu-Asp-Phe|Val-Gly-Tyr-Pro-Arg-Phe-Pro-Ala-Pro-Val-Glu-
Thermolytic peptides    Arg-Arg-Cys|Val-Glu-Ala-Cys|

                                        80                  85                  90                  95                  100
                        Phe-Ile-Ala-Ala-Val-Ile-Ala-Tyr-Tyr-Val-His-Pro-Val-Asn-Ile-Gln-Thr-Ala-Cys-Leu-Ile-Met-Glu-Gly-Ala-
Sequencer
Tryptic peptides
Chymotryptic peptides   Phe|Ile-Ala-Ala-Val-Ile-Ala-Tyr|Tyr|Val     Pro-Val-Asx-Ile-Glx-Thr     Cys-Leu|Ile(Met)Glu-Gly-Ala-
Thermolytic peptides                                     |Val-His-Pro|                           |Leu-Ile-Met-Glu-Gly-Ala-
CNBr peptides                                                                                            |Glx-Gly-Ala-

                                        105                 110                 115                 120                 125
                        Glu-Phe-Thr-Glu-Asn-Ile-Ile-Asn-Gly-Val-Glu-Arg-Pro-Val-Lys-Ala-Ala-Glu-Leu-Phe-Ala-Phe-Thr-Leu-Arg-
Sequencer
Tryptic peptides
Chymotryptic peptides   Glu-Phe|Thr-Glx-Asx-Ile-Ile-Asx-Gly-Val-Glx-Arg-Pro-Val-Lys-Ala-Ala-Glx-Leu-Phe|Ala-Phe|
Thermolytic peptides    Glu|Phe-Thr-Glx-Asx|Ile-Ile-Asx-Gly-Val-Glx-Arg(Pro)|Val-Lys-Ala|                    |Leu-Arg¦
CNBr peptides           Glx-Phe-Thr-Glx-Asx-Ile-Ile

                                        130                 135                 140                 145                 150
                        Val-Arg-Ala-Gly-Asn-Thr-Asp-Val-Leu-Thr-Asp-Ala-Glu-Glu-Asn-Val-Arg-Gln-Lys-Leu-Arg-Ala-Glu-Gly-Val-Met
Sequencer
Tryptic peptides        |Val-Arg|Ala-Gly-Asx-Thr-Asx-Val-Leu-Thr-Asx-Ala-Glx-Glx(Asx)Val-Arg|Gln-Lys-Leu-Arg|Ala-Glu-Gly-Val-Met
Chymotryptic peptides                                                                               |Arg-Ala-Glu-Gly-Val-Met
Thermolytic peptides    Val-Arg-Ala-Gly-Asx-Thr-Asx-Val-Leu-Thr(Asx,Ala,Glx,Glx,Asx)|Val-Arg-Gln-Lys|
CNBr peptides
Carboxypeptidase                                                                                                 Gly-Val-Met
```

Fig. 5 Amino acid sequence analysis of the D protein. Sequences of φX174 D protein and peptides derived by tryptic, chymotryptic, thermolytic and CNBr digestion. Solid bars indicate sites of cleavage; the broken bar shows partial cleavage. The residues shown were obtained by automatic (residues 1–17) or manual sequencing. Residues in parentheses were present in the amino acid composition but not identified in the sequencing work. Blanks indicate that the peptide was not sufficiently purified for a reliable amino acid analysis. The protein was isolated from 50 g of frozen φX174-infected *E. coli* C (grown at the Microbiological Research Establishment, Porton Down). Cells were thawed with 50 ml buffer A (20 mM Tris, 10 mM $MgCl_2$, 1 mM mercaptoethanol, 1 mM EDTA *p*H 8.0), sonicated, and the cell debris centrifuged off, extracted with buffer A and 1 M NaCl, and centrifuged again. The supernatants were dialysed against buffer A at 4 °C, centrifuged again, and the supernatant loaded on a column (3×20 cm) of Whatman DE-52 equilibrated with buffer A at 4 °C. After the unadsorbed proteins had been washed through a linear gradient was started (600 ml each A and A+0.6 M KCl). The D protein eluted about half way through the gradient as monitored by 15% acrylamide gel electrophoresis in 0.1% SDS.

Fractions containing the D protein were combined and 18.7 g $(NH_4)_2SO_4$ added per 100 ml at 4 °C. The resulting precipitate was centrifuged down, resuspended in buffer A, dialysed against buffer A and the suspension centrifuged again. The supernatant contained the D protein sufficiently pure for amino acid sequence analysis. It was dialysed extensively against water and freeze dried.

The amino acid sequencing was by the standard procedures as described previously[34,35]. The amounts of *S*-carboxymethylated protein used were 2 mg for N-terminal automatic sequencing in the Beckman sequencer, 5 mg for succinylation and tryptic digestion, 2 mg for chymotryptic digestion, 15 mg for CNBr cleavage, and 5 mg for thermolytic digestion. Purification of the insoluble and the neutral thermolytic peptides was not attempted.

the initiation codon of gene *J* in the sequence T–A–A–T–G. Overlapping initiation and termination codons have also been observed in *E. coli* tryptophan mRNA[14]. In this case the sequence was shown to be U–G–A–U–G.

Genetic mapping experiments of Benbow *et al.*[4] suggested the existence of a gene *J* in this region coding for the smallest structural component of the virion. Alignment of the DNA sequence with the amino acid sequence of a small basic protein isolated from the virion (D. Freymeyer, P. R. Shank, T. Vanaman, C. A. H. and M. H. Edgell, in preparation) clearly

Table 2 Rescue of gene *E* mutants by restriction fragments from gene *D*

Gene *E* mutant	*Hae*III fragments Z3	Z7	Z5	*Hha*I fragments H8(ab)	H4	H13
*am*3	1.2×10^3	2×10^4	8×10^2	5×10^2	2×10^4	4×10^2
*am*27	1×10^2	$>2\times10^4$	2×10^2	—	—	—
*am*34	6×10^1	$>2\times10^4$	6×10^1	6×10^1	2×10^3	2×10^1
*am*N11	3×10^1	4×10^3	4×10^1	1×10^1	1.5×10^3	2×10^1
*dE*3	8×10^1	$>2\times10^4$	6×10^1	4×10^1	1×10^4	4×10^1
*dE*5	1×10^1	2×10^3	$<10^1$	$<10^1$	1.2×10^3	$<10^1$
S13 *am*En15	2×10^2	4×10^3	2×10^2	—	—	—

Restriction fragments of wild-type RF DNA were purified by gel electrophoresis. Fragments were denatured, annealed to viral strand from each of the gene *E* mutants listed, and used to infect spheroplasts of *E. coli* as previously reported[16]. Wild-type plaque-forming phage were selectively assayed on *E. coli* C (su^-). The number of wild-type phage per ml following chloroform lysis of the spheroplasts is tabulated, and positive results are underlined. In the experiment shown here only fragments from the gene *D* region were assayed. In another experiment (data not shown) all of the *Hae*III fragments and all of the *Hha*I fragments were assayed for the *am*3 site and wild-type markers were only recovered from fragments Z7 and H4. The *am* mutants were assayed on *E. coli* HF4714. The *dE* mutants (see text) were assayed on MacConkey agar (Difco) plates. The phage sample was overlaid in a mixture of 2.5 ml of soft agar, 0.25 ml of 5 mg ml^{-1} egg white lysozyme, 10µl of 1 M $CaCl_2$, and 0.3 ml of an overnight culture of *E. coli* C (su^-). Plates were incubated at 37 °C.

demonstrates that it is coded by the gene immediately following *D*. We suggest that gene *J* be explicity defined as the gene coding for this protein since marker rescue experiments show that the mutation (*am*6) which had genetically defined *J* (ref. 4) is located in *Hae*III fragment 7 (ref. 15) which ends 179 nucleotides before the initiation codon of *J* as defined here.

The complete nucleotide sequence of gene *J* has now been determined and has been shown to extend from the end of gene *D* to within approximately 35 nucleotides of the initiation codon of gene *F* (B. G. B., unpublished). This establishes the order of genes *D*, *J* and *F* in this region, but gene *E* which mapped genetically immediately following gene *D* (ref. 4) is not accounted for.

Location of gene *E* mutants within gene *D*

It is possible to rescue genetic markers from restriction fragments[16] and thus to establish the location of genetically defined sites on the physical map of φX174. This technique has been applied to locate gene *E*. The double-stranded RF DNA of wild-type φX174 was cut with restriction endonucleases and the resulting fragments were purified by gel electrophoresis. Fragments were denatured and annealed to the viral single-stranded DNA bearing a gene *E* mutation.

Spheroplasts of *E. coli* were transfected with the resulting complexes and the progeny phage were selectively assayed for wild-type plaque-forming particles. Wild-type marker can only be rescued from fragments which span the site of the mutation. Several gene *E* mutants were tested in this way, including *am* nonsense mutants. All were rescued by the wild-type *Hae*III fragment Z7 (Table 2). Weisbeek *et al.*[15] have also mapped one of these mutants (*am*3) and the mutant (*am*6) (probably a gene *E* mutant) in fragment Z7. Surprisingly, inspection of the nucleotide and amino acid sequence data showed that this fragment is completely contained within gene *D* (Figs 2 and 4). Since gene *E am* mutants make a full length D protein, these gene *E* nonsense codons cannot be in phase with the translation of gene *D*.

Nucleotide sequence of gene *E* mutants

Sequences within *Hae*III fragment Z7 have been determined for several φX174 strains bearing mutations in gene *E*. Figure 6 compares a gel which displays the sequence synthesised on a viral strand template containing *am*3 (gene *E*) with a gel showing the corresponding wild-type sequence. The mutation clearly produces a C→T change in the synthesised minus strand which corresponds to a G→A transition at position 195 in the viral (plus) strand sequence (Fig. 4). Thus the mutation results in the minus strand sequence 5′–C–T–A–3′ which corresponds to the amber codon 5′–T–A–G–3′ on the viral strand. The phase of this nonsense codon is displaced one nucleotide to the right of that used for translation of gene *D*. In this phase a tryptophan codon (T–G–G) is changed to *am* (T–A–G). The mutation does not produce a change in the D product since one valine codon (G–T–G) is changed to another (G–T–A). Similar analysis of an (*am*3) revertant showed that the wild-type sequence was present at position 195.

An independently isolated gene *E* mutant, *am*34, was shown to contain the same altered nucleotide sequence as *am*3 (Fig. 4, data not shown). Two other independently isolated gene *E* amber mutants have also been sequenced (*am*27 and *am*N11). Both result in G→A transitions on the viral strand at position 186 (Fig. 4, data not shown). This results again in a T–G–G→T–A–G change which produces an amber codon which is in phase with that found in *am*3 and *am*34, and displaced one nucleotide to the right of the D translation phase. Again, no coding change is produced in the phase used for D translation (C–T–G→C–T–A, both coding for leucine). The sequence alterations produced by amber mutations at both these sites have been confirmed by pyrimidine tract analysis of the minus strand of *Hae*III fragment Z7. The G→A change reported here for *am* mutations induced by nitrous acid treatment of phage (plus strand) has been predicted by previous studies on the mechanism of nitrous acid mutagenesis[17,18].

One observation which seems inconsistent with the results presented here concerns the characterisation of a population of φX174 deletion mutants[19]. These deletions were originally interpreted to be located in gene *E* and presumed to be defective only in lysis function. This is inconsistent with our results which predict that any gene *E* deletion would also have a defect in gene *D*. Further analysis of the φX174 deletion population (R. L. Sinsheimer, personal communication) has shown, however, that it must contain deletions from all parts of the genome. We have also investigated this deletion population and obtained evidence that the population does not contain

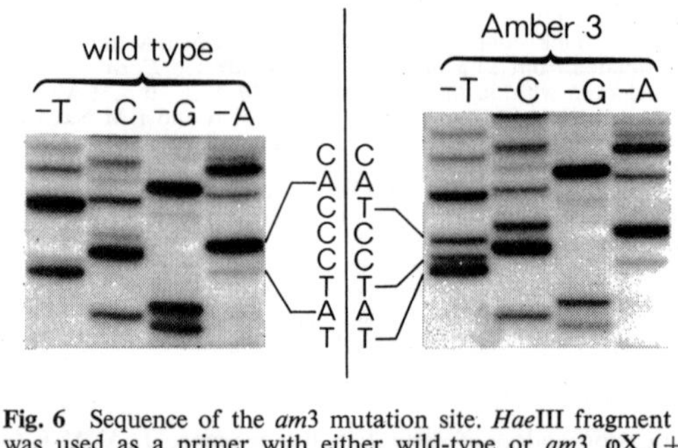

Fig. 6 Sequence of the *am*3 mutation site. *Hae*III fragment 5 was used as a primer with either wild-type or *am*3 φX (+) strand as a template. Only a portion of the autoradiograph of the minus system is shown in both cases with the sequences of the (−) strand deduced. The viral strand sequence is therefore:

wt 5′–G–T–G–G–G–A–T–A–3′
*am*3 5′–G–T–A–G–G–A–T–A–3′

any particles whose only genetic defect is a gene *E* deletion. We observed that all gene *E am* mutants tested were able to form plaques efficiently on su^- hosts when plated on special medium containing bile salts and egg white lysozyme (see legend to Table 2). If simple gene *E* deletions exist it should be possible to clone them on such plates. In fact, about 1–2% of the particles in the deletion population do form plaques on these special plates but not on normal plates on su^- or su^+ hosts. However, all 20 isolates tested reverted with frequencies reasonable for simple point mutations. Two of these isolates (dE3 and dE5) were mapped by marker rescue on to Z7 and sequenced. Both isolates contain the same missense mutation which produces a predicted Pro → Leu (C–C–G → C–T–G) change in the gene *E* translation phase as defined by sequenced gene *E* amber mutations (position 237, Fig. 4). This would again produce an unchanged D protein since one alanine codon (G–C–C) is changed to another (G–C–T).

Extent of the gene *E* coding region

The gene *E* amber mutations are suppressible in su^+ hosts indicating that gene *E* is being translated at those locations. It is therefore possible to determine the boundaries of the gene *E* coding region by looking for termination codons in the same phase as the amber codons. The termination codon at position 449–451 (Fig. 4) defines the 3′ terminus of gene *E* protein as it is the first available termination codon in this phase after the amber mutations. Initiation of the gene *E* protein must occur between the termination codon at position 107–109 and the amber mutations. There are two possible initiation codons in this region, a G–T–G at position 146–148 and an A–T–G at position 176–178. Of these the A–T–G is the only initiation codon preceded by a sequence complementary to the 3′ sequence of 16S rRNA that has been implicated in the initiation of protein synthesis[20,21]. Figure 7 shows a comparison of the predicted gene *E* initiation site with other initiation regions found in φX174. Starting at the A–T–G codon at residues 176–178 a polypeptide of 91 amino acids would be synthesised. We have been unable to identify clearly either the E protein or the *am*3 fragment *in vivo* or *in vitro*, so amino acid sequence analysis has not been possible. However, the calculated molecular weight of the E protein is in good agreement with the identification of an *E* gene product with a molecular weight of 10,000 by Burgess and Denhardt[22], although other estimates have varied (see Table 1).

Evolution of overlapping genes

The mechanism by which the gene *E* product acts to induce lysis of the infected cell is not clearly understood. There is no evidence that it functions as a lysozyme. Bile salts in the presence of egg white lysozyme are able to induce lysis of cells infected with mutants deficient in gene *E* function, although lysozyme alone is not sufficient. This suggests that the gene *E* protein may act on the cell membrane. The predicted amino acid sequence of the E protein (Fig. 4) is very hydrophobic in the N-terminal region, in particular in the sequence Leu^{13}–Leu–Leu–Leu–Ser–Lue–Leu–Leu–Pro–Ser^{23}. A large percentage of hydrophobic amino acids including tracts of leucine residues have been observed at the N terminus of precursors to several pancreatic secretory proteins[23] and in the N-terminal region of the light chain IgG precursor[24]. These N-terminal regions have been proposed to function in the transfer of the proteins across the microsomal membrane. The similarity between these sequences and the N-terminal region of the gene *E*-protein is consistent with a membrane-related activity.

An analysis of possible secondary structure in the E protein using the rules of Chou and Fasman[25] predicts that the N-terminal hydrophobic region could consist of an α-helix terminating before the proline at residue 21. The missense mutations dE3 and dE5 cause a Pro to Leu change at residue 21. This mutation, which causes a loss of lysis function, could affect the secondary structure of the E protein by preventing α-helix termination at this point.

It is interesting to speculate on the evolution of overlapping genes *D* and *E*. It is possible that the size of the φX174 genome is limited by packaging or other unknown constraints. If so, the genes required could only be fitted in either by overlapping or by reading from both strands of the DNA. The gene *D* codons contain a high proportion (40%) of T residues in the third position, as is observed in other φX174 genes[26,27]. This suggests the possibility that gene *D* arose first and that ancestors of φX174 could survive without a specific lysis gene in their natural environment. It is clear that if some mutational event led to initiation of protein synthesis in a phase one nucleotide to the right of the D reading phase the resulting codons would have a high proportion of T residues at the second position. Such codons specify hydrophobic amino acids and it seems plausible that a polypeptide could be produced which would interact with the host cell membrane. This could provide a basis for natural selection leading to an efficient lysis gene.

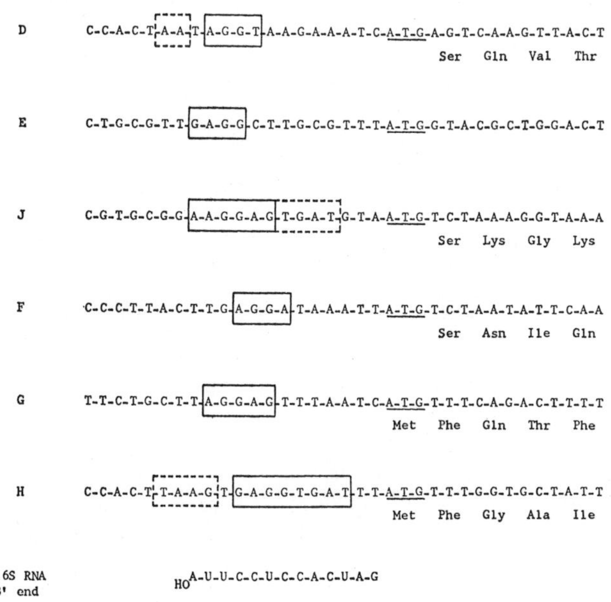

Fig. 7 Comparison of the predicted gene *E* initiation site with other known φX174 initiation sites. The boxed areas indicate sequences complementary to the 3′ end of 16S rRNA. Dashed boxes show further complementarity if some nucleotides are looped out or mismatched. Known N-terminal amino acid sequences are also shown.

It is worth noting that a gene *E* amber mutant in the closely related phage S13 is rescued by fragment Z7 from *wt* φX174 (Table 2).

It seems likely that translation of the gene *D* protein, which is made in large amounts, would obstruct binding of ribosomes at the *E* initiation site. This might be a convenient control if the E protein is only required in small amounts. On the other hand, some control may be required to facilitate initiation of the E protein, for example, mRNA processing. Initiation of the gene *J* protein poses a similar problem, since the sequence involved in gene *J* ribosome binding is being translated in both *D* and *E* phases.

The fact that genes *D* and *E* overlap alleviates to some extent the problem of fitting the known φX genes into the available DNA (Table 1). However, it is clear that if the molecular weight estimates of the A, B and C proteins are correct, they cannot be separately coded in the region between genes *H* and *D*. Thus, in spite of the severe constraints imposed on the amino acid sequence by translation in two phases, it seems possible that other examples of overlapping genes may be found.

We thank R. Kamen, N. L. Brown and S. Sato for gifts of enzymes used in this study and R. L. Sinsheimer for supplying the deletion stock. C. A. H. was supported by a grant from the US Public Health Service.

Received August 19; accepted September 14, 1976.

1 Sanger, F., and Coulson, A. R., *J. molec. Biol.*, **94**, 441–448 (1975).
2 Lee, A. S., and Sinsheimer, R. L., *Proc. natn. Acad. Sci. U.S.A.*, **71**, 2882–2886 (1974).
3 Jeppesen, P. G. N., Sanders, L., and Slocombe, P. M., *Nucleic Acids Res.*, **3**, 1323–1339 (1976).
4 Benbow, R. M., Mayol, R. F., Picchi, J. C., and Sinsheimer, R. L., *J. Virol.*, **10**, 99–114 (1972).
5 Burgess, A. B., thesis, Harvard Univ. (1969).
6 Farber, M. B., *J. Virol.*, **17**, 1027–1037 (1976).
7 Hayashi, M. N., and Hayashi, M., *J. Virol.*, **14**, 1142–1151 (1974).
8 Brown, N. L., and Smith, M., *FEBS Lett.*, **65**, 284–287 (1976).
9 Sanger, F., Donelson, J. E., Coulson, A. R., Kössel, H., and Fischer, D., *Proc. natn. Acad. Sci. U.S.A.*, **70**, 1209–1213 (1973).
10 Donelson, J. E., Barrell, B. G., Weith, H. L., Kössel, H., and Schott, H., *Eur. J. Biochem.*, **58**, 383–395 (1975).
11 Maniatis, T., Jeffrey, A., and Kleid, D. G., *Proc. natn. Acad. Sci. U.S.A.*, **72**, 1184–1188 (1975).
12 Chen, C. Y., Hutchison, C. A. III, and Edgell, M. H., *Nature new Biol.*, **243**, 233–236 (1973).
13 Smith, L. H., and Sinsheimer, R. L., *J. molec. Biol.*, **103**, 711–735 (1976).
14 Platt, T., and Yanofsky, C., *Proc. natn. Acad. Sci. U.S.A.*, **72**, 2399–2408 (1975).
15 Weisbeek, P. J., Vereijken, J. M., Baas, P. D., Jansz, H. S., and van Arkel, G. A., *Virology*, **72**, 61–71 (1976).
16 Hutchison, C. A. III, and Edgell, M. H., *J. Virol.*, **8**, 181–189 (1971).
17 Tessman, I., Poddar, R. K., and Kumar, S., *J. molec. Biol.*, **9**, 352–363 (1964).
18 Vanderbilt, A. S., and Tessman, I., *Genetics*, **66**, 1–10 (1970).
19 Zuccarelli, A. J., Benbow, R. M., and Sinsheimer, R. L., *Proc. natn. Acad. Sci. U.S.A.*, **69**, 1905–1910 (1972).
20 Shine, J., and Dalgarno, L., *Eur. J. Biochem.*, **57**, 221–230 (1975).
21 Steitz, J. A., and Jakes, K., *Proc. natn. Acad. Sci. U.S.A.*, **72**, 4734–4738 (1975).
22 Burgess, A. B., and Denhardt, D. T., *J. molec. Biol.*, **44**, 377–386 (1969).
23 Devillers-Thiery, A., Kindt, T., Scheele, G., and Blobel, G., *Proc. natn. Acad. Sci. U.S.A.*, **72**, 5016–5020 (1975).
24 Burstein, Y., and Schechter, I., *Biochem. J.*, **157**, 145–151 (1976).
25 Chou, P. Y., and Fasman, G. D., *Biochemistry*, **13**, 222–245 (1974).
26 Air, G. M., Blackburn, E. H., Sanger, F., and Coulson, A. R., *J. molec. Biol.*, **96**, 703–719 (1975).
27 Air, G. M., Blackburn, E. H., Coulson, A. R., Galibert, F., Sanger, F., Sedat, J., and Ziff, E. B., *J. molec. Biol.* (in the press).
28 Borras, M-T., Vanderbilt, A. S., and Tessman, E. S., *Virology*, **45**, 802–803 (1971).
29 Linney, E. A., Hayashi, M. N., and Hayashi, M., *Virology*, **50**, 381–387 (1972).
30 Godson, G. N., *J. molec. Biol.*, **77**, 467–477 (1973).
31 Siden, E. J., and Hayashi, M., *J. molec. Biol.*, **89**, 1–16 (1974).
32 Tessman, I., Tessman, E. S., Pollock, T. J., Borras, M-T., Puga, A., and Baker, R., *J. molec. Biol.* **103**, 583–598 (1976).
33 Van der Mei, D., Zandberg, J., and Jansz, H. S., *Biochim. biophys. Acta*, **287**, 312–321 (1972).
34 Air, G. M., *J. molec. Biol.* (in the press).
35 Air, G. M., Sanger, F., and Coulson, A. R., *J. molec. Biol.* (in the press).

(Reprinted from Nature, Vol. 265, No. 5596, pp. 702–705, February 24, 1977)

DNA sequence at the C termini of the overlapping genes *A* and *B* in bacteriophage ΦX174

M. Smith, N. L. Brown, G. M. Air, B. G. Barrell, A. R. Coulson, C. A. Hutchison III & F. Sanger

MRC Laboratory of Molecular Biology, Hills Road, Cambridge CB2 2QH, UK

Sequence determination of mutants in genes A *and* B *confirms that these constitute another pair of genes in ΦX174 which are translated from the same DNA sequence using different reading frames.*

BACTERIOPHAGE ΦX174 is known to code for nine proteins in its DNA (ref. 1) which is about 5,400 nucleotides in length (Fig. 1). The locations of the boundaries of six of the genes—*D, E, J, F, G* and *H*—have been defined by sequence determination (F.S. *et al.*, in preparation). Two of these genes (*D* and *E*) overlap on the DNA sequence[2]. To define the boundaries for genes *A, B* and *C* which lie in the remaining region between gene *H* and gene *D*, the sequence of ΦX174 *am*3 DNA in the region between the *Hae*III 6a/6b and the *Taq*I 2/7 sites (Fig. 1) has been determined (N.L.B. and M.S., in preparation). A newly recognised ribosome binding site[3] has been located in the sequence[4]. The sequence determination of a *B* gene mutant—*am*16—strongly suggested that the *A* and *B* genes are translated from the same DNA using different reading frames (N.L.B. and M.S., in preparation). We now report the 200 nucleotide sequence of ΦX174 *am*3 DNA from the *Taq* 2/7 to the *Hha* 8b/4 sites (Fig. 1). In addition, the sequences of three mutants from this region—*am*18 and *am*35 (gene *A*) and a revertant of *am*18, *ts*116 (gene *B*)—have been determined. The mutants *am*18 and *am*35 have an amber codon at the same site. The gene *B* mutant, *ts*116, results from a single base change in the termination codon of *am*18/35. These results confirm the overlap of genes *A* and *B* and establish the C-termini of the two genes.

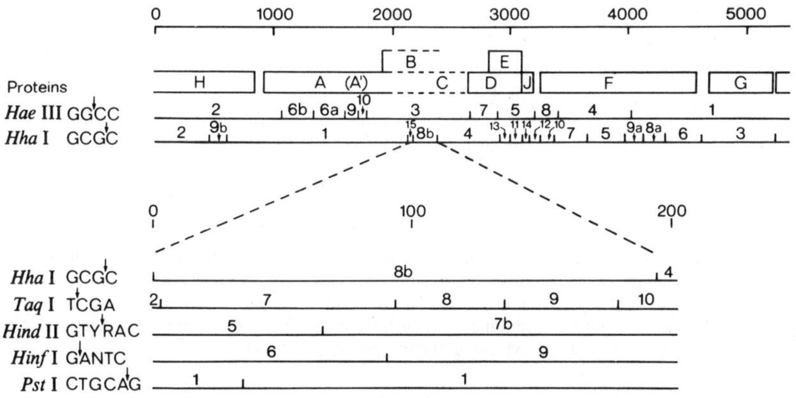

Fig. 1 Map of the ΦX174 genome showing the *Hae*III and *Hha*I fragments and the protein-coding sequences and details of other restriction cleavages within *Hha* fragment 8b.

Nucleotide sequence analysis

The cleavage sites of five restriction enzymes which cleave in the region are shown in Fig. 1. The sequence of the DNA was obtained by a slight modification of the 'plus and minus' method (ref. 5 and N.L.B. and M.S., in preparation) by priming from the *Hin*fI 6/9 site, on both viral and complementary templates and from the *Pst*I site on the complementary template (Fig. 2). The sequence was confirmed by depurination analysis[6] of *Taq*I fragments 7, 8, 9 and 10 labelled by 'nick translation'[7] of ΦX174 *am*3 RFI DNA.

Nucleotide sequence of *A* and *B* mutants

Gene *A* mutants (*am*18, *am*35) have been mapped[8] between the left end of *Hha*I fragment 8b (fragment 7b in ref. 8) and the right end of *Hin*dII fragment 5. The mutant *am*18 and its revertant *ts*116 do not recombine to product *wt* progeny at a frequency above background. These experiments set an upper limit of approximately three nucleotides on the distance separating these mutants[9]. The *B* gene mutant (*ts*116) maps close to, or in, the *Hin*dII 5/7b site[10]. Consequently, the mutant sequences can be determined very easily using the plus and minus method by priming from the *Hin*fI 6/9 site on the mutant viral strand template. The results of these determinations for *am*3 (*E* gene mutant; wild type in this region), *am*18, *am*35 and *ts*116 are shown in Fig. 3. The independently isolated *A* gene mutants, *am*18 and *am*35, show the same altered nucleotide sequence. In the viral strand, this is a C → T transition at nucleotide 64 (Fig. 2) which results in a CAG → TAG change producing an amber codon in the reading frame which has been assigned to gene *A* (N.L.B. and M.S., in preparation). No other alterations in the

```
F9(T8)/V(S)   A-T-G-C-G-C-G-C-T-T-C-G-A-T-A-A-A-A-A-T-G-A-T-T-G-G-C-G-T-A-T-C-C-A-A-C-C-T-G-C-A-G-A-G-T-T-T-T-A-T-C-G-C-T-T-C-C-A-T-G-
P1(T2)/C(A)                                                                                                                   C-A-T-G-
              ---------------------------------------------------------------------------------------------------------------------------
              A-T-G-C-G-C-G-C-T-T-C-G-A-T-A-A-A-A-A-T-G-A-T-T-G-G-C-G-T-A-T-C-C-A-A-C-C-T-G-C-A-G-A-G-T-T-T-T-A-T-C-G-C-T-T-C-C-A-T-G-
              Hha I 15↑  ↑8b 2↑7 Taq I                              Pst I 1↑1
                   10              20              30              40              50              60

T9/V(S)                                                    T-C-T-G-A-T-G-A-G-T-C-G-A-A-A-A-A-T-T-A-T-C-T-T-G-A-T-A-A-A-G-C-A-
F9(T8)/V(S)   A-C-G-C-A-G-A-A-G-T-T-A-A-C-A-*-T-T-T-C
P1(T2)/C(S)   A-C-G-C-A-G-A-A-G-T-T-A-A-C-A-C-T-T-T-C-G-G-A-T-A-T-T-T-C-T-G-A-T-G-A-G-T-C-G-A-A-A-A-A-T-T
P1(T2)/C(L)                                                         A-T-G-A-G-T-C-G-A-A-A-A-A-T-T-A-T-C-T-T-G-A-T-A-A-A-G-C-A-
              ---------------------------------------------------------------------------------------------------------------------------
              A-C-G-C-A-G-A-A-G-T-T-A-A-C-A-C-T-T-T-C-G-G-A-T-A-T-T-T-C-T-G-A-T-G-A-G-T-C-G-A-A-A-A-A-T-T-A-T-C-T-T-G-A-T-A-A-A-G-C-A-
                    Hind II 5↑7b                          Hinf I 6↑9  7↑8 Taq I
                   70              80              90              100             110             120

T9/V(S)       G-G-A-A-T-T-A-C-T-A-C-T-G-C-T-T-G-T-T-T-A-C-G-A-A-T-T-A
T10/V(S)                T-G-C-T-T-G-T-T-T-A-C-G-A-A-T-T-A-A-A-T-C-G-A-A-G-T-G-G-A-C-T-G-C-T-G-G-C-G-G-A-A-A-A-T-G-A-G-A-A-
P1(T2)/C(L)   G-G-A-A-T-T-A-C-T-A-C-T-G-C
F6/C(S)       G-G-A-A-T-T-A-C-T-A-C-T-*-C-T-T-G-*-*-T-A-C-G-A-A-T-T-A-A-A-T-C-G-A-A-G-T-G-G-A-C-T-G-C-T-G
F6/C(L)                                                                                       G-C-T-G-G-C-*-G-A-A-A-A-T-G-A-G-A-A-
              ---------------------------------------------------------------------------------------------------------------------------
              G-G-A-A-T-T-A-C-T-A-C-T-G-C-T-T-G-T-T-T-A-C-G-A-A-T-T-A-A-A-T-C-G-A-A-G-T-G-G-A-C-T-G-C-T-G-G-C-G-G-A-A-A-A-T-G-A-G-A-A-
                                                       Taq I 8↑9
                  130             140             150             160             170             180

T10/V(S)      A-A-T-T-C
F6/C(L)       A-A-T-T-C-G-A-C-C-T-A-T-C-C-T-T-G-C-G-C
              ---------------------------------------
              A-A-T-T-C-G-A-C-C-T-A-T-C-C-T-T-G-C-G-C
              Taq I 9↑10          Hha I 8b↑4
                        190
```

Fig. 2 Summary of the plus and minus sequence determination of the *Hha*I fragment 8b region. At the left of each determination is indicated the fragment used as the origin of the plus and minus priming. (The single letter designations used for restriction endonucleases are: *Taq* I (T); *Hin*fI,(F); and *Pst*I, (P).) Fragments indicated in parentheses are the primers used when these differed from the one corresponding to the origin (for example, F9(T8) indicates that T8 was the primer with the origin at the adjacent *Hin*fI site). The letters V and C indicate that the template was viral or complementary strand DNA, respectively. The designations (S) or (L) indicate the use of a short (4 h) or long (8 h) gel electrophoresis run.

*am*18 or *am*35 sequences can be seen between the *Hin*dII 5/7b site and the *Pst*I site (Fig. 2). This transition also results in an amino acid change (Ala→Val) in the reading frame assigned to gene *B* on the basis of the phasing of the ribosome binding site[3,4] and the gene *B* mutant, *am*16. This assignment of the *am*18, *am*35 (gene *A*) mutation to a specific site to the right of the *am*16 (gene *B*) as originally suggested by recombination mapping[1,9,11], and to a different reading frame, verifies the overlap of genes *A* and *B*.

The gene *B* mutant, *ts*116, was isolated as a revertant of *am*18. The mutation is a G → C change at nucleotide 66, which produces a TAG→TAC (amber→Tyr) codon change in the gene *A* reading frame. In gene *B*, the mutation results in a GAA→CAA (Glu→Gln) codon change. The resultant change in charge may be responsible for the *ts* phenotype.

The original stocks of these *am* mutants were obtained after nitrous acid treatment of viral DNA. The mutations characterised so far include G → A (*am*3,34; *am*N11,27) (ref. 2), C → T (*am*18,35) and G → T (*am*16). This variety of mutations illustrates the caution which should be exercised in predicting the consequences of chemical mutagenesis[12].

The changes induced in the second gene by mutation in its overlapping companion are not necessarily restrictive, as evidenced by the Leu→Phe change in gene *A* which accompanies the *am*16 mutation in gene *B*, the Ala→Val change in gene *B* accompanying the *am*18,35 mutation in gene *A*, and the reversion of *am*18, which by two steps induces a Gln → Tyr in gene *A*. There is no evidence for any impairment in protein function due to these changes, whereas in *ts*116 the second amino acid change (Glu→Gln) in gene *B* which follows the initial change (Ala → Val) induced by the gene *A* amber mutation evinced a phenotypic response (*ts*).

C termini of A and B proteins

The sequence determination and establishment of the *A* gene reading frame taken together with the *B* gene reading frame allow the prediction of the C termini of the two proteins (Fig. 4). The B protein is 120 amino acids long with a molecular weight of 13,845. The *A* gene extends for 85 nucleotides (28 codons) beyond the C terminus of gene *B*. If, as has been suggested (N.L.B. and M.S., in preparation), the gene *A*/gene *B* overlap arose by read-through of a previously shorter gene *A*, then a question arises about the original role of the DNA beyond the C terminus of gene *B*. The mutant *am*6 is a double amber mutant mapping in fragments F9 (gene *A*) and Z7 (gene *E*) (P. Weisbeek, personal communication). In gene *E* the mutation is identical to *am*3. The gene *A* mutation has also been sequenced and is a G → A change at nucleotide 158. This gives an amber (TAG) codon, and would result in a gene *A* product six amino acids shorter than in the wild type. In addition there is a C → T change at nucleotide 98 which would result in a missense Ser → Leu in the gene *A* protein, and preliminary results suggest yet another change in this region. The

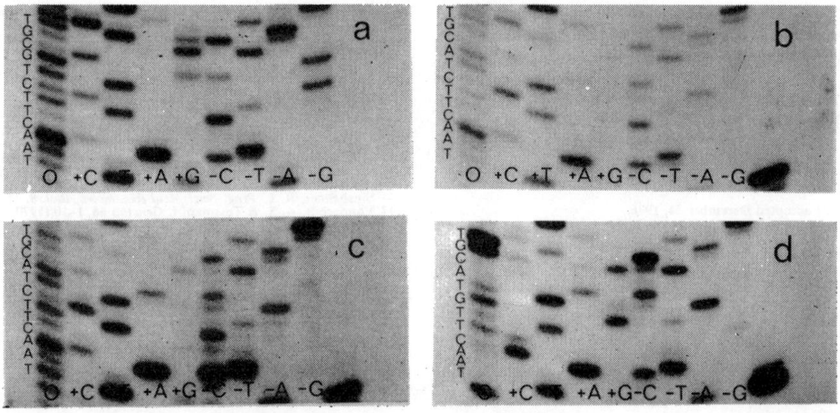

Fig. 3 Sequences of the *am*18, *am*35 and *ts*116 mutations. The *am*3 (wild type in this region) sequence is also shown. *Taq*I fragment 8 was used as primer with *Hin*fI as the datum restriction endonuclease with either (*a*) *am*3, (*b*) *am*18, (*c*) *am*35 or (*d*) *ts*116 viral DNA as template. The same portion of the autoradiogram of each plus and minus determination is shown with the sequence of the complementary strand deduced. The viral strand sequences, positions 60–74 (Fig. 2), are therefore:

```
am3 ('wt')     G-A-C-G-C-A-G-A-A-G-T-T-A-A-C
                .   ,   .   ,   .   , .   , .

am18 (A)       G-A-C-G-T-A-G-A-A-G-T-T-A-A-C
                .   ,   .   ,   .   , .   , .

am35 (A)       G-A-C-G-T-A-G-A-A-G-T-T-A-A-C
                .   ,   .   ,   .   , .   , .

ts116 (B)      G-A-C-G-T-A-C-A-A-G-T-T-A-A-C
                .   ,   .   ,   .   , .   , .
                              Hind II 5 ↑ 7b
```

Dots indicate the reading frame for gene *A* and commas that for gene *B*.

```
B protein   - Tyr - Ala - Arg - Phe - Asp - Lys - Asn - Asp - Trp - Arg - Ile - Gln - Pro - Ala - Glu - Phe - Tyr - Arg - Phe - His -

A protein    - Met - Arg - Ala - Ser - Ile - Lys - Met - Ile - Gly - Val - Ser - Asn - Leu - Gln - Ser - Phe - Ile - Ala - Ser - Met -

             A-T-G-C-G-C-G-C-T-T-C-G-A-T-A-A-A-A-A-T-G-A-T-T-G-G-C-G-T-A-T-C-C-A-A-C-C-T-G-C-A-G-A-G-T-T-T-T-A-T-C-G-C-T-T-C-C-A-T-G-
             Hha I/15   /8b  2/7 Taq I                                     Pst I 1/1
                             10                   20                  30                  40                  50                  60

             (Val)  Gln
              Val
               ↑     ↑
B protein    Asp - Ala - Glu - Val - Asn - Thr - Phe - Gly - Tyr - Phe

               Tyr
               Ter                                                        Leu
                ↑                                                          ↑
A protein    Thr - Gln - Lys - Leu - Thr - Leu - Ser - Asp - Ile - Ser - Asp - Glu - Ser - Lys - Asn - Tyr - Leu - Asp - Lys - Ala -

             A-C-G-C-A-G-A-A-G-T-T-A-A-C-A-C-T-T-T-C-G-G-A-T-A-T-T-T-C-[T-G-A]-T-G-A-G-T-C-G-A-A-A-A-A-T-T-A-T-C-T-T-G-A-T-A-A-A-G-C-A-
             am18 ↓  |                                        ↑   Hinf I 6/9   7/8 Taq I
             am35 T  |       5/7b Hind II                     |                  |
                     ↓       70                  80           90                 ↓ 100               110                 120
              (T)    C ts116                               B stop                T am6

                                                                          Ter
                                                                           ↑
A protein    Gly - Ile - Thr - Thr - Ala - Cys - Leu - Arg - Ile - Lys - Ser - Lys - Trp - Thr - Ala - Gly - Gly - Lys

             G-G-A-A-T-T-A-C-T-A-C-T-G-C-T-T-G-T-T-T-A-C-G-A-A-T-T-A-A-A-T-C-G-A-A-G-T-G-G-A-C-T-G-C-T-G-G-C-G-G-A-A-A-A-[T-G-A]-G-A-A-
                                                              8/9 Taq I    |                          ↑
                             130                 140                 150   ↓      160                 170  |               180
                                                                           A am6                         A stop

             A-A-T-T-C-G-A-C-C-T-A-T-C-C-T-T-G-C-G-C
             Taq I 9/10              Hha I 8b/4
                             190
```

Fig. 4 The sequence of ΦX174 viral DNA in the region of gene *B* and the C terminus of gene *A*. The amino acid sequences of the gene *A* and gene *B* proteins predicted from the DNA sequence are shown together with the changes resulting from the mutations *am*18 and the subsequent *ts*116, *am*35 and *am*6.

resulting gene product is impaired in function, indicating that there is some requirement for the region of the A protein coded by the DNA beyond the end of gene *B*.

We thank R. Kamen, R. J. Roberts and S. Sato for gifts of some of the enzymes used in this study. The *am*18 and *am*35 stock were kindly provided by H. S. Jansz.

Received November 25; accepted December 24, 1976.

1 Benbow, R. M., Mayal, R. F., Picchi, J. C. & Sinsheimer, R. L. *J. Virol.* **10,** 99–114 (1972).
2 Barrell, B. G., Air, G. M. & Hutchison, C. A. III *Nature* **264,** 34–41 (1976).
3 Ravetch J., Model, P. & Robertson, H. D. *Nature* **265,** 698–702 (1977).
4 Brown, N. L. & Smith, M. *Nature* **265,** 695–698 (1977).
5 Sanger, F. & Coulson, A. R. *J. molec. Biol.* **94,** 441–448 (1975).
6 Ling, V. *J. molec. Biol.* **64,** 87–102 (1972).
7 Maniatis, T., Jeffrey, A. & Kleid, D. G. *Proc. natn. Acad. Sci. U.S.A.* **72,** 1184–1188 (1975).
8 Baas, P. D., van Heusden, G. P. H., Vereijken, J. M., Weisbeek, P. J. & Jansz, H. S. *Nucl. Acids Res.* **3,** 1947–1960 (1976).
9 Hutchison, C. A. III thesis, California Institute of Technology (1969).
10 Weisbeek, P. J., Vereijken, J. M., Baas, P. D., Jansz, H. S. & van Arkel, G. G. *Virology* **72,** 61–71 (1976).
11 Sinsheimer, R. L. *Prog. Nucl. Acid Res. molec. Biol.* **8,** 115–169 (1968).
12 Vanderbilt, A. S. & Tessman, I. *Genetics* **66,** 1–10 (1970).

Printed in Great Britain by Henry Ling Ltd., at the Dorset Press, Dorchester, Dorset

(Reprinted from Nature, Vol. 265, No. 5596, pp. 687–695, February 24, 1977)

Nucleotide sequence of bacteriophage ΦX174 DNA

F. Sanger, G. M. Air*, B. G. Barrell, N. L. Brown†, A. R. Coulson, J. C. Fiddes, C. A. Hutchison III‡, P. M. Slocombe§ & M. Smith¶

MRC Laboratory of Molecular Biology, Hills Road, Cambridge CB2 2QH, UK

A DNA sequence for the genome of bacteriophage ΦX174 of approximately 5,375 nucleotides has been determined using the rapid and simple 'plus and minus' method. The sequence identifies many of the features responsible for the production of the proteins of the nine known genes of the organism, including initiation and termination sites for the proteins and RNAs. Two pairs of genes are coded by the same region of DNA using different reading frames.

THE genome of bacteriophage ΦX174 is a single-stranded, circular DNA of approximately 5,400 nucleotides coding for nine known proteins. The order of these genes, as determined by genetic techniques[2–4], is *A-B-C-D-E-J-F-G-H*. Genes *F*, *G* and *H* code for structural proteins of the virus capsid, and gene *J* (as defined by sequence work) codes for a small basic protein that is also part of the virion. Gene *A* is required for double-stranded DNA replication and single-strand synthesis. Genes *B*, *C* and *D* are involved in the production of viral single-stranded DNA: however, the exact function of these gene products is not clear as they may either be involved directly in DNA synthesis or be required for DNA packaging, which is coupled with single-strand production. Gene *E* is responsible for lysis of the host.

The first nucleotide sequences established in ΦX were pyrimidine tracts[5–7] obtained by the Burton and Petersen[8] depurination procedure. The longer tracts could be obtained pure and sequences of up to 10 nucleotides were obtained. More recently Chadwell[9] has improved the hydrazinolysis method to obtain the longer purine tracts. These results are included in the sequence given in Fig. 1.

More extensive ΦX sequences were obtained using partial degradation techniques, particularly with endonuclease IV (refs 10 and 11). Ziff *et al.*[12,13] used this enzyme in conditions of partial hydrolysis to obtain fragments 50–200 nucleotides long which were purified as ^{32}P-labelled material by electrophoresis on polyacrylamide gels. The fragments came from the same region of the genome and the sequence of a 48-nucleotide long fragment (band 6, positions 1,047–1,094) was determined using mainly further degradation with endonuclease IV and partial exonuclease digestions.

Another 50-nucleotide long fragment was obtained by Robertson *et al.*[14] as a ribosome binding site. The viral (or plus) strand DNA of ΦX has the same sequence as the mRNA and, in certain conditions, will bind ribosomes so that a protected fragment can be isolated and sequenced. Only one major site was found. By comparison with the amino acid sequence data it was found that this ribosome binding site sequence coded for the initiation of the gene *G* protein[15] (positions 2,362–2,413).

At this stage sequencing techniques using primed synthesis with DNA polymerase were being developed[16] and Schott[17] synthesised a decanucleotide with a sequence complementary to part of the ribosome binding site. This was used to prime into the intercistronic region between the *F* and *G* genes, using DNA polymerase and ^{32}P-labelled triphosphates[18]. The ribo-substitution technique[16] facilitated the sequence determination of the labelled DNA produced. This decanucleotide-primed system was also used to develop the plus and minus method[1]. Suitable synthetic primers are, however, difficult to prepare and as DNA fragments generated by restriction enzymes are more readily available these have been used for most of the work reported here.

Another approach to DNA sequencing is to make an RNA copy using RNA polymerase with α-^{32}P-labelled ribotriphosphates and then to determine the RNA sequence by more established methods. Blackburn[19,20] used this approach on intact single-stranded ΦX and on fragments obtained by digestion with endonuclease IV or with restriction enzymes. Sedat *et al.*[21] were extending their studies on the larger endonuclease IV fragments and their results, taken in conjunction with the transcription of the DNA fragments[20], amino acid sequence of the F protein[22], and the plus and minus method results, made it possible to deduce a sequence of 281 nucleotides (positions 1,016–1,296, Fig. 1) within the *F* gene[23]. Transcription of *Hin*dII fragment 10, amino acid sequence data in the *G* gene, and the plus and minus method using *Hin*dII fragments 2 and 10 as primers, gave a sequence of 195 nucleotides (positions 2,387–2,582, Fig. 1) at the N terminus of gene *G* (ref. 24).

The 'plus and minus' method

Further work on the ΦX sequence has been done using chiefly the plus and minus method primed with restriction fragments. Figure 2 shows the various restriction enzymes used and the fragment maps for each (refs 25–30 and C.A.H., submitted for publication, and N.L.B., C.A.H. and M.S., submitted for publication).

Figure 1 shows the combined results of the sequence work to date. The sequence is numbered from the single cleavage site of the restriction enzyme *Pst*I. As with other methods of sequencing nucleic acids, the plus and minus technique used by itself cannot be regarded as a completely reliable system and occasional errors may occur. Such errors and uncertainties can only

Present addresses: *John Curtin School of Medical Research, Microbiology Department, Canberra City ACT 2601, Australia, †Department of Biochemistry, University of Bristol, Bristol BS8 1TD, UK, ‡Department of Bacteriology and Immunology, University of North Carolina, Chapel Hill, North Carolina 27514, §Max-Planck-Institut für Molekulare Genetik, 1 Berlin 33, FRG, ¶Department of Biochemistry, University of British Columbia, Vancouver BC, Canada V6T 1W5.

```
P1/1                              R5/7b                             F6/9  T7/8                                                                             T8/9
GAGTTTTATC GCTTCCATGA CGCAGAAGTT AACACTTTCG GATATTTCTG ATGAGTCGAA AATTATCTT GATAAAGCAG GAATTACTAC TGCTTGTTTA CGAATTAAAT CGAAGTGGAC
        10         20         30         40         50         60         70         80         90        100        110        120
                                                 End B ↑

                                         A5/18
                     T9/10          H8b/4   T10/4  A18/6                                         F9/13
TGCTGGCGGA AAATGAGAAA ATTCGACCTA TCCTTGCGCA GCTCGAGAAG CTCTTACTTT GCGACCTTTC GCCATCAACT AACGATTCTG TCAAAAACTG ACGCGTTGGA TGAGGAGAAG
       130 ↑      140        150        160        170        180        190        200        210        220        230        240
     End A

                                                 F13/17                       F17/16a   R7b/6c                                  F16a/16b
TGGCTTAATA TGCTTGGCAC GTTCGTCAAG GACTGGTTTA GATATGAGTC ACATTTTGTT CATGGTAGAG ATTCTCTTGT TGACATTTTA AAAGAGCGTG GATTACTATC TGAGTCCGAT
       250        260        270        280        290        300        310        320        330        340        350        360
                                                                                                                       mRNA start ↑

                                 F16b/1                                        Z3/7       A6/1
GCTGTTCAAC CACTAATAGG TAAGAAATCA TGAGTCAAGT TACTGAACAA TCCGTACGTT TCCAGACCGC TTTGGCCTCT ATTAAGCTCA TTCAGGCTTC TGCCGTTTTG GATTTAACCG
       370        380        390        400        410        420        430        440        450        460        470        480
                         D start ↑

M1/7      T4/5
AAGATGATTT CGATTTTCTG ACGAGTAACA AAGTTTGGAT TGCTACTGAC CGCTCTCGTG CTCGTCGCTG CGTTGAGGCT TGCGTTTATG GTACGCTGGA CTTTGTAGGA TACCCTCGCT
       490        500        510        520        530        540        550        560        570        580        590        600
                                                                                         E start ↑

                                                       R6c/7a        Z7/5                  H4/13
TTCCTGCTCC TGTTGAGTTT ATTGCTGCCG TCATTGCTTA TTATGTTCAT CCCGTCAACA TTCAAACGGC CTGTCTCATC ATGGAAGGCG CTGAATTTAC GGAAAACATT ATTAATGGCG
       610        620        630        640        650        660        670        680        690        700        710        720

T5/3    Y1/3                                          H13/11                          M7/3
TCGAGCGTCC GGTTAAAGCC GCTGAATTGT TCGCGTTTAC CTTGCGTGTA CGCGCAGGAA ACACTGACGT TCTTACTGAC GCAGAAGAAA ACGTGCGTCA AAAATTACGT GCGGAAGGAG
       730        740        750        760        770        780        790        800        810        820        830        840
                                                                                                                              End E ↑

                                 H11/14                                                               H14/12                R7a/6b
TGATGTAATG TCTAAAGGTA AAAAACGTTC TGGCGCTCGC CCTGGTCGTC CGCAGCCGTT GCGAGGTACT AAAGGCAAGC GTAAAGGCGC TCGTCTTTGG TATGTAGGTG GTCAACAATT
       850        860        870        880        890        900        910        920        930        940        950        960
End D ↑ ↑ J start

            Z5/8                                         H12/10
TTAATTGCAG GGGCTTCGGC CCCTTACTTG AGGATAAATT ATGTCTAATA TTCAAACTGG CGCCGAGCGT ATGCCGCATG ACCTTTCCCA TCTTGGCTTC CTTGCTGGTC AGATTGGTCG
       970        980        990       1000       1010       1020       1030       1040       1050       1060       1070       1080
↑ End J                          F start ↑

                       Y3/2       F1/14b  T3/1                 H10/7                              Z8/4       F14b/2
TCTTATTACC ATTTCAACTA CTCCGGTTAT CGCTGGCGAC TCCTTCGAGA TGGACGCCGT TGGCGCTCTC CGTCTTTCTC CATTGCGTCG TGGCCTTGCT ATTGACTCTA CTGTAGACAT
      1090       1100       1110       1120       1130       1140       1150       1160       1170       1180       1190       1200

                                  Q1/3c                                                                R6b/1
TTTTACTTTT TATGTCCCTC ATCGTCACGT TTATGGTGAA CAGTGGATTA AGTTCATGAA GGATGGTGTT AATGCCACTC CTCTCCCGAC TGTTAACCAA ACTACTGGTT ATATTGACCA
      1210       1220       1230       1240       1250       1260       1270       1280       1290       1300       1310       1320

                                                                                                   H7/5
TGCCGCTTTT CTTGGCACGA TTAACCCTGA TACCAATAAA ATCCCTAAGC ATTTGTTTCA GGGTTATTTG ATATCTATAG CGTATTTTAA AGCGCCGTGG ---ATGCCTG ACCGTACCGA
      1330       1340       1350       1360       1370       1380       1390       1400       1410       1420       1430       1440

            A1/12c                                                                                            A12c/13
GGCTAACCCT AATGAGCTTA ATCAAGATGA TGCTCGTTAT GGTTTCCGTT GCTGCCATCT CAAAAACATT TGGACTGCTC CGCTTCCTCC TGAGACTGAG CTTTCTCGCC AAATGACGAC
      1450       1460       1470       1480       1490       1500       1510       1520       1530       1540       1550       1560

                                  A13/2                                                                          M3/4
TTCTACCACA TCTATTGACA TTATGGGTCT GCAAGCTGCT TATGCTAATT TGCATACTGA CCAAGAACGT GATTACTTCA TGCAGCGTTA CCATGA-GTT ATTTCTTCAT TTGGAGGTAA
      1570       1580       1590       1600       1610       1620       1630       1640       1650       1660       1670       1680

                                  H5/9a                                                                   Z4/1
AACCTCATAT GACGCTGACA ACCGTCCTTT ACTTGTCATG CGCTCTAATC TCTGGGCATC TGGCTATGAT GTTGATGGAA CTGACCAAAC GTCGTTAGGC CAGTTTTCTG GTCGTGTTCA
      1690       1700       1710       1720       1730       1740       1750       1760       1770       1780       1790       1800

                                                                  H9a/8a                                F2/11
ACAGACCTAT AAACATTCTG TGCCGCGTTT CTTTGTTCCT GAGCATGGCA CTATGTTTAC TCTTGCGCTG GTTCGTTTTC CGCCTACTGC GACTAAAGAG ATTCAGTACC TTAACGCTAA
      1810       1820       1830       1840       1850       1860       1870       1880       1890       1900       1910       1920

                                                                                                 Q3c/6 F11/7
AGGTGCTTTG ACTTATACCG ATATTGCTGG CGACCCTGTT TTGTATGGCA ACTTGCCGCC GCGTGAAATT TCTATGAAGG ATGTTTTCCG TTCTGGTGAT TCGTCTAAGA AGTTTAAGAT
      1930       1940       1950       1960       1970       1980       1990       2000       2010       2020       2030       2040

                       H8a/6                  Q6/5                                        Q5/3b
TGCTGAGGGT CAGTGGTATC GTTATGCGCC TTCGTATGTT TCTCCTGCTT ATCACCTTCT TGAAGGCTTC CCATTCATTC AGGAACCGCC TTCTGGTGAT TTGCAAGAAC GCGTACTTAT
      2050       2060       2070       2080       2090       2100       2110       2120       2130       2140       2150       2160

                                                                                                  F7/5b
TCGCAACCAT GATTATGACC AGTGTTTCAG TCGTTCAGTT GTTGCAGTGG ATAGTCTTAC CTCATGTGAC GTTTATCGCA ATCTGCCGAC CACTCGCGAT TCAATCATGA CTTCGTGATA
      2170       2180       2190       2200       2210       2220       2230       2240       2250       2260       2270     ↑ 2280
                                                                                                                           End F
```

Fig. 1 For full description see p. 690.

R1/9 H6/3
AAAGATTGAG TGTGAGGTTA TAACCGAAGC GGTAAAAATT TTAATTTTTG CCCGCTGAGG GTTGACCAAG CGAAGCGCGG TAGGTTTTCT GCTTAGGAGT TTAATCATGT TTCAGACTTT
2290 2300 2310 2320 2330 2340 2350 2360 2370 2380 2390 2400
G start ↑

A2/16 A16/15a M4/10 A15a/3 R9/10
TATTTCTCGC CACAATTCAA ACTTTTTTTC TGATAAGCTG GTTCTCACTT CTGTTACTCC AGCTTCTTCG GCACCTGTTT TACAGACACC TAAAGCTACA TCGTCAACGT TATATTTTGA
2410 2420 2430 2440 2450 2460 2470 2480 2490 2500 2510 2520

M10/9 R10/2
TAGTTTGACG GTTAATGCTG GTAATGGTGG TTTTCTTCAT TGCATTCAGA TGGATACATC TGTCAACGCC GCTAATCAGG TTGTTTCAGT TGGTGCTGAT ATTGCTTTTG ATGCCGACCC
2530 2540 2550 2560 2570 2580 2590 2600 2610 2620 2630 2640

F5b/8 M9/2
TAAATTTTTT GCCTGTTTGG TTCGCTTTGA GTCTTCTTCG GTTCCGACTA CCCTCCCGAC TGCCTATGAT GTTTATCCTT TGGATGGTCG CCATGATGGT GGTTATTATA CCGTCAAGGA
2650 2660 2670 2680 2690 2700 2710 2720 2730 2740 2750 2760

Y2/5 F8/4
CTGTGTGACT ATTGACGTCC TTCCCCGTAC GCCGGGCAAT AACGTCTACG TTGGTTTCAT GGTTTGGTCT AACTTTACCG CTACTAAATG CCGCGGATTG GTTTCGCTGA ATCAGGTTAT
2770 2780 2790 2800 2810 2820 2830 2840 2850 2860 2870 2880

Q3b/4 H3/2
TAAAGAGATT ATTTGTCTCC AGCCACTTAA GTGAGGTGAT TTATGTTTGG TGCTATTGCT GGCGGTATTG CTTCTGCTCT TGCTGGTGGC GCCATGTCTA AATTGTTTGG AGGCGGTCAA
2890 2900 2910 2920 2930 2940 2950 2960 2970 2980 2990 3000
End G ↑ H start ↑

Y5/4 Q4/7 Q7/2
AAAGCCGCCT CCGGTGGCAT TCAAGGTGAT GTGCTTGCTA CCGATAACAA TACTGTAGGC ATGGGTGATG CTGGTATTAA ATCTGCCATT CAAGGCTCTA ATGTTCCTAA CCCTGATGAG
3010 3020 3030 3040 3050 3060 3070 3080 3090 3100 3110 3120

Z1/2 A3/9
GCCGCCCCTA GTTTTGTTTC GTGTGCTATT GCTAAAGCTG GTAAAGGACT TCTTGAAGGT ACGTTGCAGG CTGGCACTTC TGCCGTTTCT GATAAGTTGC TTGATTTGGT TGGACTTGGT
3130 3140 3150 3160 3170 3180 3190 3200 3210 3220 3230 3240

A9/12d R2/6a
GGCAAGTCTG CCGCTGATAA AGGAAAGGAT ACTCGTGATT ATCTTGCTGC TGCATTTCCT GAGCTTAATG CTTGGGAGCG TGCTGGTGCT GATGCTTCCT CTGCTGGTAT GGTTGACGCC
3250 3260 3270 3280 3290 3300 3310 3320 3330 3340 3350 3360

Y4/1 F4/14a A12d/7c F14a/12
GGATTTGAGA ATCAAAAAGA GCTTACTAAA ATGCAACTGG ACAATCAGAA AGAGATTGCC GAGATGCAAA ATGAGACTCA AAAAGAGATT GCTGGCATTC AGTCGGCGAC TTCACGCCAG
3370 3380 3390 3400 3410 3420 3430 3440 3450 3460 3470 3480

F12/10
AATACGAAAG ACCAGGTATA TGCACAAAAT GAGATGCTTG CTTATTC-AC AGAAGGAGTC TACTGCTGCG TTGCGTCTAT TATGGAAAAC ACCAATCTTT CCAAGCAACA GCAGGTTTCC
3490 3500 3510 3520 3530 3540 3550 3560 3570 3580 3590 3600

H2/9b A7c/8 F10/15 R6a/4
GAGATTATGC GCCAAATGCT TACTCAAGCT CAAACGGCTG GTCAGTATTT TACCAATGAC CAAATCAAAG AAATGACTCG CAAGGTTAGT GCTGAGGTTG ACTTAGTTCA TCAGCAAACG
3610 3620 3630 3640 3650 3660 3670 3680 3690 3700 3710 3720

F15/5c M2/5 H9b/1 A8/14
CAGAATCAGC GGTATGGCTC TTCTCATATT GGCGCTACTG CAAAGGATAT TTCTAATGTC GTCACTGATG CTGCTTCTGG TGTGGTTGAT ATTTTTCATG GTATTGATAA AGCTGTTGCC
3730 3740 3750 3760 3770 3780 3790 3800 3810 3820 3830 3840

A14/7b
GATACTTGGA ACAATTTCTG GAAAGACGGT AAAGCTGATG GTATTGGCTC TAATTTGTCT AGGAAATAAC CGTCAGGATT GACACCCTCC CAATTGTATG TTTTCATGCC TCCAAATCTT
3850 3860 3870 3880 3890 3900 3910 3920 3930 3940 3950 3960
End H ↑ mRNA start ↑

T1/6
GGAGGCTTTT TTATGGTTCG TTCTTATTAC CCTTCTGAAT GTCACGCTGA TTATTTTGAC TTTGAGCGTA TCGAGGCTCT TAAACCTGCT ATTGAGGCTT GTGGCATTTC TACTCTTTCT
3970 3980 3990 4000 4010 4020 4030 4040 4050 4060 4070 4080
A start ↑

A7b/7a M5/8 F5c/3 T6/2 Q2/3a R4/3 Z2/6b
CAATCCCCAA TGCTTGGCTT CCATAAGCAG ATGGATAACC GCATCAAGCT CTTGGAAGAG ATTCTGTCTT TTCGTATGCA GGGCGTTGAG TTCGATAATG GTGATATGTA TGTTGACGGC
4090 4100 4110 4120 4130 4140 4150 4160 4170 4180 4190 4200

CATAAGGCTG CTTCTGACGT TCGTGATGAG TTTGTATCTG TTACTGAGAA GTTAATGGAT GAATTGGCAC AATGCTACAA TGTGCTCCCC CAACTTGATA TTAATAACAC TATAGACCAC
4210 4220 4230 4240 4250 4260 4270 4280 4290 4300 4310 4320

M8/6 A7a/4
CGCCCCGAAG GGGACGAAAA ATGGTTTTTA GAGAACGAGA AGACGGTTAC GCAGTTTTGC AAGCTGGCTG CTGAACGCCC TCTTAAGGAT ATTCGCGATG AGTATAATTA CCCCAAAAAG
4330 4340 4350 4360 4370 4380 4390 4400 4410 4420 4430 4440

Z6b/6a
AAAGGTATTA AGGATGAGTG TTCAAGATTG CTGGAGGCCT CCACTAAGAT ATCGCGTAGA GGCTTTGCTA TTCAGCGTTT GATGAATGCA ATGCGACAGG CTCATGCTGA TGGTTGGTTT
4450 4460 4470 4480 4490 4500 4510 4520 4530 4540 4550 4560

ATCGTTTTTG ACACTCTCAC GTTGGCTGAC GACCGATTAG AGGCGTTTTA TGATAATCCC AATGCTTTGC GTGACTATTT TCGTGATATT GGTCGTATGG TTCTTGCTGC CGAGGGTCGC
4570 4580 4590 4600 4610 4620 4630 4640 4650 4660 4670 4680

Fig. 1 (Continued).

```
          F3/5a                                                                A4/11     Z6a/9  M6/1
AAGGCTAATG ATTCACACGC CGACTGCTAT CAGTATTTTT GTGTGCCTGA GTATGGTACA GCTAATGGCC GTCTTCATTT CCATGCGGTG CACTTTATGC GGACACTTCC TACAGGTAGC
      4690       4700       4710       4720       4730       4740       4750       4760       4770       4780       4790       4800

R3/8                                                    A11/10          Z9/10
GTTGACCCTA ATTTGGTCG TCGGGTACGC AATCGCCGCC AGTTAAATAG CTTGCAAAAT ACGTGGCCTT ATGGTTACAG TATGCCCATC GCAGTTCGCT ACACGCAGGA CGCTTTTTCA
      4810       4820       4830       4840       4850       4860       4870       4880       4890       4900       4910       4920

                Z10/3                  Q3a/1        A10/12b                                                         R8/5
CGTTCTGGTT GGTTGTGGCC TGTTGATGCT AAAGGTGAGC CGCTTAAAGC TACCAGTTAT ATGGCTGTTG GTTTCTATGT GGCTAAATAC GTTAACAAAA AGTCAGATAT GGACCTTGCT
      4930       4940       4950       4960       4970       4980       4990       5000       5010       5020       5030       5040

             A12b/15b                           A15b/17                     A17/12a      F5a/6
GCTAAAGGTC TAGGAGCTAA AGAATGGAAC AACTCACTAA AAACCAAGCT GTCGCTACTT CCCAAGAAGC TGTTCAGAAT CAGAATGAGC CGCAACTTCG GGATGAAAAT GCTCACAATG
      5050       5060   ↑   5070       5080       5090       5100       5110       5120       5130       5140       5150       5160
                 B start

                                              A12a/5
ACAAATCTGT CCACGGAGTG CTTAATCCAA CTTACCAAGC TGGGTTACGA CGCGACGCCG TTCAACCAGA TATTGAAGCA GAACGCAAAA AGAGAGATGA GATTGAGGCT GGGAAAAGTT
      5170       5180       5190       5200       5210       5220       5230       5240       5250       5260       5270       5280

                        H1/15                                         H15/8b  T2/7                             P1/1
ACTGTAGCCG ACGTTTTGGC GGCGCAACCT GTGACGACAA ATCTGCTCAA ATTTATGCGC GCTTCGATAA AAATGATTGG CGTATCCAAC CTGCA
      5290       5300       5310       5320       5330       5340       5350       5360       5370  5375
```

Fig. 1 A provisional nucleotide sequence for the DNA of bacteriophage ΦX174 *am*3 *cs*70. Solid underlining indicates sequences that are fully confirmed; sequences with no underlining probably do not contain more than one mistake per 50 residues. Broken underlining indicates more uncertain sequences. Restriction enzyme recognition sites are indicated (for key to single letter enzyme code see legend to Fig. 2), as are mRNA starts and protein initiation and termination sites. Nucleotides 4,127 to 4,201 have been independently sequenced by van Mansfield *et al.*[58]. The *am*3 codon is at position 587.

be eliminated by more laborious experiments and, although much of the sequence has been so confirmed, it would probably be a long time before the complete sequence could be established. We are not certain that there is any scientific justification for establishing every detail and, as it is felt that the results may be useful to other workers, it has been decided to publish the sequence in its present form.

As template we have used both the viral (plus) and complementary (minus) strands of ΦX. Usually it is possible to determine a sequence with a single primer starting at about 15–100 nucleotides from the appropriate restriction enzyme site. In a particularly good experiment the sequence can be read out to 150–200 nucleotides but the results may become less reliable. Most sequences have been derived by priming on both strands; this allows more confidence than when only one strand could be used.

A useful method for confirming runs of the same nucleotide is depurination of ^{32}P-labelled small restriction enzyme fragments or of products of the DNA polymerase priming experiments (ref. 31 and N.L.B. and M.S., in preparation). The most satisfactory way of confirming the DNA sequences is through amino acid sequence data. As the methods used are entirely unrelated, the results of the two approaches complement each other very well and therefore complete sequences can usually be deduced from incomplete data obtained by each method. The complete sequence of genes *G* (ref. 32), *D* (ref. 33), *J* (ref. 33 and Freymeyer, unpublished) and most of *F* have been obtained in this way.

Many of the sequences in Fig. 1 have been amply confirmed and are regarded as established: these are indicated in the figure by underlining. Some sequences are considered to be reasonably accurate and probably contain no more than one mistake in every 50 nucleotides. Sequences that are particularly uncertain—either because of lack of data or conflicting results—are also indicated in Fig. 1.

In considering the sequence of ΦX174 as a functional unit it is convenient to begin in the region between the *H* and *A* genes and to continue around the DNA in the direction of transcription and translation.

A promoter and terminator

Sinsheimer *et al.*[34,35] and Axelrod[36] have determined the sequences of the 5′ end of three ΦX *in vitro* mRNA species and have located them on the restriction map. These sequences have been identified on the DNA sequence and one of them (AAATCTTGG) is found only at position 3,954 at which an *in vivo* unstable mRNA start has been located[37]. The sequence to the left of this has some characteristics of typical *E. coli* promoters[38] in that five out of the 'ideal' TATPuATPu residues are present. Nearby, to the right of this mRNA initiation, however, is the sequence TTTTTTA which is similar to sequences found at the 3′ ends of a number of mRNAs (see ref. 39) and seems a likely signal for mRNA termination. The presence of a rho-independent termination site in this approximate position has been suggested[36,37], but the relative positions of the initiating and putative termination signals is rather surprising since the terminator for one mRNA would be expected to precede the initiator for the next. One possibility is that the T_6A might be acting as an 'attenuator' involved in the control of mRNA production in a similar manner to that suggested for the tryptophan operon by Bertrand *et al.*[40]. If indeed it were acting as a transcription terminator one would expect a small RNA of 20 nucleotides to be produced, but no such product has yet been detected. Recent work, however, (Rosenberg, unpublished and ref. 41) indicates that termination may require the presence of a base-paired loop structure before the termination site. From the DNA sequence such a loop is probably present before the T_6A sequence, but in mRNA starting from the initiation site at position 3,954 this loop is not formed (Fig. 3). Therefore mRNA that had started at an earlier promoter and extended through the *H* gene would be expected to terminate here, whereas mRNA newly initiated at position 3,954 would not. This could be a way in which the phage has economised on the use of DNA—by having the ends of the two mRNAs overlapping.

The A protein

Where the amino acid sequence is available there is no problem in relating the DNA sequence to its coding properties, but it is

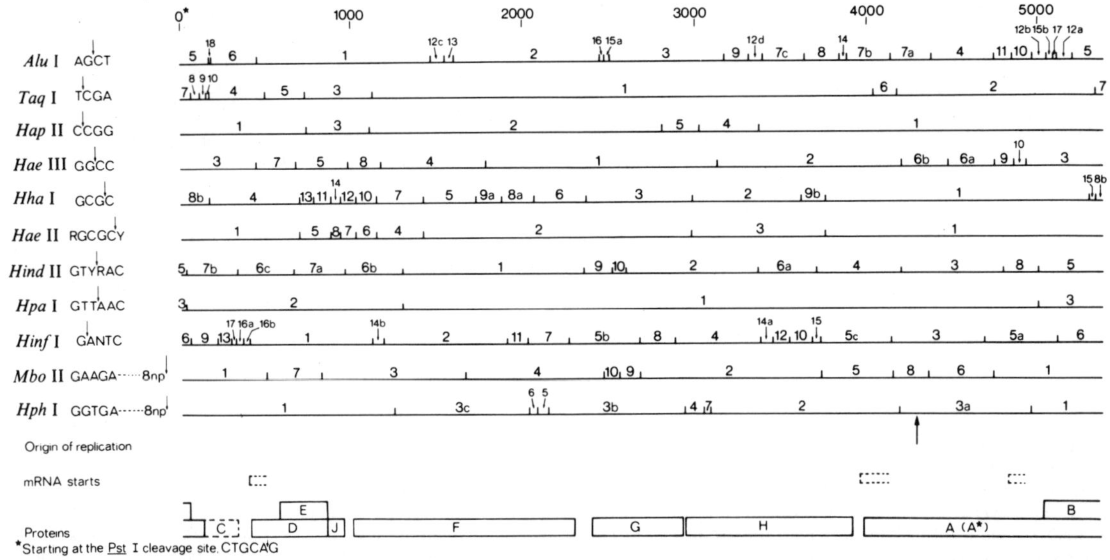

Fig. 2 Fragment maps of restriction enzymes used in the sequence analysis of ΦX174 *am*3 RFI DNA. Fragment maps of ΦX174 have been prepared for *Hin*dII (R), *Hae*III (Z) and *Hpa*I+II by Lee and Sinsheimer[25], *Hin*HI and *Hap*II (Y) by Hayashi and Hayashi[26], and for *Alu*I (A) by Vereijken *et al.*[27] and for *Pst*I (P) by Brown and Smith[30]. B.G.B., G.M.A., C.A.H. and D. Jaffe prepared the *Hin*fI (F) map, C.A.H. the *Hph*I (Q) map, and Jeppesen *et al.*[28] the *Hha*I (H), *Alu*I, *Hae*II and *Hap*II maps by using a rapid method depending on priming with DNA polymerase. A rapid two-dimensional hybridisation technique has been developed by C.A.H. (submitted for publication) and recently used for mapping *Mbo*II (M) (N.L.B., C.A.H., and M.S., submitted for publication) and *Taq*I (T)[29]. *Hha*I and *Hin*fI maps have also been prepared by Baas *et al.*[52].

more difficult to do so in the absence of such data, as is the case for the A protein. One way of identifying the reading phase of the DNA is from the distribution of nonsense codons. Over a sufficiently long sequence that is known to be coding for a single protein there is usually one phase that contains no nonsense codons, and this is identified as the reading phase. This requires completely accurate determination of the DNA sequences however: omission of a single nucleotide may give completely erroneous results. Another approach is possible in the case of ΦX. The results with the *F* and *G* genes[23,24,32] showed an unexpectedly high frequency of codons ending in T. Therefore in a coding region there is a tendency for every third nucleotide to be a T and it is then possible to define the reading phase. Figure 4 illustrates how this characteristic was used to help determine the reading phase for the A protein and to identify its initiation codon at position 3,973. In a similar way the distribution of Ts may be used to identify errors in the DNA sequence, provided that such errors occur only infrequently.

A different approach to identifying the initiation site and reading phase in a coding sequence is by looking for a characteristic 'initiation sequence'. Shine and Dalgarno have shown that a common feature of ribosome binding sites is a number of nucleotides (at least three) preceding the ATG that are capable of forming base pairs with a sequence at the 3′ end of 16S rRNA[42,43]. All of the known initiation sites in ΦX174 that have been identified by direct amino acid sequencing (for the F, G, H, J and D proteins) satisfy this criterion (see Table 2) and the fact that the sequence preceding the ATG in position 3,973 also has this characteristic supports its identification as the initiation site for the A protein.

If, as has been suggested[37], some mRNA from the previous promoter does extend beyond the hairpin structure, initiation of A protein synthesis may be controlled by the inclusion of the region complementary to the 16S rRNA in the hairpin loop. This could explain the presence of two types of mRNA covering the *A* cistron, as suggested by Hayashi *et al.*[37]—one unstable and active and the other stable but inactive. The former would be intiated at the A promoter, the latter at an earlier promoter and result from 'read-through' at the terminator. The postulated reading frame for the A protein was confirmed by sequencing amber mutants mapping in the N-terminal region of gene *A*. *am*86 proved to be a C→T change at position 4,108 and *am*33 a C→T change at position 4,372. These both result in formation of an amber codon (TAG) in the same reading frame as the proposed initiating ATG and the sequence continues to the termination codon at position 133. The A protein, which is the largest coded by ΦX174, is thus 512 amino acids long with a molecular weight of 56,000, in good agreement with SDS gel estimations (see refs 4 and 44). The A* protein, with a molecular weight of about 35,000, is believed to result from an internal translational start in the *A* gene, in the same reading phase[45]. From consideration of possible ribosome binding sequences[42,43] the ATG in position 4,657 seems to be the most likely initiation site for the A* protein.

The origin of replication

The origin of ΦX viral strand DNA synthesis has been located in gene *A*, in restriction fragment Z6b (ref. 46). This origin, while

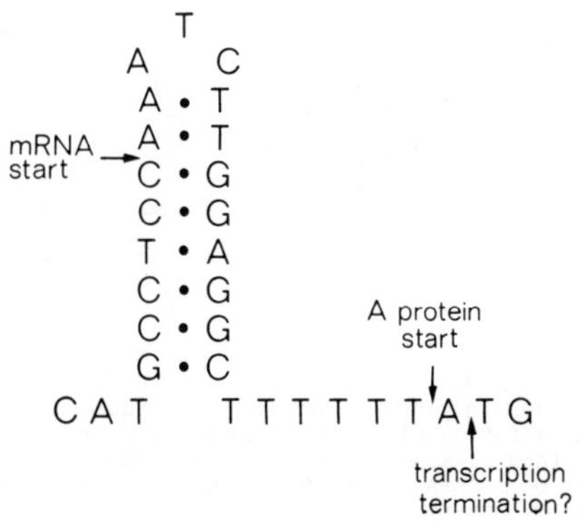

Fig. 3 Potential secondary structure at the *A* mRNA start.

		Ts in codon position (1)	(2)	(3)
3,910	CCG.TCA.GGA.TTG.ACA.CCC.TCC.CAA.TTG.TAT.	5	2	1
		3	4	3*
3,940	GTT.TTC.ATG.CCT.CCA.AAT.CTT.GGA.GGC.TTT.	2	5	5
		3	5	7*
3,970	TTT.<u>ATG</u>.GTT.CGT.TCT.TAT.TAC.CCT.TCT.GAA.	5	3	7
		5	0	7*
4,000	TCT.CAC.CCT.GAT.TAT.TTT.GAC.TTT.CAC.CGT.	4	2	7

*These figures refer to the last five codons of the previous line and the first five of the next line.

Fig. 4 Identification of the initiation codon for the A protein. Sequences of 30 nucleotides in the region in which the initiation was expected were written down and arbitrarily marked off in triplets. The number of T residues in the first position in each triplet was then counted and listed, and similarly the number of Ts in the second and third positions. The marked preference for Ts in position 3 in the last two lines, as compared with the first two lines, suggests that they are coding for protein and that the triplets are correctly marked off. The most likely initiation codon for the A protein is the ATG in position 3,973.

coding for part of the A protein, probably corresponds to the position of the plus strand nick made by the same protein[44]. Gaps in this region that are found in replicating double-stranded (RF) DNAs are probably related to the position of the nick. Eisenberg *et al.*[47] have investigated such gaps by depurination analysis and identified, in particular, the product C_6T. The sequence CTC_5 is found in position 4,285 (Fig. 1) and the location of the origin in this region agrees precisely with the results of Baas *et al.*[46]. It is not possible at present to identify the actual position of the origin nick. The region shows no apparent secondary structure or symmetrical sequences, although there is an AT-rich region (4,298–4,307) between two GC-rich regions which might be of significance. Such a region is found near the origin of replication of SV40 DNA (ref. 48).

B promoter

The second of the mRNA 5′ sequences (AUCGC)[34] has been mapped in restriction fragment R8 (Fig. 2), which starts about 300 nucleotides on from the proposed A* initiation. The sequence ATCGC is found at positions 4,832 and 4,888 in Fig. 1. The only way we can choose between them at the moment is that the second is preceded by the sequence TACAGTA (position 4,877), which is more akin to sequences found in known promoters[38] than are sequences preceding the other possible mRNA start. Irrespective of which of these sequences is used, the mRNA has a long 'leader' sequence (232 or 176 nucleotides) before the next proposed initiation codon (gene *B*).

Table 1 ΦX174 coding capacity

Gene	Protein molecular weight from SDS gels*	Number of nucleotides (Fig. 1)	Protein molecular weight from sequence information
A	55,000–67,000	1,536	56,000
(*A**)	35,000		
B	19,000–25,000	(360)†	13,845‡
C	7,000		
D	14,500	456	16,811‡
E	10,000–17,500	(273)†	9,940
J	5,000	114	4,097‡
F	48,000	1,275	46,400
G	19,000	525	19,053‡
H	37,000	984	35,800
Non-coding and *C*		485	
Total		5,375	

*See ref. 4.
†Values in parenthesis are overlapping sequences and therefore not included in the addition to obtain the total length of DNA.
‡These values are calculated from the amino acid sequence (in the case of B deduced from the nucleotide sequence). The others are derived using the formula

$$\text{Protein molecular weight} = \frac{\text{No. of nucleotides}}{3\times 0.00915}$$

The B protein

From a study of the ribonuclease T_1 digestion products of the ribosome binding sites of ΦX mRNAs[49], it was possible to identify an initiating ATG in position 5,064. From the genetic map[2,3], this would be expected to be gene *B* but, as discussed above, the A protein coding sequence extends right through this region, past the *Pst* site at residue 1 in Fig. 1, and terminates at residue 133. The initiating codon contained in the ribosome-protected sequence is, however, out of phase with the A protein reading frame. The proposed B protein coding sequence is one nucleotide to the left of the A protein phase, and continues until a termination codon occurs at position 49. Therefore the B protein coding sequence is totally contained within the *A* gene. These reading frames have been confirmed by sequencing mutants in genes *A* and *B* (*am*16, N.L.B. and M.S., in preparation; *am*18, *am*35, *ts*116 (ref. 50). Since the B protein has not been purified no protein sequence data is available. The complete amino acid sequence can be predicted from the DNA sequence however. The protein is 120 amino acids long with a molecular weight of 13,845 (including the N-terminal Met). The molecular weight estimates of the B protein obtained by SDS–gel electrophoresis are mostly greater than this (see review, ref. 4), but the electrophoretic mobility varied with gel concentration and cross linker. Such anomalous behaviour suggests that there may be, for instance, carbohydrate attached to the B protein.

The C protein

The next known gene product, protein C, maps between genes *B* and *D*. Examination of the DNA sequence in this region indicates that the most probable initiating ATG overlaps the termination codon, TGA, of gene A in the sequence ATGA at position 134. A possible termination codon for gene *C* could then be at position 391, although the sequence and phasing is not yet confirmed through this region. There is another possible protein initiation codon (position 51, overlapping the B protein termination codon) which would result in a slightly shorter gene product terminating at nucleotide 219. For the C protein, however, we favour the 'A terminator' start, since only this reading frame contains a CAA sequence, which by a C→T alteration could give the ochre 6 mutant. Ochre 6 is a gene *C* mutant produced by the decay of ^{3}H-cytosine[51] and has been mapped in fragments A6 and F9 (ref. 52); that is, between nucleotides 170 and 205 (Fig. 1).

Sequence following the *D* promoter

The mRNA 5′ sequence which maps before the *D* gene (GAUGC)[34] is found at position 358 in Fig. 1. The sequence preceding the messenger start has only four of the TATPuATPu nucleotides[38]. Thirty-two nucleotides after the mRNA initiation is the ATG (position 390) that initiates D protein synthesis. The amino acid sequence of the D protein has been determined

Table 2 Initiation sequences of ΦX174 coded proteins

D	C-C-A-C-T-A-A-T-A-G-G-T-A-A-G-A-A-A-T-C-A-T-G-A-G-T-C-A-A-G-T-T-A-C-T
	Ser Gln Val Thr
E	C-T-G-C-G-T-T-G-A-G-G-C-T-T-G-C-G-T-T-T-A-T-G-G-T-A-C-G-C-T-G-G-A-C-T
J	C-G-T-G-C-G-G-A-A-G-G-A-G-T-G-A-T-G-T-A-A-T-G-T-C-T-A-A-A-G-G-T-A-A-A
	Ser Lys Gly Lys
F	C-C-C-T-T-A-C-T-T-G-A-G-G-A-T-A-A-A-T-T-A-T-G-T-C-T-A-A-T-A-T-T-C-A-A
	Ser Asn Ile Gln
G	T-T-C-T-G-C-T-T-A-G-G-A-G-T-T-T-A-A-T-C-A-T-G-T-T-T-C-A-G-A-C-T-T-T-T
	Met Phe Gln Thr Phe
H	C-C-A-C-T-T-A-A-G-T-G-A-G-G-T-G-A-T-T-T-A-T-G-T-T-T-G-G-T-G-C-T-A-T-T
	Met Phe Gly Ala Ile
A	C-A-A-A-T-C-T-T-G-G-A-G-G-C-T-T-T-T-T-T-A-T-G-G-T-T-C-G-T-T-C-T-T-A-T
B	A-A-A-G-G-T-C-T-A-G-G-A-G-C-T-A-A-A-G-A-A-T-G-G-A-A-C-A-A-C-T-C-A-C-T
16S RNA 3' end	HOA-U-U-C-C-U-C-C-A-C-U-A-G

Where the protein start has been independently confirmed by protein sequencing data the amino acid sequences are indicated. The other initiation regions were identified as described in the text. Sequences complementary to the 3′ end of 16S rRNA (refs 42, 43) are boxed; broken lines indicate further complementarity if some nucleotides are looped out or not matched. Ribosome binding to mRNA has been demonstrated in these regions for genes *J*, *F*, *G* and *B* (ref. 49).

almost completely, and nucleotide and amino acid sequences can be correlated to the termination codon at position 846 (ref. 33). The D protein, which is involved in capsid assembly, is 151 amino acids in length, with a molecular weight of 16,811. The *D* termination codon overlaps the initiation codon for gene *J* in the sequence TAATG. A similar structure has also been found by Platt and Yanofsky[53] in the tryptophan operon. The DNA sequence following this initiation codon matches the amino acid sequence of the small basic protein (37 amino acids) of the virion determined by D. Freymeyer, P. R. Shank, T. Vanaman, C.A.H. and M. H. Edgell (personal communication). Benbow *et al.*[2,3] suggested that the mutation *am*6 was located in a gene *J*, coding for the small protein of the virion, and mapping immediately before gene *F*. Although marker rescue experiments indicate that *am*6 is not in this region[54], the DNA sequence shows that there is a gene coding for the virion protein and we

Table 3 Promoter sequences in ΦX174

A promoter	A-G-G-A-T-T-G-A-C-A-C-C-C-T-C-C-C-A-A-T-T-G-T-A-T-G-T-T-T-T-C-A-T-G-C-C-T-C-C-A-A-A-T-C-T — — — (3954; ↑ 18 nucleotides to A protein start)
D promoter	R7b/R6c G-T-T-G-A-C-A-T-T-T-T-A-A-A-A-G-A-G-C-G-T-G-G-A-T-T-A-C-T-A-T-C-T-G-A-G-T-C-C-G-A-T-G-C-T (358; ↑ 32 nucleotides to D protein start)
B promoter?	R3/R8 C-A-G-G-T-A-G-C-G-T-T-G-A-C-C-C-T-A-A-T-T-T-T-G-G-T-C-G-T-C-G-G-G-T-A-C-G-C-A-A-T-C-G-C-C (4832; ↑ 232 nucleotides to B protein start)
B promoter?	A-G-C-T-T-G-C-A-A-A-A-T-A-C-G-T-G-G-C-C-T-T-A-T-G-G-T-T-A-C-A-G-T-A-T-G-C-C-C-A-T-C-G-C-A (4888; ↑ 176 nucleotides to B protein start)

mRNA initiation sequences[34–36] are underlined. Boxed regions indicate sequences that may correspond to the TATPuATPu sequence found in other promoters[38], taking into account the distance from the mRNA starts.

have defined this as gene *J* (ref. 33). Since the *J* initiation codon overlaps the *D* termination codon we had to look elsewhere for gene *E*, which genetic mapping[2,3] had placed between them. Amber mutants in gene *E* (*am*3, *am*27, *am*34 and *am*N11) were located by the marker rescue technique and sequenced. All were found to be within the *D* coding sequence, with the mutant amber codons one nucleotide to the right of the *D* reading frame[33]. Thus the *E* coding sequence is completely contained within the *D* coding region but in a different reading frame. The proposed initiation and termination codons for the E protein are at nucleotides 568 and 840, respectively[33], giving a protein 91 amino acids in length with a molecular weight of about 9,900 (including the N-terminal methionine).

The F protein

Following the *J* gene is an intercistronic region of 39 nucleotides before initiation of the F protein. There is no known function of this apparently untranslated sequence, although the presence of a hairpin structure (positions 969–984) suggests that it could be the site of the *in vivo* messenger termination signal [37] mapped in this region. The F protein is initiated by the ATG at position 1,001. This is the capsid component of the virion, and almost all the amino acid sequence is known[22,24]. There are regions in this gene where the DNA sequence is not completely established, but the protein is about 424 amino acids in length, giving a molecular weight of ≃46,300.

The G protein region

The termination signal for the F protein (position 2,276) is followed by an unusually long untranslated sequence of 111 nucleotides until the G protein initiation codon[31]. This region contains a looped structure which was postulated to have some functional role, as yet unknown, in the single-stranded DNA or the mRNA.

Initiation of the G protein at position 2,387 is followed by a sequence of 425 nucleotides until termination at position 2,912, giving a spike protein of molecular weight 19,053. The nucleotide and amino acid sequences of this gene and product are known[24,32].

The H protein

The initiation codon for the H protein (position 2,923) was identified first on the basis of the distribution of T nucleotides between the three reading phases, and later confirmed by amino acid sequence analysis. Amino acid sequence data on the H protein is minimal but the five peptide sequences known do correspond to the amino acid sequence, deduced from the DNA sequence by using the high frequency of third position T to help in assigning a reading frame to any given region. The DNA sequence is not entirely confirmed but it is possible to write a reasonably accurate amino acid sequence for the H protein. The protein terminates at nucleotide 3,907, in agreement with carboxypeptidase results, giving a spike protein of molecular weight ≃ 35,600 (326 amino acids). The amino acid sequence at the N terminus seems to be particularly rich in hydrophobic residues, which is consistent with its suggested function as the 'pilot' protein that reacts with the bacterial membrane[55,56]. After H protein termination there are 66 nucleotides before initiation of the A protein at position 3,973.

Coding capacity of the ΦX174 genome

The most striking feature of the ΦX DNA sequence is the way in which the various functions of the genome are compressed within the 5,375 nucleotides. Since the identification of ΦX gene products[2,4] it has been clear that proteins of the accepted molecular weights could not be separately coded on the available length of DNA. However, with the presence of two pairs of overlapping genes (*B* within *A* (ref. 50), *E* within *D* (ref. 33)) the genome has more coding capacity than had been originally supposed on the assumption that each gene was physically separate. Table 1 summarises the molecular weights of the known ΦX-coded proteins. There are other potential initiation sites for polypeptide synthesis (for example, in genes *A*, *F*, *G* and *H*) and further genetic work may clarify whether there are in fact other ΦX genes as yet unidentified.

Initiation of protein synthesis

Table 2 lists the protein initiation sequences for genes *A*, *B*, *D*, *E*, *J*, *F*, *G* and *H*. It can be noted that there are no extra precursor sequences in proteins D, J, F, G or H at either the N or C terminus. There seems to be no relationship between the degree of complementarity to the 16S rRNA and the amount of protein synthesised, and we see no other features in the sequence that could explain different efficiencies of translation except where genes overlap.

Transcription of ΦX174

The sequences preceding known mRNA starts[34–36] are shown in Table 3. Other studies on promoter sequences[38] have suggested certain features that they may have in common. Although some of these features are present in the sequences preceding the ΦX transcription initiations others are not, and at present it is difficult to suggest what signal on the DNA determines a promoter site or the efficiency with which it initiates RNA synthesis. It is interesting to note that a polymerase binding site found by Chen *et al.*[57], but not associated with any *in vitro* or *in vivo* mRNA starts, mapped near the region where there is the sequence TATGATG characteristic of promoters[38] (positions 2,705–2,711).

Table 4 Codons used in ΦX174

Phe	TTT	39	Ser	TCT	35	Tyr	TAT	36	Cys	TGT	12
	TTC	26		TCC	9		TAC	15		TGC	10
Leu	TTA	19		TCA	16	Ter	TAA	3	Ter	TGA	5
	TTG	26		TCG	14		TAG	0	Trp	TGG	16
Leu	CTT	36	Pro	CCT	34	His	CAT	16	Arg	CGT	40
	CTC	15		CCC	6		CAC	7		CGC	29
	CTA	3		CCA	6	Gln	CAA	27		CGA	4
	CTG	24		CCG	21		CAG	34		CGG	8
Ile	ATT	45	Thr	ACT	40	Asn	AAT	37	Ser	AGT	9
	ATC	12		ACC	18		AAC	25		AGC	5
	ATA	2		ACA	13	Lys	AAA	47	Arg	AGA	6
Met	ATG	42		ACG	19		AAG	31		AGG	1
Val	GTT	53	Ala	GCT	64	Asp	GAT	44	Gly	GGT	38
	GTC	14		GCC	17		GAC	35		GGC	28
	GTA	10		GCA	12	Glu	GAA	27		GGA	13
	GTG	11		GCG	12		GAG	34		GGG	3

The totals are derived from sequences in Fig. 1 which are fully confirmed, that is, 377 codons in gene *A*, 120 in gene *B*, 152 in gene *D*, 91 in gene *E*, 38 in gene *J*, 344 in gene *F*, 175 in gene *G* and 49 in gene *H*. Out of a total of 1,346 codons 42.9% terminate in T. The percentages in the different genes are: *A*, 37.1 (non-overlapping region 47.1; overlapping region 15.8); *B*, 34.2; *D*, 42.1; *E*, 14.3; *J*, 47.4; *F*, 52.0; *G*, 54.3; *H*, 49.0. The initiating ATG is included in all cases.

The use of codons in ΦX174

Table 4 shows the codons used in regions where the nucleotide sequence is fully confirmed. It is clear that the pattern established by early observations on non-random use of codons[23,24] is continued now that more information is available. In particular, the preference for T at the third position of the codon is marked throughout the genome, as shown in Table 4. In regions of overlapping genes, one of the pair tends to continue the 'third T' trend (*D* and *B*), thus excluding the other (*E* and *A*). This may give some indication of the order in which overlapping genes evolved[33,50]. Another interesting feature is the very low occurrence of codons starting AG, particularly in non-overlapping regions. The base composition of the sequence of ΦX174 DNA shown in Fig. 1 is: *A*, 23.9%; *C*, 21.5%; *G*, 23.3% and *T*, 31.2%. This is in good agreement with previously determined values (see ref. 4).

We thank D. McCallum and R. Staden for carrying out the computer data storage and analysis of the sequence.

Note added in proof: J. E. Sims and D. Dressler (personal communication) have independently determined the sequence in positions 263–375 and 4,801–4,940. Their results agree with those given in Fig. 1. They have also identified the 'B' mRNA start as being at position 4,888.

Received November 30; accepted December 24, 1976.

1 Sanger, F. & Coulson, A. R. *J. molec. Biol.* **94**, 441–448 (1975).
2 Benbow, R. M., Hutchison, C. A. III, Fabricant, J. D. & Sinsheimer, R. L. *J. Virol.* **7**, 549–558 (1971).
3 Benbow, R. M., Zuccarelli, A. J., Davis, G. C. & Sinshiemer, R. L. *J. Virol.* **13**, 898–907 (1974).
4 Denhardt, D. T. *CRC Crit. Rev. Microbiol.* **4**, 161–222 (1975).
5 Hall, J. B. & Sinsheimer, R. L. *J. molec. Biol.* **6**, 115–127 (1963).
6 Ling, V. *Proc. natn. Acad. Sci. U.S.A.* **69**, 742–746 (1972).
7 Harbers, B., Delaney, A. D., Harbers, K. & Spencer, J. H. *Biochemistry* **15**, 407–414 (1976).
8 Burton, K. & Petersen, G. B. *Biochem. J.* **75**, 17–27 (1960).
9 Chadwell, H. A. Thesis, University of Cambridge (1974).
10 Sadowski, P. D. & Bakyta, I. *J. biol. Chem.* **247**, 405–412 (1972).
11 Ling, V. *FEBS Lett.* **19**, 50–54 (1971).
12 Ziff, E. B., Sedat, J. W. & Galibert, F. *Nature new Biol.* **241**, 34–37 (1973).
13 Galibert, F., Sedat, J. W. & Ziff, E. B. *J. molec. Biol.* **87**, 377–407 (1974).
14 Robertson, H. D., Barrell, B. G., Weith, H. L. & Donelson, J. E. *Nature new Biol.* **241**, 38–40 (1973).
15 Air, G. M. & Bridgen, J. *Nature new Biol.* **241**, 40–41 (1973).
16 Sanger, F., Donelson, J. E., Coulson, A. R., Kössel, H. & Fischer, D. *Proc. natn. Acad. Sci. U.S.A.* **70**, 1209–1213 (1973).
17 Schott, H. *Makromolek. Chem.* **175**, 1683–1693 (1974).
18 Donelson, J. E., Barrell, B. G., Weith, H. L., Kössel, H. & Schott, H. *Eur. J. Biochem.* **58**, 383–395 (1975).
19 Blackburn, E. H. *J. molec. Biol.* **93**, 367–374 (1975).
20 Blackburn, E. H. *J. molec. Biol.* **107**, 417–432 (1976).
21 Sedat, J. W., Ziff, E. B. & Galibert, F. *J. molec. Biol.* **107**, 391–416 (1976).
22 Air, G. M. *J. molec. Biol.* **107**, 433–444 (1976).
23 Air, G. M. *et al. J. molec. Biol.* **107**, 445–458 (1976).
24 Air, G. M., Blackburn, E. H., Sanger, F. & Coulson, A. R. *J. molec. Biol.* **96**, 703–719 (1975).
25 Lee, A. S. & Sinsheimer, R. L. *Proc. natn. Acad. Sci. U.S.A.* **71**, 2882–2886 (1974).
26 Hayashi, M. N. & Hayashi, M. *J. Virol.* **14**, 1142–1152 (1974).
27 Vereijken, J. M., van Mansfeld, A. D. M., Baas, P. D. & Jansz, H. S. *Virology* **68**, 221–233 (1975).
28 Jeppesen, P. G. N., Sanders, L. & Slocombe, P. M. *Nucl. Acids Res.* **3**, 1323–1339 (1976).
29 Sato, S., Hutchison, C. A. III & Harris, J. I. *Proc. natn. Acad. Sci. U.S.A.* (in the press).
30 Brown, N. L. & Smith, M. *FEBS Lett.* **65**, 284–287 (1976).
31 Fiddes, J. C. *J. molec. Biol.* **107**, 1–24 (1976).
32 Air, G. M., Sanger, F. & Coulson, A. R. *J. molec. Biol.* **108**, 519–533 (1976).
33 Barrell, B. G., Air, G. M. & Hutchison, C. A. III *Nature* **264**, 34–41 (1976).
34 Smith, L. H. & Sinsheimer, R. L. *J. molec. Biol.* **103**, 699–735 (1976).
35 Grohmann, K., Smith, L. H. & Sinsheimer, R. L. *Biochemistry* **14**, 1951–1955 (1975).
36 Axelrod, N. *J. molec. Biol.* **108**, 753–779 (1976).
37 Hayashi, M., Fujimura, F. K. & Hayashi, M. *Proc. natn. Acad. Sci. U.S.A.* **73**, 3519–3523 (1976).
38 Pribnow, D. *Proc. natn. Acad. Sci. U.S.A.* **72**, 784–788 (1975).
39 Rosenberg, M., de Crombrugghe, B. & Musso, R. *Proc. natn. Acad. Sci. U.S.A.* **73**, 717–721 (1976).
40 Bertrand, K. *et al. Science* **189**, 22–26 (1975).
41 Sugimoto, K., Sugisaki, H., Okamoto, T. & Takanami, M. *J. molec. Biol.* (in the press).
42 Shine, J. & Dalgarno, L. *Proc. natn. Acad. Sci. U.S.A.* **71**, 1342–1346 (1974).
43 Steitz, J. A. & Jakes, K. *Proc. natn. Acad. Sci. U.S.A.* **72**, 4734–4738 (1975).
44 Henry, T. J. & Knippers, R. *Proc. natn. Acad. Sci. U.S.A.* **71**, 1549–1553 (1974).
45 Linney, E. & Hayashi, M. *Nature* **249**, 345–348 (1974).
46 Baas, P. D., Jansz, H. S. & Sinsheimer, R. L. *J. molec. Biol.* **102**, 633–656 (1976).
47 Eisenberg, S., Harbers, B., Hours, C. & Denhardt, D. T. *J. molec. Biol.* **99**, 107–123 (1975).
48 Subramanian, K. N., Dhar, R. & Weissman, S. M. *J. biol. Chem.* (in the press).
49 Ravetch, J. V., Model, P. & Robertson, H. D. *Nature* **265**, 698–702 (1977).
50 Smith, M. *et al.* (submitted to Nature).
51 Funk, F. & Sinsheimer, R. L. *J. Virol.* **6**, 12–19 (1970).
52 Baas, P. D., van Heusden, G. P. H., Vereijken, J. M., Weisbeek, P. J. & Jansz H. S. *Nucl. Acids Res.* **3**, 1947–1960 (1976).
53 Platt, T. & Yanofsky, C. *Proc. natn. Acad. Sci. U.S.A.* **72**, 2399–2403 (1975).
54 Weisbeek, P. J., Vereijken, J. M., Baas, P. D., Jansz, H. S. & Van Arkel, G. A. *Virology* **72**, 61–71 (1976).
55 Jazwinski, S. M., Lindberg, A. A. & Kornberg, A. *Virology* **66**, 283–293 (1975).
56 Kornberg, A. *DNA Synthesis* (W. H. Freeman, San Francisco, 1974).
57 Chen, C. Y., Hutchison, C. A. III & Edgell, M. H. *Nature new Biol.* **243** 233–236 (1973).
58 van Mansfeld, A. D. M., Vereijken, J. M. & Jansz, H. S. *Nucl. Acids Res.* **3**, 2827–2843 (1976).

Reprinted from
Proc. Natl. Acad. Sci. USA
Vol. 74, No. 12, pp. 5463–5467, December 1977
Biochemistry

DNA sequencing with chain-terminating inhibitors

(DNA polymerase/nucleotide sequences/bacteriophage φX174)

F. SANGER, S. NICKLEN, AND A. R. COULSON

Medical Research Council Laboratory of Molecular Biology, Cambridge CB2 2QH, England

Contributed by F. Sanger, October 3, 1977

ABSTRACT A new method for determining nucleotide sequences in DNA is described. It is similar to the "plus and minus" method [Sanger, F. & Coulson, A. R. (1975) *J. Mol. Biol.* 94, 441–448] but makes use of the 2′,3′-dideoxy and arabinonucleoside analogues of the normal deoxynucleoside triphosphates, which act as specific chain-terminating inhibitors of DNA polymerase. The technique has been applied to the DNA of bacteriophage φX174 and is more rapid and more accurate than either the plus or the minus method.

The "plus and minus" method (1) is a relatively rapid and simple technique that has made possible the determination of the sequence of the genome of bacteriophage φX174 (2). It depends on the use of DNA polymerase to transcribe specific regions of the DNA under controlled conditions. Although the method is considerably more rapid and simple than other available techniques, neither the "plus" nor the "minus" method is completely accurate, and in order to establish a sequence both must be used together, and sometimes confirmatory data are necessary. W. M. Barnes (*J. Mol. Biol.*, in press) has recently developed a third method, involving ribo-substitution, which has certain advantages over the plus and minus method, but this has not yet been extensively exploited.

Another rapid and simple method that depends on specific chemical degradation of the DNA has recently been described by Maxam and Gilbert (3), and this has also been used extensively for DNA sequencing. It has the advantage over the plus and minus method that it can be applied to double-stranded DNA, but it requires a strand separation or equivalent fractionation of each restriction enzyme fragment studied, which makes it somewhat more laborious.

This paper describes a further method using DNA polymerase, which makes use of inhibitors that terminate the newly synthesized chains at specific residues.

Principle of the Method. Atkinson *et al.* (4) showed that the inhibitory activity of 2′,3′-dideoxythymidine triphosphate (ddTTP) on DNA polymerase I depends on its being incorporated into the growing oligonucleotide chain in the place of thymidylic acid (dT). Because the ddT contains no 3′-hydroxyl group, the chain cannot be extended further, so that termination occurs specifically at positions where dT should be incorporated. If a primer and template are incubated with DNA polymerase in the presence of a mixture of ddTTP and dTTP, as well as the other three deoxyribonucleoside triphosphates (one of which is labeled with ^{32}P), a mixture of fragments all having the same 5′ and with ddT residues at the 3′ ends is obtained. When this mixture is fractionated by electrophoresis on denaturing acrylamide gels the pattern of bands shows the distribution of dTs in the newly synthesized DNA. By using analogous terminators for the other nucleotides in separate incubations and running the samples in parallel on the gel, a pattern of bands is obtained from which the sequence can be read off as in the other rapid techniques mentioned above.

Two types of terminating triphosphates have been used—the dideoxy derivatives and the arabinonucleosides. Arabinose is a stereoisomer of ribose in which the 3′-hydroxyl group is oriented in *trans* position with respect to the 2′-hydroxyl group. The arabinosyl (ara) nucleotides act as chain terminating inhibitors of *Escherichia coli* DNA polymerase I in a manner comparable to ddT (4), although synthesized chains ending in 3′ araC can be further extended by some mammalian DNA polymerases (5). In order to obtain a suitable pattern of bands from which an extensive sequence can be read it is necessary to have a ratio of terminating triphosphate to normal triphosphate such that only partial incorporation of the terminator occurs. For the dideoxy derivatives this ratio is about 100, and for the arabinosyl derivatives about 5000.

METHODS

Preparation of the Triphosphate Analogues. The preparation of ddTTP has been described (6, 7), and the material is now commercially available. ddA has been prepared by McCarthy *et al.* (8). We essentially followed their procedure and used the methods of Tener (9) and of Hoard and Ott (10) to convert it to the triphosphate, which was then purified on DEAE-Sephadex, using a 0.1–1.0 M gradient of triethylamine carbonate at pH 8.4. The preparation of ddGTP and ddCTP has not been described previously; however we applied the same method as that used for ddATP and obtained solutions having the requisite terminating activities. The yields were very low and this can hardly be regarded as adequate chemical characterization. However, there can be little doubt that the activity was due to the dideoxy derivatives.

The starting material for the ddGTP was *N*-isobutyryl-5′-*O*-monomethoxytrityldeoxyguanosine prepared by F. E. Baralle (11). After tosylation of the 3′-OH group (12) the compound was converted to the 2′,3′-didehydro derivative with sodium methoxide (8). The isobutyryl group was partly removed during this treatment and removal was completed by incubation in NH_3 (specific gravity 0.88) overnight at 45°. The didehydro derivative was reduced to the dideoxy derivative (8) and converted to the triphosphate as for the ddATP. The monophosphate was purified by fractionation on a DEAE-Sephadex column using a triethylamine carbonate gradient (0.025–0.3 M) but the triphosphate was not purified.

ddCTP was prepared from *N*-anisoyl-5′-*O*-monomethoxytrityldeoxycytidine (Collaborative Research Inc., Waltham, MA) by the above method but the final purification on DEAE-Sephadex was omitted because the yield was very low and the solution contained the required activity. The solution was used directly in the experiments described in this paper.

An attempt was made to prepare the triphosphate of the intermediate didehydrodideoxycytidine because Atkinson *et*

Abbreviations: The symbols C, T, A, and G are used for the deoxyribonucleotides in DNA sequences; the prefix dd is used for the 2′,3′-dideoxy derivatives (e.g., ddATP is 2′,3′-dideoxyadenosine 5′-triphosphate); the prefix ara is used for the arabinose analogues.

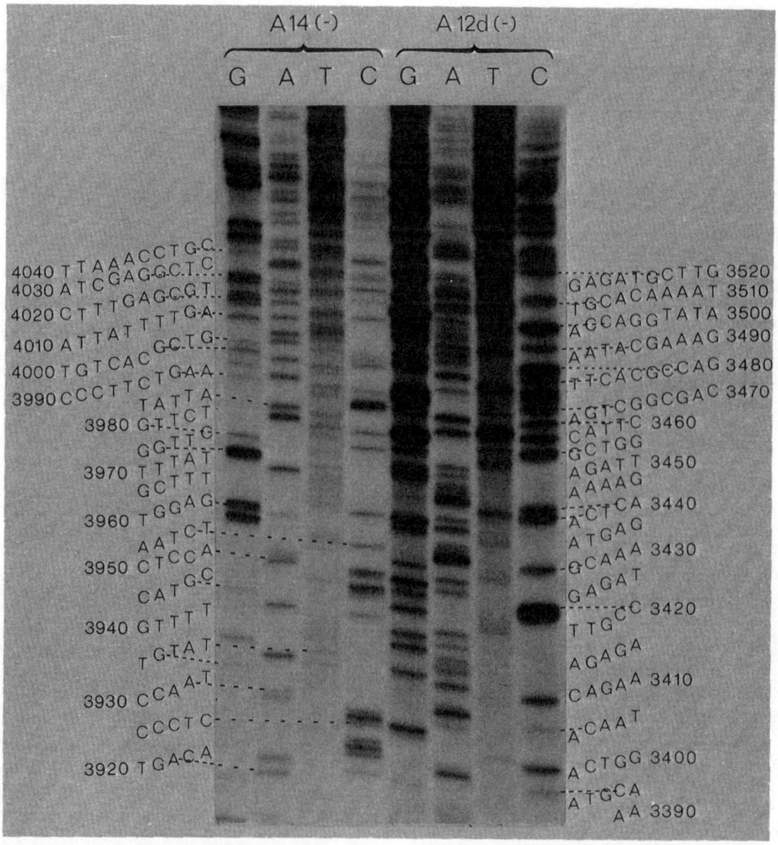

FIG. 1. Autoradiograph of the acrylamide gel from the sequence determination using restriction fragments A12d and A14 as primers on the complementary strand of ϕX174 DNA. The inhibitors used were (left to right) ddGTP, ddATP, ddTTP, and araCTP. Electrophoresis was on a 12% acrylamide gel at 40 mA for 14 hr. The top 10 cm of the gel is not shown. The DNA sequence is written from left to right and upwards beside the corresponding bands on the radioautograph. The numbering is as given in ref. 2.

al. (4) have shown that the didehydrodideoxy-TTP is also active as a terminator. However, we were unsuccessful in this. These compounds seem much less stable than the dideoxy derivatives.

araATP and araCTP were obtained from P-L Biochemicals Inc., Milwaukee, WI.

Sequencing Procedure. Restriction enzyme fragments were obtained from ϕX174 replicative form and separated by electrophoresis on acrylamide gels. The material obtained from 5 μg of ϕX174 replicative form in 5 μl of H_2O was mixed with 1 μl of viral or complementary strand ϕX174 DNA (0.6 μg) and 1 μl of H $\times$ 10 buffer (13) and sealed in a capillary tube, heated to 100° for 3 min, and then incubated at 67° for 30 min. The solution was diluted to 20 μl with H buffer and 2 μl samples were taken for each incubation and mixed with 2 μl of the appropriate "mix" and 1 μl of DNA polymerase (according to Klenow, Boehringer, Mannheim) (0.2 units). Each mix contained 1.5 $\times$ H buffer, 1 μCi of [α-^{32}P]dATP (specific activity approximately 100 mCi/μmol) and the following other triphosphates.

ddT: 0.1 mM dGTP, 0.1 mM dCTP, 0.005 mM dTTP, 0.5 mM ddTTP

ddA: 0.1 mM dGTP, 0.1 mM dCTP, 0.1 mM dTTP, 0.5 mM ddATP

ddG: 0.1 mM dCTP, 0.1 mM dTTP, 0.005 mM dGTP, 0.5 mM ddGTP

ddC: 0.1 mM dGTP, 0.1 mM dTTP, 0.005 mM dCTP, approximately 0.25 mM ddCTP

(The concentration of the ddCTP was uncertain because there was insufficient yield to determine it, but the required dilution of the solution was determined experimentally.)

araC: 0.1 mM dGTP, 0.1 mM dTTP, 0.005 mM dCTP, 12.5 mM araCTP

Incubation was at room temperature for 15 min. Then 1 μl of 0.5 mM dATP was added and incubation was continued for a further 15 min. If this step (chase) was omitted some termination at A residues occurred in all samples due to the low concentration of the [α-^{32}P]dATP. With small primers, where it was unnecessary to carry out a subsequent splitting (as in the experiment shown in Fig. 1), the various reaction mixtures were denatured directly and applied to the acrylamide gel for electrophoresis (1). If further splitting was necessary (see Fig. 2), 1 μl of the appropriate restriction enzyme was added shortly after the dATP "chase," and incubation was at 37°.

The single-site ribo-substitution procedure (N. L. Brown, unpublished) was carried out as follows. The annealing of template and primer was carried out as above but in "Mn buffer" (66 mM TrisCl, pH 7.4/1.5 mM 2-mercaptoethanol/

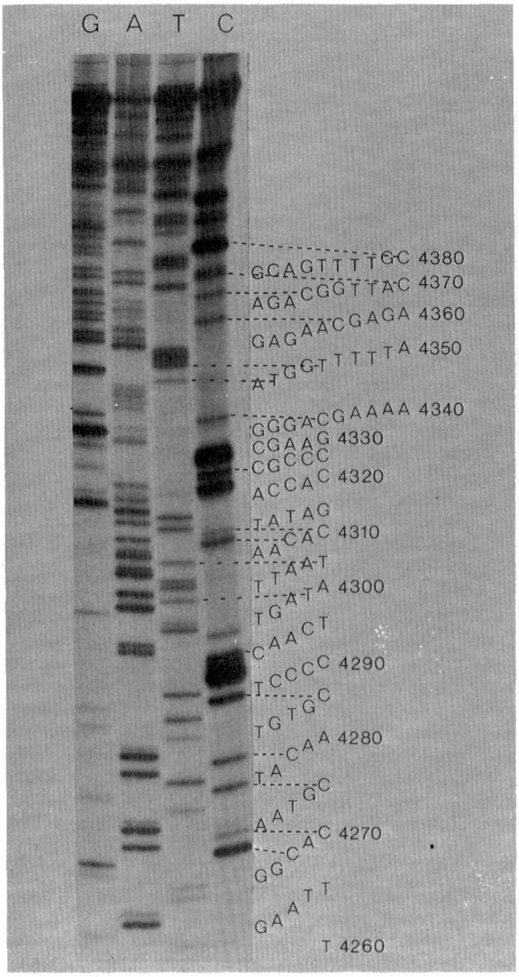

FIG. 2. Autoradiograph from an experiment using fragment R4 as primer on the complementary strand of ϕX174 DNA. Conditions were as in Fig. 1 with the following exceptions: ddCTP was used as inhibitor instead of araCTP. After incubation of the solutions at room temperature for 15 min, 1 μl of 0.5 mM dATP and 1 μl of restriction enzyme *Hae* III (4 units/μl) were added and the solutions were incubated at 37° for 10 min. The *Hae* III cuts close to the *Hin*dII site and it was used because it was more readily available. The electrophoresis was on a 12% acrylamide gel at 40 mA for 14 hr. The top 10.5 cm of the gel is not shown.

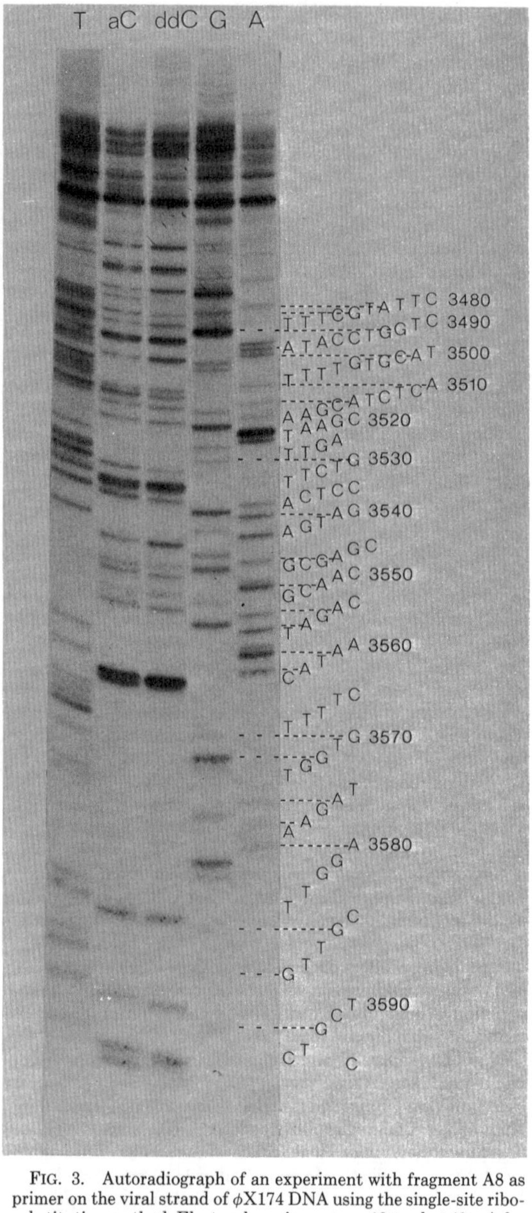

FIG. 3. Autoradiograph of an experiment with fragment A8 as primer on the viral strand of ϕX174 DNA using the single-site ribo-substitution method. Electrophoresis was on a 12% gel at 40 mA for 6 hr. The top 5.5 cm is not shown. Inhibitors used were (left to right) ddTTP, araCTP, ddCTP, ddGTP, and ddATP.

0.67 mM $MnCl_2$) rather than in H buffer. To 7 μl of annealed fragment was added 1 μl of 10 mM rCTP, 2 μl of H_2O, and 1 μl of 10 × Mn buffer. Five microcuries of dried [α-^{32}P]dTTP (specific activity approximately 1 mCi/μmol) was dissolved in this and 1 unit DNA polymerase (Klenow) was added. Incubation was for 30 min in ice. One microliter of 0.2 M EDTA was added before loading on a 1-ml Sephadex G-100 column. Column buffer was 5 mM Tris, pH 7.5/0.1 mM EDTA. The labeled fragment was followed by monitor, collected in a minimum volume (approximately 200 μl), dried down, and redissolved in 30 μl of 1 × H buffer. Samples (2 μl) of this were taken for treatment as above. Following the chase step, 1 μl of 0.1 M EDTA and 1 μl of pancreatic ribonuclease A at 10 mg/ml were added and incubated for 60 min at 37°.

RESULTS

Figs. 1–3 show examples of the use of the method for determining sequences in the DNA of ϕX174. In the experiment shown in Fig. 1 two small restriction enzyme fragments (A12d and A14, ref. 2) were used as primers on the complementary strand and there was no final digestion step to cut between the primer and the newly synthesized DNA. This is the most simple and rapid procedure, requiring only a preliminary annealing of template and primer, incubation of the four separate samples with DNA polymerase and appropriate triphosphates, followed by a chase with unlabeled dATP and application to the gel for electrophoresis. In these experiments the inhibitors used were ddGTP, ddATP, ddTTP, and araCTP. The conditions used for the "T" samples were not entirely optimal, resulting in the faster-moving bands being relatively weak.

The sequences can be read with reasonable accuracy starting at 88 nucleotides from the 5′ end of the primer for about 80 nucleotides (apart from some difficulty at position 3459 with A12d). For the next 50 nucleotides there is some uncertainty in the number of nucleotides in "runs" because bands are not actually resolved.

With longer restriction enzyme fragments as primers it is necessary to split them off from the newly synthesized DNA chains before the electrophoresis. This is normally done by digestion with a restriction enzyme. Fig. 2 shows such an experiment in which fragment R4 was used as primer on the complementary strand of ϕX174 DNA. In this experiment only dideoxynucleoside triphosphates were used as inhibitors because the results with araC were much less satisfactory when a restriction enzyme was used for the subsequent splitting. This may be due to the araC being removed by the 3′-exonuclease activity of the DNA polymerase during the incubation at 37° (which is necessary for the restriction enzyme splitting), resulting in a few C bands being either very faint or missing. Alternatively, the enzyme may be able to extend some chains beyond the araC at the higher temperature while being unable to do so at lower temperatures. araATP, which has been used only under these conditions, shows the same limitations as araCTP. These problems do not arise when ddCTP is used in this reaction.

With one exception (positions 4330–4343, see below), a sequence of 120 nucleotides, starting at a position 61 nucleotides from the restriction enzyme splitting site, could be read off; the sequence agreed with the published one. This region is believed to contain the origin of viral strand replication (2, 14). The bands beyond position 4380 indicated that there was an error in the provisional sequence (2), and further work (to be published later) has shown that the trinucleotide C-G-C should be inserted between positions 4380 and 4381.

When this technique is used the products are cut with a restriction enzyme as above, difficulties arise if there is a second restriction enzyme site close to the first one, because this will give rise to a separate pattern of bands that is superimposed on the normal one, making interpretation impossible. One way in which this can be avoided is by the single-site ribo-substitution method (N. L. Brown, unpublished). After annealing of the template and primer a single ribonucleotide is incorporated by incubation with DNA polymerase in the presence of manganese and the appropriate ribonucleoside triphosphate. Extension of the primer is then carried out with the separate inhibitors as above and the primer is split off at the ribonucleotide by ribonuclease or alkali. The method is particularly suitable for use with fragments obtained with the restriction enzyme *Alu*, which splits at the tetranucleotide sequence A-G-C-T. This enzyme is in fact inhibited by single-stranded DNA and cannot be used for the subsequent splitting of the primer from the newly synthesized DNA chain. The initial incorporation is carried out in the presence of rCTP and [α-^{32}P]dTTP. The incorporation of the ^{32}P facilitates subsequent purification on the Sephadex column.

Fig. 3 shows an example of the use of this method with fragment A8 on the viral strand of ϕX174 DNA. A sequence of about 110 nucleotides starting 33 residues from the priming site can be read off. In the provisional sequence (2) this region was regarded as very tentative. Most of it is confirmed by this experiment, but there is a clear revision required at positions 3524–3530. The sequence of the viral strand should read A-T-C-A-A-C, replacing A-T-T-C- - -A-C given in the provisional sequence. There is difficulty in reading the sequence at 3543–3550, where there is considerable variation in the distance between bands, suggesting the presence of a looped structure. Further work in which the electrophoresis was carried out at a higher temperature indicates that the sequence here is actually G-C-T-C-G-C-G (viral strand); i.e., an insertion of C between positions 3547 and 3548 in the provisional sequence.

DISCUSSION

The method described here has a number of advantages over the plus and minus methods. First, it is simpler to perform because it requires no preliminary extension, thus avoiding one incubation and purification on a Sephadex column. It requires only the commercially available DNA polymerase I (Klenow fragment). The results appear to be more clear-cut with fewer artefact bands, and can usually be read further than with the plus and minus methods. Intermediate nucleotides in "runs" show up as bands, thus avoiding a source of error in the plus and minus method—estimating the number of nucleotides in a run. Theoretically one would expect the different bands in a run to be of the same strength, but this is not always the case. Frequently, the first nucleotide is the strongest, but in the case of ddCTP the second is the strongest (see Fig. 2). The reasons for these effects are not understood, but they do not usually cause difficulties with deducing the sequences. For the longer sequences in which the separate bands in a run are not resolved, experience has shown that it is frequently possible to estimate the number of nucleotides from the strength and width of the band.

The inhibitor method can also be used on a smaller scale than the plus and minus method because better incorporation from ^{32}P-labeled triphosphates is obtained. This is presumably due to the longer incubation period used, which allows a more quantitative extension of primer chains.

In general, sequences of from 15 to about 200 nucleotides from the priming site can be determined with reasonable accuracy using a single primer. Frequently it is possible to read the gels further and, on occasions, a sequence of about 300 nucleotides from the priming site has been determined. Occasional artefacts are observed, but these can usually be readily identified. It seems likely that these are usually due to contaminants in the fragments. The most serious difficulties are due to "pile-ups" of bands, which are usually caused by the DNA forming base-paired loops under the conditions of the acrylamide gel electrophoresis. These pile-ups are seen as a number of bands in the same position or unusually close to one another on the electrophoresis. They generally occur at different positions when the priming is carried in opposite directions along the DNA over the same sequence. An example of this effect is seen in Fig. 2 at position 4330, where there is a single strong band in the G channel that in fact represents four G residues. They are presumably forming a stable loop by pairing with the four Cs at positions 4323–4326. Another example is in Fig. 3 at positions 3545–3550. This effect is likely to be found in all the rapid techniques that use gel electrophoresis.

It is felt that for an accurate determination of sequence one should not rely completely on single results obtained by this method alone but that confirmation should be obtained by some other technique or by priming on the opposite strand. This consideration probably applies to all other available methods also. The main disadvantage of the present method is the difficulty in obtaining all the inhibitors—particularly ddGTP, which is not commercially available.

We wish to thank Dr. K. Geider for a gift of ddTTP, Dr. F. E. Baralle for a gift of *N*-isobutyryl-5′-*O*-monomethoxytrityldeoxyguanosine, and Dr. M. J. Gait for useful advice on the synthetic work.

1. Sanger, F. & Coulson, A. R. (1975) *J. Mol. Biol.* **94,** 441–448.
2. Sanger, F., Air, G. M., Barrell, B. G., Brown, N. L., Coulson, A. R., Fiddes, J. C., Hutchison, C. A., Slocombe, P. M. & Smith, M. (1977) *Nature* **265,** 687–695.
3. Maxam, A. M. & Gilbert, W. (1977) *Proc. Natl. Acad. Sci. USA* **74,** 560–564.
4. Atkinson, M. R., Deutscher, M. P., Kornberg, A., Russell, A. F. & Moffatt, J. G. (1969) *Biochemistry* **8,** 4897–4904.
5. Hunter, T. & Francke, B. (1975) *J. Virol.* **15,** 759–775.
6. Russell, A. F. & Moffatt, J. G. (1969) *Biochemistry* **8,** 4889–4896.
7. Geider, K. (1974) *Eur. J. Biochem.* **27,** 555–563.
8. McCarthy, J. R., Robins, M. J., Townsend, L. B. & Robins, R. K. (1966) *J. Am. Chem. Soc.* **88,** 1549–1553.
9. Tener, G. M. (1961) *J. Am. Chem. Soc.* **83,** 159–168.
10. Hoard, D. E. & Ott, D. G. (1965) *J. Am. Chem. Soc.* **87,** 1785–1788.
11. Büchi, H. & Khorana, H. G. (1972) *J. Mol. Biol.* **72,** 251–288.
12. Robins, M. J., McCarthy, J. R. & Robins, R. K. (1966) *Biochemistry* **5,** 224–231.
13. Air, G. M., Sanger, F. & Coulson, A. R. (1976) *J. Mol. Biol.* **108,** 519–533.
14. Slocombe, P. M. (1976) Ph.D. Dissertation, University of Cambridge.

Volume 87, number 1 FEBS LETTERS March 1978

THE USE OF THIN ACRYLAMIDE GELS FOR DNA SEQUENCING

F. SANGER and A. R. COULSON
MRC Laboratory of Molecular Biology, Cambridge CB2 2QH, England

Received 13 January 1978

1. Introduction

The newer rapid methods for DNA sequencing [1–3] all use electrophoresis on acrylamide gels in denaturing conditions for the final sequence analysis, and the extent to which sequences can be read is often dependent on the degree of resolution that can be obtained with these gels. The gels are normally 1–2 mm thick and there is some spreading of the bands on the autoradiograph due to the fact that part of the radiation source is some distance from the film. One way of overcoming this is to use thinner gels and we have found that with gels of about 0.4 mm thickness sharper bands can be obtained, with a corresponding improvement in resolution. This simple modification in technique has a number of other advantages. Firstly, less material can be used so that one can economise both on the DNA used and the acrylamide. Secondly, it is possible to run the electrophoresis at higher voltages than is used with conventional gels. Using the plus and minus [1] and the terminator [3] methods, we have found serious variations in the distances between consecutive nucleotide bands in regions of dyad symmetry where base-paired loop structures can form, and this can lead to difficulties of interpretation. The effect can be overcome by running the gels at higher voltages so that the resultant higher temperatures in the gels cause complete denaturation of the loops. High temperatures may cause distortion of the bands and cracking of the glass plates, but these effects are less apparent with the thin gels than with the conventional ones.

Abbreviations: As this paper is concerned with DNA, the letters C, T, A and G are used for the deoxyribonucleotides. The prefix dd is used for the 2′,3′-dideoxy derivatives (e.g., ddATP, 2′,3′-dideoxyadenosine 5′-triphosphate) and the prefix Ara for the arabinose analogues

2. Materials and methods

The conditions for the terminator sequencing procedure were as described [3]. Spacers and well-forming templates for the thin acrylamide gels were made from 0.35 mm thick 'Plastikard' (Slater's (Plastikard) Ltd, Matlock Bath, Derbyshire) or, in earlier experiments, from two or three thicknesses of 0.2 mm polythene sheet. The electrophoresis was carried out in the apparatus [4] with minor modifications. An apparatus of this type is available from Raven Ltd (Haverhill, Suffolk).

The gels were formed between two 20 × 40 × 0.4 cm glass plates, one with a notch 16.5 × 2 cm cut out from the top to allow contact between the gel and the upper buffer compartment of the apparatus. The inside of this plate was siliconised with 'Repelcote' (Hopkin & Williams Ltd., Romford) to facilitate removal of the plate from the gel after the electrophoresis. The sides and bottom of the gel-former were sealed with waterproof electrical tape. It was found most convenient to pour the gel at an angle of about 45° from a pipette. After pouring, the well-former was inserted and the gel allowed to polymerize in a horizontal position. Slots were normally 0.75 cm deep. All solutions and buffer were as described [5].

When running samples from a reaction mixture of the chain terminator method, as in the experiments shown in fig.2, 1 μl sample was heated at 90°C for 3 min in 4 μl formamide/dye mixture [5]. This was loaded onto the gel using a finely drawn glass capillary. Different amounts of samples were used in the experiments shown in fig.1.

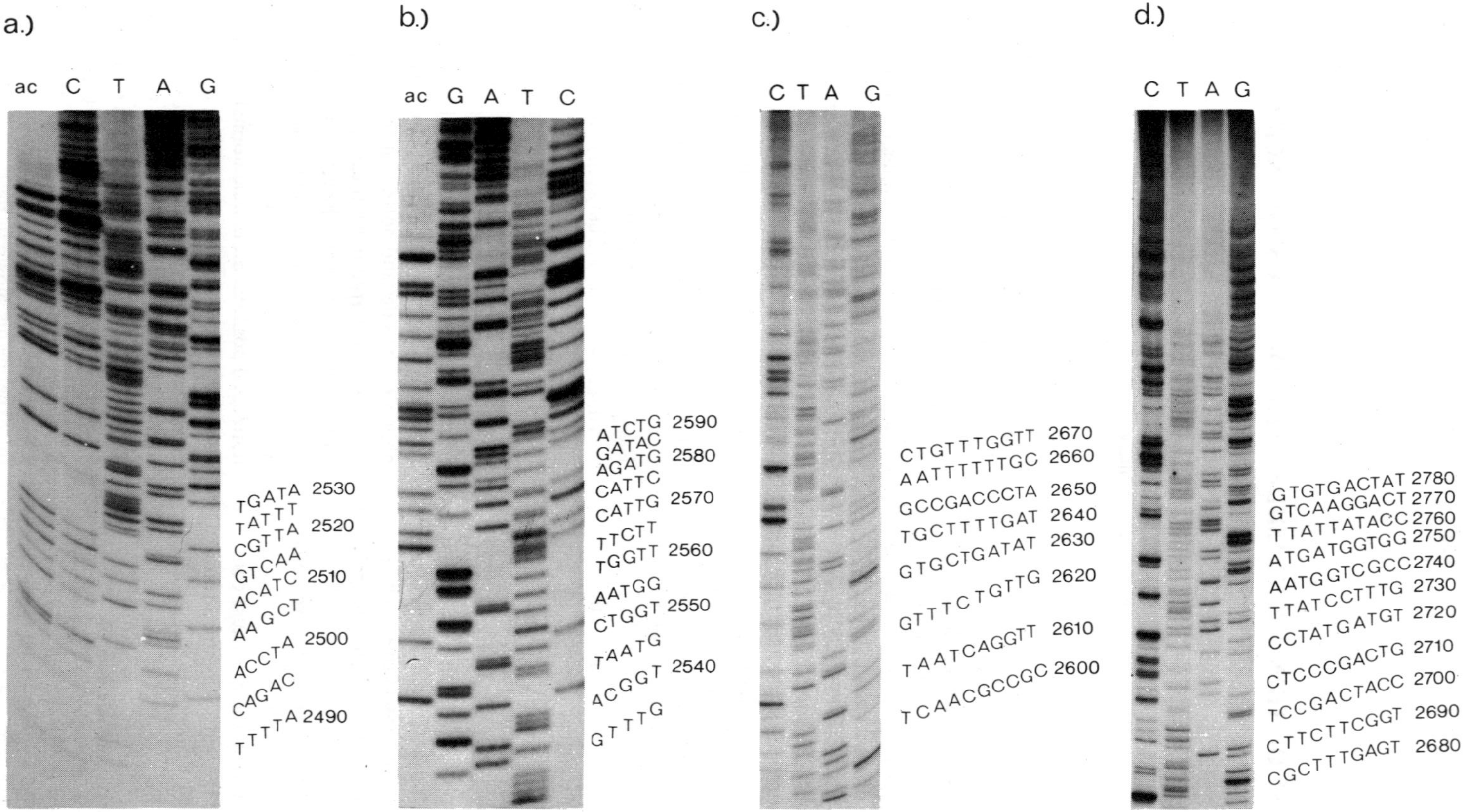

Fig.1. Autoradiographs of acrylamide gel electrophoreses from experiments using the terminator method [3] in which restriction enzyme fragment A16 [4] was used as primer on the complementary strand of ϕX174. Conditions were as in [3]. Incubations were carried out in the presence of the following concentrations of terminators: aC, 5 mM AraCTP; C, 0.1 mM ddCTP; T, 0.2 mM ddTTP; A, 0.1 mM ddATP; G, 0.2 mM ddGTP. All gels were 8% acrylamide. The top 5–9 cm of the gels are not included in the figure. Other conditions are: (a) Gel 0.4 mm thick, slots 14 mm wide. Electrophoresis for 2 h at 25 mA. (b) Gel 1.5 mm thick, slots 14 mm wide. Electrophoresis for 8 h at 45 mA. (c) Gel 0.6 mm thick, slots 8 mm wide. Electrophoresis for 6 h at 30 mA. (d) Gel 0.6 mm thick, slots 8 mm wide. Electrophoresis for approx. 10 h at 30 mA. The deduced sequences are those of the viral strand of ϕX and are written from left to right and upwards so as to come opposite the part of the gel to which they refer. The numbering is from the revised sequence (to be published) and is slightly different from that used in [6].

Volume 87, number 1 FEBS LETTERS March 1978

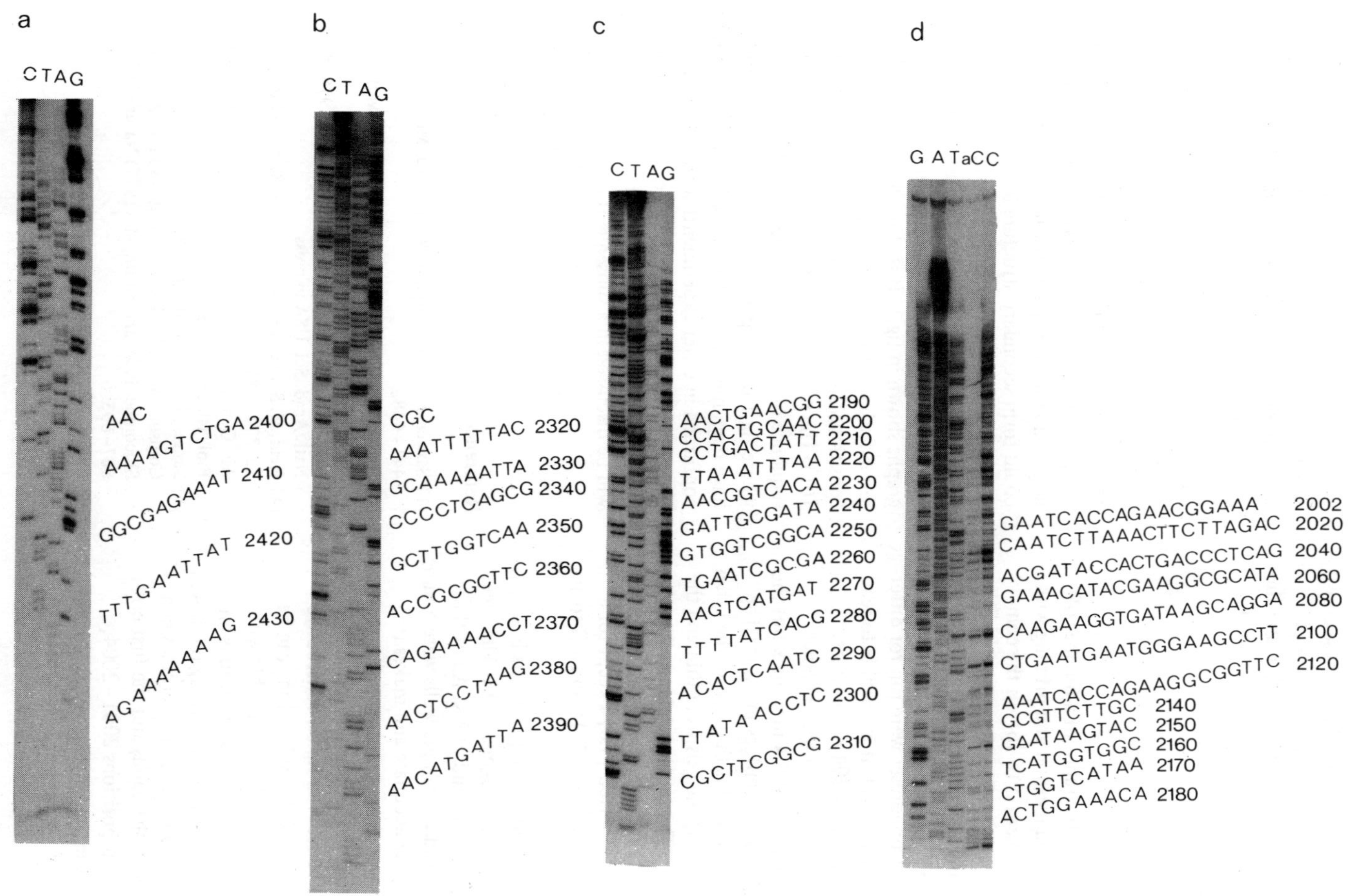

Fig.2. Autoradiographs from experiments, as in fig.1 but priming with fragment A16 on the viral strand of ϕX. All gels were 0.4 mm thick and were run at 25 mA. Slots were 5 mm wide. Concentrations of terminators were: (a) C, 0.1 mM ddCTP; T, 0.2 mM ddTTP; A, 0.1 mM ddATP; G, 0.2 mM ddGTP. (b) C, 0.1 mM ddCTP; T, 0.1 mM ddTTP; A, 0.05 mM ddATP; G, 0.2 mM ddGTP. (c), as (b) but 3 μCi [α-^{32}P]dATP (spec. act. 250 mCi/μmol) was present instead of the normal 1 μCi in the A incubation. (d) aC, 1.2 mM AraCTP; C, 0.05 mM ddCTP; T, 0.1 mM ddTTP; A, 0.05 mM ddATP and 3 μCi [α-^{32}P]dATP; G, 0.05 mM ddGTP. Other conditions were: (a) 8% acrylamide, 2 h electrophoresis. (b) 8% acrylamide, 3.5 h. (c) 8% acrylamide, 5 h. (d) 6% acrylamide, 6 h. The sequence deduced is that of the complementary strand of ϕX.

Volume 87, number 1 FEBS LETTERS March 1978

Gels were run at 25 mA (initially 1.5 kV, dropping to about 1.3 kV after heating up). Under these conditions, in an 8% gel, the bromophenol blue dye reached the bottom of the gel in about 1.5 h and the xylene cyanol FF (equivalent to an 80-base oligonucleotide) in about 3 h.

3. Results and discussion

Figure 1 shows autoradiographs of a number of gels using the termination method and priming with restriction enzyme fragment A16 [6] on the complementary strand of ϕX174. This fragment is 25 nucleotides long and was not split from the extended nucleotides after the DNA polymerase reactions. The various electrophoreses were run for different times and from the four gels it is possible to read off a consecutive sequence of 294 nucleotides. Figure 1b shows a gel of conventional thickness (1.5 mm). The gel in fig.1a was 0.4 mm thick but the loading wells were the same width as in fig.1b. The bands are clearly sharper on this thin gel. In the other two experiments (fig.1c,d) the gels were thin and the loading wells narrower. The sequence shown is in the coding region for the G gene of ϕX174 and essentially confirms the sequence reported [5,6] except for a change of an A to a T at position 2616 and a G to an A at position 2731.

Figure 2 shows the results of experiments in which the same restriction enzyme fragment (A16) has been used as primer on the viral strand of ϕX. All of these were run on thin gels and the sample wells were 5 mm wide. This width is quite adequate and makes it possible to run four to five experiments on one gel slab 20 cm wide. From the four gels a sequence of 437 nucleotides can be read. In fig.2d consecutive oligonucleotides of up to 465 residues long can be resolved – very much longer than is possible with thicker gels. Although individual bands in runs are actually overlapping one another in this region, it was possible to estimate the number of nucleotides in a run from the thickness of the bands and the first error in reading this gel was at positions 2001–2004, where A–A–A was estimated instead of the correct A–A–A–A. This was of course a particularly favourable case and it should not be assumed that an unknown sequence read out this far could be regarded as established from a single experiment. Beyond position 2000 there is still a pattern of bands from which an approximate sequence could be read and which could be used to establish overlaps with patterns from other priming experiments.

This sequence contains the complete untranslated region between the F and G genes, which was particularly difficult to work out with earlier methods [7], largely due to two symmetrical sequences which could form secondary structure loops during the electrophoresis. Using the thin gels at high temperatures these loops do not seem to form and it is possible to read the sequence correctly from the autoradiographs shown in fig.1. It is also evident that a C–G sequence had been missed out in the original sequence (positions 2310–2311 in fig.1, or position 2303 in [6]). The most extensive error in the provisional sequence of ϕX [6] was at positions 2217–2228, where fig.1 of this paper shows that six nucleotides have been missed out. The revised sequence has been confirmed in other experiments with fragment A16 and by priming with fragment Q3c on the complementary strand. Other revisions of the ϕX sequence will be discussed in a subsequent publication.

References

[1] Sanger, F. and Coulson, A. R. (1975) J. Mol. Biol. 94, 441–448.
[2] Maxam, A. M. and Gilbert, W. (1977) Proc. Natl. Acad. Sci. USA 74, 560–564.
[3] Sanger, F., Nicklen, S. and Coulson, A. R. (1978) Proc. Natl. Acad. Sci. USA in press.
[4] Reid, M. S. and Bieleski, R. L. (1968) Anal. Biochem. 22, 374–381.
[5] Air, G. M., Sanger, F. and Coulson, A. R. (1976) J. Mol. Biol. 108, 519–533.
[6] Sanger, F., Air, G. M., Barrell, B. G., Brown, N. L., Coulson, A. R., Fiddes, J. C., Hutchison, C. A., Slocombe, P. M. and Smith, M. (1977) Nature 265, 687–695.
[7] Fiddes, J. C. (1976) J. Mol. Biol. 107, 1–24.

J. Mol. Biol. (1978) **125**, 225–246

The Nucleotide Sequence of Bacteriophage ϕX174

F. Sanger, A. R. Coulson, T. Friedmann[1], G. M. Air[2],
B. G. Barrell, N. L. Brown[3], J. C. Fiddes[4], C. A. Hutchison III[5],
P. M. Slocombe[6] and M. Smith[7]

Medical Research Council Laboratory of Molecular Biology
Hills Road, Cambridge CB2 2QH, England

(Received 2 June 1978)

The complete nucleotide sequence of the DNA of bacteriophage ϕX174 has been determined. The provisional sequence (Sanger *et al.*, 1977*a*) deduced largely by the plus and minus method, has been completed and confirmed, predominantly using the terminator method (Sanger *et al.*, 1977*b*). About 30 revisions were found to be necessary in the 5386-nucleotide sequence. The amino acid sequences of the ten proteins for which the DNA codes have also been deduced.

1. Introduction

We have reported an almost complete nucleotide sequence of the DNA of bacteriophage ϕX174 from which its main biological features can be identified (Sanger *et al.*, 1977*a*). This was made possible largely by the use of the rapid and simple "plus and minus" method of sequencing (Sanger & Coulson, 1975). The sequence was not quite complete and there were a number of regions of uncertainty. Two new methods have now been used to complete the sequence and to confirm most of it.

Maxam & Gilbert (1977) have described a rapid chemical method of sequencing restriction enzyme fragments labelled with ^{32}P at their 5′ end. This method is particularly convenient for determining the sequences adjacent to the restriction enzyme cutting sites, which are not easily obtained by the enzymic methods.

An extension of the plus and minus method has recently been described which makes use of specific chain-terminating inhibitors of DNA polymerase (Sanger *et al.*, 1977*b*). The method is more rapid and simpler than either the plus or minus method and has now been used extensively on ϕX DNA.

Present addresses:
[1] Department of Pediatrics M-009, University of California, San Diego School of Medicine, La Jolla, Calif. 92039, U.S.A.
[2] John Curtin School of Medical Research, Microbiology Department, Canberra City, ACT 2601, Australia.
[3] Department of Biochemistry, University of Bristol Medical School, University Walk, Bristol BS8 1TD.
[4] Department of Biochemistry and Biophysics, University of California, San Francisco, Calif. 94143, U.S.A.
[5] Department of Bacteriology and Immunology, University of North Carolina, Chapel Hill, North Carolina 27514, U.S.A.
[6] Max-Planck Institut für Molekulaire Genetik, 1 Berlin 33, Germany.
[7] Department of Biochemistry, University of British Columbia, 2075 Wesbrook Place, Vancouver, B.C., Canada V6T 1W5.

00222836/78/300225–22 $02.00/0

The complete sequence of the F protein and of the DNA coding for it is described in the accompanying paper (Air *et al.*, 1978). This paper describes the completion of the rest of the sequence. We have not attempted to give full details for the establishment of the whole sequence. Only the results of a number of representative experiments are shown, particularly in regions where changes have had to be made to the original sequence. Some such experiments have been described in recent publications (Sanger *et al.*, 1977*b*; Sanger & Coulson, 1978).

The numbering of nucleotides in the following discussion is that of the revised sequence (Fig. 4). The one-letter code for restriction endonucleases is explained in the legend to Figure 4. The nomenclature used for the restriction enzyme fragments of ϕX is as in the previous paper (Sanger *et al.*, 1977*a*).

2. Materials and Methods

The DNA studied was obtained from bacteriophage ϕX174 *am*3 *cs*70. The *am*3 mutation is a change from the wild type of a G to an A at position 587 in the E protein coding region. *cs*70 is a cold-sensitive mutation believed to be in the *F* gene (C. A. Hutchison III, J. H. Middleton and M. H. Edgell, unpublished data). The preparation of restriction enzyme fragments and of complementary strand DNA were as previously described (Sanger & Coulson, 1975; Air *et al.*, 1976; Fiddes, 1976; Greenfield *et al.*, 1975). The sequencing experiments shown here were done by the terminator method (Sanger *et al.*, 1977*b*; Sanger & Coulson, 1978). The chemical degradation method was as described by Maxam & Gilbert (1977).

The nucleotide sequences were stored and studied using the computer programmes devised by Staden (1977).

3. Results

(a) *The region between the sequences coding for the H and A proteins*

The untranslated space between the *H* and *A* genes is of particular interest since it contains signals for the initiation and termination of messenger RNA synthesis (Sanger *et al.*, 1977*a*). The sequence was originally determined using the plus and minus method and by depurination. Although it is contained in a region of 160 nucleotides with no restriction enzyme site the sequence was confirmed when the terminator method was applied to it using as primers A14 on the minus strand (Sanger *et al.*, 1977*b*) and T6 on the plus strand (see Fig. 1). The main uncertainty was the A-G-G-A sequence at positions 3923 to 3926. When A14 was used as primer there appeared to be only one G whereas there were two when T6 was used, and the depurination results also indicated two. The only explanation seems to be that the G-G was base-pairing with the C-C at position 3918 to 3919 and causing anomalous migration of the bands when priming was on the minus strand. The single G band was observed in several experiments with A14 with the plus and minus and the terminator methods on the standard gels (1·5 mm thick), but the double band was finally observed using a thin gel (0·4 mm), which was run at a higher temperature (Sanger & Coulson, 1978).

(b) *The A protein coding region*

Much of the sequence of the A protein coding region was determined at the later stages of the ϕX work when there was considerable experience with the plus and minus method and it was possible to obtain accurate results (Brown & Smith, 1977;

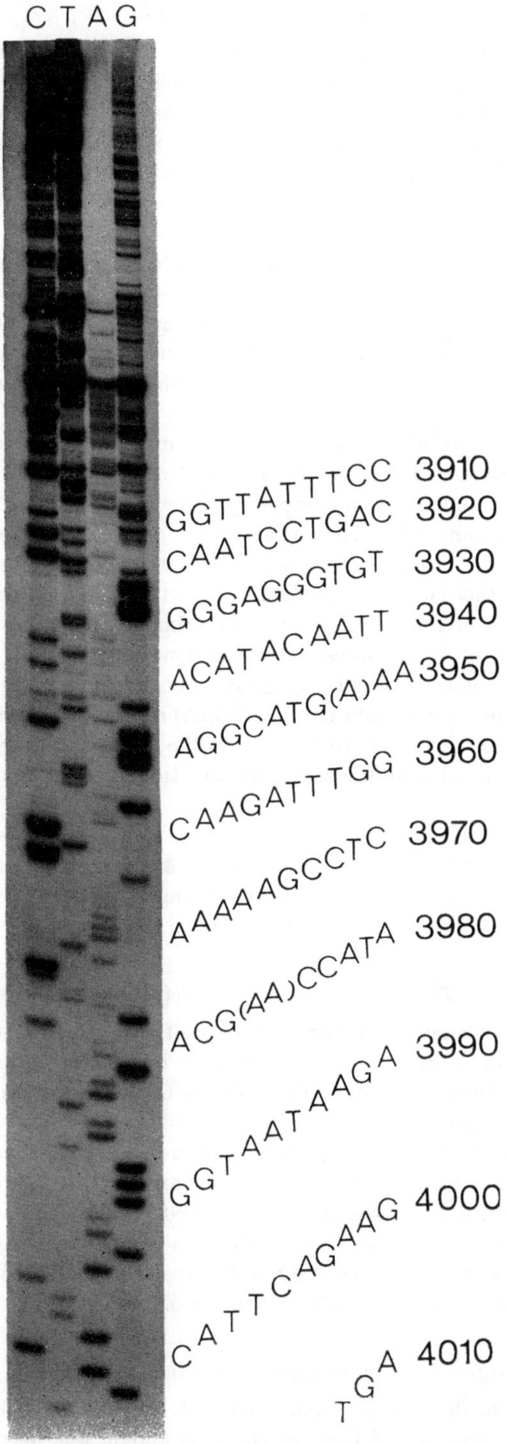

FIG. 1. Autoradiograph of an experiment using the terminator technique (Sanger & Coulson, 1978) with fragment T6 as primer on the viral (plus) strand of ϕX. The products were digested with the restriction enzyme *Taq*I before electrophoresis at 25 mA for 2 h on an 8% acrylamide gel 0·4 mm thick.

Slocombe, 1977). However, particular difficulties were encountered in one region due to two stretches of 172 and 213 nucleotides which contained no restriction enzyme sites (positions 4208 to 4379 and 4489 to 4701, respectively). Since this region contained the origin of plus strand replication it was extensively investigated (Slocombe, 1977) using the plus and minus method, the ribosubstitution method (Sanger *et al.*, 1973) and depurination. The sequence has now been confirmed by the termination method with two minor revisions.

The site of action of the A protein which initiates plus strand replication has now been shown to be between nucleotides 4305 and 4306 (Langeveld *et al.*, 1978). The sequence in this region was confirmed by priming with fragment R4 on the minus strand (Sanger *et al.*, 1977*b*). This primer also showed the presence of an extra C-G-C at positions 4389 to 4391. The initial difficulty in this region was due to the formation of a base-paired looped structure.

Figure 2 shows the results of two experiments in which fragment A11 was used as primer on the viral strand. Figure 2(a) shows three artifacts where there are strong bands in all four channels which completely obscure the reading of the sequence at positions 4642, 4626 and 4614. The one at 4642 is probably due to the excess double-stranded fragment A11 itself, which is labelled at its 3′ end during the incubation reaction, but the other two are unexplained. Apart from these artifacts a sequence of 208 nucleotides can be read from the two experiments, which shows that a revision from A-G-A-T to T-G-A-A is necessary at positions 4498 to 4501. Besides the results from the plus and minus method the original sequence was strongly supported by the identification of the depurination product $T_3C(A)$ on the complementary strand near this position (i.e. corresponding to $T(A_3G)$ on the viral strand) (Slocombe, 1977). It will, however, be noted that the same depurination product is consistent with the revised sequence.

No revisions have been made to the B protein coding region which is completely contained within that for the A protein (Brown & Smith, 1977; Smith *et al.*, 1977). Its identification has now been confirmed by studies of the N-terminal amino acid sequence of the B protein (Walker *et al.*, 1978).

(c) *The C, K, D, E and J protein coding regions*

The nucleotide sequence of the region from position 51 to 1000 was well established (Sanger *et al.*, 1977*a*; Barrell, 1978; J. E. Sims & D. Dressler, unpublished data) and no revisions have been made to it. The position of the C protein coding region (see Fig. 4) has been confirmed by determination of the N-terminal sequence of the C protein (Walker *et al.*, 1978) and by identification of a C to T change at position 196 resulting in a C-A-A (Gln) to T-A-A (term.) change in the ochre 6 mutant (B. G. Barrell, unpublished data). A hitherto unidentified protein (K) has been found in the related bacteriophage G4 (Shaw *et al.*, 1978). Its coding region overlaps those of A and C and the nucleotide sequence of ϕX in this region conserves the K protein coding sequence, suggesting the protein is also made in ϕX and coded as shown in Figure 4.

(d) *The region between the sequences coding for the F and G proteins*

Considerable difficulties were encountered with the plus and minus method in this region due to the presence of base-paired loops and it was necessary to use other techniques (ribosubstitution, endonuclease IV digestion and depurination) to establish the sequence (Fiddes, 1976). Fragment A16 was a particularly good primer and when used on the viral strand with the terminator method and thin gels the complete sequence of the region could be read (Sanger & Coulson, 1978). This confirmed the work of Fiddes but showed the presence of an additional C-G at positions 2310–2311.

(e) *The G protein coding region*

Although the original sequence of the region coding for the G protein was established both by amino acid and nucleotide sequence techniques (Air *et al.*, 1975,1976), some of the DNA results were not particularly satisfactory (e.g. priming with fragment F4; Fig. 5 of Air *et al.*, 1976) and the sequence was reinvestigated using the terminator method. Some minor changes were found (see Table 1). Most of these were in the third positions of codons and were most probably due to errors, but two had been rather

TABLE 1

Revisions to provisional sequence of ϕX174

Position in provisional sequence (Sanger *et al.*, 1977*a*)	Revised position (Fig. 4)	Change
1298–1300	1298	Delete C-A-A
1390–1391	1387–1389	Insert A
1400–1402	1398–1402	G-G-C → A-C-A-A-C
1421–1423	1421	Delete ———
1653	1650	A → G
1657	1654	T
1687	1684	A → T
1870	1867	G → T
2165	2162	A → C
2188–2189	2186	Insert C
2191–2192	2191	Insert C
2210–1	2210	Insert A
2216–2217	2217–2219	Insert A-G-G
2220–2223	2223–2228	C-C-T-C- → A-A-T-T-T-A
2230–2231	2236	Insert C
2303–2304	2310–2311	Insert C-G
2413	2421	C → T
2608	2616	A → T
2723	2731	G → A
2794	2802	C → G
2806	2814	C → T
2809	2817	C → T
3141–3142	3149–3150	G-T → T-G
3150	3158	T → G
3526	3533–3534	Delete T
3528	3535	A
3547–4548	3555	Insert C
4880–4881	4889–4891	Insert C-G-C
4487	4498	A → T
4490	4501	T → A

Changes in the F protein coding region are described in the accompanying paper (Air *et al.*, 1978).

N.B. There
are some
errors in
this Table

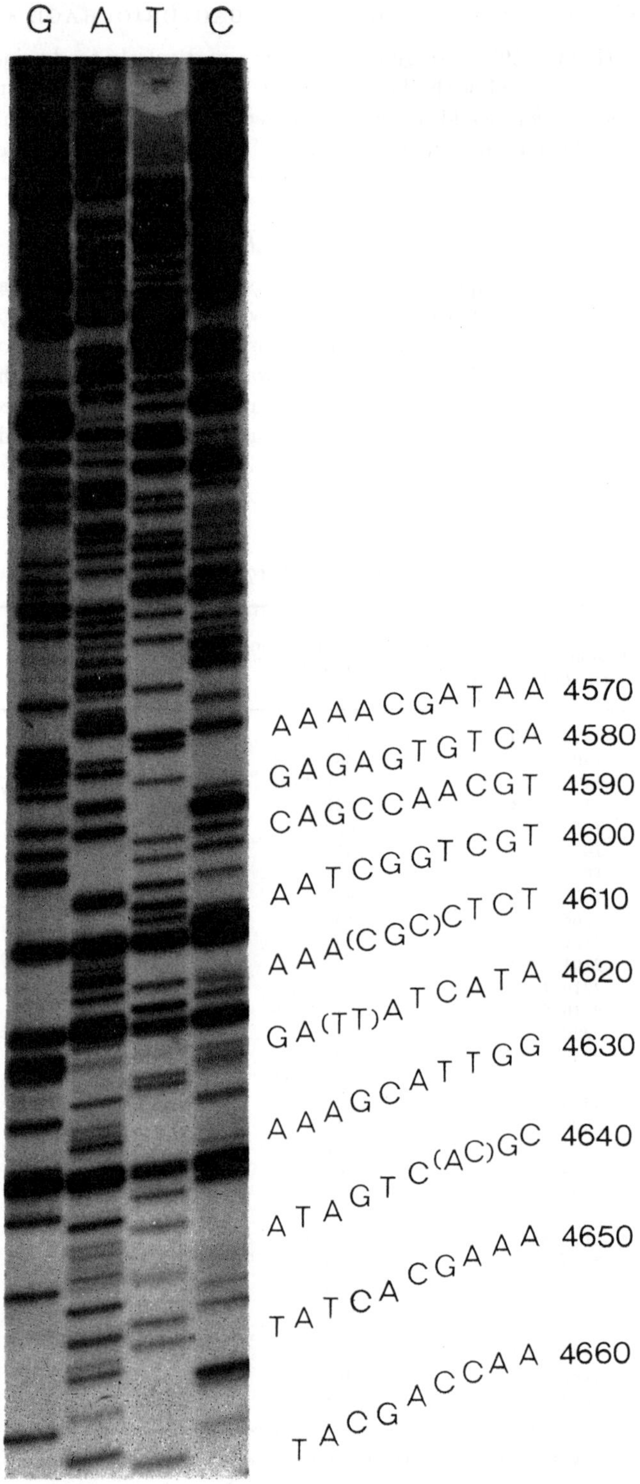

(a)

FIG. 2. Autoradiograph of an experiment using fragment A11 as primer on the viral strand of ϕX DNA using the single-site ribosubstitution method as described by Sanger *et al.* (1977*b*). Electrophoresis was on 8% acrylamide gels 1·5 mm thick at 45 mA for 10 h (a) and 16 h (b).

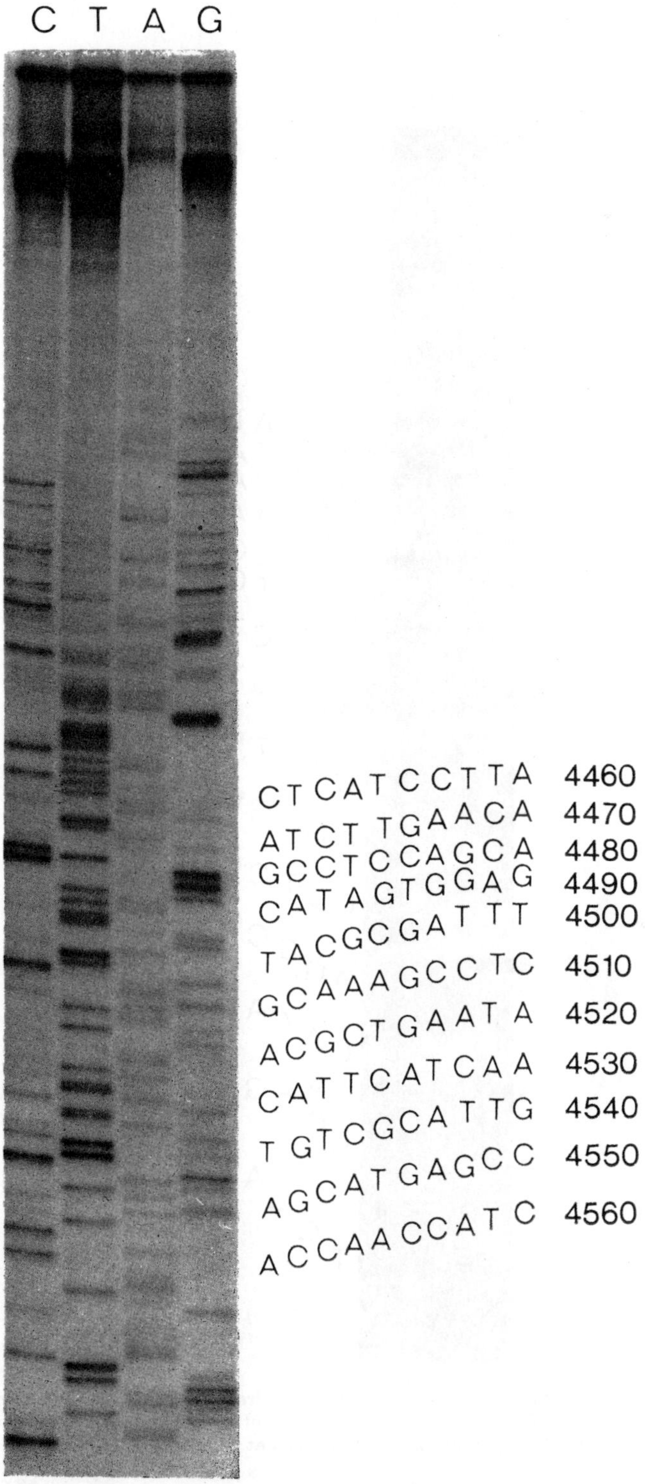
C T A G
CTCATCCTTA 4460
ATCTTGAACA 4470
GCCTCCAGCA 4480
CATAGTGGAG 4490
TACGCGATTT 4500
GCAAAGCCTC 4510
ACGCTGAATA 4520
CATTCATCAA 4530
TGTCGCATTG 4540
AGCATGAGCC 4550
ACCAACCATC 4560

(b)

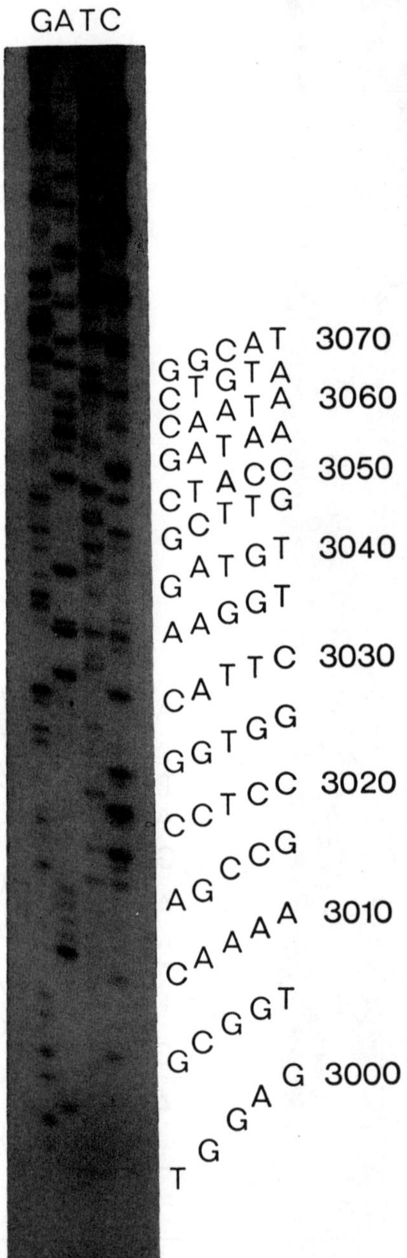

FIG. 3. Autoradiograph of an experiment in which fragment Q3b was used as primer on the complementary strand of ϕX and the products were cut by the restriction enzyme *Hha*I before electrophoresis on an 8% acrylamide gel 0·4 mm thick at 25 mA for 3 h.

```
  C C G T C A G G A T T G A C A C C C T C C C A A T T G T A T G T T T T C A T G C C T C C A A A T C T T G G A G G C T T T
                  3927                    3937                    3947                    3957    ↑               3967                    3977
                                                                                                  mRNA start

A        MET   VAL   ARG   SER   TYR   TYR   PRO   SER   GLU   CYS   HIS   ALA   ASP   TYR   PHE   ASP   PHE   GLU   ARG
  T T T A T G G T T C G T T C T T A T T A C C C T T C T G A A T G T C A C G C T G A T T A T T T G A C T T T G A G C G T
        ↑         3987                    3997                    4007                    4017                    4027                    4037
  mRNA end

A  ILE   GLU   ALA   LEU   LYS   PRO   ALA   ILE   GLU   ALA   CYS   GLY   ILE   SER   THR   LEU   SER   GLN   SER   PRO
  A T C G A G G C T C T T A A A C C T G C T A T T G A G G C T T G T G G C A T T T C T A C T C T T T C T C A A T C C C C A
     T1/6         4047                    4057                    4067                    4077                    4087                    4097

A  MET   LEU   GLY   PHE   HIS   LYS   GLN   MET   ASP   ASN   ARG   ILE   LYS   LEU   LEU   GLU   GLU   ILE   LEU   SER
  A T G C T T G G C T T C C A T A A G C A G A T G G A T A A C C G C A T C A A G C T C T T G G A A G A G A T T C T G T C T
                  4107                    4117                    4127                    4137                    4147                    4157
                                                                                          A7b/7a            M5/8      F5c/3

A  PHE   ARG   MET   GLN   GLY   VAL   GLU   PHE   ASP   ASN   GLY   ASP   MET   TYR   VAL   ASP   GLY   HIS   LYS   ALA
  T T T C G T A T G C A G G G C G T T G A G T T C G A T A A T G G T G A T A T G T A T G T T G A C G G C C A T A A G G C T
                  4167                    4177     T6/2           4187   Q2/3a            4197        R4/3      4207                        4217
                                                                                                             Z2/6b

A  ALA   SER   ASP   VAL   ARG   ASP   GLU   PHE   VAL   SER   VAL   THR   GLU   LYS   LEU   MET   ASP   GLU   LEU   ALA
  G C T T C T G A C G T T C G T G A T G A G T T T G T A T C T G T T A C T G A G A A G T T A A T G G A T G A A T T G G C A
                  4227                    4237                    4247                    4257                    4267                    4277

A  GLN   CYS   TYR   ASN   VAL   LEU   PRO   GLN   LEU   ASP   ILE   ASN   ASN   THR   ILE   ASP   HIS   ARG   PRO   GLU
  C A A T G C T A C A A T G T G C T C C C C C A A C T T G A T A T T A A T A A C A C T A T A G A C C A C C G C C C C G A A
                  4287                    4297                    4307                    4317                    4327                    4337
                                 Origin of viral strand replication ↑

A  GLY   ASP   GLU   LYS   TRP   PHE   LEU   GLU   ASN   GLU   LYS   THR   VAL   THR   GLN   PHE   CYS   ARG   LYS   LEU
  G G G G A C G A A A A A T G G T T T T T A G A G A A C G A G A A G A C G G T T A C G C A G T T T T G C C G C A A G C T G
                  4347                    4357                    4367   M8/6             4377                    4387              A7a/4 4397

A  ALA   ALA   GLU   ARG   PRO   LEU   LYS   ASP   ILE   ARG   ASP   GLU   TYR   ASN   TYR   PRO   LYS   LYS   LYS   GLY
  G C T G C T G A A C G C C C T C T T A A G G A T A T T C G C G A T G A G T A T A A T T A C C C C A A A A A G A A A G G T
                  4407                    4417                    4427                    4437                    4447                    4457

A  ILE   LYS   ASP   GLU   CYS   SER   ARG   LEU   LEU   GLU   ALA   SER   THR   MET   LYS   SER   ARG   ARG   GLY   PHE
  A T T A A G G A T G A G T G T T C A A G A T T G C T G G A G G C C T C C A C T A T G A A A T C G C G T A G A G G C T T T
                  4467                    4477                    4487  Z6b/6a          4497                    4507                    4517
```

Fig. 4

```
A   ALA   ILE   GLN   ARG   LEU   MET   ASN   ALA   MET   ARG   GLN   ALA   HIS   ALA   ASP   GLY   TRP   PHE   ILE   VAL
    G C T A T T C A G C G T T T G A T G A A T G C A A T G C G A C A G G C T C A T G C T G A T G G T T G G T T T A T C G T T
              4527                4537                4547                4557                4567                4577

A   PHE   ASP   THR   LEU   THR   LEU   ALA   ASP   ASP   ARG   LEU   GLU   ALA   PHE   TYR   ASP   ASN   PRO   ASN   ALA
    T T T G A C A C T C T C A C G T T G G C T G A C G A C C G A T T A G A G G C G T T T T A T G A T A A T C C C A A T G C T
              4587                4597                4607                4617                4627                4637

A   LEU   ARG   ASP   TYR   PHE   ARG   ASP   ILE   GLY   ARG   MET   VAL   LEU   ALA   ALA   GLU   GLY   ARG   LYS   ALA
    T T G C G T G A C T A T T T T C G T G A T A T T G G T C G T A T G G T T C T T G C T G C C G A G G G T C G C A A G G C T
              4647                4657                4667                4677                4687                4697

A   ASN   ASP   SER   HIS   ALA   ASP   CYS   TYR   GLN   TYR   PHE   CYS   VAL   PRO   GLU   TYR   GLY   THR   ALA   ASN
    A A T G A T T C A C A C G C C G A C T G C T A T C A G T A T T T T T G T G T G C C T G A G T A T G G T A C A G C T A A T
          F3/5a 4707                4717                4727                4737                4747        A4/11  4757

A   GLY   ARG   LEU   HIS   PHE   HIS   ALA   VAL   HIS   PHE   MET   ARG   THR   LEU   PRO   THR   GLY   SER   VAL   ASP
    G G C C G T C T T C A T T T C C A T G C G G T G C A C T T T A T G C G G A C A C T T C C T A C A G G T A G C G T T G A C
    Z6a/9      M6/1 4767                4777                4787                4797                4807          R3/8 4817

A   PRO   ASN   PHE   GLY   ARG   ARG   VAL   ARG   ASN   ARG   ARG   GLN   LEU   ASN   SER   LEU   GLN   ASN   THR   TRP
    C C T A A T T T T G G T C G T C G G G T A C G C A A T C G C C G C C A G T T A A A T A G C T T G C A A A A T A C G T G G
              4827                4837                4847                4857   A11/10       4867                4877
                                                                                                                  Z9/10

A   PRO   TYR   GLY   TYR   SER   MET   PRO   ILE   ALA   VAL   ARG   TYR   THR   GLN   ASP   ALA   PHE   SER   ARG   SER
    C C T T A T G G T T A C A G T A T G C C C A T C G C A G T T C G C T A C A C G C A G G A C G C T T T T T C A C G T T C T
              4887                4897 ↑              4907                4917                4927                4937
                           mRNA start

A   GLY   TRP   LEU   TRP   PRO   VAL   ASP   ALA   LYS   GLY   GLU   PRO   LEU   LYS   ALA   THR   SER   TYR   MET   ALA
    G G T T G G T T G T G G C C T G T T G A T G C T A A A G G T G A G C C G C T T A A A G C T A C C A G T T A T A T G G C T
              4947   Z10/3        4957                4967                4977  A10/12b       4987                4997
                                                      Q3a/1

A   VAL   GLY   PHE   TYR   VAL   ALA   LYS   TYR   VAL   ASN   LYS   LYS   SER   ASP   MET   ASP   LEU   ALA   ALA   LYS
    G T T G G T T T C T A T G T G G C T A A A T A C G T T A A C A A A A A G T C A G A T A T G G A C C T T G C T G C T A A A
              5007                5017                5027                5037                5047                5057
                                                  R8/5
```

```
A    GLY   LEU   GLY   ALA   LYS   GLU   TRP   ASN   ASN   SER   LEU   LYS   THR   LYS   LEU   SER   LEU   LEU   PRO   LYS
B                                      MET   GLU   GLN   LEU   THR   LYS   ASN   GLN   ALA   VAL   ALA   THR   SER   GLN   GLU
     G G T C T A G G A G C T A A A G A A T G G A A C A A C T C A C T A A A A A C C A A G C T G T C G C T A C T T C C C A A G
                     5067                5077                5087                5097                5107                5117
                     A12b/15b                                                    A15b/17

A    LYS   LEU   PHE   ARG   ILE   ARG   MET   SER   ARG   ASN   PHE   GLY   MET   LYS   MET   LEU   THR   MET   THR   ASN
B       ALA   VAL   GLN   ASN   GLN   ASN   GLU   PRO   GLN   LEU   ARG   ASP   GLU   ASN   ALA   HIS   ASN   ASP   LYS   SER
     A A G C T G T T C A G A A T C A G A A T G A G C C G C A A C T T C G G G A T G A A A A T G C T C A C A A T G A C A A A T
      A17/12a        5127  F5a/6         5137                5147                5157                5167                5177

A    LEU   SER   THR   GLU   CYS   LEU   ILE   GLN   LEU   THR   LYS   LEU   GLY   TYR   ASP   ALA   THR   PRO   PHE   ASN
B       VAL   HIS   GLY   VAL   LEU   ASN   PRO   THR   TYR   GLN   ALA   GLY   LEU   ARG   ARG   ASP   ALA   VAL   GLN   PRO
     C T G T C C A C G G A G T G C T T A A T C C A A C T T A C C A A G C T G G G T T A C G A C G C G A C G C C G T T C A A C
                     5187                5197                5207  A12a/5        5217                5227                5237

A    GLN   ILE   LEU   LYS   GLN   ASN   ALA   LYS   ARG   GLU   MET   ARG   LEU   ARG   LEU   GLY   LYS   VAL   THR   VAL
B       ASP   ILE   GLU   ALA   GLU   ARG   LYS   LYS   ARG   ASP   GLU   ILE   GLU   ALA   GLY   LYS   SER   TYR   CYS   SER
     C A G A T A T T G A A G C A G A A C G C A A A A A G A G A G A T G A G A T T G A G G C T G G G A A A A G T T A C T G T A
                     5247                5257                5267                5277                5287                5297

A    ALA   ASP   VAL   LEU   ALA   ALA   GLN   PRO   VAL   THR   THR   ASN   LEU   LEU   LYS   PHE   MET   ARG   ALA   SER
B       ARG   ARG   PHE   GLY   GLY   ALA   THR   CYS   ASP   ASP   LYS   SER   ALA   GLN   ILE   TYR   ALA   ARG   PHE   ASP
     G C C G A C G T T T T G G C G G C G C A A C C T G T G A C G A C A A A T C T G C T C A A A T T T A T G C G C G C T T C G
                     5307          H1/15 5317                5327                5337                5347  H15/8b        5357
                                                                                                                      T2/7

A    ILE   LYS   MET   ILE   GLY   VAL   SER   ASN   LEU   GLN   SER   PHE   ILE   ALA   SER   MET   THR   GLN   LYS   LEU
B       LYS   ASN   ASP   TRP   ARG   ILE   GLN   PRO   ALA   GLU   PHE   TYR   ARG   PHE   HIS   ASP   ALA   GLU   VAL   ASN
     A T A A A A A T G A T T G G C G T A T C C A A C C T G C A G A G T T T T A T C G C T T C C A T G A C G C A G A A G T T A
                     5367                5377                  1                 11                  21                  31
                                                          P1/1                                                      R5/7b

A    THR   LEU   SER   ASP   ILE   SER   ASP   GLU   SER   LYS   ASN   TYR   LEU   ASP   LYS   ALA   GLY   ILE   THR   THR
K                                         MET   SER   ARG   LYS   ILE   ILE   LEU   ILE   LYS   GLN   GLU   LEU   LEU   LEU
B          THR   PHE   GLY   TYR   PHE   ***
     A C A C T T T C G G A T A T T T C T G A T G A G T C G A A A A A T T A T C T T G A T A A A G C A G G A A T T A C T A C T
                     41                  51    F6/9  T7/8  61                  71                  81                  91
```

FIG. 4—*continued*

```
A     ALA  CYS  LEU  ARG  ILE  LYS  SER  LYS  TRP  THR  ALA  GLY  GLY  LYS  ***
K       LEU  VAL  TYR  GLU  LEU  ASN  ARG  SER  GLY  LEU  LEU  ALA  GLU  ASN  GLU  LYS  ILE  ARG  PRO  ILE
C                                                                         MET  ARG  LYS  PHE  ASP  LEU  SER
      G C T T G T T T A C G A A T T A A A T C G A A G T G G A C T G C T G G C G G A A A A T G A G A A A A T T C G A C C T A T
                101                 111                 121                 131                 141                 151
                                    T8/9                                                            T9/10

K       LEU  ALA  GLN  LEU  GLU  LYS  LEU  LEU  LEU  CYS  ASP  LEU  SER  PRO  SER  THR  ASN  ASP  SER  VAL
C         LEU  ARG  SER  SER  ARG  SER  SER  TYR  PHE  ALA  THR  PHE  ARG  HIS  GLN  LEU  THR  ILE  LEU  SER
      C C T T G C G C A G C T C G A G A A G C T C T T A C T T T G C G A C C T T T C G C C A T C A A C T A A C G A T T C T G T
                161                 171                 181                 191                 201                 211
          H8b/4     A5/18  T10/4    A18/6                                                           F9/13

K       LYS  ASN  ***
C         LYS  THR  ASP  ALA  LEU  ASP  GLU  GLU  LYS  TRP  LEU  ASN  MET  LEU  GLY  THR  PHE  VAL  LYS  ASP
      C A A A A A C T G A C G C G T T G G A T G A G G A G A A G T G G C T T A A T A T G C T T G G C A C G T T C G T C A A G G
                221                 231                 241                 251                 261                 271

C         TRP  PHE  ARG  TYR  GLU  SER  HIS  PHE  VAL  HIS  GLY  ARG  ASP  SER  LEU  VAL  ASP  ILE  LEU  LYS
      A C T G G T T T A G A T A T G A G T C A C A T T T T G T T C A T G G T A G A G A T T C T C T T G T T G A C A T T T T A A
                281                 291                 301                 311                 321                 331
                              F13/17                                F17/16a             R7b/6c

D                                                                                                          MET
C         GLU  ARG  GLY  LEU  LEU  SER  GLU  SER  ASP  ALA  VAL  GLN  PRO  LEU  ILE  GLY  LYS  LYS  SER  ***
      A A G A G C G T G G A T T A C T A T C T G A G T C C G A T G C T G T T C A A C C A C T A A T A G G T A A G A A A T C A T
                341                 351 F16a/16b  ↑     361                 371                 381                 391
                                        mRNA start

D       SER  GLN  VAL  THR  GLU  GLN  SER  VAL  ARG  PHE  GLN  THR  ALA  LEU  ALA  SER  ILE  LYS  LEU  ILE
      G A G T C A A G T T A C T G A A C A A T C C G T A C G T T T C C A G A C C G C T T T G G C C T C T A T T A A G C T C A T
      F16b/1    401                 411                 421                 431       Z3/7      441       A6/1      451

D       GLN  ALA  SER  ALA  VAL  LEU  ASP  LEU  THR  GLU  ASP  ASP  PHE  ASP  PHE  LEU  THR  SER  ASN  LYS
      T C A G G C T T C T G C C G T T T T G G A T T T A A C C G A A G A T G A T T T C G A T T T T C T G A C G A G T A A C A A
                461                 471                 481 M1/7            491                 501                 511
                                                                            T4/5

D       VAL  TRP  ILE  ALA  THR  ASP  ARG  SER  ARG  ALA  ARG  ARG  CYS  VAL  GLU  ALA  CYS  VAL  TYR  GLY
E                                                                                                     MET  VAL
      A G T T T G G A T T G C T A C T G A C C G C T C T C G T G C T C G T C G C T G C G T T G A G G C T T G C G T T T A T G G
                521                 531                 541                 551                 561                 571
```

```
D      THR   LEU   ASP   PHE   VAL   GLY   TYR   PRO   ARG   PHE   PRO   ALA   PRO   VAL   GLU   PHE   ILE   ALA   ALA   VAL
E        ARG   TRP   THR   LEU   TRP   ASP   THR   LEU   ALA   PHE   LEU   LEU   LEU   LEU   SER   LEU   LEU   LEU   PRO   SER
     T A C G C T G G A C T T T G T G G G A T A C C C T C G C T T T C C T G C T C C T G T T G A G T T T A T T G C T G C C G T
                   581                 591                 601                 611                 621                 631

D      ILE   ALA   TYR   TYR   VAL   HIS   PRO   VAL   ASN   ILE   GLN   THR   ALA   CYS   LEU   ILE   MET   GLU   GLY   ALA
E        LEU   LEU   ILE   MET   PHE   ILE   PRO   SER   THR   PHE   LYS   ARG   PRO   VAL   SER   SER   TRP   LYS   ALA   LEU
     C A T T G C T T A T T A T G T T C A T C C C G T C A A C A T T C A A A C G G C C T G T C T C A T C A T G G A A G G C G C
                   641                 651    R6c/7a   661              671             681              691
                                                                     Z7/5                              H4/13

D      GLU   PHE   THR   GLU   ASN   ILE   ILE   ASN   GLY   VAL   GLU   ARG   PRO   VAL   LYS   ALA   ALA   GLU   LEU   PHE
E        ASN   LEU   ARG   LYS   THR   LEU   LEU   MET   ALA   SER   SER   VAL   ARG   LEU   LYS   PRO   LEU   ASN   CYS   SER
     T G A A T T T A C G G A A A A C A T T A T T A A T G G C G T C G A G C G T C C G G T T A A A G C C G C T G A A T T G T T
                   701                 711                 721                 731                 741                 751
                                                              T5/3                Y1/3

D      ALA   PHE   THR   LEU   ARG   VAL   ARG   ALA   GLY   ASN   THR   ASP   VAL   LEU   THR   ASP   ALA   GLU   GLU   ASN
E        ARG   LEU   PRO   CYS   VAL   TYR   ALA   GLN   GLU   THR   LEU   THR   PHE   LEU   LEU   THR   GLN   LYS   LYS   THR
     C G C G T T T A C C T T G C G T G T A C G C G C A G G A A A C A C T G A C G T T C T T A C T G A C G C A G A A G A A A A
                   761                 771 H13/11              781                 791                 801    M7/3     811

J                                                                                MET   SER   LYS   GLY   LYS   LYS   ARG   SER
D      VAL   ARG   GLN   LYS   LEU   ARG   ALA   GLU   GLY   VAL   MET   ***
E        CYS   VAL   LYS   ASN   TYR   VAL   ARG   LYS   GLU   ***
     C G T G C G T C A A A A A T T A C G T G C G G A A G G A G T G A T G T A A T G T C T A A A G G T A A A A A A C G T T C T
                   821                 831                 841                 851                 861                 871

J    GLY   ALA   ARG   PRO   GLY   ARG   PRO   GLN   PRO   LEU   ARG   GLY   THR   LYS   GLY   LYS   ARG   LYS   GLY   ALA
     G G C G C T C G C C C T G G T C G T C C G C A G C C G T T G C G A G G T A C T A A A G G C A A G C G T A A A G G C G C T
      H11/14       881                 891                 901                 911                 921        H14/12   931

J      ARG   LEU   TRP   TYR   VAL   GLY   GLY   GLN   GLN   PHE   ***
     C G T C T T T G G T A T G T A G G T G G T C A A C A A T T T T A A T T G C A G G G G C T T C G G C C C C T T A C T T G A
                   941                 951  R7a/6b         961                 971                 981                 991
                                                                                                  Z5/8       Minor mRNA end

F                    MET   SER   ASN   ILE   GLN   THR   GLY   ALA   GLU   ARG   MET   PRO   HIS   ASP   LEU   SER   HIS
     G G A T A A A T T A T G T C T A A T A T T C A A A C T G G C G C C G A G C G T A T G C C G C A T G A C C T T T C C C A T
                  1001                1011                1021                1031                1041                1051
                                                         H12/10
```

FIG. 4—*continued*

```
F  LEU GLY PHE LEU ALA GLY GLN ILE GLY ARG LEU ILE THR ILE SER THR THR PRO VAL ILE
   C T T G G C T T C C T T G C T G G T C A G A T T G G T C G T C T T A T T A C C A T T T C A A C T A C T C C G G T T A T C
   1061 1071 1081 1091 1101 1111
   Y3/2

F  ALA GLY ASP SER PHE GLU MET ASP ALA VAL GLY ALA LEU ARG LEU SER PRO LEU ARG ARG
   G C T G G C G A C T C C T T C G A G A T G G A C G C C G T T G G C G C T C T C C G T C T T T C T C C A T T G C G T C G T
   1121 1131 1141 1151 1161 1171
   F1/14b T3/1 H10/7

F  GLY LEU ALA ILE ASP SER THR VAL ASP ILE PHE THR PHE TYR VAL PRO HIS ARG HIS VAL
   G G C C T T G C T A T T G A C T C T A C T G T A G A C A T T T T T A C T T T T T A T G T C C C T C A T C G T C A C G T T
   1181 1191 1201 1211 1221 1231
   Z8/4 F14b/2

F  TYR GLY GLU GLN TRP ILE LYS PHE MET LYS ASP GLY VAL ASN ALA THR PRO LEU PRO THR
   T A T G G T G A A C A G T G G A T T A A G T T C A T G A A G G A T G G T G T T A A T G C C A C T C C T C T C C C G A C T
   1241 1251 1261 1271 1281 1291
   Q1/3c

F  VAL ASN THR THR GLY TYR ILE ASP HIS ALA ALA PHE LEU GLY THR ILE ASN PRO ASP THR
   G T T A A C A C T A C T G G T T A T A T T G A C C A T G C C G C T T T T C T T G G C A C G A T T A A C C C T G A T A C C
   1301 1311 1321 1331 1341 1351
   R6b/1

F  ASN LYS ILE PRO LYS HIS LEU PHE GLN GLY TYR LEU ASN ILE TYR ASN ASN TYR PHE LYS
   A A T A A A A T C C C T A A G C A T T T G T T T C A G G G T T A T T T G A A T A T C T A T A A C A A C T A T T T T A A A
   1361 1371 1381 1391 1401 1411

F  ALA PRO TRP MET PRO ASP ARG THR GLU ALA ASN PRO ASN GLU LEU ASN GLN ASP ASP ALA
   G C G C C G T G G A T G C C T G A C C G T A C C G A G G C T A A C C C T A A T G A G C T T A A T C A A G A T G A T G C T
   1421 1431 1441 1451 1461 1471
   H7/5 A1/12c

F  ARG TYR GLY PHE ARG CYS CYS HIS LEU LYS ASN ILE TRP THR ALA PRO LEU PRO PRO GLU
   C G T T A T G G T T T C C G T T G C T G C C A T C T C A A A A A C A T T T G G A C T G C T C C G C T T C C T C C T G A G
   1481 1491 1501 1511 1521 1531

F  THR GLU LEU SER ARG GLN MET THR THR SER THR THR SER ILE ASP ILE MET GLY LEU GLN
   A C T G A G C T T T C T C G C C A A A T G A C G A C T T C T A C C A C A T C T A T T G A C A T T A T G G G T C T G C A A
   1541 1551 1561 1571 1581 1591
   A12c/13

F  ALA ALA TYR ALA ASN LEU HIS THR ASP GLN GLU ARG ASP TYR PHE MET GLN ARG TYR HIS
   G C T G C T T A T G C T A A T T T G C A T A C T G A C C A A G A A C G T G A T T A C T T C A T G C A G C G T T A C C A T
   1601 1611 1621 1631 1641 1651
   A13/2
```

```
F  ASP VAL ILE SER SER PHE GLY GLY LYS THR SER TYR ASP ALA ASP ASN ARG PRO LEU LEU
   G A T G T T A T T T C T T C A T T T G G A G G T A A A A C C T C T T A T G A C G C T G A C A A C C G T C C T T T A C T T
   1661 M3/4  1671  1681  1691  1701  1711

F  VAL MET ARG SER ASN LEU TRP ALA SER GLY TYR ASP VAL ASP GLY THR ASP GLN THR SER
   G T C A T G C G C T C T A A T C T C T G G G C A T C T G G C T A T G A T G T T G A T G G A A C T G A C C A A A C G T C G
   H5/9a 1721  1731  1741  1751  1761  1771

F  LEU GLY GLN PHE SER GLY ARG VAL GLN GLN THR TYR LYS HIS SER VAL PRO ARG PHE PHE
   T T A G G C C A G T T T T C T G G T C G T G T T C A A C A G A C C T A T A A A C A T T C T G T G C C G C G T T T C T T T
   Z4/1 1781  1791  1801  1811  1821  1831

F  VAL PRO GLU HIS GLY THR MET PHE THR LEU ALA LEU VAL ARG PHE PRO PRO THR ALA THR
   G T T C C T G A G C A T G G C A C T A T G T T T A C T C T T G C G C T T G T T C G T T T C C G C C C T A C T G C G A C T
   1841  1851  1861 H9a/8a  1871  1881  1891

F  LYS GLU ILE GLN TYR LEU ASN ALA LYS GLY ALA LEU THR TYR THR ASP ILE ALA GLY ASP
   A A A G A G A T T C A G T A C C T T A A C G C T A A A G G T G C T T T G A C T T A T A C C G A T A T T G C T G G C G A C
   1901 F2/11  1911  1921  1931  1941  1951

F  PRO VAL LEU TYR GLY ASN LEU PRO PRO ARG GLU ILE SER MET LYS ASP VAL PHE ARG SER
   C C T G T T T T G T A T G G C A A C T T G C C G C C G C G T G A A A T T T C T A T G A A G G A T G T T T T C C G T T C T
   1961  1971  1981  1991  2001  2011

F  GLY ASP SER SER LYS LYS PHE LYS ILE ALA GLU GLY GLN TRP TYR ARG TYR ALA PRO SER
   G G T G A T T C G T C T A A G A A G T T T A A G A T T G C T G A G G G T C A G T G G T A T C G T T A T G C G C C T T C G
   Q3c/6 F11/7 2021  2031  2041  2051  2061 H8a/6  2071

F  TYR VAL SER PRO ALA TYR HIS LEU LEU GLU GLY PHE PRO PHE ILE GLN GLU PRO PRO SER
   T A T G T T T C T C C T G C T T A T C A C C T T C T T G A A G G C T T C C C A T T C A T T C A G G A A C C G C C T T C T
   2081  2091 Q6/5  2101  2111  2121  2131

F  GLY ASP LEU GLN GLU ARG VAL LEU ILE ARG HIS HIS ASP TYR ASP GLN CYS PHE GLN SER
   G G T G A T T T G C A A G A A C G C G T A C T T A T T C G C C A C C A T G A T T A T G A C C A G T G T T T C C A G T C C
   Q5/3b 2141  2151  2161  2171  2181  2191

F  VAL GLN LEU LEU GLN TRP ASN SER GLN VAL LYS PHE ASN VAL THR VAL TYR ARG ASN LEU
   G T T C A G T T G T T G C A G T G G A A T A G T C A G G T T A A A T T T A A T G T G A C C G T T T A T C G C A A T C T G
   2201  2211  2221  2231  2241  2251
```

FIG. 4—*continued*

```
F   PRO THR THR ARG ASP SER ILE MET THR SER ***
    C C G A C C A C T C G C G A T T C A A T C A T G A C T T C G T G A T A A A A G A T T G A G T G T G A G G T T A T A A C G
    2261  F7/5b  2271  2281  2291  2301  2311

    C C G A A G C G G T A A A A A T T T A A T T T T T G C C G C T G A G G G G T T G A C C A A G C G A A G C G C G G T A G
    2321  2331  2341  2351 R1/9  2361  H6/3  2371

G   MET PHE GLN THR PHE ILE SER ARG HIS ASN SER ASN PHE
    G T T T T C T G C T T A G G A G T T T A A T C A T G T T T C A G A C T T T T A T T T C T C G C C A T A A T T C A A A C T
    2381  2391  2401  2411  2421  2431

G   PHE SER ASP LYS LEU VAL LEU THR SER VAL THR PRO ALA SER SER ALA PRO VAL LEU GLN
    T T T T T T C T G A T A A G C T G G T T C T C A C T T C T G T T A C T C C A G C T T C T T C G G C A C C T G T T T T A C
    2441  A2/16  2451  2461  2471 A16/15a  M4/10  2481  2491

G   THR PRO LYS ALA THR SER SER THR LEU TYR PHE ASP SER LEU THR VAL ASN ALA GLY ASN
    A G A C A C C T A A A G C T A C A T C G T C A A C G T T A T A T T T T G A T A G T T T G A C G G T T A A T G C T G G T A
    2501  A15a/3  2511  R9/10  2521  2531  2541  2551

G   GLY GLY PHE LEU HIS CYS ILE GLN MET ASP THR SER VAL ASN ALA ALA ASN GLN VAL VAL
    A T G G T G G T T T T C T T C A T T G C A T T C A G A T G G A T A C A T C T G T C A A C G C C G C T A A T C A G G T T G
    2561  M10/9  2571  2581  2591 R10/2  2601  2611

G   SER VAL GLY ALA ASP ILE ALA PHE ASP ALA ASP PRO LYS PHE PHE ALA CYS LEU VAL ARG
    T T T C T G T T G G T G C T G A T A T T G C T T T T G A T G C C G A C C C T A A A T T T T T T G C C T G T T T G G T T C
    2621  2631  2641  2651  2661  2671

G   PHE GLU SER SER SER VAL PRO THR THR LEU PRO THR ALA TYR ASP VAL TYR PRO LEU ASN
    G C T T T G A G T C T T C T T C G G T T C C G A C T A C C C T C C C G A C T G C C T A T G A T G T T T A T C C T T T G A
    2681 F5b/8  M9/2  2691  2701  2711  2721  2731

G   GLY ARG HIS ASP GLY GLY TYR TYR THR VAL LYS ASP CYS VAL THR ILE ASP VAL LEU PRO
    A T G G T C G C C A T G A T G G T G G T T A T T A T A C C G T C A A G G A C T G T G T G A C T A T T G A C G T C C T T C
    2741  2751  2761  2771  2781  2791

G   ARG THR PRO GLY ASN ASN VAL TYR VAL GLY PHE MET VAL TRP SER ASN PHE THR ALA THR
    C C C G T A C G C C G G G C A A T A A C G T T T A T G T T G G T T T C A T G G T T T G G T C T A A C T T T A C C G C T A
    2801 Y2/5  2811  2821  2831  2841  2851
```

```
G     LYS CYS ARG GLY LEU VAL SER LEU ASN GLN VAL ILE LYS GLU ILE ILE CYS LEU GLN PRO
      C T A A A T G C C G C G G A T T G G T T T C G C T G A A T C A G G T T A T T A A A G A G A T T A T T T G T C T C C A G C
                  2861                2871                2881                2891                2901                2911
                                                         F8/4

H                                 MET PHE GLY ALA ILE ALA GLY GLY ILE ALA SER ALA LEU ALA
G       LEU LYS ***
      C A C T T A A G T G A G G T G A T T T A T G T T T G G T G C T A T T G C T G G C G G T A T T G C T T C T G C T C T T G C
                  2921     Q3b/4      2931                2941                2951                2961                2971

H     GLY GLY ALA MET SER LYS LEU PHE GLY GLY GLY GLN LYS ALA ALA SER GLY GLY ILE GLN
      T G G T G G C G C C A T G T C T A A A T T G T T T G G A G G C G G T C A A A A A G C C G C C T C C G G T G G C A T T C A
               H3/2 2981                2991                3001                3011                3021                3031
                                                                                                Y5/4

H     GLY ASP VAL LEU ALA THR ASP ASN ASN THR VAL GLY MET GLY ASP ALA GLY ILE LYS SER
      A G G T G A T G T G C T T G C T A C C G A T A A C A A T A C T G T A G G C A T G G G T G A T G C T G G T A T T A A A T C
         Q4/7     3041                3051                3061                3071   Q7/2         3081                3091

H     ALA ILE GLN GLY SER ASN VAL PRO ASN PRO ASP GLU ALA ALA PRO SER PHE VAL SER GLY
      T G C C A T T C A A G G C T C T A A T G T T C C T A A C C C T G A T G A G G C C G C C C C T A G T T T T G T T T C T G G
                  3101                3111                3121                3131                3141                3151
                                                                             Z1/2

H     ALA MET ALA LYS ALA GLY LYS GLY LEU LEU GLU GLY THR LEU GLN ALA GLY THR SER ALA
      T G C T A T G G C T A A A G C T G G T A A A G G A C T T C T T G A A G G T A C G T T G C A G G C T G G C A C T T C T G C
                  3161     A3/9       3171                3181                3191                3201                3211

H     VAL SER ASP LYS LEU LEU ASP LEU VAL GLY LEU GLY GLY LYS SER ALA ALA ASP LYS GLY
      C G T T T C T G A T A A G T T G C T T G A T T T G G T T G G A C T T G G T G G C A A G T C T G C C G C T G A T A A A G G
                  3221                3231                3241                3251                3261                3271

H     LYS ASP THR ARG ASP TYR LEU ALA ALA ALA PHE PRO GLU LEU ASN ALA TRP GLU ARG ALA
      A A A G G A T A C T C G T G A T T A T C T T G C T G C T G C A T T T C C T G A G C T T A A T G C T T G G G A G C G T G C
                  3281                3291                3301                3311                3321                3331
                                                                              A9/12d

H     GLY ALA ASP ALA SER SER ALA GLY MET VAL ASP ALA GLY PHE GLU ASN GLN LYS GLU LEU
      T G G T G C T G A T G C T T C C T C T G C T G G T A T G G T T G A C G C C G G A T T T G A G A A T C A A A A A G A G C T
                  3341                3351                3361                3371                3381                3391
                                                      R2/6a       Y4/1              F4/14a              A12d/7c
```

Fig. 4—*continued*

```
H     THR   LYS   MET   GLN   LEU   ASP   ASN   GLN   LYS   GLU   ILE   ALA   GLU   MET   GLN . ASN   GLU   THR   GLN   LYS
     T A C T A A A A T G C A A C T G G A C A A T C A G A A A G A G A T T G C C G A G A T G C A A A A T G A G A C T C A A A A
                3401                    3411                    3421                    3431                    3441    F14a/12   3451

H     GLU   ILE   ALA   GLY   ILE   GLN   SER   ALA   THR   SER   ARG   GLN   ASN   THR   LYS   ASP   GLN   VAL   TYR   ALA
     A G A G A T T G C T G G C A T T C A G T C G G £ G A C T T C A C G C C A G A A T A C G A A A G A C C A G G T A T A T G C
                3461                    3471                    3481                    3491                    3501                    3511

H     GLN   ASN   GLU   MET   LEU   ALA   TYR   GLN   GLN   LYS   GLU   SER   THR   ALA   ARG   VAL   ALA   SER   ILE   MET
     A C A A A A T G A G A T G C T T G C T T A T C A A C A G A A G G A G T C T A C T G C T C G C G T T G C G T C T A T T A T
                3521                    3531                    3541   F12/10   3551                    3561                    3571

H     GLU   ASN   THR   ASN   LEU   SER   LYS   GLN   GLN   GLN   VAL   SER   GLU   ILE   MET   ARG   GLN   MET   LEU   THR
     G G A A A A C A C C A A T C T T T C C A A G C A A C A G C A G G T T T C C G A G A T T A T G C G C C A A A T G C T T A C
                3581                    3591                    3601                    3611            H2/9b   3621                    3631

H     GLN   ALA   GLN   THR   ALA   GLY   GLN   TYR   PHE   THR   ASN   ASP   GLN   ILE   LYS   GLU   MET   THR   ARG   LYS
     T C A A G C T C A A A C G G C T G G T C A G T A T T T T A C C A A T G A C C A A A T C A A A G A A A T G A C T C G C A A
          A7c/8 3641                    3651                    3661                    3671                    3681    F10/15    3691

H     VAL   SER   ALA   GLU   VAL   ASP   LEU   VAL   HIS   GLN   GLN   THR   GLN   ASN   GLN   ARG   TYR   GLY   SER   SER
     G G T T A G T G C T G A G G T T G A C T T A G T T C A T C A G C A A A C G C A G A A T C A G C G G T A T G G C T C T T C
                3701      R6a/4     3711                    3721                    3731    F15/5c          3741            M2/5  3751

H     HIS   ILE   GLY   ALA   THR   ALA   LYS   ASP   ILE   SER   ASN   VAL   VAL   THR   ASP   ALA   ALA   SER   GLY   VAL
     T C A T A T T G G C G C T A C T G C A A A G G A T A T T T C T A A T G T C G T C A C T G A T G C T G C T T C T G G T G T
                3761 H9b/1              3771                    3781                    3791                    3801                    3811

H     VAL   ASP   ILE   PHE   HIS   GLY   ILE   ASP   LYS   ALA   VAL   ALA   ASP   THR   TRP   ASN   ASN   PHE   TRP   LYS
     G G T T G A T A T T T T T C A T G G T A T T G A T A A A G C T G T T G C C G A T A C T T G G A A C A A T T T C T G G A A
                3821                    3831                    3841 A8/14              3851                    3861                    3871

H     ASP   GLY   LYS   ALA   ASP   GLY   ILE   GLY   SER   ASN   LEU   SER   ARG   LYS   ***
     A G A C G G T A A A G C T G A T G G T A T T G G C T C T A A T T T G T C T A G G A A A T A A
                3881 A14/7b             3891                    3901                    3911
```

FIG. 4. The nucleotide sequence of the DNA of bacteriophage ϕX174, *cs*70 and (above) the amino acid sequences of the proteins for which it codes. The sequence of the circular DNA is written starting after the termination codon of the H protein but the numbering starts at the unique *Pst*I cleavage site. The letters in the left-hand margin indicate the protein whose amino acid sequence is shown in the corresponding line. Restriction enzyme recognition sites are indicated by underlining. The single letter code for the enzymes is as follows:

A *Alu*I
F *Hinf*I
H *Hha*I
M *Mbo*II
P *Pst*I
Q *Hph*I
R *Hind*II
T *Taq*I
Y *Hap*II
Z *Hae*III

The position of the origin of viral strand replication is based on the work of Langeveld *et al.* (1978). Other references are given in our earlier paper (Sanger *et al.*, 1977*a*) and elsewhere in this paper.

The *cs*70 mutation is cold-sensitive and is believed to be a point mutation in the F protein (C. A. Hutchison III, J. H. Middleton & M. H. Edgell, unpublished data).

well established and it is possible that they may be due to mutations or changes in the batches of ϕX DNA used. These are the change of a G to an A at position 2731 which leads to an Asn-to-Asp change in the amino acid sequence and the change of C to T at position 2421. The latter is a third position change but the C was well characterized in the original work on the ribosome binding site (Robertson *et al.*, 1973).

(f) *The H protein coding region*

The provisional sequence of the region coding for the H protein (Sanger *et al.*, 1977*a*) was largely dependent on the plus and minus method, and most sequences were derived from priming done on both strands. In the initial stages considerable use was made of the previous observation that regions coding for ϕX proteins have a high frequency of T in the third positions of codons. Sequences were inspected to see if they satisfied this rule and, if they did not, errors were suspected and the region was reinvestigated. Difficulties were experienced in obtaining amino acid sequences in the H protein, which was not easy to prepare in a soluble form, but a number of peptides were finally obtained by Air (unpublished data) all of which were consistent with the DNA sequence. These are the regions with solid underlining in the provisional sequence (Sanger *et al.*, 1977*a*). The whole of the H protein coding region has now been studied by the terminator method and only six errors have been found, although this was considered to be the most uncertain region of the genome.

There was some difficulty with the sequence between residues 2996 and 3026. This was due to a symmetrical sequence which could form a stable base-paired loop affecting migration rates on the acrylamide gels. This problem could, however, be resolved by priming on both strands and the sequence was confirmed by the terminator method using Q7 as primer on the viral strand and Q3b on the complementary strand. Figure 3 shows an experiment using the latter system and cutting the product with *Hha*. Apart from artifact bands at position 3012 the sequence over the loop region can be read from this gel. Presumably the loop did not form at the high temperatures at which this thin gel was run.

Another uncertain region was at positions 3146 to 3153. By using the method of Maxam & Gilbert (1977) on fragment A3 three errors were found and the revised sequence was confirmed by the termination method.

The sequence 3525 to 3611 was initially determined by the plus and minus method using only A8 as primer on the plus strand. The only other potential priming sites in this region are *Hinf* restriction enzyme sites. *Hinf* fragments have not been found to be very good primers and the results with F12 were particularly ambiguous. These difficulties were probably due to the presence of exonuclease in the enzyme preparation used. An error was indicated in this region because digestion of ϕX174 replicative form DNA with a new restriction enzyme, *Fnu*DII, which cleaves in the centre of the sequence C-G-C-G, revealed a cleavage site near position 3560 or 3730 (A. C. P. Lui, B. C. McBride and M. Smith, unpublished data). Using the terminator method C-G-C-G was located at positions 3555 to 3558, the sequence at 3534 to 3536 was revised (see Table 1), and the rest of the sequence in this area was confirmed. (An experiment in which fragment A8 was used as primer on the plus strand was shown in a previous paper (Sanger *et al.*, 1977*b*).) Good results were also obtained with fragment R4 as primer on the plus strand and with A12d on the minus strand. An experiment with the latter system was shown in the previous paper and other gels that were run for longer times confirmed the sequence beyond position 3560.

4. Discussion

Table 1 lists the various revision that have had to be made to the provisional sequence of Sanger *et al.* (1977*a*) and Figure 4 shows the revised DNA sequence together with the amino acid sequence of the proteins for which it codes. While it is difficult to be certain that a sequence of this size contains no errors, it is believed that it is reasonably accurate. All sequences have been confirmed either by using different methods or by priming on opposite strands, or both, and errors in copying out results have been avoided by storing them in the computer (Staden, 1977).

Various biological conclusions were deduced from the provisional sequence and the present revisions do not in any way alter these. Most of the changes are in protein coding regions and some of them lead to changes in the amino acid sequences of the proteins. Otherwise the only change that may contribute to our biological understanding of ϕX is that at position 4498. This shows an A-T-G codon (4497 to 4499) in the reading phase of the A protein. It is preceded, eight residues away, by a G-G-A-G-G which is a characteristic sequence for a ribosome binding site (Shine & Dalgarno, 1974) and probably is the initiation site for the A* protein. Linney & Hayashi (1974) have shown that this protein of molecular weight approximately 35,000 results from an internal translational start in the *A* gene in the same reading phase. The start had originally been tentatively identified at position 4668 (Brown & Smith, 1977; Sanger *et al.*, 1977*a*) but from the revised sequence this position (4497) seems much more likely. The ribosomal binding site is more characteristic and it would give rise to a protein containing 341 amino acids and of molecular weight about 37,200.

The finding of non-coding intervening sequences within coding regions in eukaryotic DNA (Jeffreys & Flavell, 1977; Brack & Tonegawa, 1977) raises the question of whether they may also occur in ϕX and whether therefore one is justified in deducing the amino acid sequences of proteins solely from DNA sequence data. Our results show no evidence of insertions and there probably are none. However, we cannot rigorously exclude the possibility of their presence in regions such as that coding for the A protein where there are no amino acid data. If present, any insertions would have to be small and, fortuitously, have no "in-phase" termination codons.

REFERENCES

Air, G. M., Blackburn, E. H., Sanger, F. & Coulson, A. R. (1975). *J. Mol. Biol.* **96**, 703–719.

Air, G. M., Sanger, F. & Coulson, A. R. (1976). *J. Mol. Biol.* **108**, 519–533.

Air, G. M., Coulson, A. R., Fiddes, J. C., Friedmann, T., Hutchison, C. A. III, Sanger, F., Slocombe, P. M. & Smith, A. J. H. (1978). *J. Mol. Biol.* **125**, 247–254.

Barrell, B. G. (1978). Ph.D. Thesis, University of Cambridge.

Brack, C. & Tonegawa, S. (1977). *Proc. Nat. Acad. Sci., U.S.A.* **74**, 5652–5656.

Brown, N. L. & Smith, M. (1977). *J. Mol. Biol.* **116**, 1–30.

Fiddes, J. C. (1976). *J. Mol. Biol.* **107**, 1–24.

Greenfield, L., Simpson, L. & Kaplan, D. (1975). *Biochim. Biophys. Acta*, **407**, 365–375.

Jeffreys, A. J. & Flavell, R. A. (1977). *Cell*, **12**, 1097–1108.

Langeveld, S. A., van Mansfeld, A. D. M., Baas, P. D., Jansz, H. S., van Arkel, G. A. & Weisbeek, P. J. (1978). *Nature (London)*, **271**, 417–420.

Linney, E. & Hayashi, M. (1974). *Nature (London)*, **249**, 345–348.

Maxam, A. & Gilbert, W. (1977). *Proc. Nat. Acad. Sci., U.S.A.* **74**, 560–564.

Robertson, H. D., Barrell, B. G., Weith, H. L. & Donelson, J. E. (1973). *Nature New Biol.* **241**, 38–40.

Sanger, F. & Coulson, A. R. (1975). *J. Mol. Biol.* **94**, 441–448.

Sanger, F. & Coulson, A. R. (1978). *FEBS Letters*, **87**, 107–110.

Sanger, F., Donelson, J. E., Coulson, A. R., Kössel, H. & Fischer, D. (1973). *Proc. Nat. Acad. Sci., U.S.A.* **70**, 1209–1213.

Sanger, F., Air, G. M., Barrell, B. G., Brown, N. L., Coulson, A. R., Fiddes, J. C., Hutchison, C. A. III, Slocombe, P. M. & Smith, M. (1977*a*). *Nature (London)*, **265**, 687–695.

Sanger, F., Nicklen, S. & Coulson, A. R. (1977*b*). *Proc. Nat. Acad. Sci., U.S.A.* **74**, 5463–5467.

Shaw, D. C., Walker, J. E., Northrop, F. D., Barrell, B. G., Godson, G. N. & Fiddes, J. C. (1978). *Nature (London)*, **272**, 510–515.

Shine, J. & Dalgarno, L. (1974). *Proc. Nat. Acad. Sci., U.S.A.* **71**, 1342–1346.

Slocombe, P. M. (1977). Ph.D. Thesis, University of Cambridge.

Smith, M., Brown, N. L., Air, G. M., Barrell, B. G., Coulson, A. R., Hutchison, C. A. III & Sanger, F. (1977). *Nature (London)*, **265**, 702–705.

Staden, R. (1977). *Nucleic Acids Res.* **4**, 4037–4051.

Walker, J. E., Shaw, D. C., Northrop, F. D. & Horsnell, T. (1978). In *Solid Phase Methods in Protein Sequence Analysis* (Previero, A. & Coletti-Previero, M.-A., eds), pp. 277–285, North-Holland Inc., Amsterdam.

J. Mol. Biol. (1978) **125**, 247–254

Nucleotide Sequence of the F Protein Coding Region of Bacteriophage ϕX174 and the Amino Acid Sequence of its Product

G. M. Air[1], A. R. Coulson, J. C. Fiddes[2], T. Friedmann[3]
C. A. Hutchison III[4], F. Sanger, P. M. Slocombe[5] and A. J. H. Smith

Medical Research Council
Laboratory of Molecular Biology
Hills Road, Cambridge CB2 2QH
England

(Received 2 June 1978)

The nucleotide sequence of the region coding for the F protein of bacteriophage ϕX174 has now been completed, in conjunction with extended peptide sequence analysis of the gene *F* protein. The protein is 426 amino acids in length, with a molecular weight, calculated from the sequence, of 48,340.

1. Introduction

In the provisional nucleotide sequence of the DNA of bacteriophage ϕX174 reported by Sanger *et al.* (1977*a*), there were several regions of uncertainty within the *F* gene, which codes for the major capsid protein of the virion. Since these were also regions where the amino acid sequence data were uncertain or absent, further nucleotide sequence work has now been done, particularly using the chain-termination method of Sanger *et al.* (1977*b*). We now report the complete nucleotide sequence of the region coding for the F protein and give the amino acid sequence data which have been obtained since our preliminary reports of the *F* gene work (Air, 1976; Sedat *et al.*, 1976; Air *et al.*, 1976*a*). The numbering of the nucleotides is that of the revised sequence given in Figure 4 of the accompanying paper (Sanger *et al.*, 1978).

2. Materials and Methods

(a) *Nucleic acid sequence determination*

The methods used to determine the provisional sequence (Sanger *et al.*, 1977*a*; Fiddes, 1976; J. C. Fiddes, C. A. Hutchison III and P. M. Slocombe, unpublished data) were those

Present addresses:
[1] John Curtin School of Medical Research, Microbiology Department, Canberra City, ACT 2601, Australia.
[2] Department of Biochemistry and Biophysics, University of California, San Francisco, Calif. 94143, U.S.A.
[3] Department of Pediatrics M-009, University of California, San Diego School of Medicine, La Jolla, Calif. 92039, U.S.A.
[4] Department of Bacteriology and Immunology, University of North Carolina, Chapel Hill, North Carolina 27514, U.S.A.
[5] Max-Planck Institut für Molekulaire Genetik, 1 Berlin 33, Germany.

00222836/78/300247–09 $02.00/0

of Sanger & Coulson (1975), with modifications as previously described (Air *et al.*, 1976*b*), apart from the early work using endonuclease IV (Sedat *et al.*, 1976). More recently the chain termination method of Sanger *et al.* (1977*b*) has proved reliable, and most of the sequence has been reinvestigated by this method.

(b) *Purification of the F protein and amino acid sequence determination*

The methods used for purifying the F protein and determining its amino acid sequence were as previously described (Air, 1976; Air *et al.*, 1975). While the work was in progress it became clear that the sequencing would be completed faster and more easily using the nucleic acid sequencing techniques. Therefore the amino acid sequence work concentrated on obtaining as much peptide sequence data as possible from standard purification procedures, with no attempt to isolate insoluble peptides or to resolve complex mixtures of neutral peptides (such as are obtained in a thermolysin digest).

Basically, the strategy employed for the amino acid sequence determination was fragmentation into relatively large polypeptides using CNBr to cleave at methionine residues, and trypsin to cleave at arginine residues after blocking the ϵ-amino groups of lysine residues. Further fragmentation of the larger peptides was then by trypsin, chymotrypsin or thermolysin. The peptides were separated by 2-dimensional electrophoresis and chromatography, located with fluoroescamine, and eluted for amino acid analysis and sequencing by the manual dansyl-Edman procedure, as described previously (Air, 1976; Air *et al.*, 1975).

3. Results and Discussion

(a) *The DNA sequence*

In the provisional sequence (Sanger *et al.*, 1977*a*) there were a number of regions of uncertainty in the F protein coding region, including a possible gap of three unidentified nucleotides, and in fact the most extensive revisions in the ϕX174 DNA sequence (Sanger *et al.*, 1978) are in this gene.

By using the chemical degradation procedure (Maxam & Gilbert, 1977) on fragment A1 (for nomenclature of restriction fragments see Sanger *et al.*, 1977*a*), it was clear that there were not three extra nucleotides following position 1420 as had been indicated in the original sequence and that some revision of the sequence before this was required. The revised sequence was confirmed by the terminator method (Sanger *et al.*, 1977*b*) using fragments A12c and A2 as primers on the viral strand (Fig. 1), and fragment Q1 on the complementary strand (Fig. 2). The results in Figure 2 also showed that three extra residues had been inserted between positions 1297 and 1298. This sequence had been partly derived from early and rather provisional results with endonuclease IV digests of ϕX DNA (Sedat *et al.*, 1976). In this case the A-C-A sequence at position 1296 to 1298 had been read as A-A-A using the partial exonuclease digestion method.

The other gap in the F protein coding region at position 1654 was identified as a T, and this has been confirmed by the terminator method using fragment A13 as primer on the complementary strand (Fig. 3). Two other changes were found in this region using primer A13, although the original sequence was believed to be correct as it was based on both DNA and protein data. The most surprising difference was a change from A to G at position 1650 which would give rise to a corresponding change of His to Arg at amino acid 216 (Fig. 4) in the F protein (corresponding to amino acid C23 of Air, 1976). It is not likely that both amino acid and nucleotide sequencing results obtained earlier were in error, and since the result in Figure 3 is very clear and was confirmed by an independent method (A. J. H. Smith, unpublished data) we conclude

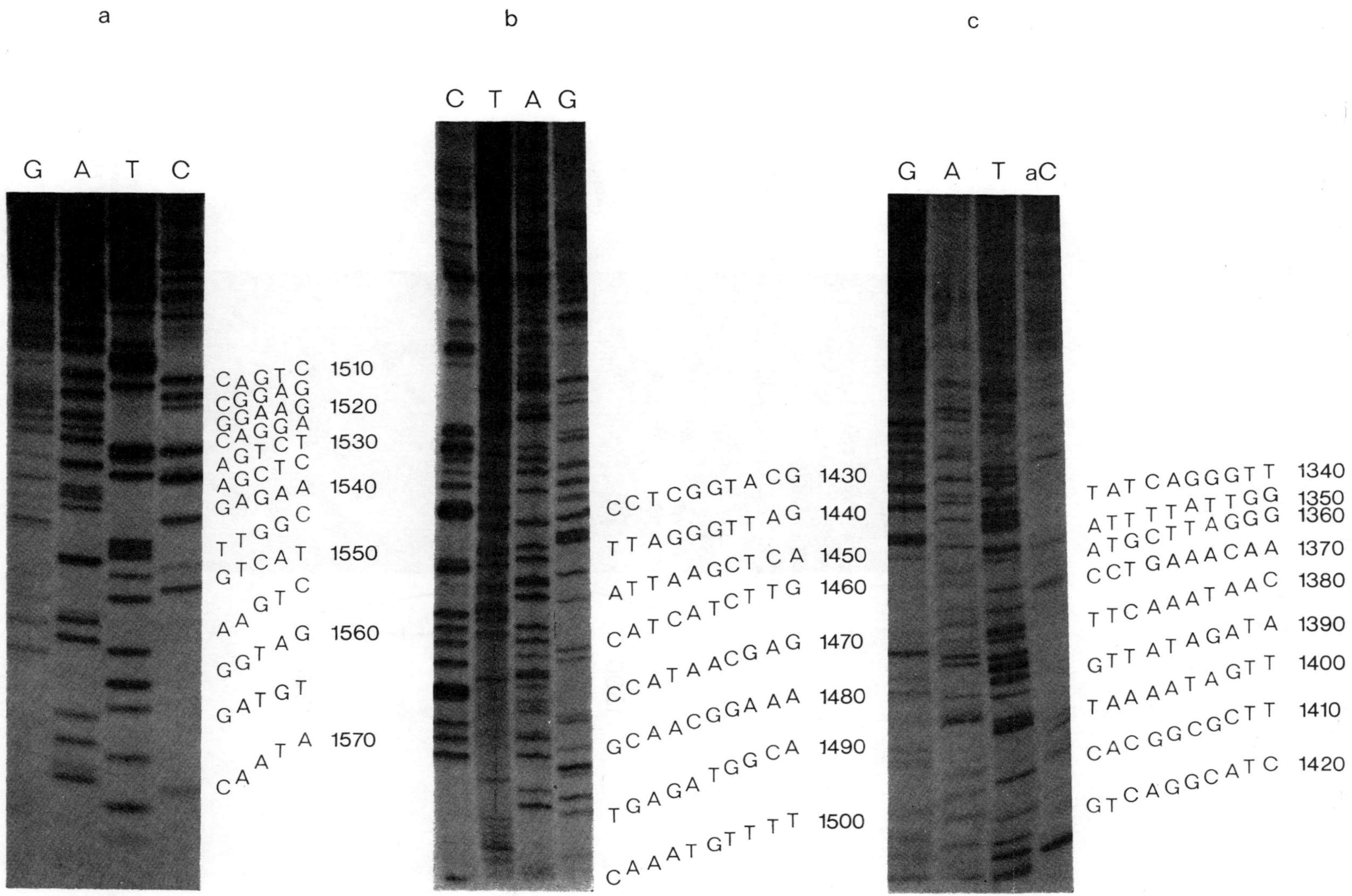

FIG. 1. Autoradiographs of experiments using the terminator method and "single-site ribosubstitution" technique (N. L. Brown, unpublished data; Sanger *et al.*, 1977*b*) with fragment A2 as primer on the viral strand of ϕX. Gels were 1·5 mm thick and were electrophoresed at 45 mA. (a) 12% acrylamide gel, 4 h; (b) 12% acrylamide, 16 h; (c) 8% acrylamide, 12 h.

[facing p. 248

that the ϕX used in the two sets of experiments was different. This could be due to mutation, or to a different stock of phage being used. One possibility is that this change corresponds to the *cs*70 mutation, which is probably in the *F* gene.

The most extensive error in the whole provisional sequence (Sanger *et al.*, 1977*a*) was between positions 2217 and 2229. This was partly due to an error in copying out the results and partly due to the fact that it was based on only one priming with one fragment (Q5 on the complementary strand). The revised sequence was determined by

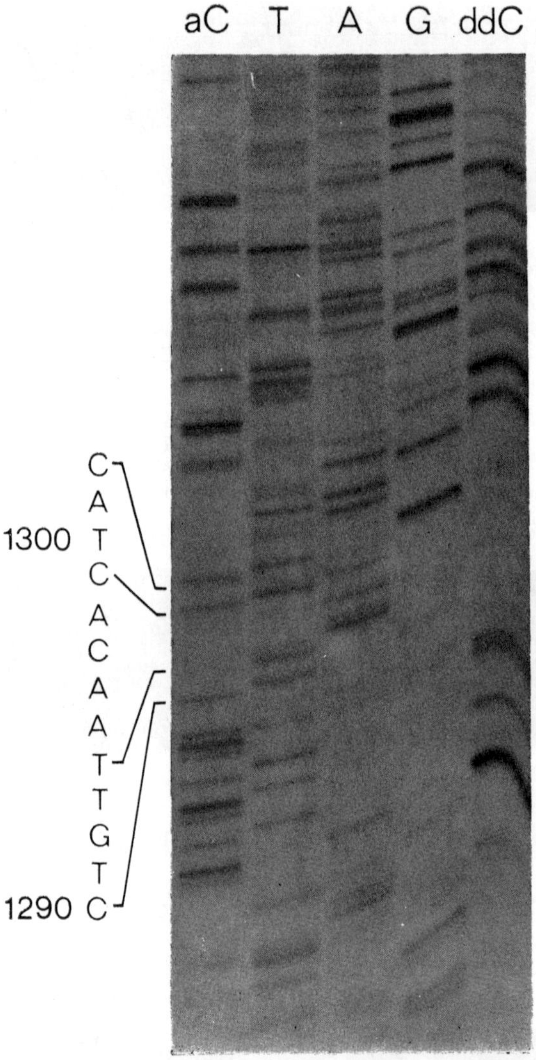

FIG. 2. Autoradiograph of an experiment using the terminator method with fragment Q1 as primer on the complementary strand of ϕX. The products were digested with the restriction enzyme *Hin*fI before electrophoresis on an 8% acrylamide gel 1·5 mm thick at 45 mA for 7 h.

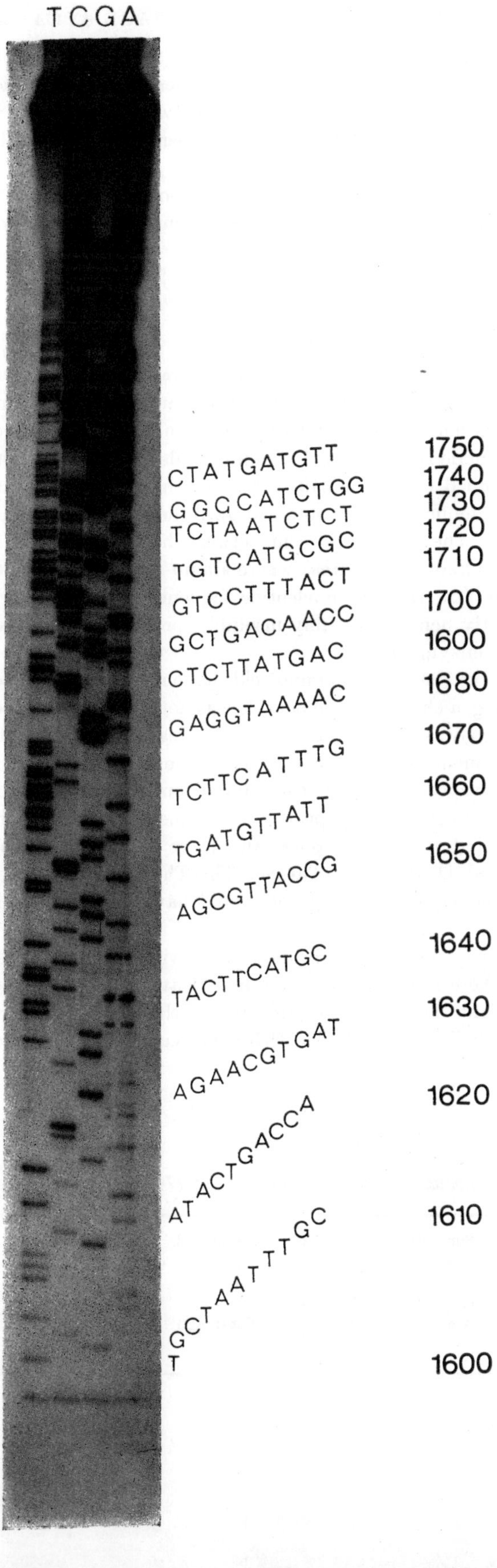

FIG. 3. Autoradiograph of an experiment using the terminator method (Sanger & Coulson, 1978) with fragment A13 as primer on the complementary strand of ϕX on an 8% acrylamide gel 0·35 mm thick at 30 mA and 1·4 kV for 2·5 h. The samples were electrophoresed without digesting away the primer.

the terminator method using fragment A16 on the viral strand (Sanger & Coulson, 1978) and fragment Q3c on the complementary strand. Three other single nucleotides have now been inserted near here and it is largely these errors between positions 2185 and 2229 that were responsible for the total nucleotide number being low in the previous paper (Sanger *et al.*, 1977*a*).

The nucleic acid sequence of the F protein coding region is given in Figure 4 of the accompanying paper and the revisions made to the original sequence are summarized in Table 1 of that paper (Sanger *et al.*, 1978).

(b) *The amino acid sequence*

The peptide sequence data obtained for the F protein are summarized in Figure 4. As mentioned in the previous report (Air, 1976), the block designated "C" (starting at amino acid 194 in Fig. 4) was derived independently of the nucleotide sequence (except for assignment of some amide residues) for 142 amino acids (residue 336), and this has subsequently been extended to residue 349. After this the peptide data are fragmentary, and no overlapping sequences were obtained until the C-terminal region is reached (residues 412 to 426). The few peptides in between have been fitted in after the nucleotide sequence was known. The areas where amino acid sequence data are entirely missing can be seen, now that the nucleotide sequence gives the amino acid sequence, to represent mostly neutral, possibly insoluble, peptides which have not been resolved in the various enzyme digests.

In the sequence preceding residue 194, several amino acids have not been identified, and a few peptides are missing (e.g. a chymotryptic peptide was not found for residues 136 to 139). One sequence which was initially treated with great scepticism until confirmed by the nucleotide sequence is from 184 to 193. The Met–Thr bond is normally resistant to CNBr cleavage (Schroeder *et al.*, 1969), but in one early experiment some partial cleavage did occur. The peptide was obtained in low yield, the amino acid analysis indicated that it was not pure, and there were considerable doubts that the sequence obtained, starting Thr–Thr–Ser–Thr–Thr–Ser, was correct. In the absence of the nucleotide sequence this region would have been extremely difficult to confirm.

Since the N- and C-terminal sequences of the protein agree with those predicted from the nucleotide sequence, there is no post-translational cleavage at either end. The amino acid composition (Table 1) is in good agreement with that obtained by hydrolysis and analysis (Suruda & Poljak, 1971; Air, 1976). The calculated molecular

TABLE 1

Amino acid composition of the F protein of ϕX174

Lys	15	Ser	28	Val	22	Phe	22
His	14	Glx	37	Met	11	Cys	3
Arg	24	Pro	28	Ile	23	Trp	6
Asx	48	Gly	28	Leu	35		
Thr	34	Ala	26	Tyr	22	Total	426

	80 90 100
Final sequence	Met-Lys-Asp-Gly-Val-Asn-Ala-Thr-Pro-Leu-Pro-Thr-Val-Asn-Thr-Thr-
CNBr	|Lys-Asp-Gly-Val-Asx-Ala-Thr-Pro-Leu- - -Val-Asn-Thr-
CNBr/trypsin	—— Residues 1–91, Air (1976) ——→
CNBr/chymotrypsin	
Arg/chymotrypsin	|Met-Lys-Asx-Gly-Val-Asx-Ala-Thr-Pro-Leu-Pro-Thr-
	110 120 125
Final sequence	Gly-Tyr-Ile-Asp-His-Ala-Ala-Phe-Leu-Gly-Thr-Ile-Asn-Pro-Asp-Thr-Asn-Lys-Ile-Pro-Lys-His-Leu-Phe-Gln-
CNBr/trypsin	-Tyr-Ile-Asx- -Ala-Ala-Phe- |Ile-Pro-Lys|
Arg/chymotrypsin	|Ile-Asx-His- - - |Leu-Gly-Thr-Ile-Asx-Pro-Asx-Thr-Asx-Lys-Ile(Pro, Lys, His, Leu)Phe|Gln-
	130 140 150
Final sequence	Gly-Tyr-Leu-Asn-Ile-Tyr-Asn-Asn-Tyr-Phe-Lys-Ala-Pro-Trp-Met-Pro-Asp-Arg-Thr-Glu-Ala-Asn-Pro-Asn-Glu-
	←—— Air (1976) (Residues B1–41) ——
Arg/chymotrypsin	Gly-Tyr|Leu-Asn-Ile-Tyr|Asn-Asn-Tyr-Phe|
	160 170 175
Final sequence	Leu-Asn-Gln-Asp-Asp-Ala-Arg-Tyr-Gly-Phe-Arg-Cys-Cys-His-Leu-Lys-Asn-Ile-Trp-Thr-Ala-Pro-Leu-Pro-Pro-
	180 190 200
Final sequence	Glu-Thr-Glu-Leu-Ser-Arg-Gln-Met-Thr-Thr-Ser-Thr-Thr-Ser-Ile-Asp-Ile-Met-Gly-Leu-Gln-Ala-Ala-Tyr-Ala-
CNBr	⋮Thr-Thr-Ser-Thr-Thr-Ser-Ile-Asx-Ile-Met|←——Air (1976) (Residues C1–73)––
	————————→
	210 220 225
Final sequence	Asn-Leu-His-Thr-Asp-Gln-Glu-Arg-Asp-Tyr-Phe-Met-Gln-Arg-Tyr(Arg)Asp-Val-Ile-Ser-Ser-Phe-Gly-Gly-Lys-
	(His)
	—————— His ——————
	230 240 250
Final sequence	Thr-Ser-Tyr-Asp-Ala-Asp-Asn-Arg-Pro-Leu-Leu-Val-Met-Arg-Ser-Asn-Leu-Trp-Ala-Ser-Gly-Tyr-Asp-Val-Asp-
	260 270 275
Final sequence	Gly-Thr-Asp-Gln-Thr-Ser-Leu-Gly-Gln-Phe-Ser-Gly-Arg-Val-Gln-Gln-Thr-Tyr-Lys-His-Ser-Val-Pro-Arg-Phe-
CNBr/trypsin	|Val-Glx-Glx-Thr-Tyr-Lys|His-Ser-Val- -Arg|Phe-
CNBr/chymotrypsin	————————→|Val-Gln-Gln-Thr-Tyr| |His-Ser-Val-Pro-Arg|Phe-
Arg	|Val-Gln-Gln-Thr-Tyr(Lys)His-Ser(Val,Pro)Arg|Phe
Arg/chymotrypsin	|Val-Gln-Gln-Thr-Tyr|Lys-His-Ser-Val-Pro-Arg|
	280 290 300
Final sequence	Phe-Val-Pro-Glu-His-Gly-Thr-Met-Phe-Thr-Leu-Ala-Leu-Val-Arg-Phe-Pro-Pro-Thr-Ala-Thr-Lys-Glu-Ile-Gln-
CNBr	|Phe-Thr-Leu-Ala-Leu-Val- -Phe-Pro-Pro-Thr-Ala- - -Glx-Ile-Glx-

```
                                280                                     290                                   306
CNBr/trypsin           Phe-Val-Pro-Glu-His-Gly(Thr)Met|                        |Phe-Pro-Pro-Thr-Ala-Thr-Lys|Glx-Ile-Glx-
CNBr/chymotrypsin      Phe-Val-Pro-Glu-His-Gly-Thr-Met|
Arg                    Phe-Val-Pro-Glu-His-Gly-Thr-                            |Phe-    -Pro-Thr-Ala-Thr-
Arg/chymotrypsin                                                               |Phe-Pro-Pro-Thr-Ala-Thr(Lys)Glx(Ile,Glx,

                                                  310                                        320
Final sequence         Tyr-Leu-Asn-Ala-Lys-Gly-Ala-Leu-Thr-Tyr-Thr-Asp-Ile-Ala-Gly-Asp-Pro-Val-Leu-Tyr-Gly-Asn-Leu-Pro-Pro-
CNBr                   Tyr-Leu-Asx-Ala-    -Gly-Ala-Leu-Thr-    -Thr-Asx-
CNBr/trypsin           Tyr(Leu,Asx,Ala)Lys|Gly-Ala-Leu-Thr-Tyr-Thr-Asx-Ile-Ala-Gly-Asx-Pro-Val-Leu-    -Gly-Asx-Leu-Pro-    -
Arg/chymotrypsin       Tyr)|Leu-Asn-Ala-Lys-Gly-Ala(Leu,Thr)Tyr|Thr-Asx-Ile-Ala-Gly-Asx(Pro,Val,Leu)Tyr|Gly-Asn-Leu-Pro-Pro-

                               330   °   °   °   °   °   °   °      ° 340              °      °   °      °       350
Final sequence         Arg-Glu-Ile-Ser-Met-Lys-Asp-Val-Phe-Arg-Ser-Gly-Asp-Ser-Ser-Lys-Lys-Phe-Lys-Ile-Ala-Glu-Gly-Gln-Trp-
CNBr                                      |Lys-Asp-Val-Phe-Arg-Ser-Gly-Asp-Ser-Ser-Lys-    -Phe-Lys-Ile-Ala-Glu-Gly-
CNBr/trypsin           Arg|Glu-Ile-Ser-Met|Lys-Asp-Val-Phe-Arg|Ser-Gly-Asx-   -   (Lys)   |Phe-Lys|Ile-Ala-Glx-Gly-
CNBr/chymotrypsin                         |Lys-Asp-Val-Phe|Arg-Ser-Gly-Asx-Ser-Ser-            |Lys-Ile-Ala-Glx-Gly-Glx-
CNBr/thermolysin                          |Lys-Asp|Val-Phe-Arg-Ser-Gly-Asx(Ser,Ser,Lys,Lys)|Phe-Lys|Ile-Ala-Glx-Gly-Glx
Arg                       |Glu-Ile-Ser-Met-Lys-Asp-Val-Phe-Arg|Ser-Gly-Asx-Ser-Ser(Lys,Lys)Phe(Lys)Ile-Ala-Glx-Gly-Glx-
Arg/chymotrypsin       Arg|

                        °                                  360                                  370           375
Final sequence         Tyr-Arg-Tyr-Ala-Pro-Ser-Tyr-Val-Ser-Pro-Ala-Tyr-His-Leu-Leu-Glu-Gly-Phe-Pro-Phe-Ile-Gln-Glu-Pro-Pro-
CNBr/chymotrypsin      Tyr|                                                                        |Ile-Glx-Glx-Pro-Pro-
CNBr/thermolysin       |Tyr-Arg|Tyr-Ala-Pro-Ser|                                                   |Ile-Glx-Glx-    -Pro-
Arg                    (Tyr)Arg
Arg/chymotrypsin           |Tyr-Ala-Pro-                                                           |Ile-Glx-Glx-Pro-Pro-

                                380                                  390                                   400
Final sequence         Ser-Gly-Asp-Leu-Gln-Glu-Arg-Val-Leu-Ile-Arg-His-His-Asp-Tyr-Asp-Gln-Cys-Phe-Gln-Ser-Val-Gln-Leu-Leu-
CNBr/trypsin                                      |Val-Leu-Ile-Arg
CNBr/chymotrypsin      Ser(Gly,Asx)Leu⁝Glx-Glx-Arg
Arg                                               |Val-Leu-Ile-Arg|
Arg/chymotrypsin       Ser-Gly-Asx-Leu

                                                   410                                        420          425
Final sequence         Gln-Trp-Asn-Ser-Gln-Val-Lys-Phe-Asn-Val-Thr-Val-Tyr-Arg-Asn-Leu-Pro-Thr-Thr-Arg-Asp-Ser-Ile-Met-Thr-Ser-
CNBr/trypsin                                                              |Asn-Leu-Pro-Thr-Thr-Arg|Asp-Ser-Ile(Met,Thr,Ser)
CNBr/thermolysin                                                  |Val-Tyr-Arg-Asx-Leu-Pro-Thr-Thr-Arg-Asx-Ser|
Arg                                                                       |Asn-Leu-Pro-Thr-Thr-Arg|Asp-Ser-Ile-Met-Thr-Ser-
```

FIG. 4. Amino acid sequence data from the F protein of ϕX174. Amino acids indicated were identified by the manual dansyl-Edman method, with amides assigned where possible by electrophoretic mobility of the peptide or identification of the PTH-amino acid. Residues marked ° were obtained from the Sequemat solid-phase machine.

A vertical bar indicates cleavage by CNBr or by trypsin after blocking lysine cleavage (designated "Arg"); partial cleavages are shown by broken bars. Residues enclosed in brackets were present in the amino acid analysis but not identified in the sequencing. Where an amino acid analysis was not conclusive the rest of the peptide is left blank.

weight is 48,340, which agrees well with the estimate obtained by polyacrylamide gel electrophoresis in the presence of sodium dodecyl sulphate (Burgess & Denhardt, 1969; and see Denhardt, 1975).

REFERENCES

Air, G. M. (1976). *J. Mol. Biol.* **107**, 433–443.

Air, G. M., Blackburn, E. H., Sanger, F. & Coulson, A. R. (1975). *J. Mol. Biol.* **96**, 703–719.

Air, G. M., Blackburn, E. H., Coulson, A. R., Galibert, F., Sanger, F., Sedat, J. W. & Ziff, E. B. (1976*a*). *J. Mol. Biol.* **107**, 445–458.

Air, G. M., Sanger, F. & Coulson, A. R. (1976*b*). *J. Mol. Biol.* **108**, 519–533.

Burgess, A. B. & Denhardt, D. T. (1969). *J. Mol. Biol.* **44**, 377–386.

Denhardt, D. T. (1975). *CRC Crit. Rev. Microbiol.* **4**, 161–222.

Fiddes, J. C. (1976). *J. Mol. Biol.* **107**, 1–24.

Maxam, A. & Gilbert, W. (1977). *Proc. Nat. Acad. Sci., U.S.A.* **74**, 560–564.

Sanger, F. & Coulson, A. R. (1975). *J. Mol. Biol.* **94**, 441–448.

Sanger, F. & Coulson, A. R. (1978). *FEBS Letters*, **87**, 107–110.

Sanger, F., Air, G. M., Barrell, B. G., Brown, N. L., Coulson, A. R., Fiddes, J. C., Hutchison, C. A. III, Slocombe, P. M. & Smith, M. (1977*a*). *Nature (London)*, **265**, 687–695.

Sanger, F., Nicklen, S. & Coulson, A. R. (1977*b*). *Proc. Nat. Acad. Sci., U.S.A.* **74**, 5463–5467.

Sanger, F., Coulson, A. R., Friedmann, T., Air, G. M., Barrell, B. G., Brown, N. L., Fiddes, J. C., Hutchison, C. A. III, Slocombe, P. M. & Smith, M. (1978). *J. Mol. Biol.*. **125**, 225–246.

Schroeder, W. A., Shelton, J. B. & Shelton, J. K. (1969). *Arch. Biochem. Biophys.* **130**, 551–556.

Sedat, J. W., Ziff, E. B. & Galibert, F. (1976). *J. Mol. Biol.* **107**, 391–416.

Suruda, A. J. & Poljak, R. J. (1971). *Virology*, **46**, 164–167.

J. Mol. Biol. (1980) **143**, 161–178

Cloning in Single-stranded Bacteriophage as an Aid to Rapid DNA Sequencing

F. Sanger, A. R. Coulson, B. G. Barrell, A. J. H. Smith and B. A. Roe†

Medical Research Council Laboratory of Molecular Biology Hills Road, Cambridge CB2 2QH, England

(Received 24 March 1980)

An approach to DNA sequencing using chain-terminating inhibitors (Sanger *et al.*, 1977) combined with cloning of small fragments of DNA in a single-stranded DNA bacteriophage is described. Random fragments from restriction enzyme digestion of the DNA are inserted into the *Eco*RI site of the modified bacteriophage M13mp2 (Gronenborn & Messing, 1978) using a linker oligonucleotide. Individual recombinant plaques are collected, 1-ml cultures grown, and the DNA isolated. A "flanking-primer" from the vector is used to determine a nucleotide sequence in each inserted DNA fragment by the chain-terminating method. This is a relatively rapid and simple method of accumulating sequence data. The 2771-nucleotide sequence of the largest *Mbo*I restriction enzyme fragment from human mitochondrial DNA was determined by this method.

1. Introduction

We have described a rapid and simple method for determining nucleotide sequences in DNA using DNA polymerase and specific chain-terminating inhibitors (Sanger *et al.*, 1977). The method requires single-stranded DNA to act as template for the DNA polymerase. It is therefore very suitable to use with DNAs, such as certain bacteriophages, that occur naturally in the single-stranded form, or for double-stranded DNAs in which the strands can be separated. However, strand separation is sometimes difficult and this limits the scope of the procedure. One method by which this problem can be overcome has been described by Smith (1979). It involves the treatment of linear DNA with a double strand-specific exonuclease, such as exonuclease III. The resultant product, which is derived from the different ends of the two strands, is largely single-stranded DNA and can be used as a template for the sequencing method.

An alternative way of preparing single-stranded templates is by cloning the DNA as a recombinant in a single-stranded bacteriophage. A number of suitable vectors have been described (Barnes, 1978,1979; Gronenborn & Messing, 1978; Herrmann *et al.*, 1978; Ray & Kook, 1978; Boeke *et al.*, 1979) which are derived from the small single-stranded filamentous bacteriophages. The size of these bacteriophage DNAs is not limited by their packaging so that fragments can be inserted and the virus is

† Present address: Department of Chemistry, Kent State University, Kent, Ohio 44242, U.S.A.

0022–2836/80/300161–18 $02.00/0

secreted by the host at high titres into the culture medium, which makes their isolation relatively easy. This paper describes a general approach to DNA sequencing using one of these vectors in conjunction with the terminator sequencing method, and its application to a fragment of human mtDNA†.

2. Principle of the Method

The vector we used had been prepared by inserting a fragment of the *lac* operon containing the β-galactosidase gene into a bacteriophage M13 (Messing *et al.*, 1977) and then introducing a single site for the restriction enzyme *Eco*RI into this gene (Gronenborn & Messing, 1978). This recombinant (M13mp2), which grows normally like M13, can be used as a vector to insert *Eco*RI fragments into the single *Eco*RI site. The M13mp2 containing the β-galactosidase gene forms blue plaques in the presence of isopropyl-β-D-thiogalactopyranoside and 5-bromo-4-chloro-3-indolyl-β-D-galactoside, but if DNA is inserted into the RI site β-galactosidase is no longer made and thus recombinants can be recognised as "white" plaques. By inserting a second *Eco*RI site in the vector, Heidecker *et al.* (1980) were able to prepare a 96-long DNA fragment that could be used as a flanking primer for sequencing into the end of an inserted DNA. Schreier & Cortese (1979) have described a simple procedure for isolating small amounts of the recombinant DNA and its use as a template for sequencing.

Our general approach is to prepare various restriction enzyme digests of the DNA to be sequenced and to insert the fragments into the M13mp2 vector using an RI linker (G-G-A-A-T-T-C-C: Heyneker *et al.*, 1976; Scheller *et al.*, 1977). White plaques are then collected at random, grown in 1 ml of culture medium and the DNA isolated. A sequence of about 200 nucleotides on each recombinant is then determined by the dideoxy method using the above flanking primer, as described by Schreier & Cortese (1979). Thus the same primer is used on different templates, whereas in the standard dideoxy method different primers are used on one template. Preliminary studies with bacteriophage lambda DNA showed this to be a rapid method of collecting sequence data, especially at the early stages of an investigation where no sequences are yet known.

3. Materials and Methods

The DNA polymerase used throughout this work was DNA polymerase I, Klenow subfragment, obtained either from Boehringer Mannheim or as a gift from A. R. MacLeod. This latter was prepared from a λ*pol*A transducing phage (Kelley *et al.*, 1977) and contained approximately 4 units/ml. Restriction enzyme *Fnu*EI was a gift from M. Smith. Other restriction enzymes were obtained from New England Biolabs.

The dideoxynucleotide triphosphates were obtained from Collaborative Research. (α-^{32}P)-labelled deoxynucleotide triphosphates (spec. act. approx. 400 mCi/μmol) were from the Radiochemical Centre, Amersham. Phenol was redistilled, saturated with water and stored in 2 to 5 ml portions at −20°C until used. Small volumes below 1 μl were normally measured using a washed 5 μl graduated Micropet.

Appropriate manipulations involving recombinant bacteriophages were carried out under category II containment conditions as advised by the Genetic Manipulation Advisory Group.

† Abbreviations used: mtDNA, mitochondrial DNA; ddTTP, dideoxythymidine triphosphate.

(a) *Preparation of the DNA*

*Mbo*I restriction enzyme fragments of human placental mtDNA were cloned in the *Bam*HI site of pBR322 in χ1776 (Drouin, 1980). The recombinant plasmid (pmt18) containing the largest *Mbo*I fragment (M1) was transfected into *Escherichia coli* K12 strain Hb101 (Boyer & Roulland-Dussiox, 1969). 3 l of overnight culture were obtained from growth of a single colony of *E. coli* HB101 carrying the pmt18 plasmid. From this 250 μg of pmt18 were isolated by the cleared lysate method (Katz *et al.*, 1973), followed by ethidium bromide/caesium chloride centrifugation (Radloff *et al.*, 1967).

The M1 fragment was cut out from the purified pmt18 recombinant plasmid by treatment with either *Fnu*EI or *Sau*3A, resolved by fractionation on RPC5 at neutral pH (Eshaghpour & Crothers, 1978) and analysed by electrophoresis through a 1% (w/v) agarose horizontal slab gel (Fig. 1). Fractions containing the M1 fragment were pooled, precipitated with ethanol and a portion restricted with *Hpa*II. The digest was resolved on RPC-5 at neutral pH and analysed by electrophoresis through a 2% agarose horizontal gel slab.

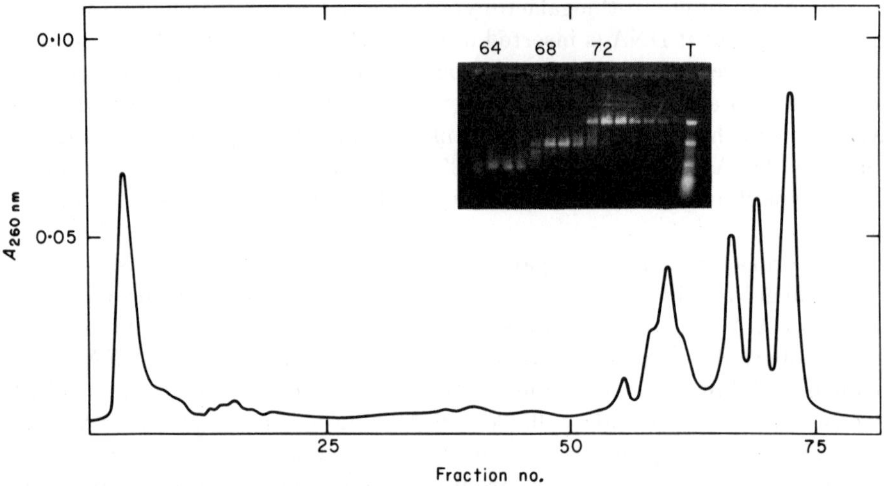

FIG. 1. Fractionation of an *Fnu*EI digest of pmt18 DNA by RPC-5 chromatography. The insert shows a 1% agarose gel analysis of fractions across the gradient. Electrophoresis was from top to bottom. Lane T is a total *Fnu*EI digest of pmt18. The slowest peak on the column and the slowest band on the gel contained fragment M1.

An alternative method for preparing the fragment was by electrophoresis on a 1% low melting agarose gel (Bethesda Research Laboratories). 10 μg of the digested plasmid DNA was applied on a 2 cm wide slot, the gel was stained with ethidium bromide (1 μg/ml) and the band corresponding to the insert visualised under ultraviolet light and excised from the gel. The agarose was melted at 70°C, an equal volume of 10 mM-Tris·HCl (pH 7·6), 0·1 mM-EDTA added and the solution then extracted with the same volume of phenol. After brief centrifugation, the aqueous layer was removed and the phenol extraction procedure repeated a further 2 to 3 times. Finally the solution was extracted with ether and the DNA precipitated overnight at −20°C by the addition of 1/10 vol. 3 M-sodium acetate (pH 5·5) and 2·5 vol. 95% ethanol, followed by centrifugation at 45,000 revs/min for 1 h in a SW60 rotor. The precipitate was washed with 95% ethanol and recentrifuged at the same speed and for the same time. The DNA precipitate was finally dried and redissolved in 10 μl water.

(b) *Digestion and labelling of fragments*

Digestion of the DNA was carried out with various restriction enzymes that recognise a sequence of 4 nucleotides. In these experiments the enzymes *Alu*I (A), *Hae*III (Z), *Hin*fI (F) and *Hpa*II (Y) were used. (The letters in parentheses refer to the symbol used as a prefix to describe the clones derived from the respective digests.) *Alu*I and *Hae*III cut the DNA to give "flush ends" that can be directly ligated to the linker. *Hin*fI and *Hpa*II give staggered ends which are filled in with DNA polymerase. The fragments were labelled with ^{32}P to facilitate purification on the Sepharose column and for the acrylamide gel electrophoreses that were used to test portions for ligation and *Eco*RI digestion (see Fig. 2). In the case of the *Hin*fI and *Hpa*II digests this was done by including a ^{32}P-labelled deoxynucleotide triphosphate in the DNA polymerase reaction. With the other 2 digests the 3′ termini were labelled by an exchange reaction using DNA polymerase and a labelled triphosphate. In this case the labelling was less efficient. To illustrate the method, the digest with *Alu*I and labelling will be described.

1 μg DNA was treated with 1 to 2 units *Alu*I in 5 μl of solution containing H buffer (6·6 mM-Tris·HCl (pH 7·4), 6·6 mM-$MgCl_2$, 6·6 mM-dithiothreitol and 0·05 M-NaCl) for 1 h at 37°C. 2 μl (2 μCi) [α-^{32}P]dGTP was dried down in a small siliconized test tube and then dissolved in the above solution. 0·5 μl of a solution prepared from equal volumes of 10 mM-dCTP, TTP and dATP was added followed by 0·25 μl DNA polymerase (0·25 unit). Incubation was for 10 min at room temperature. The reactions were terminated by the addition of 1 μl 0·2 M-EDTA (pH 7·4), 20 μl water and 20 μl phenol and vortexing. After standing to separate the layers, the aqueous layer was sucked off using a drawn-out capillary tube and the phenol layer washed with a further 10 μl water. These combined aqueous solutions were extracted twice with about 1 ml ether, and the excess ether evaporated in a stream of air. The solution was transferred to a small siliconized glass centrifuge tube (5 cm × 0·6 cm) and adjusted to 50 μl. 5 μl 3 M-sodium acetate and 125 μl ethanol were added. The solution was left at −20°C overnight and the precipitated DNA collected in an Eppendorf centrifuge (10 min), washed with 0·5 ml ethanol and dried in a stream of air.

The *Hae*III digest was treated as described above. The *Hin*fI digest was labelled by adding it to the dried down residue from 2 μCi [α-^{32}P]dATP. It was then incubated with 0·25 μl DNA polymerase for 5 min. 0·5 μl of a mixture made from equal volumes of the 4 deoxynucleotide triphosphates was added and the incubation continued for a further 20 min. The DNA was then isolated as described above. The *Hpa*II digest was treated as the *Hin*fI digest, except that labelling was with [α-^{32}P]dCTP.

(c) *Ligation with linker*

The synthetic *Eco*RI linker (G-G-A-A-T-T-C-C; Collaborative Research) was not phosphorylated on the 5′ end so was first treated with T_4 polynucleotide kinase (5 units/μl; P. L. Biochemicals). 1 μg of the linker in 9 μl of a solution containing 0·05 M-Tris·HCl (pH 7·4), 0·01 M-$MgCl_2$, 0·01 M-dithiothreitol and 0·5 mM-rATP was treated with 1 μl kinase and incubated for 1 h at 37°C.

The dried ethanol precipitate of the digested and labelled DNA was dissolved in 2 μl of the above phosphorylated linker solution using a drawn-out capillary tube. 0·5 μl of a solution containing 2 mM-rATP, 0·7 mM-spermidine and 0·1 unit of T_4 ligase (Miles) was added. The mixture was incubated in the sealed capillary tube at 14°C overnight and the enzyme inactivated by heating at 70°C for 10 min.

In order to test the efficiency of ligation, a sample (usually 0·1 vol., taken after dilution with the *Eco*RI buffer (see below)) was subjected to electrophoresis on acrylamide gel (in the absence of urea) in parallel with a sample of the unligated digest. If the ligation was successful, the bands in the digest were replaced with much larger products, which were usually unresolved (Fig. 2). Considerable differences were found with different preparations of T_4 ligase. In our experience it was necessary to test several before a satisfactory one was found. It appears that the presence of spermidine may be essential for blunt-end ligation (M. Boguski & W. M. Barnes, personal communication).

(d) *Digestion with* Eco*RI*

15 μl of a solution containing 80 mM-Tris·HCl (pH 7·5), 80 mM-NaCl, 10 mM-$MgCl_2$ and

1 mM-dithiothreitol was added to the above ligation reaction mixture, followed by 3 μl *Eco*RI (5000 units/ml; New England Biolabs). Incubation was for 3 h at 37°C. 1 μl EDTA (pH 7·4), 20 μl water and 20 μl phenol were added. After equilibration, the aqueous layer was removed and the phenol layer washed with a further 10 μl water. The combined aqueous solutions were then extracted with 1 ml ether to remove excess phenol and a small portion run on the acrylamide gel to test for complete RI digestion (see Fig. 2).

The *Eco*RI digest contained a high concentration of small fragments derived from the linker, which were readily ligated into the vector giving white plaques. These fragments were removed on a gel filtration column of Sepharose 4B, which also removed other small fragments of less than about 100 nucleotides from the digest. Small fragments appear to be incorporated into the vector more readily than larger ones, and as they give less sequence information it was an advantage to remove them.

A column of Sepharose 4B (Pharmacia) was prepared in a 1 ml disposable plastic pipette and washed well with 2 mM-Tris·HCl (pH 7·4), 0·1 mM-EDTA. The above *Eco*RI digest was applied to the column, which was monitored for radioactivity using a hand monitor, and 1-drop fractions were collected. There was usually a peak of radioactivity at the "breakthrough", followed by a long tail at a lower activity. 2 to 3 drops at the breakthrough were collected, taken to dryness in a desiccator and dissolved in water (5 μl/drop).

(e) *Incorporation into vector*

The vector used was M13mp2 described by Gronenborn & Messing (1978). The replicative form was isolated and digested with *Eco*RI. 0·5 μl (2 ng) of the vector solution was mixed with 1·2 μl of a solution containing 0·1 M-Tris·HCl (pH 7·5), 20 mM-$MgCl_2$, 10 mM-dithiothreitol and 0·4 mM-rATP. 0·5 μl of the DNA fragment with linkers attached, prepared as described above, was added, followed by 0·25 μl T_4 DNA ligase (25 units/ml). It was incubated at 15°C for 2 h and diluted with 20 μl 10 mM-Tris·HCl (pH 7·4), 0·1 mM-EDTA. Competent cells were prepared from *E. coli* strain JM101 (a tra^- mutant, Messing, 1979). The above ligation mixture was diluted into 20 μl 10 mM-Tris·HCl (pH 7·6), 0·1 mM-EDTA and added to 0·2 ml of competent cells, allowed to stand at 0°C for 40 min, heated at 45°C for 3 min and then plated out on 2 plates. The transfection yielded about 100 to 150 plaques, 25 to 30% of which were white (i.e. recombinants).

(f) *Preparation of the cloned DNA*

0·5 ml of a log culture of *E. coli* strain JM101 was added to 50 ml 2 × TY medium (1·6% tryptone, 1% yeast extract, 0·5% NaCl). 1 ml portions were pipetted into sterile glass tubes (10 cm × 1·3 cm) and the white plaques were transferred with tooth picks into the medium. It was found convenient to work up about 36 plaques at a time. The tubes were shaken at 37°C for 7 h and the solutions then transferred to capped Eppendorf tubes and centrifuged for 10 min to pellet the bacteria. The supernatant containing the bacteriophage was transferred to another centrifuge tube, but to avoid dislodging any of the precipitate no attempt was made to transfer it completely. 0·2 ml of a solution containing 2·5 M-NaCl and 20% polyethyleneglycol (M_r, 6000) was added and mixed by inverting the tube, which was then allowed to stand at room temperature for about 15 min and centrifuged for 10 min. The supernatant was poured off, taking care to get rid of the last drop without dislodging the small precipitate, which could usually be seen in the tube. After draining the tube, the inside of it was wiped with a Kleenex tissue and 0·1 ml 10 mM-Tris·HCl (pH 7·4), 0·1 mM-EDTA and 50 μl phenol were added. After vortexing and centrifuging for 1 min, the aqueous solution was sucked off using a drawn-out capillary, care being taken not to include any of the phenol layer. It was transferred to another centrifuge tube, 10 μl 3 M-sodium acetate and 0·2 ml cold ethanol were added and the mixture was left at −20°C overnight. The precipitate was collected by centrifugation for 10 min and washed with 1 ml ethanol. The tubes were allowed to drain and any residual ethanol blown off in a stream of compressed air. The DNA was dissolved in 50 μl 10 mM-Tris·HCl (pH 7·4), 0·1 mM-EDTA. This solution was used for sequencing with the flanking primer as described below.

Using the above method, the DNA in the 1 ml cultures from 36 plaques can be worked up in about 3 h. We have normally done this immediately after growing the cultures. If necessary the culture supernatant can be stored in the refrigerator after pelleting the cells; however, in this case it is advisable to centrifuge again before proceeding in case more cells have grown on standing.

(g) *Preparation of the 96-nucleotide primer*

The primer was generated by an *Eco*RI digest of M13mp2962 (Heidecker *et al.*, 1980) and separated by acrylamide gel electrophoresis or RPC-5 high pressure liquid chromatography. For gel separation, 200 μg of *Eco*RI-digested DNA in a volume of 200 μl containing 10% sucrose, 0·03% bromophenol blue and 10 mM-EDTA was loaded onto a 9-cm slot width of an 8% (w/v) polyacrylamide gel (3 mm × 20 cm × 40 cm). Electrophoresis was carried out for 4 h at 300 V in a continuous buffer system containing 0·09 M-Tris-borate (pH 8·3), 2·5 mM-EDTA. The gel was stained with ethidium bromide (1 μg/ml) for 20 min, the DNA visualised with ultraviolet light, and the gel band excised, the gel volume being kept to a minimum. The DNA was eluted by the method of Maxam & Gilbert (1977), precipitated with ethanol as described above, and transferred to a siliconised glass tube, dried and resuspended in 200 μl water before storage at −20°C.

Fractionation on RPC-5 was essentially as for plasmid fragments. The primer DNA eluted from the column at 0·6 M-NaCl.

(h) *Preliminary screening with ddT reaction*

1 μl of the DNA solution prepared as described above was mixed with 1 μl of the "R1-R1" primer and sealed in a drawn-out capillary tube. It was heated in boiling water for 3 min to denature the DNA and annealed at 67°C for about 20 min. To this was added 2 μl of a reaction solution prepared as follows: 10 μl [^{32}P]dATP (10 μCi) was dried down in a small siliconised test tube. The residue was dissolved in 40 μl of a solution containing H buffer, 0·1 mM-dGTP, 0·1 mM-dCTP, 5 μM-TTP and 0·25 mM-ddTTP. 2 μl DNA polymerase (2 units) was added immediately before addition to the above annealed primer-template solution.

Incubation was carried out at room temperature for 15 min. 1 μl 0·5 mM-dATP was added and after a further 15 min the reaction was terminated either by freezing the solution or by the addition of 5 μl deionised formamide containing 0·01 M-EDTA, 0·03% xylene cyanol FF and 0·03% bromophenol blue. This latter solution was usually added just prior to running the electrophoresis gels. Less satisfactory results were obtained if the reaction mixtures were stored at −20°C after adding the formamide solutions. After heating at 100°C for 3 min,

---

FIG. 2. Autoradiograph of a 4% non-denaturing polyacrylamide gel illustrating the preparation of *Taq*I restriction fragments with *Eco*RI "sticky ends" for cloning into the *Eco*RI site of the M13mp2 vector. The gel (40 cm × 20 cm × 0·35 mm) was run in a continuous buffer of Tris-borate, EDTA at a constant current of 20 mA until the bromophenol blue tracking dye had reached the bottom. The *Taq*I digest was of the mtDNA insert of the plasmid pmt1, which contains a $1·7 \times 10^3$ base *Bam*I-*Eco*RI fragment and the $1·1 \times 10^3$ base *Eco*RI-*Eco*RI fragment (which is also contained within the insert of pmt18). The *Taq*I fragments were blunt-ended and labelled with ^{32}P by extension of their 3′ ends with DNA polymerase and [α-^{32}P]dCTP and dGTP. Track (a) is a portion of this reaction. Track (b) is the same ^{32}P-labelled digest after blunt-end ligation with *Eco*RI linker. Track (c) is the result of an almost complete *Eco*RI digestion of the products of the linker ligation reaction. Bands 1 to 5 in track (c) are interpreted as follows: (1) *Taq*I fragment onto which no *Eco*RI linkers have been ligated. (2) *Taq*I fragment with only a single linker at one of its ends, which has been cleaved with *Eco*RI. (3) *Taq*I fragment with an *Eco*RI-cleaved linker at each end. (This is the predominant product from the reactions.) (4) *Taq*I fragment ligated to 2 *Eco*RI linkers only 1 of which is cleaved with *Eco*RI to give the *Eco*RI sticky end. (5) A band not present in the original *Taq*I digest which is probably the result of 2 of the original fragments of the digest blunt-end ligating together directly or *via* a linker(s) that has not been cleaved during the subsequent *Eco*RI digest. This is supported by the presence in track (b) of a fragment of this size with linkers added.

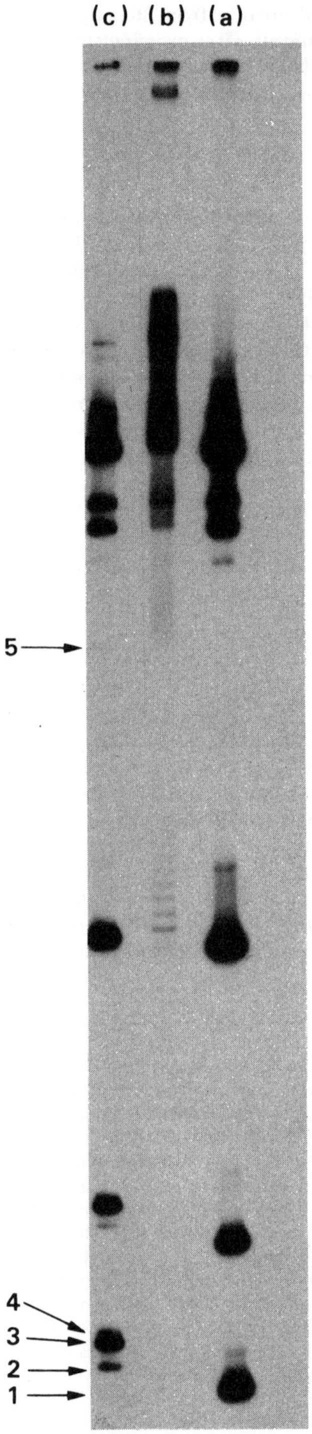

FIG. 2

about half the solution was applied to an 8% acrylamide gel for electrophoresis (Sanger & Coulson, 1978). Usually the reaction was carried out with 18 samples at a time, which were run together on a single gel.

(i) *Sequencing of the DNA*

The sequencing procedure was carried out essentially as described previously (Sanger *et al.*, 1977; Sanger & Coulson, 1978; Schreier & Cortese, 1979) using the 96-nucleotide R1-R1 primer. When this untreated fragment was used to prime, a strong artifact band was present in all 4 channels on the gel at a position about 130 nucleotides from the 5′ end of the primer, and the sequence before this could not be read satisfactorily. This effect is believed to be due to the 3′ end of the strand of the primer, which has the same sequence as the template, looping back to base-pair intramolecularly at a position where it can then act as a primer for DNA polymerase to extend by a further 40 residues. This effect could be avoided by initial treatment of the primer with exonuclease III (Schreier & Cortese, 1979). The sequence of the insert back to and including the *Eco*RI site of the linker could usually then be read unambiguously.

In a normal sequencing experiment 5 μl of the DNA solution from the clone was mixed with 5 μl of the exonuclease-treated primer in H buffer, the mixture heated and annealed, and 2 μl samples used in the 4 reaction mixtures with the appropriate dideoxynucleotide triphosphates, as described by Sanger *et al.* (1977). With different preparations of the dideoxynucleotide triphosphates we found some variation in the concentrations required to give optimum results. The concentration was usually established empirically. Also the concentrations for the G and C reactions had to be varied according to which strand of the mtDNA was being sequenced, as they have a very asymmetric G:C ratio. Normally two 6% acrylamide gel electrophoreses were run with each clone, half of the denatured reaction mixture being applied to each gel. One gel was run at 25 mA for 3 h and the other for 6 h. From the autoradiographs of these 2 gels a sequence of between 150 and 300 nucleotides could usually be determined.

4. Results

(a) *Preparation of the DNA*

In this paper we described the application of this approach to the sequence of a 2771-nucleotide long fragment of human mtDNA. The DNA was prepared from a clone made by Drouin (1980), who subjected human mtDNA to digestion with the restriction enzyme *Mbo*I and inserted the fragments into the *Bam*HI site of the plasmid pBR322. Figure 3 shows the *Mbo*I map of human mtDNA, together with some other features of the DNA that have been mapped. The fragment studied here is the largest *Mbo*I digestion product (fragment M1) and was isolated from clone pmt18 (Drouin, 1980). A certain amount of sequence data was already available on this fragment; in particular, A. J. H. Smith had determined the sequence of the *Eco*RI fragment, which is completely contained within the *Mbo*I fragment and had been isolated from a different clone (pmt1). Other sequence data had also been obtained, both on intact mtDNA and on pmt18 using the exonuclease method (Smith, 1979), but this is not described here.

The DNA of fragment M1 was isolated from the clone by fractionation on an RPC-5 column (Fig. 1). An alternative technique, which has been used with other DNAs, is to use electrophoresis on low melting agarose gels. The latter method is particularly suitable for small amounts. Advantages of the RPC-5 chromatography for the fractionation of DNA restriction enzyme digests are: (a) the fractionation

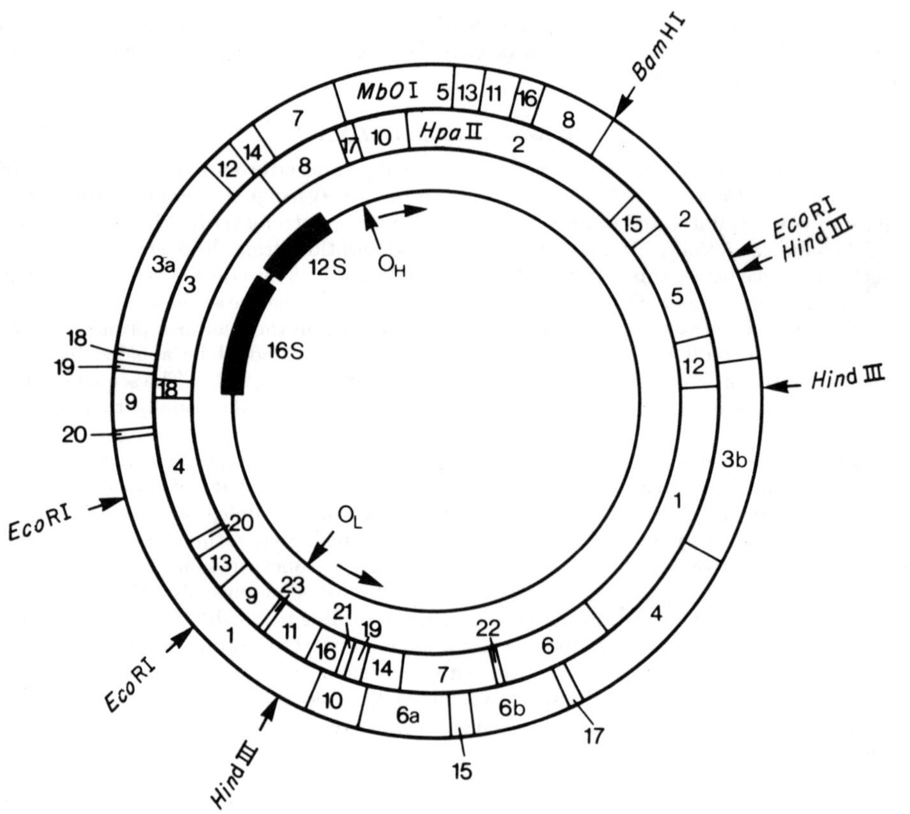

FIG. 3. Restriction enzyme digestion map of human mtDNA. Outer ring *Mbo*I (Drouin, 1980); inner ring *Hpa*II (Ojala & Attardi, 1977). The splitting sites for *Eco*RI, *Bam*HI and *Hin*dIII, the position of the ribosomal RNA genes (12 S and 16 S) and of the origins of heavy strand (O_H) and light strand (O_L) DNA replication are also shown.

step is completed in less than one hour, (b) the desired fragments can be directly precipitated with ethanol or dialysed and lyophilised, and (c) a single column can be re-used several (at least 30) times without loss of resolution. However, the resolution of RPC-5 is greatly reduced when separation of an *Eco*RI digest is attempted: thus the fragments obtained by *Eco*RI digestion of DNAs that generate oligomers similar in size to those shown in Figure 1 are not resolved. In other experiments we observed that RPC-5 can successfully resolve fragments obtained by restriction digestion with *Hin*dII, *Hin*dIII, *Hha*I, *Hae*II and *Hae*III. While these experiments were in progress similar observations were reported by Patient *et al.* (1979).

From the digest prepared with the four restriction enzymes, as described in the Materials and Methods section, we isolated 72 plaques. Cultures (1 ml) of each were grown and the phage isolated from the culture fluid by precipitation with polyethylene glycol, as described. An alternative method of isolation, which gave equally suitable DNA, was by pelleting the phage in the ultracentrifuge for six hours at 450,000 ***g*** (Schreier & Cortese, 1979). In general 1 ml cultures gave enough DNA for ten sequencing experiments.

(b) *Preliminary screening of clones*

In order to decide which clones to study, a preliminary screening was carried out by doing the normal sequencing reaction with only one dideoxynucleotide, in this case ddTTP. These ddTTP reaction mixtures were run in parallel on an acrylamide gel. An example of such a gel is shown in Figure 4. A few clones gave no pattern of bands, or were very weak, indicating that there was insufficient DNA present; these were not investigated further. Some others gave an identical pattern of bands, indicating that they contained the same fragment of DNA (e.g. samples Y6 and Y10, Fig. 4). Clones containing small inserts could also be detected by the presence of a band pattern representing the nucleotide sequence in the vector on the 3′ side of the insert. This common pattern can be seen in clones A2, F5 and Y10 (Fig. 4) and the arrows show where the insert is ligated to the vector. From the position of this pattern the size of the insert can be estimated. It was also possible to identify clones that were from a part of the DNA that had already been sequenced. From the pattern of bands it was possible to write the sequences in the forms of Ts and Xs, X representing A, G or C. Thus clone F3 could be written T-T-X-T-X-X-X-X-X-X-T-X-X-X-T-T-X-X-X-T-T-X-T-X-X-T-X-X-T-X-X-T, etc. Usually if the autoradiograph film was over-exposed a background "ladder" would show up on which the strong T bands were superimposed. In this way it was much easier to estimate the number of nucleotides between the T residues, and in the analysis it is in fact more important to know the number of nucleotides than the actual position of all the Ts. The sequence was then checked against the known sequences previously stored in the computer, using the program SEQFIT (Staden, 1977), and it could be decided which clones should be sequenced.

(c) *Sequence of fragment M1*

Figure 5 shows the location of the different cloned fragments on the physical map of the *L*-strand of fragment M1. This also shows which were subjected to the full sequencing procedure and which were only positioned using the ddT screening. For instance, most of the fragments identified as being present between the two *Eco*RI sites (positions 430, 1582) were not studied as this sequence was already known. The fragments from M1 cover the whole of the DNA except for a gap in the region between nucleotides 1582 and 1686, which contains one of the *Eco*RI sites. This site presents a particular problem since it is always digested by the *Eco*RI used in the above procedure and is never found intact in a clone. In order to obtain a clone covering this region, fragment M1 was digested with restriction endonuclease *Hpa*II and the digest fractionated by RPC-5 chromatography. A fraction containing *Hpa*II fragments 9 and 11 (Fig. 5, positions 1154 to 2050 and 2074 to 2570) was then subjected to the random M13mp2 cloning procedure. In order to obtain a sequence spanning the *Eco*RI site, the DNA was methylated with *Eco*RI methylase (Maniatis *et al.*, 1978) which renders it resistant to the restriction enzyme. Recombinants were prepared from the *Hpa*II fragments 9 and 11 without further digestion, and from

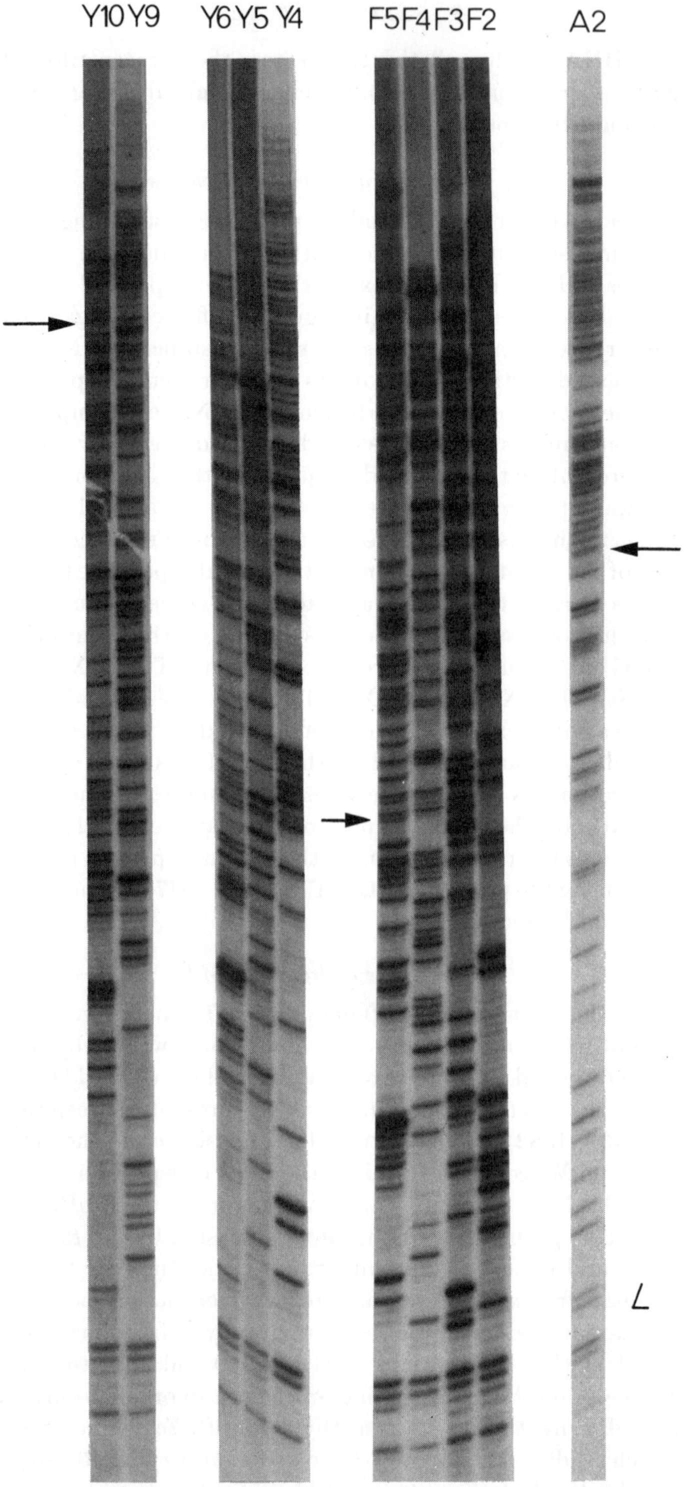

FIG. 4. Preliminary screening of some of the clones using the ddTTP reaction. (For details see text.) The origins of the gels are 5 to 10 cm off the top of the Figure. L marks the end of the 96-long R1-R1 primer. The arrows indicate the beginning of the vector sequence on the 3′ side of the insert.

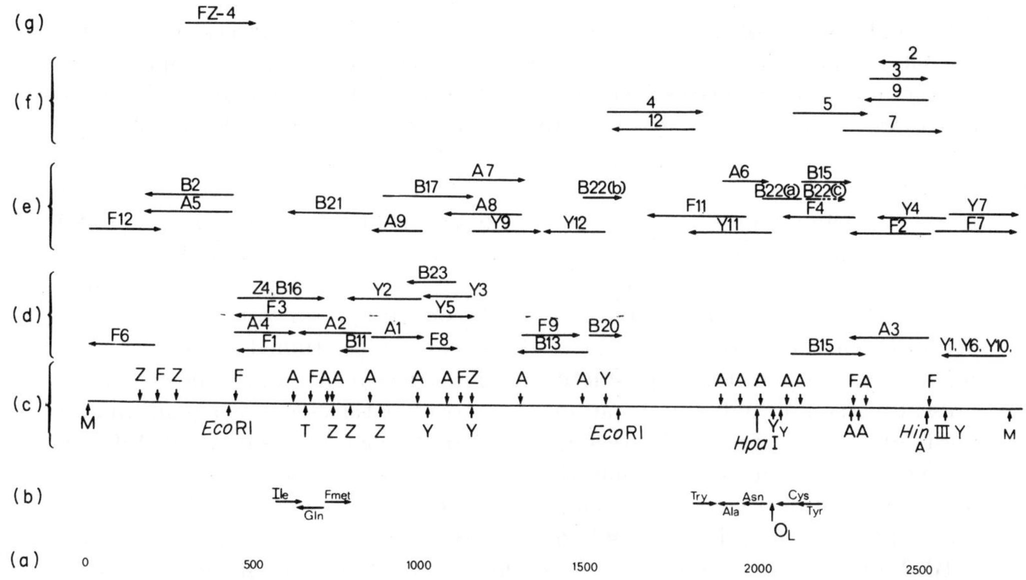

FIG. 5. Map of fragment M1 showing the location of the cloned DNA fragments. (a) Nucleotide number; (b) position of tRNA genes and origin of L-strand replication (O_L); (c) restriction enzyme sites, A = *Alu*I, F = *Hin*fI, M = *Mbo*I, Y = *Hpa*II, Z = *Hae*III; (d) location of inserts in clones that were not sequenced. (Their position was determined from the preliminary screening with ddT.) The arrows indicate the position of the 5′ end of the insert and direction of reading in the insert. In the case of the smaller clones the length of the arrow represents the length of the insert, but for the larger ones the length of the insert was not known. The fragments with prefix B were from a mixture of various restriction enzyme digests. (e) Location of inserts in clones that were sequenced. The length of the arrows shows the length of the sequence that was determined. Clone B22 contained 3 separate *Alu*I fragments ((a), (b), (c)) that had presumably been ligated together during the cloning procedure. (f) Clones derived from digests of *Hpa*II fragments 9 and 11 (see text). (g) Position of sequence previously determined by priming with fragment FZ4 on exonuclease III-treated pmt18 (Smith, 1979).

digests of them with restriction endonucleases *Alu*I and *Hin*dII. Two clones (4 and 12) covered the desired sequence and the *Eco*RI site. None of the clones from the whole M1 digests covered the other *Eco*RI site at position 430; however, this region had already been spanned in previous work by priming on the exonuclease-digested pmt18 with restriction enzyme fragment FZ4 (see Fig. 5(g)).

5. Discussion

Although extensive sequences have been determined with the methods using controlled synthesis with DNA polymerase (Sanger & Coulson, 1975; Sanger *et al.*, 1977), it has sometimes been found difficult to obtain satisfactory results with certain DNA preparations. Probably this is usually due to the presence of impurities. Small amounts of contaminating DNA can make the reading of the sequence ambiguous and the activity of DNA polymerase can be adversely affected by other contaminants, both in the template and the primers, so that some care must be taken in their preparation. In the method described here the same primer is used for each clone so that it can be prepared on a relatively large scale in a pure form, and the use of an RPC-5 column for this avoids the necessity for using gel electrophoresis, which can sometimes introduce inhibiting impurities. Furthermore, the templates prepared from the cloned DNA isolated as described above are usually sufficiently pure to give satisfactory results by the dideoxy method.

Another important advantage of the method is that it does not involve any physical fractionation of restriction enzyme fragments. In all other rapid DNA sequencing methods this is an essential step that can be quite laborious. In the present method the fractionation is replaced by a cloning procedure which necessarily yields pure DNA, and it can in theory be applied to DNA of any size. With the normal strategies the size of DNA that can be studied is limited by the resolving power of methods for fractionating restriction enzyme fragments. In the present method the only fractionation is by spreading the clones on a plate and the purity of the products is therefore independent of the complexity of the DNA digest. We have recently started to apply the method to bacteriophage λ DNA (approx. 49,000 nucleotides) and have rapidly accumulated data covering about 40,000 nucleotides. In comparison the isolation of pure restriction enzyme fragments from a DNA of this size is a formidable problem involving at least two successive digestion and fractionation steps.

Normal procedures for sequencing are usually preceded by the preparation of a restriction enzyme fragment map as a basis to a systematic approach. In the random method described here this is unnecessary since it aims to determine the whole sequence by overlapping. Because of this, however, it does depend more heavily on the use of computer analysis to interpret and store the data.

Using the random approach of collecting clones, sequence data can be obtained relatively rapidly, but towards the end of an investigation new data will accumulate more slowly along with increasing amounts of redundant data. We have found that much of this redundant data is useful for confirmation and completion of the sequence, but it may be preferable to use a more systematic approach to complete a sequence. In this investigation we have used the digest from a smaller restriction enzyme fragment to complete the sequence. Alternatively one could use hybridization probing to isolate selected clones, or prime with internal primers on the larger templates.

Ideally this method requires a completely random collection of clones in which the different regions of the DNA are equally represented. From Figure 5 it would appear that this is not entirely the case: many clones were found that covered the regions 1 to 1580 and 2070 to 2770 and only a few for region 1580 to 2070. Similarly, Drouin (1980) found that some of the *Mbo*I fragments of human mtDNA were particularly difficult to clone in the plasmid pBR322. How general this phenomenon is, and whether it will constitute a serious drawback to the method, can only be found out by further experience.

The method as described here requires digestion of the linkers with *Eco*RI endonuclease so that any *Eco*RI sites in the DNA are also split and no clones are obtained which contain them intact. This problem can be avoided by preliminary treatment of the DNA with *Eco*RI methylase (Maniatis *et al.*, 1979) which protects the sites from digestion by the endonuclease. This technique was used to obtain a

```
 I  G  A  L  R  A  V  A  Q  T  I  S  Y  E  V  T  L  A  I  I  L  L  S  T  L  L  M  S  G  S  F  N  L  S  T  L  I  T  T  Q
GATCGGCGCACTGCGAGCAGTAGCCCAAACAATCTCATATGAAGTCACCCTAGCCATCATTCTACTATCAACATTACTAATAAGTGGCTCCTTTAACCTCTCCACCCTTATCACAACACA
   10   20   30   40   50   60   70   80   90   100   110   120
URF 1 continued --->

 E  H  L  W  L  L  L  P  S  W  P  L  A  M  M  W  F  I  S  T  L  A  E  T  N  R  T  P  F  D  L  A  E  G  E  S  E  L  V  S
AGAACACCTCTGATTACTCCTGCCATCATGACCCTTGGCCATAATATGATTTATCTCCACACTAGCAGAGACCAACCGAACCCCCTTCGACCTTGCCGAAGGGGAGTCCGAACTAGTCTC
   130   140   150   160   170   180   190   200   210   220   230   240

 G  F  N  I  E  Y  A  A  G  P  F  A  L  F  F  M  A  E  Y  T  N  I  I  M  M  N  T  L  T  T  T  I  F  L  G  T  T  Y  D  A
AGGCTTCAACATCGAATACGCCGCAGGCCCCTTCGCCCTATTCTTCATAGCCGAATACACAAACATTATTATAATAAACACCCTCACCACTACAATCTTCCTAGGAACAACATATGACGC
   250   260   270   280   290   300   310   320   330   340   350   360

 L  S  P  E  L  Y  T  T  Y  F  V  T  K  T  L  L  L  T  S  L  F  L  W  I  R  T  A  Y  P  R  F  R  Y  D  Q  L  M  H  L  L
ACTCTCCCCTGAACTCTACACAACATATTTTGTCACCAAGACCCTACTTCTAACCTCCCTGTTCTTATGAATTCGAACAGCATACCCCCGATTCCGCTACGACCAACTCATACACCTCCT
   370   380   390   400   410   420   430   440   450   460   470   480

 W  K  N  F  L  P  L  T  L  A  L  L  M  W  Y  V  S  M  P  I  T  I  S  S  I  P  P  Q  T  *
ATGAAAAACTTCCTACCACTCACCCTAGCATTACTTATATGATATGTCTCCATACCCATTACAATCTCCAGCATTCCCCCTCAAACCTAAGAAATATGTCTGATAAAAGAGTTACTTTG
   490   500   510   520   530   540   550   560   570   580   590   600
                                                          tRNA^Ile --->

ATAGAGTAAATAATAGGAGCTTAAACCCCCTTATTTCTAGGACTATGAGAATCGAACCCATCCCTGAGAATCCAAAATTCTCCGTGCCACCTATCACACCCCATCCTAAAGTAAGGTCAG
   610   620   630   640   650   660   670   680   690   700   710   720
                                                          <--- tRNA^Gln     tRNA^Met_f --->

CTAAATAAGCTATCGGGCCCATACCCCGAAAATGTTGGTTATACCCTTCCCGTACTAATTAATCCCCTGGCCCAACCCGTCATCTACTCTACCATCTTTGCAGGCACACTCATCACAGCG
   730   740   750   760   770   780   790   800   810   820   830   840

                                         M  L  A  F  I  P  V  L  T  K  K  M  N  P  R  S  T  E  A  A  I  K  Y  F  L
CTAAGCTCGCACTGATTTTTTACCTGAGTAGGCCTAGAAATAAACATGCTAGCTTTTATTCCAGTTCTAACCAAAAAAATAAACCCTCGTTCCACAGAAGCTGCCATCAAGTATTTCCTC
   850   860   870   880   890   900   910   920   930   940   950   960
                                 URF 2 --->

 T  Q  A  T  A  S  M  I  L  L  M  A  I  L  F  N  N  M  L  S  G  Q  W  T  M  T  N  T  T  N  Q  Y  S  S  L  M  I  M  M  A
ACGCAAGCAACCGCATCCATAATCCTTCTAATAGCTATCCTCTTCAACAATATACTCTCCGGACAATGAACCATAACCAATACTACCAATCAATACTCATCATTAATAATCATAATAGCT
   970   980   990   1000   1010   1020   1030   1040   1050   1060   1070   1080

 M  A  M  K  L  G  M  A  P  F  H  F  W  V  P  E  V  T  Q  G  T  P  L  T  S  G  L  L  L  L  T  W  Q  K  L  A  P  I  S  I
ATAGCAATAAAACTAGGAATAGCCCCCTTTCACTTCTGAGTCCCAGAGGTTACCCAAGGCACCCCTCTGACATCCGGCCTGCTTCTTCTCACATGACAAAAACTAGCCCCCATCTCAATC
   1090   1100   1110   1120   1130   1140   1150   1160   1170   1180   1190   1200

 M  Y  Q  I  S  P  S  L  N  V  S  L  L  L  T  L  S  I  L  S  I  M  A  G  S  W  G  G  L  N  Q  T  Q  L  R  K  I  L  A  Y
ATATACCAAATCTCTCCCTCACTAAACGTAAGCCTTCTCCTCACTCTCTCAATCTTATCCATCATAGCAGGCAGTTGAGGTGGATTAAACCAGACCCAGCTACGCAAAATCTTAGCATAC
   1210   1220   1230   1240   1250   1260   1270   1280   1290   1300   1310   1320

 S  S  I  T  H  M  G  W  M  M  A  V  L  P  Y  N  P  N  M  T  I  L  N  L  T  I  Y  I  I  L  T  T  T  A  F  L  L  L  N  L
TCCTCAATTACCCACATAGGATGAATAATAGCAGTTCTACCGTACAACCCTAACATAACCATTCTTAATTTAACTATTTATATTATCCTAACTACTACCGCATTCCTACTACTCAACTTA
   1330   1340   1350   1360   1370   1380   1390   1400   1410   1420   1430   1440

 N  S  S  T  T  T  L  L  L  S  R  T  W  N  K  L  T  W  L  T  P  L  I  P  S  T  L  L  S  L  G  G  L  P  P  L  T  G  F  L
AACTCCAGCACCACGACCCTACTACTATCTCGCACCTGAAACAAGCTAACATGACTAACACCCTTAATTCCATCCACCCTCCTCTCCCTAGGAGGCCTGCCCCCGCTAACCGCCTTTTG
   1450   1460   1470   1480   1490   1500   1510   1520   1530   1540   1550   1560
```

```
P  K  W  A  I  I  E  E  F  T  K  N  N  S  L  I  I  P  T  I  M  A  T  I  T  L  L  N  L  Y  F  Y  L  R  L  I  Y  S  T  S
CCCAAATGGGCCATTATCGAAGAATTCACAAAAAACAATAGCCTCATCATCCCCACCATCATAGCCACCATCACCCTCCTTAACCTCTACTTCTACCTACGCCTAATCTACTCCACCTCA
      1570      1580      1590      1600      1610      1620      1630      1640      1650      1660      1670      1680

I  T  L  L  P  M  S  N  N  V  K  M  K  W  Q  F  E  H  T  K  P  T  P  F  L  P  T  L  I  A  L  T  T  L  L  L  P  I  S  P
ATCACACTACTCCCCATATCTAACAACGTAAAAATAAAATGACAGTTTGAACATACAAAACCCACCCCATTCCTCCCCACACTCATCGCCCTTACCACGCTACTCCTACCTATCTCCCCT
      1690      1700      1710      1720      1730      1740      1750      1760      1770      1780      1790      1800

F  M  L  M  I  L  *
TTTATACTAATAATCTTATAGAAATTTAGGTTAAATACAGACCAAGAGCCTTCAAAGCCCTCAGTAAGTTGCAATACTTAATTTCTGTAACAGCTAAGGACTGCAAAACCCCACTCTGCA
      1810      1820      1830      1840      1850      1860      1870      1880      1890      1900      1910      1920
                   tRNA^Trp -->

TCAACTGAACGCAAATCAGCCACTTTAATTAAGCTAAGCCCTTACTAGACCAATGGGACTTAAACCCACAAACACTTAGTTAACAGCTAAGCACCCTAATCAACTGGCTTCAATCTACTT
      1930      1940      1950      1960      1970      1980      1990      2000      2010      2020      2030      2040
                           <-- tRNA^Ala                                                                   <-- tRNA^Asn

CTCCCGCCGCCGGGAAAAAAGGCGGGAGAAGCCCCGGCAGGTTTGAAGCTGCTTCTTCGAATTTGCAATTCAATATGAAAATCACCTCGGAGCTGGTAAAAAGAGGCCTAACCCCTGTCT
      2050      2060      2070      2080      2090      2100      2110      2120      2130      2140      2150      2160
Origin of L strand
replication -->                                                                  <-- tRNA^Cys

                                         M  F  A  D  R  W  L  F  S  T  N  H  K  D  I  G  T  L  Y  L  L  F  G
TTAGATTTACAGTCCAATGCTTCACTCAGCCATTTTACCTCACCCCCACTGATGTTCGCCGACCGTTGACTATTCTCTACAAACCACAAAGACATTGGAACACTATACCTATTATTCGGC
      2170      2180      2190      2200      2210      2220      2230      2240      2250      2260      2270      2280
                        <-- tRNA^Tyr                CO I gene -->

A  W  A  G  V  L  G  T  A  L  S  L  L  I  R  A  E  L  G  Q  P  G  N  L  L  G  N  D  H  I  Y  N  V  I  V  T  A  H  A  F
GCATGAGCTGGAGTCCTAGGCACAGCTCTAAGCCTCCTTATTCGAGCCGAGCTGGGCCAGCCAGGCAACCTTCTAGGTAACGACCACATCTACAACGTTATCGTCACAGCCCATGCATTT
      2290      2300      2310      2320      2330      2340      2350      2360      2370      2380      2390      2400

V  M  I  F  F  M  V  M  P  I  M  I  G  G  F  G  N  W  L  V  P  L  M  I  G  A  P  D  M  A  F  P  R  M  N  N  M  S  F  W
GTAATAATCTTCTTCATAGTAATACCCATCATAATCGGAGGCTTTGGCAACTGACTAGTTCCCCTAATAATCGGTGCCCCCGATATGGCGTTTCCCCGCATAAACAACATAAGCTTCTGA
      2410      2420      2430      2440      2450      2460      2470      2480      2490      2500      2510      2520

L  L  P  P  S  L  L  L  L  L  A  S  A  M  V  E  A  G  A  G  T  G  W  T  V  Y  P  P  L  A  G  N  Y  S  H  P  G  A  S  V
CTCTTACCTCCCTCTCTCCTACTCCTGCTCGCATCTGCTATAGTGGAGGCCGGAGCAGGAACAGGTTGAACAGTCTACCCTCCCTTAGCAGGGAACTACTCCCACCCTGGAGCCTCCGTA
      2530      2540      2550      2560      2570      2580      2590      2600      2610      2620      2630      2640

D  L  T  I  F  S  L  H  L  A  G  V  S  S  I  L  G  A  I  N  F  I  T  T  I  I  N  M  K  P  P  A  M  T  Q  Y  Q  T  P  L
GACCTAACCATCTTCTCCTTACACCTAGCAGGTGTCTCCTCTATCTTAGGGGCCATCAATTTCATCACAACAATTATCAATATAAAACCCCCTGCCATAACCCAATACCAAACGCCCCTC
      2650      2660      2670      2680      2690      2700      2710      2720      2730      2740      2750      2760

F  V  W
TTCGTCTGATC
      2770
```

FIG. 6. Nucleotide sequence of fragment M1 from human mtDNA. The positions of the probable tRNA genes are indicated by boxes, the arrows showing the direction of transcription of the RNAs. The probable amino acid sequence of the proteins coded by the DNA is indicated by the single letter code above the corresponding DNA sequence. URF, unidentified reading frame. Overlining indicates the position of a probable hairpin loop that is associated with the origin of *L*-strand replication (Martens & Clayton, 1979). Hyphens omitted for clarity.

sequence spanning the site at position 1582, but it was not used for the other site (position 430) as its sequence was already known.

During the ligation reaction with the linker oligonucleotide restriction enzyme fragments present may be ligated together, leading to the formation of a new bond that was not present in the original DNA being studied. This possibility must be taken into account when interpreting the results. Because of the high concentration of linker it does not occur frequently, the only example in the present work being clone B22 (see Fig. 5), in which three *Alu* fragments were ligated together. (See also Figure 2.) In the case of restriction enzymes that give "flush ends", ligation leads to reformation of the restriction site, which can be recognised. For enzymes giving extended 5′ ends, which are filled in with DNA polymerase, the site is not necessarily restored but a predicted sequence is obtained which can be searched for. For example, *Hinf* (cutting site G-A-N-T-C) will ligate to give G-A-N-T-A-N-T-C. In the case of enzymes giving an extended 3′ end (e.g. *Hha*I or *Hph*I), the problem appears to be more serious. In fact in a single experiment in which these enzymes were used with lambda DNA almost all the clones contained more than one inserted fragment ligated together. The exact explanation of this is not clear but it was probably at least partly due to the difficulty of converting them to flush ends.

It has been observed that small fragments are more readily inserted into the vector M13mp2 than large fragments and that inserted large fragments (over about 2000 nucleotides) are frequently lost from the clones (see, for example, Cordell *et al.*, 1979). Most of the clones used in this method are for small fragments, but it does suggest the possibility that a section of an insert might be deleted and this would lead to a false result. We have not detected this effect, but as the method depends on overlapping sequences and generating a large amount of redundant data this effect would be seen if it occurred.

Figure 6 shows the complete sequence of the DNA fragment. According to the mapping experiments of Angerer *et al.* (1976), the region codes for a number of transfer RNAs. We have therefore looked for tRNA genes, using the program TRNA of Staden (1980) which identified sequences that can be folded in the cloverleaf structure. Eight such structures, arranged in two groups, have been found and identified by the sequence in the anticodon position. As with other human mt-tRNAs (Barrell *et al.*, 1979; Eperon *et al.*, 1980), they are all somewhat anomalous when compared with conventional tRNAs in that some of the so-called "invariable" features are missing. These tRNAs will be discussed in more detail in a later paper in connection with other tRNA structures from human mitochondria. Of particular interest is the sequence identified as $tRNA^{Trp}$. It has the anticodon TCA, which corresponds to the codons UGG and UGA, whereas other $tRNA^{Trp}$s that have been identified have the anticodon CCA, corresponding to UGG only. This is consistent with the identification of TGA as a codon for tryptophan in mitochondria (Barrell *et al.*, 1979; Macino *et al.*, 1979; Fox, 1979) rather than as a termination codon.

Martens & Clayton (1979) have identified the origin of *L*-strand replication (O_L) of mouse DNA and have determined a sequence of 318 nucleotides containing it. Their sequence is homologous to positions 1849 to 2167 of Figure 6. The human sequence also contains a strongly base-paired loop (positions 2038 to 2071) in the region corresponding to the loop where the origin is located in the mouse mtDNA. Thus we conclude that this is the origin in the human mtDNA. Either side of this loop we have identified tRNA genes (Fig. 6). Homologous structures can also be noted in the sequence from the mouse DNA. Indeed all five tRNA genes that we find in this region of human mtDNA seem to be present in the mouse and to have the same anticodons. As judged by the distribution of the termination codons TAG and TAA, the sequence on either side of and between the two tRNA clusters contains open "reading frames" which probably code for proteins. These also will be discussed in detail in a later paper. J. E. Walker (personal communication) has identified the reading frame starting at position 2212 as the gene for cytochrome oxidase subunit I by comparison of the DNA sequence with the amino acid sequence at the N-terminus of the bovine protein.

We thank Dr A. R. MacLeod for a gift of DNA polymerase, Dr M. Smith for a gift of restriction enzyme *Fnu*EI and Dr G. Attardi for a gift of HeLa cell mtDNA and for helpful advice. We also thank Mr R. Staden, Drs G. P. Winter and P. Schreier for useful advice and discussions in connection with this work, and Drs J. Messing and B. Gronenborn for making available to us their strains of bacteriophage and bacteria and their protocols.

One of the authors (B. A. R.) is a recipient of a National Institutes of Health Research Career Development Award (KO-4-GM-00178) and was on leave from Kent State University.

REFERENCES

Angerer, L., Davidson, N., Murphy, W., Lynch, D. & Attardi, G. (1976). *Cell*, **9**, 81–90.

Barnes, W. M. (1978). *Proc. Nat. Acad. Sci., U.S.A.* **75**, 4281–4285.

Barnes, W. M. (1979). *Gene*, **5**, 127–139.

Barrell, B. G., Bankier, A. T. & Drouin, J. (1979). *Nature (London)*, **282**, 189–194.

Boeke, J. D., Vovis, G. F. & Zinder, N. D. (1979). *Proc. Nat. Acad. Sci. U.S.A.* **76**, 2699–2702.

Boyer, H. W. & Roulland-Dussiox, D. (1969). *J. Mol. Biol.* **41**, 459–472.

Cordell, B., Bell, G., Tischer, E., DeNoto, F. M., Ullrich, A., Pictet, R., Rutter, W. J. & Goodman, H. M. (1979). *Cell*, **18**, 533–543.

Drouin, J. (1980). *J. Mol. Biol.* **140**, 15–34.

Eperon, I. C., Anderson, A. & Nierlich, D. P. (1980). *Nature (London)*, **286**, 460–467.

Eshaghpour, H. & Crothers, D. M. (1978). *Nucl. Acids Res.* **5**, 13–21.

Fox, T. D. (1979). *Proc. Nat. Acad. Sci., U.S.A.* **76**, 6534–6538.

Gronenborn, B. & Messing, J. (1978). *Nature (London)*, **272**, 375–377.

Heidecker, G., Messing, J. & Gronenborn, B. (1980). *Gene*, **10**, 69–73.

Herrmann, R., Neugebauer, K., Schaller, H. & Zentgraf, H. (1978). In *The Single-stranded DNA Phages* (Denhardt, D. T., Dressler, D. & Ray, D. S., eds), pp. 473–476, Cold Spring Harbor Laboratory, New York.

Heyneker, H. L., Shine, J., Goodman, H. M., Boyer, H. W., Rosenberg, J., Dickerson, R. E., Narang, S. A., Itakura, K., Lin, S. & Riggs, A. D. (1976). *Nature (London)*, **263**, 748–752.

Katz, L., Kingsburg, D. K. & Helinski, D. R. (1973). *J. Bacteriol.* **114**, 577.

Kelley, W. S., Chalmers, K. & Murray, N. E. (1977). *Proc. Nat. Acad. Sci., U.S.A.* **74**, 5632–5636.

Macino, G., Coruzzi, G., Nobrega, F. G., Li, M. & Tzagoloff, A. (1979). *Proc. Nat. Acad. Sci., U.S.A.* **76**, 3784–3785.

Maniatis, T., Hardison, R. C., Lacey, E., Lauer, J., O'Connell, C., Quon, D., Sim, G. K. & Efstatiadis, A. (1978). *Cell*, **15**, 687–701.

Martens, P. A. & Clayton, D. A. (1979). *J. Mol. Biol.* **135**, 327–351.

Maxam, A. M. & Gilbert, W. (1977). *Proc. Nat. Acad. Sci., U.S.A.* **74**, 560–564.

Messing, J. (1979). *Recombinant DNA Technical Bulletin*, **2**, 43–48.

Messing, J., Gronenborn, B., Müller-Hill, B. & Hofschneider, P. H. (1977). *Proc. Nat. Acad. Sci., U.S.A.* **74**, 3642–3646.
Ojala, D. & Attardi, G. (1977). *Plasmid*, **1**, 78.
Patient, R. K., Hardies, S. C., Larson, J. E., Inman, R. B., Maquat, L. Y. & Wells, R. D. (1979). *J. Biol. Chem.* **254**, 5548–5554.
Radloff, R., Bauer, W. & Vinogard, J. (1967). *Proc. Nat. Acad. Sci., U.S.A.* **57**, 1514–1520.
Ray, D. S. & Kook, K. (1978). *Gene*, **4**, 109–119.
Sanger, F. & Coulson, A. R. (1975). *J. Mol. Biol.* **94**, 441–448.
Sanger, F. & Coulson, A. R. (1978). *FEBS Letters*, **87**, 107–110.
Sanger, F., Nicklen, S. & Coulson, A. R. (1977). *Proc. Nat. Acad. Sci., U.S.A.* **74**, 5463–5467.
Scheller, R. H., Thomas, T. L., Lee, A. S., Klein, W. H., Niles, W. D., Britten, R. J. & Davidson, E. H. (1977). *Science*, **196**, 197–200.
Schreier, P. H. & Cortese, R. (1979). *J. Mol. Biol.* **129**, 169–172.
Smith, A. J. H. (1979). *Nucl. Acids Res.* **6**, 831–848.
Staden, R. (1977). *Nucl. Acids Res.* **4**, 4037–4051.
Staden, R. (1980). *Nucl. Acids Res.* **8**, 817–824.

Reprinted from Nature, Vol. 282, No. 5735, pp. 189-194, November 8 1979

A different genetic code in human mitochondria

B. G. Barrell, A. T. Bankier & J. Drouin*
MRC Laboratory of Molecular Biology, Hills Road, Cambridge, UK

Comparison of the human mitochrondrial DNA sequence of the cytochrome oxidase subunit II gene and the sequence of the corresponding beef heart protein shows that UGA is used as a tryptophan codon and not as a termination codon and suggests that AUA may be a methionine and not an isoleucine codon. The cytochrome oxidase II gene is contiguous at its 5′ end with a tRNA^{Asp} gene and there are only 25 bases at its 3′ end before a tRNA^{Lys} gene. These tRNAs are different from all other known tRNA sequences.

THE genetic code has been considered to be universal since the early work on *Escherichia coli* and mutant eukaryotic proteins. Once the code was established it was argued that any change would be lethal. This view was substantiated by the development of rapid RNA and DNA sequencing methods which have explored a wide variety of gene sequences that could be related to their protein sequences. We present evidence here, however, that mitochondria have an altered genetic code. This conclusion is based on the nucleotide sequence of a region of human mitochondrial DNA which was shown to code for an amino acid sequence homologous to the known protein sequence of the cytochrome oxidase subunit II (COII) of bovine mitochondria[1]. The homology was such that it was concluded that this region of the DNA was coding for the human COII gene. However, there were three TGA codons in the reading frame that was coding for the protein. This demonstrated that UGA was not acting as a termination codon. Moreover, the amino acid present in the bovine protein in all positions where TGA occurred in the human DNA sequence was tryptophan. Such a correspondence could only be satisfactorily explained if UGA was coding for tryptophan rather than for termination. Similar evidence is also presented that suggests that AUA is a methionine codon and not isoleucine.

Human mitochondria have a separate genetic system with a DNA genome of molecular weight 10×10^6. Figure 1 shows the organisation of the human mitochondria genome[2,3] aligned with the *Mbo*I restriction enzyme map. HeLa cell mitochondrial DNA (mtDNA) has been shown by hybridisation to code for a 12S and a 16S ribosomal RNA and at least 19 tRNAs[2-4]. HeLa cell mitochondrial translation products are less well defined and only cytochrome oxidase subunits I, II and III have been identified so far[5]. By analogy with other mitochondria systems (for review see ref. 6) genes for four proteins of the ATPase complex and one of the cytochrome bc_1 complex should also be present. However, it has become apparent recently that more proteins (at least 19) are coded for by HeLa cell mtDNA[7]. Both strands of HeLa cell mtDNA are completely transcribed symmetrically[8,9]. The rRNAs, the majority of the rRNAs and all but one of the polyadenylated RNAs hybridise to the heavy strand (H strand)[2,4,8,10]. Thus, the light strand (L strand) is defined as the main coding strand, having the same sequence as the rRNAs, tRNAs and mRNAs.

In the sequence of the COII gene region presented here we have identified two putative tRNA genes flanking the COII gene. A tRNA^{Asp} gene is 'butt-jointed' to the initiation codon of the COII gene, that is, there is no intercistronic space between them, and there are only 25 nucleotides between the termination codon of the COII gene and a tRNA^{Lys} gene. These tRNA genes are unusual in that they differ considerably in the 'D loop' and 'TψC loop' from all other known tRNA sequences.

DNA sequence analysis

*Mbo*I restriction enzyme fragments of human mtDNA obtained from a single term placenta have been cloned in the *Bam*HI site of pBR322 (ref. 42). Sequence analysis of clones pmt 110, 140 and 101, containing *Mbo*I fragments 6a, 15 and 6b, respectively (see Fig. 1), was by the chain termination method of Sanger *et al.*[11]. Single-stranded template DNA was prepared by the exonuclease III method of Smith[12]. By cleaving the duplex circular DNA at either the unique pBR322 *Hin*dIII site or the unique pBR322 *Sal*I site and digesting with exonuclease III, either strand of the region containing the inserted DNA was obtained pure for the primed synthesis DNA sequencing reaction. Small restriction fragment primers were prepared both from the pBR322 sequences flanking the insert, so as to prime into the insert from either side, and from within the insert.

The sequence determined is shown in Fig. 2. Sequence analysis of the *Alu*I restriction fragment spanning *Mbo*I fragment 15 which was cloned separately shows that there are no small *Mbo*I fragments between *Mbo*I fragments 6a, 15 and 6b (A. R. Coulson, personal communication). Two examples of chain termination sequencing gels are shown in Fig. 3. Figure 3*a* shows the sequence from −120 through the tRNA^{Asp} gene and

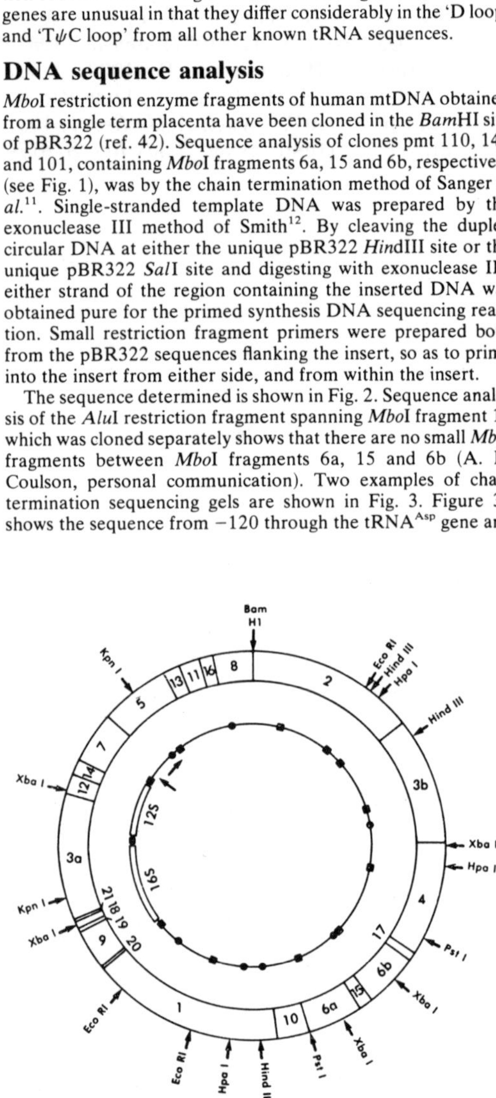

Fig. 1 The genetic map of the human mtDNA genome[2,3] is aligned with the *Mbo*I restriction enzyme map[42]. The precise order of fragments 21, 18 and 19 is not known. The sites of cleavage of some other enzymes are also shown[43,44]. The origin and direction of H-strand synthesis is shown[43,44]. The L-strand sequence 5′ to 3′ is anticlockwise and codes for the 12S and 16S rRNAs and tRNAs marked ■ (complementary to H strand). The H strand codes for the tRNAs marked ● (complementary to L strand). The COII gene is centred approximately on *Mbo*I fragment 15.

* Present address: Department of Biochemistry and Biophysics, University of California, San Francisco, California 94143.

2

```
                                                                                            tRNA^Asp
         C C C C C C A A A G C T G G T T T C A A G C C A A C C C C A T G G C C T C C A T G A C T T T T T C A A A A A G C T A T T
       -120

         A G A A A A A C C A T T T C A T A A C T T T G T C A A A G T T A A A T T A T A G G C T A A A T C C T A T A T A T C T T A
       -60
          1
Bovine   Met  Ala  Tyr  Pro  Met  Gln  Leu  Gly  Phe  Gln  Asp  Ala  Thr  Ser  Pro  Ile  Met   Glu  Glu  Leu
Human    Met  Ala  His  Ala  Ala  Gln  Val  Gly  Leu  Gln  Asp  Ala  Thr  Ser  Pro  Ile  Met?  Glu  Glu  Leu
                                                                                     (Ile)
Human    A T G G C A C A T G C A G C G C A A G T A G G T C T A C A A G A C G C T A C T T C C C C T A T C A T A G A A G A G C T T
         1
          21
         Leu  His  Phe  His  Asp  His  Thr  Leu  Met   Ile  Val  Phe  Leu  Ile  Ser  Ser  Leu  Val  Leu  Tyr
         Ile  Thr  Phe  His  Asp  His  Ala  Leu  Met?  Ile  Ile  Phe  Leu  Ile  Cys  Phe  Leu  Val  Leu  Tyr
                                                 (Ile)
         A T C A C C T T T C A T G A T C A C G C C C T C A T A A T C A T T T T C C T T A T C T G C T T C C T A G T C C T G T A T
         61                     MboI 6a/15
          41
         Ile  Ile  Ser  Leu  Met  Leu  Thr  Thr  Lys  Leu  Thr  His  Thr  Ser  Thr  Met  Asp  Ala  Gln  Glu
         Ala  Leu  Phe  Leu  Thr  Leu  Thr  Thr  Lys  Leu  Thr  Asn  Thr  Asn  Ile  Ser  Asp  Ala  Gln  Glu
         G C C C T T T T C C T A A C A C T C A C A A C A A A A C T A A C T A A T A C T A A C A T C T C A C A C G C T C A G C A A
         121
          61
         Val    Glu  Thr  Ile  Trp   Thr  Ile  Leu  Pro  Ala  Ile  Ile  Leu  Ile  Leu  Ile  Ala  Leu  Pro  Ser
         Met?   Glu  Thr  Val  Trp?  Thr  Ile  Leu  Pro  Ala  Ile  Ile  Leu  Val  Leu  Ile  Ala  Leu  Pro  Ser
         (Ile)                (***)
         A T A G A A A C C G T C T G A A C T A T C C T G C C C G C C A T C A T C C T A G T C C T C A T C G C C C T C C C A T C C
         181
          81
         Leu  Arg  Ile  Leu  Tyr  Met   Met  Asp  Glu  Ile  Asn  Asn  Pro  Ser  Leu  Thr  Val  Lys  Thr  Met
         Leu  Arg  Ile  Leu  Tyr  Met?  Thr  Asp  Glu  Val  Asn  Asp  Pro  Ser  Leu  Thr  Ile  Lys  Ser  Ile
                                  (Ile)
         C T A C G C A T C C T T T A C A T A A C A G A C G A G G T C A A C G A T C C C T C C C T T A C C A T C A A A T C A A T T
         241                                                          MboI 15/6b
          101
         Gly  His  Gln  Trp  Tyr  Trp   Ser  Tyr  Glu  Tyr  Thr  Asp  Tyr  Glu  Asp  Leu  Ser  Phe  Asp  Ser
         Gly  His  Gln  Trp  Tyr  Trp?  Thr  Tyr  Glu  Tyr  Thr  Asp  Tyr  Gly  Gly  Leu  Ile  Phe  Asn  Ser
                                  (***)
         G G C C A C C A A T G G T A C T G A A C C T A C G A G T A C A C C G A C T A C G G C G G A C T A A T C T T C A A C T C C
         301
          121
         Tyr  Met   Ile  Pro  Thr  Ser  Glu  Leu  Lys  Pro  Gly  Glu  Leu  Arg  Leu  Leu  Glu  Val  Asp  Asn
         Tyr  Met?  Leu  Pro  Pro  Leu  Phe  Leu  Glu  Pro  Gly  Asp  Leu  Arg  Leu  Leu  Asp  Val  Asp  Asn
              (Ile)
         T A C A T A C T T C C C C C A T T A T T C C T A G A A C C A G G C G A C C T G C G A C T C C T T G A C G T T G A C A A T
         361
          141
         Arg  Val  Val  Leu  Pro  Met  Glu  Met  Thr  Ile  Arg  Met   Leu   Val  Ser  Ser  Glu  Asp  Val  Leu
         Arg  Val  Val  Leu  Pro  Ile  Glu  Ala  Pro  Ile  Arg  Met?  Met?  Ile  Thr  Ser  Gln  Asp  Val  Leu
                                                              (Ile) (Ile)
         C G A G T A G T A C T C C C G A T T G A A G C C C C C A T T C G T A T A A T A A T T A C A T C A C A A G A C G T C T T G
         421
          161
         His  Ser  Trp   Ala  Val  Pro  Ser  Leu  Gly  Leu  Lys  Thr  Asp  Ala  Ile  Pro  Gly  Arg  Leu  Asn
         His  Ser  Trp?  Ala  Val  Pro  Thr  Leu  Gly  Leu  Lys  Thr  Asp  Ala  Ile  Pro  Gly  Arg  Leu  Asn
                   (***)
         C A C T C A T G A G C T G T C C C C A C A T T A G G C T T A A A A A C A G A T G C A A T T C C C G G A C G T C T A A A C
         481                                                                            HpaII 7/22
          181
         Gln  Thr  Thr  Leu  Met  Ser  Ser  Arg  Pro  Gly  Leu  Tyr  Tyr  Gly  Gln  Cys  Ser  Glu  Ile  Cys
         Gln  Thr  Thr  Phe  Thr  Ala  Thr  Arg  Pro  Gly  Val  Tyr  Tyr  Gly  Gln  Cys  Ser  Glu  Ile  Cys
         C A A A C C A C T T T C A C C G C T A C A C G A C C G G G G T A T A C T A C G G T C A A T G C T C T G A A A T C T G T
         541                                          HpaII 22/6
          201
         Gly  Ser  Asn  His  Ser  Phe  Met  Pro  Ile  Val  Leu  Glu  Leu  Val  Pro  Leu  Lys  Tyr  Phe  Glu
         Gly  Ala  Asn  His  Ser  Phe  Met  Pro  Ile  Val  Leu  Glu  Leu  Ile  Pro  Leu  Lys  Ile  Phe  Glu
         G G A G C A A A C C A C A G T T T C A T G C C C A T C G T C C T A G A A T T A A T T C C C C T A A A A A T C T T T G A A
         601
          221
         Lys    Trp  Ser  Ala  Ser  Met  Leu
         Met?   Gly  Pro  Val  Phe  Thr  Leu  ***                                                tRNA^Lys
         (Ile)                                                ------------->  ------------->
         A T A G G G C C C G T A T T T A C C C T A T A G C A C C C C C T C T A C C C C C T C T A G A G C C C A C T G T A A A G C
         661                                                                      XbaI 5/3

         T A A C T T A G C A T T A A C C T T T T A A G T T A A A G A T T A A G A G A A C C A A C A C C T C T T T A C A G T C A A
         721
```

Fig. 2 The sequence of the COII gene region showing the position of the two tRNA genes and the COII DNA sequence aligned with the bovine COII protein sequence[1]. The predicted human protein sequence is compared with the bovine sequence and homologous amino acids are shown boxed using the codons AUA as methionine and UGA as tryptophan. Their normal assignments isoleucine and termination (denoted by ***) are shown underneath. The sequence can be aligned with the genetic and restriction maps shown in Fig. 1 by the restriction enzyme sites *Mbo*I and *Xba*I shown. A 9-base repeat at the end of the COII gene is also shown.

the beginning of the COII gene to the *MboI* site at position +73, and Fig. 3*b* shows the sequence obtained using *HpaII* fragment 22 as a primer from position +521 to the *MboI* site at position +274. The positions of the TGA and ATA codons in these regions are shown.

tRNA genes

Three tRNA genes have been mapped in this region of the genome[2], two hybridising to the H strand and one to the L strand (see Fig. 1). In the sequence presented in Fig. 2 we have only been able to locate two tRNA genes, both of which are coded on the L strand, that is, the tRNA products would hybridise to the H strand. The locations of these tRNAs agree well with the hybridisation map positions[2] shown in Fig. 1, and from their positions in the sequence they probably correspond to H5 and H6 according to the nomenclature of Angerer *et al.*[2]. However, if this assignment is correct there should be a tRNA gene close to the $tRNA^{Lys}$ gene on the opposite strand that we have been unable to find. The tRNA genes were located by searching for base-paired cloverleaf structures using the computer programs HAIRPN[13] and TRNA (R. Staden, unpublished).

The sequences of the two putative tRNA genes are shown in the cloverleaf form in Fig. 4. If these tRNAs have CCA at their 3′ ends then this is not encoded in the DNA. The sequences in the D loop (left-hand loop) and the TψC loop (right-hand loop) of both tRNA genes differ considerably from the corresponding sequences found in all other tRNAs (see collection of 209 tRNA sequences compiled by Sprinzl *et al.*[14]). In the tertiary structure of yeast $tRNA^{Phe}$ (refs 15, 16) these two loops are hydrogen bonded together by conserved nucleotides not found in the two mitochondrial tRNA sequences. An examination of the primary structures of the two genes in these regions has not yet revealed how their respective D and TψC loops interact in the manner of yeast $tRNA^{Phe}$. A more detailed discussion of these and other human mitochondrial tRNA genes will be published separately.

UGA and UGG are both tryptophan codons in human mtDNA

The most interesting aspect of the sequence is in the part coding for the COII gene which is identified by the high homology with the bovine protein sequence[1]. When the human DNA sequence is aligned with the beef protein sequence as shown in Fig. 2, it becomes apparent that there are three 'in phase' UGA (*opal*) termination codons within the gene. The possibility that these are present in intervening sequences can be excluded by several criteria. First, the gene seems to be the exact length needed to code for its product, without interruption, from the AUG initiation codon to the UAG (*amber*) termination codon. Second, the UGA codons are in regions of very high homology between the beef and predicted human protein sequences. Finally, and most important, is the fact that they all align with tryptophans in the bovine protein. This suggests very strongly that UGA is a tryptophan codon in this genetic system and not a termination codon. There are five tryptophan residues in the bovine COII sequence, three of them opposite UGA codons, one opposite a UGG tryptophan codon and one opposite a GGG glycine codon. The latter is in a region where there is no homology with the bovine protein sequence at position 664 in the DNA sequence shown in Fig. 2. Tryptophan is normally coded for by a single codon, UGG, which is translated by a $tRNA^{Trp}$ with a CCA anticodon in both prokaryotes[17] and eukaryotes[18,19]. However, there is a precedent for UGA being used as a tryptophan codon. $tRNA^{Trp}$ from an su^+ UGA strain of *E. coli* (CAJ64) suppresses a UGA mutation in lysozyme mRNA during cell-free protein synthesis and synthesises polytryptophan from a repeating UGA polymer[20,21]. However, in both assays there was a low synthesis of product using $tRNA^{Trp}$ from the su^- *E. coli*CA244. This, coupled with the observation that *in vivo* UGA codons are leaky in su^- strains[22,23], suggested that the low level of UGA suppression in su^- strains was caused by the normal $tRNA^{Trp}$. Also the low level of readthrough of the Qβ coat protein UGA termination codon in su^- hosts was shown to occur by insertion of tryptophan and a functional longer protein necessary for infectivity was found in the phage particle[24,25]. Thus, there is a precedent in prokaryotes that UGA can be used as a tryptophan codon, albeit infrequently.

The codons UGA and UGG are adjacent in the genetic code (see Fig. 5) and could thus be recognised specifically by a $tRNA^{Trp}$ having the anticodon UCA according to the 'wobble hypothesis' (ref. 26) where U in the first position of the anticodon could pair with A or G in the third position of the codon. However, $tRNA^{Trp}$ from both the su^- and su^+ UGA *E. coli* strains have CCA as their anticodons and the suppressor $tRNA^{Trp}$ has a G24 → A24 in the semi-invariant base pair in the 'D arm' (ref. 17). This must somehow alter the tertiary structure of the suppressor tRNA to allow the C in the first position of the anticodon to pair with the A in the third position of the UGA codon. Although neither the human mitochondrial $tRNA^{Trp}$ nor its gene has yet been characterised there is evidence from charging experiments[4] of unfractionated mitochondrial tRNA for a mitochondrially coded $tRNA^{Trp}$.

It thus seems clear that if this gene is expressed in its entirety, UGA and UGG are both used as tryptophan codons in the genetic code of human mitochondrial DNA. The sequence of the yeast mitochondrial COII gene also has UGA codons opposite tryptophan residues when it is aligned with the bovine COII protein sequence[27,41]. The use of UGA as a tryptophan codon would preclude the use of the cytoplasmic termination factor which corresponds to all three termination codons[28]. Prokaryotes use two factors, R1, which corresponds to UAA and UAG, and R2, which corresponds to UAA and UGA[29]. Thus, it is likely that the mitochondria has a factor analogous to R1.

Does AUA code for methionine in human mtDNA?

There are two codons in the genetic code that are each the sole codon for a single amino acid. The case of UGG for tryptophan has been discussed above. The other codon is AUG, which codes for methionine, and in the same box isoleucine is encoded by AUC, AUU and AUA. All other boxes can be read using G–U wobble[25] by one tRNA that is specific for codons ending in C and U and another that is specific for codons ending in A and G. The main exception to this is when the anticodon contains modified bases that can restrict the coding properties. Thus, with the UGA/tryptophan precedent in mind, can AUA code for methionine? There are 15 internal methionine residues in the beef COII sequence and only one of these is aligned with an AUG codon but five of them are aligned with AUA codons (Fig. 2). There are eight AUA codons, five aligned with methionine, one with valine, one with leucine and one with lysine. None is aligned with isoleucine codons. Thus, there are two possibilities: either by coincidence the bovine methionine residues in these positions are being preferentially replaced by isoleucines which are all coded for by AUA, or there is another change in the code and AUA is being recognised as a methionine codon. AUA and AUG could both be recognised by a tRNA having the anticodon UAU according to the wobble hypothesis[26].

In view of the UGA/UGG precedent we think that the above data suggest that AUA and AUG are methionine codons. However, isoleucine and methionine are both hydrophobic amino acids and confirmation will depend on the determination of the human COII protein sequence.

If it is assumed that AUA is a methionine codon and UGA a tryptophan codon, the bovine protein and the predicted human protein sequences are 72% identical. The conserved amino acids are shown boxed in Fig. 2. This leaves 62 out of 227 amino acids in bovine COII that are different in the human COII. It is striking that 22 of these differences involve either methionine or serine aligning with other amino acids, eight of them threonine, using the codons ACA and ACC. However, both of these codons also align with threonine, and with the data shown in

4

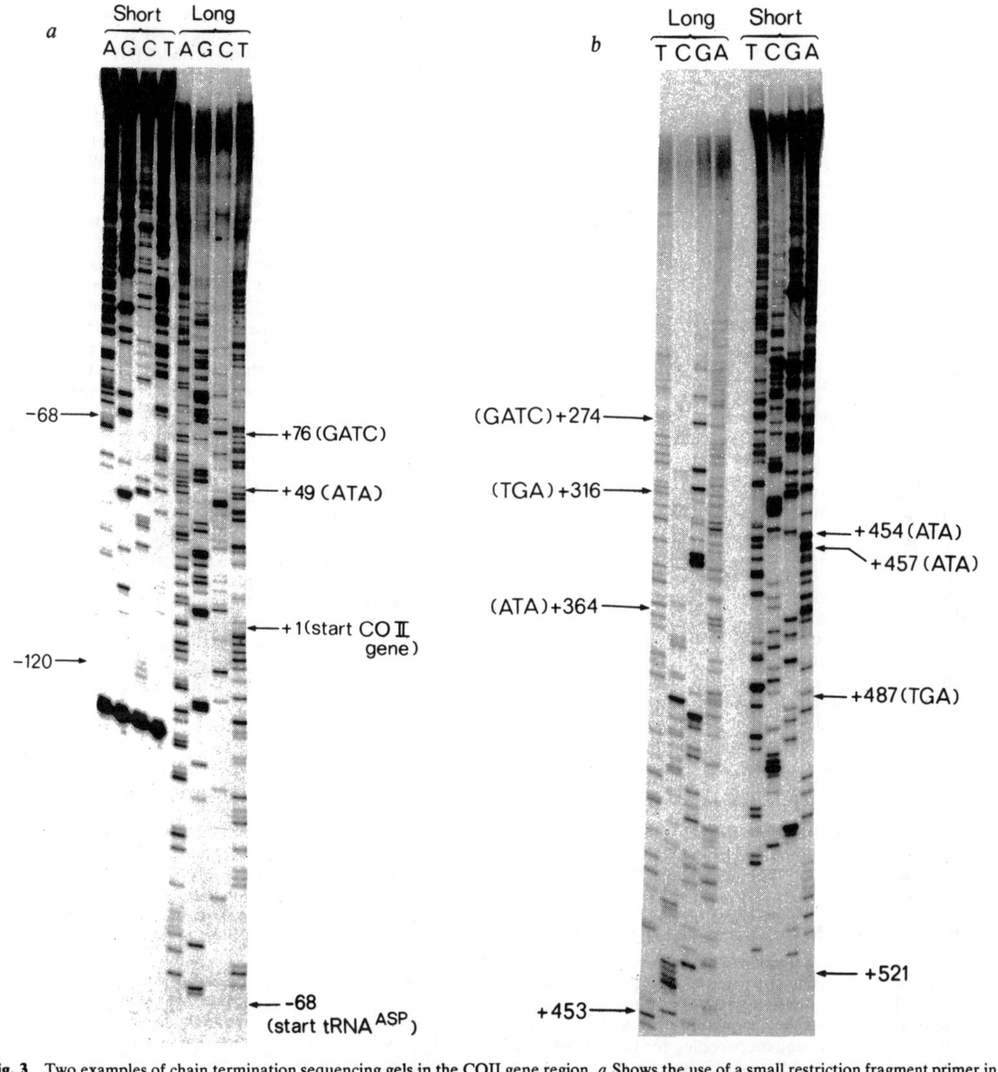

Fig. 3 Two examples of chain termination sequencing gels in the COII gene region. *a* Shows the use of a small restriction fragment primer in *Mbo*I fragment 6a primed on 'H-strand' template prepared from clone pmt 110 (see text) by the exonuclease III method of Smith[12]. The sequence corresponding to that given in Fig. 2 can be read from base −120 to the *Mbo*I site at +76 which is the junction between the clone of *Mbo*I 6a and pBR322. The starts of the $tRNA^{Asp}$ gene and the COII gene are shown along with the ATA codon at position +49. A short run and a long run of the sequencing reactions on the same 8% polyacrylamide thin gel[46] are shown. *b* Shows a similar experiment using the fragment *Hpa*II 22 as a primer on 'L-strand' template prepared from clone pmt 110 (*Mbo*I fragment 6b, see text). For ease of reading the track order A, G, C, T has been changed to the complementary order T, C, G, A so that the gel can be read as the L-strand sequence 5′ to 3′, from top to bottom. The sequence from the junction between pBR322 and *Mbo*I 6b at position +274 can be read to position +521. The positions of the UGA and AUA codons are shown.

Fig. 2 there is no clear indication of other coding changes. The one difference between the yeast mitochondrial ATPase subunit 9 protein sequence[30] and the sequence of its gene[31] is a threonine aligned with a CUA codon. In the human COII gene, 11 out of 12 CUA codons align with leucine so that this cannot be a coding change in human mtDNA. However, it is possible that other coding changes will be found in mtDNA and there may be different codes in the mitochondria of different species.

Codon usage in the COII gene

Figure 5 shows the suggested code in human mitochondria with AUA as possibly methionine and UGA as tryptophan. The number of times each codon appears in the COII gene is also shown. All amino acids are used and nearly all codons. Only one of the serine AGU/C set is used and none of the arginine AGA/G set. Excluding the UAG termination codon, only 16 out of 227 codons end in G. This reflects the low G content found in the L strand generally (unpublished results). Specifically in this region the composition is A, 29.9%; C, 30.2%; T, 25.8% and G, 14.1%. The low G content may be of significance when considering the evolution and use of the coding changes proposed here. Including AUA as methionine and UGA as tryptophan the ratio of codons ending in A to those ending in G is 5 : 1.

Fig. 4 The DNA sequences of the tRNA^{Asp} and rRNA^{Lys} genes are shown arranged in the cloverleaf conformation. The CCA sequence found at the 3′ end of all tRNAs is not encoded in the DNA sequence. The presumed codons that these tRNAs would recognise by G–U wobble[26] are indicated.

It is also interesting to consider the codon usage in relation to the number of tRNAs found to be encoded by human mtDNA. The highest number of tRNAs so far found to be coded by mammalian mtDNA is 21 (ref. 32) out of the minimum number required using G–U wobble of 32. If the AUA coding change is substantiated then a minimum of 31 would be required. On the basis of the relatively small sample of codons of the COII gene, it is probably reasonable to assume that in the whole genome all codons are used, although some may be used rarely. There is no evidence in mammalian mitochondria to suggest that any RNA is imported, and so there may be a complete set of mitochondrially coded tRNAs. Otherwise, it is difficult to see how 21 tRNAs could recognise all codons without some degree of misreading. However, it is possible that the smaller rRNAs, the novel features of the tRNAs, the possible coding changes discussed here and a limited set of tRNAs might indicate that a different protein synthesis decoding mechanism operates in the mitochondria. Selective use of a 'two out of three mechanism' (refs 33, 34) would only require 23 or 24 tRNAs, depending on the coding changes.

Organisation of genes in the COII region

Attardi *et al.*[7] discussed two models for the organisation of the tRNA genes with respect to the protein coding sequences. As can be seen in Fig. 1 the tRNA genes are scattered more or less uniformly around the genome and on both strands. These genes could either separate the sequences coding for the mRNA (the punctuation model) or could reside in them in the manner of intervening sequences (the insert model). The COII gene is flanked by two tRNA genes as shown in Fig. 2 but surprisingly, the tRNA^{Asp} gene is actually contiguous with the COII gene, with no noncoding nucleotides in between. Thus, it is possible that if there are sequences involved in ribosome recognition preceding the initiation codon these lie in the tRNA^{Asp} gene. No sequence corresponding to the *E. coli* ribosome recognition site is found[35,36]. There is a sequence similar to that postulated to have the same function in eukaryotes[37] but there is no evidence that this interaction occurs in eukaryotes and in any case mitochondria have their own rRNAs, which are much smaller than both their prokaryote and eukaryote counterparts. The possibility of a leader peptide which is subsequently processed is ruled out by in-phase UAA termination codons preceding the initiation codon.

The contiguous nature of the two genes raises the obvious difficulty in processing the primary transcript. Both strands of the mtDNA have been shown to be completely transcribed and it is not known whether there are additional promoters. If a full length transcript of the H strand is processed to give the rRNAs, and the tRNAs and mRNAs, the latter being polyadenylated at some stage, then a single cleavage between the two genes would leave no 5′-leader sequence on the COII gene mRNA. Therefore, if initiation of translation of the COII mRNA required signals in the sequence preceding the initiation codon, the tRNA would either have to be released after the cessation of translation of the mRNA or have to be transcribed separately. An intriguing possibility is that the tRNA could be spliced out of the transcript in the manner of an intervening sequence, although this seems unlikely; no evidence has yet been found for intervening sequences in human mtDNA.

There is a short noncoding region (25 bases) between the end of the COII gene and the tRNA^{Lys} gene and this contains a 9-base repeat, as shown in Fig. 2. The sequence AAUAAA is presumed to be involved in polyadenylation of eukaryotic mRNAs and the nearest is 224 bases downstream of the 3′ end of the COII gene (unpublished results). If this is the signal for polyadenylation of the COII mRNA, a similar problem to that of the tRNA^{Asp} regarding the processing of the tRNA^{Lys} arises.

There has been considerable speculation that mitochondria have evolved from bacteria that had entered into a symbiotic relationship with their host (for review see ref. 6). The economical use of the coding capacity of the DNA is more characteristic of prokaryotes, in particular bacteriophage DNAs such as ΦX (ref. 38), but the eukaryotic viruses such as SV40 (refs 39, 40) show similar economy. Thus, it is difficult to draw any conclusions as to whether this genetic system is more similar to prokaryotes or eukaryotes on this basis. If the endosymbiotic theory is correct, the different genetic code may represent a 'frozen', more primitive code, for it is unlikely that once mitochondria became essential to their 'host' the code would be able to change. Whatever their origins the novel features of the mitochondria discussed here seem to require that they be considered a separate class from the prokaryotic and eukaryotic genetic systems.

Since this paper was submitted, Macino *et al.*[41] have also demonstrated the use of UGA as tryptophan in the yeast mtDNA COII gene. Also, recent work by F. Sanger and A. R. Coulson (personal communication) has identified the sequence coding for the tryptophan tRNA gene in human mtDNA. As predicted, this structure has the anticodon UCA which would pair with UGA and UGG using G–U wobble.

We thank Dr G. Buse for communicating to us the bovine COII protein sequence before publication, and Drs F. Sanger and J. E. Walker for advice and helpful discussions.

First letter	Second letter U			Second letter C			Second letter A			Second letter G			Third letter
U	UUU	Phe	3	UCU	Ser	1	UAU	Tyr	1	UGU	Cys	1	U
U	UUC	Phe	7	UCC	Ser	4	UAC	Tyr	8	UGC	Cys	2	C
U	UUA	Leu	4	UCA	Ser	4	UAA	Ter	0	†UGA	Trp	3	A
U	UUG	Leu	1	UCG	Ser	0	UAG	Ter	1*	UGG	Trp	1	G
C	CUU	Leu	7	CCU	Pro	1	CAU	His	2	CGU	Arg	2	U
C	CUC	Leu	6	CCC	Pro	9	CAC	His	4	CGC	Arg	1	C
C	CUA	Leu	12	CCA	Pro	3	CAA	Gln	6	CGA	Arg	3	A
C	CUG	Leu	3	CCG	Pro	2	CAG	Gln	1	CGG	Arg	0	G
A	AUU	Ile	7	ACU	Thr	5	AAU	Asn	2	AGU	Ser	1	U
A	AUC	Ile	15	ACC	Thr	8	AAC	Asn	5	AGC	Ser	0	C
A	†AUA	Met	8	ACA	Thr	8	AAA	Lys	4	AGA	Arg	0	A
A	AUG	Met	2	ACG	Thr	0	AAG	Lys	0	AGG	Arg	0	G
G	GUU	Val	1	GCU	Ala	4	GAU	Asp	3	GGU	Gly	2	U
G	GUC	Val	7	GCC	Ala	5	GAC	Asp	8	GGC	Gly	4	C
G	GUA	Val	5	GCA	Ala	4	GAA	Glu	8	GGA	Gly	3	A
G	GUG	Val	0	GCG	Ala	1	GAG	Glu	3	GGG	Gly	2	G

Fig. 5 The code in human mtDNA suggested by the results presented here is shown along with the number of times each codon is used in the COII gene†AUA and UGA are normally isoleucine and termination, respectively.

Received 24 July; accepted 28 September 1979.

1. Steffens, G. J. & Buse, G. *Hoppe-Seyler's Z. physiol. Chem.* **360,** 613 (1979).
2. Angerer, L., Davidson, N., Murphy, W., Lynch, D. C. & Attardi, G. *Cell* **9,** 81 (1976).
3. Ojala, D. & Attardi, G. *Plasmid* **1,** 78 (1977).
4. Lynch, D. C. & Attardi, G. *J. molec. Biol.* **102,** 125 (1976).
5. Hare, J., Ching, E. & Attardi, G. (in preparation).
6. Borst, P. *Trends biochem. Sci.* **2,** 31 (1977).
7. Attardi, G. *et al. Proc. ICN-UCLA Meet. Extrachromosomal DNA*, March 1979 (in the press).
8. Aloni, Y. & Attardi, G. *Proc. natn. Acad. Sci. U.S.A.* **68,** 1957 (1971).
9. Murphy, W., Attardi, B., Tu, C. & Attardi, G. *J. molec. Biol.* **99,** 809 (1975).
10. Almaric, F., Merkel, C., Gelfand, R. & Attardi, G. *J. molec. Biol.* **118,** 1 (1978).
11. Sanger, F., Nicklen, S. & Coulson, A. R. *Proc. natn. Acad. Sci. U.S.A.* **74,** 5463 (1977).
12. Smith, A. J. H. *Nucleic Acids Res.* **6,** 831 (1979).
13. Staden, R. *Nucleic Acids Res.* **5,** 1013 (1978).
14. Sprinzl, M., Grüter, F. & Gauss, D. H. *Nucleic Acids Res.* Spec. Suppl., r15 (1979).
15. Ladner, J. E. *et al. Proc. natn. Acad. Sci. U.S.A.* **72,** 4414 (1975).
16. Kim, S. H. *et al. Science* **185,** 435 (1974).
17. Hirsh, D. *J. molec. Biol.* **58,** 439 (1971).
18. Keith, G., Roy, A., Ebel, J. P. & Dirheimer, G. *Biochimie* **54,** 1404 (1972).
19. Harada, F., Sawyer, R. C. & Dahlberg, J. E. *J. biol. Chem.* **250,** 3487 (1975).
20. Chan, T. S. & Green, A. *J. molec. Biol.* **49,** 231 (1970).
21. Hirsh, D. & Gold, L. *J. molec. Biol.* **58,** 459 (1971).
22. Katz, E. thesis, Cambridge Univ. (1969).
23. Roth, J. R. *J. Bact.* **102,** 467 (1970).
24. Wiener, A. M. & Weber, K. *Nature new Biol.* **234,** 206 (1971).
25. Wiener, A. M. & Weber, K. *J. molec. Biol.* **80,** 837 (1973).
26. Crick, F. H. C. *J. molec. Biol.* **19,** 548 (1966).
27. Fox, T. D. *Proc. natn. Acad. Sci. U.S.A.* (in the press).
28. Beaudet, A. L. & Caskey, C. T. *Proc. natn. Acad. Sci. U.S.A.* **68,** 619 (1971).
29. Scolnick, E., Tompkins, R., Caskey, T. & Nirenberg, M. Proc. natn. Acad. Sci. U.S.A. **61,** 768 (1968).
30. Sebald, W. & Wachter, E. in *29th Mosbacher Colloqu. Energy Conservation in Biological Membranes* (eds Schafer, G. & Klingenberg, M. 228, (Springer, Berlin, 1978).
31. Hensgens, L. A. M., Grivell, L. A., Borst, P. & Bos, J. L. *Proc. natn. Acad. Sci. U.S.A.* **76,** 1663 (1979).
32. Aujame, L. & Freeman, K. B. *Nucleic Acids Res.* **6,** 455 (1979).
33. Mitra, S. K., Lustig, F., Åkesson, B., Lagerkvist, U. & Strid, L. *J. biol. Chem.* **252,** 471 (1977).
34. Lagerkvist, U. *Proc. natn. Acad. Sci. U.S.A.* **75,** 1759 (1978).
35. Shine, J. & Dalgarno, L. *Eur. J. Biochem.* **57,** 221 (1975).
36. Steitz, J. A. & Jakes, K. *Proc. natn. Acad. Sci. U.S.A.* **72,** 4734 (1975).
37. Hagenbuchle, O., Santer, M. & Steitz, J. A. *Cell* **13,** 551 (1978).
38. Sanger, F. *et al. Nature* **265,** 687 (1977).
39. Fiers, W. *et al. Nature* **273,** 113 (1978).
40. Reddy, V. B. *et al. Science* **200,** 494 (1978).
41. Macino, G., Coruzzi, G., Nobrega, F. G., Li, M. & Tzagoloff, A. *Proc. natn. Acad. Sci. U.S.A.* **76,** 3784 (1979).
42. Drouin, J. *J. molec. Biol.* (submitted).
43. Brown, W. M. & Vinograd, J. *Proc. natn. Acad. Sci. U.S.A.* **71,** 4617 (1974).
44. Robberson, D. L., Clayton, D. A. & Morrow, J. F. *Proc. natn. Acad. Sci. U.S.A.* **71,** 4447 (1974).
45. Crews, S., Ojala, D., Posakany, J., Nishiguchi, J. & Attardi, G. *Nature* **277,** 192 (1979).
46. Sanger, F. & Coulson, A. R. *FEBS Lett.* **87,** 107 (1978).

Printed in Great Britain by Macmillan Production Ltd., Basingstoke, Hampshire

Proc. Natl. Acad. Sci. USA
Vol. 77, No. 6, pp. 3164–3166, June 1980
Biochemistry

Different pattern of codon recognition by mammalian mitochondrial tRNAs

(mitochondrial genetic code/codon recognition/two out of three reading/U·N wobble)

B. G. BARRELL*, S. ANDERSON*, A. T. BANKIER*, M. H. L. DE BRUIJN*, E. CHEN*†, A. R. COULSON*, J. DROUIN*‡, I. C. EPERON*, D. P. NIERLICH*§, B. A. ROE*†, F. SANGER*, P. H. SCHREIER*¶, A. J. H. SMITH*, R. STADEN*, AND I. G. YOUNG*‖

*Medical Research Council Laboratory of Molecular Biology, Hills Road, Cambridge CB2 2QH, England; †Chemistry Department, Kent State University, Kent, Ohio 44242; ‡Department of Biochemistry and Biophysics, School of Medicine, University of California, San Francisco, California 94143; §Department of Microbiology, College of Letters and Science, University of California, Los Angeles, California 90024; ¶Institut für Genetik der Universität, 5 Köln 41, Weyertal 121, West Germany; and ‖ Department of Biochemistry, John Curtin School of Medical Research, Australian National University, P.O. Box 334, Canberra City, A.C.T. 2601, Australia

Contributed by Frederick Sanger, March 5, 1980

ABSTRACT Analysis of an almost complete mammalian mitochondrial DNA sequence has identified 23 possible tRNA genes and we speculate here that these are sufficient to translate all the codons of the mitochondrial genetic code. This number is much smaller than the minimum of 31 required by the wobble hypothesis. For each of the eight genetic code boxes with four codons for one amino acid we find a single specific tRNA gene with T in the first (wobble) position of the anticodon. We suggest that these tRNAs with U in the wobble position can recognize all four codons in these genetic code boxes either by a "two out of three" base interaction or by U·N wobble.

Mitochondrial DNA codes for a number of tRNAs as well as ribosomal RNAs and several respiratory proteins. Nineteen tRNAs have been mapped (by hybridization) on the human mitochondrial genome (1, 2) and mitochondrial tRNAs specific for all amino acids except L-proline, glutamine, histidine, and asparagine have been identified by charging experiments (1). Similar studies on rat liver mitochondrial tRNAs have identified all species except those for glutamine and histidine (3). Assuming that all mammalian mitochondria code for the same complement of tRNAs, these studies bring the number of mammalian mitochondrial tRNAs so far identified to 21, which includes two serine tRNAs, two leucine tRNAs, a $tRNA_m^{Met}$, and a $tRNA_f^{Met}$. Studies on purified rat liver mitochondria and cytoplasmic tRNAs have failed to show any importation of cytoplasmic tRNAs into the mitochondrion (3). The mitochondrial tRNAs had different chromatographic properties on RPC-5 columns and polyacrylamide gel electrophoresis from their cytoplasmic counterparts, and no mitochondrial tRNAs hybridized to nuclear DNA.

The small number of mammalian mitochondrial tRNAs and the lack of any imported cytoplasmic tRNAs raises the problem of how these tRNAs recognize all possible codons in mitochondrial protein synthesis. On the basis of the wobble hypothesis (4), a minimum of 32 tRNAs are needed to read all possible codons of the "universal" genetic code. Recent studies have established that human and bovine mitochondria have a different genetic code in that UGA is tryptophan and not termination and AUA is methionine and not isoleucine (ref. 5; I.G.Y., unpublished results). Thus, a minimum of 31 tRNAs would be needed.

However, recently a number of *in vitro* experiments have indicated that the wobble hypothesis might not apply and that in some cases the third (wobble) position of the codon has no discriminatory function. Mitra *et al.* (6) studied the incorporation of valine in an *in vitro* protein-synthesizing system dependent upon added valine tRNA and found that valine tRNAs with anticodons VAC (V, 5-oxyacetic uridine), GAC, and IAC could each recognize all four valine codons. From these and other results, Lagerkvist (7) proposed an alternative method for codon reading called the "two out of three" method. Whereas this mechanism could operate with impunity in the genetic code boxes with four codons for one amino acid (a codon family), it would cause misreading in the genetic code boxes with four codons for two amino acids (non-family boxes). However, if it were selectively used in the four-codon family boxes, there being a mechanism to prevent misreading in the non-family boxes, then a minimum of 24 tRNAs would be needed to decode the mammalian mitochondrial genetic code. This number is consistent with the number of tRNAs found in mammalian mitochondria, and we discuss here the evidence for such a mechanism operating in human and bovine mitochondria.

RESULTS AND DISCUSSION

We have obtained extensive sequence information on human mitochondrial DNA (approximately 97% of total sequence) by using the primed synthesis chain-termination method (8). The sequence was compiled by using the computer programs of Staden (9). Fragments of human mitochondrial DNA were cloned in pBR322 (10) and single-stranded template DNA was generated by the exonuclease III technique (11). Similarly, fragments of bovine mitochondrial DNA were cloned and sequenced by using the bacteriophage M13 vector (12). These fragments span the areas in human mitochondrial DNA where we have no sequence data. Because of the high homology between human and bovine mitochondrial DNA (unpublished data), the two sets of data were combined to produce a virtually complete mammalian mitochondrial DNA sequence with few remaining ambiguous sequences. In this sequence we have identified 23 tRNA genes by visual inspection and by using the program TRNA (13) which searches for the base-paired cloverleaf structures of tRNAs. The identification of two of these genes, those for $tRNA_m^{Met}$ (AUA/G) and $tRNA^{Ser}$ (AGU/C), are considered tentative. This number confirms the 21 tRNAs identified by earlier studies (1–3) and identifies the tRNAs for glutamine and histidine which were previously not found. If additional tRNA genes lie in those areas of the sequence where there are ambiguities, then one or two more genes at the most are considered possible. The identification of several of these genes has been confirmed by sequence studies on the bovine mitochondrial tRNAs, and these studies also show that there are only 22–24 unique tRNA sequences found in bovine mitochondria (B.A.R. and E.C., unpublished results). The sequences of these tRNAs and a discussion of their unusual structures will be published elsewhere. The tRNAs have been identified by their anticodon sequences as given in Table 1.

Table 1. Anticodon sequences of human mitochondrial tRNA genes

Anticodon	Codon	Amino acid
Family boxes (codons ending in N):		
TAG	CUN	Leu
TAC	GUN	Val
TGA	UCN	Ser
TGG	CCN	Pro
TGT	ACN	Thr
TGC	GCN	Ala
TCG	CGN	Arg
TCC	GGN	Gly
Non-family boxes (codons ending in Y):		
GAA	UUY	Phe
GAT	AUY	Ile
GTA	UAY	Tyr
GTG	CAY	His
GTT	AAY	Asn
GTC	GAY	Asp
GCA	UGY	Cys
GCT*	AGY	Ser
Non-family boxes (codons ending in R):		
TAA	UUR	Leu
TAT*	AUR	Met_m
TTG	CAR	Gln
TTT	AAR	Lys
TTC	GAR	Glu
TCA	UGR	Trp
Initiation codon:		
CAT	AUG	Met_f

Sequences were deduced from the DNA sequence and assigned to specific amino acids and codons by using the modified human and bovine mitochondrial genetic code (ref. 5; I.G.Y., unpublished results). No tRNA gene for Arg (AGR) has been found. Abbreviations: Y, pyrimidine; R, purine; N, any base.

* The identification of the tRNA genes for Met_m (AUR) and Ser (AGY) is considered tentative.

If we consider the eight boxes in the genetic code where a mechanism such as "two out of three" reading can operate without causing misreading—i.e., in the four-codon fam-lilies—then a striking pattern of codon/anticodon relationships is seen. Only one tRNA for each such box is found and in each case, ignoring any possible modification, the tRNAs have U in the first position of the anticodon—i.e., the wobble position. On the basis of G·U wobble, these tRNAs should only recognize the bottom two codons in these boxes. The wobble hypothesis predicts that these boxes would be decoded by using G·U wobble as follows: the top two codons in the valine box (GUU and GUC) are recognized by the anticodon GAC and the bottom two codons (GUA and GUG) are recognized by the anticodon UAC. No mitochondrial tRNAs are found having G in the first position of the anticodon to recognize the top two codons. They are found, however, in the non-family boxes where the top two codons are used for a different amino acid than the one for the bottom two. We have found tRNAs with the predicted G·U wobble anticodons for all these boxes except arginine AGA/G, in addition to an AUG specific formylmethionine initiator tRNA. Here it is interesting to note that the two changes in the human mitochondrial genetic code are brought about by G·U wobble tRNAs for methionine codons AUA and AUG and tryptophan codons UGA and UGG.

Analysis of the codon usage in two well-characterized genes in human mitochondrial DNA, those for cytochrome oxidase subunits II and III (ref. 5; unpublished results), show that all codons are used in these genes except for Ser UCG, Lys AAG, Arg CGG, AGA, and AGG (Fig. 1). Therefore, tRNAs must exist to recognize the top two codons in the family boxes. As previously discussed, it is unlikely that the extra eight tRNAs are imported from the cytoplasm or coded for by the mitochondrial DNA. Thus, we conclude that all four codons in the family boxes are read by a tRNA with first position U in the anticodon. The predicted pattern of decoding is shown in Fig. 1. A similar conclusion has been reached for *Neurospora* and yeast mitochondria by Heckman *et al.* (14) and Bonitz *et al.* (15).

The decoding of these family boxes could operate by a two out of three base codon/anticodon pairing or the U in the first position of the anticodon could recognize any base in the third position of the codon. Whatever mechanism is being used, it must not occur in the non-family boxes; otherwise, misreading would occur. A discrimination mechanism is suggested by Heckman *et al.* (14) who studied *Neurospora crassa* mitochondrial tRNAs. *Neurospora* also codes for a limited number of tRNAs—less than the 32 needed to recognize all codons by G·U wobble—although nothing is known about the codon usage or the possible importation of cytoplasmic tRNAs into *N. crassa* mitochondria. Sequence analysis of several *N. crassa* mitochondrial tRNAs by Heckman *et al.* (14) reveals that a codon recognition pattern similar to that of human mitochondrial tRNAs is found and that the tRNAs for the family boxes have an unmodified U in the wobble position and the tRNAs specific for the bottom two codons in the nonfamily boxes have an unknown modified U in this position. Thus, the modified U could serve to prevent misreading of the top two codons in the non-family boxes, whereas the unmodified U could allow recognition of all four codons in the family boxes. The pattern of modification is not yet known for mammalian mitochondrial tRNAs but it would seem likely that a pattern similar to that described by Heckman *et al.* (14) for *N. crassa* will be found.

In this connection it is interesting that, in all published tRNA sequences (16) with U in the wobble position, the U is always found to be modified with the exception of yeast $tRNA^{Leu}_{UAG}$ (17–19). This tRNA has an unusual codon recognition pattern in that it can somehow recognize all six leucine codons. However, the situation in prokaryotes and in the cytoplasm of eukaryotes must be more complex with a greater number of tRNA species being used, and it is difficult to draw any analogy between the patterns of modification of these tRNAs and those of the mitochondria.

Does the U in the first position of the anticodon of the mitochondrial tRNAs specific for the family boxes pair with all four bases or is there a strict two out of three base codon/anticodon interaction? Crick (4) in his wobble hypothesis noted that U·U and U·C base pairs were possible but discounted them on the basis that they were too close and would cause misreading in the non-family boxes. Grosjean *et al.* (20) have examined the stability of a number of anticodon/anticodon interactions and found that the only stable nonwobble pairs involved U (or 2-thio derivatives of U) pairing with U and C. They also found by model building that U·U and U·C pairs are possible. Thus, U is the only base that can possibly form a stable base pair with A, G, C, or U and it is significant that all the tRNAs for the codon families have U in the wobble position. This would seem to suggest that this U wobbles with all four bases in the third position of the codon and that there is not a two out of three base interaction with the U remaining unpaired.

Finally, we have strongly suggestive evidence that a further simplification of the mammalian mitochondrial genetic code has occurred. The pattern of codon recognition shown in Fig.

Second letter

First letter	U			C			A			G			Third letter
U	UUU	Phe	26	UCU	Ser	2	UAU	Tyr	15	UGU	Cys	1	U
	UUC		17	UCC		11	UAC		15	UGC		3	C
	UUA	Leu	7	UCA		12	UAA	Ter	-	UGA	Trp	12	A
	UUG		1	UCG		0	UAG		-	UGG		4	G
C	CUU	Leu	10	CCU	Pro	3	CAU	His	7	CGU	Arg	2	U
	CUC		15	CCC		15	CAC		16	CGC		3	C
	CUA		29	CCA		6	CAA	Gln	15	CGA		7	A
	CUG		6	CCG		3	CAG		1	CGG		0	G
A	AUU	Ile	14	ACU	Thr	7	AAU	Asn	3	AGU	Ser	2	U
	AUC		22	ACC		18	AAC		10	AGC		3	C
	AUA	Met	16	ACA		19	AAA	Lys	7	AGA	Arg/Ter?	0	A
	AUG	F-Met	5	ACG		1	AAG		0	AGG		0	G
G	GUU	Val	3	GCU	Ala	5	GAU	Asp	5	GGU	Gly	3	U
	GUC		11	GCC		16	GAC		7	GGC		14	C
	CUA		10	GCA		7	GAA	Glu	13	GGA		7	A
	GUG		1	GCG		1	GAG		5	GGG		5	G

FIG. 1. Predicted pattern of codon recognition by mammalian mitochondrial tRNAs. Each internal box represents codons predicted to be recognized by a single tRNA using U·N wobble for the genetic code boxes with four codons for one amino acid and G·U wobble for the genetic code boxes with four codons for two amino acids. The number of codons found in the genes for cytochrome oxidase subunits II and III are listed alongside (ref. 5; unpublished results).

1 predicts that we should find a tRNA gene in the mitochondrial DNA sequence with a G·U wobble anticodon for the arginine codons AGA and AGG. No codons for arginine AGA and AGG are found in the genes for cytochrome oxidase subunits II and III, and an analysis of all significant reading frames in the rest of the sequence suggests that AGA and AGG are probably termination codons and not coding for arginine (unpublished results). Thus, mammalian mitochondria may only use 23 tRNAs to translate its genetic code.

We thank Dr. G. Attardi for a generous gift of HeLa cell mitochondria DNA. S.A. is a Research Fellow of the Cystic Fibrosis Foundation. B.A.R. was supported by a National Institutes of Health Research Career Development Award and P.H.S., by a European Molecular Biology Organization fellowship.

1. Lynch, D. C. & Attardi, G. (1976) *J. Mol. Biol.* **102,** 125–141.
2. Angener, L., Davidson, N., Murphy, W., Lynch, D. C. & Attardi, G. (1976) *Cell* **9,** 81–90.
3. Aujame, L. & Freeman, K. B. (1979) *Nucleic Acids Res.* **6,** 455–469.
4. Crick, F. H. C. (1966) *J. Mol. Biol.* **19,** 548–555.
5. Barrell, B. G., Bankier, A. T. & Drouin, J. (1979) *Nature (London)* **282,** 189–194.
6. Mitra, S. K., Lustig, F., Åkesson, B. & Lagerkvist, U. (1977) *J. Biol. Chem.* **252,** 471–478.
7. Lagerkvist, U. (1978) *Proc. Natl. Acad. Sci. USA* **75,** 1759–1762.
8. Sanger, F., Nicklen, S. & Coulson, A. R. (1977) *Proc. Natl. Acad. Sci. USA* **74,** 5463–5467.
9. Staden, R. (1979) *Nucleic Acids Res.* **6,** 2601–2610.
10. Drouin, J. (1980) *J. Mol. Biol.,* in press.
11. Smith, A. J. H. (1979) *Nucleic Acids Res.* **6,** 831–848.
12. Schreier, P. H. & Cortese, R. (1979) *J. Mol. Biol.* **129,** 169–172.
13. Staden, R. (1980) *Nucleic Acids Res.* **8,** 817–825.
14. Heckman, J. E., Sarnoff, J., Alzner-DeWeerd, B., Yin, S. & RajBhandary, U. L. (1980) *Proc. Natl. Acad. Sci. USA* **77,** 3159–3163.
15. Bonitz, S. G., Berlini, R., Coruzzi, G., Li, M., Macino, G., Nobrega, F. G., Nobrega, M. P., Thalenfeld, B. E. & Tzagoloff, A. (1980) *Proc. Natl. Acad. Sci. USA* **77,** 3167–3170.
16. Gaus, D. H., Grüter, F. & Sprinzl, M. (1979) *Nucleic Acids. Res.* **6,** Suppl. r1–r19.
17. Weissenbach, J., Dirheimer, G., Falcoff, R., Sanceau, J. & Falcoff, E. (1977) *FEBS Lett.* **82,** 71–76.
18. Randerath, K., Chia, L. S. Y., Gupta, R. C., Randerath, E., Hawkins, E. R., Brum, C. K. & Chang, S. H. (1975) *Biochem. Biophys. Res. Commun.* **63,** 157–163.
19. Randerath, E., Gupta, R. C., Chia, L. S. Y., Chang, S. H. & Randerath, K. (1979) *Eur. J. Biochem.* **93,** 79–94.
20. Grosjean, H. J., de Henau, S. & Crothers, D. M. (1978) *Proc. Natl. Acad. Sci. USA* **75,** 610–614.

Reprinted from Nature, Vol 290, No. 5806, pp. 457-465, April 9 1981

Sequence and organization of the human mitochondrial genome

S. Anderson, A. T. Bankier, B. G. Barrell, M. H. L. de Bruijn, A. R. Coulson, J. Drouin*, I. C. Eperon, D. P. Nierlich*, B. A. Roe*, F. Sanger, P. H. Schreier*, A. J. H. Smith, R. Staden & I. G. Young*

MRC Laboratory of Molecular Biology, Hills Road, Cambridge CB2 2QH, UK

The complete sequence of the 16,569-base pair human mitochondrial genome is presented. The genes for the 12S and 16S rRNAs, 22 tRNAs, cytochrome c *oxidase subunits I, II and III, ATPase subunit 6, cytochrome* b *and eight other predicted protein coding genes have been located. The sequence shows extreme economy in that the genes have none or only a few noncoding bases between them, and in many cases the termination codons are not coded in the DNA but are created post-transcriptionally by polyadenylation of the mRNAs.*

MITOCHONDRIA have a separate autonomously replicating DNA genome (for a review see ref. 1). Although most of the proteins present in the mitochondrion are encoded by nuclear DNA, a few are coded for by mitochondrial (mt) DNA and are synthesized by the separate mt translation system. The rRNAs and tRNAs of this system are mitochondrially encoded and it is likely that the other components, although nuclear coded, are distinct from those in the cytoplasmic translation system. The enzymes specific for the unique mitochondrial transcription and replication processes are probably also nuclear coded. We have recently shown that mammalian mitochondria have a different genetic code[2,3] as have other mitochondria systems that have been studied[4-7], and a different decoding mechanism[7-9].

Replication of mammalian mtDNA proceeds by initiation of heavy (H) strand synthesis at a specific origin[10,11] resulting in the formation of a displacement (D) loop with a newly synthesized H strand of ~680 bases known as 7S DNA. In mouse cells a large proportion of mtDNA exists in the D-loop form but only a few of these nascent H strands grow to full length. Initiation of light (L) strand synthesis is at a specific origin[12] and does not occur until this region has been exposed by H-strand synthesis.

Transcription of mammalian mtDNA is also unique in that both strands seem to be completely transcribed from promoters situated in the D-loop region[13,14]. These primary transcripts are then processed to give the 12S and 16S rRNAs, tRNAs and a number of presumptive mRNAs, which are not capped but are polyadenylated[15]. The small size of the ribosomal RNAs, the chloramphenicol sensitivity of mitochondrial protein synthesis and the use of formyl-methionine in initiation of translation have led to the suggestion that the mitochondrial genetic system arose by endosymbiosis of primitive bacteria within a host cell. The sequence of the rRNA genes, however, shows only low homology with present-day prokaryotic rRNAs[16].

To understand this genetic system more fully we have determined the complete DNA sequence of both the human and the bovine mitochondrial genomes. We report here the sequence of human mtDNA and an interpretation of this sequence in terms of its coding capacity, gene arrangement and gene expression. The bovine sequence will be published elsewhere.

In the accompanying articles Ojala *et al.*[17] and Montoya *et al.*[18] report the sequence analysis of the stable polyadenylated RNAs. The comparison of the DNA and RNA sequences reveals the unique nature of the precise transcript processing in mammalian mitochondria.

Organization of the genome

The DNA sequence is shown in Fig. 1 along with the predicted coding sequences. This is presented schematically in Fig. 2. The identification of the genes and their precise location in the DNA sequence are based on several unique features, the evidence for which is discussed below and in the accompanying articles[17,18]. The sequence shows extreme economy of organization in that there are none or very few noncoding bases between adjacent genes except for the D-loop region where we have been unable to assign any coding function. The sequence shown is that of the L strand which we define as the main coding strand containing the sense sequence of the rRNAs and most of the tRNAs and mRNAs. Thus, these RNAs are transcribed from (and therefore hybridize to) the H strand.

Origins of DNA replication

The 5′ end of the major 7S DNA species found in the D-loop of HeLa cell mtDNA[10] corresponds to nucleotide 191 of the DNA sequence. Other human 7S DNA species having different 5′ ends have also been reported[11,19,20], but the precise location of these termini in the sequence presented here is not known. The size of the 7S DNA (≤680 nucleotides) and mapping data[10,19,20] indicate that the 7S DNA, and presumably the D-loop, do not extend into the $tRNA^{Pro}$ gene. Although H-strand synthesis during replication initiates in the D-loop region, data from mouse cell mitochondria indicate that most of the 7S DNA molecules do not participate in DNA replication but rather are synthesized and then rapidly lost from the D-loop[21].

The ~1,100-base pair region of the human mitochondrial genome between the $tRNA^{Pro}$ and $tRNA^{Phe}$ genes contains no large open reading frames in either strand. Furthermore, except for a small 7S RNA[15], no large stable polyadenylated transcript has been found to map in this region. Unless RNA splicing is invoked it is extremely unlikely that this part of the genome codes for a major protein species. In the region covered by the D-loop, several blocks of nucleotide sequence homology exist between the human, bovine (S.A. *et al.*, unpublished results) and rat mitochondrial genomes[22]. However, the DNA separating these blocks is rather variable in length and exhibits no significant homology. The region between the 5′ end of the 7S DNA and the $tRNA^{Phe}$ gene seems to be one of the least conserved regions of the entire genome. This was initially detected by electron microscopy of heteroduplexes of sheep and goat mtDNAs[23] and has been confirmed by the DNA sequence comparisons. Very little sequence homology between the human, rat[22] or bovine mitochondrial genomes can be detected

*Present addresses: Department of Biochemistry and Biophysics, School of Medicine, University of California, San Francisco, California 94143, USA (J.D.); Department of Microbiology, College of Letters and Science, University of California, Los Angeles, California 90024, USA (D.P.N.); Chemistry Department, Kent State University, Kent, Ohio 44242, USA (B.A.R.); Institut für Genetik der Universität, 5 Koln 41, Weyertal 121, FRG (P.H.S.); Department of Biochemistry, John Curtin School of Medical Research, Australian National University, PO Box 334, Canberra City, ACT 2601, Australia (I.G.Y.).

Table 1 Human mitochondrial genetic code

Phe	UUU	77	Ser	UCU	32	Tyr	UAU	46	Cys	UGU	5
	UUC	141		UCC	99		UAC	89		UGC	17
Leu	UUA	73		UCA	83	Ter	UAA	-	Trp	UGA	93
	UUG	16		UCG	7		UAG	-		UGG	11
Leu	CUU	65	Pro	CCU	41	His	CAU	18	Arg	CGU	7
	CUC	167		CCC	119		CAC	79		CGC	25
	CUA	276		CCA	52	Gln	CAA	81		CGA	29
	CUG	45		CCG	7		CAG	9		CGG	2
Ile	AUU	125	Thr	ACU	51	Asn	AAU	33	Ser	AGU	14
	AUC	196		ACC	155		AAC	131		AGC	39
Met	AUA	167		ACA	133	Lys	AAA	85	Ter	AGA	-
	AUG	40		ACG	10		AAG	10		AGG	-
Val	GUU	30	Ala	GCU	43	Asp	GAU	15	Gly	GGU	24
	GUC	49		GCC	124		GAC	51		GGC	88
	GUA	70		GCA	80	Glu	GAA	64		GGA	67
	GUG	18		GCG	8		GAG	24		GGG	34

The genetic code of human mtDNA differs from the universal code in that UGA codes for tryptophan and not termination, AUA codes for methionine not isoleucine, and AGA and AGG are termination rather than arginine codons. In addition, AUA and possibly AUU are initiation codons as well as AUG. Boxes of four codons are each read by a single tRNA with U in the first position of the anticodon; boxes of two codons are read by tRNAs with G:U wobble anticodons. The number of methionyl tRNAs and their codon response are unclear and discussed in the text. Also shown are the total number of each amino acid codon found in the genes and predicted genes shown in Figs 1 and 2. Codons ending in A and C predominate (36.37 and 40.93%, respectively) and those ending in U and G are used less frequently (16.33 and 6.36%, respectively).

in this region. The size of the region also seems to vary much more than do intergenic spaces in other parts of the mtDNA, being 385 nucleotides in the human placenta, 291 nucleotides in rat liver and (based on alignment with the human and rat sequence) ~180 nucleotides in beef heart. This region may contain control signals for transcription of the mitochondrial genome.

L-strand synthesis during mtDNA replication initiates two-thirds of the way around the genome from the region of H-strand synthesis[24]. In mouse cell mitochondria the DNA sequence immediately surrounding the L-strand origin is a simple inverted repeat that may form a hairpin loop structure when the L-strand origin is exposed by H-strand displacement synthesis during DNA replication[12]. An analogous and highly homologous presumptive L-strand origin exists in the human mtDNA sequence (nucleotides 5,730–5,763) within a cluster of five tRNA genes. Interestingly, the differences between the human and mouse inverted repeats occur symmetrically, so that if secondary structure did exist, full base-pairing in the stem region would be maintained.

Ribosomal RNA genes

Eperon *et al.*[16] were able to align the DNA sequence with the sequences of the 5′ ends [15,25] of the 12S and 16S rRNAs and to show that there were no noncoding bases between the $tRNA^{Phe}$ gene and the 12S rRNA gene, and between the $tRNA^{Val}$ gene and the 16S rRNA gene. However, the 3′ ends of the rRNA genes could only be predicted to be very close to the 5′ ends of the flanking $tRNA^{Val}$ and $tRNA^{Leu}$ genes. It has since been shown that the mouse small subunit rRNA gene abuts the 3′ $tRNA^{Val}$ gene[26,27] and alignment of a hamster large subunit rRNA sequence with the murine gene sequence suggests that there are also no noncoding bases between the large subunit rRNA gene and the 3′ $tRNA^{Leu}$ gene[28]. Thus, the 12S rRNA and 16S rRNA and the flanking tRNAs in a primary transcript could be released by cleavage precisely at the 5′ and 3′ ends of the tRNAs without the need for further processing.

The high concentration of nascent transcripts from this region of the genome has led to the suggestion that there is a transcription attenuator in this region[20]. This is supported by the

Fig. 1 The DNA sequence of human mitochondria (pages 459–461). The sequence of the L strand is numbered arbitrarily from the *Mbo*I-5/7 boundary[56] in the D-loop region. The sequence is displayed using the program TRANMT of Staden (unpublished) and shows the position of the genes for the tRNAs (boxed) and the rRNAs (underlined). The unidentified reading frames (URF 1–6) and identified protein coding genes are translated into the one-letter amino acid code (termination codons are denoted by an asterisk). The genes for cytochrome *c* oxidase subunits I–III are abbreviated COI–III and cyt *b* denotes the gene for cytochrome *b*. Which strand contains the sense sequence of the genes is denoted by → (L strand) and ← (H strand). The ends of the presumptive mRNAs transcribed from the H strand as determined by RNA sequence analysis[17,18] are also shown below the sequence. The 5′ end of the nascent H-strand ('7S') DNA found in the D-loop is at nucleotide 191 (ref. 10). The origin of L-strand synthesis during DNA replication[12] is indicated by overlining of the possible base-pairing regions. The DNA sequence was derived mainly from a single human placenta mtDNA preparation[56] but some regions were determined on HeLa cell mtDNA. Restriction enzyme fragments of placental mtDNA were cloned in pBR322[56]. Single-stranded template DNA was prepared by the exonuclease III procedure[57] and by cloning in bacteriophage M13 (refs 58–61). The DNA sequence was determined by the chain termination method of Sanger *et al.*[62,63]. The sequence was analysed using the computer programs of Staden (refs 31, 55, 64 and unpublished). Nucleotides 10, 934–5, 14,272 and 14,365 were ambiguous and their identity assumed to be the same as in the bovine mtDNA sequence (S.A., M.H.L.B., A.R.C., I.C.E., F.S., I.G.Y., unpublished). The base composition of the L strand is 24.7% T, 31.2% C, 30.9% A and 13.1% G.

3

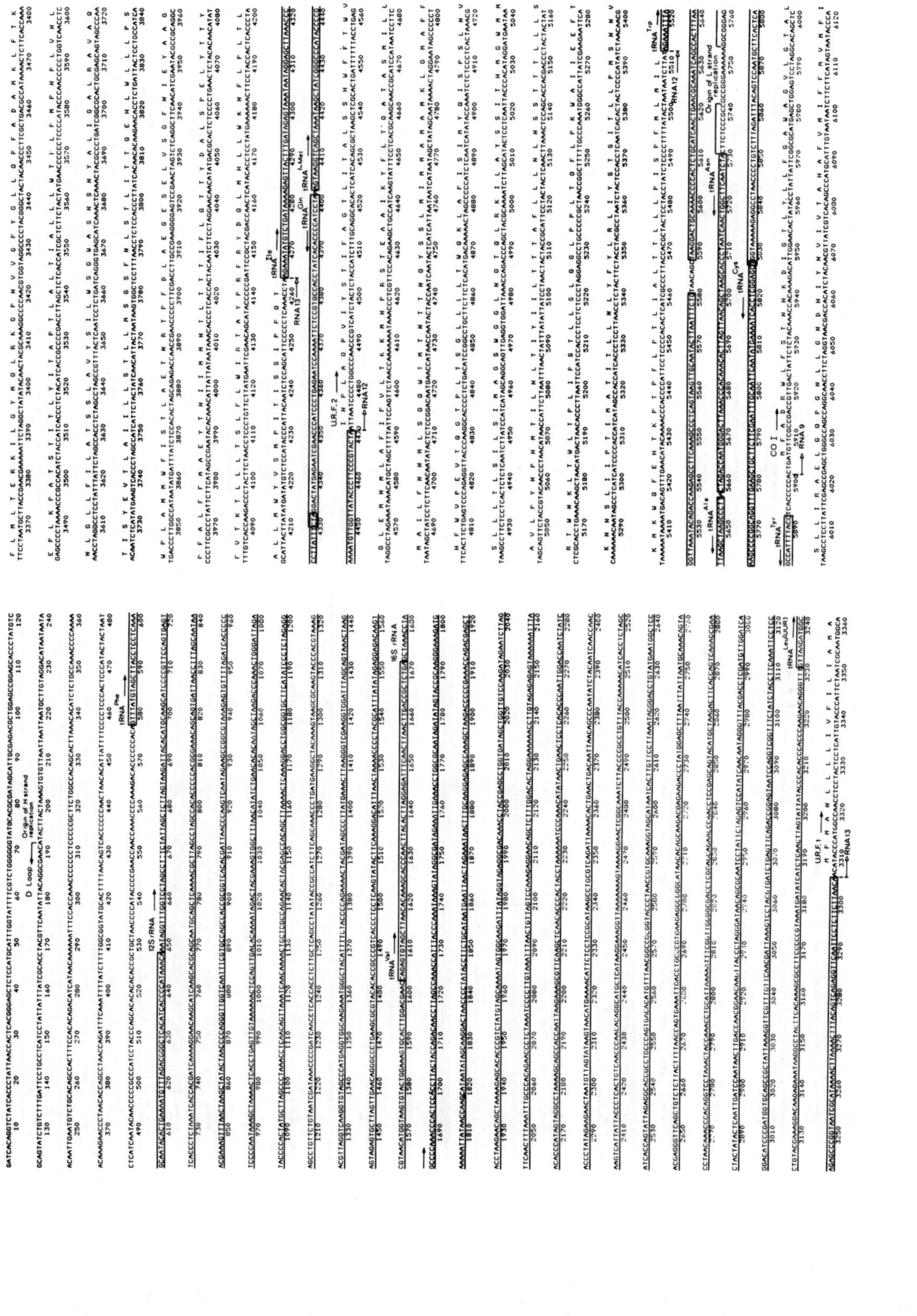

Figure 1

4

Figure 1

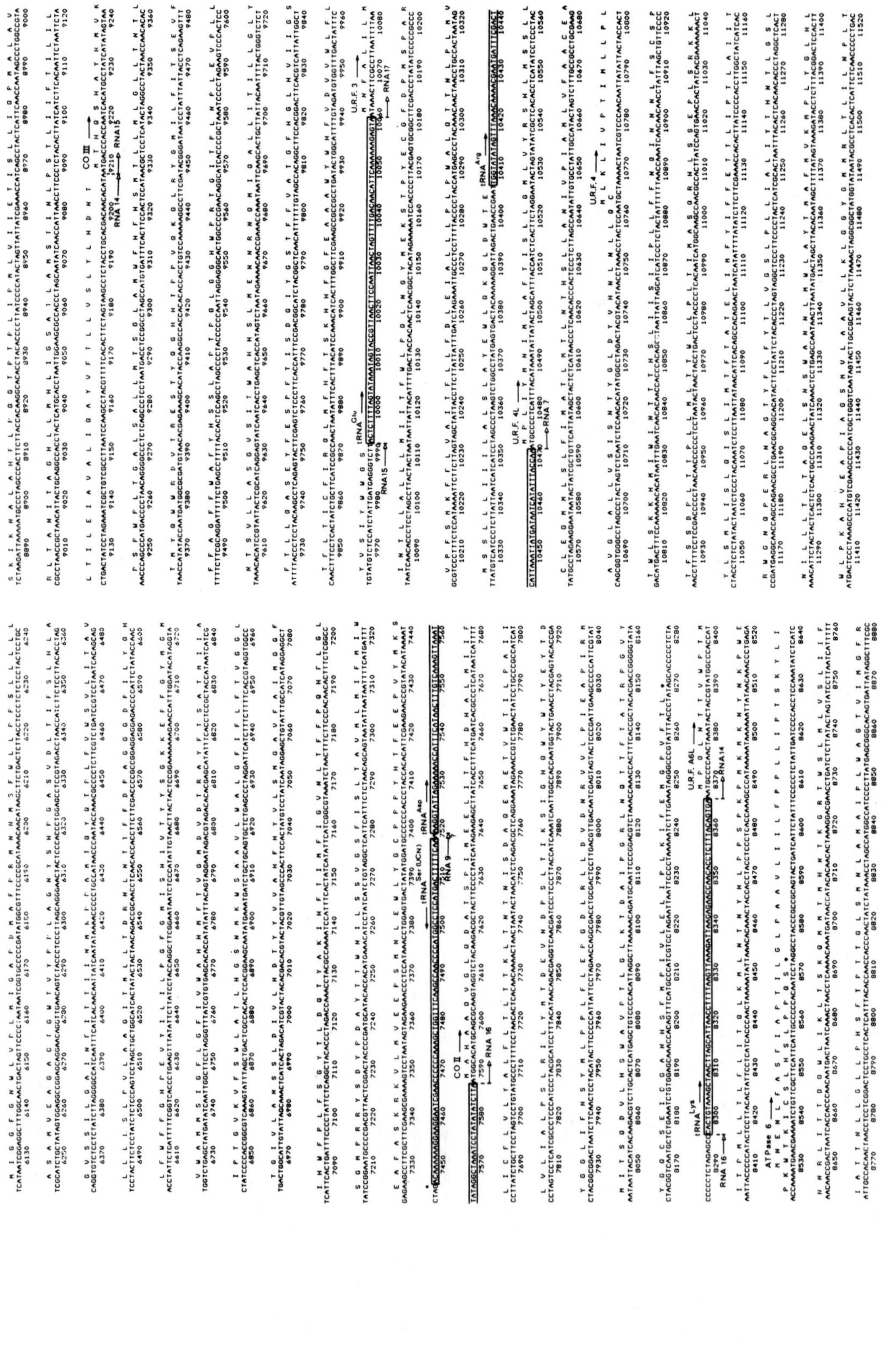

5

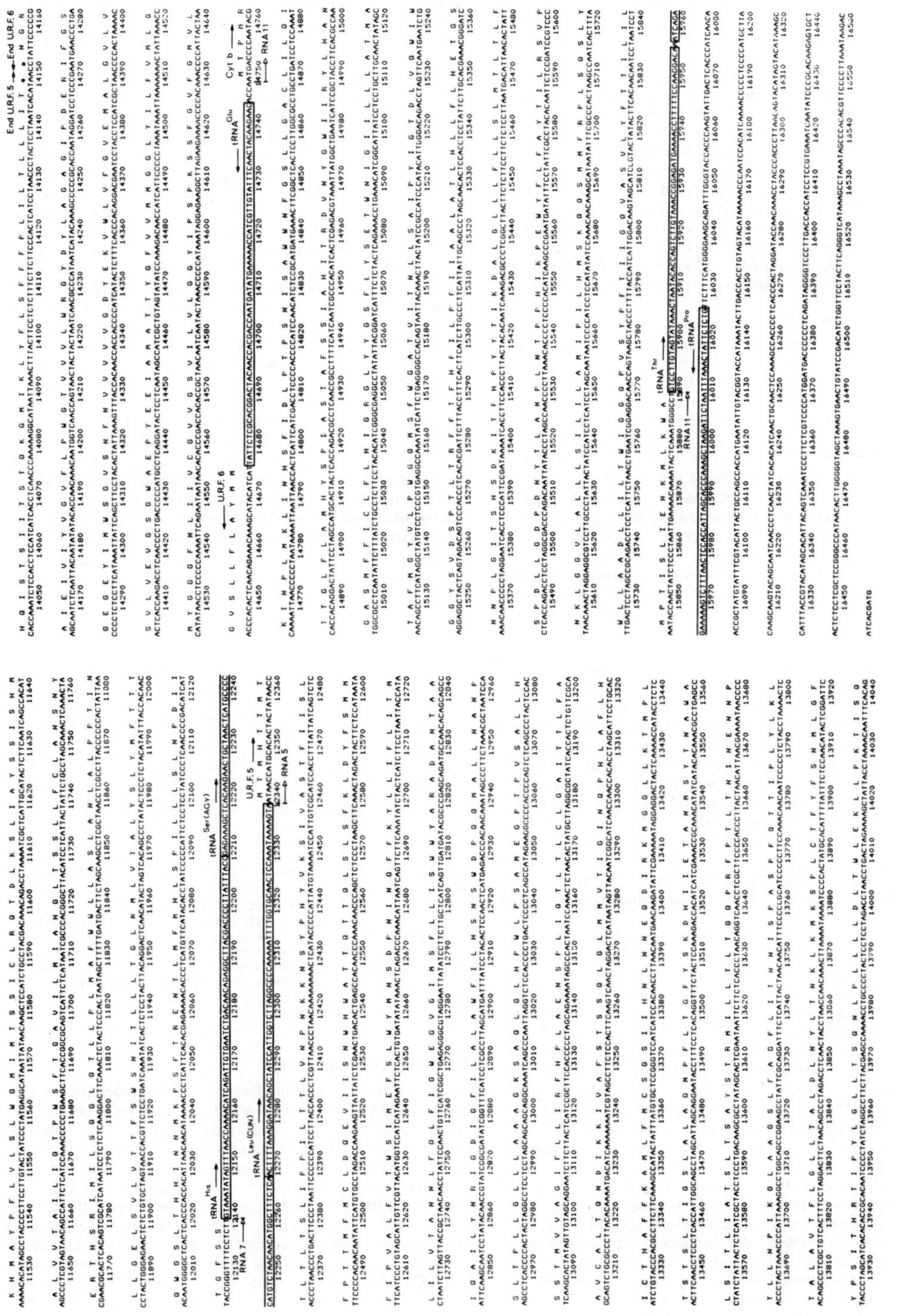

Figure 1

6

observation that the hamster large subunit rRNA ends and is polyadenylated at any of four nucleotides within the last five nucleotides of the gene[28], which contrasts with the precise termini of the 12S rRNA and mRNAs[29,30] which are presumed to be generated by processing of a primary transcript. It is possible that the 12S rRNA is also adenylated at the 3′ terminus after transcription and processing: a single addition has been proposed for the hamster small subunit rRNA by comparison with the murine gene sequence[30].

The sequences of the rRNA genes were interpreted to provide support for a model in which mt ribosomes did not have sequence-dependent signals for recognition of the correct initiation codon but rather initiated at the first initiation codon in the mRNA[16]. This is consistent with the known 5′-end sequences of the mRNAs[18] although not with the observation of single transcripts covering the URF A6L/ATPase 6 and URF 4L/URF 4 reading frames (Fig. 2) if the expected proteins are expressed.

The tRNA genes

Twenty-two tRNA genes have been identified in the human mtDNA sequence either by visual inspection or using the program TRNA[31]. Equivalent genes have been located in the bovine mtDNA sequence and most of these have been confirmed by direct sequencing of bovine mitochondrial tRNAs (B. A. Roe, E. Y. Chen, P. W. Armstrong and J. F. H. Wong, unpublished). We have been unable to substantiate a previous report[8] which tentatively identified a gene for $tRNA_m^{Met}$. Apart from the $tRNA_m^{Met}$, these 22 tRNAs are sufficient to read all codons using a mechanism unique to mitochondrial systems[7-9]. No mammalian mt tRNAs have been shown to be imported from the cytoplasm (ref. 32 and B. A. Roe and E. Y. Chen, unpublished results). It had previously been assumed that the minimum number of tRNAs required to read all codons would be 32 for the 'universal' genetic code using G:U or I:A/C/U wobble[33]. In human mitochondria the tRNAs for the two-codon families, that is, those genetic code boxes with two codons for one amino acid, have G:U wobble anticodons. However, only one tRNA is found for each of the four-codon families[8] (see Table 1). These tRNAs have a U in the first position of the anticodon although each must read all four codons in its respective family. This is probably accomplished by U:N wobble[8], although a two-out-of-three mechanism might operate[34]. Although the modification pattern of the U in the first position of the anticodon of mammalian mt tRNAs is now known, it is likely that a similar situation exists to that found in *Neurospora* mitochondria[7], whereby the tRNAs for the four-codon families have an unmodified U, but when the U is required for reading the two-codon families it is modified. This modification presumably prevents such tRNAs from reading codons ending in U or C.

Two species of mammalian mt methionyl tRNAs have been identified, a $tRNA_f^{Met}$ and a $tRNA_m^{Met}$ (refs 32, 35). Hybridization of these tRNAs to mtDNA showed partial additivity, indicating the existence of two separate genes[32]. In the DNA sequences we have only been able to identify one $tRNA^{Met}$ gene, presumably coding for the initiator tRNA, based on its anticodon sequence CAT and the close homology of its anticodon arm with other initiator tRNAs (see ref. 36). The tRNA anticodon sequence CAU should normally be specific for AUG but as discussed below we believe that AUA and possibly AUU are also initiation codons in mammalian mitochondria. To recognize AUA (and AUU?) as well as AUG, the C in the first position of the anticodon would probably have to be modified. The AUA-specific $tRNA^{Ile}$ of *Escherichia coli* and of bacteriophage T4 have modified C residues in the first position of the anticodon[37,38], and hence represent examples of specific recognition of AUA by a tRNA with a CAU anticodon. Although potential $tRNA_m^{Met}$-like structures are found in the human and bovine sequences, none of these is particularly convincing nor are they conserved in the corresponding sequence. Also, extensive sequence analysis of the bovine mt tRNAs has not revealed any candidate for a $tRNA_m^{Met}$ (B. A. Roe, E. Y. Chen, P. W. Armstrong and J. F. H. Wong, personal communication). If the gene exists in the mt genome it may have an unusual structure that is preventing its detection, for example, like that of $tRNA_{AGU/C}^{Ser}$ which completely lacks the D arm[39]. The other possibility is that one gene can give rise to both a $tRNA_f^{Met}$ and a $tRNA_m^{Met}$ by differential modification, although the hybridization data of Aujame and Freeman[32] suggest otherwise. We are at present trying to clarify this problem by hybridization of labelled methionyl tRNA to restriction fragments of mtDNA.

The sequences of the mammalian mt tRNAs are unusual in that all, except $tRNA_{UUR}^{Leu}$, lack some or all of the following features found in other tRNAs: (1) the universal sequence G–T–ψ–C–R–A, (2) the constant seven-base length of the 'TψC' loop which in mammalian mt tRNAs varies between three and nine bases, and (3) the constant bases A_{14}, G_{15} and $G_{18}G_{19}$ and the connections with U_8 and U_{48} (yeast $tRNA^{Phe}$ numbering system, see ref. 40). Thus, the mt tRNAs are apparently stabilized by fewer tertiary interactions. The most extreme case is that of $tRNA_{AGU/C}^{Ser}$, which lacks the D arm[39]. Also, considering the similarity of homologous cytoplasmic tRNA species from different animals it is surprising that substantial variations are found when specific tRNAs from human and bovine mitochondria are compared. A compilation and comparison of these tRNAs will be published elsewhere. These missing or different tertiary interactions may mean that the mammalian mitochondrial tRNAs have more freedom than cytoplasmic tRNAs to evolve in certain areas, perhaps reflecting a different interaction with the ribosome. The role of the tRNAs in processing of the primary transcript is discussed below.

Table 2 Characteristics of human mt genes and comparison with bovine mt genes

Gene	Protein length	Molecular weight	% Amino acid conservation	Initiation and termination codons			
				Human		Bovine	
URF1	318	35,600	75.79	ATA	*	ATG	*
URF2	347	38,900	62.82	ATT	*	ATA	*
COI	513	57,000	91.23	ATG	AGA	ATG	TAA
COII	227	25,500	72.69	ATG	TAG	ATG	TAA
URF A6L	68	7,900	51.52	ATG	TAG?	ATG	TAA?
ATPase 6	226	24,800	77.88	ATG	*	ATG	*
COIII	261	30,000	86.97	ATG	*	ATG	*
URF3	115	13,200	73.91	ATA	*	ATA	*
URF 4L	98	10,700	73.47	ATG	TAA?	ATG	TAA?
URF4	459	51,400	74.07	ATG	*	ATG	*
URF5	603	66,600	69.49	ATA	TAA	ATA	TAA
URF6	174	18,600	62.64	ATG	AGG	ATG	TAA
Cytochrome *b*	380	42,700	78.10	ATG	*	ATG	AGA

The predicted length and molecular weight of the expected products of the human mt genes and unidentified reading frames were calculated from the DNA sequence using the computer programs of Staden (ref. 55 and unpublished results). Also shown is the percentage identical amino acids conserved between the human and bovine sequences calculated using the program AACHAN (R. Staden, unpublished). The predicted initiation and termination codons of the human and bovine genes are shown; * indicates that the termination codon is created post-transcriptionally by polyadenylation (see text).

Protein coding genes

Thirteen significant reading frames have been identified in the human mtDNA sequence, of which five have been assigned to known mitochondrially synthesized proteins. Only reading frames of at least 40 amino acids conserved with at least 50% amino acid homology in the bovine sequence were considered. The gene for cytochrome *c* oxidase subunit I (COI) was identified from an N-terminal amino acid sequence of the bovine protein (J. E. Walker, personal communication) and confirmed by comparison with the yeast mt COI gene. COII was identified from the complete amino acid sequence of the bovine protein[41] and COIII by homology with the yeast gene[42] and by an N-terminal sequence of the bovine protein (G. Buse, personal communication). The genes for ATPase 6 and cytochrome *b*

7

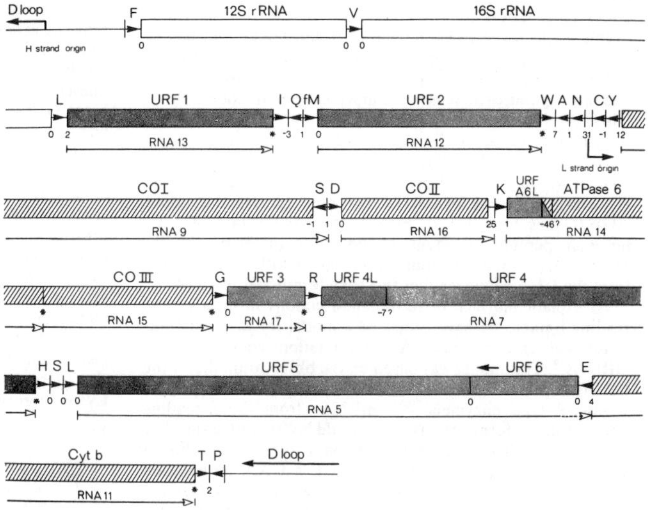

Fig. 2 Organization of the genome. A schematic representation of the sequence shown in Fig. 1. The tRNA genes are identified by the one-letter amino acid code and their sense sequence by either → (L strand) or ← (H strand). Unidentified reading frames 1–5 (URF1–5) and the identified protein genes for cytochrome *c* oxidase subunits I–III (COI–III), ATPase 6 and cytochrome *b* (cyt *b*) all have their sense sequence on the L strand; URF6, however, is H-strand coded. The number of noncoding bases between genes is given below the junctions of the genes. A minus number denotes an overlap and an asterisk indicates that there are no noncoding bases at that point and that the termination codon for the preceding gene is created in the mRNA by polyadenylation (see text). The numbered RNAs are the presumptive mRNAs transcribed from the H strand which were aligned with the DNA sequence by hybridization to restriction fragments and by sequence analysis of their 5′ and 3′ ends (see accompanying articles[17,18] for a more detailed transcription map showing the positions of the L-strand transcripts and the precursors to the H-strand transcripts).

were identified by homology with the corresponding yeast genes[43,44] and also by amino acid sequences of bovine cytochrome *b* (W. Machleidt, personal communication). We have found no homology between the bovine amino acid sequence of ATPase 9 (E. Wachter, personal communication) and any of the reading frames, so that, as in *Neurospora*[45], this protein must be nuclear coded; in yeast it is mitochondrially coded[46,47].

In regions representing identified genes the respective human and bovine sequences are highly homologous. As shown in Table 2, there is ~80% amino acid conservation, which is 5% higher than the nucleotide conservation, reflecting the degeneracy of the genetic code. Over half the codons for the conserved amino acids are different in the third (degenerate) position or, in the case of the leucine codons UUR and CUN, in the first and third positions.

We have found eight unidentified reading frames (URFs) that are conserved in human and bovine mtDNA. On average, these show ~70% nucleotide and amino acid conservation (Table 2), again with approximately half of the conserved amino acids having changed codons. This type of homology is unlikely to have occurred by chance: for example, if the reading frames are compared out of phase in the +1 and +2 frames, the amino acid homology, including all the termination codons, is 35.4% and the percentage of conserved amino acids with changed codons drops to 4.5%. Therefore, we believe that most, if not all, of these reading frames represent genes for mammalian mitochondrial proteins. Further evidence in support of this is provided by the mapping and sequence analysis of the polyadenylated presumptive mRNAs reported in the accompanying articles[17,18]. These RNAs correspond exactly to the reading frames of URF1, URF2, COI, COII, URF A6L and ATPase 6, COIII, URF3, URF 4L and URF4, URF5 and the antisense of URF6, and cytochrome *b* as shown in Figs 1 and 2. In addition, there are large L-strand transcripts covering the H-strand-coded URF6, and the number and sizes of these reading frames are in fairly good agreement with the band pattern obtained by gel electrophoresis of HeLa cell mt proteins produced in the presence of emetine, an inhibitor of cytoplasmic protein synthesis (J. E. Walker, personal communication).

Two of the H-strand transcripts contain more than one reading frame, RNA14 (URF A6L and ATPase 6) and RNA7 (URF 4L and URF4). These present an anomaly and it is not known whether the reading frames are separated by a further processing event of these RNAs. The same organization is found in bovine mtDNA, making it unlikely that the breaks in the reading frames are due to sequence errors. A possible explanation is that these frames contain short intervening sequences which, when excised, leave only a single reading frame. However, using nuclease S_1 protein experiments, Ojala *et al.*[48] found no evidence for intervening sequences in the mtDNA regions coding for the polyadenylated RNAs.

Initiation codons

Although all the identified genes (except ATPase 6) and some of the URFs have an AUG initiation codon within the first six bases of their mRNA start, others do not. The first ATG in bovine URF2 is 320 codons downstream from the RNA start in a reading frame of 347 codons. Also, in human and bovine URF3 the first AUG is, respectively, 89 and 87 codons downstream in a reading frame of 115 codons. Ignoring the cases of ATPase 6 and URF4, which are the second reading frames in their presumptive mRNAs, and also the case of human URF2, which is discussed separately below, all reading frames that do not have an AUG near the 5′ end of the RNAs have an AUA in an equivalent position. Thus, if these reading frames are translated completely, as their homology suggests they are, AUA must function as an initiation codon as well as AUG. AUA has already been shown to code for internal methionine as well as AUG[2,3]. In the case of human URF2 there is an AUU in the same position as the AUA in bovine URF2. Thus, to translate completely this reading frame we must argue that, like AUA, AUU can also be an initiation codon in mitochondrial genes.

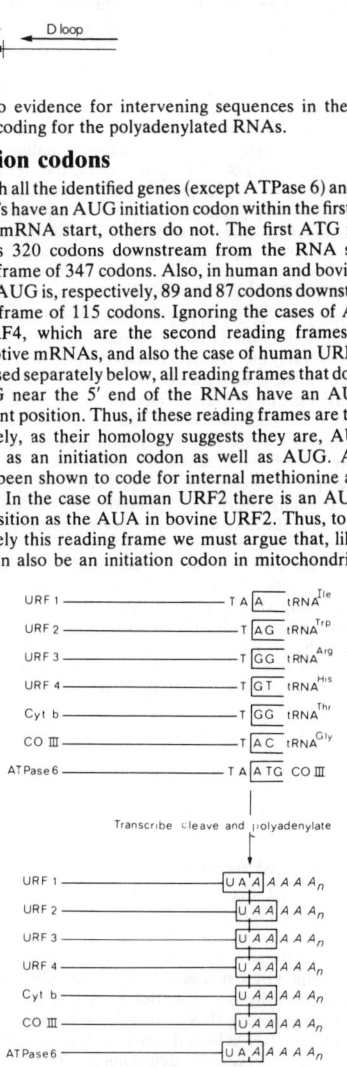

Fig. 3 The transcriptional processing and polyadenylation model for termination of translation.

Termination of translation

Over half of the presumptive mRNAs do not have UAA or UAG termination codons at the end of the reading frame (Table 2). In three cases, human COI and URF6 and bovine cytochrome *b*, termination probably occurs at AGA and AGG codons. These have been predicted to be termination codons and not arginine codons as in the universal genetic code[8]. This is based on the observation that only CGN arginine codons are used in all the reading frames, and AGA and AGG are only found at the ends of reading frames at the junction with the next gene or where the reading frame continues into the next gene. Also, no gene coding for a tRNA that could translate the codons AGA/G has been found. Finally, the composition of a C-terminal peptide fragment of bovine cytochrome *b* (W. Machleidt, personal communication) agrees with the prediction from the DNA sequence that this gene terminates in AGA.

To explain the lack of termination codons in most of the reading frames, we have proposed a model for the post-transcriptional creation of UAA termination codons in the mRNAs[49]. This model was based on the observation that if the tRNAs were precisely cleaved out of a primary transcript by an RNase P type enzyme(s)[50,51], mRNAs from those reading frames with no termination codon would be left with either a U or UA at their 3′ ends in phase with the reading frame. Thus, if these RNAs were then polyadenylated, a UAA termination codon would be created in-phase as shown in Fig. 3. Ojala *et al.*[17], in the accompanying article, have confirmed this hypothesis by sequence analysis of the 3′ ends of these mRNAs. This mechanism also provides the termination codon for ATPase 6. ATPase 6 and COIII have overlapping termination and initiation codons, that is UAAUG, and because the cleavage site is between the two A residues[18] the termination codon is destroyed by RNA processing of the primary transcript (Figs 1, 2). However, as there is no tRNA gene in this region, the signal in the RNA sequence for processing is not known.

Transcriptional processing

As discussed above, comparisons of the human DNA sequence with the polyadenylated RNA sequences[17,18] confirm the earlier predictions that the tRNA coding regions are involved in processing of the primary transcripts[15,16,49,52,53]. It is now clear that in the case of L-strand-coded tRNA genes the H-strand transcript is cleaved precisely at the 5′ and 3′ termini of the tRNA genes. Cleavage also seems to occur where the H-strand transcript has antisense, that is, H-strand-coded, tRNA genes. For example, cleavage of the H-strand transcript occurs between the antisense $tRNA^{Glu}$ gene and the cytochrome *b* gene (see Figs 1, 2). In both cases the cleavage point is a few bases downstream from the antisense tRNA gene. An additional cleavage point is between the ATPase 6 and COIII genes, where there is no known tRNA structure. Thus, except for this latter case it is reasonable to assume that in the primary H-strand transcript L-strand-coded tRNA sequences can fold up and be recognized by an RNase P-like activity. Antisense (H-strand-coded) tRNA coding regions also seem to be recognized by the RNA processing enzymes, although cleavage points have only been found on the 3′ side. With the exception the 12S rRNA and the tRNAs, the RNAs are then polyadenylated as discussed above; the tRNAs are matured by addition of CCA at their 3′ terminus. At some stage base modification of the rRNAs and tRNAs must also occur. It is likely that the H-strand-coded genes in the L-strand primary transcript are similarly processed but this has been less well studied.

Mitochondrial origins

In conclusion, it is now clear that the mammalian mitochondrial genetic system cannot generally be classified as either prokaryote-like or eukaryote-like. The mammalian mt genetic code is different from the so-called universal genetic code and is read in a unique fashion by a minimal set of mitochondrial tRNAs. The mt tRNAs generally lack a number of features previously found to be invariant among tRNAs from all other sources, and the mammalian mt rRNAs have distinctive structures only distantly related to all previously determined rRNA sequences[16]. Similarly, the economy with which the large initial mitochondrial H-strand transcripts seem to be processed is unprecedented. It is also striking that mammalian mitochondria are very different from other mitochondria. In yeast mitochondria, for example, not only is there a slightly different genetic code, but also, the genes are widely spaced and in a different order, and in some cases they contain intervening sequences[44,54]. These radical differences make it difficult to draw conclusions regarding mitochondrial evolution. Some form of endosymbiosis, involving the colonization of a primitive eukaryotic cell by a respiring bacteria-like organism, is an attractive hypothesis to explain the origins of mitochondria. However, the endosymbiont may have been no more closely related to current prokaryotes than to eukaryotes. It must also be borne in mind when making comparisons that the mechanism of selection and the selective pressures experienced by a captive cellular organelle may be quite different from those that affect a free-living organism.

We thank G. Attardi for a gift of HeLa cell mtDNA. B.A.R. was supported by NIH grants GM-21405 and GM-25962 and is a recipient of NIH Research Career Development Award GM-00124. P.H.S. was supported by the Deutsche Forschungsgemeinschaft through SFB-74 to B. Müller-Hill as well as a short-term EMBO fellowship. S.A. is a research fellow of the Cystic Fibrosis Foundation.

Received 26 January; accepted 18 February 1981.

1. Borst, P. *Trends biochem. Sci.* **2,** 31–34 (1977).
2. Barrell, B. G., Bankier, A. T. & Drouin, J. *Nature* **282,** 189–194 (1979).
3. Young, I. G. & Anderson, S. *Gene* **12,** 257–265 (1980).
4. Li, M. & Tzagoloff, A. *Cell* **18,** 47–53 (1979).
5. Macino, G., Coruzzi, G., Nobrega, F. G., Li, M. & Tzagoloff, A. *Proc. natn. Acad. Sci. U.S.A.* **76,** 3784–3785 (1979).
6. Fox, T. D. *Proc. natn. Acad. Sci. U.S.A.* **76,** 6534–6538 (1979).
7. Heckman, J. E., Sarnoff, J., Alzner DeWeerd, B., Yyn, S. & RajBhandary, U. L. *Proc. natn. Acad. Sci. U.S.A.* **77,** 3159–3163 (1980).
8. Barrell, B. G. *et al. Proc. natn. Acad. Sci. U.S.A.* **77,** 3164–3166 (1980).
9. Bonitz, S. G. *et al. Proc. natn. Acad. Sci. U.S.A.* **77,** 3167–3170 (1980).
10. Crews, S., Ojala, D., Posakony, J., Nishiguchi, J. & Attardi, G. *Nature* **277,** 192–198 (1979).
11. Gillum, A. M. & Clayton, D. A. *J. molec. Biol.* **135,** 353–368 (1979).
12. Martens, P. A. & Clayton, D. A. *J. molec. Biol.* **135,** 327–351 (1979).
13. Aloni, Y. & Attardi, G. *Proc. natn. Acad. Sci. U.S.A.* **68,** 1757–1761 (1971).
14. Murphy, W., Attardi, B., Tu, C. & Attardi, G. *J. molec. Biol.* **99,** 809–814 (1975).
15. Attardi, G. *et al.* in *Proc. ICN-UCLA Symp. Extrachromosomal DNA* (1979) (in the press).
16. Eperon, I. C., Anderson, S. & Nierlich, D. P. *Nature* **286,** 460–467 (1980).
17. Ojala, D., Montoya, J. & Attardi, G. *Nature* **290,** 470–474 (1981).
18. Montoya, J., Ojala, D. & Attardi, G. *Nature* **290,** 465–470 (1981).
19. Gillum, A. M. & Clayton, D. A. *Proc. natn. Acad. Sci. U.S.A.* **75,** 677–681 (1978).
20. Brown, W. M., Shine, J. & Goodman, H. M. *Proc. natn. Acad. Sci. U.S.A.* **75,** 735–739 (1978).
21. Bogenhagen, D. & Clayton, D. A. *J. molec. Biol.* **119,** 49–68 (1978).
22. Sekiya, T., Kobayashi, M., Seki, T. & Koike, K. *Gene* **11,** 53–62 (1980).
23. Upholt, W. B. & Dawid, I. B. *Cell* **11,** 571–583 (1977).
24. Berk, A. J. & Clayton, D. A. *J. molec. Biol.* **86,** 801–824 (1974).
25. Crews, S. & Attardi, G. *Cell* **19,** 775–784 (1980).
26. Van Etten, R. A., Walberg, M. W. & Clayton, D. A. *Cell* **22,** 157–170 (1980).
27. Baer, R. & Dubin, D. T. *Nucleic Acids Res.* **8,** 4927–4941 (1980).
28. Dubin, D. T., Timko, K. D. & Baer, R. J. *Cell* (in the press).
29. Cantatore, P. & Attardi, G. *Nucleic Acids Res.* **8,** 2605–2624 (1980).
30. Baer, R. & Dubin, D. T. *Nucleic Acids Res.* **8,** 4927–4941 (1980).
31. Staden, R. *Nucleic Acids Res.* **8,** 817–825 (1980).
32. Aujame, L. & Freeman, K. B. *Nucleic Acids Res.* **6,** 455–469 (1979).
33. Crick, F. H. C. *J. molec. Biol.* **19,** 548–555 (1966).
34. Lagerkvist, U. *Proc. natn. Acad. Sci. U.S.A.* **75,** 1759–1762 (1978).
35. Lynch, D. & Attardi, G. *J. molec. Biol.* **102,** 125–141 (1976).
36. Wrede, P., Woo, N. H. & Rich, A. *Proc. natn. Acad. Sci. U.S.A.* **76,** 3289–3293 (1979).
37. Kuchino, Y., Watanabe, S., Harada, F. & Nishimura, S. *Biochemistry* **19,** 2085–2089 (1980).
38. Fukada, K. & Abelson, J. *J. molec. Biol.* **139,** 377–391 (1980).
39. de Bruijn, M. H. L. *et al. Nucleic Acids Res.* **8,** 5213–5222 (1980).
40. Grauss, D. H., Grüter, F. & Sprinzl, M. *Nucleic Acids. Res.* **6,** r1–r19 (1979).
41. Steffens, G. J. & Buse, G. *Hoppe-Seyler's Z. physiol. Chem.* **360,** 613–619 (1979).
42. Thalenfeld, B. E. & Tzagoloff, A. *J. biol. Chem.* **255,** 6173–6180 (1980).
43. Macino, G. & Tzagoloff, A. *Cell* **20,** 507–517 (1980).
44. Nobrega, F. G. & Tzagoloff, A. *J. biol. Chem.* **255,** 9828–9837 (1980).
45. Sebald, W. & Wachter, E. in *29th Mossbacher Colloquium on Energy Conservation in Biological Membranes* (eds Schäfer, G. & Klingenberg, M.) 228–236 (Springer, Berlin, 1978).
46. Hensgens, L. A., Grivell, L. A., Borst, P. & Bos, J. L. *Proc. natn. Acad. Sci. U.S.A.* **76,** 1663–1667 (1979).
47. Macino, G. & Tzagoloff, A. *J. biol. Chem.* **254,** 4617–4623 (1979).
48. Ojala, D., Merkel, C., Gelfand, R. & Attardi, G. *Cell* **22,** 393–403 (1980).
49. Barrell, B. G. *et al.* in *31st Mosbach Colloquium* (eds Bücher, Th., Weiss, H. & Sebald, W.) (Springer, Berlin, in the press).
50. Robertson, M. D., Altman, S. & Smith, J. D. *J. biol. Chem.* **247,** 5243–5251 (1972).
51. Altman, S. & Smith, J. D. *Nature new Biol.* **233,** 35–39 (1971).
52. Attardi, G. *et al.* in *The Organisation and Expression of the Mitochondrial Genome* (eds Kroon, A. M. & Saccone, C.) (North-Holland, Amsterdam, in the press).
53. Battey, J. & Clayton, D. A. *J. biol. Chem.* **255,** 11599–11606 (1980).
54. Borst, P. & Grivell, L. A. *Cell* **15,** 705–723 (1978).
55. Staden, R. *Nucleic Acids Res.* **4,** 4037–4051 (1977); **5,** 1013–1015 (1978).
56. Drouin, J. *J. molec. Biol.* **140,** 15–34 (1980).
57. Smith, A. J. H. *Nucleic Acids Res.* **6,** 831–848 (1979).
58. Gronenborn, B. & Messing, J. *Nature* **272,** 375–377 (1978).
59. Schreier, P. H. & Cortese, R. *J. molec. Biol.* **129,** 169–172 (1979).
60. Sanger, F., Coulson, A. R., Barrell, B. G., Smith, A. J. H. & Roe, B. A. *J. molec. Biol.* **143,** 161–178 (1980).
61. Anderson, S., Gait, M. J., Mayol, L. & Young, I. G. *Nucleic Acids Res.* **8**, 1731–1743 (1980).
62. Sanger, F., Nicklen, S. & Coulson, A. R. *Proc. natn. Acad. Sci. U.S.A.* **74,** 5463–5467 (1977).
63. Sanger, F. & Coulson, A. R. *FEBS Lett.* **87,** 107–110 (1978).

Printed in Great Britain by Macmillan Production Ltd., Basingstoke, Hampshire

J. Mol. Biol. (1982) **156**, 683–717

Complete Sequence of Bovine Mitochondrial DNA

Conserved Features of the Mammalian Mitochondrial Genome

S. Anderson, M. H. L. de Bruijn, A. R. Coulson
I. C. Eperon†, F. Sanger and I. G. Young‡

Laboratory of Molecular Biology
The Medical Research Council Centre, Hills Road
Cambridge CB2 2QH, England

(Received 11 November 1981)

We present here the complete 16,338 nucleotide DNA sequence of the bovine mitochondrial genome. This sequence is homologous to that of the human mitochondrial genome (Anderson *et al.*, 1981) and the genes are organized in virtually identical fashion. The bovine mitochondrial protein genes are 63 to 79% homologous to their human counterparts, and most of the nucleotide differences occur in the third positions of codons. The minimum rate of base substitution that accounts for the nucleotide differences in the codon third positions is very high: at least 6×10^{-9} changes per position per year. The bovine and human mitochondrial transfer RNA genes exhibit more interspecies variation than do their cytoplasmic counterparts, with the "TΨC" loop being the most variable part of the molecule. The bovine 12 S and 16 S ribosomal RNA genes, when compared with those from human mitochondrial DNA, show conserved features that are consistent with proposed secondary structure models for the ribosomal RNAs. Unlike the pattern of moderate-to-high homology between the bovine and human mitochondrial DNAs found over most of the genome, the DNA sequence in the bovine D-loop region is only slightly homologous to the corresponding region in the human mitochondrial genome. This region is also quite variable in length, and accounts for the bulk of the size difference between the human and bovine mitochondrial DNAs.

1. Introduction

Mammalian mitochondrial genomes are duplex DNA circles about $16{\cdot}5 \times 10^3$ base-pairs in length. The complete DNA sequence of the human mitochondrial genome has been determined, and genes have been identified for the 12 S and 16 S ribosomal RNAs, 22 transfer RNAs, and several protein components of the inner mitochondrial membrane (Anderson *et al.*, 1981). The organization and expression of the human mitochondrial genome reflects a strikingly economical use of genetic material (Barrell *et al.*, 1980*a*; Anderson *et al.*, 1981). The genome has tRNA genes

† Present address: Department of Molecular Biophysics and Biochemistry, Yale University, New Haven, Conn., U.S.A.

‡ Present address: Department of Biochemistry, John Curtin School of Medical Research, The Australian National University, Canberra, Australia.

0022–2836/82/120683–35 $03.00/0

interspersed amongst the rRNA and protein coding sequences, and is very compact with relatively few intergenic nucleotides and no detectable introns (Ojala *et al.*, 1980; Battey & Clayton, 1980; Anderson *et al.*, 1981). The large initial mitochondrial transcripts (Aloni & Attardi, 1971; Murphy *et al.*, 1975) appear to be processed *via* precise endonucleolytic scissions to yield the messenger, ribosomal and transfer RNAs (Ojala *et al.*, 1981*a*; Montoya *et al.*, 1981). The human mitochondrial genetic code is different from the "universal" genetic code, in that UGA codes for tryptophan rather than termination, AUA for methionine and initiation rather than isoleucine, and AGA and AGG for termination rather than arginine (Barrell *et al.*, 1979,1980*a*,*b*; Anderson *et al.*, 1981). This code is composed of four termination codons and 22 codon "families", each of which is read by a single mitochondrial tRNA. For families containing two codons, the decoding mechanism relies on G·U base-pairing by the wobble position nucleotide in the tRNA. However, those tRNAs that read four-codon families all have a U in the wobble position, and hence apparently decode using U·N or "2-out-of-3" base-pairing (Barrell *et al.*, 1980*b*). The human mitochondrial tRNAs lack several "invariant" features found in all previously studied non-mitochondrial tRNAs (Anderson *et al.*, 1981), and likewise the 12 S and 16 S rRNAs exhibit only low homology to non-mitochondrial ribosomal RNAs (Eperon *et al.*, 1980).

The analysis of the human mitochondrial genome (Anderson *et al.*, 1981) relied heavily on data from the bovine mitochondrial genetic system. The amino acid sequences of the bovine cytochrome *c* oxidase subunits I, II and III, and apocytochrome *b* (J. E. Walker, personal communication; Steffens & Buse, 1979; G. Buse, personal communication; W. Machleidt, personal communication) enabled the genes for these proteins to be located in the bovine and human mtDNA† sequences. The pattern of codon conservation exhibited by the human and bovine genomes also allowed those long open reading frames ("URFs"), which did not correspond to any known mitochondrial proteins, to be identified as *bona fide* protein-coding genes (Anderson *et al.*, 1981; and see below). The existence of AUA methionine and UGA tryptophan codons in the mammalian mitochondrial genetic code was formally demonstrated using the bovine cytochrome *c* oxidase subunit II gene sequence (Young & Anderson, 1980), and the AUA start codon and AGA/AGG stop codons for the same code were inferred from a comparison of the human and bovine mtDNA sequences (Anderson *et al.*, 1981). Furthermore, although many of the mitochondrial tRNA genes could be located in the mtDNA sequence using computer methods to search for possible tRNA-like secondary structures (Staden, 1980*a*) or by visual inspection, formal proof that these represented active genes was provided by a comparison with the actual sequences of the 22 bovine mitochondrial tRNAs (B. A. Roe, personal communication). This test was especially crucial in the case of the $tRNA^{Ser}_{AGU/C}$ due to its anomalous structure (de Bruijn *et al.*, 1980; Arcari & Brownlee, 1980). Because of the high homology and virtually identical organization displayed by the human and bovine mitochondrial genomes, it was possible to combine the gene identification from the bovine genome with mRNA and rRNA mapping data from human mitochondria (Ojala *et al.*, 1980,1981*a*; Battey & Clayton, 1980; Montoya

† Abbreviations used: mtDNA, mitochondrial DNA; R, purine; Y, pyrimidine.

et al., 1981) to derive a comprehensive interpretation of gene expression in mammalian mitochondria (Anderson *et al.*, 1981).

The mammalian lines that ultimately gave rise to humans and cows diverged from each other about 80 million years ago (Wilson *et al.*, 1977). Since the rate of DNA sequence divergence is relatively fast for mammalian mitochondrial genomes (Upholt & Dawid, 1977; Brown *et al.*, 1979), those portions of the mtDNA sequence that have been conserved are likely to have been subject to function-related constraints during mammalian evolution. Functionally important higher order structures in those macromolecules coded for by mammalian mtDNA are also likely to have been conserved, so comparative analyses of mtDNA sequences can be used to derive secondary or tertiary structure models for mitochondrial rRNAs and tRNAs (Zweib *et al.*, 1981; H. F. Noller, personal communication; M. H. L. de Bruijn, unpublished results). Finally, the human and bovine mitochondrial genomes, because they code for gene products exhibiting moderate-to-high homology (Anderson *et al.*, 1981), can be used as a resource for statistical studies addressing such topics as mutation rate and genetic drift (Perler *et al.*, 1980; Kimura, 1981; Li *et al.*, 1981).

For the above reasons, we determined the complete sequence of the bovine mitochondrial genome, which is presented below. This sequence is compared with that of human mtDNA and used to identify those portions of the genome, and those parts of the various mitochondrially encoded macromolecules, that have been most conserved during mammalian evolution.

2. Materials and Methods

A clone library containing the entire bovine mitochondrial genome was prepared by digesting mtDNA from a single bovine heart with *Eco*RI and *Bam*HI and cloning the resultant fragments in multicopy plasmid vectors as described (Young and Anderson, 1980). Bovine mtDNA fragments were released from the recombinant plasmids by digestion with *Eco*RI and *Bam*HI, then purified by electrophoresis on low melting point agarose gels. Fragments recovered from the gel were subcloned into the vector M13mp2 or its derivatives (Messing *et al.*, 1977,1981; Gronenborn & Messing, 1978; Rothstein *et al.*, 1980) in order to prepare single-stranded template for sequencing by the dideoxynucleotide chain termination method (Sanger *et al.*, 1977; Sanger & Coulson, 1978). A "shotgun" DNA sequencing strategy was employed, in which M13 subclones were selected at random and the inserts sequenced using a flanking universal primer (Sanger *et al.*, 1980; Anderson *et al.*, 1980; Heidecker *et al.*, 1980; Gardner *et al.*, 1981; Anderson, 1981). The complete bovine mtDNA sequence was assembled from the individual subclone sequences using the programs BATIN, CONSEN, DBCOMP and DBUTIL (Staden, 1980*b*) to compile, edit and overlap the data. All recombinant DNA experiments were carried out in accordance with GMAG guidelines.

3. Results and Discussion

(a) *Genome organization*

The sequence of the bovine mitochondrial genome is given in Figure 1. The gene arrangement, shown schematically in Figure 2, is virtually identical to that found in the human mitochondrial genome (Anderson *et al.*, 1981).

As will be discussed below in detail, the bovine mitochondrial tRNA, rRNA and protein genes in general are highly homologous to their human counterparts; the major areas of sequence divergence between the two genomes occur in the intergenic regions. The stretch of DNA between the $tRNA^{Pro}$ and the $tRNA^{Phe}$ genes, which has no known coding function but which contains the D-loop region, is the most divergent portion of the mammalian mitochondrial genome (see below). The COII gene and the $tRNA^{Lys}$ gene are separated by three nucleotides in the bovine mitochondrial genome (Young & Anderson, 1980) but by 25 nucleotides in human mtDNA (Barrell *et al.*, 1979). In the human genome, this region contains a direct repeat of the sequence A-C-C-C-C-C-T-C-T, which is missing entirely from the bovine genome.

The human and bovine mitochondrial genomes also differ somewhat in the extent to which certain genes overlap. URF5 and URF6, which are adjacent genes encoded by opposite strands, are "butt-jointed" in the human mitochondrial genome but overlap by 17 nucleotides at their respective 3′ termini in bovine mtDNA. The $tRNA^{Cys}$ and $tRNA^{Tyr}$ genes, both encoded by the H-strand, overlap by one nucleotide in the human mitochondrial genome but are butt-jointed in the bovine mitochondrial genome. Hence, in bovine but not in human mitochondria, a single H-strand sense precursor transcript could be processed to yield both of these tRNAs. This contrasts with the arrangement of genes on the L-strand, the main coding strand, where in both bovine and human mitochondria all the tRNA, rRNA and protein-coding sequences could, in principle, be cut from a single precursor transcript (Ojala *et al.*, 1980; Anderson *et al.*, 1981). This may mean that in human mitochondria H-strand sense transcripts, which are produced at more than twice the rate that L-strand sense transcripts are produced (Cantatore & Attardi, 1980), need not be used with maximum efficiency. Alternatively, mitochondria may be able to "repair" truncated tRNAs post-transcriptionally by adding (prior to the CCA addition) one or more nucleotides to the 3′ end.

(b) *Protein genes*

(i) *Homology*

Like the human mitochondrial genome, the bovine mitochondrial genome contains 13 long open reading frames, including the five that have been identified as coding for known mitochondrially synthesized proteins (Figs 1 and 2; Anderson *et al.*, 1981). All of the L-strand sense reading frames correspond to the map positions of stable polyadenylated transcripts on HeLa cell mtDNA (Ojala *et al.*, 1980,1981*a*; Montoya *et al.*, 1981). Furthermore, when homologous human and bovine reading frames are aligned, a high proportion of the observed codon changes

---

FIG. 1. Sequence of bovine mitochondrial DNA. The sequence of the L-strand is shown, beginning in the D-loop region at a point analogous to the beginning of the previously published human mtDNA sequence (Anderson *et al.*, 1981). Major reading frames are shown translated into the 1-letter amino acid code, with asterisks designating termination codons. The genes for cytochrome *c* oxidase subunits I to III (CO I–III), ATPase 6 and apocytochrome *b* (cyt.*b*), as well as unassigned reading frames 1 to 5 (URFs 1 to 5), are L-strand sense: URF 6 is H-strand sense. The tRNA genes are shown boxed and the rRNA genes underlined. The sense of each gene is denoted by an arrow: →, L-strand; ←, H-strand. The base composition of the L-strand is 27·2% T, 25·9% C, 33·4% A and 13·5% G.

```
ACTAATGGCTAATCAGCCCATGCTCACACATAACTGTGCTGTCATACATTTGGTATTTTTTTATTTTGGGGGATGCTTGGACTCAGCTATGGCCGTCAAAGGCCCTGACCCGGAGCATCT
        10        20        30        40        50        60        70        80        90       100       110       120

ATTGTAGCTGGACTTAACTGCATCTTGAGCACCAGCATAATGATAAGCATGGACATTACAGTCAATGGTCACAGGACATAAATTATATTATATATCCCCCCTTCATAAAAATTTCCCCCT
       130       140       150       160       170       180       190       200       210       220       230       240

TAAATATCTACCACCACTTTTAACAGACTTTTCCCTAGATACTTATTTAAATTTTTCACGCTTTCAATACTCAATTTAGCACTCCAAACAAAGTCAATATATAAACGCAGGCCCCCCCCC
       250       260       270       280       290       300       310       320       330       340       350       360
 tRNA^Phe →                                                         12S rRNA →
CCCGTTGATGTAGCTTAACCCAAAGCAAGGCACTGAAAATGCCTAGATGAGTCTCCCAACTCCATAAACACATAGGTTTGGTCCCAGCCTTCCTGTTAACTCTTAATAAACTTACACATG
       370       380       390       400       410       420       430       440       450       460       470       480

CAAGCATCTACACCCCAGTGAGAATGCCCTCTAGGTTATTAAAACTAAGAGGAGCTGGCATCAAGCACACACCCTGTAGCTCACGACGCCTTGCTTAACCACACCCCACGGGAAACAGCA
       490       500       510       520       530       540       550       560       570       580       590       600

GTGACAAAAATTAAGCCATAAACGAAAGTTTGACTAAGTTATATTAATTAGGGTTGGTAAATCTCGTGCCAGCCACCGCGGTCATACGATTAACCCAAGCTAACAGGAGTACGGCGTAAA
       610       620       630       640       650       660       670       680       690       700       710       720

ACGTGTTAAAGCACCATACCAAATAGGGTTAAATTCTAACTAAGCTGTAAAAAGCCATGATTAAAATAAAAATAAATGACGAAAGTGACCCTACAATAGCCGACGCACTATAGCTAAGAC
       730       740       750       760       770       780       790       800       810       820       830       840

CCAAACTGGGATTAGATACCCCACTATGCTTAGCCCTAAACACAGATAATTACATAAACAAAATTATTCGCCAGAGTACTACTAGCAACAGCTTAAAACTCAAAGGACTTGGCGGTGCTT
       850       860       870       880       890       900       910       920       930       940       950       960

TATATCCTTCTAGAGGAGCCTGTTCTATAATCGATAAACCCCGATAAACCTCACCAATTCTTGCTAATACAGTCTATATACCGCCATCTTCAGCAAACCCTAAAAAGGAAAAAAAGTAAG
       970       980       990      1000      1010      1020      1030      1040      1050      1060      1070      1080

CGTAATTATGATACATAAAAACGTTAGGTCAAGGTGTAACCTATGAAATGGGAAGAAATGGGCTACATTCTCTACACCAAGAGAATCAAGCACGAAAGTTATTATGAAACCAATAACCAA
      1090      1100      1110      1120      1130      1140      1150      1160      1170      1180      1190      1200

AGGAGGATTTAGCAGTAAACTAAGAATAGAGTGCTTAGTTGAATTAGGCCATGAAGCACGCACACACCGCCCGTCACCCTCCTCAAATAGATTCAGTGCATCTAACCCTATTTAAACGCA
      1210      1220      1230      1240      1250      1260      1270      1280      1290      1300      1310      1320
                                                                  tRNA^Val →
CTAGCTACATGAGAGGAGACAAGTCGTAACAAGGTAAGCATACTGGAAAGTGTGCTTGGATAAATCAAGATATAGCTTAAACAAAGCATCCAGTTTACACCTAGAAGACTTCATTCATTA
      1330      1340      1350      1360      1370      1380      1390      1400      1410      1420      1430      1440
               16S rRNA →
TGAATATCTTGAACTAGACCTAGCCCAAAGATACCCTCTCGACTAAACAACCAAGATAGAATAAAACAAAACATTTAATCCCAATTTAAAGTATAGGAGATAGAAATCTAAGTACGGCGC
      1450      1460      1470      1480      1490      1500      1510      1520      1530      1540      1550      1560

TATAGAGAAAGTACCGCAAGGGAACGATGAAAGAAAAAAACTAAAAGTATAAAAAAGCAAAGATTACCCCTTGTACCTTTTGCATAATGAATTAACTAGTATAAGACTTAACAAAATGAA
      1570      1580      1590      1600      1610      1620      1630      1640      1650      1660      1670      1680

TTTTAGCTAAGCAGCCCGAAACCAGACGAGCTACTCACAAACAGTTTACCAAGAACTAACTCATCTATGTGGCAAAATAGTGAGAAGATTTGTAAGTAGAGGTGACATGCCTAACGAGCC
      1690      1700      1710      1720      1730      1740      1750      1760      1770      1780      1790      1800

TGGTGATAGCTGGTTGTCCAGAAAATGAATCTAAGTTCAGCTTTAAAGATACCAAAAATTCAAATAAACCCCACTGTAGCTTTAAAAGTTAGTCTAAAAAGGTACAGCCTTTTAGAAACG
      1810      1820      1830      1840      1850      1860      1870      1880      1890      1900      1910      1920

GATACAACCTTGACTAGAGAGTAAAATTTAACACTACCATAGTAGGCCTAAAAGCAGCCATCAATTAAGAAAGCGTTAAAGCTCAACAACAAAAATTAAATAGATTCCAACAACAAATGA
      1930      1940      1950      1960      1970      1980      1990      2000      2010      2020      2030      2040

TTAACTCCTAGCCCCAATACTGGACTAATCTATTATAGAATAGAAGCAATAATGTTAATATGAGTAACAAGAAAAATTTTCTCCTTGCATAAGTCTAAGTCAGTGCCTGATAATACTCTG
      2050      2060      2070      2080      2090      2100      2110      2120      2130      2140      2150      2160

ACCACTAACAGTCAATAAAAATAATCCAACAATAAACAATTTATTGATTATACTGTTAACCCAACACAGGAGTGCATCTAAGGAAAGATTAAAAGAAGTAAAAGGAACTCGGCAAACACA
      2170      2180      2190      2200      2210      2220      2230      2240      2250      2260      2270      2280

AACCCCGCCTGTTTACCAAAAACATCACCTCCAGCATTCCCAGTATTGGAGGCATTGCCTGCCCAGTGACAACTGTTTAACGGCCGCGGTATCCTGACCGTGCAAAGGTAGCATAATCAT
      2290      2300      2310      2320      2330      2340      2350      2360      2370      2380      2390      2400

TTGTTCTCTAAATAAGGACTTGTATGAATGGCCGCACGAGGGTTTTACTGTCTCTTACTTCCAATCAGTGAAATTGACCTTCCCGTGAAGAGGCGGGAATGCACAAATAAGACGAGAAGA
      2410      2420      2430      2440      2450      2460      2470      2480      2490      2500      2510      2520

CCCTATGGAGCTTTAACTAACCAACCCAAAGAGAATAGATTTAACCATTAAGGAATAACAACAATCTCCATGAGTTGGTAGTTTCGGTTGGGGTGACCTCGGAGAATAAAAAATCCTCCG
      2530      2540      2550      2560      2570      2580      2590      2600      2610      2620      2630      2640

AGCGATTTTAAAGACTAGACCCACAAGTCAAATCACTCTATCGCTCATTGATCCAAAAACTTGATCAACGGAACAAGTTACCCTAGGGATAACAGCGCAATCCTATTCAAGAGTCCATAT
      2650      2660      2670      2680      2690      2700      2710      2720      2730      2740      2750      2760

CGACAATAGGGTTTACGACCTCGATGTTGGATCAGGACATCCTGATGGTGCAACCGCTATCAAAGGTTCGTTTGTTCAACGATTAAAGTCCTACGTGATCTGAGTTCAGACCGGAGTAAT
      2770      2780      2790      2800      2810      2820      2830      2840      2850      2860      2870      2880

CCAGGTCGGTTTCTATCTATTACGTATTTCTCCCAGTACGAAAGGACAAGAGAAATAAGGCCAACTTTAAATCAAGCGCCTTAAGACAACCAATGATAACATCTCAACTGACAACACAAA
      2890      2900      2910      2920      2930      2940      2950      2960      2970      2980      2990      3000
                                                                                                     U.R.F. 1 →
                   tRNA^Leu(UUR) →                                                                   M  F  M  I  N  I  L
ACCCTGCCCTAGAACAGGGCTTAGTTAAGGTGGCAGAGCCCGGTAATTGCATAAAACTTAAACTTTTATATCCAGAGATTCAAATCCTCTCCTTAACAAAATGTTCATAATTAACATCTT
      3010      3020      3030      3040      3050      3060      3070      3080      3090      3100      3110      3120
```

FIG. 1 *(1st part)*

M L I I P I L L A V A F L T L V E R K V L G Y M Q L R K G P N V V G P Y G L L Q

AATACTAATTATTCCCATCCTATTGGCCGTAGCATTCCTTACGTTAGTGGAACGAAAAGTTCTAGGCTATATACAACTCCGAAAAGGTCCAAATGTCGTAGGTCCATATGGCCTACTTCA

3130 3140 3150 3160 3170 3180 3190 3200 3210 3220 3230 3240

P I A D A I K L F I K E P L R P A T S S A S M F I L A P I M A L G L A L T M W I

ACCCATCGCCGATGCAATCAAACTTTTCATTAAAGAACCACTACGACCCGCTACATCTTCAGCCTCAATATTTATCCTAGCACCTATCATAGCTTTAGGCCTAGCCTTAACCATGTGAAT

3250 3260 3270 3280 3290 3300 3310 3320 3330 3340 3350 3360

P L P M P Y P L I N M N L G V L F M L A M S S L A V Y S I L W S G W A S N S K Y

TCCCCTACCAATACCCTATCCTCTTATCAACATAAACCTAGGAGTCCTATTTATACTAGCCATATCAAGCCTAGCCGTATACTCCATTCTCTGATCAGGCTGAGCTTCCAACTCAAAATA

3370 3380 3390 3400 3410 3420 3430 3440 3450 3460 3470 3480

A L I G A L R A V A Q T I S Y E V T L A I I L L S V L L M S G S F T L S T L I T

CGCACTAATCGGAGCCCTACGAGCAGTAGCACAAACAATCTCATACGAAGTAACGCTAGCAATTATCCTGTTATCAGTACTCCTAATAAGTGGGTCCTTTACCCTCTCCACATTAATTAC

3490 3500 3510 3520 3530 3540 3550 3560 3570 3580 3590 3600

T Q E Q M W L I L P A W P L A M M W F I S T L A E T N R A P F D L T E G E S E L

TACACAAGAACAAATATGGTTAATCCTCCCAGCATGGCCTCTAGCAATAATATGATTTATCTCAACACTAGCAGAAACAAACCGAGCTCCATTTGATTTAACTGAAGGAGAATCAGAGCT

3610 3620 3630 3640 3650 3660 3670 3680 3690 3700 3710 3720

V S G F N V E Y A A G P F A L F F M A E Y A N I I M M N I F T A I L F L G T S H

AGTCTCGGGCTTCAACGTAGAATATGCAGCAGGACCATTTGCCCTCTTCTTCATAGCAGAGTACGCAAATATTATCATAATAAATATCTTTACAGCAATTTTATTCCTAGGAACATCCCA

3730 3740 3750 3760 3770 3780 3790 3800 3810 3820 3830 3840

N P H M P E L Y T I N F T I K S L L L T M S F L W I R A S Y P R F R Y D Q L M H

CAATCCACACATACCAGAACTCTACACAATCAATTTTACCATTAAATCCCTACTGCTCACAATATCCTTCCTATGAATCCGAGCATCCTACCCTCGATTTCGCTATGACCAACTAATACA

3850 3860 3870 3880 3890 3900 3910 3920 3930 3940 3950 3960

L L W K N F L P L T L A L C M W H V S L P I L T S G I P P Q T tRNA^Ile →

CTTACTATGAAAAAATTTTCTACCTCTGACACTAGCCCTGTGCATGTGACACGTATCCCTACCCATCCTTACATCAGGCATCCCACCACAAACATAAGAAATATGTCTGACAAAAGAGTT

3970 3980 3990 4000 4010 4020 4030 4040 4050 4060 4070 4080

← tRNA^Gln →

ACTTTGATAGAGTAAATAATAGAGGTTCAAACCCTCTTATTTCTAGAACTATAGGAATCGAACCTACTCCTAAGAATCCAAAACTCTTCGTGCTCCCAATTACACCAAATTCTATTAGTA

4090 4100 4110 4120 4130 4140 4150 4160 4170 4180 4190 4200

tRNA^f-Met →   U.R.F. 2 →

M N P I I F I I I L L T I M L G T I I

AGGTCAGCTAATTAAGCTATCGGGCCCATACCCCGAAAATGTTGGTTTATATCCTTCCCGTACTAATAAACCCAATTATCTTTATTATTATTCTACTAACCATTATACTAGGAACTATTA

4210 4220 4230 4240 4250 4260 4270 4280 4290 4300 4310 4320

V M I S S H W L L V W L G F E M N M L A I I P I M M K N H N P R A T E A S T K Y

TTGTCATAATCAGTTCTCACTGACTACTTGTCTGAATCGGGTTTGAAATAAATATACTCGCCATCATCCCCATCATAATAAAAAATCACAACCCACGAGCTACAGAAGCATCAACTAAAT

4330 4340 4350 4360 4370 4380 4390 4400 4410 4420 4430 4440

F L T Q S T A S M L L M M A V I I N L M F S G Q W T V M K L F N P M A S M L M T

ATTTTTTGACTCAATCAACAGCCTCAATACTACTAATAATAGCCGTCATCATTAACCTAATATTCTCAGGCCAATGAACCGTAATAAAACTATTTAACCCAATAGCCTCAATACTTATAA

4450 4460 4470 4480 4490 4500 4510 4520 4530 4540 4550 4560

M A L A M K L G M A P F H F W V P E V T Q G I P L S S G L I L L T W Q K L A P M

CGATAGCCCTAGCTATAAAACTAGGAATAGCCCCATTTCACTTCTGAGTCCCAGAAGTAACACAGGGCATCCCCCTATCCTCAGGCCTTATCCTACTGACATGACAAAAACTAGCACCTA

4570 4580 4590 4600 4610 4620 4630 4640 4650 4660 4670 4680

S V L Y Q I F P S I N L N L I L T L S V L S I L I G G W G G L N Q T Q L R K I M

TATCTGTACTTTACCAAATCTTCCCATCAATTAACCTAAACTTAATTCTAACCCTATCAGTTTTATCAATCCTAATTGGAGGCTGAGGGGGACTAAACCAAACACAACTCCGAAAAATCA

4690 4700 4710 4720 4730 4740 4750 4760 4770 4780 4790 4800

A Y S S I A H M G W M T A V L P Y N P T M T L L N L I I Y I I M T S T M F T M F

TAGCCTACTCATCAATCGCTCATATAGGCTGAATAACAGCAGTACTACCATATAACCCCACCATAACATTGCTAAACTTAATTATCTATATCATTATAACTTCCACCATATTTACCATAT

4810 4820 4830 4840 4850 4860 4870 4880 4890 4900 4910 4920

M A N S T T T T L S L S H T W N K T P I M T V L I L A T L L S M G G L P P L S G

TTATAGCCAATTCCACCACCACTACCCTGTCATTATCACACACATGAAATAAAACACCCATTATAACCGTCCTAATTCTTGCCACTCTCCTATCCATAGGAGGACTCCCTCCCCTATCTG

4930 4940 4950 4960 4970 4980 4990 5000 5010 5020 5030 5040

F M P K W M I I Q E M T K N N S I I L P T F M A I T A L L N L Y F Y M R L T Y S

GGTTTATACCAAAATGAATAATCATCCAAGAGATAACAAAAAATAACAGCATCATTCTACCCACTTTCATAGCAATCACAGCTCTACTAAACTTATATTTTTATATACGACTCACGTATT

5050 5060 5070 5080 5090 5100 5110 5120 5130 5140 5150 5160

T T L T M F P S T N N M K M K W Q F P L M K K M T F L P T M V V L S T M M L P L

CTACCACACTAACAATATTTCCCTCCACAAACAACATAAAAATAAAATGACAATTTCCCCTTATGAAAAAAATAACTTTTCTACCAACAATAGTCGTATTATCTACCATAATACTACCAC

5170 5180 5190 5200 5210 5220 5230 5240 5250 5260 5270 5280

FIG. 1 *(cont.)*

```
 T  P  M  L  S  V  L  E   tRNA^Trp →
TCACGCCAATACTATCAGTGTTAGAATAGGAATTTAGGTTAAACAGACCAAGAGCCTTCAAAGCCCTAAGCAAGTACAATTTACTTAATTCCTGATAAGGATTGCAAGACTACACCTTAC
      5290      5300      5310      5320      5330      5340      5350      5360      5370      5380      5390      5400

                         ← tRNA^Ala                                                                            ← tRNA^Asn
ATCAATTGAATGCAAATCAACCACTTTAATTAAGCTAAATCCTCACTAGACTGGTGGGCTCCACCCCCACGAAACTTTAGTTAACAGCTAAACACCCTAGCTAACTGGCTTCAATCTACT
      5410      5420      5430      5440      5450      5460      5470      5480      5490      5500      5510      5520

   Origin of L strand
      replication                                                                   ← tRNA^Cys
TCTCCCGCCGCAAGAAAAAAAAGGCGGGAGAAGCCCCGGCAGAATTGAAGCTGCTTCTCTGAATTTGCAATTCAACGTGTAAATTCACCACAGGGCTTGGTAAAAAGAGGAGTCAAACCT
      5530      5540      5550      5560      5570      5580      5590      5600      5610      5620      5630      5640

                                                COI →
                         ← tRNA^Tyr   M  F  I  N  R  W  L  F  S  T  N  H  K  D  I  G  T  L  Y  L  L  F  G  A  W
CTATCTTTAGATTTACAGTCTAATGCTTTGCTCAGCCATTTTACCCATGTTCATTAACCGCTGACTATTCTCAACCAACCATAAAGATATTGGTACCCTTTATCTACTATTTGGTGCTTG
      5650      5660      5670      5680      5690      5700      5710      5720      5730      5740      5750      5760

 A  G  M  V  G  T  A  L  S  L  L  I  R  A  E  L  G  Q  P  G  T  L  L  G  D  D  Q  I  Y  N  V  V  V  T  A  H  A  F  V  M
GGCCGGTATAGTAGGAACAGCTCTAAGCCTTCTAATTCGCGCTGAATTAGGCCAACCCGGAACTCTGCTCGGAGACGACCAAATCTACAACGTAGTTGTAACCGCACACGCATTTGTAAT
      5770      5780      5790      5800      5810      5820      5830      5840      5850      5860      5870      5880

 I  F  F  M  V  M  P  I  M  I  G  G  F  G  N  W  L  V  P  L  M  I  G  A  P  D  M  A  F  P  R  M  N  N  M  S  F  W  L  L
AATCTTCTTCATAGTAATACCAATCATAATTGGAGGATTCGGTAACTGACTTGTTCCCCTAATAATTGGTGCTCCCGATATAGCATTTCCCCGAATAAATAATATAAGCTTCTGACTCCT
      5890      5900      5910      5920      5930      5940      5950      5960      5970      5980      5990      6000

 P  P  S  F  L  L  L  L  A  S  S  M  V  E  A  G  A  G  T  G  W  T  V  Y  P  P  L  A  G  N  L  A  H  A  G  A  S  V  D  L
CCCTCCCTCATTCCTACTACTCCTCGCATCCTCTATAGTTGAAGCTGGGGCAGGAACAGGCTGAACCGTGTACCCTCCCTTAGCAGGCAACCTAGCCCATGCAGGAGCTTCAGTAGATCT
      6010      6020      6030      6040      6050      6060      6070      6080      6090      6100      6110      6120

 T  I  F  S  L  H  L  A  G  V  S  S  I  L  G  A  I  N  F  I  T  T  I  I  N  M  K  P  P  A  M  S  Q  Y  Q  T  P  L  F  V
AACCATTTTCTCTTTACACTTAGCAGGAGTTTCCTCAATTTTAGGAGCCATCAACTTCATTACAACAATTATCAACATAAAGCCCCCCGCAATGTCACAATACCAAACCCCTCTGTTCGT
      6130      6140      6150      6160      6170      6180      6190      6200      6210      6220      6230      6240

 W  S  V  M  I  T  A  V  L  L  L  L  S  L  P  V  L  A  A  G  I  T  M  L  L  T  D  R  N  L  N  T  T  F  F  D  P  A  G  G
ATGATCCGTAATAATTACCGCCGTACTACTACTACTCTCGCTCCCTGTATTAGCAGCCGGCATCACAATGCTATTAACAGACCGGAACCTAAATACAACCTTCTTCGACCCGGCAGGAGG
      6250      6260      6270      6280      6290      6300      6310      6320      6330      6340      6350      6360

 G  D  P  I  L  Y  Q  H  L  F  W  F  F  G  H  P  E  V  Y  I  L  I  L  P  G  F  G  M  I  S  H  I  V  T  Y  Y  S  G  K  K
AGGAGACCCTATTCTATATCAACACTTATTCTGATTCTTTGGACACCCCGAAGTCTATATTTTAATCTTACCTGGGTTTGGAATAATCTCTCATATCGTGACCTACTACTCAGGAAAAAA
      6370      6380      6390      6400      6410      6420      6430      6440      6450      6460      6470      6480

 E  P  F  G  Y  M  G  M  V  W  A  M  M  S  I  G  F  L  G  F  I  V  W  A  H  H  M  F  T  V  G  M  D  V  D  T  R  A  Y  F
AGAACCATTCGGATATATGGGAATAGTTTGGGCTATAATGTCAATCGGATTTCTAGGTTTCATCGTATGAGCCCACCATATATTCACTGTCGGAATAGACGTCGACACACGAGCCTACTT
      6490      6500      6510      6520      6530      6540      6550      6560      6570      6580      6590      6600

 T  S  A  T  M  I  I  A  I  P  T  G  V  K  V  F  S  W  L  A  T  L  H  G  G  N  I  K  W  S  P  A  M  M  W  A  L  G  F  I
CACATCAGCCACTATAATTATTGCTATTCCAACCGGGGTAAAAGTCTTCAGCTGATTGGCAACACTTCATGGAGGTAATATCAAATGGTCTCCTGCTATAATGTGAGCCCTAGGCTTTAT
      6610      6620      6630      6640      6650      6660      6670      6680      6690      6700      6710      6720

 F  L  F  T  V  G  G  L  T  G  I  V  L  A  N  S  S  L  D  I  V  L  H  D  T  Y  Y  V  V  A  H  F  H  Y  V  L  S  M  G  A
TTTCTTATTTACAGTAGGGGGTTTAACTGGAATTGTCTTAGCCAACTCTTCCCTCGATATTGTTCTTCACGACACATACTACGTTGTCGCACATTTCCACTATGTTTTATCAATAGGAGC
      6730      6740      6750      6760      6770      6780      6790      6800      6810      6820      6830      6840

 V  F  A  I  M  G  G  F  V  H  W  F  P  L  F  S  G  Y  T  L  N  D  T  W  A  K  I  H  F  A  I  M  F  V  G  V  N  M  T  F
TGTATTTGCTATTATAGGGGGATTTGTTCATTGATTCCCACTATTCTCAGGTTATACTCTCAACGATACATGAGCCAAAATCCACTTCGCAATTATATTTGTAGGCGTCAATATAACCTT
      6850      6860      6870      6880      6890      6900      6910      6920      6930      6940      6950      6960

 F  P  Q  H  F  L  G  L  S  G  M  P  R  R  Y  S  D  Y  P  D  A  Y  T  M  W  N  T  I  S  S  M  G  S  F  I  S  L  T  A  V
CTTCCCACAACACTTTCTAGGACTATCTGGCATGCCTCGACGATACTCCGACTACCCAGATGCATACACAATATGAAATACTATCTCATCAATAGGCTCATTCATTTCCCTAACAGCAGT
      6970      6980      6990      7000      7010      7020      7030      7040      7050      7060      7070      7080

 M  L  M  V  F  I  I  W  E  A  F  A  S  K  R  E  V  L  T  V  D  L  T  T  T  N  L  E  W  L  N  G  C  P  P  P  Y  H  T  F
TATACTAATAGTTTTCATCATCTGAGAAGCATTTGCATCTAAACGAGAAGTCTTGACTGTAGACTTAACCACGACAAATCTAGAATGATTAAACGGATGCCCTCCACCATATCACACATT
      7090      7100      7110      7120      7130      7140      7150      7160      7170      7180      7190      7200

 E  E  P  T  Y  V  N  L  K  *                                                      ← tRNA^Ser(UCN)    tRNA^Asp →
TGAAGAACCCACCTATGTTAACCTAAAATAAGAAAGGAAGGAATCGAACCCCCTACTATTGGTTTCAAGCCAACATCATAACCTCTATGTCTCTCTCAATAAACGAGGTGTTAGTAAAAC
      7210      7220      7230      7240      7250      7260      7270      7280      7290      7300      7310      7320

                                                     COII →
                                                     M  A  Y  P  M  Q  L  G  F  Q  D  A  T  S  P  I  M  E  E  L  L  H  F
ATTATATAATTTTGTCAAAGTTAAGTTACAAGTGAAAGTCCTGTACACCTCATATGGCATATCCCATACAACTAGGATTCCAAGATGCAACATCACCAATCATAGAAGAACTACTTCACT
      7330      7340      7350      7360      7370      7380      7390      7400      7410      7420      7430      7440
```

FIG. 1. *(cont.)*

```
 H D H T L M I V F L I S S L V L Y I I S L M L T T K L T H T S T M D A Q E V E T
TTCATGACCACACGCTAATAATTGTCTTCTTAATTAGCTCATTAGTACTTTACATTATTTCACTAATACTAACGACAAAGCTGACCCATACAAGCACGATAGATGCACAAGAAGTAGAGA
     7450      7460      7470      7480      7490      7500      7510      7520      7530      7540      7550      7560

 I W T I L P A I I L I L I A L P S L R I L Y M M D E I N N P S L T V K T M G H Q
CAATCTGAACCATTCTGCCCGCCATCATCTTAATTCTAATTGCTCTTCCTTCTTTACGAATTCTATACATAATAGATGAAATCAATAACCCATCTCTTACAGTAAAAACCATAGGACATC
     7570      7580      7590      7600      7610      7620      7630      7640      7650      7660      7670      7680

 W Y W S Y E Y T D Y E D L S F D S Y M I P T S E L K P G E L R L L E V D N R V V
AGTGATACTGAAGCTATGAGTATACAGATTATGAGGACTTAAGCTTCGACTCCTACATAATTCCAACATCAGAATTAAAGCCAGGGGAGCTACGACTATTAGAAGTCGATAATCGAGTTG
     7690      7700      7710      7720      7730      7740      7750      7760      7770      7780      7790      7800

 L P M E M T I R M L V S S E D V L H S W A V P S L G L K T D A I P G R L N Q T T
TACTACCAATAGAAATAACAATCCGAATGTTAGTCTCCTCTGAAGACGTATTACACTCATGAGCTGTGCCCTCTCTAGGACTAAAAACAGACGCAATCCCAGGCCGTCTAAACCAAACAA
     7810      7820      7830      7840      7850      7860      7870      7880      7890      7900      7910      7920

 L M S S R P G L Y Y G Q C S E I C G S N H S F M P I V L E L V P L K Y F E K W S
CCCTTATATCGTCCCGTCCAGGCTTATATTACGGTCAATGCTCAGAAATTTGCGGGTCAAACCACAGTTTCATACCCATTGTCCTTGAGTTAGTCCCACTAAAGTACTTTGAAAAATGAT
     7930      7940      7950      7960      7970      7980      7990      8000      8010      8020      8030      8040

                                                                                            U.R.F. A6L ——→
                                                                                             M P Q L D T S T W L T
 A S M L *  tRNA^Lys ——→
CTGCGTCAATATTATAAAAT[CACTAAGAAGCTATATAGCACTAACCTTTTAAGTTAGAGATTGAGAGCCATATACTCTCCTTGGTGA]CATGCCGCAACTAGACACGTCAACATGACTGAC
     8050      8060      8070      8080      8090      8100      8110      8120      8130      8140      8150      8160

 M I L S M F L T L F I I F Q L K V S K H N F Y H N P E L T P T K M L K Q N T P W
AATGATCTTATCAATATTCTTGACCCTTTTTATCATCTTTCAACTAAAAGTTTCAAAACACAACTTTTATCACAATCCAGAACTGACACCAACAAAAATATTAAAACAAAACACCCCTTG
     8170      8180      8190      8200      8210      8220      8230      8240      8250      8260      8270      8280

    ATPase 6 ——→
     M N E N L F T S F I T P V I L G L P L V T L I V L F P S L L F P T S N R L
 E T K W T K I Y L P L L L P L *
AGAAACAAAATGAACGAAAATTTATTTACCTCTTTTATTACCCCTGTAATTTTAGGTCTCCCTCTCGTAACCCTTATCGTACTATTCCCAAGCCTACTATTCCCAACATCAAACCGACTA
     8290      8300      8310      8320      8330      8340      8350      8360      8370      8380      8390      8400

 V S N R F V T L Q Q W M L Q L V S K Q M M S I H N S K G Q T W T L M L M S L I L
GTAAGCAATCGCTTTGTAACCCTCCAACAATGAATACTTCAACTTGTATCAAAACAAATAATGAGTATCCACAATTCTAAAGGACAAACATGAACATTAATATTAATATCTCTGATCCTA
     8410      8420      8430      8440      8450      8460      8470      8480      8490      8500      8510      8520

 F I G S T N L L G L L P H S F T P T T Q L S M N L G M A I P L W A G A V I T G F
TTTATTGGATCAACAAACCTACTAGGCCTATTACCCCATTCATTCACACCAACAACACAACTATCAATAAACCTAGGCATAGCCATCCCCCTGTGAGCAGGAGCCGTAATTACAGGATTC
     8530      8540      8550      8560      8570      8580      8590      8600      8610      8620      8630      8640

 R N K T K A S L A H F L P Q G T P T P L I P M L V I I E T I S L F I Q P M A L A
CGCAATAAAACTAAAGCATCACTTGCCCATTTCTTACCACAAGGAACACCCACTCCACTAATCCCAATACTAGTAATTATTGAAACTATCAGCCTTTTTATTCAACCTATAGCCCTCGCC
     8650      8660      8670      8680      8690      8700      8710      8720      8730      8740      8750      8760

 V R L T A N I T A G H L L I H L I G G A T L A L M S I S T T T A L I T F T I L I
GTGCGGTTAACAGCTAACATCACTGCAGGACACCTATTAATTCACCTAATCGGAGGAGCTACACTTGCACTAATAAGCATTAGCACTACAACAGCTCTAATTACATTCACCATTCTAATC
     8770      8780      8790      8800      8810      8820      8830      8840      8850      8860      8870      8880

 L L T I L E F A V A M I Q A Y V F T L L V S L Y L H D N T
                                                                                              COIII ——→
                                                                                             M T H Q T H A Y H M V
CTACTAACAATTCTAGAGTTTGCAGTAGCTATAATCCAAGCCTATGTATTCACTCTCCTAGTCAGCCTATATCTGCATGACAACACATAATGACACACCAAACTCATGCTTATCATATAG
     8890      8900      8910      8920      8930      8940      8950      8960      8970      8980      8990      9000

 N P S P W P L T G A L S A L L M T S G L T M W F H F N S M T L L M I G L T T N M
TAAACCCAAGCCCTTGACCTCTTACAGGAGCTTTGTCTGCCCTCTTAATAACATCCGGCCTAACCATGTGATTTCACTTTAACTCAATGACCCTGCTAATAATTGGCCTAACAACAAATA
     9010      9020      9030      9040      9050      9060      9070      9080      9090      9100      9110      9120

 L T M Y Q W W R D V I R E S T F Q G H H T P A V Q K G L R Y G M I L F I I S E V
TACTAACAATATACCAATGATGACGAGATGTTATCCGAGAAAGCACCTTCCAAGGGCACCATACCCCAGCTGTCCAAAAAGGCCTCCGTTATGGAATAATTCTTTTTATTATCTCCGAAG
     9130      9140      9150      9160      9170      9180      9190      9200      9210      9220      9230      9240

 L F F T G F F W A F Y H S S L A P T P E L G G C W P P T G I H P L N P L E V P L
TACTATTCTTTACCGGATTTTTCTGAGCTTTCTACCACTCAAGCCTCGCCCCCACCCCTGAACTAGGCGGCTGCTGACCCCCAACAGGCATTCACCCACTAAACCCCCTAGAAGTCCCAC
     9250      9260      9270      9280      9290      9300      9310      9320      9330      9340      9350      9360

 L N T S V L L A S G V S I T W A H H S L M E G D R K H M L Q A L F I T I T L G V
TGCTCAACACCTCTGTCCTATTGGCTTCCGGAGTTTCTATTACCTGAGCCCATCATAGTTTAATAGAAGGGGACCGAAAGCATATATTACAAGCCCTATTTATCACCATCACATTAGGAG
     9370      9380      9390      9400      9410      9420      9430      9440      9450      9460      9470      9480
```

FIG. 1. *(cont.)*

Y F T L L Q A S E Y Y E A P F T I S D G V Y G S T F F V A T G F H G L H V I I G

TCTACTTCACACTACTACAAGCCTCAGAATACTATGAAGCACCTTTTACTATCTCCGACGGAGTTTACGGCTCAACTTTTTTTGTAGCCACAGGCTTCCACGGCCTCCACGTCATCATTG

9490 9500 9510 9520 9530 9540 9550 9560 9570 9580 9590 9600

S T F L I V C F F R Q L K F H F T S N H H F G F E A G A W Y W H F V D V V W L F

GGTCCACCTTCTTAATTGTCTGCTTCTTCCGCCAATTAAAATTTCATTTTACTTCTAACCACCACTTCGGCTTTGAAGCCGGTGCCTGATACTGACATTTCGTAGACGTAGTCTGACTTT

9610 9620 9630 9640 9650 9660 9670 9680 9690 9700 9710 9720

U.R.F. 3 →

M N L M L A

L Y V S I Y W W G S tRNA^{Gly} →

TCCTCTATGTTTCTATCTATTGATGAGGCTCCTATTCTTTTAGTATTAACTAGTACAGCTGACTTCCAATCAGCTAGTTTCGGTCTAGTCCGAAAAAGAATAATAAATTTAATACTAGCC

9730 9740 9750 9760 9770 9780 9790 9800 9810 9820 9830 9840

L L T N F T L A T L L V I I A F W L P Q L N V Y S E K T S P Y E C G F D P M G S

CTCCTGACCAATTTTACACTAGCCACCCTACTCGTCATCATCGCATTCTGACTTCCCCAACTAAATGTATACTCTGAGAAAACAAGCCCATACGAATGTGGATTTGACCCCATAGGATCA

9850 9860 9870 9880 9890 9900 9910 9920 9930 9940 9950 9960

A R L P F S M K F F L V A I T F L L F D L E I A L L L P L P W A S Q T A N L N T

GCCCGCCTTCCCTTCTCTATAAAATTCTTTCTGGTAGCCATCACATTCCTCTTATTTGACCTAGAAATTGCACTCCTCCTACCACTGCCATGAGCCTCACAAACAGCAAATCTAAACACA

9970 9980 9990 10000 10010 10020 10030 10040 10050 10060 10070 10080

M L T M A L F L I I L L A V S L A Y E W T Q K G L E W T E tRNA^{Arg} →

ATGCTTACCATAGCCCTCTTCCTAATTATCCTCCTAGCTGTAAGCCTAGCCTATGAGTGAACTCAAAAAGGACTAGAATGAACCGAATATGGTACTTAGTTTAAAATAAAATAAATGATT

10090 10100 10110 10120 10130 10140 10150 10160 10170 10180 10190 10200

U.R.F. 4L →

M S M V Y M N I M M A F T V S L V G L L M Y R S H L M S

TCGACTCATTAGATTATGATTTAATTCATAATTACCAAATGTCTATAGTATACATAAACATTATAATAGCATTCACAGTATCTCTTGTAGGACTACTAATATACCGATCCCACCTAATAT

10210 10220 10230 10240 10250 10260 10270 10280 10290 10300 10310 10320

S L L C L E G M M L S L F V M A A L T I L N S H F T L A S M M P I I L L V F A A

CCTCCCTTCTATGCTTAGAAGGAATAATGCTATCCCTATTCGTTATAGCAGCCCTAACAATCCTCAACTCACATTTTACATTAGCTAGCATAATACCTATTATCCTACTAGTCTTCGCAG

10330 10340 10350 10360 10370 10380 10390 10400 10410 10420 10430 10440

U.R.F. 4 →

M L K Y I I P T I M L

C E A A L G L S L L V M V S N T Y G T D Y V Q N L N L L Q C *

CCTGTGAAGCAGCCCTAGGTCTATCTCTACTAGTAATAGTATCAAATACATATGGTACTGATTATGTACAAAACCTCAACTTACTCCAATGCTAAAATACATTATTCCAACAATTATACT

10450 10460 10470 10480 10490 10500 10510 10520 10530 10540 10550 10560

M P L T W L S K N N M I W V N S T A H S L L I S F T S L L L M N Q F G D N S L N

TATACCCCTAACCTGGTTATCAAAAAATAATATAATTTGGGTTAACTCCACAGCACACAGCCTTCTAATTAGCTTTACAAGCCTCCTCCTCATAAACCAGTTTGGCGACAACAGCCTTAA

10570 10580 10590 10600 10610 10620 10630 10640 10650 10660 10670 10680

F S L L F F S D S L S T P L L I L T M W L L P L M L M A S Q H H L S K E N L T R

TTTTTCACTACTATTTTTCTCCGACTCCCTATCCACTCCACTACTAATTTTAACCATATGGCTCCTCCCTCTAATACTAATAGCTAGCCAACATCATCTATCAAAAGAAAACCTAACCCG

10690 10700 10710 10720 10730 10740 10750 10760 10770 10780 10790 10800

K K L F I T M L I S L Q L F L I M T F T A M E L I L F Y I L F E A T L V P T L I

AAAAAAACTATTTATTACTATGCTGATCTCACTACAACTATTCCTAATTATAACCTTTACCGCCATGGAACTAATCTTATTTTATATTCTATTTGAAGCAACACTAGTCCCAACACTCAT

10810 10820 10830 10840 10850 10860 10870 10880 10890 10900 10910 10920

I I T R W G N Q T E R L N A G L Y F L F Y T L A G S L P L L V A L I Y I Q N T V

TATTATTACCCGATGAGGAAACCAAACAGAACGCCTAAACGCCGGACTCTATTTCCTATTCTATACACTAGCTGGCTCCCTACCCCTATTAGTCGCACTAATTTATATCCAAAACACAGT

10930 10940 10950 10960 10970 10980 10990 11000 11010 11020 11030 11040

G S L N F L M L Q Y W V Q P V H N S W S N V F M W L A C M M A F M V K M P L Y G

AGGATCCCTAAATTTCCTAATATTACAGTACTGAGTACAACCTGTTCATAACTCTTGATCTAATGTCTTCATATGACTAGCATGTATAATAGCTTTCATAGTAAAAATACCACTATATGG

11050 11060 11070 11080 11090 11100 11110 11120 11130 11140 11150 11160

L H L W L P K A H V E A P I A G S M V L A A V L L K L G G Y G M L R I T L I L N

CCTCCACCTTTGACTACCTAAAGCTCACGTAGAAGCCCCCATCGCAGGCTCCATAGTCCTTGCAGCAGTTCTACTAAAACTAGGGGGGTACGGTATGCTACGAATCACACTAATTCTAAA

11170 11180 11190 11200 11210 11220 11230 11240 11250 11260 11270 11280

P M T D F M A Y P F I M L S L W G M I M T S S I C L R Q T D L K S L I A Y S S V

CCCTATGACCGACTTTATAGCATACCCATTCATTATACTCTCCCTATGAGGCATAATTATAACCAGCTCAATCTGCCTCCGTCAAACGGACCTAAAATCACTCATCGCATACTCCTCTGT

11290 11300 11310 11320 11330 11340 11350 11360 11370 11380 11390 11400

S H M A L V I V A I L I Q T P W S Y M G A T A L M I A H G L T S S M L F C L A N

AAGCCACATAGCACTCGTTATCGTAGCCATCCTTATCCAGACACCTTGAAGCTACATAGGAGCAACCGCCCTTATGATTGCCCACGGCCTCACATCCTCCATACTTTTCTGTCTAGCAAA

11410 11420 11430 11440 11450 11460 11470 11480 11490 11500 11510 11520

FIG. 1 *(cont.)*

```
 S  N  Y  E  R  I  H  S  R  T  M  I  L  A  R  G  L  Q  T  L  L  P  L  M  A  T  W  W  L  L  A  S  L  T  N  L  A  L  P  P
CTCAAACTACGAACGAATCCACAGCCGAACCATAATTCTAGCTCGAGGCCTACAAACGCTCCTTCCACTAATAGCCACCTGATGACTACTAGCAAGTCTAACCAACTTAGCTCTACCCCC
     11530     11540     11550     11560     11570     11580     11590     11600     11610     11620     11630     11640

 T  I  N  L  I  G  E  L  F  V  V  M  S  T  F  S  W  S  N  I  T  I  I  L  M  G  V  N  M  V  I  T  A  L  Y  S  L  Y  M  L
AACAATCAACTTAATTGGAGAACTATTTGTAGTAATGTCAACCTTTTCATGATCTAACATTACAATTATTCTAATAGGAGTAAATATAGTAATCACCGCCCTATATTCTCTATACATGCT
     11650     11660     11670     11680     11690     11700     11710     11720     11730     11740     11750     11760

 I  M  T  Q  R  G  K  Y  T  Y  H  I  N  N  I  S  P  S  F  T  R  E  N  A  L  M  S  L  H  I  L  P  L  L  L  L  T  L  N  P
AATTATAACCCAACGAGGAAAATATACCTACCACATTAATAATATCTCGCCTTCCTTTACACGGGAAAATGCACTCATATCATTACACATCCTACCCCTACTACTCCTAACCCTAAACCC
     11770     11780     11790     11800     11810     11820     11830     11840     11850     11860     11870     11880

 K  I  I  L  G  P  L  Y   tRNA^His -->                                                          tRNA^Ser(AGY) -->
AAAAATTATTCTAGGACCTCTATACTGTAAATATAGTTTAACAAAAACATTAGATTGTGAATCTAACAATAGAAACTCATTACCTTCTTATTTACCGAAAAAGTATGCAAGAACTGCTAA
     11890     11900     11910     11920     11930     11940     11950     11960     11970     11980     11990     12000

                                                                                                              U.R.F. 5 -->
                                  tRNA^Leu(CUN) -->                                                            M  N  M  F
TTCTATGCTCCCATATCTAATAGTATGGCTTTTTCGAACTTTTAAAGGATAGTAGTTTATCCGTTGGTCTTAGGAACCAAAAAATTGGTGCAACTCCAAATAAAAGTAATAAACATATTC
     12010     12020     12030     12040     12050     12060     12070     12080     12090     12100     12110     12120

 S  S  L  S  L  V  T  L  L  L  L  T  M  P  I  M  M  M  S  F  N  T  Y  K  P  S  N  Y  P  L  Y  V  K  T  A  I  S  Y  A  F
TCCTCACTCTCACTAGTTACTTTACTCTTACTAACTATACCCATTATAATAATAAGCTTTAACACCTACAAACCTTCCAACTACCCACTCTACGTAAAAACAGCTATCTCATACGCCTTC
     12130     12140     12150     12160     12170     12180     12190     12200     12210     12220     12230     12240

 I  T  S  M  I  P  T  M  M  F  I  H  S  G  Q  E  L  I  I  S  N  W  H  W  L  T  I  Q  T  L  K  L  S  L  S  F  K  M  D  Y
ATTACCAGCATAATTCCCACAATAATATTTATCCACTCAGGCCAAGAACTAATTATTTCAAACTGACACTGACTAACCATCCAAACTCTTAAATTATCCCTCAGCTTTAAAATAGACTAT
     12250     12260     12270     12280     12290     12300     12310     12320     12330     12340     12350     12360

 F  S  M  M  F  I  P  V  A  L  F  V  T  W  S  I  M  E  F  S  M  W  Y  M  Y  S  D  P  N  I  N  K  F  F  K  Y  L  L  L  F
TTCTCAATAATATTTATCCCAGTAGCACTATTCGTCACATGATCTATTATAGAATTCTCAATATGATATATATACTCAGACCCCAATATTAACAAATTCTTCAAATATCTACTCCTATTC
     12370     12380     12390     12400     12410     12420     12430     12440     12450     12460     12470     12480

 L  I  T  M  L  I  L  V  T  A  N  N  L  F  Q  L  F  I  G  W  E  G  V  G  I  M  S  F  L  L  I  G  W  W  Y  G  R  A  D  A
CTCATTACTATGCTCATCCTTGTAACCGCAAACAACCTCTTCCAGCTATTCATTGGCTGAGAAGGCGTCGGAATCATATCATTTCTACTCATCGGATGATGATACGGACGAGCAGATGCA
     12490     12500     12510     12520     12530     12540     12550     12560     12570     12580     12590     12600

 N  T  A  A  L  Q  A  I  L  Y  N  R  I  G  D  I  G  F  I  L  A  M  A  W  F  L  T  N  L  N  T  W  D  L  Q  Q  I  F  M  L
AACACAGCAGCCCTACAAGCAATCTTATATAACCGCATCGGCGACATTGGTTTCATTTTAGCAATAGCATGGTTCCTAACAAATCTCAATACCTGAGACCTCCAACAGATCTTCATACTA
     12610     12620     12630     12640     12650     12660     12670     12680     12690     12700     12710     12720

 N  P  S  D  S  N  M  P  L  I  G  L  A  L  A  A  T  G  K  S  A  Q  F  G  L  H  P  W  L  P  S  A  M  E  G  P  T  P  V  S
AACCCAAGCGACTCAAACATACCCTTGATTGGACTAGCATTAGCTGCAACCGGAAAATCCGCCCAATTTGGCCTCCACCCGTGACTTCCCTCTGCAATAGAAGGCCCAACTCCCGTCTCA
     12730     12740     12750     12760     12770     12780     12790     12800     12810     12820     12830     12840

 A  L  L  H  S  S  T  M  V  V  A  G  I  F  L  L  I  R  F  Y  P  L  T  E  N  N  K  Y  I  Q  S  I  T  L  C  L  G  A  I  T
GCACTACTCCATTCAAGCACAATAGTGGTAGCAGGTATCTTCCTACTAATCCGTTTCTATCCCCTCACAGAAAACAATAAATACATCCAATCTATTACATTATGCTTAGGAGCCATTACC
     12850     12860     12870     12880     12890     12900     12910     12920     12930     12940     12950     12960

 T  L  F  T  A  M  C  A  L  T  Q  N  D  I  K  K  I  I  A  F  S  T  S  S  Q  L  G  L  M  M  V  T  I  G  I  N  Q  P  Y  L
ACACTATTTACAGCAATATGCGCCCTCACCCAAAATGACATTAAAAAAATCATCGCCTTCTCCACATCCAGTCAACTGGGCCTTATAATAGTAACTATTGGCATTAACCAACCTTACCTA
     12970     12980     12990     13000     13010     13020     13030     13040     13050     13060     13070     13080

 A  F  L  H  I  C  T  H  A  F  F  K  A  M  L  F  M  C  S  G  S  I  I  H  S  L  N  D  E  Q  D  I  R  K  M  G  G  L  F  K
GCTTTCCTCCACATCTGTACCCACGCCTTTTTCAAAGCTATACTATTCATATGCTCCGGTTCCATTATTCACAGCCTAAACGACGAACAAGATATTCGAAAAATAGGAGGCCTATTTAAA
     13090     13100     13110     13120     13130     13140     13150     13160     13170     13180     13190     13200

 A  M  P  F  T  T  T  A  L  I  V  G  S  L  A  L  T  G  M  P  F  L  T  G  F  Y  S  K  D  L  I  I  E  A  A  N  K  S  Y  T
GCCATGCCATTCACCACAACAGCCCTCATTGTTGGCAGTCTCGCACTAACAGGAATACCCTTCCTCACAGGATTCTACTCCAAAGACCTAATCATCGAAGCCGCCAACAAGTCTTATACC
     13210     13220     13230     13240     13250     13260     13270     13280     13290     13300     13310     13320

 N  A  W  A  L  L  M  T  L  I  A  T  S  F  T  A  I  Y  S  T  R  I  I  F  F  A  L  L  G  Q  P  R  F  P  T  L  V  N  I  N
AACGCCTGAGCCCTTCTAATAACATTAATTGCCACCTCTTTCACAGCTATTTACAGCACCCGTATTATTTTTTTCGCACTTCTAGGACAACCCCGATTCCCTACCCTAGTTAATATTAAC
     13330     13340     13350     13360     13370     13380     13390     13400     13410     13420     13430     13440

 E  N  N  P  L  L  I  N  S  I  K  R  L  L  I  G  S  L  F  A  G  Y  I  I  S  N  N  I  P  P  T  T  I  P  Q  M  T  M  P  Y
GAAAACAACCCCCTTCTGATCAACTCTATCAAACGCTTACTAATTGGAAGCCTCTTCGCAGGATACATCATTTCCAACAATATTCCTCCAACAACAATTCCCCAAATAACTATGCCCTAC
     13450     13460     13470     13480     13490     13500     13510     13520     13530     13540     13550     13560

 Y  L  K  T  T  A  L  I  V  T  I  L  G  F  I  L  A  L  E  I  S  N  M  T  K  N  L  K  Y  H  Y  P  S  N  A  F  K  F  S  T
TACCTAAAAACAACAGCCCTAATTGTTACAATCCTAGGCTTCATCTTAGCCCTAGAAATCAGTAATATAACTAAAAATCTAAAATATCACTACCCCTCAAACGCCTTCAAGTTCTCAACC
     13570     13580     13590     13600     13610     13620     13630     13640     13650     13660     13670     13680
```

FIG. 1 *(cont.)*

L L G Y F P T I M H R L A P Y M N L S M S Q K S A S S L L D L I W L E A I L P K

TTGCTAGGGTATTTCCCCACAATTATACATCGCCTAGCTCCATACATAAATTTATCAATAAGCCAAAAATCAGCATCCTCCCTTCTAGACCTAATCTGACTAGAAGCCATCCTACCAAAA

13690 13700 13710 13720 13730 13740 13750 13760 13770 13780 13790 13800

T I S L A Q M K A S T L V T N Q K G L I K L Y F L S F L I T I L I S M I L F N F

(end U.R.F. 6) * N G

ACCATCTCACTCGCCCAAATAAAAGCATCTACCCTGGTCACAAACCAAAAAGGCCTGATCAAACTATATTTCCTCTCCTTCTTAATCACAATCCTTATCAGCATAATCTTATTTAATTTC

13810 13820 13830 13840 13850 13860 13870 13880 13890 13900 13910 13920

H E * (end U.R.F. 5)

R T I E M I V V V G I L L S W G T V I V L W T G Y S Y L A A I G M A E E S F F G

CACGAGTAATTTCTATAATAACCACAACACCAATTAATAAAGACCACCCAGTTACAATAACTAATCAGGTACCATAACTGTATAAAGCCGCAATCCCTATGGCCTCTTCACTAAAGAACC

13930 13940 13950 13960 13970 13980 13990 14000 14010 14020 14030 14040

S D G T D Y I V W D G L G N F E F V V E V E K D K L V Y Y V M F F E M L L G T V

CAGAATCCCCTGTATCATAAATCACCCAATCCCCTAAACCATTAAACTCAAACACAACCTCAACTTCTTTATCCTTTAATACATAATAGACCATAAAGAACTCCATCAACAAGCCAGTAA

14050 14060 14070 14080 14090 14100 14110 14120 14130 14140 14150 14160

F A G L V A K N S L W I E P Y Q E T A M A T T Y G F V V M M G G L Y I L F V M L

CAAATGCCCCTAAAACAGCCTTATTAGAAAGCCAAATTTCAGGATACTGTTCTGTAGCCATAGCCGTTGTATAACCAAAAACTACCATCATACCTCCCAAATAAATTAAAAAGACCATCA

14170 14180 14190 14200 14210 14220 14230 14240 14250 14260 14270 14280

G L F S G G F N L V I G C G V G G S V I L G L G G Y I P S P K S S F G V F G M V

ACCCCAAAAAGGATCCACCAAAATTCAATACAATTCCACAGCCAACCCCACCACTCACAATTAACCCTAACCCCCCATAAATAGGTGAAGGTTTCGAAGAAAACCCCACAAAACCTATCA

14290 14300 14310 14320 14330 14340 14350 14360 14370 14380 14390 14400

← U.R.F. 6

F I V S L I F V I Y L M M

Cyt b →

← $tRNA^{Glu}$ M T N

CGAAAATAACGCTTAGAATAAATACAATGTATAGTATCAT[TATTCTTACATGGAATCTAACCATGACTAATGATATGAAAAACCATCGTTGTCATTCAACTACAAGAAC]ACTAATGACTA

14410 14420 14430 14440 14450 14460 14470 14480 14490 14500 14510 14520

I R K S H P L M K I V N N A F I D L P A P S N I S S W W N F G S L L G I C L I L

ACATTCGAAAGTCCCACCCACTAATAAAAATTGTAAACAATGCATTCATCGACCTTCCAGCCCCATCAAACATTTCATCATGATGAAATTTCGGTTCCCTCCTGGGAATCTGCCTAATCC

14530 14540 14550 14560 14570 14580 14590 14600 14610 14620 14630 14640

Q I L T G L F L A M H Y T S D T T T A F S S V T H I C R D V N Y G W I I R Y M H

TACAAATCCTCACAGGCCTATTCCTAGCAATACACTACACATCCGACACAACAACAGCATTCTCCTCTGTTACCCATATCTGCCGAGACGTGAACTACGGCTGAATCATCCGATACATAC

14650 14660 14670 14680 14690 14700 14710 14720 14730 14740 14750 14760

A N G A S M F F I C L Y M H V G R G L Y Y G S Y T F L E T W N I G V I L L L T V

ACGCAAACGGAGCTTCAATGTTTTTTATCTGCTTATATATGCACGTAGGACGAGGCTTATATTACGGGTCTTACACTTTTCTAGAAACATGAAATATTGGAGTAATCCTTCTGCTCACAG

14770 14780 14790 14800 14810 14820 14830 14840 14850 14860 14870 14880

M A T A F M G Y V L P W G Q M S F W G A T V I T N L L S A I P Y I G T N L V E W

TAATAGCCACAGCATTTATAGGATACGTCCTACCATGAGGACAAATATCATTCTGAGGAGCAACAGTCATCACCAACCTCTTATCAGCAATCCCATACATCGGCACAAATTTAGTCGAAT

14890 14900 14910 14920 14930 14940 14950 14960 14970 14980 14990 15000

I W G G F S V D K A T L T R F F A F H F I L P F I I M A I A M V H L L F L H E T

GAATCTGAGGCGGATTCTCAGTAGACAAAGCAACCCTTACCCGATTCTTCGCTTTCCATTTTATCCTTCCATTTATCATCATAGCAATTGCCATAGTCCACCTACTATTCCTCCACGAAA

15010 15020 15030 15040 15050 15060 15070 15080 15090 15100 15110 15120

G S N N P T G I S S D V D K I P F H P Y Y T I K D I L G A L L L I L A L M L L V

CAGGCTCCAACAACCCAACAGGAATTTCCTCAGACGTAGACAAAATCCCATTCCACCCCTACTATACCATTAAGGACATCTTAGGGGCCCTCTTACTAATTCTAGCTCTAATACTACTAG

15130 15140 15150 15160 15170 15180 15190 15200 15210 15220 15230 15240

L F A P D L L G D P D N Y T P A N P L N T P P H I K P E W Y F L F A Y A I L R S

TACTATTCGCACCCGACCTCCTCGGAGACCCAGATAACTACACCCCAGCCAATCCACTCAACACACCCCCTCACATCAAACCCGAGTGATACTTCTTATTTGCATACGCAATCTTACGAT

15250 15260 15270 15280 15290 15300 15310 15320 15330 15340 15350 15360

I P N K L G G V L A L A F S I L I L A L I P L L H T S K Q R S M M F R P L S Q C

CAATCCCCAACAAACTAGGAGGAGTACTAGCCCTAGCCTTCTCTATCCTAATTCTTGCTCTAATCCCCCTACTACACACCTCCAAACAACGAAGCATAATATTCCGACCACTCAGCCAAT

15370 15380 15390 15400 15410 15420 15430 15440 15450 15460 15470 15480

L F W A L V A D L L T L T W I G G Q P V E H P Y I T I G Q L A S V L Y F L L I L

GCCTATTCTGAGCCCTAGTAGCAGACCTACTGACACTCACATGAATTGGAGGACAACCAGTCGAACACCCATATATCACCATCGGACAACTAGCATCTGTCCTATACTTTCTCCTCATCC

15490 15500 15510 15520 15530 15540 15550 15560 15570 15580 15590 15600

FIG. 1 *(cont.)*

```
  V  L  M  F  T  A  G  T  I  E  N  K  L  L  K  W  *   tRNA(Thr) →
TAGTGCTAATACCAACGGCCGGCACAATCGAAAACAAATTACTAAAATGAAGACAGGTCTTTGTAGTACATCTAATATACTGGTCTTGTAAACCAGAGAAGGAGAACAACTAACCTCCCT
     15610      15620      15630      15640      15650      15660      15670      15680      15690      15700      15710      15720
                                                             ← tRNA(Pro)
AAGACTCAAGGAAGAAACTGCAGTCTCACCATCAACCCCCAAAGCTGAAGTTCTATTTAAACTATTCCCTGAACACTATTAATATAGTTCCATAAATACAAAGAGCCTTATCAGTATTAA
     15730      15740      15750      15760      15770      15780      15790      15800      15810      15820      15830      15840

ATTTATCAAAAATCCCAATAACTCAACACAGAATTTGCACCCTAACCAAATATTACAAACACCACTAGCTAACATAACACGCCCATACACAGACCACAGAATGAATTACCTACGCAAGGG
     15850      15860      15870      15880      15890      15900      15910      15920      15930      15940      15950      15960

GTAATGTACATAACATTAATGTAATAAAGACATAATATGTATATAGTACATTAAATTATATGCCCCATGCATATAAGCAAGTACATGACCTCTATAGCAGTACATAATACATATAATTAT
     15970      15980      15990      16000      16010      16020      16030      16040      16050      16060      16070      16080

TGACTGTACATAGTACATTATGTCAAATTCATTCTTGATAGTATATCTATTATATATTCCTTACCATTAGATCACGAGCTTAATTACCATGCCGCGTGAAACCAGCAACCCGCTAGGCAG
     16090      16100      16110      16120      16130      16140      16150      16160      16170      16180      16190      16200

GGATCCCTCTTCTCGCTCCGGGCCCATAAACCGTGGGGGTCGCTATCCAATGAATTTTACCAGGCATCTGGTTCTTTCTTCAGGGCCATCTCATCTAAAACGGTCCATTCTTTCCTCTTA
     16210      16220      16230      16240      16250      16260      16270      16280      16290      16300      16310      16320

AATAAGACATCTCGATGG
     16330
```

FIG. 1 *(final part)*

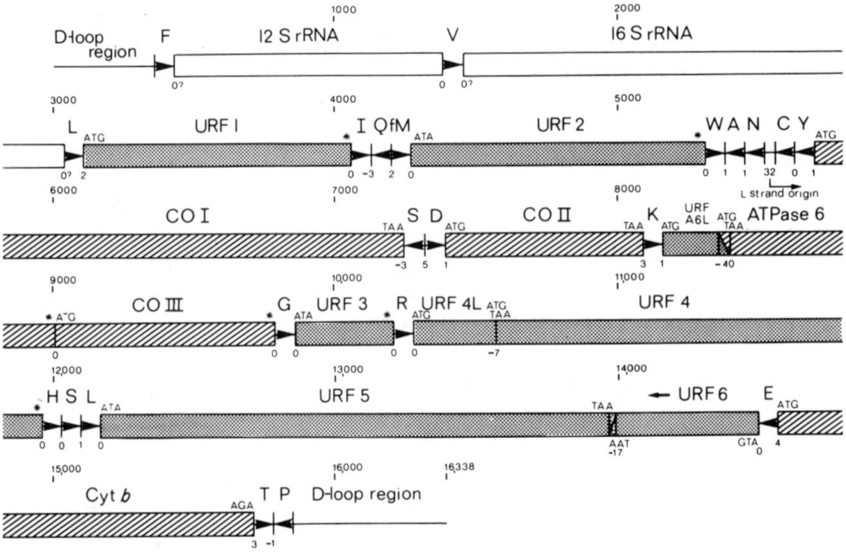

FIG. 2. Organization of the bovine mitochondrial genome. A gene map is shown, based on the sequence given in Fig. 1, of the bovine mitochondrial genome; for clarity, the map is presented in linear fashion, although the actual genome is circular. Identified protein-coding genes are represented by cross-hatched rectangles, unassigned reading frames by shaded rectangles, and rRNA genes by open rectangles. The tRNA genes are represented by ▶ (L-strand sense) or ◀ (H-strand sense) and are identified by the 1-letter amino acid code. Initiation and termination codons are shown for the protein genes; an asterisk indicates that no termination codon is present in the DNA sequence, presumably because it is generated in the transcript by polyadenylation (Anderson *et al.*, 1981). The numerals at the gene junctions denote the number of intergenic nucleotides at that point, with negative numbers signifying an overlap. The distance (in base-pairs) from the beginning of the sequence is marked above the genes.

TABLE 1

Characteristics of bovine mitochondrial genes and a comparison with the homologous human genes

Gene	Protein length	Molecular weight	% Nucleotide conservation	% Amino acid conservation	Bovine *versus* human conservative codon changes (%)		
					Frame 0	Frame +1	Frame −1
URF 1	318	35,627	73·4	78·0†	37·7	6·3	1·9
URF 2	347	39,206	65·6	62·8	34·9	7·8	4·0
CO I	514	56,965	79·4	91·2	46·4	3·9	1·6
CO II	227	25,991	70·2	72·7	43·6	3·5	4·0
URF A6L	66	7,927	63·4	51·5	22·7	6·1	12·1
ATPase 6	226	24,758	72·7	77·9	41·2	5·8	1·3
CO III	261	29,884	78·9	87·0	42·5	3·5	3·8
URF 3	115	13,039	69·9	73·9	36·5	7·0	1·8
URF 4L	98	10,784	74·8	73·5	30·6	6·1	2·0
URF 4	459	52,036	71·5	74·1	37·2	5·0	2·4
URF 5	606	68,232	71·2	69·6†	30·0	6·6	2·5
URF 6	175	19,055	65·3	62·6	33·3	5·8	2·9
Cytochrome *b*	379	42,540	72·6	78·9†	40·9	5·5	1·3

The molecular weights of the expected products of the bovine mitochondrial protein genes were calculated from the DNA sequence using the computer program MWCALC (R. Staden, unpublished). The percentage of conserved nucleotides between the bovine and human sequences was calculated with the program BSCHAN (R. Staden, unpublished), the percentage amino acid conservation with the program AACHAN (R. Staden, unpublished), and the percentage of total codons representing conservative changes with the program COCHAN (R. Staden, unpublished). For each pair of homologous genes, the high percentage of conservative codon changes in the open reading frame (frame 0) is compared with the much lower percentages that would be produced by translation of the +1 or −1 frames (assuming readthrough of the multiple termination codons in these frames). The anomalously high value for the URF A6L gene in the −1 frame is due to its extensive overlap with the downstream ATPase 6 gene, which is in this frame.

† These values differ slightly from those previously given in Table 2 of Anderson *et al.* (1981) because of several ambiguities that existed in the bovine mtDNA sequence at that time.

represent silent substitutions (Table 1). The conservative codon changes, primarily in the third position, represent additional independent evidence in favour of a protein-coding function for the reading frames. Taken together with the transcript mapping data, these results imply that all the reading frames almost certainly represent active protein genes.

Figure 3 shows the homologies that exist between the human and bovine mitochondrial proteins coded for by these reading frames. COI is the most highly conserved protein encoded by the mammalian mitochondrial genome, showing 91·2% amino acid homology between the human and bovine proteins; this includes a stretch of 140 amino acids where absolute homology exists (Fig. 3; Anderson *et al.*, 1981). URF A6L is the shortest and least conserved of the major reading frames in these genomes, having homologies of only 63·4% for the nucleotide sequences and 51·5% for the amino acid sequences (Table 1 and Fig. 3; Anderson *et al.*, 1981). The putative protein coded for by the URF A6L in either genome would have a rather unusual amino acid composition, in that only four amino acids (leucine, proline, threonine and lysine) account for more than half of the residues. The positions of the charged residues in this putative protein are also distributed asymmetrically: both the human and the bovine species have a very non-polar amino terminus containing, in each case, a run of uncharged amino acids at least 21 residues long, and both proteins have a very basic, lysine-rich carboxy terminus (Fig. 3). It is interesting to note that a gene coding for a protein having similar charge distributions has been found to precede the ATPase 6 gene in the *unc* operon of *Escherichia coli* (Gay & Walker, 1981).

For many of the human and bovine reading frames, the DNA sequence does not contain any termination codons (Fig. 1). However, when these genes are transcribed as part of a large RNA precursor and adjacent gene transcripts are separated by endonucleolytic processing, polyadenylation of the immature mRNAs creates in-phase UAA codons (Barrell *et al.*, 1980*a*; Anderson *et al.*, 1981; Ojala *et al.*, 1981*a*). The processing of these mitochondrial RNAs must operate with a precision comparable to that observed for RNA splicing. Even a small error in the site of polyadenylation would result in translation of the poly(A) tail and thus produce polylysine-terminated proteins. It is likely, however, that the mitochondrial and the nuclear polyadenylation processes are significantly different, because the A-A-U-A-A-A sequence found upstream of poly(A) addition sites in nuclear RNAs (Proudfoot & Brownlee, 1976) is absent from most of the mitochondrial mRNAs (Fig. 1; Anderson *et al.*, 1981; Ojala *et al.*, 1981*a*).

FIG. 3. Comparison of the human and bovine mitochondrially encoded proteins. Translations of the human and the bovine reading frames are shown aligned. Amino acid residues are symbolized by the 1-letter code, and asterisks between the lines mark positions of homology. For each gene, the top sequence is the human translation and the bottom is bovine. Each translation ends at the real or presumed termination codon for the reading frame, but for URF 4 the additional C-terminal extension (URF 4X) that would be produced by reading into, and terminating within, the adjacent $tRNA^{His}$ gene (see the text) is also shown. Some of the homologous translation products differ slightly in length at their respective amino or carboxy termini. However, in each pair the proteins remain in register over their entire lengths, showing that no large insertions or deletions have occurred in these genes during the course of mammalian evolution.

U.R.F. 1

```
MPMANLLLLIVPILIAMAFLMLTERKILGYMQLRKGPNVVGPYGLLQPFADAMKLFTKEPLKPATSTITLYITAPTLALTIALLLWTPLPMPNPLVNLNLGLLFILATSSLAVYSILWSG
* * * * ** *** * *** * *** ********************* *** *** **** ****     * **  **  **  * ***** ** * *** ** ** ***********
MFMINILMLIIPILLAVAFLTLVERKVLGYMQLRKGPNVVGPYGLLQPIADAIKLFIKEPLRPATSSASMFILAPIMALGLALTMWIPLPMPYPLINMNLGVLFMLAMSSLAVYSILWSG

WASNSNYALIGALRAVAQTISYEVTLAIILLSTLLMSGSFNLSTLITTQEHLWLLLPSWPLAMMWFISTLAETNRTPFDLAEGESELVSGFNIEYAAGPFALFFMAEYTNIIMMNTLTTT
***** *************************** ******* *********  ** ** **************** **** *********** *************** ******  *
WASNSKYALIGALRAVAQTISYEVTLAIILLSVLLMSGSFTLSTLITTQEQMWLILPAWPLAMMWFISTLAETNRAPFDLTEGESELVSGFNVEYAAGPFALFFMAEYANIIMMNIFTAI

IFLGTTYDALSPELYTTYFVTKTLLLTSLFLWIRTAYPRFRYDQLMHLLWKNFLPLTLALLMWYVSMPITISSIPPQT
 ****      ***** *  * ****  *****  ************************ ** ** **   * *****
LFLGTSHNPHMPELYTINFTIKSLLLTMSFLWIRASYPRFRYDQLMHLLWKNFLPLTLALCMWHVSLPILTSGIPPQT
```

U.R.F. 2

```
INPLAQPVIYSTIFAGTLITALSSHWFFTWVGLEMNMLAFIPVLTKKMNPRSTEAAIKYFLTQATASMILLMAILFNNMLSGQWTMTNTTNQYSSLMIMMAMAMKLGMAPFHFWVPEVTQ
 **     *  **  ** *    ****   * * ****** **    *  *** ***  ****** **** * **    * * *****       *   *     ** *****************
MNPIIFIIILLTIMLGTIIVMISSHWLLVWIGFEMNMLAIIPIMMKNHNPRATEASTKYFLTQSTASMLLMMAVIINLMFSGQWTVMKLFNPMASMLMTMALAMKLGMAPFHFWVPEVTQ

GTPLTSGLLLLTWQKLAPISIMYQISPSLNVSLLLTLSILSIMAGSWGGLNQTQLRKILAYSSITHMGWMMAVLPYNPNMTILNLTIYIILTTTAFLLLNLNSSTTTLLLSRTWNKLTWL
* ** *** ********* *    *** ** *    * **** ***   * ************ ***** ***** ******* ** *** **** * * *      ** **** ** ****
GIPLSSGLILLTWQKLAPMSVLYQIFPSINLNLILTLSVLSILIGGWGGLNQTQLRKIMAYSSIAHMGWMTAVLPYNPTMTLLNLIIYIIMTSTMFTMFMANSTTTTLSLSHTWNKTPIM

TPLIPSTLLSLGGLPPLTGFLPKWAIIEEFTKNNSLIIPTIMATITLLNLYFYLRLIYSTSITLLPMSNNVKMKWQFEHTKPTPFLPTLIALTTLLLPISPFMLMIL
* **  **** ****** ** *** ** * ***** * ** **    ******* ** ***   *   *   ** ******   *    ****   * *  **  *
TVLILATLLSMGGLPPLSGFMPKWMIIQEMTKNNSIILPTFMAITALLNLYFYMRLTYSTTLTMFPSTNNMKMKWQFPLMKKMTFLPTMVVLSTMMLPLTPMLSVLE
```

Cytochrome oxidase, subunit I

```
MFADRWLFSTNHKDIGTLYLLFGAWAGVLGTALSLLIRAELGQPGNLLGNDHIYNVIVTAHAFVMIFFMVMPIMIGGFGNWLVPLMIGAPDMAFPRMNNMSFWLLPPSLLLLLASAMVEA
**  *********************** **************** *** * **** ***************************************************** ****** ****
MFINRWLFSTNHKDIGTLYLLFGAWAGMVGTALSLLIRAELGQPGTLLGDDQIYNVVVTAHAFVMIFFMVMPIMIGGFGNWLVPLMIGAPDMAFPRMNNMSFWLLPPSFLLLLASSMVEA

GAGTGWTVYPPLAGNYSHPGASVDLTIFSLHLAGVSSILGAINFITTIINMKPPAMTQYQTPLFVWSVLITAVLLLLSLPVLAAGITMLLTDRNLNTTFFDPAGGGDPILYQHLFWFFGH
**************  * ******************************************* *********** *********************************************
GAGTGWTVYPPLAGNLAHAGASVDLTIFSLHLAGVSSILGAINFITTIINMKPPAMSQYQTPLFVWSVMITAVLLLLSLPVLAAGITMLLTDRNLNTTFFDPAGGGDPILYQHLFWFFGH

PEVYILILPGFGMISHIVTYYSGKKEPFGYMGMVWAMMSIGFLGFIVWAHHMFTVGMDVDTRAYFTSATMIIAIPTGVKVFSWLATLHGSNMKWSAAVLWALGFIFLFTVGGLTGIVLAN
*********************************************************************************************** * *** *  ******************
PEVYILILPGFGMISHIVTYYSGKKEPFGYMGMVWAMMSIGFLGFIVWAHHMFTVGMDVDTRAYFTSATMIIAIPTGVKVFSWLATLHGGNIKWSPAMMWALGFIFLFTVGGLTGIVLAN

SSLDIVLHDTYYVVAHFHYVLSMGAVFAIMGGFIHWFPLFSGYTLDQTYAKIHFTIMFIGVNLTFFPQHFLGLSGMPRRYSDYPDAYTTWNILSSVGSFISLTAVMLMIFMIWEAFASKR
************************************************ ********** * ***** *** *** ************************** **  ** ************ * *********
SSLDIVLHDTYYVVAHFHYVLSMGAVFAIMGGFVHWFPLFSGYTLNDTWAKIHFAIMFVGVNMTFFPQHFLGLSGMPRRYSDYPDAYTMWNTISSMGSFISLTAVMLMVFIIWEAFASKR

KVLMVEEPSMNLEWLYGCPPPYHTFEEPVYMKS
 ** *     ***** ************ *
EVLTVDLTTTNLEWLNGCPPPYHTFEEPTYVNLK
```

Cytochrome oxidase, subunit II

```
MAHAAQVGLQDATSPIMEELITFHDHALMIIFLICFLVLYALFLTLTTKLTNTNISDAQEMETVWTILPAIILVLIALPSLRILYMTDEVNDPSLTIKSIGHQWYWTYEYTDYGGLIFNS
**    * * ***********  **** *** ***   ****    * ****** *    **** ** ********* *********** ** * **** *  ****** ******   * * *
MAYPMQLGFQDATSPIMEELLHFHDHTLMIVFLISSLVLYIISLMLTTKLTHTSTMDAQEVETIWTILPAIILILIALPSLRILYMMDEINNPSLTVKTMGHQWYWSYEYTDYEDLSFDS

YMLPPLFLEPGDLRLLDVDNRVVLPIEAPIRMMITSQDVLHSWAVPTLGLKTDAIPGRLNQTTFTATRPGVYYGQCSEICGANHSFMPIVLELIPLKIFEMGPVFTL
** *   * ** **** ******** *   ***    * ********* ***************    *** ********** ********** *** **       *
YMIPTSELKPGELRLLEVDNRVVLPMEMTIRMLVSSEDVLHSWAVPSLGLKTDAIPGRLNQTTLMSSRPGLYYGQCSEICGSNHSFMPIVLELVPLKYFEKWSASML
```

U.R.F. A6L

```
MPQLNTTVWPTMITPMLLTLFLITQLKMLNTNYHLPPSPKPMKMKNYNKPWEPKWTKICSLHSLPPQS
**** *  * ***  * **** * ***     *     *    * **   * *** *****      **
MPQLDTSTWLTMILSMFLTLFIIFQLKVSKHNFYHNPELTPTKMLKQNTPWETKWTKIYLPLLLPL
```

ATPase 6

```
MNENLFASFIAPTILGLPAAVLIILFPPLLIPTSKYLINNRLITTQQWLIKLTSKQMMTMHNTKGRTWSLMLVSLIIFIATTNLLGLLPHSFTPTTQLSMNLAMAIPLWAGTVIMGFRSK
****** *** * *****   ** *** ** ***  *   **   * ***   * *****  ** ** ** *** *** **  ********************* ******** ** *** *
MNENLFTSFITPVILGLPLVTLIVLFPSLLFPTSNRLVSNRFVTLQQWMLQLVSKQMMSIHNSKGQTWTLMLMSLILFIGSTNLLGLLPHSFTPTTQLSMNLGMAIPLWAGAVITGFRNK

IKNALAHFLPQGTPTPLIPMLVIIETISLLIQPMALAVRLTANITAGHLLMHLIGSATLAMSTINLPSTLIIFTILILLTILEIAVALIQAYVFTLLVSLYLHDNT
 *  ************************ ******************** **** ****   *     ** *********** *** ******************
TKASLAHFLPQGTPTPLIPMLVIIETISLFIQPMALAVRLTANITAGHLLIHLIGGATLALMSISTTTALITFTILILLTILEFAVAMIQAYVFTLLVSLYLHDNT
```

Cytochrome oxidase, subunit III

```
MTHQSHAYHMVKPSPWPLTGALSALLMTSGLAMWFHFHSMTLLMLGLLTNTLTMYQWWRDVTRESTYQGHHTPPVQKGLRYGMILFITSEVFFFAGFFWAFYHSSLAPTPQLGGHWPRTG
**** ****** ****************** ***** ****** ** ** ********** **** ****** ************* *** ** ************** *** ** **
MTHQTHAYHMVNPSPWPLTGALSALLMTSGLTMWFHFNSMTLLMIGLTTNMLTMYQWWRDVIRESTFQGHHTPAVQKGLRYGMILFIISEVLFFTGFFWAFYHSSLAPTPELGGCWPPTG

ITPLNPLEVPLLNTSVLLASGVSITWAHHSLMENNRNQMIQALLITILLGLYFTLLQASEYFESPFTISDGIYGSTFFVATGFHGLHVIIGSTFLTICFIRQLMFHFTSKHHFGFEAAAW
* ******************************    *  * *** *** ** ********** * ******* ********************** ** *** ***** ******* **
IHPLNPLEVPLLNTSVLLASGVSITWAHHSLMEGDRKHMLQALFITITLGVYFTLLQASEYYEAPFTISDGVYGSTFFVATGFHGLHVIIGSTFLIVCFFRQLKFHFTSNHHFGFEAGAW

YWHFVDVVWLFLYVSIYWWGS
*********************
YWHFVDVVWLFLYVSIYWWGS
```

FIG. 3 *(1st part)*

U.R.F. 3

```
MNFALILMINTLLALLLMIITFWLPQLNGYMEKSTPYECGFDPMSPARVPFSMKFFLVAITFLLFDLEIALLLPLPWALQTTNLPLMVMSSLLLIIILALSLAYEWLQKGLDWTE
**  * *  *  ** ** ** ******* * **  *********  ** ***************************** ** **  *    * *** ** ****** **** ***
MNLMLALLTNFTLATLLVIIAFWLPQLNVYSEKTSPYECGFDPMGSARLPFSMKFFLVAITFLLFDLEIALLLPLPWASQTANLNTMLTMALFLIILLAVSLAYEWTQKGLEWTE
```

U.R.F. 4L

```
MPLIYMNIMLAFTISLLGMLVYRSHLMSSLLCLEGMMLSLFIMATLMTLNTHSLLANIVPIAMLVFAACEAAVGLALLVSISNTYGLDYVHNLNLLQC
*  ***** *** ** * * ******************** ** *  ** *  **   **  ********* ** ***  ***** *** *******
MSMVYMNIMMAFTVSLVGLLMYRSHLMSSLLCLEGMMLSLFVMAALTILNSHFTLASMMPIILLVFAACEAALGLSLLVMVSNTYGTDYVQNLNLLQC
```

U.R.F. 4

```
MLKLIVPTIMLLPLTWLSKKHMIWINTTTHSLIISIIPLLFFNQINNNLFSCSPTFSSDPLTTPLLMLTTWLLPLTIMASQRHLSSEPLSRKKLYLSMLISLQISLIMTFTATELIMFYI
*** * ***** *******  *** * * *** **   **  **   *    *  * ** * **** ** *****  **** *** * * ****   ******  ******* *** ***
MLKYIIPTIMLMPLTWLSKNNMIWVNSTAHSLLISFTSLLLMNQFGDNSLNFSLLFFSDSLSTPLLILTMWLLPLMLMASQHHLSKENLTRKKLFITMLISLQLFLIMTFTAMELILFYI

FFETTLIPTLAIITRWGNQPERLNAGTYFLFYTLVGSLPLLIALIYTHNTLGSLNILLLTLTAQELSNSWANNLMWLAYTMAFMVKMPLYGLHLWLPKAHVEAPIAGSMVLAAVLLKLGG
 ** ** *** ******** ****** ******* ****** ****  ** **** * *   *   *** *  ****  *****************************************
LFEATLVPTLIIITRWGNQTERLNAGLYFLFYTLAGSLPLLVALIYIQNTVGSLNFLMLQYWVQPVHNSWSNVFMWLACMMAFMVKMPLYGLHLWLPKAHVEAPIAGSMVLAAVLLKLGG

YGMMRLTLILNPLTKHMAYPFLVLSLWGMIMTSSICLRQTDLKSLIAYSSISHMALVVTAILIQTPWSFTGAVILMIAHGLTSSLLFCLANSNYERTHSRIMILSQGLQTLLPLMAFWWL
*** * ****** *  *****  *************************** ******  *********  **  ********** *********** *** ***  ********** ***
YGMLRITLILNPMTDFMAYPFIMLSLWGMIMTSSICLRQTDLKSLIAYSSVSHMALVIVAILIQTPWSYMGATALMIAHGLTSSMLFCLANSNYERIHSRTMILARGLQTLLPLMATWWL

LASLANLALPPTINLLGELSVLVTTFSWSNITLLLTGLNMLVTALYSLYMFTTTQWGSLTHHINNMKPSFTRENTLMFMHLSPILLLSLNPDIITGFSS    [CKYSLTKTSDCESDN]
**** ********** *** *   ********  * * **  ********    ** *  * ****  ******* **  *  * *** *** ** *       [******** **** *]
LASLTNLALPPTINLIGELFVVMSTFSWSNITIILMGVNMVITALYSLYMLIMTQRGKYTYHINNISPSFTRENALMSLHILPLLLLTLNPKIILGPLY    [CKYSLTKTLDCESNN]
```

U.R.F. 5

```
MTMHTTMTTLTLTSLIPPILTTLVNPNKKNSYPHYVKSIVASTFIISLFPTTMFMCLDQEVIISNWHWATTQTTQLSLSFKLDYFSMMFIPVALFVTWSIMEFSLWYMNSDPNINQFFKY
* *       **  *  **     *  *   ** ***      ** *  ** **    ** ******* * **  ****** ********************** *** ****** ****
MNMFSSLSLVTLLLLTMPIMMMSFNTYKPSNYPLYVKTAISYAFITSMIPTMMFIHSGQELIISNWHWLTIQTLKLSLSFKMDYFSMMFIPVALFVTWSIMEFSMWYMYSDPNINKFFKY

LLIFLITMLILVTANNLFQLFIGWEGVGIMSFLLISWWYARADANTAAIQAILYNRIGDIGFILALAWFILHSNSWDPQQMALLNANPSLTPLLGLLLAAAGKSAQLGLHPWLPSAMEGP
** ******************************** *** ******** **************** ***    * ** **   **   *  ** ** *** ***** *************
LLLFLITMLILVTANNLFQLFIGWEGVGIMSFLLIGWWYGRADANTAALQAILYNRIGDIGFILAMAWFLTNLNTWDLQQIFMLNPSDSNMPLIGLALAATGKSAQFGLHPWLPSAMEGP

TPVSALLHSSTMVVAGIFLLIRFHPLAENSPLIQTLTLCLGAITTLFAAVCALTQNDIKKIVAFSTSSQLGLMMVTIGINQPHLAFLHICTHAFFKAMLFMCSGSIIHNLNNEQDIRKMG
*********************** ** **   **  *********** * *********** ******************** ************************** ** ********
TPVSALLHSSTMVVAGIFLLIRFYPLTENNKYIQSITLCLGAITTLFTAMCALTQNDIKKIIAFSTSSQLGLMMVTIGINQPYLAFLHICTHAFFKAMLFMCSGSIIHSLNDEQDIRKMG

GLLKTMPLTSTSLTIGSLALAGMPFLTGFYSKDHIIETANMSYTNAWALSITLIATSLTSAYSTRMILLTLTGQPRFPTLTNINENNPTLLNPIKGLAAGSLFAGFLITNNISPASPFQT
** * ** * * *  ***** ************ *** ** ********  ****** *  **** *   * ******** ******* * * ** *  ******  * *** *     *
GLFKAMPFTTTALIVGSLALTGMPFLTGFYSKDLIIEAANKSYTNAWALLMTLIATSFTAIYSTRIIFFALLGQPRFPTLVNINENNPLLINSIKRLLIGSLFAGYIISNNIPPTTIPQM

TIPLYLKLTALAVTFLGLLTALDLNYLTNKLKMKSPLCTFYFSNMLGFYPSITHRTIPYLGLLTSQNLPLLLLDLTWLEKLLPKTISQHQISTSIITSTQKGMIKLYFLSFFFPLILTLL
* * *** *** ** **   **     *  **  *   * **  ** * * **  **  *  **      **** ***  ******  *   *      *** ********
TMPYYLKTTALIVTILGFILALEISNMTKNLKYHYPSNAFKFSTLLGYFPTIMHRLAPYMNLSMSQKSASSLLDLIWLEAILPKTISLAQMKASTLVTNQKGLIKLYFLSFLITILISMI

LIT
*
LFNFHE
```

U.R.F. 6

```
 MMYALFLLSVGLVMGFVGFSSKPSPIYGGLVLIVSGVVGCVIILNFGGGYMGLMVFLIYLGGMMVVFGYTTAMAIEEYPEAWGSGVEVLVSVLVGLAMEVGFVLWVKEYDGVVVVVNFN
 * *  * ***  ****************** ***** *** * *****   ************************ * *** * *   **     ** **   *  *     * **  **
MMLYIVFILSVIFVMGFVGFSSKPSPIYGGLGLIVSGGVGCGIVLNFGGSFLGLMVFLIYLGGMMVVFGYTTAMATEQYPEIWLSNKAVLGAFVTGLLMEFFMVYYVLKDKEVEVVFEFN

SVGSWMIYEGEGSGFIREDPIGAGALYDYGRWLVVVTGWPLFVGVYIVIEIARGN
  * * **    ***  *   *  *** ** *** **** *  **    ** ***
GLGDWVIYDTGDSGFFSEEAMGIAALYSYGTWLVIVTGWSLLIGVVVIMEITRGN
```

Cytochrome b

```
MTPMRKINPLMKLINHSFIDLPTPSNISAWWNFGSLLGACLILQITTGLFLAMHYSPDASTAFSSIAHITRDVNYGWIIRYLHANGASMFFICLFLHIGRGLYYGSFLYSETWNIGIILL
**  **  ****  *  ***** ***** ********* ****** *********  *  *****  ** *********** ************  * ********     ****** ***
MTNIRKSHPLMKIVNNAFIDLPAPSNISSWWNFGSLLGICLILQILTGLFLAMHYTSDTTTAFSSVTHICRDVNYGWIIRYMHANGASMFFICLYMHVGRGLYYGSYTFLETWNIGVILL

LATMATAFMGYVLPWGQMSFWGATVITNLLSAIPYIGTDLVQWIWGGYSVDSPTLTRFFTFHFILPFIIAALATLHLLFLHETGSNNPLGITSHSDKITFHPYYTIKDALGLLLFLLSLM
*  ********************************* ** ***** ***  ****** ********* * *  ************* ** *  *** ********* ** **  * **
LTVMATAFMGYVLPWGQMSFWGATVITNLLSAIPYIGTNLVEWIWGGFSVDKATLTRFFAFHFILPFIIMAIAMVHLLFLHETGSNNPTGISSDVDKIPFHPYYTIKDILGALLLILALM

TLTLFSPDLLGDPDNYTLANPLNTPPHIKPEWYFLFAYTILRSVPNKLGGVLALLLSILILAMIPILHMSKQQSMMFRPLSQSLYWLLAADLLILTWIGGQPVSYPFTIIGQVASVLYFT
 * ** *********** ******************** **** **********  ****** ** ** *** ********* * * * **** ********  *   *** ******
LLVLFAPDLLGDPDNYTPANPLNTPPHIKPEWYFLFAYAILRSIPNKLGGVLALAFSILILALIPLLHTSKQRSMMFRPLSQCLFWALVADLLTLTWIGGQPVEHPYITIGQLASVLYFL

TILILMPTISLIENKMLKWA
 ** ****   **** ***
LILVLMPTAGTIENKLLKW
```

FIG. 3 *(final part)*

If translation of a transcript were to begin before processing was complete, a "readthrough" product, containing C-terminal amino acid sequences coded for by the downstream gene(s), should result. We examined the possibility that such readthrough events might be producing more than one polypeptide from any of the reading frames. A comparison of the human and bovine mtDNA sequences revealed that translation from URF4 into the following $tRNA^{His}$ gene would produce, in each case, a protein 15 amino acid residues longer than the product from the polyadenylated URF4 mRNA (Fig. 3). These putative URF4 "extended" (URF4X) polypeptides from human and bovine mitochondria would have a very similar sequence, and in two positions conservative nucleotide substitutions maintain the amino acid homology. A comparison of each of the other bovine reading frames with its human counterpart showed no convincing conservation of size or sequence for any of the hypothetical translation readthrough products.

No gene homologous to any of the URFs has been reported for the yeast mitochondrial genome, but sequences homologous to URF1 and URF4 have been found in the mtDNA of the fungus *Aspergillus nidulans* (R. W. Davies, personal communication). Whether the URFs represent genes for mitochondrial proteins that are somewhat species-specific, or whether these genes are found in the mitochondrial genome of some organisms but in the nuclear genome of others, as is the case for the ATPase 9 gene (Sebald & Wachter, 1978; Hensgens *et al.*, 1979; Macino & Tzagoloff, 1979), is still uncertain.

TABLE 2

Degree of conservation for each amino acid

Amino acid	Total bovine	Total human	Conserved†	Average % conserved‡
Trp	104	104	101	97·1
Gly	219	213	198	91·7
Arg	63	63	57	90·5
Gln	87	90	77	87·1
Pro	193	219	177	85·9
Glu	95	88	77	84·2
Tyr	138	135	110	80·6
His	93	97	75	79·0
Phe	240	218	180	78·6
Asn	162	164	127	77·9
Asp	68	66	52	77·6
Lys	101	95	76	77·6
Leu	597	642	479	77·3
Cys	22	22	17	77·3
Ala	248	255	187	74·3
Ser	276	274	190	69·1
Ile	325	321	215	66·6
Met	262	207	150	64·0
Val	190	167	113	63·3
Thr	308	349	203	61·8

† The total number, for each amino acid, of homologous positions in the 2 sets of translated reading frames was determined from the alignments shown in Fig. 3.

‡ Average percentage conserved = 2 × (number conserved)/(total human + total bovine).

(ii) *Conserved amino acids*

Because the rate of sequence divergence for mammalian mitochondrial genomes is relatively fast (Upholt & Dawid, 1977; Brown *et al.*, 1979), it is likely that the conserved portions of human and bovine mitochondrially encoded proteins represent regions of structural or functional importance. In order to examine this pattern of conservation in very simple terms, the human and bovine proteins were aligned (Fig. 3) and scored for the number of times a given amino acid residue occurred in homologous positions in the two sets of sequences. A summary of the data for all 20 amino acids is shown in Table 2.

Tryptophan is the most conserved amino acid, with 101 of the total of 104 tryptophan residues in each set of mitochondrial proteins occurring in identical positions in the aligned genes (Fig. 3 and Table 2). Arginine and glycine are also highly conserved, being present in identical positions in the proteins in more than 90% of the cases. This probably means that these residues, where they occur, tend to be crucial for the function or structure of the proteins, but that their presence elsewhere in the polypeptide chains is selected against. The amino acids that have the lowest positional correlation are alanine, serine, isoleucine, methionine, valine and threonine (Table 2). These amino acids have uncharged, non-cyclic and predominantly non-polar side-groups and tend, in the amino acid sequences, to be

TABLE 3

Codon usage

A.

UUU	Phe	107(77)	UCU	Ser	51(32)	UAU	Tyr	64(46)	UGU	Cys	6(5)
UUC	Phe	133(141)	UCC	Ser	66(99)	UAC	Tyr	74(89)	UGC	Cys	16(17)
UUA	Leu	110(73)	UCA	Ser	99(83)	UAA	Term	6(2)	UGA	Trp	92(93)
UUG	Leu	16(16)	UCG	Ser	5(7)	UAG	Term	-(2)	UGG	Trp	12(11)
CUU	Leu	62(65)	CCU	Pro	42(41)	CAU	His	30(18)	CGU	Arg	7(7)
CUC	Leu	95(167)	CCC	Pro	63(119)	CAC	His	63(79)	CGC	Arg	11(25)
CUA	Leu	285(276)	CCA	Pro	85(52)	CAA	Gln	79(81)	CGA	Arg	42(29)
CUG	Leu	29(45)	CCG	Pro	3(7)	CAG	Gln	8(9)	CGG	Arg	3(2)
AUU	Ile	165(125)	ACU	Thr	44(51)	AAU	Asn	60(33)	AGU	Ser	12(14)
AUC	Ile	160(196)	ACC	Thr	96(155)	AAC	Asn	102(131)	AGC	Ser	43(39)
AUA	Met	218(167)	ACA	Thr	153(133)	AAA	Lys	90(85)	AGA	Term	1(1)
AUG	Met	44(40)	ACG	Thr	15(10)	AAG	Lys	11(10)	AGG	Term	-(1)
GUU	Val	40(30)	GCU	Ala	52(43)	GAU	Asp	22(15)	GGU	Gly	29(24)
GUC	Val	48(49)	GCC	Ala	91(124)	GAC	Asp	46(51)	GGC	Gly	62(88)
GUA	Val	87(70)	GCA	Ala	103(80)	GAA	Glu	78(64)	GGA	Gly	97(67)
GUG	Val	15(18)	GCG	Ala	2(8)	GAG	Glu	17(24)	GGG	Gly	31(34)

B.

Codon	Amino acid	Count	Codon	Amino acid	Count	Codon	Amino acid	Count	Codon	Amino acid	Count
UUU	Phe	12(8)	UCU	Ser	4(3)	UAU	Tyr	8(11)	UGU	Cys	1(1)
UUC	Phe	3(2)	UCC	Ser	1(-)	UAC	Tyr	2(-)	UGC	Cys	-(-)
UUA	Leu	10(8)	UCA	Ser	1(2)	UAA	Term	1(-)	UGA	Trp	1(3)
UUG	Leu	6(6)	UCG	Ser	1(-)	UAG	Term	-(-)	UGG	Trp	3(2)
CUU	Leu	1(-)	CCU	Pro	3(4)	CAU	His	-(-)	CGU	Arg	1(1)
CUC	Leu	-(-)	CCC	Pro	-(-)	CAC	His	-(-)	CGC	Arg	-(-)
CUA	Leu	2(-)	CCA	Pro	-(1)	CAA	Gln	-(-)	CGA	Arg	-(-)
CUG	Leu	-(3)	CCG	Pro	-(-)	CAG	Gln	1(-)	CGG	Arg	-(2)
AUU	Ile	14(12)	ACU	Thr	3(2)	AAU	Asn	4(3)	AGU	Ser	3(3)
AUC	Ile	-(-)	ACC	Thr	1(-)	AAC	Asn	-(1)	AGC	Ser	1(2)
AUA	Met	4(2)	ACA	Thr	3(1)	AAA	Lys	2(1)	AGA	Term	-(-)
AUG	Met	7(8)	ACG	Thr	1(-)	AAG	Lys	2(1)	AGG	Term	-(1)
GUU	Val	10(9)	GCU	Ala	4(4)	GAU	Asp	4(3)	GGU	Gly	8(8)
GUC	Val	2(4)	GCC	Ala	1(-)	GAC	Asp	-(-)	GGC	Gly	2(-)
GUA	Val	5(9)	GCA	Ala	1(1)	GAA	Glu	5(1)	GGA	Gly	4(6)
GUG	Val	6(9)	GCG	Ala	1(3)	GAG	Glu	4(9)	GGG	Gly	12(15)

Codon usage in the bovine mitochondrial genome is shown for all reading frames (A) and URF6 only (B). For reference, the figures in parentheses show the codon usage in human mtDNA (Anderson *et al.*, 1981). A box has been drawn around each codon family thought to be read by a single mitochondrial tRNA (Barrell *et al.*, 1980*b*).

interchangeable with each other and with leucine rather than strictly conserved. This probably means that hydrophobicity is the main chemical criterion affecting the selection of these residues, as might be the case, for example, for those portions of the polypeptide chains that penetrate lipid bilayers.

(iii) *Codon usage*

Table 3A shows the mammalian mitochondrial genetic code along with the aggregate codon usage data for all 13 of the bovine and human reading frames. Both the bovine and the human mitochondrial protein-coding genes exhibit a pronounced bias in favour of codons ending in C or A but, because of the symmetry of the code, this does not *per se* result in a discrimination against any amino acids. This codon bias reflects the overall base composition of the mtDNA strands. For example, C is the most abundant nucleotide in the human mtDNA L-strand (the main coding strand) but only the third most abundant nucleotide in the bovine L-strand (Fig. 1; Anderson *et al.*, 1981); correspondingly, the bovine genome employs fewer codons ending in C than does the human genome (Table 3A). Furthermore, in both genomes the H-strand coded URF6 uses relatively few codons ending in C, which is consistent with the low C content of this strand (Table 3B). For many organisms, the pattern of codon usage in a gene may reflect the availability of the tRNAs used to decode the message for that gene; hence, the relative concentrations of the various tRNA species could determine the basal level of expression of any given gene. Indeed, codon usage patterns have been found to correlate with gene product abundance in bacteria (Grantham *et al.*, 1980). It is unlikely, however, that the expression of URF6 relative to the other mitochondrial genes is governed by codon usage, because in mammalian mitochondria only a single tRNA species is used to decode each codon family (Barrell *et al.*, 1980*b*; B. A. Roe, personal communication).

(iv) *Multicistronic messages?*

Each of the HeLa cell mitochondrial mRNAs contains only one large open reading frame *except* RNA 14 and RNA 7 (Ojala *et al.*, 1980; Anderson *et al.*, 1981); RNA 14 contains both the URF A6L and the ATPase 6 gene, and RNA 7 contains URF 4L and URF4. Mitochondrial messages contain no recognizable ribosome binding sites and appear to be translated beginning with the first initiation codon from the 5′ end (Eperon *et al.*, 1980; Anderson *et al.*, 1981; Montoya *et al.*, 1981). How then are the reading frame pairs in each of the above mRNAs translated? The apparent reading frame pairs are unlikely to be cloning or sequencing artifacts, because they are found in both the human and the bovine mitochondrial genomes in a region of high homology. Furthermore, it is unlikely that any of the reading frames are not expressed, because of the pattern of codon conservation discussed above (Table 1).

There are several ways that RNAs 14 and 7 may be translated, which are speculative but are not inconsistent with the mitochondrial data. First, it is possible that one mode of expression may involve translating each reading frame pair as a single polypeptide. Site-specific ribosomal frameshifting during translation has been demonstrated for mutant yeast COII genes and for the overlapping coat and lysis protein genes of coliphage f2 (Fox & Weiss-Brummer, 1980; Beremand & Blumenthal, 1979). Alternatively, translation of the second reading frame may require ribosome binding and initiation of translation in the first reading frame. This has been shown to be the case for translation of the single-stranded RNA phage lysis protein gene, which requires prior initiation at and translation from the start codon of the upstream overlapping coat protein gene (Zinder & Lyons, 1968; Kirsh & Blumenthal, personal communication; R. A. Kastelein, personal communication). Finally, there is the formal possibility that additional promoters or processing sites in the URF A6L and URF 4L regions lead to the production of secondary RNA species that contain, respectively, only the ATPase 6 and the URF4 reading frames. In this regard it is noteworthy that in KB cell mitochondria there exist two stable transcripts, each having a different 5′ end, that map in the position of the ATPase 6 gene (Battey & Clayton, 1980).

(v) *Evolution of the protein-coding sequences*

The sequences of mammalian mitochondrial genomes evolve very fast. Using data based on comparisons of primate mtDNA restriction site maps, the rate of

TABLE 4

Sequence divergence in protein genes

Gene	Size (codons)†	Position 1 Obs.	Position 1 Est.	Position 2 Obs.	Position 2 Est.	Position 3 Obs.	Position 3 Est.
CO II	227	0·228	0·272	0·101	0·108	0·566	1·053
URF 5	603	0·253	0·309	0·144	0·160	0·469	0·735
URF 6	174	0·280	0·351	0·189	0·217	0·571	1·076
Average estimated divergence‡			0·311		0·162		0·955
Rate of base substitution per year ($\times 10^9$)			1·9		1·0		6·0

The differences in sequence for 3 human and bovine protein genes are shown as a function of position in the codon. The proportion of observed differences for each position (Obs.) was used to calculate an estimate of the total number of base substitutions (Est.) that have occurred since the 2 genomes diverged. This calculation was made using the formula of Jukes & Cantor (1969): estimate $= -(3/4)\ln(1-(4/3)(\text{observed}))$. Rates were calculated by dividing the average number of estimated base substitutions in each position by 2×80 million years, the approximate time of divergence for the human and bovine lines (Wilson *et al.*, 1977).

† Codons in non-overlapping regions of the genes (cf. Fig. 3) were not considered.

‡ Unweighted.

sequence drift has been estimated to be 5×10^{-9} to 10×10^{-9} base substitutions per nucleotide position per year (Brown *et al.*, 1979; Brown, 1980). This is at least twice the estimated rate for silent substitutions in mammalian globin genes (Kimura, 1981) and is comparable to the rate of evolution of globin pseudogenes (Miyata & Yasunaga, 1981; Li *et al.*, 1981), fibrinopeptides (Kafatos *et al.*, 1977) and preproinsulin C peptide (Perler *et al.*, 1980). Since the latter three cases are thought to represent sequences that are not subject to normal selective pressures, and hence are able to evolve at nearly the maximum rate for nuclear genes, it is extraordinary that the highly constrained mitochondrial genes should evolve at a similar rate.

In order to explore this question more fully, the pattern of base substitution was examined for several representative genes: COII, an assigned gene; URF5, a large unassigned gene; and URF6, the only H-strand coded protein gene. The formula of Jukes & Cantor (1969) was used to correct for multiple substitutions at single sites, thus providing an estimate of the total number of base substitutions per site that have occurred since human and bovine mtDNAs diverged from a common ancestor (Table 4). These data were calculated for each of the three codon positions and, using a value of 80 million years as the time of divergence, rates of base substitution were derived (Table 4). The results for each of the three genes were roughly the same, with the third codon position showing the highest rate of change, followed by the first then the second positions, as expected for genes coding for conserved amino acid sequences. The average rates for all three genes were $1{\cdot}9 \times 10^{-9}$, $1{\cdot}0 \times 10^{-9}$ and $6{\cdot}0 \times 10^{-9}$ changes per base-pair per year for the first, second and third codon positions, respectively; the rate averaged over all three codon positions was $3{\cdot}0 \times 10^{-9}$.

The rate of evolution calculated above is too low by at least a factor of two to account for the rate derived from restriction site comparisons (Brown *et al.*, 1979; Brown, 1980). However, the equation used to calculate the rates of divergence from

TABLE 5

Frequencies of specific base substitutions

Type of change		CO II (%)	URF 5 (%)	URF 6 (%)
Transitions:	A ↔ G	41 (20·1)	62 (11·9)	60 (33·0)
	C ↔ T	70 (34·3)	189 (36·3)	22 (12·1)
Transversions:	A ↔ C	41 (20·1)	149 (28·7)	7 (3·8)
	A ↔ T	33 (16·2)	89 (17·1)	37 (20·3)
	G ↔ C	11 (5·4)	23 (4·4)	12 (6·6)
	G ↔ T	8 (3·9)	8 (1·5)	44 (24·2)
Total		204 (100·0)	520 (99·9)	182 (100·0)

The human and bovine mitochondrial protein genes shown were aligned and the observed sequence differences tabulated according to the type of base substitution. Numbers in parentheses represent the percentages, for a given gene, of the various types of substitution.

the sequences (Jukes & Cantor, 1969) was derived by assuming that no codon usage bias existed and that the probability of any base changing to any other base was the same. That these assumptions are untrue for mammalian mtDNA is clearly shown by Tables 3 and 5. Furthermore, as discussed by Kimura (1981), comparisons of rapidly evolving sites in distantly related genes may lead to a substantial underestimate of the rate of change at those sites. This probably means that the calculated third position rates for mitochondrial DNA (Table 4) are too low. It remains to be seen whether these reservations about the rate calculations are sufficient to explain their disagreement with the divergence data derived using restriction enzyme methods. The human/bovine mtDNA sequence comparison has definitely confirmed, however, the extremely high rate at which mitochondrial genomes evolve at the nucleotide level. A more exact estimate of this rate will require a comparison between mtDNAs of more closely related species.

(c) *tRNA genes*

Twenty-two tRNA coding sequences have been identified in the human mitochondrial genome (Anderson *et al.*, 1981). Using the unique mitochondrial decoding mechanism, these tRNAs should be sufficient to read all of the non-termination codons in the mitochondrial genetic code, with the possible exception of the methionine codons (Barrell *et al.*, 1980*b*; Anderson *et al.*, 1981). The bovine mitochondrial genome also contains 22 tRNA genes that are homologous to the human ones and organized in virtually identical fashion (Figs 1 and 2). Full or partial sequences for the bovine mitochondrial tRNAs have confirmed all 22 of these gene assignments (B. A. Roe, personal communication).

(i) *Mitochondrial tRNAs compared to other tRNAs*

Although most of the mammalian mitochondrial tRNAs can form the familiar cloverleaf secondary structure, they are unusual in many other respects. Prokaryotic and non-mitochondrial eukaryotic tRNAs exhibit a number of universal or common features, both in lengths of loops and stems and in the sequence of nucleotides (Gauss & Sprinzl, 1981; Dirheimer *et al.*, 1979; Clark & Klug, 1975). Human and bovine mitochondrial tRNAs lack many of these features, notably the universal sequence T-Ψ-C-R-A, the constant seven-nucleotide size of the "TΨC" loop and a number of universal bases in the "DHU" loop and elsewhere. The most extreme case is that of $tRNA^{Ser}_{AGY}$, which lacks the entire "DHU" arm (de Bruijn *et al.*, 1980; Arcari & Brownlee, 1980). Apart from this special case, the sizes of all four stems have been conserved.

The only region to have conserved universal features, apart from the CCA at the 3′ terminus (B. A. Roe, personal communication), is the anticodon loop where the first nucleotide is always a pyrimidine, and the nucleotides immediately preceding and following the anticodon are always a U and a purine, respectively. The methionine tRNAs, however, represent a special case. Among all of the 44 known human and bovine mitochondrial tRNAs, these are unique in having a C in the first (wobble) position of the anticodon and also a C adjacent to the anticodon on the 5′ side. These sequence properties, as well as further homologies in the anticodon arm to known eukaryotic initiator tRNAs (Dirheimer *et al.*, 1979), led to the provisional identification of each of these as a $tRNA^{Met}_{f}$. However, as discussed by Anderson *et al.* (1981), the presence of a CAU anticodon would not necessarily limit these tRNAs to reading AUG initiation codons.

Some of the mitochondrial tRNAs also exhibit invariant or semi-invariant features found in non-mitochondrial tRNAs (Fig. 4). For example, in 35 tRNAs the first base in the "DHU" loop is an A, and in half of the tRNA sequences nucleotides 8 and 9 consist of UA. In the variable loop, 32 tRNAs have a purine as the second base and 35 have a pyrimidine as the last base (Fig. 4). This suggests that for most of the mitochondrial tRNAs, at least some tertiary interactions of the type found in yeast cytoplasmic $tRNA^{Phe}$ (Jack *et al.*, 1976) would be possible.

The stems of the mitochondrial tRNAs contain more than the usual number of non-standard base-pairs. The predominant non-standard pairs, A·C and G·U, are non-randomly distributed both within the individual tRNAs and within the total mitochondrial tRNA population. A·C pairs, which normally occur primarily in the aminoacyl and "DHU" stems, are found mainly in the anticodon and "TΨC" stems of the mitochondrial tRNAs. Furthermore, while the A·C pairs are found exclusively in those tRNAs produced from L-strand sense transcripts, most of the G·U pairs are found in the H-strand sense tRNAs.

A striking feature of the mammalian mitochondrial tRNAs is their high A+U content, which is in marked contrast to the base composition of all non-mitochondrial tRNAs sequenced to date (Gauss & Sprinzl, 1981). Mitochondrial tRNAs from *A. nidulans* (Köchel *et al.*, 1981) also show this A+U bias. The G+C/A+U ratio for non-mitochondrial tRNAs usually varies between one and two (Eigen & Winkler-Oswatitsch, 1981), but for the human and bovine mitochondrial tRNAs it ranges from 0·25 to 0·94. Consequently, the stem regions have a high proportion of A·U base-pairs. This, together with the absence of otherwise universally conserved features and a higher incidence of non-standard base-pairs, must result in a considerable weakening of the overall tertiary structure of these tRNAs.

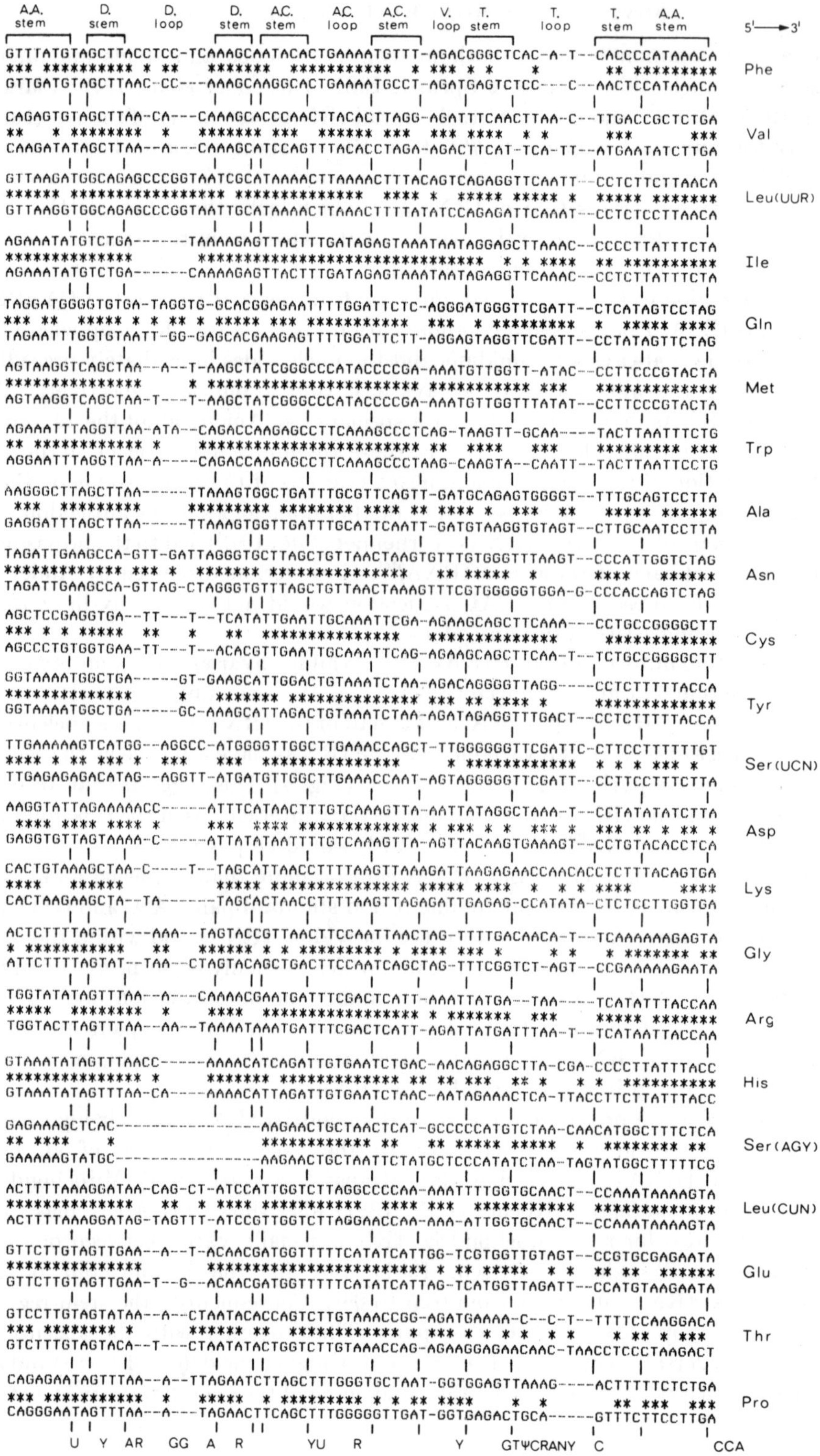
A.A. stem
D. stem
D. loop
D. stem
A.C. stem
A.C. loop
A.C. stem
V. loop
T. stem
T. loop
T. stem
A.A. stem
5'→3'
Phe
Val
Leu(UUR)
Ile
Gln
Met
Trp
Ala
Asn
Cys
Tyr
Ser(UCN)
Asp
Lys
Gly
Arg
His
Ser(AGY)
Leu(CUN)
Glu
Thr
Pro
U Y AR GG A R YU R Y GTΨCRANY C CCA

FIG. 4.

(ii) *Comparison of the human and bovine mitochondrial tRNAs*

Pairs of homologous human and bovine mitochondrial tRNA genes were aligned and examined for sequence differences (Fig. 4). This comparison revealed a variability not found in non-mitochondrial mammalian tRNAs, which show a high degree of conservation (Gauss & Sprinzl, 1981). In order to visualize the general pattern of structural variation, the differences for 21 of the 22 homologous pairs† were summed and plotted as a function of position in the tRNA structure (Fig. 5).

The "TΨC" loop was found to be the most variable region, both in size and in sequence. For example, the sequence of this loop in the human $tRNA^{Thr}$ is C-C-T, while in bovine $tRNA^{Thr}$ it is A-C-A-A-C-T-A-A. The most conserved portions, based on this pairwise comparison, were the so-called variable loop, the stem of the "DHU" arm, nucleotides 8 and 9, the tip of the aminoacyl stem, and the distal half of the anticodon stem (Fig. 5). The anticodon loop was, of course, also highly conserved. Since these structures have been conserved in the related tRNAs during the course of mammalian evolution, they may be necessary for type-specific functions such as aminoacylation. It should be stressed, however, that when sequences of mitochondrial tRNAs from more organisms become available, certain tRNAs may well exhibit subtly different patterns of variation.

The pattern of sequence divergence between the human and bovine mitochondrial tRNA genes seems to be constrained by certain structural requirements of the tRNAs. Like the sequence divergence in the protein coding regions (Table 5), A ↔ G and T ↔ C transitions are the predominant base changes, with G ↔ C and G ↔ T transversions being rare; this tendency is especially marked in the stem regions. Changes in the stem regions often conserve base-pairing (47 cases), which may explain the strong bias in favour of transitions (43 out of the 47 cases), since these allow either Watson–Crick base-pair to mutate to the other *via* a G·U or A·C intermediate. Necessary base-stacking interactions may also contribute to this bias, because in the stems purines are usually found adjacent to purines, and pyrimidines next to pyrimidines, as would be expected for helices of maximum rigidity (Dickerson & Drew, 1981). The aminoacyl acceptor stems are quite asymmetric in this regard, with one strand usually composed primarily of purines and the other of pyrimidines.

† Because of its anomalous structure, the $tRNA^{Ser}_{AGU/C}$ was omitted from this comparison.

FIG. 4. Comparison of the human and bovine mitochondrial tRNAs. A pairwise alignment of the mitochondrial tRNA genes is shown, with asterisks printed between the sequences to indicate positions of homology. The top sequence of each pair is human and the bottom bovine. All of the tRNA pairs are also aligned with each other in order to bring the sequences of the various stems and loops (marked along the top of the Figure) into vertical register. In general, the loop sizes were quite variable, so spaces were introduced in the loops to allow alignment of the other parts of the molecule. The alignments were accomplished by applying 2 criteria. First, tRNA genes were aligned so as to maximize homology in the pairs. Second, those tRNA genes that contained elements of classical tRNA structure, i.e. those that appeared to have conserved invariant or semi-invariant bases (Kim, 1979), were aligned to bring all of the "universal" nucleotides into register. These nucleotides and their positions are printed along the bottom of the Figure. Because of the extreme variability in sequence and in size of the loops, it was not possible in every case to position the nucleotides unambiguously. The somewhat arbitrary positioning of these few nucleotides did not affect the overall alignments or extents of homology, however.

FIG. 5. Regions of structural variation in mammalian mitochondrial tRNAs. The pairwise tRNA gene comparisons shown in Fig. 4 were used to construct a histogram showing sequence variability as a function of position in the tRNAs. (The $tRNA^{Ser}_{AGY}$ was not included because of its anomalous structure.) For each position in the tRNAs, the number of homologous pairs showing a mismatch at that position was plotted as a percentage of the number of pairs having at least 1 nucleotide at that position. The shaded portions designate the loop regions where in some positions fewer than 21 pairs were used as the basis for the calculation (cf. Fig. 4). A nucleotide aligned with a space at any position in a pair was counted as a mismatch.

(d) *Ribosomal RNA genes*

(i) *3′ Termini of mitochondrial ribosomal RNAs*

When the sequences of the human mitochondrial rRNA genes were published, the exact 3′ termini of the transcripts were not known (Eperon *et al.*, 1980), and homology with other rRNAs did not extend to the 3′ termini. However, the butt-jointing of the 5′ termini to tRNA genes suggested that the same arrangement could be found at the 3′ termini of the rRNA genes.

In order to establish the exact relationship of the bovine mitochondrial 12 S rRNA gene to the flanking $tRNA^{Val}$ gene, the 3′ end sequence of the bovine 12 S rRNA was determined directly (Eperon, 1981). For the bulk of the 12 S rRNA, the

3′-terminal nucleotide was found to be the U specified by the T immediately adjacent to the tRNA gene (Fig. 1); a very small proportion of the 12 S rRNA consisted of the monoadenylation product of the major component (Eperon, 1981). Thus, it appears that in bovine mtDNA, the 12 S rRNA gene is butt-jointed at its 3′ terminus to a tRNA gene. This result suggests that the rRNA precursor transcript is processed, like the less abundant mRNA precursors, by excision of tRNA sequences (Ojala *et al.*, 1980; Anderson *et al.*, 1981).

The 3′ termini of the 12 S rRNAs from mouse and hamster have also been sequenced (Baer & Dubin, 1980; Van Etten *et al.*, 1980). When the 3′ terminus of the hamster rRNA was aligned with the murine gene sequence, it appeared that the major component had been monoadenylated (Baer & Dubin, 1980). This presumption was strengthened by the discovery of a minor component that was shorter by one nucleotide (R. Baer & D. T. Dubin, personal communication). However, in the absence of hamster mtDNA sequence information, it was not formally possible to conclude that the rRNA and tRNA genes were butt-jointed. The mouse 12 S rRNA appeared to end with an A rather than the U expected from the gene sequence (Van Etten *et al.*, 1980). It is possible, therefore, that in mouse mitochondria the 12 S rRNA gene ends one nucleotide before the $tRNA^{Val}$ gene, and the 12 S rRNA itself is monoadenylated.

There is no direct sequence information available for the bovine mitochondrial 16 S rRNA, but a sequence from the 3′ terminus of a hamster 16 S rRNA has been reported (Dubin *et al.*, 1981). This terminus appeared to be ragged, with the longest species extending up to the nucleotide adjacent to the $tRNA^{Leu}$ gene in the murine mtDNA sequence, and was also oligoadenylated. The 3′ terminus of HeLa cell 16 S rRNA appears to be similar (D. T. Dubin & G. Attardi, personal communication) and, since the 3′ terminus of the bovine 16 S rRNA gene is quite homologous to that of HeLa cells, it is likely that the longest bovine transcript extends to the nucleotide adjacent to the $tRNA^{Leu}$ gene. The ragged termination (Dubin *et al.*, 1981) and the high concentration of nascent transcripts from the region of the rRNA genes (Cantatore & Attardi, 1980; Ojala *et al.*, 1980) would be consistent with a site for attenuation of transcription in this region. The small hairpin loop that might be involved (Dubin *et al.*, 1981) is found in both the human and the bovine sequences. However, the sequence is not well-conserved, and the size of the loop also varies between organisms (D. T. Dubin, personal communication). It is, therefore, not possible to confirm that such a sequence occurs in the genome only where termination is expected.

(ii) *Structure of mitochondrial rRNAs*

The sequences of the human mitochondrial rRNAs have been compared with those of the *E. coli* rRNAs (Brosius *et al.*, 1978,1980; Eperon *et al.*, 1980). The low degree of homology between the *E. coli* and human mitochondrial rRNAs made an extensive alignment of these impossible, although some sequences were identified in the mitochondrial rRNAs that are conserved amongst prokaryotes or crucial for the association of *E. coli* ribosomal subunits (Eperon *et al.*, 1980). However, it is possible to align sequences with low homology more accurately if analogous

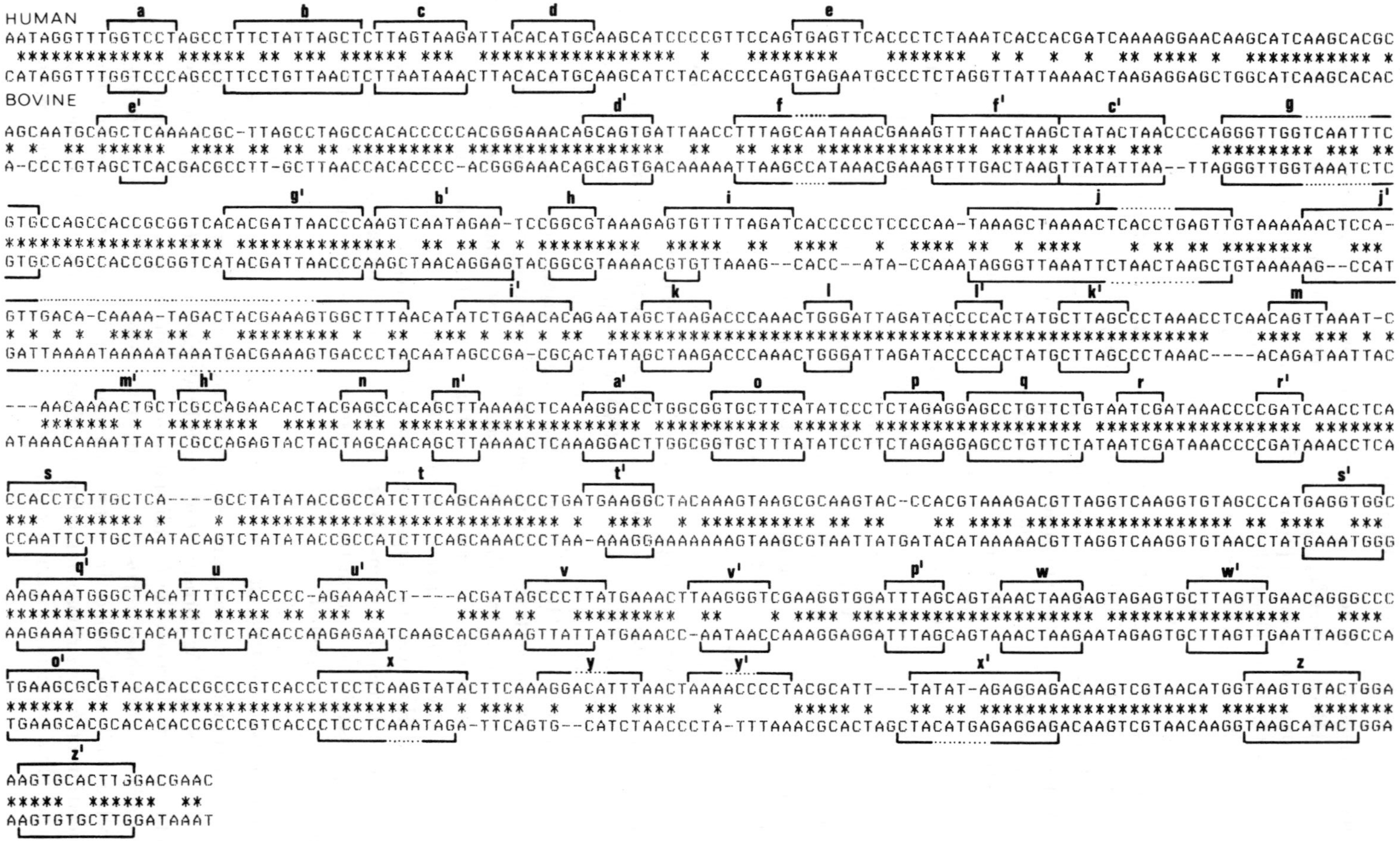

FIG. 6. Comparison of the human and bovine 12 S rRNA genes. The human and bovine 12 S rRNA gene sequences are shown aligned, with spaces introduced in order to maximize the homology (asterisks). The bracketed nucleotides represent the regions that are base-paired in the 12 S rRNA secondary structure model proposed by H. F. Noller *et al.* (unpublished results). Each of the base-paired regions (stems) is labelled with a pair of letters, 1 primed and 1 unprimed, to designate the 2 strands of the stem. The stems are lettered sequentially in the order that they occur in the primary sequence. Dotted lines indicate unpaired regions within stems.

secondary structures can be demonstrated. Recently, a substantial body of evidence from ribosomes or smaller complexes and from isolated rRNA has allowed possible secondary structures for the *E. coli* 16 S rRNA to be deduced and compared with the sequences of other bacterial rRNAs (Woese *et al.*, 1980; Noller & Woese, 1981; Glotz & Brimacombe, 1980; Stiegler *et al.*, 1981). Several possible secondary structures of human mitochondrial 12 S rRNA that were analogous to that of *E. coli* were tested for compatibility with the bovine 12 S rRNA sequence, using conservation of secondary structure as a criterion, and a single "consensus" structure was derived (H. F. Noller, personal communication). A more detailed account of this putative secondary structure for the mammalian mitochondrial 12 S rRNA will be published elsewhere (H. F. Noller, personal communication), but several points will be considered here using the structure as the basis for an accurate alignment (Fig. 6). A similar model for the mitochondrial 12 S rRNA secondary structure has been published by Zwieb *et al.* (1981).

The major differences between the mitochondrial 12 S rRNAs and the 16 S rRNA of *E. coli* include a shortening and simplification of the structure corresponding to the first 650 nucleotides of the *E. coli* sequence. The first half of this region also appears to be quite different in the 18 S rRNAs from eukaryotic cytoplasm (Rubtsov *et al.*, 1980; Salim & Maden, 1981), although the structure here is not highly variable among prokaryotes (Noller & Woese, 1981; Carbon *et al.*, 1981). The most dramatic change is in the 410 to 500 region of the *E. coli* rRNA, which is lost in mammalian mitochondrial rRNA; the 590 to 650 region, very variable in prokaryotes (Noller & Woese, 1981), also appears to have been lost in mammalian mitochondria. The other gross change exhibited by the mammalian mitochondrial 12 S rRNA is a considerable reduction in the size of the region corresponding to nucleotides 990 to 1214 of *E. coli* 16 S rRNA, wherein a considerable amount of secondary structure may have been lost, even though the sequences of some loops are still identifiable. Overall, only 11 of the 29 "universal" oligonucleotides found in prokaryotes (Woese *et al.*, 1975) were identified in human mitochondrial 12 S rRNA (Eperon *et al.*, 1980), but all of the nine proposed universal base-pairing interactions are formed (Stiegler *et al.*, 1981; Noller *et al.*, unpublished). The existence of three of the proposed hairpin loops near the 3′ terminus has been demonstrated in BHK cell mitochondrial small subunit rRNA (Baer & Dubin, 1980). The proportion of the entire 12 S rRNA that is predicted to be involved in secondary structure is approximately 40%, rather less than is found in the *E. coli* 16 S rRNA (Noller & Woese, 1981).

The most strongly conserved sequences amongst prokaryotic and eukaryotic small subunit rRNAs are found in the single-stranded regions (Noller & Woese, 1981; Stiegler *et al.*, 1981; Salim & Maden, 1981; Eperon, 1981). Similarly, in the human and bovine mitochondrial 12 S rRNAs, 64% of the 327 nucleotides homologous to the 16 S rRNA are single-stranded in the *E. coli* structure (Eperon, 1981). However, the changes between the human and bovine 12 S rRNA sequences exhibit a different tendency. Including additions and deletions as changes, the two sequences differ in 20% of the stem nucleotides and in 24% of the single-stranded nucleotides (Fig. 6). Only 6% of the nucleotides in the human sequence that are homologous to the *E. coli* sequence are different in the bovine sequence, and *vice versa*. It appears, therefore, that the 12 S rRNA sequences held in common with other organisms are necessary for the function of the mitochondrial ribosomes, whilst the remainder, whether in loops or stems, are relatively free to change.

The stems in the mitochondrial small subunit rRNAs contain a large number of non-standard base-pairs (H. F. Noller, personal communication). However, as in the tRNA genes, the changes between the human and bovine sequences are constrained. Most strikingly, 14 of the 15 cases where a G·C ⇄ A·T change has occurred represent transitions, as would be expected for base-pairs mutating *via* a G·U or A·C intermediate.

The human and bovine large subunit rRNA sequences have an overall homology of 77% and are most conserved near their 3′ ends. This is the region that is most homologous to the *E. coli* 23 S rRNA sequence (Eperon *et al.*, 1980; Brosius *et al.*, 1980). Kethoxal modification has shown that this region in the 23 S rRNA is protected during association of *E. coli* ribosomal subunits (Herr & Noller, 1979). Chloramphenicol resistance mutants of yeast, human and mouse mitochondria (Dujon, 1980; Kearsey & Craig, 1981; Blanc *et al.*, 1981) and affinity labelling of *E. coli* ribosomes (Greenwell *et al.*, 1974; Eckerman & Symons, 1978) suggest that these sequences are also involved in binding the 3′ terminus of aminoacyl tRNA in the ribosomal A site. Baer & Dubin (1981) have proposed a secondary structure for rodent mitochondrial and *E. coli* large subunit rRNAs in this region, based on a correlation between a methylated oligonucleotide from hamster mitochondrial rRNA and one of several possible positions for a methylated oligonucleotide in the *E. coli* 23 S rRNA. Both the bovine and the human 16 S rRNAs fit this structure. Recently, two models for the complete secondary structure of the *E. coli* 23 S rRNA have been proposed and used to derive similar structures for the human and mouse 16 S rRNAs (Glotz *et al.*, 1981; Branlant *et al.*, 1981). We have not attempted to reconcile the bovine 16 S rRNA sequence with either of these putative structures.

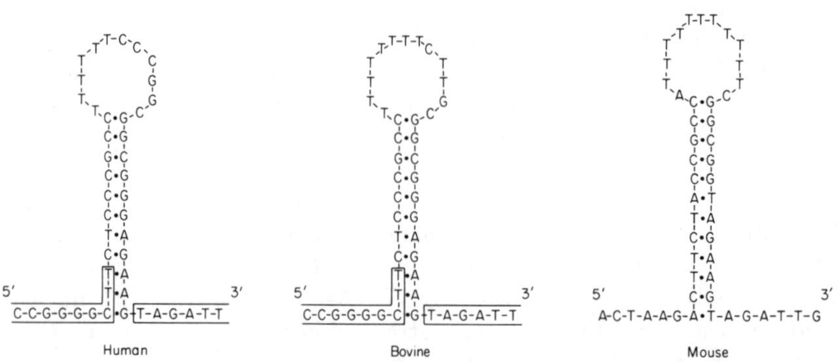

FIG. 7. Mitochondrial L-strand origin of replication. Shown is the possible secondary structure formed by the H-strand (template) at the origin of L-strand synthesis during mtDNA replication. The structures for the human, bovine and mouse (Martens & Clayton, 1979) mitochondrial genomes are compared. For the human and bovine origins sequences forming part of the flanking tRNA^{Cys} and tRNA^{Asn} genes (Fig. 1) are shown boxed.

(e) *Origins of replication*

The bovine mitochondrial L-strand origin of replication, like that for human and mouse mitochondria (Martens & Clayton, 1979; Anderson *et al.*, 1981; Tapper & Clayton, 1981), can be drawn in the form of a hairpin loop structure (Fig. 7). The genome position, the size of the stem and loop, and most of the primary sequence are all highly conserved features. It is likely, therefore, that all of these structural attributes represent elements that are important to the function of this origin during the mammalian mitochondrial DNA replication process.

Unlike the L-strand origin of replication, and indeed unlike the rest of the genome, the region containing the H-strand origin of replication (D-loop region; Figs 1 and 2) is highly variable in sequence and in size. This region between the $tRNA^{Pro}$ and $tRNA^{Phe}$ genes, where the bovine mtDNA D-loop and 7 S DNA are located (Hauswirth *et al.*, 1980), is 212 nucleotides shorter than the corresponding region in the human mitochondrial genome (Anderson *et al.*, 1981). This size variability accounts for the bulk of the total size difference between the human and bovine mtDNAs. Figure 8 shows an alignment of the human and bovine mtDNA sequences in the D-loop region. Apart from several small blocks of recognizable homology, the sequences of the two genomes have diverged markedly in this region. The homologous portions are mainly confined to the area covered by the (human) 7 S DNA in the D-loop; this is consistent with the observations reported by Upholt & Dawid (1977), who showed, using goat and sheep mtDNA heteroduplexes, that the middle of the D-loop region was more conserved than its extremities. In the region between the 5′ end of the 7 S DNA and the $tRNA^{Phe}$ gene, where the sequence variability is most extreme, the only significant homology between the two genomes occurs in a short C-rich tract (block A, Fig. 8). A similar C-rich segment has been found in the mitochondrial genomes of spontaneous yeast petite mutants and is thought to represent part of the yeast mtDNA replication origin (de Zamaroczy *et al.*, 1981).

All of the bovine/human D-loop homology blocks except H and I (Fig. 8) are also conserved in the mouse and rat mtDNA sequences (Walberg & Clayton, 1981; Sekiya *et al.*, 1980; K. Koike, personal communication). Blocks H and I comprise part of a tandemly repeated sequence (Fig. 8) that may have been deleted in rodent DNA. These blocks are located near the position of the human 7 S DNA 3′ end (Doda *et al.*, 1981), and it is intriguing to speculate that they may be involved in regulating the size of this species. Human 7 S DNA has a specific 3′ end, but in mouse mtDNA, where the repeat is missing, the 7 S DNA represents a family of molecules having different 3′ termini (Brown *et al.*, 1978; Gillum & Clayton, 1978; Bogenhagen *et al.*, 1979; Doda *et al.*, 1981).

The homology blocks A to K represent "islands" of conserved sequence in a region that otherwise has been highly divergent. This would argue that the sequences represented by the blocks are subject to functional constraints. Control signals for the initiation of H-strand synthesis during DNA replication probably exist in this region (Crews *et al.*, 1979; Bogenhagen *et al.*, 1979), and it is likely that *cis*-acting elements regulating L- and H-strand transcription are also located here (Cantatore & Attardi, 1980). Other sequences may govern the synthesis and

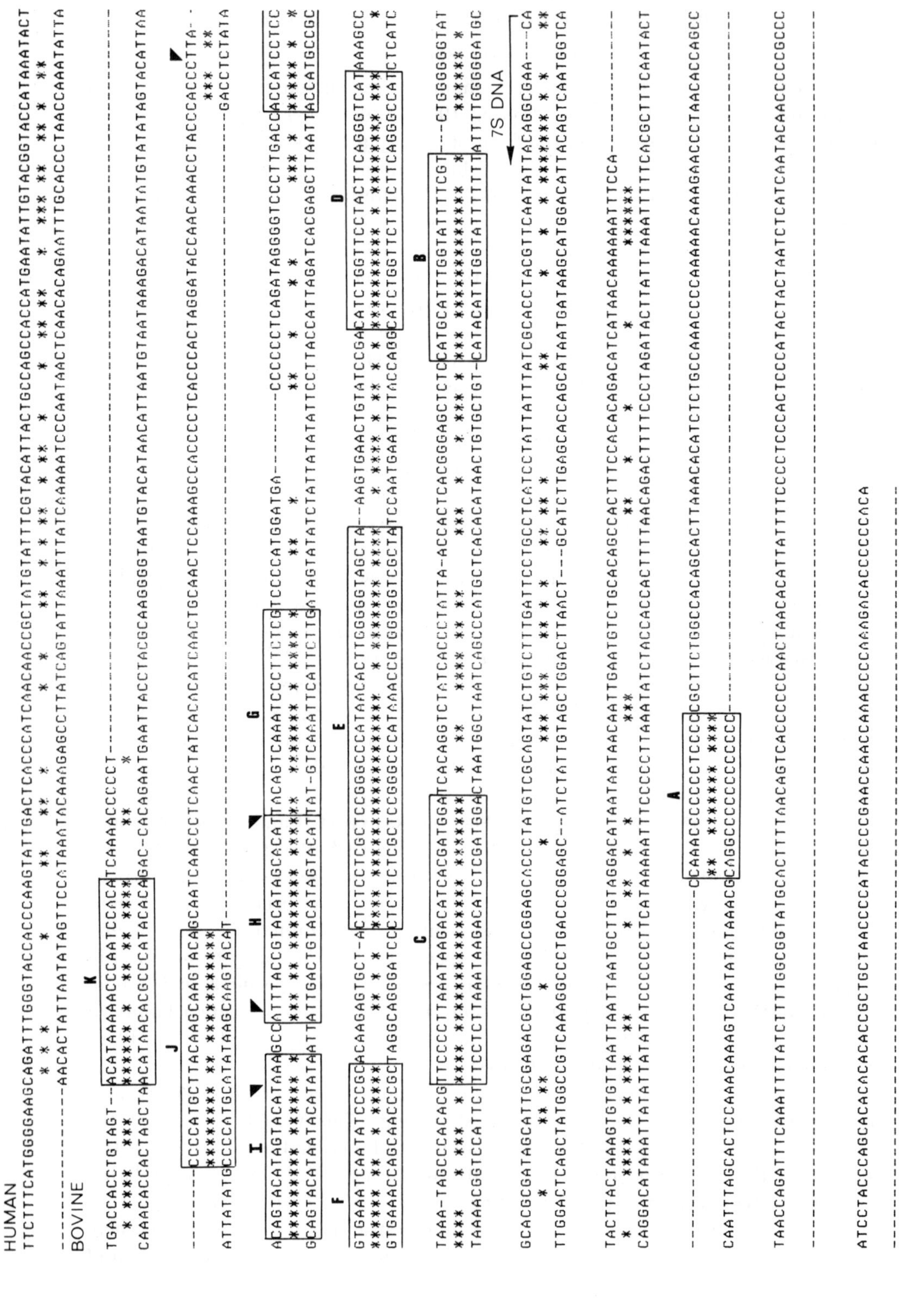

FIG. 8. Homology between the human and bovine mtDNA D-loop regions. An alignment of the human and bovine mitochondrial genomes (L-strand) in the region between the $tRNA^{Pro}$ and $tRNA^{Phe}$ genes (Fig. 1) is given. Major blocks of homology are shown in boxes. Also shown is the position of the 5′ end of the 7 S DNA in the D-loop of human mtDNA (Crews *et al.*, 1979). ◤ and ◥ indicate a tandemly repeated sequence overlapping blocks H and I (see the text).

turnover of non-replicative forms of the 7 S DNA species (Bogenhagen & Clayton, 1978). Two small transcripts, an H-strand sense 7 S RNA from HeLa cell mitochondria (Almaric *et al.*, 1978; Ojala *et al.*, 1981*b*) and an L-strand sense 4 S RNA from *Xenopus* mitochondria (Ohi *et al.*, 1978) have been mapped near the D-loop. However, none of the above phenomena has been related to any of the conserved sequences found in the D-loop region. It seems unlikely, though, that a genome so generally frugal in its use of genetic information would retain long stretches of functionally silent DNA.

The authors thank Drs B. G. Barrell and J. E. Walker for helpful discussions. We are grateful to Drs T. Blumenthal, G. Buse, D. A. Clayton, R. W. Davies, D. T. Dubin, R. A. Kastelein, K. Koike, W. Machleidt, H. F. Noller, B. A. Roe and J. E. Walker for communicating results prior to publication. We also thank Mr R. Staden for allowing us to use his unpublished sequence analysis programs. One author (S.A) was supported by fellowships from the Cystic Fibrosis Foundation and from the National Institutes of Health.

REFERENCES

Almaric, F., Merkel, C., Gelfand, R. & Attardi, G. (1978). *J. Mol. Biol.* **118**, 1–25.

Aloni, Y. & Attardi, G. (1971). *Proc. Nat. Acad. Sci., U.S.A.* **68**, 1757–1761.

Anderson, S. (1981). *Nucl. Acids Res.* **9**, 3015–3027.

Anderson, S., Gait, M. J., Mayol, L. & Young, I. G. (1980). *Nucl. Acids Res.* **8**, 1731–1743.

Anderson, S., Bankier, A. T., Barrell, B. G., de Bruijn, M. H. L., Coulson, A. R., Drouin, J., Eperon, I. C., Nierlich, D. P., Roe, B. A., Sanger, F., Schreier, P. H., Smith, A. J. H., Staden, R. & Young, I. G. (1981). *Nature (London)*, **290**, 457–465.

Arcari, P. & Brownlee, G. G. (1980). *Nucl. Acids Res.* **8**, 5207–5212.

Baer, R. J. & Dubin, D. T. (1980). *Nucl. Acids Res.* **8**, 4927–4941.

Baer, R. J. & Dubin, D. T. (1981). *Nucl. Acids Res.* **9**, 323–337.

Barrell, B. G., Bankier, A. T. & Drouin, J. (1979). *Nature (London)*, **282**, 189–194.

Barrell, B. G., Anderson, S., Bankier, A. T., de Bruijn, M. H. L., Chen, E., Coulson, A. R., Drouin, J., Eperon, I. C., Nierlich, D. P., Roe, B. A., Sanger, F., Schreier, P. H., Smith, A. J. H., Staden, R. & Young, I. G. (1980*a*). In *31st Mosbach Colloquium on Biological Chemistry of Organelle Formation* (Bucher, Th., Sebald, W. & Weiss, H., eds), pp. 11–25, Springer-Verlag, Berlin.

Barrell, B. G., Anderson, S., Bankier, A. T., de Bruijn, M. H. L., Chen, E., Coulson, A. R., Drouin, J., Eperon, I. C., Nierlich, D. P., Roe, B. A., Sanger, F., Schreier, P. H., Smith, A. J. H., Staden, R. & Young, I. G. (1980*b*). *Proc. Nat. Acad. Sci., U.S.A.* **77**, 3164–3166.

Battey, J. & Clayton, D. A. (1980). *J. Biol. Chem.* **255**, 11599–11606.

Beremand, M. N. & Blumenthal, T. (1979). *Cell*, **18**, 257–266.

Blanc, H., Wright, C. T., Bibb, M. J., Wallace, D. C. & Clayton, D. A. (1981). *Proc. Nat. Acad. Sci., U.S.A.* **78**, 3789–3793.

Bogenhagen, D. & Clayton, D. A. (1978). *J. Mol. Biol.* **119**, 49–68.

Bogenhagen, D., Gillum, A. M., Martens, P. A. & Clayton, D. A. (1979). *Cold Spring Harbor Symp. Quant. Biol.* **43**, 253–262.

Branlant, C., Krol, A., Machatt, M. A., Pouyet, J. & Ebel, J.-P. (1981). *Nucl. Acids Res.* **9**, 4303–4324.

Brosius, J., Palmer, M. L., Kennedy, P. T. & Noller, H. F. (1978). *Proc. Nat. Acad. Sci., U.S.A.* **75**, 4801–4805.

Brosius, J., Dull, T. J. & Noller, H. F. (1980). *Proc. Nat. Acad. Sci., U.S.A.* **77**, 201–204.

Brown, W. M. (1980). *Proc. Nat. Acad. Sci., U.S.A.* **77**, 3605–3609.

Brown, W. M., Shine, J. & Goodman, H. M. (1978). *Proc. Nat. Acad. Sci., U.S.A.* **75**, 735–739.

Brown, W. M., George, M. & Wilson, A. C. (1979). *Proc. Nat. Acad. Sci., U.S.A.* **76**, 1967–1971.

Cantatore, P. & Attardi, G. (1980). *Nucl. Acids Res.* **8**, 2605–2624.

Carbon, P., Ebel, J.-P. & Ehresmann, C. (1981). *Nucl. Acids Res.* **9**, 2325–2333.

Clark, B. F. C. & Klug, A. (1975). In *Proceedings of the Tenth FEBS Meeting* (Chapeville, F. & Grunberg-Manago, M., eds), vol. 39, pp. 183–189, North-Holland, Amsterdam.

Crews, S., Ojala, D., Posakony, J., Nishiguchi, J. & Attardi, G. (1979). *Nature (London)*, **277**, 192–198.

de Bruijn, M. H. L., Schreier, P. H., Eperon, I. C., Barrell, B. G., Chen, E. Y., Armstrong, P. W., Wong, J. F. H. & Roe, B. A. (1980). *Nucl. Acids Res.* **8**, 5213–5222.

de Zamaroczy, M., Marotta, R., Faugeron-Fonty, G., Goursot, R., Mangin, M., Baldacci, G. & Bernardi, G. (1981). *Nature (London)*, **292**, 75–78.

Dickerson, R. E. & Drew, H. R. (1981). *J. Mol. Biol.* **149**, 761–786.

Dirheimer, G., Keith, G., Sibler, A.-P. & Martin, R. P. (1979). In *Transfer RNA: Structure, Properties and Recognition* (Schimmel, P. R., Söll, D. & Abelson, J. N., eds), pp. 19–41, Cold Spring Harbor Laboratory Press, Cold Spring Harbor.

Doda, J. N., Wright, C. T. & Clayton, D. A. (1981). *Proc. Nat. Acad. Sci., U.S.A.* **78**, 6116–6120.

Dubin, D. T., Timko, K. D. & Baer, R. J. (1981). *Cell*, **23**, 271–278.

Dujon, B. (1980). *Cell*, **20**, 185–197.

Eckermann, D. J. & Symons, R. H. (1978). *Eur. J. Biochem.* **82**, 225–234.

Efstratiadis, A., Posakony, J. W., Maniatis, T., Lawn, R. M., O'Connell, C., Spritz, R. A., DeRiel, J. K., Forget, B. G., Weissman, S. M., Slighton, J. L., Blechl, A. E., Smithies, O., Baralle, F. E., Shoulders, C. C. & Proudfoot, N. J. (1980). *Cell*, **21**, 653–668.

Eigen, M. & Winkler-Oswatitsch, R. (1981). *Die Naturwiss.* **68**, 217–228.

Eperon, I. C. (1981). Ph.D. thesis, Cambridge University.

Eperon, I. C., Anderson, S. & Nierlich, D. P. (1980). *Nature (London)*, **286**, 460–467.

Fox, T. D. & Weiss-Brummer, B. (1980). *Nature (London)*, **288**, 60–63.

Gardner, R. C., Howarth, A. J., Hahn, R., Brown-Luedi, M., Shepherd, R. J. & Messing, J. (1981). *Nucl. Acids Res.* **9**, 2871–2888.

Gauss, D. H. & Sprinzl, M. (1981). *Nucl. Acids Res.* **9**, r1–r23.

Gay, N. J. & Walker, J. E. (1981). *Nucl. Acids Res.* **9**, 3919–3926.

Gillum, A. M. & Clayton, D. A. (1978). *Proc. Nat. Acad. Sci., U.S.A.* **75**, 677–681.

Glotz, C. & Brimacombe, R. (1980). *Nucl. Acids Res.* **8**, 2377–2395.

Glotz, C., Zwieb, C., Brimacombe, R., Edwards, K. & Kössel, H. (1981). *Nucl. Acids Res.* **9**, 3287–3306.

Grantham, R., Gautier, C. & Gouy, M. (1980). *Nucl. Acids Res.* **8**, 1893–1912.

Greenwell, P., Harris, R. J. & Symons, R. H. (1974). *Eur. J. Biochem.* **49**, 539–554.

Gronenborn, B. & Messing, J. (1978). *Nature (London)*, **272**, 375–377.

Hauswirth, W. W., Laipis, P. J., Gilman, M. E., O'Brien, T. W., Michaels, G. S. & Rayfield, M. A. (1980). *Gene*, **8**, 193–209.

Heidecker, G., Messing, J. & Gronenborn, B. (1980). *Gene*, **10**, 69–73.

Hensgens, L. A. M., Grivell, L. A., Borst, P. & Bos, J. L. (1979). *Proc. Nat. Acad. Sci., U.S.A.* **76**, 1663–1667.

Herr, W. & Noller, H. F. (1979). *J. Mol. Biol.* **130**, 421–432.

Jack, A., Ladner, J. E. & Klug, A. (1976). *J. Mol. Biol.* **108**, 619–649.

Jukes, T. H. & Cantor, C. R. (1969). In *Mammalian Protein Metabolism* (Munro, H. N., ed.), vol. 3, pp. 21–132, Academic Press, New York.

Kafatos, F. C., Efstratiadis, A., Forget, B. G. & Weissman, S. M. (1977). *Proc. Nat. Acad. Sci., U.S.A.* **74**, 5618–5622.

Kearsey, S. E. & Craig, I. W. (1981). *Nature (London)*, **290**, 607–608.

Kim, S.-H. (1979). In *Transfer RNA: Structure, Properties and Recognition* (Schimmel, P. R., Söll, D. & Abelson, J. N., eds), pp. 83–100, Cold Spring Harbor Laboratory Press, Cold Spring Harbor.

Kimura, M. (1981). *Proc. Nat. Acad. Sci., U.S.A.* **78**, 454–458.

Köchel, H. G., Lazarus, C. M., Basak, N. & Küntzel, H. (1981), *Cell*, **23**, 625–633.
Li, W.-H., Gojobori, I. & Nei, M. (1981). *Nature (London)*, **292**, 237–239.
Macino, G. & Tzagoloff, A. (1979). *J. Biol. Chem.* **254**, 4617–4623.
Martens, P. A. & Clayton, D. A. (1979). *J. Mol. Biol.* **135**, 327–351.
Messing, J., Gronenborn, B., Muller-Hill, B. & Hofschneider, P. Y. (1977). *Proc. Nat. Acad. Sci., U.S.A.* **74**, 3642–3646.
Messing, J., Crea, R. & Seeburg, P. H. (1981). *Nucl. Acids Res.* **9**, 309–321.
Miyata, T. & Yasunaga, T. (1981). *Proc. Nat. Acad. Sci., U.S.A.* **78**, 450–453.
Montoya, J., Ojala, D. & Attardi, G. (1981). *Nature (London)*, **290**, 465–470.
Murphy, W., Attardi, B., Tu, C. & Attardi, G. (1975). *J. Mol. Biol.* **99**, 809–814.
Noller, H. F. & Woese, C. R. (1981). *Science*, **212**, 403–411.
Ohi, S., Ramirez, J. L., Upholt, W. B. & Dawid, I. B. (1978). *J. Mol. Biol.* **121**, 299–310.
Ojala, D., Merkel, C., Gelfand, R. & Attardi, G. (1980). *Cell*, **22**, 393–403.
Ojala, D., Montoya, J. & Attardi, G. (1981*a*). *Nature (London)*, **290**, 470–474.
Ojala, D., Crews, S., Montoya, J., Gelfand, R. & Attardi, G. (1981*b*). *J. Mol. Biol.* **150**, 303–314.
Perler, F., Efstratiadis, A., Lomedico, P., Gilbert, W., Kolodner, R. & Dodgson, J. (1980). *Cell*, **20**, 555–556.
Proudfoot, N. J. & Brownlee, G. G. (1976). *Nature (London)*, **263**, 211–214.
Rothstein, R. J., Lau, L. F., Bahl, C. P., Narang, S. A. & Wu, R. (1980). In *Methods in Enzymology* (Colowick, S. P. & Kaplan, N. C., eds), vol. 68, pp. 101–110, Academic Press, New York.
Rubtsov, P. M., Musakhanov, M. M., Zakharyev, V. M., Krayev, A. S., Skryabin, K. G. & Bayev, A. A. (1980). *Nucl. Acids Res.* **8**, 5779–5794.
Salim, M. & Maden, B. E. H. (1981). *Nature (London)*, **291**, 205–208.
Sanger, F. & Coulson, A. R. (1978). *FEBS Letters*, **87**, 107–110.
Sanger, F., Nicklen, S. & Coulson, A. R. (1977). *Proc. Nat. Acad. Sci., U.S.A.* **74**, 5463–5467.
Sanger, F., Coulson, A. R., Barrell, B. G., Smith, A. J. H. & Roe, B. A. (1980). *J. Mol. Biol.* **143**, 161–178.
Sebald, W. & Wachter, E. (1978). In *29th Mosbacher Colloquium on Energy Conservation in Biological Membranes* (Schäfer, G. & Klingenberg, M., eds), pp. 228–236, Springer-Verlag, Berlin.
Sekiya, T., Kobayashi, M., Seki, T. & Koike, K. (1980). *Gene*, **11**, 53–62.
Staden, R. (1980*a*). *Nucl. Acids Res.* **8**, 817–825.
Staden, R. (1980*b*). *Nucl. Acids Res.* **8**, 3673–3694.
Steffens, G. J. & Buse, G. (1979). *Hoppe-Seyler's Z. Physiol. Chem.* **360**, 613–619.
Stiegler, P., Carbon, P., Zuker, M., Ebel, J.-P. & Ehresmann, C. (1981). *Nucl. Acids Res.* **9**, 2153–2172.
Tapper, D. P. & Clayton, D. A. (1981). *J. Biol. Chem.* **256**, 5109–5115.
Upholt, W. B. & Dawid, I. B. (1977). *Cell*, **11**, 571–583.
Van Etten, R. A., Walberg, M. W. & Clayton, D. A. (1980). *Cell*, **22**, 157–170.
Walberg, M. W. & Clayton, D. A. (1981). *Nucl. Acids Res.* **9**, 5411–5421.
Wilson, A. C., Carlson, S. S. & White, T. J. (1977). *Annu. Rev. Biochem.* **46**, 573–639.
Woese, C. R., Fox, G. E., Zablen, L., Uchida, T., Bonen, L., Pechman, K., Lewis, B. J. & Stahl, D. (1975). *Nature (London)*, **254**, 83–86.
Woese, C. R., Magrum, L. J., Gupta, R., Siegel, B., Stahl, D. A., Kop, J., Crawford, N., Brosius, J., Gutell, R., Hogan, J. J. & Noller, H. F. (1980). *Nucl. Acids Res.* **8**, 2275–2293.
Young, I. G. & Anderson, S. (1980). *Gene*, **12**, 257–265.
Zinder, N. D. & Lyons, L. B. (1968). *Science*, **159**, 84–86.
Zwieb, C., Glotz, C. & Brimacombe, R. (1981). *Nucl. Acids Res.* **9**, 3621–3640.

Edited by J. C. Kendrew

DETERMINATION OF NUCLEOTIDE SEQUENCES IN DNA

Nobel lecture, 8 December 1980

by

FREDERICK SANGER

Medical Research Council Laboratory of Molecular Biology,
Cambridge, England

INTRODUCTION

In spite of the important role played by DNA sequences in living matter, it is only relatively recently that general methods for their determination have been developed. This is mainly because of the very large size of DNA molecules, the smallest being those of the simple bacteriophages such as φX174 (which contains about 5,000 nucleotides). It was therefore difficult to develop methods with such complicated systems. There are however some relatively small RNA molecules – notably the transfer RNAs of about 75 nucleotides, and these were used for the early studies on nucleic acid sequences (1).

Following my work on amino acid sequences in proteins (2) I turned my attention to RNA and, with G.G. Brownlee and B.G. Barrell, developed a relatively rapid small-scale method for the fractionation of ^{32}P-labelled oligonucleotides (3). This became the basis for most subsequent studies of RNA sequences. The general approach used in these studies, and in those on proteins, depended on the principle of partial degradation. The large molecules were broken down, usually by suitable enzymes, to give smaller products which were then separated from each other and their sequence determined. When sufficient results had been obtained they were fitted together by a process of deduction to give the complete sequence. This approach was necessarily rather slow and tedious, often involving successive digestions and fractionations, and it was not easy to apply it to the larger DNA molecules. When we first studied DNA some significant sequences of about 50 nucleotides in length were obtained with this method (4,5), but it seemed that to be able to sequence genetic material a new approach was desirable and we turned our attention to the use of copying procedures.

Abbreviations The abbreviations C, A and G are used to describe both the ribonucleotides and the deoxyribonucleotides, according to context.

COPYING PROCEDURES

In the RNA field these procedures had been pioneered by C. Weissmann and his colleagues (6) in their studies on the RNA sequence of the bacteriophage Qβ. Qβ contains a replicase that will synthesize a complementary copy of the single-stranded RNA chain, starting from its 3′ end. These workers devised elegant procedures involving pulse-labelling with radioactively labelled nucleotides, from which sequences could be deduced.

For DNA sequences we have used the enzyme DNA polymerase, which copies single-stranded DNA as shown in Fig. 1. The enzyme requires a primer, which is a single-stranded oligonucleotide having a sequence that is complementary to, and therefore able to hybridize with, a region on the DNA being sequenced (the template). Mononucleotide residues are added sequentially to the 3′ end of the primer from the corresponding deoxynucleoside triphosphates, making a complementary copy of the template DNA. By using triphosphates containing ^{32}P in the α position the newly synthesized DNA can be labelled. In the early experiments synthetic oligonucleotides were used as primers, but after

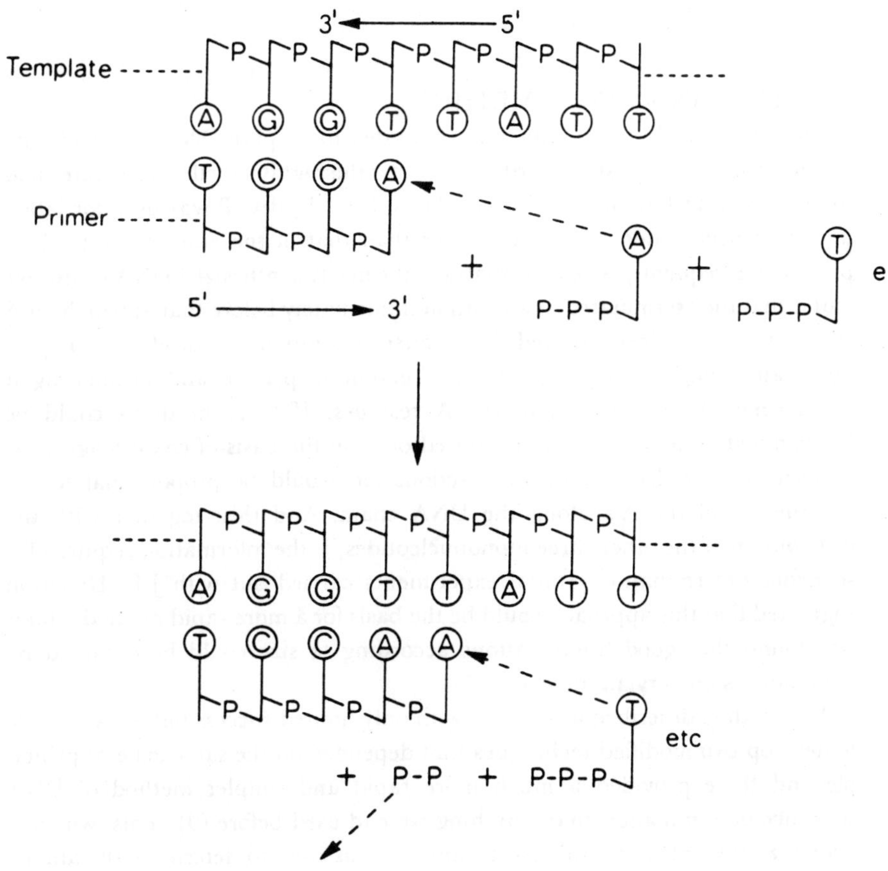

Fig. 1. Specificity requirements for DNA polymerase.

the discovery of restriction enzymes it was more convenient to use fragments resulting from their action as they were much more easily obtained.

The copying procedure was used initially to prepare a short specific region of labelled DNA which could then be subjected to partial digestion procedures. One of the difficulties of sequencing DNA was to find specific methods for breaking it down into small fragments. No suitable enzymes were known that would recognise only one nucelotide. However, Berg, Fancher & Chamberlin (7) had shown earlier that under certain conditions it was possible to incorporate ribonucleotides, in place of the normal deoxyribonucleotides, into DNA chains with DNA polymerase. Thus, for instance, if copying were carried out using riboCTP and the other three deoxynucleoside triphosphates, a chain could be built up in which the C residues were in the ribo form. Bonds involving ribonucleotides could be broken by alkali under conditions where those involving the deoxynucleotides were not, so that a specific splitting at C residues could be obtained. Using this method we were able to extend our sequencing studies to some extent (8). However extensive fractionations and analyses were still required.

THE 'PLUS AND MINUS' METHOD

In the course of these experiments we needed to prepare DNA copies of high specific radioactivity, and in order to do this the highly labelled substrates had to be present in low concentrations. Thus if $\alpha[^{32}P]$-dATP was used for labelling its concentration was much lower than that of the other three triphosphates and frequently when we analysed the newly synthesized DNA chains we found that they terminated at a position immediately before that at which an A should have been incorporated. Consequently a mixture of products was produced all having the same 5′ end (the 5′ end of the primer) and terminating at the 3′ end at the position of the A residues. If these products could be fractionated on a system that separated only on the basis of chain length, the pattern of their distribution on fractionation would be proportional to the distribution of the A's along the DNA chain. And this, together with the distribution of the other three mononucleotides, is the information required for sequence determination. Initial experiments carried out with J.E. Donelson suggested that this approach could be the basis for a more rapid method, and it was found that good fractionations according to size could be obtained by ionophoresis on acrylamide gels.

The method described above met with only limited success but we were able to develop two modified techniques that depended on the same general principle and these provided a much more rapid and simpler method of DNA sequence determination than anything we had used before (9). This, which is known as the "plus and minus" technique, was used to determine the almost complete sequence of the DNA of bacteriophage φX174 which contains 5,386 nucleotides (10).

THE 'DIDEOXY' METHOD

More recently we have developed another similar method which uses specific chain-terminating analogues of the normal deoxynucleoside triphosphates (11). This method is both quicker and more accurate than the plus and minus technique. It was used to complete the sequence of φX174 (12), to determine the sequence of a related bacteriophage, G4 (13), and has now been applied to mammalian mitochondrial DNA.

The analogues most widely used are the dideoxynucleoside triphosphates (Fig. 2). They are the same as the normal deoxynucleoside triphosphates but lack the 3′ hydroxyl group. They can be incorporated into a growing DNA chain by DNA polymerase but act as terminators because, once they are incorporated, the chain contains no 3′ hydroxyl group and so no other nucelotide can be added.

The principle of the method is summarised in Fig. 3. Primer and template are denatured to separate the two strands of the primer, which is usually a restriction enzyme fragment, and then annealed to form the primer-template complex. The mixture is then divided into four samples. One (the T sample) is incubated with DNA polymerase in the presence of a mixture of ddTTP (dideoxy thymidine triphosphate) and a low concentration of TTP, together with the other three deoxynucleoside triphosphates (one of which is labelled

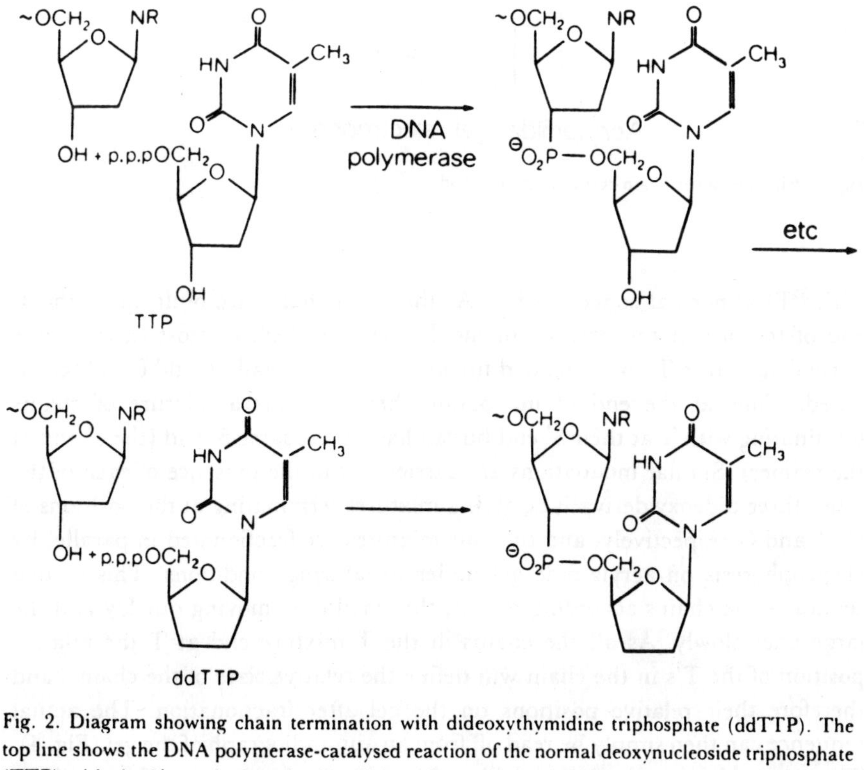

Fig. 2. Diagram showing chain termination with dideoxythymidine triphosphate (ddTTP). The top line shows the DNA polymerase-catalysed reaction of the normal deoxynucleoside triphosphate (TTP) with the 3′ terminal nucleotide of the primer: the bottom line the corresponding reaction with ddTTP.

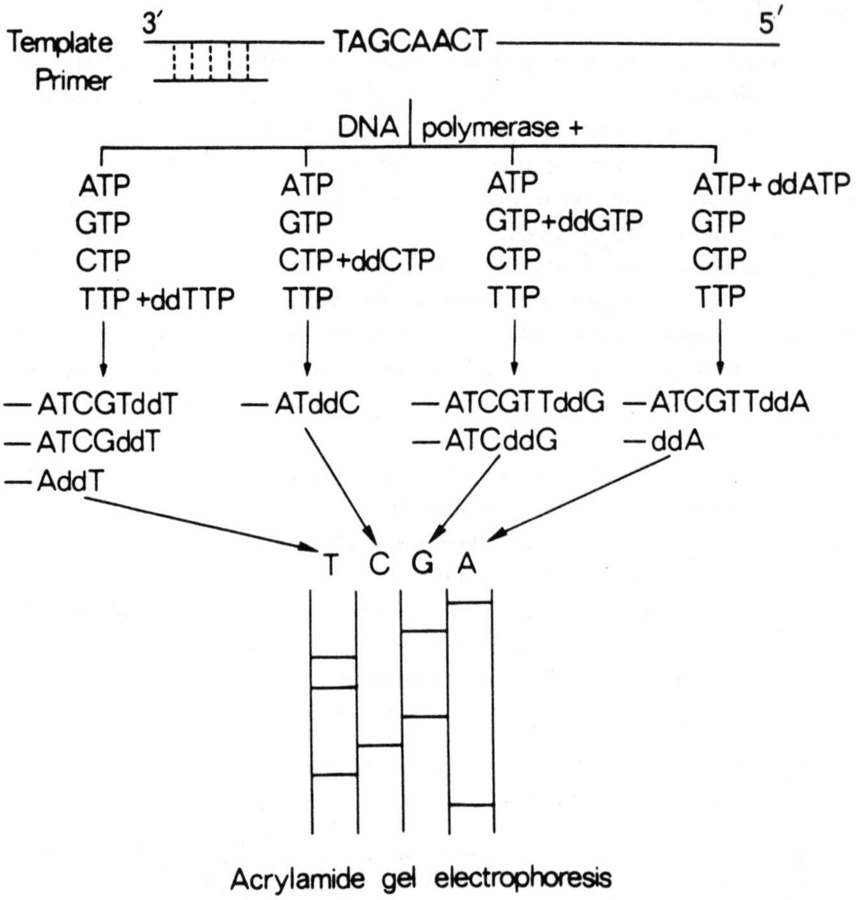

Fig. 3. Principle of the chain-terminating method.

with ^{32}P) at normal concentration. As the DNA chains are built up on the 3′ end of the primer the position of the T's will be filled, in most cases by the normal substrate T and extended further, but occasionally by ddT and terminated. Thus at the end of incubation there remains a mixture of chains terminating with T at their 3′ end but all having the same 5′ end (the 5′ end of the primer). Similar incubations are carried out in the presence of each of the other three dideoxy derivatives, giving mixtures terminating at the positions of C, A and G respectively, and the four mixtures are fractionated in parallel by electrophoresis on acrylamide gel under denaturing conditions. This system separates the chains according to size, the small ones moving quickly and the large ones slowly. As all the chains in the T mixture end at T the relative position of the T's in the chain will define the relative sizes of the chains, and therefore their relative positions on the gel after fractionation. The actual sequence can then simply be read off from an autoradiograph of the gel (Fig. 4). The method is comparatively rapid and accurate and sequences of up to about 300 nucleotides from the 3′ end of the primer can usually be determined.

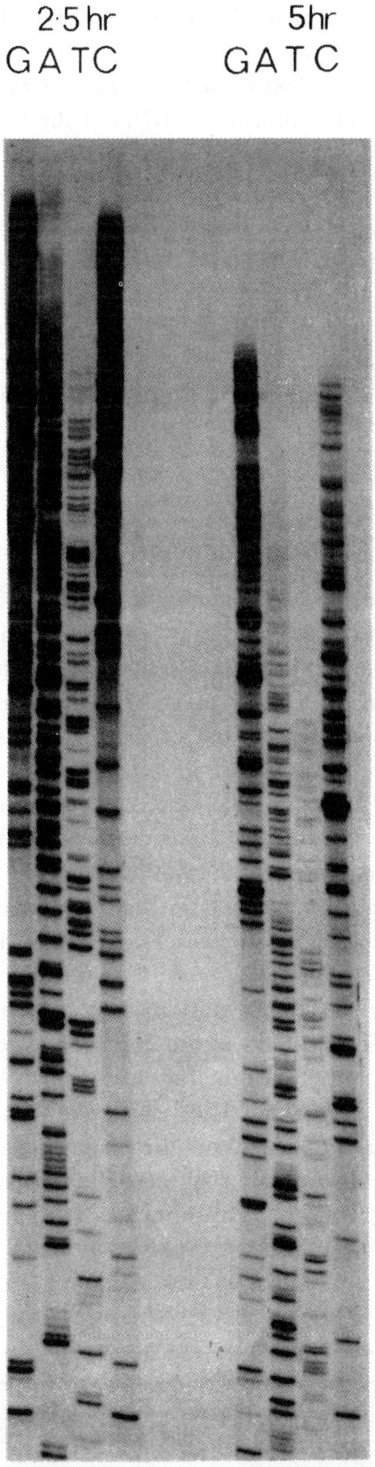

Fig. 4. Autoradiograph of a DNA sequencing gel. The origin is at the top and migration of the DNA chains, according to size, is downwards. The gel on the left has been run for 2 · 5 hr and that on the right for 5 hr with the same polymerisation mixtures.

Considerably longer sequences have been read off but these are usually less reliable.

One problem with the method is that it requires single-stranded DNA as template. This is the natural form of the DNA in the bacteriophages φX174 and G4, but most DNA is double-stranded and it is frequently difficult to separate the two strands. One way of overcoming this was devised by A.J.H. Smith (14). If the double-stranded linear DNA is treated with exonuclease III (a double-strand specific 3′ exonuclease) each chain is degraded from its 3′ end, as shown in Fig. 5, giving rise to a structure that is largely single-stranded and can be used as template for DNA polymerase with suitable small primers. This method is particularly suitable for use with fragments cloned in plasmid vectors and has been used extensively in the work on human mitochondrial DNA.

CLONING IN SINGLE-STRANDED BACTERIOPHAGE

Another method of preparing suitable template DNA that is being more widely used is to clone fragments in a single-stranded bacteriophage vector (15−17). This approach is summarised in Fig. 6. Various vectors have been described. We have used a derivative of bacteriophage M13 developed by Gronenborn & Messing (16) which contains an insert of the β-galactosidase gene with an *Eco*RI restriction enzyme site in it. The presence of β-galactosidase in a plaque can be readily detected by using a suitable colour-forming substrate (X-gal). The presence of an insert in the *Eco*RI site destroys the β-galactosidase gene, giving rise to a colourless plaque.

Besides being a simple and general method of preparing single-stranded DNA this approach has other advantages. One is that it is possible to use the same primer on all clones. Heidecker *et al.* (18) prepared a 96-nucleotide long restriction fragment derived from a position in the M13 vector flanking the *Eco*RI site (see Fig. 6). This can be used to prime into, and thus determine, a sequence of about 200 nucleotides in the inserted DNA. Smaller synthetic primers have now been prepared (19,20) which allow longer sequences to be determined. The approach that we have used is to prepare clones at random from restriction enzyme digests and determine the sequence with the flanking primer. Computer programmes (21) are then used to store, overlap and arrange the data.

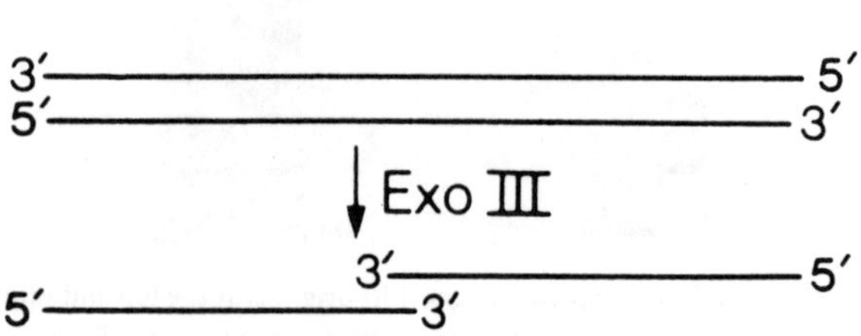

Fig. 5. Degradation of double-stranded DNA with exonuclease III.

Another important advantage of the cloning technique is that it is a very efficient and rapid method of fractionating fragments of DNA. In all sequencing techniques, both for proteins and nucleic acids, fractionation has been an important step and major progress has usually been dependent on the development of new fractionation methods. With the new rapid methods for DNA sequencing fractionation is still important and as the sequencing procedure itself is becoming more rapid more of the work has involved fractionation of the restriction enzyme fragments by electrophoresis on acrylamide. This becomes increasingly difficult as larger DNA molecules are studied and may involve several successive fractionations before pure fragments are obtained. In the

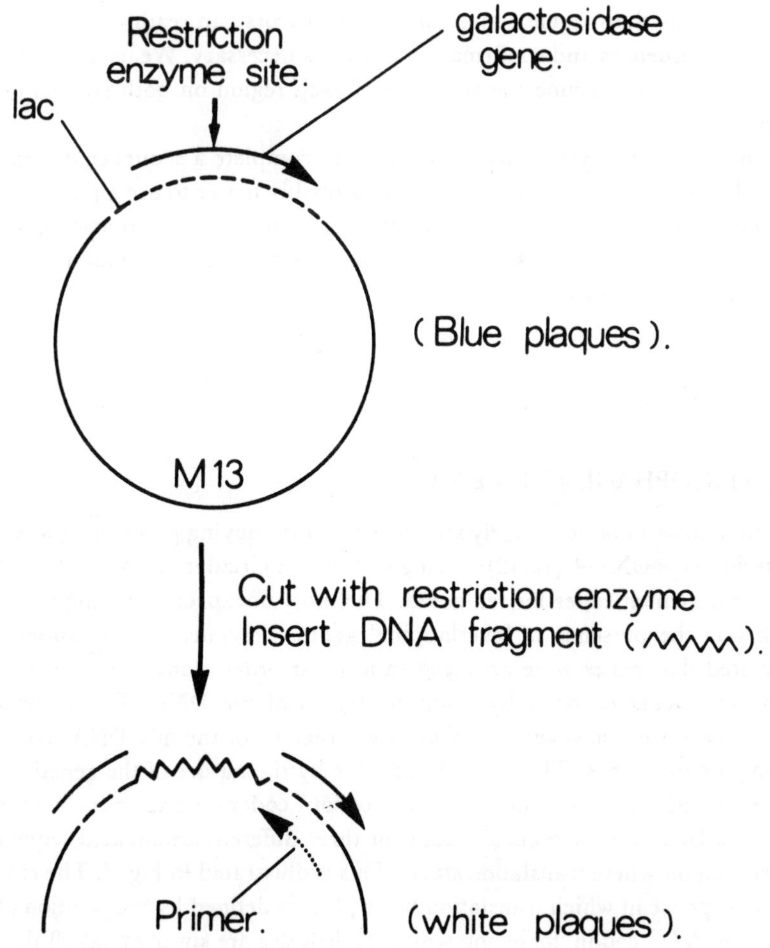

Fig. 6. Diagram illustrating the cloning of DNA fragments in the single-stranded bacteriophage vectorM13mp2 (16) and sequencing the insert with a flanking primer.

new method these fractionations are replaced by a cloning procedure. The mixture is spread on an agar plate and grown. Each clone represents the progeny of a single molecule and is therefore pure, irrespective of how complex the original mixture was. It is particulary suitable for studying large DNAs. In fact there is no theoretical limit to the size of DNA that could be sequenced by this method.

We have applied the method to fragments from mitochondrial DNA (22,23) and to bacteriophage λ DNA. Initially new data can be accumulated very quickly (under ideal conditions at about 500–1,000 nucleotides a day). However at later stages much of the data produced will be in regions that have already been sequenced, and progress then appears to be much slower. Nevertheless we find that most new clones studied give some useful data, either for correcting or confirming old sequences. Thus in the work with bacteriophage λ DNA we have about 90 % of the molecule identified in sequences and most of the new clones we study contribute some new information. In most studies on DNA one is concerned with identifying the reading frames for protein genes, and to do this the sequence must be correct. Errors can readily occur in such extensive sequences and confirmation is always necessary. We usually consider it necessary to determine the sequence of each region on both strands of the DNA.

Although in theory it would be possible to complete a sequence determination solely by the random approach, it is probably better to use a more specific method to determine the final remaining nucleotides in a sequencing study. Various methods are possible (22,24), but all are slow compared with the random cloning approach.

BACTERIOPHAGE φX174 DNA

The first DNA to be completely sequenced by the copying procedures was from bacteriophage φX174 (10,12)—a single-stranded circular DNA, 5,386 nucleotides long, which codes for ten genes. The most unexpected finding from this work was the presence of 'overlapping' genes. Previous genetic studies had suggested that genes were arranged in a linear order along the DNA chains, each gene being encoded by a unique region of the DNA. The sequencing studies indicated however that there were regions of the φX DNA that were coding for two genes. This is made possible by the nature of the genetic code. Since a sequence of three nucleotides (a codon) codes for one amino acid, each region of DNA can theoretically code for three different amino acid sequences, depending on where translation starts. This is illustrated in Fig. 7. The reading frame or phase in which translation takes place is defined by the position of the initiating ATG codon, following which nucleotides are simply read off three at a time. In φX there is an initiating ATG within the gene coding for the D protein, but in a different reading frame. Consequently this initiates an entirely

```
 Gly     His     Ala     Cys     Arg     Arg
G G C . C A C . G C A . T G C . A G G . C G G


      Ala     Thr     His     Ala     Gly
G . G C C . A C G . C A T . G C A . G G C . G G


        Pro     Arg     Met     Gln     Ala
G G . C C A . C G C . A T G . C A G . G C G . G
```

Fig. 7. Diagram illustrating how one DNA sequence can code for three different amino acid sequences. The dots indicate the positions of triplet codons coding for the amino acids.

different sequence of amino acids, which is that of the E protein. Fig. 8 shows the genetic map of φX. The E gene is completely contained within the D gene and the B gene within the A.

Further studies (25) on the related bacteriophage, G4, revealed the presence of a previously unidentified gene, which was called K. This overlaps both the A and C genes, and there is a sequence of four nucleotides that codes for part of all three proteins, A, C and K, using all of its three possible reading frames.

It is uncertain whether overlapping genes are a general phenomenon or whether they are confined to viruses, whose survival may depend on their rate of replication and therefore on the size of the DNA: with the overlapping genes more genetic information can be concentrated in a given sized DNA.

Further details of the sequence of bacteriophage φX174 DNA have been published elsewhere (10,12).

MAMMALIAN MITOCHONDRIAL DNA

Mitochondria contain a small double-stranded DNA (mtDNA) which codes for two ribosomal RNAs (rRNAs), 22–23 transfer RNAs (tRNAs) and about 10–13 proteins which appear to be components of the inner mitochondrial membrane and are somewhat hydrophobic. Other proteins of the mitochondria are encoded by the nucleus of the cell and specifically transported to the mitochondria. Using the above methods we have determined the nucleotide sequence of human mtDNA (23) and almost the complete sequence of bovine mtDNA. The sequence revealed a number of unexpected features which indicated that the transcription and translation machinery of mitochondria is rather different from that of other biological systems.

The genetic code in mitochondria

Hitherto it has been believed that the genetic code was universal. No differences were found in the *E. coli,* yeast or mammalian systems that had been studied. Our initial sequence studies were on human mtDNA. No amino acid sequence of the proteins that were encoded by human mtDNA were known. However Steffans & Buse (26) had determined the sequence of subunit II of cytochrome oxidase (COII) from bovine mitochondria, and Barrell, Bankier & Drouin (27) found that a region of the human mtDNA that they were studying had a sequence that would code for a protein homologous to this amino acid sequence – indicating that it most probably was the gene for the human COII. Surprisingly the DNA sequence contained TGA triplets in the reading frame of the homologous protein. According to the normal genetic code TGA is a termination codon and if it occurs in the reading frame of a protein the polypetide chain is terminated at that position. It was noted that in the positions where TGA occured in the human mtDNA sequence, tryptophan was found in the bovine protein sequence. The only possible conclusion seemed to be that in mitochondria TGA was not a termination codon but was coding for tryptophan. It was similarly concluded that ATA, which normally codes for isoleucine, was coding for methionine. As these studies were based on a

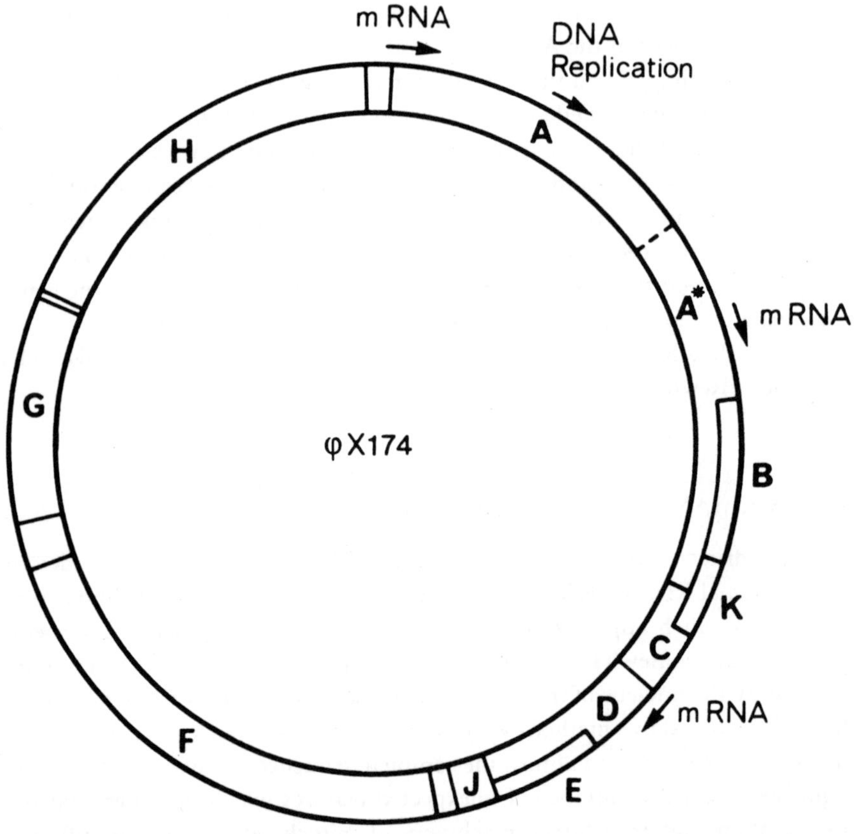

Fig. 8. Gene map of φX174 DNA.

comparison of a human DNA with a bovine protein, the possibility that the differences were due to some species variation, although unlikely, could not be completely excluded. For a conclusive determination of the mitochondrial code it was necessary to compare the DNA sequence of a gene with the amino acid sequence of the protein it was coding for. This was done by Young & Anderson (28) who isolated the bovine mtDNA, determined the sequence of its COII gene and confirmed the above differences.

Fig. 9 shows the human and bovine mitochondrial genetic code and the frequency of use of the different codons in human mitochondria. All codons are used with the exception of UUA and UAG, which are terminators, and AGA and AGG, which normally code for arginine. This suggests that AGA and AG are probably also termination codons in mitochondria. Further support for this is that no tRNA recognizing the codons has been found (see below) and that some of the unidentified reading frames found in the DNA sequence appear to end with these codons.

In parallel with our studies on mammalian mtDNA, Tzagoloff and his colleagues (29,30) were studying yeast mtDNA. They also found changes in the genetic code, but surprisingly they are not all the same as those found in mammalian mitochondria. These differences are summarised in Table 1.

SECOND LETTER

FIRST LETTER	U			C			A			G			THIRD LETTER
U	UUU	Phe	77	UCU		32	UAU	Tyr	46	UGU	Cys	5	U
	UUC		140	UCC	Ser	99	UAC		89	UGC		17	C
	UUA	Leu	73	UCA		83	UAA	Ter	-	UGA	Trp	93	A
	UUG		17	UCG		7	UAG		-	UGG		10	G
C	CUU		65	CCU		41	CAU	His	18	CGU		7	U
	CUC	Leu	167	CCC	Pro	119	CAC		79	CGC	Arg	25	C
	CUA		276	CCA		52	CAA	Gln	81	CGA		29	A
	CUG		45	CCG		7	CAG		9	CGG		2	G
A	AUU	Ile	125	ACU		51	AAU	Asn	33	AGU	Ser	14	U
	AUC		196	ACC	Thr	155	AAC		130	AGC		39	C
	AUA	Met	166	ACA		133	AAA	Lys	85	AGA	Ter	-	A
	AUG		40	ACG		10	AAG		10	AGG		-	G
G	GUU		30	GCU		43	GAU	Asp	15	GGU		24	U
	GUC	Val	49	GCC	Ala	124	GAC		55	GGC	Gly	88	C
	GUA		71	GCA		80	GAA	Glu	64	GGA		67	A
	GUG		18	GCG		8	GAG		24	GGG		34	G

Fig. 9. The human mitochondrial genetic code, showing the coding properties of the tRNAs (boxed codons) and the total number of codons used in the whole genome shown in Fig. 10. (One methionine tRNA has been detected, but as there is some uncertainty about the number present and their coding properties, these codons are not boxed.)

Table 1. Coding changes in mitochondria

Codon	Amino acid coded		
	Normal	Mammalian mitochondria	Yeast mitochondria
UGA	Term	Trp	Trp
AUA	Ile	Met	Ile
CUN	Leu	Leu	Thr
AGA, AGG	Arg	Term?	Arg
CGN	Arg	Arg	Term?

Transfer RNAs

Transfer RNAs have a characteristic base-pairing structure which can be drawn in the form of the 'cloverleaf' model. By examining the DNA sequence for cloverleaf structures and using a computer programme (31), it was possible to identify genes coding for the mt-tRNAs.

Besides the cloverleaf structure, normal cytoplasmic tRNAs have a number of so-called 'invariable' features which are believed to be important to their biological function. Most of the mammalian mt-tRNAs are anomalous in that some of these invariable features are missing. The most bizarre is one of the serine tRNA in which a complete loop of the cloverleaf structure is missing (32,33), Nevertheless it functions as a tRNA.

In normal cytoplasmic systems at least 32 tRNAs are required to code for all the amino acids. This is related to the 'wobble' effect. Codon-anticodon relationships in the first and second positions of the codons are defined by the normal base-pairing rules, but in the third position G can pair with U. The result of this is that one tRNA can recognise two codons. There are many cases in the genetic code where all four codons starting with the same two nucleotides code for the same amino acid. These are known as 'family boxes'. The situation for the alanine family box is shown in Table 2, indicating that with the normal wobble system two tRNAs are required to code for the four alanine codons.

Table 2. Coding properties of alanine tRNAs

Codon	Anticodon (wobble)	Anticodon (mitochondria)
GCU		
	GGC	
GCC		
		UGC
CA	UGC	
GCG		

Note that the first position of the codon pairs with the third position of the anticodon and vice versa; e. g.

$$^{5'}\,G\ C\ U\,^{3'}\ \text{(codon)}$$
$$_{3'}\,C\ G\ G\,_{5'}\ \text{(anticodon)}$$

Only 22 tRNA genes could be found in mammalian mtDNA, and for all the family boxes there was only one, which had a T in the position corresponding to the third position of the codon (34). It seems very unlikely that none of the other predicated tRNAs would have been detected and the most feasible explanation is that in mitochondria one tRNA can recognise all four codons in a family box and that a U in the first position of the anticodon can pair with U, C, A or G in the third position of the codon. Clearly in boxes in which two of the codons code for one amino acid and two for a different one, there must be two different tRNAs and the wobble effect still applies. Such tRNAs are found, as expected, in the mitochondrial genes. The coding properties of the mt-tRNAs are shown in Fig. 9. Similar conclusions have been reached by Heckman *et al.* (35) and by Bonitz *et al.* (36), working respectively on neurospora and yeast mitochondria.

Distribution of protein genes

Mitochondrial DNA was known to code for three of the subunits of cytochrome oxidase, probably three subunits of the ATPase complex, cytochrome b, and a number of other unidentified proteins. In order to identify the protein-coding genes, the DNA was searched for reading frames; i. e. long stretches of DNA containing no termination codons in one of the phases and thus being capable of coding for long polypeptide chains. Such reading frames should start with an initiation codon, which in normal systems is nearly always ATG, and end with a termination codon. Fig. 10 summarises the distribution of the reading frames

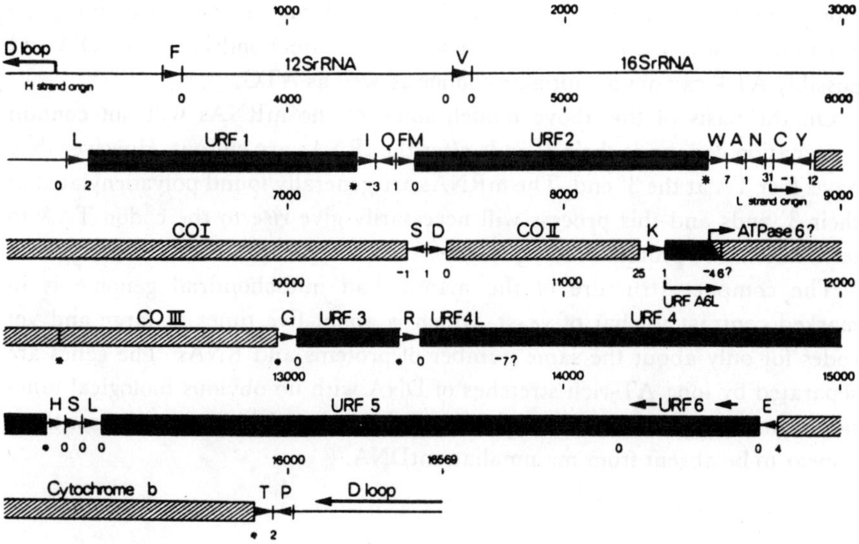

Fig. 10. Gene map of human mtDNA deduced from the DNA sequence. Boxed regions are the predicted reading frames for the proteins. URF = unidentified reading frame. tRNA genes are denoted by the one-letter amino acid code and are either L strand coded (▶) or H strand coded (◀). Numbers above the genes show the scale in nucleotides and below the predicted number between genes.

* Indicates that termination codons are created by polyadenylation of the mRNA.

on the DNA and these are believed to be the genes coding for the proteins. The gene for COII was identified from the amino acid sequence as described above, for subunit I of the cytochrome oxidase from amino acid sequence studies on the bovine protein by J. E. Walker (personal communication), and COIII, cytochrome b and, probably, ATPase 6 were identified by comparison with the DNA sequences of the corresponding genes in yeast mitochondria. Tzagoloff and his colleagues were able to identify these genes in yeast by genetic methods. It has not yet been possible to assign proteins to the other reading frames.

One unusual feature of the mtDNA is that it has a very compact structure. The reading frames coding for the proteins and the rRNA genes appear to be flanked by the tRNA genes with no, or very few, intervening nucleotides. This suggests a relatively simple model for transcription, in which a large RNA is copied from the DNA and the tRNAs are cut out by a processing enzyme, and this same processing leads to the production of the rRNAs and the messenger RNAs (mRNAs), most of which will be monocistronic. Strong support for this model comes from the work of Attardi (37,38) who has identified the RNA sequences at the 5′ and 3′ ends of the mRNAs, thus locating them on the DNA sequence. One consequence of this arrangement is that the initiation codon is at, or very near, the 5′ end of the mRNAs. This suggests that there must be a different mechanism of initiation from that found in other systems. In bacteria there is usually a ribosomal binding site before the initiating ATG codon, whereas in eucaryotic systems the 'cap' structure on the 5′ end of the mRNA appears to have a similar function and the first ATG following the cap acts as initiator. It seems that mitochondria may have a more simple, and perhaps more primitive, system with the translation machinery recognising simply the 5′ end of the mRNA. Another unique feature of mitochondria is that ATA and possibly ATT can act as initiator codons as well as ATG.

On the basis of the above model, some of the mRNAs will not contain termination codons at their 3′ ends after the tRNAs are cut out. However they have T or TA at the 3′ end. The mRNAs are generally found polyadenylated at their 3′ ends and this process will necessarily give rise to the codon TAA to terminate those protein reading frames.

The compact structure of the mammalian mitochondrial genome is in marked contrast to that of yeast, which is about five times as large and yet codes for only about the same number of proteins and RNAs. The genes are separated by long AT-rich stretches of DNA with no obvious biological function. There are also insertion sequences within some of the genes, whereas these appear to be absent from mammalian mtDNA.

REFERENCES

1. Holley, R. W., Les Prix Nobel, p. 183 (1968)
2. Sanger, F., Les Prix Nobel, p. 134 (1958)
3. Sanger, F., Brownlee, G. G. & Barrell, B. G., J. Mol. Biol. *13,* 373 (1965)
4. Robertson, H. D., Barrell, B. G., Weith, H. L. & Donelson, J. E., Nature, New Biol. *241,* 38 (1973)
5. Ziff, E. B., Sedat, J. W. & Galibert, F., Nature, New Biol. *241,* 34 (1973)
6. Billeter, M. A., Dahlberg, J. E., Goodman, H. M., Hindley, J. & Weissmann, C., Nature *224,* 1083 (1969)
7. Berg, P., Fancher, H. & Chamberlin, M., Symp. "Informational Macromolecules", pp 467–483, Academic Press: New York & London (1963)
8. Sanger, F., Donelson, J. E., Coulson, A. R., Kössel, H. & Fischer, D., Proc. Natl. Acad. Sci. USA *70,* 1209 (1973)
9. Sanger, F. & Coulson, A. R. J. Mol. Biol. *94,* 441 (1975)
10. Sanger, F., Air, G. M., Barrell, B. G., Brown, N. L., Coulson, A. R. Fiddes, J. C., Hutchison, C. A., Slocombe, P. M. & Smith, M., Nature *265,* 687 (1977)
11. Sanger, F., Nicklen, S. & Coulson, A. R., Proc. Natl. Acad. Sci. USA *74,* 5463 (1977)
12. Sanger, F., Coulson, A. R., Friedmann, T., Air, G. M., Barrell, B. G., Brown, N. L., Fiddes, J. C., Hutchison, C. A., Slocombe, P. M. & Smith, M., J. Mol. Biol. *125,* 225 (1978)
13. Godson, G. N., Barrell, B. G., Staden R. & Fiddes, J. C., Nature *276,* 236 (1978)
14. Smith, A. J. H., Nucl. Acids Res. *6,* 831 (1979)
15. Barnes, W. M., Gene *5.* 127 (1979)
16. Gronenborn, B. & Messing, J., Nature *272,* 375 (1978)
17. Herrmann, R., Neugebauer, K., Schaller, H. & Zentgraf, H:, In "The Singlestranded DNA Phages" (Eds. Denhardt, D. T., Dressler, D. & Ray, D. S.) pp 473–476, Cold Spring Harbor Laboratory, New York (1978)
18. Heidecker, G., Messing, J. & Gronenborn, B., Gene *10,* 69 (1980)
19. Anderson, S., Gait, M. J., Mayol, L. & Young, I. G., Nucl. Acids. Res. *8,* 1731 (1980)
20. Gait, M. J., Unpublished results (1980)
21. Staden, R., Nucl. Acids Res. *8,* 3673 (1980)
22. Sanger, F., Coulson, A. R., Barrell, B. G., Smith, A. J. H. & Roe, B. A., J. Mol. Biol. *143,* 161 (1980)
23. Barrell, B. G., Anderson, S., Bankier, A. T., de Bruijn, M. H. L., Chen, E., Coulson, A. R., Drouin, J., Eperon, I. C., Nierlich, D. P., Roe, B. A., Sanger, F., Schreier, P. H., Smith, A. J. H., Staden, R. & Young, I. G., 31st Mosbach Colloq. "Biological Chemistry of organelle Formation", (Eds. Bücher, T., Sebald, W. & Weiss, H.) pp 11–25, Springer-Verlag, Berlin (1980)
24. Winter, G. & Fields, S., Nucl. Acids Res. *8,* 1965 (1980)
25. Shaw, D. C., Walker, J. E., Northrop, F. D., Barrell, B. G., Godson, G. N. & Fiddes, J. C., Nature *272,* 510 (1978)
26. Steffans, G. J. & Buse, G., Hoppe-Seyler's Z. Physiol. Chem. *360,* 613 (1979)
27. Barrell, B. G., Bankier, A. T. & Drouin, J., Nature *282,* 189 (1979)
28. Young, I. G. & Anderson, S., Gene, *12,* 257 (1980)
29. Macino, G., Coruzzi, G., Nobrega, F. G., Li, M. & Tzagoloff, A., Proc. Natl. Acad. Sci. USA *76,* 3784 (1979)
30. Macino, G. & Tzagoloff, A., Proc. Natl. Acad. Sci. USA *76,* 131 (1979)
31. Staden, R., Nucl. Acids Res. *8,* 817 (1980)
32. de Bruijn, M. H. L., Schreier, P. H., Eperon, I. C., Barrell, B. G., Chen, E. Y., Armstrong, P. W., Wong, J. F. H. & Roe, B. A., Nucl. Acids Res. *8,* 5213 (1980)
33. Arcari, P. & Brownlee, G. G., Nucl. Acids. Res. *8,* 5207 (1980)
34. Barrell, B. G., Anderson, S., Bankier, A. T., de Bruijn, M. H. L., Chen, E., Coulson, A. R., Drouin, J., Eperon, I. C., Nierlich, D. P., Roe, B. A., Sanger, F., Schreier, P. H., Smith, A. J. H., Staden, R. & Young, I. G., Proc. Natl. Acad. Sci. USA *77,* 3164 (1980)

35. Heckman, J. E., Sarnoff, J., Alzner-De Weerd, B., Yin, S. & RajBhandary, U. L., Proc. Natl. Acad. Sci. USA *77*, 3159 (1980)
36. Bonitz, S. G., Berlani, R., Coruzzi, G., Li, M., Macino, G., Nobrega, F. G., Nobrega, M. P., Thalenfeld, B. E. & Tzagoloff, A., Proc. Natl. Acad. Sci. USA, *77*, 3167 (1980)
37. Montoya, J., Ojala, D. & Attardi, G., Nature *290*, 465 (1981)
38. Ojala, D., Montoya, J. & Attardi, G., Nature *290*, 470 (1981)

J. Mol. Biol. (1982) **162**, 729–773

Nucleotide Sequence of Bacteriophage λ DNA

F. SANGER, A. R. COULSON, G. F. HONG, D. F. HILL†
AND G. B. PETERSEN†

Laboratory of Molecular Biology, The Medical Research Centre
Hills Road, Cambridge CB2 2QH, England

(Received 15 September 1982)

The nucleotide sequence of the DNA of bacteriophage λ has been determined using the dideoxy chain termination method in conjunction with random cloning in M13 vectors. Various methods were studied for sequencing specific regions to complete the sequence, but all were much slower than the random approach. The DNA in its circular form contains 48,502 base-pairs. Open reading frames were identified and, where possible, ascribed to genes by comparing with the previously determined genetic map. The reading frames for 46 genes were clearly identified, though in about 20 the position of the protein initiation site could not be rigorously established. Probable positions for the *kil*, *cIII* and *lom* genes are suggested but remain uncertain. There are about 20 other unidentified reading frames that may code for proteins.

The genome is fairly compact with comparatively little non-coding DNA. In many cases the translation terminators and initiators overlap, particularly in the sequence A-T-G-A where the TGA terminates one gene and the ATG initiates the next. Such structures seem to be characterized by a purine-rich sequence, rather than by a specific "Shine and Dalgarno" sequence, before the initiator. In the whole of the left arm the codon CTA, which is normally read by a minor leucine tRNA, is absent. The distribution of other rare codons in the genes of the left arm suggests that they may have a controlling function on the relative amounts of the proteins produced.

1. Introduction

New methods for sequencing DNA (Maxam & Gilbert, 1977; Sanger *et al.*, 1977) have made it possible to determine sequences of several hundred nucleotides very rapidly. These techniques were originally applied to restriction enzyme fragments and as they improved the rate of sequencing became increasingly dependent on the isolation methods for the fragments. A major step forward was achieved by replacing standard fractionation methods with a cloning procedure using single-stranded bacteriophage as the vector. Particularly successful were the derivatives of bacteriophage M13 developed by Messing and his colleagues (Gronenborn & Messing, 1978; Messing *et al.*, 1981), which could be used in conjunction with the chain termination techniques (Sanger *et al.*, 1977,1980). During our studies on

† Present address: Department of Biochemistry, University of Otago, Box 56, Dunedin, New Zealand.

0022–2836/82/360729–45 $03.00/0

human mitochondrial DNA (Anderson *et al.*, 1981) there were difficulties in achieving the containment facilities required at that time when using M13 vectors on human DNA; consequently, we decided to try out the method on bacteriophage lambda DNA. Progress was rapid and most of the sequence was completed by the time we were able to apply the methods to mitochondrial DNA. Because of the general interest in and importance of lambda in biochemical and genetic studies, we have completed the sequence.

Bacteriophage lambda DNA in its circular form contains 48,502 base-pairs, and codes for about 60 proteins. It has been studied extensively by genetic techniques and a fairly complete gene map prepared (Echols & Murialdo, 1978; Szybalski & Szybalski, 1979; Daniels *et al.*, 1980). In many cases the proteins encoded have been identified as bands on sodium dodecyl sulphate/polyacrylamide gels, but in only a few cases has an amino acid sequence been determined.

When we started the work, a number of DNA sequences had already been established in lambda (about 6000 base-pairs), particularly in control regions. We have not rigorously re-investigated these sequences and have usually ignored clones that were found to contain them. Other sequences have been reported while our work was still in progress. In general, our results have confirmed these sequences.

2. Materials and Methods

The lambda DNA used was the standard strain λ*cI*indlts857*S*am7. It was prepared at the Centre for Applied Microbiology and Research, Porton. DNA polymerase I (Klenow subfragment) was obtained from Boehringer Mannheim. Restriction enzyme *Fnu*DII was a gift from Professor M. Smith. Other restriction enzymes were from New England Biolabs. The M13 vectors were a gift from Dr J. Messing.

(a) *DNA sequencing*

The "shotgun" sequencing was carried out initially as described previously (Sanger *et al.*, 1980) using restriction enzyme digests, the vectors mp2 (Gronenborn & Messing, 1978) or mp7 (Messing *et al.*, 1981) and priming with the 96-nucleotide R1-R1 primer, or the 30-nucleotide primer prepared by Anderson *et al.* (1980). In later work the DNA degradation was by digestion with DNase (Anderson, 1981) or by sonication (Fuhrman *et.*, 1981) and priming was with the synthetic 17-mer prepared by Duckworth *et al.* (1981). The "dideoxy" sequencing was carried out in Eppendorf polypropylene tubes rather than in capillaries. Considerable improvement in the reading of sequences could be obtained by drying the gels after fixing in acetic acid (see Garoff & Ansorge, 1981). This could be conveniently achieved by leaving the gel uncovered at room temperature for several hours and then heating at 67°C till completely dry. Alternatively, a conventional gel dryer could be used.

The main source of error in the sequencing is "compressions" on the gels. Instead of there being a regular distance between each band, some bands run closer together or occupy the same position. This appears to be caused by secondary structure effects, which are not broken by the denaturing conditions in the gels, with the result that a small region at the 3′ end of the newly synthesized DNA becomes double-stranded and the corresponding bands run abnormally fast in the electrophoresis. The best way to overcome this problem is by determining the sequence "in both directions"; i.e. on 2 clones with the sequence in opposite orientations. It is unlikely that the compression will occur in the same place in both clones. Two other techniques have been used to overcome this problem. One is to replace dGTP in the mixtures with dITP (deoxyinosine triphosphate). I·C base-pairing is weaker than G·C, so that the secondary structures are less likely to form (Mills & Kramer, 1979). Usually it is sufficient to run an extra dITP track in which 2mM-dITP is used with ddGTP (0·05 mM) next to the normal ddGTP reaction on a sequencing gel. The reaction is "chased" with a mixture of 0·5 mM-dGTP and 0·5 mM-dATP. By comparing the positions of the G bands on the 2 tracks in a region where a compression has occurred, it is usually possible to resolve the problem. The method does not work on clones prepared in the vector M13mp7. Presumably the secondary structure that can be formed due to complementary sequences in the vector inhibits the progress of the DNA polymerase in the presence of dITP. Another method of resolving compressions is by running the gels in 25% (v/v) formamide.

Another occasional source of error occurs mainly at G-T sequences. At the position of the G residue, bands appear in both the G and T channels with the T band sometimes stronger than the G. The effect is usually associated with unsatisfactory polymerases and can usually be eliminated by using a higher concentration of enzyme.

(b) *Identification of clones by hybridization*

In order to prepare clones covering regions of incomplete or doubtful sequence, ^{32}P-labelled probes were prepared from clones that we already had and that were complementary to a sequence at or near to the required position. These had usually been stored as DNA in 10 mM-Tris · HCl (pH 7·4), 0·1 mM-EDTA at −20°C. Competent cells were transfected with a small amount (0·1 to 0·5 μl) of the DNA and plated out to give new plaques. The number of plaques varied considerably, depending largely on the age of the DNA. A better method of storing clones is to set aside about 50 μl of the supernatant after centrifuging down the bacteria, and keeping it at −20°C in small capped polypropylene tubes. The low-phosphate growth medium contained 0·15% (w/v) KCl, 0·5% (w/v) NaCl, 0·1% (w/v) NH_4Cl, 0·2% (w/v) vitamin-free Casamino acids, 0·2% (w/v) Bactopeptone, 1·21% (w/v) Tris (adjusted to pH 7·4), 0·04% (w/v) $CaCl_2$, 0·5% (w/v) glucose: 0·2 ml was added to each sterile glass tube and a plaque or 5 μl of a bacteriophage supernatant was added. To the remaining low-phosphate medium, 10 μl of a log culture of *Escherichia coli* strain JM101 in 2 × TY medium and 10 μCi of [^{32}P]phosphate (carrier free) were added per 1 ml of medium, and 1 ml of this added to each of the tubes. The cells were grown for 6 to 10 h and the DNA isolated as described (Sanger *et al.*, 1980). Usually, the cultures were combined before isolation and several regions were probed for at the same time. The isolated DNA was finally dissolved in 30 to 60 ml of the hybridizing solution (Southern, 1979). This was sufficient for 5 to 10 filters.

A random library of sonicated lambda fragments (average size 500 nucleotides) was prepared (Fuhrman *et al.*, 1981) and cloned. Individual pure plaques were usually transferred to fresh plates so that there were about 50 per plate. Blots were prepared using nitrocellulose filters and hybridization and washing carried out as described by Jeffreys & Flavell (1977). Autoradiographs were developed with fluorescent screens for 1 to 3 days and DNA prepared from the plaques giving a positive response. In order to identify plaques containing the desired sequence, a preliminary screening with the ddT or ddA reaction was carried out. The results using ddA are usually rather clearer than those with ddT and were done in the same way (Sanger *et al.*, 1980), except that the reaction mixture contained H buffer, 0·1 mM-dGTP, 0·1 mM-dCTP, 0·1 mM-TTP and 0·02 mM-ddATP.

(c) *Subcloning from clones with larger inserts*

In order to digest a cloned DNA with a restriction enzyme, double-stranded DNA was prepared using DNA polymerase primed with the normal flanking primer (Hong, 1981). For clones that had been stored for a long time, fresh DNA was usually prepared by growing a 1 ml culture and dissolving the final product in 25 μl of Tris-EDTA (Sanger *et al.*, 1980). To 10 μl were added 2 μl of 10 × H buffer, 2 μl (0·4 pmol) of primer and 6 μl of water. The mixture was heated in a sealed capillary tube at 100°C for 3 min and allowed to cool slowly to room temperature. To this was added 20 μl of a solution containing 5 μCi of [α-^{32}P]dATP (spec. act. approx. 400 Ci/mmol), 0·1 mM-dGTP, 0·1 mM-dCTP, 0·1 mM-TTP and 1·5 μM-dATP in H buffer followed by 2 μl (2 units) of DNA polymerase. After incubation for 15 min at room temperature, the appropriate restriction enzyme(s) was added and incubated at 37°C for a further 15 min (65°C for *Taq*I). A 2-μl sample was usually taken at this stage and run on an acrylamide gel to determine whether the digestion was satisfactory. The ligation was done using vector DNA (M13mp8 or mp9) cut with restriction enzyme *Sma*I to generate blunt ends (Messing & Vieira, 1982), and treated with phosphatase. Therefore, if enzymes that produce staggered ends had been used, it was necessary to convert these to blunt ends. After heating at 70°C for 10 min to inactivate the enzyme, 1 μl of 0·5 mM-dATP and 1 μl of DNA polymerase were added and the mixture incubated at room temperature for 15 min: 1 μl of 0·2 M-EDTA, 50 μl of water and 50 μl of water-saturated phenol were added. After separation of the aqueous layer, the DNA was precipitated with ethanol and the product dissolved in 5 μl and applied to a 6% acrylamide gel (containing no urea). The required band was located by autoradiography and eluted by the method of Maxam & Gilbert (1977). After precipitation with ethanol it was dissolved in 5 μl of water; 2 μl of this solution was used for ligation into 20 ng of the vector and the DNA transfected into competent cells, which were then plated. Several plaques were usually screened using the ddATP reaction.

3. Results and Discussion

(a) *Determination of the sequence*

This work was largely carried out as a study of methods, so that we have used a variety of techniques rather than limiting ourselves to the most productive ones.

One of the most important attributes of the cloning approach is that it is a uniquely efficient purification procedure that is not affected by the size of the DNA being studied. Consequently, no attempt was made to carry out an initial restriction enzyme digestion and fractionation of the lambda into smaller fragments, but the whole DNA was digested and the resultant complex mixtures cloned directly. We believe that this approach is more rapid and efficient, though it may be less easy to carry out in a research environment where each collaborator would prefer to study a unique piece of the sequence.

Initially, we used restriction enzyme digests and inserted the fragments into the vector M13mp2 using RI linkers (Sanger *et al.*, 1980). The most successful of these were digests obtained with *Alu*I, *Hae*III, *Hinf*I, *Fnu*DII, *Taq*I, *Dde*I, and *Sau*96A. About 40 to 70 clones were normally sequenced using the "flanking" primer before redundant clones became too frequent. Good results were also obtained using *Sau*3A digests and cloning in a *Bam*HI vector that was an amber derivative of strain M13mWJ43 (Winter & Fields, 1980).

Because of their specificity, the use of restriction enzymes imposes a limitation on the number of useful clones that can be obtained from them, whereas from the random approach it is desirable to have as complex and varied a mixture as possible representing the whole of the DNA in equivalent amounts. Anderson (1981) has described a method of random cleavage using DNase I in the presence of manganese, and Deininger has used sonication (see Fuhrman *et al.*, 1981). Using these methods, it is possible to get many more useful clones from a single digest and these have been mainly used in the more recent stages of the work.

Initially, the method of sequencing was that described by Sanger *et al.* (1980). The DNA from each clone was sequenced by the chain terminating technique with a flanking primer, and the results entered into the data base using the computer programs of Staden (1980). To begin with, new data accumulated rapidly.

However, as the work proceeded more of the sequences were redundant and it seemed that it would be necessary to replace the random approach with a more specific one in order to find certain missing sequences. We chose to do this when about 90% of the whole sequence was in the data base in the form of about 10 to 15 contiguous sequences ("contigs"; Staden, 1980). In fact, it would probably have been better to continue the random approach farther, since it is much more rapid than any specific methods. The final stages of the sequence analysis required, firstly, joining the separate contigs and then confirming and correcting uncertain sequences. Two methods were particularly useful for the former. The ends of contigs can often be extended by reading the DNA sequence further from certain clones, so that any improvements in the electrophoresis can prove very valuable. Normally, sequences of 250 to 300 nucleotides were obtained from each clone. Using special care and techniques it is possible to read a gel out to as much as 500 nucleotides. In our experience, using lower acrylamide concentrations (5%) and drying the gels before autoradiography have proved the most successful. Although it is frequently possible to overlap contigs in this way, the sequences obtained are usually less accurate and will require subsequent confirmation by some other method.

A useful way of extending contigs is by sequencing the insert in an appropriate clone from its opposite end. (The sequence normally determined is the complement of the plus strand of the M13 recombinant.) Two methods have been described for obtaining the sequence of the opposite strand, both of which have been used in this work. In the method of Winter *et al.* (1981), the insert in the clone is cut out with a suitable restriction enzyme and re-ligated into the vector: 50% of the resultant clones should contain the insert in the opposite orientation and can be sequenced with the normal flanking primer. Another method, developed by Hong (1981), makes use of the normal flanking primer to make an unlabelled complete copy of the insert and this is sequenced using a synthetic primer corresponding to a sequence of the M13 at the opposite end of the clone.

The main method we have used for obtaining sequences from a specific region is the use of hybridization probes to select clones from random mixtures. For probes, we chose M13 clones having the complementary sequence to that required and the DNA from these was labelled by growing on medium containing [^{32}P]phosphate. Several probes could be used at the same time to screen blots from plates containing random clones. Although not all of the positively reacting clones were from the desired region, the method was relatively rapid and enabled us to complete the sequence. Towards the end of the work when the whole sequence had been studied, there were some regions that were still uncertain, either because they had only been covered by gel readings on one strand or had been read from a region far out from the priming sequence. These could frequently be studied by taking a clone containing a larger insert, digesting it with an appropriate restriction enzyme (which could be identified from the sequence we already had) and re-cloning the smaller fragment. This was a relatively slow process but was useful at this point in the work.

With a sequence the size of lambda it is difficult to ensure that there are no mistakes, but it is believed that there cannot be many. Most of the sequence has

TABLE 1

Sequences not confirmed in this work

28,462–28,807	Hoess *et al.* (1980)
35,246–35,279	Franklin & Bennett (1979)
37,710–38,040	Schwarz *et al.* (1978)
38,198–38,498	Schwarz *et al.* (1978); Roberts *et al.* (1977)
38,681–38,756	Schwarz *et al.* (1978)
39,658–39,699	Schwarz *et al.* (1980)
44,170–44,500	Daniels & Blattner (1982); Petrov *et al.* (1981)

been determined on both strands. Where this was not done there were usually several gels giving clear and unequivocal readings. The only sequences that we have not determined are listed in Table 1. They come to a total of 1454 nucleotides and were completed by others before we started sequencing.

(b) *Identification of genes*

Figure 1 shows the gene map based on both previous work and the sequence. Since transcription is in different directions in different parts of the genome, the

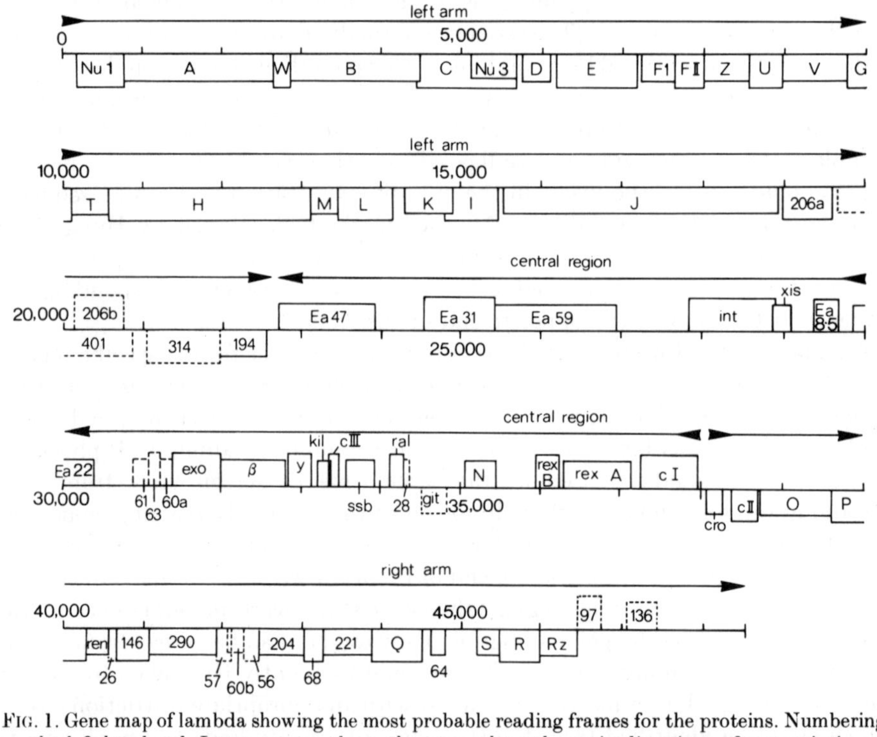

FIG. 1. Gene map of lambda showing the most probable reading frames for the proteins. Numbering is from the left-hand end. Large arrows above the genes show the main directions of transcription and translation. Genes in boxes below the line are translated from left to right; those above from right to left. Boxes with broken lines indicate the more speculative reading frames. *git* is a gene proposed by Ineichen *et al.* (1981). Where no gene or protein product has been assigned to an open reading frame (*orf*) it has been assigned a number that corresponds to the number of amino acid residues that the putative protein would contain.

TABLE 2

Sizes of the lambda proteins

Left arm			Central region			Right arm		
Gene	Mol. Wt.		Gene	Mol. Wt.		Gene	Mol. Wt.	
	from gels† (kdal)	from sequence‡ (daltons)		from gels† (kdal)	from sequence‡ (daltons)		from gels† (kdal)	from sequence‡ (daltons)
Nu1	21	20,444	*cI*	27	26,245	*cro*	7–9	7365
A	80	73,280	*rexA*	29	31,259	*cII*	11–14	11,057
W	5–10	7614	*rexB*	—	15,970	*O*	34–37	33,870
B	60	59,474	*N*	13·5	15,376	*P*	23–27	26,522
C	61	45,915	*ral*	—	7605	*ren*	—	10,613
Nu3	19	20,789	*ssb*	10–16	13,782	*orf146*	—	16,650
D	11–12	11,574	*cIII*	—	6048	*orf290*	—	33,566
E	38	38,194			or 10,067§	*orf57*	—	6964
FI	17	14,310	*kil*	16	5519	*orf60b*	—	7384
FII	11·5	12,731			or 11,646§	*orf56*	—	6321
Z	20	21,564	*γ*	16·5	11,646	*orf204*	—	24,119
U	13–16	14,651			or 16,349§	*orf68*	—	7882
V	25–31	25,814	*β*	28–30	29,693	*orf221*	—	25,222
G	33	15,610	*exo*	25–26	25,912	*Q*	23	22,476
T	16	16,061	*Ea22*	19–22	20,939	*orf64*	—	7084
H	90	92,294	*Ea8·5*	8·5	10,751	*S*	—	11,522
M	10	12,532	*xis*	—	8606	*R*	18	17,828
L	29	25,713	*int*	36–44	40,310	R_Z	11–13	17,232
K	27	23,014	*Ea59*	—	59,494			
I	—	23,127	*Ea31*	—	34,588			
J	130–140	124,440	*Ea47*	—	48,103			
orf206a	—	21,859	*orf206b*	—	20,186			
orf401	—	39,889						
orf314	—	32,024						
orf194	—	21,605						

† SDS-acrylamide gel electrophoresis. For references see Szybalski & Szybalski (1979).
‡ Calculated using programs TRANDK (Staden, 1978) and MWCALC (Staden, unpublished).
§ Ineichen *et al.* (1981).

sequence is presented in three sections, together with the proposed amino acid sequences of the protein genes (Fig. 2). Transcription and translation are shown from left to right. The numbering throughout is from the left-hand end of the genome; therefore in the central region (Fig. 2(b)) it goes backwards. In the main, genes have been identified by examining the sequence for open reading frames of the expected length (see Table 2) preceded by a suitable initiation codon and comparing with the genetic map.

The *R* gene protein was conclusively identified by comparison of the DNA sequence with the known amino acid sequence (Imada & Tsugita, 1971). In order to establish unequivocally the position of some of the structural genes, limited amino acid sequence data at the N termini were obtained (Walker *et al.*, 1982). In this way, the N-terminal sequences of the proteins of genes *D*, *E*, *V*, *B** and *H* were determined and the corresponding position in the DNA sequence identified.

(a)

```
      GGGCGGCGACCTCGCGGGTTTTCGCTATTTATGAAAATTTTCCGGTTTAAGGCGTTTCCGTTCTTCTTCGTCATAACTTAATGTTTTTAT
               10        20        30        40        50        60        70        80        90

      TTAAAATACCCTCTGAAAAGAAAGGAAACGACAGGTGCTGAAAGCGAGGCTTTTTGGCCTCTGTCGTTTCCTTTCTCTGTTTTTGTCCGT
              100       110       120       130       140       150       160       170       180

Nu1             M  E  V  N  K  K  Q  L  A  D  I  F  G  A  S  I  R  T  I  Q  N  W  Q  E  Q  G  M
      GGAATGAACAATGGAAGTCAACAAAAAGCAGCTGGCTGACATTTTCGGTGCGAGTATCCGTACCATTCAGAACTGGCAGGAACAGGGAAT
              190       200       210       220       230       240       250       260       270

Nu1     P  V  L  R  G  G  G  K  G  N  E  V  L  Y  D  S  A  A  V  I  K  W  Y  A  E  R  D  A  E  I
      GCCCGTTCTGCGAGGCGGTGGCAAGGGTAATGAGGTGCTTTATGACTCTGCCGCCGTCATAAAATGGTATGCCGAAAGGGATGCTGAAAT
              280       290       300       310       320       330       340       350       360

Nu1     E  N  E  K  L  R  R  E  V  E  E  L  R  Q  A  S  E  A  D  L  Q  P  G  T  I  E  Y  E  R  H
      TGAGAACGAAAAGCTGCGCCGGGAGGTTGAAGAACTGCGGCAGGCCAGCGAGGCAGATCTCCAGCCAGGAACTATTGAGTACGAACGCCA
              370       380       390       400       410       420       430       440       450

Nu1     R  L  T  R  A  Q  A  D  A  Q  E  L  K  N  A  R  D  S  A  E  V  V  E  T  A  F  C  T  F  V
      TCGACTTACGCGTGCGCAGGCCGACGCACAGGAACTGAAGAATGCCAGAGACTCCGCTGAAGTGGTGGAAACCGCATTCTGTACTTTCGT
              460       470       480       490       500       510       520       530       540

Nu1     L  S  R  I  A  G  E  I  A  S  I  L  D  G  L  P  L  S  V  Q  R  R  F  P  E  L  E  N  R  H
      GCTGTCGCGGATCGCAGGTGAAATTGCCAGTATTCTCGACGGGCTCCCCCTGTCGGTGCAGCGGCGTTTTCCGGAACTGGAAAACCGACA
              550       560       570       580       590       600       610       620       630

Nu1     V  D  F  L  K  R  D  I  I  K  A  M  N  K  A  A  A  L  D  E  L  I  P  G  L  L  S  E  Y  I
A                                                                                    V  N  I  S
      TGTTGATTTCCTGAAACGGGATATCATCAAAGCCATGAACAAAGCAGCCGCGCTGGATGAACTGATACCGGGGTTGCTGAGTGAATATAT
              640       650       660       670       680       690       700       710       720

Nu1     E  Q  S  G  *
A        N  S  Q  V  N  R  L  R  H  F  V  R  A  G  L  R  S  L  F  R  P  E  P  Q  T  A  V  E  W  A
      CGAACAGTCAGGTTAACAGGCTGCGGCATTTTGTCCGCGCCGGGCTTCGCTCACTGTTCAGGCCGGAGCCACAGACCGCCGTTGAATGGG
              730       740       750       760       770       780       790       800       810

A        D  A  N  Y  Y  L  P  K  E  S  A  Y  Q  E  G  R  W  E  T  L  P  F  Q  R  A  I  M  N  A  M
      CGGATGCTAATTACTATCTCCCGAAAGAATCCGCATACCAGGAAGGGCGCTGGGAAACACTGCCCTTTCAGCGGGCCATCATGAATGCGA
              820       830       840       850       860       870       880       890       900

A        G  S  D  Y  I  R  E  V  N  V  V  K  S  A  R  V  G  Y  S  K  M  L  L  G  V  Y  A  Y  F  I
      TGGGCAGCGACTACATCCGTGAGGTGAATGTGGTGAAGTCTGCCCGTGTCGGTTATTCCAAAATGCTGCTGGGTGTTTATGCCTACTTTA
              910       920       930       940       950       960       970       980       990

A        E  H  K  Q  R  N  T  L  I  W  L  P  T  D  G  D  A  E  N  F  M  K  T  H  V  E  P  T  I  R
      TAGAGCATAAGCAGCGCAACACCCTTATCTGGTTGCCGACGGATGGTGATGCCGAGAACTTTATGAAAACCCACGTTGAGCCGACTATTC
             1000      1010      1020      1030      1040      1050      1060      1070      1080

A        D  I  P  S  L  L  A  L  A  P  W  Y  G  K  K  H  R  D  N  T  L  T  M  K  R  F  T  N  G  R
      GTGATATTCCGTCGCTGCTGGCGCTGGCCCCGTGGTATGGCAAAAAGCACCGGGATAACACGCTCACCATGAAGCGTTTCACTAATGGGC
             1090      1100      1110      1120      1130      1140      1150      1160      1170

A        G  F  W  C  L  G  G  K  A  A  K  N  Y  R  E  K  S  V  D  V  A  G  Y  D  E  L  A  A  F  D
      GTGGCTTCTGGTGCCTGGGCGGTAAAGCGGCAAAAAACTACCGTGAAAAGTCGGTGGATGTGGCGGGTTATGATGAACTTGCTGCTTTTG
             1180      1190      1200      1210      1220      1230      1240      1250      1260

A        D  D  I  E  Q  E  G  S  P  T  F  L  G  D  K  R  I  E  G  S  V  W  P  K  S  I  R  G  S  T
      ATGATGATATTGAACAGGAAGGCTCTCCGACGTTCCTGGGTGACAAGCGTATTGAAGGCTCGGTCTGGCCAAAGTCCATCCGTGGCTCCA
             1270      1280      1290      1300      1310      1320      1330      1340      1350

A        P  K  V  R  G  T  C  Q  I  E  R  A  A  S  E  S  P  H  F  M  R  F  H  V  A  C  P  H  C  G
      CGCCAAAAGTGAGAGGCACCTGTCAGATTGAGCGTGCAGCCAGTGAATCCCCGCATTTTATGCGTTTTCATGTTGCCTGCCCGCATTGCG
             1360      1370      1380      1390      1400      1410      1420      1430      1440

A        E  E  Q  Y  L  K  F  G  D  K  E  T  P  F  G  L  K  W  T  P  D  D  P  S  S  V  F  Y  L  C
      GGGAGGAGCAGTATCTTAAATTTGGCGACAAAGAGACGCCGTTTGGCCTCAAATGGACGCCGGATGACCCCTCCAGCGTGTTTTATCTCT
             1450      1460      1470      1480      1490      1500      1510      1520      1530

A        E  H  N  A  C  V  I  R  Q  Q  E  L  D  F  T  D  A  R  Y  I  C  E  K  T  G  I  W  T  R  D
      GCGAGCATAATGCCTGCGTCATCCGCCAGCAGGAGCTGGACTTTACTGATGCCCGTTATATCTGCGAAAAGACCGGGATCTGGACCCGTG
             1540      1550      1560      1570      1580      1590      1600      1610      1620

A        G  I  L  W  F  S  S  S  G  E  E  I  E  P  P  D  S  V  T  F  H  I  W  T  A  Y  S  P  F  T
      ATGGCATTCTCTGGTTTTCGTCATCCGGTGAAGAGATTGAGCCACCTGACAGTGTGACCTTTCACATCTGGACAGCGTACAGCCCGTTCA
             1630      1640      1650      1660      1670      1680      1690      1700      1710

A        T  W  V  Q  I  V  K  D  W  M  K  T  K  G  D  T  G  K  R  K  T  F  V  N  T  T  L  G  E  T
      CCACCTGGGTGCAGATTGTCAAAGACTGGATGAAAACGAAAGGGGATACGGGAAAACGTAAAACCTTCGTAAACACCACGCTCGGTGAGA
             1720      1730      1740      1750      1760      1770      1780      1790      1800

A        W  E  A  K  I  G  E  R  P  D  A  E  V  M  A  E  R  K  E  H  Y  S  A  P  V  P  D  R  V  A
      CGTGGGAGGCGAAAATTGGCGAACGTCCGGATGCTGAAGTGATGGCAGAGCGGAAAGAGCATTATTCAGCGCCCGTTCCTGACCGTGTGG
             1810      1820      1830      1840      1850      1860      1870      1880      1890

A        Y  L  T  A  G  I  D  S  Q  L  D  R  Y  E  M  R  V  W  G  W  G  P  G  E  E  S  W  L  I  D
      CTTACCTGACCGCCGGTATCGACTCCCAGCTGGACCGCTACGAAATGCGCGTATGGGGATGGGGGCCGGGTGAGGAAAGCTGGCTGATTG
             1900      1910      1920      1930      1940      1950      1960      1970      1980

A        R  Q  I  I  M  G  R  H  D  D  E  Q  T  L  L  R  V  D  E  A  I  N  K  T  Y  T  R  R  N  G
      ACCGGCAGATTATTATGGGCCGCCACGACGATGAACAGACGCTGCTGCGTGTGGATGAGGCCATCAATAAAACCTATACCCGCCGGAATG
             1990      2000      2010      2020      2030      2040      2050      2060      2070

A        A  E  M  S  I  S  R  I  C  W  D  T  G  G  I  D  P  T  I  V  Y  E  R  S  K  K  H  G  L  F
      GTGCAGAAATGTCGATATCCCGTATCTGCTGGGATACTGGCGGGATTGACCCGACCATTGTGTATGAACGCTCGAAAAAACATGGGCTGT
             2080      2090      2100      2110      2120      2130      2140      2150      2160

A        R  V  I  P  I  K  G  A  S  V  Y  G  K  P  V  A  S  M  P  R  K  R  N  K  N  G  V  Y  L  T
      TCCGGGTGATCCCCATTAAAGGGGCATCCGTCTACGGAAAGCCGGTGGCCAGCATGCCACGTAAGCGAAACAAAAACGGGGTTTACCTTA
             2170      2180      2190      2200      2210      2220      2230      2240      2250

A        E  I  G  T  D  T  A  K  E  Q  I  Y  N  R  F  T  L  T  P  E  G  D  E  P  L  P  G  A  V  H
      CCGAAATCGGTACGGATACCGCGAAAGAGCAGATTTATAACCGCTTCACACTGACGCCGGAAGGGGATGAACCGCTTCCCGGTGCCGTTC
             2260      2270      2280      2290      2300      2310      2320      2330      2340

A        F  P  N  N  P  D  I  F  D  L  T  E  A  Q  Q  L  T  A  E  E  Q  V  E  K  W  V  D  G  R  K
      ACTTCCCGAATAACCCGGATATTTTTGATCTGACCGAAGCGCAGCAGCTGACTGCTGAAGAGCAGGTCGAAAAATGGGTGGATGGCAGGA
             2350      2360      2370      2380      2390      2400      2410      2420      2430

A        K  I  L  W  D  S  K  K  R  R  N  E  A  L  D  C  F  V  Y  A  L  A  A  L  R  I  S  I  S  R
      AAAAAATACTGTGGGACAGCAAAAAGCGACGCAATGAGGCACTCGACTGCTTCGTTTATGCGCTGGCGGCGCTGCGCATCAGTATTTCCC
             2440      2450      2460      2470      2480      2490      2500      2510      2520

A        W  Q  L  D  L  S  A  L  L  A  S  L  Q  E  E  D  G  A  A  T  N  K  K  T  L  A  D  Y  A  R
      GCTGGCAGCTGGATCTCAGTGCGCTGCTGGCGAGCCTGCAGGAAGAGGATGGTGCAGCAACCAACAAGAAAACACTGGCAGATTACGCCC
             2530      2540      2550      2560      2570      2580      2590      2600      2610
```

FIG. 2. Nucleotide sequence of lambda DNA and the most probable amino acid sequences of proteins it codes for. The sequence given is for λ*cI*indlts857*S*am7; mutations from the wild type are at 37,589 (*cI*indl), 37,742 (*cI*ts857) and 45,351 to 45,353 (*S*am7). The sequence is in 3 sections according to the main directions of transcription (see Fig. 1): (a) the left arm; (b) the right arm; (c) the central region. Transcription and translation are shown from left to right throughout. The numbering is from left to right of the whole DNA and consequently goes backwards in the central region. The individual gene

```
W                            M  T  R  Q  E  E  L  A  A  A  R  A  A  L  H  D  L  M  T  G  K  R  V
A          A  L  S  G  E  D  E  *
      GTGCCTTATCCGGAGAGGATGAATGACGCGACAGGAAGAACTTGCCGCTGCCCGTGCGGCACTGCATGACCTGATGACAGGTAAACGGGT
                2620      2630      2640      2650      2660      2670      2680      2690      2700

W      A  T  V  Q  K  D  G  R  R  V  E  F  T  A  T  S  V  S  D  L  K  K  Y  I  A  E  L  E  V  Q
      GGCAACAGTACAGAAAGACGGACGAAGGGTGGAGTTTACGGCCACTTCCGTGTCTGACCTGAAAAAATATATTGCAGAGCTGGAAGTGCA
                2710      2720      2730      2740      2750      2760      2770      2780      2790

B                                                     M  K  T  P  T  I  P  T  L  L  G  P  D  G  M
W      T  G  M  T  Q  R  R  R  G  P  A  G  F  Y  V  *
      GACCGGCATGACACAGCGACGCAGGGGACCTGCAGGATTTTATGTATGAAAACGCCCACCATTCCCACCCTTCTGGGGCCGGACGGCATG
                2800      2810      2820      2830      2840      2850      2860      2870      2880

B      T  S  L  R  E  Y  A  G  Y  H  G  G  G  S  G  F  G  G  Q  L  R  S  W  N  P  P  S  E  S  V
      ACATCGCTGCGCGAATATGCCGGTTATCACGGCGGTGGCAGCGGATTTGGAGGGCAGTTGCGGTCGTGGAACCCACCGAGTGAAAGTGTG
                2890      2900      2910      2920      2930      2940      2950      2960      2970

B      D  A  A  L  L  P  N  F  T  R  G  N  A  R  A  D  D  L  V  R  N  N  G  Y  A  A  N  A  I  Q
      GATGCAGCCCTGTTGCCCAACTTTACCCGTGGCAATGCCCGCGCAGACGATCTGGTACGCAATAACGGCTATGCCGCCAACGCCATCCAG
                2980      2990      3000      3010      3020      3030      3040      3050      3060

B      L  H  Q  D  H  I  V  G  S  F  F  R  L  S  H  R  P  S  W  R  Y  L  G  I  G  E  E  E  A  R
      CTGCATCAGGATCATATCGTCGGGTCTTTTTTCCGGCTCAGTCATCGCCCAAGCTGGCGCTATCTGGGCATCGGGGAGGAAGAAGCCCGT
                3070      3080      3090      3100      3110      3120      3130      3140      3150

B      A  F  S  R  E  V  E  A  A  W  K  E  F  A  E  D  D  C  C  C  I  D  V  E  R  K  R  T  F  T
      GCCTTTTCCCGCGAGGTTGAAGCGGCATGGAAAGAGTTTGCCGAGGATGACTGCTGCTGCATTGACGTTGAGCGAAAACGCACGTTTACC
                3160      3170      3180      3190      3200      3210      3220      3230      3240

B      M  M  I  R  E  G  V  A  M  H  A  F  N  G  E  L  F  V  Q  A  T  W  D  T  S  S  S  R  L  F
      ATGATGATTCGGGAAGGTGTGGCCATGCACGCCTTTAACGGTGAACTGTTCGTTCAGGCCACCTGGGATACCAGTTCGTCGCGGCTTTTC
                3250      3260      3270      3280      3290      3300      3310      3320      3330

B      R  T  Q  F  R  M  V  S  P  K  R  I  S  N  P  N  N  T  G  D  S  R  N  C  R  A  G  V  Q  I
      CGGACACAGTTCCGGATGGTCAGCCCGAAGCGCATCAGCAACCCGAACAATACCGGCGACAGCCGGAACTGCCGTGCCGGTGTGCAGATT
                3340      3350      3360      3370      3380      3390      3400      3410      3420

B      N  D  S  G  A  A  L  G  Y  Y  V  S  E  D  G  Y  P  G  W  M  P  Q  K  W  T  W  I  P  R  E
      AATGACAGCGGTGCGGCGCTGGGATATTACGTCAGCGAGGACGGGTATCCTGGCTGGATGCCGCAGAAATGGACATGGATACCCCGTGAG
                3430      3440      3450      3460      3470      3480      3490      3500      3510

B      L  P  G  G  R  A  S  F  I  H  V  F  E  P  V  E  D  G  Q  T  R  G  A  N  V  F  Y  S  V  M
      TTACCCGGCGGGCGCGCCTCGTTCATTCACGTTTTTGAACCCGTGGAGGACGGGCAGACTCGCGGTGCAAATGTGTTTTACAGCGTGATG
                3520      3530      3540      3550      3560      3570      3580      3590      3600

B      E  Q  M  K  M  L  D  T  L  Q  N  T  Q  L  Q  S  A  I  V  K  A  M  Y  A  A  T  I  E  S  E
      GAGCAGATGAAGATGCTCGACACGCTGCAGAACACGCAGCTGCAGAGCGCCATTGTGAAGGCGATGTATGCCGCCACCATTGAGAGTGAG
                3610      3620      3630      3640      3650      3660      3670      3680      3690

B      L  D  T  Q  S  A  M  D  F  I  L  G  A  N  S  Q  E  Q  R  E  R  L  T  G  W  I  G  E  I  A
      CTGGATACGCAGTCAGCGATGGATTTTATTCTGGGCGCGAACAGTCAGGAGCAGCGGGAAAGGCTGACCGGCTGGATTGGTGAAATTGCC
                3700      3710      3720      3730      3740      3750      3760      3770      3780

B      A  Y  Y  A  A  A  P  V  R  L  G  G  A  K  V  P  H  L  M  P  G  D  S  L  N  L  Q  T  A  Q
      GCGTATTACGCCGCAGCGCCGGTCCGGCTGGGAGGCGCAAAAGTACCGCACCTGATGCCGGGTGACTCACTGAACCTGCAGACGGCTCAG
                3790      3800      3810      3820      3830      3840      3850      3860      3870

B      D  T  D  N  G  Y  S  V  F  E  Q  S  L  L  R  Y  I  A  A  G  L  G  V  S  Y  E  Q  L  S  R
      GATACGGATAACGGCTACTCCGTGTTTGAGCAGTCACTGCTGCGGTATATCGCTGCCGGGCTGGGTGTCTCGTATGAGCAGCTTTCCCGG
                3880      3890      3900      3910      3920      3930      3940      3950      3960

B      N  Y  A  Q  M  S  Y  S  T  A  R  A  S  A  N  E  S  W  A  Y  F  M  G  R  R  K  F  V  A  S
      AATTACGCCCAGATGAGCTACTCCACGGCACGGGCCAGTGCGAACGAGTCGTGGGCGTACTTTATGGGGCGGCGAAAATTCGTCGCATCC
                3970      3980      3990      4000      4010      4020      4030      4040      4050

B      R  Q  A  S  Q  M  F  L  C  W  L  E  E  A  I  V  R  R  V  V  T  L  P  S  K  A  R  F  S  F
      CGTCAGGCGAGCCAGATGTTTCTGTGCTGGCTGGAAGAGGCCATCGTTCGCCGCGTGGTGACGTTACCTTCAAAAGCGCGCTTCAGTTTT
                4060      4070      4080      4090      4100      4110      4120      4130      4140

B      Q  E  A  R  S  A  W  G  N  C  D  W  I  G  S  G  R  M  A  I  D  G  L  K  E  V  Q  E  A  V
      CAGGAAGCCCGCAGTGCCTGGGGGAACTGCGACTGGATAGGCTCCGGTCGTATGGCCATCGATGGTCTGAAAGAAGTTCAGGAAGCGGTG
                4150      4160      4170      4180      4190      4200      4210      4220      4230

B      M  L  I  E  A  G  L  S  T  Y  E  K  E  C  A  K  R  G  D  D  Y  Q  E  I  F  A  Q  Q  V  R
      ATGCTGATAGAAGCCGGACTGAGTACCTACGAGAAAGAGTGCGCAAAACGCGGTGACGACTATCAGGAAATTTTTGCCCAGCAGGTCCGT
                4240      4250      4260      4270      4280      4290      4300      4310      4320

B      E  T  M  E  R  R  A  A  G  L  K  P  P  A  W  A  A  A  A  F  E  S  G  L  R  Q  S  T  E  E
      GAAACGATGGAGCGCCGTGCAGCCGGTCTTAAACCGCCCGCCTGGGCGGCTGCAGCATTTGAATCCGGGCTGCGACAATCAACAGAGGAG
                4330      4340      4350      4360      4370      4380      4390      4400      4410

B      E  K  S  D  S  R  A  A  *
C          V  T  A  E  L  R  N  L  P  H  I  A  S  M  A  F  N  E  P  L  M  L  E  P  A  Y  A  R
      GAGAAGAGTGACAGCAGAGCTGCGTAATCTCCCGCATATTGCCAGCATGGCCTTTAATGAGCCGCTGATGCTTGAACCCGCCTATGCGCG
                4420      4430      4440      4450      4460      4470      4480      4490      4500

C       V  F  F  C  A  L  A  G  Q  L  G  I  S  S  L  T  D  A  V  S  G  D  S  L  T  A  Q  E  A  L
      GGTTTTCTTTTGTGCGCTTGCAGGCCAGCTTGGGATCAGCAGCCTGACGGATGCGGTGTCCGGCGACAGCCTGACTGCCCAGGAGGCACT
                4510      4520      4530      4540      4550      4560      4570      4580      4590

C       A  T  L  A  L  S  G  D  D  D  G  P  R  Q  A  R  S  Y  Q  V  M  N  G  I  A  V  L  P  V  S
      CGCGACGCTGGCATTATCCGGTGATGATGACGGACCACGACAGGCCCGCAGTTATCAGGTCATGAACGGCATCGCCGTGCTGCCGGTGTC
                4600      4610      4620      4630      4640      4650      4660      4670      4680

C       G  T  L  V  S  R  T  R  A  L  Q  P  Y  S  G  M  T  G  Y  N  G  I  I  A  R  L  Q  Q  A  A
      CGGCACGCTGGTCAGCCGGACGCGGGCGCTGCAGCCGTACTCGGGGATGACCGGTTACAACGGCATTATCGCCCGTCTGCAACAGGCTGC
                4690      4700      4710      4720      4730      4740      4750      4760      4770

C       S  D  P  M  V  D  G  I  L  L  D  M  D  T  P  G  G  M  V  A  G  A  F  D  C  A  D  I  I  A
      CAGCGATCCGATGGTGGACGGCATTCTGCTCGATATGGACACGCCCGGCGGGATGGTGGCGGGGGCATTTGACTGCGCTGACATCATCGC
                4780      4790      4800      4810      4820      4830      4840      4850      4860

C       R  V  R  D  I  K  P  V  W  A  L  A  N  D  M  N  C  S  A  G  Q  L  L  A  S  A  A  S  R  R
      CCGTGTGCGTGACATAAAACCGGTATGGGCGCTTGCCAACGACATGAACTGCAGTGCAGGTCAGTTGCTTGCCAGTGCCGCCTCCCGGCG
                4870      4880      4890      4900      4910      4920      4930      4940      4950

C       L  V  T  Q  T  A  R  T  G  S  I  G  V  M  M  A  H  S  N  Y  G  A  A  L  E  K  Q  G  V  E
      TCTGGTCACGCAGACCGCCCGGACAGGCTCCATCGGCGTCATGATGGCTCACAGTAATTACGGTGCTGCGCTGGAGAAACAGGGTGTGGA
                4960      4970      4980      4990      5000      5010      5020      5030      5040

C       I  T  L  I  Y  S  G  S  H  K  V  D  G  N  P  Y  S  H  L  P  D  D  V  R  E  T  L  Q  S  R
      AATCACGCTGATTTACAGCGGCAGCCATAAGGTGGATGGCAACCCCTACAGCCATCTTCCGGATGACGTCCGGGAGACACTGCAGTCCCG
                5050      5060      5070      5080      5090      5100      5110      5120      5130
```

products or reading frames (*orf*) are identified in the left-hand margin. Some of the more important control regions are also marked above the sequences. The start points of the main transcription products are indicated as follows: s_R, start site for p_R RNA; s_E, for p_E RNA; $s_{R'}$, for $p_{R'}$ RNA; s_I, for *int* message; s_L, for p_L message. In some genes the position of initiation is uncertain and the most probable has been chosen (see Table 3 and the text).

```
C      M  D  A  T  R  Q  M  F  A  Q  K  V  S  A  Y  T  G  L  S  V  Q  V  V  L  D  T  E  A  A  V
      GATGGACGCAACCCGCCAGATGTTTGCGCAGAAGGTGTCGGCATATACCGGCCTGTCCGTGCAGGTTGTGCTGGATACCGAGGCTGCAGT
            5140      5150      5160      5170      5180      5190      5200      5210      5220

C      Y  S  G  Q  E  A  I  D  A  G  L  A  D  E  L  V  N  S  T  D  A  I  T  V  M  R  D  A  L  D
      GTACAGCGGTCAGGAGGCCATTGATGCCGGACTGGCTGATGAACTTGTTAACAGCACCGATGCGATCACCGTCATGCGTGATGCACTGGA
            5230      5240      5250      5260      5270      5280      5290      5300      5310

C      A  R  K  S  R  L  S  G  G  R  M  T  K  E  T  Q  S  T  T  V  S  A  T  A  S  Q  A  D  V  T
      TGCACGTAAATCCCGTCTCTCAGGAGGGCGAATGACCAAAGAGACTCAATCAACAACTGTTTCAGCCACTGCTTCGCAGGCTGACGTTAC
            5320      5330      5340      5350      5360      5370      5380      5390      5400

C      D  V  V  P  A  T  E  G  E  N  A  S  A  A  Q  P  D  V  N  A  Q  I  T  A  A  V  A  A  E  N
      TGACGTGGTGCCAGCGACGGAGGGCGAGAACGCCAGCGCGGCGCAGCCGGACGTGAACGCGCAGATCACCGCAGCGGTTGCGGCAGAAAA
            5410      5420      5430      5440      5450      5460      5470      5480      5490

C      S  R  I  M  G  I  L  N  C  E  E  A  H  G  R  E  E  Q  A  R  V  L  A  E  T  P  G  M  T  V
      CAGCCGCATTATGGGGATCCTCAACTGTGAGGAGGCTCACGGACGCGAAGAACAGGCACGCGTGCTGGCAGAAACCCCCGGTATGACCGT
            5500      5510      5520      5530      5540      5550      5560      5570      5580

C      K  T  A  R  R  I  L  A  A  A  P  Q  S  A  Q  A  R  S  D  T  A  L  D  R  L  M  Q  G  A  P
      GAAAACGGCCCGCCGCATTCTGGCCGCAGCACCACAGAGTGCACAGGCGCGCAGTGACACTGCGCTGGATCGTCTGATGCAGGGGGCACC
            5590      5600      5610      5620      5630      5640      5650      5660      5670

C D    A  P  L  A  A  G  N  P  A  S  D  A  V  N  D  L  L  N  T  P  V  *           M  T  S  K  E
      GGCACCGCTGGCTGCAGGTAACCCGGCATCTGATGCCGTTAACGATTTGCTGAACACACCAGTGTAAGGGATGTTTATGACGAGCAAAGA
            5680      5690      5700      5710      5720      5730      5740      5750      5760

D      T  F  T  H  Y  Q  P  Q  G  N  S  D  P  A  H  T  A  T  A  P  G  G  L  S  A  K  A  P  A  M
      AACCTTTACCCATTACCAGCCGCAGGGCAACAGTGACCCGGCTCATACCGCAACCGCGCCCGGCGGATTGAGTGCGAAAGCGCCTGCAAT
            5770      5780      5790      5800      5810      5820      5830      5840      5850

D      T  P  L  M  L  D  T  S  S  R  K  L  V  A  W  D  G  T  T  D  G  A  A  V  G  I  L  A  V  A
      GACCCCGCTGATGCTGGACACCTCCAGCCGTAAGCTGGTTGCGTGGGATGGCACCACCGACGGTGCTGCCGTTGGCATTCTTGCGGTTGC
            5860      5870      5880      5890      5900      5910      5920      5930      5940

D      A  D  Q  T  S  T  T  L  T  F  Y  K  S  G  T  F  R  Y  E  D  V  L  W  P  E  A  A  S  D  E
      TGCTGACCAGACCAGCACCACGCTGACGTTCTACAAGTCCGGCACGTTCCGTTATGAGGATGTGCTCTGGCCGGAGGCTGCCAGCGACGA
            5950      5960      5970      5980      5990      6000      6010      6020      6030

D      T  K  K  R  T  A  F  A  G  T  A  I  S  I  V  *
      GACGAAAAAACGGACCGCGTTTGCCGGAACGGCAATCAGCATCGTTTAACTTTACCCTTCATCACTAAAGGCCGCCTGTGCGGCTTTTTT
            6040      6050      6060      6070      6080      6090      6100      6110      6120

E                  M  S  M  Y  T  T  A  Q  L  L  A  A  N  E  Q  K  F  K  F  D  P  L  F  L  R  L
      TACGGGATTTTTTTATGTCGATGTACACAACCGCCCAACTGCTGGCGGCAAATGAGCAGAAATTTAAGTTTGATCCGCTGTTTCTGCGTC
            6130      6140      6150      6160      6170      6180      6190      6200      6210

E      F  F  R  E  S  Y  P  F  T  T  E  K  V  Y  L  S  Q  I  P  G  L  V  N  M  A  L  Y  V  S  P
      TCTTTTTCCGTGAGAGCTATCCCTTCACCACGGAGAAAGTCTATCTCTCACAAATTCCGGGACTGGTAAACATGGCGCTGTACGTTTCGC
            6220      6230      6240      6250      6260      6270      6280      6290      6300

E      I  V  S  G  E  V  I  R  S  R  G  G  S  T  S  E  F  T  P  G  Y  V  K  P  K  H  E  V  N  P
      CGATTGTTTCCGGTGAGGTTATCCGTTCCCGTGGCGGCTCCACCTCTGAATTTACGCCGGGATATGTCAAGCCGAAGCATGAAGTGAATC
            6310      6320      6330      6340      6350      6360      6370      6380      6390

E      Q  M  T  L  R  R  L  P  D  E  D  P  Q  N  L  A  D  P  A  Y  R  R  R  R  I  I  M  Q  N  M
      CGCAGATGACCCTGCGTCGCCTGCCGGATGAAGATCCGCAGAATCTGGCGGACCCGGCTTACCGCCGCCGTCGCATCATCATGCAGAACA
            6400      6410      6420      6430      6440      6450      6460      6470      6480

E      R  D  E  E  L  A  I  A  Q  V  E  E  M  Q  A  V  S  A  V  L  K  G  K  Y  T  M  T  G  E  A
      TGCGTGACGAAGAGCTGGCCATTGCTCAGGTCGAAGAGATGCAGGCAGTTTCTGCCGTGCTTAAGGGCAAATACACCATGACCGGTGAAG
            6490      6500      6510      6520      6530      6540      6550      6560      6570

E      F  D  P  V  E  V  D  M  G  R  S  E  E  N  N  I  T  Q  S  G  G  T  E  W  S  K  R  D  K  S
      CCTTCGATCCGGTTGAGGTGGATATGGGCCGCAGTGAGGAGAATAACATCACGCAGTCCGGCGGCACGGAGTGGAGCAAGCGTGACAAGT
            6580      6590      6600      6610      6620      6630      6640      6650      6660

E      T  Y  D  P  T  D  D  I  E  A  Y  A  L  N  A  S  G  V  V  N  I  I  V  F  D  P  K  G  W  A
      CCACGTATGACCCGACCGACGATATCGAAGCCTACGCGCTGAACGCCAGCGGTGTGGTGAATATCATCGTGTTCGATCCGAAAGGCTGGG
            6670      6680      6690      6700      6710      6720      6730      6740      6750

E      L  F  R  S  F  K  A  V  K  E  K  L  D  T  R  R  G  S  N  S  E  L  E  T  A  V  K  D  L  G
      CGCTGTTCCGTTCCTTCAAAGCCGTCAAGGAGAAGCTGGATACCCGTCGTGGCTCTAATTCCGAGCTGGAGACAGCGGTGAAAGACCTGG
            6760      6770      6780      6790      6800      6810      6820      6830      6840

E      K  A  V  S  Y  K  G  M  Y  G  D  V  A  I  V  V  Y  S  G  Q  Y  V  E  N  G  V  K  K  N  F
      GCAAAGCGGTGTCCTATAAGGGGATGTATGGCGATGTGGCCATCGTCGTGTATTCCGGACAGTACGTGGAAAACGGCGTCAAAAAGAACT
            6850      6860      6870      6880      6890      6900      6910      6920      6930

E      L  P  D  N  T  M  V  L  G  N  T  Q  A  R  G  L  R  T  Y  G  C  I  Q  D  A  D  A  Q  R  E
      TCCTGCCGGACAACACGATGGTGCTGGGGAACACTCAGGCACGCGGTCTGCGCACCTATGGCTGCATTCAGGATGCGGACGCACAGCGCG
            6940      6950      6960      6970      6980      6990      7000      7010      7020

E      G  I  N  A  S  A  R  Y  P  K  N  W  V  T  T  G  D  P  A  R  E  F  T  M  I  Q  S  A  P  L
      AAGGCATTAACGCCTCTGCCCGTTACCCGAAAAACTGGGTGACCACCGGCGATCCGGCGCGTGAGTTCACCATGATTCAGTCAGCACCGC
            7030      7040      7050      7060      7070      7080      7090      7100      7110

E      M  L  L  A  D  P  D  E  F  V  S  V  Q  L  A  *
      TGATGCTGCTGGCTGACCCTGATGAGTTCGTGTCCGTACAACTGGCGTAATCATGGCCCTTCGGGGCCATTGTTTCTCTGTGGAGGAGTC
            7120      7130      7140      7150      7160      7170      7180      7190      7200

FI     M  T  K  D  E  L  I  A  R  L  R  S  L  G  E  Q  L  N  R  D  V  S  L  T  G  T  K  E  E  L
      CATGACGAAAGATGAACTGATTGCCCGTCTCCGCTCGCTGGGTGAACAACTGAACCGTGATGTCAGCCTGACGGGGACGAAAGAAGAACT
            7210      7220      7230      7240      7250      7260      7270      7280      7290

FI     A  L  R  V  A  E  L  K  E  E  L  D  D  T  D  E  T  A  G  Q  D  T  P  L  S  R  E  N  V  L
      GGCGCTCCGTGTGGCAGAGCTGAAAGAGGAGCTTGATGACACGGATGAAACTGCCGGTCAGGACACCCCTCTCAGCCGGGAAAATGTGCT
            7300      7310      7320      7330      7340      7350      7360      7370      7380

FI     T  G  H  E  N  E  V  G  S  A  Q  P  D  T  V  I  L  D  T  S  E  L  V  T  V  V  A  L  V  K
      GACCGGACATGAAAATGAGGTGGGATCAGCGCAGCCGGATACCGTGATTCTGGATACGTCTGAACTGGTCACGGTCGTGGCACTGGTGAA
            7390      7400      7410      7420      7430      7440      7450      7460      7470

FI     L  H  T  D  A  L  H  A  T  R  D  E  P  V  A  F  V  L  P  G  T  A  F  R  V  S  A  G  V  A
      GCTGCATACTGATGCACTTCACGCCACGCGGGATGAACCTGTGGCATTTGTGCTGCCGGGAACGGCGTTTCGTGTCTCTGCCGGTGTGGC
            7480      7490      7500      7510      7520      7530      7540      7550      7560

FII                                                       V  A  D  F  D  N  L  F  D  A  A  I  A
FI     A  E  M  T  E  R  G  L  A  R  M  Q  *
      AGCCGAAATGACAGAGCGCGGCCTGGCCAGAATGCAATAACGGGAGGCGCTGTGGCTGATTTCGATAACCTGTTCGATGCTGCCATTGCC
            7570      7580      7590      7600      7610      7620      7630      7640      7650

FII    R  A  D  E  T  I  R  G  Y  M  G  T  S  A  T  I  T  S  G  E  Q  S  G  A  V  I  R  G  V  F
      CGCGCCGATGAAACGATACGCGGGTACATGGGAACGTCAGCCACCATTACATCCGGTGAGCAGTCAGGTGCGGTGATACGTGGTGTTTTT
            7660      7670      7680      7690      7700      7710      7720      7730      7740
```

```
FII    D  D  P  E  N  I  S  Y  A  G  Q  G  V  R  V  E  G  S  S  P  S  L  F  V  R  T  D  E  V  R
      GATGACCCTGAAAATATCAGCTATGCCGGACAGGGCGTGCGCGTTGAAGGCTCCAGCCCGTCCCTGTTTGTCCGGACTGATGAGGTGCGG
           7750      7760      7770      7780      7790      7800      7810      7820      7830

FII    Q  L  R  R  G  D  T  L  T  I  G  E  E  N  F  W  V  D  R  V  S  P  D  D  G  G  S  C  H  L
      CAGCTGCGGCGTGGAGACACGCTGACCATCGGTGAGGAAAATTTCTGGGTAGATCGGGTTTCGCCGGATGATGGCGGAAGTTGTCATCTC
           7840      7850      7860      7870      7880      7890      7900      7910      7920

FII    W  L  G  R  G  V  P  P  A  V  N  R  R  R  *
Z                                                           M  A  I  K  G  L  E  Q  A  V  E  N
      TGGCTTGGACGGGGCGTACCGCCTGCCGTTAACCGTCGCCGCTGAAAGGGGGATGTATGGCCATAAAAGGTCTTGAGCAGGCCGTTGAAA
           7930      7940      7950      7960      7970      7980      7990      8000      8010

Z       L  S  R  I  S  K  T  A  V  P  G  A  A  A  M  A  I  N  R  V  A  S  S  A  I  S  Q  S  A  S
      ACCTCAGCCGTATCAGCAAAACGGCGGTGCCTGGTGCCGCCGCAATGGCCATTAACCGCGTTGCTTCATCCGCGATATCGCAGTCGGCGT
           8020      8030      8040      8050      8060      8070      8080      8090      8100

Z       Q  V  A  R  E  T  K  V  R  R  K  L  V  K  E  R  A  R  L  K  R  A  T  V  K  N  P  Q  A  R
      CACAGGTTGCCCGTGAGACAAAGGTACGCCGGAAACTGGTAAAGGAAAGGGCCAGGCTGAAAAGGGCCACGGTCAAAAATCCGCAGGCCA
           8110      8120      8130      8140      8150      8160      8170      8180      8190

Z       I  K  V  N  R  G  D  L  P  V  I  K  L  G  N  A  R  V  V  L  S  R  R  R  R  R  K  K  G  Q
      GAATCAAAGTTAACCGGGGGGATTTGCCCGTAATCAAGCTGGGTAATGCGCGGGTTGTCCTTTCGCGCCGCAGGCGTCGTAAAAAGGGGC
           8200      8210      8220      8230      8240      8250      8260      8270      8280

Z       R  S  S  L  K  G  G  G  S  V  L  V  V  G  N  R  R  I  P  G  A  F  I  Q  Q  L  K  N  G  R
      AGCGTTCATCCCTGAAAGGTGGCGGCAGCGTGCTTGTGGTGGGTAACCGTCGTATTCCCGGCGCGTTTATTCAGCAACTGAAAAATGGCC
           8290      8300      8310      8320      8330      8340      8350      8360      8370

Z       W  H  V  M  Q  R  V  A  G  K  N  R  Y  P  I  D  V  V  K  I  P  M  A  V  P  L  T  T  A  F
      GGTGGCATGTCATGCAGCGTGTGGCTGGGAAAAACCGTTACCCCATTGATGTGGTGAAAATCCCGATGGCGGTGCCGCTGACCACGGCGT
           8380      8390      8400      8410      8420      8430      8440      8450      8460

Z       K  Q  N  I  E  R  I  R  R  E  R  L  P  K  E  L  G  Y  A  L  Q  H  Q  L  R  M  V  I  K  R
      TTAAACAAAATATTGAGCGGATACGGCGTGAACGTCTTCCGAAAGAGCTGGGCTATGCGCTGCAGCATCAACTGAGGATGGTAATAAAGC
           8470      8480      8490      8500      8510      8520      8530      8540      8550

U       M  K  H  T  E  L  R  A  A  V  L  D  A  L  E  K  H  D  T  G  A  T  F  F  D  G  R  P  A  V
Z       *
      GATGAAACATACTGAACTCCGTGCAGCCGTACTGGATGCACTGGAGAAGCATGACACCGGGGCGACGTTTTTTGATGGTCGCCCCGCTGT
           8560      8570      8580      8590      8600      8610      8620      8630      8640

U       F  D  E  A  D  F  P  A  V  A  V  Y  L  T  G  A  E  Y  T  G  E  E  L  D  S  D  T  W  Q  A
      TTTTGATGAGGCGGATTTTCCGGCAGTTGCCGTTTATCTCACCGGCGCTGAATACACGGGCGAAGAGCTGGACAGCGATACCTGGCAGGC
           8650      8660      8670      8680      8690      8700      8710      8720      8730

U      ·E  L  H  I  E  V  F  L  P  A  Q  V  P  D  S  E  L  D  A  W  M  E  S  R  I  Y  P  V  M  S
      GGAGCTGCATATCGAAGTTTTCCTGCCTGCTCAGGTGCCGGATTCAGAGCTGGATGCGTGGATGGAGTCCCGGATTTATCCGGTGATGAG
           8740      8750      8760      8770      8780      8790      8800      8810      8820

U       D  I  P  A  L  S  D  L  I  T  S  M  V  A  S  G  Y  D  Y  R  R  D  D  D  A  G  L  W  S  S
      CGATATCCCGGCACTGTCAGATTTGATCACCAGTATGGTGGCCAGCGGCTATGACTACCGGCGCGACGATGATGCGGGCTTGTGGAGTTC
           8830      8840      8850      8860      8870      8880      8890      8900      8910

U       A  D  L  T  Y  V  I  T  Y  E  M  *
V                                                        M  P  V  P  N  P  T  M  P  V  K  G  A  G  T  T
      AGCCGATCTGACTTATGTCATTACCTATGAAATGTGAGGACGCTATGCCTGTACCAAATCCTACAATGCCGGTGAAAGGTGCCGGGACCA
           8920      8930      8940      8950      8960      8970      8980      8990      9000

V       L  W  V  Y  K  G  S  G  D  P  Y  A  N  P  L  S  D  V  D  W  S  R  L  A  K  V  K  D  L  T
      CCCTGTGGGTTTATAAGGGGAGCGGTGACCCTTACGCGAATCCGCTTTCAGACGTTGACTGGTCGCGTCTGGCAAAAGTTAAAGACCTGA
           9010      9020      9030      9040      9050      9060      9070      9080      9090

V       P  G  E  L  T  A  E  S  Y  D  D  S  Y  L  D  D  E  D  A  D  W  T  A  T  G  Q  G  Q  K  S
      CGCCCGGCGAACTGACCGCTGAGTCCTATGACGACAGCTATCTCGATGATGAAGATGCAGACTGGACTGCGACCGGGCAGGGGCAGAAAT
           9100      9110      9120      9130      9140      9150      9160      9170      9180

V       A  G  D  T  S  F  T  L  A  W  M  P  G  E  Q  G  Q  Q  A  L  L  A  W  F  N  E  G  D  T  R
      CTGCCGGAGATACCAGCTTCACGCTGGCGTGGATGCCCGGAGAGCAGGGGCAGCAGGCGCTGCTGGCGTGGTTTAATGAAGGCGATACCC
           9190      9200      9210      9220      9230      9240      9250      9260      9270

V       A  Y  K  I  R  F  P  N  G  T  V  D  V  F  R  G  W  V  S  S  I  G  K  A  V  T  A  K  E  V
      GTGCCTATAAAATCCGCTTCCCGAACGGCACGGTCGATGTGTTCCGTGGCTGGGTCAGCAGTATCGGTAAGGCGGTGACGGCGAAGGAAG
           9280      9290      9300      9310      9320      9330      9340      9350      9360

V       I  T  R  T  V  K  V  T  N  V  G  R  P  S  M  A  E  D  R  S  T  V  T  A  A  T  G  M  T  V
      TGATCACCCGCACGGTGAAAGTCACCAATGTGGGACGTCCGTCGATGGCAGAAGATCGCAGCACGGTAACAGCGGCAACCGGCATGACCG
           9370      9380      9390      9400      9410      9420      9430      9440      9450

V       T  P  A  S  T  S  V  V  K  G  Q  S  T  T  L  T  V  A  F  Q  P  E  G  V  T  D  K  S  F  R
      TGACGCCTGCCAGCACCTCGGTGGTGAAAGGGCAGAGCACCACGCTGACCGTGGCCTTCCAGCCGGAGGGCGTAACCGACAAGAGCTTTC
           9460      9470      9480      9490      9500      9510      9520      9530      9540

V       A  V  S  A  D  K  T  K  A  T  V  S  V  S  G  M  T  I  T  V  N  G  V  A  A  G  K  V  N  I
      GTGCGGTGTCTGCGGATAAAACAAAAGCCACCGTGTCGGTCAGTGGTATGACCATCACCGTGAACGGCGTTGCTGCAGGCAAGGTCAACA
           9550      9560      9570      9580      9590      9600      9610      9620      9630

V  G    P  V  V  S  G  N  G  E  F  A  A  V  A  E  I  T  V  T  A  S  *              M  F  L  K
      TTCCGGTTGTATCCGGTAATGGTGAGTTTGCTGCGGTTGCAGAAATTACCGTCACCGCCAGTTAATCCGGAGAGTCAGCGATGTTCCTGA
           9640      9650      9660      9670      9680      9690      9700      9710      9720

G       T  E  S  F  E  H  N  G  V  T  V  T  L  S  E  L  S  A  L  Q  R  I  E  H  L  A  L  M  K  R
      AAACCGAATCATTTGAACATAACGGTGTGACCGTCACGCTTTCTGAACTGTCAGCCCTGCAGCGCATTGAGCATCTCGCCCTGATGAAAC
           9730      9740      9750      9760      9770      9780      9790      9800      9810

G       Q  A  E  Q  A  E  S  D  S  N  R  K  F  T  V  E  D  A  I  R  T  G  A  F  L  V  A  M  S  L
      GGCAGGCAGAACAGGCGGAGTCAGACAGCAACCGGAAGTTTACTGTGGAAGACGCCATCAGAACCGGCGCGTTTCTGGTGGCGATGTCCC
           9820      9830      9840      9850      9860      9870      9880      9890      9900

G       W  H  N  H  P  Q  K  T  Q  M  P  S  M  N  E  A  V  K  Q  I  E  Q  E  V  L  T  T  W  P  T
      TGTGGCATAACCATCCGCAGAAGACGCAGATGCCGTCCATGAATGAAGCCGTTAAACAGATTGAGCAGGAAGTGCTTACCACCTGGCCCA
           9910      9920      9930      9940      9950      9960      9970      9980      9990

G       E  A  I  S  H  A  E  N  V  V  Y  R  L  S  G  M  Y  E  F  V  V  N  N  A  P  E  Q  T  E  D
      CGGAGGCAATTTCTCATGCTGAAAACGTGGTGTACCGGCTGTCTGGTATGTATGAGTTTGTGGTGAATAATGCCCCTGAACAGACAGAGG
          10000     10010     10020     10030     10040     10050     10060     10070     10080

T                                                 V  F  D  G  E  L  S  F  A  L  K  L  A  R  E  M  G  R  P
G       A  G  P  A  E  P  V  S  A  G  K  C  S  T  V  S  *
      ACGCCGGGCCCGCAGAGCCTGTTTCTGCGGGAAAGTGTTCGACGGTGAGCTGAGTTTTGCCCTGAAACTGGCGCGTGAGATGGGGCGACC
          10090     10100     10110     10120     10130     10140     10150     10160     10170

T       D  W  R  A  M  L  A  G  M  S  S  T  E  Y  A  D  W  H  R  F  Y  S  T  H  Y  F  H  D  V  L
      CGACTGGCGTGCCATGCTTGCCGGGATGTCATCCACGGAGTATGCCGACTGGCACCGCTTTTACAGTACCCATTATTTCATGATGTTCT
          10180     10190     10200     10210     10220     10230     10240     10250     10260
```

```
T    L  D  M  H  F  S  G  L  T  Y  T  V  L  S  L  F  F  S  D  P  D  M  H  P  L  D  F  S  L  L
   GCTGGATATGCACTTTTCCGGGCTGACGTACACCGTGCTCAGCCTGTTTTTCAGCGATCCGGATATGCATCCGCTGGATTTCAGTCTGCT
       10270     10280     10290     10300     10310     10320     10330     10340     10350

T    N  R  R  E  A  D  E  E  P  E  D  D  V  L  M  Q  K  A  A  G  L  A  G  G  V  R  F  G  P  D
   GAACCGGCGCGAGGCTGACGAAGAGCCTGAAGATGATGTGCTGATGCAGAAAGCGGCAGGGCTTGCCGGAGGTGTCCGCTTTGGCCCGGA
       10360     10370     10380     10390     10400     10410     10420     10430     10440

T    G  N  E  V  I  P  A  S  P  D  V  A  D  M  T  E  D  D  V  M  L  M  T  V  S  E  G  I  A  G
   CGGGAATGAAGTTATCCCCGCTTCCCCGGATGTGGCGGACATGACGGAGGATGACGTAATGCTGATGACAGTATCAGAAGGGATCGCAGG
       10450     10460     10470     10480     10490     10500     10510     10520     10530

T    G  V  R  Y  G  *
H                 M  A  E  P  V  G  D  L  V  V  D  L  S  L  D  A  A  R  F  D  E  Q  M  A  R  V  R
   AGGAGTCCGGTATGGCTGAACCGGTAGGCGATCTGGTCGTTGATTTGAGTCTGGATGCGGCCAGATTTGACGAGCAGATGGCCAGAGTCA
       10540     10550     10560     10570     10580     10590     10600     10610     10620

H    R  H  F  S  G  T  E  S  D  A  K  K  T  A  A  V  V  E  Q  S  L  S  R  Q  A  L  A  A  Q  K
   GGCGTCATTTTTCTGGTACGGAAAGTGATGCGAAAAAAACAGCGGCAGTCGTTGAACAGTCGCTGAGCCGACAGGCGCTGGCTGCACAGA
       10630     10640     10650     10660     10670     10680     10690     10700     10710

H    A  G  I  S  V  G  Q  Y  K  A  A  M  R  M  L  P  A  Q  F  T  D  V  A  T  Q  L  A  G  G  Q
   AAGCGGGGATTTCCGTCGGGCAGTATAAAGCCGCCATGCGTATGCTGCCTGCACAGTTCACCGACGTGGCCACGCAGCTTGCAGGCGGGC
       10720     10730     10740     10750     10760     10770     10780     10790     10800

H    S  P  W  L  I  L  L  Q  Q  G  G  Q  V  K  D  S  F  G  G  M  I  P  M  F  R  G  L  A  G  A
   AAAGTCCGTGGCTGATCCTGCTGCAACAGGGGGGGCAGGTGAAGGACTCCTTCGGCGGGATGATCCCCATGTTCAGGGGGCTTGCCGGTG
       10810     10820     10830     10840     10850     10860     10870     10880     10890

H    I  T  L  P  M  V  G  A  T  S  L  A  V  A  T  G  A  L  A  Y  A  W  Y  Q  G  N  S  T  L  S
   CGATCACCCTGCCGATGGTGGGGGCCACCTCGCTGGCGGTGGCGACCGGTGCGCTGGCGTATGCCTGGTATCAGGGCAACTCAACCCTGT
       10900     10910     10920     10930     10940     10950     10960     10970     10980

H    D  F  N  K  T  L  V  L  S  G  N  Q  A  G  L  T  A  D  R  M  L  V  L  S  R  A  G  Q  A  A
   CCGATTTCAACAAAACGCTGGTCCTTTCCGGCAATCAGGCGGGACTGACGGCAGATCGTATGCTGGTCCTGTCCAGAGCCGGGCAGGCGG
       10990     11000     11010     11020     11030     11040     11050     11060     11070

H    G  L  T  F  N  Q  T  S  E  S  L  S  A  L  V  K  A  G  V  S  G  E  A  Q  I  A  S  I  S  Q
   CAGGGCTGACGTTTAACCAGACCAGCGAGTCACTCAGCGCACTGGTTAAGGCGGGGGTAAGCGGTGAGGCTCAGATTGCGTCCATCAGCC
       11080     11090     11100     11110     11120     11130     11140     11150     11160

H    S  V  A  R  F  S  S  A  S  G  V  E  V  D  K  V  A  E  A  F  G  K  L  T  T  D  P  T  S  G
   AGAGTGTGGCGCGTTTCTCCTCTGCATCCGGCGTGGAGGTGGACAAGGTCGCTGAAGCCTTCGGGAAGCTGACCACAGACCCGACGTCGG
       11170     11180     11190     11200     11210     11220     11230     11240     11250

H    L  T  A  M  A  R  Q  F  H  N  V  S  A  E  Q  I  A  Y  V  A  Q  L  Q  R  S  G  D  E  A  G
   GGCTGACGGCGATGGCTCGCCAGTTCCATAACGTGTCGGCGGAGCAGATTGCGTATGTTGCTCAGTTGCAGCGTTCCGGCGATGAAGCCG
       11260     11270     11280     11290     11300     11310     11320     11330     11340

H    A  L  Q  A  A  N  E  A  A  T  K  G  F  D  D  Q  T  R  R  L  K  E  N  M  G  T  L  E  T  W
   GGGCATTGCAGGCGGCGAACGAGGCCGCAACGAAAGGGTTTGATGACCAGACCCGCCGCCTGAAAGAGAACATGGGCACGCTGGAGACCT
       11350     11360     11370     11380     11390     11400     11410     11420     11430

H    A  D  R  T  A  R  A  F  K  S  M  W  D  A  V  L  D  I  G  R  P  D  T  A  Q  E  M  L  I  K
   GGGCAGACAGGACTGCGCGGGCATTCAAATCCATGTGGGATGCGGTGCTGGATATTGGTCGTCCTGATACCGCGCAGGAGATGCTGATTA
       11440     11450     11460     11470     11480     11490     11500     11510     11520

H    A  E  A  A  Y  K  K  A  D  D  I  W  N  L  R  K  D  D  Y  F  V  N  D  E  A  R  A  R  Y  W
   AGGCAGAGGCTGCGTATAAGAAAGCAGACGACATCTGGAATCTGCGCAAGGATGATTATTTTGTTAACGATGAAGCGCGGGCGCGTTACT
       11530     11540     11550     11560     11570     11580     11590     11600     11610

H    D  D  R  E  K  A  R  L  A  L  E  A  A  R  K  K  A  E  Q  Q  T  Q  Q  D  K  N  A  Q  Q  Q
   GGGATGATCGTGAAAAGGCCCGTCTTGCGCTTGAAGCCGCCCGAAAGAAGGCTGAGCAGCAGACTCAACAGGACAAAAATGCGCAGCAGC
       11620     11630     11640     11650     11660     11670     11680     11690     11700

H    S  D  T  E  A  S  R  L  K  Y  T  E  E  A  Q  K  A  Y  E  R  L  Q  T  P  L  E  K  Y  T  A
   AGAGCGATACCGAAGCGTCACGGCTGAAATATACCGAAGAGGCGCAGAAGGCTTACGAACGGCTGCAGACGCCGCTGGAGAAATATACCG
       11710     11720     11730     11740     11750     11760     11770     11780     11790

H    R  Q  E  E  L  N  K  A  L  K  D  G  K  I  L  Q  A  D  Y  N  T  L  M  A  A  A  K  K  D  Y
   CCCGTCAGGAAGAACTGAACAAGGCACTGAAAGACGGGAAAATCCTGCAGGCGGATTACAACACGCTGATGGCGGCGGCGAAAAAGGATT
       11800     11810     11820     11830     11840     11850     11860     11870     11880

H    E  A  T  L  K  K  P  K  Q  S  S  V  K  V  S  A  G  D  R  Q  E  D  S  A  H  A  A  L  L  T
   ATGAAGCGACGCTGAAAAAGCCGAAACAGTCCAGCGTGAAGGTGTCTGCGGGCGATCGTCAGGAAGACAGTGCTCATGCTGCCCTGCTGA
       11890     11900     11910     11920     11930     11940     11950     11960     11970

H    L  Q  A  E  L  R  T  L  E  K  H  A  G  A  N  E  K  I  S  Q  Q  R  R  D  L  W  K  A  E  S
   CGCTTCAGGCAGAACTCCGGACGCTGGAGAAGCATGCCGGAGCAAATGAGAAAATCAGCCAGCAGCGCCGGGATTTGTGGAAGGCGGAGA
       11980     11990     12000     12010     12020     12030     12040     12050     12060

H    Q  F  A  V  L  E  E  A  A  Q  R  R  Q  L  S  A  Q  E  K  S  L  L  A  H  K  D  E  T  L  E
   GTCAGTTCGCGGTACTGGAGGAGGCGGCGCAACGTCGCCAGCTGTCTGCACAGGAGAAATCCCTGCTGGCGCATAAAGATGAGACGCTGG
       12070     12080     12090     12100     12110     12120     12130     12140     12150

H    Y  K  R  Q  L  A  A  L  G  D  K  V  T  Y  Q  E  R  L  N  A  L  A  Q  Q  A  D  K  F  A  Q
   AGTACAAACGCCAGCTGGCTGCACTTGGCGACAAGGTTACGTATCAGGAGCGCCTGAACGCGCTGGCGCAGCAGGCGGATAAATTCGCAC
       12160     12170     12180     12190     12200     12210     12220     12230     12240

H    Q  Q  R  A  K  R  A  A  I  D  A  K  S  R  G  L  T  D  R  Q  A  E  R  E  A  T  E  Q  R  L
   AGCAGCAACGGGCAAAACGGGCCGCCATTGATGCGAAAAGCCGGGGGCTGACTGACCGGCAGGCAGAACGGGAAGCCACGGAACAGCGCC
       12250     12260     12270     12280     12290     12300     12310     12320     12330

H    K  E  Q  Y  G  D  N  P  L  A  L  N  N  V  M  S  E  Q  K  K  T  W  A  A  E  D  Q  L  R  G
   TGAAGGAACAGTATGGCGATAATCCGCTGGCGCTGAATAACGTCATGTCAGAGCAGAAAAAGACCTGGGCGGCTGAAGACCAGCTTCGCG
       12340     12350     12360     12370     12380     12390     12400     12410     12420

H    N  W  M  A  G  L  K  S  G  W  S  E  W  E  E  S  A  T  D  S  M  S  Q  V  K  S  A  A  T  Q
   GGAACTGGATGGCAGGCCTGAAGTCCGGCTGGAGTGAGTGGGAAGAGAGCGCCACGGACAGTATGTCGCAGGTAAAAAGTGCAGCCACGC
       12430     12440     12450     12460     12470     12480     12490     12500     12510

H    T  F  D  G  I  A  Q  N  M  A  A  M  L  T  G  S  E  Q  N  W  R  S  F  T  R  S  V  L  S  M
   AGACCTTTGATGGTATTGCACAGAATATGGCGGCGATGCTGACCGGCAGTGAGCAGAACTGGCGCAGCTTCACCCGTTCCGTGCTGTCCA
       12520     12530     12540     12550     12560     12570     12580     12590     12600

H    M  T  E  I  L  L  K  Q  A  M  V  G  I  V  G  S  I  G  S  A  I  G  G  A  V  G  G  G  A  S
   TGATGACAGAAATTCTGCTTAAGCAGGCAATGGTGGGGATTGTCGGGAGTATCGGCAGCGCCATTGGCGGGGCTGTTGGTGGCGGCGCAT
       12610     12620     12630     12640     12650     12660     12670     12680     12690

H    A  S  G  G  T  A  I  Q  A  A  A  A  K  F  H  F  A  T  G  G  F  T  G  T  G  G  K  Y  E  P
   CCGCGTCAGGCGGTACAGCCATTCAGGCCGCTGCGGCGAAATTCCATTTTGCAACCGGAGGATTTACGGAACCGGCGGCAAATATGAGC
       12700     12710     12720     12730     12740     12750     12760     12770     12780
```

```
H     A  G  I  V  H  R  G  E  F  V  F  T  K  E  A  T  S  R  I  G  V  G  N  L  Y  R  L  M  R  G
   CAGCGGGGATTGTTCACCGTGGTGAGTTTGTCTTCACGAAGGAGGCAACCAGCCGGATTGGCGTGGGGAATCTTTACCGGCTGATGCGCG
       12790     12800     12810     12820     12830     12840     12850     12860     12870

H     Y  A  T  G  G  Y  V  G  T  P  G  S  M  A  D  S  R  S  Q  A  S  G  T  F  E  Q  N  N  H  V
   GCTATGCCACCGGCGGTTATGTCGGTACACCGGGCAGCATGGCAGACAGCCGGTCGCAGGCGTCCGGGACGTTTGAGCAGAATAACCATG
       12880     12890     12900     12910     12920     12930     12940     12950     12960

H     V  I  N  N  D  G  T  N  G  Q  I  G  P  A  A  L  K  A  V  Y  D  M  A  R  K  G  A  R  D  E
   TGGTGATTAACAACGACGGCACGAACGGGCAGATAGGTCCGGCTGCTCTGAAGGCGGTGTATGACATGGCCCGCAAGGGTGCCCGTGATG
       12970     12980     12990     13000     13010     13020     13030     13040     13050

M                                                     M  K  T  F  R  W  K  V  K  P  G  M  D  V
H     I  Q  T  Q  M  R  D  G  G  L  F  S  G  G  G  R  *
   AAATTCAGACACAGATGCGTGATGGTGGCCTGTTCTCCGGAGGTGGACGATGAAGACCTTCCGCTGGAAAGTGAAACCCGGTATGGATGT
       13060     13070     13080     13090     13100     13110     13120     13130     13140

M     A  S  V  P  S  V  R  K  V  R  F  G  D  G  Y  S  Q  R  A  P  A  G  L  N  A  N  L  K  T  Y
   GGCTTCGGTCCCTTCTGTAAGAAAGGTGCGCTTTGGTGATGGCTATTCTCAGCGAGCGCCTGCCGGGCTGAATGCCAACCTGAAAACGTA
       13150     13160     13170     13180     13190     13200     13210     13220     13230

M     S  V  T  L  S  V  P  R  E  E  A  T  V  L  E  S  F  L  E  E  H  G  G  W  K  S  F  L  W  T
   CAGCGTGACGCTTTCTGTCCCCCGTGAGGAGGCCACGGTACTGGAGTCGTTTCTGGAAGAGCACGGGGGCTGGAAATCCTTTCTGTGGAC
       13240     13250     13260     13270     13280     13290     13300     13310     13320

M     P  P  Y  E  W  R  Q  I  K  V  T  C  A  K  W  S  S  R  V  S  M  L  R  V  E  F  S  A  E  F
   GCCGCCTTATGAGTGGCGGCAGATAAAGGTGACCTGCGCAAAATGGTCGTCGCGGGTCAGTATGCTGCGTGTTGAGTTCAGCGCAGAGTT
       13330     13340     13350     13360     13370     13380     13390     13400     13410

L                    M  Q  D  I  R  Q  E  T  L  N  E  C  T  R  A  E  Q  S  A  S  V  V  L  W
M     E  Q  V  V  N  *
   TGAACAGGTGGTGAACTGATGCAGGATATCCGGCAGGAAACACTGAATGAATGCACCCGTGCGGAGCAGTCGGCCAGCGTGGTGCTCTGG
       13420     13430     13440     13450     13460     13470     13480     13490     13500

L     E  I  D  L  T  E  V  G  G  E  R  Y  F  F  C  N  E  Q  N  E  K  G  E  P  V  T  W  Q  G  R
   GAAATCGACCTGACAGAGGTCGGTGGAGAACGTTATTTTTTCTGTAATGAGCAGAACGAAAAAGGTGAGCCGGTCACCTGGCAGGGGCGA
       13510     13520     13530     13540     13550     13560     13570     13580     13590

L     Q  Y  Q  P  Y  P  I  Q  G  S  G  F  E  L  N  G  K  G  T  S  T  R  P  T  L  T  V  S  N  L
   CAGTATCAGCCGTATCCCATTCAGGGGAGCGGTTTTGAACTGAATGGCAAAGGCACCAGTACGCGCCCCACGCTGACGGTTTCTAACCTG
       13600     13610     13620     13630     13640     13650     13660     13670     13680

L     Y  G  M  V  T  G  M  A  E  D  M  Q  S  L  V  G  G  T  V  V  R  R  K  V  Y  A  R  F  L  D
   TACGGTATGGTCACCGGGATGGCGGAAGATATGCAGAGTCTGGTCGGCGGAACGGTGGTCCGGCGTAAGGTTTACGCCCGTTTTCTGGAT
       13690     13700     13710     13720     13730     13740     13750     13760     13770

L     A  V  N  F  V  N  G  N  S  Y  A  D  P  E  Q  E  V  I  S  R  W  R  I  E  Q  C  S  E  L  S
   GCGGTGAACTTCGTCAACGGAAACAGTTACGCCGATCCGGAGCAGGAGGTGATCAGCCGCTGGCGCATTGAGCAGTGCAGCGAACTGAGC
       13780     13790     13800     13810     13820     13830     13840     13850     13860

L     A  V  S  A  S  F  V  L  S  T  P  T  E  T  D  G  A  V  F  P  G  R  I  M  L  A  N  T  C  T
   GCGGTGAGTGCCTCCTTTGTACTGTCCACGCCGACGGAAACGGATGGCGCTGTTTTTCCGGGACGTATCATGCTGGCCAACACCTGCACC
       13870     13880     13890     13900     13910     13920     13930     13940     13950

L     W  T  Y  R  G  D  E  C  G  Y  S  G  P  A  V  A  D  E  Y  D  Q  P  T  S  D  I  T  K  D  K
   TGGACCTATCGCGGTGACGAGTGCGGTTATAGCGGTCCGGCTGTCGCGGATGAATATGACCAGCCAACGTCCGATATCACGAAGGATAAA
       13960     13970     13980     13990     14000     14010     14020     14030     14040

L     C  S  K  C  L  S  G  C  K  F  R  N  N  V  G  N  F  G  G  F  L  S  I  N  K  L  S  Q  *
   TGCAGCAAATGCCTGAGCGGTTGTAAGTTCCGCAATAACGTCGGCAACTTTGGCGGCTTCCTTTCCATTAACAAACTTTCGCAGTAAATC
       14050     14060     14070     14080     14090     14100     14110     14120     14130

   CCATGACACAGACAGAATCAGCGATTCTGGCGCACGCCCGGCGATGTGCGCCAGCGGAGTCGTGCGGCTTCGTGGTAAGCACGCCGGAGG
       14140     14150     14160     14170     14180     14190     14200     14210     14220

K                                                        M  S  P  E  D  W  L  Q  A  E  M  Q
   GGGAAAGATATTTCCCCTGCGTGAATATCTCCGGTGAGCCGGAGGCTATTTCCGTATGTCGCCGGAAGACTGGCTGCAGGCAGAAATGCA
       14230     14240     14250     14260     14270     14280     14290     14300     14310

K     G  E  I  V  A  L  V  H  S  H  P  G  G  L  P  W  L  S  E  A  D  R  R  L  Q  V  Q  S  D  L
   GGGTGAGATTGTGGCGCTGGTCCACAGCCACCCCGGTGGTCTGCCCTGGCTGAGTGAGGCCGACCGGCGGCTGCAGGTGCAGAGTGATTT
       14320     14330     14340     14350     14360     14370     14380     14390     14400

K     P  W  W  L  V  C  R  G  T  I  H  K  F  R  C  V  P  H  L  T  G  R  R  F  E  H  G  V  T  D
   GCCGTGGTGGCTGGTCTGCCGGGGGACGATTCATAAGTTCCGCTGTGTGCCGCATCTCACCGGGCGGCGCTTTGAGCACGGTGTGACGGA
       14410     14420     14430     14440     14450     14460     14470     14480     14490

K     C  Y  T  L  F  R  D  A  Y  H  L  A  G  I  E  M  P  D  F  H  R  E  D  D  W  W  R  N  G  Q
   CTGTTACACACTGTTCCGGGATGCTTATCATCTGGCGGGGATTGAGATGCCGGACTTTCATCGTGAGGATGACTGGTGGCGTAACGGCCA
       14500     14510     14520     14530     14540     14550     14560     14570     14580

K     N  L  Y  L  D  N  L  E  A  T  G  L  Y  Q  V  P  L  S  A  A  Q  P  G  D  V  L  L  C  C  F
   GAATCTCTATCTGGATAATCTGGAGGCGACGGGGCTGTATCAGGTGCCGTTGTCAGCGGCACAGCCGGGCGATGTGCTGCTGTGCTGTTT
       14590     14600     14610     14620     14630     14640     14650     14660     14670

K     G  S  S  V  P  N  H  A  A  I  Y  C  G  D  G  E  L  L  H  H  I  P  E  Q  L  S  K  R  E  R
   TGGTTCATCAGTGCCGAATCACGCCGCAATTTACTGCGGCGACGGCGAGCTGCTGCACCATATTCCTGAACAACTGAGCAAACGAGAGAG
       14680     14690     14700     14710     14720     14730     14740     14750     14760

I                 M  A  A  T  H  T  L  P  L  A  S  P  G  M  A  R  I  C  L  Y  G  D  L  Q  R  F
K     Y  T  D  K  W  Q  R  R  T  H  S  L  W  R  H  R  A  W  R  A  S  A  F  T  G  I  Y  N  D  L
   GTACACCGACAAATGGCAGCGACGCACACACTCCCTCTGGCGTCACCGGGCATGGCGCGCATCTGCCTTTACGGGGATTTACAACGATTT
       14770     14780     14790     14800     14810     14820     14830     14840     14850

I     G  R  R  I  D  L  R  V  K  T  G  A  E  A  I  R  A  L  A  T  Q  L  P  A  F  R  Q  K  L  S
K     V  A  A  S  T  F  V  *
   GGTCGCCGCATCGACCTTCGTGTGAAAACGGGGGCTGAAGCCATCCGGGCACTGGCCACACAGCTCCCGGCGTTTCGTCAGAAACTGAGC
       14860     14870     14880     14890     14900     14910     14920     14930     14940

I     D  G  W  Y  Q  V  R  I  A  G  R  D  V  S  T  S  G  L  T  A  Q  L  H  E  T  L  P  D  G  A
   GACGGCTGGTATCAGGTACGGATTGCCGGGCGGGACGTCAGCACGTCCGGGTTAACGGCGCAGTTACATGAGACTCTGCCTGATGGCGCT
       14950     14960     14970     14980     14990     15000     15010     15020     15030

I     V  I  H  I  V  P  R  V  A  G  A  K  S  G  G  V  F  Q  I  V  L  G  A  A  A  I  A  G  S  F
   GTAATTCATATTGTTCCCAGAGTCGCCGGGGCCAAGTCAGGTGGCGTATTCCAGATTGTCCTGGGGGCTGCCGCCATTGCCGGATCATTC
       15040     15050     15060     15070     15080     15090     15100     15110     15120

I     F  T  A  G  A  T  L  A  A  W  G  A  A  I  G  A  G  G  M  T  G  I  L  F  S  L  G  A  S  M
   TTTACCGCCGGAGCCACCCTTGCAGCATGGGGGGCAGCCATTGGGGCCGGTGGTATGACCGGCATCCTGTTTTCTCTCGGTGCCAGTATG
       15130     15140     15150     15160     15170     15180     15190     15200     15210

I     V  L  G  G  V  A  Q  M  L  A  P  K  A  R  T  P  R  I  Q  T  T  D  N  G  K  Q  N  T  Y  F
   GTGCTCGGTGGTGTGGCGCAGATGCTGGCACCGAAAGCCAGAACTCCCCGTATACAGACAACGGATAACGGTAAGCAGAACACCTATTTC
       15220     15230     15240     15250     15260     15270     15280     15290     15300
```

```
I    S  S  L  D  N  M  V  A  Q  G  N  V  L  P  V  L  Y  G  E  M  R  V  G  S  R  V  V  S  Q  E
    TCCTCACTGGATAACATGGTTGCCCAGGGCAATGTTCTGCCTGTTCTGTACGGGGAAATGCGCGTGGGGTCACGCGTGGTTTCTCAGGAG
        15310     15320     15330     15340     15350     15360     15370     15380     15390

I    I  S  T  A  D  E  G  D  G  G  Q  V  V  V  I  G  P  *
    ATCAGCACGGCAGACGAAGGGGACGGTGGTCAGGTTGTGGTGATTGGTCGCTGATGCAAAATGTTTTATGTGAAACCGCCTGCGGGCGGT
        15400     15410     15420     15430     15440     15450     15460     15470     15480

J                            M  G  K  G  S  S  K  G  H  T  P  R  E  A  K  D  N  L  K  S  T  Q
    TTTGTCATTTATGGAGCGTGAGGAATGGGTAAAGGAAGCAGTAAGGGGCATACCCCGCGCGAAGCGAAGGACAACCTGAAGTCCACGCAG
        15490     15500     15510     15520     15530     15540     15550     15560     15570

J    L  L  S  V  I  D  A  I  S  E  G  P  I  E  G  P  V  D  G  L  K  S  V  L  L  N  S  T  P  V
    TTGCTGAGTGTGATCGATGCCATCAGCGAAGGGCCGATTGAAGGTCCGGTGGATGGCTTAAAAAGCGTGCTGCTGAACAGTACGCCGGTG
        15580     15590     15600     15610     15620     15630     15640     15650     15660

J    L  D  T  E  G  N  T  N  I  S  G  V  T  V  V  F  R  A  G  E  Q  E  Q  T  P  P  E  G  F  E
    CTGGACACTGAGGGGAATACCAACATATCCGGTGTCACGGTGGTGTTCCGGGCTGGTGAGCAGGAGCAGACTCCGCCGGAGGGATTTGAA
        15670     15680     15690     15700     15710     15720     15730     15740     15750

J    S  S  G  S  E  T  V  L  G  T  E  V  K  Y  D  T  P  I  T  R  T  I  T  S  A  N  I  D  R  L
    TCCTCCGGCTCCGAGACGGTGCTGGGTACGGAAGTGAAATATGACACGCCGATCACCCGCACCATTACGTCTGCAAACATCGACCGTCTG
        15760     15770     15780     15790     15800     15810     15820     15830     15840

J    R  F  T  F  G  V  Q  A  L  V  E  T  T  S  K  G  D  R  N  P  S  E  V  R  L  L  V  Q  I  Q
    CGCTTTACCTTCGGTGTACAGGCACTGGTGGAAACCACCTCAAAGGGTGACAGGAATCCGTCGGAAGTCCGCCTGCTGGTTCAGATACAA
        15850     15860     15870     15880     15890     15900     15910     15920     15930

J    R  N  G  G  W  V  T  E  K  D  I  T  I  K  G  K  T  T  S  Q  Y  L  A  S  V  V  M  G  N  L
    CGTAACGGTGGCTGGGTGACGGAAAAAGACATCACCATTAAGGGCAAAACCACCTCGCAGTATCTGGCCTCGGTGGTGATGGGTAACCTG
        15940     15950     15960     15970     15980     15990     16000     16010     16020

J    P  P  R  P  F  N  I  R  M  R  R  M  T  P  D  S  T  T  D  Q  L  Q  N  K  T  L  W  S  S  Y
    CCGCCGCGCCCGTTTAATATCCGGATGCGCAGGATGACGCCGGACAGCACCACAGACCAGCTGCAGAACAAAACGCTCTGGTCGTCATAC
        16030     16040     16050     16060     16070     16080     16090     16100     16110

J    T  E  I  I  D  V  K  Q  C  Y  P  N  T  A  L  V  G  V  Q  V  D  S  E  Q  F  G  S  Q  Q  V
    ACTGAAATCATCGATGTGAAACAGTGCTACCCGAACACGGCACTGGTCGGCGTGCAGGTGGACTCGGAGCAGTTCGGCAGCCAGCAGGTG
        16120     16130     16140     16150     16160     16170     16180     16190     16200

J    S  R  N  Y  H  L  R  G  R  I  L  Q  V  P  S  N  Y  N  P  Q  T  R  Q  Y  S  G  I  W  D  G
    AGCCGTAATTATCATCTGCGCGGGCGTATTCTGCAGGTGCCGTCGAACTATAACCCGCAGACGCGGCAATACAGCGGTATCTGGGACGGA
        16210     16220     16230     16240     16250     16260     16270     16280     16290

J    T  F  K  P  A  Y  S  N  N  M  A  W  C  L  W  D  M  L  T  H  P  R  Y  G  M  G  K  R  L  G
    ACGTTTAAACCGGCATACAGCAACAACATGGCCTGGTGTCTGTGGGATATGCTGACCCATCCGCGCTACGGCATGGGGAAACGTCTTGGT
        16300     16310     16320     16330     16340     16350     16360     16370     16380

J    A  A  D  V  D  K  W  A  L  Y  V  I  G  Q  Y  C  D  Q  S  V  P  D  G  F  G  G  T  E  P  R
    GCGGCGGATGTGGATAAATGGGCGCTGTATGTCATCGGCCAGTACTGCGACCAGTCAGTGCCGGACGGCTTTGGCGGCACGGAGCCGCGC
        16390     16400     16410     16420     16430     16440     16450     16460     16470

J    I  T  C  N  A  Y  L  T  T  Q  R  K  A  W  D  V  L  S  D  F  C  S  A  M  R  C  M  P  V  W
    ATCACCTGTAATGCGTACCTGACCACACAGCGTAAGGCGTGGGATGTGCTCAGCGATTTCTGCTCGGCGATGCGCTGTATGCCGGTATGG
        16480     16490     16500     16510     16520     16530     16540     16550     16560

J    N  G  Q  T  L  T  F  V  Q  D  R  P  S  D  K  T  W  T  Y  N  R  S  N  V  V  M  P  D  D  G
    AACGGGCAGACGCTGACGTTCGTGCAGGACCGACCGTCGGATAAGACGTGGACCTATAACCGCAGTAATGTGGTGATGCCGGATGATGGC
        16570     16580     16590     16600     16610     16620     16630     16640     16650

J    A  P  F  R  Y  S  F  S  A  L  K  D  R  H  N  A  V  E  V  N  W  I  D  P  N  N  G  W  E  T
    GCGCCGTTCCGCTACAGCTTCAGCGCCCTGAAGGACCGCCATAATGCCGTTGAGGTGAACTGGATTGACCCGAACAACGGCTGGGAGACG
        16660     16670     16680     16690     16700     16710     16720     16730     16740

J    A  T  E  L  V  E  D  T  Q  A  I  A  R  Y  G  R  N  V  T  K  M  D  A  F  G  C  T  S  R  G
    GCGACAGAGCTTGTTGAAGATACGCAGGCCATTGCCCGTTACGGTCGTAATGTTACGAAGATGGATGCCTTTGGCTGTACCAGCCGGGGG
        16750     16760     16770     16780     16790     16800     16810     16820     16830

J    Q  A  H  R  A  G  L  W  L  I  K  T  E  L  L  E  T  Q  T  V  D  F  S  V  G  A  E  G  L  R
    CAGGCACACCGCGCCGGGCTGTGGCTGATTAAAACAGAACTGCTGGAAACGCAGACCGTGGATTTCAGCGTCGGCGCAGAAGGGCTTCGC
        16840     16850     16860     16870     16880     16890     16900     16910     16920

J    H  V  P  G  D  V  I  E  I  C  D  D  D  Y  A  G  I  S  T  G  G  R  V  L  A  V  N  S  Q  T
    CATGTACCGGGCGATGTTATTGAAATCTGCGATGATGACTATGCCGGTATCAGCACCGGTGGTCGTGTGCTGGCGGTGAACAGCCAGACC
        16930     16940     16950     16960     16970     16980     16990     17000     17010

J    R  T  L  T  L  D  R  E  I  T  L  P  S  S  G  T  A  L  I  S  L  V  D  G  S  G  N  P  V  S
    CGGACGCTGACGCTCGACCGTGAAATCACGCTGCCATCCTCCGGTACCGCGCTGATAAGCCTGGTTGACGGAAGTGGCAATCCGGTCAGC
        17020     17030     17040     17050     17060     17070     17080     17090     17100

J    V  E  V  Q  S  V  T  D  G  V  K  V  K  V  S  R  V  P  D  G  V  A  E  Y  S  V  W  E  L  K
    GTGGAGGTTCAGTCCGTCACCGACGGCGTGAAGGTAAAAGTGAGCCGTGTTCCTGACGGTGTTGCTGAATACAGCGTATGGGAGCTGAAG
        17110     17120     17130     17140     17150     17160     17170     17180     17190

J    L  P  T  L  R  Q  R  L  F  R  C  V  S  I  R  E  N  D  D  G  T  Y  A  I  T  A  V  Q  H  V
    CTGCCGACGCTGCGCCAGCGACTGTTCCGCTGCGTGAGTATCCGTGAGAACGACGACGGCACGTATGCCATCACCGCCGTGCAGCATGTG
        17200     17210     17220     17230     17240     17250     17260     17270     17280

J    P  E  K  E  A  I  V  D  N  G  A  H  F  D  G  E  Q  S  G  T  V  N  G  V  T  P  P  A  V  Q
    CCGGAAAAAGAGGCCATCGTGGATAACGGGGCGCACTTTGACGGCGAACAGAGTGGCACGGTGAATGGTGTCACGCCGCCAGCGGTGCAG
        17290     17300     17310     17320     17330     17340     17350     17360     17370

J    H  L  T  A  E  V  T  A  D  S  G  E  Y  Q  V  L  A  R  W  D  T  P  K  V  V  K  G  V  S  F
    CACCTGACCGCAGAAGTCACTGCAGACAGCGGGGAATATCAGGTGCTGGCGCGATGGGACACACCGAAGGTGGTGAAGGGCGTGAGTTTC
        17380     17390     17400     17410     17420     17430     17440     17450     17460

J    L  L  R  L  T  V  T  A  D  D  G  S  E  R  L  V  S  T  A  R  T  T  E  T  T  Y  R  F  T  Q
    CTGCTCCGTCTGACCGTAACAGCGGACGACGGCAGTGAGCGGCTGGTCAGCACGGCCCGGACGACGGAAACCACATACCGCTTCACGCAA
        17470     17480     17490     17500     17510     17520     17530     17540     17550

J    L  A  L  G  N  Y  R  L  T  V  R  A  V  N  A  W  G  Q  Q  G  D  P  A  S  V  S  F  R  I  A
    CTGGCGCTGGGGAACTACAGGCTGACAGTCCGGGCGGTAAATGCGTGGGGGCAGCAGGGCGATCCGGCGTCGGTATCGTTCCGGATTGCC
        17560     17570     17580     17590     17600     17610     17620     17630     17640

J    A  P  A  A  P  S  R  I  E  L  T  P  G  Y  F  Q  I  T  A  T  P  H  L  A  V  Y  D  P  T  V
    GCACCGGCAGCACCGTCGAGGATTGAGCTGACGCCGGGCTATTTTCAGATAACCGCCACGCCGCATCTTGCCGTTTATGACCCGACGGTA
        17650     17660     17670     17680     17690     17700     17710     17720     17730

J    Q  F  E  F  W  F  S  E  K  Q  I  A  D  I  R  Q  V  E  T  S  T  R  Y  L  G  T  A  L  Y  W
    CAGTTTGAGTTCTGGTTCTCGGAAAAGCAGATTGCGGATATCAGACAGGTTGAAACCAGCACGCGTTATCTTGGTACGGCGCTGTACTGG
        17740     17750     17760     17770     17780     17790     17800     17810     17820

J    I  A  A  S  I  N  I  K  P  G  H  D  Y  Y  F  Y  I  R  S  V  N  T  V  G  K  S  A  F  V  E
    ATAGCCGCCAGTATCAATATCAAACCGGGCCATGATTATTACTTTTATATCCGCAGTGTGAACACCGTTGGCAAATCGGCATTCGTGGAG
        17830     17840     17850     17860     17870     17880     17890     17900     17910
```

```
J       A  V  G  R  A  S  D  D  A  E  G  Y  L  D  F  F  K  G  K  I  T  E  S  H  L  G  K  E  L  L
       GCCGTCGGTCGGGCGAGCGATGATGCGGAAGGTTACCTGGATTTTTTCAAAGGCAAGATAACCGAATCCCATCTCGGCAAGGAGCTGCTG
            17920     17930     17940     17950     17960     17970     17980     17990     18000

J       E  K  V  E  L  T  E  D  N  A  S  R  L  E  E  F  S  K  E  W  K  D  A  S  D  K  W  N  A  M
       GAAAAAGTCGAGCTGACGGAGGATAACGCCAGCAGACTGGAGGAGTTTTCGAAAGAGTGGAAGGATGCCAGTGATAAGTGGAATGCCATG
            18010     18020     18030     18040     18050     18060     18070     18080     18090

J       W  A  V  K  I  E  Q  T  K  D  G  K  H  Y  V  A  G  I  G  L  S  M  E  D  T  E  E  G  K  L
       TGGGCTGTCAAAATTGAGCAGACCAAAGACGGCAAACATTATGTCGCGGGTATTGGCCTCAGCATGGAGGACACGGAGGAAGGCAAACTG
            18100     18110     18120     18130     18140     18150     18160     18170     18180

J       S  Q  F  L  V  A  A  N  R  I  A  F  I  D  P  A  N  G  N  E  T  P  M  F  V  A  Q  G  N  Q
       AGCCAGTTTCTGGTTGCCGCCAATCGTATCGCATTTATTGACCCGGCAAACGGGAATGAAACGCCGATGTTTGTGGCGCAGGGCAACCAG
            18190     18200     18210     18220     18230     18240     18250     18260     18270

J       I  F  M  N  D  V  F  L  K  R  L  T  A  P  T  I  T  S  G  G  N  P  P  A  F  S  L  T  P  D
       ATATTCATGAACGACGTGTTCCTGAAGCGCCTGACGGCCCCCACCATTACCAGCGGCGGCAATCCTCCGGCCTTTTCCCTGACACCGGAC
            18280     18290     18300     18310     18320     18330     18340     18350     18360

J       G  K  L  T  A  K  N  A  D  I  S  G  S  V  N  A  N  S  G  T  L  S  N  V  T  I  A  E  N  C
       GGAAAGCTGACCGCTAAAAATGCGGATATCAGTGGCAGTGTGAATGCGAACTCCGGGACGCTCAGTAATGTGACGATAGCTGAAAACTGT
            18370     18380     18390     18400     18410     18420     18430     18440     18450

J       T  I  N  G  T  L  R  A  E  K  I  V  G  D  I  V  K  A  A  S  A  A  F  P  R  Q  R  E  S  S
       ACGATAAACGGTACGCTGAGGGCGGAAAAAATCGTCGGGGACATTGTAAAGGCGGCGAGCGCGGCTTTTCCGCGCCAGCGTGAAAGCAGT
            18460     18470     18480     18490     18500     18510     18520     18530     18540

J       V  D  W  P  S  G  T  R  T  V  T  V  T  D  D  H  P  F  D  R  Q  I  V  V  L  P  L  T  F  R
       GTGGACTGGCCGTCAGGTACCCGTACTGTCACCGTGACCGATGACCATCCTTTTGATCGCCAGATAGTGGTGCTTCCGCTGACGTTTCGC
            18550     18560     18570     18580     18590     18600     18610     18620     18630

J       G  S  K  R  T  V  S  G  R  T  T  Y  S  M  C  Y  L  K  V  L  M  N  G  A  V  I  Y  D  G  A
       GGAAGTAAGCGTACTGTCAGCGGCAGGACAACGTATTCGATGTGTTATCTGAAAGTACTGATGAACGGTGCGGTGATTTATGATGGCGCG
            18640     18650     18660     18670     18680     18690     18700     18710     18720

J       A  N  E  A  V  Q  V  F  S  R  I  V  D  M  P  A  G  R  G  N  V  I  L  T  F  T  L  T  S  T
       GCGAACGAGGCGGTACAGGTGTTCTCCCGTATTGTTGACATGCCAGCGGGTCGGGGAAACGTGATCCTGACGTTCACGCTTACGTCCACA
            18730     18740     18750     18760     18770     18780     18790     18800     18810

J       R  H  S  A  D  I  P  P  Y  T  F  A  S  D  V  Q  V  M  V  I  K  K  Q  A  L  G  I  S  V  V
       CGGCATTCGGCAGATATTCCGCCGTATACGTTTGCCAGCGATGTGCAGGTTATGGTGATTAAGAAACAGGCGCTGGGCATCAGCGTGGTC
            18820     18830     18840     18850     18860     18870     18880     18890     18900

orf     *
206a                                                                        M  R  N  V  C  I  A  V  A
       TGAGTGTGTTACAGAGGTTCGTCCGGGAACGGGCGTTTTATTATAAAACAGTGAGAGGTGAACGATGCGTAATGTGTGTATTGCCGTTGC
            18910     18920     18930     18940     18950     18960     18970     18980     18990

orf
206a    V  F  A  A  L  A  V  T  V  T  P  A  R  A  E  G  G  H  G  T  F  T  V  G  Y  F  Q  V  K  P
       TGTCTTTGCCGCACTTGCGGTGACAGTCACTCCGGCCCGTGCGGAAGGTGGACATGGTACGTTTACGGTGGGCTATTTTCAAGTGAAACC
            19000     19010     19020     19030     19040     19050     19060     19070     19080

orf
206a    G  T  L  P  S  L  S  G  G  D  T  G  V  S  H  L  K  G  I  N  V  K  Y  R  Y  E  L  T  D  S
       GGGTACATTGCCGTCGTTGTCGGGCGGGGATACCGGTGTGAGTCATCTGAAAGGGATTAACGTGAAGTACCGTTATGAGCTGACGGACAG
            19090     19100     19110     19120     19130     19140     19150     19160     19170

orf
206a    V  G  V  M  A  S  L  G  F  A  A  S  K  K  S  S  T  V  M  T  G  E  D  T  F  H  Y  E  S  L
       TGTGGGGGTGATGGCTTCCCTGGGGTTCGCCGCGTCGAAAAAGAGCAGCACAGTGATGACCGGGGAGGATACGTTTCACTATGAGAGCCT
            19180     19190     19200     19210     19220     19230     19240     19250     19260

orf
206a    R  G  R  Y  V  S  V  M  A  G  P  V  L  Q  I  S  K  Q  V  S  A  Y  A  M  A  G  V  A  H  S
       GCGTGGACGTTATGTGAGCGTGATGGCCGGACCGGTTTTACAAATCAGTAAGCAGGTCAGTGCGTACGCCATGGCCGGAGTGGCTCACAG
            19270     19280     19290     19300     19310     19320     19330     19340     19350

orf
206a    R  W  S  G  S  T  M  D  Y  R  K  T  E  I  T  P  G  Y  M  K  E  T  T  T  A  R  D  E  S  A
       TCGGTGGTCCGGCAGTACAATGGATTACCGTAAGACGGAAATCACTCCCGGGTATATGAAAGAGACGACCACTGCCAGGGACGAAAGTGC
            19360     19370     19380     19390     19400     19410     19420     19430     19440

orf
206a    M  R  H  T  S  V  A  W  S  A  G  I  Q  I  N  P  A  A  S  V  V  V  D  I  A  Y  E  G  S  G
       AATGCGGCATACCTCAGTGGCGTGGAGTGCAGGTATACAGATTAATCCGGCAGCGTCCGTCGTTGTTGATATTGCTTATGAAGGCTCCGG
            19450     19460     19470     19480     19490     19500     19510     19520     19530

orf
206a    S  G  D  W  R  T  D  G  F  I  V  G  V  G  Y  K  F  *
       CAGTGGCGACTGGCGTACTGACGGATTCATCGTTGGGGTCGGTTATAAATTCTGATTAGCCAGGTAACACAGTGTTATGACAGCCCGCCG
            19540     19550     19560     19570     19580     19590     19600     19610     19620

orf401                                M  A  V  K  I  S  G  V  L  K  D  G  T  G  K  P  V  Q  N  C  T
       GAACCGGTGGGCTTTTTTGTGGGGTGAATATGGCAGTAAAGATTTCAGGAGTCCTGAAAGACGGCACAGGAAAACCGGTACAGAACTGCA
            19630     19640     19650     19660     19670     19680     19690     19700     19710

orf401    I  Q  L  K  A  R  R  N  S  T  T  V  V  V  N  T  V  G  S  E  N  P  D  E  A  G  R  Y  S  M
       CCATTCAGCTGAAAGCCAGACGTAACAGCACCACGGTGGTGGTGAACACGGTGGGCTCAGAGAATCCGGATGAAGCCGGGCGTTACAGCA
            19720     19730     19740     19750     19760     19770     19780     19790     19800

orf401    D  V  E  Y  G  Q  Y  S  V  I  L  Q  V  D  G  F  P  P  S  H  A  G  T  I  T  V  Y  E  D  S
       TGGATGTGGAGTACGGTCAGTACAGTGTCATCCTGCAGGTTGACGGTTTTCCACCATCGCACGCCGGGACCATCACCGTGTATGAAGATT
            19810     19820     19830     19840     19850     19860     19870     19880     19890

orf401    Q  P  G  T  L  N  D  F  L  C  A  M  T  E  D  D  A  R  P  E  V  L  R  R  L  E  L  M  V  E
       CACAACCGGGGACGCTGAATGATTTTCTCTGTGCCATGACGGAGGATGATGCCCGGCCGGAGGTGCTGCGTCGTCTTGAACTGATGGTGG
            19900     19910     19920     19930     19940     19950     19960     19970     19980

orf401    E  V  A  R  N  A  S  V  V  A  Q  S  T  A  D  A  K  K  S  A  G  D  A  S  A  S  A  A  Q  V
       AAGAGGTGGCGCGTAACGCGTCCGTGGTGGCACAGAGTACGGCAGACGCGAAGAAATCAGCCGGCGATGCCAGTGCATCAGCTGCTCAGG
            19990     20000     20010     20020     20030     20040     20050     20060     20070

orf401    A  A  L  V  T  D  A  T  D  S  A  R  A  A  S  T  S  A  G  Q  A  A  S  S  A  Q  E  A  S  S
       TCGCGGCCCTTGTGACTGATGCAACTGACTCAGCACGCGCCGCCAGCACGTCCGCCGGACAGGCTGCATCGTCAGCTCAGGAAGCGTCCT
            20080     20090     20100     20110     20120     20130     20140     20150     20160

orf401    G  A  E  A  A  S  A  K  A  T  E  A  E  K  S  A  A  A  A  E  S  S  K  N  A  A  A  T  S  A
       CCGGCGCAGAAGCGGCATCAGCAAAGGCCACTGAAGCGGAAAAAAGTGCCGCAGCCGCAGAGTCCTCAAAAAACGCGGCGGCCACCAGTG
            20170     20180     20190     20200     20210     20220     20230     20240     20250

orf401    G  A  A  K  T  S  E  T  N  A  A  A  S  Q  Q  S  A  A  T  S  A  S  T  A  A  T  K  A  S  E
       CCGGTGCGGCGAAAACGTCAGAAACGAATGCTGCAGCGTCACAACAATCAGCCGCCACGTCTGCCTCCACCGCGGCCACGAAAGCGTCAG
            20260     20270     20280     20290     20300     20310     20320     20330     20340

orf401    A  A  T  S  A  R  D  A  V  A  S  K  E  A  A  K  S  S  E  T  N  A  S  S  S  A  G  R  A  A
       AGGCCGCCACTTCAGCACGAGATGCGGTGGCCTCAAAAGAGGCAGCAAAATCATCAGAAACGAACGCATCATCAAGTGCCGGTCGTGCAG
            20350     20360     20370     20380     20390     20400     20410     20420     20430
```

```
orf401   S  S  A  T  A  A  E  N  S  A  R  A  A  K  T  S  E  T  N  A  R  S  S  E  T  A  A  E  R  S
       CTTCCTCGGCAACGGCGGCAGAAAATTCTGCCAGGGCGGCAAAAACGTCCGAGACGAATGCCAGGTCATCTGAAACAGCAGCGGAACGGA
            20440     20450     20460     20470     20480     20490     20500     20510     20520

orf401   A  S  A  A  A  D  A  K  T  A  A  A  G  S  A  S  T  A  S  T  K  A  T  E  A  A  G  S  A  V
       GCGCCTCTGCCGCGGCAGACGCAAAAACAGCGGCGGCGGGAGTGCGTCAACGGCATCCACGAAGGCGACAGAGGCTGCGGGAAGTGCGG
            20530     20540     20550     20560     20570     20580     20590     20600     20610

orf401   S  A  S  Q  S  K  S  A  A  E  A  A  A  I  R  A  K  N  S  A  K  R  A  E  D  I  A  S  A  V
       TATCAGCATCGCAGAGCAAAAGTGCGGCAGAAGCGGCGGCAATACGTGCAAAAAATTCGGCAAAACGTGCAGAAGATATAGCTTCAGCTG
            20620     20630     20640     20650     20660     20670     20680     20690     20700

orf401   A  L  E  D  A  D  T  T  R  K  G  I  V  Q  L  S  S  A  T  N  S  T  S  E  T  L  A  A  T  P
       TCGCGCTTGAGGATGCGGACACAACGAGAAAGGGGATAGTGCAGCTCAGCAGTGCAACCAACAGCACGTCTGAAACGCTTGCTGCAACGC
            20710     20720     20730     20740     20750     20760     20770     20780     20790

orf401   K  A  V  K  V  V  M  D  E  T  N  R  K  A  H  W  T  V  R  H  *
       CAAAGGCGGTTAAGGTGGTAATGGATGAAACGAACAGAAAAGCCCACTGGACAGTCCGGCACTGACCGGAACGCCAACAGCACCAACCGC
            20800     20810     20820     20830     20840     20850     20860     20870     20880

       GCTCAGGGGAACAAACAATACCCAGATTGCGAACACCGCTTTTGTACTGGCCGCGATTGCAGATGTTATCGACGCGTCACCTGACGCACT
            20890     20900     20910     20920     20930     20940     20950     20960     20970

orf314                                                          M  T  N  A  L  A  G  K  Q  P  K
       GAATACGCTGAATGAACTGGCCGCAGCGCTCGGGAATGATCCAGATTTTGCTACCACCATGACTAACGCGCTTGCGGGTAAACAACCGAA
            20980     20990     21000     21010     21020     21030     21040     21050     21060

orf314 N  A  T  L  T  A  L  A  G  L  S  T  A  K  N  K  L  P  Y  F  A  E  N  D  A  A  S  L  T  E
       GAATGCGACACTGACGGCGCTGGCAGGGCTTTCCACGGCGAAAAATAAATTACCGTATTTTGCGGAAAATGATGCCGCCAGCCTGACTGA
            21070     21080     21090     21100     21110     21120     21130     21140     21150

orf314 L  T  Q  V  G  R  D  I  L  A  K  N  S  V  A  D  V  L  E  Y  L  G  A  G  E  N  S  A  F  P
       ACTGACTCAGGTTGGCAGGGATATTCTGGCAAAAAATTCCGTTGCAGATGTTCTTGAATACCTTGGGGCCGGTGAGAATTCGGCCTTTCC
            21160     21170     21180     21190     21200     21210     21220     21230     21240

orf314 A  G  A  P  I  P  W  P  S  D  I  V  P  S  G  Y  V  L  M  Q  G  Q  A  F  D  K  S  A  Y  P
       GGCAGGTGCGCCGATCCCGTGGCCATCAGATATCGTTCCGTCTGGCTACGTCCTGATGCAGGGGCAGGCGTTTGACAAATCAGCCTACCC
            21250     21260     21270     21280     21290     21300     21310     21320     21330

orf314 K  L  A  V  A  Y  P  S  G  V  L  P  D  M  R  G  W  T  I  K  G  K  P  A  S  G  R  A  V  L
       AAAACTTGCTGTCGCGTATCCATCGGGTGTGCTTCCTGATATGCGAGGCTGGACAATCAAGGGGAAACCCGCCAGCGGTCGTGCTGTATT
            21340     21350     21360     21370     21380     21390     21400     21410     21420

orf314 S  Q  E  Q  D  G  I  K  S  H  T  H  S  A  S  A  S  G  T  D  L  G  T  K  T  T  S  S  F  D
       GTCTCAGGAACAGGATGGAATTAAGTCGCACACCCACAGTGCCAGTGCATCCGGTACGGATTTGGGGACGAAAACCACATCGTCGTTTGA
            21430     21440     21450     21460     21470     21480     21490     21500     21510

orf314 Y  G  T  K  T  T  G  S  F  D  Y  G  T  K  S  T  N  N  T  G  A  H  A  H  S  L  S  G  S  T
       TTACGGGACGAAAACAACAGGCAGTTTCGATTACGGCACCAAATCGACGAATAACACGGGGGCTCATGCTCACAGTCTGAGCGGTTCAAC
            21520     21530     21540     21550     21560     21570     21580     21590     21600

orf314 G  A  A  G  A  H  A  H  T  S  G  L  R  M  N  S  S  G  W  S  Q  Y  G  T  A  T  I  T  G  S
       AGGGGCCGCGGGTGCTCATGCCCACACAAGTGGTTTAAGGATGAACAGTTCTGGCTGGAGTCAGTATGGAACAGCAACCATTACAGGAAG
            21610     21620     21630     21640     21650     21660     21670     21680     21690

orf314 L  S  T  V  K  G  T  S  T  Q  G  I  A  Y  L  S  K  T  D  S  Q  G  S  H  S  H  S  L  S  G
       TTTATCCACAGTTAAAGGAACCAGCACACAGGGTATTGCTTATTTATCGAAAACGGACAGTCAGGGCAGCCACAGTCACTCATTGTCCGG
            21700     21710     21720     21730     21740     21750     21760     21770     21780

orf314 T  A  V  S  A  G  A  H  A  H  T  V  G  I  G  A  H  Q  H  P  V  V  I  G  A  H  A  H  S  F
       TACAGCCGTGAGTGCCGGTGCACATGCGCATACAGTTGGTATTGGTGCGCACCAGCATCCGGTTGTTATCGGTGCTCATGCCCATTCTTT
            21790     21800     21810     21820     21830     21840     21850     21860     21870

orf314 S  I  G  S  H  G  H  T  I  T  V  N  A  A  G  N  A  E  N  T  V  K  N  I  A  F  N  Y  I  V
       CAGTATTGGTTCACACGGACACACCATCACCGTTAACGCTGCGGGTAACGCGGAAAACACCGTCAAAAACATTGCATTTAACTATATTGT
            21880     21890     21900     21910     21920     21930     21940     21950     21960

orf194             M  A  F  R  M  S  E  Q  P  R  T  I  K  I  Y  N  L  L  A  G  T  N  E  F  I  G
orf314 R  L  A  *
       GAGGCTTGCATAATGGCATTCAGAATGAGTGAACAACCACGGACCATAAAAATTTATAATCTGCTGGCCGGAACTAATGAATTTATTGGT
            21970     21980     21990     22000     22010     22020     22030     22040     22050

orf194 E  G  D  A  Y  I  P  P  H  T  G  L  P  A  N  S  T  D  I  A  P  P  D  I  P  A  G  F  V  A
       GAAGGTGACGCATATATTCCGCCTCATACCGGTCTGCCTGCAAACAGTACCGATATTGCACCGCCAGATATTCCGGCTGGCTTTGTGGCT
            22060     22070     22080     22090     22100     22110     22120     22130     22140

orf194 V  F  N  S  D  E  A  S  W  H  L  V  E  D  H  R  G  K  T  V  Y  D  V  A  S  G  D  A  L  F
       GTTTTCAACAGTGATGAGGCATCGTGGCATCTCGTTGAAGACCATCGGGGTAAAACCGTCTATGACGTGGCTTCCGGCGACGCGTTATTT
            22150     22160     22170     22180     22190     22200     22210     22220     22230

orf194 I  S  E  L  G  P  L  P  E  N  F  T  H  L  S  P  G  G  E  Y  Q  K  W  N  G  T  A  W  V  K
       ATTTCTGAACTCGGTCCGTTACCGGAAAATTTTACCTGGTTATCGCCGGGAGGGGAATATCAGAAGTGGAACGGCACAGCCTGGGTGAAG
            22240     22250     22260     22270     22280     22290     22300     22310     22320

orf194 D  T  E  A  E  K  L  F  R  I  R  E  A  E  E  T  K  K  S  L  M  Q  V  A  S  E  H  I  A  P
       GATACGGAAGCAGAAAAACTGTTCCGGATCCGGGAGGCGGAAGAAACAAAAAAAAGCCTGATGCAGGTAGCCAGTGAGCATATTGCGCCG
            22330     22340     22350     22360     22370     22380     22390     22400     22410

orf194 L  Q  D  A  A  D  L  E  I  A  T  K  E  E  T  S  L  L  E  A  W  K  K  Y  R  V  L  L  N  R
       CTTCAGGATGCTGCAGATCTGGAAATTGCAACGAAGGAAGAAACCTCGTTGCTGGAAGCCTGGAAGAAGTATCGGGTGTTGCTGAACCGT
            22420     22430     22440     22450     22460     22470     22480     22490     22500

orf194 V  D  T  S  T  A  P  D  I  E  W  P  A  V  P  V  M  E  *
       GTTGATACATCAACTGCACCTGATATTGAGTGGCCTGCTGTCCCTGTTATGGAGTAATCGTTTTGTGATATGCCGCAGAAACGTTGTATG
            22510     22520     22530     22540     22550     22560     22570     22580     22590

       AAATAACGTTCTGCGGTTAGTTAGTATATTGTAAAGCTGAGTATTGGTTTATTTGGCGATTATTATCTTCAGGAGAATAATGGAAGTTCT
            22600     22610     22620     22630     22640     22650     22660     22670     22680

       ATGACTCAATTGTTCATAGTGTTTACATCACCGCCAATTGCTTTTAAGACTGAACGCATGAAATATGGTTTTTCGTCATGTTTTGAGTCT
            22690     22700     22710     22720     22730     22740     22750     22760     22770

       GCTGTTGATATTTCTAAAGTCGGTTTTTTTTCTTCGTTTTCTCTAACTATTTTCCATGAAATACATTTTTGATTATTATTTGAATCAATT
            22780     22790     22800     22810     22820     22830     22840     22850     22860

       CCAATTACCTGAAGTCTTTCATCTATAATTGGCATTGTATGTATTGGTTTATTGGAGTAGATGCTTGCTTTTCTGAGCCATAGCTCTGAT
            22870     22880     22890     22900     22910     22920     22930     22940     22950

       ATCCAAATGAAGCCATAGGCATTTGTTATTTTGGCTCTGTCAGCTGCATA
            22960     22970     22980     22990     23000     23010     23020     23030     23040
```

(b)

```
                                                       M  E  Q  R  I  T  L  K  D  Y  A  M  R  F  G  Q  T
cro    CTCTGGCGGTGATAATGGTTGCATGTACTAAGGAGGTTGTATGGAACAACGCATAACCCTGAAAGATTATGCAATGCGCTTTGGGCAAAC
            38010     38020     38030     38040     38050     38060     38070     38080     38090

         K  T  A  K  D  L  G  V  Y  Q  S  A  I  N  K  A  I  H  A  G  R  K  I  F  L  T  I  N  A  D
cro    CAAGACAGCTAAAGATCTCGGCGTATATCAAAGCGCGATCAACAAGGCCATTCATGCAGGCCGAAAGATTTTTTTAACTATAAACGCTGA
            38100     38110     38120     38130     38140     38150     38160     38170     38180

        G  S  V  Y  A  E  E  V  K  P  F  P  S  N  K  K  T  T  A  *
cro    TGGAAGCGTTTATGCGGAAGAGGTAAAGCCCTTCCCGAGTAACAAAAAAACAACAGCATAAATAACCCCGCTCTTACACATTCCAGCCCT
            38190     38200     38210     38220     38230     38240     38250     38260     38270
                                                                                  S_E
                                                                                             M
cII    GAAAAAGGGCATCAAATTAAACCACACCTATGGTGTATGCATTTATTTGCATACATTCAATCAATTGTTATCTAAGGAAATACTTACATA
          38280     38290     38300     38310     38320     38330     38340     38350     38360

         V  R  A  N  K  R  N  E  A  L  R  I  E  S  A  L  L  N  K  I  A  M  L  G  T  E  K  T  A  E
cII    TGGTTCGTGCAAACAAACGCAACGAGGCTCTACGAATCGAGAGTGCGTTGCTTAACAAAATCGCAATGCTTGGAACTGAGAAGACAGCGG
            38370     38380     38390     38400     38410     38420     38430     38440     38450

         A  V  G  V  D  K  S  Q  I  S  R  W  K  R  D  W  I  P  K  F  S  M  L  L  A  V  L  E  W  G
cII    AAGCTGTGGGCGTTGATAAGTCGCAGATCAGCAGGTGGAAGAGGGACTGGATTCCAAAGTTCTCAATGCTGCTTGCTGTTCTTGAATGGG
            38460     38470     38480     38490     38500     38510     38520     38530     38540

                                                                    ------------------ oop RNA ------
         V  V  D  D  D  M  A  R  L  A  R  Q  V  A  A  I  L  T  N  K  K  R  P  A  A  T  E  R  S  E
cII    GGGTCGTTGACGACGACATGGCTCGATTGGCGCGACAAGTTGCTGCGATTCTCACCAATAAAAAACGCCCGGCGGCAACCGAGCGTTCTG
            38550     38560     38570     38580     38590     38600     38610     38620     38630

       -------- oop RNA ----------------------------S_O
O                                                                    M  T  N  T  A  K  I  L  N  F  G  R
cII      Q  I  Q  M  E  F  *
       AACAAATCCAGATGGAGTTCTGAGGTCATTACTGGATCTATCAACAGGAGTCATTATGACAAATACAGCAAAAATACTCAACTTCGGCAG
            38640     38650     38660     38670     38680     38690     38700     38710     38720

         G  N  F  A  G  Q  E  R  N  V  A  D  L  D  D  G  Y  A  R  L  S  N  M  L  L  E  A  Y  S  G
O      AGGTAACTTTGCCGGACAGGAGCGTAATGTGGCAGATCTCGATGATGGTTACGCCAGACTATCAAATATGCTGCTTGAGGCTTATTCGGG
            38730     38740     38750     38760     38770     38780     38790     38800     38810

         A  D  L  T  K  R  Q  F  K  V  L  L  A  I  L  R  K  T  Y  G  W  N  K  P  M  D  R  I  T  D
O      CGCAGATCTGACCAAGCGACAGTTTAAAGTGCTGCTTGCCATTCTGCGTAAAACCTATGGGTGGAATAAACCAATGGACAGAATCACCGA
            38820     38830     38840     38850     38860     38870     38880     38890     38900

         S  Q  L  S  E  I  T  K  L  P  V  K  R  C  N  E  A  K  L  E  L  V  R  M  N  I  I  K  Q  Q
O      TTCTCAACTTAGCGAGATTACAAAGTTACCTGTCAAACGGTGCAATGAAGCCAAGTTAGAACTCGTCAGAATGAATATTATCAAGCAGCA
            38910     38920     38930     38940     38950     38960     38970     38980     38990

         G  G  M  F  G  P  N  K  N  I  S  E  W  C  I  P  Q  N  E  G  K  S  P  K  T  R  D  K  T  S
O      AGGCGGCATGTTTGGACCAAATAAAAACATCTCAGAATGGTGCATCCCTCAAAACGAGGGAAAATCCCCTAAAACGAGGGATAAAACATC
            39000     39010     39020     39030     39040     39050     39060     39070     39080

                                                      ------------- ori mutations ---------
         L  K  L  G  D  C  Y  P  S  K  Q  G  D  T  K  D  T  I  T  K  E  K  R  K  D  Y  S  S  E  N
O      CCTCAAATTGGGGGATTGCTATCCCTCAAAACAGGGGGACACAAAAGACACTATTACAAAAGAAAAAAGAAAAGATTATTCGTCAGAGAA
            39090     39100     39110     39120     39130     39140     39150     39160     39170

         S  G  E  S  S  D  Q  P  E  N  D  L  S  V  V  K  P  D  A  A  I  Q  S  G  S  K  W  G  T  A
O      TTCTGGCGAATCCTCTGACCAGCCAGAAAACGACCTTTCTGTGGTGAAACCGGATGCTGCAATTCAGAGCGGCAGCAAGTGGGGGACAGC
            39180     39190     39200     39210     39220     39230     39240     39250     39260

         E  D  L  T  A  A  E  W  M  F  D  M  V  K  T  I  A  P  S  A  R  K  P  N  F  A  G  W  A  N
O      AGAAGACCTGACCGCCGCAGAGTGGATGTTTGACATGGTGAAGACTATCGCACCATCAGCCAGAAAACCGAATTTTGCTGGGTGGGCTAA
            39270     39280     39290     39300     39310     39320     39330     39340     39350

         D  I  R  L  M  R  E  R  D  G  R  N  H  R  D  M  C  V  L  F  R  W  A  C  Q  D  N  F  W  S
O      CGATATCCGCCTGATGCGTGAACGTGACGGACGTAACCACCGCGACATGTGTGTGCTGTTCCGCTGGGCATGCCAGGACAACTTCTGGTC
            39360     39370     39380     39390     39400     39410     39420     39430     39440

         G  N  V  L  S  P  A  K  L  R  D  K  W  T  Q  L  E  I  N  R  N  K  Q  Q  A  G  V  T  A  S
O      CGGTAACGTGCTGAGCCCGGCCAAACTCCGCGATAAGTGGACCCAACTCGAAATCAACCGTAACAAGCAACAGGCAGGCGTGACAGCCAG
            39450     39460     39470     39480     39490     39500     39510     39520     39530

P                                                               M  K  N  I  A  A  Q  M  V  N  F  D  R
O        K  P  K  L  D  L  T  N  T  D  W  I  Y  G  V  D  L  *
       CAAACCAAAACTCGACCTGACAACACAGACTGGATTTACGGGGTGGATCTATGAAAAACATCGCCGCACAGATGGTTAACTTTGACCGT
            39540     39550     39560     39570     39580     39590     39600     39610     39620

         E  Q  M  R  R  I  A  N  N  M  P  E  Q  Y  D  E  K  P  Q  V  Q  Q  V  A  Q  I  I  N  G  V
P      GAGCAGATGCGTCGGATCGCCAACAACATGCCGGAACAGTACGACGAAAAGCCGCAGGTACAGCAGGTAGCGCAGATCATCAACGGTGTG
            39630     39640     39650     39660     39670     39680     39690     39700     39710

         F  S  Q  L  L  A  T  F  P  A  S  L  A  N  R  D  Q  N  E  V  N  E  I  R  R  Q  W  V  L  A
P      TTCAGCCAGTTACTGGCAACTTTCCCGGCGAGCCTGGCTAACCGTGACCAGAACGAAGTGAACGAAATCCGTCGCCAGTGGGTTCTGGCT
            39720     39730     39740     39750     39760     39770     39780     39790     39800

         F  R  E  N  G  I  T  T  M  E  Q  V  N  A  G  M  R  V  A  R  R  Q  N  R  P  F  L  P  S  P
P      TTTCGGGAAAACGGGATCACCACGATGGAACAGGTTAACGCAGGAATGCGCGTAGCCCGTCGGCAGAATCGACCATTTCTGCCATCACCC
            39810     39820     39830     39840     39850     39860     39870     39880     39890

         G  Q  F  V  A  W  C  R  E  E  A  S  V  T  A  G  L  P  N  V  S  E  L  V  D  M  V  Y  E  Y
P      GGGCAGTTTGTTGCATGGTGCCGGGAAGAAGCATCCGTTACCGCCGGACTGCCAAACGTCAGCGAGCTGGTTGATATGGTTTACGAGTAT
            39900     39910     39920     39930     39940     39950     39960     39970     39980

         C  R  K  R  G  L  Y  P  D  A  E  S  Y  P  W  K  S  N  A  H  Y  W  L  V  T  N  L  Y  Q  N
P      TGCCGGAAGCGAGGCCTGTATCCGGATGCGGAGTCTTATCCGTGGAAATCAAACGCGCACTACTGGCTGGTTACCAACCTGTATCAGAAC
            39990     40000     40010     40020     40030     40040     40050     40060     40070

         M  R  A  N  A  L  T  D  A  E  L  R  R  K  A  A  D  E  L  V  H  M  T  A  R  I  N  R  G  E
P      ATGCGGGCCAATGCGCTTACTGATGCGGAATTACGCCGTAAGGCCGCAGATGAGCTTGTCCATATGACTGCGAGAATTAACCGTGGTGAG
            40080     40090     40100     40110     40120     40130     40140     40150     40160

         A  I  P  E  P  V  K  Q  L  P  V  M  G  G  R  P  L  N  R  A  Q  A  L  A  K  I  A  E  I  K
P      GCGATCCCTGAACCAGTAAAACAACTTCCTGTCATGGGCGGTAGACCTCTAAATCGTGCACAGGCTCTGGCGAAGATCGCAGAAATCAAA
            40170     40180     40190     40200     40210     40220     40230     40240     40250

P        A  K  F  G  L  K  G  A  S  V  *
ren                                    M  T  G  K  E  A  I  I  H  Y  L  G  T  H  N  S  F  C  A  P  D
       GCTAAGTTCGGACTGAAAGGAGCAAGTGTATGACGGGCAAAGAGGCAATTATTCATTACCTGGGGACGCATAATAGCTTCTGTGCGCCGG
            40260     40270     40280     40290     40300     40310     40320     40330     40340

          V  A  A  L  T  G  A  T  V  T  S  I  N  Q  A  A  A  K  M  A  R  A  G  L  L  V  I  E  G  K
ren    ACGTTGCCGCGCTAACAGGCGCAACAGTAACCAGCATAAATCAGGCCGCGGCTAAAATGGCACGGGCAGGTCTTCTGGTTATCGAAGGTA
            40350     40360     40370     40380     40390     40400     40410     40420     40430
```

FIG. 2 (b).

```
ren        V  W  R  T  V  Y  Y  R  F  A  T  R  E  E  R  E  G  K  M  S  T  N  L  V  F  K  E  C  R  Q
         AGGTCTGGCGAACGGTGTATTACCGGTTTGCTACCAGGGAAGAACGGGAAGGAAAGATGAGCACGAACCTGGTTTTTAAGGAGTGTCGCC
              40440     40450     40460     40470     40480     40490     40500     40510     40520

ren        S  A  A  M  K  R  V  L  A  V  Y  G  V  K  R  *
         AGAGTGCCGCGATGAAACGGGTATTGGCGGTATATGGAGTTAAAAGATGACCATCTACATTACTGAGCTAATAACAGGCCTGCTGGTAAT
              40530     40540     40550     40560     40570     40580     40590     40600     40610

orf146                                     M  K  K  L  T  F  E  I  R  S  P  A  H  Q  Q  N  A  I  H
         CGCAGGCCTTTTTATTTGGGGGAGAGGGAAGTCATGAAAAAACTAACCTTTGAAATTCGATCTCCAGCACATCAGCAAAACGCTATTCAC
              40620     40630     40640     40650     40660     40670     40680     40690     40700

orf146 A  V  Q  Q  I  L  P  D  P  T  K  P  I  V  V  T  I  Q  E  R  N  R  S  L  D  Q  N  R  K  L
         GCAGTACAGCAAATCCTTCCAGACCCAACCAAACCAATCGTAGTAACCATTCAGGAACGCAACCGCAGCTTAGACCAAAACAGGAAGCTA
              40710     40720     40730     40740     40750     40760     40770     40780     40790

orf146 W  A  C  L  G  D  V  S  R  Q  V  E  W  H  G  R  W  L  D  A  E  S  W  K  C  V  F  T  A  A
         TGGGCCTGCTTAGGTGACGTCTCTCGTCAGGTTGAATGGCATGGTCGCTGGCTGGATGCAGAAAGCTGGAAGTGTGTGTTTACCGCAGCA
              40800     40810     40820     40830     40840     40850     40860     40870     40880

orf146 L  K  Q  Q  D  V  V  P  N  L  A  G  N  G  F  V  V  I  G  Q  S  T  S  R  M  R  V  G  E  F
         TTAAAGCAGCAGGATGTTGTTCCTAACCTTGCCGGGAATGGCTTTGTGGTAATAGGCCAGTCAACCAGCAGGATGCGTGTAGGCGAATTT
              40890     40900     40910     40920     40930     40940     40950     40960     40970

orf146 A  E  L  L  E  L  I  Q  A  F  G  T  E  R  G  V  K  W  S  D  E  A  R  L  A  L  E  W  K  A
         GCGGAGCTATTAGAGCTTATACAGGCATTCGGTACAGAGCGTGGCGTTAAGTGGTCAGACGAAGCGAGACTGGCTCTGGAGTGGAAAGCG
              40980     40990     41000     41010     41020     41030     41040     41050     41060

orf146 R  W  G  D  R  A  A  *
orf290                       M  I  N  V  V  S  F  S  G  G  R  T  S  A  Y  L  L  W  L  M  E  Q  K  R
         AGATGGGGAGACAGGGCTGCATGATAAATGTCGTTAGTTTCTCCGGTGGCAGGACGTCAGCATATTTGCTCTGGCTAATGGAGCAAAAGC
              41070     41080     41090     41100     41110     41120     41130     41140     41150

orf290   R  A  G  K  D  V  H  Y  V  F  M  D  T  G  C  E  H  P  M  T  Y  R  F  V  R  E  V  V  K  F
         GACGGGCAGGTAAAGACGTGCATTACGTTTTCATGGATACAGGTTGTGAACATCCAATGACATATCGGTTTGTCAGGGAAGTTGTGAAGT
              41160     41170     41180     41190     41200     41210     41220     41230     41240

orf290   W  D  I  P  L  T  V  L  Q  V  D  I  N  P  E  L  G  Q  P  N  G  Y  T  V  W  E  P  K  D  I
         TCTGGGATATACCGCTCACCGTATTGCAGGTTGATATCAACCCGGAGCTTGGACAGCCAAATGGTTATACGGTATGGGAACCAAAGGATA
              41250     41260     41270     41280     41290     41300     41310     41320     41330

orf290   Q  T  R  M  P  V  L  K  P  F  I  D  M  V  K  K  Y  G  T  P  Y  V  G  G  A  F  C  T  D  R
         TTCAGACGCGAATGCCTGTTCTGAAGCCATTTATCGATATGGTAAAGAAATATGGCACTCCATACGTCGGCGGCGCGTTCTGCACTGACA
              41340     41350     41360     41370     41380     41390     41400     41410     41420

orf290   L  K  L  V  P  F  T  K  Y  C  D  D  H  F  G  R  G  N  Y  T  T  W  I  G  I  R  A  D  E  P
         GATTAAAACTCGTTCCCTTCACCAAATACTGTGATGACCATTTCGGGCGAGGGAATTACACCACGTGGATTGGCATCAGAGCTGATGAAC
              41430     41440     41450     41460     41470     41480     41490     41500     41510

orf290   K  R  L  K  P  K  P  G  I  R  Y  L  A  E  L  S  D  F  E  K  E  D  I  L  A  W  W  K  Q  Q
         CGAAGCGGCTAAAGCCAAAGCCTGGAATCAGATATCTTGCTGAACTGTCAGACTTTGAGAAGGAAGATATCCTCGCATGGTGGAAGCAAC
              41520     41530     41540     41550     41560     41570     41580     41590     41600

orf290   P  F  D  L  Q  I  P  E  H  L  G  N  C  I  F  C  I  K  K  S  T  Q  K  I  G  L  A  C  K  D
         AACCATTCGATTTGCAAATACCGGAACATCTCGGTAACTGCATATTCTGCATTAAAAAATCAACGCAAAAAATCGGACTTGCCTGCAAAG
              41610     41620     41630     41640     41650     41660     41670     41680     41690

orf290   E  E  G  L  Q  R  V  F  N  E  V  I  T  G  S  H  V  R  D  G  H  R  E  T  P  K  E  I  M  Y
         ATGAGGAGGGATTGCAGCGTGTTTTTAATGAGGTCATCACGGGATCCCATGTGCGTGACGGACATCGGGAAACGCCAAAGGAGATTATGT
              41700     41710     41720     41730     41740     41750     41760     41770     41780

orf290   R  G  R  M  S  L  D  G  I  A  K  M  Y  S  E  N  D  Y  Q  A  L  Y  Q  D  M  V  R  A  K  R
         ACCGAGGAAGAATGTCGCTGGACGGTATCGCGAAAATGTATTCAGAAAATGATTATCAAGCCCTGTATCAGGACATGGTACGAGCTAAAA
              41790     41800     41810     41820     41830     41840     41850     41860     41870

orf57                                                                              M  M  R  C
orf290   F  D  T  G  S  C  S  E  S  C  E  I  F  G  G  Q  L  D  F  D  F  G  R  E  A  A  *
         GATTCGATACCGGCTCTTGTTCTGAGTCATGCGAAATATTTGGAGGGCAGCTTGATTTCGACTTCGGGAGGGAAGCTGCATGATGCGATG
              41880     41890     41900     41910     41920     41930     41940     41950     41960

orf57    Y  R  C  G  E  C  K  E  D  N  R  F  R  P  N  Q  P  Y  W  N  R  W  C  L  R  C  E  R  T  P
         TTATCGGTGCGGTGAATGCAAAGAAGATAACCGCTTCCGACCAAATCAACCTTACTGGAATCGATGGTGTCTCCGGTGTGAAAGAACACC
              41970     41980     41990     42000     42010     42020     42030     42040     42050

orf60b                                          V  A  R  Q  R  R  S  I  T  D  I  I  C  E  N  C  K
orf57    T  G  V  L  P  L  P  Q  E  K  E  D  V  W  R  D  S  D  E  V  S  P  T  *
         AACAGGGGTGTTACCACTACCGCAGGAAAAGGAGGACGTGTGGCGAGACAGCGACGAAGTATCACCGACATAATCTGCGAAAACTGCAAA
              42060     42070     42080     42090     42100     42110     42120     42130     42140

orf60b Y  L  P  T  K  R  T  R  N  K  P  K  P  I  P  K  E  S  D  V  K  T  F  N  Y  T  A  H  L  W
         TACCTTCCAACGAAACGCACCAGAAATAAACCCAAGCCAATCCCAAAAGAATCTGACGTAAAAACCTTCAACTACACGGCTCACCTGTGG
              42150     42160     42170     42180     42190     42200     42210     42220     42230

orf60b D  I  R  W  L  R  R  R  A  R  K  T  R  *
orf56                                          V  I  D  Q  N  R  S  Y  E  Q  E  S  V  E  R  A  L  T
         GATATCCGGTGGCTAAGACGTCGTGCGAGGAAAAACAAGGTGATTGACCAAAATCGAAGTTACGAACAAGAAAGCGTCGAGCGAGCTTTAA
              42240     42250     42260     42270     42280     42290     42300     42310     42320

orf56    C  A  N  C  G  Q  K  L  H  V  L  E  V  H  V  C  E  H  C  C  A  E  L  M  S  D  P  N  S  S
         CGTGCGCTAACTGCGGTCAGAAGCTGCATGTGCTGGAAGTTCACGTGTGTGAGCACTGCTGCGCAGAACTGATGAGCGATCCGAATAGCT
              42330     42340     42350     42360     42370     42380     42390     42400     42410

orf204                       M  M  A  K  P  A  R  R  R  C  K  N  D  E  C  R  E  W  F  H  P  A  F  A
orf56     M  H  E  E  E  D  D  G  *
         CGATGCACGAGGAAGAAGATGATGGCTAAACCAGCGCGAAGACGATGTAAAAACGATGAATGCCGGGAATGGTTTCACCCTGCATTCGCT
              42420     42430     42440     42450     42460     42470     42480     42490     42500

orf204 N  Q  W  W  C  S  P  E  C  G  T  K  I  A  L  E  R  R  S  K  E  R  E  K  A  E  K  A  A  E
         AATCAGTGGTGGTGCTCTCCAGAGTGTGGAACCAAGATAGCACTCGAACGACGAAGTAAAGAACGCGAAAAAGCGGAAAAAGCAGCAGAG
              42510     42520     42530     42540     42550     42560     42570     42580     42590

orf204 K  K  R  R  R  E  E  Q  K  Q  K  D  K  L  K  I  R  K  L  A  L  K  P  R  S  Y  W  I  K  Q
         AAGAAACGACGACGAGAGGAGCAGAAACAGAAAGATAAACTTAAGATTCGAAAACTCGCCTTAAAGCCCCGCAGTTACTGGATTAAACAA
              42600     42610     42620     42630     42640     42650     42660     42670     42680

orf204 A  Q  Q  A  V  N  A  F  I  R  E  R  D  R  D  L  P  C  I  S  C  G  T  L  T  S  A  Q  W  D
         GCCCAACAAGCCGTAAACGCCTTCATCAGAGAAAGAGACCGCGACTTACCATGTATCTCGTGCGGAACGCTCACGTCTGCTCAGTGGGAT
              42690     42700     42710     42720     42730     42740     42750     42760     42770

orf204 A  G  H  Y  R  T  T  A  A  A  P  Q  L  R  F  N  E  R  N  I  H  K  Q  C  V  V  C  N  Q  H
         GCCGGACATTACCGGACAACTGCTGCGGCACCTCAACTCCGATTTAATGAACGCAATATTCACAAGCAATGCGTGGTGTGCAACCAGCAC
              42780     42790     42800     42810     42820     42830     42840     42850     42860

orf204 K  S  G  N  L  V  P  Y  R  V  E  L  I  S  R  I  G  Q  E  A  V  D  E  I  E  S  N  H  N  R
         AAAAGCGGAAATCTCGTTCCGTATCGCGTCGAACTGATTAGCCGCATCGGGCAGGAAGCAGTAGACGAAATCGAATCAAACCATAACCGC
              42870     42880     42890     42900     42910     42920     42930     42940     42950
```

```
orf204 H  R  W  T  I  E  E  C  K  A  I  K  A  E  Y  Q  Q  K  L  K  D  L  R  N  S  R  S  E  A  A
orf68                                                                                            M
       CATCGCTGGACTATCGAAGAGTGCAAGGCGATCAAGGCAGAGTACCAACAGAAACTCAAAGACCTGCGAAATAGCAGAAGTGAGGCCGCA
           42960     42970     42980     42990     43000     43010     43020     43030     43040

       *
orf68     T  F  S  V  K  T  I  P  D  M  L  V  E  T  Y  G  N  Q  T  E  V  A  R  R  L  K  C  S  R  G
       TGACGTTCTCAGTAAAAACCATTCCAGACATGCTCGTTGAAACATACGGAAATCAGACAGAAGTAGCACGCAGACTGAAATGTAGTCGCG
           43050     43060     43070     43080     43090     43100     43110     43120     43130

orf68     T  V  R  K  Y  V  D  D  K  D  G  K  M  H  A  I  V  N  D  V  L  M  V  H  R  G  W  S  E  R
       GTACGGTCAGAAAATACGTTGATGATAAAGACGGGAAAATGCACGCCATCGTCAACGACGTTCTCATGGTTCATCGCGGATGGAGTGAAA
           43140     43150     43160     43170     43180     43190     43200     43210     43220

orf221    M  R  Y  Y  E  K  I  D  G  S  K  Y  R  N  I  W  V  V  G  D  L  H  G  C  Y  T  N  L  M
orf68   D  A  L  L  R  K  N  *
       GAGATGCGCTATTACGAAAAAATTGATGGCAGCAAATACCGAAATATTTGGGTAGTTGGCGATCTGCACGGATGCTACACGAACCTGATG
           43230     43240     43250     43260     43270     43280     43290     43300     43310

orf221 N  K  L  D  T  I  G  F  D  N  K  K  D  L  L  I  S  V  G  D  L  V  D  R  G  A  E  N  V  E
       AACAAACTGGATACGATTGGATTCGACAACAAAAAAGACCTGCTTATCTCGGTGGGCGATTTGGTTGATCGTGGTGCAGAGAACGTTGAA
           43320     43330     43340     43350     43360     43370     43380     43390     43400

orf221 C  L  E  L  I  T  F  P  W  F  R  A  V  R  G  N  H  E  Q  M  M  I  D  G  L  S  E  R  G  N
       TGCCTGGAATTAATCACATTCCCCTGGTTCAGAGCTGTACGTGGAAACCATGAGCAAATGATGATTGATGGCTTATCAGAGCGTGGAAAC
           43410     43420     43430     43440     43450     43460     43470     43480     43490

orf221 V  N  H  W  L  L  N  G  G  G  W  F  F  N  L  D  Y  D  K  E  I  L  A  K  A  L  A  H  K  A
       GTTAATCACTGGCTGCTTAATGGCGGTGGCTGGTTCTTTAATCTCGATTACGACAAAGAAATTCTGGCTAAAGCTCTTGCCCATAAAGCA
           43500     43510     43520     43530     43540     43550     43560     43570     43580

orf221 D  E  L  P  L  I  I  E  L  V  S  K  D  K  K  Y  V  I  C  H  A  D  Y  P  F  D  E  Y  E  F
       GATGAACTTCCGTTAATCATCGAACTGGTGAGCAAAGATAAAAAATATGTTATCTGCCACGCCGATTATCCCTTTGACGAATACGAGTTT
           43590     43600     43610     43620     43630     43640     43650     43660     43670

orf221 G  K  P  V  D  H  Q  Q  V  I  W  N  R  E  R  I  S  N  S  Q  N  G  I  V  K  E  I  K  G  A
       GGAAAGCCAGTTGATCATCAGCAGGTAATCTGGAACCG●GAACGAATCAGCAACTCACAAAACGGGATCGTGAAAGAAATCAAAGGCGCG
           43680     43690     43700     43710     43720     43730     43740     43750     43760

orf221 D  T  F  I  F  G  H  T  P  A  V  K  P  L  K  F  A  N  Q  M  Y  I  D  T  G  A  V  F  C  G
       GACACGTTCATCTTTGGTCATACGCCAGCAGTGAAACCACTCAAGTTTGCCAACCAAATGTATATCGATACCGGCGCAGTGTTCTGCGGA
           43770     43780     43790     43800     43810     43820     43830     43840     43850

orf221 N  L  T  L  I  Q  V  Q  G  E  G  A  *
Q                                        M  R  L  E  S  V  A  K  F  H  S  P  K  S  P  M  M  S  D
       AACCTAACATTGATTCAGGTACAGGGAGAAGGCGCATGAGACTCGAAAGCGTAGCTAAATTTCATTCGCCAAAAAGCCCGATGATGAGCG
           43860     43870     43880     43890     43900     43910     43920     43930     43940

Q         S  P  R  A  T  A  S  D  S  L  S  G  T  D  V  M  A  A  M  G  M  A  Q  S  Q  A  G  F  G  M
       ACTCACCACGGGCCACGGCTTCTGACTCTCTTTCCGGTACTGATGTGATGGCTGCTATGGGGATGGCGCAATCACAAGCCGGATTCGGTA
           43950     43960     43970     43980     43990     44000     44010     44020     44030

Q         A  A  F  C  G  K  H  E  L  S  Q  N  D  K  Q  K  A  I  N  Y  L  M  Q  F  A  H  K  V  S  G
       TGGCTGCATTCTGCGGTAAGCACGAACTCAGCCAGAACGACAAACAAAAGGCTATCAACTATCTGATGCAATTTGCACACAAGGTATCGG
           44040     44050     44060     44070     44080     44090     44100     44110     44120

Q         K  Y  R  G  V  A  K  L  E  G  N  T  K  A  K  V  L  Q  V  L  A  T  F  A  Y  A  D  Y  C  R
       GGAAATACCGTGGTGTGGCAAAGCTTGAAGGAAATACTAAGGCAAAGGTACTGCAAGTGCTCGCAACATTCGCTTATGCGGATTATTGCC
           44130     44140     44150     44160     44170     44180     44190     44200     44210

Q         S  A  A  T  P  G  A  R  C  R  D  C  H  G  T  G  R  A  V  D  I  A  K  T  E  L  W  G  R  V
       GTAGTGCCGCGACGCCGGGGGCAAGATGCAGAGATTGCCATGGTACAGGCCGTGCGGTTGATATTGCCAAAACAGAGCTGTGGGGGAGAG
           44220     44230     44240     44250     44260     44270     44280     44290     44300

Q         V  E  K  E  C  G  R  C  K  G  V  G  Y  S  R  M  P  A  S  A  A  Y  R  A  V  T  M  L  I  P
       TTGTCGAGAAAGAGTGCGGAAGATGCAAAGGCGTCGGCTATTCAAGGATGCCAGCAAGCGCAGCATATCGCGCCGTGACGATGCTAATCC
           44310     44320     44330     44340     44350     44360     44370     44380     44390

Q         N  L  T  Q  P  T  W  S  R  T  V  K  P  L  Y  D  A  L  V  V  Q  C  H  K  E  E  S  I  A  D
       CAAACCTTACCCAACCCACCTGGTCACGCACTGTTAAGCCGCTGTATGACGCTCTGGTGGTGCAATGCCACAAAGAAGAGTCAATCGCAG
           44400     44410     44420     44430     44440     44450     44460     44470     44480

Q         N  I  L  N  A  V  T  R  *
       ACAACATTTTGAATGCGGTCACACGTTAGCAGCATGATTGCCACGGATGGCAACATATTAACGGCATGATATTGACTTATTGAATAAAAT
           44490     44500     44510     44520     44530     44530     44550     44560     44570

                      SR'---------------------------- 6S RNA------------------------------------
orf64                                                 M  K  R  G  G  A  Y  Y  R  F  R  L  V  G
       TGGGTAAATTTGACTCAACGATGGGTTAATTCGCTCGTTGTGGTAGTGAGATGAAAAGAGGCGGCGCTTACTACCGATTCCGCCTAGTTG
           44580     44590     44600     44610     44620     44630     44640     44650     44660

       ------------------------------------- 6S RNA ---------------------------------------------
orf64     H  F  D  V  S  S  G  T  P  T  I  A  G  R  E  V  C  K  M  Q  S  R  N  S  S  Q  V  I  V  R
       GTCACTTCGACGTATCGTCTGGAACTCCAACCATCGCAGGCAGAGAGGTCTGCAAAATGCAATCCCGAAACAGTTCGCAGGTAATAGTTA
           44670     44680     44690     44700     44710     44720     44730     44740     44750

       ------------ 6S RNA ---------
orf64     A  C  I  T  V  S  G  F  F  I  S  A  Q  Q  V  R  A  L  S  R  *
       GAGCCTGCATAACGGTTTCGGGATTTTTTATATCTGCACAACAGGTAAGAGCATTGAGTCGATAATCGTGAAGAGTCGGCGAGCCTGGTT
           44760     44770     44780     44790     44800     44810     44820     44830     44840

       AGCCAGTGCTCTTTCCGTTGTGCTGAATTAAGCGAATACCGGAAGCAGAACCGGATCACCAAATGCGTACAGGCGTCATCGCCGCCCAGC
           44850     44860     44870     44880     44890     44900     44910     44920     44930

       AACAGCACAACCCAAACTGAGCCGTAGCCACTGTCTGTCCTGAATTCATTAGTAATAGTTACGCTGCGGCCTTTTACACATGACCTTCGT
           44940     44950     44960     44970     44980     44990     45000     45010     45020

       GAAAGCGGGTGGCAGGAGGTCGCGCTAACAACCTCCTGCCGTTTTGCCCGTGCATATCGGTCACGAACAAATCTGATTACTAAACACAGT
           45030     45040     45050     45060     45070     45080     45090     45100     45110

S                                                                                   M  K  M  P  E
       AGCCTGGATTTGTTCTATCAGTAATCGACCTTATTCCTAATTAAATAGAGCAAATCCCCTTATTGGGGGTAAGACATGAAGATGCCAGAA
           45120     45130     45140     45150     45160     45170     45180     45190     45200

S        K  H  D  L  L  A  A  I  L  A  A  K  E  Q  G  I  G  A  I  L  A  F  A  M  A  Y  L  R  G  R
       AAACATGACCTGTTGGCCGCCATTCTCGCGGCAAAGGAACAAGGCATCGGGGCAATCCTTGCGTTTGCAATGGCGTACCTTCGCGGCAGA
           45210     45220     45230     45240     45250     45260     45270     45280     45290

S        Y  N  G  G  A  F  T  K  T  V  I  D  A  T  M  C  A  I  I  A  *  F  I  R  D  L  L  D  F  A
       TATAATGGCGGTGCGTTTACAAAAACAGTAATCGACGCAACGATGTGCGCCATTATCGCCTAGTTCATTCGTGACCTTCTCGACTTCGCC
           45300     45310     45320     45330     45340     45350     45360     45370     45380

S        G  L  S  S  N  L  A  Y  I  T  S  V  F  I  G  Y  I  G  T  D  S  I  G  S  L  I  K  R  F  A
       GGACTAAGTAGCAATCTCGCTTATATAACGAGCGTGTTTATCGGCTACATCGGTACTGACTCGATTGGTTCGCTTATCAAACGCTTCGCT
           45390     45400     45410     45420     45430     45440     45450     45460     45470
```

```
S       A  K  K  A  G  V  E  D  G  R  N  Q  *
R                         M  V  E  I  N  N  Q  R  K  A  F  L  D  M  L  A  W  S  E  G  T  D  N
     GCTAAAAAAGCCGGAGTAGAAGATGGTAGAAATCAATAATCAACGTAAGGCGTTCCTCGATATGCTGGCGTGGTCGGAGGGAACTGATAA
           45480     45490     45500     45510     45520     45530     45540     45550     45560

R       G  R  Q  K  T  R  N  H  G  Y  D  V  I  V  G  G  E  L  F  T  D  Y  S  D  H  P  R  K  L  V
     CGGACGTCAGAAAACCAGAAATCATGGTTATGACGTCATTGTAGGCGGAGAGCTATTTACTGATTACTCCGATCACCCTCGCAAACTTGT
           45570     45580     45590     45600     45610     45620     45630     45640     45650

R       T  L  N  P  K  L  K  S  T  G  A  G  R  Y  Q  L  L  S  R  W  W  D  A  Y  R  K  Q  L  G  L
     CACGCTAAACCCAAAACTCAAATCAACAGGCGCCGGACGCTACCAGCTTCTTTCCCGTTGGTGGGATGCCTACCGCAAGCAGCTTGGCCT
           45660     45670     45680     45690     45700     45710     45720     45730     45740

R       K  D  F  S  P  K  S  Q  D  A  V  A  L  Q  Q  I  K  E  R  G  A  L  P  M  I  D  R  G  D  I
     GAAAGACTTCTCTCCGAAAAGTCAGGACGCTGTGGCATTGCAGCAGATTAAGGAGCGTGGCGCTTTACCTATGATTGATCGTGGTGATAT
           45750     45760     45770     45780     45790     45800     45810     45820     45830

R       R  Q  A  I  D  R  C  S  N  I  W  A  S  L  P  G  A  G  Y  G  Q  F  E  H  K  A  D  S  L  I
     CCGTCAGGCAATCGACCGTTGCAGCAATATCTGGGCTTCACTGCCGGGCGCTGGTTATGGTCAGTTCGAGCATAAGGCTGACAGCCTGAT
           45840     45850     45860     45870     45880     45890     45900     45910     45920

Rz                                                   M  S  R  V  T  A  I  I  S  A  L  V  I  C  I
R       A  K  F  K  E  A  G  G  T  V  R  E  I  D  V  *
     TGCAAAATTCAAAGAAGCGGGCGGAACGGTCAGAGAGATTGATGTATGAGCAGAGTCACCGCGATTATCTCCGCTCTGGTTATCTGCATC
           45930     45940     45950     45960     45970     45980     45990     46000     46010

Rz    I  V  C  L  S  W  A  V  N  H  Y  R  D  N  A  I  T  Y  K  A  Q  R  D  K  N  A  R  E  L  K
     ATCGTCTGCCTGTCATGGGCTGTTAATCATTACCGTGATAACGCCATTACCTACAAAGCCCAGCGCGACAAAAATGCCAGAGAACTGAAG
           46020     46030     46040     46050     46060     46070     46080     46090     46100

Rz    L  A  N  A  A  I  T  D  M  Q  M  R  Q  R  D  V  A  A  L  D  A  K  Y  T  K  E  L  A  D  A
     CTGGCGAACGCGGCAATTACTGACATGCAGATGCGTCAGCGTGATGTTGCTGCGCTCGATGCAAAATACACGAAGGAGTTAGCTGATGCT
           46110     46120     46130     46140     46150     46160     46170     46180     46190

Rz    K  A  E  N  D  A  L  R  D  D  V  A  A  G  R  R  R  L  H  I  K  A  V  C  Q  S  V  R  E  A
     AAAGCTGAAAATGATGCTCTGCGTGATGATGTTGCCGCTGGTCGTCGTCGGTTGCACATCAAAGCAGTCTGTCAGTCAGTGCGTGAAGCC
           46200     46210     46220     46230     46240     46250     46260     46270     46280

Rz    T  T  A  S  G  V  D  N  A  A  S  P  R  L  A  D  T  A  E  R  D  Y  F  T  L  R  E  R  L  I
     ACCACCGCCTCCGGCGTGGATAATGCAGCCTCCCCCCGACTGGCAGACACCGCTGAACGGGATTATTTCACCCTCAGAGAGAGGCTGATC
           46290     46300     46310     46320     46330     46340     46350     46360     46370

Rz    T  M  Q  K  Q  L  E  G  T  Q  K  Y  I  N  E  Q  C  R  *
     ACTATGCAAAAACAACTGGAAGGAACCCAGAAGTATATTAATGAGCAGTGCAGATAGAGTTGCCCATATCGATGGGCAACTCATGCAATT
           46380     46390     46400     46410     46420     46430     46440     46450     46460

     ATTGTGAGCAATACACACGCGCTTCCAGCGGAGTATAAATGCCTAAAGTAATAAAACCGAGCAATCCATTTACGAATGTTTGCTGGGTTT
           46470     46480     46490     46500     46510     46520     46530     46540     46550

     CTGTTTTAACAACATTTTCTGCGCCGCCACAAATTTTGGCTGCATCGACAGTTTTCTTCTGCCCAATTCCAGAAACGAAGAAATGATGGG
           46560     46570     46580     46590     46600     46610     46620     46630     46640

     TGATGGTTTCCTTTGGTGCTACTGCTGCCGGTTTGTTTTGAACAGTAAACGTCTGTTGAGCACATCCTGTAATAAGCAGGGCCAGCGCAG
           46650     46660     46670     46680     46690     46700     46710     46720     46730

     TAGCGAGTAGCATTTTTTTCATGGTGTTATTCCCGATGCTTTTTGAAGTTCGCAGAATCGTATGTGTAGAAAATTAAACAAACCCTAAAC
           46740     46750     46760     46770     46780     46790     46800     46810     46820

     AATGAGTTGAAATTTCATATTGTTAATATTTATTAATGTATGTCAGGTGCGATGAATCGTCATTGTATTCCCGGATTAACTATGTCCACA
           46830     46840     46850     46860     46870     46880     46890     46900     46910

     GCCCTGACGGGGAACTTCTCTGCGGGAGTGTCCGGGAATAATTAAAACGATGCACACAGGGTTTAGCGCGTACACGTATTGCATTATGCC
           46920     46930     46940     46950     46960     46970     46980     46990     47000

     AACGCCCCGGTGCTGACACGGAAGAAACCGGACGTTATGATTTAGCGTGGAAAGATTTGTGTAGTGTTCTGAATGCTCTCAGTAAATAGT
           47010     47020     47030     47040     47050     47060     47070     47080     47090

     AATGAATTATCAAAGGTATAGTAATATCTTTTATGTTCATGGATATTTGTAACCCATCGGAAAACTCCTGCTTTAGCAAGATTTTCCCTG
           47100     47110     47120     47130     47140     47150     47160     47170     47180

     TATTGCTGAAATGTGATTTCTCTTGATTTCAACCTATCATAGGACGTTTCTATAAGATGCGTGTTTCTTGAGAATTTAACATTTACAACC
           47190     47200     47210     47220     47230     47240     47250     47260     47270

     TTTTTAAGTCCTTTTATTAACACGGTGTTATCGTTTTCTAACACGATGTGAATATTATCTGTGGCTAGATAGTAAATATAATGTGAGACG
           47280     47290     47300     47310     47320     47330     47340     47350     47360

     TTGTGACGTTTTAGTTCAGAATAAAACAATTCACAGTCTAAATCTTTTCGCACTTGATCGAATATTTCTTTAAAAATGGCAACCTGAGCC
           47370     47380     47390     47400     47410     47420     47430     47440     47450

     ATTGGTAAAACCTTCCATGTGATACGAGGGCGCGTAGTTTGCATTATCGTTTTTATCGTTTCAATCTGGTCTGACCTCCTTGTGTTTTGT
           47460     47470     47480     47490     47500     47510     47520     47530     47540

     TGATGATTTATGTCAAATATTAGGAATGTTTTCACTTAATAGTATTGGTTGCGTAACAAAGTGCGGTCCTGCTGGCATTCTGGAGGGAAA
           47550     47560     47570     47580     47590     47600     47610     47620     47630

     TACAACCGACAGATGTATGTAAGGCCAACGTGCTCAAATCTTCATACAGAAAGATTTGAAGTAATATTTTAACCGCTAGATGAAGAGCAA
           47640     47650     47660     47670     47680     47690     47700     47710     47720

     GCGCATGGAGCGACAAAATGAATAAAGAACAATCTGCTGATGATCCCTCCGTGGATCTGATTCGTGTAAAAAATATGCTTAATAGCACCA
           47730     47740     47750     47760     47770     47780     47790     47800     47810

     TTTCTATGAGTTACCCTGATGTTGTAATTGCATGTATAGAACATAAGGTGTCTCTGGAAGCATTCAGAGCAATTGAGGCAGCGTTGGTGA
           47820     47830     47840     47850     47860     47870     47880     47890     47900

     AGCACGATAATAATATGAAGGATTATTCCCTGGTGGTTGACTGATCACCATAACTGCTAATCATTCAAACTATTTAGTCTGTGACAGAGC
           47910     47920     47930     47940     47950     47960     47970     47980     47990

     CAACACGCAGTCTGTCACTGTCAGGAAAGTGGTAAAACTGCAACTCAATTACTGCAATGCCCTCGTAATTAAGTGAATTTACAATATCGT
           48000     48010     48020     48030     48040     48050     48060     48070     48080

     CCTGTTCGGAGGGAAGAACGCGGGATGTTCATTCTTCATCACTTTTAATTGATGTATATGCTCTCTTTTCTGACGTTAGTCTCCGACGGC
           48090     48100     48110     48120     48130     48140     48150     48160     48170

     AGGCTTCAATGACCCAGGCTGAGAAATTCCCGGACCCTTTTTGCTCAAGAGCGATGTTAATTTGTTCAATCATTTGGTTAGGAAAGCGGA
           48180     48190     48200     48210     48220     48230     48240     48250     48260

     TGTTGCGGGTTGTTGTTCTGCGGGTTCTGTTCTTCGTTGACATGAGGTTGCCCCGTATTCAGTGTCGCTGATTTGTATTGTCTGAAGTTG
           48270     48280     48290     48300     48310     48320     48330     48340     48350

     TTTTTACGTTAAGTTGATGCAGATCAATTAATACGATACCTGCGTCATAATTGATTATTTGACGTGGTTTGATGGCCTCCACGCACGTTG
           48360     48370     48380     48390     48400     48410     48420     48430     48440

     TGATATGTAGATGATAATCATTATCACTTTACGGGTCCTTTCCGGTGATCCGACAGGTTACG
           48450     48460     48470     48480     48490     48500
```

(c)

```
                         SR
      TTGTTCCATACAACCTCCTTAGTACATGCAACCATTATCACCGCCAGAGGTAAAATAGTCAACACGCACGGTGTTAGATATTTATCCCTT
          38040     38030     38020     38010     38000     37990     37980     37970     37960

cI                       M  S  T  K  K  K  P  L  T  Q  E  Q  L  E  D  A  R  R  L  K  A  I  Y  E
      GCGGTGATAGATTTAACGTATGAGCACAAAAAAGAAACCATTAACACAAGAGCAGCTTGAGGACGCACGTCGCCTTAAAGCAATTTATGA
          37950     37940     37930     37920     37910     37900     37890     37880     37870

cI     K  K  K  N  E  L  G  L  S  Q  E  S  V  A  D  K  M  G  M  G  Q  S  G  V  G  A  L  F  N  G
      AAAAAAGAAAAATGAACTTGGCTTATCCCAGGAATCTGTCGCAGACAAGATGGGGATGGGGCAGTCAGGCGTTGGTGCTTTATTTAATGG
          37860     37850     37840     37830     37820     37810     37800     37790     37780

cI     I  N  A  L  N  A  Y  N  A  A  L  L  T  K  I  L  K  V  S  V  E  E  F  S  P  S  I  A  R  E
      CATCAATGCATTAAATGCTTATAACGCCGCATTGCTTACAAAAATTCTCAAAGTTAGCGTTGAAGAATTTAGCCCTTCAATCGCCAGAGA
          37770     37760     37750     37740     37730     37720     37710     37700     37690

cI     I  Y  E  M  Y  E  A  V  S  M  Q  P  S  L  R  S  E  Y  E  Y  P  V  F  S  H  V  Q  A  G  M
      AATCTACGAGATGTATGAAGCGGTTAGTATGCAGCCGTCACTTAGAAGTGAGTATGAGTACCCTGTTTTTTCTCATGTTCAGGCAGGGAT
          37680     37670     37660     37650     37640     37630     37620     37610     37600

cI     F  S  P  K  L  R  T  F  T  K  G  D  A  E  R  W  V  S  T  T  K  K  A  S  D  S  A  F  W  L
      GTTCTCACCTAAGCTTAGAACCTTTACCAAAGGTGATGCGGAGAGATGGGTAAGCACAACCAAAAAAGCCAGTGATTCTGCATTCTGGCT
          37590     37580     37570     37560     37550     37540     37530     37520     37510

cI     E  V  E  G  N  S  M  T  A  P  T  G  S  K  P  S  F  P  D  G  M  L  I  L  V  D  P  E  Q  A
      TGAGGTTGAAGGTAATTCCATGACCGCACCAACAGGCTCCAAGCCAAGCTTTCCTGACGGAATGTTAATTCTCGTTGACCCTGAGCAGGC
          37500     37490     37480     37470     37460     37450     37440     37430     37420

cI     V  E  P  G  D  F  C  I  A  R  L  G  G  D  E  F  T  F  K  K  L  I  R  D  S  G  Q  V  F  L
      TGTTGAGCCAGGTGATTTCTGCATAGCCAGACTTGGGGGTGATGAGTTTACCTTCAAGAAACTGATCAGGGATAGCGGTCAGGTGTTTTT
          37410     37400     37390     37380     37370     37360     37350     37340     37330

cI     Q  P  L  N  P  Q  Y  P  M  I  P  C  N  E  S  C  S  V  V  G  K  V  I  A  S  Q  W  P  E  E
      ACAACCACTAAACCCACAGTACCCAATGATCCCATGCAATGAGAGTTGTTCCGTTGTGGGGAAAGTTATCGCTAGTCAGTGGCCTGAAGA
          37320     37310     37300     37290     37280     37270     37260     37250     37240

cI     T  F  G  *
      GACGTTTGGCTGATCGGCAAGGTGTTCTGGTCGGCGCATAGCTGATAACAATTGAGCAAGAATCTTCATCGAATTAGGGGAATTTTCACT
          37230     37220     37210     37200     37190     37180     37170     37160     37150

rexA                                       M  K  N  G  F  Y  A  T  Y  R  S  K  N  K  G  K  D  K  R
      CCCCTCAGAACATAACATAGTAAATGGATTGAATTATGAAGAATGGTTTTTATGCGACTTACCGCAGCAAAAATAAAGGGAAAGATAAGC
          37140     37130     37120     37110     37100     37090     37080     37070     37060

rexA    S  I  N  L  S  V  F  L  N  S  L  L  A  D  N  H  H  L  Q  V  G  S  N  Y  L  Y  I  H  K  I
      GCTCAATAAACCTGTCTGTTTCCTTAATTCTCTGCTGGCTGATAATCATCACCTGCAGGTTGGCTCCAATTATTTGTATATTCATAAAA
          37050     37040     37030     37020     37010     37000     36990     36980     36970

rexA    D  G  K  T  F  L  F  T  K  T  N  D  K  S  L  V  Q  K  I  N  R  S  K  A  S  V  E  D  I  K
      TCGATGGAAAAACTTTTCTCTTTACCAAAACAAATGACAAGAGTCTGGTTCAGAAGATAAATCGCTCTAAAGCTTCAGTTGAAGATATTA
          36960     36950     36940     36930     36920     36910     36900     36890     36880

rexA    N  S  L  A  D  D  E  S  L  G  F  P  S  F  L  F  V  E  G  D  T  I  G  F  A  R  T  V  F  G
      AGAACAGCCTCGCAGATGACGAATCATTGGGATTCCCATCTTTTTTGTTTGTTGAAGGCGACACCATTGGTTTTGCCAGAACTGTTTTCG
          36870     36860     36850     36840     36830     36820     36810     36800     36790

rexA    P  T  T  S  D  L  T  D  F  L  I  G  K  G  M  S  L  S  S  G  E  R  V  Q  I  E  P  L  M  R
      GGCCGACCACATCCGATCTGACAGATTTTTTAATCGGGAAAGGAATGTCATTAAGCAGTGGAGAGCGCGTTCAGATAGAGCCACTGATGA
          36780     36770     36760     36750     36740     36730     36720     36710     36700

rexA    G  T  T  K  D  D  V  M  H  M  H  F  I  G  R  T  T  V  K  V  E  A  K  L  P  V  F  G  D  I
      GGGGAACCACCAAAGACGATGTTATGCATATGCATTTCATCGGCCGAACAACGGTGAAGGTAGAAGCCAAGCTACCTGTATTTGGCGATA
          36690     36680     36670     36660     36650     36640     36630     36620     36610

rexA    L  K  V  L  G  A  T  D  I  E  G  E  L  F  D  S  L  D  I  V  I  K  P  K  F  K  R  D  I  K
      TATTAAAGGTCTTAGGGGCAACAGATATTGAAGGGGAGCTTTTTGACTCATTGGATATAGTCATTAAGCCAAAATTTAAAAGGGATATAA
          36600     36590     36580     36570     36560     36550     36540     36530     36520

rexA    K  V  A  K  D  I  I  F  N  P  S  P  Q  F  S  D  I  S  L  R  A  K  D  E  A  G  D  I  L  T
      AAAAGGTTGCCAAGGATATTATTTTTAACCCGTCACCTCAATTTTCAGACATTAGCCTGCGGGCAAAAGATGAGGCCGGAGATATTTTAA
          36510     36500     36490     36480     36470     36460     36450     36440     36430

rexA    E  H  Y  L  S  E  K  G  H  L  S  A  P  L  N  K  V  T  N  A  E  I  A  E  E  M  A  Y  C  Y
      CAGAACATTATCTATCAGAAAAAGGCCATCTCTCAGCGCCTCTGAACAAGGTCACCAATGCTGAGATAGCTGAAGAGATGGCATATTGCT
          36420     36410     36400     36390     36380     36370     36360     36350     36340

rexA    A  R  M  K  S  D  I  L  E  C  F  K  R  Q  V  G  K  V  K  D  *              M  R  N  R
      ACGCAAGAATGAAAAGTGATATACTGGAATGTTTTAAAAGGCAGGTGGGCAAAGTTAAGGATTAATTATCAGGAGTAATTATGCGGAACA
          36330     36320     36310     36300     36290     36280     36270     36260     36250

rexB    I  M  P  G  V  Y  I  V  I  I  P  Y  V  I  V  S  I  C  Y  L  L  F  R  H  Y  I  P  G  V  S
      GAATCATGCCTGGTGTTTACATAGTAATAATTCCTTACGTTATCGTAAGCATTTGCTATCTCCTTTTCCGCCACTACATTCCTGGTGTTT
          36240     36230     36220     36210     36200     36190     36180     36170     36160

rexB    F  S  A  H  R  D  G  L  G  A  T  L  S  S  Y  A  G  T  M  I  A  I  L  I  A  A  L  T  F  L
      CTTTTTCAGCTCATAGAGATGGTCTTGGGGCGACATTGTCATCATATGCAGGAACCATGATTGCAATCCTGATTGCTGCCTTGACGTTTC
          36150     36140     36130     36120     36110     36100     36090     36080     36070

rexB    I  G  S  R  T  R  R  L  A  K  I  R  E  Y  G  Y  M  T  S  V  V  I  V  Y  A  L  S  F  V  E
      TAATCGGAAGCAGAACGCGCCGACTGGCCAAGATTAGAGAGTATGGGTATATGACATCGGTAGTTATTGTCTATGCCCTTAGTTTTGTTG
          36060     36050     36040     36030     36020     36010     36000     35990     35980

rexB    L  G  A  L  F  F  C  G  L  L  L  L  S  S  I  S  G  Y  M  I  P  T  I  A  I  G  I  A  S  A
      AGCTTGGAGCTTTGTTTTTCTGCGGGTTATTGCTTCTTTCCAGCATAAGCGGCTACATGATACCCACTATCGCCATCGGCATTGCCTCTG
          35970     35960     35950     35940     35930     35920     35910     35900     35890

rexB    S  F  I  H  I  C  I  L  V  F  Q  L  Y  N  L  T  R  E  Q  E  *
      CATCGTTCATTCATATATGCATCCTTGTTTTCCAACTATATAATTTGACCAGAGAACAAGAATAACCCGGCCTCAGCGCCGGGTTTTCTT
          35880     35870     35860     35850     35840     35830     35820     35810     35800

      TGCCTCACGATCGCCCCCAAAACACATAACCAATTGTATTTATTGAAAAATAAATAGATACAACTCACTAAACATAGCAATTCAGATCTC
          35790     35780     35770     35760     35750     35740     35730     35720     35710

      TCACCTACCAAACAATGCCCCCCTGCAAAAAATAAATTCATATAAAAAACATACAGATAACCATCTGCGGTGATAAATTATCTCTGGCGG
          35700     35690     35680     35670     35660     35650     35640     35630     35620
                                               SL
      TGTTGACATAAATACCACTGGCGGTGATACTGAGCACATCAGCAGGACGCACTGACCACCATGAAGGTGACGCTCTTAAAAATTAAGCCC
          35610     35600     35590     35580     35570     35560     35550     35540     35530

      TGAAGAAGGGCAGCATTCAAAGCAGAAGGCTTTGGGGTGTGTGATACGAAACGAAGCATTGGCCGTAAGTGCGATTCCGGATTAGCTGCC
          35520     35510     35500     35490     35480     35470     35460     35450     35440

N        M  C  Q  S  R  G  V  F  V  Q  D  Y  N  C  H  T  P  P  K  L  T  D  R  R  I  Q  M  D  A  Q
      AATGTGCCAATCGCGGGGGGTTTTCGTTCAGGACTACAACTGCCACACACCACCAAAGCTAACTGACAGGAGAATCCAGATGGATGCACA
          35430     35420     35410     35400     35390     35380     35370     35360     35350
```

FIG. 2 (c).

```
N          T  R  R  R  E  R  R  A  E  K  Q  A  Q  W  K  A  A  N  P  L  L  V  G  V  S  A  K  P  V  N
       AACACGCCGCCGCGAACGTCGCGCAGAGAAACAGGCTCAATGGAAAGCAGCAAATCCCCTGTTGGTTGGGGTAAGCGCAAAACCAGTTAA
            35340     35330     35320     35310     35300     35290     35280     35270     35260

N          R  P  I  L  S  L  N  R  K  P  K  S  R  V  E  S  A  L  N  P  I  D  L  T  V  L  A  E  Y  H
       CCGCCCTATTCTCTCGCTGAATCGCAAACCGAAATCACGAGTAGAAAGCGCACTAAATCCGATAGACCTTACAGTGCTGGCTGAATACCA
            35250     35240     35230     35220     35210     35200     35190     35180     35170

N          K  Q  I  E  S  N  L  Q  R  I  E  R  K  N  Q  R  T  W  Y  S  K  P  G  E  R  G  I  T  C  S
       CAAACAGATTGAAAGCAACCTGCAACGTATTGAGCGCAAGAATCAGCGCACATGGTACAGCAAGCCTGGCGAACGCGGCATAACATGCAG
            35160     35150     35140     35130     35120     35110     35100     35090     35080

N          G  R  Q  K  I  K  G  K  S  I  P  L  I  *
       TGGACGCCAGAAAATTAAGGGAAAATCGATTCCTCTTATCTAGTTACTTAGATATTGGCCTTGGCTTTATCTCAATATTATATGGATCAT
            35070     35060     35050     35040     35030     35020     35010     35000     34990

       AGCTGGCAACTAATTCAGTCCAGTAAATATCCTCAATAGGGAATAATATATGCTTTCCATTCCATCGGGAAAAAGTTTTGTTCAACACAC
            34980     34970     34960     34950     34940     34930     34920     34910     34900

       CAAGCTCAATCAACTCACTAATGTATGGGAATTGTTTTGATGTAACCACATACTTCCTGCCTTCATTAAGGGCTGCGCACAAAACCATAG
            34890     34880     34870     34860     34850     34840     34830     34820     34810

       ATTGCTCTTCTGTAAGGTTTTGAATTACTGATCGCACTTTATCGTTTTGCATCTTAATGCGTTTTCTTAGCTTAAATCGCTTATATCTGG
            34800     34790     34780     34770     34760     34750     34740     34730     34720

       CGCTGGCAATAGCTGATAATCGATGCACATTAATTGCTAGCGAAAATGCAAGAGCAAAGACGAAAACATGCCACACATGAGGAATACCGA
            34710     34700     34690     34680     34670     34660     34650     34640     34630

       TTCTCTCATTAACATATTCAGGCCAGTTATCTGGGCTTAAAAGCAGAAGTCCAACCCAGATAACGATCATATACATGGTTCTCTCCAGAG
            34620     34610     34600     34590     34580     34570     34560     34550     34540

       GTTCATTACTGAACACTCGTCCGAGAATAACGAGTGGATCCATTTCTATACTCATCAAACTGTAGGGGTTGTAATAGTTTATCCGATTTC
            34530     34520     34510     34500     34490     34480     34470     34460     34450

       TCGCTGTAGGGGTACACGAGAACCACCGAGCCTGATGTGGTTAAAAGACAGGCACAATCTTTACTACCGCAATCCACTATTTAAGGTGAT
            34440     34430     34420     34410     34400     34390     34380     34370     34360

ral                                                                                M  T  T  T  I  D
orf28      M  E  E  E  F  E  E  F  E  E  H  P  Q  D  V  M  E  Q  Y  Q  D  Y  P  Y  D  Y  D  Y  *
       ATATGGAAGAAGAATTTGAAGAGTTCGAAGAGCATCCTCAGGATGTGATGGAACAATACCAGGACTATCCGTATGACTACGACTATTGAT
            34350     34340     34330     34320     34310     34300     34290     34280     34270

ral        K  N  Q  W  C  G  Q  F  K  R  C  N  G  C  K  L  Q  S  E  C  M  V  K  P  E  E  M  F  P  V
       AAAAATCAATGGTGTGGACAATTCAAGCGATGCAATGGATGCAAGCTGCAATCGGAATGCATGGTTAAGCCTGAAGAAATGTTTCCTGTA
            34260     34250     34240     34230     34220     34210     34200     34190     34180

ral        M  E  D  G  K  Y  V  D  K  W  A  I  R  T  T  A  M  I  A  R  E  L  G  K  Q  N  N  K  A  A
       ATGGAAGATGGGAAATATGTCGATAAATGGGCAATACGAACGACGGCAATGATTGCCAGAGAACTTGGTAAACAGAACAACAAAGCTGCC
            34170     34160     34150     34140     34130     34120     34110     34100     34090

       *
       TGATAGTGGCCTTTATTTTTGGCATAAATAACAGAATAAACACTGCACTGTGTATTCATTCCAACGAGTGAATACACGGAGCAATGTCGC
            34080     34070     34060     34050     34040     34030     34020     34010     34000

       TCGTAACTAAACAGGAGCCGACTTGTTCTGATTATTGGAAATCTTCTTTGCCCTCCAGTGTGAGGGCGATTTTTTATCTGTGAGGATATG
            33990     33980     33970     33960     33950     33940     33930     33920     33910

ssb           M  S  N  I  K  K  Y  I  I  D  Y  D  W  K  A  S  I  E  I  E  I  D  H  D  V  M  T  E  E
       AACAGATGTCAAACATCAAAAAATACATCATTGATTACGACTGGAAAGCATCAATAGAAATTGAAATCGACCATGACGTAATGACAGAGG
            33900     33890     33880     33870     33860     33850     33840     33830     33820

ssb        K  L  H  Q  I  N  N  F  W  S  D  S  E  Y  R  L  N  K  H  G  S  V  L  N  A  V  L  I  M  L
       AAAAACTTCACCAGATTAATAATTTCTGGTCAGACTCTGAATACCGACTCAATAAACACGGCTCTGTATTAAATGCTGTATTAATCATGC
            33810     33800     33790     33780     33770     33760     33750     33740     33730

ssb        A  Q  H  A  L  L  I  A  I  S  S  D  L  N  A  Y  G  V  V  C  E  F  D  W  N  D  G  N  G  Q
       TGGCGCAACATGCTCTGCTTATAGCAATTTCAAGCGACTTAAATGCATATGGTGTTGTGTGTGAGTTCGACTGGAATGATGGAAATGGTC
            33720     33710     33700     33690     33680     33670     33660     33650     33640

ssb        E  G  W  P  P  M  D  G  S  E  G  I  R  I  T  D  I  D  T  S  G  I  F  D  S  D  D  M  T  I
       AGGAAGGATGGCCTCCAATGGATGGTAGCGAAGGAATAAGAATTACCGATATCGATACATCAGGAATATTTGATTCAGATGATATGACTA
            33630     33620     33610     33600     33590     33580     33570     33560     33550

ssb        K  A  A  *                                                                          M  Q
       TCAAGGCCGCCTGAGTGCGGTTTTACCGCATACCAATAACGCTTCACTCGAGGCGTTTTTCGTTATGTATAAATAAGGAGCACACCATGC
            33540     33530     33520     33510     33500     33490     33480     33470     33460

cIII       Y  A  I  A  G  W  P  V  A  G  C  P  S  E  S  L  L  E  R  I  T  R  K  L  R  D  G  W  K  R
       AATATGCCATTGCAGGGTGGCCTGTTGCTGGCTGCCCTTCCGAATCTTTACTTGAACGAATCACCCGTAAATTACGTGACGGATGGAAAC
            33450     33440     33430     33420     33410     33400     33390     33380     33370

kil                                                  M  D  Q  T  L  M  A  I  Q  T  K  F  T  I  A  T  F
cIII       L  I  D  I  L  N  Q  P  G  V  P  K  N  G  S  N  T  Y  G  Y  P  D  *
       GCCTTATCGACATACTTAATCAGCCAGGAGTCCCAAAGAATGGATCAAACACTTATGGCTATCCAGACTAAATTCACTATCGCCACTTTT
            33360     33350     33340     33330     33320     33310     33300     33290     33280

kil       I  G  D  E  K  M  F  R  E  A  V  D  A  Y  K  K  W  I  L  I  L  K  L  R  S  S  K  S  I  H
       ATTGGCGATGAAAAGATGTTTCGTGAAGCCGTCGACGCTTATAAAAAATGGATATTAATACTGAAACTGAGATCAAGCAAAAGCATTCAC
            33270     33260     33250     33240     33230     33220     33210     33200     33190

γ          *                                                                             M  N  A  Y  Y
       TAACCCCCTTTCCTGTTTTCCTAATCAGCCCGGCATTTCGCGGGCGATATTTTCACAGCTATTTCAGGAGTTCAGCCATGAACGCTTATT
            33180     33170     33160     33150     33140     33130     33120     33110     33100

γ          I  Q  D  R  L  E  A  Q  S  W  A  R  H  Y  Q  Q  L  A  R  E  E  K  E  A  E  L  A  D  D  M
       ACATTCAGGATCGTCTTGAGGCTCAGAGCTGGGCGCGTCACTACCAGCAGCTCGCCCGTGAAGAGAAAGAGGCAGAACTGGCAGACGACA
            33090     33080     33070     33060     33050     33040     33030     33020     33010

γ          E  K  G  L  P  Q  H  L  F  E  S  L  C  I  D  H  L  Q  R  H  G  A  S  K  K  S  I  T  R  A
       TGGAAAAAGGCCTGCCCCAGCACCTGTTTGAATCGCTATGCATCGATCATTTGCAACGCCACGGGGCCAGCAAAAAATCCATTACCCGTG
            33000     32990     32980     32970     32960     32950     32940     32930     32920

γ          F  D  D  D  V  E  F  Q  E  R  M  A  E  H  I  R  Y  M  V  E  T  I  A  H  H  Q  V  D  I  D
       CGTTTGATGACGATGTTGAGTTTCAGGAGCGCATGGCAGAACACATCCGGTACATGGTTGAAACCATTGCTCACCACCAGGTTGATATTG
            32910     32900     32890     32880     32870     32860     32850     32840     32830

β                       M  S  T  A  L  A  T  L  A  G  K  L  A  E  R  V  G  M  D  S  V  D  P  Q
           S  E  V  *
γ      ATTCAGAGGTATAAAACGAATGAGTACTGCACTCGCAACGCTGGCTGGGAAGCTGGCTGAACGTGTCGGCATGGATTCTGTCGACCCACA
            32820     32810     32800     32790     32780     32770     32760     32750     32740
```

```
β          E  L  I  T  T  L  R  Q  T  A  F  K  G  D  A  S  D  A  Q  F  I  A  L  L  I  V  A  N  Q  Y
        GGAACTGATCACCACTCTTCGCCAGACGGCATTTAAAGGTGATGCCAGCGATGCGCAGTTCATCGCATTACTGATCGTTGCCAACCAGTA
            32730     32720     32710     32700     32690     32680     32670     32660     32650

β          G  L  N  P  W  T  K  E  I  Y  A  F  P  D  K  Q  N  G  I  V  P  V  V  G  V  D  G  W  S  R
        CGGCCTTAATCCGTGGACGAAAGAAATTTACGCCTTTCCTGATAAGCAGAATGGCATCGTTCCGGTGGTGGGCGTTGATGGCTGGTCCCG
            32640     32630     32620     32610     32600     32590     32580     32570     32560

β          I  I  N  E  N  Q  Q  F  D  G  M  D  F  E  Q  D  N  E  S  C  T  C  R  I  Y  R  K  D  R  N
        CATCATCAATGAAAACCAGCAGTTTGATGGCATGGACTTTGAGCAGGACAATGAATCCTGTACATGCCGGATTTACCGCAAGGACCGTAA
            32550     32540     32530     32520     32510     32500     32490     32480     32470

β          H  P  I  C  V  T  E  W  M  D  E  C  R  R  E  P  F  K  T  R  E  G  R  E  I  T  G  P  W  Q
        TCATCCGATCTGCGTTACCGAATGGATGGATGAATGCCGCCGCGAACCATTCAAAACTCGCGAAGGCAGAGAAATCACGGGGCCGTGGCA
            32460     32450     32440     32430     32420     32410     32400     32390     32380

β          S  H  P  K  R  M  L  R  H  K  A  M  I  Q  C  A  R  L  A  F  G  F  A  G  I  Y  D  K  D  E
        GTCGCATCCCAAACGGATGTTACGTCATAAAGCCATGATTCAGTGTGCCCGTCTGGCCTTCGGATTTGCTGGTATCTATGACAAGGATGA
            32370     32360     32350     32340     32330     32320     32310     32300     32290

β          A  E  R  I  V  E  N  T  A  Y  T  A  E  R  Q  P  E  R  D  I  T  P  V  N  D  E  T  M  Q  E
        AGCCGAGCGCATTGTCGAAAATACTGCATACACTGCAGAACGTCAGCCGGAACGCGACATCACTCCGGTTAACGATGAAACCATGCAGGA
            32280     32270     32260     32250     32240     32230     32220     32210     32200

β          I  N  T  L  L  I  A  L  D  K  T  W  D  D  D  L  L  P  L  C  S  Q  I  F  R  R  D  I  R  A
        GATTAACACTCTGCTGATCGCCCTGGATAAAACATGGGATGACGACTTATTGCCGCTCTGTTCCCAGATATTTCGCCGCGACATTCGTGC
            32190     32180     32170     32160     32150     32140     32130     32120     32110

exo                                                                                     M  T  P
β          S  S  E  L  T  Q  A  E  A  V  K  A  L  G  F  L  K  Q  K  A  A  E  Q  K  V  A  A  *
        ATCGTCAGAACTGACACAGGCCGAAGCAGTAAAAGCTCTTGGATTCCTGAAACAGAAAGCCGCAGAGCAGAAGGTGGCAGCATGACACCG
            32100     32090     32080     32070     32060     32050     32040     32030     32020

exo       D  I  I  L  Q  R  T  G  I  D  V  R  A  V  E  Q  G  D  D  A  W  H  K  L  R  L  G  V  I  T
        GACATTATCCTGCAGCGTACCGGGATCGATGTGAGAGCTGTCGAACAGGGGGATGATGCGTGGCACAAATTACGGCTCGGCGTCATCACC
            32010     32000     31990     31980     31970     31960     31950     31940     31930

exo       A  S  E  V  H  N  V  I  A  K  P  R  S  G  K  K  W  P  D  M  K  M  S  Y  F  H  T  L  L  A
        GCTTCAGAAGTTCACAACGTGATAGCAAAACCCCGCTCCGGAAAGAAGTGGCCTGACATGAAAATGTCCTACTTCCACACCCTGCTTGCT
            31920     31910     31900     31890     31880     31870     31860     31850     31840

exo       E  V  C  T  G  V  A  P  E  V  N  A  K  A  L  A  W  G  K  Q  Y  E  N  D  A  R  T  L  F  E
        GAGGTTTGCACCGGTGTGGCTCCGGAAGTTAACGCTAAAGCACTGGCCTGGGGAAAACAGTACGAGAACGACGCCAGAACCCTGTTTGAA
            31830     31820     31810     31800     31790     31780     31770     31760     31750

exo       F  T  S  G  V  N  V  T  E  S  P  I  I  Y  R  D  E  S  M  R  T  A  C  S  P  D  G  L  C  S
        TTCACTTCCGGCGTGAATGTTACTGAATCCCCGATCATCTATCGCGACGAAAGTATGCGTACCGCCTGCTCTCCCGATGGTTTATGCAGT
            31740     31730     31720     31710     31700     31690     31680     31670     31660

exo       D  G  N  G  L  E  L  K  C  P  F  T  S  R  D  F  M  K  F  R  L  G  G  F  E  A  I  K  S  A
        GACGGCAACGGCCTTGAACTGAAATGCCCGTTTACCTCCCGGGATTTCATGAAGTTCCGGCTCGGTGGTTTCGAGGCCATAAAGTCAGCT
            31650     31640     31630     31620     31610     31600     31590     31580     31570

exo       Y  M  A  Q  V  Q  Y  S  M  W  V  T  R  K  N  A  W  Y  F  A  N  Y  D  P  R  M  K  R  E  G
        TACATGGCCCAGGTGCAGTACAGCATGTGGGTGACGCGAAAAAATGCCTGGTACTTTGCCAACTATGACCCGCGTATGAAGCGTGAAGGC
            31560     31550     31540     31530     31520     31510     31500     31490     31480

exo       L  H  Y  V  V  I  E  R  D  E  K  Y  M  A  S  F  D  E  I  V  P  E  F  I  E  K  M  D  E  A
        CTGCATTATGTCGTGATTGAGCGGGATGAAAAGTACATGGCGAGTTTTGACGAGATCGTGCCGGAGTTCATCGAAAAAATGGACGAGGCA
            31470     31460     31450     31440     31430     31420     31410     31400     31390

exo       L  A  E  I  G  F  V  F  G  E  Q  W  R  *
orf60a                                            M  T  H  P  H  D  N  I  R  V  G  A  I  T  F  V  Y  S
        CTGGCTGAAATTGGTTTTGTATTTGGGGAGCAATGGCGATGACGCATCCTCACGATAATATCCGGGTAGGCGCAATCACTTTCGTCTACT
            31380     31370     31360     31350     31340     31330     31320     31310     31300

orf60a     V  T  K  R  G  W  V  F  P  G  L  S  V  I  R  N  P  L  K  A  Q  R  L  A  E  E  I  N  N  K
        CCGTTACAAAGCGAGGCTGGGTATTTCCCGGCCTTTCTGTTATCCGAAATCCACTGAAAGCACAGCGGCTGGCTGAGGAGATAAATAATA
            31290     31280     31270     31260     31250     31240     31230     31220     31210

orf63                M  H  K  A  S  S  V  E  L  R  T  S  I  E  M  A  H  S  L  A  Q  I  G  I  R  F
orf60a    R  G  A  V  C  T  K  H  L  L  L  S  *
        AACGAGGGGCTGTATGCACAAAGCATCTTCTGTTGAGTTAAGAACGAGTATCGAGATGGCACATAGCCTTGCTCAAATTGGAATCAGGTT
            31200     31190     31180     31170     31160     31150     31140     31130     31120

orf63     V  P  I  P  V  E  T  D  E  E  F  H  T  L  A  A  S  L  S  Q  K  L  E  M  M  V  A  K  A  E
        TGTGCCAATACCAGTAGAAACAGACGAAGAATTTCATACGTTAGCCGCATCCCTTTCACAAAAGCTGGAAATGATGGTGGCGAAAGCAGA
            31110     31100     31090     31080     31070     31060     31050     31040     31030

orf63     A  D  E  R  N  Q  V  *
orf61        M  R  E  T  R  Y  D  N  H  G  M  H  F  S  G  S  G  L  H  I  L  C  A  Y  A  C  R  H  G
        AGCAGATGAGAGAAACCAGGTATGACAACCACGGAATGCATTTTTCTGGCAGCGGGCTTCATATTCTGTGTGCTTATGCTTGCCGACATG
            31020     31010     31000     30990     30980     30970     30960     30950     30940

orf61      T  C  S  M  T  P  Q  Q  E  N  A  L  R  S  I  A  R  Q  A  N  S  E  I  K  K  S  Q  T  A  V
        GGACTTGTTCAATGACACCTCAGCAGGAAAACGCCCTTCGCAGCATTGCCCGTCAGGCTAATTCTGAAATCAAAAAAAGCCAGACAGCAG
            30930     30920     30910     30900     30890     30880     30870     30860     30850

orf61      S  G  *
        TTTCCGGATAAAAACGTCGATGACATTTGCCGTAGCGTACTGAAGAAGCACCGCGAAACGGTAACGCTGATGGGATTCACACCGACTCAT
            30840     30830     30820     30810     30800     30790     30780     30770     30760

        TTAAGCCTGGCAATCGGCATGTTAAACGGCGTCTTTAAGGAACGATGAACATGAAAAGCAAAATCATCAGGGAGCTACAGGCTCCTTTTT
            30750     30740     30730     30720     30710     30700     30690     30680     30670

        TATTATTCGCATTCACCCTCAAGCGTATTAACCAACAGTTCAGGGATTAATGAAAGATGGCAGACATCATTGATTCAGCATCAGAAATAG
            30660     30650     30640     30630     30620     30610     30600     30590     30580

        AAGAATTACAGCGCAACACAGCAATAAAAATGCGCCGCCTGAACCACCAGGCTATATCTGCCACTCATTGTTGTGAGTGTGGCGATCCGA
            30570     30560     30550     30540     30530     30520     30510     30500     30490

        TAGATGAACGAAGACGCCTGGTCGTTCAGGGTTGTCGGACTTGTGCAAGTTGCCAGGAGGATCTGGAACTTATCAGTAAACAGAGAGGTT
            30480     30470     30460     30450     30440     30430     30420     30410     30400

Ea22         V  S  E  I  N  S  Q  A  L  R  E  A  A  E  Q  A  M  H  D  D  W  G  F  D  A  D  L  F  H
        CGAAGTGAGCGAAATTAACTCTCAGGCACTGCGTGAAGCGGCAGAGCAGGCAATGCATGACGACTGGGGATTTGACGCAGACCTTTTCCA
            30390     30380     30370     30360     30350     30340     30330     30320     30310

Ea22      E  L  V  T  P  S  I  V  L  E  L  L  D  E  R  E  R  N  Q  Q  Y  I  K  R  R  D  Q  E  N  E
        TGAATTGGTAACACCATCGATTGTGCTGGAACTGCTGGATGAACGGGAAAGAAACCAGCAATACATCAAACGCCGCGACCAGGAGAACGA
            30300     30290     30280     30270     30260     30250     30240     30230     30220
```

```
Ea22       D  I  A  L  T  V  G  K  L  R  V  E  L  E  T  A  K  S  K  L  N  E  Q  R  E  Y  Y  E  G  V
         GGATATTGCGCTAACAGTAGGGAAACTGCGTGTTGAGCTTGAAACAGCAAAATCAAAACTCAACGAGCAGCGTGAGTATTACGAAGGTGT
              30210     30200     30190     30180     30170     30160     30150     30140     30130

Ea22       I  S  D  G  S  K  R  I  A  K  L  E  S  N  E  V  R  E  D  G  N  Q  F  L  V  V  R  H  P  G
         TATCTCGGATGGGAGTAAGCGTATTGCTAAACTGGAAAGCAACGAAGTCCGTGAAGACGGAAACCAGTTTCTTGTTGTTCGCCATCCTGG
              30120     30110     30100     30090     30080     30070     30060     30050     30040

Ea22       K  T  P  V  I  K  H  C  T  G  D  L  E  E  F  L  R  Q  L  I  E  Q  D  P  L  V  T  I  D  I
         GAAGACTCCTGTTATCAAGCACTGCACTGGTGACCTGGAAGAGTTTCTGCGGCAGTTAATCGAACAAGACCCGTTAGTAACTATCGACAT
              30030     30020     30010     30000     29990     29980     29970     29960     29950

Ea22       I  T  H  R  Y  Y  G  V  G  G  Q  W  V  Q  D  A  G  E  Y  L  H  M  M  S  D  A  G  I  R  I
         CATTACGCATCGCTATTACGGGGTTGGAGGTCAATGGGTTCAGGATGCAGGTGAGTATCTGCATATGATGTCTGACGCTGGCATTCGCAT
              29940     29930     29920     29910     29900     29890     29880     29870     29860

Ea22       K  G  E  *
         CAAAGGAGAGTGAGATCGGTTTTGTAAAAGATAACGCTTGTGAAAATGCTGAATTTCGCGTCGTCTTCACAGCGATGCCAGAGTCTGTAG
              29850     29840     29830     29820     29810     29800     29790     29780     29770

         TGTCAGATGATGACCGTACTCAAACATCGGGTTGAGTATTATCTTACTGTTTCTTTACATAAACATTGCTGATACCGTTTAGCTGAAACG
              29760     29750     29740     29730     29720     29710     29700     29690     29680

Ea8.5                             M  S  I  N  E  L  E  S  E  Q  K  D  W  A  L  S  M  L  C  R  S  G
         ACATACATTGCAAGGAGTTTATAAATGAGTATCAATGAGTTAGAGTCTGAGCAAAAAGATTGGGCGTTATCAATGTTGTGCAGATCCGGT
              29670     29660     29650     29640     29630     29620     29610     29600     29590

Ea8.5     V  L  S  P  C  R  H  H  E  G  V  Y  V  D  E  G  I  D  I  E  S  A  Y  K  Y  S  M  K  V  Y
         GTCTTGTCTCCATGCAGACATCACGAAGGTGTTTATGTAGATGAAGGTATAGATATAGAGTCGGCATACAAATATTCCATGAAGGTTTAT
              29580     29570     29560     29550     29540     29530     29520     29510     29500

Ea8.5     K  S  N  E  D  K  S  P  F  C  N  V  R  E  M  T  D  T  V  Q  N  Y  Y  H  E  Y  G  G  N  D
         AAGTCTAATGAAGACAAATCCCCATTCTGCAATGTGCGAGAAATGACTGATACCGTGCAAAATTATTATCACGAGTACGGTGGAAACGAT
              29490     29480     29470     29460     29450     29440     29430     29420     29410

Ea8.5     T  C  P  L  C  T  K  H  I  D  D  *
         ACTTGCCCTCTCTGTACAAAACATATAGATGATTAAACCCAATATTACATAACAATCCTCGCACTCGCGGGGATTTATTTTATCTGAACT
              29400     29390     29380     29370     29360     29350     29340     29330     29320

         CGCTACGGCGGGTTTTGTTTTATGGAGATGATAAATGCACTTCCGAGTCACAGGAGAATGGAATGGAGAGCCATTCAACAGAGTTATCGA
              29310     29300     29290     29280     29270     29260     29250     29240     29230

         AGCGGAGAACATCAACGACTGCTACGACCACTGGATGATATGGGCGCAGATAGCACATGCAGACGTAACCAATATTCGAATTGAAGAACT
              29220     29210     29200     29190     29180     29170     29160     29150     29140

                                                                                S1
xis                                                                   M  Y  L  T  L  Q  E  W  N  A
         GAAAGAACACCAAGCCGCCTGATGGCGGTTTTTCTTGCGTGTAATTGCGGAGACTTTGCGATGTACTTGACACTTCAGGAGTGGAACGC
              29130     29120     29110     29100     29090     29080     29070     29060     29050

xis        R  Q  R  R  P  R  S  L  E  T  V  R  R  W  V  R  E  C  R  I  F  P  P  P  V  K  D  G  R  E
         ACGCCAGCGACGTCCAAGAAGCCTTGAAACAGTTCGTCGATGGGTTCGGGAATGCAGGATATTCCCACCTCCGGTTAAGGATGGAAGAGA
              29040     29030     29020     29010     29000     28990     28980     28970     28960

xis        Y  L  F  H  E  S  A  V  K  V  D  L  N  R  P  V  T  G  G  L  L  K  R  I  R  N  G  K  K  A
int                                                                                 M  G  R  R  R
         GTATCTGTTCCACGAATCAGCGGTAAAGGTTGACTTAAATCGACCAGTAACAGGTGGCCTTTTGAAGAGGATCAGAAATGGGAAGAAGGC
              28950     28940     28930     28920     28910     28900     28890     28880     28870

xis        K  S  *
int         S  H  E  R  R  D  L  P  P  N  L  Y  I  R  N  N  G  Y  Y  C  Y  R  D  P  R  T  G  K  E  F
         GAAGTCATGAGCGCCGGGATTTACCCCCTAACCTTTATATAAGAAACAATGGATATTACTGCTACAGGGACCCAAGGACGGGTAAAGAGT
              28860     28850     28840     28830     28820     28810     28800     28790     28780

int         G  L  G  R  D  R  R  I  A  I  T  E  A  I  Q  A  N  I  E  L  F  S  G  H  K  H  K  P  L  T
         TTGGATTAGGCAGAGACAGGCGAATCGCAATCACTGAAGCTATACAGGCCAACATTGAGTTATTTTCAGGACACAAACACAAGCCTCTGA
              28770     28760     28750     28740     28730     28720     28710     28700     28690

int         A  R  I  N  S  D  N  S  V  T  L  H  S  W  L  D  R  Y  E  K  I  L  A  S  R  G  I  K  Q  K
         CAGCGAGAATCAACAGTGATAATTCCGTTACGTTACATTCATGGCTTGATCGCTACGAAAAAATCCTGGCCAGCAGAGGAATCAAGCAGA
              28680     28670     28660     28650     28640     28630     28620     28610     28600

int         T  L  I  N  Y  M  S  K  I  K  A  I  R  R  G  L  P  D  A  P  L  E  D  I  T  T  K  E  I  A
         AGACACTCATAAATTACATGAGCAAAATTAAAGCAATAAGGAGGGGTCTGCCTGATGCTCCACTTGAAGACATCACCACAAAAGAAATTG
              28590     28580     28570     28560     28550     28540     28530     28520     28510

int         A  M  L  N  G  Y  I  D  E  G  K  A  A  S  A  K  L  I  R  S  T  L  S  D  A  F  R  E  A  I
         CGGCAATGCTCAATGGATACATAGACGAGGGCAAGGCGGCGTCAGCCAAGTTAATCAGATCAACACTGAGCGATGCATTCCGAGAGGCAA
              28500     28490     28480     28470     28460     28450     28440     28430     28420

int         A  E  G  H  I  T  T  N  H  V  A  A  T  R  A  A  K  S  E  V  R  R  S  R  L  T  A  D  E  Y
         TAGCTGAAGGCCATATAACAACAAACCATGTCGCTGCCACTCGCGCAGCAAAATCAGAGGTAAGGAGATCAAGACTTACGGCTGACGAAT
              28410     28400     28390     28380     28370     28360     28350     28340     28330

int         L  K  I  Y  Q  A  A  E  S  S  P  C  W  L  R  L  A  M  E  L  A  V  V  T  G  Q  R  V  G  D
         ACCTGAAAATTTATCAAGCAGCAGAATCATCACCATGTTGGCTCAGACTTGCAATGGAACTGGCTGTTGTTACCGGGCAACGAGTTGGTG
              28320     28310     28300     28290     28280     28270     28260     28250     28240

int         L  C  E  M  K  W  S  D  I  V  D  G  Y  L  Y  V  E  Q  S  K  T  G  V  K  I  A  I  P  T  A
         ATTTATGCGAAATGAAGTGGTCTGATATCGTAGATGGATATCTTTATGTCGAGCAAAGCAAAACAGGCGTAAAAATTGCCATCCCAACAG
              28230     28220     28210     28200     28190     28180     28170     28160     28150

int         L  H  I  D  A  L  G  I  S  M  K  E  T  L  D  K  C  K  E  I  L  G  G  E  T  I  I  A  S  T
         CATTGCATATTGATGCTCTCGGAATATCAATGAAGGAAACACTTGATAAATGCAAAGAGATTCTTGGCGGAGAAACCATAATTGCATCTA
              28140     28130     28120     28110     28100     28090     28080     28070     28060

int         R  R  E  P  L  S  S  G  T  V  S  R  Y  F  M  R  A  R  K  A  S  G  L  S  F  E  G  D  P  P
         CTCGTCGCGAACCGCTTTCATCCGGCACAGTATCAAGGTATTTTATGCGCGCACGAAAAGCATCAGGTCTTTCCTTCGAAGGGGATCCGC
              28050     28040     28030     28020     28010     28000     27990     27980     27970

int         T  F  H  E  L  R  S  L  S  A  R  L  Y  E  K  Q  I  S  D  K  F  A  Q  H  L  L  G  H  K  S
         CTACCTTTCACGAGTTGCGCAGTTTGTCTGCAAGACTCTATGAGAAGCAGATAAGCGATAAGTTTGCTCAACATCTTCTCGGGCATAAGT
              27960     27950     27940     27930     27920     27910     27900     27890     27880

int         D  T  M  A  S  Q  Y  R  D  D  R  G  R  E  W  D  K  I  E  I  K  *
         CGGACACCATGGCATCACAGTATCGTGATGACAGAGGCAGGGAGTGGGACAAAATTGAAATCAAATAATGATTTTATTTTGACTGATAGT
              27870     27860     27850     27840     27830     27820     27810     27800     27790

                                                             ----att site----
         GACCTGTTCGTTGCAACAAATTGATAAGCAATGCTTTTTTATAATGCCAACTTAGTATAAAAAAGCTGAACGAGAAACGTAAAATGATAT
              27780     27770     27760     27750     27740     27730     27720     27710     27700

         AAATATCAATATATTAAATTAGATTTTGCATAAAAAACAGACTACATAATACTGTAAAACACAACATATGCAGTCACTATGAATCAACTA
              27690     27680     27670     27660     27650     27640     27630     27620     27610
```

```
         CTTAGATGGTATTAGTGACCTGTAACAGAGCATTAGCGCAAGGTGATTTTTGTCTTCTTGCGCTAATTTTTTGTCATCAAACCTGTCGCA
              27600     27590     27580     27570     27560     27550     27540     27530     27520

         CTCCAGAGAAGCACAAAGCCTCGCAATCCAGTGCAAAGCTTTGTGTGCCACCCACTACGACCTGCATAACCAGTAAGAAGATAGCAGTGA
              27510     27500     27490     27480     27470     27460     27450     27440     27430

         TGTCAAACGACGCAGCTGACTTCTTTTCTTTCACGACTTCCCCACACCCAGCATGCATACCTTTCCGCCATAACTGTAGTGAATGTCTGT
              27420     27410     27400     27390     27380     27370     27360     27350     27340

         TATGAGCGAGGAGCGGAAGTTAACACTTATGAAAAATGGCTACGAAGTCCGTGGCTATCTATCGGCTTATTAGTACTTGAAACGCTTCTT
              27330     27320     27310     27300     27290     27280     27270     27260     27250

         CAGAAGCCTGAAGAGCTAATCGTTCGGCGATACTATATATGCATTAATAGACTATATCGTTGGTATAAACAGTGCACCATGCAACATGAA
              27240     27230     27220     27210     27200     27190     27180     27170     27160

         TAACAGTGGGTTATCCAAAAGGAAGCAGAAAGCTAAATATGGAAAACTACAATACGATGCCCCGTTAAGTTCAATACTACTAATTTTTAG
              27150     27140     27130     27120     27110     27100     27090     27080     27070

         ATGGAAAACGTATGTAATAGAGAGTAACTTAAAAGAGAGATCCTGTGTTGCCGCCAAATAAATTGCGGTTATTTTAATAAAATTAAGGGT
              27060     27050     27040     27030     27020     27010     27000     26990     26980

Ea59           M  L  E  F  S  V  I  E  R  G  G  Y  I  P  A  V  E  K  N  K  A  F  L  R  A  D  G  W
         TACTATATGTTGGAGTTTAGTGTTATTGAAAGAGGCGGGTATATTCCTGCAGTAGAAAAAAATAAGGCATTCCTACGAGCAGATGGTTGG
              26970     26960     26950     26940     26930     26920     26910     26900     26890

Ea59     N  D  Y  S  F  V  T  M  F  Y  L  T  V  F  D  E  H  G  E  K  C  D  I  G  N  V  K  I  G  F
         AATGACTATTCCTTTGTTACAATGTTTTATCTTACTGTCTTTGATGAGCATGGTGAAAAATGCGATATCGGAAATGTTAAAATTGGTTTT
              26880     26870     26860     26850     26840     26830     26820     26810     26800

Ea59     V  G  Q  K  E  E  V  S  T  Y  S  L  I  D  K  K  F  S  Q  L  P  E  M  F  F  S  L  G  E  S
         GTAGGTCAAAAAGAAGAAGTAAGCACTTATTCATTAATAGATAAAAAATTCAGTCAACTCCCTGAAATGTTTTTTTCCTTAGGTGAAAGC
              26790     26780     26770     26760     26750     26740     26730     26720     26710

Ea59     I  D  Y  Y  V  N  L  S  K  L  S  D  G  F  K  H  N  L  L  K  A  I  Q  D  L  V  V  W  P  N
         ATTGACTACTATGTTAATCTCAGCAAATTAAGCGATGGTTTTAAACATAACCTTCTTAAAGCTATTCAGGATTTAGTAGTATGGCCAAAT
              26700     26690     26680     26670     26660     26650     26640     26630     26620

Ea59     R  L  A  D  I  E  N  E  S  V  L  N  T  S  L  L  R  G  V  T  L  S  E  I  H  G  Q  F  A  R
         CGATTAGCCGACATTGAAAATGAAAGCGTCCTTAACACCTCATTACTTAGAGGGGTAACTCTTTCAGAAATTCATGGACAGTTCGCACGT
              26610     26600     26590     26580     26570     26560     26550     26540     26530

Ea59     V  L  N  G  L  P  E  L  S  D  F  H  F  S  F  N  R  K  S  A  P  G  F  S  D  L  T  I  P  F
         GTGTTAAATGGTTTGCCAGAATTGTCAGATTTCCACTTTTCATTTAATAGAAAAAGTGCTCCCGGATTCAGTGATTTAACTATACCTTTT
              26520     26510     26500     26490     26480     26470     26460     26450     26440

Ea59     E  V  T  V  N  S  M  P  S  T  N  I  H  A  F  I  G  R  N  G  C  G  K  T  T  I  L  N  G  M
         GAGGTGACGGTTAATTCTATGCCCAGCACGAACATTCATGCTTTTATCGGGCGGAATGGGTGTGGTAAAACAACAATTTTGAATGGAATG
              26430     26420     26410     26400     26390     26380     26370     26360     26350

Ea59     I  G  A  I  T  N  P  E  N  N  E  Y  F  F  S  E  N  N  R  L  I  E  S  R  I  P  K  G  Y  F
         ATTGGTGCAATCACCAACCCAGAAAACAATGAATATTTTTTCTCTGAAAATAATAGACTTATCGAGTCAAGAATCCCAAAGGGATATTTT
              26340     26330     26320     26310     26300     26290     26280     26270     26260

Ea59     R  S  L  V  S  V  S  F  S  A  F  D  P  F  T  P  P  K  E  Q  P  D  P  A  K  G  T  Q  Y  F
         CGATCGCTTGTTTCAGTTTCGTTTAGTGCATTTGATCCTTTTACTCCTCCTAAAGAACAACCTGACCCAGCAAAAGGTACACAATACTTT
              26250     26240     26230     26220     26210     26200     26190     26180     26170

Ea59     Y  I  G  L  K  N  A  A  S  N  S  L  K  S  L  G  D  L  R  L  E  F  I  S  A  F  I  G  C  M
         TATATTGGACTCAAGAATGCTGCCAGCAATAGTTTAAAATCACTAGGCGATCTCCGCTTAGAATTCATTTCAGCATTTATTGGTTGTATG
              26160     26150     26140     26130     26120     26110     26100     26090     26080

Ea59     R  V  D  R  K  R  Q  L  W  L  E  A  I  K  K  L  S  S  D  E  N  F  S  N  M  E  L  I  S  L
         AGAGTAGATAGAAAAAGACAACTCTGGCTTGAAGCTATCAAAAAACTAAGTAGTGATGAAAACTTTTCAAATATGGAACTCATCAGCCTC
              26070     26060     26050     26040     26030     26020     26010     26000     25990

Ea59     I  S  K  Y  E  E  L  R  R  N  E  P  Q  I  Q  V  D  D  D  K  F  T  K  L  F  Y  D  N  I  Q
         ATTTCTAAATATGAAGAGTTAAGACGTAATGAACCACAGATTCAAGTGGACGATGATAAATTCACTAAATTGTTTTATGACAATATCCAG
              25980     25970     25960     25950     25940     25930     25920     25910     25900

Ea59     K  Y  L  L  R  M  S  S  G  H  A  I  V  L  F  T  I  T  R  L  V  D  V  V  G  E  K  S  L  V
         AAATATCTGCTTCGAATGAGCTCTGGACATGCAATTGTTTTATTTACTATCACAAGATTAGTAGATGTCGTTGGCGAAAAGTCATTAGTT
              25890     25880     25870     25860     25850     25840     25830     25820     25810

Ea59     L  F  D  E  P  E  V  H  L  H  P  P  L  L  S  A  F  L  R  T  L  S  D  L  L  D  A  R  N  G
         TTATTCGATGAACCAGAGGTTCATCTGCATCCACCTTTGCTCTCTGCTTTTTTACGAACATTAAGCGACTTACTCGATGCACGCAATGGT
              25800     25790     25780     25770     25760     25750     25740     25730     25720

Ea59     V  A  I  I  A  T  H  S  P  V  V  L  Q  E  V  P  K  S  C  M  W  K  V  L  R  S  R  E  A  I
         GTAGCAATAATTGCAACTCATTCCCCAGTAGTACTGCAAGAGGTTCCAAAATCCTGCATGTGGAAAGTCCTACGGTCAAGAGAAGCAATA
              25710     25700     25690     25680     25670     25660     25650     25640     25630

Ea59     N  I  I  R  P  D  I  E  T  F  G  E  N  L  G  V  L  T  R  E  V  F  L  L  E  V  T  N  S  G
         AATATTATCCGTCCGGATATTGAGACATTCGGTGAGAACTTAGGTGTTTTAACTCGTGAGGTGTTTTTACTTGAAGTGACAAATTCTGGA
              25620     25610     25600     25590     25580     25570     25560     25550     25540

Ea59     Y  H  H  L  L  S  Q  S  V  D  S  E  L  S  Y  E  T  I  L  K  N  Y  N  G  Q  I  G  L  E  G
         TACCACCACTTATTATCGCAGTCCGTTGATTCAGAGCTTTCTTATGAAACCATTCTAAAAAATTATAATGGTCAGATAGGATTAGAAGGT
              25530     25520     25510     25500     25490     25480     25470     25460     25450

Ea59     R  T  V  L  K  A  M  I  M  N  R  D  E  G  K  V  Q  *
Ea31                                                        M  K  K  L  P  L  P  A  R  T  Y  S  E  M
         CGAACCGTTTTAAAAGCGATGATAATGAACAGAGATGAAGGTAAAGTACAATGAAAAAACTACCTCTTCCAGCGAGAACTTATAGCGAAA
              25440     25430     25420     25410     25400     25390     25380     25370     25360

Ea31       L  N  K  C  S  E  G  M  M  Q  I  N  V  R  N  N  F  I  T  H  F  P  T  F  L  Q  K  E  Q  Q
         TGCTTAATAAATGCTCGGAAGGTATGATGCAGATAAATGTTAGAAATAATTTCATTACTCACTTCCCCACTTTTTTGCAGAAAGAACAAC
              25350     25340     25330     25320     25310     25300     25290     25280     25270

Ea31       Y  R  I  L  S  S  T  G  Q  L  F  T  Y  D  R  T  H  P  L  E  P  T  T  L  V  V  G  N  L  T
         AATATAGAATATTAAGCTCGACAGGTCAGTTATTTACCTACGACAGGACACACCCTCTTGAGCCTACAACCTTAGTAGTTGGTAACCTGA
              25260     25250     25240     25230     25220     25210     25200     25190     25180

Ea31       K  V  K  L  E  K  L  Y  E  N  N  L  R  D  K  N  K  P  A  R  T  Y  Y  D  D  M  L  V  S  S
         CAAAGGTTAAATTAGAAAAGCTTTATGAAAATAATCTCCGAGATAAAAACAAACCCGCTAGAACATATTACGATGACATGCTTGTTTCAT
              25170     25160     25150     25140     25130     25120     25110     25100     25090

Ea31       G  E  K  C  P  F  C  G  D  I  G  Q  T  K  N  I  D  H  F  L  P  I  A  H  Y  P  E  F  S  V
         CAGGTGAAAAAATGTCCATTTTGTGGTGATATAGGACAGACAAAAAATATAGATCATTTTCTTCCTATTGCACATTATCCTGAATTTTCGG
              25080     25070     25060     25050     25040     25030     25020     25010     25000

Ea31       M  P  I  N  L  V  P  S  C  R  D  C  N  M  G  E  K  G  Q  V  F  A  V  D  E  V  H  Q  A  I
         TGATGCCTATTAATTTAGTTCCATCGTGCCGCGACTGCAATATGGGAGAGAAAGGTCAAGTTTTCGCAGTAGATGAGGTACACCAAGCGA
              24990     24980     24970     24960     24950     24940     24930     24920     24910
```

```
Ea31       H  P  Y  I  D  K  D  I  F  F  R  E  Q  W  V  Y  A  N  F  V  S  G  T  P  G  A  I  S  F  Y
        TTCATCCCTATATCGACAAGGACATTTTTTTTCGTGAGCAATGGGTATATGCAAATTTCGTTTCCGGAACTCCGGGTGCTATCAGTTTTT
                 24900     24890     24880     24870     24860     24850     24840     24830      24820

Ea31       V  E  C  P  A  N  W  R  Q  E  D  K  H  R  A  L  H  H  F  K  L  L  N  I  A  N  R  Y  R  L
        ATGTTGAATGCCCGGCGAACTGGAGGCAGGAAGACAAACACAGAGCTCTTCATCATTTCAAGCTATTAAATATTGCTAACAGGTATCGTT
                 24810     24800     24790     24780     24770     24760     24750     24740      24730

Ea31       E  A  G  K  H  L  S  E  V  I  T  Q  R  N  S  F  V  K  V  I  R  K  Y  S  S  T  A  T  F  Q
        TGGAGGCAGGGAAGCACTTGAGTGAAGTGATTACTCAAAGAAACTCTTTCGTAAAAGTTATAAGGAAATATAGTTCAACCGCAACGTTTC
                 24720     24710     24700     24690     24680     24670     24660     24650      24640

Ea31       Q  L  Q  S  E  F  I  E  A  N  L  K  P  I  I  D  L  N  D  F  P  N  Y  W  K  R  V  M  Y  Q
        AGCAGCTACAGTCAGAATTTATTGAAGCAAATCTGAAACCTATTATAGATTTGAATGACTTCCCCAATTATTGGAAAAGAGTTATGTATC
                 24630     24620     24610     24600     24590     24580     24570     24560      24550

Ea31       C  L  A  N  S  E  D  F  F  R  G  I  *
        AGTGCCTAGCAAACTCGGAAGATTTTTTCAGAGGGATCTAGAATATGATGAAAGATAGAAAATTACGACGCTTATCGGAAGTGAACGAAT
                 24540     24530     24520     24510     24500     24490     24480     24470      24460

        ACTTTTTATATGAGGAGGGCTGTTTTTACAAAATCCGGTAGTAACTTGCTAACCAATTCCTAGGCAGGTCATTGGCAACAGTGGCATGCA
                 24450     24440     24430     24420     24410     24400     24390     24380      24370

        CCGAGAAGGACGTTTGTAATGTCCGCTCCGGCACATAGCAGTCCTAGGGACAGTGGCGTACAGTCATAGATGGTCGGTGGGAGGTGGTAC
                 24360     24350     24340     24330     24320     24310     24300     24290      24280

        AAATTCTCTCATGCAAAAAATATGTAAAATCGGTAGCAACTGGAAATCATTCAACACCCGCACTATCGGAAGTTCACCAGCCAGCCGCAG
                 24270     24260     24250     24240     24230     24220     24210     24200      24190

        CACGTTCCTGCATACGACGTGTCTGCGGCTCTACCATATCTCCTATGAGCAACGTGTTAGCAGAGCCAAGCCACAACTCTAATTTTAATA
                 24180     24170     24160     24150     24140     24130     24120     24110      24100

        CATAATGAATGATAATAATAATATTAAAAATTTCCTGTGTAACTAATTTACTATATGGTTTCTGATAAGAATCATTGCAAAGATCAAACA
                 24090     24080     24070     24060     24050     24040     24030     24020      24010

        ACTTGTATTACATTGACAGTTAAGCAGTTAATTTTATCACCTCTAAAATATATCAGCATCTAGCATGCAACCTATCAAAATGGAGAGTTT
                 24000     23990     23980     23970     23960     23950     23940     23930      23920

Ea47       M  T  K  K  P  W  E  R  R  L  K  D  L  S  H  L  L  K  C  C  I  D  T  Y  F  D  P  E  L  F
        TATGACTAAAAAACCATGGGAAAGAAGACTTAAAGATTTATCGCACTTGCTCAAATGCTGCATTGATACATATTTTGACCCTGAATTATT
                 23910     23900     23890     23880     23870     23860     23850     23840      23830

Ea47       R  L  N  L  N  Q  F  L  Q  T  A  R  T  V  T  F  I  I  Q  K  N  K  N  Q  I  I  G  Y  D  I
        TCGCTTGAATTTGAATCAATTCCTCCAAACCGCAAGAACAGTAACATTTATTATTCAAAAAAACAAAAACCAGATTATAGGATATGACAT
                 23820     23810     23800     23790     23780     23770     23760     23750      23740

Ea47       W  Y  N  N  N  V  I  E  K  W  K  N  D  P  L  M  A  W  A  K  N  S  R  N  T  I  E  K  Q  G
        TTGGTATAACAATAATGTTATTGAAAAATGGAAAAATGATCCATTAATGGCTTGGGCTAAAAATTCTCGCAATACGATAGAAAAACAAGG
                 23730     23720     23710     23700     23690     23680     23670     23660      23650

Ea47       D  L  E  M  Y  S  E  A  K  A  T  L  I  S  S  Y  I  E  E  N  D  I  E  F  I  T  N  E  S  M
        CGATTTAGAAATGTATAGCGAGGCAAAGGCTACTCTTATTTCATCTTACATTGAAGAAAATGACATTGAGTTTATTACAAATGAAAGTAT
                 23640     23630     23620     23610     23600     23590     23580     23570      23560

Ea47       L  N  I  G  I  K  K  L  V  R  L  A  Q  K  K  L  P  S  Y  L  T  E  S  S  I  I  K  S  E  R
        GTTAAACATTGGTATAAAAAAGTTAGTCAGACTTGCACAAAAGAAATTACCTTCATATTTAACTGAATCATCTATTATTAAATCAGAAAG
                 23550     23540     23530     23520     23510     23500     23490     23480      23470

Ea47       R  W  V  A  N  T  L  K  D  Y  E  L  L  H  A  L  A  I  I  Y  G  R  M  Y  N  C  C  N  S  L
        ACGATGGGTCGCTAATACGCTAAAAGATTACGAATTATTACATGCCTTAGCTATAATCTATGGCAGAATGTATAACTGCTGTAACTCTCT
                 23460     23450     23440     23430     23420     23410     23400     23390      23380

Ea47       G  I  Q  I  N  N  P  M  G  D  D  V  I  S  P  T  S  F  D  S  L  F  D  E  A  R  R  I  T  Y
        TGGCATACAAATAAACAATCCAATGGGTGACGATGTGATTTCGCCAACATCATTCGACTCTTTATTTGATGAAGCCAGGAGAATAACTTA
                 23370     23360     23350     23340     23330     23320     23310     23300      23290

Ea47       L  K  L  K  D  Y  S  I  S  K  L  S  F  S  M  I  Q  Y  D  N  K  I  I  P  E  D  I  K  E  R
        TTTAAAATTAAAAGATTACTCCATAAGCAAATTGTCATTTAGCATGATACAATATGACAATAAAATAATTCCTGAAGATATTAAAGAGCG
                 23280     23270     23260     23250     23240     23230     23220     23210      23200

Ea47       L  K  L  V  D  K  P  K  N  I  T  S  T  E  E  L  V  D  Y  T  A  K  L  A  E  T  T  F  L  K
        TCTAAAACTGGTAGATAAGCCTAAAAATATCACTTCGACAGAAGAGTTAGTTGACTATACAGCCAAGCTTGCAGAAACGACTTTTTTAAA
                 23190     23180     23170     23160     23150     23140     23130     23120      23110

Ea47       D  G  Y  H  I  Q  T  L  I  F  Y  D  K  Q  F  H  P  I  D  L  I  N  T  T  F  E  D  Q  A  D
        GGACGGTTATCACATTCAAACATTAATTTTTTATGATAAACAATTCCATCCAATTGATTTAATCAATACAACATTTGAAGATCAAGCAGA
                 23100     23090     23080     23070     23060     23050     23040     23030      23020

Ea47       K  Y  I  F  W  R  Y  A  A  D  R  A  K  I  T  N  A  Y  G  F  I  W  I  S  E  L  W  L  R  K
        TAAATATATTTTTTGGCGTTATGCAGCTGACAGAGCCAAAATAACAAATGCCTATGGCTTCATTTGGATATCAGAGCTATGGCTCAGAAA
                 23010     23000     22990     22980     22970     22960     22950     22940      22930

Ea47       A  S  I  Y  S  N  K  P  I  H  T  M  P  I  I  D  E  R  L  Q  V  I  G  I  D  S  N  N  N  Q
        AGCAAGCATCTACTCCAATAAACCAATACATACAATGCCAATTATAGATGAAAGACTTCAGGTAATTGGAATTGATTCAAATAATAATCA
                 22920     22910     22900     22890     22880     22870     22860     22850      22840

Ea47       K  C  I  S  W  K  I  V  R  E  N  E  E  K  K  P  T  L  E  I  S  T  A  D  S  K  H  D  E  K
        AAAATGTATTTCATGGAAAATAGTTAGAGAAAACGAAGAAAAAAAACCGACTTTAGAAATATCAACAGCAGACTCAAAACATGACGAAAA
                 22830     22820     22810     22800     22790     22780     22770     22760      22750

Ea47       P  Y  F  M  R  S  V  L  K  A  I  G  G  D  V  N  T  M  N  N  *
        ACCATATTTCATGCGTTCAGTCTTAAAAGCAATTGGCGGTGATGTAAACACTATGAACAATTGAGTCATAGAACTTCCATTATTCTCCTG
                 22740     22730     22720     22710     22700     22690     22680     22670      22660

        AAGATAATAATCGCCAAATAAACCAATACTCAGCTTTACAATATACTAACTAACCGCAGAACGTTATTTCATACAACGTTTCTGCGGCAT
                 22650     22640     22630     22620     22610     22600     22590     22580      22570

        ATCACAAAACGATTACTCCATAACAGGGACAGCAGGCCACTCAATATCAGGTGCAGTTGATGTATCA
                 22560     22550     22540     22530     22520     22510     22500     22490      22480
```

As a further identification of reading frames, we have used the computer method of Staden & McLachlan (1982), which calculates the frequency with which codons occur in the different reading frames. The method is now contained in a more general analytical program called ANALYSEQ. Known genes are chosen and the frequency of occurrence of the different codons calculated and used as a standard. A region of the sequence is then scanned and the frequency of codons in each reading frame calculated and compared with that of the known genes. The results are expressed as "probability" that the reading frame is coding for a protein, assuming that coding regions will have a similar codon distribution. Figure 3 shows results using ANALYSEQ on the lambda sequences. In these diagrams the probability is plotted against the position in the sequence for the three different reading frames. Thus the presence of a coding region is indicated by a peak in a particular reading frame. In most cases the results are quite clear and confirm the expected positions of the genes. We have also used these diagrams to help identify the position of protein initiation. At the beginning of the coding region there is usually a sharp rise in the probability curve and we have measured the positions where this crosses the 50% probability mark. These figures are given in Table 3. It appears that the distribution of codons differs in the different transcription units of lambda (see Table 5). Thus, when the ANALYSEQ results were calculated using as standard a gene from the left-hand end of the sequence (e.g. gene *J*), they were clear for other genes in the left-hand end but less so for those in the right-hand end; and, conversely, good results were obtained for the right-hand end when the standard was from that end (e.g. genes *O* and *P* or *exo* and *β*). Differences between the two ends of lambda have been noted previously, particularly in the nucleotide composition, the left-hand end being G+C-rich and the right relatively A+T-rich.

(c) *Protein initiation sites*

Table 3 lists the proposed start points of the gene products. In many cases they are not rigorously established and depend largely on choosing an appropriate AGT or GTG preceded by the best "Shine and Dalgarno" (SD) sequence (Shine & Dalgarno, 1974). Various attempts to define more precisely the nucleotide sequence important for initiation in the vicinity of initiation sites have met with only limited success. In an extensive study on 124 initiation sites, Stormo *et al.* (1982) have shown that other sequences both before and after the ATG may contribute to the formation of an initiation signal and have suggested additional rules that may be used to identify sites. These "Stormo" rules have also been taken into account in this work.

An interesting feature of the lambda sequence is the frequent occurrence of overlapping termination and initiation sites, particularly in the sequence A-T-G-A, where the TGA is the termination codon for one reading frame and ATG the initiation of the next. These structures are listed in Table 4. The initiators are usually preceded by an SD sequence, but more characteristic is a very purine-rich region separated from the ATG or GTG by a few nucleotides. This effect is particularly marked in the *nin* region (see below), where there is a continuous series of 13 overlapping reading frames. It is not known what the function of this

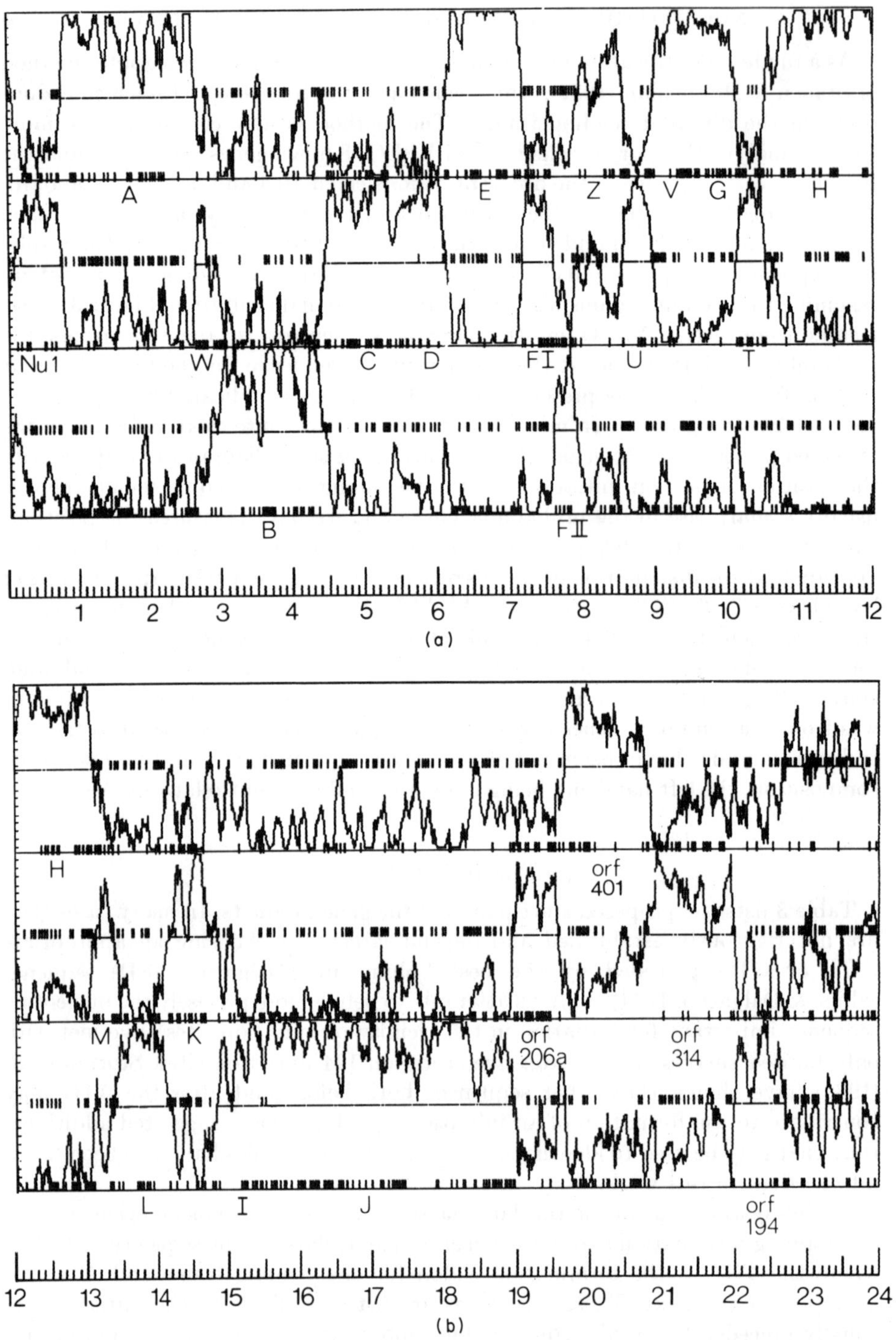

FIG. 3. Gene predictions for the DNA of lambda using the codon preference method of Staden & McLachlan (1982), which is one function of ANALYSEQ, a general sequence analysis program. The x axis represents sections of the lambda sequence (Figs 1 and 2) and the probability of coding is plotted in the y direction. The method assumes that the codon preferences for neighbouring genes are similar, and hence that the codon usage of some of the known genes can be used as standard.

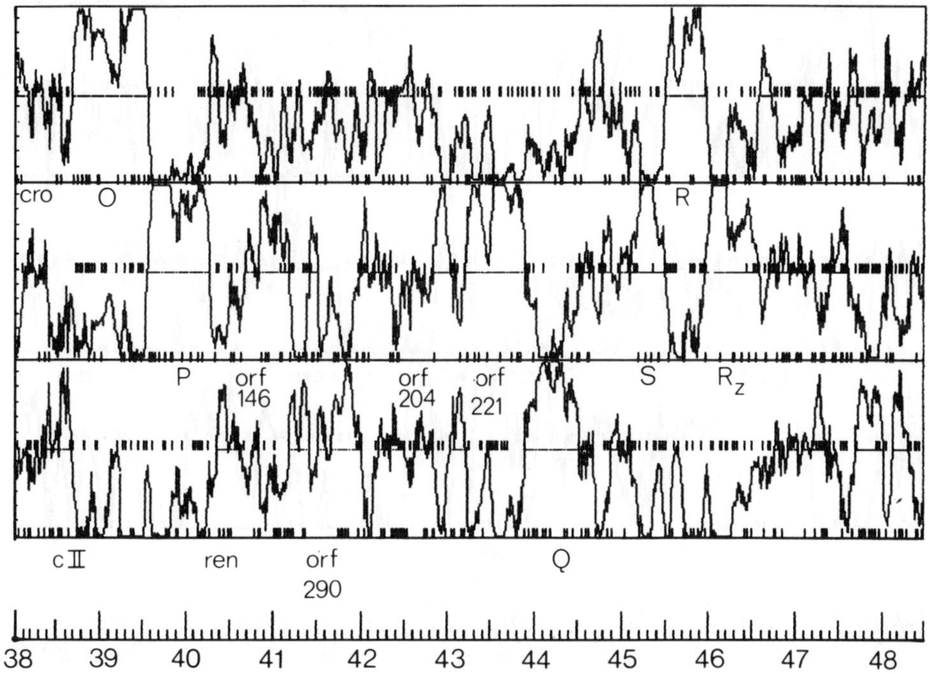

(c)

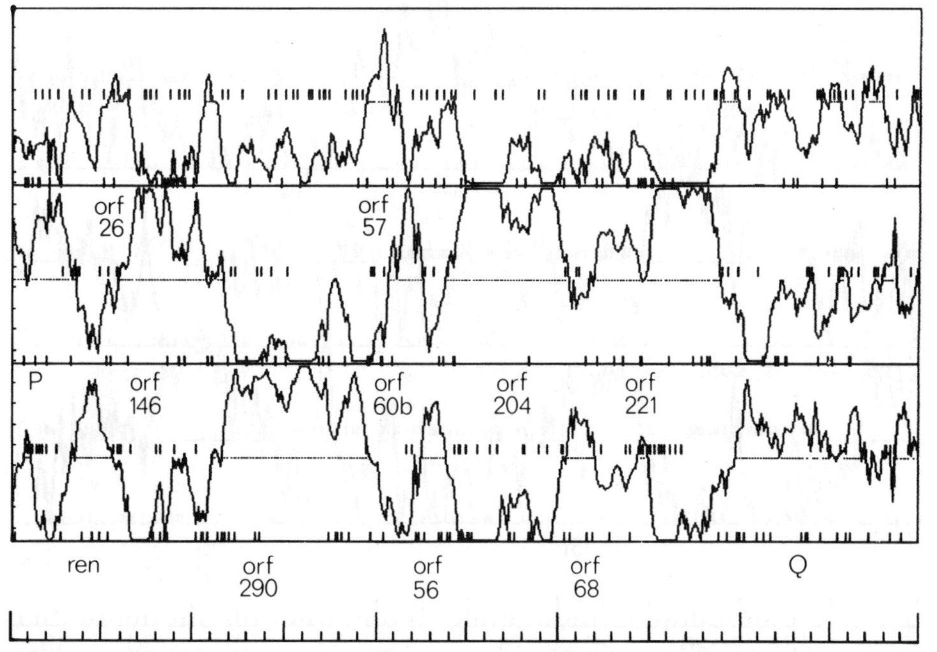

(d)

Probabilities of coding are calculated by sliding a window of 35 codons along the sequence, one codon at a time. For every position of the window, the codons found in each of the 3 reading frames are compared with those in the standard and the corresponding probabilities of coding calculated. The probabilities for each of the 3 frames have been plotted, one above the other, every 5 codons. A solid horizontal line at the mid-point of a reading frame (this is at the 50% level of probability) indicates

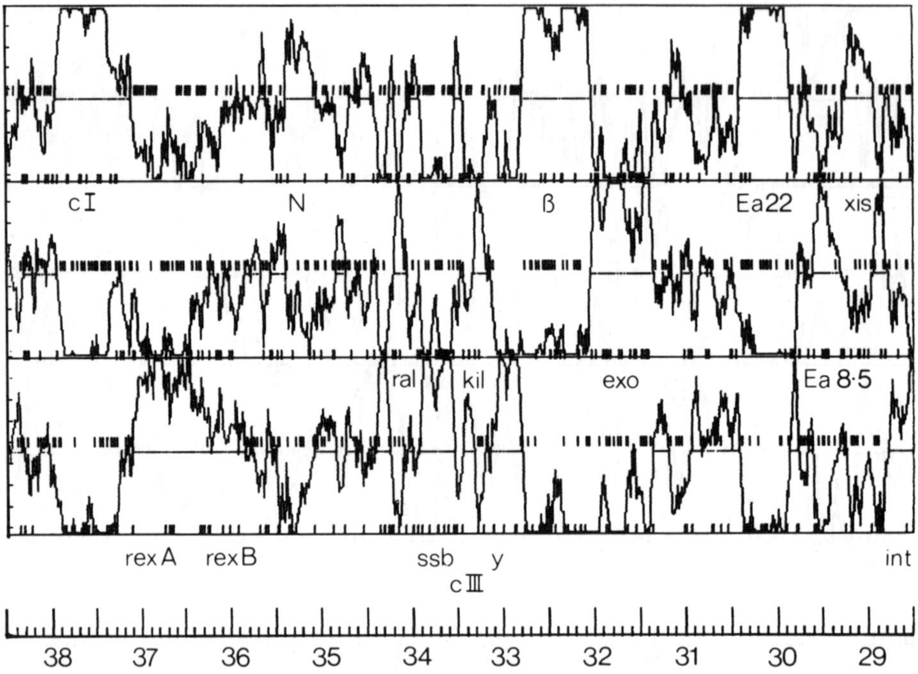

(e)

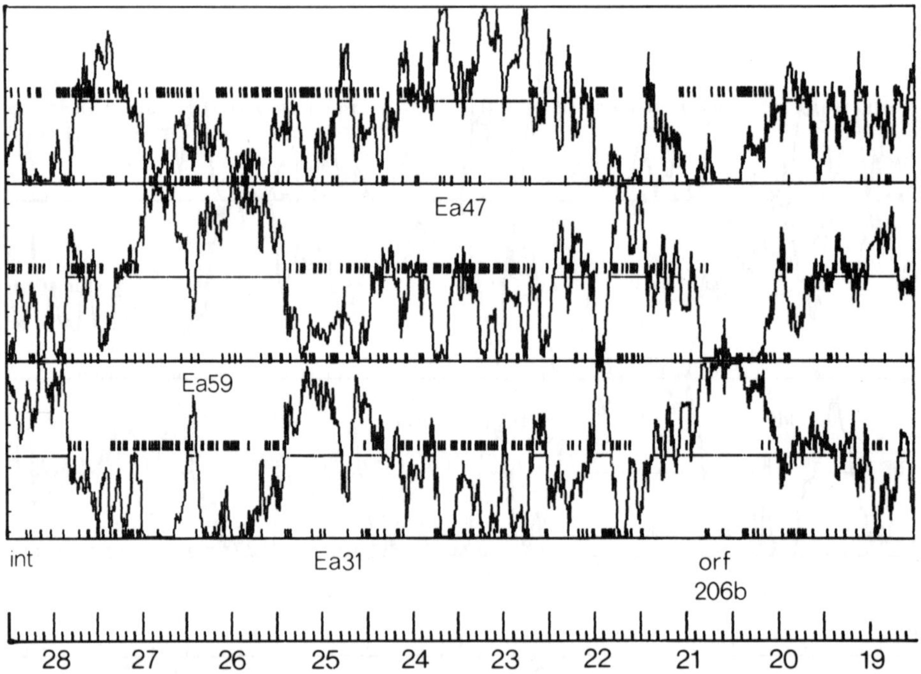

(f)

which of the 3 frames is most likely to be coding. The initiation codons ATG, GTG are marked as vertical bars along the base of each plot and the termination codons as vertical bars along the 50% level. (a) and (b) The left arm; genes *A*, *B*, *C*, *D*, *E*, *FI*, *V*, *H*, *L*, *J* used as standards; (c) the right arm, using genes *O*, *P*, *Q*, *R* as standards; (d) positions 40,000 to 45,000 of the right arm using reading frames *orf146*, *orf290* and *orf204* as standards; (e) and (f) the central region using genes *cI*, *rexA*, *ssb*, γ, β, *exo*, *Ea59* and *Ea31* as standards.

TABLE 3

Most probable initiation sites for the λ proteins

Gene	Proposed start From sequence	From ANALYSEQ	Shine & Dalgarno sequence†	Status (also see the text)
Nu1	191	210	G-G-A(7)ATG	One of 2 possible
A	711	770	G–G-G-T(8)GTG	Uncertain
W	2633	2660	G-G-A-G(8)ATG	Established by amber mutation
B	2836	2880	A-G-G-A(8)ATG	One of 2 possible
C	4418	4430	A-G-G-A-G-G(8)GTG	Best of 6 possibilities
D	5747	—	A-A-G-G(6)ATG	Amino acid sequence
E	6135	6150	G-G-A(7)AGT	Amino acid sequence
FI	7202	7170	G-G-A-G-G(7)ATG	Most probable of 2
FII	7612	7620	G-G-A-G-G(6)GTG	Most probable of 2
Z	7977	7960	A-A-G-G–G-G(6)ATG	Most probable of 3
U	8552	8550	G-G-A(13)ATG	Only possible
V	8955	8970	G-A-G-G(7)ATG	Amino acid sequence
G	9711	—	G-G-A-G(9)ATG	Most probable of 2
T	10,115	10,110	G-G(6)GTG	Uncertain
H	10,542	10,520	A-G-G-A-G-G(10)ATG	Amino acid sequence
M	13,100	13,090–13,160	G-G-A-G-G-T(8)ATG	Most probable of 5
L	13,429	13,410	A-G-G-G-T-T(9)AGT	Only probable
K	14,276	14,250	G-G-A-G-G(12)ATG	Uncertain
I	14,773	14,860	G-A-G-G-T(13)ATG	Most probable of 3
J	15,505	—	G-G-A-G(9)ATG	Most probable
orf206a	18,965	18,960	G-A-G-G-T(8)ATG	Most probable of 3
orf401	19,650	19,660	G-G–G(6)ATG	Most probable
orf314	21,029	20,845	A-G(12)ATG	Most probable (see the text)
orf194	21,937	21,990	G-A-G-G(10)ATG	Only probable
cro	38,041	37,980	A-A-G-G-A-G-G-T(6)ATG	Schwarz *et al.* (1978); Roberts *et al.* (1977)
cII	38,360	38,370	A-A-G-G-A(11)ATG	Schwarz *et al.* (1978); Roberts *et al.* (1977)
O	38,686	38,680	A-G-G-A-G(6)ATG	Schwarz *et al.* (1978)
P	39,582	39,560	G–G-G-T(9)ATG	Schwarz *et al.* (1980)
ren	40,280	40,370	A-A-G-G-A-G(8)ATG	Only probable one
nin region			See the text and Table 4	
Q	43,886	43,910	G-G-A-G(8)ATG	Daniels & Blattner (1982); Petrov *et al.* (1981)
orf64	44,621	–	G–G-G-T(9)ATG	Daniels & Blattner (1982)
S	45,186	45,130	G-G–G-G-T(8)ATG	Daniels & Blattner (1982)
R	45,493	45,490	G-G-A-G(7)ATG	Amino acid sequence (Imada & Tsugita, 1971)
R_z	45,966	45,960	G-A(7)ATG	Most probable
cI	37,940	37,950	‡	Ptashne *et al.* (1976)
rexA	37,114	37,100	G-G-A(7)ATG	Only possible Landsmann *et al.* (1982)
rexB	36,259	–	A-G-G-A-G(6)AGT	One of 2 possible Landsmann *et al.* (1982)
N	35,438	35,390	G-G-A(10)ATG	Franklin & Bennett (1979)
ral	34,287	34,210	A-G-G-A(8)ATG	Ineichen *et al.* (1981)
ssb	33,904	33,930	A-G-G-A(9)ATG	Ineichen *et al.* (1981)
cIII	33,463	33,460	A-A-G-G-A-G(7)ATG	See the text
kil	33,330	33,340	A-G-G-A-G(10)ATG	See the text
γ	33,112	33,100	A-G-G-A-G(8)ATG	See the text

TABLE 3 (*continued*)

Gene	Proposed start From sequence	From ANALYSEQ	Shine & Dalgarno sequence†	Status (also see the text)
β	32,810	32,800	G-A-G-G(12)AGT	Only probable
exo	32,028	32,040	A-A-G-G–G-G(6)ATG	Most probable
Ea22	30,395	30,430	G-A-G-G(8)GTG	Only probable
Ea8·5	29,655	29,600	A-A-G-G-A-G(8)ATG	Only probable
xis	29,078	29,250	G-G-A-G(9)ATG	Hoess *et al.* (1980)
int	28,882	28,780	A-G-G-A(6)ATG	Hoess *et al.* (1980)
Ea59	26,973	27,040	A-A-G-G(7)ATG	Only probable
Ea31	25,399	25,410	A-A-G-G(8)ATG	One of 2 possible
Ea47	23,918	23,750	G-G-A-G(7)ATG	Most probable
orf206b	20,767	20,900	A-A-G(8)GTG	See the text

† The figures in parentheses are the number of nucleotides between the first residue of the initiating ATG or GTG and the A of the G-G-A-G-G in the SD sequence, or the corresponding position in the SD sequence if there is no A there.

‡ The initiating ATG for the *cI* protein is at the 5′ terminus of the mRNA when the mRNA is transcribed from $p_{M'}$ and consequently has no SD sequence.

characteristic overlapping structure is, though it is tempting to suggest that it could allow the ribosome to read through from one frame to the next. Oppenheim & Yanofsky (1980) showed that there was "translational coupling" between the *trpD* and *trpE* genes, which are connected by an overlapping termination and initiating codon in the sequence T-G-A-T-G (i.e. that the expression of *trpD* was dependent on the expression of *trpE*) and suggested that this effect may be due to a ribosome, or a component of it that terminates on the first reading frame, being able to initiate more efficiently on the second one than a free ribosome. We have no evidence as to whether there is translational coupling between the lambda proteins that are connected by overlapping termination and initiation codons, but it is certainly an attractive hypothesis.

(d) *Sizes of the proteins*

Table 2 shows the molecular weights of the various proteins calculated from the most probable reading frames shown in Figure 2. These are compared with values found previously by sodium dodecyl sulphate/acrylamide gel electrophoresis. The agreement seems reasonably good. The most notable exception is *G*, which is discussed below. There is a similar discrepancy in the case of the *xis* gene (Hoess *et al.*, 1980), where a molecular weight of 32,000 was suggested by acrylamide gel electrophoresis (Hendrix, 1971). In this case there seems to be ample proof for the smaller value obtained from the DNA sequence.

(e) *The codon CUA*

An interesting feature of the left arm is that the codon CUA is completely absent from its 7476 codons. This is not the case in the other regions, where there are 40 CUAs in 6220 codons, or in other *E. coli* systems. CUA is a leucine codon and is

TABLE 4

Overlapping initiation and termination sites of λ proteins

Terminator	Initiator	Purine content†	Gap size‡	Structure§
A	*W*	10/11	0	ATGA
W	*B*	11/15	8	ATGA
B	*C*	14/14	0	GTG(14)TAA
Z	*U*	13/16	2	ATGA
G	*T*	6/6	0	GTG(13)TGA
T	*H*	8/8	6	ATG(2)TGA
H	*M*	8/9	2	ATGA
M	*L*	11/14	3	TGATG
K	*I*	14/19	0	ATG(97)TGA
orf314	*orf194*	7/12	0	TAATG
O	*P*	7/8	3	ATGA
P	*ren*	11/12	3	ATGA
ren	*orf26*	9/11	0	ATGA
orf26	*orf146*	14/14	2	ATGA
orf146	*orf290*	25/29	4	ATGA
orf290	*orf57*	10/10	4	ATGATG‖
orf57	*orf60b*	13/13	3	GTG(28)TAA
orf60b	*orf56*	11/12	0	GTGA
orf56	*orf204*	10/10	0	ATGATG(2)TAA‖
orf204	*orf68*	15/18	4	ATGA
orf68	*orf221*	14/16	0	ATG(17)TGA
orf221	*Q*	10/10	3	ATGA
S	*R*	16/19	0	ATG(11)TAA
R	R_Z	9/11	3	ATGA
β	*exo*	14/17	1	ATGA
exo	*orf60a*	10/12	2	ATGA
orf60a	*orf63*	21/25	4	ATG(22)TAA
orf63	*orf61*	12/14	0	ATG(14)TGA
xis	*int*	12/14	0	ATG(17)TGA
Ea59	*Ea31*	14/16	4	ATGA

† Number of purine residues/total nucleotides in purine-rich region preceding the initiator.

‡ Number of nucleotides between the purine-rich sequence and the initiator.

§ The initiator is underlined and the terminator overlined. Where appropriate, the number of nucleotides between the initiator and the terminator is shown in parentheses. The hyphens have been omitted for clarity.

‖ These sequences have 2 consecutive ATG triplets, either of which could be the initiator.

recognised by a unique tRNA, which is present in a relatively low concentration (Ikemura, 1981*b*). It seems that this unusual feature must have some biological significance. Possibly, at the late stages of infection CUA becomes a limiting codon so that the structural proteins can be synthesized at the expense of other proteins of the bacteriophage or host that contain CUA codons. Such a situation could be achieved if the CUA tRNA were inactivated by one of the late proteins of lambda and it would be interesting to know if this is indeed the case. In this connection, Yudelevich (1971) has shown that a different leucine tRNA, which recognizes the codon CUG, is split into two fragments by an enzyme produced by bacteriophage T4 when it infects the cell.

(f) *Yield of the protein products*

There are probably 25 proteins coded for by the left arm of lambda (positions 1 to 22,560). These are produced in very different amounts (Murialdo & Siminovitch, 1972; Hendrix, 1971) but are probably made from a single transcriptional unit initiated at the late promoter, p'_R. Ray & Pearson (1974,1975) have produced evidence that the various regions of the RNA are present in equal amounts during the production of the proteins; therefore, the varying yield of proteins is probably due to differences in translation. There is no obvious relationship between the nature of the initiation sites and the yield of proteins. This does not of course preclude the possibility that initiation is involved, as the characteristics of the sites are still not well-understood and it may be that secondary structures are concerned in the control of initiation (Iserentant & Fiers, 1980), although this is not apparent from the primary structure. Another possible mechanism of translational control may be the frequency of certain codons that correspond to minor transfer RNAs and may limit the rate of protein synthesis (Ikemura, 1981*a*,*b*). Table 5 lists the frequency of rare codons in the proteins of the left arm and the approximate yield of some of these proteins (Murialdo & Siminovitch, 1972). It can be seen that there is a general correlation; the codons are more frequent in proteins that are produced in small amounts than in the major components. The figures are probably not accurate enough to define an exact relationship, but it seems likely that this may be an important mechanism for translation control.

As mentioned above, the two arms have different codon distributions and the rarer codons also differ. Only CTA and ATA seem to be rare in both. In the right arm, the rare codons seem to be related mainly to its being an A+T-rich region. For instance, the rarest codon is CCC, and UUG is considerably rarer than UUA, whereas the reverse is true in the left arm.

(g) *Possible secondary structures*

Although the new sequences reported here have not been rigorously searched for possible secondary structures, it was noticed that there are a number of possible looped structures in the left arm occurring in intergenic regions before the initiation sites of certain proteins. These are shown in Figure 4. Some of these (preceding *E*, *J*, *orf401*; Fig. 1) are followed by a run of T residues, which is a characteristic of transcription termination sites. These would seem to be unlikely places for such sites, since all the proteins are believed to be translated from one long mRNA initiated at the p'_R promoter, and in any case they would probably be anti-terminated by the *Q* protein (Schechtman *et al.*, 1980). Another possibility is that these loops may be sites for RNA processing by ribonuclease III. It is interesting that four of these loops (*E*, *Z*, *J*, *orf401*) contain the sequence C-C-G-C-C.

TABLE 5

Frequency of certain rare codons in the left arm of λ

Gene	Frequency of codon per 100							Relative concentration†
	AUA	UUA	CAA	CGA	CGG	AG^{A}_{G}	UCA	
Nu1	1·10	0	0	1·65	2·75	1·10	0·55	
A	0·62	1·16	0	0·31	1·09	0·62	0·47	1
W	0	0	0	4·3	1·45	2·9	0	
B	0·56	0·38	0·19	0·56	2·4	0·38	0·93	
C	0·24	0·24	0·48	0·48	1·7	0	0·73	
Nu3‡	0	0	0·57	0·57	0	0	1·7	230
D	0	0	0	0	0·9	0	0	870
E	0	0	0·88	0	0	0	0·58	470
FI	0	0	1·5	0	0	0·76	0·76	360 (*FI* and *FII*)
FII	1·7	0	0	0	4·2	0	1·7	
Z	2·1	0	1·6	0·52	3·1	3·1	1·6	
U	0	0	2·3	0	1·5	0	2·3	375
V	0	0	0	0	0	0	0·41	265
G	0	0	0	0	2·1	0·71	2·1	18
T	0	0	0	0·69	1·4	0	1·4	
H	0·12	0	0·59	0·35	1·6	0·70	0·59	12
M	0·92	0	0	0·92	1·8	0·92	0	
L	0	0	0	0·43	0·86	0	0	44
K	0	0	0·50	1·0	3·0	0·50	1·5	14
I	0·45	1·3	0·45	0·45	1·3	0·90	2·2	
J	0·88	0·09	0·26	0·26	1·1	0·71	0·35	13
orf206a	0·49	0·49	0·97	0	0·97	0·49	0·49	
orf401	0·75	0	0·75	0·25	0·75	1·2	5·7	
orf314	0	1·3	0·32	0·32	0	0·96	1·6	
orf194	0·52	1·5	0·52	0	2·5	0·52	0·52	

	AUA	UUA	CAA	CGA	CGG	AG^{A}_{G}	UCA	CUA
Total for left arm	0·45	0·21	0·38	0·38	1·38	0·62	1·06	0
Total for central region	1·91	2·35	1·41	0·72	0·44	2·43	2·07	0·58
Total for right arm	0·65	0·77	1·77	1·50	0·88	2·27	1·12	0·73

† Relative number of copies of protein in cells 45 min after infection with lambda (Murialdo & Siminovitch, 1972).

‡ Assuming initiation at position 5132.

(h) *The reading frames*

In this section we discuss certain features of some of the individual reading frames. Other aspects of the sequence will be discussed in a separate paper (Daniels *et al.*, 1982).

Nu1

There are two possible starts at positions 191 and 269. Both have satisfactory SD sequences that satisfy Stormo rule 2D. We have chosen the former from the ANALYSEQ result.

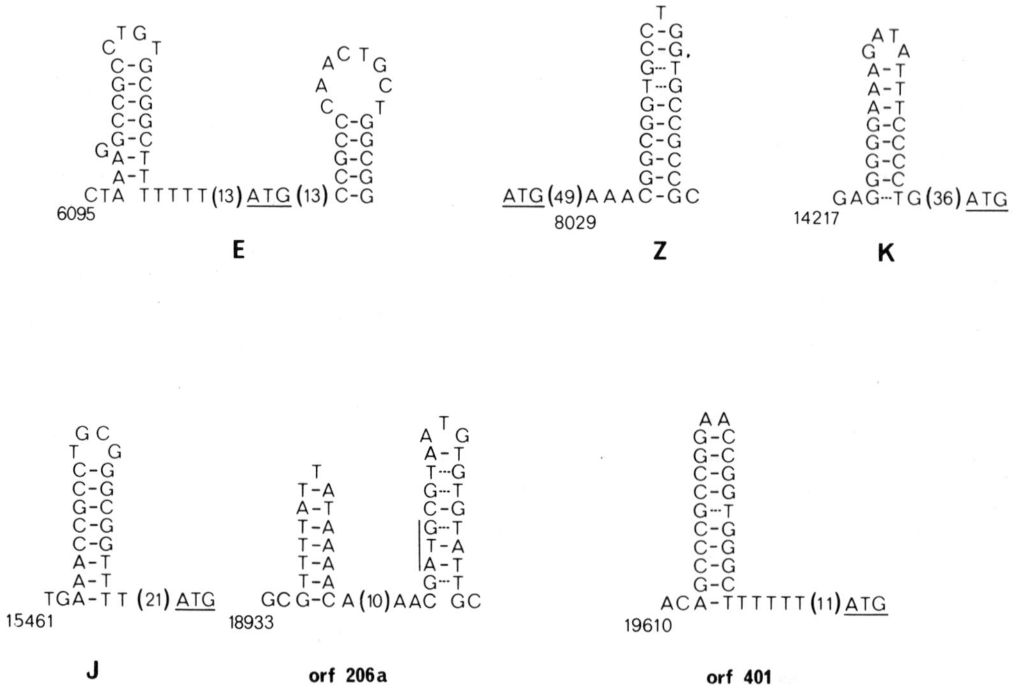

FIG. 4. Possible looped structures found near the initiating sites of some of the genes in the left arm of lambda. The initiating ATG is indicated by underlining and its relative position is shown by the numbers of nucleotides in parentheses. Base-pairing is indicated by dashes and sequence hyphens have been omitted for clarity.

A

There are five possible sites. The ATG triplets at positions 891 and 900 have very poor SD sequences, whereas the GTG triplets at 930 and 933 have good ones. However, the ANALYSEQ results suggest that this is a coding frame back to 770 and there is a GTG at position 771 preceded by G-G-G-G-T, which is probably an acceptable SD sequence involving a single G · U base-pair.

W

G. F. Hong (unpublished results) has shown that a strain of lambda with a mutation in gene *W* has T at position 2642 instead of the normal C. This establishes the ATG at 2633 as the only possible start.

B

The start of the main protein product *B** has been shown by amino acid sequencing to be at position 2902 (Walker *et al.*, 1982). This is a processed protein and so the initiation site must precede this. There are two possible ATG triplets at 2836 and 2878, both of which have SD sequences. That at 2836 is better and obeys Stormo rule 2B. It also overlaps with the termination codon of the previous protein (*W*) in the sequence A-T-G-A. This seems to be a feature of several of the lambda protein starts, though in this case it is not preceded by the usual purine-rich sequence (see Table 4).

C

The GTG at position 4418 is prefered to five other possible start sites as it has a much better SD sequence. It also overlaps the end of the *B* protein and is preceded by a run of 14 purines. It is present in the sequence G-T-G-A but, unlike most other sequences of this type (see Table 4), the TGA is not the terminator of the previous protein. The size of this protein is considerably smaller than that estimated by acrylamide gel electrophoresis (see Table 2). Even if the first possible GTG (position 4283) were used, the molecular weight would be only 48,475.

Nu3

Nu3 is believed to be a protein of 19,000 M_r initiating within the *C* gene and in the same reading frame. There are a number of ATG and GTG triplets that would give a protein of about this size, but none has a satisfactory SD sequence. Better sites are at positions 5219 and 5342: they would give proteins of 29,555 and 25,255 M_r, respectively.

E

In spite of being one of the major protein components, it has a rather poor SD sequence (GGA), but does obey Stormo rule 2C. It is preceded by an interesting structure, which has the characteristics of a transcription termination site (Fig. 4). There is also a smaller loop after the ATG containing the sequence C-C-G-C-C, which is also present in the other loop. It is tempting to suggest that in some way these structures might play a part in translation control and ensure that what appears from its primary structure to be a poor initiation site may produce relatively large amounts of protein.

Z

Although the reading frame is very clear here with only one possible start site with an SD sequence, the *Z* gene gives a rather poor peak in the ANALYSEQ program (see Fig. 3), suggesting an unusual codon composition. This may be partly due to the relatively high frequency of the rare codons (see Table 5) and to the small amount of asparagine and glutamine codons. There is a possible looped structure containing the sequence C-C-G-C-C within the coding region (see Fig. 4).

U

There is a very poor SD sequence, but the ATG overlaps the terminator of gene *Z* and is preceded by a purine-rich sequence (see Table 4).

G

Murialdo & Siminovitch (1972) identified the *G* gene product as a protein of about 33,000 M_r or approximately 330 amino acids, whereas the reading frame in this position contains only 140 amino acids. The DNA sequence has been checked carefully. If there were an extra nucleotide between positions 10,068 and 10,133 there would be a reading frame of 279 amino acids; however, there would then be no reading frame for the *T* gene.

T

There are three possible initiation sites, at positions 10,115, 10,160 and 10,184, all of which have only G-G as an SD sequence. The one at position 10,115 is chosen as it overlaps the end of the *G* gene and is preceded by a purine-rich sequence (see above).

H

The site at 10,542 corresponds to the N terminus of the *H** protein (Walker *et al.*, 1982; Hendrix & Casjens, 1974). Thus the processing of *H* to *H** does not involve removal at the N terminus.

M

The start from the ANALYSEQ program is not clear here, but there is only one initiator with a good SD sequence and this overlaps the terminator of *H*. This small protein has an unusually high tryptophan content (5 residues out of 109). This may account for the low peak in the ANALYSEQ program.

K

There are two possible start sites. The one at 14,276 has a good SD sequence but it is 12 residues from the ATG; the other one at 14,306 has AGG seven residues from the ATG. Neither obeys Stormo rules 2 or 6. The former has been chosen rather arbitrarily. The coding sequence is preceded by a gap of 148 nucleotides that contain a possible loop structure (see Fig. 4). The protein has a somewhat unusual composition with a high tryptophan and histidine content (9 and 12 residues, respectively, out of 199). This may partly account for the poor peak in the ANALYSEQ program.

I

The ATG at position 14,773 is selected as the only initiator with an SD sequence, although it is separated by 13 residues. Other possibilities are an ATG at 14,812 or a GTG at 14,872. The latter overlaps the terminator of the *K* gene but is not preceded by a run of purines, which seems to be a characteristic of such structures. If position 14,773 is correct, there is an overlap of 103 nucleotides with the *K* gene.

J

There are several GTG triplets farther downstream of the probable start site, but these have poorer SD sequences. There is a gap of 61 residues between the *I* and *J* genes containing a looped structure followed by a run of T residues, which thus resembles a transcription terminator (see Fig. 4).

orf206a

There are two GTG triplets at positions 19,118 and 19,142 that could be initiation sites, but these have poorer SD sequences than the ATG at 18,965. This site is within a possible looped structure (see Fig. 4).

orf401

There is no obvious start site. Besides the ATG at 19,650, there are several GTG triplets farther downstream, but the ANALYSEQ result makes the former more probable. Position 19,650 is preceded by a looped structure resembling a transcription termination site (see Fig. 4). An unusual feature of this reading frame is the high content (5·7%) of the codon UCA. This is normally a rare codon, no other frame on the left arm having more than 2·3% (Table 5). There are two quite long open reading frames on the opposite strand in this region. One (20,744 to 19,876) has no good initiation site and gives a trough with ANALYSEQ; the other (20,955 to 20,149), which we call *orf206b*, gives a good peak and has a possible initiation GTG at position 20,767 (Fig. 5). Translation of this reading frame would give a protein with a very unusual amino acid composition (Table 6). It would contain 26% alanine and be very hydrophobic. For this reason, it seems possible that it may be the gene (*lom*) for the outer membrane protein identified by Reeve & Shaw (1979) (see below). It is interesting that if *orf401* makes a protein it also would be very rich in alanine. This is probably due mainly to the fact that the alanine codon, GCN, will inevitably be found on both strands in a different reading phase.

orf314

This long reading frame has no really satisfactory start site. The nearest possible one to the position indicated by ANALYSEQ is at 21,029. It contains 12 nucleotides between the SD sequence (A-G) and the ATG. It may be noted that there is an ATG in a different phase at position 20,815 which contains a good SD site (A-A-G-G–G-G-T). This could be the initiation site if there were an error (insertion) in the sequence between positions 20,818 and 20,964. However, the sequence has been determined on both strands and there is no evidence of any ambiguity. One possible explanation could be that there is no protein encoded by this reading frame in the strain of lambda used. The *b* region appears to have no essential function in phage grown in the laboratory, so that a fortuitous frame-shift mutation could have taken place and been preserved in the strain without having any effect on its growth or other properties.

```
                      V  L  L  V  A  L  L  S  C  T  I  P  F  L  V  V  S  A  S  S  S  A  T
CGTTGCAGCAAGCGTTTCAGACGTGCTGTTGGTTGCACTGCTGAGCTGCACTATCCCCTTTCTCGTTGTGTCCGCATCCTCAAGCGCGAC
     20780     20770     20760     20750     20740     20730     20720     20710     20700

 A  E  A  I  S  S  A  R  F  A  E  F  F  A  R  I  A  A  A  S  A  A  L  L  L  C  D  A  D  T
AGCTGAAGCTATATCTTCTGCACGTTTTGCCGAATTTTTTGCACGTATTGCCGCCGCTTCTGCCGCACTTTTGCTCTGCGATGCTGATAC
     20690     20680     20670     20660     20650     20640     20630     20620     20610

 A  L  P  A  A  S  V  A  F  V  D  A  V  D  A  L  P  A  A  A  V  F  A  S  A  A  A  E  A  L
CGCACTTCCCGCAGCCTCTGTCGCCTTCGTGGATGCCGTTGACGCACTCCCCGCCGCCGCTGTTTTTGCGTCTGCCGCGGCAGAGGCGCT
     20600     20590     20580     20570     20560     20550     20540     20530     20520

 R  S  A  A  V  S  D  D  L  A  F  V  S  D  V  F  A  A  L  A  E  F  S  A  A  V  A  E  E  A
CCGTTCCGCTGCTGTTTCAGATGACCTGGCATTCGTCTCGGACGTTTTTGCCGCCCTGGCAGAATTTTCTGCCGCCGTTGCCGAGGAAGC
     20510     20500     20490     20480     20470     20460     20450     20440     20430

 A  R  P  A  L  D  D  A  F  V  S  D  D  F  A  A  S  F  E  A  T  A  S  R  A  E  V  A  A  S
TGCACGACCGGCACTTGATGATGCGTTCGTTTCTGATGATTTTGCTGCCTCTTTTGAGGCCACCGCATCTCGTGCTGAAGTGGCGGCCTC
     20420     20410     20400     20390     20380     20370     20360     20350     20340

 D  A  F  V  A  A  V  E  A  D  V  A  A  D  C  C  D  A  A  A  F  V  S  D  V  F  A  A  P  A
TGACGCTTTCGTGGCCGCGGTGGAGGCAGACGTGGCGGCTGATTGTTGTGACGCTGCAGCATTCGTTTCTGACGTTTTCGCCGCACCGGC
     20330     20320     20310     20300     20290     20280     20270     20260     20250

 L  V  A  A  A  F  F  E  D  S  A  A  A  A  L  F  S  A  S  V  A  F  A  D  A  A  S  A  P  E
ACTGGTGGCCGCCGCGTTTTTTGAGGACTCTGCGGCTGCGGCACTTTTTTCCGCTTCAGTGGCCTTTGCTGATGCCGCTTCTGCGCCGGA
     20240     20230     20220     20210     20200     20190     20180     20170     20160

 D  A  S  *
GGACGCTTCCTGAGCTGACGATGCAGC
     20150     20140     20130
```

FIG. 5. Nucleotide and amino acid sequence of *orf206b*.

The nin *region*

This region, which is not essential for the growth of the phage, has an unusual structure. It contains eight or nine possible reading frames, all of which start with an ATG or GTG that overlap the end of the previous reading frame. Their structures are shown in Table 4. It will be seen that there is also a very short reading frame (*orf26*) that overlaps the ends of the *ren* and *orf146* gene with similar structures.

Altogether, there is a continuous series of 13 overlapping reading frames from the beginning of the *O* gene to the end of the *Q* gene. The *O*, *P* and *Q* genes make proteins, but it is not known whether the others do. Figure 3(c) shows the application of the ANALYSEQ program to the right transcription region using genes *O*, *P*, *Q* and *R* as standards. Although there are peaks corresponding to the *nin* reading frames, they are not very clear. Figure 3(d) shows the results for the *nin* region using *orf146*, *orf290* and *orf204* as standards. Here, all the other reading frames show up as definite peaks, indicating that they all have a similar codon distribution.

orf64

This is a reading frame proposed by Daniels & Blattner (1982). It does not give a peak in the ANALYSEQ program.

S

There are two ATG triplets close together. The first is chosen as having the better SD sequence. The reading frame has a TAG at position 45,351. This is the site of the *S*7 mutation, which was present in the strain used.

The right-hand end (46,428 to 48,502)

Other than the *cos* site, no function has yet been ascribed to this region and there do not appear to be any possible rightward reading frames in it. However, there are two possible reading frames on the opposite strand with acceptable SD sequences (see Fig. 6). In the ANALYSEQ program, using standards from the left arm or from the central region, they do show as peaks, but the values are not very high. As far as we know, there is no other evidence for leftward transcription or for gene products in this region.

rexB

There are two ATG triplets with good SD sequences at 36,259 and 36,244.

orf28

In the ANALYSEQ program there is a high peak (position 34,357 to 34,271) just preceding the reading frame assigned to the *ral* gene (Ineichen *et al.*, 1981). It corresponds to a reading frame having a satisfactory possible initiating site. It would code for a protein of only 28 amino acids, of which eight would be glutamic acid, four aspartic acid and five tyrosine. Such a bizarre protein could have interesting properties.

```
                                                M  A  Q  V  A  I  F  K  E  I  F  D  Q  V  R
CGATAATGCAAACTACGCGCCCTCGTATCACATGGAAGGTTTTACCAATGGCTCAGGTTGCCATTTTTAAAGAAATATTCGATCAAGTGC
     47490     47480     47470     47460     47450     47440     47430     47420     47410

  K  D  L  D  C  E  L  F  Y  S  E  L  K  R  H  N  V  S  H  Y  I  Y  Y  L  A  T  D  N  I  H
GAAAAGATTTAGACTGTGAATTGTTTTATTCTGAACTAAAACGTCACAACGTCTCACATTATATTTACTATCTAGCCACAGATAATATTC
     47400     47390     47380     47370     47360     47350     47340     47330     47320

  I  V  L  E  N  D  N  T  V  L  I  K  G  L  K  K  V  V  N  V  K  F  S  R  N  T  H  L  I  E
ACATCGTGTTAGAAAACGATAACACCGTGTTAATAAAAGGACTTAAAAAGGTTGTAAATGTTAAATTCTCAAGAAACACGCATCTTATAG
     47310     47300     47290     47280     47270     47260     47250     47240     47230

  T  S  Y  D  R  L  K  S  R  E  I  T  F  Q  Q  Y  R  E  N  L  A  K  A  G  V  F  R  W  V  T
AAACGTCCTATGATAGGTTGAAATCAAGAGAAATCACATTTCAGCAATACAGGGAAAATCTTGCTAAAGCAGGAGTTTTCCGATGGGTTA
     47220     47210     47200     47190     47180     47170     47160     47150     47140

  N  I  H  E  H  K  R  Y  Y  Y  T  F  D  N  S  L  L  F  T  E  S  I  Q  N  T  T  Q  I  F  P
CAAATATCCATGAACATAAAAGATATTACTATACCTTTGATAATTCATTACTATTTACTGAGAGCATTCAGAACACTACACAAATCTTTC
     47130     47120     47110     47100     47090     47080     47070     47060     47050

  R  *
CACGCTAAATCATAACGTCCGGTTTCTTCCGTGTCAGCACCGGGGCGTTGGCATAATGCAATACGTGTACGCGCTAAACCCTGTGTGCAT
     47040     47030     47020     47010     47000     46990     46980     46970     46960

CGTTTTAATTATTCCCGGACACTCCCGCAGAGAAGTTCCCCGTCAGGGCTGTGGACATAGTTAATCCGGGAATACAATGACGATTCATCG
     46950     46940     46930     46920     46910     46900     46890     46880     46870

CACCTGACATACATTAATAAATATTAACAATATGAAATTTCAACTCATTGTTTAGGGTTTGTTTAATTTTCTACACATACGATTCTGCGA
     46860     46850     46840     46830     46820     46810     46800     46790     46780

                            M  K  K  M  L  L  A  T  A  L  A  L  L  I  T  G  C  A  Q  Q  T
ACTTCAAAAAGCATCGGGAATAACACCATGAAAAAAATGCTACTCGCTACTGCGCTGGCCCTGCTTATTACAGGATGTGCTCAACAGACG
     46770     46760     46750     46740     46730     46720     46710     46700     46690

 F  T  V  Q  N  K  P  A  A  V  A  P  K  E  T  I  T  H  H  F  F  V  S  G  I  G  Q  K  K  T
TTTACTGTTCAAAACAAACCGGCAGCAGTAGCACCAAAGGAAACCATCACCCATCATTTCTTCGTTTCTGGAATTGGGCAGAAGAAAACT
     46680     46670     46660     46650     46640     46630     46620     46610     46600

 V  D  A  A  K  I  C  G  G  A  E  N  V  V  K  T  E  T  Q  Q  T  F  V  N  G  L  L  G  F  I
GTCGATGCAGCCAAAATTTGTGGCGGCGCAGAAAATGTTGTTAAAACAGAAACCCAGCAAACATTCGTAAATGGATTGCTCGGTTTTATT
     46590     46580     46570     46560     46550     46540     46530     46520     46510

 T  L  G  I  Y  T  P  L  E  A  R  V  Y  C  S  Q  *
ACTTTAGGCATTTATACTCCGCTGGAAGCGCGTGTGTATTGCTCACAATAATTGCATG
     46500     46490     46480     46470     46460     46450
```

FIG. 6. Nucleotide sequence of the *r* strand (i.e. the opposite strand to that shown in Fig. 2(b)) in the region 46,452 to 47,500 and the amino acid sequence of 2 open reading frames.

cIII, *kil* and *γ*

There are various interpretations possible in this region, none of which is entirely satisfactory. The DNA sequence has been determined by Ineichen *et al.* (1981) and by us independently, so it is very unlikely that there are any mistakes in it. According to Ineichen *et al.*, the *cIII* protein starts at 33,456, *γ* at 33,232 and *kil* at 33,112 in the same reading frame as *γ*. We present here an alternative interpretation that was suggested by the ANALYSEQ results (see Fig. 3 and Table 3) and which seems to fit the results better, although giving rather small proteins.

The ATG start proposed by Ineichen *et al.* for the *cIII* gene at 33,456 is preceded by a good SD sequence at 33,475 to 33,470, 14 nucleotides upstream. If this is indeed a ribosome binding site, it would seem much more likely that it would initiate at the ATG at 33,463, which is seven nucleotides away. This would give a protein of molecular weight 6040, which is probably the *cIII*. There are no data concerning its molecular weight. There is an alternative start upstream in the same phase at 33,535, a GTG with a good SD sequence that would give a protein of molecular weight 9006. However, it would not fit so well with the ANALYSEQ result. This would leave the reading frame from 33,477 to 33,189 free and we suggest that it codes for the *kil* gene, probably starting at the ATG at 33,330, and giving a protein of 47 amino acids. The initiation site at 33,232 proposed by Ineichen *et al.* for the *γ* gene has a very poor SD site, and the ANALYSEQ program would suggest that the site at 33,112, which has a better SD sequence, may be the actual start although it would give rise to a smaller protein than predicted (Karn *et al.*, 1974). Clearly, further work is required to identify the reading frames for these genes.

exo

There are a number of possible start sites farther downstream but none has as good an SD as that at 32,028. The A-T-G-A overlap with gene β also makes this probable.

Region 31,347 to 30,395

According to Epp (1978), there should be a gene (*Ea9*) coding for a protein of 9000 M_r in this region. There does not appear to be a reading frame of this size, though there are three possible smaller frames starting at 31,351, 31,196 and 31,024 that could code for proteins of 60, 63 and 61 amino acids, respectively. They all have overlapping termination and initiation sites, the first one overlapping with *exo*, and could form a continuous series like that found in the *nin* region. They give weak peaks in the ANALYSEQ program. We refer to these reading frames as *orf60a*, *orf63* and *orf61*. Two functions have been mapped in this region by Court *et al.* (1980*a*,*b*) and are probably encoded by these reading frames. Most of this region has also been sequenced by Luk & Szybalski (1982). The results agree with ours, except at position 31,139–31,140, where they have only one T residue.

xis

The ANALYSEQ results are not very satisfactory in the region of the *xis* gene. This may be related to the unusual amino acid composition, particularly the high content of basic residues (Hoess *et al.*, 1980).

Region 27,810 to 26,972

This region, which contains the *att* site, probably does not code for any proteins. There is a short reading frame starting at 27,604, but it has an unlikely initiation codon.

The b *region*

The *b* region is not essential for lytic growth and has not, therefore, been studied in detail by genetic methods. Several proteins, identified by their size, have been shown to be encoded by this region (Hendrix, 1971; Murialdo & Siminovitch, 1972; Epp, 1978). In the DNA sequence there are eight possible reading frames; four on the left arm and four in the central region. *orf401* and *orf206b* overlap on opposite strands (Fig. 1).

There are three clear reading frames in the central region, which are clearly identified as *Ea59*, *Ea31* and *Ea47* (Fig. 1, Hendrix, 1971; Murialdo & Siminovitch, 1972; Epp, 1978). Epp *et al.* (1981) have identified a product (*Ea24*, M_r 24,000) that maps in the same region as *Ea59*. There is, however, no other reading frame of this size in this position. It may be initiated within the same reading frame and use the same terminator as *Ea59*. There is a possible, but unlikely initiating ATG at position 26,082.

Reeve & Shaw (1979) identified a lambda encoded protein of 21,000 M_r in the outer membrane of the infected cell and mapped the gene (*lom*) encoding it near the end of the *J* gene. There are two possible reading frames, *orf206a* and *orf206b*. The amino acid composition expected from frame *orf206b* (Table 6) shows it to be very

TABLE 6

Amino acid composition (number of residues) of proteins that would be coded by reading frames orf206a *and* orf206b

Amino acid	*orf206a*	*orf206b*
Phe	7	19
Leu	8	16
Ile	8	3
Met	8	0
Val	23	21
Ser	21	24
Pro	6	6
Thr	18	4
Ala	22	74
Tyr	10	0
His	5	0
Gln	4	0
Asn	3	0
Lys	9	0
Glu	8	19
Asp	8	11
Cys	1	4
Trp	3	0
Arg	10	5
Gly	24	0
Polarity index†	42	30

† Capaldi & Vanderkooi (1972).

hydrophobic, as would be expected for a membrane protein. *orf206a* is less hydrophobic but contains local regions of high hydrophobicity, which might anchor it to the membrane. Reeve & Shaw believed *lom* to be a late protein, which would favour *orf206a*, though *orf206b* could possibly be translated on a different mRNA from the other early proteins. There is thus no clear evidence as to which reading frame is coding for this protein. If *orf206b* is *lom*, *orf206a* may be *La21* (Hendrix, 1971). Hendrix identified late products (*La43*, *La40* and *La38*) that he suggested were related to one another by degradation. These probably correspond to *orf401* though *La38* could perhaps be related to *orf314* (see above). Murialdo & Siminovitch identified late products *p16* and *p17* (23,000 and 20,000 M_r, respectively), which could correspond to *orf206a* and *orf194*. They also found evidence for a late protein product, *p8*, of 79,000 M_r in the *b* region: however, there is no reading frame of this size.

Other reading frames

There are a number of fairly long open frames reading from right to left on the left arm. They are found in the same position as the coding frames on the opposite strand and are probably related to the fact that the codons UCA, CUA and UUA, which are complementary to the termination codons UGA, UAG and UAA, respectively, are absent or rare in the coding regions (Table 5). Whether or not this effect has a more general biological significance or whether it is a coincidence applying only to this region of sequence is not clear. The reading frames on the strand opposite to genes *E*, *L* and *J* show slight peaks with the ANALYSEQ program, but do not have satisfactory initiation sites. ATG and GTG are in fact rare codons in these frames, since the complementary codons CAC and CAT, which code for histidine, are rare on the coding strand. The only possible initiation site in these reading frames is an ATG at position 17,853, which could code for a protein of 472 amino acids.

Some of the sequencing was carried out by C. Howe, V. Devalia and C. J. Edge while they were summer students in this laboratory, to whom we are very grateful. We wish to thank J. Messing for gifts of the M13 vectors, F. R. Blattner and D. L. Daniels for supplying us with up-to-date lambda data and for helpful advice, K. McKenney for advice on interpretation of the data and other members of this laboratory, especially B. G. Barrell, for help and advice. The computer analyses were very much dependent on the help of R. Staden, to whom we are very grateful. D.F.H. thanks the Medical Research Council of New Zealand for post-doctoral support and G.B.P. thanks the Royal Society and Nuffield Foundation for a Commonwealth Bursary.

REFERENCES

Anderson, S. (1981). *Nucl. Acids Res.* **9**, 3015–3027.

Anderson, S., Gait, M. J., Mayol, I. & Young, I. G. (1980). *Nucl. Acids Res.* **8**, 1731–1743.

Anderson, S., Bankier, A. T., Barrell, B. G., de Bruijn, M. H. L., Coulson, A. R., Drouin, J., Eperon, I. C., Nierlich, D. P., Roe, B. A., Sanger, F., Schreier, P. H., Smith, A. J. H., Staden, R. & Young, I. G. (1981). *Nature (London)*, **290**, 457–465.

Capaldi, R. A. & Vanderkooi, G. (1972). *Proc. Nat. Acad. Sci., U.S.A.* **69**, 930–932.

Court, D., Gottesman, M. & Gallo, M. (1980*a*). *J. Mol. Biol.* **138**, 715–729.

Court, D., de Crombrugghe, B., Adhya, S. & Gottesman, M. (1980*b*). *J. Mol. Biol.* **138**, 731–743.

Daniels, D. L. & Blattner, F. R. (1982). *Virology*, **117**, 81–92.

Daniels, D. L., de Wet, J. R. & Blattner, F. R. (1980). *J. Virol.* **33**, 390–400.

Daniels, D. L., Sanger, F. & Coulson, A. R. (1982). *Cold Spring Harbor Laboratory*, in the press.

Duckworth, M. L., Gait, M. J., Goelet, P., Hong, G. F., Singh, M. & Titmus, R. C. (1981). *Nucl. Acids Res.* **9**, 1691–1706.

Echols, H. & Murialdo, H. (1978). *Microbiol. Rev.* **42**, 577–591.

Epp, C. (1978). Ph.D. thesis, University of Toronto.

Epp, C., Pearson, M. L. & Enquist, L. (1981). *Gene*, **13**, 327–337.

Franklin, N. C. & Bennett, G. N. (1979). *Gene*, **8**, 107–119.

Fuhrman, S. A., Deininger, P. L., La Porte, P., Friedmann, T. & Geiduschek, E. P. (1981). *Nucl. Acids Res.* **9**, 6439–6456.

Garoff, H. & Ansorge, W. (1981). *Anal. Biochem.* **115**, 450–457.

Gronenborn, B. & Messing, J. (1978). *Nature (London)*, **272**, 375–377.

Hendrix, R. W. (1971). In *The Bacteriophage Lambda* (Hershey, A. D., ed.), pp. 355–370, Cold Spring Harbor Laboratory Press, Cold Spring Harbor.

Hendrix, R. W. & Casjens, S. R. (1974). *Virology*, **61**, 156–159.

Hoess, R., Foeller, C., Bidwell, K. & Landy, A. (1980). *Proc. Nat. Acad. Sci., U.S.A.* **77**, 2482–2486.

Hong, G. F. (1981). *Biosci. Rep.* **1**, 243–252.

Ikemura, T. (1981*a*). *J. Mol. Biol.* **146**, 1–21.

Ikemura, T. (1981*b*). *J. Mol. Biol.* **151**, 389–409.

Imada, M. & Tsugita, A. (1971). *Nature New Biol.* **233**, 230–231.

Ineichen, K., Shepherd, J. C. W. & Bickle, T. A. (1981). *Nucl. Acids Res.* **9**, 4639–4653.

Iserentant, D. & Fiers, W. (1980). *Gene*, **9**, 1–12.
Jeffreys, A. J. & Flavell, R. A. (1977). *Cell*, **12**, 429–439.
Karn, A., Sakari, Y., Echols, H. & Linn, S. (1974). In *Mechanisms of Recombination* (Grell, R. F., ed.), pp. 95–106, Plenum Press, New York.
Landsmann, J., Kroger, M. & Hobom, G. (1982). *Gene* **20**, 11–24.
Luk, K.-C. & Szybalski, W. (1982). *Gene*, **17**, 247–258.
Maxam, A. M. & Gilbert, W. (1977). *Proc. Nat. Acad. Sci., U.S.A.* **74**, 560–564.
Messing, J. & Vieira, J. (1982). *Gene*, in the press.
Messing, J., Crea, R. & Seeburg, P. H. (1981). *Nucl. Acids Res.* **9**, 309–321.
Mills, D. R. & Kramer, F. R. (1979). *Proc. Nat. Acad. Sci., U.S.A.* **76**, 2232–2235.
Murialdo, H. & Siminovitch, L. (1972). *Virology*, **48**, 785–823.
Oppenheim, D. S. & Yanofsky, C. (1980). *Genetics*, **95**, 785–795.
Petrov, N. A., Karginov, V. A., Mikriukov, N. N., Serpinski, O. I. & Kravchenko, V. V. (1981). *FEBS Letters*, **133**, 316–320.
Ptashne, M., Backman, K., Humayun, M. Z., Jeffrey, A., Maurer, R., Meyer, B. & Sauer, R. T. (1976). *Science*, **194**, 156–161.
Ray, P. N. & Pearson, M. L. (1974). *J. Mol. Biol.* **85**, 163–175.
Ray, P. N. & Pearson, M. L. (1975). *Nature (London)*, **253**, 647–650.
Reeve, J. N. & Shaw, J. E. (1979). *Mol. Gen. Genet.* **172**, 243–248.
Roberts, T. M., Shimatake, H., Brady, C. & Rosenberg, M. (1977). *Nature (London)*, **270**, 274–275.
Sanger, F., Nicklen, S. & Coulson, A. R. (1977). *Proc. Nat. Acad. Sci., U.S.A.* **74**, 5463–5467.
Sanger, F., Coulson, A. R., Barrell, B. G., Smith, A. J. H. & Roe, B. A. (1980). *J. Mol. Biol.* **143**, 161–178.
Schechtman, M. G., Alegre, J. N. & Roberts, J. W. (1980). *J. Mol. Biol.* **142**, 269–288.
Schwarz, E., Scherer, G., Hobom, G. & Kössel, H. (1978). *Nature (London)*, **272**, 410–414.
Schwarz, E., Scherer, G., Hobom, G. & Kössel, H. (1980). *Biochem. Int.* **1**, 386–394.
Shine, J. & Dalgarno, L. (1974). *Proc. Nat. Acad. Sci., U.S.A.* **71**, 1342–1346.
Southern, E. (1979). *Methods Enzymol.* **68**, 152–176.
Staden, R. (1978). *Nucl. Acids Res.* **5**, 1013–1015.
Staden, R. (1980). *Nucl. Acids Res.* **8**, 3673–3694.
Staden, R. & McLachlan, A. D. (1982). *Nucl. Acids Res.* **10**, 141–156.
Stormo, G. D., Schneider, T. D. & Gold, L. M. (1982). *Nucl. Acids Res.* **10**, 2971–2996.
Szybalski, E. H. & Szybalski, W. (1979). *Gene*, **7**, 217–270.
Walker, J. E., Auffret, A. D., Carne, A., Gurnett, A., Hamish, P., Hill, D. & Saraste, M. (1982). *Eur. J. Biochem.* **123**, 253–260.
Winter, G. & Fields, S. (1980). *Nucl. Acids Res.* **8**, 1965–1974.
Winter, G., Fields, S. & Ratti, G. (1981) *Nucl. Acids Res.* **9**, 6907–6915.
Yudelevich, A. (1971). *J. Mol. Biol.* **60**, 20–29.

Edited by S. Brenner

Note added in proof: Kröger, M. & Hobom, G. (*Gene*, **20**, 25–38 (1982)) have recently published the DNA sequence from position 40,218 to 43,972. Our results agree with their sequence except at positions 41,978 and 43,082. At position 41,978 we originally had the sequence C-G, but have now re-examined the sequencing gels and find that they were misread and that the sequence is G-C, which agrees with Kröger & Hobom. This has now been corrected in Figure 2(b). At position 43,082 they find G whereas we find A. We believe that this may be due to a difference in the strain of lambda used.

Ann. Rev. Biochem. 1988. 57:1–28

SEQUENCES, SEQUENCES, AND SEQUENCES

Frederick Sanger

Retired from Medical Research Council Laboratory of Molecular Biology, Hills Road, Cambridge CB2 2QH, England

CONTENTS

INTRODUCTION

These prefatory chapters are usually accounts of biochemists' experiences in research, teaching, and administration. In my case the last two are easily dealt with as I have done hardly any and have indeed actively tried to avoid both teaching and administrative work. This was partly because I thought I would be no good at them, but also out of selfishness. I do not enjoy them, whereas I find research most enjoyable and rewarding.

Of the three main activities involved in scientific research, thinking, talking, and doing, I much prefer the last and am probably best at it. I am all right at the thinking, but not much good at the talking. "Doing" for a scientist implies doing experiments, and I managed to work in the laboratory as my main occupation from 1940, when I started as a PhD student with Albert Neuberger in Cambridge, until I retired in 1983. Unlike most of my scientific colleagues, I was not academically brilliant. I never won scholarships and would probably not have been able to attend Cambridge University if my parents had not been fairly rich; however, when it came to research where

0066-4154/88/0701-0001$02.00

experiments were of paramount importance and fairly narrow specialization was helpful, I managed to hold my own even with the most academically outstanding.

This article is a personal account of my research career, which has been concerned primarily with the development of sequencing methods. Most of the significant work has been summarized in a number of reviews and articles. In these there was, of necessity, a good deal of simplification and omission of detail, both for reasons of space and, sometimes, to make a good and logical story. With the passage of time even I find myself accepting such simplified accounts, with no detail, no failures recorded, and very little reference to the colleagues who have helped me. This article is perhaps an appropriate place to clarify things somewhat and to describe certain lesser known aspects of the work, some of which have never appeared in print.

When I arrived at the University of Cambridge as an undergraduate, I had to decide which three scientific subjects I should take. I had chosen two-and-a-half subjects and was looking through the list of "half" subjects when I noticed one I had not heard of before: "Biochemistry, supervisor Ernest Baldwin." The idea that biology could be explained in terms of chemistry seemed an exciting one, and this was amply confirmed when I met the enthusiastic Dr. Baldwin. My first real academic success was when I stayed on at Cambridge for an extra year (1939) to take the advanced course in Biochemistry. I left Cambridge after taking the final examination with no high hopes of having achieved a good result, so I was very surprised to read in the newspaper two weeks later that I had earned a first class degree. This raised the possibility of my becoming a PhD student and I wrote to the Biochemistry Department at Cambridge enquiring if there was any prospect of this. As I received no reply I decided to visit Cambridge to investigate and found that plenty of research work was going on in spite of the war and that there were several people who would be glad to have a PhD student (especially one who, like me, did not need any money). Furthermore, I was a conscientious objector and had been given exemption from military service. Much of the work in the department centered on enzymes and intermediary metabolism, and the main tools were Warburg manometers and methylene blue tubes. During the advanced course I had been supervised by E. Friedmann, for whom I had developed a great respect. He was a refugee from Germany and a chemist of the old school, and in biochemical studies he believed in isolation and structure and did not have much use for what he called "making nonsense with methylene blue tubes." I think it was partly his influence that made me choose to work with N. W. (Bill) Pirie, who was the protein expert in the lab. Bill's chief interest at that time was in making edible protein from grass. His method of dealing with PhD students was the fairly effective one of "throwing them in at the deep end" and letting them find out things for themselves. He presented me with a large bucket of frozen grass extract and suggested that I investigate it. Unfortunately by the time it had thawed Bill had left the lab for a new job, but not before I had assimilated some biochemical wisdom from him and enjoyed his caustic comments about the shortcomings of other members of the lab. I then needed another supervisor and was fortunate in that Albert Neuberger, who was a postdoctoral fellow in the laboratory, was prepared to take me on. I regard Albert as my main teacher. The most important thing he taught me, both by instruction and by example, was how to do research. I shall always be grateful for his kindness and patience. He also had an extremely wide knowledge of biochemistry, which I admired and used but could never emulate. The subject of my thesis was the "Metabolism of lysine": nothing very profound, but through it I gained much experience in amino acid chemistry, a good introduction to "sequences."

The Department of Biochemistry was an interesting and exciting place in which to work. It had been founded largely for F. G. Hopkins, and most of the staff members were his disciples, but also leaders in their own fields. Biochemistry was then a new subject, and there was plenty going on and a lot of enthusiasm. It seems to me that the quality of research in a laboratory is very much dependent on the atmosphere of the place. I remember being impressed with the friendliness and enthusiasm of everyone with whom I came into contact. The researchers seemed to be interested not only in what they themselves were doing, but also in what I was doing, and they were forever talking about biochemistry with no time wasted on "small talk." I liked this sort of environment and found it stimulating, and I have been fortunate in that both the laboratories in which I have worked have had this kind of atmosphere. The Laboratory of Molecular Biology, to which I moved in 1962, we built up ourselves and it could scarcely have failed to generate excitement and enthusiasm, but that initial impetus seemed to survive at least until I retired, and there were none of the major personal frictions that can have such an adverse effect on the research output of a laboratory.

INSULIN

When I had completed my PhD, in 1943, and Neuberger had left the lab, I was given the opportunity to work with Professor A. C. Chibnall and his group supported by an MRC (Medical Research Council) grant. Chibnall had just been appointed Professor of Biochemistry in succession to Hopkins and was about to move to Cambridge from Imperial College in London with a number of collaborators. Their main research interest at that time was the amino acid analysis of proteins, especially of insulin. Insulin had been chosen both for its medical importance and because it was one of the few proteins that could be bought in a pure form. It was unlikely to have been chosen because of its small size (M_r 6000), as this was not known then, but it was of course very fortunate for our later work that it was so small. At that time amino acid analysis was a painstaking undertaking, and in many cases it was difficult to get accurate results. The analyses obtained by Chibnall's group were probably the best published at that time.

My work on insulin was done about 40 years ago and, because it is associated with the beginning of protein sequencing, has already been described, with varying degrees of accuracy, in a number of reviews and books. Some (e.g. Ref. 1), give a rather romantic impression of me, all alone, resolving to determine the structure of insulin against the advice of my colleagues who believed that proteins were some ill-defined amorphous mixture that could not be studied by chemical methods. This impression, however, is entirely wrong. Although there had previously been papers suggesting this pessimistic view of proteins, it was not one held by those with whom I associated. I think we all followed the lead of Emil Fischer and his school, which depended entirely on a chemical approach to proteins. If anything, the errors were in the opposite direction, ascribing more stoichiometry to proteins than was actually the case. In Britain the chief proponent of stoichiometry was the X-ray crystallographer W. T. Astbury. On the basis of a few spots on an X-ray film he proposed detailed structures for the folding of protein chains, and these were for a while widely accepted. The theory of Bergmann & Niemann (la) was particularly popular. They suggested that residues of a particular amino acid were arranged at regular intervals along polypeptide chains and that the content of each amino acid could be expressed by the formula $2^m \times 3^n$. A useful outcome of this theory was the stimulation of interest in amino acid analysis. People would express their results in this form, and, because of the lack of accuracy of the methods and as most small numbers fit the formula, this was not too difficult. Chibnall's was probably the first group to have results accurate enough to cause the theory to be questioned (2).

So the environment in which I worked was entirely favorable, and my initial experiments were logical continuations of the group's work, and in fact were suggested to me by Chibnall. Had it not been for him I would probably have continued working on lysine metabolism. Chibnall et al (2) had found that the number of free amino groups in insulin was considerably greater than could be accounted for by the lysine content. They suggested that the excess was due to free α-amino groups, indicating that the chains were relatively short and therefore particularly suitable for chemical studies. Chibnall suggested that I try to obtain a quantitative estimation and identification of the amino acids on which these free α-amino groups were located, and that is how my work on insulin started. In fact, the work on insulin could be considered to have started somewhat earlier than that as Neuberger had done some preliminary experiments on the end groups, and probably his interest and experience were passed on to me. There was no question of my setting out to determine the complete sequence of insulin, then or for a considerable time thereafter. Probably the only time that I felt any real pressure to do so was in the final stages when only the disulfide bridges and one amide group remained to be located.

According to reviews that I and others have written, I then developed the use of fluorodinitrobenzene (FDNB) as a general reagent for determining end groups in proteins (3). Although this is so, on looking through my old notebooks I find it was by no means that simple. Several methods had already been suggested for labeling free amino groups, although in only one case was an N-terminal residue identified. In 1935 Jensen & Evans (4) had detected phenylalanine at the N-terminus of insulin, using phenylisocyanate, which is a somewhat less reactive analogue of the Edman reagent (phenylisothiocyanate) (5). This phenylalanine residue was, I think, the first amino acid to be located in any protein. For a general method what we needed was a reagent that would react under mild conditions with amino groups to form a derivative that was stable to acid hydrolysis. It was also important to have a relatively simple method for the separation and identification of the substituted amino acids. For this purpose we were anxious to use the newly discovered partition chromatography (6). It was a much better fractionation procedure than any that had been used previously. It worked well with the acetyl amino acids (6) and was in fact being used in the group by G. R. Tristram.

Besides the isocyanates the main possibilities were sulfonyl chlorides (7) or substituted benzenes (8). Benzene sulfonyl chloride had been used and was readily available, but the products did not seem to have very suitable solubility characteristics. The first reagent I tried was methane sulfonyl chloride. The main reason for this choice was that I expected the methane sulfonyl amino acids to behave similarly to the acetyl derivatives on partition chromatograms. This they did, but I was unable to get any clear results and we next turned to the substituted benzenes. 2,4 dinitrochlorobenzene had been studied by Abderhalden & Stix in 1923 (8); however it only reacted at high temperatures, where there was some hydrolysis of peptide bonds. I used it to make some dinitrophenyl (DNP) amino acids, and found they could be fractionated well on silica gel partition chromatograms. One great advantage was that the DNP derivatives were colored yellow. At that time chromatography really was "chroma"tography: there were no really reliable fraction collectors so that it was important to be able to see the bands on the columns. Clearly the chloro compound was not reactive enough, but the corresponding fluoro or nitro compounds could be expected to be more reactive. I made a small amount of the 1,2,4 trinitrobenzene and Chibnall discovered that Dr. B. C. Saunders, who worked in the nearby Chemical Laboratories, had a supply of the fluoro compound, and he kindly let us have some. Both reacted satisfactorily in the cold. We had a good supply of the FDNB and so it was used in further work. The first experiments were, however, somewhat discouraging. I tried out the reaction of FDNB with glycine, but when I applied the product to a column, instead of the expected single band of DNP-glycine, I obtained two bands in significant amounts. It seemed that the second product was diDNP-glycine, with two DNP groups on the one N atom. I had considerable difficulties getting over this problem, but it turned out that glycine was an exception: other amino acids did not give so much of the second product and a change to bicarbonate from phosphate buffer almost completely eliminated the effect.

When I applied the method to insulin, DNP-phenylalanine and DNP-glycine were identified (3). There was one residue of each per M_r 6000 (although at the time we thought the M_r was 12,000). This suggested that there were two types of chain. These were joined by disulfide cross links of the cystine residues, which could be broken by performic acid oxidation (9), and it was possible to isolate two fractions that corresponded to the two chains (10).

At this time (1947) I had an opportunity to visit the laboratory of A. Tiselius in Uppsala, Sweden. Tiselius was one of the best-known names in proteins as he had invented the analytical electrophoresis method that bore his name and was very widely used. I had shown that the cross-linking disulfide bonds could be broken by oxidation and was trying to find a way to fractionate the chains. There were really no good methods for products of that size, but Tiselius's colleagues were working on a chromatographic system based on adsorption on charcoal which looked rather hopeful. It was suggested that I spend some time working on this in Uppsala. It seemed an exciting opportunity to me, especially as England was still in the postwar period of austerity and Sweden was, comparatively, a land of plenty.

The experiments I carried out with the charcoal columns were not very satisfactory. Detection of the bands coming out of the bottom of the columns was by a rather sophisticated method depending on refractive index: one had to look through an eyepiece observing "fringes" whose positions depended on refractive index. The trouble was the fringes would disappear when the refractive index changed and only reappear in different positions when the refactive index settled down again. One thus got rather few points on the plot. The custom in the lab was to have skilled technical assistants do the practical work, and there was one technician who was considered particularly good at reading these columns, so I got her to do some readings for me. A frontal analysis method was used, so that each product should show up as a step on the plot. The technician did several experiments, and one time the plot had four sharp steps. I was quite excited by this because at the time we thought the molecular weight of insulin was 12,000 and that there were four chains. I showed this to Tiselius, and he immediately suggested we send a letter to *Nature* under both his and my name and include all the preliminary work on oxidation of the chains that I had already done in Cambridge. I was rather shocked as he had not really contributed anything and I had never seen him working in the lab. I managed to persuade him that I was not quite ready to publish the oxidation work, but being his guest and very much his junior I did not protest too hard about publishing the one promising result we had. The resultant paper (11) is fortunately one of my lesser known ones, and it is the only one of which I am ashamed. There are of course only two chains in insulin. I think the four sharp steps were due to the expert technician's way of plotting her readings; if the fringes disappeared she would assume there was no change and only record a higher reading when they reappeared, thus producing artificial steps. Tiselius was really a very kind and charming person, and I am certainly grateful for all the support he gave me both then and later, but this experience did make me realize how lucky I was to have worked with Chibnall, who had allowed me to publish my paper on the amino groups of insulin by myself even though he had actually initiated the work and could justifiably have put his name on it. One bonus from my visit to Sweden was meeting R. L. M. (Dick) Synge, who was working there and who introduced me to zone ionophoresis on starch, which was the forerunner of other more successful ionophoreses such as paper or acrylamide gel.

Complete acid hydrolysis of DNP insulin yielded DNP amino acids, but if the conditions of hydrolysis were reduced partial hydrolysis resulted and DNP peptides were obtained. By studying the DNP peptides from insulin I was able to identify short sequences of four to five amino acids near the N-terminus (12). The fact that only two sequences were obtained made it fairly certain that there were only two types of chains, and, what was probably more important, showed that specific sequences could be determined in proteins by partial hydrolysis, thus eliminating any doubts that they were pure stoichiometric compounds that could be studied by the methods of organic chemistry. In most subsequent work, both on proteins and nucleic acids, partial hydrolysis has been the main general approach to all sequencing, and it is only relatively recently in DNA sequencing that it has been largely superseded. Although my first paper on the N-terminal residues (3) has been more widely quoted (13), I was myself more excited about this peptide paper (12), and indeed in retrospect the results seem to me to be of greater significance. It was the first time that sequences were located in a protein molecule, and we could draw important conclusions about stoichiometry. Also it was done on my own initiative, whereas the first paper relied more on Chibnall's stimulus.

Apart from my two teachers, Neuberger and Chibnall, my first actual collaborator was Rodney Porter. Although a little older than I was, he was officially my PhD student. This was because he had been in the army during the war, while I had been working in the lab. Rodney was not the sort of person one could supervise, but he was very good as a collaborator and I think I learned more from him than he did from me. Probably the main thing he taught me was a certain lighthearted attitude toward research, which makes it more enjoyable and, provided standards of complete integrity and honesty are maintained, can do no harm. He joined me just after the development of the DNP method, when we were extending it to peptides and to other proteins. Rodney had just developed an interest in γ-globulin and wanted to study its end groups. I warned him that it was a complex mixture and that it was not worth trying to determine its end groups. However, this did not stop him, and to everybody's surprise he got a sensible, simple result; and from then on he was inseparable from γ-globulin, to the great benefit of science.

During this period we were consigned to a lab in the basement next to the experimental rats. In spite of these rather unsavory surroundings, it was probably the most enjoyable lab I have worked in. We shared it with Kenneth Bailey and his PhD student, S. V. (Sam) Perry, who had also been in the army and had been at the University of Liverpool with Porter. Thus Bailey and myself, two reserved and serious-minded scientists, were thrown together with these two wild boys fresh from the army. It certainly kept us awake. Both had vocal tendencies: Perry's favorite tune was "Mighty mountain" rendered after the manner of Paul Robeson, and Porter's was "Nobody knows the trouble I've seen." It was a good atmosphere and at the same time we accomplished quite a lot. The only serious black spot was when a flask exploded in front of Rodney's eye and damaged the cornea. Although the sight was not completely lost, I think he never had good vision in that eye.

Although the DNP method was an improvement on previous ones and important results were obtained using it, there were certain disadvantages, the most serious being that the DNP amino acids were not completely stable in acid and a correction factor had to be used. This factor differed with the various amino acids and for some (e.g. glycine) was fairly large; in fact in the case of hemoglobin this problem led to an error in the estimation of the free amino groups. I was therefore still interested in trying to develop a better method. The sulfonamide (SO_2NH) bond is considerably more stable than the DNP, so that it seemed a sulfonyl chloride might be an improvement, and I was anxious to try a more highly colored material to increase the sensitivity. The reagent on which I spent quite a lot of time was dimethylaminophenyl-azo-benzene sulfonyl chloride (or helianthyl chloride). The helianthyl amino acids gave beautiful red bands on the chromatograms and appeared to be stable to acid, but we did not have much success when we tried this reagent with proteins. There were probably some undesirable side reactions with the azo group. There were also complaints from other occupants of the lab when their biological preparations all became bright red. Eventually we gave it up, but more success was obtained later by Hartley & Gray (14) with another sulfonyl chloride, diamino naphthalene sulfonyl chloride (dansyl chloride). The dansyl amino acids were satisfactorily stable and their strong fluorescence made them easy to detect and assay at low concentrations.

The DNP method was fairly widely used after its development, but is now completely superseded by other methods, especially the Edman procedure (5), which has the advantage that it can be used not only for determining N-terminal residues but also for extensive sequential degradations of polypeptide chains. It is interesting that the Edman method was developed about the same time as the DNP method but at first was not used much. The main reason for this was probably the color of the DNP derivatives, which made them much easier to handle at a time when reliable fraction collectors and micro methods of assay were not available.

For sequencing purposes the DNP method was very limited as it could only be used for a few residues at the N-terminus, and the more general approach of subjecting a whole chain to partial hydrolysis seemed necessary to extend the sequence. The main technical problem was fractionation of the complex mixture of peptides produced. Indeed in almost all sequencing work fractionation has been a crucial factor, and progress in sequencing has often depended on progress in fractionation methods. With the insulin chains we were again lucky in that the method of paper chromatography had just been developed by Martin and his colleagues (15), and had been used to determine the pentapeptide sequence of gramicidin S (16). The fractionation of small peptides was far superior to anything that had been achieved previously, and it seemed something of a miracle at the time to see these hitherto intractable products separated from one another on a simple sheet of filter paper. Using this method we were able to isolate small peptides from acid hydrolysates, some of which could be fitted together to deduce larger sequences (17, 18). However, with acid hydrolysates alone we could not obtain the complete sequence of either chain, and what we clearly needed were longer peptides. A more specific method of hydrolysis was required and the obvious choice was proteolytic enzymes. Hitherto we had been reluctant to use enzymes because it was rather generally thought that they might synthesize bonds as well as hydrolyze them, and thus sequences could be rearranged. Enzyme reactions were known to be reversible, and a good deal of work was being done on the synthetic activity of proteolytic enzymes, since it was considered that this might be involved in protein biosynthesis. In spite of these warnings, we decided to risk it and to try using proteolytic enzymes, and we were able to obtain the complete sequence of both chains (19, 20). Knowing that we already had much of the sequence from the acid hydrolysis results and that there were no anomalies between the two methods convinced us that there had been no rearrangement. In all subsequent work proteolytic hydrolysis has been the method of choice, and in fact rearrangements have been shown to occur during acid hydrolysis (21).

The work on the sequencing of the two chains owes much to my two collaborators—Tuppy and Thompson. Hans Tuppy was from Vienna and was one of the hardest workers I have known. Just after the war there were only limited facilities for research in Austria, and he was delighted to come to a well-equipped lab where there was no limitation on the amount of work he could do. We were at the time collecting peptides from acid hydrolysates of the chains of insulin. He took the longer phenylalanyl chain, and I worked, among other things, on the glycyl chain. Much of the work required the use of paper chromatography; the chromatography room was at the other end of the basement corridor, and a common sight at that time was Tuppy walking at full speed along that corridor bearing chromatograms. He never ran, but to keep up with him anyone else had to run. He only stayed a year, but by the end of that year the phenylalanyl chain was practically finished and the glycyl chain had to wait for my next collaborator, E. O. P. (Ted) Thompson from Australia. I have worked with several Australians and have always found them easy to collaborate with. They seem to have a certain practical "no nonsense" view of research that takes the attitude, "If there's an experiment to be done, get on and do it." Ted was no exception and we had great fun working together. Among other things we finished up the glycyl chain, which I think really was harder than the longer phenylalanyl chain, and his persistence in tracking down the final amide group was ultimately rewarded.

Once we had determined the complete sequence of the two chains, the only remaining problem was the arrangement of the disulfide bridges. This proved particularly difficult and probably involved as many man-hours, and more frustrations, than the rest of the work put together. The idea was to hydrolyze the intact insulin, isolate cystine peptides, oxidize them, and see which cysteic acid peptides were produced. The results were, however, very equivocal, and it eventually transpired that there was an interchange reaction taking place at the disulfide bonds during acid hydrolysis (21a). One bond between the two chains could be identified using enzymic digestion, but the other two occurred in a Cys-Cys sequence in the glycyl chain, and no enzyme would split the two half-cystine residues. Clearly, in this respect insulin had been a bad protein to choose; hardly any other protein would have given us this problem. It was quite a long time before we found conditions for avoiding this interchange reaction during acid hydrolysis and were able to complete the structure (22).

Insulin was the first protein to be sequenced (22, 23), but its chains were relatively short, and it is doubtful if the methods used would have succeeded with larger, more normal, chains. Further fractionation methods were needed and the most important was that introduced by Moore & Stein (24, 25): ion-exchange chromatograms to separate the large peptides from proteolytic digests. They also used ion-exchange columns to develop an accurate method of amino acid analysis, and this, though considerably modified and improved in detail, is still the method of choice today. Using these techniques they were able to determine the complete sequence of ribonuclease (120 amino acids) (25), and subsequent work on protein sequence has more closely followed their techniques than the less quantitative methods employed for insulin.

THE LEAN YEARS

After completion of the insulin sequence around 1955, there followed a number of "lean years" when there were no major successes. I think these periods occur in most people's research careers and can be depressing and sometimes lead to disillusion. I have found that the best antidote is to keep looking ahead. When an experiment is a complete failure it is best not to spend too much time worrying about it but rather to get on with planning and becoming involved in the next one. This is always exciting and you soon forget your troubles. One disadvantage of this attitude, though, is that it makes the writing of an article of this sort difficult, because if you are always thinking about the future you tend to forget the past.

Toward the end of the insulin work people would ask me what I would do after I finished insulin. My usual reply was that knowing the structure was only a beginning and that the next thing was to find out what it meant and how the insulin worked. I have always been rather unsuccessful in predicting the future, especially in my own research, and although I did try to do some experiments toward this end, my interest began to go in other directions, and perhaps the sequencing bug had already taken hold of me.

During these "lean years" there were a number of events that are worth recording as they had an important effect on the subsequent work on nucleic acids, although they may not have led to publishable results. The first was initiated by the visit of Chris Anfinsen on sabbatical leave to our laboratory in 1954. Chris had been using, and was very enthusiastic about, the new technique of labeling with radioactive isotopes, and in discussion with him I caught some of his enthusiasm. Previously I had assumed that isotopes were in the realm of the physicists and that the apparatus and techniques would be beyond my means. But I learned that this was not the case and that already a number of radioactive substrates were available. Together we labeled a rat with ^{35}S in an attempt to prepare radioactive insulin. The experiment was not a success, but nevertheless I became a complete convert to the use of radioactive isotopes and to the autoradiographic technique, which was a very powerful and simple tool especially when used in conjunction with paper fractionation methods (paper chromatography and paper ionophoresis). After the introduction of ion-exchange chromatography for fractionating peptides (24), paper methods became less fashionable and were largely replaced by columns and fraction collectors, which were more accurate and reproducible. I was, however, never really converted to columns and retained a certain affection for the paper methods. It seemed to me much more exciting to develop a two-dimensional paper separation and to see all the spots spread out before one than to collect innumerable fractions from a column, analyze them all, and end up with what was usually the expected clear, clean data. There was more, if less clean, data on the paper, and hence more opportunity for unexpected and exciting findings. Although my preference for paper methods was to some extent sentimental, I think it was scientifically justified if only because of its subsequent successful application to RNA.

Somewhat more successful than the ^{35}S-labeling were later experiments with ^{32}P. Some proteins, notably ovalbumin (26), could be labeled with [^{32}P]-phosphate, and some enzymes, such as trypsin and chymotrypsin (27, 28), could be labeled in their active centers with [^{32}P]-diisopropyl phosphofluoridate (DFP). In this way just one or two residues in a whole protein were labeled, and if a partial hydrolysate were fractionated on a two-dimensional paper system and autoradiographed a relatively simple picture was obtained. I was particularly interested in developing methods for deducing the sequence of peptides without resorting to the standard procedure of carrying out complete amino acid analyses, which were somewhat time-consuming and tedious. The ^{32}P-labeled proteins were particularly suitable for trying this type of approach. For instance, information could be obtained by comparing autoradiographs of ionophoreses of partial acid hydrolysates of ^{32}P-labeled proteins. In this way it was shown that the sequence around the reactive serine in elastase was the same as that in trypsin and chymotrypsin (27). Each labeled residue gave a specific pattern of bands or "fingerprint," and techniques were worked out for identifying the sequences by autoradiographic methods without carrying out an amino acid analysis. The main rationale behind this approach was that the radioactive peptides were usually heavily contaminated with nonradioactive substances, but they were radiochemically pure and thus the autoradiographs gave a simpler and more interpretable fingerprint. The relationships of the different bands in a fingerprint could be determined by subjecting them to a further partial hydrolysis. Information about charged amino acids was obtained from the ionophoretic behavior of the peptide bands, and about neutral amino acids from chromatographic behavior. Treatment with phenylisothiocyanate or with periodate gave information about the free amino groups. Using this type of method it was often possible to determine a sequence of four or five amino acids around a specific labeled residue (27–31). Particularly rewarding at this time was a conversation I had with Cesar Milstein, who was working at the other end of the corridor on phosphoglucomutase. This led to a collaboration in which we labeled its active center with ^{32}P and determined a sequence of five amino acids (30). Although this was our only collaboration, my continued association with him has meant a great deal to me.

Even more ambitious but less successful was a means I devised for determining the complete sequence of a protein by autoradiographic methods. The idea was to prepare 20 samples of the protein, each labeled with a different amino acid. Ovalbumin was chosen to study as it was relatively easy to label in isolated oviduct tissue. The samples were then digested with a proteolytic enzyme and the digests fractionated in parallel by paper electrophoresis and an autoradiograph prepared. Assuming that a peptide was pure, its amino acid composition could be deduced from the darkness (radioactivity) of its band in the separate channels. The whole paper was then to be subjected to the Edman procedure and washed to remove the phenylthiohydrantoin, and another autoradiograph was to be prepared. The N-terminal residue of each peptide should then be identified as the band missing or weaker in the second autoradiograph. By doing further Edman degradation on the paper, further sequences of the individual peptides were to be determined. The idea was certainly rather wild, and it did not show much sign of working. However, if it had worked it would have been exciting and would have been an alternative to the standard partial hydrolysis approach, and it might have been quicker and simpler to carry out. We spent quite a lot of time working on this idea and the question arises whether this was really justified. At that time I was in the fortunate position of having a permanent research appointment with the (British) Medical Research Council, and was not under the usual obligation of having to produce a regular output of publishable material, with the result that I could afford to attack problems that were more "way out" and longer term: in fact, as few others could adopt this approach, I felt under some obligation to do so. This type of work also suited my temperament. I like the idea of doing something that nobody else is doing rather than racing to be the first to complete a project, and I prefer to concentrate more on the practical work itself than on its ultimate outcome. Although this work was unsuccessful, the proposed method bears some resemblance to the DNA sequencing methods we finally developed, and so probably our efforts with proteins were worthwhile, and they give some indication of how our ideas about sequencing were developing. My only regret about this work is that it seemed rather unfair to my excellent assistant, Biddy Segall, who devoted so much time and skill to this project but whose name never appeared on a paper.

During most of the work on insulin I was supported by a Beit Fellowship; this lasted seven years (a Junior Beit, a fourth year, and a Senior Beit), and I was certainly very grateful for this continued support, which allowed me to devote all my time to research in the Biochemistry Laboratory. Thereafter I was supported by the Medical Research Council—initially as an external member of the scientific staff. In 1962 I moved to the new MRC Laboratory of Molecular Biology where I joined forces with the group led by Max Perutz from the Cavendish (Physics) Laboratory; this included, among others, Francis Crick, Sydney Brenner, and John Smith. The new laboratory was divided into three more or less self-governing divisions, with Max as Chairman. One division (the Division of Protein Chemistry, which later became the Division of Protein and Nucleic Acid Chemistry) was allotted to my group. There was much more space than I had ever had previously, which meant that I could expand considerably. As I wanted to continue doing experimental work, I had no ambition to have a large team working under my direction, but this gave me a chance to bring together a number of people with similar interests in protein chemistry to make a viable group—a few collaborating directly with me and others either working independently or with a more senior member of the group. Initially the senior members were Ieuan Harris and Brian Hartley, who had been fairly closely associated with me in the Biochemistry Department, and each had his own group. We were later joined by Cesar Milstein. The laboratory rapidly acquired a good reputation so that we were able to attract plenty of good postdoctorals and PhD students, which greatly helped in maintaining the enthusiasm and momentum of the lab. At first there was plenty of room and everyone had a large bay to himself, which seemed a great luxury; however this did not last long and soon the laboratory was as overcrowded as most. In fact the work seemed to go better in the overcrowded conditions.

I think the move to the new laboratory was probably an influence in my conversion to nucleic acid sequencing. Previously I had not had much interest in nucleic acids. I used to go to Gordon Conferences on Protein and Nucleic Acids when the two subjects were bracketed together, and would sit through the nucleic acid talks waiting to get back to proteins. However, with people like Francis Crick around, it was difficult to ignore nucleic acids or to fail to realize the importance of sequencing them. An even more seminal influence was John Smith, who was the nucleic acid expert in the new laboratory and who was extremely helpful to me, so that I could turn to him for advice in this new field. He might well have felt I was "muscling in" on his area, but he did not, and was always ready to give me the benefit of his long experience.

RNA

Initially the sequencing of nucleic acids seemed much more difficult than that of proteins, and little progress had been made until the early 1960s. This was partly because of the lack of pure small substrates on which to develop methods, and partly because of their composition. For nucleic acids with only the four monomers, the interpretation of results would be expected to be more difficult than for proteins with their 20 amino acids, and larger degradation products would have to be isolated to give significant overlaps for sequence deduction. On the other hand, only having four components would make the final analyses much easier. To begin with the difficulties of interpretation were predominant, but as the techniques have improved and larger molecules have been studied the question of analysis has become more important, so that today nucleic acid sequencing is much quicker and simpler than protein sequencing. Whereas for proteins much of the time is spent in identifying and estimating the amino acid end products, it has now been possible to evolve methods for DNA in which this final analytical step is completely eliminated.

The first nucleic acid to be sequenced was in fact the first small RNA to be purified. This was the alanine transfer RNA, which was purified by Holley and his collaborators (32). They used sequencing methods that were very similar to those in use with proteins: partial hydrolysis with enzymes, and fractionation of the products on ion exchange columns. They introduced the use of the enzyme ribonuclease T_1, which, because of its unique specificity for splitting after G[1] residues, has been used as the main degradative method in most subsequent work.

Our own early efforts in RNA sequencing were directed toward the development of more rapid and simpler fractionation techniques and were influenced by our recent experience with ^{32}P-labeled proteins. Since every nucleotide contains a phosphorous atom, RNA is particularly suitable for ^{32}P-labeling in vivo, making very sensitive detection by autoradiography possible. The use of columns for fractionation was somewhat laborious, and it seemed that the simplicity and resolving power of two-dimensional techniques would offer considerable advantages for separating the complex mixtures of small nucleotides that one would obtain from partial degradation of larger RNAs. In general, oligonucleotides did not separate well by simple paper techniques (partition chromatography or ionophoresis), but we were able to develop a two-dimensional method that used ionophoresis on cellulose acetate followed by ionophoresis on ion exchange paper, and this proved considerably better than any of the previous methods (33).

I am sometimes asked such questions as "What was the most exciting moment of your scientific career?" or "What did it feel like when you discovered. . . .?" Such questions are usually prompted by the popular idea

[1]Where there is no possible ambiguity in the text the terms A, G, C are used for both the ribo and deoxyribonucleotides; otherwise rA, rG, rC, U are used for the ribo and dA, dG, dC, T for the deoxyribonucleotides.

that scientific progress depends on sudden breakthroughs or moments of inspiration (cf Archimedes jumping out of his bath, or Watson and Crick discovering the double helix). For me, however, progress does not seem to go like that, and I find it hard to remember moments of sudden exhilaration. Although there were such moments, they were probably more often associated with small and gradual advances; and these were more numerous and therefore gave more enjoyment than would have been the case if there had been sudden big leaps forward. However, I do have a clear recollection of one occasion when Bart Barrell, who usually developed the day's autoradiographs first thing in the morning, came into my lab brandishing a beautiful sheet of film with clear, round, well-separated spots on it. This was certainly exciting after the streaky, unresolved pictures we had been getting before.

Figure 1 *(left)* shows a diagram of an autoradiograph of a ribonuclease T_1 digest on ribosomal RNA. In such a digest all products have G at their 3' ends; there are three dinucleotides and nine trinucleotides, and most of these occupy unique positions so that in subsequent experiments 10 oligonucleotides could be identified simply by their position. This avoided a good deal of final analysis, and was a first step toward our ambition of eliminating the analytical step in nucleic acid sequencing. Furthermore, the composition of larger nucleotides could be determined from the "graticules," shown in the righthand side drawing. All fragments contain one G. The residue that has the greatest effect on the mobility is U, so that the pattern is divided into separate, distinct groups of spots; all oligonucleotides with no U are in the fastest

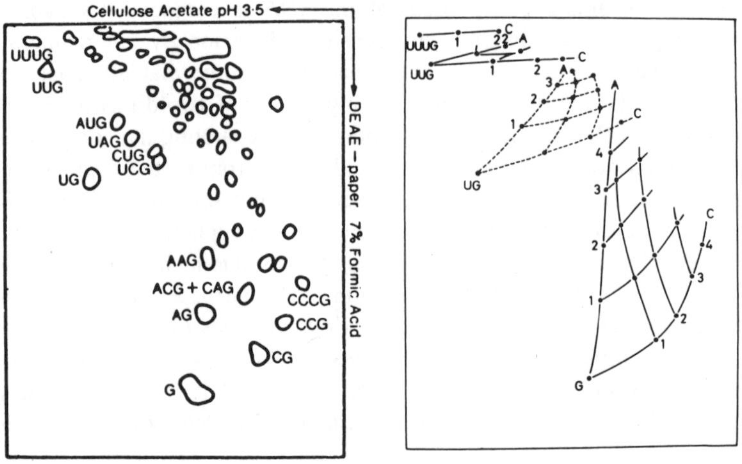

Figure 1 *Left:* Drawing of autoradiograph of two-dimensional separation of ribonuclease T_1 digest of ribosomal RNA. *Right:* Graticules showing the relation of the composition of an oligonucleotide to its position on the autoradiograph.

moving group, those with one U in the next, and so on. Within the groups it is possible to draw a graph or graticule in which one axis gives the number of C residues and the other the number of A residues.

Following this general approach we developed a number of methods for studying the isolated nucleotides. This was in a way an extension of the work on ^{32}P-labeled peptides, though much easier for RNA with only the four components to be considered. One such method that was used with oligonucleotides from pancreatic or T_1 ribonuclease digests was to subject them to partial digestion with a 5' exonuclease and to run the products on ionophoresis on DEAE-cellulose paper at pH 1.9. Sequential degradation from the 5' end will give a mixture of fragments all with the same 3' end but differing in their 5' ends. On the ionophoresis these will be arranged in size order. From the relative positions of two adjacent bands it was possible to identify the nature of the nucleotide by which they differed. Figure 2 shows an example of this approach with the oligonucleotide GAAGC from a pancreatic ribonuclease digest. We defined a value M as the distance between two adjacent oligonucleotides, *x* and *y*, divided by the distance of *x* (the larger) from the origin. If the two differ by an A the M value is 0.5–1.0, whereas in the case of G it is 1.2–3.1. In Figure 2 *a* is the unchanged material and *b* the

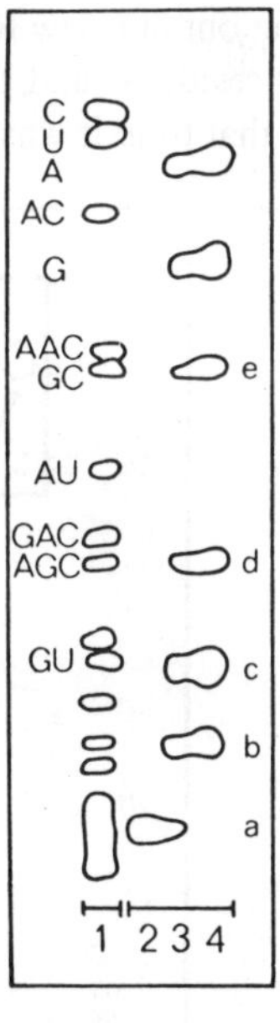

Figure 2 Sequence determination by partial digestion with 5' exonuclease of the oligonucleotide GAAAGC. 1. A pancreatic digest of ribosomal RNA used as marker. 2. Untreated GAAAGC. 3. and 4. GAAAGC digested with 5' exonuclease for 30 and 60 min, respectively.

first degradation product. The M value is 1.9 and therefore the first residue must be G. The M value for *b* and *c* is 0.55, so the second residue is A, and so is the third. Bands *d* and *e* are recognized as AGC and GC by their unique position relative to known markers.

Another useful method was the "wandering spot" technique, which was also sometimes known as "walking the graticule." This was essentially a two-dimensional extension of the above technique and was used mainly with DNA. It was first worked out by Victor Ling to sequence large depurination products (34). A two-dimensional system was devised in which the fragments from an exonuclease digest were arranged according to size, so that each spot differed from the one next to it by a single nucleotide. The system was also arranged to that the relative positions of two neighboring spots depended on the nucleotide by which they differed, as shown in the inset in Figure 3. The method was extended for use with more complex digests, but it was not possible to distinguish A and G with absolute certainty. Figure 3 shows the application of the method to a dodecamer that contained only C, T, and A. The diagram on the left shows a 3' exonuclease digest from which the sequence TTACCATT can be read, and that on the right a 5' exonuclease digest giving the sequence CTTATTAC. Combining the results gives the full sequence CTTATTACCATT.

A suitable substrate for trying out our new RNA sequencing strategies was the 5S ribosomal RNA of 120 residues, and this was completely sequenced mainly by G. G. Brownlee; at that time it was the largest RNA to have been

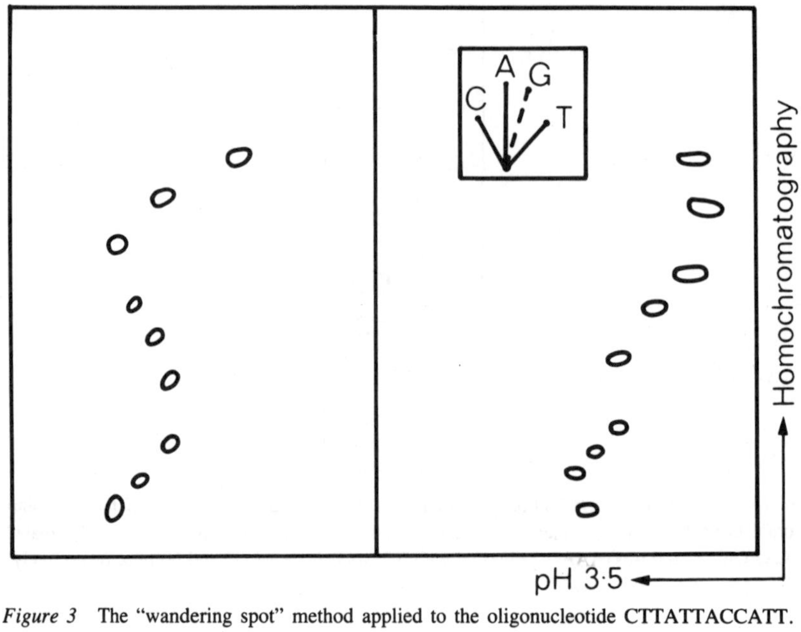

Figure 3 The "wandering spot" method applied to the oligonucleotide CTTATTACCATT.

sequenced (35). We were also able to determine sequences in the RNA bacteriophage R17 that were coding for a part of the head protein whose amino acid sequence was known (36, 37). When we started sequencing RNA one of the main objectives was to try to "break the genetic code." However this was done by entirely different in vitro translation techniques before the sequencing methods were sufficiently advanced. Nevertheless the results with R17 were a useful confirmation of its correctness.

During the work on nucleic acids I had more collaborators than previously and I have been extremely fortunate in the people who have worked with me. Many have been American postdoctorals. We used to get rather many applications, and it was always difficult to choose from such a glittering group. The main problem is that it is not only academic brilliance you are looking for. For a successful collaboration one of the most important things is to have someone you like and can get on well with, and references are often not helpful in this respect. When I was considering John Donelson, one of the referees had added a postscript saying "I think you will enjoy his sense of humour." This proved to be the case and I found the comment helpful. For successful collaborative work there must be complete understanding and trust, and a sharing of ideas and resources. If frictions arise then usually the work suffers. Usually we have had a good atmosphere in the group, but at one time when we were starting our work on DNA we had a relatively large number of people and I noticed that a certain amount of rivalry and bad feeling seemed to be building up. This worried me rather as I did not know how to deal with it, never having had the problem before, so I largely "acted ignorant" and took no notice of it. I think this was probably the best policy; by assuming that they were all friends they were shamed into recognizing the stupidity of their behavior.

The two people to whom I owe most during the nucleic acid work are my two personal assistants, B. G. (Bart) Barrell and Alan Coulson. At about the time I was changing from proteins to RNA I was looking for a new assistant and there were very few applications. The only serious possibility was B. G. Barrell, a young fellow fresh from school. He did not have the academic qualifications we were looking for, but he had plenty of enthusiasm. He came for an interview; I took to him and decided to risk it. He already knew a surprising amount about molecular biology, which he had clearly found out for himself—probably at the expense of his more orthodox school work. He came from Gloucestershire, which was also my own home county, and spoke with a nice thick Gloucestershire accent. He soon became a first-class sequencer and subsequently took charge of his own group. He gradually took over more and more of the nucleic acid sequencing work, especially the actual determinations, while I concentrated more on methods. This meant that I needed a new assistant, and I was fortunate to find Alan Coulson, who was my main collaborator in the later DNA work, and who in his own quiet way contributed to every aspect of it. Hitherto I had preferred working with the more outgoing, extroverted types whose personalities were in marked contrast to my own. I suppose I admired them and found them stimulating. With Alan it was different; we were both quiet by nature. Perhaps as I became more mature I felt less need for external stimulation, though there were always plenty of noisy people around in the lab.

DNA

The first attack on DNA in our group was made in the mid-1960s by Kenneth Murray, who developed some two-dimensional systems for fractionating and analyzing small oligodeoxyribonucleotides (38). Small sequences could be obtained, but at that time there was no way of locating them in the chains. The main problem with DNA was the very large size: the smallest pure DNAs that were available were the genomes of the single-stranded bacteriophages (such as øX174, which will be referred to as øX) of about 5000 nucleotides, and these were rather large for testing out methods. Another difficulty was the absence of suitable degradative enzymes. The sequencing of RNA depended very much on the use of ribonuclease T_1 with its specificity for one residue (G). No enzyme with an analogous specificity exists for DNA. The restriction enzymes, which have since played a very important part in DNA sequencing, had not yet been developed. We did however, around 1973, carry out some experiments using similar techniques to those employed for RNA and were able to determine a few sequences of about 50 residues (39–42). However, the methods were both slow and laborious, and it seemed that if we were really going to be able to attack the vast sequences of genetic materials, then an entirely new approach was needed. As an alternative to partial hydrolysis, some work had been done on the possibility of using copying techniques for sequencing, particularly by C. Weissmann and his colleagues (43). The RNA bacteriophage Qβ contains an RNA polymerase that copies its own RNA, and they had developed elegant pulse-labeling techniques for the RNA and for deducing sequences. An obvious enzyme to choose for copying DNA was DNA polymerase, which had already been used by Wu & Kaiser in 1969 (44) to determine the sequence of the "sticky ends" of bacteriophage λ. This dodecanucleotide sequence was in fact the first bit of DNA to be sequenced (45).

An elegant approach to specific digestion of DNA, which could be combined with a copying procedure, was suggested by Berg, Fancher, & Chamberlin in 1963 (46). In normal conditions the only substrates for DNA polymerase are the deoxyribonucleoside triphosphates, but if magnesium is replaced by manganese in the medium, ribonucleoside triphosphates can be used. Thus, for instance, if incubation is with one ribonucleoside triphosphate (say rCTP) and three deoxyribonucleoside triphosphates, a DNA chain is built up with all the dCs replaced by rCs. These bonds are labile to alkali or ribonuclease, and so it was possible to prepare a C-specific digest, and similar digests to split at other residues. The method looked very promising, and we did a good deal of work on it. DNA polymerase requires a single-stranded DNA as template and a primer that is another piece of single-stranded DNA that base-pairs with a specific region on the template. As the template we could use the single-stranded DNA of bacteriophage f1. Our intention was to use as primer an oligodeoxyribonucleotide having a sequence that could be predicted from the known amino acid sequence of the coat protein of the bacteriophage. The main question was how to make such a primer. At the time Khorana and his colleagues were developing methods for synthesizing oligodeoxyribonucleotides, but the methods were rather slow and specialized and we had no experience. Fortunately when attending a meeting at about that time I met Hans Kössel from the University of Freiburg, who had worked in Khorana's lab. I found that he had the same idea and was thinking of synthesizing the same primer, so we decided to collaborate. It took him and D. Fischer more than a year of hard work to make the octanucleotide primer. If we had waited a year or two restriction enzymes would have been available and we could have used a restriction enzyme fragment as primer. The ribo-substitution method worked reasonably well and we were able to determine a sequence of about 80 nucleotides (39, 40). Our plan was to make another primer with a sequence corresponding to the 3' end of the one we had determined and get another 80 residues, and so gradually work round the genome. This would obviously still be a slow process with many fragments to be analyzed, as well as primers to be synthesized. Fortunately, the research developed in a different direction.

In the above experiments we wanted to obtain very highly labeled DNA, and therefore used the radioactive substrate at high specific activity and in low concentrations. In such experiments (using ^{32}P-ATP) it was frequently observed that the DNA products formed ended at a position before an A would have been incorporated. Presumably the enzyme was running out of ATP. This suggested a new approach to DNA sequencing. If one could produce a mixture of chains all having the same 5' end (corresponding to the 5' end of the primer) and finishing at the 3' end at positions corresponding to the A residues, determination of the relative sizes of all these chains should give a measure for the relative positions of the A residues, and this, combined with similar data for the other three nucleotides, is all one needs for the complete sequence determination. Various methods were studied for fractionating the products of synthesis according to size, and one of these was found to be satisfactory. In fact it proved to be much more efficient than we had ever imagined would be possible. This was electrophoresis on acrylamide gel. If carried out in suitable denaturing conditions, oligonucleotides up to over 300 residues long could be separated according to size, the smaller ones migrating more rapidly on the gel than the larger ones, with each product clearly separated from its neighbors, which differed in size by only one nucleotide.

This new approach to DNA sequencing was I think the best idea I have ever had, being original and ultimately successful, so I have attempted to describe its development in some detail, but on reading it through I must confess that I am by no means certain that it really did happen like that, I certainly do not remember having the idea, whereas I do remember doing some preliminary experiments and discussing it with Alan Coulson and John Donelson, I have a feeling that the above account may have originated to some extent from my attempts to explain the method in a simple way when giving lectures, and that subsequent frequent repetition resulted in its being established as part of my "official," but perhaps not actual, memory.

The method as first conceived, using low concentrations of the substrates, was not very satisfactory, but we did not wish to give up an approach that seemed so promising and, after introducing various modifications of the conditions, two analogous techniques (which we called the "plus and minus" method [47]) were developed and proved to be a much more rapid way of sequencing DNA than anything that had gone before. Most of the sequence of øX was determined using this method. However, a much more efficient and reliable method soon superseded this. For the approach to work we needed to be able to prepare a mixture in which the various end products (all with the same residue at the 3' end) were present in about equal amounts. In the plus and minus method this was not the case, and the sizes of the products were distributed over a rather narrow range so that only relatively short sequences could be determined from one incubation. Another way of achieving the same effect was suggested by the work of Kornberg and his colleagues (48) with analogues of the normal DNA polymerase substrates, which act as chain-terminating inhibitors. One of these was 2',3'-dideoxythymidine triphosphate (dideoxy TTP), identical to the normal TTP but lacking a 3' hydroxyl group. It is a substrate for DNA polymerase, though not such a good one as TTP. Once incorporated the chain no longer has a 3' hydroxyl group and so cannot be extended. If a reaction with DNA polymerase is carried out using the four normal nucleoside triphosphates and a suitable concentration of dideoxy TTP, one would expect on long incubation to end up with a mixture of products all with dideoxy T at their 3' ends. The main problem initially was that these analogues were not available. However, at a meeting in Germany I discovered that K. Geider had made some dideoxy TTP, and he very kindly let me have some. The first experiment we did with it gave a beautiful autoradiograph with sharp bands of equal intensities extending over a long sequence. Clearly this would be much better than the plus and minus method. The problem then was to obtain the other three dideoxy NTPs, which had never been synthesized. I was told by a representative of one of the drug companies that his company was going to make them. After waiting about a year, I met him again and he casually told me that his company had decided not to, so Alan Coulson and I decided to make them ourselves. We had no experience, but we had some expert chemists in the lab to advise us, and quite enjoyed this different type of activity. Eventually we succeeded in preparing all four and the method worked well (49). It was used first to complete the øX sequence. In this case the single-stranded phage was used as template, and restriction enzyme fragments from the double-stranded (replicative) form of øX were used as primers. We naturally wanted to extend the method to other and larger DNAs; however there were two major problems. The first was the eternal one of fractionation, which has always played such an important part in sequencing studies. As the methods became more speedy the preparation of restriction fragments for primers became more and more of a limiting factor. This had been done by acrylamide gel electrophoresis. With øX reasonably pure products could be obtained from a single run, but with larger DNAs this was clearly going to be more of a problem. The other difficulty was that the method requires single-stranded DNA as a template. This was fine for øX whose normal form is single-stranded, but most DNAs are double-stranded, and it is difficult if not impossible to prepare the single strands. Various means were tried to overcome this problem, but the best was cloning the DNA in a single-stranded phage vector. Not only is this a means of obtaining single-stranded DNA, but it also solves the fractionation problem as fractionating is by cloning, which is really the ultimate method of purification and can be applied to a mixture of any complexity. The most efficient method was devised by Messing and his colleagues (50, 51) and is now the quickest method for sequencing DNA of almost any size. The vector used is the double-stranded form of the single-stranded bacteriophage M13, modified to contain a site into which fragments of the DNA to be sequenced can be inserted. The mixture of recombinants is then plated out and grown. This is essentially the fractionation method. Each plaque is derived from a single molecule and is therefore pure. The phage from the plaque is then grown and the single-stranded DNA isolated from it. This is used as the template for the dideoxy sequencing, using as a primer a synthetic oligonucleotide that is complementary to a part of the M13 vector that flanks the insert in such a way that the insert is copied by the DNA polymerase. In this way the same synthetic primer, which can be prepared pure in relatively large amounts, is utilized to prime the different recombinants.

The method is normally used as a random procedure. Originally restriction enzyme digests were used for inserting into the vector, but this was subsequently found to be unnecessary and better results are now obtained by random cleavage, usually by sonication (52).

I suppose the dideoxy method can be regarded as the climax of my research career and the fulfilment of an ambition that had gradually been forming as I became more and more involved in sequencing. It is of course very exciting and gratifying to read of the method now being used in many laboratories and of vast regions of the genomes of both simple and higher organisms becoming exposed in the form of sequences, and these are helping in the understanding of some of the fundamental problems of life. But I think I have derived even more pleasure from the development of the work—seeing the method gradually improving until we were able to read a sequence straight from an autoradiograph. Before this reading became automated, people complained that it was a tedious process, but to me it was always a delight, having in the back of my mind the way we used to do sequences one residue at a time by painstaking partial hydrolysis, fractionation, and analysis. At one stage in the work I would take the autoradiographs home with me and look forward to the pleasure of reading them in the peace of the evening.

At about the same time that the plus and minus method came out, another rapid sequencing technique was developed by Maxam & Gilbert (53, 53a). This was somewhat similar to our method in that it gave an autoradiograph of an ionophoresis from which a sequence could be read off. However, whereas we obtained the mixtures of fragments all terminating at the same nucleotide by a copying procedure using DNA polymerase, they used partial degradation by chemical methods of fragments labeled at their 5' ends. The method worked well and was widely used in preference to the plus and minus method, and in the United States probably in preference to the dideoxy technique until the introduction of the M13 cloning greatly increased the applicability of the dideoxy method. I think it would be fair to say that at present, for a long sequence, the dideoxy-M13 approach is preferable, whereas if one is interested only in a short specific sequence it is probably easier to set up the chemical method.

While the two techniques are to some extent complementary and one must welcome any scientific progress, I cannot pretend that I was altogether overjoyed by the appearance of a competitive method. However, this did not generate any sort of a "rat race," and I do not think it affected our subsequent work at all. I was by no means satisfied with the plus and minus method and certainly the dideoxy method would have been developed in any case. It may be wondered why I did not also try a similar chemical method. I had indeed considered similar ideas but suspect I rejected them because of lack of specificity. Throughout the work on proteins and RNA it had usually been found that enzymic methods were more successful than chemical ones because of their greater specificity. I was thus rather prejudiced in favor of the use of enzymes and assumed that for a rapid degradative sequencing method to work the reactions would have to be specific. In fact those used by Maxam & Gilbert are not specific, but it turns out that this does not matter and may even result in stronger data than would be obtained by completely specific reactions.

In this account I have dealt almost entirely with our work on methods for sequencing and have said little about the results obtained. This is partly for reasons of space, but also because I consider that my own personal contributions have been to the methods. In the course of the work we have determined many sequences and obtained significant results, which have been reported elsewhere, but in these I have usually been part of a team. I quite enjoyed doing the somewhat routine DNA sequencing work where it was possible to produce results without too much effort, and I probably found this a form of relaxation compared with the more original, exacting, variable, and often frustrating, work on development of methods. With the larger DNA sequences it is necessary to have someone collect and analyze the data, and this was usually done by Bart Barrell, who is something of a wizard with sequences. An exception was when we were working on bacteriophage λ (54), which was sequenced mainly by Alan Coulson and me. At first the routine work of collecting the data was good fun, but I think both of us were happy when the sequence was finally completed. Filling up the last gaps in a sequence can be both slow and frustrating. The main high spots of these sequencing results were two unexpected findings—the overlapping genes in bacteriophage øX174 (55, 56), and the altered genetic code in mitochondrial DNA (57). This work is adequately described in other articles (57–61).

Table 1 summarizes the progress made in sequencing, and lists the various landmarks discussed in this article, starting from Jensen & Evans's positioning of a single amino acid in insulin in 1935 and ending with the recent enormous acceleration in DNA sequencing. At present the number of sequences in the data banks is in the millions, and is still increasing rapidly.

Although I have dealt only with sequencing methods in this article, my interests have not been quite so narrow, and from time to time I have ventured into other fields, although usually unsuccessfully. The most successful diversion was the discovery of the initiator tRNA, formyl-methionyl tRNA (62). This was largely the work of Kjeld Marcker, who came as a visiting worker to the lab at about the time we were first trying to develop methods on RNA. The idea was to get some sequence information near the amino acid binding sites of tRNAs by labeling them with radioactive amino acids. Partial hydrolysis should then yield oligonucleotides of various lengths joined to an amino acid, and methods could be designed for sequencing them. Methionine was chosen to try out the method because it could be labeled at high specific activity with

Table 1 The progress in sequencing

Year	Protein	RNA	DNA	Number of residues	Ref.
1935	Insulin			1	4
1945	Insulin			2	3
1947	Gramicidin S			5	16
1949	Insulin			9	12
1955	Insulin			51	22
1960	Ribonuclease			120	25
1965		$tRNA_{Ala}$		75	32
1967		5S RNA		120	35
1968			Bacteriophage λ	12	45
1978			Bacteriophage ϕX 174	5,386	61
1981			Mitochondria	16,569	58
1982			Bacteriophage λ	48,502	54
1984			Epstein-Barr virus	172,282	64

^{35}S. To see if the coupling of the methionine to the tRNA had worked, the product was digested with pancreatic ribonuclease and subjected to ionophoresis. The expected adenosyl ester of methionine was identified, but there was another, unexpected, spot present. This could easily have been dismissed as an artifact, but that did not satisfy Marcker. It turned out to be the adenosyl ester of N-formyl methionine, and further experiments enabled him to show that there were two methionyl tRNAs in *Escherichia coli:* one, $tRNA_m^{Met}$, which incorporates methionine within protein chains, and the other, $tRNA_f^{Met}$, which incorporates it only in the N-terminus and is the chain initiator (63).

Unlike many scientists, I decided to retire and give up research when I reached the age of 65. This surprised my colleagues, and to some extent myself also. I had not thought about retirement until I suddenly realized that in a few years I would be 65 and would be entitled to stop work and do some of the things I had always wanted to do and had never had time for. The possibility seemed surprisingly attractive, especially as our work had reached a climax with the DNA sequencing method and I rather felt that to continue would be something of an anticlimax. The decision was I think a wise one—not only because I have greatly enjoyed the new life-style, but also because the aging process was not improving my performance in the laboratory and I think that if I had gone on working I would have found it frustrating and have felt guilty at occupying space that could have been available to a younger person. For more than 40 years I had had wonderful opportunities for research, and had been given the chance to fulfill some of my wildest dreams.

Literature Cited

1. Loewy, A. G., Siekevitz, P. 1969. *Cell Structure and Function,* pp. 193–94. New York: Holt, Rinehart & Winston
1a. Bergmann, M., Niemann, C. 1938. *J. Biol. Chem.* 122:577–96
2. Chibnall, A. C. 1942. *Proc. R. Soc. London Ser. B* 131:136–60
3. Sanger, F. 1945. *Biochem. J.* 39:507–15
4. Jensen, H., Evans, E. A. 1935. *J. Biol. Chem.* 108:1–12
5. Edman, P. 1950. *Acta Chem. Scand.* 4:283–93
6. Gordon, A. H., Martin, A. J. P., Synge, R. L. M. 1943. *Biochem. J.* 37:79–86
7. Gurin, S., Clarke, H. T. 1934. *J. Biol. Chem.* 107:395–99
8. Abderhalden, E., Stix, W. 1923. *Hoppe-Seyler's Z.* 129:143–56
9. Sanger, F. 1947. *Nature* 160:295–96
10. Sanger, F. 1949. *Biochem. J.* 44:126–28
11. Tiselius, A., Sanger, F. 1947. *Nature* 160:433–34
12. Sanger, F. 1949. *Biochem. J.* 45:563–74
13. Sanger, F. 1985. *Curr. Contents* 28(12): 23
14. Gray, W. R., Hartley, B. S. 1963. *Biochem. J.* 89:59
15. Consden, R., Gordon, A. H., Martin, A. J. P. 1944. *Biochem. J.* 38:224–32
16. Consden, R., Gordon, A. H., Martin, A. J. P., Synge, R. L. M. 1947. *Biochem. J.* 41:596–602
17. Sanger, F., Tuppy, H. 1951. *Biochem. J.* 49:463–81
18. Sanger, F., Thompson, E. O. P. 1953. *Biochem. J.* 53:353–66
19. Sanger, F., Tuppy, H. 1951. *Biochem. J.* 49:481–90
20. Sanger, F., Thompson, E. O. P. 1953. *Biochem. J.* 53:366–74
21. Sanger, F., Thompson, E. O. P. 1952. *Biochim. Biophys. Acta* 9:225–26
21a. Ryle, A. P., Sanger, F. 1955. *Biochem. J.* 60:535–40
22. Ryle, A. P., Sanger, F., Smith, L. F., Kitai, R. 1955. *Biochem. J.* 60:541–56
23. Sanger, F. 1959. *Les Prix Nobel en 1958,* pp. 134–46. Reprinted in *Science* 129:1340–45
24. Spackman, D. H., Stein, W. H., Moore, S. 1958. *Anal. Chem.* 30:1190–1206
25. Hirs, C. H. W., Moore, S., Stein, W. H. 1960. *J. Biol. Chem.* 235:633–47
26. Sanger, F., Hocquard, E. 1962. *Biochim. Biophys. Acta* 62:606–7
27. Hartley, B. S., Naughton, M. A., Sanger, F. 1959. *Biochim. Biophys. Acta* 34:243–44
28. Naughton, M. A., Sanger, F., Hartley, B. S., Shaw, D. C. 1960. *Biochem. J.* 77:149–63
29. Sanger, F., Shaw, D. C. 1960. *Nature* 187:872–73
30. Milstein, C., Sanger, F. 1961. *Biochem. J.* 79:456–69
31. Sanger, F. 1963. *Proc. Chem. Soc.* (UK), March 1963:76–83
32. Holley, R. W., Agpar, J., Everett, G. A., Madison, J. T., Marquisee, M., et al. 1965. *Science* 147:1462–65
33. Sanger, F., Brownlee, G. G., Barrell, B. G. 1964. *J. Mol. Biol.* 13:373–98
34. Ling, V. 1972. *J. Mol. Biol.* 64:87–102
35. Brownlee, G. G., Sanger, F., Barrell, B. G. 1968. *J. Mol. Biol.* 34:379–412
36. Adams, J. M., Jeppesen, P. G. N., Sanger, F., Barrell, B. G. 1969. *Nature* 223:1009–14
37. Jeppesen, P. G. N., Barrell, B. G., Sanger, F., Coulson, A. R. 1972. *Biochem. J.* 128:993–1006
38. Murray, K. 1970. *Biochem. J.* 118:831–41
39. Sanger, F., Donelson, J. E., Coulson, A. R., Kössel, H., Fischer, D. 1973. *Proc. Natl. Acad. Sci. USA* 70:1209–13
40. Sanger, F., Donelson, J. E., Coulson, A. R., Kössel, H., Fischer, D. 1974. *J. Mol. Biol.* 90:315–33
41. Ziff, E. B., Sedat, J. W., Galibert, F. 1973. *Nature New Biol.* 241:34–37
42. Robertson, H. D., Barrell, B. G., Weith, H. L., Donelson, J. E. 1973. *Nature New Biol.* 241:38–40
43. Billeter, M. A., Dahlberg, J. E., Goodman, H. M., Hindley, J., Weissmann, C. 1969. *Nature* 224:1083–86
44. Wu, R., Kaiser, A. D. 1968. *J. Mol. Biol.* 35:523–27
45. Wu, R., Taylor, E. 1971. *J. Mol. Biol.* 57:491–511
46. Berg, P., Fancher, H., Chamberlin, M. 1963. In *Symposium on Informational Macromolecules,* ed. H. Vogel, V. Bryson, J. O. Lampen, pp. 467–83. New York/London: Academic
47. Sanger, F., Coulson, A. R. 1975. *J. Mol. Biol.* 94:441–48
48. Atkinson, M. R., Deutscher, M. P., Kornberg, A., Russell, A. F., Moffatt, J. G. 1969. *Biochemistry* 8:4897–904
49. Sanger, F., Nicklen, S., Coulson, A. R. 1977. *Proc. Natl. Acad. Sci. USA* 74:5463–67
50. Gronenborn, B., Messing, J. 1978. *Nature* 272:375–77
51. Messing, J., Vierira, J. 1982. *Gene* 19: 269–76
52. Deininger, P. 1983. *Anal. Biochem.* 129:216–23
53. Maxam, A. M., Gilbert, W. 1977. *Proc. Natl. Acad. Sci. USA* 74:560–64

53a. Maxam, A. M., Gilbert, W. 1980. *Methods Enzymol.* 65(P. 1):499–560
54. Sanger, F., Coulson, A. R., Hong, G. F., Hill, D. F., Petersen, G. B. 1982. *J. Mol. Biol.* 162:729–73
55. Barrell, B. G., Air, G. M., Hutchison, C. A. III. 1976. *Nature* 264:34–41
56. Smith, M., Brown, N. L., Air, G. M., Barrell, B. G., Coulson, A. R., et al. 1977. *Nature* 265:702–5
57. Barrell, B. G., Anderson, S., Bankier, A. T., de Bruijn, M., Chen, E., et al. 1980. *Proc. Natl. Acad. Sci. USA* 77:3164–66
58. Anderson, S., Bankier, A. T., Barrell, B. G., de Bruijn, M. H. L., Coulson, A. R., et al. 1981. *Nature* 290:457–65
59. Anderson, S., de Bruijn, M. H. L., Coulson, A. R., Eperon, I. C., Sanger, F., et al. 1982. *J. Mol. Biol.* 156:683–717
60. Sanger, F. 1981. *Les Prix Nobel en 1980*, pp. 143–59. Reprinted in *Biosci. Rep.* 1:3–18; *Science* 214:1205–10
61. Sanger, F., Coulson, A. R., Friedmann, T., Air, G. M., Barrell, B. G., et al. 1978. *J. Mol. Biol.* 125:225–46
62. Marcker, K., Sanger, F. 1964. *J. Mol. Biol.* 8:835–40
63. Clark, B. F. C., Marcker, K. A. 1966. *J. Mol. Biol.* 17:394–406
64. Baer, R., Bankier, A. T., Biggin, M. D., Deininger, P. L., Farrell, P. J., et al. 1984. *Nature* 310:207–11